saur

Union of International Associations, Brussels

International Biographical Dictionary of Religion

An encyclopedia
of more than 4,000 leading personalities

Edited by Jon C. Jenkins
assisted by Cécile Vanden Bloock

K · G · Saur
München · New Providence · London · Paris 1994

The cover shows leading personalities from six international organizations
Those shown on the cover, from left to right, are:

Israel Singer
Secretary-General of the World Jewish Congress

Brahama Kumari Jayanti
European Director of the Brahma kumaris World University

Al-Hafiz B A Masri
Founder-Present of the International Muslim Association
for Animals and Nature

Lama Thubten Zopa Rinpoche
Spiritual Director of the Foundation for the Preservation
of the Mahayana Tradition

Archbishop Jan Schotte
Secretary-General of the Synod of Bishops

Rev. Dr. Milan Opocensky
Secretary-General of the World Alliance of Reformed Churches

Die Deutsche Bibliothek – CIP-Einheitsaufnahme

International biographical dictionary of religion : an
encyclopedia of more than 4,000 leading personalities / Union
of International Associations, Brussels. Ed. by Jon C. Jenkins.
Assisted by Cécile Vanden Bloock. – München ; New
Providence ; London ; Paris : Saur, 1994
ISBN 3-598-11100-2
NE: Jenkins, Jon C. [Hrsg.]; Union of International Associations

Printed on acid-free paper

A Reed Reference Publishing Company
Printed in the Federal Republic of Germany

Data preparation and automatic data processing by
Microcomposition, Wittmar

Druck / Printed by grafik & druck GmbH, München
Binden / Bound by Thomas Buchbinderei, Augsburg

ISBN 3-598-11100-2

Contents

Introduction

The *International Biographical Dictionary of Religion* has been prepared by the *Union of International Associations* and is intended to serve the following purposes:

- to provide those concerned with religious organizations with information on the principal personalities involved in the development, operation and coordination of these important institutions and networks;
- to provide an easy means of checking biographical details of significant individuals on the multinational and global levels;
- to provide biographical information about the people behind the operations of the major religious organizations of the world;
- to provide a resource for the ongoing development of relations among religions;
- to improve access to religious organizations.

International non-governmental and intergovernmental organizations, public libraries, universities, press organizations and research institutes will find this reference work indispensable. In fact, anyone concerned with prominent individuals in the religions of the world will be interested in this particularly useful tool. Quite a number of biographical reference books are available today. Some of them are well established and widely used. As far as we can tell, the *International Biographical Dictionary of Religion* is unique of its kind. The set contains nearly 4,000 entries of eminent individuals in every major religion including monks, pastors, rabbis, priests, nuns, and lay people. They range from poets to scientists, from politicians to doctors and from teachers to journalists. The major Protestant and Catholic denominations are represented as well as Hinduism, Buddhism, Judaism, Shintoism among others. Nearly every nation has representatives.

Method of preparing the International Biographical Dictionary of Religion

The concept of doing a *Who's Who in International Organizations* developed over a three-year period from 1988 to 1991. During conversations with Mr Anthony Judge, the Assistant Secretary General of the Union of International Associations (UIA), a number of proposals were discussed. In 1989 a proposal for six Who's Who type publications was sent to K. G. Saur. In the course of negotiations, we decided to start with the *Who's Who in International Organizations* to be followed by a *Who's Who in Religion.*

The practical process of developing the publication began with selecting the range of religions to be included. Of the more than 30,000 organizations in the files of the UIA, 3,000 of those related to various religions were to be contacted.

The next step was to design a questionnaire and to write an accompanying letter requesting biographical information. A mailing of approximately 1,400 letters was sent in August 1991 to individuals in all of the religious institutions. Another 700 letters were sent to recipients of the earlier mailing and to other organizations. Various directories (World Council of Churches, the Vatican, etc.) and reports were searched. More than 1,500 questionnaires and/or curricula vitae have been returned. In addition to direct contact with individuals, other biographical reference books were searched. Those are listed in the bibliography. Some 1,200 names were added and a significant number of existing entries were supplemented.

Warning

We are conscious, however, of the various shortcomings of this first edition; the most serious flaw is the actual size of a significant number of the entries. Restricted finances limited the amount of time available for the project, the type and quantity of contact with potential biographees and size of staff, all of which contributed to its defects. Other problems encountered for which the responsibility falls on the editors include the following:

- some biographees enclosed an extensive curriculum vitae, small books in fact, and the editors were forced to be selective and in being so may have made errors;
- in some cases religions are represented by individuals who do not hold the most significant positions. The editors felt that it was more important to have the traditions represented;
- questionnaires were completed in various languages (including English, French, German, Dutch, Spanish, Italian, Danish, Esperanto, etc.) and in some cases more than one language in spite of the request on the questionnaire and in the letter to use English and French, and language barriers have not always been satisfactorily overcome;
- not all of the questionnaires were filled in carefully and legibly. A great deal of effort and time was spent discerning the meaning of various elegant, beautiful, shaky or coded but illegible handwritten words. When feasible, individuals were sent proof sheets to so that corrections could be made but costs made it impossible to do this in all cases.
- some of the biographees used acronyms and in a few cases the editors were unable to discover what they meant. These are included.
- as the book was being prepared a number of new nations came into being; Estonia, Latvia, and Lithuania to name a few. While to editors were able to correct addresses due to the reunification of Germany and some others they regret that they could not correct all of the addresses due to these most recent changes.

A full entry consists of the following: full name and title; religion, nationality; personal information: date and place of birth, celibate or married, spouse's name (if applicable), number of children (boys and girls if known), religion, and languages spoken; education: post-secondary school degrees, the name of the institution and attendance dates; current position(s) in international and other organization(s); main field of work; career: position, organization and dates; publications; special interests; memberships; honors; and residential and mailing addresses. The minimum acceptable entry is Name, Organization, Position, and Religion. In these cases the religion is listed as unknown. The three indexes are by religion, by nationality and by international organization. The Nationality Index only has those names listed who indicated their citizenship on the questionnaire or a second reliable source was found. The Religion Index is based on the same criteria as the nationality with a few exceptions. Some individuals were in significant inter-religious organizations and the editors did not know the religion of the entry. Most entries are listed in the International Organization Index.

The editors are indebted to the institutions and people who provided us with material concerning the biographees. We would like to thank all the people who filled out a questionnaire or sent their curriculum vitae. We thank those who recommended others for the Who's Who. We feel obliged to express our apologies to those same people for any errors appearing in their biography due to mistranscription. We also apologize to the people who sent us their questionnaires but who have not been included because, in our opinion, they did not fit within the Who's Who's criteria. Acknowledgments must be made to those who contributed to the preparation of the publication. We would like to thank Cécile Denet and Anne Degimee for their support and advice with the computers, the network and Revelation (the text processing software). We thank Marie Aeles for her assistance with the Organizations database, Carine Faveere for her support, and all members of the staff of the *Union of International Associations*. Finally, we would like to express our gratitude to Manfred Link for his help in defining the parameters of the biographies. Manfred is with K.G. Saur Verlag GmbH & Co. KG.

The use of computers

The information in this publication is maintained in computer files. This is necessary in order to facilitate the incorporation of descriptions of entries, their interrelationships, as well as the correction and updating of such information. These tasks become very onerous when it is necessary to deal with the amounts of information processed for this publication.

Technical Information

a. System: The local area network is based on a Novell Netware 286 v2.12, using an MS-DOS 3.1 operating system. It is presently composed of 18 independent micro-computer workstations (PC compatible) linked by Arcnet. Main disk capacity is currently 540 megabytes. Backups and some communications with a UK service bureau are via a streamer tape unit for standard industry tapes (9 track, 1600 bi, odd parity, record length 180, blocking factor 20) and an Emerald 120 MB cartridge unit. Communications with a German service bureau are via 5 1/4" floppy disks using PKZIP FAST Create/Update Utility version 1.1 to compress data and MS-DOS's backup command to copy information.
b. Software: The software used for management of the relational text database is Advanced Revelation 1.16 network version. An extensive suite of programs has been specially written for UIA to take advantage of the facilities of this software for handling entries in a text database mode.

The Union of International Associations

The UIA was founded on 1 June 1907 in Brussels, as *Central Office of International Associations – Office central des institutions internationales*. It became a federation, under the present name, at the 1st World Congress of International Organizations, 1910. Activities were closely associated with *International Institute of Bibliography – Institut international de bibliographie*, set up 2 September 1895, Brussels, which is currently the *International Federation for Information and Documentation (FID)*. It also worked closely with *Institut international de la paix, Monaco*, which had been set up in 1903 and produced the 1905, 1906 and 1907 editions of the *Annuaire de la Vie internationale*; the UIA collaborated with the Institute on the 1908-1909 edition of this publication, which subsequently evolved into the Yearbook of International Organizations. The aims of the UIA are:

- to contribute to a universal order based on principles of human dignity, solidarity of peoples and freedom of communication;
- to undertake and promote research and study on transnational associative networks, considered as essential components of contemporary society;
- to collect and distribute the most comprehensive documentation possible on international organizations and associations, both governmental and non-governmental, and on new forms of transnational cooperation;
- to collect and distribute data on the various meetings organized by international bodies;
- to encourage and undertake activity aimed at promoting the development and efficiency of non-governmental networks, as well as intercommunication between people working in the international framework and in inter-associative cooperation;
- to study, categorize, analyze, compare and illuminate world problems as perceived by international organizations.

Other UIA publications:

Yearbook of International Organizations, the most universal and up-to-date source of information available on international organizations, in three volumes

Vol 1: *Organization descriptions and Index*, 1991 (28th ed.)
Vol 2: Geographic Volume: *International Organization Participation*; Country directory of secretariats and membership, 1991 (9th ed.)
Vol 3: Subject Volume: *Global Action Networks*; Classified directory by subject and region, 1991 (9th ed.)

Encyclopedia of World Problems and Human Potential in two volumes, 1990 (3rd ed.)

International Congress Calendar, 1990 (Vol 30) quarterly

International Association Statutes Series, 1988 (Vol 1)

Guides to International Organizations

No 1: *African International Organization Directory* (and African participation in other international organizations), 1984/85 (1st ed.)
No 2: *Arab-Islamic International Organization Directory*, (and Arab and Islamic participation in other international organizations), 1984/85 (1st ed.)
No 3: *International Organization Abbreviations and Addresses*, 1984/85 (1st ed.)
No 4: *Intergovernmental Organization Directory*, 1984/85 (1st ed.)

Abbreviations

The majority of abbreviations used are academic degrees or Religious communities. The degrees listed are from the questionnaires and other sources. Two different people may have attended the same university at the same period and took identical degrees but listed them differently, one translating his into an English or American equivalent degrees. For example a Doctorate in law may be listed as LL.D., Dr.Pub.Law, Doc.d'Etat en Droit Pub., Deg. Law, D.C.L., D. en D., D.law, Dr.law, etc. We made no attempt to standardize the biographees' use of abbreviations.

1st	1st degree honours
1st State Exam.	Passing the 1st State Examination
2nd State Exam.	Passing the 2nd State Examination
A.A.	Associate of Arts
A.A.	Augustinians of the Assumption (Assumptionists)
A.B.	Bachelor of Arts
A.B.A.	Associate of Business Administration
Acct.	Accounting
Accty.	Accountancy
A.C.G.I.	Associate, City and Guilds of London Institute
adm. et gest. du pers.	Personnel Administration and Development
Admin.	Administration
Admin. Crse	Adminstration Course
Admin. Eng.	Administrative Engineering
Admin. Sc.	Administration Science
adm.publ.	Public Administration
Adv.	Advertising
Adv. Eur. Stds.	Advanced European Studies
Adv. Man. Prog.	Advanced Management Program
Adv. Stds. Dipl.	Advanced Studies in Diplomacy
Adv. Trng. Crse. Cert.	Advanced Training Course Certificate
Afr. Stds.	African Studies
Agrég.	Certificate to teach
Agrég. enseign. second. inf.	Certificate to teach in Lower Secondary Schools
Agrég. enseign. second. sup.	Certificate to teach in Upper Secondary Schools
Agrég. enseign. supér.	Certificate to teach at University level
Agrég. enseign.	Certificate to teach
Agrég. Sc. Math.	Certificate to teach Mathematical Sciences
Agri., Agr.	Agriculture, Agricultural
Agri. Econ., Agr. Econ.	Agriculture Economics
Agri. Ed.	Agriculture Education
Agri. Eng.	Agricultural Engineer(ing)
Agri. Sc., Agr. Sc.	Agricultural Sciences
Agri. Soil Sc.	Agricultural Soil Science
Agri.Chem., Agr. Chem.	Agricultural Chemistry
Agron., Agro.	Agronomy
Agron. Eng.	Agronomy Engineer
A.L.A.	Associate of Library Association of London
A.M.	Master of Arts
Amer. Hist.	American History
Amer. Lit.	American Literature
Anal.-Prog.	Analysist Programmer
anatom. patho.	Pathological Anatomony
Anc. Ind. Cult.	Ancient Indian Culture
Angl.	English
Anim. Husb.	Animal Husbandry
Anim. Sc.	Animal Science
Anthr.	Anthropology
App. Math.	Applied Mathematics
App. Phys.	Applied Physics
Apprent.	Apprenticeship
app.sc., App.Sc.	Applied Sciences
Arch.	Architect(ure)
Arch. and Town Plan.	Architecture and Town Planning
Arch. Eng.	Architectural Engineer(ing)
Arch.Sc.	Archaeological Science
A.R.C.S	Associate, Royal College of Science
A.R.I.S	Associate, Royal Institute of Chemistry
Art Hist.	Art History
Arts Ed.	Degree in Arts Education
Assess.	Assessor
Assist.	Assistant
Assoc.	Associate; Associate of Professional Organization
Astro.	Astronomy, Astrology
A.T.I.	Associate, Textile Institute
Atty at Law	Attorney at Law
B.	Clerics Regular of Saint Paul (Barnabites)
B.A.	Bachelor of Arts
B.A.	Ordre Basilien Alépin
B.A. (uncompl)	Courses toward a Bachelor of Arts
Bacc.	Baccalauréat
B.Acct.	Bachelor of Accountancy
B.Ag.Chem.	Bachelor of Agricultural Chemistry
B.Agri.	Bachelor of Agriculture
B.Ag.Sc.	Bachelor of Agricultural Science
B.A.O.	Bachelor of Obstetrics
B.App.Sc.	Bachelor of Applied Science
Bar.	Certified to practice law
Bar. at Law	Barrister-at-Law
B.Arch.	Bachelor of Architecture
B.B.A.	Bachelor of Business Administration

B.B.M.	Bachelor of Business Management
B.C.E.	Bachelor of Civil Engineering
B.Chem.E.	Bachelor of Chemical Engineering
B.C.L.	Bachelor of Civil Law
B.Comm.	Bachelor of Commerce
B.D.	Bachelor of Divinity
B.D.S.	Bachelor of Divine Scriptures
B.E.	Bachelor of Engineering
B.Econ.	Bachelor of Economics
B.Ed.	Bachelor of Education
B.E.E.	Bachelor of Electrical Engineering
B.Eng.Sc.	Bachelor of Engineering Science
B.F.A.	Bachelor of Fine Arts
B.Fin. and Econ.	Bachelor of Finances and Economics
B.G.S.	Brothers of the Good-Shepard
B.H.L.	Bachelor of Hebrew Letters
B.Hum.	Bachelor of Humanities
Bib. Ed.	Bible Education
B.I.E.	Bachelor of Industrial Engineering
Bio.	Biology
Bio. Sc.	Biological Sciences
Biochem.	Biochemistry
Biophys.	Biophysics
B.Litt.	Bachelor of Letters
B.L.L.	Bachelor of Laws
B.L.S.	Bachelor of Library Science
B.M.	Bachelor of Medicine
B.Metal.	Bachelor of Metallurgy
B.Mus.	Bachelor of Music
Bot.	Botany
B.Pharm.	Bachelor of Pharmacology
B.Phil.	Bachelor of Philosophy
B.S., B.Sc.	Bachelor of Science
B.S.	Basilian Salvatorian Fathers
B.S.A.	Bachelor of Scientific Agriculture
B.S.B.A.	Bachelor of Science in Business Administration
B.Sc.A.	Bachelor of Science in Agriculture
B.S.D.	Bachelor of Didactic Science
B.S.E.E.	Bachelor of Science in Electrical Engineering
B.S.E.	Bachelor of Science in Engineering
B.S.F.	Bachelor of Science in Forestry
B.S.L.	Bachelor of Science in Law
B.Soc.Sc.	Bachelor of Social Science
B.S.T.	Bachelor of Sacred Theology
B.Surg.	Bachelor of Surgery
B.Tech.	Bachelor of Technology
B.Th.	Bachelor of Theology
B.T.S.	Certified Higher Level Technician
Bus.	Business
Bus. Admin.	Business Administration
Bus. Fin.	Business Finance
Bus. Org. and Man.	Business Organization and Management
B.V.M.	Bachelor of Veterinary Medicine
B.V.Sc.	Bachelor of Veterinary Science
C.Acct.	Chartered Accountant
Cand.	Candidate
C.A.P.	Certificate of Professional Aptitude
C.A.P.C.E.G.	Certificate of Pedagogical Aptitude to Teach in College
C.A.P.E.S.	Certificate of Pedagogical Aptitude to Teach in Secondary School
C.Chem.	Chartered Chemist
CC.RR.MM.	Clerics Regular Minor (Adorno fathers)
C.D.E.	Certified Diabetic Educator
Cert.	Certificate
C.E.S. Médecine Trop.	Certificat d'Etudes Cliniques Spéciales Mention Médecine Tropicale
C.E.S. Psy.	Certificat d'Etudes Cliniques Spéciales Mention Psychiatrie
C.E.S.M.O.	Certificat d'Etudes Cliniques Spéciales Mention Orthodontie
C.F.A.	Congregation of Cellites
C.F.C.	Congregation of Christian Brothers
C.F.I.C.	Fils de l'Immaculatée-Conception (Conceptionists)
C.F.P.	Brothers of the Poor of Saint Francis
C.F.S.	Congregation of the Sacordotal Fraternity
C.F.X.	Brothers of Saint Francis Xavier (Xaverian Brothers)
Ch.B.	Bachelor of Surgery
Ch.D.	Doctor of Chemistry
Chem.	Chemistry
Chem. Eng.	Chemical Engineer(ing)
Chem. Tech.	Chemical Technology
Chem.Sc.	Chemical Science(s)
chim.phys.	Physical Chemistry
Chin.	Chinese
C.I.	Josephite Fathers
C.I.C.M.	Congregation of the Imaculate Heart of Mary (Scheut Fathrs)
C.I.M.	Congregation of Jesus and Mary (Eudisti)
C.I.M.S.	Certificate, Institute of Management Services, London
Civ. Av. Eng.	Civil Aviation Engineer
Civ. Eng.	Civil Engineer(ing)
Civ. Law.	Civil Law
Civ. Plan.	Civil Planning
Class.	Classics
Class. Philo.	Classical Philology
C.M.	Congregation of the Mission (Lazarists)
C.M.	Master in Surgery
C.M.F.	Missionary Sons of the Immaculate Heart of Mary (Claretians)
C.M.I.	Carmalites of Mary Immaculate
C.M.M.	Congregation of Marianhill Missionaries
C.M.S.F.	Missionary Brothers of Saint Francis of Assisi
C.M.V.	Mechitarists

Comm.	Commerce
Comm. Int.	International Commerce
Comm. Law	Community Law
Comm. Off.	Officer Commission in the Armed Forces
comm.sc., Comm.Sc.	Commercial Science, Communication Science
Commun.	Communications
Comp. Law Sea	Comparative law of the Sea
Comput. Sc.	Computer Sciences
C.Op.	Congrégation des Ouvriers Chrétiens de Saint Joseph Calasanzio (Calasantins)
Cosmic ray Phys.	Cosmic ray Physics
C.P.	Congregation of the Passion (Passionists)
C.P.C.R.	Coopérateurs Paroissiaux du Christ Roi
C.P.G.E.	Classes Préparatoires aux Grandes Ecoles
C.P.M.	Congregation of the Fathers of Mercy
C.PP.S	Society of the Prescious Blood
C.P.S.	Congregation of the Sacred Stigmata (Stimmatini)
C.R.	Congregation of Regular Clerics
C.R.	Congregation of the Resurrection (Resurrectionists)
C.R.D.	Canonici Regolari della Congrégazione Ospitaliers del Gran San Bernardo
C.R.I.C.	Canonici Regolari dell' Immacolata Concezione
C.R.l.	Canonici Regolari della Congregazione del Ss.mo Salvatore Lateranenso
C.R.S.	Clerics Regular of Somasca (Somascan Fathers)
C.R.S.A.	Canons Regular of Saint Augustine
Crse.	Course(s)
C.S.	Pious Society of the Missionaries of Saint Charles (Scalabrini Fathers)
C.S.B.	Congregation of Saint Basil
C.S.C.	Congregation of the Holy Cross
C.S.C.T.	Certificat de Synthèse Clinique et Thérapeutique
C.S.D.	Doctor of Christian Science
C.Sec.	Chartered Secretary
C.S.F.	Congrégation de la Sainte Familie de Bergame
C.S.I.	Pious Congregation of Saint Joseph (Giuseppini del Murialdo)
C.S.J.	Sisters of Saint Joseph
C.S.M.A.	Congregation of the Archangel Saint Michael
C.S.P.	Congregation of the Missionary Society of Saint Paul the Apostle (Paulist Fathers)
C.S.S.	Congregation of the Sacred Stigmata (Stimmatini)
C.S.Sp.	Congregation of the Holy Spirit
C.SS.R.	Congregation of the Most Holy Redeemer (Redemptorists/Liguoriens)
C.S.V.	Clerics of Saint Viator
Cult. Anthr.	Cultural Anthropology
C.V.	Vicentian Congregation
C.V.U.O.S.B.	Congregazione Benedettina Vallombrosana
d., D.	Law
D. en D.	Doctor of Law
D. en D. Comparé	Doctor of Comparative Law
D. en D. Privé	Doctor of Private Law
D. ès L.	Doctor of Letters
D. ès Sc.Econ.	Doctor of Economic Sciences
D. ès Sc.	Doctor of Sciences
D.A.S.	Doctor of Agricultural Science
D.B.A.	Doctor of Business Administration
D.C.	Doctor of Chiropractic
D.C.	Prères de la Doctrine Chrétienne
D.Catéchèse	Doctor of Catechism
d.C.d.G.	Society of Jesus (Jesuits)
D.C.E.M.	Deuxième Cycle d'Etudes Médicales
D.C.L.	Doctor of Civil Law
D.D.	Doctor of Divinity
D.D.S.	Doctor of Dental Surgery
D.E.A.	Diplôme d'Etudes Approfondies
D.Econ.	Doctor of Economics
D.Ed.	Doctor of Education
D.E.F.A.	Diplôme d'Etudes Fondamentales en Architecture
Deg.	Degree
D.Eng.	Doctor of Engineering
Derecho	Law
D.E.S.	Diplôme d'Etudes Supérieures
D.E.S.S.	Diplôme d'Etudes Supérieures Spécialisées
D.E.U.G.	Diplôme d'Etudes Universitaires Générales
D.E.U.S.T.	Diplôme d'Etudes Universitaires Scientifiques et Techniques
D.F.A.	Doctor of Fine Arts
D.H.L.	Doctor of Hebrew Letters
D.H.P.	Doctor of Public Health
D.I.C.	Diploma of the Imperial College
Dip.	Diploma, Diplomacy
Dip. 1er Cycle	Diplome du 1er Cycle
Dip. 2ème Cycle	Diplome du 2ème Cycle
Dip. 3ème Cycle	Diplome du 3ème Cycle
Dip. A.A.	Diploma from the Agricultural Association, London
Dip. All.	Diploma in German Language
Dip. Sc.	Diplomatic Science(s)
D.Jur.	Doctor of Jurisprudence
D.Law	Doctor of Law
D.Litt.	Doctor of Letters
D.L.S.	Doctor of Library Science
D.M.	Doctor of Medicine
D.M.D.	Doctor of Medical Dentistry
D.M.Sc.	Doctor of Medical Sciences
D.Mus.	Doctor of Music
D.M.V.	Doctor of Veterinary Medicine
D.N.I.Chr.	Congregation of the Resurrection (Resurrectionists)
d.O.	Confederation of the Oratory of Saint Philip Neri (Oratorian Fathers)

Doc.	Doctorate
Doc. 3ème Cycle	Doctorat du 3ème Cycle
Doc. d'Etat	Doctorate certified by the State
D.P.A.	Doctor of Public Administration
D.Phil.	Doctor of Philosophy
D.P.H.	Doctor of Public Health
D.P.L.G.	Diplômépar le Gouvernement
D.P.M.	Diploma in Psychological Medicine
Dr.	Doctor
Dr. D.	Doctor of Dentistry
Dr. ès Lett	Doctor of Letters
Dr. ès Sc.Econ.	Doctor of Economic Sciences
Dr. ès Sc.	Doctor of Sciences
Dr.habil.	Doctor with University Teaching Qualification
Dr.h.c.	Honorary Doctor
droit, D.	Law
droit des gens	Law of Nations
droit priv.	Private Law
droit pub.	Public Law
Drs.	Doctorandus
D.Sc.	Doctor of Science
D.S.T.	Doctor of Sacred Theology
D.T.A.	Doctor of Tropical Agiculture
D.T.M.	Doctor of Tropical Medicine
D.U.T.	Diplôme Universitaire de Technologie
D.V.M.	Doctor of Veterinary Medicine
E.Asian	East Asian
E.C.M.C.	Camaldolese Hermits of the Congregation of Monte Corona
Ecol.	Ecology
Econ.	Economics, Economics Degree
Econ. Geol.	Economic Geology
Econ. Sc.	Economic Sciences
E.D.	Doctor of Education
Ed.	Education
Ed. D.	Doctor of Education
ed.phy.	Physical Education
E.E.	Electrical Engineering
Elect., Electron.	Electronics
Elect. Eng.	Graduate Electronics Engineer
Electro Tech. Eng.	Electro Technical Engineer
Electron.E.	Electronic Engineering
Eng.	Engineering, Graduate Engineer
Eng. Air	Graduate Air Engineer
Eng. and Mech.	Graduate Engineer and Mechanic
Eng. Phys.	Engineering Physics, Graduate in Engineering Physics
Engl.	English
Env. Sc.	Environmental Sciences
Esth.	Esthétique, Aesthetics
Eur.	Européennes, European
Eur. Law	European Law
Eur. Stds.	European Studies

F.B.F.	Hospitalier Order of Saint John of God (fete Bene Fratelli/Frères Hospitaliers de Saint Jean de Dieu)
F.C.	Fils de la Charité
F.d.C.C.	Sons of Charity (Canossians)
F.D.M.	Fratelli di Nostra Signora della Misericordia
F.D.P.	Little Work of Divine Providence (Don Orione)
Fed. dip.	Nationally recognized diploma
F.F.S.C.	Franciscan Brother of the Holy Cross
F.F.S.I.	Frères Fils de Saint Joseph du Rwanda (Bayozefiti)
F.I.C.	Brothers of the Immaculate Conception of the Blesses Virgin Mary
F.I.C.P.	Brothers of Christian Instruction of Saint Gabriel
Fil. Dr.	Doctor of Philosophy
Fil.Kand.	Candidate in Philosophy
Fish.	Fisheries
F.M.I.	Sons of Mary Immaculate (Fathers of Chavagnes)
F.M.I.	Sons of Mary Immaculate (Pavonians)
F.M.M.	Brothers of Mercy of Montabaur
F.M.M.A.	Frères de la Misericorde de Sainte Marie-Auxiliatrice
F.M.S.	Marist Brothers (Little Brothers of Mary)
F.N.	Congrégation de la Sainte Famille de Nazereth
F.N.D.L.	Brothers of Our Lady of Lourdes
For., Forest.	Forestry
Foreign Ser.	Foreign Service
F.P.M.	Presentation Brothers
Fr.	French
Fr. Lang. and Lit.	French Language and Literature
F.S.C.	Brothers of the Christian Schools (Christian Brothers)
F.S.F.	Institut des Frères de la Sainte Famille de Belley
F.S.G.C.	Frères de Saint Joseph-Benout Cottniengo (Cottolenghini)
F.S.M.I.	Fils de Sainte Marie Immaculée
F.S.P.	Brothers of Sainte Patrick
F.T.C.L.	Fellowship Trinity College of Music, London
Gen. Agr.	General Agriculture
Gen. Sc.	General Science
géo., Geog.	Geography
Geol.	Geology
Geol. Sc.	Geological Sciences
Ger.	German
Govt. and Int. Rel.	Government and International Relations
Grad.	Graduate
gramm.	Grammar
habil., Habil.	Certificate to Teach
hautes études eur., High Eur. St.	Graduate Studies of Europe

H.E.	Home Economics
High. Ed.	Higher Education
hist., Hist.	History
Hist. Geog.	Historical Geography
H.N.C. Mech.	Higher National Certificate in Mechanics
Hons	Honours degree
Hort.	Horticulture
Hosp. Admin.	Hospital Administration
Hotel Man.	Hotel Management
Hum.	Humanities
Hum. Bio.	Human Biology
Hum. Factors	Human Factors
Hum. Geog.	Human Geography
Hydr.	Hydraulics
Hydraulic Eng.	Hydraulic Engineering
Hydro.	Hydrology
I.C.	Institute of Charity (Rosminians)
I.E.M.E.	Spanish Institute for Foreign Missions
I.H.V.E. Dip.	Diploma, Institute of Heating and Ventilating Engineers (UK)
I.M.C.	Institute of the Consolata for Foreign Missionaries
Imm.	Immunology
Ind. Admin.	Industrial Administration
Ind. Man.	Industrial Management
Ind. Rel.	Industrial Relations
Indon. Lang. and Lit.	Indonesian Language and Literature
Info. Sc.	Information Science
ing., Ing.	Engineering, Graduate Engineer
Ing. Agri.	Agricultural Engineering
Ing. ès sc.	Sciences Engineer
Ing.agri.	Graduate in Agriculture
Ing.agro.	Graduate Agronomist
Ing.arch.	Graduate Architect
Ing.chem.	Graduate Chemist
Ing.civ.	Graduate Civil Engineer
Ing.civ. du Génie rural	Civil Engineer for Rural Engineering
Ing.comm.	Graduate in Commerce
Ing.Dr.Techn.	Doctor of Engineering Technology
Ing.Elec.	Electrical Engineer
Ing.Nucl.Power	Nuclear Power Engineer
Int. Dev. and Coop.	International Development and Cooperation
Int. Econ. Rel.	International Economic Relations
Int. Econ.	International Economics
Int. Law	International Law
Int. leg. Stds.	International Legal Studies
Int. Mark.	International Marketing
Int. Rel.	International Relations
Int. Stds.	International Studies
Int. Tax Prog.	Internal Tax Programme
Interp.	Interpretation
J.B.	Bachelor of Law

J.C.B.	Bachelor of Canon Law
J.C.D.	Doctor of Canon Law
J.C.L.	Lecturer of Canon Law
J.D.	Doctor of Jurisprudence
Jour.	Journalism
J.S.D.	Doctor of Juridical Science
J.U.D.	Doctor of both Canon and Civil Law
Juris., Jur., jur.	Jurisprudence
Juris Dr.	Doctor of Jurisprudence
L.A. Stds.	Latin American Studies
Lang.	Language
L.C.	Legionaries of Christ
Ldrshp and Org. Behav.	Leadership and Organizational Behaviour
Leg.	Legal
Legal Ed.	Legal Education
lett.	Letters
lett.class.	Classical Letters
L.H.D.	Doctor of Humane Letters
Lib.	Library, Librarianship
Lib. Arts	Liberal Arts
Lic.	Licentiate
Lic.sc., Lic. Sc.	Licentiate in Sciences
Ling.	Linguistics
Ling. App.	Applied Linguistics
Lit.	Literature
Lit. Hum.	Literary Humanities
Litt.	Letters
Litt. angl.	Littérature Anglaise, English Literature
LL.B.	Bachelor of Laws
LL.D.	Doctor of Laws
L.L.L	Licentiate in Laws
LL.M.	Master of Laws
L.M.C.C.	Licentiate, Medical Council of Canada
L.S.	Library Sciences
M. en psy.	Master of Psychology
M. es sc.	Master of Sciences
M.A.	Master of Arts
M.Afr.	Society of Missionaries of Africa (White Fathers)
M.Ag.Sc.	Master of Agricultural Science
Maître	Master of Arts
M.A.I.	Master of Engineering
Man.	Management
Man. Sc.	Management Sciences
Man. Stds.	Management Studies
Mar. Sc.	Marine Science
Math.	Mathematics
M.B.	Bachelor of Medicine
M.B.A.	Master in Business Administration
M.B.B.S.	Bachelor in Medicine and Surgery
M.C.	Order of the Discalced Mercedarians
M.C.B.S.	Congregazione Missionaria del Santissimo Sacramento

M.C.C.I.	Comboni Missionaries of the Heart of Jesus
M.C.E.	Master in Civil Engineering
M.Chem.A	Master in Chemical Analysis
M.Chem.	Master in Chemistry
M.C.L.	Master in Civil Law
M.C.P.	Master of City Planning
M.D.	Doctor of Medicine
M.Div.	Master of Divinity
M.D.S.	Master of Divine Scriptures
M.D.V.	Doctor of Veterinary Medicine
M.E. Stds.	Middle East Studies
Mech. Eng.	Mechanical Engineer(ing), Graduate Mechanical Engineer
Mech. Sc.	Mechanical Science
M.Econ.Sc.	Master of Economic Sciences
M.Econ.	Master of Economics
M.Ed.	Master of Education
med., Med.	Medicine
Med. and Surg.	Medicine and Surgery
Med. Soc.	Medical Sociology
Med. Sur. and Obs.	Medicine, Surgery and Obstetrics
Med. Vet., med.vet.	Veterinary Medicine
med.dent	Dental Medicine
Med.Sc.	Medical Sciences
M.E.P.	Paris Foreign Mission Society
Met.	Metallurgy
Meteo.	Meteorology
M.E.	Master of Engineering
M.F.A.	Master of Fine Arts
M.F.	Master of Forestry
M.G.	Missions Etrangères de Sainte Marie de Guadalupe
M.H.L.	Master of Hebrew Letters
M.I.	Order of the Servants of the Sick (Camillians)
M.I.A.G.E.	Maîtrise de Méthodes Informatiques Appliquées àla Gestion
Microbio.	Microbiology
M.I.C.	Marian Fathers of the Immaculate Conception of the Blessed Virgin Mary (Masrian Fathers and Brothers)
Mil. Off.	Military Officer
Mil.Sc.	Military Science
Min. Prosp. Expl. and Mining	Mineral Prospecting, Exploration and Mining
Min.Inf.	Order of the Servants of the Sick (Camillians)
Mining Eng.	Mining Engineer
M.J.	Master of Jurisprudence
M.J.	Missionaries of Saint Joseph of Mexico
M.L.	Master of Laws
M.L.A.	Master of Landscape Architecture
M.L.D.	Master of International Law
M.L.S.	Master of Library Science
M.M.	Catholic Foreign Mission Society of America (Maryknoll Missioners)
M.Math.	Master of Mathemathics
M.Mus.	Master of Music
M.O.	Medical Officer
M.O.	Missionaries of Workers
Mod. Chin.	Modern Chinese
Mod. Lang.	Modern Languages
M.P.A.	Master of Public Administration
M.P.H.	Master of Public Health
M.Phar.	Master of Pharmacology
M.R.E.	Master of Religious Education
M.S.	Master of Science
M.S.	Missionaries of Our Lady of LaSatto (LaSaletto Fathers)
M.Sc.	Master of Science
M.S.C.	Missionaries of the Sacred Heart of Jesus
M.S.D.	Master of Dental Surgery
M.S.F.	Congregation of the Missionaries of the Holy Family
M.S.F.S.	Missionnaries de Saint François de Sales d'Annecy
M.S.G.	Maîtrise de Sciences de Gestion
M.S.P.	Missionnaries Serviteurs des Pauvres (Boccone del Povero)
M.S.P.H.	Master of Science in Public Health
M.Sp.S.	Missionaries of the Holy Ghost
M.SS.CC.	Missionaries of the Sacred Hearts of Jesus and Mary
M.S.S.P.	Missionary Society of Saint Paul
M.S.SS.T.	Missionary Servants of the Most Holy Trinity
M.S.S.S.	Master of Science in Social Science
M.S.T.	Maîtrise de Sciences et Techniques
M.S.U.	Monaci della Regola Studita (Monaci Studiti Ucraini)
M.S.W.	Master of Social Work
M.T.D.	Midwife Teachers Diploma
M.Th.	Master of Theology
Mus.	Music
Mus. Ped.	Music Education
M.X.Y.	Foreign Missions of Yarumel
Nat. Sc., Nat.	Natural Science(s)
Nav.	Navigation
Nav. Stds.	Naval Studies
Non-Acc. Man.	Non-Accounting Managers
Not.	Notary
Nuc. Phys.	Nuclear Physics
O. de M.	Order of Our Lady of Mercy
O.A.D.	Order of the Discalced Augustines
O.A.M.	Ordine Antoniano Maronita (Antonines)
O.A.R.	Order of the Augustiniane Recollects
O.C.	Order of Brothers of the Blessed Virgin Mary of Mount Carmel (Carmelites)
O.Carm.	Order of Brothers of the Blessed Virgin Mary of Mount Carmel (Carmelites)
O.Cart.	Carthusians

Occ. Med.	Occupational Medicine
Occ. Therapy	Occupational Therapy
O.C.D.	Order of Discalced Brothers of the Blessed Virgin Mary of Mount Carmel (Discalced Carmelites)
O.Cist.	Order of Cistercians
O.C.R.	Order of Cistercians of the Strict Observance (Trappists)
O.C.S.O.	Order of Cistercians
oec.	Business Management
oec.publ.	Public Economics
O.F.M.	Order of Friers Minor (Franciscans)
O.F.M.Cap.	Order of Friers Minor Capuchin (Capuchina)
O.F.M.Conv.	Order of Friers Minor Conventural (Conventuals)
O.H.	Hospitalier Order of Saint John of God (fete Bene Fratelli/Frères Hospitaliers de Saint Jean de Dieu)
O.I.C.	Order of the Imitation of Christ
O.L.M.	Ordine Libanese Maronita (Antonines)
O.M.	Order of Minims
O.M.D.	Clerks Regular of the Mother of God
O.M.I.	Missionary Oblates of Mary Immaculate
O.M.M.	Ordine Maronita Mariamita (Antonines)
O.M.V.	Oblats de la Vierge Marie
O.P.	Order of Preachers (Dominicans)
O.Praem.	Order of the Canons Regular of Prémontré
O.R.C.	Canons Regular of the Order of the Holy Cross
Ord.Cist.	Order of Cistercians
Org. Chem.	Organic Chemistry
Org. Perf.	Organ Performance
Orient. Stds.	Oriental Studies
O.S.A.	Order of Saint Augustine (Augustinians)
O.S.B.	Order of Saint Benedict (Benedictines)
O.S.B.I.	Ordine Basiliano Italiano di Grottaferrata (Order of Saint Basil the Great; Basilians)
O.S.B.M.	Basilian Order of Saint Josaphat (Order of Saint Basil th Great; Basilina)
O.S.C.	Order of the Holy Cross (Crosiers)
O.S.F.	Order of Saint Francis (Franciscan Brothers)
O.S.F.S.	Oblates of Saint Francis de Sales
O.S.H.	Order of Saint Jarome
O.S.I.	Congregation of the Oblates of Saint Joseph (Giuseppini d'Asti)
O.S.M.	Order of the Servants of Mary (Servites)
O.S.P.	Oblate Sisters of Providence
O.S.P.P.E.	Order of Saint Paul the First Hermit
O.SS.T.	Order of the Most Holy Trinity for the Redemption of Captives (Trinitarians)
O.S.U.	Order of Saint Ursula (Ursulines)
O.T.	Ordine Teutonico
paed.	Pedagogy
P.C.	Franciscan Poor Clare Nuns (Poor Clares)
P.C.E.M.	Premier Cycle d'Etudes Médicales, 1st Cycle of Medical Studies
P.D. Fel.	Post Doctoral Fellow
P.D. Res.	Post Doctoral Research
P.D. Stds.	Post Doctoral Studies
Pd.D.	Doctor of Pedagogy
Peace Ed.	Peace Education
Ped.	Pedagogy
Ped.Sc.	Pedagogical Science(s)
P.F.M.	Marist Brothers (Little Brothers of Mary)
P.G.	Post Graduate Studies
P.G. Deg.	Post Graduate Degree
P.G. Dip.	Post Graduate Diploma
P.G.C.E., P.G.Cert. Ed.	Post Graduate Certificate in Education
Pharm. Sc.	Pharmacology
Pharm.D.	Doctor of Pharmacy
Ph.B.	Bachelor of Philosophy
Ph.D.	Doctor of Philosophy
Ph.D. (uncompl)	Uncompleted Doctor of Philosophy Program
phil., Phil.	Philosophy
Phil. Sc.	Philosophical Science
phil. Thom.	Thomistic Philosophy
Philo.	Philology
Philo.Sc.	Philological Science
Ph.L.	Licentiate in Philosophy
Ph.M.	Master of Philosophy
Phys.	Physics
Phys. Anthro.	Physical Anthropology
Phys. Ocean.	Physical Oceanography
Phy.Sc.	Physical Science
phys.ed., Phys. Ed.	Physical Education
physio., Physio.	Physiology
Phys.Ther.	Physio Therapy
phys.-med.sc.	Physical and Medical Sciences
Pilot Off.	Pilot Officer
P.I.M.E.	Pontifical Institute for Foreign Missions
Plant Physiol.	Plant Physiology
P.M.E.	Society of Foreign Missions of the Province of Quebec
P.O.C.R.	Pieux Ouvriers Catéchistes Ruraux
Pol.	Politics, Political Science
Pol. Econ.	Political Economics
Pol. Sc.	Political Science
Pol. Stds.	Political Studies
Post Doc.	Post Doctoral Studies
P.P.E.	Philosophy, Politics and Economics
Pre-Med.	Pre-Medical Studies
P.S.D.P.	Poor Servants of the Divine Providence (Don Calabria)
Ps.Dr.	Doctor of Political Science
P.S.M.	Prètres de Sainte Maria de Tinchebray
P.S.S.	Society of the Priests of Saint Sulpice (Sulpicians)
Psy.	Psychology

Pub. Admin.	Public Administration
Pub. Fin.	Public Finance
Pub. Health	Public Health
Pub. Law	Public Law
Radio and TV Dir.	Radio and Television Directing
Radio and TV Prod.	Radio and Television Production
R.C.I.	Rogationistes du Coeur de Jésus
Rel.	Religion
rel.int.	International Relations
rer.agr.	Agriculture Science
rer.nat.	Natural Science
rer.pol.	Political Science
Res. Econ.	Resource Economics
R.M.	Registered Midwife
R.N.	Registered Nurse
R.N.Dr.	Doctor of Nursing
Rom. Philo.	Romanic Philology
R.S.M.	Sisters of Charity
R.S.V.	Religieux de Saint Vincent-dePaul (Péres et Fréres)
Rural Econ.	Rural Economy
S.A.	Franciscan Friars of the Atonement
S.A.C.	Society of the Catholic Apostolate (Pallotines)
S.B.	Bachelor of Science
S.C.	Institute of the Brothers of the Sacred Heart
sc., Sc.	Science
sc.com.consul.	Commercial and Consular Sciences
sc.comm.	Commercial Sciences
Sc.D.	Doctor of Science
sc.econ., sc.écon.	Economic Science
Sch.P.	Poor Clerks Regular of the Mother of God of the Pious Schools (Piarist Fathers/Order of the Pious Schools)
S.Chr.	Società di Cristo per Gli Emigrati della Polonia
S.C.I.	Priests of the Sacred Heart of Jesus (Dehonians)
S.C.J.	Congregation of the Sacred Heart of Jesus (Pères de Timon-David)
sc.mor. et pol.	Moral and Political Sciences
sc.nat.	Natural Science
sc.phys.	Physical Sciences
sc.pol.	Political Science
sc.pol. et coloniales	Political and Colonial Sciences
S.C.S.C.	Sisters of Mercy of the Holy Cross
sc.soc	Social Sciences
S.D.B.	Society of Saint Francis de Salew (Salesians of Don Bosco)
S.d.C.	Servants of Charity (Opera Don Guanelia)
S.D.S.	Society of the Divine Savior (Salvatorians)
S.D.V.	Society of the Divine Vocations (Vocationalists Fathers)
S.F.	Sons of the Holy Family
S.F.M.	Scarboro Foreign Mission Society
S.G.	Brothers of Christian Instruction of Saint Gabriel
S.I.	Society of Jesus (Jesuits)
S.J.	Society of Jesus (Jesuits)
S.J.D.	Doctor of Juridical Science
S.M.	Master of Science
S.M.	Society of May (Marist Fathers/Brothers of Mary)
S.M.A.	Society of African Missions
S.M.B.	Bethlehem Mission Society
S.M.N.	Missionaries of the Company of Mary (Montfot Fathers)
S.M.P.	Portuguese Catholic Missions
S.N.D.	Sisters of Notre Dame de Namur
soc., Soc.	Sociology
Soc. and Phys. Geog.	Social and Physical Geography
Soc. Ldrshp.	Social Leadership
Soc. Psy.	Social Psychology
Soc. Rel.	Social Relations
Soc. Sc.	Social Sciences
Soc. Stds., Soc.St.	Social Studies
Soc. Wrk., Soc. Work	Social Work
S.O.Cist.	Cistercians of the Common Observance
S.P.	Poor Clerks Regular of the Mother of God of the Pious Schools (Piarist Fathers/Order of the Pious Schools)
S.P.	Servants of the Pareclete
S.P.S.	Saint Patrick's Society of the Foreign Missions
S.S.	Society of Saint Sulpice (Sulpicians)
SS.CC.	Congregation of the Sacred Hearts of Jesus and Mary
S.S.C.M.E.	Missionary Society of Saint Columban
S.S.C.	Sociétè Sacerdotale de Saint Joseph-Benoit Cottoiengo
S.S.E.	Society of Saint Edmund
S.S.I.	Missionaries of Saint Joseph of Mexico
S.S.J.	Saint Joseph's Society of the Sacred Heart (Josephite Fathers)
S.S.P.	Society of Saint Paul
S.S.S.	Congregation of the Blessed Sacrament (Pères du Saint-Sacrement)
S.T.	Missionary Servants of the Most Holy Trinity
Stat.	Statistics
S.T.L.	Professor of Sacred Theology
S.T.M.	Master of Sacred Theology
S.V.D.	Society of the Divine Word (Divine Word Missionaries/Verbiti)
S.X	Society of Saint Francis Xavier for the Foreign Missions (Xaverians)
T.C.	Third Order Capuchin of Our Lady of Sorrows
T.Chr.	Società di Cristo per gli Emigrati della Polonia
Tchr.	Teacher

Tchr. Cert.	Teaching Certificate
Tchr. Prim.	Primary Teacher
Tchr. Sec.	Secondary Teacher
T.D.	Congregation of the Sacred Heart of Jesus (Pères de Timon-David)
Tech.	Technology
Tech.Sc.	Technical Sciences
Th., Theol., théol.	Theology
Th. and Hist. Mus.	Theory and History of Music
Th.B.	Bachelor of Theology
Th.D.	Doctor of Theology
Theol. Sc.	Theological Sciences
Theor. Phys.	Theoretical Physics
Th.M.	Master of Theology
Thom.Phil.B.	Bachelor of Thomistic Philosophy
T.O.R.	Secular Franciscan Order
Toxi.	Toxicology
Trans.	Translation
Trop. Agri.	Tropical Agriculture
Urban/Reg.Stds.	Urban/Regional Studies
Vet.	Veterinary Medicine
Zoo.	Zoology

Bibliography

Canadian Who's Who, 1991, Volume XXVI, University of Toronto Press, Toronto, Buffalo, London, 1991.

Debrett's Distinguished People of Today, 1990, Published by Debrett's Peerage Limited, London.

The International Who's Who, 1990-91, 54th Edition 1990, Europa Publications Limited.

Who's Who in Japan, 1991-92, Asia Press Co Ltd. 1991.

Who's Who in the Arab World, 1990-1991, Tenth Edition, Publitec Publications, an Imprent for K.G. Saur, München, New York, London Paris, 1990.

Who's Who in World Jewry, A Biographical Dictionary of Outstanding Jews, 1987, Published by Who's Who in World Jewry Inc., Baltimore MD, New York NY.

The World of Learning, 1989, 39th Edition, Europa Publications Limited.

Biographies

Aagard, J, Dr
Rel: Christian.
International Organizations: Chairman for Scandinavia of the Christian Mission to Buddhists.
Mailing address: Aarhus Universitet, NDR Ringgade, DK-8000 Århus C, Denmark.

Aarflot, Andreas, Rt Rev
Rel: Christian.
International Organizations: Presiding Bishop of Church of Norway.
Mailing address: PO Box 5816 Hegdehaugen, N-0308 Oslo 3, Norway.

Aarflot, Helge
Rel: Lutheran.
International Organizations: Secretary-General of the Lutheran European Commission on the Church and the Jewish People.
Mailing address: Collettsgate 43, 2 Etg, N-0456 Oslo, Norway. T. (47 2) 46 18 58.

Aasi, Ghulam-Haider, Prof Dr
Rel: Muslim; Unordained; b. 10 SEP 1942 in Hairo-Sharqi, DG Khan, Punjab, Pakistan; m. to Zubaida Zareen; c. 2 boys and 3 girls; Citizen of Pakistan; Lang. English, Urdu, Persian, Arabic and French.
International Organizations: Chairman and Professor of Islamic Studies (since 1983) of the American Islamic College.
Member of the Council (since 1990) of the Council for a Parliament of the World's Religions.
Other organizations: Co-Chair and Founder (since 1988), Conference for Improved Christian-Muslim Relations; Lecturer at Loyola University, Chicago IL (since 1991).
Main field of work: University education, dialogue. **Educ.:** M.A.s Arabic, Islamic, Urdu from University of the Punjab, Lahore, Pakistan (1964-1970); M.A. Comp.Rel. from Temple University, Philadelphia PA, USA (1974-1976); Ph.D. Rel. from Temple University, Philadelphia PA, USA (1978-1986). **Interests:** inter-religious dialogue. **Mem.:** AAR, MESA, NASR **Pubs.:** Islam and Other Religions; Muslim Understanding of Other Religions and Iqbal and Dewey.
Mailing address: 640 W Irving Park Road, Chicago IL 60613, USA. T. (1 312) 281 4700.
Residence: 850 W Eastwood Ave, Suite 1601, Chicago IL 60640, USA. T. (1 312) 989 7206

Abbacchi, Janet Marie, SSND, Sr
Rel: Roman Catholic; Ordained; celibate.
International Organizations: Contact of the Association of Advisers on Education in International Religious Congregations.
Member of the School Sisters of Notre Dame.
Mailing address: EDUC International, School Sisters of Notre Dame, Via della Stazione Aurelia 95, I-00165 Roma, Italy. T. (36 6) 680 80 65.

Abbas, S
Rel: Muslim.
International Organizations: President of the International Islamic Society.
Mailing address: 8F ORS Islamic Cnt, 40 Oi Kwan Road, Wanchai, Hong Kong, Hong Kong. T. (852 5) 366 06 71. Fax 366 96 73.

Abb Pierre, A
See Henri Grous.

Abdin, Al Tayib Zein Al, Dr
Rel: Muslim.
International Organizations: Director of the Islamic African Centre, Khartoum.
Mailing address: PO Box 2489, Khartoum, Sudan. T. (249 11) 224430 – 224469 – 221345. C. ISLAMI KHARTOUM. Tx 24086 BADR.

Abdulnoor, Yusef, Rev
Rel: Evangelical; Ordained; b. 05 APR 1933 in Egypt; m. to Soad Abdul Noor; c. 3 girls; Citizen of Egypt.
International Organizations: Member, Executive Committee of the Middle East Council of Churches.
Main field of work: Ministry. **Career:** Pastor, The Arab Congregation. Pastor, The National Evangelical Church, Kuwait in 1959. President, The National Evangelical Church in Kuwait in 1963.
Mailing address: PO Box 80, Safat, Kuwait. T. (965) 433 230
Residence: See Mailing Address. T. (965) 438 498

Abe, Nikken
Rel: Buddhist.
International Organizations: High Priest of the Nichiren Shoshu.
Mailing address: No mailing address.

Abebaw, Yegzaw, Ato
Rel: Orthodox: Ethiopian.
International Organizations: General Secretary of the Ethiopian Orthodox Church.
Mailing address: Patriarchate Head Office, PO Box 1283, Addis Ababa, Ethiopia.

Abedin, Syed Z, Dr
Rel: Muslim.
International Organizations: Director of the Institute of Muslim Minority Affairs.
Mailing address: PO Box 8856, Jeddah, Saudi Arabia.

Abed-Rabbo, Samir
Rel: Muslim.
International Organizations: President of the Center for Arab-Islamic Studies, Brattleboro VT.
Mailing address: PO Box 678, Brattleboro VT 05302, USA. T. (1 802) 257 0872.

Abel, Lorry, Rev
Rel: Presbyterian.
International Organizations: Moderator of the Presbyterian Church of Vanuatu.
Mailing address: PO Box 150, Vila, Vanuatu.

Abeno, Ryusho
Rel: Buddhist; Ordained; b. 08 NOV 1912; Citizen of Japan.
International Organizations: Manager of the Koyasan Shingonshu.
Career: Chief, Kongobuji. **Educ.:** Koyasan University, Japan.
Mailing address: 132 Kongobu-ji Koyacho Koyasan, Ito-gun, Wakayama 648-02, Japan. T. (81 7365) 6 2011

Abernathy, Ralph David, Rev
Rel: Baptist; Ordained; b. 11 MAR 1926 in Linden AL, USA; m. to Juanita Odessa Jones; c. 4 children; Citizen of USA.
International Organizations: President Emeritus (since 1977) of the Southern Christian Leadership Conference.
Other organizations: Pastor (since 1961), West Hunter Street Baptist Church, Atlanta.
Main field of work: Civil rights. **Career:** Army service during the Second World War. Pastor, First Baptist Church of Montgomery from 1951 to 1961. Organized Montgomery Improvement Association in 1955. Helped organize the Southern Christian Leadership Conference, Atlanta in 1957. Secretary/Treasurer, then Vice-President, Southern Leadership Christian Conference from 1965 to 1968. President, Southern Leadership Christian Conference from 1968 to 1977. Jailed numerous times with late Dr Martin Luther King. **Educ.:** Alabama State University, Montgomery AL, USA; Atlanta University, GA, USA. **Hons.:** Hon. LL.D., Allen University (1960), Long Island University (1969), Alabama State University (1974); Hon. D.D., Morehouse College (1971), Kalamazoo College (1978); Peace Medallion, German Democratic Republic (1971). **Mem.:** National Association for the Advancement of Colored People (NAACP) **Pubs.:** *And The Walls Came Tumbling Down*, 1989 (autobiography).
Mailing address: 1040 Gordon Street SW, Atlanta GA 30310, USA.

Ablondi, Alberto, Mgr
Rel: Roman Catholic; Ordained on 31 MAY 1947; b. 18 DEC 1924 in Milan, Italy; celibate.
International Organizations: President of the Catholic Biblical Federation.
Member of the Pontifical Council for Promoting Christian Unity.
Other organizations: Bishop (since 1970), Livorno.
Mailing address: Via del Seminario 61, I-57100 Livorno, Italy. T. (39 586) 88 90 51

Aboa, S K, Rev
Rel: Presbyterian.
International Organizations: Secretary of the Presbyterian Church of Ghana.
Mailing address: PO Box 1800, Accra, Ghana.

Aboodi, Timotheos Aphrem, H E Archbishop
Rel: Syrian Orthodox; Ordained on 16 FEB 1953; b. 01 JAN 1930 in Mousel, Nineva, Iraq; celibate; Lang. Syriac, Arabic, English and Malayalam.
International Organizations: Archbishop Primate (since 1991) of Syrian Orthodox Church in Australia.
Other organizations: Archbishop (since 1972), Syrian Orthodox Church of Antioch and All the East.
Main field of work: Ministry. **Career:** Chairman, Oriental Orthodox Council of Churches, UK from 1988 to 1991. Editor-in-Chief, *News from Seema* magazine. **Educ.:** Dip. from Clergical School of the Syrian Orthodox Church of Antioch and All the East (1946-1951). **Interests:** Reading, writing. **Pubs.:** *Weapon of the Faithful* (in Syriac and Arabic) and various articles in newspapers and magazines in several countries.
Mailing address: PO Box 257, Lidcombe NSW 2141, Australia. T. (61 2) 649 4608
Residence: 82 Joseph Street, Lidcombe NSW 2141, Australia

Abouchacra, Muhammad, Sheikh
Rel: Druze; Citizen of Lebanon.
International Organizations: Supreme Spiritual Leader of the Druze Community.
Mailing address: Rue Abou Chacra, Beirut, Lebanon.

Aboudi, Mohamad Ibn Nassir
Rel: Muslim.
International Organizations: Secretary-General of the Islamic Call.
Mailing address: General Sec. of the Islamic Call, Oleisha, 37 Ohod Street, Riyadh, Saudi Arabia.

Abraham, Dezso, Rt Rev
Rel: Reformed.
International Organizations: Contact of the Hungarian Reformed Church in America.
Mailing address: 9901 Allen Road, Allen Park MI 48101, USA.

Abraham, K C
Rel: Protestant.
International Organizations: President of the Ecumenical Association of Third World Theologians.
Mailing address: United Theological College, 63 Millers Road, PO Box 4613, Bangalore 560 046, India.

Abrahams, Stuart N
Rel: Christian.
International Organizations: Representative of the Archbishop of Sydney's Overseas Relief and Aid Fund.
Mailing address: St Andrew's House, Sydney Square, Sydney NSW 2000, Australia. T. (61 2) 265 1509. Fax 267 7873.

Abrego de Lacy, José Maria A, Prof
Rel: Roman Catholic; Ordained on 30 JUN 1973; b. 25 MAR 1945 in Pamplona, Navarra, Spain; celibate; Citizen of Spain; Lang. Spanish, Italian, German, English and French.
International Organizations: Dean (since 1988) of the Faculty of Theology, Universidad de Deusto.
Other organizations: Vice-Rector for Students (since 1991), Universidad de Deusto, Bilbao.
Main field of work: University education. **Educ.:** Lic. Phil. from Berchmannskollege, Munich, Germany (1968); Bac. Theol. from University of Deusto, Bilbao, Spain (1974); Doc. Sac. Scripture from Pontifical Biblical Institute, Vatican City State (1982). **Interests:** Biblical research. **Mem.:** Asociación Bíblica Española (ABE); Society of Jesus **Pubs.:** Jeremías y el final del reino, Est AT, 3, Valencia, 1983; "Reflexiones sobre el pecado en el AT", Proyección 35, 1988, pp. 91-103 and "El 'resto' de Israel – La gratitud y la búsqueda, elementos fundantes de una relación con Dios", Sal Terrae, Octubre 1989, pp. 709-718.
Mailing address: Universidad de Deusto, Apartado 1, E-48080 Bilbao, Spain. T. (34 4) 445 31 00

Abril y Castelló, Santos , H E Mgr
Rel: Roman Catholic; Ordained on 19 MAR 1960; b. 21 SEP 1935 in Alfambra, Teruel, Spain; celibate; Citizen of Vatican and Spain; Lang. Spanish, English, French and Italian.
International Organizations: Apostolic Pro-Nuncio to Cameroon, Gabon and Equatorial Guinea (since 1989) of the Roman Catholic Church.
Other organizations: Titular Archbishop, Tamada.
Career: Apostolic Nuncio to Bolivia from 1985 to 1989. **Educ.:** Dip. from Pontifical Academy for Diplomats, Vatican City State; Dr. Soc.Sc.; Dr. Law. **Hons.:** Commander, Order of Isabel la Católica (Spain); Grand Cross, Civil Order of Merit (Germany); Grand Cross, Condor de los Andes (Bolivia).
Mailing address: Rue du Vatican 2, BP 210, Yaoundé, Cameroon. T. (237) 220 475. Tx 8382 NUNCAM KN
Residence: See Mailing Address. T. See Mailing Address

Abromowitz, J M, Rabbi
Rel: Jewish; Ordained.
International Organizations: Chairman, Jerusalem Office of the Agudath Israel World Organization.
Mailing address: Abraham Hirsch, Hacherut Square, POB 326, Jerusalem, Israel. T. (972 2) 384357. Fax 383 634. C. Worldaguda, Jerusalem.

Accogli, Luigi, H E Mgr
Rel: Roman Catholic; Ordained on 06 MAR 1943; b. 16 AUG 1917 in Andrano, Italy; celibate.
International Organizations: Apostolic Pro-Nuncio to Syria (since 1988) of the Roman Catholic Church.
Other organizations: Titular Archbishop, Treba.
Mailing address: Rue Masr 82, BP 2271, Damascus, Syrian AR. T. (963 11) 332 601 – 332 446. Fax 332 601 – 332 446

Acerbi, Angelo, H E Mgr
Rel: Roman Catholic; Ordained on 27 MAR 1948; b. 23 SEP 1925 in Sesta Godano, Italy; celibate.
International Organizations: Apostolic Nuncio to Hungary (since 1990) of the Roman Catholic Church.
Other organizations: Titular Archbishop, Zella.
Mailing address: Budakeszi ut. 55/D p/5, H-1021 Budapest, Hungary. T. (36 1) 761 853 – 763 334. Fax 761 815

Acharuparambil, Daniel, O.C.D., Rev Prof
Rel: Roman Catholic; Ordained on 14 MAR 1966; b. 12 MAY 1939 in Palliport, Kerala State, India; celibate; Citizen of India; Lang. Malayalam, English, Italian, Latin, Hindi, Sanskrit and French.
International Organizations: Rector Magnificus (since 1988) of the Urbanian Pontifical University.
Other organizations: Professor (since 1972), Urbanian Pontifical University; Consultant to Pontifical Council for Inter-Religious Dialogue (since 1990).
Main field of work: University education. **Career:** Professor, Lateran Pontifical University from 1973 to 1983. Dean, Faculty of Missiology, Urbanian Pontifical University from 1986 to 1988. Visiting Professor, Theresianum, Rome in 1976. Visiting Professor, Jyothir Bhavan Institute of Spirituality, India in 1980. **Educ.:** B.D. from Pontifical Seminary, Alwaye, India (1962-1966); B.A. Econ. from St Albert's College, Ernakulam, India (1966-1969); L.Ph. from Pontifical Athenaeum, Poona, India (1969-1970); M.A. Indology from Benares Hindu University, India (1970-1972); Ph.D. from Pontifical University San Tommaso d'Aquino, Vatican City State (1975-1978). **Hons.:** Two Gold Medals for M.A. in Indology from Benares Hindu University (1972). **Pubs.:** *The Destiny of Man in the Evolutionary Thought of Sri Aurobindo*, Rome, 1978; *Espiritualidad hinduista*, Madrid, BAC, 1982 and *La spiritualita dell' Induismo*, Rome, 1986.
Mailing address: Pontificia Università Urbaniana, Via Urbano VIII 16, I-00165 Roma, Italy. T. (39 6) 6548361/2. Fax 6548363
Residence: Piazza San Pancrazio 5/A, I-00152 Roma, Italy. T. (39 6) 5810140

Acker, Raymond A
Rel: Protestant.
International Organizations: Head, Secretariat of the Biblical Ministries Worldwide.
Mailing address: PO Box 464-250, Lawrenceville GA 30246, USA. T. (1 404) 339 3500.

Ackerman, William A
Rel: Protestant.
International Organizations: Director of the World Home Bible League.
Mailing address: 16801 Van Dam Road, South Holland IL 60473, USA. T. (312) 928 7022.

Adachi, Daishin
Rel: Buddhist: Zen; Ordained; b. 02 MAY 1932; Citizen of Japan.
International Organizations: Manager of the Enkakuji Sect of Rinzai Zen Buddhism.
Career: Chief, Enkakuji. **Educ.:** University of Tokyo Graduate School, Japan.
Mailing address: Enkakuji Yamanouchi, Kamakura, Kanagawa 247, Japan. T. (81 467) 22 0478

Adams, David M
Rel: Christian.
International Organizations: Chairman of the International Christian Media Commission.
Mailing address: Trans World Radio, PO Box 2020, 1200 CA Hilversum, Netherlands. T. (31 35) 2117740. Fax 234861.

Adams, Paulus Leopold Maurits, Br
Rel: Roman Catholic; Ordained; b. 07 NOV 1922 in Lubbeek, Belgium; celibate; Citizen of Belgium; Lang. Dutch, French and English.
International Organizations: Secretary-General (since 1983) of the Catholic International Education Office.
Educ.: Dr.paed.; Lic. en Orientation Professionnelle; Grad. en Sc.Rel. from Catholic University of Louvain (KUL), Belgium (1945-1949).
Mailing address: Rue des Eburons 60, B-1040 Bruxelles, Belgium. T. (32 2) 230 72 52.
Residence: See Mailing Address. T. See Mailing Address

Adcock, W O
Rel: Church of Christ.
International Organizations: President of the Associated Churches of Christ in New Zealand.
Mailing address: 90A Mount Street, Nelson, New Zealand.

Addy, Anthony J.
Rel: Christian.
International Organizations: Contact of the Work and Economy Network in the European Churches.
Programme Development Secretary of the European Contact Group on Urban Industrial Mission.
Mailing address: c/o William Temple Foundation, Manchester Business School, Manchester M15 6PB, UK. T. (44 61) 275 6534.

Adejobi, E O A, Dr
Rel: Church of the Lord (Aladura).
International Organizations: Primate of the Church of the Lord (Aladura).
Mailing address: Anthony Village, Ikorudu Road, PO Box 308, Ikeja, Lagos State, Nigeria.

Adeyemo, Tokunboh, Dr
Rel: Evangelical.
International Organizations: Secretary-General of the Association of Evangelicals of Africa and Madagascar.
Mailing address: PO Box 49332, Nairobi, Kenya. T. (254 2) 72 27 69 – (254 2) 72 02 20. Tx 22321 ITCHUS KE. Fax (254 2) 72 02 55.

Adhikary, M S
Rel: Baptist.
International Organizations: President of the Bangladesh Baptist Sangha.
Mailing address: Sadarghat, PO Box 1108, Dhaka I, Bangladesh.

Adiwinoto, Lilian E, Commr
Rel: Salvation Army, The.
International Organizations: Territorial Commander of The Salvation Army.
Mailing address: Kotak Pos 641, Bandung 40001, Java, Indonesia.

Adolf, Felipe
Rel: Christian.
International Organizations: Secretary-General of the Latin American Council of Churches.
Mailing address: Casilla 85-22, Quito, Ecuador. T. (593 2) 561 539 – (593 2) 568 373 – 239 890. Fax 504 377. Tx 21150 CLAI ED.

Ador, Samuel Nyawelo, Rev
Rel: Presbyterian; Ordained on 10 AUG 1970; b. 01 JAN 1941 in Malakal, Upper Nile, Sudan; m. to Tabitha Dak John; c. 3 boys and 3 girls; Citizen of Sudan; Lang. English and Arabic.
International Organizations: Member of the Executive Committee of the World Alliance of Reformed Churches.
Other organizations: Director, Church and Society (since 1987), Sudan Council of Churches.
Main field of work: Evangelism, education. **Educ.:** B.T.H. from Evangelical Seminary, Cairo, Egypt (1965-1969); M.Div. from Near East School of Theology, Beirut, Lebanon (1972-1974); M.Phil. from University of Birmingham, UK (1984-1986). **Pubs.:** Theological Stand on some specific issues, 1988; Islamic Faith Historical Background and Practices, July 1989 and Family and Religion, June 1989.
Mailing address: c/o Sudan Council of Churches, Box 469, Khartoum, Sudan. T. (249 11) 41 509 – 42 855 – 42 859 – 41 137. Tx 22190 ACROP (VIA HôTEL ACROPOLE)

Afonso, Manuel Castro, S.M.P., Rev
Rel: Roman Catholic; Ordained on 31 JUL 1960; b. 26 JUL 1932 in Monçao, Portugal; celibate; Citizen of Portugal; Lang. Portuguese, French, English, Spanish and Bemba.
International Organizations: Superior General (since 1990) of the Portuguese Catholic Missions.
Main field of work: Ministry and education. **Career:** Teacher at Missions' Seminary from 1960 to 1967. Novice Master, Portuguese Catholic Missions from 1967 to 1974. Superior General, Portuguese Catholic Missions from 1974 to 1980. President, Portuguese Religious Conference from 1977 to 1980. Missionary Priest in Ndola Diocese, Zambia from 1981 to 1990. **Educ.:** Dip. Phil. from Missions' Major Seminar, Cocujaes, Portugal (1952-1955); Dip. Theol. from Missions' Major Seminar, Cocujaes, Portugal (1955-1960). **Interests:** ministry, missions, education. **Pubs.:** several articles in various missions' reviews and magazines.
Mailing address: Rua da Bempostinha 30, 1100 Lisboa, Portugal. T. (351 1) 3520646 – 3520647. Fax 3524058.
Residence: See Mailing Address. T. See Mailing Address

Aga Khan, Karim A, IV, H H Prince
Rel: Muslim: Ismaili; b. 13 DEC 1936 in Creux-de-Genthod, Geneva, Switzerland; m. to Sarah Frances Croker-Poole; c. 2 boys and 1 girl.
International Organizations: Founder President (since 1967) of the Aga Khan Foundation.
Main field of work: Spiritual leader and Imam of Ismaili Muslims. **Career:** Institute of Ismaili Studies in 1977. Aga Khan Trust for Culture in 1988. Founder and Chancellor, Aga Khan University, Pakistan in 1983. **Educ.:** B.A. from Harvard University, Cambridge MA, USA. **Interests:** Breeding race-horses, yachting, golf, skiing. **Hons.:** Became Aga Khan on the death of his grandfather Sir Sultan Mohamed Shah, Aga Khan III (1957); Granted title of His Highness by Queen Elizabeth II (1957), of His Royal Highness by the Shah of Iran (1959); Aga Khan Award for Architecture (1976); Commandeur de l'Ordre du Mérite Mauritanien (1960); Grand Croix, Ordre du Prince Henry du Gouvernement Portugais (1960); Ordre National de la Côte d'Ivoire (1965), de la Haute-Volta (1965); Ordre Malgache (1966); Ordre du Croissant Vert des Comores (1966); Grand Cordon Ordre du Tadj de l'Empire d'Iran (1967); Nishan-i-Imtiaz Pakistan (1970); Cavaliere di Gran Croce dell'Ordine al Merito della Repubblica, Italy (1977); Grand Officier de l'Ordre National du Lion, Senegal (1982); Nishan-e-Pakistan, Pakistan (1983); Grand Cordon of Ouissam-al Arch, Morocco (1986); Cavaliere del Lavoro, Italy (1988); Hon. LL.D.: Peshawar University (1967), University of Sind (1970), McGill University Montreal (1983).
Mailing address: Aiglemont, F-60270 Gouvieux, France.

Agazzi, Evandro, Prof Dr
Rel: Catholic; b. 23 OCT 1934 in Bergamo, Italy; m. to Lucia Castelli; c. 1 boy and 1 girl; Citizen of Italy; Lang. Italian, English, French, German and Spanish.
International Organizations: President (since 1988) of the International Federation of Philosophical Societies.
President (since 1978) of the International Academy of Philosophy of Science. Member of the Executive Board (since 1979) of the International Society for Metaphysics.
Other organizations: Full Professor (Philosophy of Science) (since 1970), University of Genova, Italy; Full Professor (Philosophical Anthropology, Philosophy of Nature and Philosophy of Science) (since 1979), University of Fribourg, Switzerland; President (since 1987), Swiss Society for Logic and Philosophy of Science; Dean, Faculty of Humanities (since 1989), University of Fribourg, Switzerland; President, Ligurian Philosophical Association.
Main field of work: Philosophy of science. **Career:** Chairman, Dept of Philosophy, University of Genova, Italy from 1973 to 1982. Visiting Professor at the Universities, Düsseldorf, Berne, Pittsburgh and Geneva. President, Italian Society of Logic and Philosophy of Science. President, Italian Philosophical Society. President, many other learned institutions. Treasurer, International Council for Philosophy and Humanities, UNESCO. **Educ.:** M.Phil. from Catholic University of the Sacred Hearth, Milan, Italy (1953-1957); M.S. Phys. from University of Milan, Italy (1957-1961); P.G. at University of Oxford, UK (1961-1962). **Interests:** Metaphysics, logic, systems theory, ethics of science, bioethics, history of science, philosophy of language, philosophical anthropology, education, music. **Hons.:** Venia legendi in philosophy of Science (1963) and in Mathematical Logic (1966). **Mem.:** International Institute of Phylosophy (Paris), and several other Academies and learned institutions of different countries **Pubs.:** Publications include more than 20 books (author or editor) and more than 300 papers and articles, including contributions to books, anthologies, encyclopedias and journals, apart of many book reviews and newspaper articles: *Introduzione ai problemi dell'assiomatica*, 1961; *La logica simbolica*, 1964; *Temi e problemi di filosofia della fisica*, 1969; *Le geometrie non euclidee e i fondamenti della geometria*, 1978 and *Philosophie, Science, Métaphysique*, 1987.
Mailing address: Séminaire de Philosophie, Université de Fribourg, CH-1700 Fribourg, Switzerland. T. (41 37) 21 94 36. Tx 36110 TxCAB CH.
Residence: Impasse de la Forêt 24, CH-1700 Fribourg, Switzerland T. (41 37) 28 38 85

Agazzi, Mario
Rel: Roman Catholic.
International Organizations: Director of the Italian Church Centre for Latin America.
Mailing address: Bacilieri 1/A, I-37139 Verona, Italy. T. (39 45) 890 0329.

Agré, Bernard, H E Mgr
Rel: Roman Catholic; Ordained on 20 JUL 1953; b. 02 MAR 1926 in Monga, Abidjan, Ivory Coast; celibate.
International Organizations: President of the Regional Episcopal Conference of French-Speaking West Africa.
Other organizations: Bishop, Man.
Mailing address: 08 BP 22, Abidjan 08, Ivory Coast. T. (225) 49 39 43.

Aguirre Herrera, Emilio Manuel A, O. de M., Fr
Rel: Roman Catholic; Ordained; celibate.
International Organizations: Master General of the Order of Our Lady of Mercy.
Mailing address: Via Monte Carmelo 3, I-00166 Roma, Italy. T. (39 6) 680 15 98 – 680 17 16

Agustin, Obregozo Juaregui José, Fr
Rel: Roman Catholic; Ordained; celibate.
International Organizations: Superior General of the Congregation of the Passion.
Mailing address: Piazza SS Giovanni e Paolo 13, I-00184 Roma, Italy. T. (39 6) 700 5236. Fax 700 8454.

Agustoni, Gilberto, H E Mgr
Rel: Roman Catholic; Ordained on 20 APR 1946; b. 26 JUL 1922 in Schaffhausen, Basel, Switzerland; celibate.
International Organizations: Secretary of the Congregation for the Clergy.
Advisor of the Congregation for Bishops. Advisor of the Pontifical Council for the Laity. Member of the Pontifical Commission for Latin America. Member of the Pontifical Council for the Pastoral Care of Migrants and Itinerant People.
Other organizations: Titular Archbishop, Caorle.
Mailing address: Palazzo delle Congregazioni, Piazza Pio XII 3, I-00193 Roma, Italy. T. (39 6) 698 4151. Fax 698 4845
Residence: See Mailing Address. T. (39 6) 698 4137

Ahkön Norbu Lhamo, Jetsunma A, Tulku
Rel: Buddhist: Tibetan Mahayana; b. in Brooklyn NY, USA.
International Organizations: Founder and Director of the Kunzang Odsal Palyul Changchub Chöling, Poolesville MD (formerly Center for Discovery and New Life, also known as World Prayer Center).
Main field of work: Teaching. **Career:** Recognized as the incarnation of Genyenma Ahkön Lhamo in 1987.
Mailing address: 18400 River Road, Poolesville MD 20837, USA. T. (1 301) 428 8116

Ahmad, Mabruk Othman, Dr
Rel: Muslim.
International Organizations: Secretary-General of the Islamic Call College.
Mailing address: PO Box 2549, Tripoli, Libya AJ.

Ahmad, Manzoor, Prof Dr
Rel: Muslim; b. 22 SEP 1931 in Rampur, Uttar Pradesh, India; m. to Husn Ara; c. 2 boys and 1 girl; Citizen of Pakistan; Lang. English, Urdu, Persian, Arabic and Hindi.
International Organizations: Director-General of the Islamic Documentation and Information Centre.
Other organizations: Vice-Chancellor (since 1992), Hamdard University, Karachi; Director, Pakistan Centre for Democracy Studies.
Main field of work: University education. **Career:** Assistant Professor, S.M. College, Karachi from 1954 to 1955. Assistant Professor to Professor, University of Karachi from 1961 to 1991. Visiting Professor at Temple University, Philadelphia PA (1974), Columbia University New York NY (1983), University of the South, Sewanee TN (1984). **Educ.:** B.A., M.A. from University of Karachi, Pakistan (1951-1955); Ph.D. from University of London, UK (1963-1966). **Pubs.:** Ethics and Morality; God and Reason and A Short History of Philosophy.
Mailing address: Hamdard Univ., Deh Bund Murad, Karachi 74700, Pakistan. T. (92 21) 332434
Residence: B-6, Univ. of Karachi Campus, Karachi 75270, Pakistan. T. (92 21) 474269.

Aipa, Benson Nathaniel, Rt Rev
Rel: Roman Catholic; Ordained; b. 27 MAY 1937 in Malindi Mangochi; m. to Mtendere Chiwaya; c. 5 boys and 3 girls; Citizen of Malawi.
International Organizations: Bishop (since 1987) of South Malawi.
Career: Tutor, Chilema from 1969 to 1974. Vicar-General in 1974. Archdeacon, Mangochi West from 1978 to 1980. Archdeacon, Blantyre from 1985 to 1987. Vicar-General from 1986 to 1987. Canon in 1986. **Educ.:** St John's Seminary, Lusaka, Zambia; West Hill College, Birmingham, UK; Ecumenical Institute, Geneva, Switzerland. **Interests:** football, athletics, gardening.
Mailing address: Diocese of S Malawi, PO Chilema, Zomba, Malawi.

Aitken, I C, Rev
Rel: Presbyterian; Ordained.
International Organizations: General Secretary of the Presbyterian Church of Southern Africa.
Mailing address: PO Box 72057, Parkview, Johannesburg 2122, South Africa.

Akagi, Yu
Rel: New Religion; b. 05 AUG 1913; Citizen of Japan.
International Organizations: Chairman of the Daihorindaii Komyokyokai.
Educ.: Kumamoto Teacher's Course School.
Mailing address: 726 Oaza Sano Aza Haraguchi, Dazaifu, Fukuoka 818-01, Japan. T. (81 92) 922 8561

Akamatsu, Shinjo

Rel: Buddhist; Ordained; b. 18 AUG 1921; Citizen of Japan.
International Organizations: Chief of the Jodoshinshu Sonkoji.
Educ.: Kyoto University, Japan. **Hons.:** Honorary Citizen of Tarlac, Philippines.
Mailing address: 41 Ichibacho Onojimatenjin, Awa-gun, Tokushima 771-16, Japan. T. (81 883) 36 3026

Akande, Samuel Titilola Oladele, Rev Dr

Rel: Baptist; Ordained; b. 31 MAR 1926 in Awe-Oyo, Oyo State, Nigeria; m. to Comfort Olalonpe Akande; c. 1 boy and 4 girls; Citizen of Nigeria; Lang. Yoruba, English and a Ghanian language.
International Organizations: Past Secretary-General/Treasurer of the All Africa Baptist Fellowship.
Regional Secretary for Africa of the Baptist World Alliance.
Other organizations: Chief Executive (since 1979), Nigerian Baptist Convention.
Main field of work: Clergyman and administrator. **Educ.:** Th.B. from Nigerian Baptist Theological Seminary, Ogbomoso, Nigeria (1952-1955); B.A. Engl. from Wayland Baptist College, Plainview TX, USA (1959-1961); B.D. from Southern Baptist Seminary, Louisville KY, USA (1961-1962); M.S.T. from Union Theological Seminary, New York NY, USA (1969-1970); Ph.D. New Testament from Southern Baptist Seminary, Louisville KY, USA (1970-1973). **Mem.:** Faith and Order Commission, Society for New Testament Study **Pubs.:** *The Courage to Live*; *Marriage and Home Making in Nigerian Society*; *An Invitation to Become Member of God's Family (visions)*; *The Type of Life God Blesses*; *For Better and For Worse !*.
Mailing address: PO Box 16490, General Post Office, Dugbe, Ibadan, Nigeria.
Residence: 1 Ashi Road, Off Akobo Road, Bashorun, Ibadan, Nigeria T. (234 22) 71 00 34

Aker, John B, Dr

Rel: Protestant.
International Organizations: Contact of the Slavic Gospel Association.
Mailing address: PO Box 1122, Wheaton IL 60189, USA. T. (1 708) 690 8900. Fax 690 9524.

Akihito, A, Emperor

Rel: Shinto; m. to Empress Michiko; c. 3 children; Citizen of Japan; Lang. Japanese and English.
International Organizations: Emperor (since 1989) of the Japan.
Career: Assumed imperial duties in 1988. **Educ.:** Jakushuin University, Japan.

Akiya, Einosuke

Rel: Buddhist: Soka Gakkai; b. 15 JUL 1930 in Tokyo, Japan; m. to Akiko Ishida; c. 2 boys; Citizen of Japan.
International Organizations: General Director (since 1981) of the Soka Gakkai International.
Other organizations: With Soka Gakkai (since 1951); Representative Director (since 1975), Soka Gakkai; Chairman, Central Council (since 1979), Soka Gakkai; President (since 1981), Soka Gakkai.
Career: Chief, Young Men's Division, Soka Gakkai from 1956 to 1959. Chief, Youth Division, Soka Gakkai from 1959 to 1961. Director, Soka Gakkai from 1961 to 1962. Vice-General Director, Soka Gakkai from 1962 to 1967. Executive Director, Soka Gakkai from 1967 to 1970. Vice-President, Soka Gakkai from 1970 to 1981. General Editor, Seikyo Shimbun from 1970 to 1975. **Educ.:** Waseda University, Tokyo, Japan. **Interests:** reading, music.
Mailing address: 32 Shinano-machi, Shinjuku-ku, Tokyo 160, Japan. T. (81 3) 3353 7111. Fax 3353 5431. C. SOKAGAKKAI TOKYO. Tx J 33145 SKG.

Akiyama, Yuga

Rel: Buddhist; Ordained; b. 25 SEP 1895; Citizen of Japan.
International Organizations: Manager of the ShinJonshu Chizan-Sect and Chishakuin.
Career: Chief, Kongoji. **Educ.:** Chizan Kangakuin.
Mailing address: 733 Takahata, Hino, Tokyo 191, Japan. T. (81 425) 91 0032

Al-Alami, Saad Ed-din, Sheikh

Rel: Muslim; b. in Jerusalem, Israel; m. to (1st) Ismat Nuseibeh (died), (2nd) Nuzhah Khalidy; c. 3 boys and 4 girls; Citizen of Jordan.
International Organizations: Mufti (since 1953) of Jerusalem.
Other organizations: Chairman (since 1982), Supreme Muslim Council for Jerusalem; Head, Waqf; Chief Justice.
Career: Teacher, Dar el-Ulum, Jaffa in 1932. Teacher, Orphanage School, Jerusalem from 1935 to 1936. Chief Cleric, Shari'ah Courts from 1937 to 1945. Qadi in Tiberias from 1946 to 1947. Qadi in Nazareth in 1948. Qadi in Ramallah from 1949 to 1950. Qadi in Nablus from 1951 to 1952. Qadi in Jerusalem from 1974 to 1982. President, Hebrew University of Jerusalem in 1984. **Educ.:** B.A. from Al-Azhar University, Cairo, Egypt. **Interests:** reading, study. **Mem.:** League of Islamic World, Mecca (since 1984) **Pubs.:** Book of the Higher Islamic Council, 1985.
Mailing address: PO Box 20002, Jerusalem, Israel. T. (972 2) 285994

al'Alwani, Taha Jabir Fayyad, Prof

Rel: Muslim; b. 12 MAY 1935 in Al Fallujah, Al Anbar, Iraq; married; Citizen of USA; Lang. Arabic, Persian and English.
International Organizations: President (since 1986) of the International Institute of Islamic Thought.
Member of the Founding Committee of the Muslim World League.
Other organizations: Teacher of Islamic Law, College of Shari'ah (since 1975), Imam Muhammad ibn Saud Islamic University, Riyadh; Counsellor (part-time) (since 1975), International Islamic Federation for Student Organizations; Head (since 1988), Fiqh Counsel of North America.
Career: Imam, Khattib and teacher at Masjid Al Bajah Ji in Baghdad in 1953. Imam, Khattib and teacher at Masjid Al Bajah Ji in Baghdad from 1959 to 1961. Teacher of Islamic Studies at the Military College, Baghdad from 1963 to 1968. Legal Advisor for Special Law in the Home Ministry, Saudi Arabia in 1975. Assistant Professor of Islamic Law, College of Shari'ah, Imam Muhammad ibn Saud Islamic University, Riyadh from 1976 to 1979. Associate Professor of Islamic Law, College of Shari'ah, Imam Muhammad ibn Saud Islamic University, Riyadh from 1979 to 1984. Professor of Islamic Law, College of Shari'ah, Imam Muhammad ibn Saud Islamic University, Riyadh from 1984 to 1985. Professor of Islamic Culture (part-time), Institute for Security Officers, Riyadh from 1978 to 1983. Advisor (part-time) to the World Assembly of Muslim Youth, Riyadh from 1975 to 1982. Member of the Legal Committee and Legal Consultant to the National Committee for the Planning of Water Resources in Saudi Arabia from 1982 to 1984. Chairman, Committee for Research, Shari'ah Science, International Institute of Islamic Thought (IIIT) in 1980. Member, Department of Islamic Law, Imam Muhammad ibn Saud Islamic University, Riyadh from 1975 to 1985. Visiting Professor at the University of Strasbourg. Visiting Professor at the University of Brunei. Visiting Professor at the International Islamic University. Visiting Professor at Emir Abdelqadir Islamic University, Constantine. **Educ.:** B.A. Islamic Shari'ah and Law from Al-Azhar University, Cairo, Egypt (1958-1959); M.A. Islamic Shari'ah and Law from Al-Azhar University, Cairo, Egypt (1968); Ph.D. Islamic Shari'ah and Law (Hons) from Al-Azhar University, Cairo, Egypt (1973). **Mem.:** International Islamic Fiqh Academy, Jeddah **Pubs.:** "Al Fiqh wal hadarah", paper presented at the Conference of Islamic Civilization held at Kuala Lumpur, May 1984; "Al Nahy 'An al Isti'anah wal Istinsar fi Umur al Muslimin bi Ahl al Dhimmah wal Kuffar", edited manuscript, published by Dar al Jil, Cairo, 1984 and Dal al 'Ubaykan, Riyadh, 1985 and *Introduction to Logic and Preliminary Usul*, prepared for the College of Shari'ah, Imam Muhammad ibn Saud Islamic University, Riyadh, 1977.
Mailing address: 555 Grove Street, Suite 101, Box 669, Herndon VA 22070, USA. T. (1 703) 471 1133. Tx 901153 IIIT WASH. Fax (1 701) 471 3922.

Al-Askeri, Hussein Muhamad

Rel: Muslim.
International Organizations: Secretary-General of the Islamic States Broadcasting Organization.
Mailing address: PO Box 6351, Jeddah 21442, Saudi Arabia. T. (966 2) 6721121 – 6722269. Fax 672 2600. Tx 601442 ISBO SJ.

Alavi, Khalid, Prof Dr

Rel: Muslim; b. 16 NOV 1940; married; Citizen of Pakistan.
International Organizations: Director of the Sheikh Zaid Islamic Centre, University of the Punjab, Lahore.

Other organizations: Seerat Professor, Dept. of Islamic Studies, University of the Punjab, Lahore.
Career: President, Academic Staff Association, University of the Punjab from 1983 to 1984. Member, Academic Council and Senate, University of the Punjab. Member, Board of Studies, Islamia University, Islamabad. Elected twice as Member, Punjab University Syndicate. 28 years teaching/research experience at the university level. **Educ.:** M.A. Arabic (1st); M.A. Islamic Stds. (1st); M.A. Pol.Sc. (2nd); M.D.L. from University of the Punjab, Lahore, Pakistan; Ph.D. Hadith Lit. from University of Edinburgh, UK. **Pubs.:** 10 books/research studies, including: *Islam Ka Maasharti Nizam* (Social System of Islam); *Payghambrana Duaun* (Prophetic supplication) and over 20 articles in national journals, including: "Quranic concept of justice" (in Urdu), in *Minhaj*, Lahore, October 1983.
Mailing address: Institute of Islamic Studies, University of the Punjab, Quai-e-Azam Campus, Lahore, Pakistan. T. (92 42) 867993
Residence: 64-Rizwan Block, Awan Town, Lahore, Pakistan.

Al Baz, Abdulaziz A

Rel: Muslim; b. in Riyadh, Saudi Arabia.
International Organizations: General President, Islamic Research, Ifta' and Preaching Departments of the Supreme Religious Council.
Career: Lecturer, Faculty of Shari'a, King Saud University, Riyadh from 1953 to 1962. Rector, Islamic University, Medina from 1970 to 1975. Member, Constituent Council, Muslim World League (later Chairman). **Interests:** reading, writing. **Hons.:** Order of Merit, Mauritania; King Faisal Prize for Service to Islam (1982). **Pubs.:** several books dealing with fundamental matters of religion.
Mailing address: Al Badiah Street, Riyadh, Saudi Arabia.

Albert, Christel

Rel: Christian.
International Organizations: Member of the Christian Medical Commission.
Mailing address: PO Box 2100, 150 route de Ferney, CH-1211 Genève 2, Switzerland. T. (41 22) 791 61 11. Fax 791 03 61. C. Oikoumene. Tx 415 730 OIKCH.

Albert, Michel Maurice Louis Delphin, Dr

Rel: Christian; b. 25 FEB 1930 in Fontenay-le-Comte, France; m. to Claude Balland; c. 4 boys; Citizen of France.
International Organizations: President of the International Christian Union of Business Executives.
President (since 1983) of the Centre for International Prospective Studies and Information, Paris.
Other organizations: Deputy Commissioner (since 1976), Equipment and Productivity Plan; President (since 1982), Assurances Générales de France (AGF); President, Board of Directors (since 1983), Central Society of the National Group of Insurance Companies AGF; President (since 1984), Association for the Industrial Development of the West Atlantic; Director (since 1987), Crédit foncier de France.
Career: Student, Ecole Nationale d'Administration, Paris, France in 1954. Inspector of Finance in 1956. Secretary-General, Rueff-Armand Committee in 1959. Inspector-General of Finance in Morocco in 1960. Deputy Director-General, National Bank for Economic Development, Rabat from 1961 to 1963. Deputy Director, European Investment Bank, Brussels from 1963 to 1966. Director, European Investment Bank, Brussels from 1966 to 1970. Director for Structure and Economic Development, Commission of the European Communities (CEC) from 1966 to 1969. Vice-President, Express-Union from 1969 to 1970. Vice-President, Supervision Council, Groupe Express from 1970 to 1971. 1st Class Inspector of Finance in 1971. Director-General, Union d'études et d'investissements and Union de crédit pour le développement régional (Unicrédit) from 1972 to 1975. Director, Unicrédit from 1973 to 1979. Managing Director, Union financière de location de matériel (Unimat) and Union immobilière de crédit-bail pour le commerce et l'industrie (Unicomi) from 1973 to 1975. Commissaire Général au Plan from 1978 to 1981. **Educ.:** Dip. from Institut d'Etudes Politiques de Paris, France; D. en D. **Hons.:** Knight of the Legion of Honour, Commander of the National Order of Merit, Officer of the Ouissam Alaouite. **Mem.:** Titular Member, Crédit National des Assurances (since 1984) **Pubs.:** With Jean-Jacques Servan-Schreiber, *Ciel et Terre* (manifesto of the Radical Party), 1970; *Les Vaches maigres* (joint authorship), 1975; *Le Pari français*, 1982 (Prix Aujourd'hui); *Un Pari pour l'Europe*, 1983 and *Crise Krach Boom* (joint authorship), 1988.
Mailing address: c/o AGF, 87 rue de Richelieu, F-75002 Paris, France.
Residence: 54 rue de Varenne, F-75007 Paris, France

Albertz, Heinrich

Rel: Protestant; Ordained; b. 22 JAN 1915; Citizen of Germany.
International Organizations: Member (since 1946) of the Social Democratic Party (SPD).
Main field of work: Politics and ministry. **Career:** Curate in Berlin in 1937. Arrested during Second World War as active member, Protestant Church. Head, Refugee Office, City of Celle in 1946. Representative for Refugees, Lower Saxony Landtag in 1947. Member, Refugee Council, Executive Committee, Social Democrat Party (SPD) in 1947. Director, Senate Committee for People's Welfare, West Berlin from 1955 to 1959. Head, Senate Chancellery, West Berlin from 1961 to 1963. Senator for Internal Affairs, West Berlin from 1961 to 1963. Deputy Mayor and Senator for Internal Affairs from 1965 to 1966. Mayor, West Berlin from 1966 to 1967. **Educ.:** Theological Stds. in Berlin, Halle and Breslau, Germany. **Hons.:** Niedersächsische Landesmedaille, Grosses Verdienstkreuz mit Stern und Schulterband.
Mailing address: Riekestrasse 2, 28359 Bremen, Germany.

Alcamo, Louis P

Rel: Roman Catholic.
International Organizations: Contact of the Order of the Alhambra.
Mailing address: 4200 Leeds Ave, Baltimore MD 21229, USA. T. (1 301) 242 0660.

Aleksandr, A., Archbishop

Rel: Orthodox; Ordained; b. 08 AUG 1941 in Teykovo, USSR; Citizen of Russia.
International Organizations: Archbishop (since 1986) of Dmitrov.
Career: Military service from 1963 to 1966. Became a monk in 1971. Lecturer, Moscow Theological Academy in 1973. Professor, Moscow Theological Academy in 1981. Dean, Moscow Theological Academy and Seminary in 1982. Chairman, Education Committee, Holy Synod in 1986. **Educ.:** Moscow Theological Academy.
Mailing address: Moscow Theological Academy and Seminary, Moscow, Russia.

Alemany, Jose J, S.J., Prof Dr

Rel: Roman Catholic; Ordained; b. in Zaragossa, Spain; celibate.
International Organizations: Professor of Fundamental Theology (since 1984) of the Pontifical University 'Comillas', Madrid.
Member of the Society of Jesus.
Other organizations: Dean, Faculty of Theology (since 1989), Pontifical University 'Comillas', Madrid.
Main field of work: University education. **Career:** Lecturer in Fundamental Theology, Pontifical University 'Comillas', Madrid from 1973 to 1984. **Educ.:** Lic. Phil. from Universities of Zaragoza and Barcelona, Spain (1962); Dr. Theol. from Innsbruck University, Austria and Eberhard-Karls-Universität Tübingen, Germany (1970). **Mem.:** European Society for Catholic Theology **Pubs.:** Numerous publications on fundamental theology.
Mailing address: No mailing address.

Alesana, Enoka L, Rev

Rel: Congregational.
International Organizations: Contact of the Congregational Christian Church in American Samoa.
Mailing address: PO Box 1637, Pago Pago, Samoa USA.

Alexy, (Ridiger), II, Patriarch

Rel: Orthodox; Ordained; b. 23 FEB 1929 in Tallinn, Estonia, USSR; Citizen of Estonia.
International Organizations: Chairman (since 1972) of the Conference of European Churches.
Other organizations: Vice-Chairman, Dept of External Affairs (since 1962), Moscow Patriarchate, USSR; Metropolitan (since 1986), Leningrad and Novgorod, USSR.
Main field of work: Clergyman. **Career:** Chairman, Tartu Cathedral. Bishop, Tallinn and Estonia in 1961. Member, Central Committee, World Council of Churches from 1961 to 1968. Archbishop, Administrative Manager of Moscow Patriarchy, Permanent Member of Holy Synod, Chairman, Teaching Committee, Moscow Patriarchate from 1964 to 1968. Metropolitan, Tallinn and Estonia from 1968 to 1986. **Educ.:** Leningrad Theological Seminary, USSR.
Mailing address: Moscow Patriarchate, Danilovsky Val 22, SU-113191 Moskva, Russia. T. Fax (7 95) 230 26 10

Al-Faisal, Khalid, Prince
Rel: Muslim.
International Organizations: Director-General of the King Faisal Centre for Research and Islamic Studies.
Mailing address: PO Box 352, Riyadh, Saudi Arabia. T. (966 1) 465 2255. Fax (966 1) 465 6524. Tx 404694 KAHRI SJ.

Alford, Timothy, Rev
Rel: Protestant; Ordained.
International Organizations: UK Director of the Africa Inland Mission International.
Mailing address: 2 Vorley Road, London N19 5HE, UK. T. (44 71) 281 1184.

al-Ghabid, Hamid, H E Dr
Rel: Muslim; b. in Tanout, Niger; married; c. 7 children; Lang. French, English and Arabic.
International Organizations: Secretary-General (since 1988) of the Organization of the Islamic Conference.
Career: Director, Foreign Finances, Ministry of Finance from 1972 to 1974. Secretary-General, Ministry of Finance from 1974 to 1975. Executive Director Resident, Islamic Development Bank, Jeddah from 1975 to 1979. Secretary of State, Foreign and Cooperation Office from 1979 to 1981. Administrator non-resident, Islamic Development Bank from 1979 to 1981. Minister, Department of Trade from 1981 to 1982. Minister, Department of Trade and Transports from 1982 to 1983. Ministerial delegate to the Finances during 1983. Prime Minister from 1983 to 1988. **Educ.:** Lic. D. from National University of the Ivory Coast, Abidjan; Dip. from Institut International d'Administration Publique, Paris, France; Training Deg. from International Monetary Fund Institute, Washington DC, USA; Doc. from University of Paris I (Panthéon-Sorbonne), France.
Mailing address: 6 Km Makkah Al-Mukarramah Rd, PO Box 178, Jeddah, Saudi Arabia. T. (966 2) 6874858 – (966 2) 6873880. C. ISLAMIAH Jeddah. Tx 601366 ISLAMI SJ. Fax (966 2) 6876568.

Al-Ghamdi, Ali Abdullah Mughram, Dr
Rel: Muslim; b. in Saudi Arabia; married; c. 5 children; Citizen of Saudi Arabia; Lang. Arabic and English.
International Organizations: Director General (since 1981) of the London Central Mosque and Islamic Cultural Centre.
Other organizations: Patron/Trustee, several organizations (Muslim and non-Muslim); Counsellor (Islamic Affairs), Royal Embassy, Saudi Arabia, London.
Main field of work: Community representative, advisor. **Career:** University career (until 1981). Dean, King Abdulaziz University, Jeddah, Saudi Arabia (last post). **Educ.:** B.Sc. from King Abdulaziz University, Jeddah, Saudi Arabia; Ph.D. from University of Durham, UK. **Interests:** geography, environment, Middle Eastern affairs. **Mem.:** numerous
Mailing address: 146 Park Road, London NW8 7RG, UK. T. (44 71) 724 3363
Residence: See Mailing Address. T. See Mailing Address

Al-Haq, Jad Al-Haq Ali Jad, Sheikh
Rel: Muslim; Citizen of Egypt.
International Organizations: Grand Sheikh of the al-Azhar.
Mailing address: No mailing address.

Al Harakan, Mohamad Ali A
Rel: Muslim.
International Organizations: Past Secretary-General of the Muslim World League.
Mailing address: PO Box 537 – 538, Ummul Joud, Makkah, Saudi Arabia. T. (966 2) 5446700. Tx 540009 RABITA SJ. Fax (966 2) 5436202.

Ali, H A Mukti
Rel: Muslim; b. in Central Java, Indonesia; married; c. 2 boys and 1 girl; Citizen of Indonesia.
International Organizations: Member (since 1986) of the National Hijra Council, Islamabad.
Main field of work: Comparative religion specialist. **Career:** Vice-Chancellor, IAIN 'Sunan Kalijaga', Yogjakarta from 1964 to 1971. Minister of Religious Affairs from 1971 to 1978. Member, Supreme Advisory Council from 1978 to 1983. **Educ.:** Indonesia, Pakistan, Canada. **Interests:** reading novels. **Mem.:** Akademi Jakarta (since 1979) **Pubs.:** Modernization of Islamic Schools; Comparative Religion, its Method and System and Religion and Development in Indonesia.
Mailing address: Iain Sunan, Kalijaga, Yogyakarta, Indonesia.
Residence: Sagan GkI/100, Yogyakarta, Indonesia.

Ali, Muazzam
Rel: Muslim.
International Organizations: Secretary-General of the Islamic Press Union.
Mailing address: London, UK.

al-Johani, Maneh Hammad
Rel: Muslim.
International Organizations: Secretary-General of the World Assembly of Muslim Youth.
Mailing address: PO Box 10845, 30th Street Olaiya, Riyadh 11443, Saudi Arabia. T. (966 1) 464 1669. Fax 464 1710. Tx 400413 – 405220 ISLAMI SJ.

al-Khoja, Mohamed Habib ibn, Dr
Rel: Muslim.
International Organizations: Secretary-General of the Islamic Fiqh – Jurisprudence – Academy.
Mailing address: PO Box 13719, Jeddah 21414, Saudi Arabia. T. (966 2) 6609320

Allen, Cosie Lilian Widrig, N.S.T., Rev
Rel: Spiritualist; Ordained on 22 NOV 1987; b. 11 AUG 1955 in Traverse City MI, USA; m. to William Russell Allen; c. 3 boys; Citizen of USA; Lang. English.
International Organizations: Secretary of the Ministerial Association (National Spiritualist Association of Churches).
Other organizations: Teacher/Certified Medium (since 1989), National Spiritualist Association of Churches (NSAC); Missionary (since 1991), NSAC.
Main field of work: Education and ministry. **Educ.:** Morris Pratt Institute, Milwaukee WI, USA (1976-1983). **Interests:** music, reading, family. **Mem.:** NSAC Ministerial Association; NSAC Teachers Club; Morris Pratt Institute **Pubs.:** "Prayer for Peace – Healing our World", in *The National Spiritualist Magazine*, April 1991; "Just What Is A Spiritualist?", in *The National Spiritualist Magazine*, Sept 1991 and "Holiday Reflections", in *The National Spiritualist Magazine*, Dec 1991.
Mailing address: 13 Cottage Row, Lily Dale NY 14752, USA. T. (1 716) 595 2000
Residence: 1418 Hall SE, Grand Rapids MI 49506, USA. T. (1 616) 241 2761

Allen, Gary I
Rel: Christian.
International Organizations: President of the Christian Mission for the United Nations Community.
Mailing address: PO Box 159, Monroe CT 06468, USA. T. (1 203) 261 1277.

Allen, Joseph P, Rev
Rel: Roman Catholic; Ordained; celibate.
International Organizations: Director of the Confraternity of the Rosary for Sick and Handicapped Pilgrims to Lourdes.
Mailing address: 487 Michigan Avenue NE, Washington DC 20017, USA. T. (1 202) 529 5300.

Allen, Ronald James, Prof Dr

Rel: Disciples of Christ; Ordained in 1974; b. 26 APR 1949 in Poplar Bluff MO, USA; married; c. 2 boys and 3 girls; Citizen of USA; Lang. English, French, German, Greek and Hebrew.
International Organizations: Associate Professor of Preaching and New Testament (since 1982) of the Christian Theological Seminary, Indianapolis IN.
Other organizations: Secretary, Academy of Homiletics; Hymnal Development Committee, Christian Church (Disciples of Christ); Frequent lecturer on the Bible and church life at meetings of the Christian Church.
Main field of work: University teaching. **Career:** Co-pastor, First Christian Church, Grand Island NE from 1977 to 1982. **Educ.:** A.B. from Phillips University, Enid OK, USA (1971); M.Div. from Union Theological Seminary, New York NY, USA (1974); Ph.D. from Drew University, Madison NJ, USA (1977). **Mem.:** Academy of Homiletics; Society of Biblical Literature **Pubs.:** With Clark M Williamson, *The Teaching Minister*, Westminster-John Knox Press; *Contemporary Biblical Interpretation for Preaching*, Judson Press and *Preaching for Growth*, CBP Press.
Mailing address: Christian Theological Seminary, 1000 W 42nd St, Indianapolis IN 46208, USA. T. (1 317) 914 1331
Residence: 4444 N College, Indianapolis IN 46205, USA T. (1 317) 283 8306

Allsop, Ian E, Rev

Rel: Church of Christ; Ordained.
International Organizations: Secretary of the Churches of Christ in Australia.
Mailing address: 52 La Trobe Street, Melbourne 3000, Australia.

Almen, Lowell G, Rev Dr

Rel: Lutheran; Ordained on 11 JUN 1967; b. 25 SEP 1941 in Grafton ND, USA; m. to Sally Arlyn Clark; c. 1 boy and 1 girl; Citizen of USA; Lang. English.
International Organizations: Secretary (since 1987) of the Evangelical Lutheran Church in America.
Main field of work: Ministry. **Career:** Assistant Campus Pastor and Director of Communications, Concordia College, Moorhead MN from 1969 to 1974. Managing Editor, *The Lutheran Standard* from 1974 to 1978. Editor, *The Lutheran Standard* from 1979 to 1987. Pastoral Assistant at the Trinity Lutheran Church. **Educ.:** B.A. Phil. and Hist.-Pol.Sc. from Concordia College, Moorhead MN, USA (1963); M.D. from Luther Theological Seminary, St Paul MN, USA (1967); Dr. Litt. from Capital University, Columbus OH, USA (1981); D.D. from Carthage College, Kenosha WI, USA (1989). **Hons.:** Distinguished Alumnus, Concordia College, Moorhead MN (1982). **Pubs.:** Co-editor, *The Many Faces of Pastoral Ministry* and Writer/Editor, *World Religions and Christian Mission* and *Our Neighbor's Faith.*
Mailing address: 8765 West Higgins Road, Chicago IL 60631, USA. T. (1 312) 380 2800. Fax (1 312) 380 1465.
Residence: 2204 Broadway Lane, Schaumburg IL 60194, USA. T. (1 708) 843 7128

Al-Muwqqat, Subhi, Sheikh

Rel: Muslim; Citizen of Jordan.
International Organizations: Director of the Shari'a Courts in Jordan.
Mailing address: No mailing address.

Alongina, Xavi

Rel: Buddhist: Mahayana.
International Organizations: Member of the Foundation for the Preservation of the Mahayana Tradition.
Other organizations: Director, Ediciones Dharma.
Mailing address: Apartado 218, E-03660 Novelda, Alicante, Spain. T. (34 96) 560 3200

Alonso, Gustavo, C.M.F., Fr

Rel: Roman Catholic; Ordained; celibate.
International Organizations: Past Superior General of the Missionaries Sons of the Immaculate Heart of Mary.
Mailing address: Via Sacro Cuore di Maria 5, I-00197 Roma, Italy. T. (39 6) 87 00 56.

Al-Shariff, Muhammed Ahmed, Dr

Rel: Muslim; Citizen of Libya AJ.
International Organizations: Secretary-General of the World Islamic Call Society.
Contact person of the World Council for the Islamic Call.
Main field of work: Politician. **Career:** Dean, Faculty of Education, University of Libya, Tripoli. Minister of Education from 1972 to 1977. Chairman, Islamic-Christian Dialogue, Tripoli in 1976. Secretary for Education, General People's Committee from 1977 to 1981. **Educ.:** USA.
Mailing address: PO Box 2549, Tripoli, Libya AJ.

al Sidais, Abdallah Abdulaziz a

Rel: Muslim; b. in Qassim, Saudi Arabia; Citizen of Saudi Arabia.
International Organizations: Director of the Mecca Religious Institute (Ilmi).
Main field of work: Religious educator. **Career:** Teacher from 1972 to 1974. **Pubs.:** Modern Quranic Studies and several articles in local publications.
Mailing address: Al Ilmi Institute, Mecca, Saudi Arabia.

Altamemi, Arafat R, Dr

Rel: Muslim.
International Organizations: Contact of the Islamic Federation of Research Institutes.
Mailing address: c/o IFSTAD, PO Box 9833, Jeddah 21423, Saudi Arabia. T. (966 2) 6322291 – 6322273. C. ISLAMIAH JEDDAH. Tx 604081 IFSTAD SJ.

Althouse, LaVonne, Rev

Rel: Protestant.
International Organizations: Editor of the International Association of Women Ministers.
Mailing address: 5227 Castor Avenue, Philadelphia PA 19124, USA.

Altizer, Thomas

Rel: Christian; Citizen of USA.
International Organizations: Professor (since 1968) of the University of New York at Stony Brook.
Career: Taught religion and later English literature, Emory University, Atlanta GA from 1956 to 1968. **Educ.:** Ph.D. Hist. of Religions from University of Chicago, IL, USA (1955). **Pubs.:** *The Descent into Hell*, 1970; *Total Presence*, 1980 and With William Hamilton, *Radical Theology and the Death of God* (collection of essays).
Mailing address: State University of New York, Stony Brook, NY 11790, United States of America.

Alvares, Vincent

Rel: Roman Catholic.
International Organizations: Secretary-General of the Consociatio Internationalis Musicae Sacrae.
Mailing address: Via di Torre Rossa 21, I-00165 Roma, Italy. T. (39 6) 663 87 92.

Alvarez, Carmelo

Rel: Christian.
International Organizations: Executive Secretary of the Ecumenical Association of Third World Theologians.
Mailing address: 1100 E 55th Street, Chicago IL 60651, USA.

Alvárez García, Luís de la Cruz A, Fr

Rel: Roman Catholic; Ordained; celibate.
International Organizations: Superior Delegate of the Order of Brothers of Bethlehem.
Mailing address: Calle San Agustín 44, E-38201 La Laguna, Santa Cruz de Tenerife, Spain. T. (34 922) 26 40 05.

Alves, Colin, M.Litt.

Rel: Church of England; b. 19 APR 1930 in Weston-Super-Mare, Somerset, UK; m. to Peggy Evelyn Kember; c. 2 boys and 1 girl; Citizen of United Kingdom; Lang. English and German.
International Organizations: Trustee (since 1984) of the St Gabriel's Trust.
Other organizations: Trustee (since 1989), St Pierre de Thaon International Youth Trust.
Career: Teacher (Religious Education), Ripon Grammar School et al. from 1952 to 1959. Lecturer in Religious Education, King Alfred's College, Winchester from 1959 to 1968. Lecturer in Religious Education, Brighton College of Education from 1968 to 1974. Lecturer in Religious Education, St Gabriel's College, London from 1974 to 1977. Education Officer, Church House, Westminster from 1977 to 1990. **Educ.:** M.A. from University of Oxford, UK (1949-1952). **Interests:** Church, state and education (especially in Europe). **Hons.:** Member, Order of the British Empire. **Mem.:** Royal Commonwealth Society **Pubs.:** Religion and the Secondary School, SCM Press, 1968; The Christian in Education, SCM Press, 1972 and The Question of Jesus, National Society, 1987.
Mailing address: 9 Park Road, Haywards Heath RH16 4HY, UK. T. (44 444) 454496
Residence: See Mailing Address. T. See Mailing Address

Alves, Manuel Isidro, Prof

Rel: Roman Catholic; Ordained on 14 JUL 1963; b. 13 JAN 1940 in Lisbon, Portugal; celibate; Citizen of Portugal; Lang. Portuguese, Spanish, French, Italian, English and German.
International Organizations: Vice-Rector (since 1983) of the Catholic University of Portugal, Lisbon.
Other organizations: Professor (since 1979), Catholic University of Portugal; Director (since 1984), Association for Christian Theology and Culture; Board (since 1985), Scientific Society of the Catholic University of Portugal; Librarian (since 1988), Catholic University of Portugal.
Main field of work: University education. **Career:** Director of Studies, Conciliar Seminary, Braga from 1971 to 1974. Director, Higher Institute of Theology, Braga from 1980 to 1988. Director, Faculty of Philosophy, Catholic University of Portugal from 1988 to 1990. President, Faculty of Human Sciences, Catholic University of Portugal from 1990 to 1991. Director, Faculty of Theology, Catholic University of Portugal from 1988 to 1991. **Educ.:** Lic. from Pontifical Gregorian University, Vatican City State (1963-1965); Doc. from Pontifical Biblical Institute, Vatican City State (1965-1979); Dip. from Ecole Biblique et Ecole Archéologique Française, Jerusalem, Israel (1968-1970). **Mem.:** Görresgesellschaft **Pubs.:** Il Cristiano in Cristo, Braga, 1980; Ressurreiçao e Fé Pascal, Lisbon, 1991 and Comentário aos Salmos de D. Frei Bartolomeu dos Mártires – Traduçao Portuguesa Anotada, Fátima, 1991.
Mailing address: Universidade Católica Portuguesa, Palma de Cima, P-1600 Lisboa, Portugal. T. (351 1) 7265817
Residence: Praceta de S Luis, 12 4° Dto, Laranjeiro, P-2800 Almada, Portugal. T. (351 1) 2296342

Alves de Sousa, Pio G A, Prof

Rel: Roman Catholic; Ordained on 15 AUG 1968; b. 20 APR 1945 in Lanheses, Viana do Castelo, Portugal; celibate; Citizen of Portugal; Lang. Portuguese, Spanish, French, Italian, English and Latin.
International Organizations: Professor (since 1987) of the Catholic University of Portugal.
Other organizations: Deputy Director, Faculty of Theology (since 1981), Catholic University of Portugal.
Main field of work: University education. **Educ.:** Dr. from University of Navarra, Spain (1969-1974). **Mem.:** International Association for Patristic Studies **Pubs.:** El sacerdocio ministerial en los libros De Sacerdotio de S Juan Crisóstomo; Fil 2, 6-11 en los escritos de San Juan Crisóstomo and A conversao em Clemente de Roma. Metánoia uma palavra chave.
Mailing address: Faculdade de Teologia, Santa Margarida, P-4719 Braga, Portugal. T. (351 53) 613301
Residence: See Mailing Address. T. See Mailing Address

Amabe, Mistuhiko

Rel: Shinto; Ordained; b. 18 JAN 1931; Citizen of Japan.
International Organizations: Councilor of the Association of Shinto Shrines.
Career: Chief Priest, Isekago Shrine.
Mailing address: 121 Aza Ogaki, Miyazu, Kyoto 629-22, Japan. T. (81 7722) 7 0501

Amada, Bunji

Rel: Buddhist.
International Organizations: Member of the International Network of Engaged Buddhists.
Mailing address: Kanno-Ji, Gumisawa, 1335 Gotenba-Shi, Shizvoka-Ken, Japan.

Ambadiang de Mendeng, Grégoire A, Rev Dr

Rel: Presbyterian.
International Organizations: General Secretary of the Presbyterian Church of Cameroon.
Mailing address: BP 519, Yaoundé, Cameroon. T. (237) 32 42 36

Ambaum, Johannes Hendrik Theodorus Gerardus, Rev Prof Dr

Rel: Roman Catholic; Ordained on 13 DEC 1975; b. 27 NOV 1949 in Tegelen, Limburg, Netherlands; celibate; Citizen of Netherlands; Lang. Dutch, German, French, English and Italian.
International Organizations: President (since 1991) of the International Institute 'Mater Ecclesiae Domesticae'.
Member (since 1986) of the International Theological Commission.
Other organizations: Director (since 1986), BISO (Institute for correspondence courses in theology); Chief Editor (since 1990), *Communio* (international periodical), Dutch edition.
Main field of work: University education. **Career:** Study-Prefect, Priest Seminary, Roermond from 1974 to 1989. Librarian, Priest Seminary, Roermond from 1986 to 1991. **Educ.:** Dip. Theol. from Rheinische Friedrich-Wilhelms-Universität Bonn, Germany (1968-1974); Dr. Theol. from Rheinische Friedrich-Wilhelms-Universität Bonn, Germany (1974-1978). **Pubs.:** Glaubenszeichen. Schillebeeckx Auffassung von den Sakramenten, Regensburg, Pustet, 1980; Het gebed des Heren, Tegelen, Priorij Nazaret, 1985 and Gemeenschappelijk of persoonlijk gebed. Een relatiebepaling, in *Communio*, (NL) 10, 1985, pp. 285-291.
Mailing address: Rolduc, Heyendahllaan 82, NL-6464 EP Kerkrade, Netherlands. T. (31 45) 466833. Fax 465472
Residence: See Mailing Address. T. (31 45) 466888

Ambrozic, Aloysius Matthew, Most Rev Archbishop

Rel: Roman Catholic; Ordained on 04 JUN 1955; b. 27 JAN 1930 in Gabrje, Yugoslavia; celibate; Citizen of Canada.
International Organizations: Archbishop (since 1990) of Toronto.
Member of the Pontifical Council for the Pastoral Care of Migrants and Itinerant People.
Career: Professor, St Augustine's Seminary from 1960 to 1967. Professor, Toronto School of Theology from 1970 to 1976. Dean of Studies, St Augustine's Seminary from 1971 to 1976. Auxiliary Bishop, Toronto from 1976 to 1986. Coadjutor Bishop, Toronto from 1986 to 1990. Appointed to the Pontifical Council for the Pastoral Care of Migrants and Itinerant People in 1990. One of four Bishops representing Canada at the Synod on the Formation of Priests, Rome in 1990. **Educ.:** S.T.L. from Pontifical University San Tommaso d'Aquino, Vatican City State (1958); S.S.L. from Pontifical Biblical Institute, Vatican City State (1960); Th.D. from Bayerische-Julius-Maximilians-Universität Würzburg, Germany (1970). **Mem.:** Catholic Biblical Association of America; ACEBAC **Pubs.:** The Hidden Kingdom, 1972; Remarks on the Canadian Catechism, 1974 and a monthly column in *The Catholic Register*.
Mailing address: 355 Church Street, Toronto ON M5B 1Z8, Canada. T. (1 416) 977 1500

Amendt, P

Rel: Christian.
International Organizations: Contact of the Missionszentrale der Franziskaner.
Mailing address: Albertus-Magnus-Strasse 39, 53177 Bonn, Germany. T. (49 228) 33 10 65 – 33 10 66.

Ames, David Atwater, Rev Dr

Rel: Episcopal; Ordained on 06 JAN 1967; b. 12 SEP 1938 in Glendale OH, USA; m. to Dr Carol Landau; c. 2 boys; Citizen of USA; Lang. English.
International Organizations: Executive Director (since 1989) of The Association for Religion and Intellectual Life.

Other organizations: Episcopal Chaplain (since 1974), Brown University and Rhode Island School of Design, Providence RI; Board Member, Division of Ministries in Higher Education (since 1974), Rhode Island State Council of Churches; Field Education Supervisor (since 1975), Episcopal Divinity School, Cambridge MA; Parish Consultant Group (since 1982), Diocese of Rhode Island; Board of Directors (since 1983), Interfaith Health Care Ministries, Providence RI; Affinity Group (since 1987), Program in Liberal Medical Education, Brown University, Providence RI; Board of Directors (since 1988), Hospice Care of Rhode Island, Inc.; Board of Directors (since 1989), Planned Parenthood of Rhode Island; President (since 1992), Project Committee for Aging 2000.
Main field of work: Ministry. **Career:** Assistant to Personnel Director, American Laundry Machinery Industries, Cincinnati OH from 1960 to 1961. Served with US Army in Seoul, Korea from 1961 to 1963. Assistant Minister, Church of St Edward, Columbus OH from 1966 to 1968. Priest-in-Charge, Church of St Edward, Columbus OH from 1968 to 1969. Assistant Minister, Grace Church, Providence RI from 1969 to 1971. Ecumenical Chaplain, Rhode Island College from 1971 to 1974. Protestant Chaplain (part-time), Patrick I O'Rourke Children's Center from 1971 to 1974. Research and Human Rights Committee, Women and Infants Hospital of Rhode Island from 1974 to 1988. Bioethics Committee, Women and Infants Hospital of Rhode Island from 1985 to 1988. Faculty Study Group on Ethical Issues in Medicine, Brown University Program in Medicine, Providence RI from 1980 to 1989. President, Board of Directors, Interfaith Health Care Ministries, Providence RI from 1986 to 1989. Diocesan Council, Diocese of Rhode Island from 1986 to 1989. Lecturer in Christian Ethics, Episcopal Divinity School, Cambridge MA in 1988. Resource Scholar, National Conference of Christians and Jews in 1989. Convenor, Faculty Discussion Group focused on the intersection between religious faith and knowledge from 1990 to 1991. **Educ.:** B.A. from Miami University, Oxford OH, USA (1956-1960); M.Div. from Episcopal Theological School, Cambridge MA, USA (1963-1966); D.Min. from Episcopal Divinity School, Cambridge MA, USA (1981-1984). **Interests:** medical ethics, social ethics, theology. **Hons.:** Fellow, College of Preachers, Washington DC (1974); Certificate of Recognition, Brown University Program in Liberal Medical Education (1991). **Mem.:** American Civil Liberties Union; Amnesty International; Association for Religion and Intellectual Life; Episcopal Peace Fellowship; Episcopal Society for Ministry in Higher Education; National Conference of Christians and Jews; Society for Health and Human Values; The Hastings Center; The Society of Christian Ethics **Pubs.:** With S I Paradise, "The Church's Ministry in Higher Education", in *Plumbline*, Vol. 9, Issue 2, May 1981, pp. 4-10; *Good Genes?: Emerging Values for Science, Religion and Society*, Forward Movement Publications, Cincinnati OH, 1984 and "Physician Assisted Suicide: A Battle of Technology and Morality", in *Rhode Island Medical Journal*, March 1992.
Mailing address: Brown University, Box 1931, Providence RI 02912, USA. T. (1 401) 863 2956
Residence: 130 Slater Ave, Providence RI 02906, USA. T. (1 401) 751 0175

Amewowo, Wynnand, Rev Fr
Rel: Roman Catholic; Ordained; celibate.
International Organizations: Past Director of the Catholic Biblical Centre for Africa and Madagascar.
Mailing address: PO Box 24215, Karen, Nairobi, Kenya. T. (254 2) 882448

Amirtham, Samuel, Rt Rev Dr
Rel: Christian; Ordained.
International Organizations: Bishop of South Kerala Diocese.
Mailing address: Bishop's House, Thiruvananthapuram, South Kerala, India.

Amoore, F A, Rt Rev
Rel: Protestant.
International Organizations: Provincial Executive Officer of the Church of the Province of Southern Africa.
Mailing address: Bishopscourt, Claremont 7700, South Africa.

Amouzou, Ayakana, Rev
Rel: Evangelical; Ordained.
International Organizations: Synod Secretary of the Evangelical Church of Togo.
Mailing address: 1 rue Maréchal Foch, BP 2, Lomé, Togo.

Amritanand, Joseph, Rt Rev
Rel: Christian; Ordained in 1943; b. 17 FEB 1917; Citizen of India.
International Organizations: Retired Bishop of Durgapur.
Main field of work: Ecclesiastic. **Career:** Ordained Deacon in 1941. Bishop, Assam from 1949 to 1962. Bishop, Lucknow from 1962 to 1970. Bishop, Calcutta from 1970 to 1982. Bishop, Durgapur from 1972 to 1974. **Educ.:** Forman Christian College, Lahore, Pakistan; Bishop's College.
Mailing address: 182 Civil Lines, Jail Road, Bareilly, 243001 Uttar Pradesh, India. T. (91) 73049

Ananian, Paolo, Fr
Rel: Roman Catholic; Ordained; celibate.
International Organizations: Procurator General of the Mechitarists. Advisor to the Special Commission for the Liturgy of the Sacred Congregation for the Oriental Churches.
Mailing address: Via S Anselmo 10, I-00153 Roma, Italy. T. (39 6) 574 35 50

Anav, N
Rel: Jewish.
International Organizations: UNESCO Representative of the International Council of Jewish Women.
Mailing address: 62 rue de Monthieu, F-75008 Paris, France. T. (33 1) 45 62 01 47.

Andersen, Birte
Rel: Christian.
International Organizations: Contact of the Danish Missionary Council.
Mailing address: Skt Lukas Vej 13, DK-2900 Hellerup, Denmark. T. (45) 31 61 27 77.

Andersen, Herbert Earl
Rel: Christian.
International Organizations: Head, Secretariat of the Africa Inland Church – Hurri Hills Project.
Mailing address: PO Box 21028, Nairobi, Kenya. T. (254 2) 333 784.

Anderson, Hector David, Rev
Rel: Anglican; Ordained in 1931; b. 16 AUG 1906; m. to Muriel Louise Peters; c. 1 boy; Citizen of United Kingdom.
International Organizations: Ecclesiastic of the Anglican Church.
Career: Rector, Sandringham from 1942 to 1955. Rector, Swanage, Dorset from 1960 to 1969. Domestic Chaplain to H M King George VI from 1942 to 1952. Domestic Chaplain to H M The Queen from 1952 to 1955. Chaplain to H M The Queen from 1955 to 1976. **Educ.:** M.A., B.D. from University of Dublin Trinity College, Ireland. **Hons.:** Lieutenant of the Royal Victorian Order (1949).
Mailing address: Adare, The Hyde, Langton Matravers, Dorset, UK. T. (44) 423206

Anderson, M J
Rel: Mason.
International Organizations: Chairman of the Federation of Masons of the World.
Mailing address: 1017 E 11th St, Austin TX 78702, USA. T. (1 512) 477 5380.

Anderson, Philip James, Dr
Rel: Evangelical Covenant Church; Ordained on 12 JUN 1980; b. 03 NOV 1949 in Minneapolis MN, USA; m. to Karna Marie; c. 1 boy and 1 girl; Citizen of USA; Lang. English, German and Swedish.
International Organizations: Professor of Church History (since 1979) of the North Park Theological Seminary.
Other organizations: President (since 1989), Swedish-American Historical Society.
Main field of work: Theological education. **Educ.:** B.A. from University of Minnesota, USA (1967-1971); M.Div. from Bethel Theological Seminary, MN, USA (1971-1974); M.Th. from Princeton Theological Seminary, NJ, USA (1974-1975); D.Phil. from University of Oxford, UK (1975-1979). **Mem.:** American Academy of Religion; American Society of Church History **Pubs.:** Editor, with Dag Blenck, *Swedish-American Life in Chicago: Cultural and Urban Aspects of an Immigrant People, 1850-1930*, Acta Universitatis Upsaliensia, Uppsala, 1991 and University of Illinois Press, Urbana, 1992.

Mailing address: North Park Theological Seminary, 3225 W Foster Ave, Chicago IL 60625, USA. T. (1 312) 583 2700
Residence: 5846 N Kenton Ave, Chicago IL 60646, USA

Anderson, Roy W
Rel: Protestant.
International Organizations: Director of the Pocket Testament League.
Mailing address: 11 Toll Gate Rd, PO Box 800, Lititz PA 17543-7026, USA. T. (1 717) 626 1919. Fax 626 5553.

Anderson, Vinton R, Bishop
Rel: African Methodist Episcopal; Ordained.
International Organizations: Ecumenical Officer of the African Methodist Episcopal Church.
Mailing address: 2854 Grand Ave, St Louis MO 63107, USA.

Andiappan, Sundara Yogasali Asana R, Prof
Rel: Hindu; b. 07 DEC 1930 in CN Village, Tirunelveli, Tamil Nadu, India; m. to Sugatha Kumari; c. 2 boys and 4 girls; Citizen of India; Lang. Tamil, English and Malyalam.
International Organizations: President (since 1980) of the Tamil Nadu State Yoga Association.
Career: President, 10 District Yoga Conferences from 1980 to 1991. President, 4 Yoga State Conferences from 1985 to 1991. **Educ.:** Dip. Yoga from International Yoga Centre, Pondichery, India (1970). **Interests:** karnatic music. **Mem.:** International Yoga Centre, Pondichery; Rotary **Pubs.:** Yoga books: Yoga for children, for women, for better health, etc.
Mailing address: 43 East St, C.N. Village, Tirunelveli, Tamil Nadu 627001, India. T. (91) 23701
Residence: See Mailing Address. T. See Mailing Address

Andrew, Bill
Rel: Buddhist: Mahayana.
International Organizations: Member of the Foundation for the Preservation of the Mahayana Tradition.
Other organizations: Director, Dorje Chang Institute.
Mailing address: PO Box 2814, Auckland 1, New Zealand. T. (64 9) 766 509

Andrews, Elias, Rev
Rel: Methodist; Ordained in 1935; b. 29 JAN 1906 in Winterton, Newfoundland, Canada; widowed (1st) Flora Shannon Green (died), (2nd) Mildred Jean Caughey (died); c. 1 boy; Citizen of Canada.
International Organizations: University professor (retired) and Ecclesiastic of the Methodist Church.
Career: Professor of Religion, Queen's University, Kingston from 1955 to 1974. Professor of New Testament Studies, Queen's Theological College from 1955 to 1974. Principal, Queen's Theological College from 1955 to 1970. Minister, Deer Lake, Newfoundland from 1936 to 1937. Lecturer in Philosophy and Psychology, Pine Hill Divinity Hall from 1937 to 1939. Lecturer in Biblical Studies and Religion, Brothers' College, Drew University, Madison NJ from 1939 to 1941. Professor of New Testament Language and Literature, Pine Hill Divinity Hall from 1941 to 1955. Lecturer in Biblical Literature, Dalhousie University, Halifax from 1947 to 1955. President, Bay of Quinte Conference, Uniting Church of Canada from 1962 to 1963. **Educ.:** B.A. from Dalhousie University, Halifax, Canada (1933); B.D. from Pine Hill Divinity Hall, Halifax, Canada (1935); M.A. from Drew University, Madison NJ, USA (1940); Ph.D. from Drew University, Madison NJ, USA (1947); D.D. from Pine Hill Divinity Hall, Halifax, Canada (1962). **Hons.:** Alumnus of the Year, Alumni Association of Pine Hill Divinity Hall, Sackville NB (1987). **Pubs.:** Apostle of Grace, 1979; Surprised by Grace, 1985 and Contributor to *Encyclopaedia Britannica*, 15th Edn. (St Paul).
Mailing address: 115 Ordnance Street, Kingston, Ontario K7K 1G7, Canada.

Andrews, James E, Rev Dr
Rel: Presbyterian; Ordained.
International Organizations: Member of the Executive Committee of the World Alliance of Reformed Churches.
Mailing address: Presbyterian Church, Room 4408, 100 Witherspoon Street, Louisville KY 40202-1396, USA. T. (1 502) 569 5360. Fax (1 502) 569 8005
Residence: T. (1 502) 569 8005 – (1 404) 371 0724

Andriamanjato, Richard, Rev Dr
Rel: Christian; Ordained; Lang. French and English.
International Organizations: Moderator (since 1990) of the Christian Peace Conference.
President of the African Christian Peace Conference.
Mailing address: Jungmannova 9, PO Box 192, 111 21 Praha 1, Czech Rep. T. (42 2) 236 02 89 – (42 2) 236 02 90. C. EKUMRADA Praha. Tx 12 33 63 cpc c.
Residence: V A 15 Ambohitantely, 101 Antananarivo, Madagascar T. (261 2) 27548

Andrzejewski, Roman, H E Mgr
Rel: Roman Catholic; Ordained on 21 MAY 1961; b. 19 FEB 1938 in Modrzyczyn, Poland; celibate; Citizen of Poland.
International Organizations: Auxiliary Bishop (since 1981) of Wloclawek.
Other organizations: Titular Bishop, Tullia.
Career: Lecturer, Higher Theological Seminary, Wloclawek and Lublin Catholic University. Pastor, Contemplative Cloister, Zloczów (Order of the Cameldolite Sisters). Prefect, Wloclawek Theological Seminary. **Pubs.:** Translator of *The Acts of the Apostles* into Polish.
Mailing address: Kuria Diecezjalna, ul Gdanska 2/4, 87-800 Wloclawek, Poland. T. (48 541) 230 55

Angelini, Fiorenzo, H E Mgr
Rel: Roman Catholic; Ordained on 03 FEB 1940; b. 01 AUG 1916 in Rome, Italy; celibate.
International Organizations: President of the Pontifical Council for Pastoral Assistance to Health Care Workers.
Member of the Pontifical Council for the Family.
Other organizations: One of the Presidents, Pontifical Council, Pontifical Chapel; Titular Archbishop, Messene.
Mailing address: Via della Conciliazione 3, I-00193 Roma, Italy. T. (39 6) 689 68 45 – (39 6) 689 67 98. Fax (39 6) 689 68 41. Tx 2031 SANITPC VA
Residence: Via Anneo Lucano 47, I-00136 Roma, Italy.

Angelo, Girard F, Rev
Rel: Roman Catholic; Ordained on 06 JUN 1955; b. 04 SEP 1929; celibate; Lang. English and Italian.
International Organizations: Founder/Director of the National Shrine of the Sacred Heart of Jesus.
Other organizations: Pastor (since 1972), Church of the Sacred Heart.
Career: Founded Men of the Sacred Heart in 1977. Many positions in the Diocese of Scranton. **Educ.:** University of Western Ontario, London, Canada (1952); St Peter's Seminary, London, Ontario, Canada (1955). **Hons.:** Knights of Columbus (4th Degree).
Mailing address: 1 Church Place, Harleigh PA 18225, USA.
Residence: See Mailing Address.

Angeloni, Luciano, H E Mgr
Rel: Roman Catholic; Ordained on 18 AUG 1940; b. 02 DEC 1917 in Imperia, Italy; celibate.
International Organizations: Apostolic Nuncio to Portugal (since 1989) of the Roman Catholic Church.
Other organizations: Titular Archbishop, Vibo Valentia.
Mailing address: Avenida Luis Bivar 18, Lisboa, Portugal. T. (351 1) 54 71 86 – 57 11 01. Fax 53 80 75

Angha, Molana Hazrat Salaheddin Ali Nader Shah, H H E
Rel: Muslim: Sufi; b. 30 SEP 1945 in Teheran, Iran Islamic Rep.
International Organizations: President of the School of Islamic Sufism.
Hons.: Pir Oveyssi. **Pubs.:** Author of over 50 works in poetry and prose translated into Arabic, English, French, German, Italian, Hebrew, Hindi, Spanish, Swedish, Turkish, Urdu and Braille: *Peace*, 1987; *The Secret Word*; *The Wealth of Solouk*; *The Enlightenment* and *The Horizons of Awareness*.
Mailing address: PO Box 5827, Washington DC 20016, USA. T. (1 301) 654 9191
Residence: PO Box 1135, San Rafael CA 94915, USA. T. (1 415) 454 1555.

Angoya, Wycliffe, Commr
Rel: Salvation Army, The.
International Organizations: Territorial Commander of The Salvation Army.
Mailing address: Box 40575, Nairobi, Kenya.

Angulo Quilis, José A, T.O.R., Most Rev

Rel: Roman Catholic; Ordained on 30 NOV 1958; b. 05 APR 1934 in Toledo, Castilla, Spain; celibate; Citizen of Spain; Lang. Spanish, English, Italian and Catalan.
International Organizations: Minister General of the Secular Franciscan Order.
Pubs.: El libro del los Santos; Guión del monitor and Collaborator in DOXA.
Mailing address: Via dei Fori Imperiali 1, I-00186 Roma, Italy. T. (39 6) 699 1540
Residence: See Mailing Address. T. See Mailing Address

Angulo Quilis, José A

Rel: Roman Catholic.
International Organizations: Representative of the Secular Franciscan Order.
Mailing address: 1 via dei Fori Imperiali, Roma, Italy. T. (39 6) 66 79 28 00.

Anhelm, Fritz Erich, Dr

Rel: Protestant; Unordained; b. 29 NOV 1944 in Göttingen, Hannover, Germany; m. to Christiane Bank; c. 1 boy and 1 girl; Citizen of Germany; Lang. German and English.
International Organizations: Secretary-General (since 1985) of the Ecumenical Association of Academies and Laity Centres in Europe.
Member, Advisory Committee (since 1988) of the Conference of European Churches.
Other organizations: Member (since 1985), World Collaborating Committee of Centres and Academies; Member, several boards of Church agencies.
Main field of work: Education, conference planning, project development. **Career:** Teacher, High School from 1974 to 1975. National tutor, Protestant Association for Political Education from 1975 to 1979. **Educ.:** Pol. Sc., German, Ed. Stds. from Georg-August-Universität Göttingen, Germany (1967-1972). **Interests:** ecumenism, international politics. **Pubs.:** Editor, *Bausteine zum Frieden*, 5 Vols., Hannover, 1987; *Diskursives und konziliares Lernen*, Oikos, Frankfurt, 1988 and Editor, *1992-Fünfhundert Jahre danach. Vom kolonialen Blick zum interkulturellen Dialog*, Bielefeld, 1991.
Mailing address: Akademiestrasse 11, 73087 Bad Boll, Germany. T. (49 7164) 79241
Residence: Ebersbacherstrasse 76, 73095 Albershausen, Germany. T. (49 7161) 39727

Anne, Lois

Rel: Roman Catholic.
International Organizations: Contact of the Inter-Territorial Pastoral and Social Development Centre.
Mailing address: PO Box 202, Kenema, Sierra Leone.

Ansah, Paul A V, Prof

Rel: Roman Catholic.
International Organizations: Member of the Catholic Anglophone Media Association of West Africa.
Mailing address: PO Box 9712, Airport, Accra, Ghana. T. (233) 776491 – (233) 776492 – (233) 776493. Tx 2471 GH NCS.

Ansari, Zafar Ishaq, Dr

Rel: Muslim.
International Organizations: Director-General of the Islamic Research Institute.
Mailing address: PO Box 1035, Islamabad, Pakistan. T. (92 51) 851303. Fax 853360. C. ISLAMSERCH.

Antaki, Paul, H E Mgr

Rel: Greek-Melkite Catholic; Ordained on 02 JUL 1950; b. 16 JAN 1927 in Cairo, Egypt; celibate; Citizen of Egypt.
International Organizations: Vicar Patriarchal (since 1968) of Egypt and Sudan.
Other organizations: Titular Archbishop, Nubia.
Career: Vicar in Alexandria from 1951 to 1954. Principal, Patriarchal College of Cairo from 1954 to 1957. Patriarchal Vicar in Alexandria from 1957 to 1967. Superior, Rayak Institute from 1967 to 1968.
Mailing address: Greek Catholic Patriarchate, 16 Daher Street, Faggalah, Cairo, Egypt. T. (20 2) 905790

Anthony, Martin, Pastor

Rel: Seventh-Day Adventist; Ordained.
International Organizations: President, British Union Conference of the General Conference of Seventh-Day Adventists.
Mailing address: Stanborough Park, Watford WD2 6JP, UK.

Antonelli, Ferdinando Giuseppe, O.F.M., H E Card

Rel: Roman Catholic; Ordained on 25 JUL 1922; b. 14 JUL 1896 in Subbiano, Arezzo, Italy; celibate; Citizen of Italy.
International Organizations: Member (since 1973) of the College of Cardinals.
Member of the Order of Friars Minor.
Career: Titular Archbishop, Idicra in 1966. Created Cardinal by Pope Paul VI in 1973. Deacon, Church of San Sebastiano al Palatino, Rome (until 1984).
Mailing address: Piazza San Calisto 16, I-00153 Roma, Italy. T. (39 6) 698 71 24

Antonetti, Lorenzo, H E Mgr

Rel: Roman Catholic; Ordained on 26 MAY 1945; b. 31 JUL 1922 in Romagnano Sesia, Italy; celibate.
International Organizations: Apostolic Nuncio to France (since 1988) of the Roman Catholic Church.
Other organizations: Titular Archbishop, Roselle.
Mailing address: Avenue du Président Wilson 10, F-75116 Paris, France. T. (33 1) 47 23 58 34 – 47 23 62 29. Fax 47 23 65 44

Aoki, Jiun, Ven Dr

Rel: Buddhist; Ordained on 15 JUN 1972; b. 27 FEB 1929 in Yokohama, Kanagawa, Japan; celibate; Citizen of Japan; Lang. Japanese and English.
International Organizations: Founder and Great Patriarch (since 1977) of Myohozan (religious organization).
Other organizations: Honorary Chairman (since 1990), International Foundation for Art Exhibition (The Biten); Honorary Chairman (since 1990), Japan Culture Promoting Association; Founder and President (since 1992), International Cultural Peace Promotional Foundation.
Educ.: Dr. Phil. in Rel. from University of Hawaii, Honolulu, USA (1991). **Hons.:** Grand Prize of Culture; International Academy Award in Religion. **Mem.:** Advisor to Modern Art Association **Pubs.:** *The Blank – 1500 seconds*; *Bodhisattva – Savior in the Hell* and *The Great Hell on Earth.*
Mailing address: 1377 Shimoizue, Seiwa-mura, Taki-gun, Mie 519-22, Japan. T. (81 598) 49 3745

Aonuma, Jakujin

Rel: Buddhist; Ordained; b. 09 OCT 1900; Citizen of Japan.
International Organizations: Manager of the Higashi Eizan Rinoji.
Career: Chief, Kaneiji. **Educ.:** Tendaishu University.
Mailing address: 14-5 Uenokoen, Taito-ku, Tokyo 110, Japan. T. (81 3) 3821 6709

Aoyama, Shundo

Rel: Buddhist; Ordained; b. 15 JAN 1933; Citizen of Japan.
International Organizations: President of the Aichi Special Nun Academy.
Career: Buddhist nun, Shopoji Muryoji. **Educ.:** Komazawa University Graduate School, Tokyo, Japan.
Mailing address: 5926 Muryoji Kataoka, Shiojiri, Nagano 399-07, Japan. T. (81 263) 52 5750

Apartian, Barkev, Rev

Rel: Armenian Evangelical; Ordained.
International Organizations: Secretary of the Union of the Armenian Evangelical Churches in the Near East.
Mailing address: PO Box 110-377, Beirut, Lebanon.

Apfel, Martin Josef Konrad

Rel: Roman Catholic; Unordained; b. 28 NOV 1960 in Frankfurt/Main, Germany; m. to Sabine Kaul; Citizen of Germany; Lang. German, English, French and Spanish.

International Organizations: President (since 1987) of the European Federation of Christian Student Associations.

Other organizations: Personnel Manager (since 1991), Adam Opel AG Rüsselsheim HQ.

Main field of work: Higher education. **Career:** President, Federation of Catholic Student Fraternities in Germany, CV from 1985 to 1986. Member, Central Committee of Catholic Lay Persons in Germany, ZdK from 1987 to 1990. **Educ.:** 1st State Law Exam. from Johann Wolfgang Goethe-Universität Frankfurt, Germany (1980-1984); 2nd State Law Exam., Bar. from Johann Wolfgang Goethe-Universität Frankfurt, Germany (1988). **Mem.:** State Vice-President of the Junge Union Hessen

Mailing address: Gagernstrasse 19, 60385 Frankfurt/Main, Germany. T. (49 69) 443499

Residence: See Mailing Address. T. See Mailing Address

Aponte Martínez, Luis A, H E Card

Rel: Roman Catholic; Ordained on 10 APR 1950; b. 04 AUG 1922 in Lajas, Puerto Rico; celibate; Citizen of USA.

International Organizations: Member (since 1973) of the College of Cardinals.

Member of the Congregation for the Causes of Saints.

Other organizations: Archbishop (since 1964), San Juan de Puerto Rico; President (since 1966), Puerto Rican Episcopal Conference.

Career: Curate, Patillas. Pastor, Santa Isabel. Secretary to Bishop McManus from 1955 to 1957. Vice-Chancellor, Diocese of Ponce from 1955 to 1957. Pastor, Aibonito from 1957 to 1960. Chaplain to the National Guard from 1957 to 1960. Auxiliary Bishop, Ponce from 1960 to 1963. Titular Bishop, Lares from 1960 to 1963. Bishop, Ponce from 1963 to 1964. Created Cardinal by Pope Paul VI in 1973. Director of Development for the Catholic University of Puerto Rico from 1960 to 1963. **Hons.:** Hon. LL.D., Fordham University, Bronx NY (1966); Hon. S.T.D., Inter-American University of Puerto Rico (1969).

Mailing address: Arzobispado, Apartado S-1967, Calle del Cristo 50, San Juan PR 00903, Puerto Rico. T. (809) 725 4975

Apostol, Janos, Rev

Rel: Reformed Church; Ordained.

International Organizations: General Secretary of the Latin American Reformed Church.

Mailing address: Caixa Postal 1251, 01000 Sao Paulo, Brazil.

Appel, André

Rel: Lutheran; Ordained; b. 20 DEC 1921 in Strasbourg, France; m. to Marjorie Pedersen; c. 3 boys and 1 girl; Citizen of France.

International Organizations: Ecclesiastic of the Lutheran Church.

Career: Chaplain, University of Paris. Secretary-General, French Protestant Federation from 1956 to 1964. Pastor, Temple Neuf, Strasbourg from 1964 to 1965. Secretary-General, Lutheran World Federation from 1966 to 1974. President, Lutheran Church, Alsace and Lorraine from 1974 to 1987. President, Conférence des Eglises européennes from 1974 to 1986. President, National Alliance of French Lutheran Churches from 1983 to 1987. **Educ.:** D.D. from University of Strasbourg, France; L.H.D. from University of Paris, France; LL.D. from Eberhard-Karls-Universität Tübingen, Germany.

Mailing address: 4 rond-point de l'Esplanade, F-67000 Strasbourg, France.

Residence: See Mailing Address. T. See Mailing Address

Apple, Raymond, Rabbi

Rel: Jewish; Ordained in 1962; b. 27 DEC 1935 in Melbourne, Victoria, Australia; m. to Marian Unterman; c. 2 boys and 2 girls; Citizen of Australia; Lang. English and Hebrew.

International Organizations: Senior Rabbi of The Great Synagogue, Sydney.

Other organizations: Registrar (since 1975), Sydney Beth Din (ecclesiastical court); Lecturer in Judaic Studies (since 1975), University of Sydney; Senior Rabbi (since 1988), Australian Defence Force; Chairman (since 1991), NSW Council of Christians and Jews.

Main field of work: Ministry, university education. **Career:** Religious Director, Association for Jewish Youth, London from 1959 to 1963. Minister, Bayswater Synagogue, London from 1960 to 1965. Rabbi, Hampstead Synagogue, London from 1965 to 1972. **Educ.:** B.L. from University of Melbourne, Victoria, Australia (1958); B.A. from University of Melbourne, Victoria, Australia (1959); Rabbinical Dip. from Jews' College, London, UK (1970); M.Litt. from University of New England, Armidale, New South Wales, Australia (1984). **Interests:** historical research and writing, freemasonry. **Hons.:** Member, Order of Australia. **Mem.:** Association of Rabbis of Australia and New Zealand (President 1980-84, 1988-92) **Pubs.:** Making Australian Society: The Jews and The Hampstead Syngagogue, 1892-1967.

Mailing address: The Great Synagogue, 166 Castlereagh Street, Sydney NSW 2000, Australia. T. (61 2) 267 2477. Fax 264 8871

Residence: 67 Victoria Street, Potts Point NSW 2011, Australia. T. (61 2) 358 3726

Appleton, George, Rt Rev

Rel: Church of England; Ordained in 1926; b. 20 FEB 1902 in Windsor, UK; m. to Marjorie A Barrett; c. 1 boy and 2 girls; Citizen of United Kingdom.

International Organizations: Ecclesiastic of the Church of England.

Career: Ordained Deacon in 1925. Curate, Stepney Parish Church from 1925 to 1927. Missionary in charge, Society for the Propagation of the Gospel Mission, Irrawaddy Delta from 1927 to 1933. Warden, College of the Holy Cross, Rangoon from 1933 to 1941. Archdeacon, Rangoon from 1943 to 1946. Director of Public Relations, Government of Burma from 1945 to 1946. Vicar, Headstone from 1947 to 1950. Secretary, Conference of British Missionary Societies from 1950 to 1957. Rector, St Botolph, Aldgate, London from 1957 to 1962. Archdeacon, London from 1962 to 1963. Canon, St Paul's Cathedral from 1962 to 1963. Archbishop, Perth from 1963 to 1969. Anglican Archbishop in Jerusalem from 1969 to 1974. Retired in 1974. **Educ.:** M.A. from University of Cambridge, Selwyn College and St Augustine College, Canterbury, New Zealand. **Hons.:** Companion of (the Order of) St Michael and St George; Member of (the Order of) the British Empire; Buber-Rosenzweig Medal, Council of Christians and Jews (1975). **Pubs.:** *The Quiet Heart*, 1984; Editor, *Prayer in a Troubled World*, 1988 and Editor, *Paul The Interpreter*, 1989.

Mailing address: 112A, St Mary's Road, Oxford OX4 1QF, UK. T. (44 865) 248272

Apuzen, Valmike, Rev Dr

Rel: Protestant; Ordained on 20 OCT 1968; b. 26 AUG 1937 in Kota Kinabalu, Sabah, Malaysia; m. to Madeleine Lubaton; c. 4 boys and 3 girls; Citizen of Philippines; Lang. English, Filipino, Cebuano, Ilocano and Ilongo.

International Organizations: President (since 1989) of the Christian and Missionary Alliance Churches of the Philippines (CAMACOP).

Other organizations: Board Member (since 1985), Philippine Council of Evangelical Churches; Secretary (since 1988), Action International.

Main field of work: Biblical and pastoral theology. **Career:** Senior Pastor, Capital City Alliance Church from 1973 to 1989. District Commissioner, Luzon from 1975 to 1977. President, Alliance Biblical Seminary from 1982 to 1984. Founder-Director, Philippine Alliance College of Theology from 1984 to 1989. **Educ.:** B.Th. from Ebenezer Bible College, Zamboanga City, Philippines (1963); B.A. from Baguio Colleges Foundation, Baguio City, Philippines (1973); M.Div. from Philippines Baptist Theological Seminary, Baguio City, Philippines (1976). **Hons.:** D.Div. h.c., Ebenezer Bible College and Seminary, Zamboanga City (1992). **Mem.:** Fellowship of Biblical Preachers **Pubs.:** "Equipping for Ministry", in *Evangelical Thrust*; "Towards an Evangelical Theology of Missions", in *Evangelical Today* and "Brown Feet for Missions", in *Alliance Family Magazine.*

Mailing address: CAMACOP, 3rd Floor, API Bldg, 13 West Capitol Drive, Pasig, Metro Manila, Philippines. T. (63 2) 631 79 47

Residence: 21 Malinis Street, UP Village, Quezon City 1101, Philippines T. (63 2) 921 65 84

Aquilante, Sergio, Rev

Rel: Evangelical Methodist; Ordained.

International Organizations: President of the Evangelical Methodist Church of Italy.

Mailing address: Via Firenze 38, I-00184 Roma, Italy.

Arackal, George P, CV, Rev

Rel: Roman Catholic; Ordained; celibate.

International Organizations: Procurator General of the Vicentian Congregation.

Mailing address: Vicentian House, PO Box 1, Angamaly 683573, India. T. 336.

Arad, Yizhak, Dr
Rel: Jewish.
International Organizations: Chairman of the International Society for Yad Vashem.
Director of the Yad Vashem World Centre for Teaching the Holocaust.
Mailing address: Har Hazikaron, PO Box 3477, Jerusalem 91034, Israel. T. (972 2) 531202.

Aramburu, Juan Carlos, H E Card
Rel: Roman Catholic; Ordained on 28 OCT 1934; b. 11 FEB 1912 in Reducción; celibate.
International Organizations: Member (since 1976) of the College of Cardinals.
Member of the Sacred Congregation for the Oriental Churches. Member of the Sacred Congregation for the Sacraments and Divine Worship.
Other organizations: Archbishop Emeritus (since 1990), Buenos Aires.
Career: Created Cardinal by Pope Paul VI in 1976.
Mailing address: La Pampa 4022, 1430 Buenos Aires, Argentina. T. (54 1) 551 9031

Arbuckle, Gerald A, S.M., Fr
Rel: Roman Catholic; Ordained; b. in New Zealand; celibate.
International Organizations: Director of the Refounding and Pastoral Development Unit, Catholic Theological Union, Sydney.
Member of the Society of Mary.
Other organizations: Co-founder, Melanasian Institute for Pastoral Research; Lecturer.
Main field of work: Cultural anthropology. **Career:** Rector, a major seminary in New Zealand. Assistant-General, Marist Fathers, Rome. Professor of Cultural Anthropology, East Asian Pastoral Institute, Ateneo de Manila University, Philippines. **Educ.:** M.A., Ph.D. from University of Cambridge, UK. **Mem.:** Fellow, Royal Anthropological Institute **Pubs.:** Editor, with D Fleming, SJ, *Religious Life: Rebirth through Conversion*, NY, Alba House, 1990; *Change, Grief and Renewal in the Church: A Spirituality for a New Era*, Westminster MD, Christian Classics, 1991 and articles in periodicals, including: *Theological Studies, The Jurist, Missiology, Human Development, Review for Religious, World Justice.*
Mailing address: 1 Mary Street, Hunters Hill, Sydney NSW 2110, Australia. T. (61 2) 879 6348. Fax 816 4792

Arce, Sergio, Rev
Rel: Presbyterian Reformed; Ordained.
International Organizations: General Secretary of the Presbyterian-Reformed Church in Cuba.
Mailing address: Apartado 154, Matanzas, Cuba.

Archambault, Odile, DW, Sr
Rel: Roman Catholic; Ordained; celibate.
International Organizations: Secretary-General of the Daughters of Wisdom.
Mailing address: Via dei Casali di Torrevecchia 16, I-00168 Roma, Italy. T. (39 6) 627 8639 – 627 86 46. Fax 627 86 69. C. SAGESSE 00168 ROMA.

Ariarajah, Wesley, Rev Dr
Rel: Methodist; Ordained in 1968; b. 02 DEC 1941 in Jaffna, Sri Lanka; m. to Christine Shyamala Chinniah; c. 3 girls; Citizen of Sri Lanka.
International Organizations: Director, Programme on Dialogue with People of Living Faiths (since 1983) of the World Council of Churches.
Main field of work: Ecclesiastic. **Career:** Ordained in the Methodist Church from 1966 to 1968. Minister, Methodist Church of Sri Lanka, Jaffna from 1969 to 1971. Lecturer, Theological College Lanka, Pilimatalawa from 1974 to 1981. Chairman, North and East District, Methodist Church, Jaffna from 1981 to 1983. Staff, World Council of Churches Programme on Dialogue with People of Living Faiths, Geneva in 1987. Delivered Sixth Lambeth Interfaith Lecture. **Educ.:** Th.M., Ph.M., Ph.D. from United Theological College, Bangalore, India, Princeton Seminary, NY, USA, and University of London, UK. **Interests:** reading. **Pubs.:** *Dialogue*, 1980 and *The Bible and People of Other Faiths*, 1986.
Mailing address: PO Box 2100, Route de Ferney 150, CH-1211 Genève 2, Switzerland. T. (41 22) 91 63 28. C. OIKOUMENE GENEVA. Tx 415 730 OIK CH. Fax (41 22) 791 03 61.
Residence: 16 Avenue des Amazones, Chêne-Bougeries, CH-1224 Genève, Switzerland. T. (41 22) 48 51 41

Arinze, Francis, H E Card
Rel: Roman Catholic; Ordained on 23 NOV 1958; b. 01 NOV 1932 in Eziowelle, Anambra State, Nigeria; celibate; Citizen of Nigeria; Lang. Igbo, English, Italian, French, Latin and German.
International Organizations: President (since 1985) of the Pontifical Council for Inter-Religious Dialogue.
Member of the Pontifical Council for Promoting Christian Unity. Member of the Pontifical Council for Dialogue with Non-Believers. Member of the Pontifical Committee for International Eucharistic Congresses.
Other organizations: Member (since 1985), College of Cardinals; President, Commission for the Religious Relations with Islam.
Main field of work: Interreligious dialogue. **Career:** Professor of Liturgy, Logic and parts of Philosophy, Bigard Memorial Seminary, Enugu from 1961 to 1962. Regional Catholic Education Secretary for Eastern Nigeria from 1962 to 1965. Auxiliary Bishop, Onitsha from 1965 to 1967. Archbishop, Onitsha from 1967 to 1985. President, Catholic Bishops' Conference of Nigeria from 1979 to 1984. Vice-President for Africa, United Bible Societies from 1982 to 1985. Member at different times, Sacred Congregations for Doctrine of Faith, Causes of the Saints and Evangelization of Peoples, Secretariat for Non-Christians, Pontifical Commission for Justice and Peace from 1973 to 1984. Elected Patron, Society for Promotion of Igbo Language and Culture, Nigeria in 1974. Created Cardinal by Pope John Paul II in 1985. **Educ.:** Dip. Phil. from Bigard Memorial Seminary, Enugu, Nigeria (1953-1955); B.D. from Urbanian Pontifical University, Vatican City State (1955-1957); S.T.L. from Urbanian Pontifical University, Vatican City State (1959); D.D. from Urbanian Pontifical University, Vatican City State (1960); P.G. Cert. Ed. at University of London, Institute of Education, UK (1963-1964). **Hons.:** DLitt hc, University of Nigeria (1986). **Pubs.:** Sacrifice in Ibo Religion, 1970; Answering God's Call, 1986 and Church in Dialogue, 1990.
Mailing address: 00120 Città del Vaticano, Vatican. T. (39 6) 698 4321
Residence: Palazzo San Calisto, 00120 Città del Vaticano, Vatican. T. (39 6) 698 7318

Arita, Hiroshige
Rel: Shinto; b. 01 JAN 1913; Citizen of Japan.
International Organizations: Chief Priest of the Goo Shrine.
Educ.: Kokugakuin University, Tokyo, Japan.
Mailing address: 2-20-25 Myojocho, Uji, Kyoto 611, Japan. T. (81 774) 23 3234

Armeni, Antonio, Fr
Rel: Roman Catholic; Ordained; celibate.
International Organizations: Procurator General of the Congrégation des Prêtres des Ecoles de Charité.
Mailing address: Via Casilina 600, I-00177 Roma, Italy. T. (39 6) 27 16 336.

Armstrong, Beth
Rel: Spiritualist.
International Organizations: Secretary of the National Spiritual Alliance of the USA.
Mailing address: 16 Congress Street, Fitchburg MA 01426, USA.

Armstrong, J W, Rt Rev
Rel: Church of Ireland; Ordained.
International Organizations: Archbishop of Church of Ireland.
Mailing address: The See House, Cathedral Close, Armagh, UK.

Arnold, Perumalla B, Rev Dr
Rel: Mennonite; Ordained on 20 JUN 1976; b. 05 FEB 1937 in Gollapally village, Andhra Pradesh, India; m. to Sarada; c. 1 boy and 5 girls; Citizen of India; Lang. Telugu, English and Hindi.
International Organizations: Member, Executive Committee of the Mennonite World Conference.
Other organizations: President, Governing Council (since 1977), Conference of Mennonite Brethren Church of India; Treasurer (since 1976), M.B. Property Association, Pvt Ltd; Vice-President (since 1987), Asia Mennonite Conference; President (since 1989), Bible Society of India.
Main field of work: Medical ministry. **Educ.:** M.B.B.S. from Osmania Medical College, Hyderabad, India (1958-1965); M.Surgery from Ludhiana Christian Medical College, Punjab, India (1971-1973).
Mailing address: M.B. Medical Centre, Po-Town Jadcherla, AP 509 302, India. T. (91 8544) 487 – 488

Arns, Paulo Evaristo, O.F.M., H E Card

Rel: Roman Catholic; Ordained on 30 NOV 1945; b. 14 SEP 1921 in Santa Catarina, Brazil; celibate; Citizen of Brazil.
International Organizations: Member (since 1973) of the College of Cardinals.
Member of the Order of Friars Minor. Member of the Sacred Congregation for the Sacraments and Divine Worship.
Other organizations: Archbishop (since 1970), Sao Paulo.
Career: Taught theology and French at the Catholic University of Petrópolis. Pastoral work in Petrópolis. Auxiliary Bishop, São Paulo in 1966. Created Cardinal by Pope Paul VI in 1973. Grand Chancellor, Pontifícia Universidade Católica de Sao Paulo. **Hons.:** Nansen Prize, United Nations (1985); Hon. LL.D., University of Notre Dame, IN. **Mem.:** International Independent Commission on Humanitarian Issues (United Nations) **Pubs.:** Numerous works and translations on religious and racial topics, including: A Violencia em nossos Dias, 1983; Para Ser Jovem Hoje, 1983 and Santos e Heróis do Povo, 1984.
Mailing address: CP 54021, Rua Alfredo Maia 195, 01296 Sao Paulo SP, Brazil. T. (55 11) 227 3099

Arokiasamy, Arul Maria, Rev Roshi

Rel: Roman Catholic; Ordained on 19 MAR 1972; b. 21 MAY 1936 in Burma; celibate; Citizen of India; Lang. Tamil, English and German.
International Organizations: Member of the Society of Jesus.
Main field of work: Zen meditation. **Pubs.:** Leere und Fülle, Kösel.
Mailing address: Dhyana Ashram, 13 Madha Church Rd, Madras 600028, India. T. (91 44) 72821
Residence: See Mailing Address. T. See Mailing Address

Aroncena, Isabel

Rel: Buddhist: Mahayana.
International Organizations: Member of the Foundation for the Preservation of the Mahayana Tradition.
Other organizations: Director, O. Sel Ling Centro de Retiros.
Mailing address: 18412 Bubion, Granada, Spain. T. (34 958) 763 088

Arumi Blancafort, Rafael A, L.C., Fr

Rel: Roman Catholic; Ordained; celibate.
International Organizations: Procurator General of the Legionaries of Christ.
Mailing address: Via Aurelia 677, 00165 Roma, Italy. T. (39 6) 680 65 54

Asano, Hiroshi, Commr

Rel: Salvation Army, The; b. 05 MAY 1927; m. to Tomoko Ohara.
International Organizations: Territorial Commander for Japan (since 1986) of The Salvation Army.
Career: Entered Training College from Shizuoka Corps, Japan in 1949. Various Corps Appointments. Training College Staff in 1961. Divisional Commander for Kanto Area in 1968. Training Principal, Japan in 1969. Tokyo Divisional Commander in 1975. Social Secretary in 1977. Field Secretary in 1978. Lieutenant-Colonel in 1978. Chief Secretary in 1980. **Hons.:** Long Service Order (40 years).
Mailing address: 17 2-chome Kanda Jimbocho, Chiyodaku, Tokyo 101, Japan. T. (88 3) 3237 0881-5. Tx J26431 YULAW PAT

Asbil, Walter Gordon, Rt Rev

Rel: Anglican; Ordained in 1957; b. 03 OCT 1932 in Rawdon PQ, Canada; m. to Mavis Shaver; c. 3 boys and 1 girl; Citizen of Canada.
International Organizations: Coadjutor Bishop (since 1990) of Diocese of Niagara.
Career: Ordained Deacon in 1957. Incumbent, Aylwin, River Desert Parish in 1957. Incumbent, Montreal South Shore Parish in 1960. Rector, St Stephen's Westmount in 1965. Rector, St George's St Anne de Bellevue in 1965. Rector, St George's St Catharines in 1970. Dean, Ottawa, and Rector of Christ Church Cathedral in 1986. Canon, Christ Church Cathedral in 1972. Archdeacon, Diocese of Niagara in 1981. Prolocutor, General Synod, Anglican Conference Council in 1983. Priest Delegate from Canada in Nigeria, Singapore, Wales from 1984 to 1990. Board Member, St Catherines General Hospital from 1983 to 1986. Founding Director, Volunteers Bureau, St Catharines Un. Way. **Educ.:** B.A. from Concordia University, Montreal, Canada (1954); L.Th. from Montreal Diocesan Theological College, Canada (1957); B.D. from McGill University, Montreal, Canada (1957); S.T.M. from McGill University, Montreal, Canada (1967).
Mailing address: 67 Victoria Ave S, Hamilton ON L8N 2S8, Canada.
Residence: 26 Terrace Dr, Dundas ON L9H 3X2, Canada.

Aschkenazi, M, Dr

Rel: Jewish; Ordained.
International Organizations: Representative at UN of the Agudath Israel World Organization.
Mailing address: 36 General Guisan Quai, CH-8002 Zürich, Switzerland. T. (41 1) 202 17 59.

Ashraf, Syed Abu Nasr Ali, Prof Dr

Rel: Muslim: Sufi; Unordained; b. 01 JAN 1925 in Dhaka, Bangladesh; m. to Asia; c. 1 boy; Citizen of United Kingdom and Bangladesh; Lang. Bengali, English, Urdu and Arabic.
International Organizations: Director-General (since 1983) of the Islamic Academy, Cambridge.
Other organizations: Amir (Head) (since 1981), Jamaat-i-Medina (Sufi spiritual organisation); Professor (Visiting Scholar) and Member, Faculty of Education (since 1982), University of Cambridge; Editor (since 1983), *Muslim Education Quarterly*; Founder Rector (since 1989), Darul Ihsan University (Islamic), Dhaka; Educational Adviser to the World Muslim League, Makkah (since 1990); Literary critic and poet.
Main field of work: Spiritual, educational and literary islamisation of education. **Career:** Lecturer, then Reader, English Dept., University of Dhaka, Bangladesh from 1947 to 1955. Reader and Head, English Dept, University of Rajshahi, Bangladesh from 1955 to 1956. Professor and Head, English Dept., University of Karachi, Pakistan from 1956 to 1973. Professor and Head, English Dept., King Abdulaziz University, Jeddah, Saudi Arabia from 1974 to 1977. Professor of English, King Abdulaziz University, Jeddah, Saudi Arabia from 1977 to 1984. Director, World Centre for Islamic Education (WCIE), Makkah from 1980 to 1982. Visiting Professor at Harvard University, Cambridge MA, USA in 1971. Visiting Professor at New Brunswick University, Fredericton, Canada in 1974. Fellow, Clare Hall, Cambridge, UK from 1973 to 1974. Fellow, Wolfson College, Cambridge, UK from 1982 to 1984. Organizer, First World Conference on Muslim Education in Makkah in 1977. **Educ.:** B.A. (Hons), M.A. Eng. from University of Dhaka, Bangladesh (1942-1946); Eng. Tripos II from University of Cambridge, UK (1950-1952); M.A., Ph.D. from University of Cambridge, UK (1961-1964). **Interests:** spiritual training (Sufi), education, literature. **Mem.:** International Association of University Professors of English (IAUPE), International Federation for Modern Languages and Literatures (FILLM) **Pubs.:** *New Horizons in Muslim Education*, Hodder and Stoughton, 1985; *Islam, A Teacher's Manual*, Westhill College, 1989 and Editor, *T.S. Eliot Through Pakistani Eyes* (English).
Mailing address: 23 Metcalfe Road, Cambridge CB4 2DB, UK. T. (44 223) 35 09 76.
Residence: See Mailing Address. T. (44 223) 35 11 27

Askari, Hasan, Prof

Rel: Muslim; b. 04 JUL 1932 in Hyderabad, India; married; c. 3 boys; Citizen of India; Lang. English, Urdu and Arabic.
International Organizations: Director of the Seven Mirrors Publishing House Ltd.
Pubs.: *Alone to Alone: From Awareness to Vision*, Seven Mirrors; *The Darkest Hour Before Sunrise: A Critique of War and a Call to Peace*, Seven Mirrors and with Jon Avery, *Towards Spiritual Humanism*, Seven Mirrors.
Mailing address: 21 Daleside Ave, Pudsey LS28 8HB, UK. T. (44 274) 662 195

Aso, Bunno

Rel: Buddhist; Ordained; b. 28 MAR 1925; Citizen of Japan.
International Organizations: Manager of the Shigonshu Daigoha.
Career: Chief, Sohozan Daigoji Daihonzan Sanpoin. President, Shuchiin University. **Educ.:** Koyasan University, Japan.
Mailing address: Daigo Higashi Ojicho, Fishimi-ku, Kyoto, Kyoto 601-13, Japan. T. (81 75) 571 0002

As-Sadr, Sayed Moussa, Imam

Rel: Muslim: Shi'a; Ordained; Citizen of Lebanon.
International Organizations: President of the Supreme Islamic Council of the Shi'a Community of Lebanon.
Career: Missing since August 1978 (when visiting Libya).
Mailing address: Dar al-Iftaa al-Jaafari, Beirut, Lebanon.

Assaf, David

Rel: Jewish; b. 14 FEB 1956 in Tel Aviv, Israel; m. to Sharon Jean Makover; c. 1 girl; Citizen of Israel; Lang. Hebrew and English.
International Organizations: Secretary-General (since 1979) of the World Union of Jewish Studies.
Other organizations: Researcher (since 1985), Open University of Israel.
Main field of work: Ph.D. in Jewish Studies. **Educ.:** B.A. from Hebrew University of Jerusalem, Israel (1978-1980); M.A. from Hebrew University of Jerusalem, Israel (1980-1984); Ph.D. (uncompleted) from Hebrew University of Jerusalem, Israel (Beginning in 1984).
Mailing address: The Hebrew University, Givat Ram, Jerusalem 92477, Israel. T. (972 2) 526910.
Residence: Ben Azzai 8, Jerusalem 93505, Israel T. (972 2) 732627

Asseo, David, Rabbi

Rel: Jewish; b. in Istanbul, Turkey; m. to Doreta; c. 2 girls; Citizen of Turkey; Lang. French, Italian, Spanish, Hebrew and Turkish.
International Organizations: Chief Rabbi of Turkey.
Mailing address: Türkiye Hahambasiligi, Yemenici Sok 23, Beyoglu, 80050 Tünel, Turkey. T. (91 1) 14488794. Fax 1441980
Residence: Vali Konagi Cad. 16/4, Niçantaç, Istanbul, Turkey.

Assis, Raymundo Damasceno, H E Mgr

Rel: Roman Catholic; Ordained on 19 MAR 1968; b. 15 FEB 1937 in Capela Nova; celibate.
International Organizations: Secretary-General of the Latin American Episcopal Council.
Other organizations: Vicar General, Brasilia; Titular Bishop, Novapietra.
Mailing address: CELAM, Carrera 5 118-31, Usaquén, Apartado Aéreo 5278, Santa Fe de Bogotá, Colombia. T. (57) 612 1495 – 612 1620 – 612 8990. Fax 612 1929.

Assorow, Benedict Batabe

Rel: Roman Catholic; b. 22 SEP 1956 in Naurongo, Ghana; m. to Josephine Azemba Assorow; c. 1 boy; Citizen of Ghana; Lang. English, Italian and French.
International Organizations: Secretary (since 1984) of the Catholic Anglophone Media Association of West Africa.
Head, English-Speaking Africa Regional Office (since 1987) of the Unda, International Catholic Association for Radio and Television. Regional Coordinator (since 1983) of the International Catholic Union of the Press. Regional Secretary of the International Catholic Organization for Cinema and Audiovisual. Contact Person (since 1989) of the Pan African Episcopal Committee for Social Communication.
Other organizations: Executive Secretary, Dept. of Social Communications (since 1981), National Catholic Secretariat, Ghana; Editor (since 1984), National Catholic Secretariat Newsletter.
Main field of work: Diplomacy. **Career:** Editor, *Catholic Standard* in 1985. **Educ.:** Dip. from Institute of Journalism, Ghana (1976-1978); Dip. from University of Pro-Deo, ISOP, Rome, Italy (1978-1980); Dip. Radio and TV Dir., Dip. Radio and TV Prod. from Summer Program of the Catholic Radio and TV Training Center, London, UK (1979); Dip. Int. Rel., Dip. Ger. from Vienna University, Austria (1980). **Pubs.:** The Catholic Church in Ghana; Customs and Traditions Affecting Women in Ghana; Editorials in various newspapers and a number of radio commentaries.
Mailing address: PO Box 9712, Airport, Accra, Ghana. T. (233) 776491 – (233) 776492 – (233) 776493. Tx 2471 GH NCS
Residence: Kwabenya Road, Haatcho, Accra, Ghana.

Assur, M D

Rel: Evangelical Lutheran.
International Organizations: General Secretary of the Evangelical Lutheran Church in Southern Africa.
Mailing address: PO Box 5394, Johannesburg 2000, South Africa.

Astorino, Robert F

Rel: Roman Catholic.
International Organizations: President of the International Federation of Catholic News Agencies.
Executive Director of the Union of Catholic Asian News.
Mailing address: FIAC, PO Box 69626, Kwun Tong, Hong Kong. T. (852) 3496179. Fax 7727656. C. UCIPASIA HONGKONG. Tx 33067 UCANS HX.

Athenagoras, A, Rev Bishop

Rel: Orthodox; Ordained.
International Organizations: Convenor of the Study and Planning Commission of the Standing Conference of the Canonical Orthodox Bishops in the Americas.
Mailing address: 10 East 79th Street, New York NY 10021, USA. T. (212) 570 3500.

Attard, Victor Mark, O.C., Rev Dr

Rel: Roman Catholic; Ordained on 22 MAR 1969; b. 05 MAY 1945 in Hamrun, Malta; celibate; Citizen of Malta; Lang. Maltese, English, Italian, French and German.
International Organizations: Procurator General (since 1983) of the Order of Brothers of the Blessed Virgin Mary of Mount Carmel.
Other organizations: Professor (since 1971), Pontifical Gregorian University; Professor (since 1975), Pontifical Faculty Regina Mundi; Professor (since 1981), Pontifical Faculty Teresianum.
Main field of work: University education, ministry. **Career:** Professor, Pontifical Faculty Augustinianum from 1978 to 1983. Professor, Vew. Beda College from 1979 to 1985. Visiting Lecturer in Theology, University of San Francisco CA (summer of 1983, 1985). **Educ.:** S.T.L. from Pontifical Gregorian University, Vatican City State (1969-1971); D.D. from Pontifical Gregorian University, Vatican City State (1971-1976). **Pubs.:** *Compromise in Morality*, Rome, 1976, 370 pp.; "La persona umana come unità totale", in *Temi di Antropologia Teologica*, Rome, 1981, pp. 794-839 and "The Moral-Ascetical Dimensions of Sexuality and Love", in *Seminarium*, 24, 1984, pp. 83-94.
Mailing address: Via Giovanni Lanza 138, I-00184 Roma, Italy. T. (39 6) 481 41 67
Residence: See Mailing Address. T. See Mailing Address

Atugu, Samuel Oke Esono, Rev

Rel: Reformed Church.
International Organizations: Secretary-General of the Reformed Church of Equatorial Guinea.
Mailing address: Apartado Postal 195, Malabo, Bioko Norte, Equatorial Guinea.

Atwood, David C

Rel: Christian.
International Organizations: Secretary-General of the International Fellowship of Reconciliation.
Mailing address: Spoorstraat 38, NL-1815 BK Alkmaar, Netherlands. T. (31 72) 12 30 14. Fax 15 11 02.

Aubry, Gilbert, Mgr

Rel: Roman Catholic; Ordained; celibate.
International Organizations: President of the Episcopal Conference of the Indian Ocean.
Mailing address: Rivière des Pluies, 97438 Sainte-Marie, Reunion.

Audet, Léonard, C.S.V., Fr

Rel: Roman Catholic; Ordained on 29 JUN 1960; b. 26 NOV 1932 in Maria PQ, Canada; celibate; Citizen of Canada; Lang. French, English, Italian, German and Spanish.
International Organizations: Superior General (since 1988) of the Clerics of St Viator.
President, Theological Commission (since 1990) of the Union of Superiors General.
Main field of work: University education. **Career:** Professor of Holy Scripture, University of Montreal from 1965 to 1988. Director, Bible Dept., University of Montreal from 1973 to 1977. Dean, Faculty of Theology, University of Montreal from 1977 to 1985. **Educ.:** B.A. from University of Montreal, Quebec, Canada (1950-1954); D.Th. from University of Montreal, Quebec, Canada (1956-1962); L.S.S. from Pontifical Biblical Institute, Vatican City State (1962-1964). **Mem.:** Société canadienne de théologie; Association catholique des études bibliques au Canada (Treasurer 1968-77) **Pubs.:** Co-author: Jésus? De l'histoire à la foi, 1974; Après Jésus. Autorité et liberté dans le Peuple de Dieu, 1977 and Je crois en Dieu, 1989.
Mailing address: Via Padre Angelo Paoli 41, I-00144 Roma, Italy. T. (39 6) 529 1603.
Residence: See Mailing Address. T. See Mailing Address

Audrys, Juozas Backis, H E Mgr

Rel: Roman Catholic; Ordained on 18 MAR 1961; b. 01 FEB 1937 in Kaunas, Lithuania; celibate.
International Organizations: Apostolic Pro-Nuncio to the Netherlands (since 1988) of the Roman Catholic Church.
Other organizations: Titular Archbishop, Meta.
Mailing address: Carnegielaan 5, NL-2517 KH Den Haag, Netherlands. T. (31 70) 350 33 63 – 350 43 19. Fax 352 14 61

Augustijns, Louis, A.A., Fr

Rel: Roman Catholic; Ordained; celibate.
International Organizations: Procurator General of the Augustinians of the Assumption.
Mailing address: Via San Pio V 55, I-00165 Roma, Italy. T. (39 6) 62 23 998.

Aukerman, Dale H

Rel: Church of the Brethren; Ordained in 1956; b. 16 JUN 1930 in West Alexandria OH, USA; m. to Ruth Maria Seebass; c. 1 boy and 2 girls; Citizen of USA; Lang. English and German.
International Organizations: Coordinator (since 1973) of the Brethren Peace Fellowship.
Main field of work: Writing, peace work. **Career:** Coordinator, Puidoux Theological Peace Conferences from 1960 to 1962. Director of Studies, Internat. Freundschaftsheim (Germany) from 1966 to 1968. Coordinator, Brethren Action Movement from 1968 to 1970. Pastor, Church of the Brethren from 1970 to 1972. **Educ.:** A.B. from University of Chicago, IL, USA (1948-1949); University of Chicago, IL, USA (1949-1952); University of Glasgow, UK (1961-1962). **Interests:** Biblical studies, gardening. **Pubs.:** *Darkening Valley: A Biblical Perspective on Nuclear War*; Op.-Editor, "The Execution of Ronnie Dunkins", in *Washington Post* and *Reckoning with Apocalypse: Terminal Politics and Messianic Hope* (forthcoming).
Mailing address: Brethren Peace Fellowship, Box 455, Union Bridge MD 21791, USA. T. (1 410) 775 2254
Residence: 191 Stem Road, Union Bridge MD 21791, USA. T. See Mailing Address

Aurenche, Guy

Rel: Roman Catholic; Unordained; b. 14 JUN 1946 in Le Creusot, France; m. to Blandine Theoleyre; c. 1 boy and 2 girls; Citizen of France; Lang. French, English and Italian.
International Organizations: President (since 1975) of the International Federation of the ACAT – Action of Christians for the Abolition of Torture.
Other organizations: Counsel (since 1967), Court of Paris; Member (since 1978), French Catholic Commission 'Justice et Paix'.
Main field of work: Human rights. **Career:** President, Action of Christians for the Abolition of Torture (ACAT)-France from 1975 to 1983. **Educ.:** Lic. Droit, D.E.S. Droit Privé from Faculty of Paris, France (1969). **Interests:** justice, human rights. **Mem.:** International Movement of Catholic Lawyers **Pubs.:** *L'Aujourd'hui des Droits de l'Homme*, Ed. Nouvelle Cité, Paris, 1980; *Bonne Nouvelle à un Monde Torturé*, Ed. Le Centurion, Paris, 1986 and contributor to several collective works, including: *Les Enjeux des Droits de l'Homme*, Ed. Larousse, 1988.
Mailing address: 252 rue Saint-Jacques, F-75005 Paris, France. T. (33 1) 43 29 88 52. Tx 206661 F. Fax (33 1) 40 46 01 83.
Residence: 1 rue Paul Strauss, F-75020 Paris, France T. (33 1) 43 61 56 89

Avadhuta, Shambhushivananda, Dr

Rel: Ananda Marga; Ordained on 11 MAY 1964; b. 14 JAN 1949 in India; single; Citizen of USA; Lang. English, Hindi, Bengali, Sanskrit, Punjabi and German.
International Organizations: Member, Central Committee and Sectorial Secretary for Europe of the Ananda Marga Pracaraka Samgha.
Main field of work: Education, ministry, yoga, research. **Career:** Speaker on several occasions on 'Yoga and contemporary problems'. **Educ.:** M.Com. from University of Pennsylvania, Philadelphia, USA (1966-1968); M.A., Ph.D. from University of Pennsylvania, Philadelphia, USA (1969-1979). **Mem.:** Prout Research Institute (fouding member) **Pubs.:** Prout-Neohumanistic Economics; Economics of Malnutrition in Developing Countries and Legal History of Ananda Marga.
Mailing address: Weisenauer Weg 4, 55129 Mainz, Germany. T. (49 6131) 83 29 95.
Residence: Koburgerstrasse 5/1, 10825 Berlin, Germany. T. (49 30) 7825217

Avi, Dick, Rev

Rel: United Church of Papua New Guinea and Solomon Islands; Ordained.
International Organizations: Assembly Secretary of the United Church in Papua, New Guinea and the Solomon Islands.
Mailing address: Assembly Office, PO Box 3401, Port Moresby, Papua, New Guinea.

Awasom, Henry Anye, Rev

Rel: Presbyterian; Ordained.
International Organizations: Moderator of the Presbyterian Church in Cameroon.
Mailing address: BP 19, Buea (SW Province), Cameroon. T. (237) 32 42 36

Ayarzagüena, Antonio, O.SS.T., Fr

Rel: Roman Catholic; Ordained; celibate.
International Organizations: Procurator General of the Order of the Most Holy Trinity for the Redemption of Captives.
Mailing address: Via Massimi 114-c, I-00136 Roma, Italy. T. (39 6) 342 05 29, (39 6) 342 07 26. Fax (39 6) 34 16 73

Ayik, Arop Yor, Prof

Rel: Presbyterian Evangelical; b. 01 JAN 1935 in Khartoum, Sudan; m. to Elizabeth Arek; c. 2 boys and 4 girls; Citizen of Sudan; Lang. Shilluk, Arabic and English.
International Organizations: Elder of the Presbyterian Church in Sudan.
Other organizations: Professor of Arabic, University of Khartoum; Centre Coordinator, University of Khartoum (School of Extramural Studies), as Supervisor of Teaching Arabic as a Second Foreign Language (TAFL).
Career: Teaching Assistant, Faculty of Arts (Dept of Arabic), University of Khartoum from 1962 to 1964. Assistant Student's Warden, University of Khartoum in 1964. Lecturer in Arabic, University of Khartoum from 1965 to 1970. Minister of Works, Government of Sudan from 1966 to 1967. Director of Education, Ministry of Education from 1970 to 1972. Director of Education and Information, Southern Government from 1970 to 1973. Director of Information and Culture, Southern Government from 1972 to 1973. Principal (Registrar), University of Khartoum in 1974. Deputy Speaker, People's National Assembly (Parliament) from 1974 to 1977. Commissioner for Census (with a status of Deputy Minister) in Southern Sudan in 1981. Dean of Students, University of Juba from 1982 to 1983. Member of the Council for the South and Minister, Education in 1988. Chairman, National Committee for the re-organization of education of the displaced pupils and students from the Southern Sudan in 1989. **Educ.:** B.A. from University of Khartoum, Sudan (1958-1962); P.G. Cert. at University of London, UK (1962-1963); M.A. (TAFL) from American University in Cairo, Egypt (1980-1985). **Hons.:** Education Ladder (1969-70). **Mem.:** Sudanese Red Crescent **Pubs.:** Teaching Arabic as a Second Language, AUC, 1986 and Mechanism of Medium of Instruction: Teachers' Handbook (forthcoming).
Mailing address: School of Extramural Studies, University of Khartoum, PO Box 321, Khartoum, Sudan.
Residence: 46 Sharia Al Fasher, Hai al Matar, Khartoum 321, Sudan.

Ayivi, Eli Kofi, Rev

Rel: Evangelical.
International Organizations: Moderator of the Evangelical Church of Togo.
Mailing address: 1 rue Maréchal Foch, BP 2, Lomé, Togo.

Ayub, K Suyaga, Rev

Rel: Protestant.
International Organizations: General Secretary of the Protestant Christian Church in Bali.
Mailing address: PO Box 72, Jalan Debes 6, Denpasar, Bali, Indonesia.

Ayyamcolil, Joseph, C.V., Fr

Rel: Roman Catholic; Ordained; celibate.
International Organizations: Past Procurator General of the Vicentian Congregation.
Mailing address: Vicentian House, PO Box 1, Angamaly 683573 South, Kerala State, India. T. (91 4856) 336.

Azar, Zahi

Rel: Christian.
International Organizations: Director, Middle East Regional Office of the World Student Christian Federation.
Mailing address: Makhoul Street, Rebeiz Bldg, PO Box 1375, Beirut, Lebanon. T. (961) 34 19 02. Tx OIKLE Beirut 22662.

Azariah, Khushnud Mussarat

Rel: Christian; b. 25 JUL 1949 in Pattoki, Lahore, Punjab, Pakistan; m. to Rt Rev Samuel Robert Azariah; c. 3 girls; Citizen of Pakistan; Lang. English, Urdu and Punjabi.
International Organizations: President for West Asia (since 1986) of the World Federation of Methodist Women.
Member, Women's Committee (since 1990) of the Christian Conference of Asia.
Other organizations: Principal (since 1980), Dar-ul-Khushnud and Dar-ul-Mussarat, Centres for mentally disabled children; Director (since 1988), Women Development and Service Programme, Raiwind Diocese, Church of Pakistan; Coordinator (since 1990), Synodical of Women's Fellowship, Church of Pakistan; Chairperson (since 1990), Women's Concerns Commission, National Council of Churches of Pakistan.
Main field of work: Secondary education and social work. **Career:** Teacher, Cathedral High School, Lahore, Pakistan from 1971 to 1974. Hostel Warden, Kinnaird College for Women, Lahore, Pakistan from 1977 to 1978. Founder-Principal, Dar-ul-Khushnud Centre (for mentally handicapped children), Karachi from 1981 to 1987. **Educ.:** B.Sc. from Kinnaird College, Lahore, Pakistan (1965-1969); M.A. Ed. from University of the Punjab, Lahore, Pakistan (1969-1971); B.D. from Trinity Theological College, Singapore (1974-1977); Dip. Special Educ. from Shiyo Gakuen, Japan (1979-1980). **Interests:** Feminist theology, work with the mentally handicapped. **Mem.:** Human Rights Commission, Lahore, Pakistan **Pubs.:** "Education for the Handicapped in Pakistan"; "Community of Women and Men" and "Jesus Christ, Friend of the Poor".
Mailing address: Women Devpt. and Service Prog., 15A Warris Road, Lahore, Punjab, Pakistan. T. (92 42) 477 255
Residence: 17A Warris Road, Lahore, Pakistan T. (92 42) 489 857

Azariah, M, Rev

Rel: Church of South India; Ordained.
International Organizations: General Secretary of the Church of South India.
Mailing address: PO Box No 4906, Cathedral Road, Madras 600 086, India.

Azizullah, Mohammed

Rel: Muslim.
International Organizations: Secretary of the Islamic Cultural Centre, Hyderabad.
Mailing address: 21/5 Azamabad, Hyderabad 20, India. T. (91) 62342.

Aznárez, Marco Pedro Maria, M.SS.CC., Fr

Rel: Roman Catholic; Ordained; celibate.
International Organizations: Superior General of the Missionaries of the Sacred Hearts of Jesus and Mary.
Mailing address: Vicolo del Curato 12, I-00186 Roma, Italy. T. (39 6) 686 20 45.

Azrak, Mohamad Yosouf

Rel: Muslim.
International Organizations: Member, Islamic Research Center of the Ministry of Islamic Affairs, Afghanistan.
Mailing address: Ministry of Islamic Affairs and Pious Foundation Organisation, Kabul, Afghanistan.

Azzam, Salem

Rel: Muslim.
International Organizations: Secretary-General of the Islamic Council of Europe.
Mailing address: 16 Grosvenor Crescent, London SW1 7EP, UK. T. (44 71) 235 9832. C. Islamiah, London SW1. Tx 894240 Islami G.

Az-Zawi, Taher Ahmad, Sheikh

Rel: Muslim; Citizen of Libya.
International Organizations: Chief Mufti of Libya.
Mailing address: No mailing address.

Babingui, Daniel, Col

Rel: Salvation Army, The.
International Organizations: Territorial Commander of The Salvation Army.
Mailing address: BP 20, Brazzaville, Congo.

Babu, K Jayakar, Dr

Rel: Christian.
International Organizations: Vice-Chairperson of the Association of Christian Institutions for Social Concern in Asia.
Mailing address: Community Services Centre, 17 Balfour Road, Kilpauk, Madras 600 010, India. T. (91 44) 612 684.

Bacon, Daniel W

Rel: Protestant.
International Organizations: Director-General, Singapore of the Overseas Missionary Fellowship.
Mailing address: 10 W Dry Creek Circle, Littleton CO 80120-4427, USA. T. (1 303) 730 4160. Fax 730 4165.

Bacq, Philippe, Fr

Rel: Roman Catholic; Ordained; celibate.
International Organizations: Director and President of the International Centre for Studies in Religious Education.
Mailing address: Rue Washington 184-186, B-1050 Bruxelles, Belgium. T. (32 2) 344 18 82.

Bacqué, François, H E Mgr

Rel: Roman Catholic; Ordained on 01 OCT 1966; b. 02 SEP 1936 in Bordeaux, France; celibate.
International Organizations: Apostolic Pro-Nuncio to Sri Lanka (since 1988) of the Roman Catholic Church.
Other organizations: Titular Archbishop, Gradisca.
Mailing address: 1 Gower Street, Colombo 5, Sri Lanka. T. (94 1) 58 25 54 – 58 60 99. Fax 58 09 06

Bacque, Harvey G

Rel: Roman Catholic.
International Organizations: Director of the Columbian Squires.
Mailing address: PO Drawer 1670, New Haven CT 06507-0901, USA. T. (1 203) 772 2130.

Badawi, Jamal Abdeen, Prof

Rel: Muslim; b. 10 NOV 1939 in Giza, Egypt; m. to Azza Ibrahim El-Shazly; c. 2 boys and 3 girls; Citizen of Canada; Lang. English and Arabic.
International Organizations: Chairman/Founder (since 1981) of the Islamic Information Foundation.
Other organizations: Imam (since 1970), Islamic Association of Maritime Provinces; Associate Professor (since 1974), Saint Mary's University, Halifax; Member, Legal Council (since 1980), Islamic Society of North America; Member, Consultative Council (since 1988), Islamic Society of North America; Member, Board of Directors (since 1990), North American Islamic Trust (NAIT).
Main field of work: University education, public programs on Islam. **Career:** Department Head, El-Nasar Export and Import Co., Egypt from 1960 to 1963. Assistant Professor, University of Maine, USA from 1968 to 1969. Assistant Professor, Saint Mary's University, Halifax from 1970 to 1974. **Educ.:** B.Com. from Ain Shams University, Cairo, Egypt (1956-1960); M.B.A. from Indiana University, Bloomington, USA (1963-1965); Ph.D. from Indiana University, Bloomington, USA (1965-1970). **Interests:** seminars, dialogues and TV programs on Islam and other religions. **Mem.:** numerous local, regional, national and international Islamic and other religious organizations **Pubs.:** "Muhammad in the Bible", in T Querishi, Ed., *Some Facts about Prophet Muhammad's Life*, NAIT, Indianapolis IN, 1983; "Muhammad's Prophethood: An Analytical View", in T Querishi, Ed., *Islam: A Way of Life and a Movement*, NAIT, Indianapolis IN, 1984 and "Islam: A Brief Look" and "The Earth and Humanity", in John Hick and E Meltzer, Eds., *Three Faiths – One God*, Macmillan, 1989.
Mailing address: Saint Mary's University, Robie Street, Halifax NS, Canada. T. (1 902) 420 5776
Residence: 8 Laurel Lane, Halifax NS B3M 2P6, Canada. T. (1 902) 445 2494. FAX 445 2494

Bafile, Corrado, H E Card
Rel: Roman Catholic; Ordained on 11 APR 1936; b. 04 JUL 1903 in L'Aquila; celibate.
International Organizations: Member (since 1976) of the College of Cardinals.
Career: Privy Chamberlain to Pope John XXIII from 1958 to 1960. Papal Nuncio to Germany from 1960 to 1975. Titular Archbishop, Antiocha in Pisidia from 1960 to 1976. Created Cardinal by Pope Paul VI in 1976. Prefect, Sacred Congregation for the Causes of the Saints.
Mailing address: Via P Pancrazio Pfeiffer 10, I-00193 Roma, Italy. T. (39 6) 698 42 43

Baggio, Sebastiano, H E Card
Rel: Roman Catholic; Ordained on 21 DEC 1935; b. 16 MAY 1913 in Rosà, Italy; celibate.
International Organizations: Cardinalis Patronus (since 1984) of the Sovereign Military Order of Malta.
Other organizations: Cardinal-Bishop (since 1974), Velletri; Camerlengo (since 1985), Holy Roman Church; Member (since 1969), College of Cardinals; Subdean (since 1986), College of Cardinals.
Career: Secretary, Nunciatures in El Salvador, Bolivia, Venezuela from 1938 to 1946. With Secretariat of State from 1946 to 1948. Chargé d'Affaires, Colombia from 1948 to 1950. Sacra Congregazione Concistoriale from 1950 to 1953. Apostolic Nuncio, Chile from 1953 to 1959. Apostolic Delegate to Canada from 1959 to 1964. Apostolic Nuncio, Brazil from 1964 to 1969. Created Cardinal by Pope Paul VI in 1969. Archbishop, Cagliari from 1969 to 1973. Prefect, Sacred Congregation for the Bishops from 1973 to 1984. President, Pontifical Commission for the Vatican City State in 1984. Chamberlain, High Roman Church in 1985. **Hons.:** Orders from Bolivia, Brazil, Chile, Colombia, Ecuador, Malta, Portugal and Venezuela. **Mem.:** Rotary International (Honorary Member, since 1980)
Mailing address: Piazza della Città Leonina 9, I-00193 Roma, Italy. T. (39 6) 654 01 37

Baginski, Jan, Mgr
Rel: Roman Catholic; Ordained on 17 JUN 1956; b. 31 MAY 1932 in Kamionka, Poland; celibate; Citizen of Poland.
International Organizations: Auxiliary Bishop (since 1985) of Opole.
Other organizations: Titular Bishop, Tagarata.
Career: Parish work at Pyskowice and Nowa Wies Królewska near Opole from 1956 to 1959. Prefect, then Vice-Rector, Higher Theological Seminary, Opole. Parish Priest, Prelate in Kluczbork from 1976 to 1985.
Mailing address: ul Grunwaldzka, 45-054 Opole, Poland.

Bagna, Mario, D.C., Fr
Rel: Roman Catholic; Ordained; celibate.
International Organizations: Procurator General of the Prêtres de la Doctrine Chrétienne.
Mailing address: S Maria in Monticelli 28, I-00186 Roma, Italy. T. (39 6) 65 42 292.

Bahk, Yong Whee, Prof Dr
Rel: Roman Catholic.
International Organizations: President of the Asian Federation of Catholic Medical Associations.
Mailing address: Catholic University Graduate School, 505 Banpodong, Socho-Ku, Seoul 137 701, Korea Rep. T. (82 2) 593 5141 Ext 1257. Fax (82 2) 532 3112.

Baird, Isabel D
Rel: United Free Church of Scotland.
International Organizations: General Secretary of the United Free Church of Scotland.
Mailing address: 11 Newton Place, Glasgow G3 7PR, Scotland, UK.

Baird, Robert Dean, Dr
Rel: Baptist; Ordained on 19 NOV 1959; b. 12 AUG 1933 in Hereford TX, USA; m. to Margaret Ann Roberts; c. 1 boy and 1 girl; Citizen of USA; Lang. English.
International Organizations: Missions Director (since 1986) of the Baptist Bible Fellowship International.
Main field of work: Ministry. **Career:** Missionary in Jamaica (for 5 years). Assistant Pastor to Dr David A Cavin at the High Street Baptist Church, Springfield, MO (for 6 years). Assistant Missions Director under Dr Carl Boonstra, Baptist Bible Fellowship International from 1977 to 1981. Pastor, Hallmark Baptist Church, Fort Worth, TX from 1981 to 1986. Vice-President, Baptist Bible Fellowship International. **Educ.:** Th.B. from Baptist Bible College, Sprinfield MO, USA (1966-1970). **Interests:** Golf, stamp collecting.
Mailing address: PO Box 191, Springfield MO 65803, USA. T. (1 417) 862 5001.
Residence: Route 2, Box 326, Springfield MO 65803, USA T. (1 417) 833 3992

Bajardi, Elia
Rel: Buddhist: Mahayana; m. to Rodolfo Barjardi.
International Organizations: Member of the Foundation for the Preservation of the Mahayana Tradition.
Other organizations: Director, Muni Ghiana Centre.
Mailing address: Piazza Pallavicino 20, 90146 Palermo, Italy. T. (39 91) 454 839

Bajardi, Rodolfo
Rel: Buddhist: Mahayana; m. to Elia Bajardi.
International Organizations: Member of the Foundation for the Preservation of the Mahayana Tradition.
Other organizations: Director, Muni Ghiana Centre.
Mailing address: Piazza Pallavicino 20, 90146 Palermo, Italy. T. (39 91) 454 839

Baker, Elisabeth, Ms
Rel: Quaker.
International Organizations: Secretary of the Quaker Council for European Affairs.
Mailing address: Quaker House, Square Ambiorix 50, B-1040 Bruxelles, Belgium. T. (32 2) 230 49 35.

Baker, John Austin, Rt Rev
Rel: Church of England; Ordained in 1955; b. 11 JAN 1928; m. to Gillian Mary Leach; Citizen of United Kingdom.
International Organizations: Bishop (since 1982) of Salisbury.
Career: Official Fellow, Chaplain and Lecturer in Divinity, Corpus Christi College, Oxford from 1959 to 1973. Also Lecturer in Theology, Brasenose College and Lincoln College, Oxford from 1959 to 1973. Dorrance Visiting Professor, Trinity College, Hartford CT in 1967. Canon, Westminster from 1973 to 1982. Visiting Professor, King's College, London from 1974 to 1976. Sub-Dean, Westminster and Lector Theologiae from 1978 to 1982. Rector, St Margaret's, Westminster from 1978 to 1982. Chaplain to Speaker, House of Commons from 1978 to 1982. Member, Church of England Doctrine Commission from 1967 to 1981. Member, C. of E. Doctrine Commission from 1984 to 1987. Chairman, C. of E. Doctrine Commission from 1985 to 1987. Member, Standing Committee, World Council of Churches Faith and Order Commission from 1984 to 1987. **Educ.:** Marlborough College, Oxford, UK; University of Oxford, Oriel College, UK; Cuddesdon Theological College, UK. **Hons.:** Emeritus Fellow, Corpus Christi College, Oxford (since 1977). **Mem.:** Clubs: United Oxford and Cambridge **Pubs.:** The Foolishness of God, 1970; Travels in Oudamovia, 1976 and The Whole Family of God, 1981.
Mailing address: South Canonry 71, The Close, Salisbury SP1 2ER, UK. T. (44 722) 334031

Bakhsh, J S Qadir, Rt Rev
Rel: Church of Pakistan.
International Organizations: Moderator of the Church of Pakistan.
Mailing address: Lal Kothi, Barah Patthar, Sialkot-2, Pakistan.

Bakkevig, Trond, Rev Dr
Rel: Church of Norway; Ordained.
International Organizations: General Secretary, Council on Foreign Relations of the Church of Norway.
Mailing address: PO Box 5816, Hegdehaugen, N-0308 Oslo 3, Norway.

Bakula, Kushok, Rev Rinpoche

Rel: Buddhist; Ordained in 1925; b. 19 MAY 1917 in Matho, Ladakh, India; single; Citizen of India; Lang. Ladakhi, Tibetan and Hindi.
International Organizations: President (since 1990) of the Asian Buddhist Conference for Peace.
Other organizations: Head Lama of Ladakh (since 1948), Ladakh Gonpa Association; Ambassador (since 1989), India to Mongolia.
Main field of work: Education. **Career:** Member, Indian Parliament from 1967 to 1977. Member, Minorities Commission of India from 1978 to 1990. Minister, Government of Jammu and Kashmir (India). Head, World Fellowship of Buddhists (Ladakh Centre). **Educ.:** Gehse Lharamba Deg. from Drespung Monastery, Lhasa, Tibet (1940). **Hons.:** Medal of Honour, Asian Buddhist Conference for Peace; Padma Bhushan, India. **Mem.:** Maha Bodhi Society of India (Life Member) **Pubs.:** author of commentaries on Buddhist teachings.
Mailing address: Embassy of India, 26 Peace Street, PO Box 691, Ulan Bator, Mongolia. T. 53167 – 50772
Residence: Samkar Gonpa, PO Leh-194101, Ladakh, India

Balasuriya, Stanislaus Tissa, Rev

Rel: Roman Catholic; Ordained; b. 29 AUG 1924 in Kahatagasdigiliya, North Central Province, Sri Lanka; celibate; Citizen of Sri Lanka; Lang. English, Sinhala, French, Latin and Italian.
International Organizations: Director (since 1971) of the Centre for Society and Religion.
Main field of work: Author and theologian. **Career:** Rector, Aquinas College, Sri Lanka from 1964 to 1971. Asian Chaplain, Catholic Students' Movement from 1969 to 1979. Asian Co-ordinator, Ecumenical Association of Third World Theologican (EATWOT) from 1976 to 1986. Editor, *Logos*, *Quest*, *Voices of the Third World*, and *Social Justice*. **Educ.:** B.A. Econ. from University of Colombo, Sri Lanka (1945); Lic. Phil. from Pontifical Gregorian University, Vatican City State (1949); Lic. Theol. from Pontifical Gregorian University, Vatican City State (1952); Dip. Ag.Econ. from University of Oxford, UK (1963). **Interests:** writing on socio-economic, political and theological issues. **Hons.:** Khan Memorial Gold Medal for Economics. **Mem.:** Citizens Committee for National Harmony; Ecumenical Association of Third World Theologians (EATWOT) **Pubs.:** *Jesus Christ and Human Liberation*, CSR, 1976; *Eucharist and Human Liberation*, Orbis, New York, 1977 and *Indicators of Social Justice*, CSR, 1991.
Mailing address: 281 Deans Road, Colombo 10, Sri Lanka. T. (94 1) 69 54 25
Residence: See Mailing Address. T. See Mailing Address

Balcells Xuriach, José Maria B, S.P., Fr

Rel: Roman Catholic; Ordained; celibate.
International Organizations: Superior General of the Clerks Regular Poor of the Mother of God of the Pious Schools.
Mailing address: Piazza dei Massimi 4, I-00186 Roma, Italy. T. (39 6) 654 1494 – (39 6) 686 8456.

Baldelli, Fortunato, H E Mgr

Rel: Roman Catholic; Ordained on 18 MAR 1961; b. 06 AUG 1935 in Valfabbrica, Italy; celibate.
International Organizations: Apostolic Delegate to Angola (since 1983) of the Roman Catholic Church.
Other organizations: Apostolic Pro-Nuncio to Sao Tomé-Principe (since 1985); Titular Archbishop, Bevagna.
Mailing address: CP 1030, Luanda, Angola. T. (244 2) 33 05 32 – 33 62 89. Fax 33 23 78

Balducelli, Roger, O.S.F.S., Fr

Rel: Roman Catholic; Ordained; celibate.
International Organizations: Superior General of the Oblates of St Francis de Sales.
Mailing address: Via Dandolo 49, I-00153 Roma, Italy. T. (39 6) 58 23 89.

Baldzhian, L K

See Vazken.

Bali, Dev Dutta, Pundit

Rel: Hindu: Vedic; b. 07 MAR 1924 in Hoshiarpur, Punjab, India; m. to Sarla; c. 1 boy and 2 girls; Citizen of India; Lang. Hindi, English, Urdu and Punjabi (spoken).
International Organizations: Trustee/Secretary of the Vedic Sadhan Ashram.
Other organizations: Partner (since 1940), Bali Watch Co, Dehradun; Editor (since 1989), *Pavamaan* (monthly); District President, Arya Samaj.
Main field of work: Lecturing and writing. **Educ.:** B.A. from Agra University, India (1961). **Interests:** Indian classical music. **Hons.:** Several awards in declamation. **Mem.:** Doon Watch Merchants Association
Mailing address: Bali Niwas, 119/II, Chukhuwala, Dehradun 248001, Uttar Pradesh, India. T. (91) 25453

Bálint, Lajos, Rt Rev

Rel: Roman Catholic; Ordained on 28 APR 1957; b. 06 JUL 1929 in Dolmita, Romania; celibate.
International Organizations: Bishop (since 1990) of Alba-Julia.
Career: Parish Priest, Odorheiu Secuiesc from 1977 to 1981. Titular Bishop, Nova from 1981 to 1990.
Mailing address: Palatul Episcopiei, Str Mihai Viteazul 21, 2500 Alba-Julia, Romania. T. (40 968) 11 689

Baliozian, Aghan, His Grace Bishop

Rel: Armenian Apostolic Church; Ordained on 28 JUL 1968; b. 18 AUG 1946 in Aleppo, Syrian AR; celibate; Citizen of Australia; Lang. Armenian, English, Arabic and Turkish.
International Organizations: Primate (since 1981) of the Armenian Apostolic Church Diocese of Australia.
Other organizations: Pontifical Legate (since 1991), Far East Armenians.
Main field of work: Ministry. **Career:** Assistant Dean/Lecturer, Seminary of the Armenian Patriarchate of Jerusalem from 1968 to 1971. Lecturer, Seminary of the Armenian Church of Holy See in Etchmiadzin from 1971 to 1974. Dean/Lecturer, Seminary of the Armenian Patriarchate of Jerusalem from 1974 to 1975. **Educ.:** Non-degree student Dip. from Yerevan State University, Armenia (1971-1974). **Interests:** sports.
Mailing address: 10 MacQuarie Street, Chatswood NSW 2067, Australia. T. (61 2) 4198056
Residence: 468 Pittwater Road, North Manly, Sydney NSW 2100, Australia. T. (61 2) 9384816

Ball, Bryan William, Dr

Rel: Seventh-Day Adventist; Ordained on 11 SEP 1962; b. 11 JUL 1935 in Bere Ferrers, Devon, UK; m. to Dawn Gwynydd Macey; c. 1 boy and 2 girls; Citizen of United Kingdom; Lang. English.
International Organizations: President, South Pacific Division (since 1990) of the General Conference of Seventh-Day Adventists.
Other organizations: Chairman, Board of Governors (since 1990), Avondale College, Australia; Chairman, Board of Trustees (since 1990), Sydney Adventist Hospital.
Main field of work: Ecclesiastical history. **Career:** President, North British Conference, Seventh-Day Adventists from 1971 to 1975. Chairman, Theology Dept., Newbold College from 1975 to 1984. Principal, Avondale College, Australia from 1984 to 1990. **Educ.:** M.A. from Andrews University, Berrien Springs MI, USA (1966); Ph.D. from University of London, UK (1971). **Interests:** music, golf, photography. **Pubs.:** A Great Expectation: Eschatological Thought in English Protestantism 1640-1660; The English Connection: The Puritan Roots of SDA Belief and contribution to *Encyclopedia of World Religions*.
Mailing address: 148 Fox Valley Rd, Wahroonga NSW 2076, Australia. T. (61 2) 489 7122
Residence: 5 Pringles Rd, Martinsville NSW 2265, Australia. T. (61 49) 771 220

Ball, John M, Canon

Rel: Protestant; Ordained.
International Organizations: Secretary-General of the Bible Churchmen's Missionary Society.
Mailing address: 251 Lewisham Way, London SE4 1XF, UK. T. (44 81) 691 61 11.

Ballentine, Rudolph Miller, Dr

Rel: Lutheran; b. 16 AUG 1941 in Ballentine SC, USA; m. to Pennell Whitney; c. 3 boys and 1 girl; Citizen of USA; Lang. English, French and Spanish.

International Organizations: President-Administrator (since 1981) of the Himalayan International Institute of Yoga Science and Philosophy.

Other organizations: Private practice in Psychosomatic Medicine at Center for Holistic Medicine (since 1974); Director (since 1975), Combined Therapy Program.

Main field of work: Physician. **Career:** Instructor in Psychiatry, Louisiana State University Medical School, New Orleans in 1971. Assistant Clinical Professor of Psychiatry, LSU Medical School in 1972. **Educ.:** B.S. from Duke University, Durham NC, USA (1962); M.D. from Duke University, Medical School, Durham NC, USA (1967). **Mem.:** World Medical Association; American Institute of Homeopathy; American Medical Association; New York Academy of Sciences **Pubs.:** Editor, *Theory and Practice of Meditation*, 1975; Co-author, *Science of Breath*, 1979 and Author, *Transition to Vegetarianism*, 1987.

Mailing address: RR 1, Box 400, Honesdale PA 18431, USA. T. (1 717) 253 5551. Tx 5106001805.

Residence: 1379 Bridge Street, Honesdale PA 18431, USA. T. See Mailing Address

Ballestrero, Anastasio Alberto, O.C.D., H E Card

Rel: Roman Catholic; Ordained on 06 JUN 1936; b. 03 OCT 1913 in Genoa, Italy; celibate; Citizen of Italy.

International Organizations: Member (since 1979) of the College of Cardinals.

Member of the Order of Discalced Brothers of the Blessed Virgin Mary of Mount Carmel. Member of the Congregation for Religious and Secular Institutes.

Other organizations: Archbishop Emeritus (since 1989), Turin; Member, Council of Cardinals and Bishops.

Career: Bishop, Bari from 1974 to 1977. Created Cardinal by Pope John Paul II in 1979. Archbishop, Turin from 1977 to 1989. **Hons.:** entitled S Maria sopra Minerva. **Mem.:** Affari Pubblici della Chiesa

Mailing address: Fortino Santa Maria, CP 3, I-19030 Bocca di Magra (La Spezia), Italy. T. (39 187) 60 00 65

Balslev-Olsen, Christian

Rel: Christian.

International Organizations: Secretary-General of the Danchurchaid.

Mailing address: Skt Peders Straede 3, DK-1453 København K, Denmark. T. (45) 32 15 28 00. C. DANCHURCHAID COPENHAGEN. Tx 16217 danaid dk.

Baltasar Marcelino, Antonio B, Rt Rev

Rel: Roman Catholic; Ordained on 09 JUL 1955; b. 21 SEP 1930 in Lousa, Portugal; celibate.

International Organizations: Contact of the European Episcopal Committee on the Media.

Other organizations: Bishop (since 1988), Aveiro, Portugal.

Mailing address: Curia Diocesana, Rua Cándido Dos Reis 107, Apartado 541, P-3808 Aveiro, Portugal. T. (351 34) 23 354, (351 34) 27 605. Fax (351 34) 38 14 02

Balume, M

Rel: Christian.

International Organizations: Contact of the Organisme chrétien de coopération internationale.

Mailing address: Montée St Barthélémy 42, F-69005 Lyon, France. T. (33) 78 25 41 65.

Banana, Canaan Sodindo, Rev Dr

Rel: Methodist; Ordained in 1966; b. 05 MAR 1936 in Esiphezini, Matabeleland; m. to Janet Mbuyazwe; c. 3 boys and 1 girl; Citizen of Zimbabwe.

International Organizations: Ecclesiastic of the Methodist Church.

Career: Chaplain, Tegwani High School from 1965 to 1966. Chairman, Bulawayo Council of Churches from 1969 to 1970. Chairman, Southern Africa Urban Industrial Mission from 1970 to 1973. Founder Member and Vice-President, African National Council (ANC) from 1971 to 1973. ANC Representative in USA and UN from 1973 to 1975. Chaplain, American University from 1974 to 1975. Detained from 1975 to 1976. Press Spokesman for ANC in 1976. Founder Member and Publicity Secretary, People's Movement in 1976. In detention several times and publications banned in Rhodesia. President, Republic of Zimbabwe from 1980 to 1987. Founded Mushandira Pamure Project in 1980. Founded Kushinga-Phikelela Agricultural Institute in 1981. Chancellor, University of Zimbabwe from 1983 to 1987. **Educ.:** Tegwani Training Institute; Epworth Theological College; Kansai Industrial Center, Japan; Wesley Theological Seminary, USA; University of South Africa, Pretoria, South Africa. **Interests:** soccer, table tennis, lawn tennis. **Hons.:** Hon. LL.D., American University and University of Zimbabwe. **Pubs.:** Theology of Promise; The Woman of My Imagination and The Ethos of Socialism, 1987.

Mailing address: c/o State House, Harare, Zimbabwe.

Bang, Thakurdas

Rel: Hindu.

International Organizations: Chairman of the Sarva Seva Sangh.

Mailing address: Gopuri, Wardha 442001, India.

Banks, William, Lt-Col

Rel: Salvation Army, The.

International Organizations: Officer Commanding of The Salvation Army.

Mailing address: PO Box 70129, Kowloon Central Post Office, Kowloon, Hong Kong.

Baquete, Grace

Rel: Methodist.

International Organizations: President for South and East Africa of the World Federation of Methodist Women.

Mailing address: Ireje Metodiste, CP 2640, Maputo, Mozambique.

Barbagallo, Maria, Sr

Rel: Roman Catholic; Ordained; celibate.

International Organizations: Superior General of the Missionary Sisters of the Sacred Heart.

Mailing address: 269 Viale Cortina d'Ampezzo, I-00135 Roma, Italy.

Barbarito, Luigi, H E Mgr

Rel: Roman Catholic; Ordained on 20 AUG 1944; b. 19 APR 1922 in Atripalda, Avellino, Italy; celibate.

International Organizations: Apostolic Pro-Nuncio to the United Kingdom (since 1986) of the Roman Catholic Church.

Other organizations: Titular Archbishop, Fiorentino.

Career: Served in the Diocese, Avellino from 1944 to 1952. Entered Diplomatic Service, Holy See in 1953. Secretary, Apostolic Delegation, Australia from 1953 to 1959. Secretary of State, Vatican (Council for Public Affairs of the Church) from 1959 to 1967. Counsellor, Apostolic Nunciature, Paris from 1967 to 1969. Archbishop and Papal Nuncio to Haiti and Delegate to the Antilles from 1969 to 1975. Pro-Nuncio to Senegal, Burkina Faso, Niger, Mali, Mauritania, Cape Verde Islands, and Guinea-Bissau from 1975 to 1978. Apostolic Pro-Nuncio to Australia from 1978 to 1986. **Educ.:** J.C.D. from Pontifical Gregorian University, Vatican City State; Dip. from Pontifical Academy for Diplomats, Vatican City State. **Interests:** music, walking, snooker. **Hons.:** Grand Cross, National Order of Haiti (1975); Grand Cross, Order of the Lion, Senegal (1978); Knight Commander, Order of Merit, Italy (1966), Portugal (1967). **Mem.:** Mexican Academy of International Law

Mailing address: 54 Parkside, Wimbledon SW19 5NF, UK. T. (44 81) 946 1410. Fax 9472 494

Barberis, Bruno, Prof

Rel: Christian.

International Organizations: President of the International Centre of Sindonology.

Mailing address: Via San Domenico 28, I-10122 Torino, Italy.

Barbour of Bonskeid, Robert (Robin) Alexander Stewart B, Very Rev Prof

Rel: Church of Scotland; Ordained in 1954; b. 11 MAY 1921 in Edinburgh, Scotland, UK; m. to Margaret Isobel Pigot; c. 3 boys and 1 girl; Citizen of United Kingdom; Lang. English, French, German and Italian.

International Organizations: Chaplain-in-Ordinary to H M The Queen (since 1976) of Church of Scotland.

Other organizations: Prelate, Priory of Scotland (since 1977), Order of St John; Dean (since 1981), Chapel Royal in Scotland.

Main field of work: University education. **Career:** Military service from 1940 to 1945. Chairman, Edinburgh Christian Council for Overseas Students from 1953 to 1955. Lecturer, Faculty of Divinity, University of Edinburgh from 1955 to 1959. Senior Lecturer, Faculty of Divinity, University of Edinburgh from 1959 to 1971. Professor of New Testament Exegesis, University of Aberdeen from 1971 to 1986. Master, Christ's College, Aberdeen from 1977 to 1982. Moderator, General Assembly, Church of Scotland from 1979 to 1980. Chairman, Scottish Churches Council from 1982 to 1986. **Educ.:** University of Oxford, Balliol College, UK (1939-1940); M.A. from University of Oxford, Balliol College, UK (1945-1947); B.D. from University of St Andrews, St Mary's College, UK (1949-1952); S.T.M. from Yale University, Divinity School, New Haven CT, USA (1952-1953). **Interests:** New Testament studies, ecumenical affairs, music, forestry, walking. **Hons.:** Military Cross (1945); Hon. D.D., St Andrews (1979). **Mem.:** Studiorum Novi Testamenti Societas **Pubs.:** The Scottish Horse 1939-45, 1950; Traditio-Historical Criticism of the Gospels, 1972 and What is the Church for?, 1973.
Mailing address: Fincastle, Pitlochry, Perthshire PH16 5RJ, Scotland, UK. T. (44 796) 3209
Residence: See Mailing Address. T. See Mailing Address

Barker, Eileen Vartan, Dr
Rel: Unknown; b. 21 APR 1938 in Edinburgh, Scotland, UK; m. to Peter Johnson Barker; c. 2 girls; Citizen of United Kingdom; Lang. English and French.
International Organizations: Reader in Sociology with special reference to Religion (since 1990) of the London School of Economics.
Honorary Director (since 1988) of the Information Network Focus on Religious Movements.
Other organizations: President (since 1991), Society for the Scientific Study of Religion.
Main field of work: University education. **Educ.:** B.Sc. Soc. (1st) from University of London, London School of Economics and Political Science, UK (1967-1970); Ph.D. from University of London, London School of Economics and Political Science, UK (1984). **Hons.:** Distinguished Book Award of SSSR; Hobhouse Memorial Prize (1970). **Pubs.:** Editor, *New Religious Movements: A Perspective for Understanding Society*; *New Religious Movements: A Practical Introduction*, HMSO, London, 1989 and *The Making of a Moonie: Brainwashing or Choice?*, Blackwell, Oxford, 1994.
Mailing address: Dept of Sociology, London School of Economics, Houghton St, London WC2A 2AE, UK. T. (44 71) 242 0392

Barlage, Heinrich, S.V.D., Fr
Rel: Roman Catholic; Ordained; celibate.
International Organizations: Superior General of the Society of the Divine Word.
Mailing address: Via dei Verbiti 1, CP 5080, I-00154 Roma, Italy. T. (39 6) 575 40 21.

Barlow, Bernard Francis, Rev
Rel: Roman Catholic; Ordained on 28 AUG 1973; b. 14 OCT 1941 in Glasgow, Scotland, UK; celibate; Citizen of United Kingdom; Lang. English, French, Dutch, Italian and German.
International Organizations: General Secretary (since 1989) of the Order Secular of Servants of Mary.
General Councillor (since 1989) of the Order of the Servants of Mary.
Main field of work: Consultant in ministry and mission. **Career:** Teacher, Tayside Regional Council from 1981 to 1984. Parish Priest, Servites – Bognor Regis from 1984 to 1989. Assistant Provincial, Servites – English Province from 1978 to 1980. Assistant Provincial, Servites – English Province from 1987 to 1989. **Educ.:** Ph.B. from Heythrop College, Oxford, UK (1964-1967); M.A. from Catholic University of Louvain, Belgium (1967-1970); Drs. Medieval Studies from Catholic University of Louvain, Belgium (1970-1971); L.Th. from Catholic University of Nijmegen, The Netherlands (1971-1973). **Interests:** Church history, ecumenism. **Mem.:** Ecumenical Society of the Blessed Virgin; Liberal Democrats **Pubs.:** The Conversations at Malines and Louvain Studies, 1973.
Mailing address: Servi di Maria, Piazza S Marcello, al Corso 5, I-00187 Roma, Italy. T. (39 6) 678 08 88.
Residence: See Mailing Address. T. See Mailing Address

Barnes, Geoffrey Leon, Rev Dr
Rel: Uniting Church in Australia; Ordained in 1951; b. 26 JAN 1926 in Melbourne, Victoria, Australia; m. to Annie Louisa Matear; c. 1 boy and 1 girl; Citizen of Australia; Lang. English.
International Organizations: Secretary of the Australian and New Zealand Association of Theological Schools.
Other organizations: Lecturer in Church History (since 1975), United Theological College; President (since 1977), Church Records and Historical Society, NSW; Convenor (since 1990), Uniting Church Assembly Historical Committee; Vice-Principal (since 1991), United Theological College.
Main field of work: Theological education. **Career:** Principal, Camden Theological College from 1968 to 1975. President, Congregational Union of Australia from 1973 to 1975. Moderator, New South Wales Synod Uniting Church from 1987 to 1988. **Educ.:** B.A. from University of Melbourne, Victoria, Australia (1947-1949); B.D. from Melbourne College of Divinity, Victoria, Australia (1950-1952); S.T.M. from Hartford Seminary, CT, USA (1959-1961); Ph.D. from MacQuarie University, Sydney, New South Wales, Australia (1982-1987). **Interests:** tennis, gardening. **Mem.:** Uniting Church in Australia (Standing Committee) **Pubs.:** "The Australian Life-Style: Spirituality 'Down Under'", in *Studies of the Church in History*, H Davies, Editor; "Congregationalists", in *Many Faiths: One Nation*, I Gillman, Editor and *The Forgotten Factor: The Story of Lay People in the Church.*
Mailing address: United Theological College, 16 Masons Drive, North Parramatta NSW 2151, Australia. T. (61 2) 683 3655
Residence: 62 Downes Street, North Epping NSW 2121, Australia. T. (61 2) 876 6666

Barnett, Joe
Rel: Jewish: Progressive.
International Organizations: Secretary of the European Board of the World Union for Progressive Judaism.
Honorary Secretary of the World Union for Progressive Judaism.
Mailing address: 109 Whitfield Street, London W1P 5RP, UK. T. (44 71) 636 7627.

Barnett, Ted, Dr
Rel: Protestant.
International Organizations: US Director of the Africa Inland Mission International.
Mailing address: PO Box 178, Pearl River NY 10965, USA. T. (1 914) 735 4014. Tx 137015.

Baroi, R N, Rev
Rel: Baptist; Ordained.
International Organizations: Executive Secretary of the Bangladesh Baptist Sangha.
Mailing address: Sadarghat, PO Box 1108, Dhaka I, Bangladesh.

Baron, Salo Wittmayer
Rel: Jewish; Ordained in 1920; Citizen of Austria.
International Organizations: Retired of the Center of Israel and Jewish Studies, Columbia University, New York NY.
Main field of work: Historian and sociologist. **Career:** Taught at Columbia University, New York NY from 1930 to 1963. Director, Center of Israel and Jewish Studies, Columbia University, New York NY from 1950 to 1968. Occupied the first American chair on Jewish history. **Educ.:** Dr. Phil. from Vienna University, Austria (1917); Dr. Pol.Sc. from Vienna University, Austria (1922); Dr. Law from Vienna University, Austria (1923). **Pubs.:** *Social and Religious History of the Jews*, 18 Vols., 1952-1983 and wrote a history of the Jewish community in the American revolution, 3 Vols., 1942.
Mailing address: No mailing address.

Barquon, Jean Paul, Pastor
Rel: Seventh-Day Adventist; Ordained on 10 MAY 1981; b. 05 NOV 1949 in Casablanca, Morocco; m. to Lilyane Jublin; c. 1 girl; Citizen of France; Lang. French and English.
International Organizations: UNESCO Representative and National Secretary of the French Section (since 1982) of the International Association for the Defence of Religious Liberty.
Other organizations: Journalist for *Signes des Temps* (since 1982); Member of the Bureau (since 1982), French Committee of Non-Governmental Organizations associated with the United Nations Division of Public Information (NGO/DPI).

Main field of work: Public relations, defence of religious liberty, journalism. **Career:** Reporter at Radio FR3 for the Adventist Mission, French Polynesia from 1974 to 1977. Pastor, Adventist Federation from 1977 to 1982. Director, Communications Department, Adventist Federation in 1982. **Educ.:** Lic. Theol. from Faculté Adventiste de Théologie, France (1970-1974); Human Rights and Int. Law from International Institute of Human Rights, Strasbourg, France (1988-1989). **Interests:** tolerance, human rights, international law. **Mem.:** AJIR (Association professionnelle des journalistes de l'information religieuse) **Pubs.:** "Sectes et esprit sectaire"; "Fondamentalisme et intégrisme religieux en Occident" and "La liberté religieuse: un droit de l'homme".
Mailing address: 680 avenue de la Libération, F-77350 Le Mee sur Seine, France. T. (33 1) 60 68 62 32
Residence: 3 allée François Villon, F-95100 Argenteuil, France. T. (33 1) 34 11 20 36

Barrett, Ronald
Rel: Bahá'í.
International Organizations: Past Secretary-General of the Bahá'í International Community.
Mailing address: Bahá'í World Centre, PO Box 155, 31001 Haifa, Israel. T. (972 4) 372 433 – 510 344. Fax 372 425. C. BAHAIFAITH, HAIFA. Tx 46626 BAYT IL.

Barrington-Ward, Simon, Rt Rev
Rel: Church of England; Ordained in 1957; b. 27 MAY 1930 in London, UK; m. to Jean Caverhill; c. 2 girls; Citizen of United Kingdom.
International Organizations: Bishop (since 1985) of Coventry.
Other organizations: Chairman (since 1986), Partnership for World Mission International Affairs and Development Committee; Chairman (since 1987), Lee Abbey Council (Lee Abbey Community).
Career: National Service with the Royal Air Force from 1949 to 1950. Pilot Officer in 1949. Lecture, Free University of Berlin from 1953 to 1954. Ordained Deacon in 1956. Chaplain, Magdalene College, Cambridge from 1956 to 1960. Lecturer, Ibadan University, Nigeria from 1960 to 1963. Fellow and Dean of Chapel, Magdalene College, Cambridge from 1963 to 1970. Principal, Crowther Hall College and Selly Oak College from 1970 to 1975. General Secretary, Church Missionary Society from 1975 to 1985. Chaplain to H M The Queen from 1984 to 1985. **Educ.:** Eton College, UK; B.A., M.A. from University of Cambridge, Magdalene College, UK; Westcott House, Cambridge, UK. **Interests:** mission in modern world, walking, bicycling. **Hons.:** Prelate to the Order of St Michael and St George (since 1989); Hon. D.D., Wycliffe College, Toronto (1983); Honorary Fellow, Magdalene College, Cambridge. **Mem.:** Fellow, Royal Anthropological Institute; Member, House of Lords (1991) **Pubs.:** Contributor to *Christianity in Independent Africa*, Ed. Fashole Luke and Others, London, 1978; Contributor to *The World's Religions*, 1988 and *Love Will Out*, 1988 (anthology of CMS Newsletters 1975-85).
Mailing address: The Bishop's House, Davenport Road, Coventry CV5 6PW, UK. T. (44 203) 672244. Fax 713271

Barth, Jörg, Rev
Rel: Hutterian Brethren; Ordained on 25 AUG 1968; b. 14 JAN 1929 in Veitsteinbach, Hessen, Germany; m. to (Maria) Renata Zimmermann; c. 8 boys and 2 girls; Citizen of Germany; Lang. German, English and Spanish.
International Organizations: Preacher (since 1988) of the Hutterian Brethren.
Main field of work: Ministry and education. **Educ.:** Maestro Normal Superior from Normal School, Asunción, Paraguay (1945-1948).
Mailing address: Michaelshof, Auf der Höhe, 57612 Birnbach, Germany. T. (49 2681) 6250
Residence: See Mailing Address. T. See Mailing Address

Barth, Norman E
Rel: Lutheran.
International Organizations: Executive Director of the Lutheran World Relief.
Mailing address: 390 Park Avenue S, New York NY 10016, USA. T. (1 212) 532 63 50. Fax 213 6081.

Bartha, Tibor
Rel: Reformed Church in Hungary; Ordained; b. 13 JUL 1912 in Magyarkapud, Transylvania, Hungary; m. to Szende G Szabó; c. 1 boy and 1 girl; Citizen of Hungary.
International Organizations: President of the General Synod of the Reformed Church in Hungary.
Other organizations: Board Member (since 1958), National Peace Council; Vice-President (since 1974), National Committee for European Security and Co-operation.
Career: Professor, Theological Academy, Debrecen from 1953 to 1958. Bishop, Transtibiscan Reformed Diocese from 1958 to 1986. Chairman, Reformed Church Synod of Hungary from 1962 to 1986. Chairman, Ecumenical Church Council of Hungary from 1959 to 1986. Member, Central Board, World Council of Churches from 1958 to 1983. Member, Executive Committee, Reformed World Federation from 1961 to 1977. Vice-President, Christian Peace Conference from 1964 to 1978. Honorary President, Christian Peace Conference from 1978 to 1986. President, Ecumenical Peace Committee, Hungary's National Peace Council from 1983 to 1986. Board Member, World Peace Council from 1968 to 1986. Member, Parliament from 1958 to 1986. Member, Presidential Council from 1963 to 1986. **Educ.:** D.Div. from Debrecen Academy, Hungary (1936). **Hons.:** Dr. h.c., Evangelic Lutheran Theology Seminary of Budapest and United Theology Seminary of Kolozsvár, Cluj (Romania); Labour Order of Merit; Banner of the Order of the Hungarian People's Republic (3rd cl. 1958, 2nd cl. 1962, 1st cl. 1972); St Vladimir Order (Russian). **Pubs.:** Ige, egyház, nép (Logos, Church, People), Vols. I-II, 1972; Diaconal Manual, 1980 and Divine Service Procedural Order, 1985.
Mailing address: General Synod of the Reformed Church in Hungary, Abonyi utca 21, H-1146 Budapest, Hungary. T. (36 1) 122 7870

Basarke, Alice
Rel: Sikh.
International Organizations: Executive Assistant of the World Sikh Organization.
Mailing address: PO Box 15830, Merivale Depot, Nepean, Ottawa ON K2C 3S7, Canada. T. (1 613) 723 2026. Fax 723 0290.

Bass, Joe
Rel: Christian.
International Organizations: President of the International Christian Aid.
Mailing address: 5189 Verdugo Way, Camarillo CA 93010, USA. T. (1 805) 987 8888. Tx 4720231 INTERAID. Fax (1 805) 484 8829.

Bassotti, Giuseppe M, CRSP, Fr
Rel: Roman Catholic; Ordained; celibate.
International Organizations: Superior General of the Clerics Regular of St Paul.
Mailing address: Via Giacomo Medici 15, I-00153 Roma, Italy. T. (39 6) 58 23 39.

Basterrechea, José Pablo, F.S.C., Fr
Rel: Roman Catholic; Ordained; celibate.
International Organizations: Secretary of the Union of Superiors General.
Member, Committee for Reporting to the Congregation, the International Union of Superior Generals, and the Union of Superior Generals of the Congregation for the Evangelization of Peoples. Advisor, Section for Secular Institutes of the Congregation for Religious and Secular Institutes. Member, Committee for Reporting to the Congregation, the International Union of Superior Generals, and the Union of Superior Generals of the Congregation for Religious and Secular Institutes. Member of the Brothers of the Christian Schools.
Mailing address: Via dei Penitenzieri 19, I-00193 Roma, Italy. T. (39 6) 686 82 29. Fax (39 6) 687 43 17

Bastiaens, Willy, F.I.C., Fr
Rel: Roman Catholic; Ordained; celibate.
International Organizations: Secretary-General (since 1988) of the Brothers of the Immaculate Conception of the Blessed Virgin Mary.
Mailing address: Brusselsestraat 38, 6211 PG Maastricht, Netherlands. T. (31 43) 21 21 14.

Bastian, Donald N, Bishop
Rel: Free Methodist; Ordained.
International Organizations: Bishop of Free Methodist World Fellowship.
Mailing address: 4315 Village Centre Court, Mississauga ON L4Z 1S2, Canada.

Bastiani, Elio, Fr
Rel: Roman Catholic; Ordained; celibate.
International Organizations: Superior General of the Congregatio Filiorum Amoris misericordis.
Mailing address: c/o Congregatio Filiorum Amoris Misericordis, I-06050 Collevalenza Perugia, Italy. T. (39 75) 88 71 35

Bastos de Avila, Fernando B, Rev
Rel: Roman Catholic; Ordained; b. 17 MAR 1918; celibate; Citizen of Brazil.
International Organizations: Ecclesiastic of the Roman Catholic Church.
Other organizations: Social Director (since 1954), National Catholic Immigration Commission; Professor of Sociology (since 1957), Pontifícia Universidade Católica do Rio de Janeiro; Member, Council (since 1960), National Federation of Trade.
Main field of work: Sociology. **Career:** Director, Instituto Brasileiro de Desenvolvimento (IBRADES). **Educ.:** Universidade do Nova Friburgo, Brazil; Catholic University of Louvain, Belgium; Pontifical Gregorian University, Vatican City State. **Pubs.:** Introduçao a Sociologia, 1962; Solidarismo, 1965 and Pequena Enciclopédia de Moral Ecivismo, 1967.
Mailing address: 115 rua Bambina-Botatogo, Rio de Janeiro RJ, Brazil.

Basu, Samar, Sri
Rel: Hindu; b. 20 AUG 1922 in Gopalnagar, West Bengal, India; Citizen of India; Lang. English and Bengali.
International Organizations: Secretary-General/Treasurer of the World Union.
Educ.: University of Calcutta, India. **Pubs.:** *Social and Political Evolution of Man as envisioned by Sri Aurobindo*; *A Handbook of World Union*; *To read Sri Aurobindo: How and Why ?*; *How far is Pondicherry ?* and *Educational Philosophy of Sri Aurobindo* (in bengali).
Mailing address: Sri Aurobindo Ashram, Pondicherry 605002, India. T. (91 413) 24834. C. Aurobindo Pondicherry-2.
Residence: Mettu Street, Kuruchikuppam, Pondicherry 605012, India

Bates, Gerald E, Bishop
Rel: Free Methodist; Ordained; b. 12 SEP 1933 in Caldwell OH, USA; m. to Marlene Rachel Parsons; c. 2 boys and 1 girl.
International Organizations: President (since 1989) of the Free Methodist World Fellowship.
Other organizations: Bishop (since 1985), Free Methodist Church of North America.
Career: Missionary to Central Africa (Burundi, Zaire) with the General Missionary Board from 1957 to 1985. Area Administrative Assistant for Burundi, Rwanda, Zaire from 1971 to 1985. Chairman, Board of Editors, *Book of Discipline* in 1985. **Educ.:** A.A. from Spring Arbor College, MI, USA (1953); A.B. from Greenville College, IL, USA (1955); M. Div. from Asbury Theological Seminary, Wilmore KY, USA (1958); Th.M. from Western Theological Seminary, Holland MI, USA (1964); Ph.D. from Michigan State University, East Lansing, USA (1975); D.D. from Roberts Wesleyan College, Rochester NY, USA (1986). **Hons.:** Alumnus of the Year, Spring Arbor College, MI (1974); Goodwill Ambassador, Noble County, Chamber of Commerce, OH (1988); Distinguished Alumnus, Asbury Theological Seminary, Wilmore KY (1991). **Mem.:** Phi Kappa Phi (since 1975); Rochester Fights Back-AntiDrug Campaign, NY (Advisory Committee, since 1990); American Bible Society (Denomination Representative); Rotary (since 1987) **Pubs.:** *Soul Afire*, 1981 (biography of pioneer missionary J W Haley).
Mailing address: PO Box 535002, Indianapolis IN 46253, USA. T. (1 317) 244 3660. Fax (1 317) 244 1247
Residence: 6715 Oak Lake Drive, Indianapolis IN 46214, USA

Bates, Robert
Rel: Roman Catholic.
International Organizations: Executive Secretary of the International Catholic Deaf Association.
Mailing address: 814 Thayer Avenue, Silver Spring MD 20910, USA. T. (1 301) 588 4009.

Bath, Robert Edward, Commr
Rel: Salvation Army, The; Ordained on 06 JAN 1947; b. 26 JUN 1928 in Sydney NSW, Australia; m. to Vida Joy; c. 1 boy and 1 girl; Citizen of Australia; Lang. English and Hindi.
International Organizations: Territorial Commander for Eastern Australia and Papua New Guinea (since 1988) of The Salvation Army.
Other organizations: Chairman (since 1988), Salvation Army Trusts, Australia; Director (since 1988), Christian Children's Fund, Australia.
Main field of work: Administration, evangelism and social services. **Career:** With The Salvation Army (since 1947). International Secretary, The Salvation Army from 1985 to 1988. **Mem.:** Sydney Rotary Club (1988)
Mailing address: PO Box A435, Sydney South NSW 2000, Australia. T. (61 2) 264 1711
Residence: 1/60 Solander Street, Monterey NSW 2217, Australia. T. (61 2) 588 6775

Bathellier, A, Ms
Rel: Roman Catholic.
International Organizations: Contact of the Focolare Movement.
Mailing address: CCIC, 9 rue Clear, F-75007 Paris, France. T. (33 1) 39 71 89 01.

Battista, Armando P Gregorio, O.Cist, Fr Abbot
Rel: Roman Catholic; Ordained on 17 JUL 1938; b. 22 APR 1915 in Fontana Liri, Frosinone, Italy; celibate; Citizen of Italy; Lang. Italian, French, German, Spanish and Latin.
International Organizations: Procurator General (since 1953) of the Order of Cistercians.
Educ.: Studied Phil. and Theol. from Pontifical University San Tommaso d'Aquino, Vatican City State (1932-1938).
Mailing address: Piazza del Tempio di Diana 14, I-00153 Roma, Italy. T. (39 6) 574 36 94.
Residence: See Mailing Address. T. See Mailing Address

Bauer, Hartmut
Rel: Protestant.
International Organizations: Director of the Protestant Association for Cooperation in Development.
Mailing address: Mittelstrasse 37, 53175 Bonn, Germany. T. (49 228) 810 10. Fax 810 11 60. C. Zentralstelle Bonn 2. Tx 8861133 eze d.

Baum, Gregory G
Rel: Christian; b. 20 JUN 1923 in Berlin, Germany; Citizen of Canada.
International Organizations: Professor of Religious Studies of the McGill University.
Main field of work: University education, writing. **Career:** Came to Canada from England in 1940. Editor, *The Ecumenist*. Member, Editorial Committee, *Concilium*. Member, Editorial Committee, *Relations*. **Educ.:** B.A. Math. and Phys. from McMaster University, Hamilton, Ontario, Canada (1946); M.A. Math. from Ohio State University, Columbus, USA (1947); Th.D. from University of Fribourg, Switzerland (1956). **Hons.:** Hon. Doctorates: Huron College, St Francis Xavier, Ohio Wesleyan, Lafayette, Waterloo Lutheran, McMaster and Dubuque; Officer of the Order of Canada. **Mem.:** Canadian Theological Society; Canadian Society for the Study of Religion; Catholic Theological Society of America **Pubs.:** Ethics and Economics, 1984; Theology and Society, 1987 and Compassion and Solidarity, 1988.
Mailing address: Religious Studies, McGill University, Montreal, Québec H3A 2A7, Canada.

Baum, William Wakefield, H E Card
Rel: Roman Catholic; Ordained on 12 MAY 1951; b. 21 NOV 1926 in Dallas TX, USA; celibate; Citizen of USA.
International Organizations: Prefect (since 1980) of the Congregation for Catholic Education.
Member of the Congregation for Bishops. Member of the Sacred Congregation for the Oriental Churches. Member of the Congregation for the Causes of Saints. Member of the Congregation for Bishops. Member of the Congregation for the Evangelization of Peoples. Member of the Congregation for Religious and Secular Institutes. Member of the Pontifical Council for the Laity. Member of the Pontifical Council for the Interpretation of Legislative Texts.
Other organizations: Chancellor (since 1973), Catholic University; Member (since 1976), College of Cardinals; Penitenziere Maggiore (since 1990), Penitenziera Apostolica; Member, Congregation for the Doctrine of the Faith.

Career: Associate Pastor, St Aloysius, St Therese's and St Peter's parishes from 1951 to 1956. Instructor and Professor, Avila College, Kansas City from 1954 to 1956. Instructor and Professor, Avila College, Kansas City from 1958 to 1963. Administrator, St Cyril's Parish, Sugar Creek MO from 1960 to 1961. Honorary Chaplain, His Holiness the Pope in 1961. Peritus (Expert Adviser), Second Vatican Council from 1962 to 1965. First Executive Director, Bishops' Commission for Ecumenical and Interreligious Affairs, Washington from 1964 to 1967. Chairman, Bishops' Commission for Ecumenical and Interreligious Affairs, Washington in 1972. Member, Joint Working Group of Representatives, Catholic Church and World Council of Churches from 1965 to 1969. Member, Mixed Commission of Representatives, Catholic Church and Lutheran World Federation from 1965 to 1966. Chancellor, Diocese of Kansas City, St Joseph from 1967 to 1970. Honorary Prelate, His Holiness the Pope in 1968. Pastor, St James Parish, Kansas City from 1968 to 1970. Bishop, Diocese of Springfield-Cape Girardeau in 1970. Member, Synod of Bishops in 1971. Archbishop, Washington from 1973 to 1980. Created Cardinal by Pope Paul VI in 1976. **Interests:** reading, music. **Hons.:** Hon. D.D., Muhlenberg College, Allentown PA (1967), Georgetown University WA, St John's University, Brooklyn NY. **Pubs.:** The Teaching of Cardinal Cajetan on the Sacrifice of the Mass, 1958 and Considerations Toward the Theology on the Presbyterate, 1961.
Mailing address: Via Rusticucci 13, I-00193 Roma, Italy.

Bauswein, Jean-Jacques

Rel: Reformed Church.
International Organizations: Director of the John Knox International Reformed Centre.
Mailing address: Chemin des Crets de Pregny 27, Grand-Saconnex, CH-1218 Genève, Switzerland. T. (41 22) 798 91 61.

Bavin, Timothy John, Rt Rev

Rel: Anglican; Ordained; b. 17 SEP 1935 in Northwood, UK; Citizen of United Kingdom.
International Organizations: Bishop (since 1985) of Portsmouth.
Career: National Service with Royal Army Service Corps (now RCT) from 1957 to 1959. Commissioned in 1958. Platoon Officer (2 Lt) 90 Co. Aden from 1958 to 1959. Curate, St Alban's Cathedral, Pretoria from 1961 to 1963. Chaplain, St Alban's College, Pretoria from 1963 to 1969. Curate, Uckfield with Litte Horsted, Sussex from 1969 to 1971. Vicar, Parish of the Good Sheperd, Brighton from 1971 to 1972. Dean, Johannesburg and Rector of the Cathedral Parish from 1971 to 1974. Bishop, Johannesburg from 1974 to 1984. **Educ.:** Brighton College, UK; University of Oxford, Worcester College, UK; Cuddesdon College, Oxford, UK. **Interests:** music, walking, gardening, Victoriana. **Mem.:** Oratory of the Good Shepherd (since 1987); Clubs: Athenaeum, Royal Yacht Squadron (Cowes), Royal Naval (Portsmouth) **Pubs.:** Editor, Deacons in the Ministry of the Church, 1988.
Mailing address: Bishopwood, Fareham, Hants PO14 1NT, UK. T. (44 329) 280247

Bax, Martin

Rel: Protestant.
International Organizations: Liaison Officer of the Association of Protestant Development Organizations in Europe.
Mailing address: Rue Joseph II 174, B-1040 Bruxelles, Belgium. T. (32 2) 231 01 02. Fax (32 2) 231 14 13.

Baxendale, David Arthur, Col

Rel: Salvation Army, The; Ordained on 22 JUN 1955; b. 23 APR 1930 in New York NY, USA; m. to Alice Chamberlain; c. 1 boy and 2 girls; Citizen of USA; Lang. English.
International Organizations: Territorial Commander for Chile, Peru, Bolivia and Ecuador (since 1990) of The Salvation Army.
Main field of work: Ecclesiastical administration. **Career:** President, School for Officers' Training, The Salvation Army USA East from 1977 to 1981. Personnel Director, The Salvation Army USA East from 1981 to 1984. Chief Secretary, The Salvation Army USA West from 1984 to 1986. Territorial Commander, The Salvation Army for Caribbean Territory (15 countries) from 1986 to 1990. **Educ.:** B.Sc. from Springfield College, MA, USA (1948-1952); Bible Stds from Pittsburgh Xenia, PA, USA (1952-1953); Ordination from The Salvation Army Training College, Bronx NY, USA (1954-1955); M.A. from Columbia University, New York NY, USA (1963-1964). **Pubs.:** *Growing Together* (book on marriage – 2 chapters); "These Are My People", in *The Officer* (magazine), The Salvation Army and "Family Devotions", in *The Officer* (magazine), The Salvation Army.
Mailing address: Casilla 3225, Santiago 1, Chile. T. (56 2) 671 8237
Residence: Avda España 46, Santiago 1, Chile. T. See Mailing Address

Baycroft, John Arthur, Rt Rev

Rel: Anglican; Ordained in 1955; b. 02 JUN 1933 in Redcar, Yorkshire, UK; m. to Joan Lake; c. 1 boy and 2 girls.
International Organizations: Suffragan Bishop (since 1985) of Ottawa.
Career: Rector, Loughborough from 1955 to 1957. Assistant Rector, St Matthew's Church, Ottawa from 1957 to 1962. Rector, Perth, Ontario from 1962 to 1967. Rector, St Matthias Church, Ottawa from 1967 to 1984. Rector, Christchurch Cathedral and Dean of Ottawa from 1984 to 1986. Formerly Archdeacon, Ottawa West. Canon Theological, Diocese of Ottawa. Member, Anglican/Roman Catholic International Commission. Anglican Co-Chairman, Anglican/Roman Catholic Dialogue of Canada. Lecturer, Carleton University. Member, Faculty of Theology, St Paul University. **Educ.:** B.A. from University of Cambridge, Christ's College, UK (1954); M.A. from University of Cambridge, Christ's College, UK (1958); G.O.E. from Ripon Hall, Oxford, UK (1955); B.D. from University of Toronto, Trinity College, Canada (1959). **Interests:** theatre, art, ballet. **Hons.:** Hon. D.D., Montreal Diocesan Theological College. **Mem.:** American Academy of Religion; Society of Biblical Literature; Canadian Church History Society; Club: National Press **Pubs.:** The Eucharist Way, 1982; The Way of Prayer, 1983 and others as well as numerous scholarly articles.
Mailing address: 71 Bronson Ave, Ottawa ON K1R 6G6, Canada.
Residence: 97 Java St, Ottawa ON K1Y 3L5, Canada.

Bayer, Oswald

Rel: Evangelical; Ordained; b. 30 SEP 1939 in Nagold, Germany; m. to Eva; Citizen of Germany.
International Organizations: Ecclesiastic of the Evangelical Church.
Other organizations: Director (since 1979), Institute of Christian Sociology; Professor of Systematic Theology (since 1979), University of Tübingen.
Career: Vicar, Evangelische Landeskirche, Würtemberg in 1964. Assistant, University of Tübingen from 1965 to 1968. Evangelical Monastery, Tübingen from 1968 to 1971. Priest, Tübingen from 1972 to 1974. Professor of Systematic Theology, University of Bochum from 1974 to 1979. **Educ.:** Tübingen, Bonn and Rome. **Pubs.:** several books on theological topics.
Mailing address: Evangelisch-theologische Fakultät der Universität Tübingen, Institut für Christliche Gesellschaftslehre, Hölderlinstrasse 16, 72074 Tübingen, Germany. T. (49 7071) 292591
Residence: Am unteren Herrlesberg 36, 72074 Tübingen, Germany. T. (49 7071) 81897

Bayzelon, Jean-Paul, M.E.P., Fr

Rel: Roman Catholic; Ordained; celibate.
International Organizations: Superior General of the Paris Foreign Mission Society.
Mailing address: 128 rue du Bac, F-75341 Paris CEDEX 07, France. T. (33) 45 48 19 92

Bazyli, B, His Beatitude

Rel: Orthodox.
International Organizations: Metropolitan of Warsaw and All Poland.
Mailing address: ul Gen K Swierczewkiego 52, Warszawa, Poland.

Beach, Bert Beverly, Dr

Rel: Seventh-Day Adventist; Ordained in 1956; b. 15 JUN 1928 in Gland, Vaud, Switzerland; m. to Eliane Palange; c. 2 girls; Citizen of USA; Lang. English, French, German and Italian.
International Organizations: Director of Public Affairs and Secretary-General of Council of Inter-Church Relations (since 1980) of the General Conference of Seventh-Day Adventists.
Secretary-General (since 1980) of the International Religious Liberty Association. President (since 1991) of the International Commission for the Prevention of Alcoholism and Drug Dependency.
Other organizations: Secretary (since 1970), Conference of Secretaries of Christian World Communions.

Main field of work: Ministry and education. **Career:** Principal, Italian Union School, Florence from 1952 to 1958. Associate Professor, Columbia Union College, Takoma Pard MD from 1958 to 1960. Director of Education, Northern Europe – West Africa Division of Seventh-Day Adventists (SDA) from 1960 to 1975. Secretary, Northern Europe – West Africa Division of SDAs from 1973 to 1980. **Educ.:** B.A. from Pacific Union College, Angwin CA, USA (1946-1948); Stanford University, CA, USA (1948-1951); Dr. from University of Paris, France (1952-1958). **Hons.:** Th.D. h.c., Christian Theological Academy, Warsaw. **Pubs.:** Vatican II – Bridging the Abyss; Ecumenism – Boon or Bane? and Bright Candle of Courage.
Mailing address: 12501 Old Columbia Pike, Silver Spring MD 20904, USA. T. (1 301) 680 6680. Fax 680 6635. Tx 440186.
Residence: 14508 Cutstone Way, Silver Spring MD 20905, USA. T. (1 301) 384 2271

Beam, Fred, Dr

Rel: Protestant.
International Organizations: Director, International Headquarters of the Africa Inland Mission International.
Mailing address: CEO, 37 Alexandra Park, Bristol BS6 6QB, UK. T. (44 272) 429 771. Fax 444134.

Beaulieu, Omer, F.D.M., Br

Rel: Roman Catholic; Ordained; celibate.
International Organizations: Superior General of the Frères de Notre-Dame de la Miséricorde.
Mailing address: Via Bogliasco 34, I-00163 Roma, Italy. T. (39 6) 680 8552.

Becher, Paul, Dr

Rel: Roman Catholic.
International Organizations: Contact of the European Forum of National Laity Committees.
Mailing address: Zentralkomitee der deutschen Katholiken, Hochkreuzallee 246, D-53175 Bonn, Germany. T. (49 228) 382970.

Beck, Brian Edgar, Rev

Rel: Methodist; Ordained in 1960; b. 27 SEP 1933 in London, UK; m. to Margaret Elizabeth Christie Ludlow; c. 3 girls; Citizen of United Kingdom.
International Organizations: Member (since 1981) of the World Methodist Council.
Other organizations: Co-Chairman (since 1976), Oxford Institute of Methodist Theological Studies; Secretary (since 1984), Methodist Conference of GB.
Career: Assistant Tutor, Handsworth College from 1957 to 1959. Circuit Minister, Suffolk from 1959 to 1962. St Paul's United Theological College, Limuru, Kenya from 1962 to 1968. Tutor, Wesley House, Cambridge from 1968 to 1980. Principal, Wesley House, Cambridge from 1980 to 1984. Secretary, East African Church Union Consultation Worship and Liturgy Committee from 1963 to 1968. Member, World Methodist Council from 1966 to 1971. **Educ.:** M.A. from University of Cambridge, Corpus Christi College and Wesley House, UK. **Interests:** walking. **Pubs.:** Reading the New Testament Today, 1977; Christian Character in the Gospel of Luke, 1989 and contributor to: Suffering and Martyrdom in the New Testament, 1981.
Mailing address: 1 Central Buildings, Westminster, London SW1H 9NH, UK. T. (44 71) 222 8010. Fax (44 71) 930 5355
Residence: 76 Beaumont Road, Purley CR2 2EG, UK. T. (44 81) 645 9162

Beckford, James Arthur, Prof

Rel: Anglican; b. 01 DEC 1942 in Enfield, Middlesex, UK; m. to Julia Carolyn Hanson; c. 1 boy and 2 girls; Citizen of United Kingdom; Lang. English and French.
International Organizations: Governor (since 1990) of the Information Network Focus on Religious Movements.
Other organizations: Professor (since 1989), University of Warwick.
Main field of work: University education. **Career:** Lecturer, University of Reading from 1966 to 1973. Lecturer, University of Durham from 1973 to 1988. Professor, Loyola University, Chicago IL, USA from 1988 to 1989. **Educ.:** B.A. from University of Reading, UK (1961-1965); Ph.D. from University of Reading, UK (1966-1972); D.Litt. from University of Reading, UK (1973-1985). **Interests:** new religious movements. **Mem.:** Association for the Sociology of Religion **Pubs.:** *The Trumpet of Prophecy*, Oxford, Blackwell, 1975; *Cult Controversies*, London, Tavistock, 1986 and *Religion and Advanced Industrial Society*, London, Unwin-Hyman, 1989.
Mailing address: Dept of Sociology, University of Warwick, Coventry CV4 7AL, UK. T. (44 203) 523156
Residence: 4 Dunvegan Close, Kenilworth CV8 ZPH, UK. T. (44 926) 851252

Bedard, Margaret, Dr

Rel: Roman Catholic.
International Organizations: Representative at UN, New York of the International Catholic Child Bureau.
Mailing address: 300 Pelham Road, Apt 8A, New Rochelle NY 10805, USA.

Bedini, Ignazio, S.D.B., Most Rev Archbishop

Rel: Roman Catholic; Ordained on 21 DEC 1968; b. 27 JUN 1939 in Prignano sulla Secchia; celibate; Citizen of Iran.
International Organizations: Archbishop (since 1990) of Isfahan. Member of the Society of St Francis of Sales.
Mailing address: Consolata Church, POB 11365-445, 75 France Ave, Teheran, Iran Islamic Rep. T. (98 21) 673210

Begeman, G

Rel: Christian.
International Organizations: Contact of the Oekumenisches Begegnungszentrum.
Mailing address: Weinberg 10, 31675 Bückeburg, Germany. T. (49 5722) 2 50 47.

Behrend, Wilbur, Rt Rev

Rel: Moravian; Ordained.
International Organizations: President of the Moravian Church in America (Northern Province).
Mailing address: PO Box 1245, Bethlehem PA 18018, USA.

Bélisle, Gilles, H E Most Rev

Rel: Roman Catholic; Ordained on 02 FEB 1950; b. 07 OCT 1923 in Clarence ON, Canada; celibate; Citizen of Canada.
International Organizations: Auxiliary Bishop (since 1977) of Ottawa.
Other organizations: Titular Bishop, Uccula.
Career: Co-ordinator of pastoral works, French sector, Ottawa archdiocese. **Educ.:** Pontifical University San Tommaso d'Aquino, Vatican City State; Pontifical Gregorian University, Vatican City State. **Mem.:** Ontario Conference of Catholic Bishops; Conference of Catholic Bishops of Canada
Mailing address: 1247 Kilborn Ave, Ottawa ON K1H 6K9, Canada.
Residence: 95 Parent, Apt 3, Ottawa ON K1N 7B2, Canada.

Belkhodja, Muhammad Habir

Rel: Muslim; Citizen of Tunisia.
International Organizations: Grand Mufti of Tunisia.
Mailing address: No mailing address.

Bell, Randall

Rel: Protestant.
International Organizations: Contact of the American Association of Bible Colleges.
Mailing address: PO Box 1523, Fayetteville AR 72701, USA. T. (1 501) 521 8164.

Bellantonio, Alfredo, O.M., Fr

Rel: Roman Catholic; Ordained; celibate.
International Organizations: Procurator General of the Order of Minims.
Mailing address: Pizza S Francesco di Paola 10, 00184 Roma, Italy. T. (39 6) 488 02 50

Bellini, Pietro, O.S.A., Fr

Rel: Roman Catholic; Ordained; celibate.
International Organizations: Procurator General of the Order of St Augustine.
Mailing address: Via del S Uffizio 25, 00193 Roma, Italy. T. (39 6) 686 45 14

Bellotto, Giuseppe, C.S.I., Fr
Rel: Roman Catholic; Ordained; celibate.
International Organizations: Procurator General of the Pious Congregation of St Joseph.
Mailing address: Via del Belvedere Mondello 77, I-00166 Roma, Italy. T. (39 6) 624 28 51 – 624 34 00. Fax 62 44 64.

Belopopsky, Alexander
Rel: Orthodox.
International Organizations: Secretary-General of the World Fellowship of Orthodox Youth.
Mailing address: Résidence des Pépinières, 12 bd Desgranges, Esc 6, F-92330 Sceaux, France. T. (33 1) 46 60 17 74. Fax 46 60 45 54.

Benda, Kálmán, Prof
Rel: Calvinist; b. 27 NOV 1913 in Nagyvárad, Hungary; m. to Marie Szabó; c. 2 boys and 1 girl; Citizen of Hungary; Lang. Hungarian, German, French and English.
International Organizations: Director of the Scientific Collection of the Danubian District (since 1980) of the Calvinist Church of Hungary.
Other organizations: Titular Professor (since 1984), University of Budapest; President, Historical Committee (since 1990), Hungarian Academy; President (since 1990), World Association of Hungarian Historians.
Career: Military Service from 1940 to 1942. Fellow, Paul Teleki Scientific Institute from 1943 to 1945. Vice-Director, Paul Teleki Scientific Institute from 1945 to 1948. Archivist, Ráday Scientific Collection of the Danubian District, Hungarian Calvinist Church from 1953 to 1956. Vice-Director, Institute of Historical Research of the Hungarian Academy of Sciences from 1957 to 1958. Fellow worker, Institute of Historical Research from 1958 to 1983. Chief of Section, Institute of Historical Research from 1983 to 1985. **Educ.:** B.A. Hist. and Geog. (Hons) from Pázmány Péter University, Budapest, Hungary (1937). **Hons.:** Hon. D.D., Calvinist Theological Academy (1988). **Mem.:** Hungarian Academy of Sciences (1989) **Pubs.:** A magyar nemzeti hivatástudat története, 1937 (History of the Hungarian national consciousness); A magyar jacobinus mozgalom története, 1957 (History of the Hungarian Jacobin movement) and Habsburg abszolutizmus és a magyar rendi ellenállás a XVI-XVII században, 1984 (Habsburg absolutism and Hungarian feudal opposition in the XVI-XVII centuries).
Mailing address: Institut de sciences historiques de l'académie hongroise, I Uri U 53, H-1250 Budapest, Hungary. T. (36 1) 160 160 – 160 080
Residence: 5 Molnár u, H-1056 Budapest, Hungary. T. (36 1) 371 424

Bender, Karpol, Rabbi
Rel: Jewish; Ordained on 08 APR 1957; b. 11 FEB 1930 in Slonim, Poland; m. to Rena Joyce Zambrowsky; c. 1 boy and 1 girl; Citizen of Israel and Canada; Lang. English, Hebrew, Yiddish and Afrikaans.
International Organizations: Executive Vice-Chairman (since 1969) of the International Board of Overseers.
Main field of work: University education. **Career:** Served as a Rabbi in Brooklyn NY. While a student at Yeshiva University, served as National President, Religious Zionist Youth of America. Rabbi, Beth Israel Congregation, Kingston, Ontario from 1960 to 1969. Director, B'nai B'rith Hillel Foundation, Queen's University, Kingston, Ontario from 1960 to 1969. Chaplain, Federal Penitentiary Service, Canada from 1960 to 1969. Member, National Committee for Information and Education, Zionist Organization of Canada from 1964 to 1969. Member, Joint Committee, Canadian Jewish Congress and B'nai B'rith on Jewish Affairs. **Educ.:** B.A. Pol.Sc. from Yeshiva University, New York NY, USA (1949-1953); Rabb. Dip. from Yeshiva University, Rabbi Isaac Elchanan Theological Seminary, New York NY, USA (1957). **Hons.:** Hon. Fellow, Great Synagogue, Tel Aviv; Torah Fellow, Isaac Elchanan Theological Seminary; Recipient of numerous awards. **Mem.:** Rabbinical Council of America; Rabbinical Council of Canada (Past Vice-Chairman) **Pubs.:** Tribute to the late President John F Kennedy in *Sunset at Midday* and articles in various journals.
Mailing address: Bar Ilan University, Ramat Gan 52900, Israel. T. (972 3) 349333
Residence: Keren Hayesod 7, Givat Shmuel, Ramat Ilan 51905, Israel. T. (972 3) 356076

Bender, Ross T
Rel: Mennonite.
International Organizations: Professor of Christian Education of the Goshen Biblical Seminary.
Mailing address: 3003 Benham Ave, Elkhart IN 46517, USA.

Benedek, Jack
Rel: Roman Catholic.
International Organizations: President of the Catholic Alumni Clubs International.
Mailing address: 3523 W 78th St, Chicago IL 60652, USA. T. (1 312) 776 9476.

Benedyktowicz, Witold Kazimierz, Rev
Rel: Methodist; Ordained in 1943; b. 25 JUN 1921 in Cracow, Poland; c. 3 boys; Citizen of Poland.
International Organizations: Member, Working Committee (since 1968) of the Christian Peace Conference.
Other organizations: Clergyman (since 1943), Methodist Church in Poland; Lecturer and Researcher at the Christian Theology Academy, Warsaw (since 1967); Head, Systematic Theology Dept (since 1967), Christian Theology Academy, Warsaw; Member (since 1969), Churches Commission for International Matters, Geneva; Ordinary Professor (since 1980), Christian Theology Academy, Warsaw; Honorary President (since 1983), Polish Ecumenical Council.
Main field of work: Theologian. **Career:** Custodian, Warsaw University Library from 1956 to 1967. Assistant Professor, Christian Theology Academy, Warsaw from 1967 to 1972. Extraordinary Professor, Christian Theology Academy, Warsaw from 1972 to 1980. Chief Superintendent, Methodist Church in Poland from 1969 to 1983. President, Polish Ecumenical Council from 1975 to 1983. Member, Consultative Council attached to the Chairman of the Council of State from 1986 to 1989. **Educ.:** Grad. from University of Warsaw, Poland (1950); Dr. Theol. from University of Warsaw, Poland (1954); Dr.hab. from University of Warsaw, Poland (1965). **Interests:** reading novels and memoirs, opera. **Hons.:** Commander's, Officer's and Knight's Cross, Order of Polonia Restituta; Gold Cross of Merit (1964); and other decorations. **Pubs.:** "Die Interpretation des Neuen Testaments in Konfrontation mit dem Sozialismus", in *Begegnung mit Polen*, 1974; *Co powinnismy czynic – Zarys ewangelickiej etyki teologicznej*, 1975 and *Ekumenia, pokój, pojednanie*, 1988.
Mailing address: ul Mokotowska 12, 00-561 Warsaw, Poland. T. (48 28) 53-28
Residence: T. (48 29) 74-25

Bennett, Asia, Ms
Rel: Quaker.
International Organizations: Director, Americas Section Office of the Friends World Committee for Consultation.
Mailing address: 1506 Race Street, Philadelphia PA 19102, USA. T. (1 215) 241 7250. Fax 864 0104.

Bennett, Jim L, Dr
Rel: Assembly of God.
International Organizations: Contact of the General Council of the Assemblies of God.
Mailing address: Continental Theological Seminary, Kasteelstraat 48, B-1600 St-Pieters-Leeuw, Belgium. T. (32 2) 378 00 77.

Bennett, John Coleman, Dr
Rel: Protestant; Ordained; b. 22 JUL 1902 in Kingston ON, Canada; m. to Anne McGrew; c. 2 boys and 1 girl; Citizen of USA.
International Organizations: Participant in Department on Church and Society of the World Council of Churches.
Main field of work: Theologian, ethicist and ecumenical leader. **Career:** Held teaching posts at Auburn Theological Seminary, New York from 1931 to 1938. Held teaching posts at Pacific School of Religion, Berkeley CA from 1938 to 1943. Professor of Christian Theology and Ethics, Union Theological Seminary, New York in 1943. Dean of Faculty, Union Theological Seminary, New York from 1955 to 1963. William E Dodge Jr Professor of Applied Christianity, Union Theological Seminary, New York from 1957 to 1961. Reinhold Niebuhr Professor of Social Ethics, Union Theological Seminary, New York from 1961 to 1970. President, Union Theological Seminary, New York from 1963 to 1970. Visiting Professor of Christian Ethics, Pacific School of Religion, Berkeley CA from 1970 to 1977. Lecturer, Claremont School of Theology from 1980 to 1981. President, American Theological Society in 1954. President, American Society for Christian Social Ethics in 1961. Delegate to the first assembly, World Council of Churches – WCC, served as secretary of the section on the Church and the Disorder of Society from 1943 to 1970. Active participant in the work of the Department on Church and Society, WCC in 1948. **Educ.:** Williams

College, Williamstown MA, USA; D.D. from University of Oxford, Mansfield College, UK; S.T.D. from Union Theological Seminary, New York NY, USA. **Hons.:** 22 honorary degrees, mainly D.D. **Pubs.:** Senior Contributing Editor, *Christianity and Crisis* (founding sponsor) and numerous books and articles on specific topics such as nuclear weapons, Communism, US foreign policy and the relationship between Christians and the State.
Mailing address: 620 Plymouth Road, Claremont CA 91711, USA. T. (1 714) 624 3752

Ben-Simhon, Shaul

Rel: Jewish.
International Organizations: Chairman of the World Union of North African Jews.
Mailing address: PO Box 6521, Tel Aviv, Israel. T. (972 3) 268859 – 223918.

Benson, Ezra Taft

Rel: Church of Jesus Christ of Latter-Day Saints; b. 04 AUG 1899; Citizen of USA.
International Organizations: Leader (since 1985) of the Mormon Church.
Career: Missionary, Church of Jesus Christ of Latter-Day Saints in UK from 1921 to 1923. County Agricultural Agent, University of Idaho Extension Service in 1929. Extension Economist and Marketing Specialist from 1930 to 1939. Head, Department of Agricultural Economics and Marketing in 1931. Secretary, Idaho Co-operative Council from 1933 to 1938. Executive Secretary, National Council of Farmer Cooperatives from 1939 to 1943. Served on several advisory committees and national boards in the field of agriculture (since 1939). Member, Executive Committee, American Institute of Co-operation from 1942 to 1952. Chairman, Board of Trustees, American Institute of Co-operation in 1952. Member, National Agricultural Advisory Committee (during Second World War). Member, National Farm Credit Committee from 1940 to 1943. US Delegate to the International Conference of Farm Organisations, London in 1946. US Secretary of Agriculture from 1953 to 1961. Member, Council of Twelve, Church of Jesus Christ of Latter-Day Saints (since 1943). President, Mormon Church European Mission in 1963. **Educ.:** Brigham Young University, Hawaii Campus, Oahu, USA; M.S. from Iowa State University of Science and Technology, Ames IA, USA; University of California, Berkeley, USA. **Mem.:** Farm Foundation (1946, 1950)
Mailing address: Church of Jesus Christ Latter-Day Saints, 50 EN Temple Street, Salt Lake City UT 84150, USA.

Bentley, Philip Jay, Rabbi

Rel: Jewish; Ordained on 03 JUN 1973; b. 29 MAY 1945 in New York City NY, USA; c. 2 boys; Citizen of USA; Lang. English, Hebrew (some) and Spanish (some).
International Organizations: President (since 1987) of the Jewish Peace Fellowship.
Other organizations: Rabbi (since 1978), Temple Or-Elohim, Jericho NY; Vice-President, Long Island Board of Rabbis; Member, Board of Governors, Synagogue Relations Commission of the Federation of Jewish Philantropies of New York; Member, Social Action Commission of Reform Judaism.
Career: Student Rabbi, Temple Sinai, New Bedford MA from 1970 to 1973. Director, B'nai B'rith Hillel Foundation, University of Denver, CO from 1973 to 1975. Rabbi, Congregation Mikve Israel-Emanu-El, Curacao from 1975 to 1978. Member, Central Conference of American Rabbis Committee on Justice and Peace from 1982 to 1987. President, Long Island Association of Reform Rabbis from 1985 to 1987. **Educ.:** A.B. from Shimer College, Mount Carroll IL, USA (1963-1966); Judaic Stds. from University of Manitoba, Winnipeg, Canada (1966-1967); B.H.L. from Hebrew Union College-Jewish Institute of Religion, Cincinnati OH, USA (1968-1972); M.H.L. from Hebrew Union College-Jewish Institute of Religion, Cincinnati OH, USA (1973). **Pubs.:** "With Nickels and Dimes", in *Compass*, Spring 1981; "Rabbinic Sources on Economic Policy", in *CCAR Yearbook*, 1982 and "The Blessing and The Sun", in *Moment*, March-April 1981, and many other brief articles published in a variety of Jewish publications.
Mailing address: PO Box 358, Santa Fé NM 87504, USA.
Residence: See Mailing Address. T. (1 505) 983 7446

Bepari, S K, Rev

Rel: Baptist; Ordained.
International Organizations: Executive Secretary of the Bengal-Orissa-Bihar Baptist Convention.
Mailing address: Sepoy Bazar, Midnapore, 721101 West Bengal, India.

Berg, David Brandt

Also know as Moïse David; Rel: Protestant; Citizen of USA.
International Organizations: Founder of the Jesus Revolution.
Mailing address: No mailing address.

Berger, Anton, Fr

Rel: Roman Catholic; Ordained; celibate.
International Organizations: Secretary of the Council of European Conferences of Priests.
Mailing address: Stephansplatz 3, A-1010 Wien, Austria. T. (43 1) 51552 – 566 232.

Berloco, Giacinto, H E Mgr

Rel: Roman Catholic; Ordained on 19 MAR 1966; b. 31 AUG 1941 in Altamura, Bari, Italy; celibate; Citizen of Vatican and Italy; Lang. Italian, French, Spanish, English and Portuguese.
International Organizations: Apostolic Delegate to Mozambique and Apostolic Pro-Nuncio to Zimbabwe (since 1990) of the Roman Catholic Church.
Other organizations: Titular Archbishop, Fidene.
Main field of work: Ministry, diplomacy. **Career:** Served in Apostolic Nunciatures in Costa Rica, The Netherlands and Spain. **Educ.:** B.Ph. from Lateran Pontifical University, Vatican City State (1960-1962); D.Theol. from Lateran Pontifical University, Vatican City State (1962-1968); Lic. Cn.L. from Lateran Pontifical University, Vatican City State (1968-1971). **Interests:** art, journey.
Mailing address: Avenida Julius Nyerere 882, CP 2738, Maputo, Mozambique. T. (258 1) 49 11 44. Fax 49 22 17
Residence: See Mailing Address. T. See Mailing Address

Bernard, S Francis, Sr

Rel: Roman Catholic; Ordained; celibate.
International Organizations: Superior General of the Sisters of Holy Cross.
Mailing address: St Mary's, Bertrand Hall, Notre Dame IN 46556, USA. T. (1 219) 284 5550.

Bernardin, Joseph Louis, H E Card

Rel: Roman Catholic; Ordained on 26 APR 1952; b. 02 APR 1928 in Columbia, USA; celibate; Citizen of USA.
International Organizations: Member (since 1983) of the College of Cardinals.
Member of the Sacred Congregation for the Sacraments and Divine Worship. Member of the Pontifical Council for Promoting Christian Unity.
Other organizations: Archbishop (since 1982), Chicago.
Career: Titular Bishop, Lugara in 1966. Archbishop, Cincinnati from 1972 to 1982. Created Cardinal by Pope John Paul II in 1983.
Mailing address: PO Box 1979, Chicago IL 60690, USA. T. (1 312) 337 5952

Bernier, Paul J

Rel: Roman Catholic.
International Organizations: Director for USA of the Priests Eucharistic League.
Mailing address: 194 E 76th Street, New York NY 10021, USA. T. (1 212) 861 1076.

Bernstein, Louis, Rabbi

Rel: Jewish; Ordained in 1950; b. 02 APR 1927 in New York NY, USA; m. to Pearl Moshel; c. 2 boys and 1 girl; Citizen of USA.
International Organizations: President (since 1984) of the Rabbinical Council of America.
Executive (since 1976) of the World Zionist Organization.
Other organizations: Editor (since 1954), *Rabbinical Council Record*; Chairman (since 1961), Yeshivath Hadarom; Associate Professor (since 1980), Yeshiva University, New York NY.

Career: Rabbi, Glenwood Jewish Center, Brooklyn from 1950 to 1952. Served with Chaplains' Corps, US Army from 1952 to 1954. Instructor-Assistant Professor, Yeshiva University, New York from 1954 to 1980. Instructor, Yeshiva University High School, New York from 1954 to 1968. President, Yeshiva College Alumni Association from 1960 to 1962. President, Rabbinical Council of America from 1973 to 1974. President, Religious Zionists of America from 1975 to 1980. Chairman of Board, Religious Zionists of America from 1980 to 1984. **Educ.:** B.A. from Yeshiva University, New York NY, USA (1947); Ph.D. from Yeshiva University, New York NY, USA (1976). **Interests:** tennis. **Hons.:** Bernard Revel Memorial Award, Yeshiva College Alumni Association. **Mem.:** Rabbinical Council of America; Religious Zionists of America; Yeshiva College Alumni Association; Yeshivath Hadarom; Rabbinical Alumni Yeshiva University; Hapoel Mizrachi America; Jewish Chaplains' Association; World Zionist Organization; Vaad Harabbonim, Queens
Mailing address: Religious Zionists of America, 15 West 26th Street, New York NY 10010, USA.

Berra, Francesco, I.C., Rev
Rel: Roman Catholic; Ordained; celibate.
International Organizations: Procurator-General of the Institute of Charity.
Mailing address: Via di Porta Latina 17, I-00179 Roma, Italy. T. (39 6) 70491777.

Berrigan, Daniel, S.J.
Rel: Roman Catholic; Ordained in 1952; b. 09 MAY 1921 in Virginia MN, USA; celibate; Citizen of USA.
International Organizations: Member of the Society of Jesus.
Main field of work: Social Activist. **Career:** Spent a year in France after ordination and became involved in social causes. Taught French and Theology in Brooklyn NY (for 3 years). Taught French and Theology at LeMoyne College, Syracuse NY (for 6 years). **Pubs.:** To Dwell in Peace (autobiography); The Trial of the Catonsville Nine (play) and many other plays and essays.
Mailing address: No mailing address.

Berrigan, Philip, S.S.J.
Rel: Roman Catholic; Ordained in 1955; b. 05 OCT 1923 in Two Harbours MN, USA; celibate; Citizen of USA.
International Organizations: Member of the St Joseph's Society of the Sacred Heart.
Main field of work: Social Activist.
Mailing address: No mailing address.

Berry, Robert Edward Fraser, Rt Rev
Rel: Anglican; Ordained in 1954; b. 21 JAN 1926 in Ottawa, Ontario, Canada; m. to Margaret Joan Trevorrow; c. 1 boy and 1 girl; Citizen of Canada.
International Organizations: Chaplain (since 1988) of Royal Canadian Air Force Association.
Career: Ordained Deacon in 1953. Rector, St Margaret's, Hamilton, Ontario in 1955. Rector, St Mark's, Orangeville, Ontario in 1961. Rector, St Luke's, Winnipeg, Manitoba in 1963. Rector, St Michael and All Angels, Kelowna, BC, and Supervising Pastor, Central Okanagan Region in 1967. Elected Bishop and consecrated in 1971. Served with the RCAF (Royal Canadian Air Force), Air Crew from 1943 to 1945. Chaplain, BC Dragoons from 1970 to 1972. Bishop, Kootenay (retired 1989). **Educ.:** B.A. from Sir George Williams College, Canada (1950); L.Th. from Montreal Diocesan College, Canada (1953); B.D. from McGill University, Montreal, Canada (1953). **Interests:** boating, fishing, computer communications. **Hons.:** Hon. D.D., McGill University (1972). **Mem.:** Clubs: Vancouver, Kelowna Yacht
Mailing address: 1857 Maple Street, Kelowna BC V1Y 1H4, Canada.
Residence: See Mailing Address.

Berry, Thomas, C.P., Fr
Rel: Roman Catholic; Ordained; celibate.
International Organizations: Member of the Congregation of the Passion.
Mailing address: 5801 Palisade Ave, Bronx NY 10471, USA.

Berry, Victoria
Rel: Christian.
International Organizations: Member of Staff of the World Council of Churches.
Mailing address: PO Box 2100, Route de Ferney 150, CH-1211 Genève 2, Switzerland. T. (41 22) 791 61 11. C. OIKOUMENE GENEVA. Tx 415 730 OIK CH. Fax (41 22) 791 03 61.

Bertello, Giuseppe, H E Mgr
Rel: Roman Catholic; Ordained on 20 JUN 1966; b. 01 OCT 1942 in Foglizzo; celibate.
International Organizations: Apostolic Nuncio to Rwanda (since 1991) of the Roman Catholic Church.
Other organizations: Titular Archbishop, Urbisaglia.
Mailing address: BP 261, Kigali, Rwanda. T. (250) 75 293 – 75 181. Fax 75 041

Bertels, Marcel
Rel: Buddhist: Mahayana.
International Organizations: Contact of the International Mahayana Institute.
Mailing address: Nalanda Monastery, Rouzegas, Labastide Saint-Georges, F-81500 Lavaur, France.

Bertels, Marcel
See Thubten Palgye.

Bertoli, Paolo, H E Card
Rel: Roman Catholic; Ordained on 15 AUG 1930; b. 01 FEB 1908 in Poggio Garf (Lucca), Italy; celibate.
International Organizations: Member (since 1969) of the College of Cardinals.
Other organizations: Suburbicarian Bishop (since 1979), Frascati.
Main field of work: Diplomat. **Career:** Secretary, Apostolic Nunciature, Belgrade from 1933 to 1938. Secretary, Apostolic Nunciature, France from 1938 to 1942. Chargé d'affaires a.i., Antilles Apostolic Nunciature, Port-au-Prince, Haiti from 1942 to 1946. Counsellor, Apostolic Nunciature, Berne from 1946 to 1952. Head, Mission of Emigration to South America. Titular Archbishop, Nicodemia in 1952. Apostolic Delegate in Turkey from 1952 to 1953. Apostolic Nuncio in Colombia from 1953 to 1959. Apostolic Nuncio in Lebanon from 1959 to 1960. Apostolic Nuncio in France from 1960 to 1969. Prefect, Congregation for the Causes of Saints from 1969 to 1973. Created Cardinal by Pope Paul VI in 1969. Camerlengo, Roman Catholic Church from 1979 to 1985.
Mailing address: Piazza della Città Leonina 1, I-00193 Roma, Italy. T. (39 6) 687 73 12

Bertrand, Hector-L, Rev
Rel: Roman Catholic; Ordained in 1939; b. 10 MAR 1907 in Warren ON, Canada; Citizen of Canada.
International Organizations: Vice-President-Administrator (since 1965) of the University of Sudbury.
Other organizations: Publisher and Editor (since 1975), *Le Voyageur* (weekly French paper).
Career: Professor of Apologetics and Dean of Discipline (Prefectus Disciplinae), Sacred Heart College of Sudbury (now present university) in 1941. Founder, 1st Convention and Exhibition for Hospitals, Province of Quebec in 1948. Founder and Director, L'Ecole Supérieure d'Administration Hospitalière in 1948. Founder, *L'Hôpital d'Aujourd'hui* (journal) in 1955. Founder, Medical Directors course in 1957. Founder, Hospital and Administration courses (evening) in 1961. Regent, College of Medicine, Bangalore from 1963 to 1965. Consultant in hospital administration. Served as Chaplain, Parachutists Corps (Canadian) in 1943. Major and Commanding Chaplain in 1944. President and Executive Director, Catholic Hospital Association from 1945 to 1952. Founder, President and 1st Executive Director, Comité des Hôpitaux du Québec in 1947. Governor, Canadian Hospital Association from 1947 to 1963. Founder, Association Medical Directors, Province of Quebec in 1957. Co-founder, Canadian Council on Hospital Accreditation. **Educ.:** B.A. from Sacred Heart College, Sudbury, Ontario, Canada (1928); D.Sc. from Georgetown University, Washington DC, USA (1928). **Hons.:** George Findlay Stephens Memorial Award for 'Outstanding services to Hospitals of Canada'; Citations from various professional associations. **Mem.:** Association of Hospital Administrators, Province of Quebec; Association of Practical Nurses, Province of

Quebec (Honorary Member); Association of Baby-Nurses, Province of Quebec; Association of Medical Record Librarians, Province of Quebec (1960); American College of Hospital Administrators (Honorary Fellow); Quebec Association of Hospital Administrators (Honorary Member); Canadian Medical Record Librarians; Association of Consultants in Industrial Relations (Life Member); Academy of Hospital Counsellors (Member)
Mailing address: University of Sudbury, Sudbury ON P3E 2C6, Canada.

Betts, Glancia
Rel: Christian.
International Organizations: Regional Secretary for Latin America and the Caribbean of the World Association for Christian Communication.
Mailing address: CAVE, CP 5002, Rudge Ramos SBC-SP 09735, Brazil.

Beukenhorst, Martin J, Rev
Rel: Protestant; Ordained.
International Organizations: President of the United Protestant Church of Belgium.
Mailing address: 5 rue du Champ de Mars, B-1050 Bruxelles, Belgium.

Beya, Marie-Bernadette Mbuy, Sr
Rel: Roman Catholic; Ordained; b. 27 SEP 1946 in Zaire; celibate; Lang. French and English.
International Organizations: Regional Coordinator for Africa (since 1986) of the Ecumenical Association of Third World Theologians.
Main field of work: Women's meetings. **Career:** Director, Institut Lubusha (Luishia) from 1968 to 1979. Director, Institut Supérieur des Sciences Religieuses from 1982 to 1985. Superior, Soeurs Ursulines from 1982 to 1990. **Educ.:** Institutrice (D6A) from Luishia (1962-1968); D.Catéchèse from Rusanda, Butare, Rwanda (1972-1973); Grad. en Pastorale et Catéchèse (G3) from Lumen Vitae, Brussels, Belgium (1980-1982). **Interests:** Women's liberation in Africa. **Hons.:** Brevet de Mérite de la Région du Shaba pour l'encadrement de la femme (1989). **Pubs.:** *La réalité de l'oppression et la lutte des femmes en Afrique*; *L'expérience féminine de Dieu*; *Faire la théologie dans la perspective des femmes africaines* and *La femme africaine et la sexualité*.
Mailing address: BP 765, Lubumbashi, Zaire.
Residence: Avenue Kasavubu 535, Lubumbashi, Zaire T. (243) 22 3257

Bharadwaj, Ishwar, Dr
Rel: Yoga.
International Organizations: Head, Yoga Department of the Gurukul Kangri University, Hardwar.
Mailing address: Yoga Dept., Gurukul Kangri University, Hardwar UP 249 404, India.

Bichkov, Alexei M, Rev
Rel: Baptist; Ordained.
International Organizations: General Secretary of the Union of Evangelical Christian Baptists.
Mailing address: GPO Box 520, Moscow, Russia.

Bickel, Jean-François
Rel: Roman Catholic.
International Organizations: Secretary-General of the International Young Catholic Students.
Mailing address: 171 rue de Rennes, F-75006 Paris, France. T. (33 1) 45 48 14 72.

Bidawid, Raphael I, I, His Beatitude
Rel: Chaldean Catholic; Ordained on 22 OCT 1944; b. 17 APR 1922 in Mosul, Iraq; celibate; Citizen of Iraq; Lang. Fluent in 12 written and spoken Oriental and European languages.
International Organizations: Member of the Sacred Congregation for the Oriental Churches.
Other organizations: Patriarch (since 1989), Babylon.
Career: Teacher at the patriarcal College, Mosul in 1947. Consecrated Bishop, Amadiya in Iraq from 1957 to 1966. Transferred to Bishopric, Beirut from 1966 to 1989. **Pubs.:** Author of several books, including: *History of the Christians under the Abbasides.*
Mailing address: Patriarcat Chaldéen Catholique, Baghdad, Iraq. T. (964 1) 537 8511

Biffi, Giacomo, H E Card
Rel: Roman Catholic; Ordained on 23 DEC 1950; b. 13 JUN 1928 in Milan, Italy; celibate; Citizen of Italy.
International Organizations: Member (since 1985) of the College of Cardinals.
Member of the Sacred Congregation for the Sacraments and Divine Worship. Member of the Congregation for the Clergy.
Other organizations: Archbishop (since 1984), Bologna.
Career: Consecrated Bishop in 1975. Created Cardinal by Pope John Paul II in 1985.
Mailing address: Arcivescovado, Via Altabella 6, I-40126 Bologna, Italy. T. (39 51) 22 25 66

Biggio, Piero, H E Mgr
Rel: Roman Catholic; Ordained on 05 AUG 1962; b. 29 JUN 1937 in Calasetta, Sardinia, Italy; celibate.
International Organizations: Apostolic Pro-Nuncio to Bangladesh (since 1988) of the Roman Catholic Church.
Other organizations: Titular Archbishop, Otricoli.
Mailing address: Lake Road 2, Baridhara Model Town, Diplomatic Enclave, Dhaka 12, Bangladesh. T. (880 2) 88 20 18 – 88 21 43. Fax 88 35 74

Bilheimer, Robert Sperry, Rev
Rel: Presbyterian; Ordained; b. 28 SEP 1917 in Denver CO, USA; m. to Dorothy Dodge; c. 3 boys; Citizen of USA.
International Organizations: Ecclesiastic of the Presbyterian Church.
Career: Minister, Westminster Presbyterian Church from 1946 to 1948. Secretary, The Inter-Seminary Movement from 1945 to 1948. Programme Secretary, World Council of Churches (WCC), New York from 1948 to 1954. Associate General Secretary and Director, Division of Studies, WCC, Geneva from 1954 to 1963. Senior Minister, Central Presbyterian Church, Rochester NY from 1963 to 1966. Director, International Affairs Programs, The National Council of Churches of Christ in the USA from 1966 to 1974. Executive Director, Institute for Ecumenical and Cultural Research, St John's University from 1974 to 1984. **Educ.:** Yale University, New Haven CT, USA; Yale University, Divinity School, New Haven CT, USA. **Interests:** trout fishing, photography. **Hons.:** Hon. D.D., Chicago Theological Seminary (1954), Butler University (1954), Hamilton College (1980). **Pubs.:** Editor, Faith and Ferment, 1983; A Spirituality for the Long Haul, 1984 and Breakthrough, the Emergence of the Ecumenical Tradition, 1989.
Mailing address: 15256 Knightwood Road, Cold Spring MN 56320, USA.

Binder, Elaine Kotell
Rel: Jewish; b. 12 OCT 1938 in Boston MA, USA; m. to Richard A Binder; c. 2 boys; Citizen of USA.
International Organizations: Executive Director (since 1985) of the B'nai B'rith Women.
Main field of work: Philanthropy. **Career:** Director, Wider Opportunities for Women from 1971 to 1976. Consultant, BEMW from 1976 to 1978. Project Director, Women's Equity Action League from 1978 to 1980. Administrative Director, American Association of University Women from 1980 to 1985. **Educ.:** A.B. from Radcliffe College, Cambridge MA, USA (1956-1960); M.A. from University of Maryland at College Park, USA (1971-1975).
Mailing address: 1828 L Street NW, Suite 250, Washington DC 20036, USA. T. (1 202) 857 1310.
Residence: 6704 Bradley Boulevard, Bethesda MD 20817, USA. T. (1 301) 365 2466

Birmingham, Kieran, Fr
Rel: Roman Catholic; Ordained; celibate.
International Organizations: Superior General of the St Patrick's Society for the Foreign Missions.
Mailing address: Saint Patrick's, Kiltegan, Ireland.

Birnbaum, Jacob
Rel: Jewish.
International Organizations: Contact of the Center for Russian and East European Jewry.
Mailing address: 210 W 91st St, New York NY 10004, USA. T. (1 212) 928 7451.

Bisai, Ana, Rev
Rel: Methodist; Ordained.
International Organizations: Member of the World Federation of Methodist Women.
Mailing address: Malmaluan, PO Box 2070, Rabaul, Papua, New Guinea.

Bisar, Muhammad Abdul Rahman
Rel: Muslim; Citizen of Egypt.
International Organizations: Grand Sheikh of the Al-Azhar.
Main field of work: Religious Leader. **Career:** Secretary-General, Academy of Islamic Research, Cairo. Director, Islamic Cultural Centre, Washington DC. Minister, Waqfs and Minister of State charged with Al-Azhar affairs from 1978 to 1979. **Educ.:** Ph.D. from University of London, UK.
Mailing address: c/o Office of the Grand Sheikh, Al-Azhar Administration, Cairo, Egypt.

Bishoi, Amba, Metropolitan
Rel: Coptic Orthodox; Ordained.
International Organizations: Co-Chairman of the Joint International Commission for Dialogue between the Roman Catholic Church and the Coptic Orthodox Church.
Mailing address: Sec. of the Holy Synod, Coptic Orthodox Church, Amba Rueiss Building, Ramses Street, Cairo, Egypt. T. (20 2) 283 0247. Fax 32 70 53.

Bishop, Cecil, Bishop
Rel: African Methodist Episcopal Zion; Ordained.
International Organizations: Secretary of the African Methodist Episcopal Zion Church.
Mailing address: 1200 Windermere Drive, Pittsburgh PA 15218, USA.

Bisimwa, Ethienne
Rel: Roman Catholic.
International Organizations: UNESCO Representative of the Pax Romana, International Movement of Catholic Students.
Member of the Pontifical Council for the Laity.
Mailing address: 171 rue de Rennes, F-75006 Paris, France. T. (33 1) 45 44 70 75.

Biyrje, Paul
Rel: Buddhist: Mahayana.
International Organizations: Member of the Foundation for the Preservation of the Mahayana Tradition.
Other organizations: Director, Wisdom Publications Australia.
Mailing address: PO Box 1326, Chatswood, Sydney NSW 2067, Australia. T. (61 2) 417 1897. Fax (61 2) 417 1857

Black, J Bernard, Rev
Rel: Roman Catholic; Ordained; b. 31 MAY 1926 in Toronto, Ontario, Canada; celibate; Citizen of Canada.
International Organizations: Member of the Congregation of St Basil.
Main field of work: Librarian. **Career:** Joined the Basilean Fathers in 1946. Counsellor, Emmanuel Convalescent Centre from 1983 to 1987. Librarian, St Michael's College, University of Toronto from 1961 to 1985. Member, Toronto Public Library Board from 1979 to 1984. Chairman, Toronto Public Library Board (in 1981, 1982). Instructor, University of Toronto Library School from 1963 to 1965. Instructor, University of Western Ontario School of Library and Information Science in 1971. Assistant Librarian, St Basil's Seminary from 1953 to 1956. Librarian and Assistant Professor of English, St Thomas More College in University of Saskatchewan from 1956 to 1959. Member, Advisory Council, Saskatchewan Library Association from 1958 to 1959. Served in Royal Canadian Infantry from 1944 to 1945. **Educ.:** B.A. from University of Toronto, St Michael's College, Canada (1949); B.L.S. from University of Toronto, St Michael's College, Canada (1952); S.T.B. from University of Toronto, St Michael's College, Canada (1955); M.L.S. from University of Toronto, St Michael's College, Canada (1956). **Mem.:** Canadian College of Teachers
Mailing address: 95 St Joseph St, Toronto, Ontario M5S 2R9, Canada.

Blajot, Victor
Rel: Roman Catholic; b. 11 AUG 1910 in Spain; single; Citizen of Spain and Bolivia; Lang. Spanish, Catalan, Italian, French and English.
International Organizations: International Coordinator (since 1973) of the Latin American Federation 'Faith and Joy'.
Other organizations: President (since 1985), Associación Boliviana de Educación Católica.
Educ.: Lic. from Faculty of Law, Barcelona, Spain (1926-1929); Lic. from Faculty of Philosophy, Angliana (1932-1936); Lic. from Faculty of Theology, Barcelona, Spain (1939-1943).
Mailing address: Casilla de Correo 672, Santa Cruz, Bolivia. T. (591 33) 21814
Residence: Avda España 21, Santa Cruz 319, Bolivia. T. (591 33) 23695

Blanch, Stuart Yarworth, Rt Rev and Rt Hon Lord
Rel: Church of England; Ordained; b. 02 FEB 1918; m. to Brenda Gertrude Coyte; c. 1 boy and 4 girls; Citizen of United Kingdom.
International Organizations: Ecclesiastic of the Church of England.
Other organizations: Chairman, Council (since 1967), Wycliffe Hall, Oxford.
Career: Served during World War II as navigator with Royal Air Force (RAF). Transport Command, India. Vicar, Eynsham Oxon from 1952 to 1957. Tutor and Vice-Principal, Wycliffe Hall, Oxford from 1957 to 1960. Oriel Canon, Rochester and Warden of Rochester Theological College from 1960 to 1966. Bishop, Liverpool from 1966 to 1975. Archbishop, York from 1975 to 1983. Retired in 1983. Pro-Chancellor, University of Hull from 1975 to 1983. Pro-Chancellor, University of York from 1977 to 1983. **Educ.:** D.D. from University of Oxford, St Catherine's College and Wycliffe Hall, UK. **Interests:** squash, walking, music, meteorology. **Hons.:** Baron (Life Peer, 1983) of Bishopthorpe in the Co. of North Yorks; Privy Councillor (1975); Hon. D.D., Toronto; Hon. Dr., York. **Mem.:** Roy Cwlth Society **Pubs.:** Living by Faith, 1983; Way of Blessedness, 1985 and Encounters with Jesus, 1988.
Mailing address: Little Garth, Church Street, Bloxham, nr Banbury, Oxfordshire, UK.

Blanquet Gas, José Maria B, S.F., Fr
Rel: Roman Catholic; Ordained; celibate.
International Organizations: Superior General of the Sons of the Holy Family.
Mailing address: Via Giovanni Giolitti 154, I-00185 Roma, Italy. T. (39 6) 73 13 362.

Bleakley, David Wylie, Rt Hon
Rel: Church of Ireland; b. 11 JAN 1925 in Belfast, N Ireland, UK; m. to Winifred Wason; c. 3 boys; Citizen of United Kingdom; Lang. English.
International Organizations: Secretary-General (since 1980) of the Irish Council of Churches.
President (since 1983) of the Church Missionary Society, UK. Member, Central Executive of the Conference of European Churches.
Main field of work: Mission (worldwide), peacemaking in Ireland. **Career:** Member, Northern Ireland Parliament (Labour) from 1958 to 1975. Minister, Northern Ireland Government. Lecturer in Industrial Studies, Kivukoni College, Dar es Salaam. Visiting Senior Lecturer in Peace Studies, University of Bradford. **Educ.:** Dip. Econ. and Pol.Sc. from University of Oxford, UK (1946-1948); B.A. from Queen's University of Belfast, UK (1948-1951); M.A. from Queen's University of Belfast, UK (1951-1955). **Interests:** theology of work, religion and ageism, reading, writing, travelling. **Hons.:** Privy Councillor (1971); Commander of the Order of the British Empire (1984); Hon. Deg., Open University. **Mem.:** International Fellowship of Reconciliation (IFOR); Fellowship of St John the Evangelist **Pubs.:** Peace in Ireland, Mowbray Press, UK and A Study of Work, 3 Vols., Scar Press, UK.
Mailing address: CMS, 8 Thornhill, Bangor, Co. Down BT19 1RD, N Ireland, UK. T. (44 247) 454898
Residence: See Mailing Address. T. See Mailing Address

Blei, Karel, Rev Dr
Rel: Reformed Church; Ordained.
International Organizations: Moderator, Department of Theology of the World Alliance of Reformed Churches.
Mailing address: PO Box 405, NL-2260 AK Leidschendam, Netherlands. T. (31 70) 313 1131. Fax (31 70) 313 1202
Residence: T. (31 23) 250 199

Bloch, Georges M
Rel: Jewish: Progressive.
International Organizations: Chairman of the B'nai B'rith International.
Representative at UNESCO of the World Union for Progressive Judaism.
Mailing address: 23 Bd de l'Orangerie, F-67000 Strasbourg, France. T. (33) 88 35 49 01.

Bloem, Renate
Rel: Methodist.
International Organizations: Contact of the World Federation of Methodist Women.
Mailing address: N103 bis Rt De Thonon, CH-1222 Vesenaz, Switzerland.

Bloom, Alfred, Dr
Rel: Buddhist; Unordained; b. 09 NOV 1926 in Philadelphia PA, USA; m. to Dorothy Nell; c. 1 boy and 1 girl; Citizen of USA; Lang. English and Japanese.
International Organizations: Dean (since 1988) of the International Association of Buddhist Studies.
Other organizations: Professor Emeritus (since 1988), University of Hawaii, Honolulu.
Career: Proctor for Center for the Study of World Religions, Harvard Divinity School, Cambridge MA from 1959 to 1961. Teaching Fellow in History of Religion, Harvard Divinity School, Cambridge MA from 1959 to 1961. Lecturer in History of Religion, Newton Junior College in 1961. Associate Professor of Religion, University of Oregon, Eugene from 1961 to 1970. Associate Professor of Religion, Dept. of Religion, University of Hawaii at Manoa from 1970 to 1974. Professor of Religion, Dept. of Religion, University of Hawaii at Manoa from 1974 to 1986. On loan to Institute of Buddhist Studies, Berkeley CA from 1986 to 1988. **Educ.:** A.B., Th.B. from Eastern Baptist Theological Seminary, USA (1947-1951); B.D., S.T.M. from Andover Newton Theological School, USA (1951-1953); Ph.D. from Harvard University, Divinity School, Cambridge MA, USA (1963). **Mem.:** American Academy of Religion, Society for Values in Higher Education, International Association of Shin Buddhist Studies **Pubs.:** *Tannisho as a Resource for Modern Man*, Buddhist Study Center, Honpa Hongwanji of Hawaii, Spring 1981; *Shoshinge: Heart of Shin Buddhism*, BSC Press, 1986 and "The Unfolding of the Lotus: Developments in Shin Buddhism in the West", in *Buddhist Christian Studies*, 1990.
Mailing address: c/o Inst of Buddhist Studies, 1900 Addison Street, Berkeley CA 94704, USA. T. (1 510) 849 2383.
Residence: 1312 Josephine Street, Suite 8, Berkeley CA 94703, USA.

Bloom, William, Dr
Rel: New Age; Unordained; b. 08 FEB 1948 in London, UK; m. to Sabrina Dearborn; Citizen of United Kingdom.
International Organizations: Fellow of the Findhorn Foundation.
Other organizations: Director (since 1989), Alternatives of St James's; Research Fellow (since 1990), Stiftung Imago Coelestis.
Main field of work: Spiritual, psychological, political research and education. **Educ.:** Ph.D. from University of London, London School of Economics and Political Science, UK. **Mem.:** Fellow, Findhorn Foundation **Pubs.:** The Sacred Magician, Paladin, 1973; Sacred Times, Findhorn Press, 1990 and Gothic Image, 1992.
Mailing address: St James's Church, 197 Piccadily, London W1V 9RF, UK. T. (44 71) 734 4511

Bluyssen, Johannes Wilhelmus Maria, Most Rev
Rel: Roman Catholic; Ordained; b. 10 APR 1926 in Nijmegen, Netherlands; celibate; Citizen of Netherlands.
International Organizations: Ecclesiastic of the Roman Catholic Church.
Career: Curate, Veghel from 1950 to 1952. Vice-Prefect, Seminary St Michielsgestel from 1952 to 1954. Studied in Rome from 1954 to 1957. Spiritual Director, Seminary St Michielsgestel from 1957 to 1961. Joint Bishop, 's-Hertogenbosch from 1961 to 1966. Bishop, 's-Hertogenbosch from 1966 to 1986. **Educ.:** St Michielsgestel and Haaren Episcopal Seminaries, Netherlands.
Mailing address: Parade 11, 's-Hertogenbosch, Netherlands. T. (31 73) 125488

Blyweert, Vincent
Rel: Christian.
International Organizations: Secretary-General of the International Independent Christian Youth.
Mailing address: Elorriaga, 11 rue Martin Bernard, F-75013 Paris, France. T. (33 1) 45 89 00 42.

Boaga, Emmanuele, O.Carm., Rev Prof
Rel: Roman Catholic; Ordained; b. 30 MAR 1934 in Padova, Italy; celibate; Citizen of Italy; Lang. Italian, Latin, Portuguese, Spanish, English and French.
International Organizations: President (since 1985) of the Institutum Carmelitum.
Member of the Carmelite Third Order.
Other organizations: Archivist (since 1984), Archivo Generale Carmelitani, Rome; Professor (since 1985), Pontifical Faculty of Theology 'Marianum', Vatican City State.
Main field of work: Research in history and theology. **Career:** Professor of History, Liceo-Filosofico S Piertommajo, Rome from 1963 to 1968. Secretary-General, Curia Generalizia Carmelitani from 1971 to 1983. Prior, Centrum Internat. S Alberto, Rome from 1985 to 1990. **Educ.:** Dr. Hist. Eccl. from Pontifical Gregorian University, Vatican City State. **Interests:** History, archivism. **Pubs.:** *La soppressione innocenziana dei piccoli conventi*, Ed. Storia e Letteratura, Rome, 1971; *Como pedras vivas* .., Rome, 1989 and *Guida degli archivi diocesani d'Italia*, Rome, 1989 (collective work).
Mailing address: Via Sforza Pallavicini 10, I-00193 Roma, Italy. T. (39 6) 654 3513
Residence: See Mailing Address. T. See Mailing Address

Boayen, Boonratna, Rev Dr
Rel: Church of Christ; Ordained.
International Organizations: General Secretary of the Church of Christ in Thailand.
Mailing address: 14 Pramuan Road, Bangkok 10500, Thailand.

Bocken, Dirk
Rel: Christian.
International Organizations: Contact of the Association of the International Christian Youth Exchange in Europe.
Mailing address: Groenstraat 35A, B-3001 Heverlee, Belgium.

Böckle, Franz, Fr
Rel: Roman Catholic; Ordained in 1945; b. 18 APR 1921 in Glarus; celibate; Citizen of Switzerland.
International Organizations: Ecclesiastic and academic of the Roman Catholic Church.
Other organizations: Professor Emeritus (since 1986), University of Bonn.
Career: Professor of Moral Theology, Theologische Hochschule, Chur from 1953 to 1963. Professor of Moral Theology, University of Bonn from 1963 to 1986. Rector, University of Bonn from 1983 to 1985. **Educ.:** Stds. in Chur, Rome and Munich. **Interests:** sport. **Hons.:** Das Ehrenzeichen der Deutschen Arzteschaft. **Pubs.:** Grundbegriffe der Moral, 1966; Fundamentalmoral, 1977 and Christlicher Glaube in moderner Gesellschaft, 1980-82.
Mailing address: Am Kottenforst 46, 53125 Bonn, Germany. T. (49 228) 25 15 78

Bøckman, Peter Wilhelm
Rel: Christian.
International Organizations: Chairman, Norwegian Branch of the Nordic Christian Mission to Buddhists.
Mailing address: Syrenveien 12, Box 74, Tasen, N-0801 Oslo 8, Norway.

Bodhi, Bhikkhu, Ven
Rel: Buddhist: Theravada; Ordained on 18 NOV 1972; b. 10 DEC 1944 in New York NY, USA; celibate; Citizen of USA; Lang. Pali and Sanskrit.
International Organizations: President (since 1988) of the Buddhist Publication Society.
Other organizations: Editor (since 1984), Buddhist Publication Society.

Main field of work: Scholarship, publishing. **Educ.:** B.A. from Brooklyn College, New York, USA (1961-1966); Ph.D. from Claremont Graduate School, CA, USA (1966-1972). **Pubs.:** The Discourse on the All-Embracing Net of Views; The Great Discourse on Causation and The Noble Eightfold Path.
Mailing address: Buddhist Publication Society, 54 Sangharaja Mawatha, PO Box 61, Kandy, Sri Lanka. T. (94 8) 23679
Residence: Forest Hermitage, Udawattakele Reserve, Kandy, Sri Lanka.

Boesak, Allan, Rev

Rel: Reformed Church; Ordained; b. 23 FEB 1946 in Kakamas, South Africa; m. to Dorothy Rose Martin; c. 1 boy and 3 girls; Citizen of South Africa.
International Organizations: Past President of the World Alliance of Reformed Churches.
Other organizations: President (since 1984), Association of Christian Students in South Africa.
Main field of work: Clergyman. **Career:** Co-founder, United Democratic Front in 1983. President, South African Council of Churches from 1984 to 1987. President, World Alliance of Reformed Churches in 1982. **Hons.:** Hon. D.D.: Victoria (1983), Yale (1984), Interdenominational Theological Centre Atlanta (1985); Hon. D.Iur., Warwick (1989); Third World Prize (1989); numerous other awards. **Pubs.:** *A Call for an End to Unjust Rule*, 1986 and *If This is Treason, I am Guilty* (speeches), 1988.
Mailing address: 150 route de Ferney, PO Box 2700, CH-1211 Genève 20, Switzerland. T. (41 22) 791 62 37 – 791 62 38. C. WARC GENEVA. Tx 415 730 OIK CH (WARC). Fax (41 22) 791 65 05 WARC REFORMIERTER WELTBUND.

Boff, Leonardo Genezio Darci, Prof

Rel: Roman Catholic; Ordained; b. 14 DEC 1938 in Concórdia, SC; celibate; Citizen of Brazil.
International Organizations: Theologian, leading advocate of the Liberation Theology.
Other organizations: Advisor to National Conference of Brazilian Bishops (CNBB) (since 1971); Member, Editorial Board (since 1971), *Revista Eclesiástica Brasileira*; Member, Board of Directors (since 1971), Vozes publishing house; President, Board of Editors (since 1985), Theology and Liberation collection.
Career: Taught systematic theology at the Petropólis Institute for Philosophy and Theology, Brazil. Adviser to Latin American Conference of Religions (CLAR) from 1971 to 1980. Condemned by the Vatican and ordered to keep one year of public silence in 1985. **Educ.:** Stds. from University of Würzburg, Germany, University of Louvain, Belgium, and University of Oxford, UK; Dr. Theol. from Ludwig-Maximilians-Universität München, Germany (1972). **Interests:** gardening, social work at the 'favelas', child-minding. **Hons.:** Paz y Justicia Award, Barcelona; Menschenrechte in der Kirche Award; Herbert Haag Foundation, Germany and Switzerland. **Pubs.:** *St Francis: A Model for Human Liberation* (English translation 1982); With Clodovis Boff, *Salvation and Liberation* (English translation 1984) and *Church, Charism, and Power: Liberation Theology and the Instituional Church* (English translation 1985).
Mailing address: Frei Luís 100, 25 689 Petrópolis, Rio de Janeiro, Brazil.

Bohdanowicz, Piotr, Fr

Rel: Roman Catholic; Ordained; celibate.
International Organizations: Procurator-General of the Fils de la Mère de Dieu Douloureuse.
Mailing address: Aleja Przyszlosci 8, PL-05 420 Warszawa, Poland.

Bojaxhiu, Agnes Gonxha

See Mother Teresa.

Bolonek, Janusz, H E Mgr

Rel: Roman Catholic; Ordained on 17 DEC 1961; b. 06 DEC 1938 in Huta Dlutowska, Lodz, Poland; celibate; Citizen of Vatican and Poland; Lang. Polish, Italian, Spanish, French, Russian, German, English and Latin.
International Organizations: Apostolic Nuncio to the Ivory Coast and Apostolic Pro-Nuncio to Burkina Faso and Niger (since 1989) of the Roman Catholic Church.
Other organizations: Titular Archbishop, Madauro.
Educ.: D.Theol. from Pontifical Gregorian University, Vatican City State (1962-1969); D.Cn.L. from Lateran Pontifical University, Vatican City State (1967-1971); Dip. from Pontifical Academy for Diplomats, Vatican City State (1967-1971). **Pubs.:** Many articles on religious subjects.
Mailing address: Nonciature Apostolique, Rue Mgr René Kouassi, BP 1347, Abidjan 08, Ivory Coast. T. (225) 44 38 35. Fax 44 72 40. Tx 26182 NUNCIO CI
Residence: See Mailing Address. T. See Mailing Address

Bolton, Brenda

Rel: Christian.
International Organizations: Secretary of the International Commission of Comparative Church History.
Mailing address: 44 King's College Court, Primrose Hill Road, London NW3 3EA, UK.

Bomers, Hendrik Joseph Alois, C.M., Rt Rev

Rel: Roman Catholic; Ordained on 19 MAR 1964; b. 19 APR 1936 in Groenlo, Utrecht, The Netherlands; celibate; Citizen of Netherlands.
International Organizations: Bishop (since 1983) of Haarlem.
Other organizations: Member (since 1967), Lazarist Fathers.
Main field of work: Ministry. **Career:** Novitiate, Lazarist Fathers in 1957. Lecturer in Philosophy, Interdiocesan Seminary, Addis Ababa from 1967 to 1973. Regional Superior, Lazarist Fathers in Ethiopia from 1973 to 1978. Secretary-General for Food Aid from 1973 to 1974. Apostolic Vicar, Gimma, Ethiopia in 1977.
Mailing address: Nieuwe Gracht 80, PB 1053, 2001 NJ Haarlem, Netherlands. T. (31 23) 319450

Bonata, Anna

Rel: Roman Catholic.
International Organizations: Member of the Arche Internationale.
Mailing address: Communauté de l'Arche, Via Umberto I N° 1, I-10080 Lugnacco (Torino), Italy.

Bonello, James (Anthony), M.S.S.P., Rev

Rel: Roman Catholic; Ordained on 19 MAR 1972; b. 23 AUG 1943 in Rabat, Malta; celibate; Citizen of Malta; Lang. Maltese, English, Italian, Spanish and French.
International Organizations: Superior General (since 1982) of the Missionary Society of St Paul.
Other organizations: Member (since 1977), Presbyterial Council; President (since 1990), Council of Major Superiors.
Career: Teacher at St Paul's Missionary College from 1968 to 1982. Moderator for Students, Missionary Society of St Paul from 1973 to 1982. Chairman, Commission for Renewal of Constitution of Missions, Missionary Society of St Paul from 1979 to 1982. President, meeting for Directors of Formation, Missionary Society of St Paul in 1991. **Educ.:** B.A., B.D., Lic.D. from University of Malta, Valletta (1963-1972). **Interests:** reading, languages. **Mem.:** Children's Home Commission (1991)
Mailing address: St Agatha's Lane, Rabat RBT 07, Malta. T. (356) 456926
Residence: See Mailing Address. T. See Mailing Address

Bonfil, Robert, Rabbi Prof

Rel: Jewish; Ordained in 1959; b. 30 APR 1937 in Karditsa, Greece; m. to Eva Rechtschaffen; c. 1 boy and 2 girls; Citizen of Israel; Lang. Hebrew, Italian, English, French, Greek, Portuguese and Latin.
International Organizations: Director of the Institute for Research on the Sephardi and Oriental Jewish Heritage.
Other organizations: Professor of Jewish History (since 1980), The Hebrew University of Jerusalem.
Main field of work: University education. **Career:** Rabbi, Jewish Community of Milan from 1959 to 1968. **Educ.:** Collegio Rabbinico Italiano, Rome/Turin, Italy (1949-1959); Doc. Phys. from University of Turin, Italy (1955-1960); Ph.D. from Hebrew University of Jerusalem, Israel (1971-1976). **Pubs.:** Rabbis and Jewish Communities, Oxford University Press, 1990; Gli ebrei in Italia nell'epoca del rinascimento, Firenze, Sansoni, 1991 and The Writings of Azariah de Rossi, Jerusalem, Mossad Bialik, 1991 (in Hebrew).
Mailing address: Dept. of Jewish History, The Hebrew University, Jerusalem, Israel.
Residence: Ben Zion 20, Jerusalem 95423, Israel. T. (972 2) 522853

Bontrager, David, Rev
Rel: Christian; Ordained.
International Organizations: Head, Secretariat of the Christian Missions to the Communist World International.
Mailing address: PO Box 938, Middlebury IN 46540, USA. T. (1 219) 825 21 17 – (1 219) 825 52 41.

Bonvin, Jean-Marc, O.R.C., Fr
Rel: Roman Catholic; Ordained; celibate.
International Organizations: Superior General of the Canons Regular of the Order of the Holy Cross.
Mailing address: Via Casale Vecchio di Aguzzano 85, I-00137 Roma, Italy. T. (39 6) 889 67 70

Booth, William Bramwell, Col
Rel: Salvation Army, The; Ordained on 20 MAY 1953; b. 19 SEP 1928 in London, UK; m. to Ruth Lilian Fiskaa; c. 1 boy and 2 girls; Citizen of United Kingdom; Lang. English, French and Italian.
International Organizations: Territorial Commander (since 1991) of The Salvation Army.
Main field of work: Ministry. **Educ.:** Officer Comm. from International Training College, London, UK (1952-1954). **Interests:** Salvation Army history.
Mailing address: Die Heils Armee, Salierring 23-27, 50677 Köln, Germany. T. (49 221) 208 19 21
Residence: Peter Dedenbach Strasse 2, 50677 Köln, Germany T. (49 221) 52 15 42

Booth-Clibborn, Stanley Eric Francis, Rt Rev
Rel: Church of England; Ordained in 1952; b. 20 OCT 1924 in London, UK; m. to Anne Roxburgh Forrester; c. 2 boys and 2 girls; Citizen of United Kingdom.
International Organizations: Bishop (since 1979) of Manchester.
Other organizations: President (since 1979), St Ann's Hospice, Cambridge; Chairman (since 1983), Namibia Communications Centre; Member (since 1985), House of Lords.
Main field of work: Ministry. **Career:** Served during World War II with Royal Artillery (RA). Curate, Heeley, Diocese of Sheffield from 1952 to 1954. Curate, Attercliffe from 1954 to 1956. Training Officer, Christian Council of Kenya from 1956 to 1963. Editor, East African Venture Newspaper, Kenya from 1963 to 1967. Leader, Lincoln Team Ministry from 1967 to 1970. Vicar, Great St Mary, Cambridge from 1970 to 1979. Moderator, Movement for the Ordination of Women in the Church of England (until 1982). **Educ.:** B.A. from University of Oxford, Oriel College, UK; M.A. from University of Oxford, Oriel College, UK. **Interests:** tennis, photography, music. **Hons.:** Honorary Fellow, Manchester Polytechnic (1989). **Mem.:** Royal Commonwealth Society **Pubs.:** Taxation – Burden or Blessing?.
Mailing address: Bishops Court, Bury New Road, Manchester M7 OLE, UK. T. (44 61) 792 2096/1799

Borgegård, Gunnel
Rel: Christian.
International Organizations: Director of the Nordic Ecumenical Council.
Mailing address: Nathan Säderblomsplan 1, Box 640, S-75127 Uppsala, Sweden. T. (46 18) 16 95 00. Fax (46 18) 133178.

Borges, Armando Rodriguez, Bishop
Rel: Methodist; Ordained.
International Organizations: Bishop of Methodist Church in Cuba.
Mailing address: Calle 58, No 4305, Havana, Cuba.

Borgomeo, Pasquale, S.J., Rev
Rel: Roman Catholic; Ordained in 1963; b. 20 MAR 1933 in Naples, Italy; celibate; Citizen of Italy.
International Organizations: Director General (since 1985) of the Radio Vaticana.
Member, Bureau (since 1976) of the International Radio and Television University. President, International Broadcasting Working Party (since 1983) of the European Broadcasting Union. Adviser of the Pontifical Council for Social Communications. Member of the Society of Jesus.
Other organizations: Board of Directors (since 1984), Centro Televisivo Vaticano.
Career: Entered Society of Jesus in 1948. Editor-in-Chief, Radio Vaticana from 1970 to 1978. Programme Director, Radio Vaticana from 1978 to 1983. Assistant Director General, Radio Vaticana from 1983 to 1985. **Pubs.:** L'Eglise de ce temps dans la prédication de Saint Augustin, 1972.
Mailing address: Vatican Radio, 00120 Vatican City, Vatican. T. (39) 698 3945
Residence: Via dei Penitenzieri 20, I-00193 Roma, Italy. T. (39 6) 68 69 357

Born, Ethel W
Rel: Methodist.
International Organizations: President for North America of the World Federation of Methodist Women.
Mailing address: 6995 Malinda Road, Salem VA 84153, USA.

Born, Horst, Dr
Rel: Christian.
International Organizations: Coordinator of the European Evangelical Accrediting Association.
Mailing address: Bettingerstrasse 112, Wyhlen 1, 79639 Groenzach, Germany.

Boro, Wanjiku
Rel: Christian.
International Organizations: Secretary of the Pan-African Christian Broadcast Association.
Mailing address: NCCK, Box 45009, Nairobi, Kenya.

Borocz, Isaac, O.S.B., Fr
Rel: Roman Catholic; Ordained; celibate.
International Organizations: Procurator General of the Benedictine Confederation.
Mailing address: Piazza dei Cavalieri di Malta 5, I-00153 Roma, Italy. T. (39 6) 578 02 47

Borunda, Luis Díaz, Fr
Rel: Roman Catholic; Ordained; celibate.
International Organizations: Procurator General of the Missionaries of the Holy Ghost.
Mailing address: Piazza S Salvatore in Campo 57, I-00186 Roma, Italy. T. (39 6) 654 20 41.

Bory, Jean-Pierre
Rel: Christian.
International Organizations: President of the Association de soutien des missions des assemblées de France.
Mailing address: 20 quai Saint-Nicolas, F-67000 Strasbourg, France.

Bos, Wytze
Rel: Bahá'í.
International Organizations: Head, European Office of the Bahá'í International Community.
Mailing address: Route des Morillons 15, CH-1290 Genève, Switzerland. T. (41 22) 98 54 00.

Bosa, Siliana
Rel: Buddhist: Mahayana.
International Organizations: Member of the Foundation for the Preservation of the Mahayana Tradition.
Other organizations: Director, Centro Studi Cenresig.
Mailing address: Via F Acri 2/a, Bologna, Italy. T. (39 51) 226 953

Boscardin, Augusto, F.d.C.C., Fr
Rel: Roman Catholic; Ordained; celibate.
International Organizations: Superior General of the Sons of Charity – Canossians.
Mailing address: Via Marcantonio Dalla Torre 5, I-37131 Verona, Italy. T. (39 45) 52 88 57

Boseto, Leslie
Rel: United Church.
International Organizations: Chairman of the Pacific Conference of Churches.
Mailing address: United Church Solomon Is. Region, Munda, Solomon Is.

Boshoff, Carel Willem Hendrik, Rev Dr

Rel: Dutch Reformed; b. 09 NOV 1927 in Nylstroom, South Africa; Citizen of South Africa.

International Organizations: Chairman (since 1972) of the South African Bureau of Racial Affairs.

Other organizations: Chairman (since 1976), N.G. Kerkboekhandel; Chairman, Council (since 1978), Institute for Missiological Research; Executive Chairman (since 1988), Afrikaner Vryheidstigting.

Main field of work: Theologian. **Career:** Missionary, Dutch Reformed Church from 1953 to 1963. Secretary of Missions, Dutch Reformed Church from 1963 to 1966. Professor and Head, Dept. of Theology, Missiology and Science of Religion, University of Pretoria from 1967 to 1988. Dean, Theology Faculty, University of Pretoria from 1978 to 1980. **Educ.:** M.A. from University of Pretoria, South Africa; D.D. from University of Pretoria, South Africa. **Interests:** small farming, breeding Nooitgedracht ponies. **Mem.:** numerous theological and scholarly committees **Pubs.:** Uit God Gebore, 1968; Die Nuwe Sendingsituasie, 1978 and Swart Teologie van Amerika Tot in Suid-Afrika, 1980.

Mailing address: 55 Kesselaar Ave, Brummeria, 0184 Pretoria, South Africa. T. (27 12) 436051

Residence: T. (27 12) 865225

Bouchard, Giorgio, Rev

Rel: Waldensian; Ordained on 30 AUG 1958; b. 01 AUG 1929 in S Germano Chisone, Torino, Italy; separated; c. 2 boys and 1 girl; Citizen of Italy; Lang. Italian, French, German, English and Spanish.

International Organizations: President (since 1988) of the Federation of Evangelical Churches in Italy (FCEI).

Other organizations: Pastor (since 1987), Napoli Waldensian Church.

Main field of work: Ministry. **Career:** Editor, *Gioventù Evangelica* from 1962 to 1971. Moderator, Evangelical Waldensian Church from 1979 to 1986. **Educ.:** Doc. from University of Turin, Italy (1948-1952); Doc. from Facoltà Valdese di Teologia, Rome, Italy (1952-1957). **Pubs.:** I Valdesi e l'Italia, Torino, 1988; La Scritta di Pilato, Torino, 1989 and Spirito Protestante e Etica del Socialismo, Roma, 1991.

Mailing address: Via Firenze 38, I-00184 Roma, Italy. T. (39 6) 48 25 120

Residence: Via dei Cimbri 8, I-80138 Napoli, Italy. T. (39 81) 26 45 10

Bouchaud, Constant, P.S.S., Fr

Rel: Roman Catholic; Ordained; celibate.

International Organizations: Procurator-General of the Society of the Priests of St Sulpice.

Advisor to the Administration of the Congregation for the Evangelization of Peoples. Advisor to Office I of the Congregation for the Clergy.

Other organizations: Superior, Procura St Sulpizio (Convitto Ecclesiastico).

Mailing address: Via dei Fratelli Bandiera 17/a, I-00152 Roma, Italy. T. (39 6) 589 14 20

Boucher, René, S.C., Br

Rel: Roman Catholic; Ordained; celibate.

International Organizations: Procurator-General of the Institute of the Brothers of the Sacred Heart.

Mailing address: Piazza del Sacro Cuore 3, I-00152 Roma, Italy. T. (39 6) 53 31 29.

Boukakaur, Dalil, Dr

Rel: Muslim.

International Organizations: Secretary-General of the International Muslim Union.

Mailing address: Grande Mosquée de Paris, 2 bis pl du Puits de l'Ermite, F-75005 Paris, France.

Bourdeau, Gilles, O.F.M., Rev

Rel: Roman Catholic; Ordained on 20 MAY 1967; b. 21 NOV 1941 in St Chrysostôme PQ, Canada; celibate; Citizen of Canada; Lang. French, English, Italian and Spanish.

International Organizations: Vicar-General/Procurator-General (since 1991) of the Order of Friars Minor.

Career: Entered the Order of Friars Minor in 1962. Delegate, Quebec to Student Christian Movement from 1966 to 1968. Founder/President, Association pour la spiritualité des jeunes du Québec from 1976 to 1979. Theological Counselor to Union des religieuses contemplatives du Canada from 1978 to 1981. Co-Director, Fonds Jean Mouroux, Dijon (France) from 1973 to 1980. Taught Catechism at Ecole Arcand from 1968 to 1969. Assistant Professor of Theology, University of Montreal from 1969 to 1971. Tutor, Master's degree in Pastoral Ministry, University of Montreal from 1970 to 1971. Professor of Spirituality (part-time), University of Montreal from 1974 to 1976. Professor, Institut de Pastorale des Dominicains from 1976 to 1981. Provincial Minister, St Joseph's Province of Canada, Order of Friars Minor from 1987 to 1991. **Educ.:** B.Th. from Scolasticat OFM de Rosemont, and Grand Séminaire de Montréal, Canada (1963-1967); M.A. Theol. from University of Toronto, St Michael's College, Canada (1967-1968); Ph.D. Theol. from University of Montreal, Quebec, Canada (1969-1975).

Mailing address: Via S Maria Mediatrice 25, I-00165 Roma, Italy. T. (39 6) 68 49 19.

Residence: See Mailing Address. T. See Mailing Address

Bourdeaux, Michael Alan, Rev Canon

Rel: Church of England; Ordained on 21 DEC 1960; b. 19 MAR 1934 in Praze, Cornwall, UK; m. to (1st) Gillian Mary Davies (died), (2nd) Lorna Elizabeth Waterton; c. 2 boys and 2 girls; Citizen of United Kingdom; Lang. English, Russian, French and German.

International Organizations: Founder/Director (since 1969) of the Keston College, Oxford.

Other organizations: Honorary Canon (since 1990), Rochester Cathedral.

Main field of work: Research on religion in the former Soviet Union and Eastern Europe. **Career:** Parish work in London from 1960 to 1964. Three further visits to USSR from 1961 to 1964. Correspondent researcher on Church in USSR for Centre de Recherches, Geneva from 1965 to 1968. Visiting Fellow at London School of Economics and Political Science, University of London from 1968 to 1969. Visiting Professor at St Bernard's Seminary, Rochester NY in 1969. Founded the CSRC-Centre for the Study of Religion and Communism (now Keston College) in 1969. Delivered the Dawson Lectures on Church and State at Baylor University, Waco TX in 1972. Lecture Tour in Africa in 1973. Oversaw th move of the CSRC to its new home at Keston in 1974. Visits to USSR (in 1975, 1977, 1979). Chavasse Memorial Lectures at Oxford University in 1976. Visit to Romania in 1978. Visit to Poland in 1979. Visiting Professor at the Summer School at Fullet Theological Seminary, Pasadena CA in 1981. Kathryn W Davi Visiting Professor of Slavic Studies at Wellesley College, MA (autumn term) in 1981. Lecture tours in Australasia (in 1982, 1983). Moorhouse Lectures (Melbourne), tour of Australia and New Zealand, visit to Albania in 1987. Sabbatical at Oxford University, St Anthony's and St Edmund Hall from 1989 to 1990. **Educ.:** B.A. (Hons) Russian and French from University of Oxford, St Edmund Hall, UK (1954-1957); B.A. (Hons) Theol. from Wycliffe Hall, Oxford, UK (1959); M.A. from University of Oxford, St Edmund Hall, UK (1960); Studied Russian History from Moscow M V Lomonosov State University, USSR (1959-1960); B.D. by Thesis from University of Oxford, St Edmund Hall, UK (1969). **Hons.:** Templeton Prize for Progress in Religion (1984); Award from Leverhulme Trust to write a book at Oxford (1988). **Mem.:** Philarmonia Chorus (1961-1991); British Tennis Umpires Association (1974); Club: Athenaeum **Pubs.:** Land of Crosses: The Catholic Church in Lithuania Today, Augustine Publishing Company, Chumleigh, Devon, 1979; With Lorna Bourdeaux, Ten Growing Soviet Churches, MARC Europe, Bromley, 1987 and Gorbachev, Glasnost and the Gospel, Hodder and Stoughton, London, 1990.

Mailing address: Keston Research, 33a Canal Street, Oxford OX2 6BQ, UK. T. (44 865) 311022.

Residence: 101 Church Way, Iffley Village, Oxford OX4 4EG, UK.

Bourel de la Roncière, Philippe B

Rel: Roman Catholic.

International Organizations: President of the European Catholic Home.

Mailing address: Ave E Lacomblé 6, B-1040 Bruxelles, Belgium.

Bourne, Joan, Sr

Rel: Roman Catholic; Ordained; celibate.

International Organizations: President of the Association of Contemplative Sisters.

Main field of work: Contemplation.

Mailing address: c/o Association of Contemplative Sisters, Monastery Rd, Elysburg PA 17824, USA. T. (1 717) 672 2935

Boutaleb, Abdelhadi
Rel: Muslim.
International Organizations: Director-General of the Islamic Educational, Scientific and Cultural Organization.
Secretary-General of the Federation of the Universities of the Islamic World.
Mailing address: 16 bis Avenue Omar Ben Khattab, BP 755, AGDAL Rabat, Morocco. T. (212 7) 77 24 33. Fax 77 74 59 – 77 20 58. Tx 326 45 M – 318 44 M.

Bovon, François, Prof Dr
Rel: Protestant; Ordained in 1964; b. 13 MAR 1938 in Lausanne, Vaud, Switzerland; divorced; c. 2 boys; Citizen of Switzerland; Lang. French, German and English.
International Organizations: Professor of New Testament, Faculty of Theology (since 1967) of the University of Geneva.
Main field of work: University education. **Career:** President, Société suisse de théologie from 1973 to 1977. Dean, Faculty of Theology, University of Geneva from 1976 to 1979. President, Association pour l'étude de la littérature apocryphe chrétienne from 1981 to 1987. **Educ.:** Lic. Theol. from University of Lausanne, Switzerland (1961); Dr. Theol. from University of Basel, Switzerland (1965). **Interests:** Gospel of Luke, Acts of Philippe.
Mailing address: Faculté de Théologie, Université de Genève, 3 place de l'Université, CH-1211 Genève 4, Switzerland. T. (41 22) 705 77 75
Residence: Rue François Grast, CH-1208 Genève, Switzerland. T. (41 22) 786 13 26

Bowen, Michael George, Most Rev
Rel: Roman Catholic; Ordained on 06 JUL 1958; b. 23 APR 1930 in Gibraltar; celibate; Citizen of United Kingdom.
International Organizations: Archbishop and Metropolitan (since 1977) of Southwark.
Career: Wine trade from 1951 to 1952. Venerable English College, Rome from 1952 to 1959. Curate, Earlsfield and Walworth, Diocese of Southwark from 1959 to 1963. Teacher of Theology, Pontifical Beda College, Rome from 1963 to 1966. Chancellor, Diocese of Arundel and Brighton from 1966 to 1970. Coadjutor Bishop, Arundel and Brighton from 1970 to 1971. Bishop, Arundel and Brighton from 1971 to 1977. **Educ.:** S.T.L. from University of Cambridge, Trinity College, UK; Ph.L. from Pontifical Gregorian University, Vatican City State. **Interests:** golf, tennis. **Hons.:** Freeman of City of London (1984).
Mailing address: Archbishop's House, 150 St George's Road, London SE1 6HX, UK. T. (44 71) 928 5592/2495

Bowman, John
Rel: Christian; Ordained in 1954; b. 13 MAY 1916 in Ayr, Scotland, UK; m. to Margaret Fanny Stanton; c. 2 boys and 5 girls; Citizen of Australia.
International Organizations: Professor Emeritus (since 1978) of the University of Melbourne.
Main field of work: Ecclesiastic. **Career:** Lecturer in Hebrew, Faculties of Arts and Divinity, University of Glasgow from 1945 to 1946. Senior Lecturer, Head of Dept. of Semitic Languages and Literature, University of Leeds from 1946 to 1959. Professor, Dept. of Middle Eastern Studies, University of Melbourne from 1959 to 1978. Became Australian citizen in 1961. UNESCO Fellow in 1964. **Educ.:** M.A. from University of Glasgow, UK; D.Phil. from University of Oxford, Christ Church, UK. **Interests:** planting trees, walking. **Mem.:** Fellow, Australian Academy of the Humanities **Pubs.:** Founder and Editor, Abr Nahrain 1961-80, Milla wa-Milla.
Mailing address: 15 Haines Street, North Melbourne, Victoria 3051, Australia.

Boyd, Robert Hugh Steele (Robin)
Rel: Presbyterian; Ordained; b. 14 MAY 1924 in Belfast, UK; m. to Frances Amy Paton; c. 2 girls; Citizen of United Kingdom.
International Organizations: Ecclesiastic of the Presbyterian Church.
Other organizations: Parish Minister (since 1988), Wesley Church.
Career: Secretary, Theological Colleges Dept., Student Christian Movement, London from 1951 to 1953. Missionary, Gujarat State, India from 1954 to 1974. Lecturer, Gujarat United School of Theology from 1961 to 1974. Presbyter, Church of North India from 1970 to 1974. Associate Minister, Presbyterian Church, Toorak, Melbourne from 1974 to 1977. Associate Minister, Uniting Church in Australia from 1977 to 1980. Director, Irish School of Ecumenics, Dublin from 1980 to 1987. **Educ.:** B.A. from University of Dublin Trinity College, Ireland; Ph.D. from University of Edinburgh, UK; D.D. from University of Basel, Switzerland. **Interests:** railways, industrial archaeology, local history and music. **Pubs.:** India and the Latin Captivity of the Church, 1975; Church History of Gujarat, 1981 and Ireland: Christianity Discredited or Pilgrim's Progress?, 1988.
Mailing address: Wesley Church, 148 Lonsdale Street, Melbourne, Victoria 3000, Australia. T. (61 3) 662 2355
Residence: 41 Kawarren Street, North Balwyn, VIC 3104, Australia. T. (61) 857 5378

Boyle, James M, C.F.X., Br
Rel: Roman Catholic; Ordained; celibate.
International Organizations: Superior General of the Brothers of St Francis Xavier.
Mailing address: 32 Strawberry Hill Road, Twickenham TW1 4PU, UK. T. (44 81) 892 3347.

Brabbe, Richard
Rel: Christian.
International Organizations: Director of the Africa Christian Press.
Mailing address: PO Box 30, Achimota, Ghana.

Bradfield, J T, Pastor
Rel: Seventh-Day Adventist; Ordained.
International Organizations: President, South African Union of the General Conference of Seventh-Day Adventists.
Mailing address: PO Box 468, Orange Free State, Bloemfontein 9300, South Africa.

Bradley, Patrick, CC.SS., Fr
Rel: Roman Catholic; Ordained; celibate.
International Organizations: Superior General of the Congregation of the Sacred Hearts of Jesus and Mary.
Member, Council for Reporting between the Congregation and the International Union of Superiors General and Superiors General of the Commission on the Catechism and Catechists of the Congregation for the Evangelization of Peoples.
Mailing address: Via Rivarone 85, I-00166 Roma, Italy. T. (39 6) 680 55 38, (39 6) 680 55 83

Brady, John Joseph, S.P.S., Fr
Rel: Roman Catholic; Ordained; celibate.
International Organizations: Procurator-General of the St Patrick's Society for the Foreign Missions.
Mailing address: Piazza di Villa Carpegna 42, I-00165 Roma, Italy. T. (39 6) 662 17 97

Brakemeier, Gottfried, Rev Dr
Rel: Lutheran; Ordained.
International Organizations: President of the Lutheran World Federation.
Other organizations: President, Evangelical Church of the Lutheran Confession in Brazil.
Mailing address: Rua Senhor dos Passos 202, Caixa Postal 2876, 90 001 Porto Alegre RS, Brazil. T. (55 512) 213433. Fax 257244. C. ECLESIA PORTOALEGRE. Tx 2332 LUTE BR

Brambilla, Franco, H E Mgr
Rel: Roman Catholic; Ordained on 29 AUG 1947; b. 27 NOV 1923 in Brugherio, Italy; celibate.
International Organizations: Apostolic Pro-Nuncio to Australia (since 1986) of the Roman Catholic Church.
Other organizations: Titular Archbishop, Viminacio.
Mailing address: 2 Vancouver Street, Red Hill, Canberra ACT 2603, Australia. T. (61 62) 95 38 76. Fax 95 36 90

Brandt, D Gerhard
Rel: Evangelical.
International Organizations: Contact of the Evangelical Church in Rhineland.
Mailing address: Hans-Böckler-Strasse 7, 40476 Düsseldorf, Germany.

Brannon, Ronal Roy, Rev Dr

Rel: Wesleyan Church; Ordained on 10 JUN 1951; b. 16 APR 1928 in Aberdeen SD, USA; m. to Rosalee V Carry; c. 2 boys and 1 girl; Citizen of USA; Lang. English.

International Organizations: Secretary-General (since 1982) of the Wesleyan World Headquarters.

Other organizations: Member and Secretary, General Board of Administration, The Wesleyan Church, Central Wesleyan College Board of Trustees, Hephzibah Children's Home Local Board of Managers, Wesleyan Investment Foundation Board of Directors; Co-founder and coordinator, police chaplaincy, Wichita KS.

Main field of work: Ministry and administration. **Career:** Pastor, Heber Wesleyan, Miltonvale KS from 1949 to 1952. Pastor, First Wesleyan, Wichita KS from 1952 to 1968. District Superintendent, The Kansas District, The Wesleyan Church from 1968 to 1983. Past President, Kansas District of Wesleyan Youth, Western Area of Wesleyan Youth, North American General Conference of Wesleyan Youth, Kansas Association of Evangelicals. **Educ.:** B.Rel. from Bartlesville Wesleyan College, OK, USA (1946-1950); D.D. from Central Wesleyan College, SC, USA (1987). **Pubs.:** Contributing author to *Program Pathways for Young Adults*, Baker Book House, 1964; *A Guide for Training Christians to be Soul-Winners* and other Wesleyan Church evangelism materials.

Mailing address: PO Box 50434, Indianapolis IN 46250-0434, USA. T. (1 317) 576 8154.

Residence: 1412 Marlin Drive, Marion IN 46952, USA. T. (1 317) 664 1389

Branover, Herman, Prof

Rel: Jewish.

International Organizations: Chairman of the Association of Jewish Religious Professionals from the Soviet Union and Eastern Europe.

Mailing address: 6 David Yellin Street, PO Box 5749, Jerusalem, Israel. T. (972 2) 22 37 02. Fax (972 2) 38 51 18.

Brasch, Rudolph, Rabbi Dr

Rel: Jewish; Ordained in 1938; b. 06 NOV 1912 in Berlin, Germany; m. to Liselotte Buchbinder; Citizen of Australia; Lang. English, German, Hebrew and French.

International Organizations: Rabbi and author of Sydney, Australia.

Main field of work: Ministry, education and literature. **Career:** Rabbi, North London Liberal Synagogue, UK from 1938 to 1948. Rabbi/Founder, Dublin Progressive Synagogue, Ireland from 1946 to 1947. Rabbi, Springs and District Reform Synagogues, South Africa from 1948 to 1949. Chief Minister, Temple Emanuel, Sydney, Australia from 1949 to 1979. Rabbi, Temple Etz Ahayem, Montgomery AL, USA in 1980. Visiting Professor, University of Manoa, Hawaii in 1981. **Educ.:** Ph.D. from Bayerische-Julius-Maximilians-Universität Würzburg, Germany (1936); Rabbi from Hochschule für die Wissenschaft des Judentums, Berlin, Germany (1938). **Interests:** origins of customs, traditions, superstitions and phrases. **Hons.:** DD hc, Hebrew Union College, Cincinnati OH (1959); Officer of the Order of the British Empire (1967); Silver Jubilee Medal (1977); Order of Australia (1979); Peace Prize, United Nations Association of Australia (1979). **Mem.:** CCAR, USA **Pubs.:** *Thank God I'm an Atheist – The Religious Origins of Expressions, Customs and Institutions*, 207 pp., Fontana/Collins (paperback edition), 1987, 1988, Collins Australia (paperback edition), 1988, Tynron Press, Scotland, 1990; *The Book of the Year – Special Days and Their Meanings*, 170 pp., Collins/Angus and Robertson, Sydney, London (fully illustrated), 1991 (2 Eds.) and *A Book of Comfort – To Help and Inspire in Troubled Times*, 56 pp., Collins/Angus and Robertson, Sydney, London, 1991.

Mailing address: 14 Derby Street, Vaucluse, Sydney NSW 2030, Australia. T. (61 2) 337 2521

Residence: See Mailing Address. T. See Mailing Address

Brauer, Jerald Carl, Prof Dr

Rel: Lutheran; Unordained; b. 16 SEP 1921 in Fond du Lac WI, USA; m. to Muriel Nelson; c. 2 boys and 1 girl; Citizen of USA.

International Organizations: Professor of the History of Christianity (since 1959) of the University of Chicago, IL.

Other organizations: Associate Member, History Department (since 1959), University of Chicago, IL; Editor (with Professors Robert Grant and Martin Marty) (since 1962), *Church History*; Naomi Shenstone Donnelley Professor Emeritus (since 1991); Member, Board of Governors, International House of the University of Chicago; Member, Board of Theological Education, Lutheran Church of America; Trustee, Carnegie Council on Ethics and International Affairs.

Career: Instructor of Church History and History of Christian Thought, Union Theological Seminary, New York City from 1948 to 1950. Assistant Professor of the History of Christianity, University of Chicago, IL from 1950 to 1954. Associate Professor of the History of Christianity, University of Chicago from 1954 to 1959. Dean, Federated Theological Faculty, University of Chicago from 1955 to 1960. Dean, Divinity School, University of Chicago from 1960 to 1970. Member, Executive Committee, American Association of Theological Schools from 1956 to 1961. Member, Executive Committee, Rockfeller Brothers Theological Schools from 1956 to 1968. Member, Board of Trustees, Carthage College from 1958 to 1962. President, American Society of Church History in 1961. President, Board of Theological Education, Lutheran Church in America from 1961 to 1969. Editor, *Criterion* from 1961 to 1970. Member, Board of Trustees, Augustana College from 1964 to 1966. Consultant to the State of New York Education Department from 1969 to 1978. Naomi Shenstone Donnelley Professor, Divinity School, University of Chicago from 1969 to 1991. Chairman, Carnegie Council on Ethics and International Affairs from 1972 to 1983. Chairman, Board of Governors, International House of the University of Chicago from 1973 to 1991. National Endowment of the Humanities Fellow from 1977 to 1978. Visiting Professor, University of Frankfurt, Germany in 1961. Visiting Lecturer, University of Tokyo and University of Kokugakuin, Japan in 1966. Visiting Fellow, Center for the Study of Democratic Institutions, Santa Barbara (1972 and 1974). **Educ.:** A.B. from Carthage College, Kenosha WI, USA (1943); B.D. from Northwestern Lutheran Theological Seminary, USA (1945); Ph.D. from University of Chicago, IL, USA (1948). **Interests:** painting, music, drama. **Hons.:** Hon. D.D., Miami University, Oxford OH; Hon. LL.D., Carthage College; Hon. S.T.D., Ripon College; Hon. L.H.D., Gettysburg College. **Mem.:** American Society of Church History; Midwestern British Historical Association; American History Association **Pubs.:** Author/Editor of several books, including: *Images of Religion in America*, Fortress Press, Philadelphia, 1967 (author); *Westminster Dictionary of Church History*, Westminster Press, Philadelphia, 1971 (editor) and monographs and chapters in books, articles in various magazines and journals.

Mailing address: Swift Hall 207, The Divinity School, University of Chicago, Chicago IL 60637, USA. T. (1 312) 702 8232

Residence: 5620 South Blackstone Ave, Chicago IL 60637, USA. T. (1 312) 493 6886

Braun, Cornelis, M.S.C., Fr

Rel: Roman Catholic; Ordained; celibate.

International Organizations: Superior General of the Missionaries of the Sacred Heart of Jesus.

Member, Council for Reporting to the Congregation and to the International Union of Superior Generals and the Union of Superior Generals of the Congregation for the Evangelization of Peoples.

Mailing address: Via Asmara 11, I-00199 Roma, Italy. T. (39 6) 831 37 41, (39 6) 839 29 75

Braun, Ingrid

Rel: Buddhist: Mahayana.

International Organizations: Member of the Foundation for the Preservation of the Mahayana Tradition.

Other organizations: Director, Do-ngag Zung-jug Ling.

Mailing address: PO Box 16, Paros 84400, Greece. T. (30 284) 91343

Braybrooke, Marcus Christopher Rossi, Rev

Rel: Church of England; Ordained; b. 16 NOV 1938; m. to Mary Elizabeth Walker; c. 1 boy and 1 girl.

International Organizations: Vice-President and Chairman, International Committee (since 1986) of the World Congress of Faiths.

Other organizations: Honorary Priest-in-Charge (since 1984), Christ Church, Bath; Editor (since 1987), *Common Ground*; Chairman (since 1990), International Interfaith Organizations Coordinating Committee; Prebendary (since 1990), Wells Cathedral.

Career: Curate, St Michael's Highgate from 1964 to 1967. Vicar, Strood Clergy Team from 1967 to 1973. Rector, Swainswick, Langridge, Woolley from 1973 to 1979. Director of Training, Diocese of Bath and Wells from 1979 to 1984. Chairman, World Congress of Faiths from 1978 to 1983. Executive Director, Council of Christians and Jews from 1984 to 1987. Examining Chaplain to the Bishop of Bath and Wells from 1984 to 1988. **Educ.:** B.A., M.A. from University of Cambridge, Magdalene College, UK; Madras Christian College, India; Wells Theological College, UK; M.Phil. from University of London, King's College, UK; Stds. from Ecumenical Institute, Jerusalem, Israel (1977).

Interests: gardening, home decorating, tennis, swimming, travel, photography. **Hons.:** Named as a winner of the Sir Sigmund Sternberg Award for a Distinguished Contribution to Christian-Jewish Relations (1992). **Pubs.:** *Interfaith Organizations: A Historical Directory*, 1980; *Time to Meet*, 1990 and *Pilgrimage of Hope: A History of the Interfaith Movement*, SCM Press, 1991.
Mailing address: 2 The Bassetts, Box, Corsham SN14 9ER, UK. T. (44 225) 742827. Fax 742954

Bressan, Luigi, H E Mgr Dr

Rel: Roman Catholic; Ordained on 28 JUN 1964; b. 09 FEB 1940 in Sarche, Trento, Italy; celibate; Citizen of Vatican and Italy; Lang. Italian, Latin, German, English, French, Spanish and Portuguese.
International Organizations: Apostolic Pro-Nuncio to Pakistan (since 1989) of the Roman Catholic Church.
Other organizations: Titular Archbishop (since 1989), Severiana.
Main field of work: Ministry, international relations. **Career:** Parish Assistant, Archdiocese of Trent from 1964 to 1967. Secretary, Apostolic Nunciature in the Republic of Korea from 1971 to 1973. Secretary, Apostolic Nunciature in the Ivory Coast from 1974 to 1976. Officer, Vatican Secretariat of State from 1976 to 1978. Secretary, Permanent Mission of the Holy See to UN/Geneva from 1978 to 1982. Secretary, Apostolic Nunciature in Brazil from 1982 to 1983. Special Envoy of Holy See to Council of Europe/Strasbourg from 1983 to 1989. **Educ.:** Ph.D. Canon Law from Pontifical Gregorian University, Vatican City State (1967-1971); Dip. Diplomacy from Pontifical Ecclesiastical Academy, Vatican City State (1969-1971). **Interests:** international relations, religious freedom. **Pubs.:** Author of many articles and several books, including: *Divorzio per adulterio*, Roma, 1973; *Il divorzio nelle Chiese orientali*, Bologna, 1976 and *Libertà religiosa nel diritto internazionale*, Padova, 1989.
Mailing address: Apostolic Nunciature, Diplomatic Enclave, PO Box 1106, Islamabad 44000, Pakistan. T. (92 51) 210 491.
Residence: See Mailing Address. T. (92 51) 210 490

Bressanelli, Virginio Domingo, S.C.J., Rev

Rel: Roman Catholic; Ordained on 17 DEC 1966; b. in Berabevu, Santa Fé, Argentina; celibate; Citizen of Argentina; Lang. Spanish and Italian.
International Organizations: Superior General (since 1991) of the Priests of the Sacred Heart of Jesus.
Educ.: Pontifical Gregorian University, Vatican City State (1963-1967).
Mailing address: Via Casale S Pio V 20, I-00165 Roma, Italy. T. (39 6) 6638809 – 6638825. Fax 66000915.
Residence: See Mailing Address. T. See Mailing Address

Briddell, David

Rel: Christian.
International Organizations: Director of the Intermedia, New York NY.
Mailing address: 475 Riverside Drive, Room 670, New York NY 10115, USA. T. (1 212) 870 2376. C. Dominicus.

Brierley, Peter William

Rel: Church of England; Unordained; b. 30 OCT 1938 in London, UK; m. to Cherry Antoinette Goatman; c. 3 boys and 1 girl; Citizen of United Kingdom; Lang. English.
International Organizations: European Director (since 1983) of the MARC Europe.
Research Associate for Europe (since 1984) of the Lausanne Committee for World Evangelization. Member, Statistics Working Party (since 1989) of the Lausanne Committee for World Evangelization. Member, Executive Committee (since 1985) of the Evangelical Missionary Alliance.
Other organizations: Member (since 1969), Christ Church, Bromley, Kent; Various studies for churches (since 1969); Various studies for missionary societies (since 1970); Member (since 1981), Parochial Church Council; Chairman (since 1985), Southeast Asia Outreach; Council of Reference (since 1987), Spinnaker Trust; Co-opted Governor (since 1989), Bullers Wood Comprehensive School for Girls.
Main field of work: Church research and leadership training. **Career:** Actuarial Clerk, Prudential Assurance Co. Ltd. from 1957 to 1958. Scientific Officer, War Office, Whitehall, London from 1961 to 1962. Teacher, Southampton Education Authority from 1965 to 1967. Statistician, Ministry of Defence, Whitehall, London from 1967 to 1970. Statistician, Cabinet Office, Central Statistical Office, London from 1960 to 1978. Programme Director, Bible Society, London from 1978 to 1983. Member, England and Wales Auxiliary of Leprosy Mission from 1970 to 1976. Member, EUSA Finance Sub-Committee from 1971 to 1982. Member, Doughty Street International Students Centre Management Committee from 1971 to 1985. Member of both General and Executive Committees, British and Foreign Bible Society from 1976 to 1978. Lecturer on form design at Civil Service College from 1969 to 1988. Statistical Consultant to Opinion Research Centre from 1977 to 1981. Statistical Consultant to Urban Church Project from 1978 to 1981. Social Section, Royal Statistical Society from 1980 to 1982. Convener, Religious Researchers Group, Social Research Association from 1980 to 1983. Organising several conferences on the Presentation of Survey Reports from 1980 to 1984. Superintendent, Primary Sunday School, Christ Church from 1972 to 1975. Organiser, Overseas Students Weekends, Christ Church from 1975 to 1978. Member, Diocesan Stewardship and Resources Committee from 1981 to 1988. Director, Christian Booksellers Association from 1984 to 1987. Member, Mission Council, Christ Church, Bromley from 1988 to 1989. **Educ.:** B.S. Stat. (Hons) from University of London, University College, UK (1958-1961); Dip. Theol. from Bible Training Institute, Glasglow, UK (1962-1964); Dip. Bookselling (by correspondence) (1966-1969). **Mem.:** Social Research Association (1978, member of Executive Committee 1978-82), Market Research Society (1979, full member since 1984), British Institute of Management (1981) **Pubs.:** *UK Christian Handbook*, Part 1 in 1972, 1976, 1980, Part 2 in 1977, Parts 3 and 4 in 1980, all 4 Parts integrated in 1982, and other Christian Handbooks; *Vision Building*, Hodder and Stoughton, 1989 and *Priorities, Planning and Paperwork*.
Mailing address: Vision Building, 4 Footscray Road, Eltham, London SE9 2TZ, UK. T. (44 81) 294 1989. Fax 294 0014.

Briggs, Betty J

Rel: Mason.
International Organizations: Contact of the General Grand Chapter, Order of the Eastern Star.
Mailing address: 1618 New Hampshire Ave NW, Washington DC 20009, USA. T. (1 202) 667 4737.

Bright, William (Bill) Rohl

Rel: Protestant; b. in Coweta OK, USA.
International Organizations: Founder and President (since 1951) of the Campus Crusade for Christ International.
Career: Established Bright's California Confections, Los Angeles CA. Became associated with Hollywood's First Presbyterian Church. Made a promise to follow Jesus Christ in 1945. Studied at the Princeton Theological Seminary, NJ. Left Princeton and returned to California, where he entered the first class, Fuller Theological Seminary in 1947. Departed from Fuller without graduating in 1951. Sold his confection business and established Campus Crusade for Christ. Participated in the Lausanne International Congress on World Evangelism in 1974. **Educ.:** B.A. from Northeastern State College, OK, USA (1943). **Hons.:** several honorary degrees from colleges in the US and Korea. **Pubs.:** Author of nineteen books and booklets, including: *Four Spiritual Laws*.
Mailing address: 44-50 Arrowhead Springs, San Bernardino CA 92414, USA. T. (1 714) 886 5224 EXT 3337.

Bringmann, Jürgen

Rel: Roman Catholic; b. 04 OCT 1937 in Hannover, Niedersachsen, Germany; m. to Marie Luise; c. 1 boy and 1 girl; Citizen of Germany; Lang. German, English, French, Italian and Spanish.
International Organizations: President (since 1991) of the International Military Apostolate.
Mailing address: Katholisches Militärbischofsamt, Adenauerallee 115, 53113 Bonn, Germany. T. (49 221) 22 10 15
Residence: Hauptstrasse 148, 53347 Alfter, Germany. T. (49 221) 64 56 43

Brinkhues, Josef, Rev

Rel: Old Catholic; Ordained in 1937; b. 21 JUN 1913 in Aachen, Germany; m. to Dr Ilse Volckmar; c. 1 boy and 1 girl; Citizen of Germany; Lang. German, English and French.
International Organizations: Bishop Emeritus of Old-Catholic Church in Germany.
Career: Consecrated Bishop, Old-Catholic Church in 1966. **Educ.:** Johann Wolfgang Goethe-Universität Frankfurt, Germany (1932-1934); Rheinische Friedrich-Wilhelms-Universität Bonn, Germany (1934-1937). **Interests:** Modern philosophy, history of religions. **Mem.:** International Old Catholics Bishops' Conference of Utrecht Union (1966)

Mailing address: Alt-Katholische Kirche, Ordinariat, Gregor Mendel Strasse 28, 53115 Bonn, Germany.
Residence: Oberdorf 18, 53347 Impekoven/Alfter, Germany. T. (49 228) 64 33 01

Brister, George W S
Rel: Christian.
International Organizations: President of the International Alliance of Catholic Churches.
Mailing address: 1836 N W 15th Street, Oklahoma City OK 73106, USA. T. (1 405) 588 4009.

Brooker, Art
Rel: Christian.
International Organizations: Director of the Christian Blind Mission International.
Mailing address: 1506 E Roosevelt Rd, Wheaton IL 60187, USA. T. (1 312) 690 0300. Fax 690 0328.

Brookes, Graham F, Rev
Rel: Uniting Church in Australia; Ordained.
International Organizations: Secretary of the Uniting Church World Mission.
Mailing address: PO Box E 266, St James NSW 2000, Australia. T. (61 2) 287 0900. Fax 287 0999.

Brookes, Leonard
Rel: Protestant.
International Organizations: Contact of the Full Gospel Business Men's Fellowship International Europe.
Mailing address: Mechelse Steenweg 30, B-3000 Leuven, Belgium. T. (32 16) 20 79 33. Fax 20 79 31. Belgium. T. (32 2) 734 79 08 – 734 20 77. Fax 734 84 83.

Brose, Susanne
Rel: Methodist.
International Organizations: Vice-President for Europe of the World Federation of Methodist Women.
Mailing address: Dijonstrasse 10, 28211 Bremen, Germany.

Brotman, Manuel
Rel: Jewish.
International Organizations: President of The Messianic Jewish Movement International.
Mailing address: PO Box 30313, Bethesda MD 20814, USA. T. (1 301) 656 7575.

Brown, Alan Scott
Rel: Church of England; Unordained; b. 27 JUN 1944 in Newport, Gwent, Wales, UK; m. to Erica Annette Lawry; c. 1 boy and 1 girl; Citizen of United Kingdom; Lang. English.
International Organizations: President (since 1988) of the Inter-European Commission on Church and School.
Chairman (since 1988) of the United Society for Christian Literature.
Other organizations: Schools Officer (since 1980), Church of England Board of Education; Director (since 1985), National Society's Religious Centre; Director (since 1989), The Chichester Project; Chairman (since 1990), Sharp Working Party on World Religions in Education.
Main field of work: Religious education in schools. **Educ.:** B.A. Theol. from University of Leeds, UK (1962-1965); M.Phil. from University of Leeds, UK (1965-1967). **Pubs.:** The Christian World; Religious Education and the Pupil with Learning Difficulties and Religions.
Mailing address: The National Society's Religious Centre, 23 Kensington Square, London W8 5HN, UK. T. (44 71) 937 4241
Residence: 7 Elyham, Purley-on-Thames, Reading RG8 8EN, UK. T. (44 734) 843 664

Brown, Arthur Durrant, Rt Rev
Rel: Anglican; Ordained in 1950; b. 07 MAR 1926 in Guelph ON, Canada; m. to Norma Inez Rafuse; c. 3 girls; Citizen of Canada.
International Organizations: Bishop (since 1981) of York-Scarborough.
Other organizations: Suffragan Bishop (since 1981), Diocese of Toronto; Member (since 1961), Corporation of Huron College; Chairman (since 1982), Canadian Friends to West Indian Christians, Ontario; Chairman, Pension Committee (since 1983), Anglican Church of Canada; Honorary Chairman (since 1983), Canadian Foundation on Compulsive Gambling, Ontario; Member, Advisory Committee (since 1985), Ontario Homes for the Aging; Patron and Member, Advisory Committee (since 1984), Canadian Diabetes Association (Annual Fund Raising Campaign).
Career: Ordained Deacon in 1949. Rector, Paisley Pinkerton Cargill from 1948 to 1950. Rector, Glanworth and St Stephen's Church from 1950 to 1953. Rector, St John's Church from 1953 to 1963. Rector, St Michael and All Angels from 1963 to 1980. Rural Dean, Essex from 1960 to 1962. Canon, St James Cathedral in 1973. Archdeacon, York, Diocese of Toronto from 1974 to 1980. Broadcaster, 'Moments of Meditation' CFRB from 1964 to 1985. Columnist, Toronto *Sunday Sun* from 1974 to 1986. Member, Judicial Council, Province of Ontario from 1979 to 1985. Member, Ontario Press Council from 1986 to 1987. Chairman, Royal Visit Children's Fund in 1984. **Educ.:** B.A. from University of Western Ontario, London, Canada (1949); L.Th. from University of Western Ontario, Huron College, London, Canada (1949); D.D. from University of Western Ontario, Huron College, London, Canada (1979); D.D. from University of Toronto, Wycliffe College, Canada (1980). **Hons.:** Humanitarian Award (1990).
Mailing address: 3333 Finch Ave E, Agincourt ON M1W 2R9, Canada.
Residence: 5 Gossamer Ave, Willowdale ON M2M 2X1, Canada.

Brown, Carol, Rev
Rel: Presbyterian; Ordained; b. 20 MAY 1950 in Pittsburgh PA, USA; single; Citizen of USA; Lang. English and German.
International Organizations: Treasurer (since 1979) of the International Association of Women Ministers.
Other organizations: Associate Pastor (since 1990), First Presbyterian Church, Stroudsburg, PA.
Main field of work: Religion. **Career:** Pastor, First Presbyterian Church, Manchester and Garrard, KY from 1975 to 1983. Chair, Education Committee, Transylvania Presbytery from 1979 to 1983. Assistant Pastor, First Presbyterian Church, Stroudsburg, PA from 1983 to 1989. **Educ.:** B.A. from Slippery Rock University, PA, USA (1968-1972); M.Div. from Pittsburgh Theological Seminary, PA, USA (1972-1975).
Interests: Camping, canoeing.
Mailing address: 579 Main Street, Stroudsburg PA 18360, USA. T. (1 717) 421 7751.
Residence: See Mailing Address. T. See Mailing Address

Brown, F Reed, Rev
Rel: Spiritualist; Ordained.
International Organizations: President of the Spiritual Churches of Science and Revelation.
Mailing address: 5121 N 33rd Street, Arlington VA 22207, USA.

Brown, Florence
Rel: Protestant.
International Organizations: Executive Director of the Evangelical and Ecumenical Women's Caucus.
Mailing address: PO Box 209, Hadley NY 12835, USA. T. (1 518) 696 2406.

Brown, George W, Pastor
Rel: Seventh-Day Adventist; b. 11 JAN 1924 in Santo Domingo, Dominican Republic; m. to Carla Brown; c. 4 girls; Citizen of USA; Lang. English and Spanish.
International Organizations: President, Inter-American Division (since 1980) of the General Conference of Seventh-Day Adventists.
Main field of work: Administration. **Educ.:** B.Th. from University of the West Indies, Mona Campus, Kingston, Jamaica (1958-1960); M.Th. from Andrews University, Berrien Springs MI, USA (1960-1961). **Hons.:** Doctor of Divinity. **Pubs.:** Contributing Editor to: *Adventist Review*, International Edition; *El Centinela* and "Let's Talk it Over" (monthly column), in *Adventist Review*, Inter-American Edition.
Mailing address: SDA-Inter-American Division, 760 Ponce de Leon Blvd, Coral Gables FL 33134, USA. T. (1 305) 443 7471
Residence: 8450 SW 27th Lane, Miami FL 33155, USA T. (1 309) 221 3056

Brown, Harold Ogden Joseph, Rev Prof

Rel: Evangelical Reformed; Ordained on 07 DEC 1958; b. 06 JUL 1933 in Tampa FL, USA; m. to Grace Winifred Hancox; c. 1 boy and 1 girl; Citizen of USA; Lang. English, German, French, Italian and Swedish.
International Organizations: Professor of Theology and Ethics (since 1976) of the Trinity Evangelical Divinity School.
Other organizations: Chairman (since 1975), Christian Action Council; Director (since 1989), Rockford Institute Center on Religion and Society.
Main field of work: Seminary education. **Career:** Associate Editor, *Christianity Today* from 1972 to 1974. Pastor, Evangelical Reformed Church – Synod of the Grisons from 1983 to 1987. **Educ.:** A.B. from Harvard College, Cambridge MA, USA (1949-1953); S.T.B., S.T.M. from Harvard University, Divinity School, Cambridge MA, USA (1955-1959); Ph.D. from Harvard University, Cambridge MA, USA (1961-1967). **Interests:** ethics, skiing, rowing, moutaineering. **Pubs.:** Death before Birth, Thomas Nelson, 1976; The Reconstruction of the Republic, Arlington, 1977 and Heresies, Doubleday, 1984.
Mailing address: Trinity Evangelical Divinity School, 2065 Half Day Rd, Deerfield IL 60015, USA. T. (1 708) 317 8062
Residence: 5306 Amby Lane, Mc Henri IL 60050, USA. T. (1 708) 497 9013

Brown, John P E

Rel: Christian.
International Organizations: Secretary-General of the International Association for the Promotion of Democracy under God.
Mailing address: Via Pola 12, I-00198 Roma, Italy.

Brown, Kerry

Rel: Unknown.
International Organizations: Contact of the International Sacred Literature Trust.
Mailing address: Manchester Polytechnic, 9A Didsbury Park, Wilmslow Road, Manchester M20 OHL, UK. T. (44 61) 434 0828. Fax 434 8374.

Brown, Lindsay

Rel: Evangelical; Unordained; b. 27 DEC 1953 in Merthyr Tydfil, Mid-Glamorgan, Wales, UK; m. to Ann; c. 1 boy; Citizen of United Kingdom; Lang. English and French.
International Organizations: Secretary-General (since 1991) of the International Fellowship of Evangelical Students.
Other organizations: Member of the Board (since 1989), Evangelical Alliance of Wales.
Main field of work: University students. **Educ.:** M.A. from University of Oxford, UK (1973-1976); Theol. Stds. from Faculté Libre de Théologie, Vaux-sur-Seine, Hauts de Seine, France (1981-1982).
Mailing address: c/o IFES Admin Office, 55 Palmerston Road, Wealdstone, Harrow HA3 7RR, UK. T. (44 81) 863 8688. Fax 863 8229. C. FELINTER HARROW. Tx FELINTER HARROW.

Brown, Margrethe B J, Rev

Rel: Reformed Church; Ordained.
International Organizations: Secretary, Caribbean and North American Area Council of the World Alliance of Reformed Churches.
Mailing address: Room 4412, 100 Witherspoon Street, Louisville KY 40202-1396, USA. T. (1 502) 569 5431. Fax (1 502) 569 8005
Residence: T. (1 502) 423 9846 – (1 716) 288 3161

Brown, Robert McAfee

Rel: Protestant; Ordained; b. 28 MAY 1920 in Carthage IL, USA; m. to Sydney Elise Thomson; c. 3 boys and 1 girl; Citizen of USA.
International Organizations: Professor Emeritus of Theology and Ethics (since 1986) of the Pacific School of Religion.
Main field of work: Clergyman. **Career:** US Navy Chaplain, Pacific from 1945 to 1946. Assistant Chaplain, Instructor in Religion, Amherst College, MA from 1946 to 1948. Instructor, Union Theological Seminary, NY from 1950 to 1951. Professor and Head of Dept. of Religion, Macalester College, MN from 1951 to 1953. Auburn Professor of Systematic Theology, Union Theological Seminary, New York NY from 1953 to 1962. Professor of Religious Studies, Stanford University, CA from 1962 to 1976. Professor of Ecumenics and World Christianity, Union Theological Seminary, New York NY from 1976 to 1979. Professor of Theology and Ethics, Pacific School of Religion from 1979 to 1986. Protestant Observer at Second Vatican Council (1963 and 1965), representing the World Alliance of Presbyterian and Reformed Churches. Delegate to 4th Assembly, World Council of Churches, Uppsala in 1968. Speaker, General Assembly, Presbyterian Church of Southern Africa, Durban in 1972. Delegate to Conference of the Fund for Reconstruction and Reconciliation in Indochina (World Council of Churches), Vientiane in 1975. Advisor and keynote speaker at 5th Assembly, World Council of Churches, Nairobi in 1975. Member, Board, Institute for Advanced Religious Studies, Tantur, Jerusalem from 1965 to 1973. Member, United States Holocaust Memorial Council from 1978 to 1986. **Educ.:** B.A. from Amherst College, MA, USA (1943); M.Div. from Union Theological Seminary, New York NY, USA (1945); Ph.D. from Columbia University, New York NY, USA (1951). **Interests:** cello, carpentry, grandchildren. **Hons.:** Hon. D.D., Amherst College (1958), Pacific School of Religion (1967), Kenyon College (1981), Hebrew Union College (1982), Lehigh University (1988); Hon. D.Litt., University of San Francisco (1964); Hon. L.H.D., Lewis and Clark College (1964), St Lewis University (1966); Hon. LL.D., Loyola University (1965), University of Notre Dame (1965), Boston College (1965), St Mary's College (1968). **Pubs.:** *Saying Yes and Saying No: On Rendering to God and Caesar*, Westminster, 1986; *Gustavo Gutierrez: An Introduction to Liberation Theology*, Orbis Books, 1990 and contributor to various symposia, journals, etc.
Mailing address: 2090 Columbia Street, Palo Alto CA 94306, USA. T. (1 415) 857 0870

Brown, Stewart Jay, Prof

Rel: Christian; b. 08 JUL 1951; m. to Teri Beth Hopkins; c. 1 boy and 1 girl.
International Organizations: Professor of Ecclesiastical History (since 1988) of the University of Edinburgh.
Main field of work: University education. **Career:** Whiting Fellow in Humanities, University of Chicago, IL from 1979 to 1980. Assistant to the Dean and Lecturer in History, Northwestern University, Evanston IL from 1980 to 1982. Associate Professor and Assistant Head of Dept. of History, University of Georgia, Athens from 1982 to 1988. **Educ.:** B.A. from University of Illinois, USA; M.A., Ph.D. from University of Chicago, IL, USA. **Interests:** swimming, hill-walking. **Mem.:** Scotland Church History Society (Council) **Pubs.:** Thomas Chalmers and the Godly Commonwealth in Scotland, 1982.
Mailing address: University of Edinburgh, New College, Mound Place, Edinburgh EH1 2LU, UK. T. (44 31) 225 8400
Residence: 160 Craigleith Hill Ave, Edinburgh EH4 2NB, UK. T. (44 31) 343 1712

Brown, Stuart E, Dr

Rel: Protestant.
International Organizations: Secretary-General of the Inter-Church Committee for Refugees.
Mailing address: 40 St Clair Avenue East, Suite 201, Toronto ON M4T LM9, Canada. T. (1 416) 921 41 52. Fax 921 7478. C. CANOIKOS TORONTO FOR ICCR. Tx 6524128 (for iccr).

Brown, Wendy

Rel: Christian.
International Organizations: Head, Secretariat of the Central American Mission Partners.
Mailing address: 2576 San Pablo Ave, Oakland CA 94612, USA. T. (1 415) 451 0919.

Brown, Winifred, Sr

Rel: Roman Catholic; Ordained; celibate.
International Organizations: Secretary-General of the Sisters of Our Lady of the Missions.
Mailing address: Via di Bravetta 628, I-00164 Roma, Italy. T. (39 6) 6615 8400 – (39 6) 6615 8348.

Browne, Bartholomew, F.P.M., Br

Rel: Roman Catholic; Ordained; celibate.
International Organizations: Procurator-General of the Presentation Brothers.
Mailing address: Mount St Joseph, Blarney Street, Cork, Ireland. T. (353 21) 43160.

Browne, George D, Most Rev
Rel: Church of the Province of West Africa; Ordained.
International Organizations: Archbishop of Church of the Province of West Africa.
Mailing address: Bishopscourt, PO Box 277, Monrovia, Liberia.

Browning, Edmond L, Most Rev
Rel: Episcopal; Ordained; b. 11 MAR 1929 in Texas, USA; m. to Patricia Sparks; c. 4 boys and 1 girl; Citizen of USA.
International Organizations: Presiding Bishop (since 1985) of Episcopal Church, USA.
Career: First missionary Bishop, Okinawa from 1968 to 1971. Bishop-in-Charge, Convocation of American Churches in Europe from 1971 to 1974. Executive for National and World Mission, Episcopal Church Center from 1974 to 1976. Bishop, Hawaii from 1976 to 1985. **Educ.:** University of the South, Sewanee TN, USA; B.D. from University of the South, School of Theology, Sewanee TN, USA; Japanese Language School, Okinawa, Japan.
Mailing address: 815 Second Ave, New York NY 10017, USA. T. (1 212) 867 8400

Brubacher, Ray
Rel: Mennonite.
International Organizations: Director of Overseas Services of the Mennonite Central Committee.
Mailing address: 21 South 12th Street, Akron PA 17501, USA.

Brubaker, A Graybill, Rev
Rel: Christian; Ordained.
International Organizations: Past Executive Director of the Brethren in Christ World Missions.
Mailing address: 500 S Angle St, PO Box 390, Mount Joy PA 17552, USA. T. (1 717) 653 8067. Fax 653 9370. C. BICMISSION.

Bruce, Florence
Rel: Roman Catholic.
International Organizations: Representative at UN, Geneva of the International Catholic Child Bureau.
Mailing address: 65 rue de Lausanne, CH-1202 Genève, Switzerland.

Brunk, George R, III, Dr
Rel: Mennonite.
International Organizations: Dean of the Eastern Mennonite Seminary.
Mailing address: Harrisonburg VA 22801-2462, USA.

Brunot, Jean-Marie
Rel: Roman Catholic.
International Organizations: President of the International Catholic Union of the Press.
Member of the Pontifical Council for Social Communications.
Mailing address: 'Bayard Presse', 3-5 rue Bayard, F-75393 Paris CEDEX 08, France.

Brunson, Douglas
Rel: Christian.
International Organizations: General Manager of the Ecumenical Development Cooperative Society.
Mailing address: P C Hooftlaan 3, NL-3818 EB Amersfoort, Netherlands. T. (31 33) 63 31 22. Fax 65 03 36. Tx 20010 PMS NL – EDCS.

Brusco, Angelo, M.I., Fr
Rel: Roman Catholic; Ordained; celibate.
International Organizations: Superior General of the Order of the Servants of the Sick.
Mailing address: Piazza della Maddalena 53, I-00186 Roma, Italy. T. (39 6) 679 77 96 – 679 77 97.

Brutosky, Veronica, Sr
Rel: Roman Catholic; Ordained; celibate.
International Organizations: Director of the International Registry for Religious Women Artists.
Mailing address: 1315 N Van Ness Ave, Fresno CA 93728-1937, USA. T. (1 209) 266 3812.

Buchanan, George Duncan, Rt Rev
Rel: Anglican; Ordained on 23 MAY 1959; b. 23 APR 1935 in Delmore, Transvaal, South Africa; m. to Diana Margaret Dacombe; c. 2 girls; Citizen of South Africa; Lang. English, Afrikaans and Xhosa.
International Organizations: Bishop (since 1986) of Johannesburg.
Other organizations: Chairman (since 1989), Church Unity Commission.
Main field of work: Ministry. **Career:** Assistant Priest, St Paul's Church, Durban from 1961 to 1962. Rector, Parish of Kingsway from 1962 to 1965. Sub-Warden, St Paul's Theological College, Grahamstown from 1966 to 1975. Warden, St Paul's Theological College, Grahamstown from 1976 to 1986. Dean, Johannesburg in 1986. Bishop, Johannesburg in 1986. **Educ.:** B.A. from Rhodes University, Grahamstown, South Africa (1954-1956); General Ord. Exams from Rippon Hall, Oxford, UK (1957-1959); M.Div. from General Theological Seminary, NY, USA (1959-1961). **Interests:** carpentry, music, English literature, tennis, hockey. **Hons.:** Hon. D.D. **Mem.:** Rand Club **Pubs.:** The Counselling of Jesus (The Jesus Library), Hodder and Stoughton, 1985.
Mailing address: PO Box 1131, Johannesburg 2000, South Africa. T. (27 11) 29 8724
Residence: 4 Crescent Drive, Westcliff, Johannesburg 2193, South Africa. T. (27 11) 486 1014

Bucherl, Gisela
Rel: Buddhist: Mahayana.
International Organizations: Member of the Foundation for the Preservation of the Mahayana Tradition.
Other organizations: Manager, Aryatara Institut.
Mailing address: Holzhamerstrasse 5, 94424 Arnstorf, Germany. T. (49 8723) 2396 OR (49 8723) 8382

Buckley, Francis Joseph, S.J., Rev
Rel: Roman Catholic; Ordained on 18 JUN 1958; b. 31 AUG 1928 in Los Angeles CA, USA; celibate; Citizen of USA; Lang. English, Latin, Greek, French, Spanish and German.
International Organizations: Professor of Systematic and Pastoral Theology (since 1960) of the University of San Francisco, CA.
Member of the Society of Jesus.
Other organizations: Chaplain (since 1963), St Elizabeth Hospital; Associate Director, Graduate Programs in Religious Education and Pastoral Ministry (since 1991), University of San Francisco, CA.
Main field of work: University teaching and research. **Career:** Chairman, Dept. of Theology, University of San Francisco, CA from 1971 to 1973. President, College Theology Society from 1972 to 1974. Director, Graduate Programs in Religious Education, USF from 1974 to 1975. Trustee, Loyola Marymount University from 1974 to 1989. Chairman, Dept. of Theology, University of San Francisco, CA from 1978 to 1979. Director, Graduate Programs in Religious Education, USF from 1979 to 1982. Trustee/Chairman, Pacific Graduate School of Psychology from 1984 to 1986. Director, Graduate Programs in Religious Education, USF from 1986 to 1987. Chairman, Dept. of Theology, University of San Francisco, CA from 1988 to 1991. Director, Graduate Programs in Religious Education, USF from 1992 to 1993. **Educ.:** University of Notre Dame, IN, USA (1944-1945); Santa Clara University, CA, USA (1945-1949); B.A. from Gonzaga University, Spokane WA, USA (1949-1951); M.A. from Gonzaga University, Spokane WA, USA (1951-1952); S.T.L. from Alma College, CA, USA (1955-1959); S.T.M. from Santa Clara University, CA, USA (1955-1959); S.T.D. from Pontifical Gregorian University, Vatican City State (1961-1964); Pontifical Biblical Institute, Vatican City State (1961-1963). **Interests:** religious education, confirmation and penance, computers. **Hons.:** Hon. D.H.L., Pacific Graduate School of Psychology (1988). **Mem.:** College Theology Society; Jesuit Ecumenists; Catholic Theological Society **Pubs.:** Children and God: Communion, Confession, Confirmation, Corpus, 1970; I Confess: The Sacrament of Penance Today, Ave Maria, 1972 and Deepening Christian Life, Harper and Row, 1987.
Mailing address: University of San Francisco, 650 Parker Ave, San Francisco CA 94118, USA. T. (1 415) 666 0123
Residence: See Mailing Address. T. See Mailing Address

Buconyori, Elie, Dr
Rel: Christian.
International Organizations: Managing Director of the Christian Learning Materials Centre, Nairobi.
Mailing address: PO Box 24345, Karen, Nairobi, Kenya. T. (254 2) 882667 – 882048. Fax 710254.

Bugin, Fiorenzo, O.C.D., Fr
Rel: Roman Catholic; Ordained; celibate.
International Organizations: Procurator-General of the Order of Discalced Brothers of the Blessed Virgin Mary of Mount Carmel.
Mailing address: Corso d'Italia 38, I-00198 Roma, Italy. T. (39 6) 841 65 78 – 884 09 58.

Bugotu, Ambrose H
Rel: Church of Melanesia.
International Organizations: Provincial Secretary of the Church of Melanesia.
Mailing address: PO Box 19, Honiara, Solomon Is.

Bukovsky, John, S.V.D., H E Mgr
Rel: Roman Catholic; Ordained on 03 DEC 1950; b. 18 JAN 1924 in Cerova; celibate.
International Organizations: Apostolic Nuncio to Romania (since 1990) of the Roman Catholic Church.
Member of the Society of the Divine Word.
Other organizations: Titular Archbishop, Tabalta.
Mailing address: Str Pictor C Stahi 5-7, 70764 Bucuresti, Romania. T. (40 0) 13 94 90

Bulaitis, Giovanni, H E Mgr
Rel: Roman Catholic; Ordained on 02 FEB 1958; b. 26 JUN 1933 in London, UK; celibate.
International Organizations: Apostolic Pro-Nuncio to Iran (since 1987) of the Roman Catholic Church.
Other organizations: Titular Archbishop, Narona.
Mailing address: Rue Razi 97, Carrefour Av de France, BP 11365-178, Teheran, Iran Islamic Rep. T. (98 21) 640 35 74

Bulgaria, Maxim of, H H Patriarch
Rel: Bulgarian Orthodox; Ordained.
International Organizations: Patriarch of Bulgarian Orthodox Church.
Mailing address: Bulgarian Patriarchate, The Holy Synod, 4 Oborishte St, PO Box 376, 1090 Sofia, Bulgaria.

Bulka, Reuven Pinchas, Rabbi Dr
Rel: Jewish; Ordained in 1965; b. 06 JUN 1944 in London, UK; m. to Naomi Jakobovits; c. 3 boys and 2 girls; Citizen of Canada; Lang. English, Hebrew and Yiddish.
International Organizations: Rabbi (since 1967) of Congregation Machzikei Hadas, Ottawa.
Other organizations: Vice-President (since 1990), Rabbinical Council of America (RCA); Chairman, RCA Publications Committee.
Career: Chairman, Rabbinical Council of America Family and Marriage Committee from 1976 to 1988. Editor, *Family and Marriage Newsletter*, RCA from 1976 to 1988. Section Editor, *RCA Sermon Manual* from 1977 to 1982. Section Editor, *RCA Sermon Manual* in 1986. Founder/Editor, *Journal of Psychology and Judaism*. Member, Editorial Boards, *Tradition, Journal of Religion and Health, International Forum for Logotherapy, Pastoral Psychology*. Chairman, Rabbinic Cabinet, State of Israel Bonds of Canada. National Chairman, Jews in Arab/Moslem Lands Committee. Co-Chairman, Ottawa Soviet Jewry Committee. Advisor to Bereaved Families of Ontario. Member, Pastoral Care Advisory Committee, Children's Hospital of Eastern Ontario. Member, League for Human Rights, B'nai Brith Canada. **Educ.:** B.A. from City University of New York, USA (1961-1965); M.A. from University of Ottawa, Ontario, Canada (1968-1969); Ph.D. from University of Ottawa, Ontario, Canada (1970-1971). **Hons.:** Man of the Year, Israel Bonds (1971); Distinguished Service Award, United Jewish Appeal (1980). **Pubs.:** *What You Thought You Knew About Judaism: 341 Common Misconceptions about Jewish Life*, Northvale, NJ, Jason Aronson, 1989; *Uncommon Sense for Common Problems*, Toronto, Lugus Productions, 1990 and contributed scholarly and popular articles to various journals, including: *AMHC Forum, Analecta Frankliana, Chronicle Review, Hadarom, Humanitas, International Forum for Logotherapy, Journal of Ecumenical Studies, Journal of Halacha and Contemporary Society, Journal of Humanistic Psychology, Journal of Religion and Health, Judaism, L'Eylah, Midstream, Pastoral Psychology, Tradition,* and *Voices*.
Mailing address: Congregation Machzikei Hadas, 2310 Virginia Drive, Ottawa, Ontario K1H 6S2, Canada. T. (1 613) 521 9700
Residence: 1747 Featherston Drive, Ottawa, Ontario K1H 6P4, Canada. T. (1 613) 731 9119

Bunchoem, Prasong
Rel: Buddhist.
International Organizations: Vice-President and Member of the Executive Council of the World Fellowship of Buddhists.
Mailing address: 33 Sukhumvit Road, Bangkok 10110, Thailand. T. (66 2) 251 1188. C. WORLBUDIST BANGKOK. Fax (66 2) 255 7363.

Bunskoek, L J
Rel: Roman Catholic.
International Organizations: Director of the Centre Contact between the Continents.
Mailing address: Amersfoortsestraat 20, 3769 AS Soesterberg, Netherlands. T. (31 3463) 1755.

Buonaiuto, G
Rel: Roman Catholic.
International Organizations: President of the Groupe de laïcs pour le Tiers-monde.
Mailing address: Via Marechiaro 42/46, I-80123 Napoli, Italy. T. (39 81) 769 52 36 – (39 81) 769 18 52.

Burden, Jasper Jacobus, Prof
Rel: Dutch Reformed; Ordained on 10 OCT 1959; b. 02 JAN 1933 in Pretoria, South Africa; m. to Annette; c. 1 boy and 4 girls; Citizen of South Africa; Lang. Afrikaans, English and Xhosa.
International Organizations: Dean (since 1987) of the Faculty of Theology, University of South Africa.
Other organizations: Professor of Old Testament (since 1978), University of South Africa; Treasurer (since 1985), Southern Africa Biblical Studies Society; Editor (since 1986), Old Testament Essays.
Main field of work: Ministry, university education. **Career:** Minister, Dutch Reformed Church from 1960 to 1967. Professor, University of South Africa (UNISA) from 1967 to 1975. Senior Lecturer, UNISA from 1976 to 1977. Associate Professor, UNISA from 1977 to 1978. Head, Old Testament Dept., UNISA from 1982 to 1985. **Educ.:** B.A. from University of Stellenbosch, South Africa (1952-1954); M.A. from University of Stellenbosch, South Africa (1955-1957); B.D. from University of Stellenbosch, South Africa (1955-1958); D.Litt. from University of Stellenbosch, South Africa (1970-1971). **Interests:** Old Testament, missiology. **Hons.:** Andrew Murray Prize. **Mem.:** Old Testament Society of South Africa (OTSSA); Southern Africa Biblical Studies Society (SABSS); SAMS **Pubs.:** An ABC of Biblical Exegesis, Pretoria, Van Schaik; Reconsidering Parallelism in the Old Testament, OTE 4, pp. 141-176 and Psalms 101-119: 'n Verklaring, Kaapstad, Lux Verbi.
Mailing address: University of South Africa, PO Box 392, Pretoria 0001, South Africa. T. (27 12) 429 4567
Residence: 521 Richard Street, Pretoria 0083, South Africa. T. (27 12) 43 57 26

Burger, Stephen E, Rev
Rel: Protestant; Ordained; b. 23 AUG 1940 in St Paul MN, USA; m. to Delores Prokofsky; c. 1 boy and 1 girl; Citizen of USA; Lang. English.
International Organizations: Executive Director and Past President (since 1989) of the International Union of Gospel Missions.
Other organizations: Board Member, Christian Ministry Management Association; Board Member, Evangelical Council for Financial Accountability.
Career: Director, Ober Boys Club and Savage Boys Club, operated by the Union Gospel Mission of St Paul MN from 1959 to 1964. Youth Director, City Rescue Mission, New Castle PA from 1964 to 1969. Superintendent, York Rescue Mission, York PA from 1969 to 1974. Executive Director, Union Gospel Mission, Seattle WA from 1974 to 1989. Board of Directors, Rotary Club of Seattle WA from 1987 to 1989. Community Service Chair, Rotary Club of Kansas City MO from 1990 to 1991. Board Member, York Hospital, York PA. Board Member, Seattle Christian School. President, Pioneer Square Association, Seattle WA. President, Seattle Association of Evangelicals. President, North Pacific District, International Union of Gospel Missions. President, Council of Community Services, New Castle PA. **Educ.:** Hamline University, St Paul MN, USA; Bethel College and Seminary, St Paul MN, USA. **Interests:** travel, volleyball. **Hons.:** Clergyman of the Year, American Rescue Workers, Washington DC (1990), Service Award, Rotary Club of Seattle WA (1989), 100 Future Newsmakers, TIME Magazine and Seattle Chamber of Commerce (1978), St Matthews Award, Northwest Harvest. **Mem.:** Rotary Club

Mailing address: 1045 Swift, North Kansas City MO 64116, USA. T. (1 816) 471 8020
Residence: 4023 N Bennington, Suite 102, Kansas City MO 64117, USA. T. (1 816) 454 8190

Burghardt, Walter J

Rel: Roman Catholic; b. 10 JUL 1914 in New York, USA; Citizen of USA.
International Organizations: Professor of Theology of the Georgetown University, Washington DC.
Other organizations: Co-Editor (since 1958), *Ancient Christian Writers*; Editor (since 1967), *Theological Studies*; Research Associate (since 1974), Woodstock Theological Center, Washington DC; Professor Emeritus (since 1978), The Catholic University of America, Washington DC; Member, Board of Patrons (since 1981), Ecumenical Institute for Advanced Theological Studies in Jerusalem (Tantur); Chairman, Committee on Theology and Ethics (since 1985), The Catholic Health Association of the US; Member (since 1986), Corporation Jesuit International Volunteers Inc.
Main field of work: Education. **Career:** Professor of Patristic (Historical) Theology, Woodstock College from 1946 to 1974. Lecturer, Union Theological Seminary, New York City from 1971 to 1974. Professor of Patristic Theology, The Catholic University of America from 1974 to 1978. Visiting Lecturer in Theology, Princeton Theological Seminary from 1972 to 1973. Managing Editor, *Theological Studies* from 1946 to 1967. **Educ.:** M.A., Ph.L. from Woodstock College; S.T.L., S.T.D. from Catholic University of America, Washington DC, USA. **Hons.:** Mariological Award (1985); Cardinal Spellman Award (1962); several honorary degrees. **Mem.:** American Theological Society; International Association for Patristic Studies; Catholic Commission on Intellectual and Cultural Affairs; Catholic Theological Society of America; Mariological Society of America; North American Academy of Ecumenists; North American Patristic Society **Pubs.:** Musings on the Human Journey, 1983; Preaching: The Art and Craft, 1987 and author of many booklets, radio and TV programmes, lectures.
Mailing address: Office of Theologian in Residence, Georgetown Univ., Washington DC 200057, USA.

Burghoorn, A, Rev

Rel: Evangelical Lutheran; Ordained.
International Organizations: President of the Evangelical Lutheran Church in the Kingdom of the Netherlands.
Mailing address: Jan de Bakkerstraat 13-15, NL-3441 ED Woerden, Netherlands.

Burkholder, Mary

Rel: Mennonite.
International Organizations: Executive Secretary of the Mennonite Church Eastern Conference.
Mailing address: Apt 201, 1414 King St E, Kitchener ON N2G 2N7, Canada.

Burnett, Bill Bendyshe, Most Rev Dr

Rel: Anglican; Ordained on 21 DEC 1946; b. 31 MAY 1917 in Grahamstown, Cape Province, South Africa; m. to Sheila Fulton Trollip; c. 2 boys and 1 girl; Citizen of South Africa; Lang. English.
International Organizations: Ecclesiastic (retired) of the Anglican Church.
Main field of work: Preaching, lecturing, ministering. **Career:** Army Service from 1940 to 1945. Chaplain, Michaelhouse School, Balgowan, Natal from 1950 to 1954. Rector, All Saints Church, Ladysmith from 1954 to 1957. Bishop, Bloemfontein from 1957 to 1966. General Secretary, South African Council of Churches from 1966 to 1970. Bishop, Grahamstown from 1970 to 1974. Archbishop, Cape Town from 1974 to 1981. **Educ.:** M.A. Hist. from Rhodes University, Grahamstown, South Africa (1936-1939); Dip. Theol. from St Paul's College, Grahamstown, South Africa (1945-1946); P.G. Stds. at Queen's Theological College, Birmingham, UK (1949-1950). **Hons.:** Hon. D.D., Rhodes University (1980). **Pubs.:** *By My Spirit*, Hodder and Stoughton, 1988; The Seven Letters, 1989-1990 and article in *Bishops Move*, 1978.
Mailing address: 20 Milner Street, Grahamstown 6140, South Africa. T. (27 461) 27703

Burniak, Marian, T.Chr., Fr

Rel: Roman Catholic; Ordained; celibate.
International Organizations: Procurator-General of the Società di Cristo per gli Emigrati della Polonia.
Mailing address: Via delle Botteghe Oscure 15, I-00186 Roma, Italy. T. (39 6) 678 82 28, (39 6) 679 53 47

Burnier, Radha, Dr

Rel: Theosophist.
International Organizations: President of the Theosophical Society.
Mailing address: Adyar, Madras 600 020, India. T. (91 44) 412815. C. Olcott, Madras.

Burns, John Henri

Rel: Christian.
International Organizations: Secretary of the Non-Denominational Bible Prophecy Study Association.
Mailing address: 339 E Laguna Drive, Tempe AZ 85282, USA. T. (1 602) 967 30 66.

Burrows, Eva Evelyn, Gen

Rel: Salvation Army, The; b. 15 SEP 1929 in Newcastle NSW, Australia; Citizen of Australia; Lang. English.
International Organizations: General and World Leader (since 1986) of The Salvation Army.
World President of the Salvation Army Medical Fellowship. World President of the Salvation Army Women's Organizations. President of the Salvation Army Home League.
Career: Missionary Educator, Howard Institute from 1952 to 1957. Consultant on curriculum, Government Education Authority (until 1967). Principal, Usher Institute (Girl's Secondary School and Domestic Science Teacher Training College) from 1967 to 1969. Vice-Principal, International College for Officers from 1970 to 1973. Principal, International College for Officers from 1974 to 1975. Leader, Women's Social Services in Great Britain and Ireland from 1975 to 1977. Territorial Commander (National Leader), The Salvation Army, Sri Lanka from 1977 to 1979. Territorial Commander (National Leader), The Salvation Army, Scotland from 1979 to 1982. Territorial Commander (National Leader), The Salvation Army, Australia from 1982 to 1986. **Educ.:** B.A. Hist. and Engl. from University of Queensland, Brisbane, Australia (1950); Theol. training leading to Ordination from William Booth Memorial Training College, London, UK (1951); P.G.Cert. Ed. at University of London, UK (1952); M.Ed. from University of Sydney, New South Wales, Australia (1959). **Interests:** classical music, reading, travelling. **Hons.:** Officer of the Order of Australia for service to humanity (1986); Hon. Degree of Doctor of Liberal Arts, EWHA Women's University of Seoul, Korea (1988); Hon. LL.D., Asbury College, USA (1988).
Mailing address: 101 Queen Victoria Street, PO Box 249, London EC4P 4EP, UK. T. (44 71) 236 5222. C. SALVATION LONDON EC4. Tx 8954847. Fax (44 71) 236 4981

Burtchaell, James Tunstead, Rev

Rel: Roman Catholic; Ordained; b. 31 MAR 1934 in Portland OR, USA; celibate; Citizen of USA.
International Organizations: Ecclesiastic of the Roman Catholic Church.
Other organizations: Member, Faculty (since 1966), University Notre Dame, IN; Professor of Theology (since 1975), University Notre Dame, IN.
Career: Chairman, Dept. of Theology, University of Notre Dame, IN from 1968 to 1970. Provost, Trustee, University of Notre Dame, IN from 1970 to 1977. President, American Academy of Religion from 1970 to 1971. Visiting Fellow, Princeton University, NJ from 1980 to 1981. Sr Fulbright Fellow, University of Cambridge from 1985 to 1986. **Educ.:** S.T.L. from University of Notre Dame, IN, USA; S.S.L. from Pontifical Gregorian University, Vatican City State; Ph.D. from Catholic University of America, Washington DC, USA; Ecole Biblique et Ecole Archéologique Française, Jerusalem, Israel; University of Cambridge, UK. **Interests:** classical music, travel, contention. **Hons.:** Honorary L.H.D., St Mary's College, Rose-Hulman Institute. **Pubs.:** For Better for Worse, 1985; There is No More Just War: The Teaching and Trial of Don Lorenzo Milani, 1988 and The Giving and Taking of Life: Essays Ethical, 1989.
Mailing address: Dept. of Theology, University of Notre Dame, Notre Dame IN 46556, USA. T. (1 219) 239 7662

Burton, Ernest James, Rt Rev

Rel: Liberal Catholic; Ordained on 21 DEC 1932; b. 13 OCT 1908 in Crayford, Kent, UK; m. to Jean Margaret Allingham; c. 1 boy and 2 girls; Citizen of United Kingdom; Lang. English, French and Latin.
International Organizations: Regionary Bishop Emeritus (since 1968) of Liberal Catholic Church Province of Great Britain and Ireland.
Main field of work: Ministry, university examiner and consultant, liturgical scholar. **Career:** Divinity Teacher at Chelmsford Grammar School from 1931 to 1942. Divinity Teacher at Latimer's School, Edmonton from 1945 to 1952. Public Preacher, Chelmsford Diocese from 1950 to 1961. Principal Lecturer, Trent Park College of Education from 1952 to 1968. Lecturer, University of London from 1952 to 1968. **Educ.:** B.A. (1st) from University of London, UK (1928); M.A. from University of London, UK (1930); Anglican Ord. Exam. (1931). **Interests:** comparative religion, diction, voice and speech in the church. **Hons.:** Quain Research Essay Prize (1930). **Mem.:** Fellow, Royal Society of Arts; Society of Teachers of Speech and Drama (Hon. Member) **Pubs.:** The Communication of Religious Experience, St Alban Press and Calgary University Press; The Liberal Catholic Church and the Apostolic Succession, St Radigund's Press and Reality and Realization – An Introduction to a Philosophy.
Mailing address: Liberal Catholic Church, Drayton House, 30 Gordon Street, London WC1H OBE, UK. T. (44 71) 387 3260
Residence: The Roseries, Monks Horton, Sellindge, Ashford TN25 6EA, UK. T. (44 303) 81 2138.

Burton-Stibben, Arthur

Rel: Buddhist.
International Organizations: Member of the Buddhist Union of Europe.
Mailing address: 58 Eccleston Square, London SW1Y 1PH, UK. T. (44 71) 828 9242.

Burua, Albert, Rev

Rel: United Church in Papua New Guinea and Solomon Islands; Ordained.
International Organizations: Moderator of the United Church in Papua Guinea and the Solomon Islands.
Mailing address: Assembly Office, PO Box 3401, Port Moresby, Papua New Guinea.

Buschlüter, Ursula

Rel: Old Catholic.
International Organizations: President of the Federation of the Old Catholic Church in the ex-GDR.
Mailing address: Georgstrasse 7, 38889 Blankenburg/Harz, Germany.

Buscicchio, Mavi

Rel: Buddhist: Mahayana.
International Organizations: Member of the Foundation for the Preservation of the Mahayana Tradition.
Other organizations: Director, Universal Education – Italy.
Mailing address: Via San Bernardino 16, I-38100 Trento, Italy. T. (39 461) 233 080

Buss, Peter Martin, Rt Rev

Rel: Swedenborgian; Ordained on 19 JUN 1964; b. 08 OCT 1940 in Springs, Transvaal, South Africa; m. to Elisabeth Sandstrom; c. 2 boys and 3 girls; Citizen of USA; Lang. English, Zulu, Swedish and Afrikaans.
International Organizations: Executive Bishop (since 1991) of General Church of the New Jerusalem.
Other organizations: Chancellor (since 1991), Academy of the New Church, Bryn Athyn PA; President (since 1991), General Church in Canada; President, General Church International, Inc.
Main field of work: Ministry. **Career:** Assistant to Pastor, Olivet Church, Toronto in 1964. Superintendent, South African Mission in 1965. Visiting Pastor to isolated members and groups in South Africa in 1965. Pastor, Durban Society in 1971. Pastor, Glenview (IL) Society in 1976. President, Midwestern Academy of the New Church in 1976. Headmaster, Immanuel Church School in 1976. Bishop's Representative in the Midwestern District in 1976. President, Midwestern Academy of the New Church, in charge of the Central Western District in 1977. Bishop's Representative in the Midwestern District and Central Western District in 1977. President, Academy of the New Church in 1983. Assistant Bishop, General Church of the New Jerusalem in 1987. Head, Evangelization in 1987. Vice-President, General Church of the New Jerusalem Corporation in 1987. **Educ.:** B.A. from University of Natal, Durban, South Africa (1957-1961); M.Div. from Academy of the New Church Theological School, USA (1961-1964). **Pubs.:** Before Marriage and After Marriage – Notes on Conjugial Love for Senior Boys Religion and Junior Girls Religion; "Principles of Government" – an address to the Academy, October 1990, published in *New Church Life*, March-June 1991 and "Forces Which Influence and Order the Human Mind".
Mailing address: General Church of the New Jerusalem, 801 Tomlinson Road, Box 711, Bryn Athyn PA 19009, USA. T. (1 215) 938 2620
Residence: 3079 Huntingdon Pike, Bryn Athyn PA 19009, USA. T. (1 215) 947 5679

Bustros, Cyrille Selim, M.S.S.P., H E Mgr

Rel: Greek Catholic; Ordained on 29 JUN 1962; b. 29 JAN 1939 in Ain-Bourday, Lebanon; celibate.
International Organizations: Archbishop (since 1988) of Baalbek. Member of the Missionary Society of St Paul.
Mailing address: Greek-Catholic Bishopric, Baalbek, Lebanon. T. (961) 870 200

Butcher, Pauline

Rel: Christian.
International Organizations: Contact of the Women's Inter-Church Council.
Mailing address: 27 Tavistock Square, London WC1H 9HH, UK.

Buthelezi, Manas, Rt Rev

Rel: Evangelical Lutheran; Ordained; b. 10 FEB 1935 in Mahlabathini, South Africa; m. to Grace Mhlungu; c. 2 boys and 2 girls; Citizen of South Africa.
International Organizations: Bishop (since 1977) of Central Diocese. **Career:** High school teacher in 1957. Visiting Professor, University of Heidelberg in 1972. Wesley Seminary in 1975. President, South African Council of Churches from 1970 to 1977. Member, Commission on Studies, Lutheran World Federation from 1975 to 1983. Member, Commission on World Mission and Evangelism, World Council of Churches. **Educ.:** S.T.M. from Yale University, New Haven CT, USA; Ph.D. from Drew University, Madison NJ, USA. **Interests:** music, photography. **Hons.:** several honorary degrees.
Mailing address: PO Box 32413, Braamfontein 2017, South Africa.

Butler, John C

Rel: Druidism.
International Organizations: Secretary of the International Grand Lodge of Druidism.
Mailing address: Belair 5052, Australia.

Byblow, Frances Eudoxia, S.S.M.I., Sr

Rel: Byzantine-Ukrainian Catholic; Ordained on 12 JUL 1942; b. 09 MAY 1924 in Springside, Saskatchewan, Canada; celibate; Citizen of Canada; Lang. English, Ukrainian, Italian and French.
International Organizations: Superior General (since 1980) of the Sisters Servants of Mary Immaculate.
Main field of work: Administration. **Career:** Taught in elementary schools in Edmonton and Ottawa. High school staffs at Sacred Heart Academy in Yorkton and Mount Mary Academy in Ancaster ON serving in turn as classroom teacher, guidance counsellor and principal. Served in the Congregation of Sisters Servants of Mary Immaculate in Canada as local Superior, Directress of Novices, Provincial Secretary and Provincial Superior. **Educ.:** B.A. from University of Saskatchewan, Saskatoon, Canada (1956); Dip. Theol. from Regina Mundi, Rome, Italy (1965). **Interests:** History, literature, religion, biography, writing. **Hons.:** Pro Ecclesia et Pontifice (1983). **Mem.:** International Union of Superiors General
Mailing address: Via Cassia Antica 104, I-00191 Roma, Italy. T. (39 6) 329 57 46.
Residence: See Mailing Address. T. See Mailing Address

Byrne, Damian Aloysius, O.P., Fr
Rel: Roman Catholic; Ordained; celibate.
International Organizations: Master General of the Order of Preachers.
Master General of the Third Order Secular of St Dominic.
Other organizations: Grand Chancellor, Pontifical University of Thomas Aquinas.
Mailing address: Convento S Sabina, Piazza Pietro d'Illiria 1, I-00153 Roma, Italy. T. (39 6) 579 45 55 – 575 06 75

Byu, Esther
Rel: Christian.
International Organizations: Women's Section of the Christian Conference of Asia.
Mailing address: 879 Epifanio de los Santos Avenue, West Triangle, 1104 Quezon City, Metro Manila, Philippines.

Cabra, Pier Giordano, F.N., Fr
Rel: Roman Catholic; Ordained; celibate.
International Organizations: Superior General of the Congrégation de la Sainte Famille de Nazareth.
Mailing address: Via Giovanni Piamarta 6, I-25121 Brescia, Italy. T. (39 30) 29 58 91

Cacchione, Ciro, Dr
Rel: Roman Catholic.
International Organizations: Executive Secretary of the World Union of Jesuit Alumni.
Mailing address: Istituto Leone XIII, Via Leone XIII 16, I-20145 Milano, Italy.

Caccia, Sisto, C.S., Fr
Rel: Roman Catholic; Ordained; celibate.
International Organizations: Superior General of the Pious Society of the Missionaries of St Charles.
Mailing address: Via Calandrelli 11, I-00153 Roma, Italy. T. (39 6) 589 88 27 - (39 6) 589 88 28.

Cacciavillan, Agostino, H E Mgr
Rel: Roman Catholic; Ordained on 26 JUN 1949; b. 14 AUG 1926 in Novale; celibate.
International Organizations: Apostolic Pro-Nuncio to the United States of America and Permanent Observer to the Organization of American States (since 1990) of the Roman Catholic Church.
Other organizations: Titular Archbishop, Amiterno.
Mailing address: 3339 Massachusetts Ave NW, Washington DC 20008-3687, USA. T. (1 202) 333 7121. Fax 337 4036. Tx 440117 APNU UI

Cáceda, Carlos Chipoco
Rel: Roman Catholic.
International Organizations: Director of the Latin American Network of Catholic Lawyers.
Mailing address: Av 6 de Agosto 589, Of 306 Jesús María, Lima, Peru.

Cadier, Gérard
Rel: Protestant.
International Organizations: Secretary of the Conference of Protestant Churches in the Latin Countries of Europe.
Mailing address: Le Rastel, F-26460 Bordeaux, France.

Cadier, Gérard
Rel: Protestant.
International Organizations: Secretary of the Conference of Protestant Churches in the Latin Countries of Europe.
Mailing address: Le Rastel, F-26460 Bourdeaux, France.

Cagnasso, Franco, P.I.M.E., Fr
Rel: Roman Catholic; Ordained; celibate.
International Organizations: Superior General of the Pontifical Institute for Foreign Missions.
Mailing address: Via F D Guerrazzi 11, I-00152 Roma, Italy. T. (39 6) 58 97 941.

Cagni, Giuseppe, B., Fr
Rel: Roman Catholic; Ordained; celibate.
International Organizations: Procurator-General of the Clerics Regular of St Paul.
Mailing address: Via Giacomo Medici 15, I-00153 Roma, Italy. T. (39 6) 58 23 39.

Caird, Donald Arthur Richard, Most Rev Dr
Rel: Church of Ireland; Ordained on 29 SEP 1950; b. 11 DEC 1913 in Dublin, Ireland; m. to Nancy Ballantine Sharp; c. 1 boy and 2 girls; Citizen of Ireland; Lang. English and Irish.
International Organizations: Archbishop (since 1985) of Dublin.
Other organizations: Chairman (since 1975), Secondary Education Committee; Chairman of the Board (since 1985), Alexandra College, Dublin; Chairman (since 1985), Church of Ireland College of Education; Primate, Ireland and Metropolitan.
Main field of work: Ministry, universal education. **Career:** Curate, St Mark's Parish, Dundela, Belfast from 1950 to 1953. Chaplain and Master, The Royal School, Portora from 1953 to 1957. Lecturer in Philosophy, St David's University College, Wales from 1957 to 1958. Rector, Rathmichael Parish, Dublin from 1960 to 1969. Dean, Ossary, St Canice's Cathedral, Kilkenny from 1969 to 1970. Bishop, Limerick, Ardfert and Aghadoe from 1970 to 1976. Bishop, Meath and Kildar from 1976 to 1985. **Educ.:** B.A., M.A. Phil. (1st) from University of Dublin Trinity College, Ireland (1944-1950); B.D. from University of Dublin Trinity College, Ireland (1955); Higher Dip.Ed. from University of Dublin Trinity College, Ireland (1955). **Interests:** tennis, walking, swimming. **Hons.:** Hon. D.D., Trinity College, Dublin (1988). **Mem.:** Fellow, St Columbus College, Dublin (1971) **Pubs.:** The Predicament of Natural Theology After the Criticism of Kant.
Mailing address: The See House, 17 Temple Road, Milltown, Dublin 6, Ireland. T. (353 1) 977849
Residence: See Mailing Address. T. See Mailing Address

Cairnscross, G
Rel: Jewish.
International Organizations: UNESCO Representative of the Consultative Council of Jewish Organizations.
Mailing address: 6 Sussex House, Glenilla Road, London NW3 4AR, UK.

Cakpo, Barthélémy
Rel: Roman Catholic.
International Organizations: President of the Catholic Union of the Press in Africa.
Mailing address: La Croix du Bénin, Boîte Postale 105, Cotonou, Benin.

Calabrese, Vito, O.S.I., Fr
Rel: Roman Catholic; Ordained; celibate.
International Organizations: Superior General of the Congregation of the Oblates of St Joseph.
Mailing address: Via Boccea 364, I-00166 Roma, Italy. T. (39 6) 622 33 47

Calabresi, Ubaldo, H E Mgr
Rel: Roman Catholic; Ordained on 27 MAR 1948; b. 02 JAN 1925 in Sezze Romano, Italy; celibate.
International Organizations: Apostolic Nuncio to Argentina (since 1981) of the Roman Catholic Church.
Other organizations: Titular Archbishop, Fondi.
Mailing address: Avenida Alvear 1605, 1014 Buenos Aires, Argentina. T. (54 1) 42 96 97. Fax 742 1109. Tx 17406 NUNAP AR

Calderón Polo, Cipriano C, H E Mgr
Rel: Roman Catholic; Ordained on 19 MAR 1953; b. 01 DEC 1927 in Plasencia, Spain; celibate.
International Organizations: Vice-President of the Pontifical Commission for Latin America.
Advisor of the Pontifical Council for Social Communications. Member of the Pontifical Committee for International Eucharistic Congresses.
Other organizations: Titular Bishop, Tagora.
Mailing address: Palazzo di San Paolo, Via della Conciliazione 1, I-00193 Roma, Italy. T. (39 6) 687 1138. Fax 654 8733
Residence: Via di Porta Angelica 63, I-00193 Roma, Italy T. (39 6) 687 9834

Caldwell, Happy

Rel: Protestant.

International Organizations: President of the International Convention of Faith Ministers.

Mailing address: 3840 S 103 E Ave, Suite 132, Tulsa OK 74146, USA.

Callaghan, Brendan Alphonsus, S.J., Rev

Rel: Roman Catholic; Ordained on 17 JUN 1978; b. 29 JUL 1948 in Liverpool, UK; celibate; Citizen of United Kingdom; Lang. English and French.

International Organizations: Principal (since 1985) of the Heythrop College, University of London.

Chairman, Advisory Board (since 1991) of the Centre for the Study of Communication and Culture. Member of the Society of Jesus.

Other organizations: Lecturer (since 1980), Heythrop College, University of London; Governing Body (since 1989), Institute of Medical Ethics; Ethics Committee (since 1990), West Lambeth Health Authority.

Main field of work: University education. **Career:** Curate, St Aloysius' Parish, Glasgow from 1979 to 1980. Lecturer, Archdiocesan Seminary, Westminster from 1981 to 1987. **Educ.:** M.A. Experimental Psy. from University of Oxford, UK (1971-1974); M.Phil. Clinical Psy. from University of Glasgow, UK (1974-1976); M.Th. Pastoral Theol. from University of London, UK (1978-1979). **Interests:** psychology and spirituality, medical ethics. **Mem.:** Fellow, Royal Society of Medicine; Associate Fellow, British Psychological Society **Pubs.:** Co-author, Life before Birth, SPCK, 1986.

Mailing address: Heythrop College, 11-13 Cavendish Square, London W1M OAN, UK. T. (44 71) 580 6941. Fax 580 5031

Residence: 25 Primrose Road, London E18 1DD, UK.

Callaghan, William Sydney, Rev

Rel: Methodist; Ordained on 18 JUN 1951; b. 20 MAR 1926 in Dublin, Ireland; m. to Edith Brenda Leesley; c. 1 boy and 1 girl; Citizen of Ireland; Lang. English.

International Organizations: General Secretary, Council on Social Welfare of the Methodist Church in Ireland.

Other organizations: Member (since 1991), Eastern Health and Social Services Council; Founder Member and Chaplain, Northern Ireland Hospice.

Main field of work: Ministry. **Career:** Full-time Lay Worker, Carlisle Road, Londonderry from 1944 to 1945. Charlemont Circuit, Co. Tyrone from 1945 to 1947. Dalkey, Co. Dublin from 1947 to 1948. With Dublin Central Mission from 1950 to 1955. Superintendent of Evangelism from 1955 to 1962. Resident Chaplain and Tutor, Edgehill College, Belfast from 1961 to 1962. Appointed to Agnes Street Circuit – in the Shankill area of Belfast from 1962 to 1968. Donegall Square from 1968 to 1977. Belfast (South) Circuit – Lisburn Road, Osborne Park and University Road from 1977 to 1984. Belfast Central Mission – Sandy Row and Springfield Road in 1984. Founder Member/Deputy Director, Belfast Branch of the Samaritans from 1961 to 1974. Director, Belfast Branch of the Samaritans from 1974 to 1987. President, Methodist Church in Ireland from 1980 to 1981. Has served on nearly all the committees and departments of the Church. Former Governor, Edgehill Theological College. Member, Northern Ireland Mental Health Review Tribunal (for 13 years). College Lecturer and Faculty Member, Queen's University, Belfast. **Educ.:** B.A. from University of Dublin Trinity College, Ireland (1947-1953); Full Div. Crse. from Edgehill Theological College, Belfast, UK (1948-1950); Dip.Ed. from Queen's University of Belfast, UK (1960-1963). **Interests:** work, music, arts. **Hons.:** People of the Year Award (Ireland); D.D., De Pauw University, IN; Officer of the Order of the British Empire. **Mem.:** Fellowship of Reconciliation **Pubs.:** The Agony of Ireland, Published by the Division of Social Responsibility of the Methodist Church in England (An interpretation of the Irish Situation), 1981, 2nd Edn. Revised, 1984; Good Grief, 1990 and "Insights from Ireland", in *The World Outlook*, 1984.

Mailing address: 22 Bawnmore Road, Belfast BT9 6LA, N Ireland, UK.

Callagy, Dennis

Rel: Roman Catholic.

International Organizations: Head of Vienna Office of the Conference of International Catholic Organizations.

Representative at UN of the International Catholic Child Bureau. Consultor of the Pontifical Council for the Laity.

Mailing address: Gonzagagasse 13, A-1010 Wien, Austria. T. (43 1) 633 09 35.

Calver, Clive Richard, Rev

Rel: Free Church; Ordained in 1971; b. 04 JAN 1949 in Ipswich, Suffolk, UK; m. to Ruth; c. 2 boys and 2 girls; Citizen of United Kingdom; Lang. English and French (some).

International Organizations: General Director (since 1983) of the Evangelical Alliance.

Other organizations: Vice-President (since 1983), British Youth for Christ; President (since 1988), Crusaders Union; Vice-President (since 1990), Universities and Colleges Christian Fellowship.

Main field of work: Preaching, co-ordinating evangelical initiatives. **Career:** Associate, London Bible College from 1968 to 1971. Team Leader, In the Name of Jesus from 1973 to 1975. National Director, British Youth for Christ from 1976 to 1982. Programme Director, Billy Graham 'Mission England' from 1982 to 1983. **Educ.:** B.D. (Hons) from University of London, UK (external student) (1968-1971). **Pubs.:** Sold Out, Marshalls, 1980; The Holy Spirit, Scripture Union, 1984 and Spring Harvest Seminar Notes 1986-1991.

Mailing address: Whitefield House, 186 Kennington Park Road, London SE11 4BT, UK. T. (44 71) 582 0228. Fax (44 71) 582 6221.

Residence: 13 Pearfield Road, London SE23 2LP, UK.

Camba, Eugenia Lily D, Ms

Rel: United Church of Christ.

International Organizations: Secretary of the Association of Christian Institutions for Social Concern in Asia.

Mailing address: Inst of Religion and Culture, c/o United Church of Christ in Philippines, 875 EDSA, Quezon City, Philippines.

Campbell, P

Rel: Roman Catholic.

International Organizations: Oceania Regional Secretary (since 1990) of the International Catholic Committee of Nurses and Medico-social Assistants.

Mailing address: PO Box 2018S, Melbourne 3001, Australia.

Campbell, Robert C, Rev Dr

Rel: Baptist: American.

International Organizations: General Secretary of the American Baptist Churches in the USA.

Mailing address: PO Box 851, Valley Forge PA 19482-0851, USA.

Campeau, Lucien, S.J., Fr

Rel: Roman Catholic; Citizen of Canada.

International Organizations: Professor Emeritus of History (since 1980) of the University of Montreal.

Member of the Society of Jesus.

Mailing address: Maison des Jésuites, CP 130, 175 blvd des Hauteurs, St Jérôme, Québec J7Z 5T8, Canada.

Campos Leite, Nelson Luis C, Bishop

Rel: Methodist; Ordained.

International Organizations: President of the Council of Bishops and of the General Council of Methodist Church in Brazil.

Mailing address: Rua Vigario Joao de Pontes, 766, CEP 04748, Santo Amaro, Sao Paulo, Brazil.

Camps, Carlos M, Rev Prof Dr

Rel: Presbyterian; Ordained on 12 SEP 1952; b. 01 JUN 1929 in Cardenas, Matanzas, Cuba; m. to Norca Iglesias Zuñigas; c. 1 boy and 6 girls; Citizen of Cuba; Lang. Spanish and English.

International Organizations: General Secretary (since 1952) of the Reformed Presbyterian Church in Cuba.

President (since 1980) of the Church and Society in Latin America. Member, Executive Committee and Moderator, Council Area for North America and Caribbean (since 1990) of the World Alliance of Reformed Churches.

Other organizations: Vice-Principal (since 1989), Theological Seminary, Matanzas.

Main field of work: University education. **Educ.:** Th.B. from Evangelical Seminary, Matanzas, Cuba (1949-1952); Ph.D. from University of Havana, Cuba (1950-1955); Th.M. from Princeton Theological Seminary, NJ, USA (1958-1959). **Mem.:** Rotary International; Masonic Lodge **Pubs.:** Nueva Espiritualidad en Cuba Hoy and Peace from the Perspective of Theology of Liberation.

Mailing address: Box 205, Matanzas MTZs 40100, Cuba. T. (53) 53 4250. Tx 512350 CECUM

Residence: 63 St 29805, Matanzas MTZs 40100, Cuba. T. (53) 52 4125

Canalini, Francesco, H E Mgr
Rel: Roman Catholic; Ordained on 18 MAR 1961; b. 23 MAR 1936 in Osimo, Italy; celibate.
International Organizations: Member of the Roman Catholic Church.
Other organizations: Titular Archbishop, Valeria.
Career: Apostolic Pro-Nuncio to Indonesia from 1986 to 1991.
Mailing address: Jalan Merdeka Timur 18, PO Box 4227, Jakarta-Pusat, Indonesia. T. (62 21) 384 11 42 – 381 07 36. Fax 384 11 43

Candy, D J, Sr
Rel: Roman Catholic; Ordained; celibate.
International Organizations: Superior General of the Congregation of La Retraite.
Mailing address: 16 av du Belvédère, F-78100 Saint-Germain-en-Laye, France. T. (33 1) 34 51 96 96.

Canestri, Giovanni, H E Card
Rel: Roman Catholic; Ordained on 12 APR 1941; b. 30 SEP 1918 in Castelpina, Italy; celibate; Citizen of Italy.
International Organizations: Member (since 1987) of the College of Cardinals.
Member of the Sacred Congregation for the Sacraments and Divine Worship. Member of the Congregation for the Clergy.
Other organizations: Archbishop (since 1987), Genova.
Career: Elected to the titular Church, Tenedo in 1961. Consecrated Bishop in 1961. Transferred to Tortona in 1971. Prefect at the titular Church, Monterano in 1975. Transferred to Cagliari with title, Archbishop in 1984. Created Cardinal by Pope John Paul II in 1987.
Mailing address: Arcivescovado, Piazza Matteotti 4, I-16123 Genova, Italy. T. (39 10) 296 469

Cano Manrique, Francisco C, M.D., Fr
Rel: Roman Catholic; Ordained; celibate.
International Organizations: Superior General of the Order of Discalced Mercedarians.
Mailing address: Santa María de la Merced, Urbanización Club de Golf, Calle Cabo Mayor 1, 28290 Las Matas (Madrid), Spain. T. (34 91) 630 30 81.

Capetola, Nicola, CC.RR.MM., Fr
Rel: Roman Catholic; Ordained; celibate.
International Organizations: Procurator-General of the Clerics Regular Minor.
Mailing address: Via Tribuna Campitelli 6/a, I-00186 Roma, Italy. T. (39 6) 654 18 19

Capon, Anthony Charles, Rev Canon
Rel: Anglican; Ordained in 1953; b. 05 OCT 1926 in Chislehurst, UK; m. to Elizabeth Mowll; c. 3 boys and 1 girl; Citizen of Canada.
International Organizations: Theological College Principal (retired) (since 1991) of the Montreal Diocesan Theological College, Anglican Church of Canada.
Career: Principal, Montreal Diocesan Theological College, Anglican Church of Canada from 1978 to 1991. Canon in 1978. Assistant Priest, St Paul's Church, Portman Square, London, UK from 1953 to 1956. Canadian Director, Scripture Union from 1956 to 1970. Regional Director, Americas Scripture Union from 1970 to 1975. Director of Development, Wycliffe College, Toronto from 1975 to 1978. Senior Scholar, Trinity College, Cambridge in 1951. Chairman, Heads of Anglican Colleges, Canada in 1988. **Educ.:** B.A. from University of Cambridge, Trinity College, UK (1951); M.A. from University of Cambridge, Trinity College, UK (1955); G.O.E. from Oak Hill College, London, UK (1953); B.D. from University of Toronto, Wycliffe College, Canada (1965); D.D. from University of Toronto, Wycliffe College, Canada (1982). **Pubs.:** Co-author, The Church and the Child, 1965.
Mailing address: 5 Loradean Cres, Kingston, Ontario K7K 6X9, Canada.

Capougi, Hilarion Bashir, H E Mgr
Rel: Roman Catholic; Ordained on 20 JUL 1947; b. 02 MAR 1922 in Aleppo, Syria; celibate; Citizen of Syria.
International Organizations: Archbishop of Roman Catholic Church.
Career: Professor, Deir-Ach-Chir Small Seminary (for 6 years). Principal, An-Nahdat Al-Wataniya School. Attached to Deir Ach-Chir Convent. In charge of village parish priests, churches and schools, Aleppo from 1953 to 1962. Raised to the Titular Archbishopric, Cesareum (Palestine). Vicar-General, Patriarchate of Jerusalem in 1965. Imprisoned and released by the Israeli Government in 1977.
Mailing address: The Vatican, Rome, Italy.

Cappello, Mario
Rel: Roman Catholic.
International Organizations: International Director of the International Catholic Programme of Evangelization.
Mailing address: Allerheiligen, 77728 Oppenau, Germany.

Capps, Carol A, Ms
Rel: Lutheran.
International Organizations: Director of the Church World Service / Lutheran World Relief.
Mailing address: 110 Maryland Ave NE, Building Mailbox No 45, Washington DC 20002-5694, USA. T. (1 202) 543 3663. Fax 543 1297.

Capraro, Giacomo, S.D.V., Rev
Rel: Roman Catholic; Ordained; celibate.
International Organizations: Superior General of the Society of Divine Vocations.
Mailing address: Via Cortina d'Ampezzo 138, I-00135 Roma, Italy. T. (39 6) 328 66 04.

Caprio, Giuseppe, H E Card
Rel: Roman Catholic; Ordained on 17 DEC 1938; b. 15 NOV 1914 in Lapio, Avellino, Italy; celibate; Citizen of Italy.
International Organizations: Member (since 1979) of the College of Cardinals.
Grand Master of the Equestrian Order of the Holy Sepulchre of Jerusalem. Member of the Congregation for Bishops. Member of the Sacred Congregation for the Oriental Churches. Member of the Congregation for the Causes of Saints. Member of the Congregation for Bishops. Member of the Congregation for the Evangelization of Peoples.
Career: Attaché, Secretariat of State, Vatican City from 1943 to 1947. Secretary, Apostolic Nunciature, Nanking, China from 1947 to 1951. Auditor, Apostolic Nunciature, Brussels from 1952 to 1956. Apostolic Visitor to Republic, Viet-Nam and later Regent of Apostolic Delegation in Saigon from 1956 to 1959. Apostolic Nuncio to China, serving in Taiwan from 1959 to 1967. Titular Archbishop, Apollonia in 1961. Apostolic Pro-Nuncio to India from 1967 to 1969. Secretary of Administration of Patrimony, Holy See from 1969 to 1977. Substitute, the Secretariat of State from 1977 to 1979. President of Administration of Patrimony, Holy See from 1979 to 1981. Created Cardinal by Pope John Paul II in 1979. Prefect, Vatican's Economic Affairs Office in 1981.
Mailing address: 00120 Città del Vaticano, Vatican. T. (39 6) 698 35 72

Captain, Shahpur F
Rel: Zoroastrian.
International Organizations: President of the World Zoroastrian Organization.
Mailing address: 135 Tennyson Road, South Norwood, London SE25 5NF, UK.

Capulong, Emilio, Jr, Atty
Rel: Reformed Church.
International Organizations: Member of the Executive Committee of the World Alliance of Reformed Churches.
Mailing address: Calamba, Laguna, Philippines. T. (63 91) 545 1245

Carberry, John Joseph, H E Card

Rel: Roman Catholic; Ordained on 28 JUL 1929; b. 31 JUL 1904 in Brooklyn NY, USA; celibate; Citizen of USA.
International Organizations: Member (since 1969) of the College of Cardinals.
Other organizations: Archbishop Emeritus (since 1979), Saint Louis MO.
Career: Titular Bishop, Elis in 1956. Bishop, Lafayette IN from 1957 to 1965. Bishop, Columbus OH from 1965 to 1968. Archbishop, St Louis MO from 1968 to 1979. Created Cardinal by Pope Paul VI in 1969.
Mailing address: 4445 Lindell Blvd, Saint Louis MO 63108, USA.

Carbonnel, Jeannine Marie Léonie

Rel: Roman Catholic; b. 14 OCT 1942 in Wasquenal, Nord, France; single; Citizen of France; Lang. French, English and Spanish.
International Organizations: Member (since 1963) of the Missionary Workers of the Immaculate.
Main field of work: Engineering. **Educ.:** Ing. from Ecole Polytechnique Féminine, Sceaux, France; Maîtrise en Informatique from University of Paris, France.
Mailing address: BP 4669, Nouméa, New Caledonia.
Residence: 20 RT 13, Nouméa, New Caledonia.

Cardenal, Ernesto

Rel: Roman Catholic; Ordained on 15 AUG 1965; b. 20 JAN 1925 in Granada, Nicaragua; celibate; Citizen of Nicaragua.
International Organizations: Founder (since 1966) of the Christian commune in Nicaragua: Solentiname.
Main field of work: Poet of Revolutionary literature. **Career:** Joined Unidad Nacional de Accion Popular (UNAP) (illegal revolutionary group). Had a spiritual conversion and entered the Monastery of Our Lady Gethsemani to study under the Order of the Trappist Monks in Kentucky in 1956. Moved to the Benedictine community at Guernavaca in Mexico in 1959. Moved to the Seminary of La Ceja in Columbia in 1961. Was Minister of Culture for the Sandinista regime after the overthrow of Samosa in 1979 (until the early 1980s). **Educ.:** Deg. Phil. and Letters from National Autonomous University of Mexico, Mexico; Dr. Phil. from Columbia University, New York NY, USA (1949). **Pubs.:** *Gethsemani, Kentucky*, 1960; *Homage to the American Indians*, 1969 and *Zero Hour and Other Documentary Poems*, 1980.
Mailing address: No mailing address.

Carew, William Aquin, H E Mgr

Rel: Roman Catholic; Ordained on 15 JUN 1947; b. 23 OCT 1922 in St John's, Newfoundland, Canada; celibate.
International Organizations: Apostolic Pro-Nuncio to Japan (since 1983) of the Roman Catholic Church.
Other organizations: Titular Archbishop, Telde.
Mailing address: 9-2 Sanban-cho, Chiyoda-ku, Tokyo, Japan. T. (81 3) 263 6851. Fax 263 6060

Carey, George Leonard, Most Rev and Rt Hon

Rel: Anglican; Ordained in 1962; b. 13 NOV 1935 in Bow, London, UK; m. to Eileen Harmsworth; c. 2 boys and 2 girls; Citizen of United Kingdom.
International Organizations: President (since 1991) of the Lambeth Conference of Bishops of the Anglican Communion.
Other organizations: Archbishop (since 1991), Canterbury.
Career: Clerk, London Electricity Board from 1950 to 1958. Wireless Operator with the Royal Air Force – National Service from 1954 to 1956. Assistant Curate, St Mary's, Islington, North London from 1962 to 1966. Lecturer in Theology, Oak Hill Theological College, Southgate from 1966 to 1970. Lecturer in Theology and Chaplain, St John's Theological College, Nottingham from 1970 to 1975. Vicar, St Nicholas' Church, Durham from 1975 to 1982. Chaplain, a Youth Custody Prison from 1975 to 1982. Chaplain, Durham branch, RAF Association from 1975 to 1982. Principal, Trinity College, Bristol from 1982 to 1987. Elected Member, General Synod from 1982 to 1987. Chairman, Faith and Order Advisory Group, and Member, Board for Mission and Unity from 1982 to 1987. Council Member, Advisory Council for the Church's Ministry (ACCM) from 1982 to 1987. Member, Theological and Religious Studies Board, Council for National Academic Award from 1982 to 1987. Bishop, Bath and Wells from 1988 to 1991. **Educ.:** B.D. (Hons) from London College of Divinity, UK (1962); M.Th. from London College of Divinity, UK (1962); Ph.D. from University of London, King's College, UK (1962). **Interests:** football (supporter of Arsenal Football Club), reading, walking, listening to music. **Pubs.:** Author of 8 books on theological issues, including: *The Church in the Market Place*, 1984; *The Message of the Bible*, 1986; *I Believe*, 1991 and contributor to numerous journals and periodicals.
Mailing address: Lambeth Palace, London SE1 7JU, UK. T. (44 71) 928 8282

Carisio, Mario, S.P./C.Op., Fr

Rel: Roman Catholic; Ordained; celibate.
International Organizations: Procurator-General of the Clerks Regular Poor of the Mother of God of the Pious Schools.
Procurator General of the Congrégation des Ouvriers Chrétiens de Saint Joseph Calasanz.
Mailing address: Piazza dei Massimi 4, I-00186 Roma, Italy. T. (39 6) 654 1494 – (39 6) 686 8456.

Carlebach, Felix, Rabbi

Rel: Jewish; Ordained; b. 15 APR 1911 in Lübeck, Germany; widowed Babette Kohn (died); c. 3 girls; Citizen of United Kingdom; Lang. English, German and Hebrew.
International Organizations: Senior Rabbi of South Manchester Synagogue.
Other organizations: Chairman, Manchester Council of Christians and Jews; Chaplain to 5 Manchester Lord Mayors; Chaplain to Christie Cancer Hospital.
Career: Came to Great Britain in 1939. Rabbi, South Manchester Synagogue from 1945 to 1986. **Educ.:** University and Musikhochschule, Cologne, Germany (1929-1933); M.A. from Victoria University of Manchester, UK (1950-1953); Yeshiva Diaspora, Jerusalem Rabbinical Seminary, Israel (1953-1956). **Interests:** music. **Hons.:** Freeman of the City of Lübeck (Germany); Felix Carlebach Concert by Halle Orchestra. **Pubs.:** Synagogue pamphlets.
Mailing address: 2A Elm Road, Didsbury, Manchester 20, UK. T. (44 61) 445 5716
Residence: See Mailing Address. T. See Mailing Address

Carlson, Alexandre

Rel: Roman Catholic.
International Organizations: Secretary-General of the Catholic International Union for Social Service.
Mailing address: Rue de la Citronelle 31, B-1348 Louvain-la-Neuve, Belgium. T. (32 10) 45 25 13.

Carmody, Joy, Sr

Rel: Roman Catholic; Ordained; celibate.
International Organizations: Contact of the Missionary Sisters of St Columban.
Mailing address: 209 Quebec Drive, East Kilbride, Glasgow G758 BB, UK.

Carnley, Peter Frederick, Most Rev Dr

Rel: Anglican; Ordained in 1962; b. 17 OCT 1937 in New Lambton, New South Wales, Australia; m. to Carol Ann Dunstan; c. 1 boy and 1 girl; Citizen of Australia; Lang. English.
International Organizations: Archbishop (since 1981) of Perth.
Other organizations: Metropolitan (since 1981), Province of Western Australia.
Career: Curate, Parkes, Diocese of Bathurst in 1966. Chaplain, Mitchell College, Diocese of Bath, UK from 1970 to 1972. Research Fellow at St John's College, Cambridge, UK from 1971 to 1972. Warden, St John's College, St Lucia, Brisbane from 1972 to 1981. **Educ.:** Th.L. (1st) from St John's Theological College, Morpeth NSW, Australia (1962); B.A. (1st) from University of Melbourne, Victoria, Australia (1966); Ph.D. from University of Cambridge, UK (1969). **Interests:** gardening, music. **Hons.:** DD hc, General Theological Seminary, USA (1984). **Pubs.:** *The Structure of Resurrection Belief*, Clarendon Press, Oxford, 1987.
Mailing address: GPO Box W2607, Perth 6001 WA, Australia. T. (61 9) 325 7455
Residence: 52 Mount Street, West Perth 6005, Australia. T. (61 9) 322 1777

Carpino, Francesco, H E Card

Rel: Roman Catholic; Ordained on 14 AUG 1927; b. 18 MAY 1905 in Palazzolo Acreide, Sicily, Italy; celibate; Citizen of Italy.
International Organizations: Member (since 1967) of the College of Cardinals.
Career: Titular Archbishop, Nicodemia in 1951. Titular Archbishop, Sardica in 1961. Archbishop, Palermo from 1967 to 1970. Titular Bishop, Albano. Created Cardinal by Pope Paul VI in 1967.
Mailing address: Piazza San Calisto 16, I-00153 Roma, Italy.

Carr, Burgess, Rev

Rel: Episcopal.
International Organizations: Executive Director of the Episcopal Migration Ministries.
Mailing address: 815 2nd Avenue, New York NY 10017, USA. T. (1 212) 867 8400. Fax 972 0860. C. FENALONG NEW YORK. Tx 971271.

Carr, Gregory Allen

Rel: Buddhist: Tibetan; b. 08 NOV 1952 in Perth, New South Wales, Australia; single; Citizen of Australia; Lang. English.
International Organizations: Member of the Foundation for the Preservation of the Mahayana Tradition.
Other organizations: Director (since 1988), Hayagriva Buddhist Centre, Perth; WA Committee Member (since 1991), 1992 H H Dalai Lama Tour of Australia.
Main field of work: Buddhist study and meditation. **Career:** Electrician (self-employed) from 1972 to 1992. **Interests:** personal development, sport, movies.
Mailing address: PO Box 302, Como, Perth WA 6152, Australia. T. (61 9) 367 1641
Residence: 80 McDonald St, Como, Perth WA 6152, Australia. T. See Mailing Address

Carrara, Giovannamaria, F.S.P., Sr

Rel: Roman Catholic; Ordained; celibate.
International Organizations: Superior General of the Pious Society of the Daughters of St Paul.
Advisor of the Pontifical Council for Social Communications.
Mailing address: Via San Giovanni Eudes 25, I-00163 Roma, Italy. T. (39 6) 626 10 00. Fax 625 72 08.

Carraro Flavio, Roberto C, O.F.M. Cap., Fr

Rel: Roman Catholic; Ordained; celibate.
International Organizations: Minister-General of the Order of Friars Minor Capuchin.
Member of the Congregation for Religious and Secular Institutes.
Mailing address: Via Piemonte 70, I-00187 Roma, Italy. T. (39 6) 474 06 43.

Carrier, Hervé, S.J., Rev Prof

Rel: Roman Catholic; Ordained; b. 26 AUG 1921 in Grand-Mère, Quebec, Canada; celibate; Citizen of Canada; Lang. French, English and Italian.
International Organizations: Secretary (since 1982) of the Pontifical Council for Culture.
Member of the Pontifical Commission for Latin America. Member of the Pontifical Council for the Pastoral Care of Migrants and Itinerant People. Consultant of the Pontifical Council for Justice and Peace. Consultant of the Pontifical Council for Dialogue with Non-Believers. Member of the Society of Jesus.
Other organizations: Professor of Sociology (since 1959), Pontifical Gregorian University, Vatican City State.
Main field of work: Sociology of religion and culture. **Career:** Rector, Pontifical Gregorian University, Vatican City State from 1966 to 1978. President, International Federation of Catholic Universities (IFCU) from 1970 to 1980. Director, Centre for Coordination of Research, IFCU from 1978 to 1982. **Educ.:** B.A. from University of Montreal, Quebec, Canada (1944); M.A. Soc. from Catholic University of America, Washington DC, USA (1953); Dr. Soc. from University of Paris IV (Paris-Sorbonne), France (1959). **Hons.:** Dr. h.c.: Sogang University (Seoul, Korea), Fu-Jen Catholic University (Taiwan); Legion of Honour (France). **Mem.:** Académie des Lettres et Sciences Humaines, Royal Society of Canada **Pubs.:** *Evangile et cultures*, LEV/Mediasp., Vatican City State/Paris, 1987; *Psico-sociologia dell'appartenenza religiosa*, LDC, Turin, 1988 and *Evangélisation et développement des cultures*, Gregorian University, Rome, 1990.
Mailing address: I-00120 Città del Vaticano, Vatican. T. (39 6) 698 7321 – (39 6) 698 7342. Tx 2013 CULTVAT VA. Fax (39 6) 698 7368.
Residence: 4 Piazza della Pilotta, I-00187 Roma, Italy. T. (39 6) 67011

Carroll, Philip, Rev

Rel: Roman Catholic; Ordained on 18 DEC 1971; b. 30 NOV 1946 in Newcastle upon Tyne, UK; celibate; Citizen of United Kingdom; Lang. English and Italian.
International Organizations: General Secretary (since 1992) of the Catholic Bishops' Conference of England and Wales.
Career: Pastoral priest, Parish of Our Lady, Washington, UK from 1972 to 1978. Director, Hexham and Newcastle Diocesan Catechetical Centre from 1979 to 1988. Assistant Secretary-General, Bishops' Conference of England and Wales from 1988 to 1991. **Educ.:** Lic. Phil. from Pontifical Gregorian University, Vatican City State (1965-1968); Lic. Theol. from Pontifical Gregorian University, Vatican City State (1968-1972); Dip. Rel.Ed. from Mount Oliver Institute of Religious Education, Dundalk, UK (1978-1979).
Mailing address: Bishop's Conference of England and Wales, 39 Eccleston Square, London SW1V 1PD, UK. T. (44 71) 630 8220
Residence: Our Lady of Victories, Kensington, London, UK.

Carson, Eugene

Rel: Presbyterian; Citizen of USA.
International Organizations: Clergyman of the Presbyterian Church of America.
Main field of work: Clergyman. **Career:** President, National Council of Churches from 1954 to 1957. Secretary-General, World Council of Churches from 1966 to 1972. Chief Executive, the Presbyterian Church during the 1950s and 1960s. Leader, the ecumenical movement.
Mailing address: No mailing address.

Carter, Alexander, Most Rev

Rel: Roman Catholic; Ordained on 06 JUN 1936; b. 16 APR 1909 in Montreal, Quebec, Canada; celibate; Citizen of Canada.
International Organizations: Bishop Emeritus (since 1985) of Sault Sainte Marie.
Career: Vice-Chancellor, Montreal from 1940 to 1946. Chancellor, Winnipeg from 1946 to 1947. Vice-official, Matrimonial Tribunal, Montreal from 1948 to 1953. Parish Priest, Holy Family, Montreal from 1953 to 1967. President, Canadian Catholic Conference of Bishops from 1967 to 1969. Member, Secretariat for Non-Believers from 1967 to 1973. Member, Vanier Institute of the Family. National Director, Pontifical Mission-Aid Societies for Canada from 1971 to 1977. President, Ontario Catholic Conference of Bishops from 1977 to 1981. Chairman, CCCB Episcopal Commission for the Laity from 1978 to 1983. Bishop, Sault Sainte Marie from 1958 to 1985. **Educ.:** B.A. from Seminary of Philosophy, Canada (1932); L.Th. from Grand Seminary of Theology, Canada (1936); J.C.L. from Appolinaris Law School, Rome, Italy (1939). **Interests:** fishing, hunting. **Hons.:** Hon. LL.D., Laurentian University of Sudbury, Ontario (1962); Officer of the Order of Canada (1989). **Pubs.:** Answers'.
Mailing address: 250 Silver Lady Lane, R R 3, North Bay ON P1B 8G4, Canada. T. (1 705) 474 9230

Carter, Gerald Emmett, H E Card

Rel: Roman Catholic; Ordained on 22 MAY 1937; b. 11 MAR 1912 in Montreal, Canada; celibate; Citizen of Canada.
International Organizations: Member (since 1979) of the College of Cardinals.
Other organizations: Archbishop Emeritus (since 1990), Toronto.
Career: Supervisor, Montreal Catholic School Commission from 1937 to 1939. Founder and Principal, St Joseph's Teachers' College, Montreal from 1939 to 1961. Chaplain, Newman Club, McGill University, Montreal from 1941 to 1956. Charter Member and first President, Thomas More Institute for Adult Education from 1945 to 1961. English Commissioner, Montreal Catholic School Commission from 1948 to 1961. Rector, St Lawrence College, Quebec in 1961. Titular Bishop, Altiburo in 1961. Consecrated Bishop in 1962. Bishop, London, Ontario from 1964 to 1978. Archbishop, Toronto from 1978 to 1990. Created Cardinal by Pope John Paul II in 1979. Canon, Law Commission, Holy See in 1981. Chairman, Episcopal Commission on Canadian Liturgy from 1966 to 1973. Member, Consilius of Liturgy, Rome in 1965. Vice-President, Doctrine and Faith Dep, Canadian Conference of Catholic

Bishops (CCCB) in 1969. Chairman, International Committee for English in Liturgy in 1971. Vice-President, Catholic Conference of Ontario from 1971 to 1973. Vice-President, CCCB from 1973 to 1975. President, CCCB from 1975 to 1977. Council Member, Synod of Bishops from 1977 to 1983. **Educ.:** B.A. from University of Montreal, Quebec, Canada (1933); B.Th. from University of Montreal, Quebec, Canada (1936); M.A. from University of Montreal, Quebec, Canada (1940); Ph.D. from University of Montreal, Quebec, Canada (1947); D.H.L. from Duquesne University, Pittsburgh PA, USA (1962). **Interests:** fishing, swimming. **Hons.:** Companion, Order of Canada (1983); Hon. Canon, Basilica of Our Lady Queen of the World, Montreal (1953); Commander of Order of Scholastic Merit of Quebec (1958); Conventual Chaplain, Knights of Malta (1960); Hon. LL.D., Western Ontario (1966), Concordia (1976), Windsor (1977), McGill (1980), Notre Dame IN (1981); Hon. D.Lit., St Mary's, Halifax NS (1980); Hon. D.D., Huron College, University of Western Ontario (1978). **Pubs.:** The Catholic Public Schools of Quebec, 1957; Psychology and the Cross, 1959 and The Modern Challenge to Religious Education, 1961.
Mailing address: 355 Church Street, Toronto ON M5B 1Z8, Canada. T. (1 416) 977 1500

Carteri, Giuseppe, Fr
Rel: Roman Catholic; Ordained; celibate.
International Organizations: Past Procurator-General of the Company of Mary for the Education of Deaf-Mutes.
Mailing address: Via Antonio Provolo 20, I-37123 Verona, Italy. T. (39 45) 8007419. Fax 575396.

Carton, Jacques
Rel: Roman Catholic.
International Organizations: Europe/Mediterranean Regional Secretary (since 1990) of the International Catholic Conference of Scouting.
Mailing address: 54 avenue Jean Jaures, F-75019 Paris, France.

Caruana, Edmund A, Fr
Rel: Roman Catholic; Ordained; celibate.
International Organizations: Contact of the Latinitas Foundation.
Mailing address: Via della Tipografia, I-00120 Città del Vaticano, Vatican. T. (39 6) 698 84648. Fax 698 84716.

Carvelli, Giuseppe, P.O.C.R., Fr
Rel: Roman Catholic; Ordained; celibate.
International Organizations: Superior General of the Pieux Ouvriers Catéchistes Ruraux.
Mailing address: 87046 Montalto Uffugo, Cosenza, Italy. T. (39 984) 93 10 20

Casaroli, Agostino, H E Card
Rel: Roman Catholic; Ordained on 27 MAY 1937; b. 24 NOV 1914 in Castel S Giovanni, Piacenza, Italy; celibate; Citizen of Italy.
International Organizations: President of the Fondazione Pio XII per l'Apostolato dei Laici.
Member of the Congregation for Bishops. Member of the Pontifical Council for the Interpretation of Legislative Texts.
Other organizations: Titular Archbishop (since 1967), Carthage; Member (since 1979), College of Cardinals; Member, Congregation for the Doctrine of the Faith; Member, Cardinal's Commission for the supervision, Institute for Religious Works.
Career: Created Cardinal by Pope John Paul II in 1979. Secretary of State and Prefect, Council for the Public Affairs of the Church. **Hons.:** Entitled Ss XII Apostoli (1985). **Pubs.:** Der Heilige Stuhl und die Völkergemeinschaft (1981).
Mailing address: 00120 Città del Vaticano, Vatican.

Casey, Eamonn, Most Rev
Rel: Roman Catholic; Ordained on 17 JUN 1951; b. 24 APR 1927 in Firies, Co Kerry, Ireland; celibate; Citizen of Ireland.
International Organizations: Bishop (since 1976) of Galway and Kilmacduagh.
Career: Curate and teacher from 1951 to 1955. Attached to St John's Cathedral, Limerick from 1955 to 1960. Chaplain to the Irish, Slough, Bucks. from 1960 to 1963. National Director, Catholic Housing Aid Society from 1963 to 1969. Bishop, Kerry from 1969 to 1976. **Educ.:** University and Seminary at Maynooth. **Interests:** music, theatre, concerts, films, conversation, motoring. **Mem.:** numerous commissions and societies **Pubs.:** With Adam Ferguson, A Home of Your Own.
Mailing address: The Diocesan Office, The Cathedral, Galway, Ireland. T. (353 91) 63566
Residence: Mount St Mary's, Taylor's Hill, Galway, Ireland.

Casey, John W
Rel: Protestant.
International Organizations: Secretary-General of the World Alliance of Young Men's Christian Associations.
Mailing address: John R Mott House, Quai Wilson 37, CH-1201 Genève, Switzerland. T. (41 22) 732 31 00. Fax 738 40 15. C. FLEMGO Genève. Tx 412332 YMCA CH.

Casilla, Ercilla Palomino
Rel: Methodist.
International Organizations: President for Latin America of the World Federation of Methodist Women.
Mailing address: 4814 Valparaiso, Chile.

Casimir, Stewart
Rel: Christian.
International Organizations: Executive Secretary of the TOC H.
Mailing address: 1 Forest Close, Wendover, Aylesbury HP22 6BT, UK. T. (44 296) 623911.

Casoria, Giuseppe, H E Card
Rel: Roman Catholic; Ordained on 21 DEC 1930; b. 01 OCT 1908 in Acerra, Italy; celibate; Citizen of Italy.
International Organizations: Member (since 1983) of the College of Cardinals.
Other organizations: Titular Archbishop (since 1972), Vescovia; Dean (since 1984), San Giuseppe in Via Trionfale.
Career: Made Bishop in 1972. Created Cardinal by Pope John Paul II in 1983. Prefect, the Sacred Congregation of the Sacraments. Consultant to the Sacred Congregation of the Oriental Churches.
Mailing address: Via P Pancrazio Pfeiffer 10, I-00193 Roma, Italy. T. (39 06) 698 42 45

Casserly, Patrick, S.M., Rev
Rel: Roman Catholic; Ordained; celibate.
International Organizations: Regional Secreatary of the International Catholic Organization for Cinema and Audiovisual.
Member of the Society of Mary. Advisor of the Pontifical Council for Social Communications. Contact person of the OCIC/Oceania.
Mailing address: PO Box 109, Suva, Fiji.

Cassidy, Edward Idris, H E Mgr
Rel: Roman Catholic; Ordained on 23 JUL 1949; b. 05 JUL 1924 in Sydney, Australia; celibate; Citizen of Australia; Lang. English, Italian, Spanish and French.
International Organizations: President (since 1989) of the Pontifical Council for Promoting Christian Unity.
President (since 1989) of the Commission for Religious Relations with the Jews.
Other organizations: Titular Archbishop (since 1970), Amantia.
Main field of work: Ecumenism. **Career:** Assistant Priest, Parish of Yenda NSW from 1950 to 1952. Attaché, Second Secretary, Apostolic Internunciature in India from 1955 to 1962. First Secretary, Apostolic Nunciature in Ireland from 1962 to 1967. Counsellor, Apostolic Nunciature in El Salvador from 1967 to 1969. Counsellor, Apostolic Nunciature in Argentina in 1970. Apostolic Pro-Nuncio to Taiwan (Rep. of China) from 1970 to 1979. Apostolic Pro-Nuncio to Bangladesh and Apostolic Delegate to Burma from 1973 to 1979. Apostolic Delegate to Southern Africa (including: South Africa, Botswana, Swaziland, Namibia) from 1979 to 1984. Apostolic Pro-Nuncio to Lesotho from 1979 to 1984. Apostolic Pro-Nuncio to the Netherlands from 1984 to 1989. Substitute Secretary of State from 1988 to 1989. Elevated by Pope John Paul II to the College of Cardinals in 1991. **Educ.:** St Columba Seminary, Springwood, Australia (1943); St Patrick's College, Manly, Australia (1944-1949); D.C.L. from Lateran Pontifical University, Vatican City State (1952-1955); Dip. Ecc. Diplomacy from Pontifical Ecclesiastical Academy, Vatican City State (1953-1955). **Interests:** Tennis, golf, music. **Hons.:** Commendador en la Orden Nacional José Matias Delgado, El Salvador (1969); Order of the Brilliant Star with Gran Cordon, China (1979); Grootkruis in de Orde van Oranje-Nassau, Netherlands (1988); Cavaliere di Gran Croce dell'Ordine Al Merito, Italy (1988); Companion in the General Division of the Order of Australia (1990); Commandeur de la Légion d'Honneur, France (1991); Commander Grand Cross of the Royal Order of the Polar Star, Sweden (1991).
Mailing address: Via dell'Erba 1, I-00193 Roma, Italy. T. (39 6) 698 41 81. FAX 698 53 65. C. SECRUNIT VATICAN CITY.
Residence: Piazza del S Uffizio 11, I-00193 Roma, Italy

Castellanos, Carlos
Rel: Christian.
International Organizations: Executive Secretary of the Centre for Ecumenical Studies, Mexico.
Mailing address: Yosemite 45, Col Nápoles, 03810 Mexico 18 DF, Mexico.

Castillo Lara, Rosalio José C, S.D.B., H E Card
Rel: Roman Catholic; Ordained on 04 SEP 1949; b. 04 SEP 1922 in San Casimiro, Maracay, Venezuela; celibate; Citizen of Venezuela.
International Organizations: Member (since 1985) of the College of Cardinals.
Member of the Society of St Francis of Sales. Member of the Congregation for Bishops. Member of the Congregation for Religious and Secular Institutes. Member of the Congregation for Catholic Education. Member of the Pontifical Council for Promoting Christian Unity.
Other organizations: President (since 1989), Amministrazione del Patrimonio della Sede Apostolica; President (since 1990), Pontifical Commission for the Vatican City State; Member, Supremo Tribunale della Segnatura Apostolica; Member, Institute for Religious Works; Member, Council of Cardinals and Bishops.
Career: Consecrated Bishop (Titular See of Praecausa) in 1973. Archbishop in 1982. Created Cardinal by Pope John Paul II in 1985. **Hons.:** Entitled Nostra Signora di Coromoto in S. Giovanni di Dio.
Mailing address: 0120 Città del Vaticano, Vatican.

Castles, Margaret
Rel: Buddhist: Mahayana.
International Organizations: Australia/New Zealand Regional Office of the Foundation for the Preservation of the Mahayana Tradition.
Mailing address: PO Box 1326, Chatswood, Sydney NSW 2067, Australia. T. (61 2) 419 8170. Fax (61 2) 417 1857

Castro, Emilio Enrique, Rev Dr
Rel: Methodist; Ordained in 1948; b. 02 MAY 1927 in Montevideo, Uruguay; m. to Gladys Nieves; c. 1 boy and 1 girl; Citizen of Uruguay; Lang. Spanish, English, French, German, Italian and Portuguese.
International Organizations: Past General Secretary (since 1985) of the World Council of Churches.
Career: Pastor, Central Methodist Church, Montevideo from 1957 to 1965. Coordinator, Commission for Evangelical Unity in Latin America from 1965 to 1972. Director, Commission on World Mission and Evangelism (WCC) from 1973 to 1983. Rector (President), Union Theological Seminary (Isedet) in 1984. Visiting Professor, Iliff Theological Seminary (USA) in 1984. Visiting Professor, Montpellier Institute of Theology (France) in 1984. **Educ.:** Th.L. from Union Theological Seminary, Buenos Aires, Argentina (1944-1950); P.G. work at University of Basel, Switzerland (1953-1954); Doc. from University of Lausanne, Switzerland (1984). **Hons.:** Hon. D.H.L., Westmar College, LeMars IA (1984). **Pubs.:** Freedom in Mission: The Perspective of the Kingdom of God – An Ecumenical Enquiry, WCC Publications, 1985; When We Pray Together, WCC Publications, 1989 and Editor, To the Wind of God's Spirit, WCC Publications, 1990.
Mailing address: PO Box 2100, Route de Ferney 150, CH-1211 Genève 2, Switzerland. T. (41 22) 791 61 11. C. OIKOUMENE Geneva. Tx 415 730 OIK CH. Fax (41 22) 791 03 61.

Castro, Erna
Rel: Quaker.
International Organizations: Honorary President of the Friends Peace Center.
Mailing address: Apartado 1507, 1000 San José, Costa Rica. T. (506) 33 61 68.

Castro, George F, Bishop
Rel: Evangelical Methodist; Ordained.
International Organizations: General Superintendent of the Evangelical Methodist Church in the Philippines.
Mailing address: 1240 General Luna Street, Ermita, Metro Manila, Philippines.

Catchings, Rose, Ms
Rel: Methodist.
International Organizations: UN Representative of the World Federation of Methodist Women in New York.
Mailing address: 5 High Terrace, Montclair NJ 07042, USA.

Catoir, John, Rev
Rel: Roman Catholic; Ordained.
International Organizations: Director of The Christophers.
Mailing address: 12 East 48th Street, New York NY 10017, USA. T. (1 212) 759 4050.

Catto, Charles, Rev
Rel: United Church of Canada; Ordained; b. 07 JUN 1929 in Toronto, Ontario, Canada; m. to Barbara Jean Loveys; c. 2 boys and 2 girls; Citizen of Canada.
International Organizations: Executive Director (since 1968) of the Frontiers Foundation.
Career: Missionary with Cree Indians, God's Lake, Manitoba from 1954 to 1957. Missionary, Board of World Mission, United Church of Canada under United Church of Zambia from 1957 to 1962. Minister, Hampton Charge from 1963 to 1968. Founded Operation Beaver under Canadian Council of Churches in 1963. Founded Operation Causepeace in 1968. Member, United Church of Canada Committee on International Affairs from 1963 to 1979. Cross-country speaker, sponsored by Association of Canadian Clubs from 1986 to 1987. Senator, Victoria University in 1990. Produced 2 LP albums ('Zambezi Valley Songs', 'The Third World Sings'). **Educ.:** B.A. from University of Toronto, Ontario, Canada (1951); M.Div. from University of Toronto, Emmanuel College, Ontario, Canada (1955). **Interests:** skating, model building, banjo playing, table tennis. **Hons.:** Order of Canada (1979); Guest of Honour, Native Council of Canada Banquet (1981); Aboriginal Order of Canada (1985).
Mailing address: 2615 Danforth Ave, Toronto, Ontario M4C 1L6, Canada. T. (1 416) 690 3930. Fax (1 416) 698 9846.
Residence: 712 Kingfisher Dr, Pickering, Ontario L1W 1X5, Canada.

Caudron, Marc
Rel: Roman Catholic.
International Organizations: Secretary-General of the International Federation of Catholic Universities.
Mailing address: 51 rue Orfila, F-75020 Paris, France. T. (33 1) 47 97 26 60. Fax 47 97 29 42.

Cavallari, Eugenio, O.A.D., Fr
Rel: Roman Catholic; Ordained; celibate.
International Organizations: Prior-General of the Order of Discalced Augustines.
Mailing address: Piazza Ottavilla 1, I-00152 Roma, Italy. T. (39 6) 589 63 45

Cavanaugh, Ellen
Rel: Roman Catholic.
International Organizations: Executive Director of the International Liaison of Lay Volunteers in Mission.
Mailing address: 4121 Harewood Rd NE, Washington DC 20017, USA. T. (1 202) 529 11000.

Cé, Marco, H E Card
Rel: Roman Catholic; Ordained on 27 MAR 1948; b. 08 JUL 1925 in Izano; celibate.
International Organizations: Member (since 1979) of the College of Cardinals.
Member of the Sacred Congregation for the Sacraments and Divine Worship. Member of the Congregation for the Clergy.
Other organizations: Patriarch, Venice.
Career: Created Cardinal by Pope John Paul II in 1979.
Mailing address: S Marco 318, I-30124 Venezia, Italy. T. (39 41) 522 56 94

Ceirano, Giovanni, H E Mgr
Rel: Roman Catholic; Ordained on 01 JUL 1951; b. 20 JUL 1927 in Lagnasco; celibate.
International Organizations: Apostolic Pro-Nuncio to Papua New Guinea (since 1989) of the Roman Catholic Church.
Other organizations: Apostolic Pro-Nuncio to the Solomon Islands (since 1990); Titular Archbishop, Tagase.
Mailing address: PO Box 98, Port Moresby, Papua New Guinea. T. (675) 25 60 21. Fax 25 28 44

Celata, Pier Luigi, H E Mgr
Rel: Roman Catholic; Ordained on 08 OCT 1961; b. 23 JAN 1937 in Pitigliano, Italy; celibate.
International Organizations: Apostolic Nuncio to Malta (since 1985) of the Roman Catholic Church.
Other organizations: Apostolic Nuncio to San Marino (since 1988); Titular Archbishop, Doclea.
Mailing address: Villa Cor Jesu, Pitkali Road, Attard, Malta.

Celeste, M, Mother
Rel: Roman Catholic; Ordained; celibate.
International Organizations: Superior General of the Sisters of the Holy Family of Nazareth.
Mailing address: Via Nazareth 400, I-00166 Roma, Italy. T. (39 6) 624 09 29. C. CAP Roma.

Cenkner, William, Prof
Rel: Roman Catholic; Ordained on 05 JUN 1958; b. 25 OCT 1930 in Cleveland OH, USA; celibate; Citizen of USA; Lang. English, Latin, French and Sanskrit.
International Organizations: Professor (since 1969) of the School of Religious Studies, Catholic University of America.
Other organizations: Dean, (since 1985), School of Religious Studies, Catholic University of America; Executive Board (since 1987), American Conference on New Religious Movements.
Main field of work: History of religions. **Career:** President, College Theology Society from 1978 to 1980. Visiting Professor, Nanzam Institute of Religion and Culture in 1983. **Educ.:** B.A. from Providence College, RI, USA (1949-1953); S.T.B., S.T.L. from Pontifical Faculty of Theology, Washington DC, USA (1954-1959); Ph.D. from Fordham University, Bronx NY, USA (1965-1969). **Interests:** religious liberty, Asian religions. **Mem.:** American Academy of Religion; Association for Asian Studies; Catholic Academy of Sciences in the US (1985); Knights of the Holy Sepulchre (1989) **Pubs.:** The Hindu Personality in Education: Tagore, Gandhi, Aurobindo; A Tradition of Teachers: Sankara and the Jagadgurus Today and Creative Moral Protest: Teachings of M K Gandhi.
Mailing address: School of Religious Studies, Catholic University of America, Washington DC 20064, USA. T. (1 202) 319 5683
Residence: 8019 Glenside Drive, Takoma Park MD 20912, USA. T. (1 301) 439 4495

Cerullo, Morris, Rev
Rel: Christian; Ordained.
International Organizations: President of the Morris Cerullo World Evangelism.
Mailing address: PO Box 700, San Diego CA 92138, USA. T. (1 619) 232 0161.

Cespites, Gennaro, C.PP.S., Fr
Rel: Roman Catholic; Ordained; celibate.
International Organizations: Procurator-General of the Society of the Precious Blood.
Mailing address: Viale di Porta Ardeatina 66, I-00154 Roma, Italy. T. (39 6) 574 16 56.

Chabot, Marie-Emmanuel, Rev Sister
Rel: Roman Catholic; Ordained; celibate; Citizen of Canada.
International Organizations: Member of the Ursulines de l'Union Canadienne.
Mailing address: Monastère des Ursulines, 2 rue de Parloir, Québec City, Québec G1R 4M5, Canada.

Chakaipa, Patrick Fani, Most Rev Archbishop
Rel: Roman Catholic; Ordained on 15 AUG 1965; b. 25 JUN 1932 in Mhondoro Communal Land, Zimbabwe; celibate; Citizen of Zimbabwe; Lang. Shona and English.
International Organizations: Archbishop (since 1976) of Harare.
Main field of work: Ministry. **Career:** Assistant Priest, Makumbi Mission from 1967 to 1969. Priest-in-Charge, All Souls Mission, Mutoko from 1969 to 1972. Episcopal Vicar, Mutoko, Murewa from 1970 to 1972. Auxiliary Bishop, Harare from 1973 to 1976. **Educ.:** Teacher Trng. from Kutama College, Zimbabwe (1956-1957); Seminary, Chishawasha, Zimbabwe (1958-1966). **Interests:** chess. **Pubs.:** *Karikoga Gumiremiseve*, 1958; *Pfumo Reropa*, 1961 and *Garandichauya*, 1963.
Mailing address: PO Box 8060, Causeway, Harare, Zimbabwe. T. (263 4) 727386
Residence: Archbishop's House, 66 Fifth Street, Harare, Zimbabwe. T. (263 4) 792125

Chan, Marc
Rel: Christian.
International Organizations: Contact of the Asian Institute of Christian Communication.
Mailing address: PO Box 581, Marine Parade, Singapore 9144, Singapore. T. (65) 748 056. Fax 746 0452.

Chance, Hugh Emery, Dr
Rel: Bahá'í; b. 28 DEC 1911 in Winfield KS, USA; m. to Margaret Chamberlin; c. 1 girl; Citizen of USA; Lang. English.
International Organizations: Member, Universal House of Justice (since 1963) of the Bahá'í International Community.
President of the International Association of Y's Men's Clubs.
Main field of work: Law and administration. **Career:** Private Practice (Law) from 1961 to 1963. US Navy from 1946 to 1961. Legal Counsel, International Chiropractors Association. Secretary, National Spiritual Assembly of the Baha'is of the United States. **Educ.:** B.A. from Cornell College, Mount Vernon IA, USA (1929-1933); J.D. from University of Iowa, Iowa City, USA (1932-1935). **Mem.:** Rotary Club of Haifa (President) **Pubs.:** With Eunice Braun, *A Crown of Beauty*, George Ronald, 1982 and *Iowa Annotations, Restatement of Conflicts of Law*.
Mailing address: Bahá'í World Centre, PO Box 155, 31001 Haifa, Israel. T. (972 4) 358 358. C. BAHAIFAITH, HAIFA. Tx 46626 BAYT IL. Fax (972 4) 372 425
Residence: 37 Harofe, Haifa, Israel. T. (972 4) 245 012

Chandrananda Kurukkal, P C, Sri
Rel: Hindu.
International Organizations: Member of the Mamanga Pillayar Kovil.
Mailing address: Mamanga Pillayar Kovil, Amirtha Kali, Sri Lanka.

Chandrasekara Kurukkal, S C, Sivasri
Rel: Hindu.
International Organizations: Chief Priest of the Sivasubramaniya Swamy Temple, Colombo.
Mailing address: Sivasubramaniya Swamy Temple, 131 Kew Road, Colombo 2, Sri Lanka.

Chang, Peter Hei Dong, Commr
Rel: Salvation Army, The.
International Organizations: Territorial Commander of The Salvation Army.
Mailing address: CPO Box 1192, Seoul 100 611, Korea Rep.

Chang, Sang, Prof Dr
Rel: Reformed Church.
International Organizations: Moderator, Department of Co-operation and Witness of the World Alliance of Reformed Churches.
Mailing address: Korean Cultural Research Inst, Ewha Women's Univ, Chin Chon, Sudaemoon-ku, Seoul, Korea Rep. T. (82 2) 362 6252 EXT 708. Fax (82 2) 393 5903
Residence: T. (82 2) 646 7650

Chang-Him, French Kitchener, Most Rev
Rel: Anglican; Ordained; b. 10 MAY 1938 in Seychelles; m. to Susan Talma; c. 2 girls; Citizen of Seychelles.
International Organizations: Bishop (since 1979) of Seychelles.
Other organizations: Archbishop (since 1984), Province of Indian Ocean.
Career: Primary school teacher in 1958. Manager, schools from 1973 to 1977. Chairman, Board of Governors, Teacher Training College from 1976 to 1977. Vicar-General, Diocese of Seychelles from 1973 to 1979. Archdeacon, Seychelles from 1973 to 1979. Dean, Province of Indian Ocean from 1983 to 1984. **Interests:** reading, international affairs, cooking, gardening, fishing. **Pubs.:** The Seychellois: In Search of an Identity, 1975.
Mailing address: Bishop's House, PO Box 44, Victoria, Seychelles. T. (248) 24242

Chan Soon Cheong, James C, Rt Rev

Rel: Roman Catholic; Ordained on 09 AUG 1959; b. 26 JUL 1926 in Selama, Perak, Malaysia; celibate.

International Organizations: Secretary-General/Treasurer (since 1975) of the Catholic Bishops' Conference of Malaysia-Singapore-Brunei.

Other organizations: Appointed 1st Bishop (since 1972), Melaka-Johor Diocese.

Career: Director, Catholic Information Service, Penang from 1959 to 1973. Assistant Priest, Church of Seven Sorrows, Penang from 1959 to 1963. Acting Parish Priest, St Anne's Church, Bukit Mertajam from 1963 to 1964. Assistant Priest, St Francis Xavier's Church, Penang from 1965 to 1966. Appointed Finance Assistant to late Bishop Francis Chan, Penang in 1967. Appointed Secretary to Bishop Gregory Yong, Penang from 1968 to 1970. Parish Priest, Holy Spirit Church, Penang from 1970 to 1973. Ordained Bishop in 1973. Vice-President, Christian Federation of Malaysia (CFM). Episcopal Chairman, Peninsular Malaysian Catechetical Commission. Episcopal Chairman, National Commission for Social Communications. **Educ.:** Studied from Minor Seminary of St Francis Xavier, Singapore (1950-1952); Studied Phil. and Theol. from College General, Penang, Malaysia (1953-1958). **Pubs.:** Living Our Faith, Bk 1 to 6 (Malaysian Series for school children) and Teaching Aids; Religious Correspondence Courses for Adults and Children, Bible Correspondence Course and Diocesan Magazine *News and Views* (English) and *The Bridge*.

Mailing address: Melaka-Johor Diocesan Secretariat, PO Box 319, 80730 Johor Bahru, Johor, Malaysia. T. (60 7) 248500.

Residence: 2101 Jalan Masai, 80730 Johor Bahru, Johor, Malaysia.

Chapellin, E Hernandez, Sr

Rel: Roman Catholic; Ordained.

International Organizations: Regional Secretary for America of the Catholic International Education Office.

Mailing address: CIEC, Apartado Aéreo 90036, Bogotá 8 DE, Colombia. T. (57 1) 55 36 76.

Chapman, Paul K, Dr

Rel: Protestant.

International Organizations: Executive Secretary of the North America Maritime Ministry Association.

Mailing address: PO Box 10, South Strafford VT 05070, USA. T. (1 802) 765 4130. Fax 765 4510.

Chapper, Adolfo, Fr

Rel: Roman Catholic; Ordained; celibate.

International Organizations: International Assistant of the International Movement for the Apostolate in the Independent Social Milieux.

Mailing address: Palazzo San Calisto 16, I-00120 Roma, Italy. T. (39 6) 698 71 83.

Chargueraud, Marc André

Rel: Christian.

International Organizations: Chairman of the International Churches-Enterprises Dialogue Steering Committee of the International Christian Union of Business Executives.

Mailing address: c/o Société Générale de Surveillance Holding SA, 34 Grand-Rue, CH-1297 Founex, Switzerland.

Charles, Nelson H, Rev

Rel: Methodist.

International Organizations: President of Conference of the Methodist Church Sierra Leone.

Mailing address: Wesley House, George Street, PO Box 64, Freetown, Sierra Leone.

Charlop, Zevulun, Rabbi

Rel: Jewish; Ordained in 1954; b. 14 DEC 1929 in New York NY, USA; m. to Judith Rosner; c. 2 boys and 6 girls; Citizen of USA; Lang. English, Hebrew and Yiddish.

International Organizations: Dean (since 1971) of the Rabbi Isaac Elchanan Theological Seminary, New York NY.

Other organizations: Rabbi (since 1954), Young Israel of Mosholu Parkway; Lecturer in American History (since 1966), Yeshiva College, New York NY; Dean, Mazer Yeshiva Program, Mazer School of Talmudic Studies, Yeshiva University, New York NY.

Main field of work: Education. **Educ.:** B.A. from Yeshiva College, New York NY, USA (1947-1951); Rabbinical Ord. from Rabbi Isaac Elchanan Theological Seminary, New York NY, USA (1951-1954); M.A. Amer.Hist. from Columbia University, New York NY, USA (1954-1959). **Pubs.:** *Chof Yamin on Rezan and Chagiga*, 2 Vols, 1980-1985; "The Making of a Rabbi", in *Encyclopedia Judaica*, 1986 and "G-D in History and Halakha from the Perspective of American History", in *The Torah U Mada Journal*, Vol. 1.

Mailing address: 2540 Amsterdam Ave, New York NY 10033, USA. T. (1 212) 960 5344

Residence: Apartment 2B, 100 East Mosholu Parkway South, Bronx NY 10458, USA T. (1 212) 365 6810

Charny, Israel W, Prof

Rel: Jewish.

International Organizations: Executive Director of the Institute on the Holocaust and Genocide.

Mailing address: PO Box 10311, Jerusalem 91102, Israel. T. (972 2) 720424. Fax 713606, Ext 7118. Tx 26144, BXJM IL, Ext 7118.

Charria, Manuelita, O.P. d. Pres., Sr

Rel: Roman Catholic; Ordained; celibate.

International Organizations: Assistant Secretary of the Latin American Confederation of Religious Orders.

Mailing address: Calle 64 Nr 10-45 – 5° piso, Apartado Aéreo 90710 Bogotá, Columbia. T. (57 91) 211 66 56, (57 91) 211 66 86

Charzat, J

Rel: Roman Catholic.

International Organizations: UNESCO Representative of the Catholic International Education Office.

Mailing address: 32 rue du Maréchal Joffre, F-78000 Versaille, France.

Chau, Thich Tam, Ven

Rel: Buddhist.

International Organizations: President of the World Buddhist Social Service.

Mailing address: 32 avenue Docteur Ménard, F-06000 Nice, France.

Chaudhry, Saeed Ahmad

Rel: Muslim.

International Organizations: Contact of the Moslem Mission.

Mailing address: Brienner Strasse 7-8, D-10713 Berlin, Germany. T. (49 30) 87 57 03.

Chavez Campos, Enrique C, Bishop

Rel: Pentecostal; Ordained.

International Organizations: Bishop of Pentecostal Church of Chile.

Mailing address: Calle Pena 1103, Casilla de Correo 2, Curico, Chile.

Chazelle, M E

Rel: Christian.

International Organizations: President of the International Missionary Benefit Society.

Mailing address: Rue Barra 7, F-42000 Saint-Etienne, France.

Checkley, David

Rel: Protestant.

International Organizations: Australian Director of the Africa Inland Mission International.

Mailing address: 37 Hercules Street, Chatswood NSW 2067, Australia.

Cheli, Giovanni, Mgr

Rel: Roman Catholic; Ordained on 21 JUN 1942; b. 04 OCT 1918 in Turin, Italy; celibate; Citizen of Italy; Lang. Italian, English, French and Spanish.

International Organizations: President (since 1989) of the Pontifical Council for the Pastoral Care of Migrants and Itinerant People.
Member of the Pontifical Commission for Latin America. Member of the Pontifical Council 'Cor Unum'. Member of the Pontifical Council for Inter-Religious Dialogue.
Main field of work: Diplomat. **Career:** Second Secretary, Apostolic Nunciature in Guatemala from 1952 to 1955. First Secretary, Madrid from 1955 to 1962. Counsellor, Nunciature in Rome from 1962 to 1967. Served on Council for Public Affairs of the Church, Vatican City from 1967 to 1973. Permanent Observer to UN from 1973 to 1986. Named Archbishop, Santa Giusta (Apostolic Nuncio), Sardinia in 1978. **Educ.:** Lic. Theol. from Lateran Pontifical University, Vatican City State (1946-1948); J.C.D. from Lateran Pontifical University, Vatican City State (1948-1952); Dip. from Pontifical Academy for Diplomats, Vatican City State (1950-1952). **Interests:** mountain climbing, reading, classical music. **Hons.:** Knight Commander, Orden de Isabel la Católica (Spain), Ordine al Merito della Repubblica Italiana; Verdienstkreuz der Bundesrepublik Deutschland. **Pubs.:** L'applicazione delle Riforme Tridentine nella diocesi do Asti, 1952.
Mailing address: Palazzo San Calisto 16, 00120 Città del Vaticano, Vatican. T. (39 6) 698 7131
Residence: See Mailing Address. T. See Mailing Address

Chenouda, Nasri Gayeb, III, H H

Rel: Coptic Orthodox; Ordained; b. in Egypt.

International Organizations: Pope and Patriarch (since 1971) of Coptic Orthodox Church.
Career: Priest in Dair as-Surian in 1954. Secretary, late Pope and Patriarch Cyrille VI from 1958 to 1971. Bishop, the Religious Education in 1963. Visited Greek and Russian Orthodox Patriarchs, Istanbul in 1971. Visited the Pope, Vatican in 1973. **Pubs.:** Author of several books and booklets.
Mailing address: Coptic Orthodox Church, Azbakia, Cairo, Egypt.

Cher, Arturo Palma, Rev

Rel: Pentecostal; Ordained.

International Organizations: Secretary of the Pentecostal Mission Church.
Mailing address: Casilla 618, Los Angeles, Chile.

Chernoff, David

Rel: Jewish.

International Organizations: Chairman of the International Alliance of Messianic Congregations and Synagogues.
Mailing address: PO Box 417, Wynnewood PA 19096, USA. T. (1 215) 896 5812.

Chernoff, Joel

Rel: Jewish: Messianic; Ordained on 15 APR 1976; b. 20 AUG 1950 in Atlanta GA, USA; m. to Mindy Lee; c. 1 boy and 2 girls; Citizen of USA; Lang. English.

International Organizations: President (since 1991) of the Messianic Jewish Alliance of America.
Other organizations: Member, Messianic Jewish Alliance of America.
Career: President, Messianic Record Inc from 1973 to 1992. Vice-President, International Messianic Jewish Alliance from 1989 to 1992. **Educ.:** B.A. Pol.Sc. from University of Cincinnati, OH, USA (1968-1972). **Mem.:** National Association of Recording Artists; Messianic Jewish Alliance of America
Mailing address: 50 E Wynnewood Road, Suite 217, Wynnewood PA 19096, USA. T. (1 215) 896 5812.
Residence: 7605 Brook Haven Road, Philadelphia PA 19151, USA T. (1 215) 477 7342

Chevrier, Jôdô André, Rev Prof Dr

Rel: Buddhist; Ordained on 29 OCT 1975; b. 19 OCT 1924 in Geneva, Switzerland; sep. from Fabienne Rossier; c. 3 boys and 1 girl; Citizen of Switzerland; Lang. French, Latin, German, English, Italian, Spanish, Russian, Dutch, Romanian and Japanese.

International Organizations: European Representative to Nishi Hongwanji (Kyoto) (since 1973) of the Jôdo Shinshû Hongwanji-ha.
Other organizations: Full Professor of Psychology (since 1988), La Jolla University, San Diego CA, European Campus (Lugano and Lausanne, Switzerland).
Main field of work: Ministry. **Career:** With SBC and SVB (banks) from 1951 to 1972. **Educ.:** Lic. ès sc.écon. et soc. from University of Geneva, Switzerland (1944-1947); Institute of Buddhist Studies, Berkeley CA, USA (1974-1978). **Interests:** psychology. **Hons.:** Ph.D. h.c., University of Oriental Studies, Los Angeles CA (1980). **Mem.:** Buddhist Union of Europe; Union bouddhique suisse **Pubs.:** Translations into French of the Shôshin Ge and Tanni Shô and publisher of *The Link* (newsletter of the Nen-Butsu followers in Europe) for 6 years.
Mailing address: Nen-Butsu An, PO Box 25, CH-1400 Yverdon-les-Bains, Switzerland. T. (41 24) 21 07 29
Residence: Rue des Chaînettes 25, CH-1400 Yverdon-les-Bains, Switzerland. T. See Mailing Address

Chianghnuna, C, Col

Rel: Salvation Army, The.

International Organizations: Territorial Commander of The Salvation Army.
Mailing address: PO Box 4510, Bombay 400 008, India.

Chikane, Frank, Rev Dr

Rel: Apostolic Faith Mission; Ordained in 1980; b. 03 JAN 1951 in Johannesburg, South Africa; m. to Kagiso Oglobry; c. 3 boys; Citizen of South Africa; Lang. English, Tswana, Zulu and S. Sotho.

International Organizations: Secretary-General (since 1987) of the South African Council of Churches.
Main field of work: Ministry. **Career:** Worked with Christ for All Nations from 1975 to 1976. Placed in a congregation as a probationary pastor under supervision from 1976 to 1979. Part-time research officer at Institute of Contextual Theology in 1981. Appointed Co-ordinator, of the Institute in 1982. Appointed Secretary-General, of the Institute in 1983. Deputy President, United Democratic Front (UDF) from 1983 to 1985. Deputy President, Soweto Civic Association in 1987. Detained several times. **Educ.:** B.Sc. (uncompleted) from University of the North, Pietersburg, South Africa (1972-1974); B.Sc. (uncompleted) from University of South Africa, Pretoria, South Africa (1972-1974); B.Th. from University of South Africa, Pretoria, South Africa (1984-1985). **Interests:** reading, keeping fit (mentally, spiritually, physically). **Hons.:** Diakonia Peace Prize (1986); Star Crystal Award (1987); Third World Prize (1989). **Mem.:** Ecumenical Association of Third World Theologians (EATWOT) **Pubs.:** "The Incarnation in the Life of the People in Southern Africa", in *Journal of Theology for Southern Africa*, June 1985, Cape Town UCT; "Doing Theology in a Situation of Conflict", in *Resistance and Hope: South African Essays in Honour of Beyers Naude*, Charles Villa-Vicencio and John W de Gruchy, Eds., Cape Town, JHB: David Phillip, 1985 and "Prophetic Witnessing" in *The Relevance of Evangelism in South Africa Today*: Papers of the seminar held at Hammanskraal Transvaal, January 1986, Mark Hestenes, Johann Botha, Bongani Mazibuko and Andrew Jossias, Eds., JHB, SACC.
Mailing address: SACC, PO Box 4921, Johannesburg 2000, South Africa. T. (27 11) 492 1380/9
Residence: 435 Diepkloof Extension, Phase Two, Soweto, South Africa. T. (27 11) 985 4988

Childs, Derrick Greenslade, Most Rev

Rel: Church of Wales.

International Organizations: Archbishop of Church of Wales.
Mailing address: c/o Religious Education and Communication Centre, Woodland Place, Penarth, South Glamorgan CF6 2EX, UK.

Chilstrom, Herbert W, Rt Rev

Rel: Lutheran.

International Organizations: Vice-President of the Lutheran World Federation.
Member, Program Committee for Theology and Studies of the Lutheran World Federation.
Other organizations: Presiding Bishop, Evangelical Lutheran Church in America.
Mailing address: 8765 West Higgins Road, Chicago IL 60631, USA. T. (1 312) 380 2700. Fax 380 1465. Tx 4900009321 ELC

Ching, Peter P S
Rel: Christian.
International Organizations: President of the American Mission to the Chinese and Asian.
Mailing address: 144-25 Roosevelt Ave, Flushing NY 11354, USA. T. (1 212) 461 5756.

Ching, Peter P S, DD, Dr
Rel: Christian.
International Organizations: Contact of the International Clergy Council.
Mailing address: 399 Pendleton St, Suite 705, Alexandria VA 22314, USA. T. (1 713) 549 2962.

Chipenda, José Belo, Rev
Rel: Evangelical Congregational; Ordained in 1955; b. 09 OCT 1929 in Bailundo, Huambo, Angola; m. to Eva Joao Miguel de C.; c. 1 boy and 1 girl; Citizen of Angola; Lang. Portuguese, English and French.
International Organizations: Secretary-General (since 1987) of the All Africa Conference of Churches.
Main field of work: Coordination, representation, ministry. **Career:** Pastor, Evangelical Church in Lobito from 1955 to 1960. Secretary for Africa, World Student Christian Federation from 1969 to 1974. Executive Secretary, World Council of Churches from 1974 to 1980. General Secretary, Anglican Council of Evangelical Churches from 1985 to 1987. **Educ.:** B.D. from Seminario Presbiteriano, Carcavelos, Portugal (1951-1955); M.A. from Hartford Foundation, CT, USA (1960-1961).
Mailing address: Waiyaki Way, Westland, PO Box 14205, Nairobi, Kenya. T. (254 2) 742281. Fax 742352. C. CHURCHCON. Tx 22175.
Residence: Olenervone Road, Lavington, Nairobi, Kenya. T. (254 2) 48588

Chiphangwi, Saindi, Rev Dr
Rel: Reformed Church.
International Organizations: Coordinator, Southern Africa Alliance of Reformed Churches of the World Alliance of Reformed Churches.
Mailing address: PO Box 1728, Gaborone, Botswana. T. (267) 313 697. Fax (267) 375 245

Chisanga, Joel, Rev
Rel: United Church of Zambia; Ordained.
International Organizations: General Secretary of the United Church of Zambia.
Mailing address: PO Box 50122, Nationalist Road at Burma Road, Lusaka, Zambia.

Chiwanga, Simon, Rev
Rel: Church of the Province of Tanzania; Ordained.
International Organizations: Provincial Secretary of the Church of the Province of Tanzania.
Mailing address: PO Box 35, Korogwe, Tanzania UR.

Cho, David J
Rel: Evangelical.
International Organizations: Secretary-General of the Asia Missions Association.
Mailing address: PO Box 2732, Seoul, Korea Rep. T. (82 813) 1129.

Cho, Paul Yonggi, Rev Dr
Rel: Christian; Ordained on 18 MAY 1958; b. 14 FEB 1936 in Eulju-Gun, Kyungnam, Korea Rep; m. to Grace Sunghae Kim; c. 3 boys; Citizen of Republic of Korea; Lang. English, Japanese and Korean.
International Organizations: Chairman (since 1976) of the Church Growth International.
Other organizations: Senior Pastor (since 1958), Yoido Full Gospel Church; Chairman (since 1982), Full Gospel World Missions Association; Professor (since 1985), Full Gospel Theological Seminary.
Main field of work: Ministry. **Career:** Pioneered Full Gospel Tent Church in 1958. General Superintendent, Korea Assemblies of God in 1966. Invited by the Secretary of State as the main speaker, Revival Crusade in the USA in 1966. Executive Member, Pentecostal World Conference in 1967. Chairman, Korea Assemblies of God in 1970. Chairman, the 10th Pentecostal World Conference in 1973. Chairman, Hoseo University in 1978. Invited as the main speaker, the 40th Inauguration Ceremony of President Reagan in the USA in 1981. Invited as the main speaker, NRB and NAE in the USA in 1983. Invited as the main speaker, The Centiannual Ceremony of Christianity in Korea in 1985. **Educ.:** Grad. Theol. from Full Gospel Theological Seminary, Seoul, Korea Rep (1956-1958); Grad. Law from Kukmin University, Seoul, Korea Rep (1962-1966). **Interests:** world mission. **Hons.:** Honorary degrees (all in the USA): Doctor of Theology, Bethany Bible College (1968), Doctor of Literature, California Graduate School of Theology (1978), Doctor in Ministry, Oral Roberts University, Tulsa OK (1989), Doctor in Ministry, Regent University, Virginia Beach (1990). **Pubs.:** *Church Growth*; *The Fourth Dimension*, I and II and *The Holy Spirit, My Senior Partner*.
Mailing address: c/o Yoido Full Gospel Church, PO Box 7, Seoul 150-600, Korea Rep. T. (82 2) 783 9920. Tx FGGCKAG K28200. Fax (82 2) 784 9429.
Residence: 11 Yoido-Dong, Seoul 150-010, Seoul, Korea Rep. T. (82 2) 783 4135

Choi, Stephen, Rev
Rel: Protestant.
International Organizations: Contact of the Korea International Seamen's Mission.
Mailing address: PO Box 84, Inchon, Korea Rep.

Chon, Pal Khn, Prof
Rel: Buddhist.
International Organizations: Member of the Executive Council of the World Fellowship of Buddhists.
Mailing address: 344-2 Sin Yong-Dong, Iri, Chun-Buk, Korea DPR.

Choné, Monique, FMM, Sr
Rel: Roman Catholic; Ordained; celibate.
International Organizations: Secretary-General of the Franciscaines Missionnaires de Marie.
Mailing address: 12 via Giusti, I-00185 Roma, Italy. T. (39 6) 731 6741

Chong, Richard Lai Pong
Rel: Roman Catholic.
International Organizations: International President of the International Catholic Committee of Nurses and Medico-social Assistants.
Mailing address: PO Box 352, Jalan Sultan, Petaling Jaya, Malaysia.

Choo, Teh Thean
Rel: Buddhist.
International Organizations: Member of the Executive Council of the World Fellowship of Buddhists.
Mailing address: 252 Jalan Seragam, United Garden, Jalan Klang, Petaling Jaya, Kuala Lumpur, Malaysia.

Chooi, Theresa Ee
Rel: Roman Catholic.
International Organizations: President of the Southeast Asian Catholic Press Association.
Mailing address: Catholic Reseach Center, 528 Jalan Bukit Nana, Kuala Lumpur, Malaysia.

Chrapek, Jan, C.S.M.A., Fr
Rel: Roman Catholic; Ordained; celibate.
International Organizations: Superior General of the Congregation of the Archangel St Michael.
Other organizations: President, Konferencja Przelozonych Wyzszych Zakonów Meskich w Polsce.
Mailing address: ul Swierczewskiego 248/252, 05-261 Warzawska, Poland. T. (48 22) 30 04 50

Christano, Peter Charles, Rev
Rel: Mennonite; Ordained on 21 OCT 1971; b. 28 APR 1939 in Pati, Central Java, Indonesia; m. to Elisabeth Naftalih; c. 2 boys and 1 girl; Citizen of Indonesia; Lang. Indonesian, Javanese and English.
International Organizations: Advisor to Mennonite Church Conference (since 1992) of the Indonesia.
Other organizations: Pastor (since 1976), Kudus Mennonite Church; Vice-Chairman (since 1989), Satya Wacana Christian University.
Main field of work: Ministry. **Career:** Pastor-at-large, Mennonite Churches of Indonesia from 1971 to 1976. Council Member, Indian National Council of Churches from 1971 to 1984. President, Indian Mennonite Conference (GKMI) from 1971 to 1978. President, Mennonite World Conference from 1978 to 1984. Vice-Chairman, Asia Mennonite Conference from 1971 to 1978. **Educ.:** B.A. from Sanata Dharma, Yogyakarta, Indonesia (1961-1964); M.Div. from Jakarta Seminary, Jakarta, Indonesia (1964-1969); Dip. Discpl. from Discipleship Training Centre, Singapore (1969-1971); Dip. Judaism from Jerusalem Institute of Biblical Research and Studies, Israel (1989-1990). **Interests:** Evangelism, pastoral, dialogue, Islam. **Pubs.:** Author of a series on anabaptist history and teachings, baptism, foundations to the Christian faith (all in Indonesian).
Mailing address: Jln Kha Wahid Hasyim 78, Kudus 59317, Indonesia. T. (62 291) 21524
Residence: Jln Kha Wahid Hasyim 74, Kudus 59317, Indonesia

Christensen, Birthe J
Rel: Christian.
International Organizations: Secretary of the Churches Third World Information.
Mailing address: Nørregade 11, 2 Sal, DK-1165 København K, Denmark. T. (45 33) 15 59 27.

Christiansen, Egon
Rel: Protestant.
International Organizations: Secretary-General of the Nordic Seamen's Church Council.
Mailing address: Dansk Kirke i Udlandet, Norrevaenget 43B, DK-5000 Odense, Denmark. T. (45 9) 66 13 95 31.

Christie, William Douglas McLaren, Ven
Rel: Anglican; Ordained in 1936; b. 12 NOV 1910 in St Johnsbury VT, USA; m. to Pamela Kerrigan Henderson; c. 1 boy and 1 girl; Citizen of Canada.
International Organizations: Ecclesiastic (retired) of the Anglican Church.
Career: Ordained Deacon in 1935. Curate at Bury, Quebec from 1935 to 1936. Incumbent, Montage and Franktown (Diocese of Ottawa) from 1936 to 1939. Incumbent, Vankleek Hill, Ontario from 1939 to 1943. Rector, Renfrew from 1943 to 1946. Rector, All Saints, Westboro, Ottawa from 1946 to 1958. Rector, Trinity Church, and Archdeacon of Cornwall in 1958. Diocesan Archdeacon and Director of Programme for Diocese of Ottawa from 1967 to 1972. Diocesan Archdeacon and Rector, St Bartholomew's Church, Ottawa from 1972 to 1979. Chaplain to the Governor General from 1967 to 1980. Rural Dean, Ottawa from 1951 to 1957. Part-time Chaplain, RCAF (Royal Canadian Air Force) from 1941 to 1943. Chaplain, Gov.-Gen. Foot Guards from 1947 to 1958. Retired in 1979. **Educ.:** B.A. Theol. from Bishop's University, Lennoxville, Quebec, Canada (1935). **Hons.:** Order of St Lazarus of Jerusalem; Efficiency Decoration.
Mailing address: 23 MacKay St, Apt 2, Ottawa ON K1M 2B1, Canada.
Residence: See Mailing Address.

Christinat, Ann-Marie, Mme
Rel: Unknown.
International Organizations: Secretary of the Meditation Group of United Nations Staff Members, Geneva.
Mailing address: Bureau E-8051, Palais des Nations, CH-1211 Genève 10, Switzerland.

Chrysostomos, C, Archbishop
Rel: Orthodox; Citizen of Cyprus.
International Organizations: Archbishop of Nova Justiniana and all Cyprus.
Other organizations: Metropolitan, Paphos and Kitium.
Mailing address: POB 1130, Arch. Kyprianos St, Nicosia, Cyprus. T. (357 2) 474411. Fax (357 2) 474180

Chuba, Bwalya S, Rev
Rel: United Church of Zambia.
International Organizations: Secretary-Treasurer of the Africa Association for Liturgy, Music and Arts.
Mailing address: c/o United Church of Zambia, PO Box 50122, Lusaka, Zambia.

Ciappi, Mario Luigi, O.P., H E Card
Rel: Roman Catholic; Ordained on 26 MAR 1932; b. 06 OCT 1909 in Florence, Italy; celibate; Citizen of Italy.
International Organizations: Member (since 1977) of the College of Cardinals.
Member of the Order of Preachers.
Other organizations: President, Pontificia Accademia Romana di S. Tommaso D'Aquino e di Religione Cattolica.
Career: Titular Bishop, Miseno in 1977. Created Cardinal by Pope Paul VI in 1977. Deacon, Nostra Signora del S Cuore. **Mem.:** Entitled S. Cuore di Gesù agonizzante a Vitinia
Mailing address: Via di Porta Angelica 63, I-00193 Roma, Italy. T. (39 6) 654 73 66

Cifuni, Pietro, R.C.I., Fr
Rel: Roman Catholic; Ordained; celibate.
International Organizations: Superior General of the Rogationistes du Coeur de Jésus.
Mailing address: Via Tuscolana 167, I-00182 Roma, Italy. T. (39 6) 702 0751.

Cinà, Giuseppe, M.I., Fr
Rel: Roman Catholic; Ordained; celibate.
International Organizations: Procurator-General of the Order of the Servants of the Sick.
Other organizations: President, Institute of Holy "Camillianum" Pastoral Theology.
Mailing address: Piazza della Maddalena 53, I-00186 Roma, Italy. T. (39 6) 679 77 96 – 679 77 97.

Citak, Audrey
Rel: Jewish: Conservative.
International Organizations: President of the Women's League for Conservative Judaism.
Mailing address: 48 East 74th Street, New York NY 10021, USA. T. (1 212) 628 1600. Fax (1 212) 772 3507.

Clancy, Edward Bede, H E Card
Rel: Roman Catholic; Ordained on 23 JUL 1949; b. 13 DEC 1923 in Lighgow NSW, Australia; celibate; Citizen of Australia.
International Organizations: Member (since 1988) of the College of Cardinals.
Member of the Synod of Bishops. Member of the Congregation for Bishops. Member of the Pontifical Council for Social Communications.
Other organizations: Archbishop (since 1983), Sydney; President, Australian Catholic Bishops' Conference; Member, Council of Cardinals for the Study of Organizational and Economic Problems of the Holy See; Member, Council of Cardinals and Bishops.
Career: Auxiliary Bishop, Archdiocese of Sydney from 1974 to 1978. Archbishop, Archdiocese of Canberra and Goulburn from 1979 to 1983. Created Cardinal by Pope John Paul II in 1988. **Interests:** golf. **Hons.:** Officer of the Order of Australia.
Mailing address: Chancery Office, Polding House, 276 Pitt Street, Sydney NSW 2000, Australia. T. (61 2) 232 8073

Clark, Andrew C
Rel: Quaker.
International Organizations: Secretary-General of the Religious Society of Friends.
Secretary-General of the Quaker Peace and Service.
Mailing address: Friends House, Euston Road, London NW1 2BJ, UK. T. (44 71) 387 36 01. C. ENHEARTEN LONDON NW1. Fax (44 71) 388 1977.

Clark, R Norris
Rel: Protestant.
International Organizations: Chairman of the International Christian Youth.
Mailing address: 756 Haddon Ave, Collingswood NJ 08108, USA. T. (609) 858 0700.

Clarke, Eric St C, Rev

Rel: Methodist.
International Organizations: President of the Conference of the Methodist Church in the Caribbean and the Americas.
Mailing address: Belmont, PO Box 9, St John's, Antigua-Barbuda. T. (1 809) 46 21234.

Clavel, Pierre, Fr

Rel: Roman Catholic; Ordained; celibate.
International Organizations: Procurator-General of the Prêtres de l'Oratoire de France.
Mailing address: 2 bis quai des Célestins, F-75004 Paris, France. T. (33 1) 42 78 40 67.

Claverie, Pierre Lucien, O.P., Mgr

Rel: Roman Catholic; Ordained on 04 JUL 1965; b. 08 MAY 1938 in Algiers, Algeria; celibate.
International Organizations: Bishop of Oran.
Member of the Order of Preachers.
Mailing address: Evêché, 2 rue Saad Ben Rebbi, Oran, Algeria. T. (213 6) 35 61 13

Clearfield, Sidney M

Rel: Jewish.
International Organizations: Past Contact of the B'nai B'rith Youth Organization.
Mailing address: 1640 Rhode Island Ave NW, Washington DC 20036, USA. T. (1 202) 857 6585. Fax 857 1099.

Clément, Pierre

Rel: Christian.
International Organizations: European Regional Coordinator of the Every Home for Christ International.
Mailing address: 39 rue Trébois, F-92300 Levallois-Perret, France.

Clements, Geoffrey, Dr

Rel: Hindu.
International Organizations: Vice-Chancellor of the Maharishi University of Natural Law.
Mailing address: Mentmore Towers, Mentmore, Leighton Buzzard LU 7 OQH, UK. T. (44 296) 661 881 – 662 186. Tx 9312130390 MT G. Fax (44 296) 662 049.

Clemis, Abraham Mar, Metropolitan

Rel: Malankara Jacobite Syrian Orthodox Church of India; Ordained.
International Organizations: Past Co-Chairman of the Joint International Commission for Dialogue between the Roman Catholic Church and the Malankara Jacobite Syrian Church.
Mailing address: Mor Aphrem Seminary, 686531 Chingavanam, Kerala, India.

Clímaco, López Montes, OAR, Fr

Rel: Roman Catholic; Ordained; celibate.
International Organizations: Procurator-General of the Order of the Augustinian Recollects.
Mailing address: Viale dell'Astronomia 27, I-00144 Roma, Italy. T. (39 6) 592 65 34.

Cline, Philip A, Rev

Rel: United Church of Canada; Ordained.
International Organizations: Secretary of the General Council of the United Church of Canada.
Mailing address: 85 St Clair Ave East, Toronto, Ontario M4T 1M8, Canada.

Clivaz-Nielsen, Isabelle

Rel: Christian.
International Organizations: Head, Secretariat of the Action Children Aid, Mother Theresa's Co-workers.
Mailing address: Ryparken 98, DK-2100 København Ø, Denmark. T. (45) 31 20 36 10.

Cloete, Glynn

Rel: Roman Catholic; Unordained; b. 04 SEP 1960 in Cape Town, South Africa; married; Citizen of South Africa; Lang. English, French and Afrikaans.
International Organizations: Secretary-General (since 1987) of the International Young Christian Workers.
Main field of work: Training and co-ordination.
Mailing address: Rue Plantin 11, B-1070 Bruxelles, Belgium. T. (32 2) 521 69 83. C. JOCINT BRUXELLES
Residence: See Mailing Address. T. See Mailing Address

Cobb, John Boswell, Jr, Prof

Rel: United Methodist; Ordained on 21 JUN 1955; b. 09 FEB 1925 in Kobe, Hyogo, Japan; m. to Jean Loftin; c. 4 boys; Citizen of USA.
International Organizations: Co-Director (since 1973) of the Center for Process Studies.
Other organizations: Part-time Professor (since 1990), Claremont Graduate School.
Career: Assistant Professor, Emory University, Atlanta GA from 1953 to 1958. Professor, School of Theology at Claremont from 1958 to 1990. **Educ.:** M.A., Ph.D. from University of Chicago, IL, USA (1947-1950). **Interests:** process theology, ecological theology, Buddhist-Christian dialogue. **Hons.:** Woodrow Wilson Center Fellow; Fulbright Professor. **Mem.:** American Academy of Religion; American Metaphysical Society **Pubs.:** With Charles Birch, The Liberation of Life, Cambridge University Press; With Herman Daly, For the Common Good, Beacon Press and Matters of Life and Death, Westminster/John Knox Press.
Mailing address: 777 N Cambridge Way, Claremont CA 91711, USA. T. (1 714) 626 2686
Residence: See Mailing Address. T. See Mailing Address

Cochrane, Charles Clarke, Rev

Rel: Presbyterian; Ordained; b. 30 NOV 1910 in Belleville ON, Canada; m. to Isobel Jean Cameron; c. 3 boys; Citizen of Canada.
International Organizations: Ecclesiastic of the Presbyterian Church.
Career: Minister, Presbyterian Tri-Congregations, Toronto from 1976 to 1979. Minister, St Andrew's, Geraldton, Ontario from 1939 to 1941. Minister, Knox., Georgetown, Ontario from 1941 to 1948. Minister, Melville, Westmount, Quebec from 1948 to 1976. Member, Senate, Presbyterian College, Montreal from 1949 to 1952. Member, Board, Presbyterian College, Montreal from 1952 to 1958. Lecturer, Presbyterian College, Montreal from 1964 to 1966. Member, Board, Westmount YMCA from 1950 to 1958. Member, Board, Montreal Council for Social Order from 1952 to 1953. Member, Board, Prot. Hungarian Service Association from 1957 to 1959. Member, Board, Greater Montreal Council of Churches from 1953 to 1960. President, Greater Montreal Council of Churches from 1957 to 1960. Moderator, Presbytery of Montreal (1955 and 1965). Clerk, Presbytery of Montreal from 1957 to 1958. President, Westmount Min. Association from 1959 to 1960. Advisor to Montreal Christian Youth Council from 1957 to 1960. Senior Teacher, Religious Instructor, Westmount High School from 1952 to 1966. Vice-President, North American Area Council, World Alliance of Reformed Churches (WARC) in 1960. President, North American Area Council, WARC in 1965. Quebec Correspondent, *The Christian Century* from 1958 to 1966. Member, Board, Children's Service Centre, Society for the Protection of Women and Children, Elizabeth House. Trustee, Fund to Contest the 'Padlock Law'. **Educ.:** B.A. from Queen's University at Kingston, Ontario, Canada (1938); Dip. from University of Toronto, Knox College, Canada (1939); B.D. from University of Toronto, Emmanuel College, Canada (1944); D.D. from University of Toronto, Knox College, Canada (1969). **Pubs.:** The Gospel According to Genesis, 1984; Contending With God, 1990 and occasional contributor to religious journals.
Mailing address: 300 Regina St N, Richmond One, Apt 212, Waterloo ON N2J 3B8, Canada.

Coertzen, Pieter, Prof

Rel: Dutch Reformed; Ordained on 07 NOV 1970; b. 20 APR 1943 in Boksburg, Transvaal, South Africa; m. to Maria Aletta Scheepers; c. 3 girls; Citizen of South Africa; Lang. Afrikaans, English, German, French and Zulu.
International Organizations: Actuarius of Synod (Member of Moderamen) (since 1991) of the Dutch Reformed Church of South Africa.
Other organizations: Professor (since 1988), University of Stellenbosch; Chairman, Board of Trustees (since 1988), Huguenot Museum.

Main field of work: University education, ministry. **Educ.:** B.A.; M.A. from Potchefstroom University for Christian Higher Education, South Africa (1961-1964); B.Th.; M.Th. from University of Stellenbosch, South Africa (1965-1969); D.Th. from University of Stellenbosch, South Africa (1969-1976). **Mem.:** Full Member, SA Academy for Science and Arts **Pubs.:** Vanaf Genève na Suidpunt van Afrika, 1984; The Huguenots of South Africa, 1988 and Gepas en Ordelik, 1991.
Mailing address: University of Stellenbosch, 171 Dorp Street, Stellenbosch 7600, South Africa. T. (27 2231) 773576
Residence: 4 Ackermann Street, Stellenbosch 7600, South Africa T. (27 2231) 72619

Coff, Pascaline, OSB, Sr

Rel: Roman Catholic; Ordained; celibate.
International Organizations: Member of the Benedictine Nuns.
Other organizations: Member, Association of Contemplative Sisters.
Mailing address: Osage Monastery, 18701 W Monastery Rd, Sand Springs OK 74063, USA.

Coffin, Richard Clifford, Rev

Rel: Baptist; Ordained; b. 19 DEC 1936 in Boston MA, USA; m. to Jean MacKinnon; c. 1 boy and 1 girl; Citizen of Canada.
International Organizations: General Secretary-Treasurer (since 1985) of the Canadian Baptist Federation.
Career: Held pastorates in Jeddor-Oyster Pond Un. Baptist Churches from 1964 to 1966. Held pastorate in New Minas Baptist Church from 1966 to 1970. Pastor, First Baptist Church, Dartmouth, NS from 1970 to 1976. Pastor, Calgary First Baptist Church from 1976 to 1978. Pastor, Blythwood Road Baptist Church, Toronto from 1979 to 1985. Executive Member, Baptist World Alliance and BWA Council. Vice-President and Executive Member, North American Baptist Fellowship. Ex-Officio Member, Canadian Baptist Overseas Mission Board. **Educ.:** B.A. from Acadia University, Wolfville, Nova Scotia, Canada (1966); M.Div. from Acadia Divinity College, Wolfville, Canada (1969); D.Min. from American Baptist Seminary of the West, Covina CA, USA (1974). **Interests:** Old Timers hockey.
Mailing address: Canadian Baptist Place, 7185 Milcreek Dr, Mississauga, Ontario L5N 5R4, Canada.
Residence: 532 Soudan Ave, Toronto ON M4S 1X4, Canada.

Coffin, William Sloane, Jr

Rel: Presbyterian; Ordained in 1956; b. 01 JUN 1924 in New York NY, USA; m. to (1st) Eva Anna Rubinstein (divorced), (2nd) Harriet Gibney (divorced), (3rd) Virginia Randolph Witson; c. 2 boys and 1 girl; Citizen of USA.
International Organizations: Head (since 1988) of the SANE/FREEZE: Campaign for Global Security, Washington DC.
Main field of work: Clergyman and social activist. **Career:** Served with US Army (served in Europe as an infantry officer, and as a liaison officer with the French Army and the Soviet Army) from 1943 to 1947. Worked abroad on Russian Affairs for Central Intelligence Agency (CIA) from 1950 to 1953. Acting Chaplain, Phillips Academy, Andover MA from 1956 to 1957. Chaplain, Williams College, Williamstown MA from 1957 to 1958. Chaplain, Yale University, New Haven CT from 1958 to 1975. Senior minister, New York City's progressive interdenominational Riverside Church from 1977 to 1987. **Educ.:** Attended from Union Theological Seminary, New York NY, USA (1949-1950); B.A. from Yale University, New Haven CT, USA (1949); B.D. from Yale University, New Haven CT, USA (1956). **Hons.:** Connecticut Valley B'nai B'rith awards for Americanism; D.D., Wesleyan College. **Mem.:** American Freedom of Residence Fund (Director); National Association for the Advancement of Colored People (Director of Legal Defense Fund); Operation Crossroads Africa **Pubs.:** With Charles E Whittaker, *Law, Order, and Civil Disobedience*, American Enterprise Institute for Public Policy Research, 1967; With Morris L Leibman, *Civil Disobedience: Aid or Hindrance to Justice?*, American Enterprise Institute for Public Policy Research, 1972 and *Once to Every Man: A Memoir*, Athenium, 1977.
Mailing address: No mailing address.

Coggan, (Frederick) Donald, Rt Rev and Rt Hon Lord

Rel: Church of England; Ordained in 1934; b. 09 OCT 1909 in London, UK; m. to Jean Strain; c. 2 girls; Citizen of United Kingdom.
International Organizations: Life President (since 1981) of the Church Army.
Honorary President of the International Council of Christians and Jews.
Career: Assistant Lecturer in Semitic Languages and Literature, University of Manchester from 1931 to 1934. Curate, St Mary, Islington from 1934 to 1937. Professor of New Testament, Wycliffe College, Toronto from 1937 to 1944. Principal, London College of Divinity from 1944 to 1956. Bishop, Bradford from 1956 to 1961. Chairman, Liturgical Commission from 1960 to 1964. Archbishop, York from 1961 to 1974. Archbishop, Canterbury from 1974 to 1980. Pro-Chancellor, University of York from 1962 to 1974. Pro-Chancellor, University of Hull from 1963 to 1974. President, Society for Old Testament Studies from 1967 to 1968. Prelate, Order of St John of Jerusalem from 1967 to 1980. Chaplain, the Merchant Taylors' Co. from 1980 to 1981. **Educ.:** M.A. from University of Cambridge, St John's College, UK; Wycliffe Hall, Oxford, UK. **Interests:** gardening, motoring, music. **Hons.:** Baron (Life Peer), created 1980, of Canterbury and Sissinghurst in the County of Kent; Freeman of City of Canterbury (1976); Hon. D.D., Cambridge, Leeds, Aberdeen, Tokyo, Saskatoon, Huron, Hull, Manchester, Moravian Theological Seminary, Virginia Theological Seminary; Hon. LL.D., Liverpool; Hon. H.H.D., Westminster Choir College, Princeton; Hon. D.Litt., Lancaster; S.T.D. h.c., General Theological Seminary, NY; Hon. D.C.L., Kent. **Mem.:** Fellow, King's College, London; Club: Athenaeum **Pubs.:** *The Sacrament of the Word*, 1987; *Cuthbert Bardsley: Bishop, Evangelist, Pastor*, 1989 and *God of Hope*, 1991.
Mailing address: 28 Lions Hall, St Swithun Street, Winchester SO23 9HW, UK. T. (44 962) 864289

Coggins, Wade T

Rel: Evangelical.
International Organizations: Executive Director of the Evangelical Foreign Missions Association.
Mailing address: 1430 K Street NW, Washington DC 20005, USA. T. (202) 628 7911.

Cohen, Bruce

Rel: Jewish.
International Organizations: Member of The Messianic Jewish Movement International.
Other organizations: Member, Messianic Jewish Alliance of America.
Mailing address: PO Box 417, Wynnewood PA 19096, USA. T. (1 215) 896 5812

Cohen, Mattie

Rel: Mason.
International Organizations: Executive Secretary of the Modern Free and Accepted Masons of the World.
Mailing address: PO Box 1072, Columbus GA 31902, USA. T. (404) 322 3326.

Colaco, Marie

See Mariella.

Colasuonno, Francesco, H E Mgr

Rel: Roman Catholic; Ordained on 28 SEP 1947; b. 02 JAN 1925 in Grummo Appula, Italy; celibate.
International Organizations: Apostolic Nuncio to the USSR (since 1990) of the Roman Catholic Church.
Other organizations: Titular Archbishop, Tronto.
Mailing address: 00120 Città del Vaticano, Vatican. T. (39 6) 698 54 11

Colbert, James D, Rev

Rel: Protestant; Ordained; b. in San Antonio TX, USA; Citizen of USA.
International Organizations: Chairman of the Christian Anti-Communism Crusade.
Career: Served as pastor in churches in Oregon and California. Served as Executive Director, Youth for Christ in Long Beach CA. Appointed Director of Missions, Christian Anti-Communism Crusade in 1956. Has travelled extensively and studied under Dr Fred C Schwarz (who is considered to be one of the world's foremost authorities on the philosophy of Communism). Has lectured for many years to and consulted with rulers, diplomats, statesmen and politicians in many countries of the world. Had debated successfully with Communist university professors and students about Communism, Capitalism and recently Perestroika and Glasnost. **Educ.:** Biola University, La Mirada CA, USA; B.A. and Masters degrees from The Masters Seminary, USA; P.G. Stds. in Pol.Sc. at California State University, Long Beach, USA.
Mailing address: PO Box 890, 227 E Sixth St, Long Beach CA 90801, USA. T. (213) 437 0941.

Cole, Owen

Rel: Buddhist: Mahayana.
International Organizations: Member of the Foundation for the Preservation of the Mahayana Tradition.
Other organizations: Director, The Maitreya Project.
Mailing address: 1 Grove Lane, Lilyfield NSW 2040, Australia. T. (61 2) 909 1330

Coleman, John W, Fr

Rel: Roman Catholic; Ordained; celibate.
International Organizations: Secretary-General of the Inter-Regional Meeting of Bishops of Southern Africa.
Mailing address: 4 Bayswater Road, Highlands, PO Box 50, Harare, Zimbabwe. T. (263 10) 47479. Tx 2390 GENSEC ZW Attn IMBISA.

Collaço, Blasco Francisco, H E Mgr

Rel: Roman Catholic; Ordained on 02 MAY 1954; b. 16 MAY 1931 in Raia, Goa, India; celibate; Lang. Portuguese, Italian, Latin, French, English, German, Spanish and Konkani.
International Organizations: Apostolic Pro-Nuncio to Madagascar and Mauritius (since 1991) of the Roman Catholic Church.
Other organizations: Titular Archbishop (since 1961), Octava; Apostolic Delegate to Comoros and Reunion (since 1991).
Career: Apostolic Nuncio to Panama from 1977 to 1982. Apostolic Nuncio to the Dominican Republic from 1982 to 1991. **Educ.:** D.C.L. from Urbanian Pontifical University, Vatican City State (1954-1957); Dip. Soc.Sc. from International University of Social Studies, Rome, Italy (1957-1959); Dip. Diplomacy from Pontifical Ecclesiastical Academy, Vatican City State (1959-1961). **Interests:** Music, photography, reading. **Hons.:** Knight Commander of the Order of Francisco Morazan (Honduras); Grand Cross Officer of the Order of Vasco Nunez de Balboa (Panama); Grand Cross Officer of the Order of Duarte, Sanchez e Mella (Dominican Rep).
Mailing address: Nonciature Apostolique, BP 650, Antananarivo 101, Madagascar. T. (261 2) 423 76

Collet, Adrien, Fr

Rel: Roman Catholic; Ordained; celibate.
International Organizations: Secretary-General of the Gospel and Mission Groups.
Mailing address: 202 avenue du Maine, Pavillon 4, F-75014 Paris, France. T. (33 1) 45 40 45 51.

Collins, (John) Grenville

Rel: Muslim; b. 24 JAN 1943; m. to Susan Elizabeth Carr; c. 1 girl.
International Organizations: Chairman (since 1971) of the Beshara Trust.
Other organizations: 1984, With G and A Kelly Ltd.
Career: Head of Islamic Studies, Beshara School of Intensive from 1975 to 1978. Chairman, Esoteric Education, Sherborne. Glos. Muhyiddin Ibn Arabic. Society (Oxford, Istanbul, San Francisco, Sydney). **Interests:** travelling, publishing. **Pubs.:** Author of papers on esoteric subjects.
Mailing address: 82 St George's Sq, London SW1, UK.
Residence: Kingham House, Kingham, Oxford OX7 6YA, UK.

Collins, J

Rel: Christian.
International Organizations: Co-Director of the Clergy and Laity Concerned.
Mailing address: 198 Broadway, Room 302, New York NY 10038, USA. T. (1 212) 964 6730.

Colombo, Giovanni, H E Card

Rel: Roman Catholic; Ordained on 29 MAY 1926; b. 06 DEC 1902 in Milan, Italy; celibate; Citizen of Italy.
International Organizations: Member (since 1965) of the College of Cardinals.
Other organizations: Archbishop Emeritus (since 1979), Milan.
Career: Consecrated Titular Bishop, Philippopolis in Arabia in 1960. Archbishop, Milan from 1963 to 1979. Created Cardinal by Pope Paul VI in 1965.
Mailing address: Corso Venezia 11, I-20121 Milano, Italy. T. (39 2) 78 39 43

Combrink, Hans Jacob Bernardus, Prof

Rel: Dutch Reformed; Ordained on 26 JUN 1968; b. 23 OCT 1940 in Lydenburgn, Transvaal, South Africa; m. to Anna Lucia Antonites; c. 2 boys and 2 girls; Citizen of South Africa; Lang. Afrikaans and English.
International Organizations: Dean (since 1992) of the Faculty of Theology, University of Stellenbosch.
Other organizations: Professor and Head, New Testament Dept. (since 1988), University of Stellenbosch; Member, Council (since 1991), South African Academy of Arts and Sciences.
Main field of work: University education. **Career:** Minister, Dutch Reformed Church, Wonderboom, Pretoria from 1968 to 1970. Senior Lecturer, Biblical Studies Dept., Rand Afrikaans University from 1970 to 1973. Senior Lecturer, New Testament Dept., University of Pretoria from 1974 to 1975. Senior Lecturer, New Testament Dept., University of Stellenbosch from 1976 to 1982. Associate Professor, New Testament Dept., University of Stellenbosch from 1982 to 1987. **Educ.:** B.A. (cum laude) from University of Pretoria, South Africa (1958-1960); B.D. (cum laude) from University of Pretoria, South Africa (1961-1963); B.A. Greek (Hons) from University of Pretoria, South Africa (1961-1964); Th.D. from Free University, Amsterdam, The Netherlands (1964-1968). **Interests:** literary criticism, rhetoric, gospel of Matthew, sociological exegesis. **Mem.:** Society of New Testament Studies; Society of Biblical Literature; New Testament Society of South Africa; Tyndale Fellowship **Pubs.:** "The Synoptic Problem – the Gospel according to Matthew", in A B du Toit, Editor, *The Synoptic Gospels and Acts*, 1983; "The Structure of the Gospel of Matthew as narrative", in *Tyndale Bulletin*, 34, pp. 61-90 and "Petrus en Judas", in *Handleiding bij die NT*, VI, 1990.
Mailing address: Faculty of Theology, 171 Dorp Street, Stellenbosch 7600, South Africa. T. (27 2231) 77 3255
Residence: 1 Meerlust Ave, Stellenbosch 7600, South Africa. T. (27 2231) 83 3920

Comiskey, Brendan Oliver, SS.CC., Rt Rev

Rel: Roman Catholic; Ordained on 25 JUN 1961; b. 13 AUG 1935 in Tasson, Clontibret, Co. Monaghan, Ireland; celibate; Citizen of Ireland.
International Organizations: Bishop (since 1984) of Ferns.
Other organizations: Chairman, Bishops' Commission for Communications; President, Catholic Communications Institute of Ireland; Chairman, Irish Churches' Council for TV and Radio Affairs (ICCTRA).
Career: Teacher of Latin and English and Dean, Damien High School, La Verne CA from 1961 to 1964. Taught Moral Theology and served as Chairman of Moral Theology Dept., Washington DC Theological Union. Elected Provincial, Anglo-Irish Province of the Congregation of the Sacred Hearts in 1969. Elected Secretary-General, Conference of Major Religious Superiors in 1974. Auxiliary Bishop, Dublin from 1979 to 1984. Ordained Bishop in 1980. **Educ.:** Sacred Hearts Novitiate, Cootehill, Ireland (1954-1955); Phil. Stds. from Sacred Hearts Seminary, Wareham MA, USA (1955-1956); Theol. Stds. from Queen of Peace Mission Seminary, Jaffrey Centre, NH, USA (1956-1961); Lic. Theol. from Lateran Pontifical University, Vatican City State (1956-1961); M.Sc. from University of Dublin Trinity College, Ireland (1979). **Hons.:** Freeman of Wexford Town (1980). **Mem.:** National Episcopal Conference; Bishop's Commission for Ecumenism; Joint Commission of Bishops and Religious Superiors; Commission for Youth; Academic Council of Irish School of Ecumenics; Standing Committee of Glenstal Ecumenical Conference
Mailing address: Bishop's House, Wexford, Ireland. T. (353 53) 22177. Fax 23436

Cone, James Hal

Rel: Methodist Episcopal; b. in Frodyce AR, USA; Citizen of USA.
International Organizations: Charles A Briggs Professor of Systematic Theology of the Union Theological Seminary, NY.
Main field of work: Theologian and originator of the contemporary expression of black theology. **Career:** Academic appointments at Philander Smith College, Little Rock AR from 1964 to 1966. Academic appointments at Adrian College, MI from 1966 to 1969. Joined the faculty, Union Theological Seminary, NY in 1969. **Educ.:** B.A. from Philander Smith College, Little Rock AR, USA (1958); B.D. from Garrett Theological Seminary (1961); M.A. from Northwestern University, Evanston IL, USA (1961); Ph.D. from Northwestern University, Evanston IL, USA (1965). **Mem.:** African Methodist Episcopal Church **Pubs.:** *Black Theology and Black Power*, 1969; *A Black Theology of Liberation*, 1970 and *God of the Oppressed*, 1975.
Mailing address: No mailing address.

Conklin, George
Rel: Christian.
International Organizations: Regional Secretary for North America of the World Association for Christian Communication.
Mailing address: 2319 Parker Street, Berkeley CA 94704, USA.

Connolly, Jerome
Rel: Roman Catholic.
International Organizations: Secretary of the European Conference Justitia et Pax.
Mailing address: 169 Booterstown Avenue, Blackrock, Dublin, Ireland. T. (353 1) 885 021. Fax 834 161. Tx 8869438 SDBK D.

Connolly, Michael Augustine, C.F.C., Br
Rel: Roman Catholic; Ordained; celibate.
International Organizations: Procurator-General of the Congregation of Christian Brothers.
Mailing address: Via della Maglianella 375, I-00166 Roma, Italy. T. (39 6) 696 02 53. C FRATRES ROMA.

Conrad, Juana C
Rel: Bahá'í; Ordained in 1988; b. 04 JUL 1939 in Callaway NB, USA; m. to Sam Conrad; c. 2 boys; Citizen of USA; Lang. English.
International Organizations: Assistant Secretary (since 1989) of the National Spiritual Assembly of the Bahá'ís of the US.
Other organizations: Founder/President (since 1985), Women for International Peace and Arbitration (WIPA); Working on a Five-Year Plan with the All China Women's Federation in the areas of health, education, family and child care, management and training (since 1991).
Main field of work: Development, women's education and equality. **Career:** Court Administrator, East Los Angeles Municipal Court from 1982 to 1988. Appointed to the California Judicial Council Advisory Committee on Gender Bias in the Courts in 1987. Member, United States Delegation to the World Congress of Women, Moscow in 1987. Served on the Steering Committee, First Sino-American Conference on Women's Issues, Beijing in 1990. Led a delegation of educators to Beijing, Dalian, Shanghai and Hangzhou, Rep. of China in 1991. Elected to the Board of Officers, Bahá'í Justice Society in 1988. **Interests:** women's international development and peace. **Hons.:** Fellow, Institute of Court Management (1984). **Pubs.:** *Peacemaker* (quarterly newsletter of WIPA).
Mailing address: Bahá'í National Center, 536 Sheridan Rd, Wilmette IL 60091, USA. T. (1 708) 869 9039

Conring, Warner
Rel: Protestant.
International Organizations: Contact of the Churches' Development Service.
Mailing address: Herrenhäuser Strasse 12, 30419 Hannover, Germany. T. (49 511) 27 96-0. Fax 276 67 07. Tx 09-23 445 ekd-d.

Conroy, Donald B
Rel: Unknown.
International Organizations: Contact of the Consortium on Religion and Ecology International.
Mailing address: 5 Thomas Circle NW, Washington DC 20005, USA. T. (1 202) 462 2591. Fax 234 0307.

Conti, Luigi, H E Mgr
Rel: Roman Catholic; Ordained on 29 SEP 1954; b. 02 MAR 1929 in Ceprano, Italy; celibate.
International Organizations: Apostolic Nuncio to Ecuador (since 1987) of the Roman Catholic Church.
Other organizations: Titular Archbishop, Graziana.
Mailing address: Avenida Orellana 692 y 702, Apartado Postal N. 4543-A-Norte, Quito, Ecuador. T. (593 2) 564 938 – 528 783. Tx 2053 NUNQ ED

Conti, Mario Giuseppe, Rt Rev
Rel: Roman Catholic; Ordained on 26 OCT 1958; b. 20 MAR 1934 in Elgin, Moray, Scotland, UK; celibate; Citizen of United Kingdom; Lang. English and Italian.
International Organizations: Bishop (since 1977) of Aberdeen.
Vice-Chairman, Aberdeen Branch of the United Nations Children's Fund.
Other organizations: Vice-President (since 1980), Scottish Catholic Heritage Commission; President (since 1985), National Commission for Christian Doctrine and Unity; Member, Bishops' Joint Committee on Bio-Ethical Issues; Member, Management Committee, Aberdeen and North East Association for Mental Health; Convener, Central Council, Action of Churches Together in Scotland (ACTS).
Career: Curate, St Mary's Cathedral, Aberdeen (for 3 years). Joint parish priest, St Joachim's, Wick and St Anne's, Thurso (until 1977). President-Treasurer, Scottish Catholic International Aid Fund (SCIAF) from 1978 to 1985. President, National Liturgy Commission from 1981 to 1985. Vice-President, Commission for Migrant Workers and Tourism from 1978 to 1984. Member, International Commission for English in the Liturgy from 1978 to 1987. **Educ.:** Ph.L. from Pontifical Scots College, Vatican City State (1952-1955); S.T.L. from Pontifical Gregorian University, Vatican City State (1955-1959). **Interests:** walking, travel, swimming, music, the arts. **Hons.:** Order of Merit of the Italian Republic (1981); D.D. h.c., University of Aberdeen (1989); Knight Commander of the Holy Sepulchre (1989); Conventual Chaplain Ad Honorem, Sovereign Military Order of St John of Jerusalem, Rhodes and Malta (1991). **Mem.:** Aberdeen Branch of the Order of Christian Unity (Founder Member); Scottish Steering Committee of the Inter-Church Process; 'Not Strangers but Pilgrims'; Pontifical Council for the Promotion of Christian Unity; Bishop's Joint Committee on Bio-Ethical Issues **Pubs.:** Occasional articles and letters in national and local press.
Mailing address: Bishop's House, 156 King's Gate, Aberdeen AB2 6BR, Scotland, UK. T. (44 224) 319154
Residence: See Mailing Address. T. See Mailing Address

Cook, David, Rev
Rel: Evangelical.
International Organizations: Secretary of the South Pacific Association of Bible Colleges.
Mailing address: PO Box 83, Croydon NSW 2132, Australia. T. (61 2) 747 4780.

Cook, Edward David, Dr
Rel: Baptist; b. 03 MAR 1947 in Newcastle, UK; m. to Kathleen; c. 2 boys; Citizen of United Kingdom; Lang. English, French, German, Greek and Latin.
International Organizations: Director (since 1987) of the Whitefield Institute, Oxford.
Other organizations: Fellow/Chaplain (since 1983), Green College, Oxford; Member, Medical Ethics Committee (since 1989), British Medical Association.
Main field of work: Medical and Christian ethics, university teaching and supervision. **Career:** Lecturer, St John's College, Nottingham from 1973 to 1979. Lecturer, Trent Polytechnic from 1977 to 1979. Head, Theology Dept., Westminster College, Oxford from 1979 to 1987. Lecturer in Christian Ethics, Regent's Park College from 1979 to 1992. **Educ.:** B.A. Phil. of Rel. from Arizona State University, Tempe, USA (1967-1968); M.A. Mental Phil. from University of Edinburgh, UK (1965-1970); Ph.D. from University of Edinburgh, New College, UK (1970-1973); M.A. from University of Oxford, UK (1984). **Hons.:** Outstanding Forensic Student (ASU). **Pubs.:** The Moral Maze, SPCK; Dilemmas of Life, IVP and Living in the Kingdom, Hodder.
Mailing address: Whitefield Institute, 3c Norham Gardens, Oxford OX2 6PS, UK. T. (44) 310869
Residence: 31 Simpson's Way, Kennington, Oxford OX1 5RZ, UK. T. (44) 739870

Cook, Henry George, Rt Rev
Rel: Anglican; Ordained; b. 12 OCT 1906 in London, UK; m. to Opal May Thompson; c. 2 boys and 1 girl; Citizen of Canada.
International Organizations: Ecclesiastic of the Anglican Church.
Career: Member, Anglican General Synod from 1944 to 1974. Served as Missionary at Fort Simpson, NW Territories from 1935 to 1943. Canon, Diocese of Athabasca in 1940. Senior Curate, St Paul's Bloor Street Church, Toronto from 1943 to 1944. Rector, South Porcupine

from 1944 to 1945. Archdeacon, James Bay from 1945 to 1947. Principal, Bishop Horden School, Moose Factory from 1945 to 1947. Superintendent, Indian School Administration, Canada from 1948 to 1963. Suffragan Bishop, the Arctic from 1963 to 1966. Bishop, MacKenzie from 1966 to 1974. Co-ordinator of Historical Programs, Government of NW Territories from 1974 to 1979. Served some of a 5-year apprenticeship in the trade of toolmaking. Served 4 years with COTC. **Educ.:** B.A., L.Th. from University of Western Ontario, Huron College, London, Canada (1935). **Interests:** mechanical models. **Hons.:** Hon. D.D., University of Western Ontario, London (1948).
Mailing address: 15 Plainfield Court, Stittsville ON K2S 1B9, Canada.

Cooke, William Dennis Davison, Rev Dr

Rel: Methodist; Ordained on 15 JUN 1965; b. 03 DEC 1938 in St George's, Grenada, West Indies; m. to Joan; c. 2 boys and 1 girl; Citizen of United Kingdom; Lang. English and French.
International Organizations: Principal (since 1984) of the Edgehill Theological College.
Other organizations: Director of Christian Education (since 1984), Edgehill Christian Education Centre.
Main field of work: Lecturer in Church history. **Career:** Circuit Minister, Methodist Church in Ireland from 1966 to 1980. Senior Tutor, Edgehill Theological College from 1980 to 1984. **Educ.:** B.A. from Queen's University of Belfast, UK (1957-1960); B.D. from Queen's University of Belfast, UK (1962-1965); M.Th. from Lexington Theological Seminary, Lexington KY, USA (1965-1966); Ph.D. from Queen's University of Belfast, UK (1975-1980). **Interests:** ecumenism, evangelism, gardening. **Mem.:** Glencree Centre for Reconciliation (Dublin) **Pubs.:** "The Religious Dimension in the Northern Ireland Problem", in *Lexington Theological Quarterly*, July 1981 and Authority and Dissent (public lecture), Union Theological College, October 1981.
Mailing address: Edgehill Theological College, 9 Lennoxvale, Belfast BT9 5BY, UK. T. (44 232) 665870
Residence: See Mailing Address. T. See Mailing Address

Cooper, Owen

Rel: Baptist.
International Organizations: President of the Pan-American Union of Baptist Men.
Mailing address: P O Box 388, Yazoo City MS 39194, USA. T. (601) 746 4131.

Cooper, Roswitha

Rel: Roman Catholic.
International Organizations: Executive Secretary of the Christian Life Community.
Mailing address: Borgo Santo Spirito 8, Casella Postale 6139, I-00195 Roma, Italy. T. (39 6) 686 80 79 FAX 687 9203

Copley, Derek Bryan, Rev Dr

Rel: Evangelical; Ordained on 01 OCT 1969; b. 09 AUG 1940 in Accrington, Lancashire, UK; m. to Nancey; c. 1 boy and 1 girl; Citizen of United Kingdom; Lang. English.
International Organizations: Principal (since 1970) of the Moorlands Bible College.
Other organizations: Chairman (since 1989), Association of Bible College Principals; Chairman (since 1991), Evangelical Alliance, UK.
Main field of work: Training for Christian service. **Career:** Fulbright Scholar, University of North Carolina, Chapel Hill from 1964 to 1966. Research Associate, University of Manchester from 1966 to 1970. **Educ.:** B.Sc. (Hons) from University of Manchester Institute of Science and Technology, UK (1958-1961); Ph.D. from University of Manchester Institute of Science and Technology, UK (1961-1964). **Interests:** human behaviour and relationships. **Mem.:** Orchid Society **Pubs.:** Building with Bananas; Guide to Evangelism and Taking a Lead.
Mailing address: Moorlands College, Sopley, Christchurch, Dorset BH23 7AT, UK. T. (44 425) 72369
Residence: Croft Rd, Neacroft, Christchurch, Dorset BH23 8JS, UK. T. (44 425) 72791

Coppa, Giovanni, H E Mgr

Rel: Roman Catholic; Ordained on 02 JAN 1949; b. 09 NOV 1925 in Alba, Italy; celibate.
International Organizations: Apostolic Nuncio to Czechoslovakia (since 1990) of the Roman Catholic Church.
Other organizations: Titular Archbishop, Serta.
Mailing address: Vorsilská 12, Praha I, Czech Rep. T. (42 2) 20 37 67 – 20 46 31. Fax 20 63 07

Corbon, Richard, R.S.V., Fr

Rel: Roman Catholic; Ordained; celibate.
International Organizations: Superior General of the Pères et Frères Religieux de Saint Vincent-de-Paul.
Mailing address: Via Palestro 26, I-00185 Roma, Italy. T. (39 6) 488 27 40

Corcoran, Theresa, Sr

Rel: Roman Catholic; Ordained; celibate.
International Organizations: Contact of the Sisters of Charity of St Vincent-de-Paul.
Mailing address: Mount St Vincent, 150 Bedford Highway, Halifax B3M 3J5, Canada. T. (1 902) 443 4620.

Cordeiro, Joseph, H E Card

Rel: Roman Catholic; Ordained on 24 AUG 1946; b. 19 JAN 1918 in Bombay, India; celibate; Citizen of Pakistan.
International Organizations: Member (since 1973) of the College of Cardinals.
Member of the Congregation for the Evangelization of Peoples. Member of the Congregation for Religious and Secular Institutes. Member of the Pontifical Council for Inter-Religious Dialogue.
Other organizations: Archbishop (since 1958), Karachi; President, Pakistan Episcopal Conference; Member, Council of Cardinals for the Study of Organizational and Economic Problems of the Holy See.
Career: Created Bishop by Pope Paul VI in 1973.
Mailing address: St Patrick's Cathedral, Shahrah-Iraq, Karachi 74400, Pakistan. T. (92 21) 51 58 70 – 51 41 26

Cordero Lanza di Montezemolo, Andrea C, H E Mgr

Rel: Roman Catholic; Ordained on 13 MAR 1954; b. 27 AUG 1925 in Torino, Italy; celibate.
International Organizations: Apostolic Pro-Nuncio to Cyprus and Apostolic Delegate to Jerusalem and Palestine (since 1990) of the Roman Catholic Church.
Other organizations: President, Union des Religieuses de Terre Sainte (URTS); Titular Archbishop, Anglona.
Mailing address: Holy Cross Catholic Church, PO Box 1964, Nicosia, Cyprus. T. (357 2) 46 21 32

Coriaty, George M, Mgr

Rel: Melkite; Ordained in 1956; b. 01 JAN 1933 in Avaré, Sao Paolo, Brazil.
International Organizations: Rector (since 1964) of the St Sauveur Cathedral.
Career: Assistant Pastor, O.L. of Annunciation, Boston from 1957 to 1960. Assistant Pastor, St Sauveur Church, Montreal from 1960 to 1964. Elevated to Monsignr in 1965. Nominated Apostolic Visitor to Melkites in Canada in 1972. Patriarchal Vicar in 1973. Founder-President, Middle East Immigrant Aid Society of Canada in 1963. Founder/Co-President, Amitiés Québec-Proche Orient in 1964. Founder/President, Centre Communautaire Bois-de-Boulogne in 1972. Founder, Vancouver Mission of St George in 1974. Founder, St Ann's Mission, Quebec (now St John Chrysostomos Mission, Quebec) in 1968. Founder, O.L. of Assumption Mission Toronto (now St Joseph Melkite Church Mission) in 1966. Founder, First Byzantine Museum in North America. **Educ.:** L.Th. from Séminaire St Sauveur, Lebanon (1956); Ph.D. in Social Psy. and Pol.Sc. from Columbia Pacific University (1981); Ph.D. in Canon Law (not yet completed) from University of Strasbourg, France (1981). **Hons.:** Named First Citizen of Montreal (1973); Member, Order of Canada (1982); Hon. Ph.D. in Byzantine Studies, International University Foundation (1985); Order of Merit First Class, Egypt (1986); Commander, Military and Hospitaller Order of St Lazarus of Jerusalem. **Mem.:** Musée des Beaux-Arts, Montreal; Chamber of Commerce, Montreal; Canadian and American Catholic Historical Associations; American and Canadian Management Association; Canadian Ethnic Studies Association (CESA); Canadian Sociological and Anthropological Association; International Political Science Association; Learned Societies; Canadian Canon Law Society (1987); Lebanese Canadian Society; Alumni Association, Columbia Pacific University; International Tennis Hall of Fame (Honorary Life Member); National Center of Oecumenism; Association of the Refugees; Canadian Council of Multiculturalism (1973-80), and other organizations
Mailing address: PO Box 578, Stn C, Montréal PQ H2L 4K4, Canada.

Corradi, Carol
Rel: Buddhist: Mahayana.
International Organizations: Member of the Foundation for the Preservation of the Mahayana Tradition.
Other organizations: Director, Tsechen Ling.
Mailing address: 909 Clayton St, San Francisco CA 94117, USA. T. (1 415) 664 9046

Corripio Ahumada, Ernesto C, H E Card
Rel: Roman Catholic; Ordained on 25 OCT 1942; b. 29 JUN 1919 in Tampico, Mexico; celibate; Citizen of Mexico.
International Organizations: Member (since 1979) of the College of Cardinals.
Member of the Sacred Congregation for the Sacraments and Divine Worship. Member of the Congregation for the Clergy. Member of the Congregation for Catholic Education.
Other organizations: Archbishop (since 1977), Mexico City; Member, Council of Cardinals for the Study of Organizational and Economic Problems of the Holy See.
Career: Titular Bishop, Zapara from 1953 to 1956. Bishop, Tampico from 1956 to 1967. Archbishop, Antequera from 1967 to 1976. Archbishop, Puebla de los Angeles from 1976 to 1977. Created Cardinal by Pope John Paul II in 1979. **Hons.:** Entitled L'Immacolata al Tiburtino.
Mailing address: Apolinar Nieto 40, Tetlameya, 04730 Mexico DF, Mexico. T. (52 5) 573 06 42

Corriveau, Ernest, M.S., Fr
Rel: Roman Catholic; Ordained; celibate.
International Organizations: Superior General of the Missionaries of Our Lady of La Salette.
Mailing address: Piazza Madonna della Salette 3, I-00152 Roma, Italy. T. (39 6) 53 04 51

Cors, Marie
Rel: Yoga.
International Organizations: Secretary of the Institut international de yoga.
Mailing address: Rue du Parc 47, B-7100 La Louvière, Belgium.

Cortesi, Filipo, Dr
Rel: Roman Catholic; married; Citizen of Italy; Lang. Italian, German and English.
International Organizations: Secretary-General of the International Catholic Rural Association.
Mailing address: Palazzo San Calisto 16, I-00153 Città del Vaticano, Vatican. T. (39 6) 698 71 23.

Cory, Robert
Rel: Quaker.
International Organizations: President Clerk of the Friends Coordinating Committee on Peace.
Mailing address: 706 E Capitol St, Washington DC 20003, USA.

Coscia, Luis, O.F.M.Cap., Fr
Rel: Roman Catholic; Ordained; celibate.
International Organizations: President of the Latin American Confederation of Religious Orders.
Member of the Order of Friars Minor Capuchin.
Mailing address: Esquiú 974, 1437 Buenos Aires, Argentina. T. (54 1) 91 12 41

Costa, Joaquín
Rel: Spiritualist.
International Organizations: Head, Secretariat of the Universal Spiritual School.
Mailing address: Ent la, Barcelona, Spain.

Cotterill, Edward, Col
Rel: Salvation Army, The.
International Organizations: Territorial Commander of The Salvation Army.
Mailing address: PO Box 320, 6 Osu Avenue, Accra, Ghana.

Cottier, Georges Marie Martin, O.P., Rev
Rel: Roman Catholic; Ordained; celibate.
International Organizations: Secretary-General of the International Theological Commission.
Member of the Order of Preachers. Consultor of the Pontifical Council for Dialogue with Non-Believers.
Other organizations: Consultor, Congregation for the Doctrine of the Faith; Theologian, Pontifical Household.
Mailing address: Palazzo del S Uffizio, I-00120 Città del Vaticano, Vatican. T. (39 6) 698 4753.

Coudreau, François, Rev Prof Dr
Rel: Roman Catholic; Ordained on 20 MAR 1943; b. 14 DEC 1916 in Poitiers, France; celibate; Citizen of France; Lang. French.
International Organizations: Permanent Representative at UNESCO of the International Catholic Child Bureau.
Other organizations: Founder and Director, Institut Supérieur de la Pastorale Catéchétique (since 1950), Institut Catholique de Paris; Pastoral Advisor to the National Confederation of Catholic Family Associations.
Main field of work: Ministry. **Career:** Professor, Université Laval, Québec from 1955 to 1980. Professor, Faculty of Letters and Human Sciences, Metz from 1975 to 1985. **Educ.:** Dr. Theol. from Université Catholique de Paris, France (1944-1945). **Pubs.:** La Foi s'enseigne-t'elle ?; La Foi des Catholiques and Le Catéchuménat des Adultes.
Mailing address: Compagnie des Prêtres de St Sulpice, Rue du Regard N°6, F-75006 Paris, France. T. (33 1) 43 54 38 51
Residence: Passage Rimbaut N°9, F-75014 Paris, France. T. (33 1) 45 40 66 08

Couloubaly, Joseph
Rel: Roman Catholic.
International Organizations: Regional Secretary of the International Catholic Organization for Cinema and Audiovisual.
Mailing address: BP 298, Bamako, Mali.

Couve de Murville, Maurice Noël Léon C, Most Rev
Rel: Roman Catholic; Ordained on 29 JUN 1957; b. 27 JUN 1929 in St Germain-en-Laye, France; celibate; Citizen of United Kingdom.
International Organizations: Archbishop (since 1982) of Birmingham.
Career: Curate, St Anselm's, Dartford, Kent from 1957 to 1960. Priest-in-charge, St Francis, Moulsecoomb from 1961 to 1964. Catholic Chaplain, University of Sussex from 1961 to 1977. Senior Catholic Chaplain, University of Cambridge from 1977 to 1982. **Educ.:** M.A. from University of Cambridge, Trinity College, UK; S.T.L. from Institut Catholique de Paris, France; M.Phil. from University of London, School of Oriental and African Studies, UK. **Interests:** walking, gardening, local history. **Pubs.:** With Philip Jenkins, Catholic Cambridge, 1983; John Milner 1752-1826, 1986 and Le Monde en Face, 1989.
Mailing address: Archbishop's House, St Chad's Queensway, Birmingham B4 6EX, UK. T. (44 21) 236 5535
Residence: 57 Mearse Lane, Barnt Green, Birmingham B45 8HJ, UK. T. (44 21) 445 1467

Coveney, Patrick, H E Mgr
Rel: Roman Catholic; Ordained on 21 FEB 1959; b. 29 JAN 1934 in Cork, Ireland; celibate.
International Organizations: Apostolic Pro-Nuncio to Ethiopia (since 1990) of the Roman Catholic Church.
Other organizations: Titular Archbishop, Satriano.
Mailing address: PO Box 588, Addis Ababa, Ethiopia. T. (251 1) 71 21 00 – 71 23 01. Tx 21815 VATIC ET

Coventry, John Seton, S.J., Rev
Rel: Roman Catholic; Ordained on 10 SEP 1947; b. 21 JAN 1915 in Deal, Kent, UK; celibate; Citizen of United Kingdom; Lang. English and French.
International Organizations: Member (since 1932) of the Society of Jesus.
Other organizations: Lecturer in Theology (since 1987), Jesuit Noviceship.
Main field of work: Theology. **Career:** Headmaster, Beaumont College from 1951 to 1958. Provincial Superior, Society of Jesus from 1958 to 1964. Lecturer in Theology, Heythrop College from 1964 to 1976. Master, St Edmund's College (then House), Cambridge from 1976 to 1985.

Lecturer Emeritus and Fellow, Heythrop College from 1985 to 1987. **Educ.:** Lic. Phil. from Heythrop College, Oxon, UK (1935-1938); M.A. from Campion Hall, Oxford, UK (1938-1942); S.T.L. from Heythrop College, Oxon, UK (1946-1950). **Interests:** ecumenism, music, birds. **Hons.:** MA, Cambridge; Honorary Fellow, St Edmund's College, Cambridge. **Pubs.:** Christian Truth, 1975; Faith in Jesus Christ, 1980 and Reconciling, 1985.
Mailing address: Manresa House, 10 Albert Road, Birmingham B17 OAN, UK. T. (44 21) 427 2638
Residence: See Mailing Address. T. See Mailing Address

Cragg, D G L, Rev Dr

Rel: Methodist.
International Organizations: Secretary for Ecumenical Affairs of the Methodist Church of Southern Africa.
Mailing address: PO Box 2256, Durban 4000, South Africa.

Craig, James Hannington, Rev Canon

Rel: Anglican; Ordained; b. in Montreal, Quebec, Canada; m. to Diane V; c. 2 boys and 1 girl; Citizen of Canada.
International Organizations: Ecclesiastic of the Anglican Church.
Career: Taught school in Vernon, BC from 1926 to 1928. Dean of Residence and Lecturer in Practical Theology, Huron College, London, Ontario from 1930 to 1935. Rector, St Luke's Church, London, Ontario from 1930 to 1935. Incumbent, St Mary's Church, Vancouver, BC and Lecturer at Anglican Theological College there from 1935 to 1944. Dean, Algoma and Rector of St Luke's Cathedral, Sault St Marie from 1944 to 1951. Dean, Calgary and Rector of Cathedral Church of the Redeemer from 1951 to 1953. Rector, Grace Church On-the-Hill, Toronto from 1953 to 1972. **Educ.:** B.A. from University of British Columbia, Vancouver, Canada (1925); B.D., M.A., D.D. from University of Toronto, Trinity College, Canada (1928-1930).
Mailing address: Suite 5, C 19, RR Suite 1, Naramata BC V0H 1N0, Canada.

Craig, Maxwell, Rev

Rel: Church of Scotland; Ordained on 21 SEP 1966; b. 25 DEC 1931 in Halifax, Yorkshire, UK; m. to Janet Margaret MacGregor; c. 1 boy and 3 girls; Citizen of United Kingdom; Lang. English, German and French.
International Organizations: General Secretary of the Action of Churches Together in Scotland.
Educ.: M.A. from University of Oxford, Oriel College, UK (1950-1954); B.D. from University of Edinburgh, UK (1961-1964); Th.M. from Princeton Theological Seminary, NJ, USA (1964-1965).
Mailing address: Scottish Churches House, Dunblane, Pertshire FK15 0AJ, Scotland, UK. T. (44 786) 823588. Fax 825844
Residence: 9 Kilbryde Crescent, Dunblane, Perthshire FK15 9BA, UK. T. (44 786) 823147

Craig, Robert, Very Rev Prof

Rel: Church of Scotland; Ordained on 11 MAY 1943; b. 22 MAR 1917 in Markinch, Fife, Scotland, UK; m. to Olga Wanda Strzelec; c. 1 boy and 1 girl; Citizen of United Kingdom; Lang. English, French and German.
International Organizations: Ecclesiastic of the Church of Scotland.
Career: Assistant Minister, St John's Kirk, Perth from 1941 to 1942. British Army Infantry Chaplain (in UK, NW Europe, Egypt, Palestine) from 1942 to 1947. Mentioned in Despatches, Normandy in 1944. Professor of Divinity, Natal University, South Africa from 1950 to 1957. Dean, College from 1953 to 1954. Professor of Religion, Smith College, Northampton MA from 1958 to 1963. Professor of Theology, University College of Rhodesia from 1963 to 1980. Vice-Principal, University College of Rhodesia in 1966. Acting Principal, University College of Rhodesia in 1967. Acting Principal, University College of Rhodesia in 1969. Principal and Vice-Chancellor, University of Rhodesia (now Zimbabwe) from 1969 to 1980. Emeritus in 1980. Minister, St Andrew's Scots Memorial Church, Jerusalem from 1980 to 1985. Emeritus in 1985. Council, Interfaith Committee from 1981 to 1985. Board of Directors, International YMCA from 1982 to 1985. Chairman, International YMCA from 1984 to 1985. Executive Committee, Spafford Community Centre from 1982 to 1985. Chairman, Church of Scotland Israel Council from 1981 to 1985. Moderator, Presbytery of Jerusalem from 1982 to 1984. Chairman, Ecumenical Theological Research Fraternity from 1983 to 1985. Moderator, General Assembly, Church of Scotland from 1986 to 1987. **Educ.:** M.A., B.D. from University of St Andrews, UK (1935-1941); S.T.M. from Union Theological Seminary, New York NY, USA (1947-1948); Ph.D. from University of St Andrews, UK (1948-1950). **Interests:** cinema, theatre, recent and contemporary history, light classical music, listening and talking to people. **Hons.:** Commander, Order of the British Empire (1981); Honorary Fellow, Zimbabwe Institute of Engineers (1976); Hon. D.D., St Andrews (1967); Hon. LL.D., Witwatersrand (1979), Birmingham (1980), Natal (1981); Hon. D.Litt., Zimbabwe (1981); Honorary Chaplain to the Forces (1947); City of Jerusalem Medal (1985); Golden Jubilee Medal, Witwatersrand (1977). **Mem.:** Jerusalem Foundation (1989) **Pubs.:** Social Concern in the Thought of William Temple, 1963; Religion: Its Reality and its Relevance, 1965 and Politics and Religion: A Christian View, 1972.
Mailing address: West Port, Falkland KY7 7BL, UK. T. (44 337) 57238
Residence: See Mailing Address. T. See Mailing Address

Crespelle, Jean-Pierre

Rel: Roman Catholic.
International Organizations: President of the Association Our Lady of the Nations: Group of Catholic International Civil Servants.
Mailing address: 147 Rue de Lausanne, CH-1202 Genève, Switzerland. T. (41 22) 732 74 87.

Crossman, Gerald Peter

Rel: Mennonite; b. 23 AUG 1953 in Glen Ridge NJ, USA; single; Citizen of USA; Lang. English, French, Dutch and Swahili.
International Organizations: Secretary-General (since 1990) of the European Ecumenical Organization for Development.
Main field of work: Coordination of development research and political actions of the churches. **Career:** Projects Coordinator, Mennonite Central Committee from 1978 to 1981. Professor of Theology and Religions, Catholic University of Louvain (KUL), Belgium from 1985 to 1989. Scientific Assistant, Catholic University of Louvain (KUL), Belgium from 1987 to 1990. **Educ.:** B.A. Hist. from King's College, Briancliff Manor NY, USA (1975); M.Th. from Wheaton College, IL, USA (1978); B.A. Phil. from Catholic University of Louvain (KUL), Belgium (1984); S.T.L. from Catholic University of Louvain (KUL), Belgium (1988). **Pubs.:** *Social Sin in the Social Gospel*, Thesis for License in Theology, 170 pp.
Mailing address: Rue Joseph II 174, B-1040 Bruxelles, Belgium. T. (32 2) 230 61 05 – (32 2) 230 4173. Fax (32 2) 231 14 13.

Crous Boatella, Filomena C

Rel: Roman Catholic.
International Organizations: Director-General of the Magdalena Aulina Institute.
Mailing address: 2 Calle San Pedro Claver, Barcelona 08017, Spain. T. (34 3) 203 90 83 – (34 3) 204 43 58.

Crouse, J B

Rel: Protestant.
International Organizations: Member of the OMS International.
Mailing address: PO Box A, Greenwood IN 46142, USA. T. (1 317) 881 6751.

Crow, Paul A, Jr, Rev Dr

Rel: Disciples of Christ; b. 17 NOV 1931 in Birmingham AL, USA; m. to Mary Evelyn Matthews; c. 1 boy and 2 girls; Citizen of USA; Lang. English.
International Organizations: Secretary-General of the Disciples Ecumenical Consultative Council.
Member, Central Committee of the World Council of Churches. Member of the Faith and Order.
Other organizations: President (since 1974), Council on Christian Unity of the Disciples of Christ; Co-Chairperson, Disciples of Christ/Roman Catholic International Dialogue; Co-Chairperson, Disciples of Christ/Russian Orthodox International Dialogue; Co-chairperson, Disciples of Christ/Reformed International Dialogue.
Main field of work: Religion, christian unity. **Career:** Campus Minister for Disciples students, University of Alabama, Tuscaloosa from 1953 to 1954. Minister, Congregations in Alabama, Kentucky, Massachusetts from 1955 to 1961. Professor, Lexington Theological Seminary, Kentucky from 1961 to 1968. Secretary-General, Consultation on Church Union from 1968 to 1974. **Educ.:** B.Sc. from University of Alabama, Tuscaloosa, USA (1955); B.D. from Lexington Theological Seminary, KY, USA (1957); S.T.M. from Hartford Theological Seminary, CT, USA (1958); Ph.D. from Hartford Seminary Foundation, CT, USA

(1962); D.D. from Yale University, Divinity School, New Haven CT, USA (1986). **Mem.:** Indianapolis Athletic Club **Pubs.:** Christian Unity: Matrix for Mission; The Anatomy of a Nineteenth Century Church; Editor of *Mid-Stream* (Ecumenical Journal) and The History of the Faith and Order Movement, 1910-1990 (in process).
Mailing address: 222 South Downey Ave, PO Box 1986, Indianapolis IN 46 206, USA. T. (1 317) 353 1491
Residence: 7215 Vauxhall Road, Indianapolis IN 46 250, USA T. (1 317) 849 7742

Crowe, Frederick Ernest, S.J., Rev

Rel: Roman Catholic; Ordained in 1949; b. 05 JUL 1915 in Jeffries Corner, New Brunswick, Canada; celibate; Citizen of Canada.
International Organizations: Professor Emeritus (since 1980) of the Regis College, Toronto.
Member (since 1936) of the Society of Jesus.
Career: Teacher at St Mary's College, Halifax from 1943 to 1946. Teacher of Theology at Regis College, Toronto in 1953. Research Professor at Regis College in 1975. President, Regis College from 1969 to 1972. Visiting Professor, Pontifical Gregorian University, Rome (in 1964 and 1984). **Educ.:** B.Sc. from University of New Brunswick, Fredericton, Canada (1934); B.A. from University of Montreal, Loyola College, Quebec, Canada (1943); S.T.D. from Pontifical Gregorian University, Vatican City State (1953); Lic. Phil. from Collège de l'Immaculée-Conception, Montreal, Canada (1962). **Hons.:** Hon. D.Litt., St Mary's, Halifax (1971); Hon. D.D., Trinity College, Toronto (1977), St Michael's College, University of Toronto (1986); Hon. LL.D., St Thomas, Fredericton (1982); John Courtney Murray Award (1977); Honoured with volume of essays by colleagues, Trinification of the World: *A Festschrift in Honour of Frederick E Crowe in Celebration of his 60th Birthday*, Regis College Press, 1978. **Mem.:** Canadian Theological Society; Catholic Theological Society of America; Jesuit Philosophical Association; American Catholic Philosophical Association **Pubs.:** Old Things and New: A Strategy for Education, 1985; Appropriating the Lonergan Idea, 1989 and contributor to *New Catholic Encyclopedia*, 1967, 1974, 1979.
Mailing address: Lonergan Research Institute, 10 St Mary St, Toronto ON M4Y 1P9, Canada.

Crowley, Regina, OSB, Sr

Rel: Roman Catholic; Ordained; celibate.
International Organizations: Contact of the Benedictine Sisters of the Federation of Saint Scholastica.
Other organizations: Member, Order of St Benedict.
Mailing address: St Scholastica Priory, 7430 Ridge Blvd, Chicago IL 60645, USA.

Crumley, James Robert, Jr, Rt Rev

Rel: Lutheran; Ordained; b. 30 MAR 1925 in Bluff City TN, USA; m. to Annette Bodie; c. 1 boy and 2 girls.
International Organizations: Co-Chairperson of the Lutheran-Roman Catholic Joint Commission.
Board of the Institute for Ecumenical Research, Strasbourg.
Other organizations: Distinguished Visiting Professor of Ecumenism (since 1988), Lutheran Theological Southern Seminary, Columbia SC. **Career:** Pastor, Greene County Lutheran Parish, Greenville TN (St James and Solomon Churches) from 1951 to 1953. Pastor, Grace Lutheran Church, Oak Ridge TN from 1953 to 1966. Pastor, Lutheran Church of the Ascension, Savannah GA from 1966 to 1974. Secretary, Lutheran Church in America from 1974 to 1978. President, Lutheran Church in America from 1978 to 1980. Bishop, Lutheran Church in America from 1980 to 1987. Founding Director, Center for Ethical Development, Newberry College, SC from 1988 to 1990. Executive Committee, Chairman of the Constitution Committee, and Chairman of the Committee on Restructuring, Lutheran World Federation (LWF). Central Committee, World Council of Churches (WCC). Governing Board, Executive Committee, and Second Vice-President, National Council of the Churches of Christ in the USA. Delegate to the 6th and 7th Assemblies, LWF. Delegate to the 6th Assembly, WCC. President, Lutheran World Ministries. Chairperson, LCA Committee on Affirmation of Faith. Various synodical offices and appointments in the Virginia Synod and in the Southeastern Synod, the United Lutheran Church in America. **Educ.:** B.A. from Roanoke College, Salem VA, USA (1948); M.Div. from Lutheran Theological Southern Seminary, Columbia SC, USA (1951). **Pubs.:** God and Science (a Vacation Church School Course); Editor, with H George Anderson, Promoting Unity and various articles for professional journals.
Mailing address: c/o LWF, Secretariat for Interconfess Dialogue, PO Box No 2100, Route de Ferney 150, CH-1211 Genève 2, Switzerland. T. (41 22) 791 61 11. C. LUTHERWORLD GENEVA. Tx 415 730 OIK CH.

Cuesta, Julio, F.D.P., Rev

Rel: Roman Catholic; Ordained; celibate.
International Organizations: Procurator-General of the Little Work of Divine Providence.
Mailing address: Via Etruria 6, 00183 Roma, Italy. T. (39 6) 757 55 37, (39 6) 77 85 88

Cullmann, Oscar, Prof Dr

Rel: Protestant; Unordained; b. 25 FEB 1902 in Strasbourg, France; single; Citizen of France.
International Organizations: Professor of Theology and Religious Sciences (retired) of the Universities of Basle and Paris.
Other organizations: Professor Emeritus (since 1972), Universities of Basle and Paris.
Main field of work: University education. **Career:** Director, Protestant Seminary, Strasbourg from 1925 to 1930. Professor, University of Strasbourg from 1930 to 1938. Professor, Universities of Basle and Paris from 1938 to 1972. Rector, University of Basle from 1968 to 1969. Professor, Ecole des Hautes Etudes, Paris from 1948 to 1972. Professor, Faculty of Letters, Sorbonne, Paris from 1955 to 1972. **Educ.:** Doc. from University of Strasbourg, France (1921-1930); Doc. from University of Paris IV (Paris-Sorbonne), France (1925-1926). **Interests:** New Testament, ecumenism, early Christianity. **Hons.:** Dr. h.c., Universities of Lausanne (1945), Edinburgh (1952), Lund (1953), Manchester (1949), Debrecen (1982); Dr. Phil., University of Basle (1972); Commander, Legion of Honour. **Mem.:** Titular Member: Académie des sciences morales et politiques, Institut de France (1972); Corresponding Member: British Academy, Akademie Lit. Mainz, Royal Netherlands Academy **Pubs.:** Over 100 titles, including: Die Christologie des Neuen Testaments, 1957; Der Johanneische Kreis, 1975 and Einheit Durch Vielheit, 1986.
Mailing address: Birmannsgasse 10A, CH-4055 Basel, Switzerland. T. (41 61) 261 15 66
Residence: See Mailing Address. T. See Mailing Address

Cumberbatch, Edward R

Rel: Protestant.
International Organizations: Secretary of the Caribbean Conference of Churches.
Mailing address: PO Box 616, Bridgetown, Barbados. T. (1 809) 427 2681 to 427 2683. Fax 429 2075. C. CHRISTOS. Tx 2335 CADEC WB.

Cumella, Elena, Dr

Rel: Roman Catholic.
International Organizations: Secretary-General of the Latin American Council of Catholic Women.
Mailing address: CLAMUC, Gelly y Obes 2213, 1425 Buenos Aires, Argentina. T. (54 1) 803 4901 – 824 1078.

Cunegatti, Pietro, P.S.D.P., Fr

Rel: Roman Catholic; Ordained; celibate.
International Organizations: Superior General of the Poor Servants of Divine Providence.
Mailing address: Via S Zeno in Monte 23, I-37129 Verona, Italy. T. (39 45) 800 40 35

Cunha, Carlos, Rev

Rel: Presbyterian; Ordained.
International Organizations: Moderator of the United Presbyterian Church of Brazil.
Mailing address: Rua Sete de Setembro 421, Vitoria ES, Brazil.

Cupitt, Don, Rev Dr

Rel: Church of England; Ordained in 1960; b. 22 MAY 1934 in Oldham, Lancashire, UK; m. to Susan Marianne Day; c. 1 boy and 2 girls; Citizen of United Kingdom; Lang. English.
International Organizations: Ecclesiastic of the Church of England.
Other organizations: Fellow and Dean (since 1966), Emmanuel College, Cambridge; Lecturer (since 1973), University of Cambridge.
Main field of work: University education. **Career:** Ordained Deacon in 1959. Assistant Curate, St Philip's Church, Salford, Lancashire from 1959 to 1962. Vice-Principal, Westcott House, Cambridge from 1962 to 1965. University Proctor, University of Cambridge from 1966 to 1969. Stanton Lecturer in the Philosophy of Religion, University of Cambridge from 1968 to 1971. Assistant Lecturer in Divinity, University of Cambridge from 1968 to 1973. Chairman, Faculty Board of Divinity, University of Cambridge from 1983 to 1985. Has held a variety of College and University appointments. Has lectured over most areas in the philosophy of religion, ethics and Christian ethics, and religious studies. **Educ.:** B.A. from University of Cambridge, Trinity Hall, UK (1955-1955); M.A. from Westcott House, Cambridge, UK (1957-1959). **Interests:** writing. **Hons.:** Hon. D.Litt., Bristol (1984). **Pubs.:** 20 books, including the following most recent: *Creation out of Nothing*, 1990; *What is a Story?*, 1991 and *The Time Being*, 1992.
Mailing address: Emmanuel College, Cambridge CB2 3AP, UK. T. (44 223) 334 200
Residence: See Mailing Address. T. See Mailing Address

Curioni, Cesare, Mgr

Rel: Roman Catholic; Ordained; b. 06 JAN 1923 in Asso, Italy; celibate; Citizen of Italy; Lang. Italian, French and Spanish.
International Organizations: President of the International Commission of General Prison Chaplains.
Main field of work: Religion. **Educ.:** Seminario cattolico. **Interests:** Pastoral assistance to prisoners.
Mailing address: Treublaan 2, NL-2555 HHS Den Haag, Netherlands.
Residence: Via Giulia 52, I-00186 Roma, Italy T. (39 6) 686 9331

Curis, Carlo, H E Mgr

Rel: Roman Catholic; Ordained on 13 JUL 1947; b. 02 NOV 1923 in La Maddalena, Sardaigne, Italy; celibate; Citizen of Italy; Lang. Italian, English, French and Spanish.
International Organizations: Apostolic Pro-Nuncio to Canada (since 1990) of the Roman Catholic Church.
Other organizations: Titular Archbishop, Medeli.
Mailing address: 724 Manor Avenue, Ottawa K1M OE3, Canada. T. (1 613) 746 4914. Fax 746 4786. Tx 533380 NONCIA OTT
Residence: See Mailing Address. T. See Mailing Address

Curran, Charles E, Rev Dr

Rel: Roman Catholic; Ordained on 13 JUL 1958; b. 30 MAR 1934 in Rochester NY, USA; celibate; Citizen of USA; Lang. English, Latin, Italian and French.
International Organizations: Elizabeth Scurlock Professor of Human Values (since 1991) of the Southern Methodist University, Dallas TX.
Other organizations: Member, Editorial Boards, *Eglise et Théologie*, *Horizons*, and *Christian Digest*.
Main field of work: University education. **Career:** Professor of Moral Theology, St Bernard's Seminary, Rochester NY from 1961 to 1965. Assistant Professor, Associate Professor and Ordinary Moral Theology, Catholic University of America, Washington DC from 1965 to 1987. Senior Research Scholar, Kennedy Center for Bio-Ethics at Georgetown University, Washington DC in 1972. External Examiner in Christian Ethics, University of the West Indies from 1982 to 1986. Visiting Kaneb Professor of Catholic Studies, Cornell University, Ithaca NY from 1987 to 1988. Visiting Brooks Professor of Religion, University of Southern California, Los Angeles from 1988 to 1989. Visiting Firestone Professor of Religion, University of Southern California, Los Angeles from 1989 to 1990. Visiting Goodwin-Philpott Eminent Scholar in Religion, Auburn University, AL from 1990 to 1991. President, The Catholic Theological Society of America from 1969 to 1970. President, Society of Christian Ethics from 1971 to 1972. President, American Theological Society from 1989 to 1990. **Educ.:** B.A. from St Bernard's College, Rochester NY, USA (1955); S.T.L. from Pontifical Gregorian University, Vatican City State (1959); S.T.D. from Pontifical Gregorian University, Vatican City State (1961); S.T.D. from Academia Alfonsiana, Rome, Italy (1961). **Interests:** reading, golf. **Hons.:** First recipient of John Courtney Murray Award of the Catholic Theological Society for distinguished achievement in theology (1972); ABC Person of the Week (August 1986); Hon. Doctorate, University of Charleston (1987). **Mem.:** Society of Christian Ethics; American Theological Society; Catholic Theological Society of America; College Theology Society **Pubs.:** Author of numerous articles in scholarly publications, author and editor of numerous books on moral theology, ethics, and dissent within the Catholic church, including: *Faithful Dissent*, Kansas City MO, Sheed and Ward, 1986; *Catholic Higher Education, Theology, and Academic Freedom*, Notre Dame IN, University of Notre Dame Press, 1990 and *The Living Tradition of Moral Theology*, Notre Dame IN, University of Notre Dame Press, 1992.
Mailing address: 317 Dallas Hall, Southern Methodist University, Dallas TX 75275-0235, USA. T. (1 214) 692 4073
Residence: 4125 Woodcreek Drive, Dallas TX 75220, USA. T. (1 214) 352 8974

Currie, David, Pastor

Rel: Seventh-Day Adventist.
International Organizations: Ministerial Director, Trans-European Division of the General Conference of Seventh-Day Adventists.
Mailing address: 119 St Peter's Street, St Albans AL1 3EY, UK.

Cuthbert, Rod

Rel: Protestant.
International Organizations: South African Director of the Africa Inland Mission International.
Mailing address: Box 109, Plumstead 7800, South Africa.

Cwajgenbaum, Serge

Rel: Jewish.
International Organizations: Secretary-General of the European Jewish Congress.
Mailing address: 78 ave des Champs-Elysees, F-75008 Paris, France. T. (33 1) 43 59 94 63. Tx 650320. Fax (33 1) 42 25 45 28.

Czaja, Tadeusz, O.M.I., Fr

Rel: Roman Catholic; Ordained; celibate.
International Organizations: Procurator-General of the Missionary Oblates of Mary Immaculate.
Mailing address: Via Aurelia 290, Casella Postale 9061, I-00165 Roma, Italy. T. (39 6) 637 02 51. FAX 637 53 22.

Dabrowski, Ray, Pastor

Rel: Seventh-Day Adventist; b. 25 SEP 1948 in Tomaszów Mazowiecka, Poland; Citizen of Poland.
International Organizations: Communication Director, Trans-European Division (since 1985) of the General Conference of Seventh-Day Adventists.
Career: Editor and Manager, Znaki Czasu Publishing House, Warsaw from 1972 to 1985. Vice-President, Society for Prevention of Drug Abuse from 1982 to 1984. **Pubs.:** Narkomania znakiem czasu (The Drug Habit as a Sign of the Time); Przestepcy prawa (Law Breakers) and Biblia mówi (The Holy Bible Says).
Mailing address: 119 St Peter's Street, St Albans AL1 3EY, UK.

Da Campo, Lino D, F.S.F., Br

Rel: Roman Catholic; Ordained; celibate.
International Organizations: Superior General of the Institut des Frères de la Sainte-Famille de Belley.
Mailing address: Viale Aurelio Saffi 24, I-00152 Roma, Italy. T. (39 6) 581 38 41

da Cruz Policarpo, José, H E Mgr

Rel: Roman Catholic; Ordained on 15 AUG 1961; b. 26 FEB 1936 in Alvorninha, Portugal; celibate.
International Organizations: Titular Bishop (since 1978) of Caliabria. Member of the Congregation for Catholic Education. Consultant of the Pontifical Council for Dialogue with Non-Believers.
Other organizations: Auxiliary, Lisbon; Rector, Catholic University of Portugal.
Mailing address: Universidade Católica Portuguesa, Palma de Cima, P-1600 Lisboa, Portugal. T. (351 1) 726 55 50

Daffern, Thomas

Rel: Eclectic; Unordained; b. 08 MAY 1956 in Montreal, Quebec, Canada; separated; c. 2 girls; Citizen of Canada and United Kingdom; Lang. English and French.
International Organizations: Secretary-General for UK and Ireland (since 1990) of the World Conference on Religion and Peace.
Convenor (since 1990) of the International Peace Research Association.
Other organizations: Director (since 1990), University of London Peace and World Order Studies Network; Co-President (since 1990), International Philosophers for Global Responsibility.
Main field of work: University education. **Career:** Research Development Officer, Peace Studies Initiative from 1989 to 1990. **Educ.:** Phil. from University of Bristol, UK (1976-1977); B.A. Hist. from University of London, UK (1985-1988); M.Phil., Ph.D. from University of London, UK (Beginning in 1989). **Interests:** study, travel, thought, friendship. **Mem.:** International Peace Research Association, International Philosophers for Global Responsibility **Pubs.:** Editor, *In Formation* (journal); *Europax* (journal) and various articles and papers on peace, world order studies and philosophy of religion.
Mailing address: Peace Studies, Institute of Education, 20 Bedford Way, London WC1H 0AA, UK. T. (44 81) 801 9981

D'Agostino, Federico

Rel: Unknown; b. 13 JUN 1939; Citizen of Italy.
International Organizations: Member of the Council of the International Society for the Sociology of Religions.
Other organizations: Full Professor of General Sociology, University of Naples.
Educ.: Dr. Sociology from University of California, Berkeley, USA.
Mailing address: Via Sprovieri 6, I-00152 Roma, Italy. T. (39 6) 580 1755 – (39 81) 285 526

Dagpo Rinpoche, D

Rel: Buddhist.
International Organizations: Member of the Foundation for the Preservation of the Mahayana Tradition.
Mailing address: 107 av Floquet, F-94240 Hay Les Roses, France.

Dahm, Gene

Rel: Mason.
International Organizations: Secretary of the High Twelve International.
Mailing address: 11155 S Towne Sq, B2, St Louis MO 63123, USA. T. (1 314) 487 3387.

Daily, Thomas Vose, Bishop

Rel: Roman Catholic; Ordained on 10 JAN 1952; b. 23 SEP 1927 in Belmont MA, USA; celibate.
International Organizations: Bishop (since 1990) of Brooklyn.
Member of the Pontifical Commission for Latin America.
Other organizations: Serves on a number of boards, local and national organizations, including Catholic Relief Services, the Florida Catholic Conference, Catholic University, St John Vianney College Seminary of Miami, Society of St James the Apostle, Palm Beach Round Table, Fellowship of Catholic Scholars, Eternal Word Television Network, Papal Foundation, and National Catholic Office for Persons with Disabilities.
Career: Curate, St Ann's parish, Wollaston MA from 1952 to 1960. Joined Missionary Society of St James the Apostle and served in Peru from 1960 to 1965. Assistant Pastor from 1965 to 1971. Appointed Secretary to Cardinal Humberto Medeiros, Archbishop, Boston in 1971. Named Auxiliary Bishop, Boston in 1973. Named Vicar-General, Boston in 1974. Appointed First Bishop, Palm Beach FL in 1984. Served as Moderator, Florida Catholic Conference Task Force for Human Sexuality. Served as Chairman of the Board of Directors, St Vincent de Paul Regional Seminary in Boynton Beach FL. Co-Founder, Por Cristo missionary organization. **Educ.:** Attended Boston College, Chestnut Hill MA, USA; Attended St John's Seminary.
Mailing address: 75 Greene Ave, PO Box C, Brooklyn NY 11202-3604, USA. T. (1 718) 638 5500. Fax 638 5552

Dalai Lama, D, H H The

Also know as Tenzin Gyatso; Rel: Buddhist: Tibetan; b. 06 JUL 1935 in Taktser, Amdo, Eastern Tibet; Citizen of Tibet.
International Organizations: Spiritual and temporal leader of the Tibet.
Other organizations: Supreme Head, all Buddhist sects in Tibet (Xizang).
Career: Recognised as the fourteenth Dalai Lama in 1937. Enthroned at Lhasa in 1940. Rights exercised by regency from 1934 to 1950. Assumed political power in 1950. Fled to Chumbi in South Tibet after abortive resistance to Chinese in 1950. Negotiated agreement with China in 1951. Vice-Chairman, Standing Committee, and Member, National Committee, CPPCC (Chinese People's Political Consultative Conference) from 1951 to 1959. Honorary Chairman, Chinese Buddhist Association from 1953 to 1959. Delegate to National People's Congress from 1954 to 1959. Chairman, Preparatory Committee for the 'Autonomous Region of Tibet' from 1955 to 1959. Fled Tibet to India after suppression, Tibetan national uprising in 1959. Since exile, has travelled widely in Asia, Europe and the United States. **Educ.:** Educated privately by Buddhist monks; Attained the Geshe Lharampa degree (equivalent to a doctorate in Buddhist Philosophy) (1959). **Interests:** gardening, mechanics. **Hons.:** Albert Schweitzer Humanitarian Award (1987); Memory Prize (1989); Congressional Human Rights Award (1989); Nobel Peace Prize (1989). **Pubs.:** *My Land and My People* (autobiography), 1962, 1987; *Key to the Middle Way*, 1971 and *A Human Approach to World Peace*, 1984.
Mailing address: Thekchen Choeling, Mcleod Ganj 176219, Dharamsala, Himachal Pradesh, India.

Daly, Edward Kevin, Rt Rev

Rel: Roman Catholic; Ordained in 1957; b. 05 DEC 1933 in Co. Donegal, Ireland; single; Citizen of Ireland.
International Organizations: Bishop (since 1974) of Derry.
Career: Curate, Castlederg, Co. Tyrone from 1957 to 1962. St Eugene's Cathedral, Derry from 1962 to 1974. **Educ.:** B.Phil., S.T.L., D.D. **Interests:** theatre, angling.
Mailing address: Bishop's House, St Eugene's Cathedral, Derry BT48 9AP, Ireland. T. (353 504) 262302

Daly, Gabriel Conor, OSA, Fr

Rel: Roman Catholic; Ordained in 1951; b. 18 NOV 1927 in Dublin, Ireland; celibate; Citizen of Ireland.
International Organizations: Member (since 1944) of the Order of St Augustine.
Other organizations: Part-time Lecturer in Theology (since 1970), Irish School of Ecumenics; Lecturer in Theology (since 1985), Trinity College, Dublin.
Career: Senior History Master and Housemaster, Austin Friars School, Carlisle, UK in 1957. Tutor in History, University College, Dublin from 1967 to 1966. Lecturer in Theology, Milltown Institute, Dublin during 1969. Chairman, Irish Theological Association from 1979 to 1985. **Educ.:** Pontifical Gregorian University, Vatican City State; University of Oxford, UK. **Interests:** walking, music. **Pubs.:** Asking the Father: A Study of the Prayer of Petition, 1982; Creation and Redemption, 1988 and articles and reviews in various theological journals.
Mailing address: St Augustine's, Taylor's Lane, Ballyboden, Dublin 16, Ireland. T. (353) 944966

Daly, Mary

Rel: Roman Catholic; Citizen of USA.
International Organizations: Feminist and theologian of the Roman Catholic Church.
Career: Taught at Boston College, Chestnut Hill MA. **Educ.:** Doc. from Saint Mary's College, Notre Dame IN, USA (1954); Doc. from University of Fribourg, Switzerland (1965). **Pubs.:** *Beyond God the Father*, 1973; *Gyn/Ecology: The Metaethics of Radical Feminism*, 1978 and *Pure Lust: Elemental Feminist Philosophy*, 1984.
Mailing address: No mailing address.

Damiba, Bernard

Rel: Roman Catholic; b. 29 SEP 1953 in Koupela, Burkina Faso; m. to Lucienne Compaore; c. 1 boy and 2 girls; Citizen of Burkina Faso; Lang. French and English.
International Organizations: Africa Regional Secretary (since 1989) of the International Catholic Conference of Scouting.
Other organizations: General Commissioner (since 1989), Scouts of the Burkina Faso; Director General (since 1989), Society for Informatics Services.

Main field of work: Education. Career: Head of Informatics Dept., National Society of Electricity from 1983 to 1988. Educ.: D.U.T. Génie Mécanique from University of Bordeaux, France (1975-1978); M.I.A.G.E. Informatique from University of Montpellier, France (1978-1981).
Mailing address: BP 729, Ouagadougou, Burkina Faso. T. (226) 30 74 09
Residence: BP 1041, Ouagadougou, Burkina Faso T. See Mailing Address

Dammertz, Viktor Joseph, Rt Rev
Rel: Roman Catholic; Ordained on 21 SEP 1957; b. 08 JUN 1929 in Schaephuysen, Nordrhein Westfalen, Germany; celibate; Citizen of Germany; Lang. German, Italian, English, French and Spanish.
International Organizations: Abbot Primate (since 1977) of the Benedictine Confederation.
Career: Secretary-General, Benedictine Congregegation of St Otilien from 1960 to 1975. Abbot, St Otilien Abbey from 1975 to 1977. **Educ.:** Innsbruck University, Austria (1950-1953); Lic. Theol. from Pontifical St Anselmo Athenaeum, Vatican City State (1954-1957); Dr.iur.can. from Ludwig-Maximilians-Universität München, Germany (1957-1961). **Hons.:** Distinguished Service Cross (1st Class) (BVK); Bavaria Distinguished Service Cross (BayVO); Knight of the Legion of Honour. **Pubs.:** Das Verfassungsrecht der benediktinischen Mönchskongregationen.
Mailing address: Piazza Cavalieri di Malta 5, I-00153 Roma, Italy. T. (39 6) 5791319
Residence: See Mailing Address. T. See Mailing Address

Danesh, H B, Dr
Rel: Bahá'í.
International Organizations: Chairman of the Association for Bahá'í Studies.
Mailing address: 34 Copernicus Street, Ottawa ON K1N 7K4, Canada. T. (1 613) 233 19 03.

Danker, William J, Dr
Rel: Christian.
International Organizations: President of the Center for World Christian Interaction.
Mailing address: 5842 Stony Island Ave, Chicago IL 60637, USA. T. (1 312) 947 90 96.

Danneels, Godfried, H E Card
Rel: Roman Catholic; Ordained on 17 AUG 1957; b. 04 JUN 1933 in Kanegem, Belgium; celibate; Citizen of Belgium; Lang. Dutch, French, English, German and Italian.
International Organizations: Member (since 1983) of the College of Cardinals.
Permanent Member of the Synod of Bishops. Member of the Sacred Congregation for the Sacraments and Divine Worship. Member of the Congregation for Bishops. Member of the Congregation for the Evangelization of Peoples. Member of the Congregation for Catholic Education. Member of the Pontifical Council for Dialogue with Non-Believers.
Other organizations: Archbishop (since 1980), Malines-Brussels; Castrene Bishop (since 1980), Belgium; President (since 1980), Belgian Bishops' Conference; Member, Council of Cardinals and Bishops; Member, Congregation for the Doctrine of the Faith.
Career: Professor, Diocesan Seminary, Bruges from 1959 to 1977. Professor, Catholic University of Louvain (KUL) from 1969 to 1977. Bishop, Antwerp from 1977 to 1979. Created Cardinal by Pope John Paul II in 1983. **Educ.:** Lic. Phil.Thom. from Catholic University of Louvain (KUL), Belgium; Dr. Theol. from Pontifical Gregorian University, Vatican City State. **Pubs.:** 21 brochures, *Parole de Vie* Series; Qui est Dieu pour vous? and Familles, Dieu vous aime.
Mailing address: Aartsbisdom, Wollemarkt 15, B-2800 Mechelen, Belgium. T. (32 15) 21 65 01
Residence: See Mailing Address. T. See Mailing Address

Dansokho, Samuel, Rev
Rel: Reformed Church.
International Organizations: Member of the Executive Committee of the World Alliance of Reformed Churches.
Mailing address: BP 847, 65 rue Wagane Diouf, Dakar, Senegal. T. (221) 215 564. Fax (221) 238 579. Tx 61310 PRICE SG

D'Arcy Wood, Harold D, Rev Dr
Rel: Uniting Church in Australia; Ordained on 20 OCT 1966; b. 09 DEC 1936 in Nuku' Alofa, Tonga; m. to Janet Rae McInnis; c. 1 boy and 1 girl; Citizen of Australia; Lang. English and French.
International Organizations: President (since 1991) of the Uniting Church in Australia.
Other organizations: Ex-President (since 1990), Australian Academy of Liturgy; Vice-President (since 1991), Australian Council of Churches; Parish Minister (since 1991), Canberra City Uniting Church.
Main field of work: Parish ministry. **Career:** Parish Minister, Methodist Church from 1966 to 1968. Studies Secretary, Australian Council of Churches from 1969 to 1973. Lecturer in Theology, Parkin-Wesley College from 1974 to 1988. **Educ.:** M.A. from University of Melbourne, Victoria, Australia (1955-1959); B.D. (Hons) from Melbourne College of Divinity, Australia (1959-1961); Ph.D. from Princeton Theological Seminary, NJ, USA (1962-1966). **Interests:** music, liturgy, social justice. **Mem.:** Australian Academy of Liturgy; Royal School of Church Music **Pubs.:** What Future for the Australian Churches?, 1974 and Building on a Solid Basis, 1987.
Mailing address: City Uniting Church, GPO Box 1526, Canberra ACT 2601, Australia.
Residence: 7 Line Place, Weetangera ACT 2614, Australia. T. (61 6) 257 4600

Dardene, Myriam, O.S.C., Sr
Rel: Roman Catholic; Ordained; celibate.
International Organizations: Member of the Cistercian Order of the Strict Observance.
Other organizations: Member, Association of Contemplative Sisters.
Mailing address: Redwoods Trappistine Monastery, Whilethorn, CA 95489, USA.

Dargan, Joseph, S.J., Rev
Rel: Roman Catholic; Ordained; celibate.
International Organizations: Secretary-General of the Union of European Conferences of Superiors General.
Member of the Society of Jesus.
Other organizations: Secretary, Conference of Major Superiors, Ireland.
Mailing address: Milltown Park, Dublin 6, Ireland. T. (353 1) 698011. Fax (353 1) 698887.

Dargan, Joseph, S.J., Fr
Rel: Roman Catholic; Ordained; celibate.
International Organizations: Secretary-General of the Union of European Conferences of Superiors General.
Mailing address: Milltown Park, Dublin 6, Ireland. T. (353 1) 2698011. Fax 2698887.

Dario, Antonio, d.O., Fr
Rel: Roman Catholic; Ordained; celibate.
International Organizations: Procurator-General of the Confederation of the Oratory of St Philip Neri.
Mailing address: Via Parione 33, I-00186 Roma, Italy. T. (39 6) 687 52 89.

Darmo, Mar Thomas, H B
Rel: Nestorian; Ordained; b. in Iraq; Citizen of Iraq.
International Organizations: Patriarch (since 1968) of Nestorian Church.
Mailing address: Mar Zayya Church, Baghdad, Iraq.

Darmojuwono, Justin, H E Card
Rel: Roman Catholic; Ordained on 25 MAY 1947; b. 02 NOV 1914 in Godean, Yogyakarta, Indonesia; celibate; Citizen of Indonesia.
International Organizations: Member (since 1967) of the College of Cardinals.
Other organizations: Archbishop Emeritus (since 1981), Semarang.
Career: Teacher, Minor Seminary from 1947 to 1948. Priest, local parish, Yogyakarta from 1948 to 1950. Priest, local parish, Klaten from 1950 to 1954. Priest, local parish, Surakarta from 1956 to 1962. Vicar General, Semarang from 1962 to 1963. Archbishop, Semarang from 1963 to 1981. Created Cardinal by Pope Paul VI in 1967. Chairman, Conference of Indonesian Bishops from 1964 to 1980. Bishop, Indonesian Armed Forces from 1964 to 1981. Member, Central Committee,

Federation of Asian Bishops' Conferences from 1970 to 1979. **Educ.:** Dip. from Teacher School, Muntilan, Indonesia (1935); Dip. from Minor Seminary, Yogyakarta, Indonesia (1942); Dip. from Major Seminary, Yogyakarta, Indonesia (1947); Dip. from Pontifical Gregorian University, Vatican City State (1955).
Mailing address: Jl Kamfer Raya 49, Banyumanik, Semarang 50237, Indonesia.

da Silva Lima, José, Prof

Rel: Roman Catholic; Ordained on 23 JUL 1978; b. 11 JUL 1954 in Milhoes, Vila de Punhe, Viana do Castelo, Portugal; celibate; Citizen of Portugal; Lang. Portuguese and French.
International Organizations: Professor of Theology (since 1986) of the Catholic University of Braga.
Other organizations: Professor (since 1988), Graduate School of Education, Viana do Castelo; Director (since 1991), Graduate School of Theology, Viana do Castelo.
Main field of work: University education. **Educ.:** Doc. Hist. of Religions from University of Paris IV (Paris-Sorbonne), France (1986); Doc. Theol. from Institut Catholique de Paris, France (1986); Doc. Cult. Anthropology from New University of Lisbon, Portugal (1987). **Pubs.:** Dieu, je n'ai rien contre Lui (doctoral thesis), Lille, 1986; Religion, Traditions Populaires, Versailles, 1990 and Portugal, Gentes e Culturas – Les Européens, Vol. II, Gallimard, Paris, 1992.
Mailing address: Povoença-Areosa, P-4900 Viana do Castelo, Portugal. T. (351 58) 829 845
Residence: See Mailing Address. T. See Mailing Address

da Silva Sebastiáo, Manuel Martins d, C.S.Sp., Rev

Rel: Roman Catholic; Ordained on 19 DEC 1964; b. 04 JUN 1939 in Proença-a-Nova, Beira Baixa, Portugal; celibate; Citizen of Portugal; Lang. Portuguese, English, French, Spanish, Italian, Latin, Greek and Lumkhumbi.
International Organizations: Procurator-General (since 1989) of the Congregation of the Holy Spirit.
Main field of work: Ministry. **Career:** Taught at lyceeum from 1967 to 1976. Taught at high school from 1975 to 1981. **Educ.:** Studied Theol. from Holy Ghost Missionary College, Dublin, Ireland (1962-1965); Studied Pedagogy from University of Sao Paulo, Brazil (1983-1987). **Interests:** pedagogy.
Mailing address: Clivo di Cinna 195, I-00136 Roma, Italy. T. (39 6) 34 82 47. Fax 345 06 76.
Residence: See Mailing Address. T. See Mailing Address

Dass, Garib, Col

Rel: Salvation Army, The.
International Organizations: Territorial Commander of The Salvation Army.
Mailing address: High Ground Road, Maharaja Nagar PO, Tirunelveli 627 011, India.

Dastur, Meherjirana Kaikobad, Vada Dasturji

Rel: Jarthosti; b. 09 JAN 1935 in Navsari, Gujrat, India; m. to Mehru Manekji Kotwal; Citizen of India; Lang. English, Gujrati, Hindi, Avesta, Pahlavi and Persian.
International Organizations: Vada Dasturji (High Priest) (since 1960) of the Parsi Community.
Other organizations: President (since 1960), Samast Parsi Jarthosti Anjuman, Navsari; President (since 1960), Bhagarsath Anjuman, Navsari; President and trustee, several other religious and social trust funds.
Main field of work: Social service. **Educ.:** B.A. from Bombay University, India (1957). **Interests:** study and understanding of Zoróastrianism. **Pubs.:** various articles on Zoroastrianism in newspapers.
Mailing address: 8/1226, Dastur Street, Navsari 396 445, India.

Daun, Christel, Sr

Rel: Roman Catholic; Ordained; celibate.
International Organizations: Contact of the Missionary Sisters Servants of the Holy Spirit.
Mailing address: Via Cassia 645, I-00189 Roma, Italy. T. (39 6) 366 02 47. Fax 365 2148. C. Missiasteyl 00189 Roma.

D'Avack, P A, Prof

Rel: Roman Catholic.
International Organizations: President of the International Association for the Study of Canon Law.
Mailing address: Università degli Studi, Città Universitaria, I-00100 Roma, Italy.

Davenport, Mikail

Rel: Muslim: Sufi.
International Organizations: Contact of the Sufi Order International.
Mailing address: PO Box 85569, Seattle WA 98145, USA. T. (1 206) 323 2944.

David, S Immanuel, Rev Prof Dr

Rel: Lutheran; Ordained on 15 JUL 1975; b. 01 APR 1949 in Coimbatore, Tamilnadu, Madras, India; m. to Beulah David; c. 1 boy and 1 girl; Citizen of India; Lang. Tamil, Hindi, English, French, Greek, Hebrew and Latin.
International Organizations: Professor of Ecclesiastical History (since 1988) of the Faculty of Protestant Theology, Yaoundé.
Member, West African Theological Commission (since 1988) of the Lutheran World Federation.
Other organizations: Member, Editorial Committee (since 1991), *Flambeau.*
Main field of work: University Education. **Career:** Member, National Committee on Lutheran-Roman Catholic Dialogue, India from 1983 to 1988. Member, National Committee on Lutheran-Orthodox Dialogue, India from 1983 to 1988. Editor, *Indian Church History Review* from 1985 to 1988. Member, Executive Committee, Church History Association of India from 1985 to 1988. Member, South India History Congress from 1985 to 1988. One of the leaders, Theological Institute for Lay Women and Men, BTESSC from 1985 to 1988. Pastor, Tranquebar Lutheran Church, South India. Assistant Professor, United Theological College, Bangalore, South India. **Educ.:** B.A. from Government Arts College, Coimbatore, India; B.D. from University of Serampore, United Theological College, Calcutta, India; M.Th. from University of Serampore, Calcutta, India (1973-1975); M.A. from University of Wisconsin – Madison, USA (1978-1980); D.Th. from Lutheran School of Theology, Chicago IL, USA (1983). **Mem.:** South Indian History Congress; Indian Church History Association **Pubs.:** *American Arcot Mission Missionaries in South India, 1839-1938: An Analytical Study*, Bangalore, Asian Trading Corporation, 1987; *Christianity and the Encounter with Other Religions*, Bangalore, United Theological College, 1988 and *Women, Church and Feminist Theology: A Bibliographical Guide*, Yaoundé, Faculty of Protestant Theology, 1992.
Mailing address: Faculty of Protestant Theology, BP 4011, Yaoundé, Cameroon. T. (237) 20 73 12
Residence: See Mailing Address. T. See Mailing Address

Davids, Leo, Rabbi

Rel: Jewish: Orthodox; Ordained in 1960; b. 22 JUL 1938 in Amsterdam, Netherlands; m. to Faigie Shields; c. 4 children.
International Organizations: Associate Professor (since 1972) of the Atkinson College, York University.
Other organizations: Founding and Current Co-Chairman, Jewish Marriage Education Committee, Toronto.
Career: Rabbi, Congregation Agudat Achim, Bridgeport, CT from 1962 to 1964. Sociology Teacher, City University, NY from 1961 to 1966. Sociology Teacher, Stern College for Women, Yeshiva University, NY from 1966 to 1968. Assistant Professor, York University from 1968 to 1972. Visiting Professor, Bar-Ilan University from 1974 to 1975. Visiting Professor, Tel Aviv University from 1981 to 1982. Visiting Researcher at Demographic Centre, Jerusalem from 1981 to 1982. Research Consultant to B'nai B'rith Canada (Divorce Project) from 1986 to 1987. Lecturer, University of Toronto, Brandeis University (Jerusalem Centre), Trent University and others. **Educ.:** B.A. (magna cum laude) from City University of New York, City College, USA (1959); M.A. from City University of New York, City College, USA (1961); Ph.D. from New York University, NY, USA (1968). **Interests:** Torah-scroll reading. **Hons.:** Phi Beta Kappa; Kappa Delta Pi, Alpha Kappa Delta; Nominee for Teaching Excellence Award, OCUFA several times. **Mem.:** Canadian Professor for Peace in the Middle East; Canadian Sociology and Anthropology Association; York University Faculty Association; Association for Jewish Demography and Statistics; etc. **Pubs.:** Defining Sociology/Anthropology, 1982; Author of scholarly articles and contributing editor to *Viewpoints Quarterly*.
Mailing address: 338, Atkinson College, 4700 Keele St, North York ON M3J 1P3, Canada.
Residence: 242 Armour Blvd, Downsview, Ontario, Canada.

Davidson, Hugh, Rev

Rel: Reformed Church.
International Organizations: Member of the Executive Committee of the World Alliance of Reformed Churches.
Mailing address: 43 Inverleith Gardens, Edinburgh EH3 5PR, UK. T. (44 31) 225 5722
Residence: T. (44 31) 552 3874

Davie, Grace R C, Dr

Rel: Anglican; Unordained; b. 02 SEP 1946 in Cambridge, UK; m. to Dr Mark Davie; c. 3 boys; Citizen of United Kingdom; Lang. English and French.
International Organizations: Member of the Council of the International Society for the Sociology of Religions.
Other organizations: Lecturer, Dept. of Sociology (since 1990), University of Exeter; Short-term research projects.
Main field of work: University education. **Educ.:** B.A. from University of Exeter, UK (1964-1967); Ph.D. from University of London, London School of Economics and Political Science, UK (1967-1971). **Interests:** sociology of religion. **Mem.:** International Sociological Association, British Sociological Association **Pubs.:** With G Ahern, *Inner City God – The Nature of Belief in the Inner City*, Hodder and Stoughton, 1987; "Believing without Belonging – Is this the Future of Religion in Britain?", in *Social Compass*, 1990/4 and "'An Ordinary God': The Paradox of Religion in Contemporary Britain", in *British Journal of Sociology*, 1990/3.
Mailing address: Dept. of Sociology, University of Exeter, Exeter EX4 4RJ, UK. T. (44 392) 263276
Residence: Sylvan Road 17, Exeter EX4 6EW, UK. T. (44 392) 75637

Davies, Noel Anthony, Rev

Rel: Congregational; Ordained on 12 SEP 1968; b. 26 DEC 1942 in Dowlais, Glamorgan, Wales, UK; m. to Patricia; Citizen of United Kingdom; Lang. Welsh and English.
International Organizations: General Secretary (since 1990) of the CYTUN: Churches Together in Wales.
Other organizations: Chairman (since 1990), Welsh Congregational Union Council; Moderator (since 1991), Churches Commission on Mission; Moderator (since 1991), WCC Consultation for National Council of Churches.
Main field of work: Ecumenical ministry. **Career:** Minister, Welsh Congregational Church, Glanaman, Dyfed, Wales from 1968 to 1977. General Secretary, Council of Churches for Wales from 1977 to 1990. General Secretary, Commission of Covenanted Churches from 1977 to 1990. **Educ.:** B.Sc. from University of Wales, University College of North Wales, Bangor, UK (1962-1965); B.A. from University of Oxford, Mansfield College, UK (1965-1968). **Interests:** gardening, cooking, hi-fi, classical music. **Pubs.:** Collections of Prayers, etc. and a number of articles in Christian magazines.
Mailing address: Churches Together in Wales, 21 St Helens Rd, Swansea SA1 4AP, UK. T. (44 792) 460876
Residence: 16 Mapple Crescent, Uplands, Swansea SA2 0QD, UK. T. (44 792) 203693

Davies, Stanley

Rel: Evangelical.
International Organizations: Secretary-General of the Evangelical Missionary Alliance.
Mailing address: Whitefield House, 186 Kennington Park Rd, London SE11 4BT, UK. T. (44 71) 735 0421.

Davis, Brian Newton, Most Rev

Rel: Anglican; Ordained on 05 FEB 1961; b. 28 OCT 1934 in Stratford, Taranaki, New Zealand; m. to Marie Lynette Waters; c. 4 girls; Citizen of New Zealand; Lang. English.
International Organizations: Primate and Archbishop (since 1986) of New Zealand.
Other organizations: Bishop (since 1986), Wellington.
Main field of work: Ministry. **Career:** Ordained Deacon in 1960. Assistant Curate, St Mark's, Wellington from 1960 to 1962. Assistant Curate, Karori and Makara, Wellington from 1962 to 1964. Vicar, Karori and Makara from 1964 to 1967. Vicar, Dannevirke from 1967 to 1973. Vicar and Dean, St John's Cathedral, Napier from 1973 to 1980. Vicar-General, Waiapu from 1979 to 1980. Bishop, Waikato from 1980 to 1986. **Educ.:** B.A. from Victoria University of Wellington, New Zealand (1957); M.A. Geog. (1st) from Victoria University of Wellington, New Zealand (1959); L.Th. from Christchurch Theological College, New Zealand (1965). **Interests:** tennis, water colour painting, wood turning, swimming. **Mem.:** Wellington Rotary Club **Pubs.:** Contributor to *An Encyclopaedia of New Zealand*, 1966; Columnist for *Church Scene* (Australian Anglican Newspaper) and *Guidelines for Church Membership* (booklet).
Mailing address: PO Box 12-046, Wellington, New Zealand. T. (64 4) 472 1057
Residence: Bishopcourt, 28 Eccleston Hill, Thorndon, Wellington, New Zealand. T. (64 4) 472 3183

Davis, F J

Rel: Druidism.
International Organizations: Secretary-General of the United Ancient Order of Druids.
Mailing address: Druids Hall, 8 Perry Road, Bristol BS1 5BQ, UK. T. (44 272) 292009.

Davis, Moshe, Prof

Rel: Jewish; Ordained in 1942; b. 12 JUN 1916 in Brooklyn NY, USA; m. to Lottie Keiser; c. 1 boy and 1 girl; Citizen of Israel and USA; Lang. English, Hebrew, Yiddish, French and Spanish.
International Organizations: Chairman (since 1980) of the International Centre for University Teaching of Jewish Civilization.
Other organizations: Project Director and General Editor, the America-Holy Land Series; Professor Emeritus, Institute of Contemporary Jewry, Hebrew University of Jerusalem.
Main field of work: University education, research. **Career:** Provost, Jewish Theological Seminary of America, New York NY from 1950 to 1963. Founding Head, Hebrew University of Jerusalem in 1959. Chairman, Israel President's Study Circle on World Jewry from 1966 to 1983. **Educ.:** B.S. from Columbia University, Teachers' College, New York NY, USA (1937); Pd.B. from Jewish Theological Seminary of America, New York NY, USA (1937); M.H.L. from Jewish Theological Seminary of America, New York NY, USA (1942); Ph.D. from Hebrew University of Jerusalem, Israel (1946). **Hons.:** Israel Knesset Speaker's Award (1987); Samuel Rothberg Prize for World Jewish Education; Jewish Theological Seminary Medal (1991). **Mem.:** Rabbinical Assembly, American Jewish Historical Association, World Union of Jewish Studies **Pubs.:** Shaping of American Judaism, 1951; From Dependence to Mutuality: The American Jewish Community and World Jewry, 1970 and Editor, Publications of Study Circle in Home of President of Israel, Series I, 1967 and Series XIII, 1983.
Mailing address: PO Box 4234, Jerusalem 91042, Israel. T. (972 2) 699 032 – 699 036. Fax 232 693 – 244 737 – 242 810 EXT 7680. Tx 26144 BXJM II EXT 7680.
Residence: Balfour Street 14, Jerusalem 92102, Israel. T. (972 2) 630454

Davis, William Emrys

Rel: Bahá'í; b. 13 OCT 1942 in Lexington KY, USA; m. to Constance Ann; c. 2 girls; Citizen of USA; Lang. English and Spanish.
International Organizations: Member/Treasurer (since 1986) of the National Spiritual Assembly of the Bahá'ís of the US.
Other organizations: Board of Directors (since 1991), Judicature, Chicago IL.
Main field of work: Law and judicial administration. **Career:** Program Officer, Peace Corps, Chile from 1968 to 1971. Director, Administrative Office of Courts, State Judiciary, Kentucky from 1975 to 1979. Circuit Executive, the US Courts in the Ninth Circuit from 1981 to 1986. Director, Administrative Office of Courts, Judicial Council of California from 1987 to 1991. **Educ.:** A.B. from University of Kentucky, Lexington, USA (1960-1964); J.D. from Transylvania University, Lexington KY, USA (1967-1972). **Interests:** travel, current affairs, social and economic development. **Mem.:** Kentucky Bar Association; Kentucky Judges Association; Judges of North Circuit
Mailing address: Bahá'í National Center, 536 Sheridan Rd, Wilmette IL 60091, USA.
Residence: 414 Bay Road, Menlo Park CA 94025, USA. T. (1 415) 324 9055

Davitti, Michele, Rev Fr

Rel: Roman Catholic; Ordained on 27 SEP 1970; b. 23 JUN 1943 in Reggello, Firenze, Italy; celibate; Citizen of Italy; Lang. Italian, German and English.

International Organizations: Secretary-General (since 1991) of the Inter-Territorial Catholic Bishops' Conference.
Main field of work: University education. **Career:** Lecturer in Cultural Anthropology, Istituto Saveriano, Parma from 1979 to 1989. Visiting Lecturer, St Paul's Major Seminary, Gbarnga, Liberia from 1979 to 1989. **Educ.:** B.Th. from Istituto Saveriano, Parma, Italy (1966-1970); M.Th. from Pontifical Gregorian University, Vatican City State (1971-1973); M.A. from University of London, School of Oriental and African Studies, UK (1973-1975). **Interests:** cultural anthropology, comparative religion. **Pubs.:** African Traditional Religion; Phenomenology of Religion and Christian-Muslim Dialogue.
Mailing address: 10 Howe St – Santanno House, PO Box 893, Freetown, Sierra Leone. T. (232 22) 24590. C. SANTANNO FREETOWN. Tx 3311 WASLTD SL (attn Santanno/ITCABIC) – 3689 CATSEC SL.
Residence: See Mailing Address. T. See Mailing Address

Dawson, Alan
Rel: Christian.
International Organizations: President of the International Religious Fellowship.
Mailing address: 1A Ingram Street, Huntington, Cambs PE18 6QG, UK.

Day, Garnett
Rel: Disciples of Christ.
International Organizations: Executive Officer of the Disciples Peace Fellowship.
Mailing address: PO Box 1986, Indianapolis IN 46206, USA. T. (1 317) 353 1491.

Daya Mataji, D, Sri Sri
Rel: Hindu.
International Organizations: Spiritual Head and President of the Self Realization Fellowship.
Other organizations: Spiritual Head and President, Yogoda Satsanga Society of India.
Mailing address: 3880 San Rafael Avenue, Los Angeles CA 90065, USA. T. (1 213) 225 2471. C. Selfreal.

Dayton, Edward, Dr
Rel: Protestant.
International Organizations: Vice-President of the World Vision International.
Mailing address: 919 West Huntington Drive, Monrovia CA 91016, USA. T. (1 818) 303 8811. C. WORVIS. Tx 275 335 WORV UR.

De Angelis, Nicola, C.F.I.C., Fr
Rel: Roman Catholic; Ordained; celibate.
International Organizations: Superior General of the Sons of the Immaculate Conception.
Mailing address: Via della Luce 46, I-00153 Roma, Italy. T. (39 6) 58 23 31

de Araújo Sales, Eugénio d, H E Card
Rel: Roman Catholic; Ordained on 21 NOV 1943; b. 08 NOV 1920 in Acari, Rio Grande do Norte, Brazil; celibate; Citizen of Brazil.
International Organizations: Member (since 1969) of the College of Cardinals.
Member of the Congregation for Bishops. Member of the Sacred Congregation for the Oriental Churches. Member of the Sacred Congregation for the Sacraments and Divine Worship. Member of the Congregation for Bishops. Member of the Congregation for the Evangelization of Peoples. Member of the Congregation for the Clergy. Member of the Congregation for Catholic Education. Member of the Pontifical Council for Culture. Member of the Pontifical Council for Social Communications.
Other organizations: Archbishop (since 1971), Sao Sebastiao do Rio de Janeiro; Primate, Brazil; Member, Council of Cardinals for the Study of Organizational and Economic Problems of the Holy See.
Career: Bishop in 1954. Apostolic Administrator, See of Sao Salvador da Bahia (until 1968). Created Cardinal by Pope Paul VI in 1969. **Interests:** reading.
Mailing address: Rua da Glória 446, CP 1362, 20001 Rio de Janeiro RJ, Brazil. T. (55 21) 222 2308

Deas, Alberta
Rel: Bahá'í.
International Organizations: Member of the National Spiritual Assembly of the Bahá'ís of the US.
Career: Director, Louis Gregory Bahá'í Institute in Hemingway, SC. Assistant Professor of Education, South Carolina State College. **Educ.:** Ed.D. from University of Massachusetts at Amherst, USA.
Mailing address: Bahá'í National Center, 536 Sheridan Rd, Wilmette IL 60091, USA.

De Boer, Stewart W
Rel: Christian.
International Organizations: President of the Asian Theological Seminary.
Mailing address: PO Box 461, Manila, Philippines. T. (63 2) 98 68 26 – (63 2) 98 29 21.

De Bonis, Donato, H E Mgr
Rel: Roman Catholic; Ordained on 05 APR 1953; b. 13 APR 1930 in Potenza, Basilicata, Italy; celibate; Citizen of Italy.
International Organizations: Prelate (since 1990) of the Institute for Religious Works.
Other organizations: General Treasurer (since 1990), Apostolic Administration; Professor of Ecclesiastic Administration, Lateran Pontifical University; Protonotary Apostolic, the Holy See.
Main field of work: University education. **Educ.:** Deg. Phil. from Lateran Pontifical University, Vatican City State; Deg. Theol. from Lateran Pontifical University, Vatican City State; Deg. Canon Law from Lateran Pontifical University, Vatican City State.
Mailing address: 00120 Città del Vaticano, Vatican.
Residence: See Mailing Address.

De Brandt, Edward
Rel: Roman Catholic.
International Organizations: President of the International Catholic Migration Commission.
Director of the Secours international de Caritas Catholica.
Mailing address: 37-39 rue de Vermont, CP 96, CH-1211 Genève 20, Switzerland. T. (41 22) 733 41 50. C. Cathmicom. Tx 41 41 22 ICMC CH. Fax (41 22) 734 79 29.

de Broucker, José
Rel: Roman Catholic.
International Organizations: President of the International Federation of Catholic Journalists.
Mailing address: 'Malesherbes Publications', 163 Bd Malesherbes, F-75859 Paris CEDEX 17, France.

de Castro Cruz, Jorio Roberto
Rel: Roman Catholic.
International Organizations: Co-Secretary-General of the International Movement of Apostolate of Children.
Mailing address: 8 rue Duguay-Trouin, F-75006 Paris, France. T. (33 1) 42 22 84 81.

Decencière, Françoise, Sr
Rel: Roman Catholic; Ordained; celibate.
International Organizations: Superior General of the Religious of Nazareth.
Mailing address: Via Caterina Fieschi 6, I-00151 Roma, Italy.

de Champigny, Yolande, Sr
Rel: Roman Catholic; Ordained; celibate.
International Organizations: Contact of the Augustinians of the Assumption.
Mailing address: 6 Sentier Henri-Dupuis, F-94230 Cachan, France.

Dechant, Virgil C
Rel: Roman Catholic.
International Organizations: Supreme Knight of the Knights of Columbus.
Member of the Pontifical Council for the Family. Consultor of the Pontifical Council for Social Communications.
Other organizations: Honorary Consultor, Pontifical Commission for the Vatican City State; Member, Institute for Religious Works.
Mailing address: Columbus Plaza, New Haven CT 06507, USA. T. (1 203) 772 2130. Fax 773 3000.

De Concini, Barbara, Dr

Rel: Unknown; b. 15 FEB 1944 in Philadelphia PA, USA; married; Citizen of USA; Lang. English and French.
International Organizations: Executive Director (since 1991) of the American Academy of Religion.
Other organizations: Professor, Emory University, Atlanta GA.
Main field of work: Administration. **Educ.:** A.B. from Rosemont College, PA, USA (1968); M.A. from Bryn Mawr College, PA, USA (1973); Ph.D. from Emory University, Atlanta GA, USA (1980). **Pubs.:** Narrative Remembering, University Press of America.
Mailing address: 1703 Clifton Rd NE, Atlanta GA 30329-4017, USA. T. (1 404) 727 7920

de Corneille, Roland, Rev

Rel: Anglican; Ordained in 1953; b. 19 MAY 1927 in Switzerland; m. to Elizabeth Cleland; c. 1 boy and 1 girl.
International Organizations: Incumbent (since 1991) of the St Paul's Runnymede, Toronto.
Career: Elected to House, Commons as Liberal for Eglinton Lawrence from 1979 to 1988. Official Opposition Critic for Secretary, State from 1984 to 1986. Official Opposition Critic External Relations and for Human Rights from 1986 to 1988. Statistician for Time Inc., NY from 1946 to 1948. Advertising and Brand Promotion for Proctor and Gamble, OH from 1948 to 1951. Curate and Rector, Parish Churches, Montreal and Toronto from 1952 to 1962. Founder and Director, 1st Jewish-Christian Dialogue Program from 1962 to 1970. National Director, League for Human Rights B'nai B'rith from 1971 to 1979. Parliamentary Secretary to Ministry of Veterans' Affairs from 1981 to 1983. Chairman, Toronto and Region Liberal Caucus from 1984 to 1985. Chairman, Canada-Italy Parliamentary Friendship Group. Founding Chairman, Canada-Israel Parliamentary Friendship Group and other parliamentary associations including Commonwealth, Canada-Europe, Canada-US and NATO. **Educ.:** B.A. from Amherst College, MA, USA (1946); L.Th. from University of Toronto, Trinity College, Canada (1953); S.T.B. from University of Toronto, Trinity College, Canada (1954); M.Th. from University of Toronto, Trinity College, Canada (1961). **Hons.:** Shared Brotherhood Book Award, National Conference of Christians and Jews with Martin Luther King Jr (1966); Recipient, First Claude Uzielli Award for pioneering Christian-Jewish Dialogue (1982); Certificate of Appreciation, Jamaican-Canadian Association (1983); Spiritual Father of the Year Award, Pioneer Woman (1984); Phi Beta Kappa. **Mem.:** many Church and Inter-Religious Committees, including World Council of Churches, National Council of Churches, Canadian Council of Churches, Anglican Diocese, Toronto **Pubs.:** Christians and Jews, 1966.
Mailing address: 9 Bernick Rd, Willowdale ON M2H 1E3, Canada.
Residence: See Mailing Address.

de Cosse, Pierre

Rel: Roman Catholic.
International Organizations: Grand Master of the Military and Hospitaller Order of St Lazarus of Jerusalem.
Mailing address: Duc de Brissac, 36 Cours Albert 1er, F-75008 Paris, France.

de Coulomme-Labarthe, G

Rel: Roman Catholic.
International Organizations: Representative at UNESCO of the International Catholic Child Bureau.
Mailing address: 19 rue de Varennes, F-75007 Paris, France. T. (33 1) 42 22 00 01.

Decourtray, Albert, H E Card

Rel: Roman Catholic; Ordained on 29 JUN 1947; b. 09 APR 1923 in Wattignies, France; celibate; Citizen of France; Lang. Italian and English.
International Organizations: Member (since 1985) of the College of Cardinals.
Member (since 1985) of the Pontifical Council for Promoting Christian Unity. Member (since 1988) of the Pontifical Council for Dialogue with Non-Believers.
Other organizations: Archbishop (since 1981), Lyon.
Main field of work: Ministry. **Career:** Director, Grand Séminaire de Lille from 1952 to 1962. Vicar-General, Lille from 1966 to 1971. Consecrated Auxiliary Bishop, Dijon and Bishop (titular church of Ippona Zanto) in 1971. Bishop, Dijon from 1974 to 1981. Prelate, Mission de France from 1982 to 1988. Created Cardinal by Pope John Paul II in 1985. Member, Pontifical Council for Inter-Religious Dialogue from 1985 to 1991. President, Episcopal Conference of France from 1987 to 1990. Member, World Synod of Bishops, Rome (in 1983, 1985, 1987). **Educ.:** Dr. Theol. from Pontifical Gregorian University, Vatican City State (1948-1951); Lic. Sc.Bibliques from Pontifical Biblical Institute, Vatican City State (1948-1951). **Hons.:** Officer, Legion of Honour (1986); Grand Cordon de l'Ordre du Cèdre du Liban; Prix des Droits de l'Homme (1987).
Mailing address: 1 place de Fourvière, F-69005 Lyon, France. T. (33) 78 25 12 27

Decraene, Michel, C.I.C.M., Rev Fr

Rel: Roman Catholic; Ordained; b. 04 OCT 1935 in Mouscron, Belgium; celibate; Citizen of Belgium; Lang. French, English, Dutch, Italian and Kiyombe.
International Organizations: Superior General (since 1987) of the Congregation of the Immaculate Heart of Mary.
Member, Council for Reporting to the Congregation and to the International Union of Superior Generals and the Union of Superior Generals of the Congregation for the Evangelization of Peoples.
Educ.: Lic. philologie classique from Catholic University of Louvain (UCL), Belgium (1961-1965).
Mailing address: Via S Giovanni Eudes 95, I-00163 Roma, Italy. T. (39 6) 661 585 52
Residence: See Mailing Address. T. See Mailing Address

de Diesbach, G

Rel: Roman Catholic.
International Organizations: UNESCO Representative of the World Union of Catholic Women's Organizations.
Mailing address: Hendecourt, F-62175 Boileux, France.

de Ferreira, Maria Victoria

Rel: Roman Catholic.
International Organizations: Head, Latin America Regional Secretariat of the International Movement for the Apostolate in the Independent Social Milieux.
Mailing address: Grtal Bernardin a Caballero 271, Asunción, Paraguay.

de Furstenberg, Maximilian, H E Card

Rel: Roman Catholic; Ordained; celibate.
International Organizations: Grand Master of the Equestrian Order of the Holy Sepulchre of Jerusalem.
Mailing address: I-00120 Città del Vaticano, Vatican. T. (39 6) 65 76 32.

Degenhardt, Johannes Joachim, Most Rev Archbishop

Rel: Roman Catholic; Ordained in 1951; b. 31 JAN 1926 in Schwelm, Germany; celibate; Citizen of Germany.
International Organizations: Archbishop (since 1974) of Paderborn.
Career: Student chaplain at Pädagogische Hochschule, Paderborn, and Bezirksdekan, Hochstift Paderborn. Bishop, Paderborn in 1968. Chairman, Education Commission, Conference of German Bishops. **Pubs.:** Lukas, Evangelist der Armen, 1966.
Mailing address: Kamp 38, 33098 Paderborn, Germany. T. (49) 20 73 27

De Jager, John

Rel: Reformed Church.
International Organizations: Contact of the Christian Reformed Church World Literature Ministries.
Mailing address: c/o CRC Publications, 2850 Kalamazoo Ave SE, Grand Rapids MI 49560, USA. T. (1 616) 246 0728. Fax 246 0834.

de Jonghe, Etienne Nathalie Honoré
Rel: Roman Catholic; b. 22 DEC 1946 in Sint-Niklaas, Belgium; m. to Magda Van Damme.
International Organizations: International Secretary of the Pax Christi – International Catholic Peace Movement.
Educ.: Lic. Pol. and Soc. Sc. from Catholic University of Louvain (KUL), Belgium. **Pubs.:** With Mieke Van Haegendoren, *Roergangers in de Mist*, Standaard Boekhandel Antwerpen and articles in various reviews.
Mailing address: Rue Marché aux Grains 21, B-1000 Bruxelles, Belgium. T. (32 2) 502 55 50. Fax 502 46 26. Tx 32675 BAVEL B.
Residence: Transvaalstraat 22, B-2600 Berchem, Belgium T. (32 3) 239 43 72

del Aguila, Jorge Mario Avila
Rel: Roman Catholic.
International Organizations: Secretary-General of the Episcopal Secretariat of Central America and Panama.
Mailing address: 7 Avenida O 42, Zona 1, 21001 Jalapa, Guatemala. T. (502 9) 361 022.

Delcourt, France, S.A., Sr
Rel: Roman Catholic; Ordained; celibate.
International Organizations: President of the Union of European Conferences of Superiors General.
Mailing address: 16 rue Saint-Jean Baptiste de la Salle, F-75006 Paris, France. T. (33) 45 67 90 90

Del Hoyo, Félix D, F.S.C., Fr
Rel: Roman Catholic; Ordained; celibate.
International Organizations: Procurator-General of the Brothers of the Christian Schools.
Mailing address: Casa Generalizia, Via Aurelia 476, I-00165 Roma, Italy. T. (39 6) 662 15 91 – 662 15 94 – 62 01 01. C. LASALLE-ROMA. Tx 622429 TELSAL I.

Delk, Robert C
Rel: Muslim.
International Organizations: Past President of the American Islamic College.
Mailing address: 640 W Irving Park Road, Chicago IL 60613, USA. T. (312) 281 4700.

Delp, Lee
Rel: Mennonite.
International Organizations: President of the MEDA.
Mailing address: 12 Greenfield Road, Lancaster PA 17602, USA.

del Portillo, Alvaro , Rev
Rel: Roman Catholic; Ordained; celibate.
International Organizations: Prelate of the Prelature of the Holy Cross and Opus Dei.
Mailing address: c/o Viale Bruno Buozzi 73, I-00197 Roma, Italy. T. (39 6) 807 87 41.

Del Zanna, Luisa D
Rel: Roman Catholic.
International Organizations: Secretary of the Focolare Movement.
Mailing address: Centro dell'Opera, Via di Frascati 306, I-00040 Rocca di Papa (Roma), Italy. T. (39 6) 949 386. Fax (39 6) 949 320.

de Madrid, Ignacio, O.S.H., Br
Rel: Roman Catholic; Ordained; celibate.
International Organizations: Prior-General of the Order of St Jerome.
Mailing address: Monasterio de San Jerónimo de Yuste, 10430 Cuacos, Spain. T. (34 927) 46 05 30.

de Mallerais, Bernard Tissier, Fr
Rel: Roman Catholic; Ordained; celibate.
International Organizations: Secretary-General of the International Sacerdotal Brotherhood Saint Pius X.
Mailing address: Solothurnerstrasse 35, CH-4613 Rickenbach SO, Switzerland. T. (41 62) 46 18 18.

Demetrios Coucouzis, D
See Iakovos.

de Mille Cameron, Angus, Rev
Rel: Unitarian; Ordained in 1937; b. 09 JUN 1913 in Sussex Corner, New Brunswick, Canada; m. to Esther Cary Horner; c. 2 boys and 2 girls; Citizen of Canada.
International Organizations: Ecclesiastic (retired) of the Unitarian Church.
Career: Minister, Adams Memorial Church, Dunkirk, NY from 1937 to 1941. Minister, Church of the Messiah, Montreal, Quebec from 1941 to 1959. Minister, First Unitarian Church of Philadelphia from 1963 to 1967. Served on Board of Governors, Montreal Council of Social Agencies from 1942 to 1946. Vice-President, American Unitarian Association from 1948 to 1951. **Educ.:** B.A. from Acadia University, Wolfville, Nova Scotia, Canada (1934); University of Chicago, IL, USA (1935-1937); B.D. from Meadville Theological School, Chicago IL, USA (1937). **Interests:** tennis, music, hunting, fishing. **Mem.:** Club: Montreal Indoor Tennis **Pubs.:** Contributing Editor of *The Christian Register* and author of chapter in *Voices of the Liberalism 1*, 1947.
Mailing address: 'Lochiel', Clifton Royal NB E0G 1N0, Canada.
Residence: See Mailing Address.

Demke, Christoph, Bishop Dr
Rel: Evangelical; Ordained.
International Organizations: Bishop of Evangelical Church of the Church Province of Saxony.
Mailing address: Am Dom 2, 39104 Magdeburg, Germany.

de Montfort, Marie, Sr
Rel: Roman Catholic; Ordained; celibate.
International Organizations: Past Secretary-General of the International Association of Logopedics and Phoniatrics.
Mailing address: School of Clinical Speech and Language Studies, Trinity College, Dublin 2, Ireland. T. (353 1) 772941 Ext 1588. Fax (353 1) 77 26 94.

De Nicoló, Pier Giacomo, H E Mgr
Rel: Roman Catholic; Ordained on 12 APR 1952; b. 11 MAR 1920 in Cattolica, Italy; celibate.
International Organizations: Apostolic Nuncio to Costa Rica (since 1984) of the Roman Catholic Church.
Other organizations: Titular Archbishop, Martana.
Mailing address: Apartado Postal 992, Centro Colón, San José de Costa Rica, Costa Rica. T. (506) 32 21 28. Fax 31 25 57

Denis, Ernes
Rel: Roman Catholic.
International Organizations: Director of the European Centre for Democratic and Laity Action.
Mailing address: Bd Saint-Germain 133, F-75006 Paris, France. T. (33 1) 43 26 81 38.

De Nittis, Francesco, H E Mgr
Rel: Roman Catholic; Ordained on 15 JUL 1956; b. 03 JUL 1933 in Vieste, Italy; celibate.
International Organizations: Apostolic Nuncio to Uruguay (since 1990) of the Roman Catholic Church.
Other organizations: Titular Archbishop, Tunes.
Mailing address: Bulevar Artigas 1270, CP 1503, 11300 Montevideo, Uruguay. T. (598 2) 411 410 – 411 947. Fax 48 89 49. Tx 26670 NUNTIUS UY

Denniston, Robin Alastair, Rev
Rel: Anglican; Ordained in 1979; b. 25 DEC 1926; m. to (1st) Anne Alice Kyffin Evans (died), (2nd) Rosa Susan Penelope Beddington; c. 1 boy and 2 girls; Citizen of United Kingdom.
International Organizations: Publisher and ecclesiastic of the Anglican Church.
Other organizations: Non-executive Chairman (since 1974), A R Mowbray and Co.; Priest-in-charge (since 1987), Great and Little Tew.

Career: Editor, Collins from 1950 to 1959. Managing Director, Faith Press from 1959 to 1960. Editor, Prism from 1959 to 1961. Promotion Manager, Hodder and Stoughton Ltd. from 1960 to 1964. Editorial Director, Hodder and Stoughton Ltd. in 1966. Managing Director, Hodder and Stoughton Ltd., also Director of Mathew Hodder Ltd. (and subsidiary cos.) from 1968 to 1972. Deputy Chairman, George Weidenfeld and Nicolson (and subsidiary cos.) in 1973. Chairman, Sphere Books from 1975 to 1976. Chairman, Thomas Nelson and Sons (and subsidiary cos.), Michael Joseph Ltd., George Rainbird Ltd. from 1975 to 1978. Director, Thomson Publishers Ltd., Hamish Hamilton Ltd. from 1975 to 1978. Academic Publisher, Oxford University Press in 1978. Senior Deputy Secretary to the Delegates, Oxford University Press from 1984 to 1988. Oxford Publisher from 1984 to 1988. Ordained Deacon in 1978. **Educ.:** B.Lit.Hum. from University of Oxford, Christ Church, UK. **Interests:** music, farming, squash. **Hons.:** Hon. Curate, Parish of Clifton-on-Teme (1978); New with S Hinksey (1985); Priest-in-Charge, Great and Little Tew (1987). **Mem.:** Clubs: Oxford and Cambridge **Pubs.:** The Young Musicians, 1956; Partly Living, 1967 and Editor, Part-Time Priests?, 1960.
Mailing address: The Vicarage, Great Tew, Oxfordshire OX7 4AG, UK. T. (44 60) 883 293

Denoyelle, Alfred

Rel: Roman Catholic.
International Organizations: President of the Archangel Saint Michael Cultural Foundation for a Catholic Europe.
Mailing address: Avenue Brigade Piron 84, B-1080 Bruxelles, Belgium. T. (32 2) 410 03 54.

Denton, John

Rel: Anglican.
International Organizations: General Secretary of the Anglican Church of Australia.
Mailing address: St Andrew's House, Sydney Square NSW 2000, Australia.

de Olivera Soares, Isman

Rel: Roman Catholic.
International Organizations: President of the Catholic Latin American Press Union.
Mailing address: R Major Newton Feliciano dos Santos 33, 04118 Sao Paulo, Brazil.

De Paoli, Ambrose B D, H E Mgr

Rel: Roman Catholic; Ordained on 18 DEC 1960; b. 19 AUG 1934 in Jeannette PA, USA; celibate.
International Organizations: Apostolic Delegate to Southern Africa and Apostolic Pro-Nuncio to Lesotho (since 1988) of the Roman Catholic Church.
Other organizations: Titular Archbishop, Lares.
Mailing address: 800 Pretorius Street, Arcadia, Pretoria 0083, South Africa. T. (27 12) 344 3815 – 344 3816. Fax 344 3595

Derde, L

Rel: Roman Catholic.
International Organizations: Contact of the European Forum of the Apostolate of Laity.
Mailing address: Mutsaerstraat 32, B-2000 Antwerpen, Belgium.

Derderian, Yeghnishe, H B

Rel: Armenian Church; Ordained; b. in Jordan.
International Organizations: Patriarch of Armenian Church.
Mailing address: Armenian Patriarch, Old City, Jerusalem, Israel.

DeRevere, David

Rel: Christian.
International Organizations: Executive Director of the International Conference of Police Chaplains.
Mailing address: 101 Rainbow Drive, Suite 82, Livingston TX 77351, USA. T. (409) 327 2332.

de Rios, Maria Marta

Rel: Roman Catholic.
International Organizations: Latin America Regional Secretary of the International Movement for the Apostolate in the Independent Social Milieux.
Mailing address: Algeciras 856, Las Condes, Santiago, Chile.

Der Nersessian, Nerses D, Fr

Rel: Roman Catholic; Ordained; celibate.
International Organizations: Superior General of the Mechitarists.
Mailing address: Isola di San Lazzaro, I-30126 Venezia, Italy. T. (39 41) 526 01 04

De Roo, Remi Joseph, Most Rev

Rel: Roman Catholic; Ordained on 08 JUN 1950; b. 24 FEB 1924 in Swan Lake MB, Canada; celibate; Citizen of Canada.
International Organizations: Bishop (since 1962) of Victoria, BC.
Other organizations: Member (since 1987), CCCB Theology Commission.
Career: Pastor, Holy Cross Parish, St Boniface, Manitoba. Chairman, Human Rights Commission, Province of British Columbia from 1974 to 1977. Member, CCCB Social Action Commission from 1973 to 1987. President, Western Conference of Bishops from 1984 to 1988. **Educ.:** S.T.D. from Dominican Pontifical University, Vatican City State (1952). **Hons.:** LLD hc, University of Antigonish, NS (1983), University of Brandon, Manitoba (1987); Honorary Fellow, Ryerson Polytechnical Institute (1987). **Mem.:** American Catholic Theological Association; Canadian Catholic Mariological Association
Mailing address: Suite 1, 4044 Nelthorpe St, Victoria BC V8X 2A1, Canada. T. (1 604) 479 1331

de Rooy-Janse, Nelly, Dr

Rel: Roman Catholic.
International Organizations: President of the International Association of Liberal Religious Women.
Mailing address: Chezeeweg 18, Postbus 42, NL-4424 ZG Wemeldinge, Netherlands. T. (31 1192) 2382.

Derrien, Enrico, S.M.M., Fr

Rel: Roman Catholic; Ordained; celibate.
International Organizations: Procurator-General of the Missionaries of the Company of Mary.
Mailing address: Viale dei Monfortani 65, I-00135 Roma, Italy. T. (39 6) 305 23 32.

Dery, Shimon

Rel: Jewish.
International Organizations: Secretary-General of the World Sephardi Federation.
Mailing address: 13 rue Marignac, CH-1206 Genève, Switzerland. T. (41 22) 47 33 13 – 47 33 14. Fax 47 28 39. C. KOLSEFARAD Geneva. Tx 427 569.

de Silva, Lily, Prof Dr

Rel: Buddhist; b. 16 JUL 1928 in Colombo, Sri Lanka; m. to Dayananda de Silva; c. 2 boys and 2 girls; Citizen of Sri Lanka; Lang. Singhalese and English.
International Organizations: Professor of the Department of Pali and Buddhist Studies, University of Peradeniya.
Main field of work: Education. **Career:** Assistant Lecturer, University of Peradeniya from 1959 to 1967. Lecturer, University of Peradeniya from 1967 to 1970. Senior Lecturer, University of Peradeniya from 1970 to 1980. Associate Professor (Merit Promotion), University of Peradeniya from 1980 to 1981. **Educ.:** B.A. (Ceylon) 1st (1955); M.A. (Ceylon) (1961); Ph.D (Ceylon) (1967). **Pubs.:** Dozens of papers published or presented to conferences and 6 books including: "The role of women in the development of society from a Buddhist perspective", accepted for publication; "The Scope and Significance of the Five Precepts", *Buddhist Ethics and Modern Society*, Greenwood Press, New York, 1991; *One Foot in the World: Buddhist Approaches to Present-day Problems*, BPS, Kandy, 1986 and *Mental Culture in Buddhism (Based on the Mahāsatipatthāna sutta) Dona Alpina Ratnayake Trust, Sixteenth Lecture, 1983.*
Mailing address: 28/3 Old Galaha Rd, Perdeniya, Sri Lanka. T. (94 88) 8601 EXT 374
Residence: See Mailing Address. T. See Mailing Address

Deskur, Andrzej Maria, H E Card

Rel: Roman Catholic; Ordained on 20 AUG 1950; b. 29 FEB 1924 in Sancygniów, Kielce, Poland; celibate; Citizen of Poland.

International Organizations: President Emeritus of the Pontifical Council for Social Communications.

Member of the Congregation for the Causes of Saints. Member of the Pontifical Council for Pastoral Assistance to Health Care Workers.

Other organizations: Member (since 1985), College of Cardinals; Member, Pontifical Commission for the Vatican City State; President, Academic Council, Pontifical Academy of the Immaculate.

Career: Consecrated Bishop (Titular See of Thenae) in 1974. Archbishop in 1980. Created Cardinal by Pope John Paul II in 1985. **Educ.:** Master's Deg. from Jagiellonian University, Cracow, Poland (1945); Dr. Theol. from University of Fribourg, Switzerland (1952).

Mailing address: 00120 Città del Vaticano, Vatican. T. (39 6) 698 37 73

de Smedt, Aemilius Josephus, Rt Rev

Rel: Roman Catholic; Ordained; b. 30 OCT 1909 in Opwijk, Belgium; celibate; Citizen of Belgium.

International Organizations: Ecclesiastic of the Roman Catholic Church.

Career: Auxiliary Bishop, Malines in 1950. Bishop, Bruges from 1952 to 1985. **Hons.:** Knight, Order of Leopold. **Mem.:** Secretariat for Christian Unity, Rome **Pubs.:** Il les fit Homme et Femme; Orientations éthiques pour une éducation sexuelle and Rencontrer.

Mailing address: c/o 4 H Geeststraat, B-8000 Bruges, Belgium. T. (32 50) 335906

De Smet, Hubert

Rel: Roman Catholic.

International Organizations: Secretary-General of the European Organization of the International Group of Catholic Parents' Associations.

Mailing address: Rue Guimard 1, B-1040 Bruxelles, Belgium. T. (32 2) 511 65 05.

Desmet, Mariette, Sr

Rel: Roman Catholic.

International Organizations: Contact of the Missionary Sisters of the Immaculate Heart of Mary.

Mailing address: Via di Villa Troili 30, I-00163 Roma, Italy. T. (39 6) 6641 71 56. C. ICMAR – 00163 Roma.

Desmond, Bernadette

Rel: Christian.

International Organizations: Director of the Ecumenical Committee on the Andes.

Mailing address: 198 Broadway No 302, New York NY 10038, USA. T. (1 212) 964 6730.

de Sousa, Desmond, Fr

Rel: Roman Catholic; Ordained; celibate.

International Organizations: Executive Secretary of the Ecumenical Coalition on Third World Tourism.

Mailing address: PO Box 24 Chorakhebua, Bangkok 10230, Thailand. T. (66 2) 519 2606. Fax 510 7287. C. Countours Bangkok.

de Sousa, José, Fr

Rel: Roman Catholic; Ordained; celibate.

International Organizations: Procurator-General of the Congregation of the Most Holy Redeemer.

Mailing address: Via Merulana 31, CP 2458, I-00100 Roma, Italy. T. (39 6) 731 58 41.

de Sousa Campos, Manuel

Rel: Evangelical Presbyterian.

International Organizations: Secretary of the Evangelical Presbyterian Church of Portugal.

Mailing address: Avenida do Brasil, 92-2° Dto, Lisbon 5, Portugal.

de Souza, Ernest Henriques, Rev

Rel: Jewish; b. 28 SEP 1933 in Kingston, Jamaica; m. to Judith Mary Bate; c. 2 boys; Citizen of Jamaica; Lang. English.

International Organizations: Spiritual Leader (since 1982) of the Jewish Community in Jamaica.

Other organizations: Lay Reader and Religion School Teacher (since 1946); Professional photographer (since 1954); Member, Island Council and Executive (since 1961), Girl Guides Association; Honorary Treasurer and Accountant (since 1967), Girl Guides Association; Member, Island Council and Executive (since 1967), Scout Association of Jamaica; Life Senator (since 1969), Junior Chamber International; Member, Board of Governors (since 1972), Hillel Academy; Secretary (since 1975), United Congregation of Israelites and Synagogue Trust Ltd.; Marriage Officer (since 1976); Honorary Vice-President (since 1980), Scout Association of Jamaica; Secretary (since 1984), Hillel Academy; Vice-Chairman (since 1989), Hillel Academy.

Career: Clerk, H M Brandon and Co. in 1950. Clerk, Stanley Motta Ltd. from 1950 to 1953. Clerk, Lascelles de Mercado and Co. from 1953 to 1954. Director, United Congregation of Israelites from 1960 to 1975. President, United Congregation of Israelites from 1966 to 1967. Trustee, Hillel Academy from 1978 to 1982. **Interests:** photography, coins, stamps and souvenir collecting, preserving historical records. **Hons.:** Justice of the Peace; Prime Minister's Medal of Appreciation for Religion (1983); Girl Guides Association: Thanks Badge (1962), The Torch (1967) and the Silver Bee (1975), 75th Anniversary Plaque and Citation for over 30 years service (1990); The Scout Association of Jamaica's Silver Pine Apple (1986); Certificates of Merit from the Jaycees of Jamaica (1963, 1965, 1967); B'nai B'rith: Ben Brit of the Year (1969), Award of Merit (1972), and Treasurers' Plaque (1980); Pro Mundi Beneficio Medal for community service from the Brazilian Academy of Humanities (1975); Gevaert Gold Medal and Blue Ribbon Award for scenic photographic competition to mark Jamaica's Tercentenary's year (1955); Kiwanis Club's Laymans' Award (1983, 1986); Certificate of Merit from International Who's Who in community service (1980); Silver Tray from the Jewish Community to mark 40 years as a Lay Reader and Religion School Teacher (1986). **Mem.:** Masonic Fraternity (Active Member since 1959); Jamaica Camera Club and Professional Photographers Association, and B'nai B'rith Jewish Service Club (Founder Member) **Pubs.:** Prayers, Meditations and Order of Services for the Synagogue, 1984 and Pictorial History of the Jewish Community in Jamaica, 1986.

Mailing address: 2A King's Drive, Kingston 6, Jamaica. T. (1 809) 927 7948

Residence: See Mailing Address. T. See Mailing Address

de Souza, Teotoni R, Rev

Rel: Roman Catholic; Ordained; celibate.

International Organizations: Contact of the Association des historiens de l'Eglise d'Asie.

Mailing address: Xavier Centre of Historical Research, Alto Porvorim, Goa 403 521, India.

De Stefano, Lucio, M.S.C., Rev

Rel: Roman Catholic; Ordained on 08 JUL 1956; b. 17 DEC 1929 in Sezze Romano, Italy; celibate; Citizen of Italy; Lang. English.

International Organizations: Procurator-General (since 1970) of the Missionaries of the Sacred Heart of Jesus.

Other organizations: General Postulator (since 1983), Missionaries of the Sacred Heart of Jesus.

Main field of work: Ministry, canonical work. **Career:** Director of Students, MSC for Italian Province from 1960 to 1965. Official, Congregation for the Causes of Saints from 1968 to 1982. **Educ.:** Deg. Cn.L. from Lateran Pontifical University, Vatican City State (1958-1962); Dip. Roman Rota Lawyer (1960-1963). **Interests:** Causes of canonization. **Hons.:** Papal Medal 'Pro Ecclesia et Pontifice'.

Mailing address: Via Asmara 11, I-00199 Roma, Italy. T. (39 6) 831 3741

Residence: See Mailing Address. T. See Mailing Address

Destree, A

Rel: Muslim.

International Organizations: Director of the Study Centre for Problems of the Contemporary Muslim World, Brussels.

Mailing address: 44 ave Jeanne, B-1050 Bruxelles, Belgium. T. (32 2) 642 33 59.

de Tarrazi, Amin A
Rel: Roman Catholic.
International Organizations: International President of the Society of St Vincent-de-Paul.
Mailing address: 5 rue du Pré-aux-Clercs, F-75007 Paris, France. T. (33 1) 42 61 50 25. C. Vincenpaul Paris. Tx 264918 TRACE F – Réf 545.

de Thoury, C
Rel: Roman Catholic.
International Organizations: UNESCO Representative of the World Union of Catholic Women's Organizations.
Mailing address: 2 rue Pierre Bizos, F-92330 Sceaux, France.

Devamitra, D, Dharmachari
Rel: Buddhist; Ordained on 12 JAN 1974; b. 02 APR 1948 in Stokesley, Yorkshire, UK; single; Citizen of United Kingdom; Lang. English.
International Organizations: Responsible for the 'mitras' of the Friends of the Western Buddhist Order.
Other organizations: President (since 1991), Norwich Buddhist Centre, UK; President (since 1991), Leeds Buddhist Centre, UK; President (since 1991), Melbourne Buddhist Centre, Australia.
Career: Chairman, Norwich Buddhist Centre, UK from 1976 to 1979. conducted many retreats and seminars. Gave hundreds of talks and lectures.
Mailing address: 12 Vivian Road, London E3 5RF, UK. T. (44 81) 981 5120
Residence: See Mailing Address. T. See Mailing Address

Devananda, Vishnu, Swami
Rel: Hindu; Ordained in 1948; b. 31 DEC 1927 in Nemnara, Kerala, India; single; Citizen of Canada; Lang. English, Malayalam and Hindi.
International Organizations: Founder-Director (since 1957) of the International Sivananda Yoga Vedanta Center.
Main field of work: Yoga teacher, peace missions. **Educ.:** Yoga-Acharya (Master of Yoga) from Yoga Vedanta Forest Academy, Rislinkesh, India. **Pubs.:** *The Complete Illustrated Book of Yoga*; *Meditation and Mantras* and *The Sivananda Companion to Yoga*.
Mailing address: Sivananda Ashram Yoga Camp HQ, 8th Avenue Val Morin, Quebec JOT 2RO, Canada. T. (1 819) 322 3226. Fax (1 819) 322 5876.
Residence: See Mailing Address. T. See Mailing Address

Devestel, Waldebert, F.C., Br
Rel: Roman Catholic; Ordained; celibate.
International Organizations: Superior General of the Brothers of Charity of Ghent.
Other organizations: Member, Commission for the Catechism and the Catechists.
Mailing address: Via G B Pagano 35, CP 9082, I-00167 Roma, Italy. T. (39 6) 62 30 476.

Deville, Raymond, P.S.S., Fr
Rel: Roman Catholic; Ordained; celibate.
International Organizations: Superior General of the Society of the Priests of St Sulpice.
Mailing address: 6 rue du Regard, F-75006 Paris, France. T. (33 1) 42 22 38 45.

de Visser, Jan Piet
Rel: Christian.
International Organizations: Contact of the Youth with a Mission.
Mailing address: Avenue des Celtes 19, B-1040 Bruxelles, Belgium. T. (32 2) 736 33 63. Fax 732 52 36.

De Wijs-Koolkin, Paula Beth
Rel: Buddhist: Mahayana; b. 29 MAR 1948 in New York NY, USA; m. to Matheus W G M De Wijs; Citizen of USA; Lang. English, Dutch and French.
International Organizations: Member, Board of Directors (since 1988) of the Foundation for the Preservation of the Mahayana Tradition.
Other organizations: Secretary (since 1976), Tibetan Affairs Coordination Office; Director (since 1979), Maitreya Institute; Chairperson (since 1990), Dutch Foundation for Tibetan Medicine; Editor (since 1991), *Tibet Nieuws*, Tibet Support Group.
Educ.: University of Arizona, Tucson, USA (1964-1965). **Interests:** Central Asia, Tibetan dogs. **Pubs.:** Author of numerous articles on Tibetan Buddhism and the Tibetan political situation.
Mailing address: Maitreya Institute, Heemhoeveweg 2, NL-8166 HA Emst, Netherlands. T. (31 5787) 1450
Residence: Ouderkerkerdijk 40, NL-1096 CR Amsterdam, Netherlands. T. (31 20) 692 0629

De Zutter, Raf
Rel: Roman Catholic.
International Organizations: Regional Secretary for Europe of the Catholic International Education Office.
Secretary-General of the Centre of Promotion of Catholic Education in Europe. Secretary-General of the European Committee for Catholic Education.
Mailing address: Rue de l'Industrie 42, B-1040 Bruxelles, Belgium. T. (32 2) 511 75 01 – 511 82 32. Fax 514 59 89.

Dezza, Paolo, S.I., H E Card
Rel: Roman Catholic; Ordained on 25 MAR 1928; b. 13 DEC 1901 in Parma, Italy; celibate; Citizen of Italy; Lang. Italian, Latin, French, Spanish, English and German.
International Organizations: Cardinal of the Roman Catholic Church. Member of the Society of Jesus.
Career: Professor of Philosophy, Pontifical Gregorian University, Rome from 1929 to 1962. Provincial Superior, Jesuit Province of Milan from 1935 to 1939. Rector, Pontifical Gregorian University, Rome from 1941 to 1951. President, Pontifical Institute Regina Mundi from 1954 to 1966. General Assistant, Society of Jesus, Rome from 1965 to 1975. Papal Delegate for the Society of Jesus, Rome from 1981 to 1983. Peritus at Vatican Council II from 1962 to 1965. Consultor for various Congregations, the Holy See from 1929 to 1991. **Educ.:** Dr. Phil. from Colegio de San Ignacio, Barcelona, Spain (1921-1924); Dr. Theol. from Innsbruck University, Austria and Facoltà di San Luigi, Naples, Italy (1925-1929). **Interests:** philosophy. **Hons.:** Hon. Dr.Phil., Catholic University of Santiago de Chile; Commander, Order of Alfonso the Wise (Spain). **Mem.:** Pontifical Academy San Tommaso d'Aquino (Fellow) **Pubs.:** Alle origini del Neotomismo, 1940; Metaphysica Generalis, 1962 and Filosofia, sintesi scolastica, 1988.
Mailing address: Borgo S Spirito 4, CP 6139, I-00195 Roma, Italy. T. (39 6) 689 771
Residence: See Mailing Address. T. See Mailing Address

Dhandhania, Kishorilal
Rel: Hindu.
International Organizations: Secretary-General of the Sri Aurobindo Society.
Mailing address: Society House, Pondicherry 605 001, India. T. 24382 – 24668 – 26396 – 26397 – 26398.

Dhargyey, D, Geshe
Also know as Yar-Chag Pal-Bum-Tsang Ngawang Dhargyey; Rel: Buddhist; Ordained in 1942; b. in Yar-Chag, Trehor, Kham, Tibet; celibate; Citizen of Tibet and New Zealand; Lang. Tibetan.
International Organizations: Spiritual/Religious Instructor/Teacher (since 1985) of the Dhargyey Buddhist Centre.
Main field of work: Spiritual/Religious instruction. **Career:** Scriptural Teacher, Sera Monastery, Tibet and India from 1942 to 1970. Head Teacher, Library of Tibetan Works and Archives, Dharamsala, India from 1970 to 1984. **Educ.:** Lharampa Geshe from Sera-Je Monastic University, Lhasa, Tibet (now Buxa, Assam, India) (1939-1964). **Pubs.:** Tibetan Tradition of Mental Development, Library of Tibetan Works and Archives, Dharamsala, 1974; Anthology of Well-Spoken Advice, LTWA, 1982 and Kalacakra Tantra, LTWA, 1985.
Mailing address: 22 Royal Terrace, Dunedin, New Zealand. T. (64 24) 4778374
Residence: See Mailing Address. T. See Mailing Address

Dharmacharini Sanghadevi, D
See Christine Janet Seymour.

Dharmasakti, Sanya, H E Prof
Rel: Buddhist.
International Organizations: President and Member of the Executive Council of the World Fellowship of Buddhists.
Mailing address: 33 Sukhumvit Road, Bangkok 10110, Thailand. T. (66 2) 251 1188. C. WORLBUDIST BANGKOK. Fax (66 2) 255 7363.

D'Heedene, Walter, C.I.C.M., Fr
Rel: Roman Catholic; Ordained; celibate.
International Organizations: Secretary-General of the Congregation of the Immaculate Heart of Mary.
Mailing address: Via S Giovanni Eudes 95, I-00163 Roma, Italy. T. (39 6) 625 85 52

Diam, Gabriel
Rel: Roman Catholic.
International Organizations: Regional Secretary for Africa of the Catholic International Education Office.
Mailing address: SRAM, BP 276, Kinshasa 1, Zaire.

Diamond, James, Rabbi
Rel: Jewish.
International Organizations: President of the Association of Hillel-Jewish Campus Professionals.
Mailing address: 6300 Forsyth Blvd, Clayton MO 63105, USA. T. (1 314) 726 6177.

Diangienda-Kuntima, Joseph, H E
Rel: Church of Christ on Earth by the Prophet Simon Kimbangu.
International Organizations: Spiritual Head of the Church of Christ on Earth by the Prophet Simon Kimbangu.
Mailing address: BP 7069, Kinshasa I, Zaire.

Dias, Ivan, H E Mgr
Rel: Roman Catholic; Ordained on 08 DEC 1958; b. 14 APR 1936 in Bombay, India; celibate.
International Organizations: Apostolic Pro-Nuncio to the Republic of Korea (since 1987) of the Roman Catholic Church.
Other organizations: Titular Archbishop, Rusubisir.
Mailing address: PO Box 393, Kwang Hwa Moon, Seoul, Korea Rep. T. (82 2) 736 5725 – 739 2310. Fax 736 5738

Diaz, Roberto, Rev
Rel: Evangelical Methodist.
International Organizations: President of the Evangelical Methodist Church of Costa Rica.
Mailing address: Apdo 5481, San José, Costa Rica.

di Battista, Myriam
Rel: Roman Catholic.
International Organizations: Secretary-General of the International Catholic Committee of the Blind.
Mailing address: 8 rue François-Elie Dujenne, St Satur, F-18300 Sancerre, France. T. (33) 48 54 23 89.

Di Berardino, Angelo , Rev Prof Dr
Rel: Roman Catholic; Ordained on 15 JUL 1962; b. 10 OCT 1936 in Furci, Chieti, Italy; celibate; Citizen of Italy; Lang. Italian, Spanish, French, English, German, Latin and Greek.
International Organizations: Secretary-General of the International Association for Patristic Studies.
Other organizations: Rector, Santa Monica College; Professor of Patristics at Augustinianum, Rome.
Main field of work: Fathers of the Church. **Educ.:** Dr. Theol. from Facoltà San Luigi, Naples, Italy; Dr. Phil. from Università La Sapienza, Rome, Italy. **Hons.:** Catholic Book Award, Theology Category (1986). **Mem.:** German Academy of Sciences **Pubs.:** *Patrology*, Vol. IV, Christian Classics, 1986; *Encyclopedia of the Early Church*, Cambridge, 1991 and others.
Mailing address: Istituto Augustinianum, Via S Uffizio 25, I-00193 Roma, Italy. T. (39 6) 686 89 26.
Residence: See Mailing Address. T. See Mailing Address

Dick, Ignace, Rev Archimandrite
Rel: Greek-Melkite Catholic; Ordained on 29 JUN 1949; b. 15 JAN 1926 in Aleppo, Syria; celibate; Citizen of Syria; Lang. Arabic, French, Greek, Latin and English (some).
International Organizations: Vicar-General (since 1986) of the Greek-Melkite Catholic Bishopric of Aleppo.
Other organizations: Judge, Intercommunal tribunal of Aleppo; Director, Revue diocésaine.
Main field of work: Administration and ministry. **Career:** Professor, Petit Séminaire Ste Anne, Lebanon from 1953 to 1957. Professor, Grand Séminaire Ste Anne, Jerusalem from 1962 to 1963. Curate, St George and various parishes in Aleppo. **Educ.:** Lic. Oriental Philology; Dr. Phil. from Catholic University of Louvain (UCL), Belgium (1957-1960). **Pubs.:** Sens et vicissitudes de l'uniatisme (in French); L'Orient chrétien (in Arabic) and Les Melkites, Coll. Fils d'Abraham (in French), forthcoming.
Mailing address: Greek-Catholic Bishopric, BP 146, Place Farhat, Aleppo, Syria. T. (963 21) 213 218

Dicker, Helen Ruth
Rel: Protestant.
International Organizations: Director, Australia Region of the International Association of Women Ministers.
Mailing address: 420 Liverpool Road, Enfield NSW 2136, Australia.

Didi, Dibalu, Rev Fr
Rel: Roman Catholic; Ordained; celibate.
International Organizations: Secretary-General of the Regional Secretariat of Catholic Education for Africa and Madagascar.
Mailing address: BP 276, Kinshasa 1, Zaire.

Diggs, Ronald J, Bishop
Rel: Lutheran; Ordained.
International Organizations: Bishop of Lutheran Church in Liberia.
Mailing address: PO Box 1046, Monrovia, Liberia.

Dilanni, Albert, S.M., Fr
Rel: Roman Catholic; Ordained; celibate.
International Organizations: Member of the Society of Mary.
Mailing address: Padri Maristi, Via Alessandro Poerio 63, I-00152 Roma, Italy.

Dill, Edwin, S.T., Fr
Rel: Roman Catholic; Ordained; celibate.
International Organizations: Custodian General of the Missionary Servants of the Most Holy Trinity.
Mailing address: 1215 North Scott St, Arlington VA 22209-3097, USA. T. (1 703) 276 8646

Dillon, Therese, Sr
Rel: Roman Catholic; Ordained; celibate.
International Organizations: Superior General of the Missionary Sisters of Our Lady of the Holy Rosary.
Mailing address: 23 Cross Avenue, Blackrock, Ireland. T. (353 1) 88 17 08.

Dilschneider, Gerhard
Rel: Reformed Church.
International Organizations: Assistant to the Secretary-General of the World Alliance of Reformed Churches.
Mailing address: 150 route de Ferney, PO Box 2100, CH-1211 Genève 2, Switzerland. T. (41 22) 791 62 39. C. WARC GENEVA. Tx 415 730 OIK CH (WARC). Fax (41 22) 791 65 05 WARC REFORMIERTER WELTBUND.
Residence: See Mailing Address. T. (41 22) 738 5417

Dimitrios, D, I
Also know as Dimitrios Papadopoulos; Rel: Orthodox; Ordained in 1942; b. 08 SEP 1914 in Istanbul, Turkey.
International Organizations: Ecumenical Patriarch and Archbishop (since 1972) of Constantinople.
Career: Ordained Deacon in 1937. Preacher in Edessa, Greece from 1937 to 1938. Preacher, Parish, Feriköy from 1939 to 1945. Priest, the Orthodox Community, Teheran from 1945 to 1950. Head Priest, Feriköy from 1950 to 1964. Bishop, Elaia, Auxiliary Bishop of the Patriarch Athenagoras in Istanbul from 1964 to 1972. Metropolitan, Imvros and Tenedos in 1972.
Mailing address: Rum Patrikhanesi, Fener, Istanbul, Turkey. T. (90 1) 23 98 50

Dinan, Jacques
Rel: Roman Catholic.
International Organizations: Regional Secretary of the International Catholic Organization for Cinema and Audiovisual.
Mailing address: 4 Avenue Fl Nightingale, Rose Hill, Mauritius.

Di Nello, Filomena , Sr
Rel: Roman Catholic; Ordained; celibate.
International Organizations: Superior General of the Comboni Missionary Sisters.
Mailing address: Via Boccea 506, I-00166 Roma, Italy. T. (39 6) 696 0273. Fax 696 1963.

Ding Guangxun, K H D, Bishop
Rel: Protestant; Ordained in 1942; b. 20 SEP 1915; Citizen of China.
International Organizations: Deputy Chairman (since 1989) of the Chinese People's Political Consultative Conference.
Main field of work: Theologian and church leader. **Career:** Secretary, Student Christian Movement of Canada, Student World Christian Federation, Geneva. Returned to China in 1951. President, Nanking Theological Seminary in 1952. Consecrated Bishop in 1955. Vice-Chairman, Three-Self Patriotic Committee, Movement of Protestant Churches of China (later Chairman of 4th Committee) in 1961. President, Second Christian Council of China.
Mailing address: Nanjing Theological Seminary, Nanjing 210005, China.

Dinitz, Simcha
Rel: Jewish; b. 23 JUN 1929 in Tel Aviv, Israel; m. to Vivian Dinitz; c. 1 boy and 2 girls; Citizen of Israel.
International Organizations: Chairman (since 1988) of the World Zionist Organization.
Chairman, Executive (since 1988) of the Jewish Agency for Israel.
Main field of work: Diplomat. **Career:** Israel Defence Forces from 1948 to 1950. Assistant to Director of Information at the Embassy, Israel in Washington, USA from 1954 to 1958. Dept. of Information, Ministry for Foreign Affairs, Jerusalem, Israel from 1958 to 1961. Director of the Bureau of the Director-General, Ministry for Foreign Affairs from 1961 to 1963. Director of the Bureau and Political Secretary to the Foreign Minister, Israel in 1963. Member, Israel Delegation to the United Nations from 1963 to 1966. Minister to Rome from 1966 to 1968. Minister-in-charge, Information, Washington, USA from 1968 to 1969. Political Adviser to the Prime Minister, Israel from 1969 to 1973. Director-General, Prime Minister's Office from 1972 to 1973. Ambassador, Israel to the USA from 1973 to 1978. Vice-President, Hebrew University of Jerusalem, Israel from 1979 to 1984. Member, Knesset from 1984 to 1988. **Educ.:** M.Sc. from University of Cincinnati, OH, and Georgetown University, Edmund A Walsh School of Foreign Service, Washington DC, USA. **Interests:** Music, sports, reading.
Mailing address: PO Box 91920, Jerusalem, Israel. T. (972 2) 39261. C. Zionit Jerusalem.
Residence: 40 Nayot, Jerusalem, Israel T. (972 2) 632314

Dinkha, D, IV, H H
Rel: Apostolic Catholic Assyrian Church of the East.
International Organizations: Catholicos Patriarch of Apostolic Catholic Assyrian Church of the East.
Mailing address: 8908 Birch Avenue, Morton Grove IL 60053, USA.

Diodoros, D, I, His Beatitude
Also know as Damianos George Karivalis; Rel: Greek Orthodox; Ordained in 1947; b. 14 AUG 1923 in Island of Chios, Greece; celibate; Citizen of Israel.
International Organizations: Patriarch (since 1981) of Jerusalem.
Career: Became a monk, ordained deacon and assigned to serve at the Church of the Nativity, Bethlehem in 1944. Appointed Superior, Church of the Praetorium, Jerusalem in 1947. Named Secretary, Committee of Lands and other properties of the Patriarchate in 1947. Appointed teacher in the High School, the Patriarchate and served also as Librarian and archivist in 1957. Appointed President, Ecclesiastical Court of the Patriarchate in 1958. Acted as Draguman from 1959 to 1962. Consecrated titular Archbishop, Hierapolis and appointed Patriarchal Vicar in Amman, Jordan in 1962. Elected Patriarch, Jerusalem by the Holy Synod in 1981. **Educ.:** Deg. Theol. from National Capodistrian University of Athens, Greece (1952-1957). **Hons.:** Recipient of numerous decorations from civil and religious authorities; Hon. Dr., Theological Academy, Leningrad (1981).
Mailing address: Greek Orthodox Patriarchate, PO Box 19632-633, Jerusalem, Israel. T. (972 2) 284917

Ditta, S A, Pastor
Rel: Seventh-Day Adventist.
International Organizations: President, Pakistan Union Section of the General Conference of Seventh-Day Adventists.
Mailing address: PO Box 32, Lahore, Pakistan.

DiUlio, Albert Joseph, S.J., Rev
Rel: Roman Catholic; Ordained on 04 MAY 1974; b. 14 FEB 1943 in Lanoa WI, USA; celibate; Citizen of USA.
International Organizations: President (since 1990) of the Marquette University, Milwaukee WI.
Member of the Society of Jesus.
Other organizations: Chairman, Greater Milwaukee Committee Task Force to review governance of the Milwaukee County Medical Complex; Director, Wisconsin Association of Independent Colleges and Universities; Member, Wisconsin Higher Educational Aids Board; Member, Greater Milwaukee Committee; Chairman, Mayor's Committee on Police and Community Relations.
Career: Faculty Member, Campion Jesuit High School, Prairie du Chien WI from 1969 to 1970. Faculty Member, Campion Jesuit High School, Prairie du Chien WI from 1973 to 1974. Acting Principal, Campion Jesuit High School, Prairie du Chien WI from 1970 to 1971. Principal, Campion Jesuit High School, Prairie du Chien WI in 1975. Assistant Dean for Arts and Sciences, Marquette University, Milwaukee WI from 1978 to 1980. Assistant to President, Loyola Marymount University, Los Angeles CA from 1980 to 1981. Assistant Professor, College of Business Administration, Marquette University, Milwaukee WI from 1983 to 1984. Associate Dean, College of Business Administration, Marquette University, Milwaukee WI from 1984 to 1986. President, Xavier University, Cincinnati OH from 1986 to 1990. **Educ.:** B.S. Business Econ. from Marquette University, Milwaukee WI, USA (1965); A.M. Econ. from Marquette University, Milwaukee WI, USA (1969); M.Div. from Weston School, Cambridge MA, USA (1974); M.A. Ed. from Stanford University, CA, USA (1976); Ph.D. Higher Ed. from Stanford University, CA, USA (1979); M.B.A. Fin. from Stanford University, CA, USA (1983). **Interests:** California wines, gardening, reading. **Hons.:** Outstanding Business Leadership Award, Marquette University; Equestrian Order of the Holy Sepulchre of Jerusalem; Alpha Sigma Nu; Beta Gamma Sigma; Beta Alpha Psi; Phi Delta Kappa. **Mem.:** Wisconsin Foundation of Independent Colleges; Newcomen Society of the United States
Mailing address: 615 North 11th Street, Milwaukee WI 53233-2305, USA. T. (1 414) 288 7223
Residence: See Mailing Address. T. See Mailing Address

Divarkar, Parmananda Roque, Dr
Rel: Roman Catholic; Ordained on 21 DEC 1952; b. 04 OCT 1922 in Benaulim, Goa, India; celibate; Citizen of India; Lang. English, Spanish, Italian, Portuguese, French, Latin, Konkani and Hindi.
International Organizations: Promoter of Ecumenism and Inter-Religious Dialogue (since 1984) of the Archdiocese of Bombay.
Main field of work: University education. **Career:** Rector, St Xavier's College, Bombay from 1967 to 1971. Vice-Chancellor, Jesuit Ecclesiastical Faculties in India from 1971 to 1975. Assistant to Superior General, Society of Jesus, Rome from 1975 to 1983. **Educ.:** M.Th. from Jnana Deepa, Pune, India (1953); Dr. Phil. from University of Bombay, India (1965). **Interests:** psychology of religion, spirituality. **Mem.:** Society of Jesus **Pubs.:** The Act of Being, Examiner Press, Bombay, 1973; The Path of Interior Knowledge, GSP, Anand, 1983 and Ablaze with God, GSP, Anand, 1991.
Mailing address: St Mary's, Mazagaon, Bombay 400010, India.
Residence: 10 Dhondy Marg, Bombay 400010, India. T. (91 22) 868265

Dizon, Rolando Ramos, Br
Rel: Roman Catholic; Ordained on 20 JUN 1970; b. 31 OCT 1944 in Bacolod City, Negros Occidental, Philippines; celibate; Citizen of Philippines; Lang. English, Filipino, Spanish and French.
International Organizations: Regional Secretary for Asia Pacific (since 1990) of the Catholic International Education Office.
Other organizations: Director (since 1983), Catholic Educational Association of the Philippines; Board Chairman (since 1987), Gregorio Araneta University Foundation; Vice-Chairman (since 1987), UNESCO Philippine Commission on Education.

Main field of work: University education. **Career:** President, La Salle Green Hills from 1979 to 1986. President, University of St La Salle, Bacolod from 1986 to 1992. President, Catholic Educational Association of the Philippines from 1983 to 1987. Founder, La Salle, Antipolo, Rizal in 1985. Founder, St Benilde School, Bacolod in 1987. **Educ.:** M.A. from De La Salle University, Manila, Philippines (1970-1973); Dr. Phil. from Stanford University, CA, USA (1974-1978). **Interests:** community development. **Hons.:** Ten Outstanding Young Men (1984). **Pubs.:** *Educational Alternatives for the Future*, Phoenix-Alemar's, Manila, 1983; *Visions and Revisions: Essays on Education*, Phoenix Publishing Co., Manila, 1991 and "Education for Evangelization", International Mission Congress, Manila, 1979.
Mailing address: c/o University of St La Salle, La Salle Avenue, Bacolod City 6100, Philippines. T. (63 34) 20577
Residence: See Mailing Address. T. See Mailing Address

Doga, Lobsang Dorje, Ven Geshe

Rel: Buddhist: Tibetan; Ordained in 1942; b. 13 JUL 1935 in Khamze, Tibet; celibate; Citizen of Australia; Lang. Tibetan and English.
International Organizations: Resident Teacher and Lama, Tara Institute (since 1984) of the Foundation for the Preservation of the Mahayana Tradition.
Main field of work: Teaching philosophy. **Career:** Teached at Atisha Centre, Bendigo. **Educ.:** Sera Monastery, Lhasa, Tibet; Acharya Degree from Varanasi University, India; Geshe Degree from Sera Monastery, Karnataka, India. **Pubs.:** Contributes to *Tara Institute News* (monthly).
Mailing address: 3 Mavis Ave, Brighton VIC 3187, Australia. T. (61 3) 596 8900. FAX 596 4856
Residence: See Mailing Address. T. See Mailing Address

Domingos, A L, Rev

Rel: United Evangelical.
International Organizations: General Secretary of the United Evangelical Church of Angola.
Mailing address: BP 122, Uige, Angola.

Domingos Caetano, José D, Rev

Rel: Evangelical Pentecostal.
International Organizations: Contact of the Evangelical Pentecostal Church of Angola.
Mailing address: CP 219, Porto Amboim, Angola.

Donahue, John A

Rel: Roman Catholic.
International Organizations: Executive Director of the Serra International.
Mailing address: 22 W Monroe Street, Chicago IL 60603, USA. T. (1 312) 782 2163 – 782 1300. Fax 782 2358.

do Nascimento, Alexandre, H E Card

Rel: Roman Catholic; Ordained on 20 DEC 1952; b. 01 MAR 1925 in Malanje, Angola; celibate; Citizen of Angola.
International Organizations: Member (since 1983) of the College of Cardinals.
Member of the Congregation for the Evangelization of Peoples.
Other organizations: Archbishop (since 1986), Luanda; President, Episcopal Conference of Angola and Sao Tomé-Principe.
Career: Bishop, Malanje in 1975. Archbishop, Lubango from 1977 to 1986. Created Cardinal by Pope John Paul II in 1983. Apostolic Administrator, Onjiva. Preacher, Retreat of the Holy Father and the Roman Curia. **Interests:** reading, music.
Mailing address: Arcebispado, CP 87, 1230-C Luanda, Angola. T. (244) 34 640

Donato, Cauzo, Fr

Rel: Roman Catholic; Ordained; celibate.
International Organizations: Secretary of the International Institute for the Pastoral Theology of Health Care.
Mailing address: CAMILLIANUM, Largo Ottorino Respighi 6, I-00135 Roma, Italy. T. (39 6) 328 86 08 – 328 42 28.

Donlon, Kevin Hastings, CSSR, Rev Fr

Rel: Roman Catholic; Ordained on 17 AUG 1952; b. 13 SEP 1925 in Roscrea, Co. Tipperary, Ireland; celibate; Citizen of Ireland; Lang. Irish, English and French.
International Organizations: President (since 1986) of the International Federation of Church Press Associations.
Other organizations: Editor, *Reality*.
Educ.: University of Santo Tomas, Manila, Philippines.
Mailing address: Redemptorist Publications, 75 Orwell Rd, Dublin 6, Ireland. T. (353 1) 961688. Fax 961654.
Residence: See Mailing Address. T. See Mailing Address

Doom, John

Rel: Evangelical.
International Organizations: General Secretary of the Evangelical Church of French Polynesia.
Mailing address: BP 113, Papeete, Tahiti, Polynesia Fr.

Dorey, Gordon, Dr

Rel: Protestant.
International Organizations: Director for Canada of the Africa Inland Mission International.
Mailing address: 1641 Victoria Park, Scarborough ON M1R 1P8, Canada.

Dorgan, Margaret, Sr

Rel: Roman Catholic; Ordained; b. 02 JUL 1927 in Providence RI, USA; celibate; Citizen of USA; Lang. English.
International Organizations: Member (since 1965) of the John of the Cross Monastery Hermitage (Carmelites of Maine).
Other organizations: Member (since 1946), Carmelite Nuns; Lectures and workshops in the America, Canada and the USA on prayer and mysticism (since 1979); Collaboration with the Carmelite Forum (since 1989).
Main field of work: reflection on prayer. **Educ.:** B.A. Phil. from Radcliffe College, Cambridge MA, USA (1944-1946); Radcliffe College, Cambridge MA, USA (1974-1975). **Interests:** prayer, spiritual guidance, feminine experience. **Hons.:** Phi Beta Kappa. **Mem.:** Carmelite Communities Associated; Association of Contemplative Sisters **Pubs.:** Simplification in Prayer: When is it the Next Step?; Thérèse of Lisieux: Mystic of the Ordinary and St Teresa of Avila: a Guide for Travel Inward.
Mailing address: Monastery Hermitage, RFD #1 Box 1760, Corland-Bucksport ME 04416, USA. T. (1 207) 469 3456
Residence: See Mailing Address. T. See Mailing Address

Doroszkiewicz, Bazyli, H E M

Rel: Orthodox; Ordained; b. 15 MAR 1914 in Cisy, Poland; Citizen of Poland.
International Organizations: Metropolitan (since 1970) of Warsaw.
Other organizations: Head (since 1970), Polish Autocephalic Orthodox Church.
Career: Parish priest, Polish Autocephalous Orthodox Church, Michalowo from 1940 to 1946. Parish priest at Gródek near Bialystok from 1946 to 1960. Bishop, Wroclaw and Szczecin Diocese from 1960 to 1970. **Interests:** collecting old prints and icons, building and restoration of sacred buildings including Cathedral in Wroclaw. **Hons.:** Commander's Cross with Star, Officer's Cross, Knight's Cross, Order of Polonia Restituta; 30th Anniversary People's Republic of Poland Medal; Badge of Poland's Millenium; St Cyril and St Methody Orders (Czechoslovakia and Bulgaria); St Paul Apostle Order and Grand Cross of the Saviour, Greek Orthodox Church; St Nina's Order (Georgia); Badges of 50th and 60th Anniversary of Patriarchate Revival, Russian Orthodox Church; Badge of 50th Anniversary of Patriarchate Revival, Romanian Church; Order of the Holy Agnus Dei (Finnish Orthodox Church); and numerous other Polish and foreign awards. **Mem.:** Soldiers' Friends Society (1946); National Unity Front (1960) **Pubs.:** religious and historical articles in ecclesiastical press, including Cerkiewny Wiestnik, Wiadomosci and pastoral letters and occasional messages to congregation.
Mailing address: Al Swierczewskiege 52 m 6, 03 402 Warszawa, Poland. T. (48 22) 19 34 09

Dorotej, D, H E

Rel: Orthodox.
International Organizations: Metropolitan of Prague and All Czechoslovakia.
Mailing address: V Jame 6, Praha 1, Czech Rep.

dos Santos, Alexander d, O.F.M., H E Card

Rel: Roman Catholic; Ordained on 25 JUN 1953; b. 18 MAR 1924 in Inhambane, Portugal; celibate; Citizen of Portugal.
International Organizations: Member (since 1988) of the College of Cardinals.
Member of the Order of Friars Minor. Member of the Congregation for the Evangelization of Peoples. Member of the Pontifical Council for Dialogue with Non-Believers.
Other organizations: Archbishop, Maputo.
Career: Elected to Church in Lourenço Marques (now Maputo) in 1974. Consecrated Bishop in 1975. Created Cardinal by Pope John Paul II in 1988.
Mailing address: Paço Arquiepiscopal, Ava Eduardo Mondiane 1448, CP 258, Maputo, Mozambique. T. (258 1) 421 873

Dossena, Luigi, H E Mgr

Rel: Roman Catholic; Ordained on 25 MAR 1951; b. 28 MAY 1925 in Campagnola Cremasca; celibate.
International Organizations: Apostolic Nuncio to Peru (since 1985) of the Roman Catholic Church.
Other organizations: Titular Archbishop, Carpi.
Mailing address: Avenida Salaverry, 6a cuadra, Apartado 397, Lima 100, Peru. T. (51 14) 31 94 36 – 31 94 73. Fax 32 32 36. Tx 25751 NUNAPLIM PE

Douek, Maurice-Haim, Rabbi

Rel: Jewish; b. in Cairo, Egypt.
International Organizations: Chief Rabbi of Egypt.
Hons.: Middle East and foreign decorations.
Mailing address: Big Temple, Adly Pasha Street, Cairo, Egypt.

Douget, Yves, Fr

Rel: Roman Catholic; Ordained; celibate.
International Organizations: Superior General of the Prêtres de Sainte Marie de Tinchebray.
Mailing address: Sainte-Marie, F-61800 Tinchebray, France. T. (33) 64 91 48.

Douglas, James Dixon, Rev Dr

Rel: Church of Scotland; Ordained in 1957; b. 23 DEC 1922 in Glasgow, Scotland, UK; single; Citizen of United Kingdom.
International Organizations: Lecturer (since 1988) of the Singapore Bible College.
Main field of work: Church history and journalism. **Career:** Lecturer, St Andrews University from 1955 to 1956. Librarian, Tyndale House Cambridge from 1958 to 1961. Editor-at-Large, *Christianity Today* Magazine from 1961 to 1983. **Educ.:** M.A., B.D. from University of St Andrews, UK (1947-1953); S.T.M., Ph.D. from Hartford Theological Seminary, CT, USA (1953-1955). **Pubs.:** Editor: *New Bible Dictionary*, 1962; *New International Dictionary of the Christian Church*, 1974 and *20th Century Encyclopedia of Religion*, 1991.
Mailing address: 9-15 Adam Road, Singapore 1128, Singapore. T. (65) 466 5189
Residence: 2 Doocot Road, St Andrews K416 8QP, UK. T. (44 334) 74876

Douglas, Jane Dempsey, Prof Dr

Rel: Presbyterian; Ordained; b. 22 MAR 1933 in Wilmington DE, USA; m. to Gordon K Douglas; c. 2 boys and 1 girl; Citizen of USA; Lang. English, French, German (some) and Spanish (some).
International Organizations: President (since 1990) of the World Alliance of Reformed Churches.
Other organizations: Ruling Elder, the Presbyterian Church (USA); Editorial Board (since 1966), Studies in Medieval and Reformation Thought, Leiden, Brill; Vice-Chair (since 1984), Special Committee to Prepare a Brief Statement of Reformed Faith, Presbyterian Church (USA); Hazel Thompson McCord Professor of Historical Theology (since 1985), Princeton Theological Seminary; Council on Theological Scholarship and Research (since 1985), Association of Theological Schools; Editorial Council (since 1985), *Theology Today*; Editorial Board (since 1986), *Church History*; Book Review Editor (since 1989), *Princeton Seminary Bulletin*.
Career: Associate Director, Presbyterian Student Center, Columbia from 1955 to 1958. Teaching Fellow in Church History, Harvard Divinity School from 1959 to 1962. Professor of Church History, School of Theology at Claremont from 1963 to 1985. Professor of Religion, Claremont Graduate School from 1963 to 1985. Member-at-large, General Council, Presbytery of San Gabriel from 1973 to 1976. Member of the Editorial Board, *Journal of the American Academy of Religion* from 1973 to 1977. Member of the Faculty Fellowships Panel, Association of Theological Schools from 1976 to 1978. Member of the Caribbean and North American Area Council, World Alliance of Reformed Churches from 1978 to 1985. Member of the Committee on Cooperation and Witness, World Alliance of Reformed Churches from 1979 to 1985. Member of the Personnel Committee, Presbytery of San Gabriel from 1979 to 1980. Member of the Council, American Society of Church History from 1980 to 1982. Chair, Worship and Music Commission, Claremont Presbyterian Church, CA (USA) from 1980 to 1982. Member of the Advisory Committee, Women's Studies in Religion, Harvard Divinity School from 1980 to 1986. Co-Chair, Lutheran-Reformed Dialogue III in North America from 1981 to 1983. Member of the Committee on Ministry, Presbytery of San Gabriel from 1981 to 1983. Member of the American Sub-Commission, International Commission for the Comparative Study of Ecclesiastical History from 1983 to 1984. President, American Society of Church History in 1983. Chair, Nominating Committee, American Society of Church History in 1984. Moderator, Presbytery of San Gabriel in 1984. Member, International Lutheran-Reformed Dialogue between World Alliance of Reformed Churches and Lutheran World Federation from 1985 to 1988. Vice-President, World Alliance of Reformed Churches from 1989 to 1990. **Educ.:** B.A. Soc. and Mod. Lang. from Syracuse University, NY, USA (1954); University of Geneva, Switzerland (1954-1955); A.M. from Radcliffe College, Cambridge MA, USA (1961); Ph.D. Hist. and Phil. of Religion from Harvard University, Cambridge MA, USA (1963). **Hons.:** Ordained Ruling Elder in the Presbyterian Church (USA). **Pubs.:** *Justification in Late Medieval Preaching: A Study of John Geiler of Keisersberg*, Leiden, Brill, 1966 and 1989; *Women, Freedom and Calvin*, Philadelphia, Westminster, 1985; Editor, with J L Stotts, *To Confess the Faith Today*, Philadelphia, Westminster/John Knox Press, 1990; "A Study of the Congregation in History", in *Beyond Clericalism: The Congregation as a Focus for Theological Education*, Ed. J C Hough, Jr. and B G Wheeler, Atlanta, Scholars Press, 1988, pp. 63-75 and With G K Douglas, "Creation, Reformed Faith, and Sustainable Food Systems", in *Reformed Faith and Economics*, Ed. Robert Stivers, New York, University Press of America, 1989, pp. 117-144.
Mailing address: Princeton Theological Seminary, CN 821, Princeton NJ 08542, USA. T. (1 609) 497 7798. Fax (1 609) 924 2973
Residence: 26 Library Place, Princeton NJ 08540, USA T. (1 609) 497 0082

Douguet, Yves, P.S.M., Fr

Rel: Roman Catholic; Ordained; celibate.
International Organizations: Superior General of the Prêtres de Sainte Marie de Tinchebray.
Mailing address: Sainte Marie, Boulevard du Nord, 61800 Tinchebray, France. T. (33) 33 64 91 48

Dowdell, Pamela, F.M.M., Sr

Rel: Roman Catholic; Ordained; celibate.
International Organizations: Secretary General of the International Union of Superiors General.
Member of the Congregation for Religious and Secular Institutes. Member of the Franciscaines Missionnaires de Marie.
Mailing address: Piazza di Ponte S Angelo 28, I-00186 Roma, Italy. T. (39 6) 687 59 21.

Downes, Stanley E, Rev

Rel: Methodist; Ordained.
International Organizations: General Secretary of the Methodist Church in India.
Mailing address: Methodist Centre, 21 YMCA Road, Bombay 400 008, India.

Doyle, A C

Rel: Christian.
International Organizations: Founder of the International Partners in Prayer.
Mailing address: PO Box 570122, Houston TX 77257, USA.

Doyle, June
Rel: Spiritualist.
International Organizations: Secretary of the International Spiritualist Federation.
Mailing address: Bosboomstraat 45, NL-1702 ZH Heerhuyswaard, Netherlands.

Doyle, Marge
Rel: Seventh-Day Adventist.
International Organizations: President of the Seventh-Day Adventist Kinship International.
Mailing address: PO Box 3840, Los Angeles CA 90078, USA. T. (1 213) 876 2076.

Doyle, Wilfred Emmett, Most Rev
Rel: Roman Catholic; Ordained on 05 JUN 1938; b. 18 FEB 1913 in Calgary, Alberta, Canada; celibate; Citizen of Canada.
International Organizations: Bishop Emeritus (since 1990) of Nelson. **Career:** Assistant, St Joshep's Cathedral, Edmonton, Alberta from 1938 to 1943. Assistant Chancellor, Archdiocese of Edmonton from 1938 to 1943. Chancellor, Archdiocese of Edmonton from 1943 to 1946. Chancellor, Archdiocese of Edmonton from 1949 to 1958. Bishop, Nelson from 1958 to 1989. Chancellor, Notre Dame University, Edmonton, BC from 1963 to 1968. Chairman, Board of Governors, Notre Dame University from 1963 to 1974. **Educ.:** B.A. from University of Alberta, Edmonton, Canada (1935); J.C.D. from University of Ottawa, Ontario, Canada (1949).
Mailing address: 10661 82nd Ave, Edmonton AB T6E 2A6, Canada. T. (1 403) 433 8936

Driedger, Florence
Rel: Mennonite.
International Organizations: Moderator of the General Conference of the Mennonite Church.
Mailing address: 3833 Montague St, Regina SK S4S 3J6, Canada.

Droste, Rolf, Rev
Rel: Evangelical Lutheran; Ordained.
International Organizations: General Secretary of the Evangelical Church of Lutheran Confession in Brazil.
Mailing address: Caixa Postal 2876, 90 000 Porto Alegre RGS, Brazil.

Drouin, Pierre Joseph Donat, C.I.M., Rev
Rel: Roman Catholic; Ordained on 09 MAR 1963; b. 22 APR 1937 in Ville St Georges PQ, Canada; celibate; Citizen of Canada; Lang. French, English, Spanish, Italian, Portuguese and German.
International Organizations: Superior General (since 1989) of the Congregation of Jesus and Mary.
Other organizations: Professor of Holy Scripture (since 1973), Universidad Javeriana, Bogotà; Professor of Holy Scripture (since 1979), Centro ITER, Caracas; Professor of Holy Scripture (since 1980), Instituto Pastoral del Celam (Colombia).
Main field of work: University education. **Career:** Provincial, Congregation of Jesus and Mary from 1979 to 1987. **Educ.:** Lic. Théol. from Pontifical University San Tommaso d'Aquino, Vatican City State (1963-1964); Lic. Ecriture Sainte from Pontifical Biblical Institute, Vatican City State (1964-1966). **Hons.:** Dr h.c., Université Sainte Anne (1990). **Pubs.:** Una lectura del Apocalipsis, Col. Pneuma, Bogotà, 2nd edn, 1987.
Mailing address: Via dei Querceti 15, I-00184 Roma, Italy. T. (39 6) 704 506 43. Fax 709 23 74
Residence: See Mailing Address. T. (39 6) 704 506 47

Droutman, J, Mrs
Rel: Quaker.
International Organizations: UNESCO Representative of the Friends World Committee for Consultation.
Mailing address: 8 rue de Rémusat, F-75016 Paris, France. T. (33 1) 45 27 57 30.

Drukier, Elisabeth
Rel: Buddhist: Mahayana.
International Organizations: Europe Regional Office of the Foundation for the Preservation of the Mahayana Tradition.
Mailing address: 7 Rue Labat, 75018 Paris, France. T. (33 1) 4255 0328. Fax (33 1) 4252 5063

Drummond, Norman Walker, Rev
Rel: Church of Scotland; Ordained on 15 AUG 1976; b. 01 APR 1952 in Greenock, Scotland, UK; m. to Lady Elizabeth Helen Kennedy; c. 2 boys and 2 girls; Citizen of United Kingdom; Lang. English and German.
International Organizations: Headmaster (since 1984) of the Loretto.
Other organizations: Member, the Queen's Bodyguard for Scotland (Royal Company of Archers); Chaplain to the Governor, Edinburgh Castle; Chairman, Musselburgh and District Council of Social Service; Member, Citizen's Advice Bureau, Musselburgh; Sponsor, Joint Loretto/Musselburgh Grammar School Community Service Committee; Member, Scottish Committee, Duke of Edinburgh's Award Scheme; Trustee, Scottish Silver Jubilee and Children's Bursary Fund; Chairman, Ronald Selby Wright Christian Leadership Trust; Governor, Crawfordton House School; Governor, St Mary's School, Melrose; Governor, Croftinloan School, Pillochry; Trustee, Arthur Smith Memorial Trust; Member of the Court, Heriot Watt University; Member, Scottish Council, ISCO (Independent Schools Careers Organisation).
Main field of work: Education. **Career:** Chaplain, Depot, The Parachute Regiment from 1976 to 1978. Chaplain, 1st Battalion, The Black Watch from 1978 to 1982. Chaplain, Fettes College from 1982 to 1984. **Educ.:** B.A. from University of Cambridge, UK (1970-1973); B.D. from University of Edinburgh, New College, UK (1973-1976). **Interests:** rugby football, cricket, golf, curling, traditional jazz, Isle of Skye. **Mem.:** Clubs: New (Edinburgh), Hawks (Cambridge), MCC, Free Foresters **Pubs.:** The First Twenty Five Years – The Official History of the Kirk Session of the Black Watch (Royal Highland Regiment).
Mailing address: Headmaster's House, Loretto, Musselburgh-Midlothian EH21 7RE, UK. T. (44 31) 665 2567
Residence: See Mailing Address. T. See Mailing Address

D'Souza, Eugene, M.S.F.S., Most Rev
Rel: Roman Catholic; Ordained on 29 SEP 1944; b. 15 NOV 1917 in Nagpur, India; celibate; Citizen of India.
International Organizations: Archbishop (since 1963) of Bhopal.
Member (since 1935) of the Missionnaires de Saint François de Sales d'Annecy.
Other organizations: Member, Standing Committee (since 1953), Catholic Bishops' Conference of India.
Career: Consecrated first native Bishop, Nagpur in 1951. First Archbishop, Nagpur in 1953. Established St Francis de Sales College of Arts and Science and St Charles Seminary, Nagpur in 1956. Established Bhopal School of Social Sciences, University College, University of Bhopal in 1965. Established Asha Niketan, Rehabilitation Centre for handicapped and deaf children in 1967. Member, Governing Board, St John's Medical College, Bangalore. Active participant in Vatican Council II and several international conventions. **Hons.:** National Award for Distinguished Service in Education and Rehabilitation of the Handicapped (1976).
Mailing address: Archbishop's House, 33 Ahmedabad Palace Road, Bhopal MP 462 001, India. T. (91) 73619

D'Souza, Henry Sebastian, Most Rev Archbishop
Rel: Roman Catholic; Ordained on 24 AUG 1948; b. 20 JAN 1926 in Igatpuri, Poona, India; celibate; Citizen of India; Lang. English, Bengali, Hindi and Italian.
International Organizations: Secretary-General of the Federation of Asian Bishops' Conferences.
Co-Head, Asia Regional Office of the Unda, International Catholic Association for Radio and Television. Member of the Congregation for the Evangelization of Peoples. Member of the Pontifical Council for Inter-Religious Dialogue.
Other organizations: Archbishop (since 1986), Calcutta.
Educ.: S.T.L.; J.C.D. from Urbanian Pontifical University, Vatican City State (1945-1959).
Mailing address: 16 Caine Road, PO Box 2984, Hong Kong, Hong Kong. T. (852 5) 25 80 21
Residence: 32 Park Street, Calcutta 700016, India.

Dubois, Yvette
Rel: Christian.
International Organizations: Secretary-Treasurer of the Fraternity International Apostolic Team.
Mailing address: 41 rue Boileau, F-75016 Paris, France.

Ducor, Jérôme, Rev Prof

Rel: Buddhist; Ordained on 28 OCT 1977; b. 18 MAR 1954 in Geneva, Switzerland; Citizen of Switzerland; Lang. French, English, Japanese and Tibetan.
International Organizations: Secretary/Vice-President Shingyôji (temple) (since 1972) of the Société bouddhique suisse Jôdo-Shinshú.
Other organizations: Assistant Professor of Japanese Religions (since 1992), McGill University, Montreal.
Career: Tokudo, Jôdoshinshú-Honganji-Ha, Kyoto in 1977. Kyôshi (master), Jôdoshinshú-Honganji-Ha, Kyoto in 1987. **Educ.:** University of Lausanne, Switzerland (1975-1978); Lic. ès Lettres from University of Geneva, Switzerland (1978-1981); Doc. ès Lettres from University of Geneva, Switzerland (1990). **Pubs.:** Tannishô (notes déplorant les divergences), Kyoto, 1983; Shinshú Gongyôshú, Kyoto, 1990 and La vie de Zonkaku (1290-1373), Genève, 1991.
Mailing address: BP 2139, CH-1211 Genève 1, Switzerland.
Residence: 2 Place de la Taconnerie, CH-1204 Genève, Switzerland. T. (41 22) 28 01 53

Duda, Betty

Rel: Lutheran.
International Organizations: President of the International Lutheran Women's Missionary League.
Mailing address: 3558 S Jefferson Ave, St Louis MO 63118-3810, USA. T. (1 314) 268 1531.

Dudit, Camille, F.M.I., Fr

Rel: Roman Catholic; Ordained; celibate.
International Organizations: Superior General of the Sons of Mary Immaculate, Fathers of Chavagnes.
Mailing address: BP 5, Ste Marie Chavagnes-en-Paillers, F-85250 Saint-Fulgent, France. T. (33) 51 42 21 28

Dugniolle, Laurence

Rel: Roman Catholic.
International Organizations: Secretary of the Euro-Caritas.
Mailing address: Rue du Commerce 70-72, B-1040 Bruxelles, Belgium. T. (32 2) 511 42 55. Tx 25 173.

Dujardin, Jean, Fr

Rel: Roman Catholic; Ordained; celibate.
International Organizations: Superior General of the Prêtres de l'Oratoire de France.
Mailing address: 2 bis quai des Célestins, F-75004 Paris, France. T. (33 1) 42 78 40 67.

Dulczin, A

Rel: Jewish.
International Organizations: President of the Jewish Agency for Israel.
Mailing address: PO Box 92, 91920, Jerusalem, Israel. T. (972 2) 39261. C. Zionit Jerusalem.

Duminuco, Vincent J, Dr

Rel: Roman Catholic.
International Organizations: Delegate of the World Union of Jesuit Alumni.
Mailing address: Borgo San Spirito 5, CP 6139, I-00195 Roma, Italy. T. (39 6) 68 69 841. C. GISA-ROMA-VATICANO. Tx 504 2018 GISAROMA VA.

Dumke, Isolde, Dr

Rel: Protestant; b. 07 AUG 1944 in Berlin, Germany; single; Citizen of Germany; Lang. German, English, French, Latin and Ancient Greek.
International Organizations: Secretary (since 1983) of the International Council of Associations of Theological Libraries.
Other organizations: Director (since 1979), Evangelische Bibliothek, Köln; Delegate to Arbeitsgemeinschaft der Archive und Bibliotheken in der Evangelischen Kirche International Council (since 1989), Verband Kirchlich-Wissenschaftlicher Bibliotheken.
Main field of work: Theological and ecclesiastical libraries. **Career:** Archivist for medieval manuscripts, Württembergische Landesbibliothek, Stuttgart from 1975 to 1979. Member of the Board, Delegate to the Council, Verband Kirchlich-Wissenschaftlicher Bibliotheken from 1980 to 1987. President, Verband Kirchlich-Wissenschaftlicher Bibliotheken from 1986 to 1987. **Educ.:** Ph.D. from Free University of Berlin, Germany (1963-1972); Assistant Librarian Exam. from Staatsbibliothek Preussischer Kulturbesitz, Berlin and Fachhochschule für Bibliotheks- und Dokumentationswesen, Köln, Germany (1973-1975). **Interests:** French, Latin, Asian culture and literature, embroidery (gobelin). **Pubs.:** *Voltaire als Religionskritiker im Spiegel der Forschung (1956-1969)*, thesis, 1972 and various articles in reviews.
Mailing address: Postfach 25 01 04, Kartäusergasse 9, 50678 Köln, Germany. T. (49 221) 3382 110.
Residence: Peter Bauer Strasse 7, 50823 Köln, Germany T. (49 221) 52 47 48

Dunbar, Hooper C

Rel: Bahá'í.
International Organizations: Member, Universal House of Justice of the Bahá'í International Community.
Mailing address: PO Box 155, 31001 Haifa, Israel. T. (972 4) 372 433 – 510 344. C. BAHAIFAITH HAIFA. Tx 46626 BAYT IL. Fax (972 4) 372 425.

Duncan, Alistair

Rel: Muslim.
International Organizations: Director of the World of Islam Festival Trust.
Mailing address: 33 Thurloe Place, London SW7 2HQ, UK. T. (44 71) 581 3522. Fax 584 1977. Tx 896988 SCOOPS G. C. Islamtrust London SW7.

Dunoyer, D Emilio, C.R.L., Fr Abbot

Rel: Roman Catholic; Ordained; celibate.
International Organizations: Abbot-General of the Cononici Regolari della Congregazione del Ss.mo Salvatore Lateranense.
Mailing address: Piazza S Pietro in Vincoli 4/a, I-00184 Roma, Italy. T. (39 6) 481 89 69, (39 6) 488 28 65

du Plessis, Paul d, Col

Rel: Salvation Army, The.
International Organizations: Territorial Commander of The Salvation Army.
Mailing address: PO Box 453, Vepery, Madras 600 007, India.

Du Preez, Jannie D, Prof

Rel: Dutch Reformed; Ordained in 1952; b. 16 AUG 1927 in Joubertina, Cape Province, South Africa; m. to Hendrika Elizabeth Burger; c. 1 boy and 3 girls; Citizen of South Africa; Lang. Afrikaans, English, Dutch, German and Xhosa.
International Organizations: Professor and Head, Department of Theology (since 1982) of the Faculty of Theology, University of Stellenbosch.
Other organizations: Member, Editorial Committee (since 1975), *Ned. Geref. Teologiese Tijdskrif*; Member, Editorial Committee (since 1982), *Missionalia*.
Main field of work: University education. **Career:** Minister of Religion, Dutch Reformed Church Mission from 1952 to 1957. Principal, Evangelist School, Umtata, Transkei from 1954 to 1957. Principal, Decoligny Theological School, Transkei from 1958 to 1975. Senior Lecturer, Faculty of Theology, University of Stellenbosch from 1975 to 1982. **Educ.:** B.A. from University of Stellenbosch, South Africa (1945-1947); M.A. Phil. from University of Stellenbosch, South Africa (1948-1950); Dip. Theol. from Theological Seminary, Stellenbosch, South Africa (1948-1951); B.D. from University of Stellenbosch, South Africa (1961-1962); D.Th. from University of Stellenbosch, South Africa (1964-1967). **Mem.:** Southern African Missiological Society; Association for the Study of Religion (South Africa); Theological Society of South Africa; South African Academy for Science and Art **Pubs.:** Die koms van die Koninkrijk volgens die boek Openbaring, 1979 (The coming of the Kingdom according to the book of Revelation); Die Briewe aan die Tessalonisense, 1979 (The Letters to the Thessalonians) and Sendingsperspektiewe in die erediens, 1992 (Mission perspectives in the divine service).
Mailing address: University of Stellenbosch, 171 Dorp Street, Stellenbosch 7600, South Africa. T. (27 2231) 773261
Residence: Oranjelaan 11, Stellenbosch 7600, South Africa. T. (27 2231) 70896

Duprey, Pierre, M.Afr., H E Mgr

Rel: Roman Catholic; Ordained on 01 FEB 1950; b. 26 NOV 1922 in Croix, France; celibate.

International Organizations: Secretary of the Pontifical Council for Promoting Christian Unity.

Co-Chairman of the Joint International Commission for Dialogue between the Roman Catholic Church and the Malankara Orthodox Syrian Church. Co-Chairman of the Joint International Commission for Dialogue between the Roman Catholic Church and the Coptic Orthodox Church. Advisor of the Sacred Congregation for the Oriental Churches. Member of the Pontifical Commission for Latin America. Member of the Pontifical Council for the Pastoral Care of Migrants and Itinerant People. Member of the Pontifical Committee for International Eucharistic Congresses. Member of the Society of Missionaries of Africa. Secretary-General of the Christian World Communion Executives Conference.

Other organizations: Titular Bishop, Thibaris; Member, Pontifical Commission for Russia; Advisor, Pontifical Commission for the Doctrine of the Faith.

Mailing address: Pontifical Council for Promoting Christian Unity, 00120 Vatican City, Vatican. T. (39 6) 698 43 84. Fax 698 53 65.

Duranti, Angelo, Prof

Rel: Christian.

International Organizations: Director of the Tutmonda Ekumena Ligo.

Mailing address: Odos Rodou 5, Athinai 11252, Greece. T. (30 1) 8674873.

Durel, Michel

Rel: Roman Catholic.

International Organizations: President of the Catholic Relief.

Unesco Representative of the Caritas Internationalis.

Mailing address: 106 rue du Bac, F-75341 Paris CEDEX 07, France. T. (33 1) 43 20 14 14. C. SECATHO PARIS. Tx 270184 Secatho F. Fax (33 1) 45 49 94 50.

Durersingh, Christopher, Rev Dr

Rel: Protestant.

International Organizations: Director of the Commission on World Mission and Evangelism of the World Council of Churches.

Mailing address: CP 66, 150 route de Ferney, CH-1211 Genève 2, Switzerland. T. (41 22) 791 61 11. C. OIKUMENE GENEVA. Tx 415 730 OIK CH. Fax (41 22) 791 03 61.

Durodola, V O

Rel: Baptist.

International Organizations: Member of the All-Africa Baptist Fellowship.

Mailing address: PMB 5113, Ibadan, Nigeria.

Dussel, Enrique

Rel: Christian.

International Organizations: President of the Commission of Studies for Latin American Church History.

Mailing address: Celaya 21-402, Colonia Hipódromo, 06100 Mexico DF, Mexico. T. (52 5) 574 5661.

Düsterfeld, Peter, Rev Prof Dr

Rel: Roman Catholic; Ordained; celibate.

International Organizations: Co-Head, Europe Regional Office of the Unda, International Catholic Association for Radio and Television.

Advisor of the Pontifical Council for Social Communications.

Mailing address: Zentralstelle Medien der Deutschen Bishofskonf, Kaiserstrasse 163, D-5300 Bonn, Germany.

Dusu, Barnaba

Rel: Christian.

International Organizations: Chairman of the Rural Development Counsellors for Christian Churches in Africa.

Mailing address: NBTC, Old Airport Road, PO Box 6617, Jos, Nigeria. T. (234 73) 57560. Fax 57863. Tx 81350 NITEL JS Box 011 (Attn RURCON).

Duteil, Armel, Rev Fr

Rel: Roman Catholic; Ordained; b. 31 MAR 1940 in Lorient, France; celibate; Citizen of France; Lang. French, English, Spanish, German, Latin, Lingala, Kikongo and Ouolof.

International Organizations: President (since 1969) of the Recherches pour l'éducation affective des jeunes en Afrique.

National Chaplain (since 1981) of the European Young Christian Workers. Responsible Pan-African person for young people education (since 1989) of the International Federation for Family Life Promotion.

Other organizations: In charge of leading (since 1980) young people, Saint Louis Diocese; Responsible Regional Person (since 1982), Association Sénégalaise pour la Promotion de la Famille.

Main field of work: Education. **Career:** Vicar, Brazaville Diocese from 1966 to 1975. Mission organizer, Paris from 1976 to 1979. Curate, Tamba Counda, Senegal from 1979 to 1980. Educator, Catholic Mission, Saint Louis in 1980. **Educ.:** Institut Catholique de Paris, France (1958-1960); Bacc.theol. from University of Fribourg, Switzerland (1960-1962); Bacc.phil. from Institut Catholique de Paris, France (1964-1966); Fin d'études Past. from ICAO (ISCR), Abidjan, Ivory Coast (1975-1976). **Interests:** Young people education in Africa. **Hons.:** Academic Palms, Senegal. **Mem.:** Amnesty International, Alliance Franco-Sénégalaise, Association Sénégalaise pour la Promotion de la Famille **Pubs.:** *La Sorcellerie, Pourquoi ?*; *Education Traditionnelle et Education Moderne*; *Une Sexualité Libérée*; *Témoins de l'Evangile en Terre d'Islam* and *Apprenons à nous connaître*.

Mailing address: c/o Mission Catholique, REDAJA, 8 rue Duret, BP 160, St Louis, Senegal. T. (221) 61 10 04 – (221) 61 10 27.

Residence: See Mailing Address. T. See Mailing Address

Duval, Ayle

Rel: Roman Catholic.

International Organizations: Head, Indian Ocean Regional Secretariat of the International Movement for the Apostolate in the Independent Social Milieux.

Mailing address: Clairfonds 2, Av Brown Square, Vacoas, Mauritius.

Duval, Léon-Etienne, H E Card

Rel: Roman Catholic; Ordained on 18 DEC 1926; b. 09 NOV 1903 in Chenex, Haute-Savoie, France; celibate; Citizen of Algeria.

International Organizations: Member (since 1965) of the College of Cardinals.

Other organizations: Archbishop Emeritus (since 1988), Algiers.

Career: Professor, Grand Séminaire Annecy from 1930 to 1942. Vicar-General and Director of works, Diocese of Annecy from 1942 to 1946. Consecrated Bishop, Constantine and Hippo in 1946. Archbishop, Algiers from 1954 to 1988. Created Cardinal by Pope Paul VI in 1965. **Hons.:** Officer of the Legion of Honour. **Pubs.:** Messages de Paix 1955-1962, 1962; Laïcs, prêtres, religieux dans l'église, 1967 and Evêque en Algérie (in collaboration), 1984.

Mailing address: 4 avenue Ali Ourak, Bologhine-Algiers, Algeria. T. (213 1) 62 35 89

Dylykov, S D, Prof

Rel: Buddhist.

International Organizations: Vice-President of the World Fellowship of Buddhists.

Mailing address: Religious Board of the Buddhist of USSR, 49 Ostozhenka Street, Moscow 119034, USSR.

Dzobo, N K, Rt Rev Prof

Rel: Evangelical Presbyterian.

International Organizations: Moderator of the Evangelical Presbyterian Church, Ghana.

Mailing address: PO Box 18, HO, Volta Region, Ghana.

Eames, Robert Henry Alexander, Most Rev

Rel: Church of Ireland; Ordained in 1964; b. 27 APR 1937 in Belfast, N. Ireland, UK; m. to Ann Christine Daly; c. 2 boys; Citizen of United Kingdom.

International Organizations: Archbishop (since 1986) of Armagh.

Other organizations: Primate (since 1986), All-Ireland and Metropolitan.

Career: Research Scholar and Tutor, Faculty of Laws, Queen's University, Belfast from 1960 to 1963. Ordained Deacon in 1963. Curate, Bangor Parish Church from 1963 to 1966. Rector, Gilnahirk from 1966 to 1974. Rector, Dundela from 1974 to 1975. Domestic Chaplain to Bishop, Down and Dromore from 1970 to 1972. Examining Chaplain to Bishop, Down and Dromore from 1973 to 1975. Consecrated Bishop in 1975. Bishop, Derry and Raphoe from 1975 to 1980. Bishop, Down and Dromore from 1980 to 1986. Chairman, Church of Ireland Priorities Committee from 1977 to 1979. Irish Representative to the Anglican Consultative Council in 1984. Member, Standing Committee, Anglican Consultative Council in 1985. Select Preacher, University of Oxford from 1986 to 1987. Chairman, the Archbishop of Canterbury's Commission on Communion and Women in the Episcopate from 1988 to 1990. **Educ.:** LL.B. (Hons) from Queen's University of Belfast, UK (1960); Ph.D. from Queen's University of Belfast, UK (1963); University of Dublin Trinity College, Ireland (1963). **Hons.:** History and Pastoral Theology Prizes; Hon. LL.D., Queen's University, Belfast (1989). **Pubs.:** Through Suffering, 1973; Thinking Through Lent, 1978 and contributed articles to *Irish Legal Quarterly, Northern Ireland Legal Quarterly, Criminal Law Review, New Divinity, University Review, The Furrow.*
Mailing address: The See House, Cathedral Close, Armagh BT61 7EE, UK. T. (44 861) 527144. Fax 527823
Residence: T. (44 861) 522851

Earwicker, John
Rel: Evangelical.
International Organizations: Member of the Evangelical Alliance.
Mailing address: Whitefield House, 186 Kennington Park Rd, London SE11 4BT, UK. T. (44 71) 582 02 28. Fax 582 62 21.

Easter, Clive J
Rel: Christian.
International Organizations: Secretary of the Church Monuments Society.
Mailing address: 18 Cotmore Way, Chillington, Kingsbridge, UK. T. (44 548) 58 00 01.

Eastman, Dick, Rev
Rel: Protestant.
International Organizations: President of the Every Home for Christ International.
Mailing address: 20232 Sunburst Street, Chatsworth CA 91311, USA. T. (1 818) 341 7870.

Ebacher, Roger, Mgr
Rel: Roman Catholic; Ordained on 27 MAY 1961; b. 06 OCT 1936 in Amos, Quebec, Canada; celibate; Citizen of Canada.
International Organizations: Archbishop Metropolitan (since 1990) of Gatineau-Hull.
Other organizations: Member of Board, Bishops' Conference of Canada.
Career: Professor of Philosophy from 1961 to 1967. Pastoral work, Diocese, Amos and parish priest. Vicar-General, Diocese, Amos. Consecrated Bishop in 1979. **Educ.:** B.A. from Université Laval, Québec, Canada (1957); L.Th. from University of Ottawa, Ontario, Canada (1961); Ph.D. from Institut Catholique de Paris, France (1966). **Mem.:** Member of various episcopal committees **Pubs.:** Education de la foi (un diocèse engage ses laics), 1979; Marie et notre pélerinage de foi, 1988 and numerous religious and philosophical articles in journals and magazines.
Mailing address: 119 rue Carillon, Hull, Québec J8X 2P8, Canada.

Eckardt, Ralph W
Rel: Christian.
International Organizations: Executive Director of the INTERSERVE.
Mailing address: Box 418, 239 Fairfield Ave, Upper Darby PA 19082, USA. T. (1 215) 352 0581. Fax 352 4394.

Eda, Koten
Rel: Buddhist; Ordained; b. 29 OCT 1915; Citizen of Japan.
International Organizations: Head Manager of the Tendaishu.
Career: Chief, Chotokuji. **Educ.:** Taisho University.
Mailing address: 3-31-23 Chotokuji Senbamachi, Kawagoe, Saitama 350, Japan. T. (81 492) 22 2031

Edelby, Neophytos Elias, H E Mgr
Rel: Greek-Melkite Catholic; Ordained in 1944; b. 10 NOV 1920 in Aleppo, Syria; Citizen of Syria.
International Organizations: Metropolitan (since 1968) of Aleppo.
Career: Professor, St Anne Seminary of Jerusalem from 1950 to 1953. Principal, St John the Apostle College of Beirut from 1953 to 1959. Assistant, Basilean Sisters of Aleppo in 1953. Secretary to the Synod, the Melkite Church in 1951. Consecrated Bishop in 1962. Appointed Advisor to His Eminence Cardinal Maximos IV Saïgh, Patriarch, the Melkite Church from 1963 to 1967. **Hons.:** Grand Cross, Order of Saint Lazarus; Officer, Order of the Merit, Lebanon (1968). **Pubs.:** Author of *Liturgicon* (French missal of the Byzantine rite), 1960; *Prayer Book* (in Arabic), 1963 and contributor to various encyclopedias and reviews.
Mailing address: Melkite Greek-Catholic Archbishopric, PO Box 146, Aleppo, Syria.

Edgar, W R
Rel: Christian.
International Organizations: General Director of The Leprosy Mission International.
Mailing address: 80 Windmill Road, Brentford TW8 0QH, UK. T. (44 81) 569 7292.

Edirisinghe, Albert
Rel: Buddhist; b. 01 SEP 1913 in Baddegama, Galle, Sri Lanka; m. to Yasawathie; c. 2 boys and 3 girls; Citizen of Sri Lanka; Lang. Singhalese and English.
International Organizations: Chairman of the Third Committee (Dhammaduta Activities) (since 1981) of the World Fellowship of Buddhists.
Other organizations: Treasurer (since 1966), Somawathi Chaityawardena; Honorary Member (since 1973), All Ceylon Buddhist Congress; President (since 1977), Mahatma Gandhi Trust; Vice-President (since 1982), India-Ceylon Society; Honorary Royal Nepalese Consul General (since 1984); Honorary Member (since 1990), Supreme Advisory Council of the Ministry of Buddha Sasana.
Main field of work: Optometry. **Career:** Director, Industrial Development Board of Sri Lanka from 1965 to 1970. Member, Buddhist Advisory Council in 1969. Director, Bureau of Standards in 1972. Director, National Metrication Board in 1972. Secretary, India-Ceylon Society from 1973 to 1977. Vice-President, World Fellowship of Buddhists (WFB) from 1975 to 1986. Secretary, WFB Sri Lanka Regional Centre from 1964 to 1987. President, WFB Sri Lanka Regional Centre from 1975 to 1985. President, All-Ceylon Buddhist Congress from 1975 to 1977. Treasurer, All-Ceylon Buddhist Congress from 1979 to 1983. President, All-Ceylon Buddhist Congress from 1983 to 1984. Chairman, Albert Edirisinghe Opticians Ltd., Lanka Optical Industries Ltd., Albert Edirisinghe Songdoh (Pvt) Ltd., Albert Edirisinghe Investments Ltd., Tajit (Pvt) Ltd. Director, Sampath Bank, Albert Edirisinghe Electronics Ltd., Sparkling Travels Ltd., Tea House Ceylon Ltd., Metro Printers. Chairman/Managing Director, Janaka Ltd. President, Target International Ltd. Director, Ceylon Eye Care Society. Managing Trustee, World Fellowship of Buddhists Dhammaduta Fund Trust. Managing Trustee, Yasodhara Foundation Trust. Secretary/Treasurer, Malalasekera Institute of Buddhist Education and Culture. Vice Patron, Sri Lanka-Nepal Friendship Society. President, Ganegama Dutugemunu Raja Maha Vihara Dayaka Sabha. Member of the Supreme Advisory Board, Buddha Sasana Ministry. Trustee, Sambodhi Trust, Galle. **Educ.:** Dip. from London Course of Optometry. **Hons.:** Appointed Justice of the Peace (1964); Certificate of Commendation, World Fellowship of Buddhists, Bangkok (1973); Appreciation Plaque presented by the Prime Minister for the services rendered to the Sri Lanka Chamber of Small Industry as Founder President (1988); Plaque of Appreciation presented by the President of the World Society for Zen Academy and the President of the Korea Fellowship of Buddhists (1984); Desabandhu Title conferred by the President of the Government of Sri Lanka (1987); Gorkha Dakshina Bahu conferred by His Majesty the King of Nepal (1991); Buddhasasana Mahasura Sri Lanka Ramanna Nikaya Senapathi. **Mem.:** Buddhist Theosophical Society, Maha-Bodhi Society (India and Sri Lanka), Japan-Sri Lanka Friendship Society, YMBA (Colombo, Nugegoda, Dehiwala, Wellawatta), Sarvodaya (Sri Lanka), Sasana Sevaka Society, Singhalese Sports Club, Saukyadana Movement, Sri Lanka Eye Donation Society, Royal Asiatic Society, Sinhala Subasadhaka Aramudala, Sambodhi Society (Galle), Anuradhapura Sri Maha Bodhi Ranweta Sasanaraksana Mandalaya, National Council for Non-Smoking, Ceylon Cancer Society, Ceylon Anti-Tuberculosis Society, Prevention of Blindness Society, Sri Lanka-Pakistan Friendship Society, Sri Lanka-Korea Friendship Society, Sri Lanka Chamber of Small Industry (Founder

President), Ceylon Optometric Association (Past President), Rotary Club of Colombo West (Past President and Charter Member)
Mailing address: Sri Lanka Regional Centre of WFB, 52 Vision House, Galle Road, Colombo 4, Sri Lanka. T. (94 1) 580987
Residence: 290 R A de Mel Mawatha, Colombo 3, Sri Lanka. T. (94 1) 574717

Edmone, M
Rel: Unknown.
International Organizations: Secretary of the Omkarananda International Ashram.
Mailing address: 41 Anton Graffstrasse, CH-8400 Winterthur, Switzerland.

Edsor, Albert W, Pastor
Rel: Protestant.
International Organizations: Member of the World Revival Crusade.
Mailing address: 17 Sumburgh Rd, London SW12 8AJ, UK. T. (44 71) 228 7363.

Edwards, David, Col
Rel: Salvation Army, The.
International Organizations: Territorial Commander of The Salvation Army.
Mailing address: PO Box 378, Kingston 10, Jamaica.

Edwards, Olive Eva
Rel: Uniting Church in Australia; b. 30 AUG 1929 in Launceston, Tasmania, Australia; single; Citizen of Australia; Lang. English.
International Organizations: Corresponding Secretary (since 1980) of the World's Woman's Christian Temperance Union.
Other organizations: National President and State Treasurer (since 1991), Woman's Christian Temperance Union; Elder, Uniting Church in Australia.
Main field of work: Temperance. **Career:** Farm worker from 1947 to 1986. **Educ.:** Correspondence Crse. (London) (1951-1955). **Interests:** children, church. **Mem.:** Horticulture Society; St John Ambulance Brigade; Country Women's Association; World, National, State, USA and Canada Woman's Christian Temperance Union (Life Member); Order of St John of Jerusalem (Serving Sister)
Mailing address: Cloverlea, Branxholm 7261 TAS, Australia. T. (61 03) 54 6172.
Residence: See Mailing Address. T. See Mailing Address

Efefe, Elonda, Rev Dr
Rel: Disciples of Christ; Ordained.
International Organizations: General Secretary of the Church of Christ in Zaire (Community of Disciples of Christ).
Mailing address: BP 178, Mbandaka, Zaire.

Efthimiou, Milton B, Dr
Rel: Orthodox.
International Organizations: Contact of the Standing Conference of the Canonical Orthodox Bishops in the Americas.
Contact of the Greek Orthodox Archdiocesan Council of North and South America.
Mailing address: 10 East 79th Street, New York NY 10021, USA. T. (1 212) 570 3500.

Egaña, Francisco Javier, S.J., Fr
Rel: Roman Catholic; Ordained; celibate.
International Organizations: Procurator-General of the Society of Jesus.
Advisor, Section on Secular Institutes of the Congregation for Religious and Secular Institutes.
Mailing address: Borgo S Spirito 5, I-00193 Roma, Italy. T. (39 6) 686 98 41.

Egede, Daniel, Lt-Col
Rel: Salvation Army, The.
International Organizations: Territorial Commander of The Salvation Army.
Mailing address: Box 215, Lagos, Nigeria.

Egger, Carlo, C.R.S.A., Fr Abbot
Rel: Roman Catholic; Ordained; celibate.
International Organizations: Abbot Primate of the Canons Regular of Saint Augustine.
President of the Latinitas Foundation.
Mailing address: Casella Postale 91, Tor Lupara, I-00010 Roma, Italy. T. (39 6) 905 90 38

Eguiguren, José V, Msgr
Rel: Roman Catholic; Ordained; celibate.
International Organizations: Head, Latin American Secretariat of the Caritas Internationalis.
Mailing address: Apartado 1389-A, Quito, Ecuador.

Ehnes, Herbert, Dr
Rel: Reformed Church; b. 18 JAN 1937 in Büderich, Germany; m. to Gudrun Ehnes; c. 3 girls; Citizen of Germany; Lang. German and English.
International Organizations: Moderator, Department of Finance (since 1989) of the World Alliance of Reformed Churches.
Other organizations: Church Council and Member of the Board (since 1978), Lippische Landeskirche.
Main field of work: Church law. **Career:** Lecturer, Kirchliche Hochschule Bethel in 1989. **Educ.:** Dr.jur. from Westfälische Wilhelmsuniversität Münster, and Eberhard-Karls-Universität Tübingen, Germany (1957-1961). **Pubs.:** *Die rechtlichen Grundlagen der Kriminaltherapie im Strafvollzug* (thesis) and articles in *Reformierte Kirchenzeitung, Evangel. Kommentare Zeitschrift für evangel. Kirchenrecht,* and others.
Mailing address: Matthias-Claudius-Weg 15, 32756 Detmold, Germany. T. (49 5231) 740 318. Fax 740 345
Residence: See Mailing Address. T. (49 5231) 67 581

Einaudi, Giulio, H E Mgr
Rel: Roman Catholic; Ordained on 16 FEB 1934; b. 11 FEB 1928 in San Damiano Marca, Italy; celibate.
International Organizations: Apostolic Nuncio to Chile (since 1988) of the Roman Catholic Church.
Other organizations: Titular Archbishop, Villamagna di Tripolitania.
Mailing address: Calle Nuncio Sotero Sanz 200, Casilla 507, Santiago, Chile. T. (56 2) 231 72 40 – 231 20 20. Fax 231 08 68. Tx 241035 APOST CL

Eisen, Sylvia
Rel: Jewish.
International Organizations: Treasurer of the International Society of Jewish Librarians.
Mailing address: 690 Anderson Avenue, Franklin Square NY 11010, USA. T. (516) 486 3748.

Ekandem, Dominic Ignatius, H E Card
Rel: Roman Catholic; Ordained on 07 DEC 1947; b. in Obio Ibiono, Cross River State, Nigeria; celibate; Citizen of Nigeria.
International Organizations: Member (since 1976) of the College of Cardinals.
Other organizations: Bishop, Abuja.
Career: Assistant Priest, Ifuho and Afaha Obong. Rector, Minor Seminary. Consecrated Auxiliary Bishop, Calabar in 1954. Apostolic Administrator, Port Harcourt Diocese from 1970 to 1973. President, Episcopal Conference of Nigeria from 1973 to 1979. Created Cardinal by Pope Paul VI in 1976. President, Association of Episcopal Conferences of Anglophone West Africa in 1977. **Interests:** reading, cultural exercises such as drama. **Hons.:** Hon. D.D. (1954); Honorary Officer of the Order of the British Empire (1960); Commander, Order of the Niger (1965). **Pubs.:** Sheperd among Sheperds, 1979 and Annual Addresses to Episcopal Conferences 1973-78.
Mailing address: PO Box 286, Garki, Abuja, Federal Capital Territory, Nigeria. T. (234 9) 234 1553

El Awad, Abdalla Suliman, Dr

Rel: Muslim; Unordained; b. 01 JAN 1935 in El Maseed, El Kamlin, Sudan; married; Lang. Arabic and English.
International Organizations: Founder and Director-General (since 1981) of the Islamic Relief Agency.
Vice-President of the International Council of Voluntary Agencies. Vice-President of the Forum of African Voluntary Development Organizations.
Main field of work: Primary health care work. **Educ.:** M.B.B.S. from University of Khartoum, Sudan (1954-1960); D.L.H. from University of Dundee, UK (1969-1970); M.Sc. Med. from University of London, UK (1970). **Pubs.:** Primary Health Care in Refugee Camps, 1983; Occupational Health in Industrially Developing Nations, 1979 (Arabic) and Interaction between Official and Voluntary Efforts in Social Work (Arabic).
Mailing address: PO Box 3372, Khartoum, Sudan. T. (249 11) 73741 – (249 11) 78766. Tx 22883 (IARA) SO. Fax (249 11) 72433.
Residence: Khartoum North, Sudan. T. (249 11) 613830

Eldad, Oded

Rel: Jewish.
International Organizations: Executive Director of the European Council of Jewish Community Services.
Mailing address: 4 bis rue de Lota, F-75116 Paris, France. T. (33 1) 44 53 31 26. Tx 611068 HIAS.

Eldred, Carl C

Rel: Protestant.
International Organizations: Executive Director of the Chosen.
Mailing address: 3642 West 26th Street, Erie PA 16506, USA. T. (1 814) 833 3023.

Eleblu, Lydia Aku

Rel: Reformed Church.
International Organizations: Member of the Executive Committee of the World Alliance of Reformed Churches.
Mailing address: PO Box 188 HO, V/R, Ghana.

El-Hachem, Basile, Rev

Rel: Maronite Catholic; Ordained on 07 JUN 1943; b. 18 OCT 1925 in Richmaya, Casa Alay, Lebanon; celibate; Citizen of Lebanon; Lang. Arabic, French, Italian and English.
International Organizations: Superior General (since 1986) of the Maronite Order of Lebanon.
Main field of work: Ministry, administration. **Educ.:** Lic. Theol. and Cn.L. from University of Strasbourg, France (1956); Lic. Soc.Sc. from Institut Catholique de Paris et Sorbonne, France (1959); Lic. for Teaching from Lebanese University, Beirut, Lebanon (1961). **Pubs.:** Obéissance et Liberté; La Sainteté de rencontre and Le Nous religieux.
Mailing address: Couvent Saint Antoine, Ghazir, Lebanon. T. (961) 906950/1/2/3/4

El-Hachem, Mounged, Mgr

Rel: Roman Catholic; Ordained; celibate.
International Organizations: Secretary of the Fondazione Pio XII per l'Apostolato dei Laici.
Other organizations: Minutanti (1st Class), General Affairs Section, Secretariat of State, Vatican City.
Mailing address: I-00120 Città del Vaticano, Vatican.

Eliahu, Mordechai, The Chief Rabbi

Rel: Jewish: Sephardi; Citizen of Israel.
International Organizations: Head of the Sephardic Community.
Mailing address: No mailing address. T. (972 2) 244785

Elias, Nabeel Radi, Prof

Rel: Christian; b. 01 JAN 1947 in Khartoum, Sudan; m. to Thérèse Adly Attalla; c. 1 boy; Citizen of Sudan; Lang. English and Arabic.
International Organizations: Chairman (since 1987) of the Evangelical Community Council.
Other organizations: Professor (since 1979), University of Khartoum; Chairman (since 1983), Evangelical Educational Board; Secretary (since 1983), Endeavor Union.
Main field of work: Administration, policy making. **Career:** Teaching Assistant, University of Khartoum from 1968 to 1972. Lecturer, University of Khartoum from 1972 to 1979. **Educ.:** B.Sc. from University of Khartoum, Sudan (1962-1968); Ph.D. from University of London, UK (1968-1972). **Interests:** stamp collecting. **Hons.:** many scientific awards. **Mem.:** Institute of Mechanical Engineers; American Society of Mechanical Engineers; Institute of Electrical and Electronics Engineers; Institute of Measurement and Control; MASEE **Pubs.:** Integrated Controller for Ships, Institute of Measurement and Control, Nov. 1978 and Discontinuous Control Strategy, PhD thesis, University of London.
Mailing address: Khartoum Tutorial College, PO Box 921, Khartoum, Sudan.
Residence: 34 Street 47, Khartoum 2, Sudan.

Eliasen, Samuel, Lt-Col

Rel: Salvation Army, The.
International Organizations: Officer Commanding of The Salvation Army.
Mailing address: Benito Castro, 7-5°c, 28028, Spain.

Eliopoulos, N

Rel: Greek Orthodox.
International Organizations: Secretary-General of the Pan-American Council for the Preservation of the Hellenic Orthodox Church And the Hellenic Language.
Mailing address: PO Box 65, Oak Park IL 60303, USA. T. (312) 725 1960.

Elizondo, Adán

Rel: Christian.
International Organizations: Vice-President for Latin America of the International Christian Union of Business Executives.
Mailing address: c/o Celulosas y Derivados, Av Santa Engracia 325, CP 624, Garza García, NL México, Mexico.

Elizondo, Juana, DC, Sr

Rel: Roman Catholic; Ordained; celibate.
International Organizations: Superior General of the Company of the Daughters of Charity of St Vincent-de-Paul.
Mailing address: 140 rue du Bac, F-75340 Paris CEDEX 07, France. T. (33 1) 45 48 10 13.

Ellerbee, Gene

Rel: Protestant.
International Organizations: Executive Vice-President of the Full Gospel Business Men's Fellowship International.
Mailing address: 3150 Bear Street, Costa Mesa CA 92626, USA. T. (1 714) 754 1400. Fax (1 714) 757 9916.

Elliott, Charles Middleton, Rev Prebendary Dr

Rel: Anglican; Ordained on 29 SEP 1964; b. 09 JAN 1939 in Wakefield, Yorkshire, UK; m. to Hilary Margaret Hambling; c. 3 boys; Citizen of United Kingdom; Lang. English, French, German, Spanish and Portuguese.
International Organizations: Chairman (since 1987) of the Feed the Minds.
Other organizations: Dean (since 1989), Trinity Hall, Cambridge; Affiliated Lecturer (since 1991), Faculty of Divinity, Cambridge.
Main field of work: University education. **Career:** Lecturer in Economics, University of Nottingham from 1962 to 1964. Reader in Economics, and Head of Dept., University of Zambia from 1964 to 1969. Assistant Secretary, Committee on Society, Development and Peace, World Council of Churches and the Pontifical Commission Justice and Peace (SODEPAX) from 1969 to 1972. Senior Lecturer in Economics, School of Development Studies, University of East Anglia from 1972 to 1977. Professor of Development Studies, University of Wales from 1977 to 1982. Director, Christian Aid from 1982 to 1984. Benjamin Meaker Professor, University of Bristol from 1985 to 1986. Visiting Professor, King's College, London from 1987 to 1988. **Educ.:** M.A. from University of Oxford, Lincoln College, UK (1957-1960); D.Phil. from University of Oxford, Nuffield College, UK (1960-1962). **Interests:** economics of developing countries. **Hons.:** Collins Religious Book Prize (1985) for *Praying the Kingdom.* **Pubs.:** Patterns of Poverty, 1973; Praying the Kingdom, 1985 and Sword and Spirit: Christianity in a Divided World, 1989.
Mailing address: Trinity Hall, Cambridge CB2 1TJ, UK. T. (44 223) 332525
Residence: 11 Perowne St, Cambridge CB2 1AY, UK. T. (44 223) 69233

Elliott, J K, Dr
Rel: Christian.
International Organizations: Secretary of the Society for the Study of the New Testament.
Mailing address: Department of Theology, University of Leeds, Leeds LS2 9JT, UK.

Ellis, Wilma M, Dr
Rel: Bahá'í.
International Organizations: Representative at UN of the Bahá'í International Community.
Mailing address: 866 United Nations Plaza, No 120, New York NY 10017, USA. T. (1 212) 486 0560. C. BAHAINTCOM NEW YORK. Tx 666363.

Ellor, James W, Rev Dr
Rel: Presbyterian; Ordained on 20 DEC 1983; b. 23 APR 1951 in Abington PA, USA; m. to Janet R; c. 2 girls; Citizen of USA; Lang. English, German, Spanish, French, Greek and Hebrew.
International Organizations: Associate Professor (since 1988) of the National-Louis University (formerly National College of Education).
Other organizations: Member of the Faculty, Division of Allied Health and Human Services (since 1983), National-Louis University, IL; Director of Gerontology, Director of Graduate Research, Human Service Dept. (since 1985), National-Louis University, IL; Parish Associate (part-time) (since 1989), Southminster United Presbyterian Church, Waukesha WI; Staff Therapist (since 1990), Exodus Pastoral Counseling Center; Chair, Forum on Religion and Aging (since 1991), American Society on Aging (formerly WGS).
Main field of work: University education, research. **Career:** Executive Director, Vista Volunteer (first Director of the Housing Rehabilitation Program), Revenna OH from 1972 to 1973. Executive Director (part-time), North Shore Interfaith Housing Council, Wilmette IL from 1974 to 1975. Social Worker/Assistant Coordinator, Augustana Hospital, Chicago IL from 1976 to 1977. Social Worker/Special Projects Coordinator, Johnston R Bowman Health Center/RUSH-Presbyterian-St Luke's Hospital, Chicago IL from 1977 to 1980. Chaplain, The Methodist Home, Chicago IL from 1978 to 1980. Consultant to the Oncology Dept., Thorek Hospital, Chicago IL from 1979 to 1981. Project Director at the University of Chicago IL from 1980 to 1984. Parish Associate, Lumen Christi United Presbyterian Church, Naperville IL from 1983 to 1989. Human Service Program Coordinator, National-Louis University, IL from 1983 to 1988. Chair of the Human Service Dept., National-Louis University, IL from 1986 to 1989. **Educ.:** B.A. Soc. from Kent State University, OH, USA (1970-1973); M.S.W from University of Chicago, IL, USA (1974-1976); M.Div. from McCormick Theological Seminary, Chicago IL, USA (1973-1978); Post M.S.W. Cert. Group Psychotherapy from Smith College, School for Social Work, Northampton MA, USA (1979-1980); D.Min. from Chicago Theological Seminary, IL, USA (1981-1983); Ph.D. Candidate from Chicago Theological Seminary, IL, USA (Beginning in 1991). **Interests:** clinical assessment of spiritual wholism. **Hons.:** The Manchester Cup Graduation Award from Kent State University (1973); Outstanding Young Men of America (1980); Men of Achievement (1981). **Mem.:** Academy of Certified Social Workers (1978) **Pubs.:** Articles in professional publications, chapters in books, book reviews, professional papers, books and monographs, including: Elaine Thomopoulos, Christine Roos, James Ellor, Eds., *Organizing a Volunteer Program Serving Seniors In Chicagoland*, Chicago, Concordia Mutual, 1985; With S S Tobin, J W Ellor, S M Anderson-Ray, *Enabling The Elderly: Religious Institutions Within The Community Service System*, Albany, State University of New York Press, 1986 and Editor, with M Kimble, S McFadden, J Ellor, J Seeber, *Handbook on Religion, Spirituality, and Aging* (currently under contract with Fortress Press, expected publication date, Fall 1993).
Mailing address: National-Louis University, 1 South 331 Grace Street, Lombard IL, USA. T. (1 708) 691 9390
Residence: 2609 Algoma Terrace, Waukesha WI 53188, USA. T. (1 414) 549 5752

Elmquist, Karl-Axel
Rel: Christian.
International Organizations: Secretary-General of the Diakonia.
Mailing address: Alvsjö Gardsväg 3, 125 30 Alvsjö, Sweden. T. (46 8) 749 15 00. Fax 99 86 19. C. DIAKONIA. Tx 813 6153.

Elom, Maarten E, Dr
Rel: Christian.
International Organizations: Secretary of the International Council for Pastoral Care and Counselling.
Mailing address: Jan van Nassaustraat 108, 2596 BW Den Haag, Netherlands. T. (31 70) 32 82565.

Elsener, Josef, Rev Fr
Rel: Roman Catholic; Ordained on 25 MAR 1956; b. 26 MAY 1929 in Rorschach, St Gallen, Switzerland; celibate; Citizen of Switzerland; Lang. German, English, French and Spanish.
International Organizations: Superior General (since 1981) of the Bethlehem Mission Society.
Main field of work: Missionary work. **Career:** Vicar-General, Diocese of Gweru, Zimbabwe from 1972 to 1977. Chancellor, Diocese of Gweru, Zimbabwe from 1977 to 1981. **Educ.:** M.A. Soc. from Fordham University, Bronx NY, USA (1956-1958); Cert. Soc. Leadership from Coady International Institute, Antigonish, Canada (1963-1964); Dip. Adult Ed. from University College, Harare, Zimbabwe (1965-1966).
Mailing address: Missionhaus Bethlehem, CH-6405 Immensee, Switzerland. T. (41 41) 81 51 81. Fax 81 69 96.
Residence: See Mailing Address. T. See Mailing Address

Eltink, Merle
Rel: Yoga.
International Organizations: Director of the International Yoga Teachers Association.
Mailing address: 17 King Street, Waverton 2060 NSW, Australia.

Eltscher, Susan M
Rel: Methodist.
International Organizations: Member of the World Methodist Historical Society.
Mailing address: PO Box 127, Madison NJ 07940, USA. T. (1 201) 822 2787.

Emerton, John Adney, Rev Prof
Rel: Anglican; Ordained in 1953; b. 05 JUN 1928 in Winchmore Hill, UK; m. to Norma Elizabeth Bennington; c. 1 boy and 2 girls; Citizen of United Kingdom.
International Organizations: Past Secretary (since 1971) of the International Organization for the Study of the Old Testament.
Other organizations: Regius Professor of Hebrew (since 1968), University of Cambridge; Fellow (since 1970), St John's College, Cambridge; Editor (since 1976), *Vetus Testamentum*.
Career: Ordained deacon in 1952. Assistant Lecturer in Theology, University of Birmingham from 1952 to 1953. Curate, Birmingham Cathedral from 1952 to 1953. Lecturer in Hebrew and Aramaic, University of Durham from 1953 to 1955. Lecturer in Divinity, University of Cambridge from 1955 to 1962. Visiting Professor of Old Testament and Near Eastern Studies, Trinity College, Toronto in 1960. Fellow, St Peter's College and Reader in Semitic Philology, University of Oxford from 1962 to 1968. Visiting Fellow, Institute for Advanced Studies, Hebrew University of Jerusalem in 1983. Visiting Professor of Old Testament, United Theological College, Bangalore in 1986. **Educ.:** B.A. from University of Oxford, Corpus Christi College, UK; M.A. from University of Oxford, Wycliffe Hall, UK; M.A., B.D., D.D. from University of Cambridge, UK. **Hons.:** Honorary Canon, St George's Cathedral, Jerusalem (since 1984); Hon. D.D., Edinburgh (1977). **Mem.:** Fellow, British Academy (1979) **Pubs.:** The Peshitta of the Wisdom of Solomon, 1959; The Old Testament in Syriac – The Song of Songs, 1966 and articles in journals.
Mailing address: St John's College, Cambridge CB2 1TP, UK.
Residence: 34 Gough Way, Cambridge CB3 9LN, UK.

Emmerson, P R
Rel: Unknown.
International Organizations: Executive Secretary of the Bible Lands Society.
Mailing address: PO Box 50, High Wycombe HP15 7QU, UK. T. (44 494) 521351. Fax 462171.

Emonson, Sue
Rel: Buddhist: Mahayana.
International Organizations: Member of the Foundation for the Preservation of the Mahayana Tradition.
Other organizations: Director, Vajrayana Institute.
Mailing address: PO Box 597, Neutral Basy NSW 2089, Australia. T. (61 2) 909 1330. Fax (61 2) 904 1247

Ender, Erwin Josef, H E Mgr

Rel: Roman Catholic; Ordained in 1965; b. 07 SEP 1937 in Steingrund, Breslau, Germany; celibate.

International Organizations: Apostolic Pro-Nuncio to Sudan (since 1990) of the Roman Catholic Church.

Other organizations: Apostolic Delegate to the Red Sea Region (since 1990); Titular Archbishop, Germania di Numida.

Mailing address: Kafouri Belgravia, PO Box 623, Khartoum, Sudan. T. (249 11) 74 692. Tx 26032 NUNC SD

Engelhardt, Klaus, H E

Rel: Protestant; Ordained; b. 11 MAY 1932 in Schillingstadt, Germany; m. to Dorothea Schlink; c. 2 boys and 1 girl; Citizen of Germany.

International Organizations: Bishop (since 1980) of Protestant Church of Baden.

Career: Curate, Evangelische Landeskirche (Protestant Church) from 1960 to 1962. Student priest, Karlsruhe from 1962 to 1965. Professor of Theology, Päd. Hochschule from 1965 to 1980.

Mailing address: Blumenstrasse 1, 76133 Karlsruhe, Germany.

Engstrom, Ted, Dr

Rel: Protestant.

International Organizations: Chairman of the Youth for Christ International.

Mailing address: c/o World Vision, 919 West Huntington Dr, Monrovia CA 91016, USA. T. (1 818) 358 2311. Fax 357 0841.

Enoch, Elizabeth, OSC, Sr

Rel: Roman Catholic; Ordained; celibate.

International Organizations: Member of the Cistercian Order of the Strict Observance.

Other organizations: Member, Association of Contemplative Sisters.

Mailing address: Monastery of St Clare, 142 Hollywood Ave, Bronx NY 10465, USA.

Enrique y Tarancón, Vicente E, H E Card

Rel: Roman Catholic; Ordained on 01 NOV 1929; b. 14 MAY 1907 in Burriana, Castellón, Spain; celibate; Citizen of Spain.

International Organizations: Member (since 1969) of the College of Cardinals.

Other organizations: Archbishop Emeritus (since 1983), Madrid.

Career: Administrative Assistant, Vinaroz from 1930 to 1933. Archpriest, Vinaroz from 1938 to 1943. Archpriest, Villarreal from 1943 to 1946. Bishop, Solsona from 1946 to 1964. Secretary-General, Spanish Bishopric from 1956 to 1964. Archbishop, Ovopie from 1964 to 1969. Archbishop, Toledo, Primate of Spain from 1969 to 1971. Archbishop, Madrid from 1971 to 1983. Created Cardinal by Pope Paul VI in 1969. **Educ.:** Universidad Pontificia Valencia, Spain. **Interests:** musical composition, listening to classical music (particularly Bach). **Pubs.:** Unidad y pluralismo en la Iglesia, 1969; Liturgia y lengua del pueblo, 1970 and El magisterio de Santa Teresa, 1970.

Mailing address: Apartado 333, Mayor de San Jaime 24, 12540 Villareal de los Infantes (Castellón de la Plana), Spain. T. (34 964) 52 19 42

Ens, Harold

Rel: Mennonite.

International Organizations: Associate Secretary of the Mennonite Brethren Missions and Services.

Mailing address: 315 S Lincoln, Hillsboro KS 67063, USA.

Epps, Dwain, Rev

Rel: Christian.

International Organizations: Director, New York Office of the Commission of the Churches on International Affairs of the World Council of Churches.

Mailing address: 777 United Nations Plaza, R9D, New York NY 10017, USA. T. (212) 967 5890. Fax 967 7462.

Eracle, Jean, Rev

Rel: Buddhist: Jôdo-Shinshú; Ordained on 16 JAN 1970; b. 11 DEC 1930 in Geneva, Switzerland; single; Citizen of Switzerland; Lang. French.

International Organizations: President/Founder (since 1970) of the Société bouddhique suisse Jôdo-Shinshú.

Other organizations: Curator, Asia Dept. (since 1969), Museum of Ethnography of Geneva; Rector (Jûshoku), Shingyôji (temple).

Main field of work: Spreading of Buddhism. **Career:** Canon Regular, Saint-Augustine at Royal Abbey of Saint-Maurice. Ordained Roman Catholic Priest in 1955. Professor, Saint-Maurice's College. Vicar, Bagnes. **Educ.:** Studied Phil. and Theol. from Abbaye de Saint-Maurice, Valais, Switzerland. **Interests:** history of fine arts, study of religions. **Mem.:** Swiss Committee of UNESCO Associated Schools (former President); Swiss Committee of UNESCO Associated Clubs (former President); Swiss National Commission for UNESCO (former Member-Expert) **Pubs.:** *La doctrine bouddhique de la Terre Pure*, Paris, Dervy, 1973; *Trois soûtras et un Traité sur la Terre Pure*, Genève, Aquarius, 1984 (introduction and translation) and *Civilisation des Samurai*, Ivrea, Prurli et Verlucca, 1991.

Mailing address: Case Postale 2139, CH-1211 Genève 1, Switzerland.

Residence: Shingyôji, 9 rue de Fribourg, CH-1201 Genève, Switzerland.

Erem, Shimon

Rel: Jewish.

International Organizations: Co-Chairman of the B'nai B'rith International.

Mailing address: 930 Westbourne Drive, Apt 100, Los Angeles CA 90069, USA.

Erickson, Fred Harvey, Rev

Rel: Disciples of Christ; Ordained on 13 JUN 1965; m. to Sally.

International Organizations: Vice-President for Communication (since 1987) of the Creative Consultants, Inc.

Other organizations: Bivocational Pastor (since 1989), Dripping Springs Christian Church (Disciples of Christ), Columbia MO; Training Materials Producer (since 1989), Orscheln Industries, Moberly MO.

Career: Youth Pastor, Alvador Church of Christ, OR from 1963 to 1964. Minister, First Christian Church, Port Orford, OR from 1964 to 1965. Associate Minister, University Park Christian Church, Indianapolis IN from 1965 to 1968. Minister, First Christian Church, Veedersburg IN from 1968 to 1970. Consecutively: Studio Cameraman, Film Editor, and Producer-Director, WISH Television, Channel 8, Indianapolis IN from 1970 to 1973. Director of Communications, Church Federation of Greater Indianapolis, IN from 1973 to 1978. Occasional Lecturer in Communication, Christian Theological Seminary, Indianapolis IN from 1974 to 1986. Manager, Electronic Communication, Office of Communication, Christian Church (Disciples of Christ) from 1978 to 1987. Minister, Arrow Rock Federated Church from 1987 to 1989. Central Committee, World Association for Christian Communication (WACC) from 1984 to 1986. Secretary, North American Regional Association, WACC from 1983 to 1987. Board of Directors, Sandcastles International (radio consortium) from 1983 to 1987. President, Association of Regional Religious Communicators from 1976 to 1978. Board of Managers, Communication Commission, National Council of Churches from 1976 to 1987. General Board, Christian Church (Disciples of Christ) from 1969 to 1974. Board of Directors, Council on Christian Unity from 1973 to 1977. Board of Directors, PuppetVision, Inc. from 1974 to 1978. Board of Directors, New World Communications from 1976 to 1977. **Educ.:** B.Th. from Northwest Christian College, Eugene OR, USA (1965); M.Div. (Hons. in Hebrew) from Christian Theological Seminary, Indianapolis IN, USA (1969); S.T.M. Communication (Magna Cum Laude) from Christian Theological Seminary, Indianapolis IN, USA (1971). **Interests:** photography, camping, SCUBA, canoeing, sailing, Macintosh Computers, telecommunication, home repairs, woodworking, travel. **Hons.:** numerous awards for radio and television spot announcements, and national satellite-delivered cable programs. **Pubs.:** "More than Advertising: Developing a Media Campaign", in *Vangard*, Jan/Feb/March 1984, Vol. 15, No 4, p. 4; "Headlines and Interstate Highways: writing newsletter headlines", in *The Newsletter Newsletter*, October 1988, Vol. 10, No 10, 'How to ..', pp. 1-4 and "Network of Hope", in *The Disciple*, September 1988, Vol. 15, No 9, p. 22.

Mailing address: Creative Consultants Inc, PO Box 86, Hallsville MO 65255, USA. T. (1 314) 696 3373

Residence: 15500 Fernwood Drive, Hallsville MO 65255, USA. T. (1 314) 696 3715

Erickson, John David, Rev Dr

Rel: Lutheran; Ordained in 1959; b. 28 APR 1933 in Wesleyville PA, USA; m. to Nancy Ann; c. 2 boys and 2 girls; Citizen of USA; Lang. English and Japanese.

International Organizations: Secretary-General (since 1991) of the United Bible Societies.

Career: Secretary for Asia, American Bible Society (ABS) from 1965 to 1967. Associate General Secretary, ABS from 1961 to 1968. Executive Secretary, ABS from 1969 to 1976. Deputy General Secretary, ABS from 1976 to 1978. World Service Officer, United Bible Societies (UBS) from 1976 to 1988. General Secretary, ABS from 1978 to 1991. Head of International Relations and Promotion Division, ABS from 1988 to 1991. **Educ.:** B.A. Phil. from Augustana College, Rock Island IL, USA (1955-1957); M.Div. from Augustana Theological Seminary, IL, USA (1957-1959); Master's Program in East and Southeast Asian Stds. from University of Minnesota, Minneapolis, USA (1959-1960); D.D. from Virginia Theological Seminary, Lynchburg, USA (1973). **Hons.:** Outstanding Achievement Award, Augustana College alumni association, Rock Island, IL, USA (1978).

Mailing address: Reading Bridge House, 7th Floor, Reading RG1 8DP, UK. T. (44 734) 500200. C. BIBLES READING. Tx 848541. Fax (44 734) 500857.

Residence: 'Rumsey', Old Bath Road, Sonning RG4 0TA, UK. T. (44 734) 693260

Eriksson, Throstur

Rel: Protestant.

International Organizations: Executive Officer of the Nordic Council for Church Music.

Mailing address: Kleppsvegur 4, 105 Reykjavik, Iceland.

Estades, Aurora, MMB, Sr

Rel: Roman Catholic; Ordained; celibate.

International Organizations: Secretary-General of the Sisters Mercedarian Missionaries of Bérriz.

Mailing address: Viale Pola 10, I-00198 Roma, Italy.

Esteban Denuncio, Raul E, Rev

Rel: Evangelical Lutheran.

International Organizations: President of the United Evangelical Lutheran Church.

Mailing address: José Pedro Varela 3035, 1417 Buenos Aires, Argentina.

Estep, John H (Jack), Dr

Rel: Baptist; Ordained in 1956; b. 30 JUN 1930 in Bellwood PA, USA; m. to Dorothy Nash; c. 1 boy and 1 girl; Citizen of USA; Lang. English.

International Organizations: General Director (since 1980) of the Conservative Baptist Home Mission Society (CBHMS).

Main field of work: Administration and ministry. **Educ.:** B.A. from Wheaton College, IL, USA (1953); M.Div. from Denver Conservative Baptist Seminary, CO, USA (1953-1956). **Hons.:** Hon. D.D., Denver Conservative Baptist Seminary, CO (1980).

Mailing address: PO Box 828, 25W560 Geneva Road, Wheaton IL 60189, USA. T. (1 708) 653 4900

Residence: 34 Brian Drive, Glendale Heights IL 60139, USA.

Etchegaray, Roger, H E Card

Rel: Roman Catholic; Ordained on 13 JUL 1947; b. 25 SEP 1922 in Espelette, France; celibate; Citizen of France.

International Organizations: Past President of the Pontifical Council for Justice and Peace.

President (since 1984) of the Pontifical Council 'Cor Unum'. Member of the Congregation for the Evangelization of Peoples. Member of the Congregation for Catholic Education. Member of the Pontifical Council for Promoting Christian Unity. Member of the Pontifical Council for Social Communications.

Other organizations: Member (since 1979), College of Cardinals.

Career: Served diocese, Bayonne from 1947 to 1960. Assistant Secretary, then Secretary-General, French Episcopal Conference from 1961 to 1970. President, French Episcopal Conference from 1975 to 1981. Archbishop, Marseilles from 1970 to 1984. President, Council of European Episcopal Conferences from 1971 to 1979. Prelate, Mission of France from 1975 to 1981. Created Cardinal by Pope John Paul II in 1979. **Hons.:** Officer of the Legion of Honour. **Pubs.:** Dieu à Marseille, 1976; J'avance comme un âne, 1984 and L'évangile aux couleurs de la vie, 1987.

Mailing address: Piazza San Calisto 16, I-00153 Roma, Italy. T. (39 6) 698 72 68. Fax 698 72 05.

Eugenis, Gay Montalvo

Rel: Roman Catholic.

International Organizations: Secretary of the International Movement of Catholic Lawyers.

Mailing address: 115 Roger de Lliuza, E-Barcelona, Spain.

Euzen, Robert, S.G., Br

Rel: Roman Catholic; Ordained; celibate.

International Organizations: Procurator-General of the Brothers of Christian Instruction of St Gabriel.

Mailing address: Via Trionfale 12840, I-00135 Roma, Italy. T. (39 6) 376 50 49. Fax 376 24 07.

Even-Chen (Edelstein), Yaacov E, Rabbi

Rel: Jewish; Ordained in 1946; b. 23 DEC 1927 in Warsaw, Poland; m. to Nechama; c. 2 boys and 2 girls; Citizen of Israel; Lang. Hebrew, Yiddish and English.

International Organizations: Chairman of the Association of Religious Writers.

Main field of work: Religious literature. **Career:** *Hatzom* (daily newspaper of National Religious Party). **Hons.:** Prime Minister's Award for Literature. **Pubs.:** Editor of annual Yearbook of Religious Writers *Mabua*.

Mailing address: Association of Religious Writers, POB 7440, Jerusalem, Israel.

Residence: Mevo Yorum 4, Jerusalem 92346, Israel.

Evrard, Jean, Fr

Rel: Roman Catholic; Ordained; celibate.

International Organizations: Administrator Delegate of the Societies of African Missions.

Mailing address: Rue Camille Lemonnier 16, B-1060 Bruxelles, Belgium.

Ewing, Tod

Rel: Bahá'í.

International Organizations: Member of the National Spiritual Assembly of the Bahá'ís of the US.

Other organizations: Partner, Hanna Ewing and Associates (management consultants), Columbia SC.

Career: Director of minority affairs at the University of St Cloud MN. Counselor for juvenile and adolescent offenders.

Mailing address: Bahá'í National Center, 536 Sheridan Rd, Wilmette IL 60091, USA.

Eyre, Charles G, Rev

Rel: Methodist; Ordained.

International Organizations: Secretary of the Methodist Church in Ireland.

Mailing address: 3 Upper Malone Road, Belfast BT9 6TD, N Ireland, UK.

Ezaz, Cyrus

Rel: Muslim: Sufi.

International Organizations: Member of the School of Islamic Sufism.

Mailing address: PO Box 5827, Washington DC 20016, USA. T. (1 301) 654 91 91.

Faaalo, Puafitu, Rev

Rel: Church of Tuvalu; Ordained.

International Organizations: General Secretary of the Church of Tuvalu.

Mailing address: Funafuti, Tuvalu.

Fabbri, Fabio, Mgr

Rel: Roman Catholic; Ordained on 22 MAR 1970; b. 03 MAY 1942 in Scarperia, Firenze, Italy; celibate; Citizen of Italy; Lang. Italian, French and Spanish.
International Organizations: Secretary (since 1985) of the International Commission of General Prison Chaplains.
Other organizations: Assistant (since 1977), General Chaplaincy of the Italian Prisons.
Main field of work: Ministry. **Educ.:** Studied Theol. from Pontifical Gregorian University, Vatican City State (1964-1970). **Interests:** music. **Hons.:** Chaplain of His Holiness. **Mem.:** social affiliations **Pubs.:** La presenza della Chiesa in carcere; I Fondamenti spirituali del 'mandato' del Vescovo and Dietro le sbarre per una nuova frontiera.
Mailing address: Ispettorato, Via Giulia 52, I-00186 Roma, Italy. T. (39 6) 686 9331.
Residence: Piano dei Mantellini 60, I-53100 Siena, Italy. T. (39 577) 44 290

Faccani, Clemente, H E Mgr

Rel: Roman Catholic; Ordained on 10 APR 1943; b. 19 OCT 1920 in Lugo, Ravenna, Italy; celibate; Citizen of Vatican and Italy; Lang. Italian, English, French, Latin, Spanish and German.
International Organizations: Apostolic Pro-Nuncio to Kenya (since 1983) of the Roman Catholic Church.
Other organizations: Apostolic Pro-Nuncio to the Seychelles (since 1985); Titular Archbishop, Serra.
Main field of work: Diplomacy. **Educ.:** Dip. Eccl. Dip. from Pontifical Ecclesiastical Academy, Vatican City State; D.S.T., D.C.L. and D.Cn.L. **Hons.:** Commander of the Order of Leopold II (Belgium).
Mailing address: Manyani Road West, PO Box 14326, Nairobi, Kenya. T. (254 2) 442975
Residence: See Mailing Address. T. (254 2) 442975/6/7

Fackenheim, Emil Ludwig, Rabbi Prof

Rel: Jewish; Ordained in 1939; b. 22 JUN 1916 in Halle, Germany; m. to Rose Komlodi; c. 3 boys and 1 girl; Citizen of Canada and Israel; Lang. English, German and Hebrew.
International Organizations: Visiting Professor of Jewish Philosophy, School for Overseas Students (since 1981) of the Hebrew University of Jerusalem.
Main field of work: Philosophy and Jewish thought after the Holocaust. **Career:** Rabbi, Congregation Anshe Sholom, Hamilton, Ontario from 1943 to 1948. Member of the Faculty of Philosophy, University of Toronto from 1948 to 1983. University Professor of Philosophy, University of Toronto from 1979 to 1983. **Educ.:** Rabb. Dip. from Hochschule für die Wissenschaft des Judentums, Berlin, Germany (1935-1939); Student from Martin-Luther-Universität Halle-Wittenberg, Halle, Germany (1937-1938); Student from University of Aberdeen, UK (1939-1940); Ph.D. from University of Toronto, Ontario, Canada (1942-1945). **Interests:** impact of the Holocaust on philosophy and theology. **Hons.:** 5 hon. doctorates; President's Medal, University of Western Ontario (1954); Jewish Welfare Board award for Best Book in Jewish Thought (1969). **Mem.:** Royal Society of Canada; Institute of Contemporary Jewry, Jerusalem (Fellow); Jerusalem Institute of Public Affairs (Fellow) **Pubs.:** The Religious Dimension in Hegel's Thought, 1968, 1970, 1982; Encounters Between Judaism and Modern Philosophy, 1973, 1980 and To Mend the World: Foundations of Post-Holocaust Jewish Thought, 1982.
Mailing address: 3 Alroi St, Apt 7, Jerusalem 92108, Israel. T. (972 2) 635 363
Residence: See Mailing Address. T. See Mailing Address

Fadden, Marie Celeste, Sr

Rel: Roman Catholic; Ordained; celibate.
International Organizations: Member of the Discalced Carmelite Nuns.
Other organizations: Member, Association of Contemplative Sisters.
Main field of work: Artist.
Mailing address: Carmelite Monastery, 1950 La Fond Dr, Reno, NV 89509, USA.

Fagiolo, Vincenzo, H E Mgr

Rel: Roman Catholic; Ordained on 06 MAR 1943; b. 05 FEB 1918 in Segni, Italy; celibate.
International Organizations: President of the Pontifical Council for the Interpretation of Legislative Texts.
Member of the Pontifical Council for the Pastoral Care of Migrants and Itinerant People.
Other organizations: Archbishop Emeritus (since 1984), Chieti-Vasto; One of the Presidents, Pontifical Council, Pontifical Chapel; President, Disciplinary Commission of the Roman Curia.
Career: Consecrated Bishop in 1971.
Mailing address: Palazzo delle Congregazioni, Piazza Pio XII 10, I-00193 Roma, Italy. T. (39 6) 698 4235 – 698 3928
Residence: Via Rusticuci 13, I-00193 Roma, Italy T. (39 6) 698 4032

Fahd Ibn Abdul Aziz, F, King

Rel: Muslim; b. in Riyadh, Saudi Arabia; Citizen of Saudi Arabia.
International Organizations: Custodian (since 1986) of the Two Holy Mosques (Mecca and Medina).
Other organizations: King, Saudi Arabia; Prime Minister (since 1982).
Career: Minister of Education in 1953. Minister of the Interior from 1962 to 1975. Second Deputy Prime Minister from 1967 to 1975. First Deputy Prime Minister in 1975. Became Crown Prince in 1975. Succeeded to the throne on the death of his brother in 1982.
Mailing address: Royal Diwan, Riyadh, Saudi Arabia.

Fahim, Amin, Maître

Rel: Coptic Catholic; Unordained; b. 08 APR 1925 in Cairo, Egypt; m. to Thérèse H Doss; c. 3 boys; Citizen of Egypt; Lang. Arabic, French, English and Italian.
International Organizations: President (since 1989) of the Conference of International Catholic Organizations.
President (since 1989) of the International Catholic Child Bureau. Member (since 1986) of the Pontifical Council for Culture.
Other organizations: President Director-General (since 1965), Christian Association of Upper Egypt for Schools and Social Promotion; Consultant to the Pontifical Council of Laity (since 1990).
Main field of work: Education and development. **Career:** Chief, Middle East desk, *La Bourse Egyptienne* (a daily French-language Egyptian newspaper) from 1953 to 1956. President, Syndicate, Société Orientale de Publicité from 1953 to 1956. Private Lawyer from 1956 to 1966. Editor-in-Chief, *Le Rayon d'Egypte* (a French-language Egyptian Catholic newspaper) in 1956. **Educ.:** B.A. Law from Ecole Française de Droit du Caire, Egypt (1953); B.A. Law from Cairo University, Egypt (1956). **Interests:** politics of the Middle East, dialogue and mutual understanding between religions, children's rights (especially in the deprived areas where poverty and illiteracy prevail). **Hons.:** Great Cross of St Gregory the Great (Vatican); Commander of St Silvester with Plaque (Vatican). **Mem.:** International Commission of Dialogue between the Catholic and Coptic Orthodox Churches (1974) **Pubs.:** *Egyptians AND Christians*, History of a Movement (under preparation) and several unpublished lectures in various languages.
Mailing address: 85A Ramsis Ave, Cairo 11599, Egypt. T. (20 2) 77 22 42
Residence: 50 Nazih Kalifa Street, Heliopolis, Cairo, Egypt. T. (20 2) 258 06 23

Fain Binda, Luigi F, F.S.M.I., Fr

Rel: Roman Catholic; Ordained; celibate.
International Organizations: Superior General of the Fils de Sainte Marie Immaculée.
Mailing address: Via del Mascherone 55, I-00186 Roma, Italy. T. (39 6) 687 59 15

Fair, Ian A, Prof Dr

Rel: Church of Christ; Ordained on 10 JAN 1959; b. 03 OCT 1934 in Pietermaritzburg, Natal, South Africa; m. to Moira June Stant; c. 3 boys; Citizen of USA; Lang. English, Afrikaans, Dutch and Zulu.
International Organizations: Professor of New Testament/Dean of College of Biblical Studies (since 1985) of the Abilene Christian University, TX.
Other organizations: Distinguished Scholar (since 1992), R S Bell Chair for Biblical Studies.
Main field of work: University administration, New Testament theology. **Educ.:** B.A. Bible from Abilene Christian University, TX, USA (1965-1968); B.A. (Hons) from University of Natal, Durban, South Africa (1969-1970); M.A. from University of Natal, Durban, South Africa (1971-1972); Ph.D. from University of Natal, Durban, South Africa (1972-1974). **Mem.:** Society of Biblical Literature; American Academy of Religion
Mailing address: Abilene Christian University, PO Box 8400, Station ACU, Abilene TX 79699, USA. T. (1 915) 674 3700
Residence: 901 Green Valley Drive, Abilene TX 79601, USA. T. (1 915) 672 7694

Fairbairn, William D

Rel: Protestant.
International Organizations: Director of the Inter-Church Committee on Human Rights in Latin America.
Mailing address: Suite 201, 40 St Clair Ave E, Toronto M4T 1M9, Canada. T. (1 416) 921 41 52.

Fairweather, Eugene Rathbone, Rev

Rel: Anglican; Ordained in 1944; b. 02 NOV 1920 in Ottawa, Ontario, Canada; Citizen of Canada.
International Organizations: Canon (since 1969) of the St James Cathedral, Toronto.
Other organizations: Sessional Lecturer in Anglican Studies, etc. (since 1988), Atlantic School of Theology.
Career: Ordained Deacon in 1943. Assistant Curate, St Matthew's Church, Toronto, Ontario from 1943 to 1947. Fellow and Tutor in Divinity, Trinity College, Toronto from 1944 to 1947. Assistant, Cathedral of St John the Divine, New York from 1948 to 1949. Associate Professor of Dogmatic Theology and Ethics, Trinity College, Toronto from 1949 to 1964. Bishop Paddock Lecturer, General Theological Seminary from 1955 to 1956. Visiting Lecturer, St Augustine's College, Canterbury in 1955. Member, World Faith and Order Commission, World Council of Churches from 1960 to 1975. Editor, *Canadian Journal of Theology* from 1960 to 1970. Anglican Delegate/Observer to Vatican Council II from 1964 to 1965. Keble Professor of Divinity, Trinity College, Toronto from 1964 to 1986. Dean of Divinity, Trinity College, Toronto from 1983 to 1985. Examining Chaplain to Bishop, Toronto from 1965 to 1986. Visiting Professor of Anglican Studies, General Theological Seminary in 1987. **Educ.:** B.A. from McGill University, Montreal, Canada (1941); M.A. from University of Toronto, Ontario, Canada (1943); B.D. from University of Toronto, Trinity College, Canada (1945); S.T.M. from Union Theological Seminary (1948); Th.D. from Union Theological Seminary (1949). **Interests:** music, swimming, travel. **Hons.:** Hon. D.D., McGill University (1971), Huron College (1973), University of King's College (1973), Montreal Diocesan Theological College (1985), Regis College (1986), St Michael's College (1988). **Mem.:** Canadian Society of Biblical Studies (President 1954-55); Canadian Theological Society (President 1957-58); American Theological Society (President 1973-74); Canadian Society of Patristic Studies (Honorary President 1977-79) **Pubs.:** Episcopacy Re-Asserted, 1955; Contributor to Early Christian Fathers, 1953 and articles in various religious journals.
Mailing address: RR 3, Wolfville NS B0P 1X0, Canada.

Falcao, José Freire, H E Card

Rel: Roman Catholic; Ordained on 19 JUN 1949; b. 23 OCT 1925 in Ererê, Brazil; celibate.
International Organizations: Member (since 1988) of the Administrative Hierarchy of the Roman Catholic Church.
Second Vice-President of the Latin American Episcopal Council. Member of the Pontifical Commission for Latin America. Member of the Pontifical Council for Promoting Christian Unity. Member of the Pontifical Council for Pastoral Assistance to Health Care Workers.
Other organizations: Archbishop, Brasilia.
Career: Created Cardinal by Pope John Paul II in 1988.
Mailing address: QL 12-Cj12, Lote 1, Lago Sul, 71600 Brasília DF, Brazil. T. (55 61) 248 3455

Falck, Alberto

Rel: Christian.
International Organizations: Vice-President of the International Christian Union of Business Executives.
Mailing address: c/o Acciaierie e Ferriere Lombarde Falck, Corso Venezia 48, I-20121 Milano, Italy.

Falconer, Alan David, Rev Prof

Rel: Church of Scotland; Ordained on 12 DEC 1972; b. 12 DEC 1945 in Edinburgh, Scotland, UK; m. to Marjorie Ellen Walters; c. 2 boys and 1 girl; Citizen of United Kingdom; Lang. English.
International Organizations: Director (since 1990) of the Irish School of Ecumenics.
Other organizations: Chairman (since 1986), British and Irish Churches Human Rights Forum; Consultant to Reformed-Roman Catholic International Dialogue (since 1986).
Main field of work: University education, ecumenics. **Career:** Assistant Minister, St Machar's Cathedral, Aberdeen from 1972 to 1974. Lecturer, Irish School of Ecumenics from 1974 to 1987. Dean of Studies, Irish School of Ecumenics from 1987 to 1990. President, European Society for Ecumenical Research from 1986 to 1990. Consultant to World Council of Churches Assembly, Canberra in 1991. **Educ.:** M.A. from University of Aberdeen, UK (1964-1967); B.D. from University of Aberdeen, UK (1967-1970); Cert. Ecum. from University of Geneva, Switzerland (1970-1971). **Interests:** ecumenical theology. **Mem.:** Irish Theological Association; Irish Council of Churches (Board of Interchurch Affairs) **Pubs.:** Editor, Understanding Human Rights, Dublin, Irish School of Ecumenics, 1980; Editor, Reconciling Memories, Dublin, Columba Press, 1987 and A Man Alone, Dublin, Columba Press, 1986.
Mailing address: Irish School of Ecumenics, Milltown Park, Sandford Rd, Dublin 6, Ireland. T. (353 1) 269 8607
Residence: 25 Orwell Park Rise, Dublin 6W, Ireland. T. (353 1) 509 325

Faley, Roland James, TOR, Fr

Rel: Roman Catholic; Ordained on 09 JUN 1956; celibate.
International Organizations: Spiritual Director (since 1991) of the Diocese of Austin, TX.
Member of the Third Order Regular of St Francis.
Career: Professor of Sacred Scripture and Biblical Languages, St Francis Seminary, Loretto PA from 1961 to 1971. Professor of Sacred Scripture and Biblical Languages, St Thomas University (Angelicum), Rome from 1973 to 1976. Visiting Professor, Catholic University of America, Washington DC from 1966 to 1969. Visiting Professor, LaSalle College, Philadelphia PA in 1966. Visiting Professor, Pittsburgh Theological Seminary, PA in 1969. Rector, St Francis Seminary, Loretto PA in 1970. Vicar-General, Third Order Regular of St Francis from 1971 to 1977. Superior General, Third Order Regular of St Francis from 1977 to 1983. Associate Pastor, St Michael, Bedford TX from 1983 to 1985. Executive Director, Conference of Major Superiors of Men from 1985 to 1990. **Educ.:** B.A. from Saint Francis College, Loretto PA, USA (1952); S.T.L. from Catholic University of America, Washington DC, USA (1957); S.S.L. from Pontifical Biblical Institute, Vatican City State (1959); S.T.D. from Pontifical University San Tommaso d'Aquino, Vatican City State (1960). **Hons.:** Dr. of Humanities h.c., St Francis College (1985). **Pubs.:** *The Cup of Grief* (sin, suffering, death), Alba House, c. 1977; *To Come and See* (Religious life today), Franciscan Herald Press, Fall Publication, 1984 and articles in *Thought*, *Theology Digest*, *The Bible Today*, *Review for Religious*, *Analecta T.O.R.*, *America*.
Mailing address: Emmaus, 3635 Ranch Road 620 South, Austin TX 78734, USA.

Falkenberg, James R

Rel: Protestant.
International Organizations: President of the Bible Literature International.
Mailing address: 625 E North Broadway, Columbus OH 43214, USA. T. (1 614) 267 3116.

Falowski, Jan, Br

Rel: Roman Catholic; Ordained; celibate.
International Organizations: Superior General of the Serviteurs de Marie-Immaculée.
Mailing address: ul Browarna 14, PL-26 420 Nowe Miasto nad Pilica, Poland.

Falquet, Claude

Rel: Roman Catholic.
International Organizations: Secretary-General of the Institute of the Heart of Jesus.
Mailing address: 202 avenue du Maine, F-75014 Paris, France. T. (33 1) 45 40 45 51.

Falwell, Jerry, Rev

Rel: Baptist; Ordained; b. 11 AUG 1933 in Lynchburg VA, USA; m. to Macel Pate; c. 2 boys and 1 girl; Citizen of USA.
International Organizations: Founder of the Moral Majority, Inc.
Other organizations: Senior Pastor, Thomas Road Baptist Church, Lynchburg VA; President, Old Time Gospel Hour, Inc.; Chancellor, Liberty University, Liberty Baptist Theological Seminary, and Liberty Home Bible Institute.

Main field of work: Fundamentalist pastor, television evangelist, political activist. **Career:** Converted to the Christian faith in 1952. Founded the Thomas Road Baptist Church, Lynchburg VA in 1956. Began weekly radio broadcasts of church services. Founded the Lynchburg Christian Academy in 1967. Began taping church services for television (programs later called 'The Old-Time Gospel Hour') in 1968. Founded Liberty University (formerly Liberty Baptist College) in 1971. Founded Liberty Baptist Theological Seminary in 1973. Founded Home Bible Institute in 1976. Founded the Moral Majority, Inc. (included a legislative lobby, educational activities and a political-action agency for promoting conservative moral and political concerns) in 1979. Remained a leading spokesman for conservative views in both church and state throughout the 1980s. Resigned from the presidency, the Moral Majority in 1987. Announced the dissolution, the Moral Majority in 1989. **Educ.:** Th.G. from Baptist Bible College, Springfield MO, USA (1956). **Hons.:** Hon. D.D., Tennessee Temple Theological Seminary; Hon. D.Litt., California Graduate School of Theology; Hon. LL.D., Central University, Seoul, Korea; Clergyman of the Year Award, Religious Heritage (1979); Jabotinsky Centennial Medal (1980); and numerous other awards. **Pubs.:** *Listen, America!*, Doubleday, 1980; *Stepping Out on Faith*, Tyndale, 1984 and *Strength for the Journey*, Simon and Schuster, 1987 (autobiography).
Mailing address: Thomas Road Baptist Church, Lynchburg VA 24514, USA.

Fanchette, Philippe, Fr

Rel: Roman Catholic; Ordained on 29 JUL 1962; b. 30 JAN 1937 in Beau-Bassin, Mauritius; celibate; Citizen of Mauritius; Lang. Creole, French, English, Italian and Spanish.
International Organizations: Executive Secretary for Adult Education (since 1988) of the World Council of Churches.
Main field of work: Education. **Educ.:** Lic. Phil. and Theol. from Pontifical Gregorian University, Vatican City State. **Interests:** Peoples movements, culture, social analysis for grass-roots. **Mem.:** Fellow, Ecumenical Institute for Development of Peoples (1975-1987)
Mailing address: 150 route de Ferney, PO Box 2100, CH-1211 Genève 2, Switzerland. T. (41 22) 791 61 11
Residence: 60 rue de Berne, CH-1201 Genève, Switzerland T. (41 22) 738 00 84

Fang, C N, Bishop

Rel: Methodist.
International Organizations: Bishop of Methodist Church in Malaysia.
Mailing address: 23 Jalan Mayang, Kuala Lumpur, Malaysia.

Fani, Italo, C.S.F., Fr

Rel: Roman Catholic; Ordained; celibate.
International Organizations: Procurator-General of the Congrégation de la Sainte Famille de Bergame.
Mailing address: Via Filippo Tajani 50, 00149 Roma, Italy. T. (39 6) 558 46 65

Fániran, Joseph Oládèyo, Fr

Rel: Roman Catholic; Ordained; celibate.
International Organizations: Executive Secretary of the Pan African Episcopal Committee for the Press.
Mailing address: SECAM Secretariat, 4 Senchi Street, PO Box 9156, Airport Residential Area, Accra, Ghana. T. (233 21) 777706 – 774724. C. SECAM ACCRA. Tx 2587 SECAM GH. Fax 772548.

Fantini, Augusto

Rel: Roman Catholic.
International Organizations: Secretary of the Pontifical Institute of Sacred Music.
Mailing address: Via di Torre Rossa 21, I-00165 Roma, Italy. T. (39 6) 663 87 92.

Farah, Augustin Joseph, H E Mgr

Rel: Catholic; Ordained in 1940; b. in Qara, Syria; Citizen of Syria.
International Organizations: Archbishop (since 1957) of Tripoli.
Career: Entered the Order, the Chouerites Basilian Fathers in 1922. Director, the Seminarists in 1940. Superior, Saint Jan Seminary and Professor of French and Latin from 1940 to 1946. Vice-Principal and Professor of Philosophy, Oriental College of Zahle in 1946. Superior, the Saint Elie Convent of Zahle from 1947 to 1948. Elected Superior, Order of the Chouerites Basilian Fathers in 1949. Visited France and the Vatican in 1951. Patriarchal Vicar at Irbed in 1951. Re-elected General Superior, His Order in 1956. Apostolic visitor in America in 1957. Founder and Director, many schools and convents in Lebanon, Syria and America in 1957. **Pubs.:** *Le Saint Esprit d'après l'Enseignement des Saint Basile* (thesis) and author of many articles on theology.
Mailing address: Archbishop's House of Tripoli, Christians Street, Tripoli, Lebanon.

Färber, Egon, M.S.F., Fr

Rel: Roman Catholic; Ordained; celibate.
International Organizations: Superior General of the Congregation of the Missionaries of the Holy Family.
Mailing address: Via di Villa Troili 56, I-00163 Roma, Italy. T. (39 6) 625 85 46.

Farhat, Edmond, H E Mgr

Rel: Roman Catholic; Ordained on 28 MAR 1959; b. 20 MAY 1933 in Ain Kfaa, Lebanon; celibate.
International Organizations: Apostolic Delegate to Libya and Apostolic Pro-Nuncio to Algeria and Tunisia (since 1989) of the Roman Catholic Church.
Other organizations: Titular Archbishop, Biblo.
Mailing address: 1 rue Noureddine Mekiri, 16090 Bologhine-Algiers, Algeria. T. (213 2) 62 34 30. Fax 57 23 75

Farina, Raffaele, Rev Prof Dr

Rel: Roman Catholic; Ordained on 01 JUL 1958; b. 24 SEP 1933 in Buonalbergo, Benevento, Italy; celibate; Citizen of Italy; Lang. Italian, German and English.
International Organizations: Rector (since 1992) of the Salesian Pontifical University.
Other organizations: President, Editorial Commission (since 1991), Libreria Editrice Vaticana; Director (since 1992), *Salesianum* (four-month revue).
Main field of work: University education. **Career:** Moderator, General Chapter of the Salesian Congregation from 1977 to 1978. Rector, Salesian Pontifical University from 1977 to 1983. Secretary, Pontifical Committee of Historical Sciences from 1981 to 1990. Director, Archivio Salesiano Centrale from 1984 to 1990. Undersecretary, Pontifical Council for Culture from 1986 to 1991. **Educ.:** Dip. Hist. and Phil. from University of Naples, Italy (1952-1956); L.Th. from Pontificio Ateneo Salesiano, Turin, Italy (1954-1958); Dr. Hist. of the Church from Pontifical Gregorian University, Vatican City State (1958-1965). **Interests:** history of the Ancient Church. **Mem.:** Pontifical Committee of Historical Sciences **Pubs.:** L'Impero e l'Imperatore cristiano in Eusebio di Cesarea, Zürich, Pas-Verlag, 1966; Bibliografia origeniana, SEI, Turin, 1970 and Metodologia, Libreria Ateneo Salesiano, Rome, 1986, 4th Edn.
Mailing address: Università Pontificia Salesiana, Piazza Ateneo Salesiano 1, I-00139 Roma, Italy. T. (39 6) 881 20 41

Farinch, Manuel

Rel: Theosophist.
International Organizations: President of the Inter-American Theosophical Federation.
Mailing address: Florencio Balcarce 71, 1405 Buenos Aires, Argentina.

Farkas, Ervin

Rel: Jewish.
International Organizations: Executive Vice-President of the World Federation of Hungarian Jews.
Mailing address: 136 E 39th Street, New York NY 10016, USA. T. (1 212) 683 5377.

Farmad, F, Dr

Rel: Muslim.
International Organizations: Head, Secretariat of the Islamic Thought Foundation.
Mailing address: PO Box 14155-3987, Teheran, Iran Islamic Rep. T. (98 21) 844 09 26.

Farrell, Seamus, Fr
Rel: Roman Catholic; Ordained; celibate.
International Organizations: Contact of the Irish Missionary Union.
Mailing address: IMU, Orwell Park, Rathgar, Dublin 6, Ireland. T. (353 1) 97 17 70 – 96 54 33. Fax (353 1) 96 50 29.

Farrelly, J P
Rel: Christian.
International Organizations: Membership and Conference Secretary of the Ecumenical Society of the Blessed Virgin Mary.
Mailing address: 11 Belmont Road, Wallington, UK. T. (44 81) 647 5992.

Farris, Stephen Charles, Rev Prof
Rel: Presbyterian; Ordained on 20 SEP 1981; b. 21 JUN 1951 in Brampton ON, Canada; m. to Patricia Jean White; c. 2 boys; Citizen of Canada; Lang. English, French and German (some).
International Organizations: Member of the Executive Committee (since 1989) of the World Alliance of Reformed Churches.
Theological Committee, Caribbean/North American Area Council (since 1989) of the World Alliance of Reformed Churches.
Other organizations: Board Member (since 1984), various Canadian social agencies; Professor of Preaching and Worship (since 1986), Knox College, Toronto School of Theology, University of Toronto; Chair (since 1990), Toronto School of Theology.
Main field of work: University education. **Career:** Church Doctrine Committee, Presbyterian Church in Canada from 1983 to 1989. Convenor, Committee on Human Sexuality, Presbyterian Church in Canada from 1988 to 1991. Faith and Order Commission, Canadian Council of Churches from 1986 to 1990. **Educ.:** B.A. from University of Toronto, Ontario, Canada (1969-1979); D.Min. from Union Theological Seminary, Richmond VA, USA (1973-1977); Th.M. from Union Theological Seminary, Richmond VA, USA (1977-1978); Ph.D. from University of Cambridge, UK (1978-1982). **Mem.:** Academy of Homiletics, Canadian Society of Biblical Studies **Pubs.:** The Hymns of Luke's Infancy Narratives, Sheffield, JSOT, 1985; "It's About Time", in B G White, *A New Teaching, New Preaching*, Toronto, United Church Publishing House, 1991 and "Worship", "The Song of Zechariah", "The Song of Simeon", in *Dictionary of Jesus and the Gospels*, Downer's Grove, IVP, 1991.
Mailing address: c/o Knox College, 59 St George Street, Toronto ON M5S 2E6, Canada. T. (1 416) 978 2787
Residence: 70 Livingston Rd, West Hill ON M1E 1L4, Canada. T. (1 416) 261 7441

Fata, Mohamad Walis
Rel: Muslim.
International Organizations: President, Publication Department of the Ministry of Islamic Affairs, Afghanistan.
Mailing address: Ministry of Islamic Affairs and Pious Foundation Organisation, Kabul, Afghanistan.

Fathea'zam, Hushmand
Rel: Bahá'í; b. 08 JAN 1924 in Teheran, Iran Islamic Rep; m. to Shafigheh Fathea'zam; c. 3 boys; Citizen of Canada; Lang. Persian and English.
International Organizations: Member, Universal House of Justice (since 1963) of the Bahá'í International Community.
Main field of work: Bahá'í administration. **Career:** Curator, Library of the Faculty of Art, University of Teheran, Iran from 1950 to 1952. Lecturer in Persian Literature, Camp College, Punjab University, New Delhi from 1959 to 1963. Staff Artist, All-India Radio, New Delhi from 1955 to 1963. Managing Director, Bahá'í Publishing Trust, New Delhi from 1955 to 1963. Secretary, National Spiritual Assembly of the Bahá'ís of India from 1955 to 1963. **Educ.:** M.A. from University of Teheran, Iran (1942-1945); M.A. from Visva-Bharati University, Santiniketan, India (1952-1954). **Pubs.:** *The New Garden* (Introduction to the Bahá'í Faith), first published in India, 1962, translated into 95 languages (54 of which are published to date) and *Namnam-i-Baran* (Collection of poems), published in India, 1986.
Mailing address: PO Box 155, 31001 Haifa, Israel. T. (972 4) 358 358
Residence: 7 Abraham Lincoln, 31001 Haifa, Israel T. (972 4) 244 834

Faul, Denis O'Beirne, Rev
Rel: Roman Catholic; Ordained in 1956; b. 14 AUG 1932 in Dundalk, Co. Louth, Ireland; celibate; Citizen of Ireland.
International Organizations: Principal (since 1983) of the St Patrick's Academy, Dungannon.
Career: Teacher at St Patrick's Academy, Dungannon, Co. Tyrone from 1958 to 1983. **Interests:** books, modern languages. **Pubs.:** 15 books and 20 leaflets on Northern Ireland problems and papers on Patristics and history of Irish Catholic church.
Mailing address: St Patrick's Academy, Dungannon, Co. Tyrone, N Ireland, UK. T. (44) 22165 – 22668

Fay, Patrick
Rel: Roman Catholic.
International Organizations: Secretary of the Legion of Mary.
Mailing address: De Montfort House, Brunswick Street, Dublin 7, Ireland. T. (353 1) 723153 – (353 1) 725093. C. LEGION, DUBLIN.

Felderer, Ditlieb
Rel: Jehovah's Witnesses.
International Organizations: Executive Director of the Jehovah's Witnesses Information.
Mailing address: Marknadsvagen 289, S-183 34 Täby, Sweden. T. (46 8) 768 13 98.

Feldkämper, Ludger Bernhard, Fr Dr
Rel: Roman Catholic; Ordained on 02 FEB 1963; b. 16 MAY 1937 in Mesum, Nordrhein-Westfalen, Germany; celibate; Citizen of Germany; Lang. German, English, Italian, Spanish and French.
International Organizations: Secretary-General (since 1984) of the Catholic Biblical Federation.
Main field of work: Teaching scripture, biblical-pastoral ministry. **Career:** Professor of Sacred Scripture, Immaculate Conception School of Theology, Vigan, Philippines from 1967 to 1972. Professor of Sacred Scripture, Immaculate Conception School of Theology from 1978 to 1983. Dean of Studies, Immaculate Conception School of Theology from 1968 to 1969. Spiritual Director, Immaculate Conception School of Theology from 1969 to 1972. Founder and Director, John Paul I Biblical Center, Vigan from 1979 to 1983. **Educ.:** S.T.L. from Pontifical Gregorian University, Vatican City State (1963-1964); S.S.L. from Pontifical Biblical Institute, Vatican City State (1964-1967); S.S.D. from Pontifical Biblical Institute, Vatican City State (1973-1977). **Mem.:** Catholic Biblical Association of America **Pubs.:** "Der betende Jesus als Heilsmittler nach Lukas", Veröffentl. des Missionspriesterseminars St Augustin 29, 1978, pp. 370; "Bibeltheologische Überlegungen zum Begriff der Evangelisierung anhand von LK 4, 16-30", in *Verbum*, svd 22, 1981, pp. 153-174 and *Basic Bible Seminar* – Handbook for core teams.
Mailing address: Mittelstrasse 12, Postfach 10 52 22, 70180 Stuttgart, Germany. T. (49 711) 60 40 99. Fax (49 711) 640 56 44.
Residence: Neckarstrasse 9, 70736 Fellbach-Öffingen, Germany. T. (49 711) 51 26 72

Feldmann, Adrian Roy
See Thubten Chodron.

Felici, Angelo, H E Card
Rel: Roman Catholic; Ordained on 04 APR 1942; b. 26 JUL 1919 in Segni, Italy; celibate; Citizen of Italy.
International Organizations: Prefect (since 1988) of the Congregation for the Causes of Saints.
Member of the Sacred Congregation for the Oriental Churches. Member of the Congregation for Bishops. Member of the Congregation for the Evangelization of Peoples. Member of the Congregation for the Clergy. Member of the Pontifical Council for Promoting Christian Unity.
Other organizations: Member (since 1988), College of Cardinals; Member, Council of Cardinals and Bishops.
Career: Elected Archbishop, Cesariana, Numidia in 1967. Consecrated Bishop in 1967. Created Cardinal by Pope John Paul II in 1988.
Mailing address: Piazza della Città Leonina 9, I-00193 Roma, Italy. T. (39 6) 689 2969

Felicitas, Mary, Sr

Rel: Roman Catholic; Ordained in 1934; b. 18 JAN 1916 in Fife Lake, Sakskatchewan, Canada; celibate; Citizen of Canada.

International Organizations: Member of the Sisters of Providence of St Vincent de Paul.

Main field of work: Counsellor and spiritual director. **Career:** Joined Sisters of Providence of St Vincent de Paul. Appointed Director of Nursing, St Mary's Hospital, Montreal and its Nursing School in 1945. 1st Vice-President, Association of Nurses of the Province of Quebec from 1950 to 1953. Chairman, Editorial Board, Canadian Nurses Journal from 1960 to 1962. Chairman, Canadian Conference of Catholic Schools of Nursing from 1960 to 1964. President, Canadian Nurses' Association from 1967 to 1970. Served as General Councillor, Sisters of Providence of St Vincent de Paul from 1971 to 1977. **Educ.:** B.Sc. from University of Ottawa, Ontario, Canada; M.Sc. from Catholic University of America, Washington DC, USA (1953); LL.D. from Queen's University at Kingston, Ontario, Canada (1974). **Interests:** music, crochet, crossword puzzles, reading. **Mem.:** Pi Gamma Mu (USA); Sigma Theta Tau (Kappa Chapter, Washington DC)

Mailing address: Providence Mother House, Box 427, Kingston, Ontario K7L 4W4, Canada.

Feliú Loriente, María del Pilar F, STJ, Sr

Rel: Roman Catholic; Ordained; b. 31 JAN 1931 in Zaragoza, Spain; celibate; Citizen of Spain; Lang. Spanish, Italian, Portuguese and French.

International Organizations: Superior General (since 1981) of the Society of St Teresa of Jesus.

Member (since 1951) of the Society of St Teresa of Jesus.

Other organizations: Professor at the schools (since 1959), Society of St Teresa of Jesus; Director (since 1965), *Jesus Maestro* (magazine), Ediciones STJ, Society of St Teresa of Jesus; General Coordinator for Education (since 1975), Society of St Teresa of Jesus.

Main field of work: Educational activities. **Educ.:** Lic. Phil. from University of Barcelona, Spain (1954-1959); Lic. Jour. from University School, Madrid, Spain (1965-1968). **Interests:** education, writing. **Pubs.:** numerous articles published in *Jesus Maestro* (twice-monthly magazine), Teresian Educational Community (1954-75).

Mailing address: Via Valcannuta 134, I-00166 Roma, Italy. T. (39 6) 663 7053.

Fell, Arnold A, Rev

Rel: Roman Catholic; Ordained; celibate.

International Organizations: Secretary of the Holy Cross Foreign Mission Society.

Mailing address: PO Box 543, Notre Dame IN 46556, USA. T. (1 219) 239 5477.

Femenia, Marisa

Rel: Buddhist: Mahayana.

International Organizations: Member of the Foundation for the Preservation of the Mahayana Tradition.

Other organizations: Director, Nagarjuna C.E.T. Madrid.

Mailing address: Costanilla de los Angeles, N.2/escalera izda./3-dcha, 28013 Madrid, Spain. T. (34 91) 541 3755. Fax (34 91) 429 6708

Feng, Allan C J, Prof Dr

Rel: Methodist; b. 07 JUL 1933 in Sunchang, Fukien, China; m. to Chi Hwa Liang; c. 1 girl; Citizen of USA; Lang. English and Chinese.

International Organizations: Secretary-General (since 1983) of the Association of Christian Universities and Colleges in Asia.

Other organizations: Dean, School of Law (since 1985), Tunghai University, Taichung, Taiwan.

Main field of work: University education. **Career:** Dean, School of Social Sciences, Pennsylvania State University, University Park from 1972 to 1981. **Educ.:** LL.B. from National Taiwan University, Taipei, Taiwan (Rep. of China) (1950-1954); M.A. from University of California, Los Angeles, USA (1958-1960); Ph.D. from New York University, USA (1962-1965). **Pubs.:** From Developing to Developed – The Taiwan Model; N/ES and American Foreign Policy and The Limitation of Power: America's Dilemma in Asia.

Mailing address: Box 5-371, Taichung, Taiwan (Rep. of China). T. (886 4) 359 2824.

Residence: 47 Tunghai Road, Taichung, Taiwan (Rep. of China). T. (886 4) 359 0423.

Fenske, Elizabeth DeLoris Wall (Pat), Dr

Rel: United Church; Ordained on 30 OCT 1960; b. 14 SEP 1929 in Savannah GA, USA; m. to Paul Brecht Fenske; Citizen of USA; Lang. English.

International Organizations: Executive Director and President Emeritus (since 1986) of the Spiritual Frontiers Fellowship International. President (since 1987) of the International Association for Near-Death Studies. Representative to the Non-Governmental Organizations Section (since 1990) of the United Nations.

Other organizations: Private Practice of Psychotherapy (since 1974); Program Chairperson (since 1979), Academy of Religion and Psychical Research, Bloomfield CT; Advisory Council (since 1986), Paul Brunton Philosophic Foundation, Hector NY; Board Member (since 1987), Center for International Cooperation, Washington DC.

Career: Assistant in College Work, Presbyterian Campus Ministry, Iowa State University, Ames from 1952 to 1953. Faculty Member at Chung Chi College, Hong Kong from 1956 to 1959. Visited Universities and Colleges under the auspices, National Student Christian Federation from 1959 to 1960. Protestant Minister to Drexel University, Philadelphia PA from 1960 to 1969. Independent study in the fields of Psychology, Parapsychology and Religion with special emphasis on the work of the Swiss Psychiatrist Carl G Jung, M.D. from 1969 to 1974. President, Spiritual Frontiers Fellowship International from 1980 to 1986. International Board of Advisors, Foundation for Universal Unity, Loveland CO from 1982 to 1984. President, USA Chapter, International Institute for Integral Human Sciences, Burlington VT from 1983 to 1984. Member, World Unity Forum, Wyoming MI from 1984 to 1988. President, Center for International Cooperation, Washington DC from 1987 to 1990. Co-President, Human Unity Conference, Philadelphia PA in 1987. **Educ.:** B.Sc. Psy. and Soc. from Georgia College, Milledgeville GA, USA (1947-1951); M.Div. from Eden Theological Seminary, Webster Groves MO, USA (1953-1956); Ph.D. Psy. from Golden State University, Los Angeles CA, USA (1982-1987). **Interests:** travel, reading, collecting, growing plants and flowers. **Hons.:** Award, Drexel University Chapter of the American Association of University Professors (1968). **Mem.:** International Association for Near-Death Studies, UK (Advisory Board); Scientific and Medical Network (UK); The Association for Humanistic Psychology; The Association for Transpersonal Psychology; Round Table Associates (Analytical Psychology) **Pubs.:** Author of numerous articles and author/editor of *Spiritual Insights for Daily Living.*

Mailing address: 3310 Baring Street, Philadelphia PA 19104, USA. T. (1 215) 222 0916

Residence: See Mailing Address. T. See Mailing Address

Fenzl, Friedrich R, Shaku

Rel: Buddhist: Pure Land; Ordained on 15 DEC 1960; b. 04 NOV 1932 in Kierberg/M., Bohemia, Czechoslovakia; single; Citizen of Austria; Lang. German, English and Japanese.

International Organizations: Chairman (since 1977) of the Buddhist Society of Salzburg.

Other organizations: Representative (since 1965), Jôdo Shinshû Austria; Publishes *Amida* and *Rundbriefe* (German Buddhist papers).

Career: Deputy Chairman, Austrian Buddhist Association from 1988 to 1990. **Educ.:** Shin Buddhism Stds. from Ryukoku University, Kyoto, Japan (1968-1970). **Interests:** Buddhist social ethics. **Mem.:** World Congress on Religion and Peace (WCRP); Buddhist Union of Europe (BUE) **Pubs.:** Some Aspects of Shin Ethics in a Modern European Society; Problems of Buddhist Education in Austria – A Model Case in a Catholic Country and Der Buddhismus in Russland in Vergangenheit und Gegenwart.

Mailing address: Merianstrasse 29-4-52, A-5020 Salzburg, Austria. T. (43 662) 79 09 52

Residence: See Mailing Address. T. See Mailing Address

Ferguson, William Everett, Prof

Rel: Church of Christ; Ordained on 26 APR 1981; b. 18 FEB 1933 in Montgomery TX, USA; m. to Nancy Lewis; c. 1 boy and 2 girls; Citizen of USA; Lang. English, Greek (reading), French (reading) and German (reading).

International Organizations: Professor (since 1962) of the Abilene Christian University, TX.

Other organizations: Editor (since 1981), *The Second Century*; Editor (since 1987), *Restoration Quarterly.*

Main field of work: University education. **Career:** Council, American Society of Church History from 1983 to 1985. **Educ.:** B.A., M.A. from Abilene Christian University, TX, USA (1949-1954); S.T.B. from Harvard University, Divinity School, Cambridge MA, USA (1954-1956); Ph.D. from Harvard University, Cambridge MA, USA (1956-1960).

Interests: early church history and patristics. **Mem.:** North American Patristic Society (President 1990-92); Society of Biblical Literature; Ecclesiastical History Society; International Association for Patristic Studies **Pubs.:** *Backgrounds of Early Christianity*, Eerdmans, 1987; Rev. Editor, *Early Christians Speak*, ACU, 1987 and Editor, *Encyclopedia of Early Christianity*, Garland, 1990.
Mailing address: Abilene Christian University, 1600 Campus Court, Abilene TX 79699, USA. T. (1 915) 674 3734
Residence: 609 E N 16th Street, Abilene TX 79601, USA. T. (1 915) 677 3893

Ferlisi, Gabriele, O.A.D., Fr

Rel: Roman Catholic; Ordained; celibate.
International Organizations: Procurator-General of the Order of Discalced Augustines.
Mailing address: Piazza Ottavilla 1, I-00152 Roma, Italy. T. (39 6) 589 63 45

Fernandes, Angelo Innocent, Most Rev

Rel: Roman Catholic; Ordained on 29 AUG 1937; b. 28 JUL 1913 in Karachi, Pakistan; celibate; Citizen of India.
International Organizations: President (since 1990) of the Catholic International Education Office.
President Emeritus (since 1984) of the World Conference on Religion and Peace.
Other organizations: Archbishop Emeritus (since 1991), Delhi.
Career: Engaged in various areas, Church service in Bombay including the fields of education and social welfare from 1938 to 1959. Archbishop, New Delhi from 1959 to 1990. Attended all the sessions, Second Vatican Council from 1962 to 1965. Member, Vatican Justice and Peace Academy from 1966 to 1976. Attended five World Synods, Catholic Bishops. Member, Synod Secretariat from 1971 to 1974. Member, Synod Secretariat from 1980 to 1983. Secretary-General, Catholic Bishops' Conference of India from 1960 to 1972. Chairman, Justice Development and Peace Commission, Catholic Bishops' Conference of India from 1968 to 1976. Chairman, Justice Development and Peace Commission, Catholic Bishops' Conference of India from 1986 to 1990. Executive President, World Conference on Religion and Peace from 1970 to 1984. Chairman, Office of Ecumenical and Inter-religious Affairs, Federation of Asian Bishops' Conference from 1985 to 1989. Has lectured on religion and philosophy, and on educational, social and cultural matters in India, the Far East, Australia, New Zealand, Britain, Ireland, in most countries of Europe and in all parts of the USA. **Educ.:** Grad. from Papal University, Kandy, Sri Lanka (1937). **Pubs.:** *God's Rule and Man's Role*, Ghandi Memorial Lecture, 1982; *Summons to Dialogue*, Paper No 34 of the Federation of Asian Bishops' Conferences, 1983 and *The Christian Way Today*, 12 lectures delivered in the eighties on salient topics to all members of the human community, 1987.
Mailing address: Archbishop's House, Ashok Place, New Delhi 110 001, India. T. (91 11) 343457

Fernández-Cueto, José María, C.P.C.R., Fr

Rel: Roman Catholic; Ordained; celibate.
International Organizations: Superior General of the Coopérateurs Paroissiaux du Christ Roi.
Mailing address: Via Castel di Leva 254, I-00134 Roma, Italy. T. (39 6) 795 43 77

Fernando, Harold A, Rev

Rel: Methodist; Ordained.
International Organizations: President of the Methodist Church in Sri Lanka.
Mailing address: Methodist Headquarters, Colombo 3, Sri Lanka.

Fernando, Londoño Bernal, Rev Fr

Rel: Roman Catholic; Ordained; celibate.
International Organizations: Contact of the Latin American Pontifical College.
Mailing address: Via Aurelia Antica 408, Roma, Italy. T. (39 6) 662 4641. Fax 662 4645.

Fernando, Nicholas Marcus, Most Rev Dr

Rel: Roman Catholic; Ordained on 20 DEC 1958; b. 06 DEC 1932 in Negombo, Western Province, Sri Lanka; celibate; Citizen of Sri Lanka; Lang. Sinhala, Tamil, English, Italian and Latin.
International Organizations: Archbishop (since 1977) of Colombo.
Member (since 1989) of the Congregation for the Evangelization of Peoples.
Other organizations: President (since 1989), Catholic Bishops' Conference of Sri Lanka.
Career: Rector, St Aloysius Minor Seminary from 1965 to 1973. **Educ.:** B.A. from University of London, UK (external) (1951-1953); Lic. Phil. from University of Propaganda Fide, Vatican City State (1955-1956); Dr. Theol. from University of Propaganda Fide, Vatican City State (1974-1976).
Mailing address: Archibshop's House, Gnanartha Pradeepaya Mawatha, Colombo 8, Sri Lanka. T. (94 1) 695471-3
Residence: See Mailing Address. T. See Mailing Address

Fernando, Swithin, Rt Rev

Rel: Christian.
International Organizations: Bishop of Colombo.
Mailing address: Bishop's House, 368/2 Bauddhaloka Mawatha, Colombo 7, Sri Lanka.

Ferranti, Andriana

Rel: Buddhist: Mahayana.
International Organizations: Member of the Foundation for the Preservation of the Mahayana Tradition.
Other organizations: Director, Maitri Leprosy Centre.
Mailing address: PO Bodhgaya (Gaya Dist), Bihar 824231, India. T. (91 631) 20704

Ferrari, Edoardo, S.C.J.;T.D., Fr

Rel: Roman Catholic; Ordained; celibate.
International Organizations: Procurator-General of the Congrégation du Sacré-Coeur de Jésus.
Mailing address: 20 rue de la Bibliothèque, F-13001 Marseille, France. T. (33) 91 48 02 21

Ferraris, Battistina, Sr

Rel: Roman Catholic; Ordained; celibate.
International Organizations: Secretary-General of the Sisters of Charity of St Jeanne-Antide Thouret.
Mailing address: Via Santa Maria in Cosmedin 5, I-00153 Roma, Italy. T. (39 6) 57 46 328.

Ferré, André, Fr

Rel: Roman Catholic; Ordained; celibate.
International Organizations: Director of the Pontifical Institute for Arabic and Islamic Studies.
Mailing address: Viale di Trastevere 89, I-00153 Roma, Italy. T. (39 6) 588 26 76 – 686 15 92.

Ferrell, Patricia, SCN, Sr

Rel: Roman Catholic; Ordained; celibate.
International Organizations: Secretary of the Sisters of Charity of Nazareth.
Mailing address: PO Box 172, Nazareth KY 40048, USA. T. (1 502) 348 1555.

Ferris, Elizabeth G, Dr

Rel: Quaker; b. 28 JUL 1950 in Quanah TX, USA; m. to Barry Childers; c. 1 boy and 1 girl; Citizen of USA; Lang. English, Spanish, French and Portuguese.
International Organizations: Research Director (since 1991) of the Life and Peace Institute.
Main field of work: Peace research. **Career:** Fulbright Professor, National Autonomous University of Mexico from 1981 to 1982. Assistant Professor, Lafayette College, Easton PA from 1982 to 1985. Refugee Secretary, World Council of Churches from 1985 to 1991. **Educ.:** B.A. from Duke University, Durham NC, USA (1968-1971); M.A. from University of Florida, Gainesville, USA (1971-1972); Ph.D. from University of Florida, Gainesville, USA (1972-1976). **Interests:** refugees, women's spirituality. **Pubs.:** Refugees and World Politics, New York, Praeger, 1985 and The Central American Refugees, New York, Praeger, 1987.

Mailing address: Sysslomansgatan, PO Box 297, S-751 05 Uppsala, Sweden. T. (46 18) 16 95 47.
Residence: Vinkelvägen 8, S-741 41 Knivsta, Sweden. T. (46 18) 34 33 52

Ferro, Anthony, C.F.C., Br
Rel: Roman Catholic; Ordained; celibate.
International Organizations: Member of the Congregation of Christian Brothers.
Other organizations: Editor of *Brothers*, National Assembly of Religious Brothers.
Mailing address: 1337 West Ohio Street, Chicago IL 60622-6490, USA. T. (1 312) 829 8525

Ferstl, Franz
Rel: Roman Catholic.
International Organizations: Contact of the Third World Working Group of Vicars.
Mailing address: Antonsgasse 20a, A-2500 Baden, Austria. T. (43 2252) 42 562.

Fichtner, Hans-Eberhard
Rel: Evangelical.
International Organizations: Contact of the Evangelical Church of the Goerlitz Church Territory.
Mailing address: Berliner Strasse 62, 02826 Görlitz, Germany.

Field, Frank, Prof
Rel: Unitarian.
International Organizations: Representative at UN, Geneva of the International Association for Religious Freedom.
Mailing address: 55 rue de Moillebeau, Flat 34, CH-1209 Genève, Switzerland.

Filho, Arnaldo Sisson
Rel: Theosophist.
International Organizations: President of the World Theosophical Youth Federation.
Mailing address: Caixa Postal 10360, Porto Alegre 90 000 RS, Brazil.

Filippucci, Mauro, S.M., Fr
Rel: Roman Catholic; Ordained; celibate.
International Organizations: Procurator-General of the Society of Mary.
Mailing address: Villa S Maria, Via Alessandro Poerio 63, I-00152 Roma, Italy. T. (39 6) 58 99 041.

Fine, Charles, S.C.J.;T.D., Fr
Rel: Roman Catholic; Ordained; celibate.
International Organizations: Superior General of the Congrégation du Sacré-Coeur de Jésus.
Mailing address: 20 rue de la Bibliothèque, F-13001 Marseille, France. T. (33) 91 48 02 21

Finkelstein, Louis
Rel: Jewish: Conservative; Ordained in 1919; b. in Cincinnati OH, USA.
International Organizations: Rabbinic scholar and leader of Conservative Judaism.
Career: President, Jewish Theological Seminary of America from 1940 to 1951. Chancellor, Jewish Theological Seminary of America from 1951 to 1972. Founder, Institute for Religious and Social Studies. **Educ.:** City University of New York, City College, USA; Doc. from Columbia University, New York NY, USA (1918). **Pubs.:** *Sifré*, 1939 (first critical edition on Deuteronomy).
Mailing address: No mailing address.

Finnegan, Peter, S.P.S., Fr
Rel: Roman Catholic; Ordained; celibate.
International Organizations: Superior General of the St Patrick's Society for the Foreign Missions.
Mailing address: Saint Patrick's, Kiltegan, Co. Wicklow, Ireland. T. (353 508) 73 233

Finzel, Hans W, Dr
Rel: Baptist: Conservative; Ordained.
International Organizations: General Director of the Conservative Baptist Foreign Mission Society.
Mailing address: PO Box 5, Wheaton IL 60189-0005, USA. T. (1 708) 665 1200. Fax 665 1418.

Fiolet, Herman Antonius
Rel: Roman Catholic; Ordained in 1946; b. 15 AUG 1920 in Amsterdam, Netherlands; celibate; Citizen of Netherlands.
International Organizations: Writer and Ecclesiastic of the Roman Catholic Church.
Career: Director, Institute for Lay Education from 1953 to 1959. Professor of Dogmatical and Ecumenical Theology from 1959 to 1973. Secretary-General, Council of Churches in the Netherlands from 1970 to 1985. Member of the Board, Ecumenical Development Cooperative Society from 1974 to 1983. Member, many national and international study commissions. **Interests:** music, chess, skiing. **Hons.:** Coebergh Prize for best ecumenical book in the Netherlands (1968). **Pubs.:** Vreemde Verleiding, 1968; Tweede Reformatie, 1969 and many articles in national and international theological periodicals.
Mailing address: 's Gravelandseweg 86 A 21, NL-1217 EW Hilversum, Netherlands. T. (31 35) 12530

Fiora, Luigi, S.D.B., Rev
Rel: Roman Catholic; Ordained; celibate.
International Organizations: Procurator-General of the Society of St Francis of Sales.
Mailing address: Casa Generalizia Salesiana, POB 9022, I-00163 Roma, Italy. T. (39 6) 65 92 915. Fax 65 92 929

Fiorini, Luigi, P.S.D.P., Br
Rel: Roman Catholic; Ordained; celibate.
International Organizations: Procurator-General of the Poor Servants of Divine Providence.
Mailing address: Via G B Soria 9, I-00168 Roma, Italy. T. (39 6) 627 03 90 – 627 96 11.

Fiorini, Luigi
Rel: Roman Catholic.
International Organizations: Procurator-General of the Poor Servants of Divine Providence.
Mailing address: Via Monti di Primavalle 3, I-00167 Roma, Italy. T. (39 6) 627 02 63.

Fischer, Jean Eugène
Rel: Reformed Church; Unordained; b. 08 JUL 1933 in Tunis, Tunisia; m. to Nicole Duchâble; c. 2 boys and 1 girl; Citizen of Switzerland; Lang. French, English, German and Lozi.
International Organizations: Secretary-General (since 1987) of the Conference of European Churches.
Other organizations: Member, Third World Commission (since 1969), National Protestant Church of Geneva.
Main field of work: Ecumenism and churches. **Career:** Served under the auspices of the Paris Missionary Society in the Church of Barotseland, later the United Church of Zambia from 1956 to 1963. Africa Secretary, Commission on Inter-Church Aid, Refugee and World Services (CICARWS) from 1964 to 1974. Deputy Director, CICARWS from 1974 to 1977. Director, CICARWS from 1977 to 1982. Executive Secretary, Swiss Inter-Church Aid from 1982 to 1986. President, Third World Commission, National Protestant Church of Geneva from 1976 to 1979. **Educ.:** Mech. Eng. from School of Engineering, ETS Winterthur, Switzerland (1952-1956). **Mem.:** National Protestant Church of Geneva; Swiss anti-apartheid movement (former founding member)
Mailing address: 150 route de Ferney, PO Box 2100, CH-1211 Genève 2, Switzerland. T. (41 22) 791 61 11. C. OIKOUMENE. Tx 415730 OIK CH. Fax (41 22) 791 03 61.
Residence: Grand Saconnex, CH-1218 Genève, Switzerland.

Fischer, Jean
Rel: Christian.
International Organizations: Secretary-General of the Conference of European Churches.
Mailing address: 150 route de Ferney, PO Box 2100, CH-1211 Genève 2, Switzerland. T. (41 22) 791 62 27. Fax 791 03 61. C. OIKOUMENE. Tx 415730 OIK CH.

Fisher, Avri
Rel: Jewish.
International Organizations: Secretary-General of the World Union of Mapam Parties.
Mailing address: Ytamar Ben-Avi 4, PO Box 1777, Tel Aviv, Israel. T. (972 3) 26 62 44.

Fishman, Samuel, Rabbi
Rel: Jewish; Ordained in 1957; b. 21 MAR 1932 in St Louis MO, USA; m. to Tamar Neuhart; c. 4 children; Citizen of USA.
International Organizations: Acting Director of the B'nai B'rith Hillel Foundations.
Career: Served with Chaplains Corps, US Army from 1957 to 1959. Director, Hillel Council, University of California at Los Angeles from 1959 to 1969. **Educ.:** B.A. from Washington University, Saint Louis MO, USA (1952); M.A. from Hebrew Union College-Jewish Institute of Religion, Cincinnati OH, USA (1957); Ph.D. from University of California, Berkeley, USA (1969). **Mem.:** Association of Jewish Studies; Central Conference of American Rabbis
Mailing address: 1640 Rhode Island Ave NW, Washington DC 20036, USA. T. (1 202) 857 6561.

Fisk, Terry D
Rel: Deist; b. 05 APR 1955 in St Croix Falls WI, USA; m. to Lora J Boyce; c. 2 boys and 1 girl; Citizen of USA; Lang. English.
International Organizations: Founder and Executive Director (since 1986) of the World Deist Society.
Main field of work: Religion. **Pubs.:** *Journal of the World Deist Society.*
Mailing address: PO Box 178, Webster WI 54893-0178, USA. T. (1 715) 866 8964
Residence: 7357 Cedar St E, Webster WI, USA. T. See Mailing Address

Fitzer, Margaret
Rel: Roman Catholic.
International Organizations: Executive Director of the Religious Formation Conference.
Mailing address: 8820 Cameron St, Silver Spring MD 20910-4113, USA.

Fitzgerald, Constance, OCD, Sr
Rel: Roman Catholic; Ordained; celibate.
International Organizations: Member of the Discalced Carmelite Nuns.
Other organizations: Member, Association of Contemplative Sisters.
Mailing address: Carmelite Forum, 1318 Dulany Valley Rd, Baltimore MD 21204-1399, USA.

Fjeld, Bjorn Oyvind, Rev
Rel: Free Evangelical.
International Organizations: Secretary-General of the International Federation of Free Evangelical Churches.
Mailing address: Mollergate 26, N-0179 Oslo 1, Norway. T. (47 2) 11 33 30.

Flavio, Carraro, Fr
Rel: Roman Catholic; Ordained; celibate.
International Organizations: Procurator-General of the Order of Friars Minor Capuchin.
Mailing address: Via Piemonte 70, I-00187 Roma, Italy. T. (39 6) 474 06 43.

Fleming, (William) Launcelot Scott, Rt Rev
Rel: Church of England; Ordained in 1933; b. 07 AUG 1906 in Edinburgh, Scotland, UK; m. to Jane Hildegarde Sabine Agutter Machen; c. 1 boy and 1 girl; Citizen of United Kingdom.
International Organizations: Explorer and Ecclesiastic of the Church of England.
Other organizations: Governor (since 1960), United World College of the Atlantic.
Career: Fellow and Chaplain, Trinity Hall, Cambridge from 1933 to 1949. Dean, Trinity Hall, Cambridge from 1937 to 1949. Expeditions to Iceland and Spitzbergen from 1932 to 1933. Chaplain and Geologist, British Graham Land Expedition from 1934 to 1937. Chaplain, RNVR (Royal Naval Volunteer Reserve) from 1940 to 1944. Director, Service Ordination Candidates from 1944 to 1946. Director, Scott Polar Research Institute, Cambridge from 1947 to 1949. Bishop, Portsmouth from 1949 to 1959. Bishop, Norwich from 1959 to 1971. Dean, Windsor, Domestic Chaplain to HM Queen Elizabeth II, Register Order of the Garter from 1971 to 1976. Chairman, Church of England Youth Council from 1950 to 1961. Chairman, Archbishops' Commission on Needs and Resources of the Church of England from 1963 to 1973. Vice-Chairman, Parliamentary Group on World Government from 1969 to 1971. Member, Royal Commission on Environmental Pollution from 1970 to 1973. President, Young Explorers Trust from 1976 to 1979. Honorary Chaplain, RNVR (now RNV) in 1950. Honorary Vice-President, Royal Geographical Society in 1961. Trustee, The Prince's Trust from 1974 to 1985. **Educ.:** M.A. from University of Cambridge, Trinity Hall, UK.
Interests: reading, writing, meeting friends, music. **Hons.:** Knight Commander of the Royal Victorian Order (1976); Hon. D.C.L., East Anglia (1976); Polar Medal (1936-37); Honorary Fellow, Trinity Hall, Cambridge (1956). **Mem.:** Fellow, Royal Society of Edinburgh
Mailing address: Tithe Barn, Poyntington, nr. Sherborne, Dorset DT9 4LF, UK. T. (44 096) 322 479

Fleurot, Robert, Rev
Rel: Roman Catholic; Ordained; celibate.
International Organizations: Indian Ocean Islands Coordinator of the African Caritas.
Contact person of the Caritas inter-îles.
Mailing address: Cure de Mon Loisir, Rivière du Rempart, Port Louis, Mauritius. T. (230) 54 75 03. Tx 4330 DEVELOP IW (POUR CARITAS).

Florent, Bugin, Br
Rel: Roman Catholic; Ordained; celibate.
International Organizations: Procurator-General of the Order of Discalced Brothers of the Blessed Virgin Mary of Mount Carmel.
Mailing address: Corso d'Italia 38, I-00198 Roma, Italy. T. (39 6) 841 6578. Fax 844 0758.

Florio, A Angela, Sr
Rel: Roman Catholic; Ordained; celibate.
International Organizations: Contact of the Daughters of St Anne.
Mailing address: Via Merulana 177, I-00185 Roma, Italy. T. (39 6) 73 87 39.

Flynn, Charles
Rel: Christian.
International Organizations: Past President of the Travellers for Christ.
Mailing address: 39/39 Upper Gardiner Street, Dublin, Ireland. T. (353 1) 72 80 27 – 74 93 46.

Flynn, Patricia, SSND, Sr
Rel: Roman Catholic; Ordained; celibate.
International Organizations: Superior General of the School Sisters of Notre Dame.
Mailing address: Via della Stazione Aurelia 95, I-00165 Roma, Italy. T. (39 6) 6641 8065 – 6641 8075.

Fogh-Hansen, Carl V, Rev
Rel: Christian.
International Organizations: Chairman, Danish Branch of the Christian Mission to Buddhists.
Mailing address: Isagervej 8, DK-6950 Ringkøbing, Denmark. T. (45 7) 32 06 29.

Foley, John J, C.S.P., Fr
Rel: Roman Catholic; Ordained; celibate.
International Organizations: Procurator-General of the Paulist Fathers.
Mailing address: Via XX Settembre 14, I-00187 Roma, Italy. T. (39 6) 475 15 10.

Foley, John Patrick, Most Rev

Rel: Roman Catholic; Ordained on 19 MAY 1962; b. 11 NOV 1935 in Sharon Hill PA, USA; celibate; Citizen of USA; Lang. English, Italian, French, Spanish and Latin.

International Organizations: President (since 1984) of the Pontifical Council for Social Communications.

Other organizations: President (since 1984), Vatican Film Library; Titular Archbishop, Neapolis in Proconsulari.

Main field of work: Communications. **Career:** Assistant Editor, *The Catholic Standard and Times*, Philadelphia in 1963. Rome Correspondent, *The Catholic Standard and Times* from 1963 to 1965. Co-producer, Philadelphia Catholic Hour (local radio) from 1966 to 1976. Assistant Editor, *The Catholic Standard and Times* from 1967 to 1970. Professor of Philosophy, Saint Charles' Seminary, Philadelphia from 1967 to 1984. Member, Regional Board of Directors, National Conference of Christians and Jews from 1968 to 1983. Editor, *The Catholic Standard and Times* from 1970 to 1984. Member, Board of Governors, 41st International Eucharistic Congress from 1975 to 1976. Member, Communications Committee, US Catholic Conference from 1978 to 1981. Vice-Chairman, Pennsylvania State Ethics Commission from 1979 to 1984. Vice-President, Catholic Press Association from 1981 to 1984. President, Vatican Television Centre from 1984 to 1989. Ordained Archbishop in 1984. **Educ.:** B.A. Hist. and Phil. from Saint Joseph's University, Philadelphia PA, USA (1953-1957); B.A. Phil. from Saint Charles' Seminary, Philadelphia PA, USA (1957-1962); Ph.L.; Ph.D. from Pontifical University San Tommaso d'Aquino, Vatican City State (1963-1965); M.S. Journalism from Columbia University, New York NY, USA (1965-1966). **Hons.:** Honorary Prelate of His Holiness (1976); St Francis de Sales Award, Catholic Press Association of USA and Canada (1984); Doctorate of Humane Letters, St Joseph's University, Philadelphia PA (1985); Sourin Award, Catholic Philopatrian Literary Institute, Philadelphia PA (1990); Doctorate of Laws, Allentown College of St Francis de Sales (1990); Knight Commander with Grand Cross, Order of the Holy Sepulchre (1991); Knight Commander with Grand Cross, Order of the Northern Star, Kingdom of Sweden (1991). **Mem.:** Knights of Columbus; American Catholic Philosophical Association; American Catholic Historical Society; Catholic League for Religions and Civil Rights; Alpha Sigma Nu (Jesuit Honor Society); Fellowship of Catholic Scholars **Pubs.:** *Natural Law, Natural Right and the Warren Court*, 1965; *The Eucharist in relation to the other Sacraments*, 1976 and *The Eucharist and the hungers of the human family*, 1976.

Mailing address: Palazzo San Carlo, I-00120 Città del Vaticano, Vatican. T. (39 6) 698 3197 – (39 6) 698 3597. Tx 2019 PCCS VA. Fax (39 6) 698 5373.

Residence: See Mailing Address. T. See Mailing Address

Foley, Michael E, Fr

Rel: Roman Catholic.

International Organizations: Superior General of the Servants of the Paraclete.

Mailing address: Jemez Springs NM 87024, USA. T. (1 505) 829 3586.

Foley, Nadine, Sr

Rel: Roman Catholic; Ordained; celibate.

International Organizations: Contact of the Dominican Sisters of the Congregation of the Most Holy Rosary.

Mailing address: 1257 E Siena Heights Drive, Adrian MI 49221, USA. T. (1 517) 265 5135.

Folkenberg, Robert S

Rel: Seventh-Day Adventist; Ordained on 24 DEC 1966; b. 01 JAN 1941 in Santurce, Puerto Rico; m. to Anita I Emmerson; c. 1 boy and 1 girl.

International Organizations: President, North American Division (since 1990) of the General Conference of Seventh-Day Adventists.
Member, Board of Trustees of the International Children's Care.

Other organizations: Member, Board of Trustees, Versacare Corporation Board; Member, Board of Trustees, Hialeah Hospital.

Career: Pastoral Intern, Battle Creek Tabernacle, MI from 1963 to 1964. Singing Evangelist with the Roger Holley Evangelistic team working in the Columbia Union Team from 1964 to 1966. Pastor, Colon, Panama from 1966 to 1968. Director, Stewardship Dept. and Conference Evangelist, Panama Conference of SDA from 1968 to 1969. President, Honduras Mission, Honduras from 1970 to 1973. Secretary, Central American Union from 1974 to 1975. President, Central American Union from 1975 to 1980. Assistant to the President, Inter-American Division, Miami FL from 1980 to 1985. President, Carolina Conference of SDA, Charlotte NC from 1985 to 1990. **Educ.:** Atlantic Union College, South Lancaster MA, USA (1954-1956); Newbold College, UK (1956-1958); Andrews University, Berrien Springs MI, USA (1958-1959); M.A. N.T. Theol. from Andrews University, Berrien Springs MI, USA (1962-1963). **Hons.:** Hon. D.D., Southwestern Adventist College, TX (1990), Andrews University, Berrien Springs MI (1991).

Mailing address: 12501 Old Columbia Pike, Silver Spring MD 20904, USA. T. (1 301) 680 6000.

Forck, Gottfried, Bishop Dr

Rel: Evangelical; Ordained.

International Organizations: Bishop of Evangelical Church in Berlin-Brandenburg.

Mailing address: Neue Grünstrasse 19-22, 10179 Berlin, Germany.

Foreman, Kenneth

Rel: Pentecostal.

International Organizations: President of the International Pentecostal Press Association.

Mailing address: PO Box 850, Joplin MO 64802, UK.

Forero, Jaime, Fr

Rel: Roman Catholic; Ordained; celibate.

International Organizations: Secretary-General of the Third Order Capuchin of Our Lady of Sorrows.

Mailing address: Via Bernardo Blumenstihl 28-36, I-00135 Roma, Italy. T. (39 6) 305 59 31.

Forgues, Réal, R.S.V., Fr

Rel: Roman Catholic; Ordained; celibate.

International Organizations: Procurator-General of the Pères et Frères Religieux de Saint Vincent-de-Paul.

Mailing address: Via Palestro 26, 00185 Roma, Italy. T. (39 6) 488 27 40

Fornari, Franco, Fr

Rel: Roman Catholic; Ordained; celibate.

International Organizations: Superior General of the Congregazione di Gesù Sacerdote.

Mailing address: Via dei Giardini 36, I-38100 Trento, Italy. T. (39 461) 98 38 63.

Forrest, Tom

Rel: Roman Catholic.

International Organizations: Director of the Evangelization 2000.

Mailing address: Città del Vaticano, Vatican.

Forrester, Duncan Baillie, Rev Prof

Rel: Church of Scotland; Ordained on 06 JUN 1962; b. 10 NOV 1933 in Scotland, UK; m. to Rev Margaret R McDonald; c. 1 boy and 1 girl; Citizen of United Kingdom; Lang. English, French and Tamil.

International Organizations: Professor of Christian Ethics and Practical Theology (since 1978) of the University of Edinburgh.
Member, Faith and Order Commission (since 1983) of the World Council of Churches.

Other organizations: Director (since 1984), Edinburgh University Centre for Theology and Public Issues; Principal (since 1986), New College, University of Edinburgh; President (since 1991), Society for the Study of Theology; President (since 1991), Society for the Study of Christian Ethics.

Main field of work: University education. **Career:** Part-time Assistant in Politics, University of Edinburgh from 1957 to 1958. Assistant Minister of Hillside Church and Leader, St James Mission from 1960 to 1961. Missionary, Church of Scotland to South India in 1962. Lecturer, then Professor of Politics, Madras Christian College, Tambaram from 1962 to 1970. Part-time Lecturer in Politics, University of Edinburgh from 1966 to 1967. Chaplain and Lecturer in Politics and Religious Studies, School of African and Asian Studies, University of Sussex from 1970 to 1978. Chairman, Edinburgh Council of Social Service from 1983 to 1987. **Educ.:** M.A. Mod. Hist. and Pol.Sc. (Hons) from University of St Andrews, UK (1955); Grad. from University of Chicago, IL, USA (1955-1956); B.D. from University of Edinburgh, UK (1957-1960); D.Phil. from University of Sussex, UK (1976-1978). **Interests:** hill-walking, ornithology. **Pubs.:** *Caste and Christianity*, London, 1980; *Theology and Politics*, Oxford, 1988 and *Beliefs, Values and Politics: conviction politics in a secular age*, 1989.

Mailing address: New College, University of Edinburgh, Mound Place, Edinburgh EH1 2LU, UK. T. (44 31) 225 8400
Residence: 25 Kingsburgh Road, Edinburgh EH12 6DZ, UK. T. (44 31) 337 5646

Forsyth, Charles Harkness, Rev

Rel: United Church of Canada; Ordained in 1950; b. 20 APR 1926 in Winnipeg MB, Canada; m. to Myrna Lee George; c. 2 boys and 2 girls; Citizen of Canada.
International Organizations: Consultant on Church and Urban Development (since 1990) of the Housing Organizations.
Career: Served at Sioux Lookout, Ontario from 1950 to 1954. John Black Memorial Church from 1954 to 1960. Superintendent, Central Winnipeg Parish from 1960 to 1964. Sackville Uniting Church, New Brunswick from 1964 to 1966. Executive Assistant to Premier, New Brunswick from 1966 to 1968. Member, Canadian Council from 1965 to 1966. Regent, United College from 1963 to 1964. Served on various civic and philanthropic boards from 1954 to 1966. Secretary, Board of Evangelism and Social Service, United Church of Canada from 1968 to 1970. Minister, St Andrew's Church from 1970 to 1978. Minister, First Pilgrim United Church, Hamilton from 1978 to 1990. Director, Jubilee Consultant Services Inc., Hamilton. Founder and Director, 'Renew' Community Projects, Hamilton. Chairman, Corktown Cooperative Homes. **Educ.:** B.A. from University of Manitoba, University College, Winnipeg, Canada (1947); B.D. from University of Manitoba, University College, Winnipeg, Canada (1950); D.D. from University of Winnipeg, Manitoba, Canada (1969). **Interests:** music, gardening. **Hons.:** Recipient Centennial Medal (1967). **Mem.:** Secretariat on Aging; Ontario Welfare Council; Sudbury Regional Development Corporation; Hamilton City Central Area Planning Group
Mailing address: 175 Delaware Ave, Hamilton, Ontario L8M 1V9, Canada.

Fortier, Jean-Marie, Most Rev

Rel: Roman Catholic; Ordained on 16 JUN 1944; b. 01 JUL 1920 in Quebec, Canada; celibate; Citizen of Canada.
International Organizations: Archbishop (since 1968) of Sherbrooke.
Other organizations: President (since 1985), A.E.Q.; Member (since 1985), Comité du Laïcat.
Career: History Teacher at Grand Séminaire de Québec from 1950 to 1960. Auxiliary Bishop, Ste-Anne-de-la-Pocatière in 1960. Bishop, Gaspé in 1965. Member, Vatican's Sacred Congregation of Sacraments and Divine Worship from 1975 to 1984. Vice-President, A.E.Q. from 1981 to 1985.
Mailing address: 130 ave de la Cathédrale, Sherbrooke PQ J1H 4M1, Canada. T. (1 819) 563 9934

Foster, Robert G, Rev

Rel: Moravian; Ordained.
International Organizations: President of the Moravian Church in Jamaica.
Mailing address: 3 Hector St, Kingston 5, Jamaica.

Foucher, Joseph, SCJ, Père

Rel: Roman Catholic; Ordained; b. 30 OCT 1932 in Frossay, France; celibate; Citizen of France and Canada; Lang. French and Spanish.
International Organizations: Secretary (since 1984) of the International Catholic Federation of Young People's Missionary Organizations. Member of the Priests of the Sacred Heart of Jesus.
Main field of work: Group leading and missions. **Career:** Assistant Director, Service missionnaire des jeunes from 1964 to 1969. National Director, Jeunesse du monde from 1969 to 1983. Vice-Provincial, Père des Sacrés-Coeurs from 1973 to 1980. **Interests:** education to international solidarity.
Mailing address: 920 rue Richelieu, Québec G1R 1L2, Canada. T. (1 418) 694 1222
Residence: 945 Rue St Olivier, Québec G1R 1H8, Canada T. (1 418) 692 1649

Foulkes, Garry

Rel: Buddhist: Mahayana.
International Organizations: Member of the Foundation for the Preservation of the Mahayana Tradition.
Other organizations: Director, Chenrizig Institute.
Mailing address: PO Box 41, Eudlo QLD 4554, Australia. T. (61 74) 459 047. Fax (61 74) 459 720

Fourie, Enid

Rel: Methodist.
International Organizations: Member of the World Federation of Methodist Women.
Mailing address: 11 Budree Crescent, Austerville 4052, Durban, South Africa.

Fourré, Paulette, Sr

Rel: Roman Catholic; Ordained; celibate.
International Organizations: Superior General of the Missionary Sisters of the Holy Ghost.
Mailing address: 16 rue de Billancourt, F-92100 Boulogne-Billancourt, France. T. (33 1) 46 05 51 62.

Fox, Jean

Rel: Roman Catholic.
International Organizations: Contact of the Madonna House Apostolate.
Mailing address: Madonna House, Combermere ON KOJ 1LO, Canada. T. (1 613) 756 3713.

Fox, Marjorie Joyce

Rel: Christian; Unordained; b. 05 FEB 1940 in Barnsley, Yorkshire, UK; m. to David Henry Fox; c. 1 girl; Citizen of United Kingdom; Lang. English and French.
International Organizations: International Coordinator (since 1986) of the Girls' Brigade International.
Main field of work: Ministry and coordination. **Mem.:** British Institute of Management, Society Executive Associations
Mailing address: Brigade House, Old Bridewell, 3 Burgh Road, Aylsham, Norwich NR11 6AJ, UK. T. (44 263) 734917. Fax 734917.
Residence: See Mailing Address. T. See Mailing Address

Fox, Ruth, Sr

Rel: Roman Catholic; Ordained; celibate.
International Organizations: President of the Benedictine Sisters of the Federation of Saint Gertrude.
Mailing address: Sacred Heart Monastery, PO Box 364, Richardton ND 58652, USA. T. (1 701) 974 2121.

Foxall, George M, Rev

Rel: Christian; b. 20 JUN 1930 in Slettler, Alberta, Canada; m. to Ruth; c. 1 boy; Citizen of Canada; Lang. English and Hausa.
International Organizations: Administrator (since 1979) of the Accrediting Council for Theological Education in Africa.
Main field of work: Theological education. **Educ.:** M.R.E. from Canadian Theological Seminary (1972-1974).
Mailing address: PMB 2049, Kaduna, Nigeria.
Residence: 8 A Ali Akilu Road, Kaduna, Nigeria

Francelle, Jean

Rel: Roman Catholic.
International Organizations: Head, Europe Regional Secretariat of the International Movement for the Apostolate in the Independent Social Milieux.
Mailing address: c/o ACI, 3 bis François-Ponsard, F-75116 Paris, France.

Frances, N

Rel: Roman Catholic.
International Organizations: UNESCO Representative of the Catholic International Union for Social Service.
Mailing address: 69 avenue de Ségur, F-75007 Paris, France. T. (33 1) 45 66 69 31.

Francesconi, Emanuelle, F.D.M., Fr

Rel: Roman Catholic; Ordained; celibate.
International Organizations: Procurator-General of the Frères de Notre-Dame de la Miséricorde.
Mailing address: Via Bogliasco 34, I-00163 Roma, Italy. T. (39 6) 680 8552.

Francis, Daisy

Rel: Christian.
International Organizations: Contact of the Canada-Asia Working Group.
Mailing address: 11 Madison Ave, Toronto ON M5R 2S2, Canada. T. (1 496) 921 5626. Fax 924 5356. Tx 065 24128 CA.

Francis, Michael Kpalala, Mgr
Rel: Roman Catholic; Ordained on 04 AUG 1963; b. 12 FEB 1936 in Kakata, Liberia; celibate.
International Organizations: President of the Association of Episcopal Conferences of Anglophone West Africa.
Other organizations: Archbishop (since 1981), Monrovia.
Mailing address: PO Box 10-502, 1000 Monrovia, Liberia. T. (231) 26 24 47. C. CATHOLIC MISSION. Tx 44529 CM LI.

Francisco, Julio, Rev
Rel: Evangelical Congregational; Ordained.
International Organizations: General Secretary of the Evangelical Congregational Church in Angola.
Mailing address: CP 551, Huambo, Angola.

Françoise, George P
Rel: Roman Catholic.
International Organizations: Pan-African Secretariat of the International Young Christian Workers.
Mailing address: Panafrican YCW, PO Box 46528, Nairobi, Kenya.

Franklin, Rose Marie, Sr
Rel: Roman Catholic; Ordained; celibate.
International Organizations: Superior General of the Maryknoll Sisters of Saint Dominic.
Mailing address: Maryknoll Sisters Center, Maryknoll NY 10545, USA. T. (1 914) 941 7575.

Fraser, Roy
Rel: Buddhist: Mahayana.
International Organizations: Member of the Foundation for the Preservation of the Mahayana Tradition.
Other organizations: Director, Mahamudra Centre.
Mailing address: Colville RD, Coromandel, New Zealand. T. FAX AND TEL (64 843) 56851

Fraser, Sally
Rel: Buddhist: Mahayana.
International Organizations: Member of the Foundation for the Preservation of the Mahayana Tradition.
Other organizations: Director, Mahamudra Centre.
Mailing address: Colville RD, Coromandel, New Zealand. T. FAX AND TEL (64 843) 56851

Fredericks, Donald G
Rel: Christian.
International Organizations: Director-General of the United Indian Missions International.
Mailing address: PO Box 3600, Flagstaff AZ 86003, USA. T. (1 602) 774 0651.

Freeman, James Darcy, H E Card
Rel: Roman Catholic; Ordained on 14 JAN 1951; b. 19 NOV 1907 in Sydney NSW, Australia; celibate; Citizen of Australia.
International Organizations: Member (since 1973) of the College of Cardinals.
Other organizations: Archbishop Emeritus (since 1983), Sydney.
Career: Private Secretary to H E Cardinal Gilroy, Archbishop, Sydney from 1941 to 1946. Auxiliary Bishop, Sydney in 1957. Bishop, Armidale in 1968. Archbishop, Sydney during 1983. Created Cardinal by Pope Paul VI in 1973. **Hons.:** Knight of the Holy Sepulchre; Knight Commander of the Order of the British Empire.
Mailing address: PO Box 246, Saint John Vianney Villa, Clovelly Rd, Randwick NSW 2031, Australia. T. (61 2) 399 85 64

Frei Betto, F
See Carlos Alberto Libanio Christo.

Freise, Josef, Dr
Rel: Roman Catholic; b. 29 JUN 1951 in Billerbeck, Germany; m. to Elisabeth Eichholz; c. 1 boy and 1 girl; Citizen of Germany; Lang. German, French and English.
International Organizations: Secretary-General (since 1988) of the International Christian Service for Peace.
Main field of work: International coordination of services for peace and development. **Educ.:** Dip. Theol. from Westfälische Wilhelmsuniversität Münster, Germany (1970-1972); Dr. Päd. from Westfälische Wilhelmsuniversität Münster, Germany (1973-1978); Catholic University of Louvain, Belgium (1972-1973); Eberhard-Karls-Universität Tübingen, Germany (1973-1973). **Interests:** intercultural exchanges, ecumenical movement for justice, peace, safeguard of the creation. **Pubs.:** Interkulturelles Lernen in Begegnungen – Eine neue Möglichkeit Entwicklungs – Politischer Bildung?, Saarbrücken, Fort Lauderdale, 1982.
Mailing address: Engerser Strasse 74b, 56564 Neuwied, Germany. T. (49 2631) 22 011
Residence: See Mailing Address. T. (49 2631) 28 460

Freitag, Erling
Rel: Druidism.
International Organizations: President of the International Grand Lodge of Druidism.
Mailing address: Asveien 16, Teie, 3100 Tønsberg, Norway.

Frend, William Hugh Clifford, Rev Prof
Rel: Church of England; Ordained on 11 JUL 1982; b. 11 JAN 1916 in Shuttermill, Surrey, UK; m. to Mary Grace Crook; c. 1 boy and 1 girl; Citizen of United Kingdom; Lang. English, French, German and Italian (reading).
International Organizations: Clergyman of the Church of England.
Other organizations: Professor Emeritus (since 1984), University of Glasgow.
Main field of work: Teaching and research in early Christianity. **Career:** Assistant Principal, War Cabinet Offices (WO) from 1941 to 1942. Intelligence Officer, Political Welfare Executive (PWE) from 1942 to 1946. Editor, German Foreign Ministry Documents Project from 1947 to 1951. Research Fellow, University of Nottingham from 1951 to 1952. Assistant Lecturer, then Lecturer in Church History and Doctrine, University of Cambridge from 1953 to 1969. Fellow, Caius College, Cambridge in 1956. Director of Studies in Archaeology and Anthropology, Caius College, Cambridge from 1961 to 1969. Professor of Ecclesiastical History, University of Glasgow from 1969 to 1984. Dean of Divinity, University of Glasgow from 1972 to 1975. Chairman, Association of University Teachers, Scotland from 1976 to 1978. President, International Commission of Comparative Church History from 1980 to 1983. Vice-President, International Association for Patristic Studies from 1983 to 1987. Priest-in-charge, Barnwell, Peterborough from 1984 to 1990. **Educ.:** Modern Hist. (1st) from University of Oxford, Keble College, UK (1934-1940). **Interests:** early Christian archaeology, golf, collecting stamps and coins, gardening. **Hons.:** Hon. D.D., Edinburgh University (1974); Gold Cross of Merit with Swords, Polish Government in exile (1950); Territorial Officers' Decoration. **Mem.:** Fellow, Society of Antiquaries (1952); Fellow, Royal Historical Society (1954); Fellow, Royal Society of Edinburgh (1979); Fellow, British Academy (1983) **Pubs.:** The Donatist Church, Oxford, 1951, 3rd Edn., 1985; Martyrdom and Persecution in the Early Church, 1965 and The Rise of Christianity, 1984.
Mailing address: The Clerks Cottage, Little Wilbraham, Cambridge CB1 5LB, UK. T. (44 223) 811 731
Residence: See Mailing Address. T. See Mailing Address

Fresno Larraín, Juan Francisco F, H E Card
Rel: Roman Catholic; Ordained on 18 DEC 1937; b. 26 JUL 1914 in Santiago de Chile, Chile; celibate; Citizen of Chile.
International Organizations: Member (since 1985) of the College of Cardinals.
Member of the Congregation for Catholic Education.
Other organizations: Archbishop Emeritus (since 1990), Santiago de Chile.
Career: Consecrated Bishop, Copiapó in 1958. Archbishop, La Serena in 1967. Archbishop, Santiago de Chile from 1983 to 1990. Created Cardinal by Pope John Paul II in 1985.
Mailing address: Laura Tromben 5675, Vitacura, Santiago, Chile. T. (56 2) 211 7916

Freund-Rosenthal, Miriam Kottler, Dr

Rel: Jewish; b. 17 FEB 1906 in New York City NY, USA; widowed (1st) Milton Freund (died), (2nd) Harry Rosenthal; c. 2 boys; Citizen of USA; Lang. Hebrew, English and French (some).

International Organizations: Trustee (since 1978) of the American Friends of the Hebrew University in the USA.

Life Member, Actions Committee (since 1956) of the World Zionist Organization. Member, Assembly (since 1971) of the Jewish Agency for Israel. Life Member, Zionist General Council (since 1978) of the World Zionist Organization.

Other organizations: Chairman, Governing Board of the Women's Division (since 1983), State of Israel Bonds and Member of the Cabinet of Bonds; Honorary Vice-President, Hadassah and Representative to UN; Member, Board, Mercaz; National Officer, Keren Kayemeth.

Main field of work: Zionist leader. **Career:** Teacher and lecturer in New York City Schools (until 1943). National Youth Activities Chairman, Hadassah from 1943 to 1948. National Vocational Education Chairman, Hadassah from 1948 to 1953. National Youth Aliyah Chairman, Hadassah from 1953 to 1956. Vice-President, Hadassah from 1953 to 1956. President, Hadassah from 1956 to 1960. Editor, Hadassah Magazine from 1966 to 1971. Member, Board, United Israel Appeal in 1960. Co-Chairman, American Zionist Youth Commission from 1944 to 1949. Member, World Zionist Presidium from 1960 to 1964. Chairman, Executive Committee, American Zionist Council from 1960 to 1965. Elected to the Board of Directors, World Council of Synagogues in 1966. Elected National Secretary, Herzliah Hebrew Teachers College in 1968. Member, National Council, National Planning Association in 1970. Vice-President, American Zionist Federation from 1970 to 1972. Founding Vice-President, Mercaz from 1978 to 1983. Served as National Associate Chairman for the Women's Division, State of Israel Bonds (until 1980). Founder, Inter-Collegiate Zionist Youth Federation of America. Founding Director, Brandeis Youth Foundation. National Vice-President, Women's Division of Brandeis University, Keren Kayemeth. Founding Member, Manhattan Chapter, Keren Kayemeth. Served on the Board of Governors, Hebrew University of Jerusalem. **Educ.:** Doc. American Hist. from New York University, NY, USA. **Interests:** travel, painting, writing. **Hons.:** Recipient citation of Jewish Book Council of America in recognition of her contributions to the literature on American Jewish History (1954); Hall of Fame of the Alumni Association of Hunter College (1972). **Mem.:** American Association of University Women **Pubs.:** Jewish Merchants in Colonial America; Jewels for a Crown, 1963 and My Lifetime.

Mailing address: 11 Island Ave, Miami Beach FL 33139, USA. T. (1 305) 672 8878

Residence: New York City and Miami FL, USA.

Frezzato, Matteo, F.S.G.C., Br

Rel: Roman Catholic; Ordained; celibate.

International Organizations: Superior General of the Frères de Saint Joseph-Benoît Cottolengo.

Mailing address: Via Cottolengo 14, I-10152 Torino, Italy. T. (39 11) 522 51 11.

Friant, Jean, S.G., Br

Rel: Roman Catholic; Ordained; celibate.

International Organizations: Superior General of the Brothers of Christian Instruction of St Gabriel.

Mailing address: Via Trionfale 12840, I-00135 Roma, Italy. T. (39 6) 376 50 49. Fax 376 24 07.

Friedlander, Albert H, Rabbi Prof Dr

Rel: Jewish: Progressive; Ordained in 1952; b. 10 APR 1927 in Berlin, Germany; m. to Evelyn Philipp; c. 3 girls; Citizen of USA; Lang. English, German, Hebrew, Spanish and Yiddish.

International Organizations: Vice-President (since 1970) of the World Union for Progressive Judaism.

Co-chairman (UK) (since 1988) of the World Conference on Religion and Peace.

Other organizations: Dean (since 1968), Leo Baeck College, London; Minister (since 1971), Westminster Synagogue, London; Co-editor (since 1988), Littman Library for Oxford University Press.

Main field of work: Ministry, university work. **Career:** Rabbi, congregations in Arkansas and Pennsylvania from 1956 to 1961. Counselor/chaplain, Columbia University, New York NY from 1961 to 1966. Rabbi, Wembley Liberal Synagogue from 1966 to 1971. **Educ.:** B.Phil. from University of Chicago, IL, USA (1944-1946); B.H.L.; M.H.L. from Hebrew Union College-Jewish Institute of Religion, Cincinnati OH, USA (1946-1952); Ph.D. from Columbia University, New York NY, USA (1966). **Hons.:** D.Div. h.c., Hebrew Union College, Cincinnati OH (1976); Sternberg Medal for Interfaith (1990). **Mem.:** PEN; Athenaeum Club; CCAR; ULPS; RSGB **Pubs.:** *Leo Baeck, Teacher of Theresienstadt*, 1967, 1991; *Out of the Whirlwind*, 1968; With E Wiesel, *The Six Days of Destruction*, 1988, 1992 and *The Thread of Gold*, SCM Press, 1991.

Mailing address: Kent House, Rutland Gardens, London SW7 1BX, UK. T. (44 71) 584 3953

Residence: See Mailing Address. T. See Mailing Address

Friend, Stephen

Rel: Christian.

International Organizations: Secretary of the International Association for the Study of Maritime Mission.

Mailing address: Univ. Coll. of Ripon and York St John, College Road, Ripon HG4 2QX, UK. T. (44 765) 602691. Fax 600516.

Frink, Ferdinand, F.M.M., Br

Rel: Roman Catholic; Ordained; celibate.

International Organizations: Superior General of the Brothers of Mercy of Montabaur.

Mailing address: Horresser Strasse 16, 56412 Niederelbert, Germany. T. (49 2302) 4832

Fritz, Edward, Col

Rel: Salvation Army, The.

International Organizations: Territorial Commander of The Salvation Army.

Mailing address: Apartado Postal 12-668, 03020 México DF, Mexico.

Frost, Francis, Rev Prof Dr

Rel: Roman Catholic; Ordained on 20 APR 1958; b. 03 MAR 1933 in London, UK; celibate; Citizen of United Kingdom; Lang. English and French.

International Organizations: Professor (since 1989) of the Ecumenical Institute Bossey.

Other organizations: Full Professor of Theology, Faculty of Theology (since 1976), Fédération Universitaire et Polytechnique de Lille, France.

Main field of work: University education. **Career:** Professor at an Institute for Pedagogical Education, Liverpool (for 4 years). Lecturer, Faculty of Theology. **Educ.:** Lic. Lettres from University of Lille III, France (1952-1954); Dr. Theol. from Fédération Universitaire et Polytechnique de Lille, France (1954-1964). **Pubs.:** 5 articles (including 3 major) in *Dictionnaire de Spiritualité* and over 20 major articles in *Le Dictionnaire du Catholicisme*.

Mailing address: Château de Bossey, CH-1298 Céligny, Switzerland. T. (41 22) 776 25 31. C. INSTITUTCELIGNY GENEVA. Fax (41 22) 776 01 69.

Residence: See Mailing Address. T. See Mailing Address

Fuchs, Eric V, Prof Dr

Rel: Protestant; Ordained on 15 JUN 1958; b. 28 DEC 1932 in Geneva, Switzerland; m. to Rose-Marie Nicolas; c. 1 boy and 2 girls; Citizen of Switzerland; Lang. French, English, Italian, Greek, Latin and Hebrew.

International Organizations: Professor of Ethics and Dean, Faculty of Theology (since 1987) of the University of Geneva.

Main field of work: University education. **Career:** Director, Centre Protestant d'Etudes from 1960 to 1979. Director, Atelier Oecuménique de Théologie from 1973 to 1980. Ordinary Professor, University of Lausanne from 1981 to 1987. **Educ.:** Lic. Theol. from University of Geneva, Switzerland (1951-1957); Dr. Theol. from University of Geneva, Switzerland (1979). **Interests:** ethics. **Pubs.:** La Morale selon Calvin, 1986; L'Ethique Protestante, 1990 and Le Droit de Résister, 1990.

Mailing address: Université de Genève, CH-1211 Genève 4, Switzerland. T. (41 22) 705 74 21

Residence: Tabazan 9, CH-1204 Genève, Switzerland. T. (41 22) 310 47 45

Fugere, Robert

Rel: Protestant.

International Organizations: Executive Secretary of the Inter-Church Fund for International Development.

Mailing address: 85 St Clair E, Room 204, Toronto M4T 1MB, Canada. T. (1 416) 968 1411. Fax 925 3394. C. ICFIDTOR. Tx 06524128.

Fuji, Koei
Rel: Buddhist; Ordained; b. 03 JAN 1931; Citizen of Japan.
International Organizations: Manager of the Shinshu Izumojiha.
Career: Chief, Gusetsuji. **Educ.:** Otani University Graduate School.
Mailing address: 2-9 Shimizugashira, Takefu, Fukui 915, Japan. T. (81 778) 27 1108

Fujii, Jitsuo
Rel: Buddhist; Ordained; b. 20 OCT 1898; Citizen of Japan.
International Organizations: Manager of the Jodo Shu.
Main field of work: Monk. **Educ.:** Shukyo University.
Mailing address: 400 Rinkacho, Higashi Yama-ku, Kyoto, Kyoto 605, Japan. T. (81 75) 531 2111

Fukui, Kojun
Rel: Buddhist; Ordained; b. 27 APR 1898; Citizen of Japan.
International Organizations: Chief of the Tendaishu Myohoin.
Career: President, Taisho University. **Educ.:** Waseda University, Tokyo, Japan.
Mailing address: Higashi Yama 7-jo Agaru, Higashiyama-ku, Kyoto 605, Japan. T. (81 75) 561 1744

Fukutomi, Settei
Rel: Buddhist: Zen; Ordained; b. 23 SEP 1921; Citizen of Japan.
International Organizations: Chief of the Daitokuki Sect.
Career: Manager, Daitokuji Sect of Rinzai Zen Buddhism. **Educ.:** Taisho University, Japan.
Mailing address: 53 Murasakino Daitokujicho, Kita-ku, Kyoto 603, Japan.

Fumagalli, Pier Francesco, Rev
Rel: Roman Catholic; Ordained on 10 JUN 1978; b. 09 OCT 1948 in Luino, Italy; celibate.
International Organizations: Secretary (since 1986) of the Commission for Religious Relations with the Jews.
Member of the Pontifical Council for Promoting Christian Unity.
Other organizations: Consultant, Center for Hebrew Bibliography and Jewish Heritage, Italy; Member, Executive Committee, Italian Association of Ecclesiastical Librarians.
Career: Scriptor, Ambrosian Library in Milan from 1978 to 1981. Doctor, Ambrosian Library in Milan from 1981 to 1986. **Educ.:** Studied Classics and Oriental Lang. and Lit. (Doc. Class. Stds.) from Catholic University of the Sacred Hearth, Milan, Italy; Studied Hebrew Paleography, Lit. and Lang. from Hebrew University of Jerusalem, Israel; Studied Phil. and Theol. (Bacc. Theol.) from Milan's major seminary, Italy. **Pubs.:** Author of many scholarly articles on Hebrew manuscripts and typography, Italian illuminated Ketubbot, and Christian-Jewish relations in history, liturgy and theology, including: "Ebrei e cristiani: dal disprezzo all'amore per i 'Fratelli maggiori'", in *Ambrosius*, 62, 4, 1986, pp. 316-329; "The Embrace of the Pope and the Rabbi: Two Years Later", in *Journal of Ecumenical Studies*, 25:3, Summer, 1988, pp. 499-501 and "La Biblia, animatrice di dialogo, di comunione e di evangelizzazione per il Terzo Millennio", *L'Osservatore Romano*, 26 January 1990, p. 5 (La settimana di preghiera per l'unità dei cristiani).
Mailing address: c/o Pontifical Council for Promoting Christian Unity, I-00120 Città del Vaticano, Vatican. T. (39 6) 698 4386 &h- 698 3071. Fax 698 5365. C/ SECRUNIT – I-00120 VATICAN CITY. Tx 2024 DIRGENTEL VA.

Funamoto, Sakao
Rel: Christian; b. 20 FEB 1906; m. to Ayako; Citizen of Japan.
International Organizations: Chairman of the Osaka Jogakuin.
Career: Honorary Minister, Osaka Johoku Church. **Educ.:** Aoyama Gakuin University, Tokyo, Japan.
Mailing address: 17-8 Hanayashiki-Matsugaoka, Takarazuka, Hyogo 665, Japan. T. (81 727) 87 6754

Fürer, Ivo, Rev Dr
Rel: Roman Catholic; Ordained; celibate.
International Organizations: Secretary-General of the Council of European Bishops' Conferences.
Mailing address: Klosterhof 6 b, CH-9000 St Gallen, Switzerland. T. (41 71) 22 81 06.

Furio, Angelo
Rel: Buddhist: Mahayana.
International Organizations: Member of the Foundation for the Preservation of the Mahayana Tradition.
Other organizations: Director, Nagarjuna C.E.T. Valencia.
Mailing address: Calle Joaquin Costa 10 Pta 9, E-46005 Valencia, Spain. T. (34 96) 333 1259

Furman, L G Wayne
Rel: Christian.
International Organizations: Contact of the International Order of the King's Daughters and Sons.
Mailing address: PO Box 1017, Chautauqua NY 14722, USA. T. (1 716) 357 4951.

Furman, Marilyn
Rel: Christian.
International Organizations: Contact of the International Order of the King's Daughters and Sons.
Mailing address: PO Box 1017, Chautauqua NY 14722, USA. T. (1 716) 357 4951.

Furno, Carlo, H E Mgr
Rel: Roman Catholic; Ordained on 25 JUN 1944; b. 02 DEC 1921 in Bairo Canavese, Italy; celibate.
International Organizations: Apostolic Nuncio to Brazil (since 1982) of the Roman Catholic Church.
Other organizations: Titular Archbishop, Abari.
Mailing address: Avenida das Naçoes, lote n 1, CP 07-0153, 70359 Brasilia-DF, Brazil. T. (55 61) 223 0794. Fax 224 93 65. Tx 612125 NUAP BR

Furuse, Nichiu
Rel: Buddhist; Ordained; b. 30 MAR 1917; Citizen of Japan.
International Organizations: Manager of the Kenpon Hokkashu.
Career: Chief, Myomanji, Kuonji. **Educ.:** Waseda University, Tokyo, Japan.
Mailing address: 16 Nisoninmonzen Chojincho, Saga, Ukyo-ku, Kyoto 616, Japan. T. (81 75) 861 0249

Furutan, Ali-Akbar
Rel: Bahá'í.
International Organizations: Member of the Bahá'í International Community.
Mailing address: Bahá'í World Center, PO Box 155, 31 001 Haifa, Israel.

Futaba, Kenko
Rel: Buddhist; Citizen of Japan.
International Organizations: President of the Kyoto Joshi Gakuen.
Career: President, Ryukoku University (Japanese Buddhism history). **Educ.:** Ryukoku University, Kyoto, Japan.
Mailing address: 1-26 Katagihara Sumitacho, Nishikyo-ku, Kyoto 615, Japan. T. (81 75) 381 5067

Gabriel, R N
Rel: Roman Catholic.
International Organizations: Director, Near and Middle East of the Catholic International Education Office.
Mailing address: Collège de la Ste Famille, BP 73, Fagallah, 11000 Cairo, Egypt.

Gaétan, Mutembe
Rel: Lutheran.
International Organizations: Executive Secretary of the All-Africa Lutheran Churches Information and Coordination Centre.
Other organizations: Editor, *Pan-African Lutheran Information.*
Mailing address: ELCT-Bldg, PO Box 314, Arusha, Tanzania UR. T. (255) 2318. C. AFRILUTHCO. Tx 42054 LUTH T2.

Gafah, Pierre-Elise, Rev
Rel: Roman Catholic; Ordained; celibate.
International Organizations: Executive Secretary of the Pan-African Episcopal Committee for Social Communication.
Mailing address: c/o SECAM Secretariat, 4 Senchi Street, PO Box 9156, Airport Residential Area, Accra, Ghana. T. (233) 777706 – 774724. Fax 772548. C. SECAM ACCRA. Tx 2587 SECAM GH.

Gagnon, Edouard, P.S.S., H E Card
Rel: Roman Catholic; Ordained on 15 AUG 1940; b. 15 JAN 1918 in Port-Daniel, Gaspé, Canada; celibate; Citizen of Canada.
International Organizations: President (since 1991) of the Pontifical Committee for International Eucharistic Congresses.
Member of the Society of the Priests of St Sulpice. Member of the Sacred Congregation for the Sacraments and Divine Worship. Member of the Congregation for the Causes of Saints.
Other organizations: Member (since 1985), College of Cardinals; Member, Supreme Court of Appeal, Vatican City State.
Career: Consecrated Bishop, Saint Paul, Alberta in 1969. Renounced position in 1972. Archbishop (Titular See of Iustiniana prima) in 1983. Created Cardinal by Pope John Paul II in 1985. President, Pontifical Council for the Family.
Mailing address: Via di Porta Angelica 31, I-00193 Roma, Italy.

Gailani, Sayed Ahmad
Rel: Muslim.
International Organizations: Head of the National Islamic Front, Peshawar.
Career: Served as religious adviser to King Zahir Shah and President Daud.
Mailing address: No mailing address.

Gaitán, Amparo Ramírez
Rel: Roman Catholic.
International Organizations: Director, Latin America Regional Office of the International Federation for Catholic Associations of the Blind.
Mailing address: Avenida Caracas 34-47, Bogotá, Colombia.

Gaitán, Jorge
Rel: Evangelical.
International Organizations: Director-General of the Latin American Evangelical Center for Pastoral Studies.
Mailing address: Apartado 1307, San José, Costa Rica.

Galichet, Jean Paul, Fr
Rel: Roman Catholic; Ordained; celibate.
International Organizations: Procurator-General of the Carthusian Order.
Mailing address: Via Caprareccia, I-02030 Bocchignano, Italy. T. (39 765) 29 033.

Galignano, Eugenio, MI, Fr
Rel: Roman Catholic; Ordained; celibate.
International Organizations: Director-General of the Knights of the Immaculate.
Mailing address: Convento P Kolbe, Via S Teodoro 42, I-00186 Roma, Italy. T. (39 6) 679 3828 – (39 6) 679 49 74.

Gallagher, Clarence, Fr
Rel: Roman Catholic; Ordained; celibate.
International Organizations: Dean, Faculty of Oriental Canon Law and Rector of the Pontifical Oriental Institute.
Mailing address: Piazza S Maria Maggiore 7, I-00185 Roma, Italy. T. (39 6) 446 55 89.

Gallagher, John C, C.S.B., Fr
Rel: Roman Catholic; Ordained; celibate.
International Organizations: Superior General of the Congregation of St Basil.
Mailing address: Basilwood, 15 Side Road RR3, Milton L9T 2X7, Canada. T. (1 416) 875 0222. Fax 875 1920.

Gallagher, Joseph V, C.S.P., Fr
Rel: Roman Catholic; Ordained; celibate.
International Organizations: Superior General of the Paulist Fathers.
Mailing address: 86 Dromore Rd, Scarsdale NY 10583, USA. T. (1 914) 472 4800

Gallimore, Gerry O
Rel: Protestant.
International Organizations: President of the Youth for Christ International.
Mailing address: 1 Colombo Court 08-14, Singapore 0617, Singapore. T. (65) 336 8776. Tx 24557 Leelim.

Galten, John W
Rel: Roman Catholic.
International Organizations: Contact of the Catholic Evidence Guild.
Mailing address: 4228 23rd Street, San Francisco CA 94114, USA. T. (1 415) 282 0511.

Galuzzi, Alessandro M, O.M., Fr
Rel: Roman Catholic; Ordained; celibate.
International Organizations: Superior General of the Order of Minims.
Advisor of the Congregation for the Causes of Saints.
Mailing address: Pizza S Francesco di Paola 10, I-00184 Roma, Italy. T. (39 6) 488 02 50

Gamarra Mayor, José G, O.SS.T., Fr
Rel: Roman Catholic; Ordained; celibate.
International Organizations: Minister-General of the Order of the Most Holy Trinity for the Redemption of Captives.
Mailing address: Via Massimi 114-c, I-00136 Roma, Italy. T. (39 6) 342 05 29, (39 6) 342 07 26. Fax (39 6) 34 16 73

Gamedze, A B, Dr
Rel: Evangelical; Ordained in 1956; b. 03 APR 1921 in Shiselweni District, Swaziland; married; c. 3 boys and 3 girls; Citizen of Swaziland.
International Organizations: Director of the various companies.
Main field of work: Theologian and diplomat. **Career:** Head Teacher, Jerusalem School in 1942. Transferred to Franson Christian High School in 1943. Head Teacher, Makhonza School from 1946 to 1947. Studied education, Christian education and theology at Wheaton College, IL from 1947 to 1951. Lecturer, Evangelical Teacher Training College, Natal from 1951 to 1956. Founder and Editor, Africa's Hope from 1955 to 1961. President, African Teachers' Christian Fellowship in 1956. Assisted organizing inspection of religious education in secondary schools, Transvaal in 1958. Lecturer in Divinity and Education, and Chaplain, University of Fort Hare from 1960 to 1961. Joined staff, Franson Christian High School in 1962. Superintendent (Grantee), all Evangelical Church Schools from 1962 to 1965. Vice-President, Evangelical Church. Vice-President, Swaziland Conference of Churches in 1964. President, Swaziland Conference of Churches from 1979 to 1980. Senior Liaison Officer, Swaziland from 1965 to 1971. Member, Senate in 1967. Minister, Education, Swaziland from 1967 to 1971. Editor, Imbokodvo Bulletin from 1969 to 1970. High Commissioner in UK, concurrently Ambassador to France, Belgium and Federal Republic of Germany from 1971 to 1972. Secretary to Swazi Nation, Protocol, Dept. of Foreign Affairs from 1973 to 1974, then Private Secretary to King Sobhuza II, then Chief, Imbokodvo Bulletin. Editor, Association of Evangelicals from 1969 to 1970. President, Swaziland Bible Society from 1975 to 1980. Chairman, Church Extension in northern Swaziland from 1980 to 1981. Director, various companies. **Interests:** writing, walking, indoor games. **Hons.:** Hon. LL.D., Wheaton College IL, USA (1968).
Mailing address: PO Box 868, Mbabane, Swaziland. T. (268) 2661
Residence: T. (268) 2427

Gandolfi, Stefania
Rel: Christian.
International Organizations: President of the International Voluntary Christian Service.
Mailing address: Via Gavazzeni 3, I-24100 Bergamo, Italy. T. (39 35) 243022.

Gandolfo, Guido, SSP, Rev
Rel: Roman Catholic; Ordained; celibate.
International Organizations: Procurator-General of the Society of St Paul.
Mailing address: Via della Fanella 39, I-00148 Roma, Italy. T. (39 6) 522 2927.

Gantin, Bernardin, H E Card

Rel: Roman Catholic; Ordained on 14 JAN 1951; b. 08 MAY 1922 in Toffo, Benin; celibate; Citizen of Benin.
International Organizations: Prefect (since 1984) of the Congregation for Bishops.
President (since 1984) of the Pontifical Commission for Latin America. Member of the Sacred Congregation for the Oriental Churches. Member of the Sacred Congregation for the Sacraments and Divine Worship. Member of the Congregation for the Causes of Saints. Member of the Congregation for the Evangelization of Peoples. Member of the Congregation for Religious and Secular Institutes. Member of the Congregation for Catholic Education. Member of the Pontifical Council for the Interpretation of Legislative Texts. Member of the Pontifical Council for Inter-Religious Dialogue. Member of the Pontifical Council for Social Communications.
Other organizations: Member (since 1977), College of Cardinals; Titular Bishop (since 1986), Palestrina; Member, Council of Cardinals and Bishops; Member, Congregation for the Doctrine of the Faith; Member, Institute for Religious Works.
Career: Elected to Titular Church, Tipasa, Mauritania in 1956. Consecrated Bishop in 1957. Appointed to Cotonou, Benin in 1960. Created Cardinal by Pope Paul VI in 1977.
Mailing address: Piazza San Calisto 16, I-00153 Roma, Italy.

Garbi, Ermanno, Pastor

Rel: Seventh-Day Adventist.
International Organizations: President, Israel Field of the General Conference of Seventh-Day Adventists.
Mailing address: PO Box 592, Jerusalem, Israel.

Garcia, Raúl Oscar, Prof

Rel: Evangelical Mennonite; b. 26 AUG 1930 in Pehuajo, Buenos Aires, Argentina; m. to Anita Edna Swartzentruber; c. 3 boys; Citizen of Argentina; Lang. English, Spanish and French.
International Organizations: President (since 1992) of the Mennonite World Conference.
Main field of work: Education. **Career:** Rector, Colegio Nacional de Pehuajo from 1962 to 1990. Secretary of Education, Municipalidad de Pehuajo from 1965 to 1966. President of the Board, Argentine Mennonite Church (several times). **Pubs.:** *I am a Christian, Evangelical, Anabaptist*; "The Argentine Mennonite Church: an Interpretation", in *The Role of Society* and "El Contexto Hispanoamericano", in *Los Niños y el Reino.*
Mailing address: Godoy 448, Pehuajo, Buenos Aires 6450, Argentina. T. (54 1) 396 2660

García del Cerro, Cesáreo G, Rev

Rel: Roman Catholic; Ordained on 15 JUL 1960; b. 07 MAR 1927 in Sevilla, Spain; celibate; Citizen of Spain; Lang. Spanish, French, English and Latin.
International Organizations: Director (since 1963) of the Latin American Bureau of Society of Jesus.
Member (since 1945) of the Society of Jesus.
Other organizations: Director (since 1963), Secretariado Latinoamericano; Director (since 1980), Viajes Selasi, Madrid; Vice-President (since 1985), Fe y Alegría, Madrid; Council (since 1986), Via Maria.
Main field of work: Manager. **Career:** Professor, Colegio San José, Arequipa, Peru from 1954 to 1957. **Educ.:** Lic. Phil. from Universidad Complutense de Madrid, Spain (1951-1954); Lic. Theol. from University of Granada, Spain (1957-1961); M.Soc. and Ind.Rel. from Loyola University, Chicago IL, USA (1962-1963); Técnico en Turismo from Escuela Central, Madrid, Spain (1981). **Interests:** promotion and development of Third World management.
Mailing address: Almagro 6, E-28010 Madrid, Spain. T. (34 1) 319 7581 – 308 0254. Fax 200 9140. C. SELASI. Tx 47642 E.
Residence: See Mailing Address. T. See Mailing Address

Gardinier, Meg

Rel: Roman Catholic.
International Organizations: Representative at UN, New York of the International Catholic Child Bureau.
Mailing address: ICCB, c/o ICO Center, 323 East 47th Floor, New York NY 10017, USA.

Garmo, Georges, Most Rev

Rel: Chaldean Catholic; Ordained on 08 DEC 1945; b. 08 DEC 1921 in Telfek, Iraq; celibate; Citizen of Iraq.
International Organizations: Archbishop of Mosul.
Mailing address: Archevêché Chaldéen, Mayassa, Mosul, Iraq. T. (964 60) 764733

Garnsworthy, Lewis Samuel, Most Rev

Rel: Anglican; Ordained in 1946; b. 18 JUL 1922 in Edmonton, Alberta, Canada; m. to Jean V Allen; c. 1 boy and 1 girl; Citizen of Canada.
International Organizations: Ecclesiastic of the Anglican Church.
Career: Assistant Curate, St John the Baptist in 1946. Rector, St Nicholas' Church, Birch Cliff in 1948. Rector, Church of the Transfiguration, Toronto in 1956. Rector, St John, York Mills in 1960. Suffragan Bishop, Diocese of Toronto in 1968. Bishop, Toronto in 1972. Archbishop, Toronto from 1979 to 1988. Metropolitan, the Ecclesiastical Province of Ontario from 1979 to 1985. **Interests:** gardening. **Hons.:** Hon. D.D., Wycliffe (1969), Trinity (1973), Huron (1976).
Mailing address: Suite 201, 1210 Don Mills Road, Don Mills, Ontario M3B 2N9, Canada.

Garrido, Pablo

Rel: Roman Catholic.
International Organizations: Member of the Carmelite Third Order.
Mailing address: Via Sforza Pallavicini 10, I-00193 Roma, Italy.

Garrone, Gabriel Marie, H E Card

Rel: Roman Catholic; Ordained on 11 APR 1925; b. 12 OCT 1901 in Aix-les-Bains, France; celibate; Citizen of France.
International Organizations: Member (since 1967) of the College of Cardinals.
Career: Professor and Superior, Grand Seminaire de Chambéry from 1925 to 1939. Officer and Prisoner of War from 1940 to 1945. Archbishop Coadjutor, Toulouse from 1947 to 1956. Archbishop, Toulouse from 1956 to 1966. Vice-President, Permanent Council of Plenary Assembly of French episcopate from 1964 to 1966. Pro-Prefect, Congregation of Seminaries and of Universities, Rome from 1966 to 1967. Prefect, Congregation for Catholic Education from 1968 to 1978. Created Cardinal by Pope Paul VI in 1967. In charge of organizing the dialogue between the Church and culture from 1978 to 1988. **Educ.:** D.E.S. Phil. from Institut Notre Dame de la Villette, France; Dr. Phil. Scolastique from University of Grenoble, France; Dr. Théol. from Pontifical Gregorian University, Vatican City State. **Hons.:** Grand Cross, Legion of Honour; Military Cross (1939-45). **Pubs.:** Author of over 50 religious works, including: Que faut-il croire; Morale chrétienne et Valeurs humaines and Le Concile. Orientations.
Mailing address: Piazza San Pietro in Vincoli 6, I-00184 Roma, Italy. T. (39 6) 47 45 410
Residence: See Mailing Address. T. See Mailing Address

Gartenhaus, Jacob, Mrs

Rel: Jewish.
International Organizations: Contact of the International Board of Jewish Missions.
Mailing address: 1805 Balley Ave, Chattanooga TN 37404, USA. T. (1 615) 698 3417.

Garuana, Edmund

Rel: Roman Catholic.
International Organizations: Member of the Carmelite Third Order.
Mailing address: Via Sforza Pallavicini 10, I-00193 Roma, Italy.

Garvey, Helen Maher, Sr

Rel: Roman Catholic; Ordained; celibate.
International Organizations: Superior General of the Sisters of Charity of the Blessed Virgin Mary.
Mailing address: BVM Center, Mount Carmel, 1100 Drive Carmel, Dubuque IA 52001, USA.

Gasparotto, Rinaldo, D.C., Fr

Rel: Roman Catholic; Ordained; celibate.
International Organizations: Superior General of the Prêtres de la Doctrine Chrétienne.
Mailing address: S Maria in Monticelli 28, I-00186 Roma, Italy. T. (39 6) 65 42 292.

Gaspoz, Jean-Marc, O.F.M.Cap., Fr

Rel: Roman Catholic; Ordained on 21 JUN 1964; b. 09 JAN 1939 in Sion, Valais, Switzerland; celibate; Citizen of Switzerland; Lang. French and English.
International Organizations: Guide (since 1989) of the International Young Christian Workers.
Member of the Order of Friars Minor Capuchin.
Main field of work: Coordination, training. **Educ.:** Lic. Theol. from Ecole de Théologie OFM CAP, Sion, Switzerland (1961-1965); ISPC, Paris, France (1965-1966).
Mailing address: Rue Plantin 11, B-1070 Bruxelles, Belgium. T. (32 2) 521 69 83. C. JOCINT BRUXELLES.
Residence: See Mailing Address. T. See Mailing Address

Gassmann, Günther, Rev

Rel: Lutheran; Ordained; b. 15 AUG 1931 in Frankenhausen, Germany; m. to Ursula Kähler; c. 3 boys; Citizen of Germany.
International Organizations: Director, Commission on Faith and Order (since 1984) of the World Council of Churches.
Career: Assistant Professor, University of Heidelberg from 1962 to 1969. Research Professor, Institute for Ecumenical Research, Strasbourg from 1969 to 1976. President, Central Office, Lutheran Churches in Germany, Hanover from 1976 to 1982. Ecumenical Officer, Lutheran World Federation, Geneva from 1982 to 1983. **Interests:** trains, stamps. **Pubs.:** Concepts of Unity in the Faith and Order Movement, 1977; The Unity of the Church, 1983 and numerous articles in periodicals.
Mailing address: PO Box 2100, Route de Ferney 150, CH-1211 Genève 2, Switzerland. T. (41 22) 791 63 37. C. OIKOUMENE GENEVA. Tx 415 730 OIK CH. Fax (41 22) 791 03 61.

Gast, G

Rel: Roman Catholic.
International Organizations: UNESCO Representative of the Pax Christi – International Catholic Peace Movement.
Mailing address: Pax Christi-France, 18 rue Cousté, F-94230 Cachan, France.

Gastel, Mark, C.F.P., Br

Rel: Roman Catholic; Ordained; celibate.
International Organizations: Superior General of the Brothers of the Poor of St Francis.
Mailing address: Johannes-Höver-Haus, Rotbendenstrasse 62, 52076 Aachen, Germany. T. (49 241) 62 044

Gatto, Santi, Dr

Rel: Unknown.
International Organizations: Secretary of the International Institute for the Psychology of Religious Feelings.
Mailing address: Corso XXII Marzo 57, I-20129 Milano, Italy. T. (39 2) 742 64 89. Fax (39 2) 749 10 51.

Gatwa, Tharcisse

Rel: Presbyterian.
International Organizations: General Secretary of the Presbyterian Church of Rwanda.
Mailing address: BP 56, Kigali, Rwanda.

Gaudeul, Bernard, F.I.C.P., Br

Rel: Roman Catholic; Ordained; celibate.
International Organizations: Past Superior General of the Prêtres de Sainte Marie de Tinchebray.
Mailing address: Via della Divina Provvidenza 44, I-00166 Roma, Italy. T. (39 6) 680 56 18.

Gauthier, Aquin, S.C., Br

Rel: Roman Catholic; Ordained; celibate.
International Organizations: Member of the Institute of the Brothers of the Sacred Heart.
Other organizations: Treasurer, National Assembly of Relgious Brothers.
Mailing address: 1337 West Ohio Street, Chicago IL 60622-6490, USA. T. (1 312) 829 8525

Gauthier, Léon, Bishop

Rel: Old Catholic.
International Organizations: Bishop of Old Catholic Church of Switzerland.
Mailing address: St Niklausstrasse 61, CH-4500 Solothurn, Switzerland.

Gavai, R S

Rel: Buddhist.
International Organizations: Vice-President, Member of the Executive Council, and Chairman of the Fourth Committee (Humanitarian Services) of the World Fellowship of Buddhists.
Mailing address: 902 Purna Bldg, Worli Sagar Co-op. Housing Society, Sir Pochkhanawala Road, Worli, Bombay 400 025, India.

Gawa, Aka

Rel: Unknown.
International Organizations: Director, European Branch of the World Conference on Religion and Peace.
Mailing address: Château Pictet, 14 chemin Auguste Vilbert, Grand Saconnex, CH-1218 Genève, Switzerland. T. (41 22) 798 51 62. Fax 791 00 34. C. RELPEACE, GENEVA.

Gay, Arthur E, Dr

Rel: Evangelical.
International Organizations: Executive Director of the World Relief Corporation.
Mailing address: PO Box WRC, Wheaton IL 60189, USA. T. (1 708) 665 0235. Tx 9102524402. Fax (1 708) 653 8023.

Gay, John Dennis, Rev Dr

Rel: Christian; Ordained in 1966; b. 18 JAN 1943 in Stockport, Cheshire, UK; married; c. 1 boy and 1 girl; Citizen of United Kingdom; Lang. English.
International Organizations: Director (since 1980) of the Culham College Institute, Abingdon.
Other organizations: Member (since 1982), Diocese of Oxford Board of Education; Member (since 1983), Oxfordshire Education Committee.
Main field of work: Church educational research. **Career:** Parish Priest, Parish Church of Paddingdon from 1967 to 1972. Chaplain/Senior Lecturer, Culham College of Education from 1972 to 1978. Lecturer, University of Oxford from 1978 to 1980. **Educ.:** B.A. from University of Oxford, UK (1964); D.Phil. from University of Oxford, UK (1969); M.Sc. from University of Oxford, UK (1978). **Pubs.:** Geography of Religion in England, 1971; The Final Report of the Church Colleges, Research Project, 1986 and Religious Education and Collective Worship in Primary Schools, 1992.
Mailing address: Culham College Institute, 60 East St Helen St, Abingdon OX14 5EB, UK. T. (44 235) 502458

Gear, John A., Rev

Rel: Protestant.
International Organizations: Secretary-General of the International Bible Reading Association.
Mailing address: Robert Denholm House, Nutfield, Redhill RH1 4HW, UK. T. (44 73) 782 2411.

Gebs, Simone

Rel: Roman Catholic; b. 22 FEB 1924 in Meudon, France; m. to Charles Gebs; Citizen of Switzerland; Lang. French and English.
International Organizations: International President (since 1988) of the International Movement for the Apostolate in the Independent Social Milieux.
Main field of work: Apostolate. **Educ.:** Ecole de Secretariat de Direction.
Mailing address: Palazzo San Calisto 16, I-00120 Roma, Italy. T. (39 6) 698 71 83
Residence: Chemin du Perreret, CH-1261 Trélex, Switzerland T. (41 22) 69 17 20

Geelen, W, Rev

Rel: Roman Catholic.
International Organizations: Rector of the International Union of Prayer of Banneux Notre-Dame.
Mailing address: Rue de l'Esplanade 57, B-4141 Louveigné, Belgium.

Geertz, Armin W

Rel: Unknown.

International Organizations: Honorary Treasurer of the International Association for the History of Religions.

Mailing address: Dept. of Study of Religion, University of Århus, Building 415, NDR Ringgade, DK-8000 Århus C, Denmark. T. (45 86) 13 67 11. Fax 13 04 90.

Geffré, Claude, O.P., Prof

Rel: Roman Catholic; Ordained on 12 JUL 1953; b. 23 JAN 1926 in Niort, Deux-Sèvres, France; celibate; Citizen of France; Lang. French, English, German and Italian.

International Organizations: Professor of Theology of Religions and Theological Hermeneutics (since 1968) of the UER of Theology and Religious Sciences, Institut Catholique de Paris.

Member (since 1948) of the Order of Preachers. Member, International Council of the World Conference on Religion and Peace.

Other organizations: Director of Series (*Cogitatio fidei*), Editions Du Cerf (since 1975); Vice-President, Association de théologie catholique européenne; Member, Editorial Board, *Le Supplément, Revue d'éthique et théologie morale*.

Main field of work: University education. **Career:** Professor of Dogmatic Theology, Faculty of Theology, Saulchoir from 1957 to 1968. Rector, Facultés Dominicaines du Saulchoir from 1965 to 1968. Ordinary Professor of Fundamental Theology, UER of Theology and Religious Sciences, Institut Catholique de Paris from 1968 to 1988. Director, Cycle des Etudes du Doctorat (CED), Institut Catholique de Paris from 1973 to 1984. Co-Director, Spirituality Dept., then Fundamental Theology Dept., *Concilium* from 1965 to 1991. Theologian member of the founding, *Concilium* from 1975 to 1987. Member, Philosophy Commission, Centre national des lettres from 1987 to 1991. Visiting Professor, Ecole des Sciences Philosophiques et Religieuses, Facultés Universitaires Saint-Louis, Brussels from 1972 to 1975. Visiting Professor, Faculty of Theology, University of Sherbrooke, Quebec in 1979. Visiting Professor, Faculty of Theology, University of Fribourg from 1984 to 1985. Visiting Professor, Faculty of Theology, Laval University, Quebec in 1988. Visiting Professor, Université inter-âges de la Sorbonne from 1984 to 1989. **Educ.:** Phil. and Theol. Stds. from Facultés Dominicaines du Saulchoir, Corbeil, France (1949-1955); D.Theol. from Pontifical University San Tommaso d'Aquino, Vatican City State (1957); Maître en Théol. from Institut Catholique de Paris, France (1989). **Interests:** history of religions. **Hons.:** Knight of the National Order of Merit. **Mem.:** Groupe de recherches islamo-chrétien (GRIC) **Pubs.:** Un nouvel âge de la théologie; Le Christianisme au risque de l'interprétation and Passion de l'homme, Passion de Dieu.

Mailing address: Institut Catholique de Paris, 21 rue d'Assas, F-75270 Paris Cedex 06, France.

Residence: 29 bd Latour-Maubourg, F-75340 Paris Cedex 07, France.

Gega Lama, G

Rel: Buddhist.

International Organizations: Master of Tibetan painting of the Institut Yeunten Ling.

Other organizations: Official holder, the typically Kagyu tradition of Tibetan art (style known as Karma Gardri).

Pubs.: *Principles of Tibetan Art.*

Mailing address: Institut Yeunten Ling, Château de Fond l'Evêque, 4 promenade Saint Jean l'Agneau, B-4500 Huy, Belgium. T. (32 85) 21 48 20

Geiser, Daniel

Rel: Mennonite.

International Organizations: Ecumenical Contact Person of the United German Mennonite Congregations.

Mailing address: Torneystrasse 90a, 56567 Neuwied, Germany.

Gelegjamts, Lubsantseren, Dr

Rel: Buddhist; b. 10 JUL 1935 in Dornogobi, Mongolia; m. to Gombo Tuja; c. 2 boys and 1 girl; Citizen of Mongolia; Lang. Mongolian, Russian and English (reading).

International Organizations: Secretary-General (since 1986) of the Asian Buddhist Conference for Peace.

Main field of work: Philosophy research. **Educ.:** Dip. from Mongolian State University, Ulan Bator, Mongolia (1952-1956). **Interests:** Buddhist philosophy. **Hons.:** Mongolian orders and medals. **Mem.:** International Council, World Conference on Religion and Peace (1973); Council of Researchers, Academy of Science, Mongolia (1973) **Pubs.:** Ideas of the Nagarjuna Philosophy, Ulan Bator, 1981 (in Mongolian); Main Principles of the Madhyamaka Philosophy, Ulan Bator, 1987 and Philosophical Analysis of Nagarjuna's Treatise 'Shunyatasaptatikarika', Novosibirsk, 1989 (Istochnikovedenye i Tekstologiya pamyatnikov .. – in Russian).

Mailing address: ABCP Headquarters, Ulan Bator 51, Mongolia. T. 53497

Residence: Peace St, I-40000 Bld 7, Flat 13, Ulan Bator, Mongolia. T. 24000

Gemayel, Boutros, Most Rev

Rel: Maronite Catholic; Ordained on 12 APR 1959; b. 29 JUN 1932 in Ain el Kharroubé, Cyprus; celibate; Citizen of Cyprus.

International Organizations: Archbishop of Cyprus.

Member, Special Commission for the Liturgy of the Sacred Congregation for the Oriental Churches.

Mailing address: POB 2249, Maronite Archbishop's House, 8 Favierou St, Nicosia, Cyprus. T. (357 2) 427966. Fax 499155. Tx 2933 LITSCO CY

Gemello, Francesco, S.S.C., Fr

Rel: Roman Catholic; Ordained; celibate.

International Organizations: Superior General of the Société Sacerdotale de Saint Joseph-Benoît Cottolengo.

Mailing address: Via di Villa Alberici 14, I-00165 Roma, Italy. T. (39 6) 63 09 03

Gendrot, Marcel, O.M.V., Fr

Rel: Roman Catholic; Ordained; celibate.

International Organizations: Pontifical Commissioner of the Oblats de la Vierge Marie.

Mailing address: Viale Trenta Aprile 17, I-00153 Roma, Italy. T. (39 6) 589 10 71, (39 6) 589 05 36

Genheimer, Ron

Rel: Evangelical.

International Organizations: International Director of the Africa Evangelical Fellowship.

Mailing address: 35 Kingfisher Court, Hambridge Road, Newbury RG14 5SJ, UK. T. (44 734) 583848. C. Fellowship Reading England.

Genoud, Jocelyne

Rel: Roman Catholic.

International Organizations: Secretary-General of the International Committee of the Order of Malta for Leprosy Relief.

Mailing address: 3 Place Claparède, CH-1205 Genève, Switzerland. T. (41 22) 346 86 87. Fax 347 08 61.

Gensichen, Hans-Werner, Prof Dr

Rel: Lutheran; Ordained in 1943; b. 10 MAR 1915 in Lintorf, Germany; m. to Dr E Anneliese Rudert; Citizen of Germany; Lang. German and English.

International Organizations: Professor Emeritus of History of Religions and Missiology of the University of Heidelberg.

Main field of work: University education. **Career:** Chairman, Deutsche Gesellschaft für Missionwissenschaft from 1966 to 1989. Chairman, International Association for Mission Studies from 1972 to 1974. **Educ.:** M.Th. from Princeton Theological Seminary, NJ, USA (1938); Dr. Theol. from Georg-August-Universität Göttingen, Germany (1942). **Pubs.:** Glaube für die Welt, 1971; Missionsgeschichte der neuen Zeit, 1976 and Weltreligionen und Weltfriede, 1985.

Mailing address: Eckenerstrasse 1, 69121 Heidelberg, Germany. T. (49 6221) 480935

Residence: See Mailing Address. T. See Mailing Address

George, Kondothra M, Prof Dr

Rel: Protestant.

International Organizations: Member of the Ecumenical Institute Bossey.

Mailing address: Château de Bossey, CH-1298 Céligny, Switzerland. T. (41 22) 776 25 31. C. INSTITUTCELIGNY GENEVA. Fax (41 22) 776 01 69.

Gerada, Emanuele, H E Mgr

Rel: Roman Catholic; Ordained on 01 AUG 1943; b. 18 MAY 1920 in Zejtun, Malta; celibate.
International Organizations: Apostolic Nuncio to Ireland (since 1989) of the Roman Catholic Church.
Other organizations: Titular Archbishop, Nomento.
Mailing address: 183 Navan Road, Dublin 7, Ireland. T. (353 1) 30 93 44. Fax 30 45 61

Geraisy, Sami F, Dr

Rel: Christian.
International Organizations: Chairman of the International Christian Committee for Israel.
Mailing address: PO Box 304, Nazareth, Israel.

Gérard, Yves, Mgr

Rel: Roman Catholic; Ordained; celibate.
International Organizations: General Responsible of the Gospel and Mission Groups.
Other organizations: Vicar-General, Verdun.
Mailing address: Evêché, BP 90, 2 Place Mgr Ginisty, F-55103 Verdun CEDEX, France. T. (33) 29 86 01 72.

Géraud, A

Rel: Roman Catholic.
International Organizations: UNESCO Representative of the International Catholic Union of the Press.
Other organizations: Editor-in-Chief, *La Croix*.
Mailing address: 3-5 rue Bayard, F-75393 Paris CEDEX 08, France.

Gerber, Samuel

Rel: Mennonite.
International Organizations: Member, Executive Committee of the Mennonite World Conference.
Mailing address: Liestalerstrasse 18, CH-4412 Nuglar, Switzerland.

Gerig, Zenas, Dr

Rel: Evangelical.
International Organizations: Contact of the Evangelical Association of the Caribbean.
Mailing address: PO Box 121, Constant Spring, Kingston 8, Jamaica.

Gerling, Helene, Rev

Rel: Spiritualist.
International Organizations: Director and Secretary of the Universal Harmony Foundation.
Mailing address: 5903 Seminole Blvd, Seminole FL 33542, USA.

German, G, H H

Rel: Orthodox: Serbian.
International Organizations: Patriarch of Serbia.
Mailing address: PO Box 182, 1101 Belgrade, Yugoslavia.

Gerö, Julia

Rel: Baptist.
International Organizations: President of the European Baptist Women's Union.
Mailing address: c/o Baptist Union of Hungary, Jarno utca 5, H-1155 Budapest, Hungary.

Gervais, Christiane, PDDM, Sr M

Rel: Roman Catholic; Ordained; celibate.
International Organizations: Superior General of the Sister Disciples of the Divine Master.
Mailing address: Via Gabriele Rossetti 17, I-00152 Roma, Italy. T. (39 6) 588 0872

Gervais, Marc, S.J., Rev

Rel: Roman Catholic; Ordained; b. 03 DEC 1929 in Sherbrooke PQ, Canada; celibate; Citizen of Canada.
International Organizations: Professor of Film, Aesthetics, etc., Department of Communication Studies of the Concordia University, Montreal.
Member of the Society of Jesus.
Other organizations: Founding Director and Board Chairman (since 1988), The Loyola Institute for Studies in International Peace.
Main field of work: University education. **Career:** Commissioner, CRTC from 1981 to 1986. Board of Directors, Canadian Film Institute from 1986 to 1989. Member, Board of Judges, Canadian Film Awards. Member, film juries (Cannes, Venice, Oxford, etc.). TV and radio work including TV Specials on Ingmar Bergman for CBC and CTV. **Educ.:** B.A. from Loyola College (1950); L.Ph. from L'Immaculée Conception, Montreal, Canada (1956); M.F.A. Drama from Catholic University of America, Washington DC, USA (1960); L.Th. from Regis College, Toronto, Canada (1964); M.A. from Saint Mary's University, Halifax, Canada (1964); Ph.D. Film Aesth. from Sorbonne University, Paris, France (1973). **Interests:** sports, arts. **Pubs.:** Pasolini, 1972 and other writings including film articles in journals and newspapers in UK, France, USA, Canada, Spain, Sweden, Italy and Australia.
Mailing address: 7141 Sherbrooke St W, Montréal PQ H4B 1R6, Canada.

Gese, H, Prof Dr

Rel: Evangelical.
International Organizations: Professor of Old Testament of the University of Tubingen.
Mailing address: Evangelisch-theologische Fakultät, Universität Tübingen, Whilhemstrasse 7, 72074 Tübingen, Germany.

Geuns, André Jozef, Dr

Rel: Catholic; b. 30 SEP 1936 in Mol, Belgium; m. to Anna Paola Mundula; c. 1 boy and 1 girl; Citizen of Belgium; Lang. Dutch, English, French, Italian, German, Spanish, Latin, Greek and Chichewa.
International Organizations: President (since 1989) of the International Council of Associations of Theological Libraries.
Other organizations: Member (since 1973), Vereniging van het Theologisch Bibliothecariaat, Netherlands.
Main field of work: Coordination of library work. **Career:** Professor, International Seminary Malawi and Zambia from 1966 to 1969. Professor, Theologicum, Heverlee from 1971 to 1973. Librarian, Faculty of Theology, Tilburg University, Netherlands from 1973 to 1990. President, Vereniging van het Theologische Bibliothecariaat, Netherlands from 1979 to 1989. **Educ.:** Lic. Theol. from International Theological College, Totteridge, London, UK (1958-1962); J.C.D. from Pontifical Gregorian University, Vatican City State (1962-1966); Post Doc. Soc. Anthro. from Catholic University of Louvain (UCL), Belgium (1971-1973); Librarianship from University of Amsterdam, Netherlands (1973-1974). **Interests:** Librarianship, archeology. **Pubs.:** *Le sacerdoce diocesain et la consécration à Dieu dans les Instituts séculiers de prêtres, Rome, 1966, 530 pp.; Bibliographie commentée du prophétisme*, Congo, 1973, 81 pp.; *Chronologie des mouvements religieux indépendants au Bas-Zaïre*, Leiden, 1974, 35 pp.; *De bibliotheek van het voormalig seminarie*, Haaren, 1981 and *Card Index Catalogue on the Patrologia Latina et Graeca*, by J P Migne, Brepols, Turnhout, 1984.
Mailing address: Via Muzio Clementi 71C, Valcanneto di Cerveteri, I-00050 Roma, Italy. T. (39 6) 991 6907
Residence: See Mailing Address. T. See Mailing Address

Ghebray, Kidane, Abba

Rel: Orthodox: Ethiopian.
International Organizations: Abbot of the Ethopian Orthodox Church.
Mailing address: PO Box 2454, Addis Ababa, Ethiopia.

Ghilan, Maxim

Rel: Jewish.
International Organizations: Secretary-General of the International Jewish Peace Union.
Mailing address: BP 130, F-75463 Paris CEDEX 10, France. T. (33 1) 42 46 82 47. Fax 42 46 51 77.

Ghirardo, Danilo
Rel: Buddhist: Mahayana.
International Organizations: Member of the Foundation for the Preservation of the Mahayana Tradition.
Other organizations: Director, Centro Lama Tzong Khapa.
Mailing address: Ouazza V Emanuele 18, I-31050 Villorba, Italy. T. (39 522) 928 079

Ghirotto, Domenico, M.C.C.I., Fr
Rel: Roman Catholic; Ordained; celibate.
International Organizations: Procurator-General of the Comboni Missionaries of the Heart of Jesus.
Mailing address: Via Luigi Lilio 80, CP 10733, I-00143 Roma, Italy. T. (39 6) 519 1224.

Ghose, Chinmoy K, Sri
Rel: Spiritualist.
International Organizations: Director of the Sri Chinmoy Centre.
Mailing address: 85-45 149th St, Jamaica Hills NY 11435, USA. T. (1 212) 523 3471.

Giacomo, Capraro D, SDV, Rev
Rel: Roman Catholic; Ordained; celibate.
International Organizations: Superior General of the Society of Divine Vocations.
Mailing address: Via Cortina D'Ampezzo, I-00135 Roma, Italy. T. (39 6) 328 66 04.

Giacon, Modesto, F.d.C.C., Fr
Rel: Roman Catholic; Ordained; celibate.
International Organizations: Procurator-General of the Sons of Charity – Canossians.
Mailing address: Lg San Giorgio 4, I-00125 Acilia, Italy. T. (39 6) 605 00 52, (39 6) 605 13 48

Giampiccoli, Franco, Rev
Rel: Waldensian Reformed; Ordained on 25 AUG 1959; b. 12 SEP 1934 in Milan, Italy; m. to Danielle Rollier; c. 3 girls; Citizen of Italy; Lang. English and French.
International Organizations: Moderator (since 1986) of the Waldensian National Board (Tavola Valdese).
Main field of work: Ministry. **Career:** Director, Ecumenical Centre Agape, Italy from 1966 to 1971. Editor, *La Luce* (Waldensian-Methodist weekly newspaper) from 1976 to 1986. **Educ.:** L.Th. from Waldensian Theological Seminary, Rome, Italy (1954-1957); M.Th. from Princeton Theological Seminary, NJ, USA (1958). **Pubs.:** a biography on Dag Hammarskjöld; a book on mixed marriages in Italy and an essay on suffering.
Mailing address: Tavola Valdese, Via Firenze 38, I-00184 Roma, Italy. T. (39 6) 474 55 37
Residence: Via Marianna Dionigi 57, I-00193 Roma, Italy. T. (39 6) 320 49 26

Giannini, Paolo, O.S.B.I., Most Rev.
Rel: Roman Catholic; Ordained on 21 NOV 1943; b. 11 JUL 1920 in Albania; celibate.
International Organizations: Superior General of the Order of St Basil the Great.
Advisor of the Sacred Congregation for the Oriental Churches.
Other organizations: Archbishop (since 1972), Territorial Abbacy Santa Maria di Grottaferrata.
Mailing address: Badia Greca, I-00046 Grottaferrata, Italy. T. (39 6) 945 93 09

Gibran, Gabriel, Most Rev Archbishop
Rel: Antiochian Orthodox.
International Organizations: Archbishop of Antiochian Orthodox Church.
Mailing address: PO Box 120, Randwick NSW 2031, Australia.

Gibson, John Clark Love, Rev Prof
Rel: Protestant; Ordained; b. 28 MAY 1930; m. to Agnes Gilmour Russell; c. 4 boys and 1 girl.
International Organizations: Professor of Hebrew and Old Testament Studies (since 1987) of the University of Edinburgh.
Career: Minister, New Machar, Aberdeenshire from 1959 to 1962. Lecturer in Hebrew and Semitic Languages, University of Edinburgh from 1962 to 1973. Reader, University of Edinburgh from 1973 to 1987. **Educ.:** M.A., B.D. from University of Glasgow, UK; D.Phil. from University of Oxford, Magdalen College, UK. **Interests:** Scottish literature, spy novels, golf. **Pubs.:** Job: Daily Study Bible, 1985; Contributor to: The Bible in Scottish Life and Literature, 1988 and Ascribe to the Lord: In Memory of Peter C Craigie, 1988.
Mailing address: New College, Mound Place, Edinburgh EH1 2LX, UK. T. (44 31) 225 8400
Residence: 10 South Morton Street, Edinburgh EH15 2NB, UK. T. (44 31) 669 3635

Gibsone, Craig John Kenneth
Rel: New Age; b. 25 OCT 1941 in Harvey, Western Australia, Australia; single; Citizen of Australia; Lang. English.
International Organizations: Roving Ambassador of the Findhorn Foundation.
Main field of work: Meditation and personal growth. **Career:** Foculiser, Findhorn Foundation (for the past 5 years). **Interests:** practical spirituality, windsurfing.
Mailing address: The Park, Findhorn, Forres IV36 OTZ, UK. T. (44 309) 69 03 11. Fax 69 09 33.
Residence: See Mailing Address. T. See Mailing Address

Gienke, Horst, Bishop Dr
Rel: Evangelical; Ordained.
International Organizations: Bishop of Evangelical Church of Greifswald.
Mailing address: Bahnhofstrasse 35/36, 17489 Greifswald, Germany.

Gigacz, Stefan Robert
Rel: Greek Catholic; Unordained; b. 02 SEP 1955 in Melbourne, Victoria, Australia; single; Citizen of Australia; Lang. English and French.
International Organizations: Treasurer (since 1990) of the International Young Christian Workers.
Main field of work: Coordination, training. **Career:** Australian Young Christian Workers from 1978 to 1985. Solicitor from 1986 to 1987. International Young Christian Workers, Asia Pacific from 1988 to 1989. **Educ.:** LL.B. from University of Melbourne, Victoria, Australia (1973-1977). **Pubs.:** *Moscow-Sydney Line* (bimonthly newsletter).
Mailing address: Rue Plantin 11, B-1070 Bruxelles, Belgium. T. (32 2) 521 69 83. C. JOCINT BRUXELLES.
Residence: See Mailing Address. T. See Mailing Address

Giglio, Paolo, H E Mgr
Rel: Roman Catholic; Ordained on 12 APR 1952; b. 20 JAN 1927 in Valletta, Malta; celibate; Citizen of Vatican; Lang. Italian, English, French, Spanish, Maltese, Portuguese and Latin.
International Organizations: Apostolic Nuncio to Nicaragua (since 1986) of the Roman Catholic Church.
Other organizations: Titular Archbishop, Tindari.
Educ.: S.T.L. from Pontifical Gregorian University, Vatican City State (1954); D.Cn.Law from Pontifical Gregorian University, Vatican City State (1958).
Mailing address: Apostolic Nunciature, Apartado Postal 506, Managua DN, Nicaragua. T. (505 2) 658 657
Residence: See Mailing Address. T. See Mailing Address

Giglioni, Paolo, Rev Fr
Rel: Roman Catholic; Ordained; celibate.
International Organizations: Director of the Institute of Missionary Catechesis.
Advisor of the Congregation for the Clergy.
Other organizations: Member, Commission for the Catechism and the Catechists.
Mailing address: Pontificia Universitas Urbaniana, 16 via Urbano VIII, I-00165 Roma, Italy. T. (39 6) 68 75 992 – (39 6) 68 68 640.

Gijsen, Johannes Baptist Matthijs, Rev

Rel: Roman Catholic; Ordained on 06 APR 1957; b. 07 OCT 1932 in Oeffelt, Netherlands; celibate.
International Organizations: Past Administrator of the International Institute 'Mater Ecclesiae Domesticae'.
Other organizations: Bishop (since 1972), Roermond.
Mailing address: Rolduc, Heyendahllaan 82, NL-6464 EP Kerkrade, Netherlands.

Gilbert, Arthur Joseph, Most Rev

Rel: Roman Catholic; Ordained on 03 JUN 1943; b. 26 OCT 1915 in Hedley BC, Canada; celibate; Citizen of Canada.
International Organizations: Bishop Emeritus (since 1986) of St John, NB.
Career: Served as Curate, St Andrews Parish, NB. Secretary to Bishop and Chancellor, Diocese of Saint John from 1943 to 1949. Director, Silver Falls Orphanage from 1949 to 1955. Pastor, St Pius X Parish from 1955 to 1969. Pastor, St Joseph's, Loch Lomond from 1969 to 1971. Pastor, St Joachim's, Saint John in 1974. Bishop, Saint John, NB from 1974 to 1986. Former Chancellor and Chairman, Board of Governors, St Thomas University, Fredericton, NB. Former Chairman, Board of Directors, *The New Freeman* (diocesan newspaper). **Educ.:** B.A. from Saint Francis Xavier University, Antigonish, Nova Scotia, Canada; B.D. from Holy Heart Seminary. **Hons.:** Knight, Order of Canada.
Mailing address: 100 Villa Madonna Road, Renforth, Saint John NB E2H 2T2, Canada. T. (1 506) 847 4012
Residence: See Mailing Address.

Gilbert, Bruce

Rel: Protestant.
International Organizations: Director, North American Regional Office of the World Student Christian Federation.
Mailing address: SCM Canada, 310 Danforth Ave, C3, Toronto M4K 1N6, Canada. T. (1 416) 463 4312. Fax 778 8938.

Gilbert, Patrick Nigel Geoffrey

Rel: Church of England; b. 12 MAY 1934.
International Organizations: Secretary-General (since 1971) of the Society for Promoting Christian Knowledge.
Other organizations: Governor (since 1971), St Martin-in-the-Fields School; Trustee (since 1971), Richards Trust; Trustee (since 1973), Overseas Bishoprics Fund; Governor (since 1978), Roehampton Institute of Higher Education; Chairman (since 1984), Nikaean Club; Governor (since 1985), Pusey House; Member of Court (since 1987), City University; Director (since 1988), Surrey Building Society; Member of Board (since 1988), National Youth Dance; Member of Council (since 1990), The Publishers Association.
Career: Lecturer in Further Education at South Berks College from 1959 to 1962. Personal Assistant to Sir Edward Hulton from 1962 to 1964. With Oxford University Press from 1964 to 1969. Managing Director, Linguaphone Group, Westinghouse from 1969 to 1971. Chairman, Camden Arts Council from 1970 to 1974. Chairman, Standing Conference of London Arts Council from 1975 to 1978. Trustee, World Association for Christian Communication from 1975 to 1987. Chairman (and Founder), National Association of Local Arts Councils from 1976 to 1980. Chairman, Embroiders' Guild from 1977 to 1978. Chairman, Greater London Arts Association from 1980 to 1984. Chairman, Concord Multicultural Arts Trust from 1980 to 1989. Vice-President, National Association of Local Arts Councils from 1980 to 1989. Chairman, Working Party on Arts Funding (set up by Minister for the Arts) in 1981. Member of Executive, Anglican Centre in Rome from 1981 to 1990. Governor, Contemporary Dance Trust from 1981 to 1990. Trustee, Anglican Consultative Council Research Project from 1982 to 1984. Trustee, Dancers Resettlement Fund from 1982 to 1990. Chairman, Dancers Resettlement Trust from 1987 to 1990. **Educ.:** University of Oxford, Merton College, UK. **Interests:** reading, writing, travel (visited 120 countries), arts, golf. **Hons.:** Freeman of the City of London (1966), City of Savannah; Order of St Vladimir (1977); Hon. D.Litt., Columbia Pacific University (1982); Institute of Directors Diploma in Company Administration (1984). **Mem.:** Honorary Life Member: Greater London Arts Association (1978), Association for Development in the Arts (1989); Member: National Economic Research Council (1982), Worshipful Company of Paris Clercks, Guild of Freemen of the City of London; Fellow: Royal Society of Arts (1978), Institute of Directors (1982), British Institute of Management (1982); Honorary Fellow, Columbia Pacific University (1982); Clubs: Athenaeum, Walton Heath Golf, City Livery Club, Bread Street Ward Club, Coleman Street Ward Club, United Wards Club; and member of a large number of societies.
Mailing address: Holy Trinity Church, Marylebone Road, London NW1 4DU, UK. T. (44 71) 387 5282. Fax 388 2352
Residence: 3 The Mount Square, London NW3 6SU, UK T. (44 71) 794 8807

Gilby, Peter S

Rel: Christian.
International Organizations: President of the International Secretariat of Christian Artists.
Mailing address: Rosehill House, Lydiate L31 4JF, UK.

Gill, David Muir, Rev

Rel: Uniting Church in Australia; Ordained on 27 MAY 1965; b. 19 JUN 1939 in Sydney NSW, Australia; single; Citizen of Australia; Lang. English and French.
International Organizations: General Secretary (since 1988) of the Australian Council of Churches.
Main field of work: Church administration. **Career:** Travelling Secretary, Australian Student Christian Movement from 1963 to 1964. Secretary, Church and Society, World Council of Churches (WCC) from 1968 to 1973. Executive Secretary, Fifth Assembly, WCC from 1973 to 1976. Director, Renewal and Congregational Life, WCC from 1976 to 1979. General Secretary, Uniting Church in Australia from 1980 to 1988. **Educ.:** B.A. from University of Sydney, New South Wales, Australia (1957-1960); B.D. from Melbourne College of Divinity, Australia (1961-1965); M.S.T. from Hartford Seminary Foundation, USA (1965-1967). **Pubs.:** Editor, Gathered for Life, WCC, 1983 (the official report of the sixth assembly of the World Council of Churches).
Mailing address: POB C199, Clarence St, Sydney NSW 2000, Australia. T. (61 2) 299 2215. Fax 262 4514. Tx 171715
Residence: 20/114 North Steyne, Manly NSW 2095, Australia. T. (61 2) 977 6593

Gillespie, John M, Rev

Rel: Protestant; Ordained.
International Organizations: Director-General of the World Missions Fellowship.
Mailing address: PO Box 5148, 729 Molalla Ave 3, Oregon City OR 97045, USA. T. (1 503) 655 5152.

Gillon, Luc-Pierre A, Mgr

Rel: Roman Catholic; Ordained; b. 15 SEP 1920 in Rochefort, Belgium; celibate; Citizen of Belgium.
International Organizations: Scientist and Ecclesiastic of the Roman Catholic Church.
Career: Research physicist, Institut Interuniversitaire des Sciences Nucléaires in 1948. Guest staff member, Brookhaven National Laboratory, New York in 1953. Rector, Lovanium University in 1954. Professor of Nuclear Engineering, Catholic University of Louvain in 1972. Trustee, Nuclear Research Centre, Mol. **Interests:** flying (private aeroplane).
Mailing address: 23 rue du Blanc Try, B-1301 Bierges, Belgium.

Gilmore, Alec, Rev

Rel: Baptist; Ordained on 01 JUL 1952; b. 23 JAN 1928 in Bacup, Lancashire, UK; m. to Enid Ruth Batten; c. 2 boys; Citizen of United Kingdom; Lang. English, French, German and Spanish.
International Organizations: Secretary-General (since 1976) of the United Society for Christian Literature.
Chairman, Books and Translation Commission (since 1987) of the European Baptist Federation.
Other organizations: Member (since 1959), Baptist Union Council; Director (since 1983), Feed the Minds; Chairman (since 1983), Baptist Times, Ltd.; Trustee (since 1984), Psalms and Hymns Trust, Chairman of new hymn book Editorial Committee; Member of the Council (since 1985), Bible Reading Fellowship.
Main field of work: Ministry. **Career:** Governor, Northern Baptist College from 1956 to 1985. Minister, Kingsthorpe Baptist Church, Northampton from 1952 to 1962. Minister, West Worthing Baptist Church from 1962 to 1974. Editor, Carey Kingsgate Press (part-time) from 1962 to 1967. Associate Editor, Lutterworth Press (part-time) from 1967 to 1974. Editor, Lutterworth Press from 1975 to 1983. Visiting Minister, First Baptist Church, Philadelphia PA in 1974. Visiting

Professor, Louisville Baptist Seminary, KY in 1979. Member of the Court, University of Sussex in 1990. Regular contributor to *Partners in Learning* from 1968 to 1975. Regular contributor to *Upper Room Disciplines* (USA) from 1968 to 1977. Regular contributor to *Pulpit Digest* (USA). Regular contributor over thirty years to *Baptist Times*, *British Weekly*, *Baptist Quarterly*, *Christian*, *New Christian*, and local papers in Northampton and Worthing. Creator, several series of books. Extra-mural Lectures under Leicester University, Southampton University and the WEA. Regular contributor over a period of ten years to religious programmes (ITV). Regular contributor over a period of ten years to religious programmes (BBC/TV), especially 'Seeing and Believing' and 'Meeting Place'. Regular contributor over eights years to 'Pause for Thought' (BBC/Radio 2). Regular contributor for nine years to 'Prayer for the Day' (BBC/Radio 4). For ten years a member of Radio Brighton's Religious Advisory Panel, a regular contributor to religious programmes and originator of the programme 'Insight'. **Educ.:** B.A. Theol., B.D., M.A. from Manchester Baptist College and Victoria University of Manchester, UK (1946-1952). **Interests:** writing, publishing, broadcasting. **Mem.:** Society for Old Testament Study (1953), Society for the Study of Theology (1974), Athenaeum Club, London (1979) **Pubs.:** Editor and contributor to *Christian Baptism*, Lutterworth Press (UK) and Judson Press (USA), 1959; *Tomorrow's Pulpit*, Lutterworth Press (UK) and Judson Press (USA), 1975 and *Have a Good Day*, Hodder and Stoughton, 1975.
Mailing address: Rannoch Mount Way, North Lancing BN15 0QD, UK. T. (44 483) 577877
Residence: See Mailing Address. T. (44 903) 767661

Gilmour, Daphne Elizabeth

Rel: Reformed Church.
International Organizations: Member of the Executive Committee of the World Alliance of Reformed Churches.
Mailing address: 18 Ballyhannon Park, Portadown, UK. T. (44 762) 333 593

Gindi, Nathan, Supt

Rel: Free Methodist.
International Organizations: Member of the Free Methodist World Fellowship.
Mailing address: PO Box 36, Asyut, Egypt.

Gingrich, Paul M

Rel: Mennonite.
International Organizations: President of the Mennonite Board of Missions.
Mailing address: Box 3470, Elkhart IN 46515-0370, USA. T. (1 219) 294 75 23.

Gintings, E P, Rev

Rel: Protestant; Ordained.
International Organizations: General Secretary of the Karo Batak Protestant Church.
Mailing address: Jln Kapt Pala Banun, 90 Kabanjahe, North Sumatra, Indonesia.

Giordano, Michele, H E Card

Rel: Roman Catholic; Ordained on 05 JUL 1953; b. 26 SEP 1930 in Tursi-Lagonegro, Italy; celibate; Citizen of Italy.
International Organizations: Member (since 1988) of the College of Cardinals.
Member of the Congregation for Bishops. Member of the Congregation for the Evangelization of Peoples. Member of the Congregation for the Clergy. Member of the Pontifical Council for Pastoral Assistance to Health Care Workers.
Other organizations: Archbishop, Naples; Member, Council of Cardinals and Bishops.
Career: Elected to the Titular Church, Lari Castello in 1971. Consecrated Bishop in 1972. Prefect at Matera e Irsina in 1974. Transferred to Naples in 1987. Created Cardinal by Pope John Paul II in 1988.
Mailing address: Largo Donnaregina 23, I-80138 Napoli, Italy. T. (39 81) 44 91 18

Giorgio, Giuseppe, M.S.P., Fr

Rel: Roman Catholic; Ordained; celibate.
International Organizations: Superior General of the Missionnaires Serviteurs des Pauvres.
Mailing address: Via Pindemonte 1/E, I-90129 Palermo, Italy. T. (39 91) 42 20 63.

Girardi Oliviero, Giuseppe G, S.C.J., Fr

Rel: Roman Catholic; Ordained; celibate.
International Organizations: Procurator-General of the Lay Dehonians.
Procurator-General of the Priests of the Sacred Heart of Jesus.
Mailing address: c/o Sacerdoti del Sacro Cuore, Via Casale S Pio V 20, I-00165 Roma, Italy. T. (39 6) 66 38 809 – 66 38 825.

Giri, Chidda, Prof Dr

Rel: Hindu; b. 15 JUL 1932 in Bulandshar, Uttar Pradesh, India; m. to Rajwati; c. 5 boys; Citizen of India; Lang. English, Hindi, Sanskrit and Punjabi.
International Organizations: Professor of Yoga (since 1963) of the Higher Education Department of Punjab, Government of India.
Main field of work: Research, administration and teaching of yoga. **Career:** Has organized many seminars, conferences, workshops on yoga. **Educ.:** M.A. Sanskrit from Agra University, Uttar Pradesh, India (1961-1962); M.A. Physical Educ. from Punjabi University, Patiala, India (1965-1966); Ph.D. Physical Educ. from Punjabi University, Patiala, India (1973-1977). **Interests:** Yoga, physical education in yoga. **Mem.:** Many academic bodies of the various universities of India **Pubs.:** Author of 2 books and about 30 research papers and articles on yoga, physical education and religion.
Mailing address: Quarter No 467, School Street, Anand Nagar, Tripuri 147 001.

Giri, Rakesh, Dr

Rel: Hindu; b. 02 MAY 1956 in Khurja, Uttar Pradesh, India; m. to Jaishree Kaushal; c. 2 boys; Citizen of India; Lang. Hindi, Punjabi and English.
International Organizations: Head, Yoga Institution as State Yoga Organiser (since 1982) of the Government Yoga Health Organisation.
Main field of work: Yoga. **Educ.:** M.Sc. from Punjabi University, Patiala, India (1974-1976); Dip. Yoga Educ. from Government College of Education, Chandigarh (1976-1977); Ph.D. from Panjab University, Chandigarh, India (1982-1986). **Mem.:** Divine Life Society **Pubs.:** Research papers on diabetes and yoga; articles on mental stress and yoga and article on critical evaluation of yoga.
Mailing address: Govt. Yoga Health Organisation, Sector 23-A, Chandigarh 160023, India.
Residence: 3271 B, Sector 24-D, Chandigarh 160023, India

Giroux, Gilberte, Sr

Rel: Roman Catholic; Ordained; celibate.
International Organizations: Superior General of the Missionary Sisters of Our Lady of the Angels.
Mailing address: 323 rue Queen, Lennoxville PQ J1M 1K8, Canada. T. (1 819) 569 9248.

Girsang, H M, Rev

Rel: Protestant; Ordained.
International Organizations: General Secretary of the Simalungun Protestant Christian Church.
Mailing address: Jln Jend. Sudirman 14, Kotak Pos 2, Pematang Siantar, Sumatra Utara, Indonesia.

Gjerdi, Ivar

Rel: Christian.
International Organizations: Executive Officer of the Christian Temperance Council for the Nordic Countries.
Mailing address: c/o KE, Keysersgate 1 4, N-Oslo 1, Norway. T. (47 2) 20 77 06.

Glazemaker, Antonius Jan, Mgr

Rel: Old Catholic; Ordained on 08 DEC 1979; b. 19 APR 1931 in Hilversum, Netherlands; m. to G G de Groot; c. 2 girls; Citizen of Netherlands; Lang. Dutch, English, German, French and Latin.
International Organizations: President of the International Old Catholic Bishops' Conference.
Other organizations: Archbishop of Utrecht (since 1982), Old-Catholic Church of the Netherlands.
Career: Parish priest in Leiden from 1956 to 1963. Parish priest in Ijmuiden from 1963 to 1980. Dean, Harlem from 1976 to 1980. Bishop, Arnhem from 1980 to 1982. **Educ.:** Theological Stds. from Old-Catholic Seminary. **Interests:** Music, liturgy. **Hons.:** Dr. h.c. **Pubs.:** *5 jaar Seminarie – Old-Catholic Seminarie*, 1975 and *Communion and episcopacy: an Old-Catholic Perspective*, Ripon College Cuddendon Publ., 1988.
Mailing address: Kon Wilhelminalaan 3, NL-3818 HN Amersfoort, Netherlands. T. (31 33) 620875.
Residence: See Mailing Address. T. See Mailing Address

Glebe-Møller, Jens, Prof Dr

Rel: Lutheran; b. 21 AUG 1933 in Hull, UK; m. to Kirsten; c. 3 boys and 1 girl; Citizen of Denmark; Lang. Danish, English, German and French.
International Organizations: Professor of Dogmatics (since 1980) of the Faculty of Theology, University of Copenhagen.
Other organizations: Dean, Faculty of Theology (since 1991), University of Copenhagen.
Main field of work: University education. **Career:** Professor, University of Roskilde from 1972 to 1980. **Educ.:** Cand. Theol. from University of Copenhagen, Denmark (1954-1959). **Pubs.:** Wittgenstein 06 Religionen, 1968; A Political Dogmatic, 1987 and Jesus and Theology, 1989.
Mailing address: Institut for Systematisk Teologi, Købmagergade 44-46, DK-1150 København, Denmark. T. (45 33) 152811
Residence: Borups Allé 144, 4th, DK-2000 Frederiksberg, Denmark. T. (45 38) 343556

Glemp, Józef, H E Card

Rel: Roman Catholic; Ordained on 25 MAY 1956; b. 18 DEC 1929 in Inowroclaw, Poland; celibate; Citizen of Poland.
International Organizations: Member (since 1983) of the College of Cardinals.
Member of the Sacred Congregation for the Oriental Churches.
Other organizations: Metropolitan Archbishop (since 1981), Gniezno and Warsaw and Primate of Poland; President (since 1981), Bishops' Conference of Poland; President (since 1981), Bishops' Committee for Pastoral Care Abroad; Ordinary for the Armenian-rite communities in Poland (since 1981).
Career: Educational work with young and sick, in Mielzyn and Witkowo, and catechistic work in Wagrowiec, and Miasteczko Krajenskie from 1956 to 1958. Advocate, Rota Romana Tribunal in 1964. Various positions, Gniezno Curia Tribunal and Lecturer, Major Theological Seminary, Gniezno from 1964 to 1967. Secretary, Primate's Office in Warsaw, Primate's Chaplain from 1967 to 1979. Roman Law Lecturer, Catholic Theological Academy, Warsaw. Former member, Bishops' Committee for Review of Canon Law. Secretary, Bishops' Committee for Polish Institutions in Rome. Consultant and protector, the marriage tie in the Primate's Tribunal from 1965 to 1967. Bishop, Warmia in Olsztyn from 1979 to 1981. Co-Chairman, the Legislative Team set up by Joint Commission of the Polish Government and the Episcopate from 1979 to 1981. President, Pontifical Council on Justice and Peace in 1979. Created Cardinal by Pope John Paul II in 1983. **Educ.:** Theological Seminary in Gniezno, Poland (1950-1956); Dr. Canon and Civil Law from Lateran Pontifical University, Vatican City State (1958-1964). **Hons.:** Dr. h.c., Catholic Academy of Theology, Warsaw (1982), Villanova University, USA (1985), Lublin Catholic University (1985), St Thomas University, Manila (1988); Culture for Peace Prize, Sardinia (1986). **Pubs.:** Naucza pasterskie, 1988; Let my call come to you, 1988 and Nauczaine spoleczne 1981-1986, 1989.
Mailing address: Miodowa 17/19, 00-246 Warszawa, Poland. T. (48 22) 312157
Residence: See Mailing Address. T. See Mailing Address

Glisky, Joan, Sr

Rel: Roman Catholic; Ordained; celibate.
International Organizations: Superior General of the Sisters, Servants of the Immaculate Heart of Mary of Monroe.
Mailing address: 610 W Elm Avenue, Monroe MI 48161, USA. T. (1 313) 241 0092.

Gnanadason, Jonathan N

Rel: Christian.
International Organizations: Secretary-General of the International Christian Federation for the Prevention of Alcoholism and Drug Addiction.
Mailing address: No 1 First Street, Haddows Road, Nungambakkan, Madras 600 006, India.

Goetschmann, M

Rel: Salvation Army, The.
International Organizations: UNESCO Representative of The Salvation Army.
Mailing address: 7 chemin Bayard, CH-2024 St-Aubin NE, Switzerland. T. (41 38) 55 33 80.

Gogne, René Joseph, F.N.D.L., Br

Rel: Roman Catholic; Ordained on 08 SEP 1951; b. 06 MAR 1926 in Loppem (Bruges), West Flanders, Belgium; celibate; Citizen of Belgium; Lang. Dutch, French, Italian and English.
International Organizations: Procurator-General (since 1989) of the Brothers of Our Lady of Lourdes.
Other organizations: Advisory Member (since 1991), chapters of Congregation of Brothers of Our Lady of Lourdes; Editor-in-Chief, *Vriendengroet* (periodical of Glorieux-Instituut, Oostakker).
Main field of work: Administration. **Career:** Teacher (secondary education) from 1951 to 1983. Headmaster (secondary education) (until 1990). **Educ.:** Tchr. Cert. (primary) from Normaalschool, Oostakker, Belgium (1942-1946); Tchr. Cert. (secondary) from Normaalschool, Brussels, Belgium (1949-1951); Tchr. Cert. French from Ecole Normale, Arlon, Belgium (1948); Rel. Crse. from Catholic University of Louvain (KUL), Belgium (1956-1957); Italian Crse. from Volkshogeschool, Ghent, Belgium (1972-1974). **Interests:** social and religious matters, history, arts.
Mailing address: Sint Jozefstraat 1, B-9041 Oostakker, Belgium. T. (32 91) 51 01 85
Residence: See Mailing Address. T. (32 91) 51 12 81

Gohil, Jamnadas Zaver Ragha, Prof

Rel: Hindu; Ordained on 02 FEB 1968; b. 02 DEC 1943 in Moshi, Tanzania; m. to Godelieve Smets; c. 1 boy and 1 girl; Citizen of Belgium; Lang. English, German, Dutch, Swahili, Hindi, Gujarati, Sanskrit, Prakrit, Avestan and Persian.
International Organizations: Professor of Hinduism of the Faculty for the Comparative Study of Religions, Antwerp.
Other organizations: Board of Directors (since 1988), Interfilm Academy.
Main field of work: University education. **Educ.:** Lic. Criminology from Catholic University of Louvain (KUL), Belgium (1968-1972); Lic. Fam. and Sex.Sc. from Catholic University of Louvain (KUL), Belgium (1970-1975); Lic. Oriental Stds. from Catholic University of Louvain (KUL), Belgium (1988-1991); Drs. from Maharaja Sayajirao University of Baroda, India (Beginning in 1992); Drs. from Catholic University of Louvain (KUL), Belgium (Beginning in 1992). **Interests:** African religions, Jainism. **Hons.:** Appointed by the Ministry of Justice to visit Hindus in the Belgian penitentiary institutions. **Mem.:** Brahma Kumaris (supporting member) **Pubs.:** The Hindus of England; Gods and goddesses in Jainism and Sex in the Indian thought.
Mailing address: Jan Van Rijswijcklaan 73, B-2018 Antwerpen, Belgium. T. (32 3) 237 84 20
Residence: See Mailing Address. T. See Mailing Address

Gohla, Hans Peter

Rel: Roman Catholic.
International Organizations: Executive Director of the Catholic Media Council.
Mailing address: Anton-Kurze-Allee 2, Postfach 1912, D-52074 Aachen, Germany. T. (49 241) 73081. Fax 73462. C. MEDIENPLANUNG. Tx 0832719 mira d.

Goldman, Ralph I

Rel: Jewish; b. 01 SEP 1914 in Russia; m. to Helen Goldberg; c. 1 boy and 2 girls; Citizen of USA.
International Organizations: President of the World Confederation of Jewish Community Centers.
Career: Came to US in 1925. Served with US Army during World War II. Personal aide to Prime Minister David Ben-Gurion on 1st (1951) and last (1967) visits to US. In Prime Minister's Office from 1953 to 1957. Executive Vice-President, American-Israel Cultural Fund from 1958 to 1964. Founding Executive Director, Israel Education Fund of United Jewish Appeal from 1964 to 1969. Associate Director-General, JDC-MALBEN from 1969 to 1976. World Director, American Jewish Joint Distribution Committee (JDC) from 1976 to 1985. President, International Conference of Jewish Communal Service in 1981. Executive Committee, International Council of Social Welfare in 1984. **Educ.:** B.J.E. from Boston Hebrew College, MA, USA (1936); B.S. from Boston University, MA, USA (1937); B.S. from Harvard University, Cambridge MA, USA (1938); M.S. Social Work from Boston University, MA, USA (1941). **Hons.:** Philip L Lown Distinguished Service Award, Boston Hebrew College (1978); French Legion of Honour.
Mailing address: 15 E 26th Street, New York NY 10010, USA. T. (1 212) 532 4949.

Goldstein, Joseph

Rel: Buddhist; b. 20 MAY 1944 in Jersey City NJ, USA; single; Citizen of USA; Lang. English.
International Organizations: Guiding Teacher (since 1976) of the Insight Meditation Society.
Other organizations: Co-Founder (since 1990), Barre Center for Buddhist Studies.
Main field of work: Teaching meditation. **Educ.:** B.A. from Columbia University, New York NY, USA (1961-1965). **Pubs.:** The Experience of Insight and Co-Author, Seeking the Heart of Wisdom.
Mailing address: Insight Meditation Society, 1230 Pleasant Street, Barre MA 01005, USA. T. (1 508) 355 4378
Residence: See Mailing Address. T. See Mailing Address

Golish, Robert W, M.M., Fr

Rel: Roman Catholic; Ordained; celibate.
International Organizations: Procurator-General of the Catholic Foreign Missionary Society of America.
Mailing address: Via Sardegna 83, I-00187 Roma, Italy. T. (39 6) 488 5700. Fax 487 4158.

Gombrich, Richard Francis, Prof

Rel: None; Unordained; b. 17 JUL 1937 in London, UK; m. to (1st) Dorothea Amanda Friedrich (divorced), (2nd) Sanjukta Gupta; c. 1 boy and 1 girl; Citizen of United Kingdom; Lang. Sanskrit, Pali, Prakrit, Sinhala, Latin, Ancient Greek, Italian, French, German and Dutch.
International Organizations: Regional Secretary for Europe of the International Association of Buddhist Studies.
Honorary Secretary and Treasurer (since 1981) of the Pali Text Society.
Other organizations: Member (since 1982), Shap working party on world religions in education; Adviser on religious studies to the National Council for Academic Awards (since 1983).
Main field of work: University education. **Career:** University Lecturer in Sanskrit and Pali, University of Oxford from 1965 to 1976. Benjamin Meaker Visiting Professor, University of Bristol from 1981 to 1982. Directeur d'études invité, Ecole des Hautes Etudes en Sciences Sociales, Paris during 1982. Stewart Visiting Fellow in the Humanities, Princeton University, NJ from 1986 to 1987. **Educ.:** B.A. from University of Oxford, Magdalen College, UK (1957-1961); A.M. from Harvard University, Cambridge MA, USA (1961-1963); D.Phil. from University of Oxford, Magdalen College, UK (1963-1970). **Interests:** history and anthropology of Buddhism. **Hons.:** Hon. D.Litt., Kalyani University. **Mem.:** Academia Europea, Society for South Asian Studies, Royal Asiatic Society **Pubs.:** *Theravada Buddhism: a social history from ancient Benares to modern Colombo*, 1988; With G Obeyesekere, *Buddhism Transformed: Religious Change in Sri Lanka*, Princeton, 1988 and *Precept and Practice*, Oxford, 1971, 2nd Ed. *Buddhist Precept and Practice*, Delhi, 1991.
Mailing address: Balliol College, Oxford OX1 3BJ, UK. T. (44 865) 277777
Residence: 15 Church Street, Kidlington OX5 2BA, UK. T. (44 8675) 6032

Gomes, Rui

Rel: Christian.
International Organizations: Secretary of the Christian Movement for Peace.
Mailing address: Secrétariat européen, Rue Stévin 92, B-1040 Bruxelles, Belgium. T. (32 2) 230 84 99. Fax 230 82 38.

Gómez, Luis, Prof

Rel: Unknown.
International Organizations: Regional Secretary for the Americas of the International Association of Buddhist Studies.
Mailing address: Dept. of Far Eastern Languages and Literatures, Frieze Bldg, The University of Michigan, Ann Arbor MI 48104, USA.

Gómez, Eugenio Julio, C.R., Fr

Rel: Roman Catholic; Ordained; celibate.
International Organizations: Preposito Generale of the Congregation of Regular Clerics.
Member, Committee for Reporting to the Congregation, the International Union of Superior Generals, and the Union of Superior Generals of the Congregation for Religious and Secular Institutes.
Mailing address: Piazza Vidoni 6, I-00186 Roma, Italy. T. (39 6) 68 61 339.

Gonçalves, Jaime Pedro, H E Mgr

Rel: Roman Catholic; Ordained on 17 DEC 1967; b. 26 NOV 1936 in Manica, Mozambique; celibate.
International Organizations: President of the Inter-Regional Meeting of Bishops of Southern Africa.
Member of the Congregation for the Evangelization of Peoples.
Other organizations: Archbishop, Beira.
Mailing address: 4 Bayswater Road, Highlands, PO Box 50, Harare, Zimbabwe. T. (263 10) 47479. Tx 2390 GENSEC ZW ATTN IMBISA.

Gondry, J

Rel: Roman Catholic.
International Organizations: Contact of the Commission on Justice and Peace.
Mailing address: Av d'Auderghem 12, B-1040 Bruxelles, Belgium. T. (32 2) 736 54 13.

Gontard, Françoise, Dr

Rel: Roman Catholic.
International Organizations: Secretary of the European Federation of Catholic Medical Associations.
Mailing address: 5 av de l'Observatoire, F-75006 Paris, France. T. (33 1) 46 34 59 15.

González Gómez, Santiago G, C.M.F., Fr

Rel: Roman Catholic; Ordained; celibate.
International Organizations: Procurator-General of the Missionaries Sons of the Immaculate Heart of Mary.
Procurator-General of the Congregation of St Peter in Chains.
Mailing address: Via Sacro Cuore di Maria 5, I-00197 Roma, Italy. T. (39 6) 87 00 56.

González Martín, Marcelo G, H E Card

Rel: Roman Catholic; Ordained on 29 JUN 1941; b. 16 JAN 1918 in Villanubla, Valladolid, Spain; celibate; Citizen of Spain.
International Organizations: Member (since 1973) of the College of Cardinals.
Member of the Congregation for the Clergy.
Other organizations: Archbishop (since 1971), Toledo and Primate of Spain.
Career: Consecrated Bishop, Astorga in 1961. Titular Archbishop, Case Mediane in 1966. Auxiliary Archbishop, Barcelona in 1967. Created Cardinal by Pope Paul VI in 1973.
Mailing address: Arco de Palacio 3, E-45002 Toledo, Spain. T. (34 925) 22 34 39

González-Montes, Adolfo, Rev Prof Dr

Rel: Roman Catholic; Ordained on 29 JUN 1972; b. 13 NOV 1946 in Salamanca, Spain; celibate; Citizen of Spain; Lang. Spanish, French, German, Italian and English (reading).

International Organizations: Director (since 1982) of the John XXIII Centre of Ecumenical Studies and Oriental Studies.

Member of the Pontifical Council for Promoting Christian Unity.

Other organizations: Professor of Theology (since 1977), Pontifical University of Salamanca; Director (since 1982), *Diálogo Ecuménico*; Consultant to Pontifical Council for Promoting Christian Unity (since 1989).

Main field of work: University education. **Career:** Professor of Theology, 'Gaudium et Sepes' Institute of Philosophy and Theology, Salamanca from 1975 to 1980. Professor of Theology, 'Pius X' Higher Institute of Catechetical Studies, Salamanca from 1976 to 1977. **Educ.:** Lic. Phil. and Litt. from Pontifical University of Salamanca, Spain (1963-1968); Dr. Theol. from Pontifical University of Salamanca, Spain (1977) (1968-1973); Theol. Stds. from Eberhard-Karls-Universität Tübingen, Germany (1973-1975); Doc. Crse. from Universidad Complutense de Madrid, Spain (1979-1980). **Interests:** ecumenical theology. **Mem.:** Societas Oecumenica – European Society for Ecumenical Research **Pubs.:** *Religión y nacionalismo*, 1982; *Enchiridion oecumenicum*, 1986 and *Reforma luterana y tradición católica*, 1987.

Mailing address: Universidad Pontificia, Calle Companía 5, E-37008 Salamanca, Spain. T. (34 23) 21 59 80. Fax (34 923) 21 34 50.

Residence: c/o Padre Cámara, 18-20, 10° C, E-37003 Salamanca, Spain. T. (34 23) 23 82 35

Goodsell, James Nelson, Dr

Rel: Church of Christ, Scientist; b. 07 JUN 1929 in Evanston IL, USA; m. to Rhoda Merle Ford; c. 1 boy and 2 girls; Citizen of USA; Lang. English, Spanish and Portuguese.

International Organizations: Latin America Editor, *The Christian Science Monitor* (since 1957) of the Church of Christ, Scientist.

Main field of work: Journalism. **Career:** Editor, El Monitor de Hoy (TV) from 1957 to 1989. **Educ.:** B.A. from Principia College, Elsah IL, USA (1947-1951); M.A. from Mexico City College, Mexico (1951-1953); Ph.D. from Harvard University, Cambridge MA, USA (1956-1966). **Interests:** Latin America history. **Mem.:** American Historical Association, Conference on Latin America History, Sigma Del Chi **Pubs.:** *The Quest for Change in Latin America*; *Fidel Castro's Personal Revolution in Cuba*; *Nicaragua: A Tortured History* and *The Press in Latin America.*

Mailing address: 330 Dartmouth Street 2-S, Boston MA 02116, USA. T. (1 617) 247 4811

Residence: See Mailing Address. T. See Mailing Address

Goody, Jeremy

See Lokamitra.

Gorai, Dinesh Chandra, Rt Rev

Rel: Methodist; Ordained in 1962; b. 15 JAN 1934 in West Bengal, India; m. to Binapani; c. 2 boys; Citizen of India.

International Organizations: Bishop (since 1982) of Calcutta.

Career: Consecrated Bishop, Church of North India in 1970. Bishop, Barrackpore from 1970 to 1982. Moderator, Church of North India from 1983 to 1986.

Mailing address: Bishop's House, 51 Chowringhee Road, Calcutta 700 071, India.

Gorder, John Martin, Rev

Rel: Christian; b. in Wisconsin, USA; m. to Gordeen Gorder; Citizen of USA; Lang. English, French, Sangu and Neo-Melanesian.

International Organizations: Secretary-General (since 1981) of the Joint Christian Ministry in West Africa.

Main field of work: Networking. **Career:** Intern, Elcong, Papua New Guinea from 1969 to 1971. Pastor, Evangelical Lutheran Church, Central African Republic from 1973 to 1980. Visiting Scholar, Saint Olaf College, Northfield, MN, USA during 1981. **Educ.:** B.A. from Saint Olaf College, Northfield MN, USA (1963-1967); M.Div. from Luther Seminary, St Paul MN, USA (1967-1972); M.Th. from Luther Northwestern Seminary, St Paul MN, USA (1980-1981). **Pubs.:** *Tensions in Resource Sharing*, thesis M.Theol.

Mailing address: BP 495, Navaguta Av 29, Jos, Nigeria.

Residence: Jos, Plateau State, Nigeria T. (234 73) 52532

Gordillo, Ismael Gomez, Fr

Rel: Roman Catholic; Ordained; celibate.

International Organizations: Contact of the Missionaries of the Holy Ghost.

Mailing address: Avenida Universidad 1702, 04010 México DF, Mexico.

Gordon, Margaret

Rel: Methodist.

International Organizations: President for South Pacific of the World Federation of Methodist Women.

Mailing address: 3 Bleakhouse Road, Howick, Auckland, New Zealand.

Gorekar, Nizamuddin S, Prof Dr

Rel: Muslim; b. 12 AUG 1921 in Bhiwandi, Maharashtra State, India; m. to Haseena; c. 1 boy; Citizen of India; Lang. Urdu, English, Hindi, Marathi, Persian and Arabic.

International Organizations: Director (since 1975) of the Anjuman-i-Islam Urdu Research Institute and Anjuman-i-Islamic Research Association (Bombay).

Other organizations: Professor of Persian, Urdu and Islamics (since 1944), St Xavier's College, Bombay; Member, Executive Council (since 1970), Anjuman-i-Islam, Bombay; President (since 1970), University Students Association of Kalyan; Conducts certificate and diploma courses in Urdu and Persian at the St Xavier's College, Bombay (since 1970); Editor (since 1976), *Nawa-e-Adab* (biannual research journal), Anjuman-i-Islam Urdu Research Institute, Bombay; Assistant Director (since 1982), Heras Institute of Indian History and Culture, Bombay; Organizes guidance classes for the Jamai Urdu courses at the Anjuman-i-Islam Urdu Research Institute, Bombay (since 1984); Principal (since 1989), I/C KME Society's Womens College, Bhiwandi; Member, Executive Council, PEN Centre, Bombay.

Main field of work: Teaching and research guidance for Ph.D. degrees. **Career:** Editor, *Karvan-e-Adab* (annual Urdu journal), St Xavier's College, Bombay from 1945 to 1968. Founder-Secretary, Educational Uplift Society of Kalyan from 1956 to 1961. President, Educational Uplift Society of Kalyan from 1974 to 1978. Member, Bureau for Promotion of Urdu, Government of India, New Delhi from 1976 to 1983. Member, General Council, Jamia Urdu, Aligarh from 1981 to 1986. Vice-President, Asiatic Society of Bombay (until 1983). Special Executive Magistrate, Thane District (for 10 years). Member, Board of Studies in Urdu and Persian, Bombay, Poona, Shivaji, Baroda Universities. **Educ.:** M.A. (Hons) from Bombay University, India (1944); Ph.D. from Bombay University, India (1969); D.Litt. from Maithili Vishva-Vidhyapith, Bihar, India (1979). **Hons.:** Kalyan Bhushan, Kalyan Municipal Corporation (1979); Mahamahopadhaya, Maithili Vishva Vidhyapith (1982); Eminent Persian Teacher, VI All-India Persian Teachers Conference (1983); President of India's Certificate of Honour in Persian (1985). **Mem.:** Kalyan Rotary Club and Giants Group (Honorary Member, 1986) **Pubs.:** *Glimpses of Urdu Literature*, 1964; Textbooks in Urdu, Persian and Arabic for secondary and university students in the Maharashtra State like *Kaif-e-Adab* and *Suroor-e-Adab* in Urdu, *Amuzgar-e-Farsi* and *Armugha-e-Farsi* in Persian, and *Talimul-Arabiyah* and *Faraidul-Adab* in Arabic and contributed articles to journals like *Indo-Iranica* (Bombay), *Sabras* (Hyderabad), *Indica* (Bombay), *Shair* (Bombay), *Tehzibul-Akhlaq* (Aligarh), *Nawa-e-Adab* (Bombay), journals of the universities of Kerala, Punjab and Bombay, *India Magazine* (Delhi), etc.

Mailing address: 92 Dr D.N. Rd, Bombay 400 001, India. T. (91) 2620177 (BOM)

Residence: Bait-Ush-Sharaf, Bunder Rd, Kalyan-Bombay 421 301, India. T. (91) 23479 (KYN)

Gormit Singh, G, Dr

Rel: Sikh.

International Organizations: President of the International Sikh Organization.

Mailing address: 2025 I St NW, Suite 901, Washington DC 20006, USA. T. (1 202) 833 3262.

Gottschalk, Alfred, Dr

Rel: Jewish; Ordained in 1957; b. 07 MAR 1930 in Oberwesel, Germany; m. to Deanna Zeff; c. 1 boy and 1 girl; Citizen of USA.

International Organizations: Member of the Executive Committee, American Section of the World Jewish Congress.

International Council of the World Union of Jewish Studies. Governing Body of the World Union for Progressive Judaism. Executive Committee of the Memorial Foundation for Jewish Culture. Chairman, North American Advisory Committee of the International Centre for University Teaching of Jewish Civilization.
Other organizations: President (since 1971), Hebrew Union College-Jewish Institute of Religion; National Chairman of Academic Relations, State of Israel Bonds National Campaign Cabinet; Chairman, Academic Committee, United States Holocaust Memorial Council; Executive Committee, Center for the Study of Ethics and Contemporary Moral Problems (UC-HUC); Executive Committee, Central Conference of American Rabbis; Executive Committee, Union of American Hebrew Congregations; Executive Committee, United States Holocaust Memorial Council; Editorial Advisory Board, *Dictionary of American Jewish Biography*; Editorial Board, *Hebrew Union College Annual*; Honorary Editorial Advisory Board, *Holocaust and Genocide Studies*; Publications Committee, Jewish Publication Society of America; Editorial Committee, US Holocaust Memorial Museum; other governance responsibilities.
Career: Emigrated to the US in 1939. US citizen in 1944. Director, Hebrew Union College-Jewish Institute of Religion (HUC-JIR), California School from 1957 to 1958. Acting Dean, HUC-JIR, California School from 1958 to 1959. Dean, HUC-JIR, California School from 1959 to 1971. Instructor (Jewish History), HUC-JIR from 1957 to 1959. Assistant Professor of Bible and Jewish Religious Thought, HUC-JIR from 1959 to 1962. Associate Professor of Bible and Jewish Religious Thought, HUC-JIR from 1962 to 1965. Professor of Bible and Jewish Religious Thought, HUC-JIR in 1965. Founder, HUC-JIR Center for the Study of Ethics and Contemporary Moral Problems. Founder, Irwin Daniel School of Jewish Communal Service in Los Angeles. **Educ.:** A.B. from City University of New York, Brooklyn College, USA (1952); B.H.L. from Hebrew Union College-Jewish Institute of Religion, Cincinnati OH, USA (1954); M.A. (Hons) from Hebrew Union College-Jewish Institute of Religion, Cincinnati OH, USA (1956); Ph.D. from University of Southern California, Los Angeles, USA (1965). **Hons.:** American Jewish Committee Human Relations Award (1971); Los Angeles City Council Award for contribution to Education (1965, 1971); Alumnus of the Year Award, Brooklyn College (1972); Tower of David Award for Cultural Contributions to Israel and America, Israel Government (1972); Gold Medallion Award of the Jewish National Fund (1972); Man of the Year Award, West Coast Chapter, Boys High School Alumni Association (1976); Myrtle Wreath Award, Southern Pacific Coast Region of Hadassah (1977); Louis Dembitz Brandeis Award, Zionist Organization for America (1977); Establishment of Dr Alfred Gottschalk Department of Judaica at Kfar Silver Youth and Education Center, Israel (1977); Establishment of the Alfred Gottschalk Chair in Jewish Communal Service by the Board of Governors of Hebrew Union College-Jewish Institute of Religion (February 1979); National Brotherhood Award, National Conference of Christians and Jews, New York (October 1979); Israel Bonds Man of the Year, Cincinnati (1982); Jerusalem/City Peace Award, Jerusalem (1988); Defender of Jerusalem Award honoree, Jabotinsky Foundation (1990); The Isaac Mayer Wise Award, Congregation Emanuel-Temple Sinai, Denver CO (March 1991); Honorary degrees: University of Southern California (S.T.D., 1968), University of Judaism (D.Hu.L., 1971), Dropsie University (Litt.D., 1974), University of Cincinnati (LL.D., 1976), University of Southern California (LL.D., 1976), The Hebrew University (Honorary Fellow, 1976), Loyola Marymount University (D.Relig.Ed., 1977), Xavier University (LL.D., 1981), St Thomas Institute (Lit.D., 1982), New York University (D.D., 1985), The Jewish Theological Seminary of America (D.Hu.L., 1986), Brooklyn College (D.Hu.L., 1991). **Mem.:** American Academy of Religion, American Association for Higher Education, American Association of Presidents of Independent Colleges and Universities, American Philosophical Society, Conference of Presidents of Major American Jewish Organizations, International Conference of Jewish Communal Service, International Organization for Scientific Co-operation in Oriental and Jewish Studies, Israel Exploration Society, National Education Association, Society for Biblical Literature and Exegesis, World Institute for Sephardic Studies, Zalman Shazar Centre for the Furtherance of the Study of Jewish History; Honorary Societies: Delta Sigma Rho, Phi Beta Kappa; Honorary Memberships: Commercial Club, Greater Cincinnati Interfaith Holocaust Foundation, Losantiville Country Club, Lotos Club, New York Board of Rabbis (Honorary Vice-President), Southern California Jewish Historical Society (Honorary President) **Pubs.:** "Ahad Ha'Am: Confronting the 'Plight of Judaism'", in *Journal of Reform Judaism*, Summer 1987; *To Learn and To Teach*, revised, 1988 and over 200 essays, articles and translations.
Mailing address: 1 West 4th Street, New York NY 10012, USA.
Residence: 17 Belsaw Place, Cincinnati OH 45220, USA. T. (1 513) 281 5880

Goudreault, Henri, O.M.I., Rt Rev

Rel: Roman Catholic; Ordained on 17 JUN 1956; b. 30 APR 1928 in Casey Twp, Ontario, Canada; celibate; Citizen of Canada.
International Organizations: Bishop (since 1987) of Labrador City-Schefferville.
Member of the Missionary Oblates of Mary Immaculate. Member of the Congregation for the Evangelization of Peoples.
Career: Professor of Theology, University of Ottawa (now St Paul University) from 1957 to 1962. Professor of Theology, University of Ottawa from 1965 to 1966. Professor of Theology, University of Ottawa from 1968 to 1971. Director, Institute of Mission Studies, St Paul University from 1971 to 1977. Rector, St Paul University from 1977 to 1985. Superior Provincial, the Oblate St Joseph's Province from 1985 to 1987. Member, Executive, Pastoral Council, Archdiocese of Ottawa from 1970 to 1972. Vice-President, Priests' Synod from 1972 to 1973. Board, L'Entraide Missionnaire, Montreal from 1971 to 1977. Board, National Missionary Council from 1972 to 1975. Board, Association canadienne d'éducation des adultes dans les universités de langue française from 1979 to 1982. Board, Le Centre Saint Pierre Apôtre, Montreal from 1979 to 1982. Provincial Council, O.M.I. (Saint Joseph's Province) from 1973 to 1979. Provincial Council, O.M.I. (Saint Joseph's Province) from 1982 to 1985. Observer, Holy See to 5th meeting of World Food Council in 1979. Delegate, Holy See to Centennial Forum (100th session of International Wheat Council) in 1984. Delegate, Roman Catholic Church to Anglican National Meeting 'Partners in Mission Consultation' in 1979. Member, Canadian Oblate Conference of Mission from 1976 to 1980. Board, Le Centre Missionnaire Oblat, Montreal from 1974 to 1977. Co-founder and Member of Board, Medical Aid Foundation Canada from 1981 to 1984. Official Preacher, Ecumenical Service Canadian Week in 1978. Vice-President, Oblate Conference of Canada from 1985 to 1987. Teaching resource person in Lesotho, Malawi, Liberia, Haiti, USA. Delegate to the 1990 Roman Synod on the formation of future priests and on-going formation of priests. **Educ.:** B.A., B.Ph. from University of Ottawa, Ontario, Canada (1951); L.Ph. from University of Ottawa, Ontario, Canada (1952); L.Th. from University of Ottawa, Ontario, Canada (1957); L.S.Sc. from Pontifical Biblical Institute, Vatican City State (1965); D.Th. from Institut Catholique de Paris, France (1968). **Interests:** swimming, volleyball, billiards. **Mem.:** L'Association Jacques Maritain (1979); Société Canadienne de Théologie; Association Catholique des Etudes Bibliques au Canada **Pubs.:** Author of numerous articles.
Mailing address: Bishop's Residence, EE 318 Elizabeth Ave, Box 545, Labrador City, Labrador, Newfoundland A2V 2K7, Canada.

Gough, Hugh Rowlands, Rt Rev

Rel: Anglican; Ordained in 1929; b. 19 SEP 1905 in Pakistan; m. to Madeline Elizabeth Kinnaird; c. 1 girl; Citizen of United Kingdom.
International Organizations: Ecclesiastic of the Anglican Church.
Career: Deacon in 1928. Curate, St Mary Islington from 1928 to 1931. Perpetual Curate, St Paul, Walcot, Bath from 1931 to 1934. Vicar, St James Carlisle from 1934 to 1939. Vicar, St Matthew Bayswater from 1939 to 1946. Vicar, Islington and Rural Dean from 1946 to 1948. Prebendary, St Paul's Cathedral in 1948. Archdeacon, West Ham from 1948 to 1958. Suffragan Bishop, Barking from 1948 to 1959. Archbishop, Sydney from 1959 to 1966. Primate, Australia from 1959 to 1966. Retired in 1972. Deputy Lieutenant, Co. of Essex from 1952 to 1959. War service from 1939 to 1945. **Hons.:** Companion of the Order of St Michael and St George; Officer of the Order of the British Empire, mentioned in despatches; Honorary Chaplain to the Forces 2nd Class; Hon. D.D., Lambeth and Wycliffe College, Toronto; Hon. Th.D., Australia; Sub-Prelate, Order of St John (1959).
Mailing address: The Forge, Over Wallop, Stockbridge, Hants, UK. T. (44 264) 781315

Gouyon, Paul, H E Card

Rel: Roman Catholic; Ordained on 13 MAR 1937; b. 24 OCT 1910 in Bordeaux, France; celibate; Citizen of France.
International Organizations: Member (since 1969) of the College of Cardinals.
Other organizations: Archbishop Emeritus (since 1985), Rennes.
Career: Consecrated Bishop in 1957. Bishop, Bayonne in 1957. Titular Archbishop, Pessinonte in 1963. Archbishop, Rennes from 1964 to 1985. Created Cardinal by Pope Paul VI in 1969. National President, Pax Christi from 1966 to 1982. **Hons.:** Lauréat de l'Académie

Française; Officer, Legion of Honour; Commander, National Order of Merit; Military Cross (1939-45). **Pubs.:** Commentaire de la Déclaration conciliaire sur l'éducation chrétienne, 1966; L'intercommunion est-elle possible?, 1976 and Marcel Callo, témoin d'une génération, 1981.
Mailing address: 181 rue Judaïque, F-33081 Bordeaux CEDEX, France. T. (33) 56 96 08 30

Gowans, John, Commr
Rel: Salvation Army, The.
International Organizations: Territorial Commander of The Salvation Army.
Mailing address: BP 284 08, F-75364 Paris Cedex 08, France.

Grace, P J, Fr
Rel: Roman Catholic; Ordained; celibate.
International Organizations: Director of the Worldwide Marriage Encounter.
Mailing address: 1908 E Highland, San Bernardino CA 92494, USA. T. (1 714) 881 3456.

Grady, Laureen, OCD, Sr
Rel: Roman Catholic; Ordained; celibate.
International Organizations: Member of the Discalced Carmelite Nuns.
Other organizations: Member, Association of Contemplative Sisters.
Mailing address: Carmelite Monastery, 1950 La Fond Dr, Reno NV 89509, USA.

Grady, Thomas, Br
Rel: Roman Catholic; Ordained; celibate.
International Organizations: Superior General of the Franciscan Brothers of the Third Order Regular of Brooklyn.
Mailing address: 135 Remsen St, Brooklyn NY 11201-4296, USA. T. (1 718) 858 8217

Graham, William (Billy) Franklin
Rel: Baptist: Southern; Ordained in 1940; b. 07 NOV 1918 in Charlotte NC, USA; m. to Ruth McCue Bell; c. 2 boys and 3 girls; Citizen of USA.
International Organizations: International evangelist of the Billy Graham Crusade.
Career: Converted to Christ in 1934. Served as a lay preacher during student days from 1936 to 1940. Pastor, First Baptist Church, Western Springs IL from 1943 to 1945. Following a brief pastorate, became an evangelist for Youth for Christ from 1944 to 1945. Charter Vice-President, Youth for Christ International, Chicago IL from 1945 to 1950. Made two trips to UK for evangelistic tours from 1946 to 1947. Served as President, Northwestern Schools, Minneapolis MN, consisting of three institutions: liberal arts college, Bible school and theological seminary from 1947 to 1952. Launched an evangelical campaign that brought him into public view in 1949. Founded the Billy Graham Evangelistic Association and began weekly 'Hour of Decision' radio program in 1950. Ministered with his team during a crusade in UK that brought him to international attention in 1954. Contributed to the establishment, *Christianity Today* in 1956. Became an unofficial spiritual advisor and confidant to many US presidents, beginning with Dwight D Eisenhower. Involved in launching a number of important congresses on evangelism: Berlin (1966), Lausanne (1974) and Itinerant Evangelists (1983, 1986). Traveled widely, including tours of Korea (1973) and the Soviet Union (1982, 1984). **Educ.:** Grad. from Florida Bible Institute (now Trinity College), Tampa, USA (1937-1940); B.A. from Wheaton College, IL, USA (1940-1943). **Interests:** swimming, aerobics, walking, golf, tennis. **Hons.:** Numerous awards, including: Gold Medal Award, National Institute of Social Science, New York (1957); Annual Gutenberg Award, Chicago Bible Society (1962); Speaker of the Year Award (1964); Golden Plate Award, American Academy of Achievement (1965); Ten Most Admired Men in the World from the Gallup Poll (since 1955); Big Brother of the Year Award at the White House, Washington DC (1966), for contribution to the welfare of children; International Brotherhood Award, National Conference of Christians and Jews (1971); Distinguished Service Award, National Association of Broadcasters (1972); Franciscan International Award (1972); Sylvanus Thayer Award, United States Military Academy Association of Graduates at West Point (1972); Philip Award, Association of United Methodist Evangelists (1976); First National Interreligious Award, American Jewish Committee (1977); Jabotinsky Centennial Medal, The Jabotinsky Foundation (1980); Religious Broadcasting Hall of Fame Award (1981); Templeton Foundation Prize for Progress in Religion Award (1982); Presidential Medal of Freedom (1983); National Religious Broadcasters Award of Merit (1986); Many honorary degrees. **Pubs.:** Author of numerous books, including: *Revival in Our Times*, 1950; *Peace with God*, 1953 (revised edition, 1985) and *The Secret of Happiness*, 1955.
Mailing address: PO Box 9313, Minneapolis MN 55440-9313, USA. T. (1 612) 338 0500. Fax 338 6362. Tx 290259. C. BGTEAM

Grams, Louis
Rel: Christian.
International Organizations: President of the Christians in Commerce.
Mailing address: 2300 East 88th St, Bloomington MN 55425-2185, USA.

Gramzow, Oberin Margret
Rel: Christian.
International Organizations: Secretary of the World Federation of Diaconal Associations and Sisterhoods.
Mailing address: Glockenstrasse 8, 14163 Berlin, Germany. T. (49 30) 8018091.

Grangy, G, Ms
Rel: Roman Catholic.
International Organizations: UNESCO Representative of the International Association of Charities.
Mailing address: 9 rue d'Assas, F-75006 Paris, France.

Grant, Michael
Rel: Hindu.
International Organizations: Contact of the International Society for Krishna Consciousness.
Mailing address: 1030 Grand Ave, San Diego CA 92109, USA. T. (1 619) 272 8334. Fax 483 6068.

Grants, Alnis
Rel: Buddhist: Mahayana.
International Organizations: Member of the Foundation for the Preservation of the Mahayana Tradition.
Other organizations: Director, Aryatara City Centre.
Mailing address: Lucille-Grahnstrasse 47, 81675 München, Germany. T. (49 89) 471 415

Graus, Lambert, O.S.C., Fr
Rel: Roman Catholic; Ordained; celibate.
International Organizations: Master-General of the Order of the Holy Cross.
Mailing address: Via del Velabro 19, I-00186 Roma, Italy. T. (39 6) 679 33 35

Gray, Betty
Rel: Episcopal; Unordained; b. 21 AUG 1926 in Iron City GA, USA; widowed Rev William Boyce Gray; c. 2 girls; Citizen of USA; Lang. English, French, Japanese and Spanish.
International Organizations: Executive Director (since 1983) of the Japan International Christian University Foundation.
Executive Director of the International Christian University, New York NY.
Other organizations: Chair, Board of Directors (since 1988), Society for the Arts, Religion and Contemporary Culture; President/Officer, New York Chapter (since 1988), Religious Public Relations Council.
Main field of work: University education, development. **Career:** General Assignment Reporter, Atlanta (GA) Constitution from 1951 to 1953. Copy Editor, Associated Press, Topeka KS from 1955 to 1956. Legislative Reporter, *Episcopal New Yorker*, New York NY from 1956 to 1958. Editor, *Response* – Methodist Women's Magazine from 1973 to 1976. Associate Editor, The Episcopal Church from 1976 to 1981. National Executive for the Ministries With Women from 1981 to 1982. **Educ.:** B.A. from Florida State University, Tallahassee, USA (1949); M.A. from University of Northern Iowa, Cedar Falls, USA (1965-1966). **Interests:** birding, hiking, swimming, reading. **Hons.:** Associated Press Feature Prize. **Mem.:** Church Club of New York, Metropolitan Museum (Fellow) **Pubs.:** *The Episcopal Church Welcomes You*, Seabury Press, 1974; New York Times Sunday Book Review of Mary Daly's *Beyond God the Father* and New World Outlook Series on Japan: Discrimination, Status of Women, Religion in Japan, 1991.
Mailing address: 70 LaSalle Street, Apt 21-C, New York NY 10027, USA. T. (1 212) 865 3440
Residence: See Mailing Address. T. See Mailing Address

Gray, Gordon Joseph, H E Card

Rel: Roman Catholic; Ordained on 15 JUN 1935; b. 10 AUG 1910 in Leith, Edinburgh, Scotland, UK; celibate; Citizen of United Kingdom.
International Organizations: Member (since 1969) of the College of Cardinals.
Other organizations: Archbishop Emeritus (since 1985), Saint Andrews and Edinburgh.
Career: Assistant Priest, St Andrews from 1935 to 1941. Parish Priest, Hawick from 1941 to 1947. Rector, Blairs College, Aberdeen (National Junior Seminary for Scotland) from 1947 to 1951. Archbishop, St Andrews and Edinburgh from 1951 to 1985. Created Cardinal by Pope Paul VI in 1969. **Educ.:** M.A. from University of St Andrews, UK. **Interests:** gardening, carpentry, liturgy, local historical research. **Hons.:** Honorary Fellow, Educational Institute, Scotland (1970); Hon. D.D., St Andrews (1967); Hon. D.Univ., Heriot-Watt (1981).
Mailing address: The Hermitage, St Margaret's Convent, Whitehouse Loan, Edinburgh EH9 1BB, UK. T. (44 31) 447 6210

Gray, Hugh, Dr

Rel: Theosophist.
International Organizations: Past President of the Theosophical Society.
Mailing address: Adyar, Madras 600 020, India. T. (91 44) 415552. C. OLCOTT, MADRAS.

Gray, Patrick Trevor Robert, Prof

Rel: Anglican; Ordained; b. 16 OCT 1940 in Toronto ON, Canada; m. to Allison McNaught; c. 4 boys; Citizen of Canada.
International Organizations: Associate Professor, Humanities Department (since 1984) of the Atkinson College, York University.
Canadian Committee (since 1976) of the International Association for Byzantine Studies.
Career: Divinity Tutor, Trinity College from 1967 to 1969. Associate Rector, Church of St Simon-the-Apostle from 1969 to 1971. Instructor and Visiting Assistant Professor, York University from 1972 to 1975. Senior Tutor, Stong College, York University from 1973 to 1976. Resident Tutor, Stong College from 1974 to 1976. Research Associate at Pontifical Institute of Medieval Studies from 1976 to 1978. Assistant and Associate Professor of History and Philosophical Theology, McMaster Divinity College from 1978 to 1984. Director, Canadian Federation for the Humanities from 1984 to 1987. Convenor, 'Outside the Nuclear Club' Conference, York University in 1987. President, Canadian Society of Patristic Studies from 1977 to 1979. Honorary President, Canadian Society of Patristic Studies from 1979 to 1983. **Educ.:** B.A. from University of Toronto, Canada (1962); S.T.B. from University of Toronto, Trinity College, Canada (1965); Th.D. from University of Toronto, Trinity College, Canada (1973); S.T.M. from Yale University, New Haven CT, USA (1966). **Interests:** music, gardening, winemaking. **Mem.:** Canadian Society of Patristic Studies **Pubs.:** The Defense of Chalcedon in the East (451-453), 1979.
Mailing address: 702 Atkinson, York University, 4700 Keele St, North York ON M3J 1P3, Canada.
Residence: 'Dove Farm', RR 1, Beeton ON L0G 1A0, Canada.

Graz, Franz Friedrich

Rel: Lutheran; b. 16 MAY 1939 in Königsberg, Germany; m. to Brigitte Maria Graz; c. 2 boys and 2 girls; Citizen of Germany; Lang. German, English and Afrikaans.
International Organizations: Secretary-General (since 1988) of the Federation of Evangelical Lutheran Churches in Southern Africa.
Other organizations: Pastor (since 1976), Evangelical Lutheran Church in Southern Africa (ELCSA), Natal-Transvaal; President (since 1990), ELCSA.
Main field of work: Theology and church work. **Career:** Pastor, Lutheran Church, Hannover from 1968 to 1976. Deputy Dean, Lutheran Church, Hannover from 1973 to 1975. Dean, Evangelical Lutheran Church in Southern Africa (ELCSA) from 1981 to 1990. Vice-President, ELCSA from 1983 to 1990. **Educ.:** Philipps-Universität Marburg, Germany (1960-1961); 1st and 2nd Theol. Exam. from Westfälische Wilhelmsuniversität Münster, Germany (1961-1965). **Interests:** theology, systematics. **Pubs.:** FELCSA Sunday School material and Publications on church unity and the political responsibility of the church.
Mailing address: PO Box 7170, Bonaero Park 1622, South Africa. T. (27 11) 973 1851. Fax (27 11) 973 1873.
Residence: 5 Schonefeld Ave, Bonaero Park, Kempton Park, South Africa T. (27 11) 973 1461

Grebe Lopez, Ronald G

Rel: Roman Catholic.
International Organizations: Co-Head, Latin America Regional Office of the Unda, International Catholic Association for Radio and Television.
Mailing address: Educación Radiofonica de Bolivia (ERBOL), Casilla 5946, La Paz, Bolivia.

Grebe Lopez, Ronald G

Rel: Catholic.
International Organizations: Regional Representative for Latin America of the Unda, International Catholic Association for Radio and Television.
Mailing address: ERBOL, Casilla 4313, La Paz, Bolivia.

Grech, Pierre

Rel: Roman Catholic.
International Organizations: Secretary-General of the Conference of Latin Bishops in Arab Regions.
Mailing address: CELRA, Patriarchatus latinus, PO Box 14152, Jerusalem, Israel. T. (972 2) 282323.

Green, Bernard, Rev

Rel: Baptist; Ordained in 1952; b. 11 NOV 1925 in Walgrave, Northants, UK; m. to Joan Viccars; c. 2 boys and 1 girl; Citizen of United Kingdom.
International Organizations: Secretary-General (since 1982) of the Baptist Union of Great Britain.
Other organizations: Director (since 1982), Baptist Times Ltd.; Director (since 1982), Baptist Holiday Fellowship Ltd.; Director (since 1982), Baptist Union Corporation Ltd.; Director (since 1982), Baptist Insurance PLC; Director (since 1982), London Baptist Property Board.
Career: National service in coal mines from 1944 to 1947. Minister, Yardley Baptist Church, Birmingham from 1952 to 1961. Minister, Mansfield Road Baptist Church, Nottingham from 1961 to 1976. Minister, Horfield Baptist Church, Bristol from 1976 to 1982. Baptist Chaplain, Nottingham University from 1961 to 1966. Moderator, Free Church Federal Council from 1988 to 1989. Frequent broadcaster and member, religious advisory panels, BBC Radio Nottingham and Bristol from 1970 to 1982. Chairman, Mansfield Road Baptist Housing Association from 1968 to 1976. Chairman, Horfield Housing Association from 1976 to 1982. Director, Baptist House Ltd. in 1989. **Interests:** reading, music, gardening, walking.
Mailing address: Baptist Union of Great Britain, Baptist House, PO Box 44, 129 Brodway, Didcot OX11 8RT, UK. T. (44 235) 512077

Green, Ian Russell

Rel: Buddhist: Tibetan Mahayana; b. 26 JAN 1946 in Bendigo VIC, Australia; m. to Judy; c. 3 boys; Citizen of Australia; Lang. English.
International Organizations: Member of the Foundation for the Preservation of the Mahayana Tradition.
Other organizations: Creative Director (since 1986), Cook Green and Moor; Director (since 1990), Atisha Centre.
Main field of work: Administration, meditation. **Career:** Director, Atisha Centre from 1981 to 1985. **Educ.:** B.A. from Monash University, Clayton, Victoria, Australia (1983-1986).
Mailing address: RMB 1530, Eaglehawk VIC 3556, Australia. T. (61 54) 41 2705
Residence: 16 Brodie Street, Bendigo VIC 3550, Australia

Green, Judy

Rel: Buddhist: Mahayana.
International Organizations: Member of the Foundation for the Preservation of the Mahayana Tradition.
Other organizations: Director, Atisha Centre.
Mailing address: RMB 1530, Eaglehawk VIC 3556, Australia. T. (61 54) 42 2866. Fax (61 54) 42 5301

Green, Warren

Rel: Jewish.
International Organizations: Secretary-General of the Consultative Council of Jewish Organizations.
Mailing address: 420 Lexington Avenue, Suite 1733, New York NY 10170, USA. T. (1 212) 808 5437. Fax 983 0094.

Greenberg, Janet

Rel: Jewish.

International Organizations: Past Executive Director of the Action for Soviet Jewry.

Mailing address: 24 Crescent Street, Suite 3B, Waltham MA 02154, USA. T. (1 617) 893 2331.

Greene, Cheryl

See Thubten Chodron.

Greenspoon, Leonard Jay, Prof Dr

Rel: Jewish; Unordained; b. 05 DEC 1945 in Richmond VA, USA; m. to Eliska Morsel; c. 2 girls; Citizen of USA; Lang. English, French, Italian, German, Hebrew, Greek, Latin and Aramaic.

International Organizations: Secretary/Treasurer (since 1982) of the International Organization for Septuagint and Cognate Studies.

Other organizations: Professor of Religion (since 1987), Clemson University, SC; Chair (since 1987), Textual Criticism Society of Biblical Literature; President (since 1988), Congregation Beth Israel Synagogue, Greenville SC; President (since 1989), South Carolina Academy of Religions.

Main field of work: Research and university teaching. **Career:** Instructor of Religion, Clemson University, SC from 1975 to 1977. Assistant Professor of Religion, Clemson University, SC from 1977 to 1983. Associate Professor of Religion, Clemson University, SC from 1983 to 1987. **Educ.:** B.A. from University of Richmond, VA, USA (1964-1967); University of Rome, Italy (1967-1968); M.A. from University of Richmond, VA, USA (1968-1970); Ph.D. from Harvard University, Cambridge MA, USA (1970-1977); Jewish Theological Seminary, NY, USA (1978); University of California, Santa Barbara, USA (1978-1979). **Interests:** Bible translations, public service and uniting. **Hons.:** Grants, Fellowships and Awards from: Clemson University, National Endowment for the Humanities, American Council of Learned Societies, American Jewish Archives, American Schools of Oriental Research, Hebrew University of Jerusalem, Annenberg Research Institute, Endowment for Biblical Research. **Mem.:** American Schools of Oriental Research, Association for Jewish Studies, B'nai B'rith, Catholic Biblical Association, International Organization for Septuagint and Cognate Studies, Society for Values in Higher Education, South Carolina Academy of Religion, the Democratic Party, Planned Parenthood **Pubs.:** *Textual Studies in the Book of Joshua*, Harvard Semitic Monographs, 1983; *Max Leopold Mangolis. A Scholar's Scholar*, Biblical Scholarship in North America, 1987 and "The Warrior God", in *Religion and Politics in the Modern World.*

Mailing address: Dept. of Philosophy and Religion, Clemson University, Clemson SC 29634-1508, USA. T. (1 803) 656 5358.

Residence: 300 Hunting Hill Circle, Greer SC 29650, USA. T. (1 803) 268 9312

Greet, Kenneth Gerald, Rev Dr

Rel: Methodist; Ordained in 1947; b. 17 NOV 1918 in Bristol, Avon, UK; m. to Mary Eileen Edbrooke; c. 1 boy and 2 girls; Citizen of United Kingdom; Lang. English.

International Organizations: Chairman of the Methodist Peace Fellowship.

President (since 1989) of the World Disarmament Campaign.

Other organizations: World Methodist Council (since 1957); Chairman of Governors (since 1986), Southlands College.

Main field of work: Ministry. **Career:** Cwm and Kingstone Methodist Church from 1940 to 1942. Ogmore Vale Methodist Church from 1942 to 1945. Tonypandy Central Hall from 1947 to 1954. Secretary, Dept. of Christian Citizenship, Methodist Church and Social Responsibility Division from 1954 to 1971. Secretary, Methodist Conference from 1971 to 1984. President, Methodist Conference from 1980 to 1981. Chairman, Executive Committee, World Methodist Council from 1976 to 1981. Moderator, Free Church Federal Council from 1981 to 1982. Member, British Council of Churches from 1955 to 1984. Chairman, Executive, British Council of Churches from 1977 to 1980. Chairman, Executive, Temperance Council of Christian Churches from 1961 to 1971. Co-Chairman, World Disarmament Campaign from 1982 to 1986. **Educ.:** Handsworth College, Birmingham, UK (1946-1947). **Hons.:** Hon. D.D., Ohio. **Pubs.:** The Art of Moral Judgement, 1970; The Big Sin: Christianity and the Arms Race, 1982 and What Shall I Cry?, 1986.

Mailing address: 'Redcroft', 89 Broadmark Lane, Rustington, West Sussex BN16 2JA, UK. T. (44 903) 773326

Residence: See Mailing Address. T. See Mailing Address

Grégoire, Paul, H E Card

Rel: Roman Catholic; Ordained on 22 MAY 1937; b. 24 OCT 1911 in Verdun, Montreal, Canada; celibate; Citizen of Canada.

International Organizations: Member (since 1988) of the College of Cardinals.

Member of the Sacred Congregation for the Oriental Churches. Member of the Pontifical Council for the Pastoral Care of Migrants and Itinerant People. Member of the Congregation for Catholic Education.

Other organizations: Archbishop Emeritus (since 1990), Montreal.

Career: Elected to the Titular Church, Curubi in 1961. Consecrated Bishop in 1961. Prefect to Montreal in 1968. Created Cardinal by Pope John Paul II in 1988.

Mailing address: 1071 rue de la Cathédrale, Montréal PQ H3B 2V4, Canada. T. (1 514) 866 1661

Gregorios, Paulos Mar, Metropolitan Dr

Rel: Eastern Orthodox; Ordained in 1961; b. 09 AUG 1922 in Tripunithura, Kerala, India; single; Citizen of India; Lang. English, French, Malayalam, Amharic, Hindi and Tamil.

International Organizations: Co-President of the Joint International Commission for Dialogue between the Roman Catholic Church and the Malankara Orthodox Syrian Church.

Member, Central Committee (since 1968) of the World Council of Churches. Member, Executive Committee (since 1975) of the World Council of Churches. Member, Presidium (since 1983) of the World Council of Churches. Patron of the Council for the World's Religions. President, Indian National Chapter of the International Society for Neoplatonic Studies.

Other organizations: Principal (since 1967), Orthodox Theological Seminary, Kottayam, Kerala; Vice-President (since 1968), Kerala Philosophical Congress; Metropolitan (since 1976), Delhi, Orthodox Syrian Church of the East; Member, Senate (since 1984), Serampore University; Member, Standing Committee, Orthodox Episcopal Synod; President, Synodal Committee for Inter-Church Relations, Inter-Religious Federation for World Peace (New York), PMGM Charitable Trust, Thalacode (Kerala); Director, Delhi Orthodox Centre (New Delhi), Sarvadharma Nilaya, A Home for All Faiths (New Delhi), Niti Santi Kendra, A Centre for Justice and Peace Initiatives (New Delhi); Chairman, Divyabodhanam, a Programme for Formation of Orthodox Lay Men and Women, St Thomas Orthodox Vaidika Sanghom (Orthodox Clergy Association), Federated Faculty for Research in Religion and Culture (Kerala), Oriental Orthodox Churches' Curriculum Committee, Commission on National Affairs of the Orthodox Church of India; Vice-President, MOC Medical Mission Hospital, Kolencheri (Kerala); Chief Editor, *Star of the East* (quarterly, New Delhi), *Purohitan* (quarterly, Kottayam).

Career: Freelance journalist, India from 1937 to 1942. Secretary, Public Library, Tripunithura from 1939 to 1942. Post and Telegraph Dept., Government of India from 1942 to 1947. Associate Secretary, All-India Post and Telegraph Union for Travancore-Cochin from 1945 to 1947. Teacher, Government Schools, Ethiopia from 1947 to 1950. Honorary Lecturer in Religion, Union Christian College, Alwaye from 1954 to 1956. Secretary-General, Orthodox Christian Student Movement of India from 1955 to 1957. Personal Aide and Advisor to the Emperor Haile Sellassie I, Ethiopia from 1956 to 1959. Advisor to the Ministry of Education, Ethiopia from 1957 to 1959. Founder, Haimanote Abew Orthodox Student Association, Ethiopia in 1957. Director, Division of Ecumenical Action and Associate Secretary-General, World Council of Churches (WCC) from 1962 to 1967. Member, Joint Working Group WCC-Roman Catholic Church from 1963 to 1975. Member, Faith and Order Commission, WCC from 1968 to 1975. Moderator, Church and Society Working Group, WCC from 1975 to 1983. Observer, II Vatican Council from 1963 to 1965. Vice-President, Christian Peace Conference from 1970 to 1990. Member, Senate, Serampore University from 1970 to 1974. Member, Senate, Kerala University from 1972 to 1976. Lectured in numerous universities all over the world. **Educ.:** B.A. from Goshen College, IN, USA (1950-1952); M.Div. from Princeton University, NJ, USA (1952-1954); S.T.M. from Yale University, New Haven CT, USA (1959-1960); D.Th. from Serampore University, India (1971-1975). **Hons.:** Dr. Theol. h.c., Leningrad Theological Academy, Lutheran Theological Academy, Budapest, and Jan Hus Faculty, Prague; Hall of Fame Award for Extraordinary Service to Peace and Human Unity (USA); Certificate of Merit for Distinguished Service and Inspired Leadership of the World Church, Dictionary of International Biography, Cambridge (UK); Order of St Vladimir (USSR); Order of St Sergius (USSR); Order of St Mary Magdalen (USSR); Order of Bishop Fransiszek Hodur, First Class (Poland); Otto Nuschke Prize for Peace (Germany, 1988); Soviet Land Nehru Award (India, 1988); The International Biographical Roll of Honour for Distinguished Service to World

Unity and Understanding among Religions (USA); The Distinguished Leadership Award for Extraordinary Service to Peace and Human Unity (USA); Man of the Year Award, American Biographical Institute (1990); Bhai Param Vir Singh International Award (India, 1990). **Mem.:** Indian Council of Philosophical Research (Government of India); Council of Serampore College, West Bengal; Council for Research in Values and Philosophy, Washington DC; Pro Oriente Foundation (Honorary Member); The Research Board of Advisors, American Biographical Institute (Honorary Member) **Pubs.:** Author of several books, including: *Enlightenment – East and West*, Indian Institute of Advanced Study, Shimla, 1989; *Too Bright A Light*, State University of New York Press, Albany NY, 1991 and hundreds of periodical articles, symposia and encyclopedia contributions published in a dozen countries.
Mailing address: The Delhi Orthodox Centre, 2 Tughlakabad Institution Area, New Delhi 1110 062, India. T. (91 11) 643 6417 – 647 4975. Fax 647 4975. C. DELHORTH, NEW DELHI
Residence: See Mailing Address. T. See Mailing Address

Greiveldinger, P

Rel: Roman Catholic.
International Organizations: UNESCO Representative of the World Union of Catholic Teachers.
Mailing address: 36 avenue des Ternes, F-75017 Paris, France.

Griffer, Andrew

Rel: Jewish.
International Organizations: Executive Director of the American Jewish World Service.
Mailing address: 1290 Ave of the Americas, New York NY 10104, USA. T. (1 212) 468 7380. Fax (1 212) 468 7398.

Griffin, David Ray, Prof

Rel: Disciples of Christ; Ordained in 1962; b. 08 AUG 1939 in Wilbur WA, USA; m. to (2nd) Ann Jagua; c. 1 girl; Citizen of USA; Lang. English and German.
International Organizations: Professor (since 1973) of the School of Theology at Claremont, CA.
Other organizations: Executive Director (since 1973), Center for Process Studies; Founding President (since 1983), Center for a Postmodern World.
Main field of work: Seminary education. **Career:** Assistant Professor, University of Dayton, OH from 1968 to 1973. **Educ.:** B.A. from Northwest Christian College, Eugene OR, USA (1956-1962); M.A. from University of Oregon, Eugene, USA (1957-1963); Ph.D. from Claremont Graduate School, CA, USA (1963-1970). **Pubs.:** Editor, The Reenchantment of Science, 1988; Varieties of Postmodern Theology, 1989 and Evil Revisited: Responses and Reconsiderations, 1991.
Mailing address: School of Theology, 1325 N College, Claremont CA 91711, USA. T. (1 714) 626 3521
Residence: 6891 Del Playa, Isla Vista CA 93117, USA. T. (1 805) 968 3825

Griffin, Victor Gilbert Benjamin, Rev

Rel: Anglican; Ordained; b. 24 MAY 1924 in Carnew, Co. Wicklow, Ireland; m. to Daphne Elizabeth Mitchell; c. 2 boys; Citizen of Ireland.
International Organizations: Ecclesiastic of the Anglican Church.
Other organizations: Dean (since 1969), St Patrick's Cathedral, Dublin.
Career: Curate, St Augustine's, Londonderry from 1947 to 1950. Curate, Christchurch, Londonderry from 1950 to 1957. Rector, Christchurch, Londonderry from 1957 to 1969. Lecturer in Philosophy, Magee University College, Londonderry from 1950 to 1969. **Interests:** golf, walking, music. **Hons.:** Irish People of the Year Award (1979). **Pubs.:** Trends in Theology (1870-1970), 1970; Anglican and Irish, 1976 and Pluralism and Ecumenism, 1983.
Mailing address: The Deanery, St Patrick's Cathedral, Dublin 8, Ireland. T. (353 1) 543428 – 754817

Grimes, Edward, Rev

Rel: Roman Catholic; Ordained; b. 14 JAN 1940 in Cork, Ireland; celibate; Citizen of Ireland; Lang. English.
International Organizations: Former Secretary-General of the Inter-Territorial Catholic Bishops' Conference.
Other organizations: Co-Director (since 1981), Caritas, Gambia; Director (since 1982), Pontifical Mission Societies.
Main field of work: Administration. **Career:** Secretary-General, Inter-Territorial Catholic Bishops' Conference from 1981 to 1991. **Educ.:** B.Sc. from National University of Ireland, University College, Dublin, Ireland (1960-1962); S.T.L. from Pontifical Gregorian University, Vatican City State (1967-1969). **Interests:** Environment.
Mailing address: 10 Howe St – Santanno House, PO Box 893, Freetown, Sierra Leone. T. (232 22) 24590. C. SANTANNO FREETOWN. Tx 3311 WASLTD SL (attn Santanno/ITCABIC).
Residence: See Mailing Address. T. See Mailing Address

Gripkey, R Joseph

Rel: Christian.
International Organizations: President of the Children International.
Mailing address: 2000 E Red Bridge Rd, Kansas City MO 64131, USA. T. (1 816) 942 2000.

Grisez, Germain, Dr

Rel: Roman Catholic; Unordained; b. 30 SEP 1929 in University Heights OH, USA; m. to Jeannette Selby; c. 4 children; Citizen of USA; Lang. English, French and Latin.
International Organizations: The Bishop Harry J Flynn Professor of Christian Ethics (since 1979) of Mount St Mary's College, Emmitsburg MD.
Main field of work: Teaching, research and writing in moral theology. **Career:** Clerical work at Federal Reserve Bank, Chicago IL from 1951 to 1956. Assistant Professor to Full Professor, Georgetown University, Washington DC from 1957 to 1972. Lecturer in Medieval Philosophy (part-time), University of Charlottesville, VA from 1961 to 1962. Special assistant to Patrick Cardinal O'Boyle, Archbishop, Washington from 1968 to 1969. Consultant (part-time), Archdiocese, Washington from 1969 to 1972. Professor of Philosophy, Campion College, University of Regina, Canada from 1972 to 1979. Executive Council, American Catholic Philosophical Association from 1968 to 1970. Vice-President, American Catholic Philosophical Association from 1982 to 1983. President, American Catholic Philosophical Association from 1983 to 1984. **Educ.:** B.A. Phil. (magna cum laude) from John Carroll University, Cleveland OH, USA (1947-1951); M.A., Ph.L. (summa cum laude) from Dominican College of St Thomas Aquinas, River Forest IL, USA (1950-1951); Ph.D. from University of Chicago, IL, USA (1951-1959). **Hons.:** Pro ecclesia et pontifice medal (1972); Fellowship of Catholic Scholars: Special award in recognition of scholarly work (1981), the Cardinal Wright Award for service to the Church (1983). **Mem.:** Catholic Theological Society of America; American Philosophical Association; Fellowship of Catholic Scholars (founding member); American Philosophical Association **Pubs.:** With Russell Shaw, *Beyond the New Morality: The Responsibilities of Freedom*, 3rd Ed., Notre Dame, University of Notre Dame Press, 1988, xi and 256; *The Way of the Lord Jesus*, Vol. 1, *Christian Moral Principles*, Chicago, Franciscan Herald, 1983, xxxiii and 971 and With Joseph Boyle and John Finnis, "Practical Principles, Moral Truth, and Ultimate Ends", in *American Journal of Jurisprudence*, 32, 1987, pp. 99-151.
Mailing address: Mount Saint Mary's College, Emmitsburg MD 21727-7799, USA. T. (1 301) 447 5771

Groen, H James

Rel: Christian.
International Organizations: Chairman of the Youth for Christ International.
Mailing address: Suite 205, 6890 South Tucson Way, Englewood CO 80112-3923, USA. T. (1 303) 790 4477. Tx 910 931-0599 PROTEX OVR. Fax (1 303) 790 4486.

Groër, Hans Hermann, O.S.B., H E Card

Rel: Roman Catholic; Ordained on 12 APR 1942; b. 13 OCT 1919 in Vienna, Austria; celibate; Citizen of Austria.
International Organizations: Member (since 1988) of the College of Cardinals.
Member of the Benedictine Confederation. Member of the Sacred Congregation for the Oriental Churches. Member of the Sacred Congregation for the Sacraments and Divine Worship. Member of the Congregation for Religious and Secular Institutes. Member of the Congregation for Catholic Education.
Other organizations: Archbishop, Vienna; President, Bishops' Conference of Austria.
Career: Consecrated Bishop in 1986. Created Cardinal by Pope John Paul II in 1988. Ordinand for the faithful in the Byzantine Rite resident in Austria.
Mailing address: Wollzeile 2, A-1010 Wien, Austria. T. (43 222) 51 552

Gross, Iris

Rel: Jewish.

International Organizations: Executive Director of the National Council of Jewish Women.

Career: Served as Commissioner on the Montgomery County (MD) Commission for Women from 1980 to 1981. Regional Director, B'nai B'rith Women in Rockville, MD. **Educ.:** B.A. from American University, Washington DC, USA. **Hons.:** Certified Association Executive, American Society of Association Executives; Named Outstanding Chairperson, New York Society of Association Executives; Honored by the March of Dimes and by the City of Rockville MD for her work with senior citizens.

Mailing address: 53 West 23rd Street, New York NY 10010, USA. T. (1 212) 645 4048. Fax 645 7466

Grou, Claude, C.S.C., Fr

Rel: Roman Catholic; Ordained; celibate.

International Organizations: Superior General of the Congregation of Holy Cross.

Mailing address: Via Framura 85, I-00168 Roma, Italy. T. (39 6) 627 69 20 – (39 6) 627 76 44.

Grouès, Henri, Abbe

Also know as Abbé Pierre; Rel: Roman Catholic; Ordained in 1938; b. 05 AUG 1912 in Lyon, France; celibate; Citizen of France.

International Organizations: Philanthropist and Ecclesiastic of the Roman Catholic Church.

Career: Entered Capuchin Order in 1930. Left for health reasons in 1938. Almoner at the hospital, La Mure and in charge of the Groupements de Jeunesse and the Orphanage of the Côte St André in 1940. Vicar, Grenoble Cathedral in 1941. Founded an escape organization through the Alps and the Pyrenees. Founded the committee against forced labour. Joined Free French Forces in Algiers as Almoner to the Fleet in 1944. Deputy for Meurthe-et-Moselle from 1946 to 1951. Organized help for the destitute and the homeless in France and abroad, and created the Centre d'Emmaüs. Founded the revue *Faims et Soifs* in 1954. **Hons.:** Officer, Legion of Honour (1981); Military Cross (2 citation avec palmes); Médaille de la Résistance; Médaille des Evadés; Médaille des Combattants Volontaires; Médaille des Maquisards Belges; Médaille Albert Schweitzer (1975). **Pubs.:** Feuilles éparses (poems); L'Abbé Pierre vous parles, Emmaüs, 1959 and Le scandale de la faim interpelle l'église.

Mailing address: Monastère de Saint Wandrille, F-76490 Caudebec en Caux, France. T. (33) 35 96 23 11

Gruenwald, Oskar, Dr

Rel: Christian.

International Organizations: President of the International Christian Studies Association.

Mailing address: 2828 Third Street 11, Santa Monica CA 90405, USA. T. (1 310) 396 0517.

Gruer, David, Col

Rel: Salvation Army, The.

International Organizations: Territorial Commander of The Salvation Army.

Mailing address: Exercito de Salvacao, Caixa Postal 8631, 01051 Sao Paulo, Brazil.

Grünewald, Hans Isaak, Rabbi

Rel: Jewish; Ordained on 30 MAR 1960; b. 15 MAR 1914 in Frankfurt/Main, Germany; m. to Margot; c. 2 boys; Citizen of Germany and Israel; Lang. English, German and Hebrew.

International Organizations: Member of Standing Committee/Coordinator (since 1961) of the Conference of European Rabbis.

Main field of work: Ministry. **Career:** Rabbi, Jüdische Gemeinde, Hamburg from 1960 to 1963. Rabbi, Israelitische Kultusgemeinde, Munich from 1963 to 1982. President, Bne Brith Bialik Lodge, Tel Aviv from 1955 to 1956. President, Bne Brith Hebraica Lodge, Munich from 1976 to 1980. Jewish Chairman, Council of Christians and Jews, Munich from 1968 to 1982. Retired from ministry in 1982. **Educ.:** Johann Wolfgang Goethe-Universität Frankfurt, Germany (1932-1933); Talmudical College, Frankfurt, Germany (1933-1936); Hebrew University of Jerusalem, Israel (1936-1940); Jews' College, London, UK (1957-1960). **Pubs.:** Die Lehre Israels, Munich, 1970 and Einblicke, Frankfurt, 1989.

Mailing address: 36 Monarch Court, Lyttelton Road, London N2 0RA, UK. T. (44 81) 455 0811

Residence: See Mailing Address. T. See Mailing Address

Gryn, Hugo Gabriel, Rabbi Dr

Rel: Jewish: Reform; Ordained in 1957; b. 25 JUN 1930 in Berehovo, Czechoslovakia; m. to Jacqueline; c. 1 boy and 3 girls; Citizen of USA; Lang. English, Hebrew, Yiddish, Czech, Russian and Hungarian.

International Organizations: President of the Reform Synagogues of Great Britain.

Other organizations: Senior Rabbi (since 1964), West London Synagogue; Joint Chair (since 1987), Interfaith Network for UK.

Main field of work: Ministry. **Career:** Rabbi, Jewish Religious Union, Bombay from 1957 to 1959. Executive Director, World Union for Progressive Judaism from 1959 to 1962. Executive, American Jewish Joint Distribution Committee from 1962 to 1964. **Educ.:** B.A. from University of Cincinnati, OH, USA (1951-1952); M.A.H.L., Rabb. Dip. from Hebrew Union College-Jewish Institute of Religion, Cincinnati OH, USA (1952-1957); D.D. from Hebrew Union College-Jewish Institute of Religion, Cincinnati OH, USA (1982).

Mailing address: 33 Seymour Place, London W1H 6AT, UK. T. (44 71) 723 4404

Residence: See Mailing Address. T. See Mailing Address

Guastavino, Jorge

Rel: Christian.

International Organizations: Area Executive of the Caribbean Alliance of YMCAs.

Mailing address: PO Box 433, Oranjestad, Aruba. T. (297 8) 47157. Tx 5024 MARTN NA.

Gudmundsson, Bernhardur

Rel: Protestant.

International Organizations: Executive Officer of the Nordic Council for Church Studies.

Mailing address: Kirkens Undervisningskomite, Biskopsstofu, Sudergota 22, 150 Reykjavik, Iceland. T. (354 1) 911 24 45.

Gudmundsson, J Eric, Pastor

Rel: Seventh-Day Adventist; Ordained; b. 25 JUL 1949 in Reykjavik, Iceland; m. to Laila Panduro-Olesen; c. 1 boy and 1 girl; Citizen of Iceland; Lang. Icelandic, Danish and English.

International Organizations: President, Iceland Conference (since 1988) of the General Conference of Seventh-Day Adventists.

Main field of work: Church administration and ministry. **Career:** Chaplain, Masanga Leprosy Hospital, Sierra Leone from 1980 to 1982. Pastor, Reykjavik Church of Seventh-Day Adventists from 1984 to 1989. **Educ.:** B.A. Theol. from Newbold College, UK (1973-1976); M.Div. from Andrews University, Berrien Springs MI, USA (1982-1984).

Mailing address: 36 Sudurhlid, 105 Reykjavik, Iceland. T. (354 1) 91 679260

Residence: 43 Tveras, 110 Rejkjavik, Iceland T. (354 1) 91 673626

Gudmundsson, Karen

Rel: Buddhist.

International Organizations: Secretary of the Foundation for the Preservation of the Mahayana Tradition.

Mailing address: PO Box 1778, Soquel CA 95073, USA. T. (1 408) 476 8435. Fax 476 4823.

Guerrero, Bernardita

Rel: Spiritualist.

International Organizations: Secretary-General of the International Spiritual Alliance.

Mailing address: Calle Sazié 2078, Casilla 2941, Santiago, Chile. T. (56 2) 695 1170.

Guerri, Sergio, H E Card

Rel: Roman Catholic; Ordained on 30 MAR 1929; b. 25 DEC 1905 in Tarquinia, Italy; celibate; Citizen of Italy.

International Organizations: Member (since 1969) of the College of Cardinals.

Career: Created Cardinal by Pope Paul VI in 1969. Pro-President, Pontifical Commission for the State of the Vatican City.

Mailing address: 00120 Città del Vaticano, Vatican. T. (39 6) 698 34 43

Guggenbühl, Ad, Dr

Rel: Christian.
International Organizations: President of the Christian Business Men's Association International.
Mailing address: Seefeldstrasse 45, CH-8008 Zürich, Switzerland. T. (41 1) 252 79 12.

Guillerm, Vincent, F.I.C.P., Br

Rel: Roman Catholic; Ordained; celibate.
International Organizations: Procurator-General of the Prêtres de Sainte Marie de Tinchebray.
Procurator-General of the Brothers of Christian Instruction of Ploërmel.
Mailing address: Via della Divina Provvidenza 44, I-00166 Roma, Italy. T. (39 6) 680 56 18.

Guillet, Jean-Paul, Fr

Rel: Roman Catholic; Ordained; celibate.
International Organizations: Director of the OCIC Missionary Service.
Mailing address: Piazza San Calisto 16, I-00120 Città del Vaticano, Vatican. T. (39 6) 698 72 55. Tx (504) 2014 CI VA for OCIC. Fax (39 6) 698 7335.

Gulbinowicz, Henryk Roman, H E Card

Rel: Roman Catholic; Ordained on 18 JUN 1950; b. 17 OCT 1928 in Szukiszki, Vilna, Poland; celibate; Citizen of Poland.
International Organizations: Member (since 1985) of the College of Cardinals.
Member of the Sacred Congregation for the Oriental Churches. Member of the Congregation for the Clergy.
Other organizations: Archbishop, Wroclaw, Breslavia.
Career: Vicar in Szudzialowo parish from 1950 to 1951. Academic Curate in Bialystok from 1955 to 1959. Staff Member, Olsztyn Curia from 1959 to 1970. Titular Bishop, Acco, Apostolic Administrator of Archdiocese of Bialystok from 1970 to 1976. Created Cardinal by Pope John Paul II in 1985. **Educ.:** Dr. Theol. from Catholic University of Lublin, Poland (1951-1955).
Mailing address: ul Katedralna 11, 50-328 Wroclaw, Poland. T. (48 71) 22 42 14

Gunaratana Mahathera, Henepola G, Rev Dr

Rel: Buddhist: Theravada; Ordained in 1940; b. 07 DEC 1927 in Henepola, Sri Lanka; Citizen of USA; Lang. Sinhalese, English, Pali, Sanskrit, Hindi, German, Tamil and Japanese.
International Organizations: President (since 1982) of the Bhavana Society.
Other organizations: Patron (since 1978), Ottawa Buddhist Association; Patron (since 1989), Buddhist Association in British Columbia.
Main field of work: Teaching, preaching, writing. **Career:** Teacher, Buddhist Institute, Malaysia from 1958 to 1968. Assistant Professor, American University, Washington DC from 1975 to 1983. Assistant Professor, Georgetown University, Washington DC from 1981 to 1982. Assistant Professor, Bucknell University, Lewisburg PA from 1974 to 1983. Assistant Professor, University of Maryland from 1976 to 1980. **Educ.:** B.A. from American University, Washington DC, USA (1975); M.A. from American University, Washington DC, USA (1977); Ph.D. from American University, Washington DC, USA (1980). **Interests:** Meditation, writing. **Pubs.:** Come and See; Path of Serenity and Insight; The Jhanas in Theravada Buddhist Meditation and Mindfulness in Plain English.
Mailing address: Bhavana Society Meditation Center, Rt 1, Box 218-3, High View WV 26808, USA. T. (1 304) 856 3241

Gundiaev, Kirill (Vladimir), H E

Rel: Orthodox: Russian; Ordained on 01 JUN 1969; b. 20 NOV 1946 in Leningrad (now St Petersburg), USSR; celibate; Citizen of Russia; Lang. Russian, English and German (some).
International Organizations: Metropolitan (since 1991) of Smolensk and Kaliningrad.
Other organizations: Head, Smolensk Diocese (since 1986), Russian Orthodox Church; Chairman, Dept. for External Church Relations Permanent member (since 1989), Moscow Patriarchate; Permanent Member (since 1989), Holy Synod of Russian Orthodox Church; Chairman, Synodal Commission on the Revival of Charity and Religious Education (since 1990), Russian Orthodox Church.
Career: Representative, Moscow Patriarchate to World Council of Churches, Geneva from 1971 to 1974. Rector, Leningrad Theological Academy of the Russian Orthodox Church from 1974 to 1984. Elevated to the rank of Archbishop in 1977. **Educ.:** Dip. Theol. from Leningrad Theological Academy, St Petersburg, USSR (1965-1970). **Hons.:** Dr. h.c., Reformed Theological Academy, Budapest; Honorary Member, Theological Academy of St Petersburg. **Pubs.:** "Church vis-à-vis Society in Perestroika", in *Journal of Moscow Patriarchate* (JMP), 2/90; Interview for the JMP in connection with the appointment to the position of the Head of the DECR, JMP, 6/90 and "On the Ecology of the Spirit", in *JMP*, 5-6/91.
Mailing address: Dept. for External Church Relations, Moscow Patriarchate, Danilovsky val 22, Danilovsky Val 22, Moskva 113191, Russia. T. (7 95) 2350708
Residence: See Mailing Address. T. See Mailing Address

Gunter, Nina G

Rel: Nazarene.
International Organizations: Director-General of the Nazarene World Mission Society.
Mailing address: 6401 The Paseo, Kansas City MO 64131, USA. T. (1 816) 333 7000. Fax 363 3100.

Günther, Rolf-Dieter, Rev

Rel: Evangelical.
International Organizations: Secretary of the Ecumenical Workshop for Information in Europe.
Other organizations: Press Secretary, Federation of Evangelical Churches.
Mailing address: Auguststrasse 80, 10117 Berlin, Germany. T. (37 2) 886 116

Gurugé, Ananda Weihena Palliya, Dr

Rel: Buddhist; Unordained; b. 28 DEC 1928 in Galle, Southern Province, Sri Lanka; m. to Sujata Nalini Jayasinghe; c. 1 boy; Citizen of Sri Lanka; Lang. Sinhala, English, French, Tamil, Hindi, Pali and Sanskrit.
International Organizations: Vice-President (since 1988) of the World Fellowship of Buddhists.
Other organizations: Ambassador and Permanent Delegate (since 1985), Sri Lanka to UNESCO; Ambassador (since 1989), Sri Lanka to France, Spain and Algeria; Patron (since 1988), European Buddhist Union.
Main field of work: Diplomat and research scholar. **Career:** Officer, Sri Lanka Civil Service from 1952 to 1967. Professor, State University of New York at Buffalo from 1967 to 1968. Staff Member, UNESCO from 1968 to 1985. **Educ.:** B.A. Sanskrit (Hons) from University of Colombo, Sri Lanka (1947-1950); Ph.D. from University of Colombo, Sri Lanka (1950-1952). **Interests:** history of Asian culture. **Hons.:** Kalakirthi (national award). **Pubs.:** several publications, including: From the Living Fountains of Buddhism, 1984; Mahavamsa – New Translation, 1990 and Voices from Ancient Sri Lanka, 1991.
Mailing address: Sir Lanka Embassy, 15 rue d'Astorg, F-75008 Paris, France. T. (33 1) 42 66 38 56
Residence: 76 rue Lecourbe, F-75015 Paris, France T. (33 1) 45 67 11 30

Guske, Hubertus

Rel: Roman Catholic; b. 21 MAR 1930 in Wroclaw, Poland; married; c. 2 boys and 1 girl; Citizen of Germany; Lang. German and English.
International Organizations: Secretary-General (since 1987) of the Berlin Conference of European Catholics.
Member of the Presidium (since 1982) of the Berlin Conference of European Catholics.
Other organizations: Journalist (since 1951), various papers.
Main field of work: Peace research and church affairs. **Career:** Editor, *Neue Zeit*, Berlin (daily) from 1951 to 1961. Editor-in-Chief, *Begegnung*, Berlin (monthly) from 1961 to 1987. **Educ.:** Dip. Hist. from Humboldt University of Berlin, Germany (1957-1963). **Mem.:** Christian Democratic Union of Germany **Pubs.:** Some booklets about people of importance and many articles in papers and magazines.
Mailing address: Poststrasse 4-5, 10178 Berlin, Germany. T. (37 2) 212 54 92
Residence: Am Friedrichshain 4, 10407 Berlin, Germany T. (37 2) 439 86 43

Gustavo, Mejia P

Rel: Roman Catholic; Ordained; celibate.
International Organizations: Superior General of the Foreign Missions of Yarumal.
Mailing address: Carrera 81 No 52 B-120, Medellín, Colombia. T. (57 4) 234 09 21.

Gutiérrez, Gustavo

Rel: Roman Catholic; Ordained in 1959; b. 08 JUN 1928 in Lima, Peru; celibate; Citizen of Peru.
International Organizations: Ecclesiastic and theologian of the Roman Catholic Church.
Other organizations: Professor (since 1960), Catholic University of Lima; Adviser to the National Union of Catholic Students (since 1960); Board (since 1974), Instituto Bartolomé Las Casas-Rímac.
Career: Member, Pastoral-Theological team, Latin American Conference of Catholic Bishops (CELAM) from 1967 to 1968. Worked among slum dwellers. Developed Liberation Theology (during the 1960s). **Educ.:** Studied Medicine (for 5 years) from National University of Lima, Peru; M.A. Phil. and Psy. from Catholic University of Louvain, Belgium (1955); Dr. Theol. from University of Lyon, France (1959). **Interests:** swimming, literature. **Hons.:** Dr. h.c., Nijmegen (1979), Tübingen (1985). **Mem.:** Ecumenical Association of Third World Theologians (EATWOT) **Pubs.:** *A Theology of Liberation*, 1971 (English translation 1973); *The Power of the Poor in History*, 1979 (English translation 1983) and *We Drink from Our Own Wells: The Spiritual Journey of a People*, 1983 (English translation 1984).
Mailing address: Instituto Bartolomé Las Casas-Rímac, Apartado 3090, Lima 100, Peru. T. (51 14) 814704 – 814663

Gutiérrez, Isaias, Bishop

Rel: Methodist; Ordained.
International Organizations: Bishop of Methodist Church of Chile.
Mailing address: Sargento Aldea 1041, Casilla 67, Santiago de Chile, Santiago.

Gyaltsab Rinpoche, G, H E

Rel: Buddhist: Kagyupa.
International Organizations: One of the main leaders of the Kagyupa tradition.
Mailing address: Monastery of Rumtek, PO Ranipul, Gangtok, Sikkim, India.

Gyatso, Tenzin

See Dalai Lama.

Gyatso, Thubten, Ven Dr

Also know as Adrian Roy Feldmann; Rel: Buddhist: Mahayana; Ordained on 18 NOV 1975; b. 21 OCT 1943 in Melbourne, Victoria, Australia; celibate; Citizen of Australia; Lang. English.
International Organizations: Member of the Foundation for the Preservation of the Mahayana Tradition.
Other organizations: Director (since 1991), Jingxu Dharma Centre (Taiwan), FPTM.
Main field of work: Education (Buddhist philosophy). **Career:** Spiritual Director, Tara Institute (Australia), FPTM from 1979 to 1981. Director, Nalanda Monastery (France), FPTM from 1981 to 1988. Spiritual Teacher, FPTM (Australia) from 1988 to 1990. **Educ.:** M.B.B.S. from University of Melbourne, Victoria, Australia (1963-1968); D.T.M. and H. from University of London, School of Hygiene and Tropical Medicine, UK (1971). **Interests:** Buddhist psychology.
Mailing address: Jingxu Dharma Centre, 4F No 5, Alley 1, Lane 329, Fulin Road, Shihlin, Taipei, Taiwan (Rep of China). T. (886 2) 833 8754 (SAME FOR Fax)
Residence: 3 Mavis Ave, Brighton East 3187, Australia.

Gyaw, Gregory Hla, Most Rev

Rel: Christian.
International Organizations: Archbishop of Burma.
Mailing address: Bishopscourt, 140 Pyidaungsu Yeiktha Road, Rangoon 11191, Myanmar (Burma).

Gyürki-kis, Adam, C.Op., Fr

Rel: Roman Catholic; Ordained; celibate.
International Organizations: Superior General of the Congrégation des Ouvriers Chrétiens de Saint Joseph Calasanz.
Mailing address: Gebrüder Langgase 7, A-1550 Wien 15, Austria. T. (43 1) 83 14 94

Haan, Sheri D

Rel: Christian.
International Organizations: Director of the Christian Schools International.
Mailing address: 3350 E Paris Avenue SE, Grand Rapids MI 49508, USA. T. (1 616) 957 10 70.

Haarbeck, Ako, Dr

Rel: Church of Lippe.
International Organizations: Landessuperintendent of the Church of Lippe.
Mailing address: Leopoldstrasse 27, 32756 Detmold, Germany.

Haas, Pierre, C.S.Sp., Fr

Rel: Roman Catholic; Ordained; celibate.
International Organizations: Superior General of the Congregation of the Holy Spirit.
Member of the Congregation for Religious and Secular Institutes.
Mailing address: Clivo di Cinna 195, I-00136 Roma, Italy. T. (39 6) 34 82 47. Fax 345 06 76.

Habgood, John Stapylton, Most Rev and Rt Hon Dr

Rel: Church of England; Ordained in 1954; b. 23 JUN 1927 in Stoney Stratford, Bucks, UK; m. to Rosalie Mary Anne Boston; c. 2 boys and 2 girls; Citizen of United Kingdom; Lang. English.
International Organizations: Archbishop (since 1983) of York.
Other organizations: Patron, President of Vice-President, numerous voluntary organizations.
Main field of work: Ministry. **Career:** Demonstrator in Pharmacology, University of Cambridge from 1950 to 1953. Fellow, King's College, Cambridge from 1952 to 1955. Curate, St Mary Abbott's Church, Kensington from 1954 to 1956. Vice-Principal, Westcott House, Cambridge from 1956 to 1962. Rector, St John's Church, Jedburgh from 1962 to 1967. Principal, Queen's College, Birmingham from 1967 to 1973. Honorary Canon, Birmingham Cathedral from 1971 to 1973. Bishop, Durham from 1973 to 1983. **Educ.:** B.A., M.A., Ph.D. from University of Cambridge, King's College, UK (1945-1952); Gen. Ord. Exam from University of Oxford, Cuddesdon College, UK (1953-1954). **Interests:** science and religion, ethics, ecumenism, carpentry, painting. **Hons.:** Privy Councillor (1983); 5 Hon. D.D.'s; Hon. Fellow, King's College, Cambridge (1984). **Mem.:** Club: Athenaeum **Pubs.:** Religion and Science, 1964; Church and Nation in a Secular Age, 1983 and Confessions of a Conservative Liberal, 1988.
Mailing address: Bishopthorpe Palace, Bishopthorpe, York YO2 1QE, UK. T. (44 904) 707 021
Residence: See Mailing Address. T. See Mailing Address

Habib, Gabriel Gergi

Rel: Orthodox, Church of Antioch; b. 14 FEB 1937 in Wadi El-Olia, Beirut, Lebanon; m. to Kathryn Huenemann; c. 2 girls; Citizen of Lebanon; Lang. French, English and Arabic.
International Organizations: Secretary-General (since 1977) of the Middle East Council of Churches.
Main field of work: Ecumenical relations. **Career:** General Secretary, Orthodox Youth Movement from 1967 to 1974. General Secretary, Syndesmos from 1967 to 1974. Middle East Secretary, World Student Christian Federation from 1974 to 1977. Middle East Secretary, World Council of Churches Youth Movement from 1974 to 1977. **Educ.:** Legal Stds. from Saint Joseph's University, Beirut, Lebanon (1958-1962). **Hons.:** Ted Johnson Award 1988, Canada; High Commander of Saint Peter and Saint Paul 1990, Church of Antioch.
Mailing address: PO Box 4259, 1 Petrarchi Street, Limassol, Cyprus. T. (357 51) 326022
Residence: See Mailing Address. T. (357 51) 32 64 89

Habib, Rafic
Rel: Christian.
International Organizations: Regional Secretary for the Middle East of the World Association for Christian Communication.
Mailing address: PO Box 1029, Limassol, Cyprus.

Habib, Samuel, Rev Dr
Rel: Protestant.
International Organizations: General Secretary of the Protestant Churches of Egypt.
Mailing address: PO Box 1304, Cairo, Egypt.

Habiby, Samir J, Rev Canon
Rel: Episcopal; Ordained.
International Organizations: Presiding Bishop of Fund for World Relief.
Career: Came to the US in 1954. Served two tours of duty in Vietnam as a navy chaplain. Former Director, Episcopal Church's global relief agency. Worked closely with Terry Waite to further the Anglican Church's humanitarian efforts for the release of American hostages held in Lebanon.
Mailing address: St Paul's Episcopal Church, Riverside CT, USA.

Haddad, Heskel M, Dr
Rel: Jewish.
International Organizations: President of the World Organization of Jews from Arab Countries.
Mailing address: 1125 Park Ave, New York NY 10128, USA. T. (1 212) 427 1246.

Haddad, Pierre Naaman, H E
Rel: Greek Catholic; Ordained in 1950; b. 14 OCT 1923 in Cairo, Egypt; celibate.
International Organizations: Bishop of Greek Catholic Chruch.
Mailing address: Greek Catholic Patriarchate, 16 Daher Street, Cairo, Egypt.

Haddick, Donald H, Rev
Rel: Spiritualist; Ordained.
International Organizations: Secretary of the National Spiritualist Teachers Club.
Mailing address: 870 Junipero Serra Blvd, San Francisco CA 94127-2850, USA.

Hadfield, Anne
Rel: Reformed Church.
International Organizations: Member of the Executive Committee of the World Alliance of Reformed Churches.
Mailing address: PO Box 9049, Wellington, New Zealand. T. (64 4) 801 6000. Fax (64 4) 801 6001

Hadjeddin, S A
Rel: Unknown.
International Organizations: UNESCO Representative of the World Conference on Religion and Peace.
Mailing address: 89 quai Gabriel Péri, F-94340 Joinville-le-Pont, France. T. (33 1) 48 89 53 91.

Hafiz, Omar Z A
Rel: Muslim; b. in Medina, Saudi Arabia; Citizen of Saudi Arabia.
International Organizations: Deputy Director of the Centre for Research in Islamic Economics.
Other organizations: Assistant Professor at the Faculty of Economics and Administration and Member (since 1981), King Abdulaziz University, Jeddah, Saudi Arabia.
Main field of work: Academic. **Career:** Demonstrator from 1975 to 1978. Lecturer from 1978 to 1981. Chairman, King Abdulaziz University Cultural Committee from 1981 to 1982. **Educ.:** M.A., Ph.D. Econ. **Interests:** Reading, writing. **Mem.:** King Faisal Welfare Society **Pubs.:** *Public Expenditure of Saudi Arabia, Structure and Growth* (M.A. Thesis) and *A Foreign Trade Model for Saudi Arabia: an Econometric Approach* (Ph.D. Thesis).
Mailing address: King Abdulaziz University, PO Box 16711, Jeddah 21474, Saudi Arabia. T. (966 2) 687 94 04 – (966 2) 687 92 02. C. JAMEATABDULAZIZ. Tx 601141 KAUNI SJ.
Residence: Jeddah, Saudi Arabia T. (966 2) 671 63 57

Haggart, Alastair Ian MacDonald, Rt Rev
Rel: Episcopal; Ordained in 1942; b. 10 OCT 1915 in Glasgow, Scotland, UK; m. to (1st) Margaret Agnes Trundle (died), (2nd) Mary Elizabeth Scholes; c. 2 girls; Citizen of United Kingdom.
International Organizations: Ecclesiastic (retired) of the Episcopal Church.
Career: Ordained Deacon in 1941. Curate, St Mary's Cathedral, Glasgow from 1941 to 1945. Canon, St Mary's Cathedral, Glasgow from 1958 to 1959. Curate, St Mary's, Hendon from 1945 to 1948. Precentor, St Ninian's Cathedral, Perth from 1948 to 1957. Rector, St Oswald's, Glasgow from 1951 to 1959. Synod Clerk from 1958 to 1959. Provost, St Paul's Cathedral, Dundee from 1959 to 1971. Principal, Edinburgh Theological College from 1971 to 1975. Bishop, Edinburgh from 1975 to 1986. Primus, Scottish Episcopal Church from 1977 to 1986. **Interests:** walking, reading, music. **Hons.:** Hon. LL.D., Dundee (1970).
Mailing address: 19 Eglinton Crescent, Edinburgh EH12 5BY, Scotland, UK. T. (44 31) 337 8948

Haimanot, Abune Tekle, H H
Rel: Orthodox: Ethiopian.
International Organizations: Patriarch of Ethiopian Orthodox Church.
Mailing address: Patriarchate Head Office, PO Box 1283, Addis Ababa, Ethiopia.

Hajek, Miloslav, Rev
Rel: Evangelical Brethren; Ordained.
International Organizations: Moderator of the Evangelical Church of Czech Brethren.
Mailing address: Jungmannova 9, Praha 1, Czech Rep.

Hakenewerth, Quentin, Fr
Rel: Roman Catholic; Ordained; celibate.
International Organizations: Superior General of the Society of Mary.
Mailing address: Via Latina 22, I-00179 Roma, Italy. T. (39 6) 757 58 92.

Hakim, Maximos V, His Beatitude
Rel: Greek-Melkite Catholic; Ordained on 20 JUL 1930; b. 18 MAY 1908 in Tanta, Egypt; celibate; Citizen of Lebanon; Lang. Arabic, French, English, Italian, Latin, Greek and Hebrew.
International Organizations: Patriarch (since 1967) of Antioch and All the East (Alexandria and Jerusalem).
Member of the Sacred Congregation for the Oriental Churches.
Other organizations: Head, Greek-Melkite Catholic Patriarchal Synod; President, Assembly of Diocesan Bishops of the Syrian AR.
Main field of work: Ministry. **Career:** Teacher at Patriarchal College, Cairo from 1930 to 1943. Archbishop, Acre and All Galilee from 1943 to 1967. Founder, *Le Lien* (French magazine). **Educ.:** Dr. Theol. **Hons.:** Grand Cross of the Legion of Honour; Dr. h.c., Laval University, Algiers University and many US universities (including Loyola). **Pubs.:** Pages d'Evangile lues en Galilée (translated into English, Dutch and Spanish), 1954; Life of Jesus (Arabic), 1980 and Paroissien Byzantin (Arabic and French), 1980.
Mailing address: Melkite Greek Catholic Patriarchate, PO Box 50076, Beirut, Lebanon or PO Box 22249, Damascus, Syria. T. BEIRUT (961 1) 413111 – DAMASCUS (963 11) 433129

Halberstram, Naftali, Rabbi
Rel: Jewish.
International Organizations: President of the World Jewish Genealogy Organization.
Mailing address: PO Box 420, Brooklyn NY 11219, USA. T. (1 718) 435 4400.

Hale, Joseph Rice, Dr

Rel: Methodist; Ordained; b. 25 MAR 1935 in Texakarna TX, USA; m. to Mary Richey; c. 1 boy.

International Organizations: Secretary-General (since 1976) of the World Methodist Council.

Main field of work: Church organization executive. **Career:** Ordained to ministry in the Methodist Church in 1958. Pastor, Methodist Church, Sunset, TX from 1958 to 1960. Evangelist from 1960 to 1966. Associate Director for the Dept. of Evangelism, Board of Evangelism, Board of Evangelism, Methodist Church from 1966 to 1968. Director for ecumenical evangelism from 1968 to 1974. Member, Executive Committee, Key 73 from 1970 to 1973. Secretary, working group for evangelism, National Council of Churches in 1972. Executive Committee, Evangelization Forum from 1973 to 1975. President, Communications Foundation, Inc. from 1974 to 1975. Registrar, World Methodist Evangelism Convocation, Jerusalem in 1974. World ambassador, International Prayer Fellowship in 1974. Director for evangelization development, Board of Discipleship, United Methodist Church from 1975 to 1976. Chairman, Christian World Communions from 1983 to 1986. Producer, a film and several videos. **Educ.:** B.A. from Asbury College, Wilmore KY, USA (1957); B.D. from Southern Methodist University, Dallas TX, USA (1960); D.D. from Asbury Theological Seminar (1978). **Hons.:** Recipient Key to the City of Daytona Beach, FL (1963, 1964); Named Kentucky Colonel (1977); Recipient Asbury College Alumni Award (1977). **Pubs.:** Author: *Design for Evangelism*, 1970; *Christ Matters !*, 1971; *God's Moment*, 1972; Also articles and chapters in books and encyclopedias and Editor, proceedings of the 13th, 14th, 15th World Methodist Conferences.

Mailing address: PO Box 518, Lake Junaluska NC 28745, USA. T. (1 704) 456 9432.

Residence: 301 Forest Park Drive, Waynesville NC 28786, USA

Halliden, William H, S.S.C.M.E., Fr

Rel: Roman Catholic; Ordained; celibate.

International Organizations: Procurator-General of the Missionary Society of St Columban.

Mailing address: Corso Trieste 57, I-00198, Italy. T. (39 6) 841 66 03

Halme, Kaija Liisa

Rel: Christian.

International Organizations: Contact of the Study Committee of the Nordic Missions Council.

Mailing address: Church Mission Centre, PL 185, SF-00161 Helsinki, Finland.

Halonen, Anna-Liisa

Rel: Seventh-Day Adventist.

International Organizations: Treasurer, Trans-European Division of the General Conference of Seventh-Day Adventists.

Mailing address: 119 St Peter's Street, St Albans AL1 3EY, UK.

Hamani, Ahmad

Rel: Muslim; Citizen of Algeria.

International Organizations: President of the Superior Islamic Council.

Mailing address: Place Cheik Abd al-Hamid ibn Badis, Algiers, Algeria.

Hambidge, Douglas Walter, Most Rev

Rel: Anglican; Ordained in 1954; b. 06 MAR 1927 in London, UK; m. to Denise Colvill; c. 2 boys and 1 girl; Citizen of United Kingdom.

International Organizations: Metropolitan (since 1981) of Ecclesiastical Province of British Columbia.

Other organizations: Archbishop (since 1981), New Westminster.

Career: Curate, St Mark and St Bartholomew, UK from 1953 to 1956. Curate, Cassiar, BC, Canada from 1956 to 1958. Curate, Smithers, BC from 1958 to 1964. Curate, Fort St John, BC from 1964 to 1969. Rural Dean, Tweedsmuir from 1960 to 1964. Canon, St Andrew's Cathedral in 1965. Rural Dean, Peace River from 1965 to 1969. Bishop, Caledonia in 1969. **Interests:** cycling, badminton, squash.

Mailing address: 302-814 Richards Street, Vancouver BC V6B 3A7, Canada. T. (1 604) 684 6306

Residence: 5887 Highbury Street, Vancouver BC V6N 1Y9, Canada.

Hamdani, Viqar Ahmad, Dr

Rel: Muslim.

International Organizations: Permanent Representative to the UN, New York of the World Muslim Congress.

Mailing address: President Representative of WMC, 5 Stuyvessant Oval, Apt 4M, New York NY 10009, USA.

Hamelin, Jean-Guy, Most Rev

Rel: Roman Catholic; Ordained on 11 JUN 1949; b. 08 OCT 1925 in St Severin PQ, Canada; celibate; Citizen of Canada.

International Organizations: Bishop (since 1974) of Diocese of Rouyn-Noranda.

Other organizations: Advisor to International Cooperation for Development and Solidarity (CIDSE) (since 1988); Member, Executive Committee (since 1989), Canadian Conference of Bishops.

Career: Teacher at Theology and Social Science's Seminary, Trois-Rivières, Quebec from 1955 to 1958. Chaplain for Social Movements, Diocese of Trois-Rivières, Quebec from 1958 to 1964. Director, Social Action Dept., Canadian Conference of Catholic Bishops from 1964 to 1968. Secretary-General, Conference of Bishops of Quebec from 1968 to 1974. Member of Board, Vice-President, Conference of Bishops of Quebec from 1975 to 1985. **Educ.:** B.A. from Université Laval, Québec, Canada (1945); L.Th. from Pontifical University San Tommaso d'Aquino, Vatican City State (1953); L.Sc.Soc. from Pontifical Gregorian University, Vatican City State (1955).

Mailing address: 515 Cuddihy, Rouyn, Québec J9X 5W9, Canada. T. (1 819) 764 4660

Hamer, Jean Jérôme, O.P., H E Card

Rel: Roman Catholic; Ordained on 03 AUG 1941; b. 01 JUN 1916 in Brussels, Belgium; celibate; Citizen of Belgium.

International Organizations: Prefect (since 1985) of the Congregation for Institutes of Consecrated Life and for Societies of Apostolic Life. Member of the Order of Preachers. Member of the Sacred Congregation for the Sacraments and Divine Worship. Member of the Congregation for Bishops. Member of the Congregation for the Evangelization of Peoples. Member of the Congregation for Religious and Secular Institutes. Member of the Congregation for Catholic Education. Member of the Pontifical Council for the Interpretation of Legislative Texts.

Other organizations: Member (since 1985), College of Cardinals; Deacon, Saba; Member, Council of Cardinals and Bishops; Member, Congregation for the Doctrine of the Faith.

Career: Consecrated Titular Bishop, Lorium with personal title of Archbishop in 1973. Created Cardinal by Pope John Paul II in 1985.

Mailing address: Piazza del S Uffizio 11, I-00193 Roma, Italy. T. (39 6) 698 4121

Hamilton, Susan

Rel: Roman Catholic.

International Organizations: President of the Christian Family Movement.

Mailing address: 505 East Huron Street, Apt 607, Ann Arbor MI, USA.

Hamilton, Wayne

Rel: Roman Catholic.

International Organizations: President of the Christian Family Movement.

Mailing address: 505 East Huron Street, Apt 607, Ann Arbor MI, USA.

Hammer, Reuven, Rabbi Prof

Rel: Jewish; Ordained on 08 JUN 1958; b. 30 JUN 1933 in Syracuse NY, USA; m. to Rahel; c. 2 boys and 3 girls; Citizen of USA and Israel; Lang. English and Hebrew.

International Organizations: Professor (since 1991) of the Oranim College.

Main field of work: Writing, lecturing. **Career:** Came to Israel in 1973. Dean, Jewish Theological Seminary in Jerusalem from 1974 to 1989. Director, Seminary of Judaic Studies from 1985 to 1988. **Educ.:** B.A. from Yeshiva University, New York NY, USA (1949-1953); Rabb. Ord., M.H.L. from Jewish Theological Seminary of America, New York NY, USA (1953-1958); D.H.L. from Jewish Theological Seminary of America, New York NY, USA (1965-1968); Ph.D. from Northwestern University, Evanston IL, USA (1971-1974). **Mem.:** Rabbinical Assembly of Israel **Pubs.:** The Other Child in Jewish Education; Sifre Deuteronomy and The Classical Midrash.

Mailing address: 31 Adam Street, Jerusalem, Israel.

Hammerling, Mary Anselm, Sr
Rel: Roman Catholic; Ordained; celibate.
International Organizations: President of the Benedictine Sisters of the Federation of Saint Gertrude.
Mailing address: 620 East Summit Avenue, Crookston MN 56716, USA.

Hammond, Robert B, Dr
Rel: Christian.
International Organizations: President of the Voice of China and Asia Missionary Society.
Mailing address: PO Box 15-M, Pasadena CA 91102, USA. T. (1 818) 441 0640.

Hampton, J E
Rel: Pentecostal.
International Organizations: Administrator of the Pentecostal Assemblies of the World.
Mailing address: 3944 Meadows Drive, Indianapolis IN 46205, USA. T. (1 317) 547 9541.

Hancock, Tinai
Rel: Bahá'í.
International Organizations: Head, Pacific Office of the Bahá'í International Community.
Mailing address: United Nations Office, PO Box 15888, Suva, Fiji.

Handl, Mathilde, Sr
Rel: Roman Catholic; Ordained; celibate.
International Organizations: Contact of the Benedettine Missionarie.
Mailing address: Via dei Bevilacqua 60, I-00163 Roma, Italy. T. (39 6) 680 9002. Fax 627 1347 CICM.

Handschin, Hanna
Rel: United Methodist; b. 05 NOV 1933 in Rorbas, Zurich, Switzerland; m. to Paul Handschin; c. 4 boys and 1 girl; Citizen of Switzerland; Lang. German, English, French and Swahili.
International Organizations: President for Europe (since 1986) of the World Federation of Methodist Women.
Other organizations: Member, Mission Board (since 1976), United Methodist Church in Switzerland; Member of the Board (since 1972), Frauendienst der Evangelisch-Methodistischen Kirche.
Main field of work: Editor of church periodical. **Career:** Vice-President for Europe, World Federation of Methodist Women from 1981 to 1986. **Educ.:** Cert. Tchr. Prim. from Oberseminar, Zürich, Switzerland (1953-1954); Cert. Tchr. Sec. from University of Zürich, Switzerland (1962-1965). **Pubs.:** Monthly periodical, *Der Missionsbote.*
Mailing address: Rosenbergstrasse 4, Postfach 289, CH-8820 Wadenswil, Switzerland. T. (41 1) 780 30 95

Hanegraaf, Hendrik
Rel: Christian.
International Organizations: President of the Christian Research Institute International.
Mailing address: Box 500, San Capistrano CA 92693, USA. T. (714) 855 9926.

Hanna, Samuel, Rev
Rel: Evangelical; Ordained.
International Organizations: Moderator of the National Evangelical Synod of Syria and Lebanon.
Mailing address: PO Box 110-235, Beirut, Lebanon.

Hannevik, Edward, Col
Rel: Salvation Army, The.
International Organizations: Territorial Commander of The Salvation Army.
Mailing address: Frederiksberg Allé 9, DK-1621 København 5, Denmark.

Hanselmann, Johannes, Dr
Rel: Evangelical Lutheran.
International Organizations: Bishop of Evangelical Lutheran Church in Bavaria.
Mailing address: Meiserstrasse 11-13, 80333 Munich, Germany.

Hanssens, Jan, Rev
Rel: Roman Catholic; Ordained; celibate.
International Organizations: Contact of the Congregation of the Immaculate Heart of Mary.
Mailing address: Via San Giovanni Eudes 95, I-00163 Roma, Italy. T. (39 6) 661 585 52. Fax 661 613 47. C. Corimac.

Hansulrich, Gerber
Rel: Mennonite.
International Organizations: Member of the Mennonite Central Committee.
Mailing address: Les Fraises 1, CH-2722 Les Reusilles, Switzerland.

Haq, Abdul
Rel: Muslim; Citizen of Afghanistan.
International Organizations: Field Commander of the the underground network in Kabul and the rebel forces surrounding the city.
Career: Joined an Islamic Movement opposing the growing Communist influence in the Afghan government in 1975. Imprisoned by President Daud after participating in an unsuccessful antigovernment uprising. Visited the United States several times. **Mem.:** moderate wing of the fundamentalist Islamic party led by Khalis.
Mailing address: No mailing address.

Harbans Singh, H, Dr
Rel: Sikh; b. 02 JAN 1942 in Chak Bhai Ka Dist. Sangrur, Punjab, India; m. to Surinder Pal Kaur; c. 1 boy and 2 girls; Citizen of India; Lang. Punjabi.
International Organizations: Lecturer (since 1972) of the SGPC Mata Gujri College Fatehgarh Sahib.
Main field of work: University education. **Educ.:** A.C.T. Fine Arts from Panjab University, Chandigarh, India (1965); IG.D. from University of Bombay, India (1965); M.A.; Ph.D. from Panjab University, Chandigarh, India (1985). **Interests:** Comparative religion studies. **Mem.:** Guru Gobind Singh Study Circle
Mailing address: Hamayunpur, Sirhind, Punjab 140406, India.

Harbuziuk, Olexa R, Rev
Rel: Baptist.
International Organizations: President of the All-Ukrainian Evangelical Baptist Fellowship.
Mailing address: 6751 Riverside Dr, Berwyn IL 60402, USA. T. (1 312) 788 09 99 – (1 312) 832 07 09.

Harcombe, E G, Pastor
Rel: Seventh-Day Adventist.
International Organizations: Member of the Adventist Development and Relief Agency International.
Mailing address: c/o PO Box 66150, Broadway 2020, South Africa.

Hardig, Warren G
Rel: Protestant.
International Organizations: Executive Director of the Men for Missions International.
Mailing address: Box A, Greenwood IN 46142, USA. T. (1 317) 881 6752.

Hare Duke, Michael Geoffrey H, Rt Rev
Rel: Anglican; Ordained in 1953; b. 28 NOV 1925 in Calcutta, India; m. to Grace Lydia Frances McKean Dodd; c. 1 boy and 3 girls; Citizen of United Kingdom.
International Organizations: Bishop (since 1969) of St Andrew's, Dunkeld and Dunblane.
Other organizations: President for the British Region (since 1982), Christian Peace Conference.
Career: Sub-Lieutenant, RNVR (Royal Naval Volunteer Reserve) from 1944 to 1946. Ordained Deacon in 1952. Curate, St John's Wood, London from 1952 to 1956. Vicar, St Mark's, Bury from 1956 to 1962. Pastoral Director, Clinical Theology Association from 1962 to 1964. Vicar, St Paul's, Daybrook from 1964 to 1969. Officiating Chaplain to the Forces, East Midlands District Headquarters from 1968 to 1969. Chairman, Scottish Association for Mental Health from 1978 to 1985. Chairman, Scottish Pastoral Association from 1970 to 1974. Vice-President, Scottish Institute of Human Relations from 1974 to 1976. **Educ.:** M.A. from Westcott House, Cambridge, UK. **Interests:** writing, broadcasting, journalism, walking. **Pubs.:** Freud, 1972; Good News, 1976 and Stories, Signs and Sacraments in the Emerging Church, 1982.
Mailing address: Bishop's House, Fairmount Road, Perth PH2 7AP, UK. T. (44 738) 21580

Hargis, Billy James

Rel: Evangelical; Ordained in 1943; b. in Texarkana TX, USA.
International Organizations: Publisher of *The Christian Crusade Newspaper* (monthly periodical).
Main field of work: Fundamentalist evangelist. **Career:** Pastor, several churches. Founded, The Church of the Christian Crusade, Tulsa OK in 1948. Founded, Billy James Hargis Evangelists Association in 1974. **Pubs.:** author of several books on Communism.
Mailing address: No mailing address.

Hark, Edgar, H E

Rel: Evangelical Lutheran; Ordained.
International Organizations: Archbishop of Estonian Evangelical Lutheran Church.
Mailing address: Raamatukogu 8, 200103 Tallinn, Estonia.

Harkavy, Angela

Rel: Unknown.
International Organizations: Contact of the South American Conference on Religion and Ecology.
Mailing address: 9422 Goshen VA, Burke VA 22015, USA.

Harkianakis, Stylianos, Most Rev Dr

Rel: Greek Orthodox; Ordained in 1970; b. 29 DEC 1935 in Rethymnon, Crete, Greece; single; Citizen of Australia; Lang. Greek, German and English.
International Organizations: Primate of Australia (since 1975) of the Greek Orthodox Archdiocese of Australia.
Other organizations: Co-Chairman (since 1980), Joint Official Theological Dialogue with the Roman Catholic Church; Dean (since 1986), St Andrew's Greek Orthodox Theological College.
Main field of work: Ministry. **Career:** Abbot, Patriarchal Monastery of Vlatadon from 1965 to 1975. Lecturer, University of Regensburg in 1973. Lecturer, Aristotelian University of Thessaloniki from 1969 to 1975. Titular Metropolitan, Miletoupolis from 1970 to 1975. **Educ.:** B.Div. (Hons) from Theological School of Halki, Constantinople, Greece (1958); D.D. from National Capodistrian University of Athens, Greece (1965). **Interests:** reading, walking, swimming. **Pubs.:** The Infallibility of the Church in Orthodox Theology, 1965 (Greek); Orthodoxy and Catholicism, 1975 (German and English) and *En Ge Allotria* – Collected Works of Poetry, Vols. 1 and 2.
Mailing address: 242 Cleveland Street, Redfern NSW 2016, Australia. T. (61 2) 698 5066
Residence: See Mailing Address. T. See Mailing Address

Harms, Hans Heinrich, Bishop Emer

Rel: Lutheran; Ordained on 14 MAY 1939; b. 04 JUL 1914 in Scharmbeck, Hannover, Germany; m. to Marianne Kiel; c. 2 girls; Citizen of Germany; Lang. German, English and French.
International Organizations: Bishop Emeritus (since 1986) of Evangelical Lutheran Church in Oldenburg.
Main field of work: Ministry, university education. **Career:** Pastor in Duderstadt in 1939. Lecturer, University of Göttingen from 1946 to 1950. Lecturer, University of Hamburg from 1960 to 1967. Senior Pastor in Hamburg from 1960 to 1967. Oberkirchenrat, Foreign Office, Evangelical Church in Germany from 1950 to 1952. Associate Director, Division of Studies, World Council of Churches (WCC) from 1952 to 1960. Member, Council, Evangelical Church in Germany from 1973 to 1985. Member, Central Committee, WCC from 1961 to 1968. Member, Central Committee and Executive Committee, WCC from 1975 to 1983. Bishop, Evangelical Lutheran Church in Oldenburg from 1967 to 1985. **Educ.:** Georg-August-Universität Göttingen, Germany (1932-1933); Rheinische Friedrich-Wilhelms-Universität Bonn, Germany (1933-1934); Georg-August-Universität Göttingen, Germany (1934-1936); Th.M. from Princeton Theological Seminary, NJ, USA (1936-1937); Dr. Theol. from Georg-August-Universität Göttingen, Germany (1941). **Hons.:** Dr. Theol. h.c., University of Göttingen (1963); Winner of the Niedersachsen Landesmedaille (1985). **Pubs.:** Bekenntnis und Kircheneinheit bei den Jungen Kirchen, Berlin, 1962; "Der Ökumenische Rat der Kirchen", in Beilage zu *Das Parlament*, 1962 and editor of series *Die Kirchen der Welt*, 1958-1988.
Mailing address: Beethovenstrasse 9, 26135 Oldenburg, Germany. T. (49 441) 14384
Residence: See Mailing Address. T. See Mailing Address

Harper, C, Rev

Rel: Protestant; Ordained.
International Organizations: Acting Director of the Commission of the Churches on International Affairs of the World Council of Churches.
Mailing address: PO Box 2100, Route de Ferney 150, CH-1211 Genève 20, Switzerland. T. (41 22) 791 61 11. Fax 791 03 61. C. OIKOUMENE Geneva. Tx 415730 OIK CH.

Harries, Richard Douglas, Rt Rev

Rel: Church of England; Ordained; b. 02 JUN 1936; m. to Josephine Bottomley; c. 1 boy and 1 girl; Citizen of United Kingdom.
International Organizations: Bishop (since 1987) of Oxford.
Career: Lieutenant, Royal Corps of Signals from 1955 to 1958. Curate, Hampstead Parish Church from 1963 to 1969. Chaplain, Westfield College from 1966 to 1969. Lecturer, Wells Theological College from 1969 to 1972. Warden, Wells, Salisbury and Wells Theological College from 1971 to 1972. Vicar, All Saints, Fulham from 1972 to 1981. Dean, King's College, London from 1981 to 1987. GOE (General Ordination Examination) Examiner in Christian Ethics from 1972 to 1976. Director, Post Ordination Training for Kensington Jurisdiction from 1973 to 1979. Vice-Chairman, Council of Christian Action from 1979 to 1987. Vice-Chairman, Council for Arms Control from 1982 to 1987. Member, Home Office Advisory Committee for reform of law on sexual offences from 1981 to 1985. Chairman, Southwark Ordination Course, Shalom and End Loans to South Africa (ELSTA) from 1982 to 1987. President, Johnson Society from 1988 to 1989. **Educ.:** Royal Military Academy, Sandhurst, UK; M.A. from University of Cambridge, Selwyn College, UK (1965); Cuddesdon College, Oxford, UK (1965). **Interests:** theatre, literature, sport. **Hons.:** Sir Sigmund Steinberg Award (1989). **Mem.:** Fellow, King's College, London (1983) **Pubs.:** *C S Lewis: the man and his God*, 1987; *Christ is Risen*, 1988 and articles in *Theology*, *The Times*, *The Observer* and various other periodicals.
Mailing address: Diocesan Church House, North Hinksey, Oxford OX2 ONB, UK. T. (44 865) 244566. Fax 790470

Harrington, Patrick, S.M.A., Fr

Rel: Roman Catholic; Ordained; celibate.
International Organizations: Superior General of the Society of African Missions.
Other organizations: Member of the Council for Reporting to the Congregation and the International Union of Superior Generals and the Union of Superior Genreals, Commissione per la Catechesi e per i Catechisti.
Mailing address: Via della Nocetta 111, I-00164 Roma, Italy. T. (39 6) 625 77 39, (39 6) 625 77 35

Harris, Augustine, Rt Rev

Rel: Roman Catholic; Ordained in 1942; b. 27 OCT 1917 in Liverpool, UK; celibate; Citizen of United Kingdom.
International Organizations: Bishop (since 1978) of Middlesbrough.
Career: Curacies from 1942 to 1952. Prison Chaplain from 1952 to 1966. Principal Roman Catholic Prison Chaplain from 1958 to 1966. Auxiliary Bishop, Liverpool from 1966 to 1978. **Interests:** golf, walking.
Mailing address: Bishop's House, 16 Cambridge Road, Middlesbrough, Cleveland TS5 5NN, UK.

Harris, Barbara Clementine

Rel: Episcopal; Ordained; b. in Philadelphia PA, USA; divorced.
International Organizations: Bishop (since 1989) of Boston, Massachusetts.
Career: Staff Member, Joseph V Baker and Associates (public relations firm), Philadelphia PA from 1948 to 1958. President, Joseph V Baker and Associates, Philadelphia PA from 1958 to 1968. Became Head of Community Relations Dept., Sun Oil Co. Active in leadership, Church of the Advocate, North Philadelphia PA, serving on the vestry, working with prisoners, and participating in a campaign to open the church to female clergy. Ordained Deacon, Episcopal Church in 1979. Served as Priest, St Augustine of Hippo Church, Philadelphia from 1980 to 1984. Served as Chaplain, Philadelphia County Prison from 1980 to 1984. Served as Executive Director, Episcopal Church Publishing Co. and Writer, Editor and Publisher of Church journal *The Witness* (until 1988). Served as interim Rector, Church of the Advocate (until 1988). Suffragan (assistant) Bishop, Episcopal Diocese of Boston, MA in 1988. **Educ.:** Attended Metropolitan Collegiate Center (1976); Attended from Villanova University, PA, USA (1977-1979).
Mailing address: Episcopal Diocese of Massachusetts, 138 Tremont St, Boston MA 02111, USA.
Residence: Boston MA, USA.

Harris, Myrtle W

Rel: Protestant.
International Organizations: Secretary of the Atkinson Foundation.
Mailing address: 10 West Orange Avenue, San Francisco CA 94080, USA. T. (1 415) 876 1559. Fax 876 1143.

Harris, Virginia

Rel: Church of Christ, Scientist.
International Organizations: Clerk of the Church of Christ, Scientist.
Mailing address: First Church of Christ, 175 Huntington Avenue, Boston MA 02115, USA. T. (1 617) 450 2000. C. SCIENCE, Boston.

Harris, Wesley, Commr

Rel: Salvation Army, The.
International Organizations: Territorial Commander of The Salvation Army.
Mailing address: PO Box 4021, Postal Station A, Toronto M5W 2B1, Canada.

Hart, Shirley Alexander, Dr

Rel: Protestant.
International Organizations: Executive Office of the International Ministers' Wives and Widows Association.
Mailing address: 305 Dexter St E, Chesapeake VA 23324, USA. T. (1 804) 543 0427.

Haryono, Anwar, Dr

Rel: Muslim.
International Organizations: Director of the Islamic Centre for Research and Development.
Mailing address: PO Box 3067, Jakarta, Indonesia. T. (62 21) 32 71 18.

Hasan, Munawar Syed, Syed

Rel: Muslim; Unordained; b. 05 AUG 1942 in Karachi, Sind, Pakistan; m. to Kaneez Aisha; c. 1 boy and 1 girl; Citizen of Pakistan; Lang. Urdu and English.
International Organizations: Secretary-General (since 1978) of the Islamic Research Academy.
Other organizations: President (Amir), Karachi Division (since 1988), Jamaat-e-Islami; Elected Member, Parliament, Pakistan.
Main field of work: Social Sciences Research. **Career:** Research Scholar, Islamic Research Academy in 1963. **Educ.:** B.Sc. from Government College, Karachi, Pakistan (1956-1960); M.A. Sociology from University of Karachi, Pakistan (1961-1966); M.A. Islamic Stds. (1961-1966). **Pubs.:** Polygamy vs. the Modernists; Charms of Islam and Introducing Islam.
Mailing address: D-35, Block 5, Federal 'B' Area, Karachi 75950, Pakistan. T. (92 21) 68 11 57.
Residence: Karachi, Pakistan T. (92 21) 42 33 69 – 41 83 10

Hasan, Okuyucu

Rel: Muslim.
International Organizations: Secretary of the Fondation religieuse islamique turque en Belgique.
Mailing address: Rue Montoyer 4, B-1040 Bruxelles, Belgium.

Hashem, Muhammad Abdo, Sheikh

Rel: Muslim; Citizen of Jordan.
International Organizations: Mufti of Hashemite Kingdom of Jordan.
Mailing address: No mailing address.

Hashmi, Shahid

Rel: Muslim.
International Organizations: Director of the Islamic Research Academy.
Mailing address: D-35, Block 5, Federal 'B' Area, Karachi 75950, Pakistan. T. (92 21) 68 11 57.

Hassel, David John, Prof

Rel: Roman Catholic; Ordained on 15 JUN 1955; b. 13 JUL 1923 in Chicago IL, USA; celibate; Citizen of USA; Lang. English, Latin, Greek (Ancient) and French.
International Organizations: Co-Editor (since 1988) of the *Review for Religious* and *Studies in Spirituality of Jesuits*.
Main field of work: Teaching, research, writing in philosophy and theology. **Career:** Chairman, Dept. of Philosophy, St Mary of the Lake Seminary from 1964 to 1966. President, Jesuit Philosophical Association from 1975 to 1976. Tertian Director, Chicago Province, Society of Jesus from 1975 to 1976. **Educ.:** B.Lit. from Xavier University, Cincinnati OH, USA (1945-1947); M.Ph. from Loyola University, Chicago IL, USA (1947-1952); L.Th. from West Baden College, IN, USA (1952-1955); Ph.D. from Saint Louis University, MO, USA (1957-1963). **Interests:** secularization theory. **Mem.:** Jesuit Philosophical Association **Pubs.:** *Searching the Limits of Love*, Loyola University Press, Chicago; *City of Wisdom. A Christian Vision of American University*, Loyola University Press and *Radical Prayer*, Paulist Press, Mahwah NJ.
Mailing address: Jesuit Residence, Loyola University, 6525 N Sheridan Road, Chicago IL 60626, USA. T. (1 312) 508 8800
Residence: See Mailing Address. T. (1 312) 508 2012

Hastings, Adrian Christopher, Rev Prof

Rel: Roman Catholic; Ordained on 21 JUN 1955; b. 23 JUN 1929 in Kuala Lumpur, Malaysia; m. to Elizabeth Ann Spence; Citizen of United Kingdom; Lang. English, Italian and French.
International Organizations: Professor of Theology (since 1985) of the University of Leeds.
Other organizations: Editor (since 1985), Journal of Religion in Africa; Director (since 1988), SCM Press.
Main field of work: University education. **Career:** Priest, Diocese, Masaka, Uganda from 1958 to 1966. Priest, Tanzania and Zambia from 1966 to 1970. Consultant to Anglican archbishops in Africa on marriage problems from 1971 to 1972. Research Officer at the School of Oriental and African Studies, London from 1973 to 1976. Lecturer in Religious Studies, University of Aberdeen from 1976 to 1982. Professor of Religious Studies, University of Zimbabwe, Harare from 1982 to 1985. **Educ.:** M.A. from University of Oxford, UK (1946-1949); D.Th. from Urbanian Pontifical University, Vatican City State (1953-1958); Dip. Ed. from University of Cambridge, UK (1957-1958). **Interests:** African Christian history, the Church in the 20th century world. **Pubs.:** A History of African Christianity (1950-1975), 1979; A History of English Christianity (1920-1990), 1986, 3rd Ed., 1991 and Robert Runcie, 1991.
Mailing address: Dept. of Theology and Religious Studies, University of Leeds, Leeds LS2 9JT, UK. T. (44 532) 333641. Fax 336017. Tx 556473 UNILDS G
Residence: 3 Hollin Hill House, 219 Oakwood Lane, Leeds LS8 2PE, UK. T. (44 532) 400154

Hasuike, Nitto

Rel: Buddhist; Ordained; b. 13 DEC 1914; Citizen of Japan.
International Organizations: Chief of the Kochoji.
Career: Manager, Hoke-Shu, Honmon-Sect. **Educ.:** Chuo University, Tokyo, Japan.
Mailing address: 1-12-16 Shitaya, Taito-ku, Tokyo 110, Japan. T. (81 3) 3841 2569

Hasuo, Zenryu

Rel: Buddhist; Ordained; b. 28 JAN 1915; Citizen of Japan.
International Organizations: Manager of the Shingonshu Zentsujiha.
Career: Chief, Znetsuji Kyoto Zuishinin. **Educ.:** Nihon University, Tokyo, Japan.
Mailing address: 3-3-1 Zentsujicho, Zentsuji, Kagawa 765, Japan. T. (81 877) 62 0111

Hatano, Motoi, Bishop

Rel: Free Methodist.
International Organizations: Member of the Free Methodist World Fellowship.
Mailing address: 7-9 Sembon-Naka, 2-chome Nishinari-ku, Osaka 557, Japan.

Hauschildt, Lennart

Rel: Christian.
International Organizations: Contact of the Joint Board of Nordic Clergymen's Associations.
Mailing address: c/o Svenska Kyrkans, Personalförbund, Box 19609, S-104 32 Stockholm, Sweden. T. (08) 15 10 50.

Havea, S A, Rev Dr

Rel: Free Wesleyan.
International Organizations: President of the Free Wesleyan Church of Tonga.
Mailing address: PO Box 57, Nuku'alofa, Tonga.

Havens, Lani J

Rel: Protestant.
International Organizations: Executive of the Church World Service.
Mailing address: 475 Riverside Drive, Room 678, New York NY 10115-0050, USA. T. (1 212) 870 2257. Fax 870 2055.

Hawter, Ken

Rel: Buddhist: Mahayana.
International Organizations: Member of the Foundation for the Preservation of the Mahayana Tradition.
Other organizations: Director, Karuna Centre.
Mailing address: 38 Lamont Road, Wilston QLD 4051, Australia. T. (61 7) 356 7100

Hayek, Ignace Antoine, II, His Beatitude

Rel: Syrian Catholic; Ordained on 10 JUN 1933; b. 14 SEP 1910 in Aleppo, Syria; celibate; Citizen of Syria.
International Organizations: Patriarch (since 1968) of Antioch.
Member of the Sacred Congregation for the Oriental Churches.
Other organizations: Head, Syrian Patriarchal Synod.
Career: Successively or concurrently Director, School, Curate and Vicar-General, Aleppo. Archbishop, Aleppo from 1959 to 1968. **Hons.:** Lebanese and foreign awards.
Mailing address: Syrian Catholic Patriarchate, Damascus Street, PO Box 116-5087, Beirut, Lebanon. T. (961 1) 387 132

Hayes, Eugen Joseph, O.Praem., Very Rev Dr

Rel: Roman Catholic; Ordained on 29 JUN 1977; b. 21 MAR 1951 in Brooklyn NY, USA; celibate; Citizen of USA; Lang. English, Italian, French (reading) and Latin.
International Organizations: Procurator-General (since 1988) of the Order of the Canons Regular of Prémontré.
Other organizations: Canonical Consultant, Canonesses (since 1989), Holy Sepulcher.
Main field of work: Administration, teaching. **Career:** Campus Minister, Mater Dei High School, Santa Ana CA from 1983 to 1985. Religion Teacher, Mater Dei High School from 1980 to 1985. Priests Council Member, Diocese of Orange, CA from 1982 to 1985. **Educ.:** B.Phil. from Cathedral College, Douglaston NY, USA (1968-1972); S.T.B.; S.T.L. from Pontifical University San Tommaso d'Aquino, Vatican City State (1976-1978); J.C.L. from Catholic University of America, Washington DC, USA (1985-1987); J.C.D. from Catholic University of America, Washington DC, USA (1987-1989). **Interests:** Canon law, comparative religions. **Mem.:** Canon Law Society of the United States **Pubs.:** The Rightful Autonomy of Life and Charism in Proper Law, 1990 (Doctoral dissertation, CUA, Washington DC).
Mailing address: Viale Giotto 27, I-00153 Roma, Italy. T. (39 6) 574 69 79
Residence: See Mailing Address. T. See Mailing Address

Hayes, James Martin, Most Rev

Rel: Roman Catholic; Ordained on 15 JUN 1947; b. 27 MAY 1924 in Halifax NS, Canada; celibate; Citizen of Canada.
International Organizations: Archbishop Emeritus (since 1991) of Halifax.
Career: Consecrated Bishop in 1965. President, Canadian Conference of Catholic Bishops from 1987 to 1989. Archbishop, Halifax from 1967 to 1990. **Educ.:** B.A. from Saint Mary's University, Halifax, Canada (1943); J.C.D. from Pontifical University San Tommaso d'Aquino, Vatican City State (1957). **Hons.:** Hon. D.D., University of King's College (1967); Hon. D.Litt., Mt St Vincent University (1985), St Francis Xavier University, Antigonish (1987), St Thomas University, Fredericton (1989); Hon. LL.D., St Mary's University (1985); Hon. D.D., Atlantic School of Theology (1986).
Mailing address: PO Box 1527, Halifax NS B3J 2Y3, Canada. T. (1 902) 429 9800. Fax 423 5201

Hazim, Ignace, Mgr

Rel: Greek Orthodox; Ordained; b. in Mharde, Syria; Citizen of Syria.
International Organizations: Greek Orthodox Patriarch (since 1979) of Antioch and All the East.
Career: Director, a secondary Theological Institute, Beirut. Rector, Theological Institute, Antioch, and then elected Bishop of Latakia in 1966. Took up post in 1970. President, Middle East Ecumenical Council. Member, Central Committee, Ecumenical Council of Geneva. **Pubs.:** The Telling of Your Word Enlightens (in Arabic); The Church in the Middle East (translation of the work by Père Corbon) and God's Design (translation of the work by Suzanne de Dietrich).
Mailing address: Box 9, Damascus, Syria. T. (963 11) 6329 – 7403

Headley, Thomas A

Rel: Transcendental Meditation.
International Organizations: Director US Office of the Association for the World Government of the Age of Enlightenment.
Mailing address: 2021 N Main Street, Fairfield IA 52556, USA. T. (1 515) 472 0108.

Healy, Timothy Stafford, S.J.

Rel: Roman Catholic; Ordained in 1953; b. 25 APR 1923 in New York NY, USA; celibate.
International Organizations: President (since 1989) of the New York Public Library.
Member of the Society of Jesus.
Career: Taught English and Latin at Fordham Preparatory School, New York City from 1947 to 1959. Began as an Instructor at Fordham University, Bronx NY in 1956. Became Professor of English in 1968. Served as Director of Alumni Relations from 1957 to 1962. Served as Executive Vice-President from 1965 to 1969. Professor of English and Vice-Chancellor of Academic Affairs, City University, New York NY from 1969 to 1976. Professor of English and President, Georgetown University, Washington DC from 1976 to 1989. Served as Director and Chairman, American Council on Education and National Association of Independent Colleges and Universities. Served as Chairman, National Advisory Committee on Accreditation and Institutional Eligibility. Served as Secretary, State's Advisory Committe on South Africa. Trustee, LeMoyne College, Xavier University, Regis High School, and the Shakespeare Theater at the Folger. Member, National Collegiate Athletics Association Presidents Commission, Consortium on Financing Higher Education, National Commission for the Study of Migration and Cooperative Economic Development, Folger Library Committee, President's Commission on Foreign Languages and International Studies, Middle States Commission on Higher Education. **Educ.:** A.B. from Woodstock College, USA (1946); Ph.L. from Woodstock College, USA (1947); M.A. from Woodstock College, USA (1948); S.T.L. from Catholic University of Louvain, Belgium (1954); M.A. from Fordham University, Bronx NY, USA (1958); D.Phil. from University of Oxford, UK (1966). **Hons.:** more than two dozen honorary degrees. **Pubs.:** Co-Editor, *John Donne: Selected Prose*, 1967 and *John Donne: Ignatius His Conclave*, 1969.
Mailing address: New York Public Library, Fifth Ave and 42nd St, New York NY 10018, USA.

Heaslewood, Jeanne

Rel: Mason.
International Organizations: Grand Secretary of the International Co-Freemasonry.
Mailing address: 10 Ladbroke Terrace, Notting Hill Gate, London W11 3PQ, UK. T. (44 71) 727 6111.

Heath, David Gordon, Rev

Rel: Roman Catholic; Ordained; b. 20 FEB 1930 in Detroit MI, USA; celibate; Citizen of Canada.
International Organizations: Treasurer-General (since 1991) of the Basilean Fathers of Toronto.
Main field of work: Education. **Career:** High School Teacher and Administrator from 1953 to 1965. University Professor and Administrator from 1965 to 1979. President, Assumption University from 1979 to 1991. Former Chairman, Board of Governors and Member, Senate, University of Windsor. **Educ.:** B.A. from University of Western Ontario, London, Canada (1953); S.T.B. from University of Toronto, St Michael's College, Canada (1958); M.S. from Indiana University, Bloomington, USA (1961); D.Ed. from Indiana University, Bloomington, USA (1967). **Mem.:** Congregation of St Basil (CSB), Toronto
Mailing address: Basilwood, 15th Side Rd, RR 3, Milton ON L9T 2X7, Canada.

Heaton, Eric William, Very Rev

Rel: Anglican; Ordained in 1945; b. 15 OCT 1920; m. to Rachel Mary Dodd; c. 2 boys and 2 girls; Citizen of United Kingdom.
International Organizations: Ecclesiastic of the Anglican Church.
Other organizations: Dean (since 1979), Christ Church, Oxford; Pro-Vice-Chancellor (since 1984), University of Oxford.
Career: Curate, St Oswald's, Durham from 1944 to 1945. Chaplain, Gonville and Caius College, Cambridge from 1945 to 1946. Dean and Fellow, Gonville and Caius College, Cambridge from 1946 to 1953. Tutor, Gonville and Caius College, Cambridge from 1951 to 1953. Canon Residentiary, Salisbury Cathedral from 1953 to 1960. Chancellor, Salisbury Cathedral from 1956 to 1960. Tutor in Theology, Official Fellow and Chaplain, St John's College, Oxford from 1960 to 1964. Senior Tutor, St John's College, Oxford from 1967 to 1973. Dean, Durham from 1974 to 1979. Honorary Lecturer, University of Durham from 1975 to 1979. **Educ.:** M.A. from University of Cambridge, Christ's College, UK. **Hons.:** Honorary Fellow, Champlain College, University of Trent, Ontario (1973), St John's College, Oxford (1979), Christ's College, Cambridge (1983). **Pubs.:** The Hebrew Kingdoms, 1968; Solomon's New Men, 1974 and articles in theological journals.
Mailing address: The Deanery, Christ Church, Oxford OX1 1DP, UK. T. (44 865) 276161
Residence: See Mailing Address. T. (44 865) 276162

Hebler, Klaus

Rel: Christian.
International Organizations: Contact of the International Ecumenical Centre.
Mailing address: c/o BK, Friedrichstrasse 169/170, 10117 Berlin, Germany.

Hedberg, Lennart, Commr

Rel: Salvation Army, The.
International Organizations: Territorial Commander of The Salvation Army.
Mailing address: Box 5090, S-102 42 Stockholm, Sweden.

Hees, Karl-Wilhelm

Rel: Evangelical.
International Organizations: Past Secretary-General of the European Evangelical Alliance.
Mailing address: Sehrtenbachstrasse 11, 57610 Altenkirchen, Germany. T. (49 2681) 3022. Fax (49 2681) 7630.

Hegedüs, Lóránt, Bishop, Dr

Rel: Reformed Church in Hungary; Ordained; Citizen of Hungary.
International Organizations: President of the Synod of the Reformed Church in Hungary.
Other organizations: President, "Collegium Doctorum" of the Reformed Church in Hungary.
Mailing address: Ráday u. 28, Budapest H 1092, Hungary.

Hegel, Eduard

Rel: Christian; b. 28 FEB 1911 in Wüppertal-Barmen, Germany; Citizen of Germany.
International Organizations: Professor Emeritus, (since 1976) of the Univerity of Bonn.
Main field of work: Theologian. **Career:** Professor for Middle and New Church History, Trier in 1949. Professor, Münster in 1953. Professor, Bonn from 1966 to 1976. Member, Historical Commission for Westphalia in 1958. **Mem.:** Rheinisch-Westfälischen Akademie der Wissenschafte (1973) **Pubs.:** Geschichte der Katholisch-Theologischen Fakultät Münster, 2 Vols., 1966, 1971; Geschichte des Enzbistums Köln, Vols. 4-5, 1979, 1987 and Ecclesiastica Rhenana, 1986.
Mailing address: Gregor-Mendel-Strasse 29, 53115 Bonn, Germany. T. (49 228) 23 22 73

Heidkamp, Martha

Rel: Roman Catholic.
International Organizations: Trustee of US Branch of the International Grail Movement.
Mailing address: c/o Grailville, 932 O'Bannonville Road, Loveland OH 45140, USA. T. (1 513) 683 2340.

Heil, Hans-Joachim

Rel: Protestant.
International Organizations: President of the Family Life Mission.
Mailing address: Postfach 1965, 77694 Kehl, Germany. T. (49 7851) 78330. •

Heintze, Gerhard

Rel: Lutheran; Ordained in 1938; b. 14 NOV 1912 in Wehre, Kreis Goslar, Germany; m. to (1st) Ilse Hoppe (died), (2nd) Renate Wigand; c. 2 boys and 3 girls; Citizen of Germany.
International Organizations: Ecclesiastic (retired) of the Lutheran Church.
Career: Pastor, Hollern-Tweilenfleth, Netherlands from 1942 to 1946. Inspector, Mission, Hermannsburg from 1946 to 1950. Director, seminary, Hannover from 1950 to 1957. Senior Minister, Hildesheim from 1957 to 1961. Senior Minister, Lower Saxony, Hildesheim from 1961 to 1965. Bishop, Protestant-Lutheran Church, Brunswick from 1965 to 1982. **Interests:** music. **Pubs.:** Luthers Predigt von Gesetz und Evangelium, 1958.
Mailing address: Herdweg 100, 70193 Stuttgart, Germany.

Heisey, Nancy

Rel: Mennonite.
International Organizations: Associate Executive Secretary of the Mennonite Central Committee.
Mailing address: 21 S 12th Street, Akron PA 17501, USA.

Hekmatyar, Gulbuddin

Rel: Muslim; Citizen of Afghanistan.
International Organizations: Head of the one of the largest rebel groups.
Main field of work: Fundamentalist leader. **Career:** Became active in a fundamentalist student group in the mid-1970s. Launched an unsuccessful uprising against President Daud. **Educ.:** University of Kabul, Afghanistan.
Mailing address: No mailing address.

Held, Heinz Joachim

Rel: Evangelical; b. 16 MAY 1928 in Wesseling/Rhein, Germany; m. to Anneliese Novak; c. 1 boy and 3 girls; Citizen of Germany.
International Organizations: Moderator, Central Committee and Executive Committee (since 1983) of the World Council of Churches.
Member, Central Committee (since 1968) of the World Council of Churches.
Career: Research Assistant at Wüppertal Theological Seminary from 1953 to 1956. Clergyman, Friedrichsfeld/Niederrhein from 1957 to 1964. Professor of Theology, Buenos Aires Lutheran Seminary from 1964 to 1968. President, River Plate Evangelical Church, Buenos Aires from 1968 to 1974. Chairman, National Council of Churches, Federal Republic of Germany from 1982 to 1988. **Interests:** stamp collecting, amateur music (piano), photography. **Hons.:** Hon. Dr.Theol., Lutheran Theological Academy, Budapest (1985); Hon. D.D., Academy of Ecumenical Indian Theology and Church Administration, Madras (1988).
Mailing address: Herrenhäuser Strasse 12, 30419 Hannover, Germany. T. (49 511) 71 11 125

Hell, Theo

Rel: Roman Catholic.
International Organizations: Secretary of the International Catholic Union of the Press – Europe Region.
Mailing address: Arbeitsgemeinschaft Katolische Presse, Adenauerallee 176, D-5300 Bonn 1, Germany.

Hempel, Johannes, Rt Rev

Rel: Evangelical Lutheran; Ordained; b. 23 MAR 1929 in Zittau, Saxony, Germany; m. to Dorothea Schönbach; c. 2 boys and 1 girl; Citizen of Germany.
International Organizations: President of the World Council of Churches.
Other organizations: Bishop (since 1971), Evangelical-Lutheran Church in Dresden, Saxony.
Career: Parish priest in Karl-Marx-Stadt district from 1955 to 1957. Supervisor from 1957 to 1963. Students' Pastor in Leipzig from 1963 to 1967. Director, St Paul Theological Seminary, Leipzig from 1967 to 1971. **Hons.:** Hon. D.Theol., Karl-Marx University, Leipzig (1983).
Mailing address: Tauscherstrasse 44, 01277 Dresden, Germany.

Henau, Ernest, Rev

Rel: Roman Catholic; Ordained; celibate.
International Organizations: Head, Europe Office of the Unda, International Catholic Association for Radio and Television.
Mailing address: KTRO, Roodebeeklaan 44, B-1040 Bruxelles, Belgium.

Henderson, Gregor, Rev

Rel: Uniting Church in Australia; Unordained; Citizen of Australia.
International Organizations: Secretary of the Uniting Church in Australia.
Main field of work: Church Administration.
Mailing address: POB E266, St James NSW 2000, Australia. T. (61 2) 287 0900, Fax (61 2) 287 0999.

Henderson, Robert C, Dr

Rel: Bahá'í.
International Organizations: Secretary-General of the National Spiritual Assembly of the Bahá'ís of the US.
Career: Managing Director, Tarkenton and Co. Executive Officer, Air Atlanta. **Educ.:** Ed.D. from University of Massachusetts at Amherst, USA.
Mailing address: Bahá'í National Center, 536 Sheridan Rd, Wilmette IL 60091, USA.

Hendrickse, Helenard Joe (Alan), Rev

Rel: Congregational; Ordained; b. 22 OCT 1927 in Uitenhage, Cape Province, South Africa; m. to Patricia Heber (Terry); c. 2 boys and 2 girls; Citizen of South Africa.
International Organizations: Leader (since 1978) of the Labour Party of South Africa.
Main field of work: Religious official and politician. **Career:** Founded Blacks School, Uitenhage in 1951. Elected Head, United Congregational Church of Southern Africa in 1972. Teacher from 1951 to 1969. Elected to Coloured Persons Representative Council in 1969. Elected to Executive, responsible for Education in 1975. Elected to the House of Representatives (Coloured Chamber) in 1984. Chairman, Ministers' Council from 1984 to 1986. Minister Without Portofolio and Chairman, Ministers' Council, House of Representatives from 1986 to 1987.
Mailing address: Private Bag 9068, Cape Town, South Africa.

Hengel, M, Prof Dr

Rel: Evangelical.
International Organizations: Professor of New Testament and Classical Jewish Studies of the University of Tubingen.
Mailing address: Evangelisch-theologische Fakultät, Universität Tübingen, Whilhemstrasse 7, 72074 Tübingen, Germany.

Hengsbach, Franz, H E Card

Rel: Roman Catholic; Ordained on 13 MAR 1937; b. 10 SEP 1910 in Velmede, Paderborn, Germany; celibate; Citizen of Germany.
International Organizations: Member (since 1988) of the College of Cardinals.
Other organizations: Bishop, Essen.
Career: Elected to the Titular Church, Cantano in 1953. Consecrated bishop in 1953. Transferred to Essen in 1957. Created Cardinal by Pope John Paul II in 1988.
Mailing address: Burgplatz 2, 45127 Essen, Germany. T. (49 201) 2204 201

Henihan, Mary, SMR, Sr

Rel: Roman Catholic; Ordained; celibate.
International Organizations: Superior General of the Society of Mary Reparatrix.
Mailing address: Via dei Lucchesi 3, I-00187 Roma, Italy. T. (39 6) 679 20 17.

Henjo, Daiyu

Rel: Buddhist; Ordained; b. 28 FEB 1908; Citizen of Japan.
International Organizations: Manager of the Koyasan Shingonshu, Kongobuji.
Main field of work: Monk. **Educ.:** Koyasan University, Japan.
Mailing address: 132 Takanocho Koyasan, Ito-gun, Wakayama 648-02, Japan. T. (81 736) 6 2011

Henri, Harry Y, Rev

Rel: Methodist.
International Organizations: President of the Protestant Methodist Church in the People's Republic of Benin.
Mailing address: Carré 206, BP 34, 54 Ave Sékou Touré, Cotonou, Benin. T. (229) 312520

Henry, Brian, Very Rev

Rel: Episcopal; Ordained.
International Organizations: Provincial Secretary of the Episcopal Church in Jerusalem and the Middle East.
Mailing address: PO Box 2075, Nicosia 118, Cyprus.

Henry, Carl Ferdinand Howard, Rev Prof Dr

Rel: Baptist; Ordained in 1941; b. 22 JAN 1913 in New York City NY, USA; m. to Helga Bender; c. 1 boy and 1 girl; Citizen of USA; Lang. English.
International Organizations: Lecturer-at-Large (since 1990) of the Prison Fellowship Ministries.
Other organizations: Visiting Professor of Theology (since 1974), Trinity Evangelical Divinity School; Member, Board of Directors (since 1981), Prison Fellowship Ministries; Member, Board (since 1982), Institute on Religion and Democracy.
Main field of work: Lecturing and writing. **Career:** Editor, *Smithtown Star* in 1932. Made a commitment to Jesus Christ in 1933. Assistant Professor of Theology, Northern Baptist Theological Seminary, Chicago IL from 1940 to 1942. Professor of Theology, Northern Baptist Theological Seminary, Chicago IL from 1942 to 1947. Acting Dean, Fuller Theological Seminary in 1947. Professor, Fuller Theological Seminary from 1947 to 1956. Editor, *Christianity Today* from 1956 to 1968. Editor-at-Large, *Christianity Today* from 1968 to 1974. Professor, Eastern Baptist Theological Seminary from 1969 to 1974. Lecturer-at-Large, World Vision International from 1974 to 1986. Co-Chairman, Rose Bowl Easter Sunrise Service from 1950 to 1956. Member, Board of Administration, National Association of Evangelicals from 1956 to 1970. Member, Board of Trustees, Gordon College from 1962 to 1968. Chairman, World Congress on Evangelism, Berlin in 1966. President, Evangelical Theological Society from 1969 to 1970. Program Chairman, Jerusalem Conference on Biblical Prophecy in 1971. President, Board of Directors, Institute for Advanced Christian Studies from 1971 to 1974. Member, Board, Ethics and Public Policy Center in 1980. President, American Theological Society from 1980 to 1981. Co-Chairman, Evangelical Affirmations Conference in 1989. **Educ.:** B.A. from Wheaton College, IL, USA (1938); M.A. from Wheaton College, IL, USA (1940); B.D. from Northern Baptist Seminar, Chicago IL, USA (1941); Th.D. from Northern Baptist Theological Seminary, Chicago IL, USA (1942); Ph.D. from Boston University, MA, USA (1949). **Interests:** writing, antiques. **Hons.:** Freedoms Foundation Award (1954 and 1966); Hon. D.Litt., Seattle Pacific College (1963), Wheaton College (1968); Hon. L.H.D., Houghton College (1973); Religious Heritage of America 'Faith and Freedom Award' (1975); Hon. D.D., Northwestern College (1979), Gordon-Conwell Theological Seminary (1984); Hon. LL.D., Hillsdale College (1989). **Mem.:** member of a dozen theological, philosophical, and literary societies **Pubs.:** Editor of 11 books and author of 35 books, including: *Confessions of a Theologian*; *Twilight of a Great Civilization* and *God, Revelation and Authority*, 6 Vols.
Mailing address: 3824 N 37th Street, Arlington VA 22207, USA. T. (1 703) 528 2401
Residence: See Mailing Address. T. See Mailing Address

Hensley, Kirby J

Rel: Protestant.
International Organizations: Contact of the Universal Life Church.
Mailing address: 601 3rd St, Modesto CA 95351, USA. T. (1 209) 527 81 11.

Heon, Céline, PSA, Sr

Rel: Roman Catholic; Ordained; celibate.
International Organizations: Superior General of the Little Sisters of the Assumption.
Mailing address: 57 rue Violet, F-75015 Paris, France. T. (33 1) 45 79 29 90.

Heptner, Elisabeth, S.A.C., Sr

Rel: Roman Catholic; Ordained; celibate.
International Organizations: Superior General of the Missionary Sisters of the Catholic Apostolate.
Other organizations: Member, Commission for the Catechism and the Catechists.
Mailing address: Viale delle Mura Aurelie 7b, I-00165 Roma, Italy. T. (39 6) 638 36 35.

Herkenrath, Norbert, Mgr

Rel: Roman Catholic; Ordained; celibate.
International Organizations: Director of the Misereor, Action Against Hunger and Disease in the World – Bishops' Relief Fund.
Mailing address: Mozartstrasse 9, Postfach 1450, D-52064 Aachen, Germany. T. (49 241) 44 20. Fax 44 21 88. Tx 832370 Misa D.

Hermaniuk, Maxim, Most Rev

Rel: Ukrainian Catholic; Ordained in 1938; b. 30 OCT 1911 in Nove Selo, Western Ukraine, Ukraine; Citizen of Ukraine.
International Organizations: Archbishop Metropolitan (since 1956) of Ukrainian Catholics in Canada.
Career: Joined Redemptorist Congregation in 1933. Co-founder, Ukrainian Relief Committee, Belgium from 1942 to 1948. Professor of Moral Theology, Sociology and Hebrew, Beauplateau, Belgium from 1943 to 1945. Organizer, Ukrainian University Students Organization (Obnova), Belgium from 1946 to 1948. Co-founder and first President, Ukrainian Cultural Society, Belgium in 1947. Came to Canada in 1948. Professor of Moral Theology and Holy Scripture, Redemptorist Seminary, Waterford, Ontario from 1949 to 1951. Superior, Vice-Provincial, Canada and USA from 1948 to 1951. Consecrated Bishop in 1951. Auxiliary Bishop, Winnipeg in 1951. Organizer, Ukrainian University Students Organization (Obnova), Canada in 1953. Apostolic Administrator, Exarchate of Manitoba in 1956. Member, Vatican II Council from 1962 to 1965. Member, Secretariat for Promoting Christian Unity, Rome in 1963. Member, Joint Working Group of Catholic Church and World Council of Churches, Geneva in 1969. Member, Council to Secretariat, Synod of Bishops in 1977. Member, Council to Secretariat, Synod of Bishops in 1983. Member, Pontifical Commission for Revision of the Kodex of Oriental Canon Law in 1983. **Educ.:** Dr. Theol. from Catholic University of Louvain, Belgium (1943); Maître Agrégé Théol. from Catholic University of Louvain, Belgium (1947). **Hons.:** Officer, Order of Canada (1982); D.D. h.c., St Michael's College, University of Toronto (1988). **Mem.:** World Congress of Free Ukrainians (1967); Taras Shevchenko Scientific Society (1951); Pontifical Commission for the Revision of the Kodex of Oriental Canon Law (1983); Mark Twain Society (Honorary Member, 1972); Ukrainian History Association **Pubs.:** La Parabole Evangélique, 1947 and Our Duty, 1960.
Mailing address: 235 Scotia Street, Winnipeg, Manitoba R2V 1V7, Canada.

Hermans, A M

Rel: Roman Catholic.
International Organizations: President of the Lay Cooperation Service for Development.
Mailing address: Rue des Fripiers 15-17, Galerie du Centre, Bloc 1, B-1000 Bruxelles, Belgium. T. (32 2) 217 72 50. Fax 223 02 17.

Hermans, Louis, SS.CC., Fr

Rel: Roman Catholic; Ordained; celibate.
International Organizations: Contact of the Congregation of the Sacred Hearts of Jesus and Mary.
Mailing address: Via Rivarone 85, I-00166 Roma, Italy. T. (39 6) 680 55 83 – 630 55 38.

Hermiz, Thomas H, Dr

Rel: Protestant.
International Organizations: President of the World Gospel Mission. President of the World Gospel Crusades.
Mailing address: Box 948, 3783 State Road 18 E, Marion IN 46952, USA. T. (1 317) 664 7331.

Hernandez, Andrés Delgado, Fr

Rel: Roman Catholic.
International Organizations: Secretary-General of the Catholic International Education Office.
Mailing address: Rue des Eburons 60, B-1040 Bruxelles, Belgium. T. (32 2) 230 72 52. Fax 230 97 45.

Herranz Casado, Julián H, H E Mgr

Rel: Roman Catholic; Ordained on 07 AUG 1955; b. 31 MAR 1930 in Baena, Spain; celibate.
International Organizations: Secretary of the Pontifical Council for the Interpretation of Legislative Texts.
Advisor of the Congregation for Bishops.
Other organizations: Titular Bishop, Vertara.
Mailing address: Palazzo delle Congregazioni, Piazza Pio XII 10, I-00193 Roma, Italy. T. (39 6) 698 4008
Residence: Viale Bruno Buozzi 75, I-00197 Roma, Italy T. (39 6) 87 9042

Herrmann, Siegfried

Rel: Evangelical; b. 15 MAY 1926 in Dresden, Germany; m. to Ruth; c. 1 boy and 1 girl; Citizen of Germany.
International Organizations: Professor of Old Testament, Evangelical Theology Dept. (since 1966) of the University of Bochum.
Main field of work: Theologian. **Career:** Assistant in Old Testament Studies, University of Leipzig from 1951 to 1959. Dozent, Humboldt University of Berlin from 1960 to 1966. **Pubs.:** Die prophetischen Heilserwartungen, 1965; A History of Israel in Old Testament Times, 1980 and Jeremia, 1986.
Mailing address: Paracelsusweg 14, 44801 Bochum, Germany. T. (49 234) 70 17 75

Herron, Andrew, Very Rev

Rel: Church of Scotland; Ordained on 20 MAY 1934; b. 29 SEP 1909 in Glasgow, Scotland, UK; m. to Joanna Fraser Neill; c. 4 girls; Citizen of United Kingdom; Lang. English.
International Organizations: Editor (since 1961) of the *Church of Scotland Year Book.*
Career: Minister, Linwood from 1936 to 1940. Minister, Houston and Killellan from 1940 to 1959. Clerk to the Presbytery, Paisley from 1953 to 1959. Clerk to the Presbytery, Glasgow from 1959 to 1981. Special Lecturer in Practical Theology, University of Glasgow from 1968 to 1985. Moderator, General Assembly, Church of Scotland from 1971 to 1972. Convener of Business Committee, General Assembly, Church of Scotland from 1972 to 1976. Baird Lecturer in 1985. Barclay Lecturer in 1990. **Educ.:** M.A. Mental Phil. (2nd) from University of Glasgow, UK (1932); B.D. Syst. Theol. (Dist.) from University of Glasgow, UK (1935); B.L. from University of Glasgow, UK (1956). **Hons.:** Hon. D.D., St Andrews (1975), Glasgow (1979); Hon. LL.D., Strathclyde (1983). **Pubs.:** Record Apart, 1972 and Guides: to the General Assembly (1976), to Congregational Affairs (1979), to the Presbytery (1982), to the Ministry (1987), to Ministerial Income (1987).
Mailing address: 36 Darnley Road, Glasgow GH1 4NE, Scotland, UK. T. (44 41) 423 6422
Residence: See Mailing Address. T. See Mailing Address

Hertzberg, Arthur, Rabbi Prof Dr

Rel: Jewish; Ordained in 1943; b. 09 JUN 1921 in Lubaczow, Poland; m. to Phyllis Cannon; c. 2 girls; Citizen of USA; Lang. English, Hebrew, German, French and Yiddish.
International Organizations: Vice-President (since 1975) of the World Jewish Congress.
Other organizations: Adjunct Professor of History at the Columbia University, New York NY (since 1960); President (since 1978), American Jewish Policy Foundation; Rabbi Emeritus (since 1985), Temple Emanu-El, Englewood NJ; Professor of Religion at Dartmouth College, Hannover NH (since 1985); Visiting Professor of the Humanities at New York University (since 1991).
Career: Came to the US in 1926. Hillel Director at Smith College from 1943 to 1944. Rabbi, Ahavat Israel, Philadelphia PA from 1944 to 1947. Rabbi, West End Syngagogue, Nashville TN from 1947 to 1956. On leave, serving as Chaplain in the US Air Force from 1951 to 1953. Rabbi, Temple Emanu-El, Englewood NJ from 1956 to 1985. President, Conference of Jewish Social Studies from 1967 to 1972. Member, Board of Governors, Jewish Agency for Israel from 1969 to 1978. Member of the Executive, World Zionist Organization from 1969 to 1978. President, American Jewish Congress from 1972 to 1978. Visiting Professor of Jewish Studies at Rutgers University, NJ from 1966 to 1968. Lecturer in Religion at Princeton University, NJ from 1968 to 1969. Visiting Professor of History at the Hebrew University of Jerusalem from 1970 to 1971. Visiting Professor at the Jewish Theological Seminary, NY in 1980. Visiting Professor at the Ecole des Hautes Etudes, Paris in 1989. Visiting Scholar at St Antony's College, Oxford in 1989. Consulting Editor to the Encyclopedia Judaica (16 Vols), published in January

1972. **Educ.:** A.B. from Johns Hopkins University, Baltimore MD, USA (1940); M.H.L. from Jewish Theological Seminary of America, NY, USA (1943); Ph.D. Hist. from Columbia University, New York NY, USA (1966). **Hons.:** Hon. D.D., Lafayette College, Easton PA (1970); Hon. D.H.L., Baltimore Hebrew College (1974). **Pubs.:** *The Jews in America*, Simon and Schuster, November 1989; *Judaism*, 1961 (published in French, Spanish, Portuguese, and English), 2nd Ed. Revised, Simon and Schuster, 1991 and *Jewish Polemics: Selected Essays by Arthur Hertzberg*, Columbia University Press (forthcoming 1992).
Mailing address: 83 Glenwood Road, Englewood NJ 076231, USA. T. (1 201) 568 3259
Residence: See Mailing Address. T. See Mailing Address

Hes, Jan

Rel: Christian.
International Organizations: Secretary-General of the International Inter-Church Film Centre.
Mailing address: Steynlaan 8, Postbus 515, 1200 AM Hilversum, Netherlands. T. (31 35) 217645. Fax 235906.

Hess, Robert, Rev

Rel: Roman Catholic.
International Organizations: President of the Sacred Heart League.
Mailing address: Walls MS 38686, USA. T. (1 601) 781 1360.

Heubach, Joachim, Bishop Dr

Rel: Evangelical Lutheran.
International Organizations: Bishop of Evangelical Lutheran Church of Schaumburg-Lippe.
Mailing address: Landeskirchenamt, Herderstrasse 27, 31675 Bückeburg, Germany.

Heye, Bekele, Pastor

Rel: Seventh-Day Adventist; Ordained on 15 SEP 1967; b. 02 DEC 1936 in Addis Ababa, Ethiopia; m. to Haregewine Gamachu Bekele; c. 3 boys and 1 girl; Citizen of Ethiopia; Lang. Amharic, Oromegna and English.
International Organizations: President, Eastern Africa Division (since 1980) of the General Conference of Seventh-Day Adventists.
Chairman of Executive Committee, Eastern Africa Division (since 1982) of the General Conference of Seventh-Day Adventists.
Other organizations: Chairman, Board of Trustees (since 1984), Soluxi College; Chancellor (since 1990), University of Eastern Africa.
Main field of work: Ministry, teaching. **Career:** Teacher, Ethiopian Adventist College from 1964 to 1967. Department Director, Ethiopian Union Mission, SDA from 1967 to 1969. Field Secretary, Afro-Mideast Division, SDA from 1970 to 1975. President, Ethiopian Union Mission, SDA from 1975 to 1980. President, Afro-Mideast Division, SDA from 1980 to 1982. **Educ.:** Dip.Ed. from Newbold College, UK (1960-1962); B.A. from Newbold College, UK (1960-1964); M.A. from Andrews University, Berrien Springs MI, USA (1969-1970). **Interests:** Church history. **Pubs.:** Master's thesis on the History of the Sabbath in Ethiopia.
Mailing address: PO Box HG 100, Highlands, Harare, Zimbabwe. T. (263 4) 732531/2/3. Fax 739284. Tx 26462
Residence: 6 Sunridge Road, Colne Valley, Harare, Zimbabwe. T. (263 4) 884 102

Heymann, Michael

Rel: Jewish.
International Organizations: Contact of the World Council on Jewish Archives.
Mailing address: c/o Israel Archives Asn, PO Box 1149, Jerusalem 91010, Israel. T. (972 2) 635716.

Hibbs, William E, III, Mgr

Rel: Roman Catholic; Ordained on 25 JUN 1975; b. 28 SEP 1950 in Boston MA, USA; celibate; Citizen of USA; Lang. English and French.
International Organizations: President of the Interfaith Council for Human Rights.
Other organizations: Chairman, Co-Executive Director (since 1976), National Ecumenical Coalition, Inc.; President, National AIDS Bereavement Center; Executive Producer, *The AIDS Digest.*
Main field of work: Research, education, religions. **Career:** Roman Catholic Priest, involved extensively in congressional legislation, community outreach programs, and national AIDS foundations from 1975 to 1982. Business Manager, United Press International (UPI), at the World Headquarters, Washington DC from 1985 to 1986. Controller (employed by the National Forest Products Association), the American Forest Council and the American Forest Foundation, Inc. Served as delegate, the National Ecumenical Coalition (NEC) to the United Nations High Commissioner for Refugees (UNHCR) and the Social Economic Council. Served in an advisory capacity to several Latin American governments and five US administrations on civil rights and social welfare issues. Appeared nationally on network television as well as local programs. National/international public speaker. **Educ.:** Grad. from Harvard University, Cambridge MA, USA; Grad. from St Thomas School of Theology. **Pubs.:** Marriage – The Sacrament and Contract and many published articles.
Mailing address: 4300 Old Dominion Dr, Suite 502, Arlington VA 22207-3246, USA. T. (1 703) 522 9758.
Residence: See Mailing Address. T. See Mailing Address

Hickey, James Aloysius, H E Card

Rel: Roman Catholic; Ordained on 15 JUN 1946; b. 11 OCT 1920 in Midland MI, USA; celibate; Citizen of USA.
International Organizations: Member (since 1988) of the College of Cardinals.
Member (since 1989) of the Pontifical Council for the Family. Member (since 1989) of the Congregation for Catholic Education. Member (since 1989) of the Congregation for Religious and Secular Institutes. Member, Council of the General Secretariat (since 1990) of the Synod of Bishops. Member (since 1990) of the Congregation for the Causes of Saints. Member of the Congregation for the Clergy.
Other organizations: Archbishop (since 1980), Washington DC; Chancellor (since 1980), Catholic University of America, Washington DC; Chairman, Board of Trustees (since 1980), Basilica of the National Shrine of the Immaculate Conception, Washington DC; Chairman, Board of Governors (since 1989), Pontifical North American College.
Main field of work: Ministry, education, administration. **Career:** Served as Associate Pastor, St Joseph Church, Saginaw MI (following ordination). Served as Secretary to the Bishop, Saginaw in 1951. Named first Rector, St Paul Seminary, Saginaw MI in 1959. Named Auxiliary Bishop, Saginaw in 1967. Consecrated as Titular Bishop, Taraqua in 1967. Servec briefly as Administrator, Diocese of Saginaw. Appointed Rector, North American College, Rome in 1969. Named Bishop, Cleveland in 1974. Prefect in 1980. Created Cardinal by Pope John Paul II in 1988. Chairman of Priestly Formation, National Conference of Catholic Bishops (NCCB) from 1968 to 1969. Chairman of Pastoral Research and Practices, NCCB from 1974 to 1977. Chairman, Committee for Doctrine, NCCB from 1979 to 1981. Chairman, Committee on Human Values, NCCB from 1984 to 1987. **Educ.:** S.T.L. from Catholic University of America, Washington DC, USA (1942-1946); S.T.D. from Pontifical University San Tommaso d'Aquino, Vatican City State (1947-1949); J.C.D. from Lateran Pontifical University, Vatican City State (1947-1950). **Hons.:** Honorary Doctorates: Saginaw Valley College, MI (1971), John Carroll University, Cleveland OH (1975), Ursuline College, Cleveland OH (1977), Cleveland State University (1981), Georgetown University, Washington DC (1981), Xavier University, Cincinnati OH (1981), Mount St Mary's College, Emmitsburg MD (1982), Providence College, RI (1983). **Pubs.:** Mary at the Foot of the Cross, 1989.
Mailing address: Archidioce of Washington, 5001 Eastern Ave, PO Box 29260, Washington DC 20017, USA. T. (1 301) 853 3800
Residence: See Mailing Address. T. See Mailing Address

Hickey, Thomas A, M.M., Br

Rel: Roman Catholic; Ordained; celibate.
International Organizations: Member of the Catholic Foreign Missionary Society of America.
Other organizations: Executive Secretary, National Assembly of Religious Brothers.
Mailing address: 1337 West Ohio Street, Chicago IL 60622-6490, USA. T. (1 312) 829 8525

Hicks, Murray L

Rel: Protestant.
International Organizations: Executive Director of the International Ministries to Israel.
Mailing address: 3323 N Ridge Avenue, Arlington Heights IL 60004, USA. T. (1 708) 394 4405. Fax 394 4408.

Hiestand, Galen, Rev

Rel: Evangelical; Ordained on 04 JAN 1982; b. 15 JUN 1945 in Lancaster PA, USA; m. to Carol Longenecker; c. 3 boys; Citizen of USA; Lang. English.

International Organizations: North American Director (since 1987) of the World Evangelical Fellowship.

Other organizations: President (since 1985), Development-Communications Inc.

Main field of work: Executive. **Career:** Pastor, Evangelical Free Church of America from 1973 to 1977. Marketing Manager, Hiestand Distributors from 1977 to 1981. Director of Development, Inter City Impact from 1981 to 1982. Account Executive, Management Development Association from 1982 to 1985. **Educ.:** B.A. Soc.Sc. from Wheaton College, IL, USA (1967-1969); M.Div. from Trinity Evangelical Divinity School, Deerfield IL, USA (1969-1973); M.A. Commun. from Wheaton Graduate School, Wheaton IL, USA (1981-1986).

Mailing address: 450 Gundersen Drive, Carol Stream IL 60189, USA. T. (1 708) 668 0440

Residence: 1099 Rose Avenue, Des Plaines IL 60016, USA. T. (1 708) 635 7718

Higgins, George Gilmary, Mgr

Rel: Roman Catholic; Ordained in 1940; b. 21 JAN 1916 in La Grange IL, USA; celibate; Citizen of USA.

International Organizations: Ecclesiastic of the Roman Catholic Church.

Member of the Pontifical Council for Justice and Peace.

Other organizations: Adjunct Lecturer, Dept. of Economics, School of Social Science (since 1981), Catholic University of America.

Career: Papal Chamberlain in 1953. Domestic Prelate in 1959. Former teacher, Dept. of Economics, School of Social Science, Catholic University of America. Appointed to staff, Social Action Dept., National Catholic Welfare Conference in 1944. Assistant Director, Social Action Dept. in 1946. Director, Social Action Dept. in 1954. Appointed Secretary for Research, US Catholic Conference in 1972. Appointed Secretary for Special Concerns, US Catholic Conference in 1979. Retired from the Conference in 1980. Chairman, Public Review Board, United Auto Workers of America AFL-CIO. Chairman, Bishops' Committee for Catholic-Jewish Relations. Adviser to Chairman, US Delegation to Belgrade Conference on Human Rights. Consultant to Bishops' Committee on Farm Labor. Member, Preparatory Commission on Lay Apostolate, Vatican Council II and Consultant to Council. Executive Committee, Leadership Conference on Human Rights. Bishops' Committee on Catholic Social Teaching and US Economy. Dunlop Commission on Farm Labor. Advisory Board, National Council on Foundations. Former member, American Arbitration Association. **Interests:** reading, theology, history, politics, swimming, golf. **Hons.:** Honorary degrees from eight universities; Murray-Green-Meany Award AFL-CIO (1980). **Pubs.:** syndicated column in *The Yardstick* and several journal articles.

Mailing address: 220 Curley Hall, Catholic University of America, Washington DC 20064, USA. T. (1 202) 635 5660

Higman, Francis Montgomery, Prof

Rel: Anglican; b. 17 SEP 1935 in Georgetown, Guyana; separated; c. 1 boy and 2 girls; Citizen of United Kingdom; Lang. English, French and German.

International Organizations: Director (since 1988) of the Institut d'Histoire de la Réformation, Geneva.

Treasurer (since 1990) of the International Federation of Societies and Institutes for the Study of the Renaissance.

Main field of work: Research in Reformation history. **Career:** Lecturer, University of Bristol from 1962 to 1969. Professor, University of Saskatchewan from 1969 to 1970. Professor, University of Dublin, Trinity College from 1970 to 1979. Professor, University of Nottingham from 1979 to 1988. **Educ.:** B.A. from University of Oxford, UK (1956-1959); B.Litt. from University of Oxford, UK (1959-1964). **Interests:** vocal music. **Hons.:** Hon. D.Th., University of Lausanne. **Pubs.:** The Style of John Calvin, OUP, 1967; Censorship and the Sorbonne, Droz, Geneva, 1979 and G Farel, Ed., Pater Noster et le Credo, Droz, Geneva, 1982.

Mailing address: 3 place de l'Université, CH-1211 Genève 4, Switzerland. T. (41 22) 705 71 28

Residence: 7 rue Monnier, CH-1206 Genève, Switzerland. T. (41 22) 789 10 39

Hilborn, Harry, Rev

Rel: Spiritualist.

International Organizations: President of the Independent Spiritualist Association.

Mailing address: 5130 West 25th Street, Cicero IL 60650, USA.

Hilchey, Harry St C, Ven

Rel: Anglican.

International Organizations: General Secretary of the Anglican Church of Canada.

Mailing address: 600 Jarvis Street, Toronto, Ontario M4Y 2J6, Canada.

Hill, Christopher John, Rev Canon

Rel: Church of England; Ordained on 24 MAY 1970; b. 10 OCT 1945 in Sedgley, Staffordshire, UK; m. to Hilary Ann Whitehouse; c. 3 boys and 1 girl; Citizen of United Kingdom; Lang. English, French and Italian.

International Organizations: Precentor and Canon Residentiary (since 1989) of the St Paul's Cathedral, London.

Other organizations: Chaplain to HM the Queen (since 1987); Co-Chairman (since 1990), London Society for Jews and Christians.

Main field of work: Music, liturgy and choir school. **Career:** Ordained Deacon in 1969. Assistant Curate, St Michael Tividale, Lichfield Diocese from 1969 to 1973. Assistant Curate, St Nicholas, Codsall, Lichfield Diocese from 1973 to 1974. Co-Secretary, Anglican-Roman Catholic International Commission from 1974 to 1981. Secretary, Anglican-Lutheran European Commission from 1981 to 1982. Assistant Chaplain to Archbishop, Canterbury for Foreign Relations from 1974 to 1981. Secretary to Archbishop, Canterbury for Ecumenical Affairs from 1982 to 1989. Canon, Canterbury Cathedral from 1982 to 1989. **Educ.:** B.D. (Hons) from University of London, King's College, UK (1964-1967); M.Th. from University of London, King's College, UK (1967-1968). **Interests:** music, walking, reading. **Mem.:** Church of England Nordic/Baltic Conversations (1989); Clubs: Athenaeum, Nikaean **Pubs.:** Miscellaneous ecumenical articles.

Mailing address: 3 Amen Court, London EC4M 7BU, UK. T. (44 71) 236 4532

Residence: See Mailing Address. T. See Mailing Address

Hill, Kent R

Rel: Evangelical; b. 24 MAY 1949 in Nampa ID, USA; m. to Janice Hurn; c. 1 boy and 1 girl.

International Organizations: Member, National Executive Board of the World Without War Council.

Other organizations: President (since 1986), Institute on Relgion and Democracy; Board of Directors, Keston College, USA; National Advisory Board for 'Peace, Freedom and Security Program', National Association of Evangelicals; US Board, Operation Hunger, South Africa.

Career: Russian Translator for US Army from 1971 to 1974. Associate Professor of History at Seattle Pacific University from 1980 to 1986. Actively involved in the 'Siberian Seven' emigration case from 1978 to 1983. **Educ.:** B.A. Hist. from Northwest Nazarene College, Nampa ID, USA (1971); M.A. Russian Studies from University of Washington, Seattle, USA (1976); Ph.D. Hist. from University of Washington, Seattle, USA (1980). **Hons.:** Pushkin Award for Outstanding Scholarship, Defense Language Institute, Monterey CA (1972); 'Citation of Merit', Northwest Nazarene College Alumni Association (1985); Baccalaureate Address, Seattle Pacific University (1985); Professor of the Year, Seattle Pacific University (1985-86); 'Alumnus of the Year', Northwest Nazarene College (1988). **Mem.:** American Association for the Advancement of Slavic Studies; American Historical Association; Conference on Faith and History **Pubs.:** *The Puzzle of the Soviet Church: An Inside Look at Christianity and Glasnost*, Portland OR, Multnomah Press, 1989, 417 pp.; *The Soviet Union on the Brink*, Portland OR, Multnomah Press, 1991, 520 pp. and articles/reviews in: *New Oxford Review*, *Fides et Historia*, *The Review of Metaphysics*, *Religion in Communist Dominated Areas*, *Christian Scholar's Review*, *This World*, *Christianity Today*, *Missiology*, *The National Catholic Register*, *Christian Century*, *Slavic Review*, and *Christian Science Monitor*.

Mailing address: 1331 H Street NW, Suite 900, Washington DC 20005, USA. T. (1 202) 393 3200. Fax 638 4948

Hill, Peter Allen, Rev

Rel: Anglican; Ordained; b. 23 AUG 1940 in Guelph, Ontario, Canada; m. to Nancy Morrow; Citizen of Canada.

International Organizations: Chaplain (since 1989) of St George's College.
Career: Assistant Curate, St John's Church, Thunder Bay, Ontario from 1965 to 1968. Rector, St Joseph Island Parish, Richard Landing, Ontario from 1968 to 1976. Teacher at Central Algoma Secondary School from 1968 to 1976. Chaplain, Trinity College School, Port Hope, Ontario from 1976 to 1985. Headmaster, Queen Margaret's School from 1985 to 1989. **Educ.:** B.A. from McMaster University, Hamilton, Ontario, Canada (1962); B.Th. from University of Western Ontario, Huron College, London, Canada (1965).
Mailing address: 638 Manning Ave, Toronto ON M6G 2V9, Canada.

Hill, Richard, Rev
Rel: Protestant.
International Organizations: President of the International Ecumenical Fellowship.
Mailing address: All Saints Vicarage, Brackenfield Road, Newcastle-upon-Tyne NE3 4DX, UK.

Hills, Christopher, Dr
Rel: Yoga.
International Organizations: Secretary of the World Yoga Society.
Mailing address: PO Box 644, Boulder Creek CA 95006, USA.

Hilton, David, Dr
Rel: Protestant; b. 23 AUG 1931 in Plymouth WI, USA; m. to Laveta Smith; c. 2 boys and 2 girls; Citizen of USA; Lang. English, Hausa and French.
International Organizations: Associate Director (since 1986) of the Christian Medical Commission.
Main field of work: Health consultant. **Career:** In charge of primary health care, Protestant Church from 1959 to 1979. Health Director, Seminole Tribe of Florida from 1979 to 1985. **Educ.:** B.A. (cum laude) from Indiana University, Indianapolis, USA (1949-1953); M.D. from Indiana University, Bloomington, USA (1953-1957). **Mem.:** Certified Fellow: American Board of Family Practice, American Academy of Family Practice
Mailing address: PO Box 2100, 150 route de Ferney, CH-1211 Genève 2, Switzerland. T. (41 22) 791 62 14. C. OIKOUMENE. Tx 415 730 OIKCH. Fax (41 22) 791 03 61.

Himbeault, Claire, R.N.D.M., Sr
Rel: Roman Catholic; Ordained; celibate.
International Organizations: Superior General of the Sisters of Our Lady of the Missions.
Other organizations: Member, Commission for the Catechism and the Catechists.
Mailing address: Via di Bravetta 628, I-00164 Roma, Italy. T. (39 6) 6615 8400 – (39 6) 6615 8348.

Hincq, L
Rel: Roman Catholic.
International Organizations: Secretary of the Oeuvres sociales et éducatives des jésuites au Tiers-Monde.
Mailing address: Rue A Fauchille 12, B-1150 Bruxelles, Belgium. T. (32 2) 762 43 20.

Hirayama, Shoji, Rev
Rel: United Church of Christ; Ordained on 13 OCT 1940; b. 04 OCT 1907 in Yama-Uchi Cho, Saga Prefecture, Japan; m. to Akiko; c. 2 boys and 4 girls; Citizen of Japan; Lang. Japanese and English (some).
International Organizations: President of the Asian-Pacific Christian Peace Conference.
Other organizations: Chief Pastor (since 1948), Tokyo Yamate Church of the United Church of Christ in Japan; Representative, Japan Religious Peace Conference; Representative, Japan Christian Peace Association; Representative, Innovation Conference all over the country.
Main field of work: Ministry. **Educ.:** Deg. Theol. from Doshisha University, Kyoto, Japan (1931-1934). **Pubs.:** In Japanese only: AI to heiwa no kikyu (In search of love and peace); Nakumono to tomoni nake (Weep with those in sorrow) and Korekara no kristokyo (Christianity in the future).
Mailing address: Tokyo Yamate Church, 19-5-1001 Udagawa-Cho, Shibuya, Tokyo 150, Japan. T. (81 3) 3463 2971. Fax 3463 1344
Residence: See Mailing Address. T. (81 3) 3461 5758 (SAME FOR FAX)

Hirose, Takashi
Rel: Buddhist; b. 22 JUN 1924; Citizen of Japan.
International Organizations: Professor of the Otani University.
Career: President, Otani University (Buddhism). **Educ.:** Otani University, Japan.
Mailing address: 1-67 Iwakura Nagatanicho, Sakyo-ku, Kyoto 606, Japan. T. (81 75) 711 3786

Hirsch, Abraham
Rel: Jewish.
International Organizations: Secretary-General of the Agudath Israel World Organization.
Mailing address: Hacherut Square, POB 326, Jerusalem, Israel. T. (972 2) 384357. Fax 383 634. C. Worldaguda, Jerusalem.

Hirsch, Richard G, Rabbi Dr
Rel: Jewish: Progressive; Ordained.
International Organizations: Executive Director of the World Union for Progressive Judaism.
Mailing address: 838 Fifth Avenue, New York NY 10021, USA. T. (1 212) 249 0100. Fax 517 3940.

Hirschler, Horst, Landesbischof
Rel: Lutheran.
International Organizations: Vice-President of the Lutheran World Federation.
Member, Program Committee for Theology and Studies of the Lutheran World Federation.
Other organizations: Landesbischof, Evangelical Church of Hannover.
Mailing address: Haarstrasse 6, 30449 Hannover, Germany. T. (49 511) 80 01 88/89

Ho, David, Rev
Rel: Evangelical.
International Organizations: Executive Secretary of the Evangelical Association of the Caribbean.
Mailing address: 9 West Avenue, Kingston 8, Jamaica. T. (1 809) 924 1378.

Hoare, Liam, S.P., Fr
Rel: Roman Catholic; Ordained; celibate.
International Organizations: Superior General of the Servants of the Paraclete.
Mailing address: Jemez Springs NM 87025, USA. T. (1 505) 829 3586.

Hochbaum, Jerry, Dr
Rel: Jewish.
International Organizations: Executive Director of the Memorial Foundation for Jewish Culture.
Mailing address: 15 E 26th Street, New York NY 10010, USA. T. (1 212) 679 40 74. C. MECULTFUND New York.

Hodakov, M A, Rabbi
Rel: Jewish; Ordained.
International Organizations: Executive Officer of the Lubavitch Movement.
Mailing address: 770 Eastern Parkway, Brooklyn NY 11213, USA. T. (1 718) 774 4000.

Hodder, Kenneth L, Commr
Rel: Salvation Army, The.
International Organizations: Territorial Commander of The Salvation Army.
Mailing address: 1424 Northeast Expressway, Atlanta GA 30329, USA.

Hodge, Stephen
Rel: Buddhist.
International Organizations: President of the Buddhist Union of Europe.
Mailing address: 58 Eccleston Square, London SW1Y 1PH, UK. T. (44 71) 828 9242.

Hodosy, Imre, Bishop
Rel: Reformed Church.
International Organizations: Bishop of Reformed Church in Yugoslavia.
Mailing address: Ul Bratstva 26, 24323 Feketic, Yugoslavia.

Hoeffel, Michel, Pastor
Rel: Lutheran.
International Organizations: Chairperson, Program Committee for Communication Services of the Lutheran World Federation.
Other organizations: President, Church of the Augsburg Confession of Alsace and Lorraine.
Mailing address: 1A quai Saint-Thomas, F-67081 Strasbourg CEDEX, France. T. (33) 88 25 90 05. Fax 88 25 90 99. Tx MESTRA 890156
Residence: T. (33) 88 22 25 77

Hoekema, Alle
Rel: Mennonite.
International Organizations: Pastor of the Haarlem Mennonite Church.
Mailing address: Verspronckweg 11, NL-2023 BA Haarlem, Netherlands.

Hofman, David
Rel: Bahá'í; Unordained; b. 26 SEP 1908 in Poona, India; m. to Marion Elizabeth Holley; c. 1 boy and 1 girl; Citizen of United Kingdom; Lang. English and French.
International Organizations: Vice-President (since 1988) of the World Congress of Faiths.
Other organizations: Founder and Chairman (since 1948), George Ronald, Publisher.
Main field of work: Administration. **Career:** Announcer, BBC Television and Radio from 1937 to 1940. Member, National Spiritual Assembly, Bahá'ís of the British Isles from 1936 to 1963. Member, Universal House of Justice, Bahá'í International Community from 1963 to 1988. **Pubs.:** *The Renewal of Civilization*; *George Townshend* (life of Archdeacon G Townshend) and *Bahá'u'lláh, The Prince of Peace*.
Mailing address: 'The Bivvy', Heyford Road, Steeple Aston, Bicester OX6 3SH, UK. T. (44 869) 47342

Hofmann, Nadi Angelika
Rel: Bahá'í; b. 08 APR 1970 in München, Germany; single; Citizen of Germany; Lang. German, English, Persian and Esperanto.
International Organizations: Member, Youth Section (since 1987) of the Bahá'í Esperanto League.
Member (since 1989) of the European Bahá'í Youth Council.
Main field of work: Psychology. **Educ.:** Technische Hochschule Darmstadt, Germany (Beginning in 1989). **Interests:** Comparison of religions, music, psychology, esperanto. **Mem.:** Bahá'í Esperanto League, Bahá'í Youth Committees, European Bahá'í Youth Council, Universal Esperanto Association (UEA)
Mailing address: Henry-Dunantstrasse 30, 64807 Dieburg, Germany.
Residence: Am Schloss Stockau 15, 64807 Dieburg, Germany T. (49 6071) 1532

Hogan, William Francis, C.S.C., Rev
Rel: Roman Catholic; Ordained on 08 JUN 1956; b. 12 APR 1930 in Tewkesbury MA, USA; celibate; Citizen of USA; Lang. English, French and Italian.
International Organizations: Procurator-General (since 1984) of the Congregation of Holy Cross.
Consultant (since 1984) of the Congregation for Religious and Secular Institutes.
Main field of work: Ministry, adult education. **Career:** Provincial Superior, Congregation of Holy Cross from 1970 to 1974. Procurator-General, Congregation of Holy Cross from 1974 to 1984. **Educ.:** A.B. from University of Notre Dame, IN, USA (1950-1952); J.C.D. from Pontifical Gregorian University, Vatican City State (1957-1960). **Interests:** renewal of religious life. **Mem.:** Religious Formation Conference **Pubs.:** "Focus: The Mission", in *Review for Religious*, May-June 1991; "Les chapitres généraux", in *La vie des communautés religieuses*, Sept-Oct 1991 and "Evangelizing Community", in *Review for Religious*, Jan-Feb 1992.
Mailing address: Via Framura 85, I-00168 Roma, Italy. T. (39 6) 627 69 20
Residence: See Mailing Address. T. See Mailing Address

Holeton, David, Rev Canon Prof Dr
Rel: Christian.
International Organizations: Secretary of the Societas Liturgica – International Society for Liturgical Study and Renewal.
Mailing address: Trinity College, 6 Hoskin Ave, Toronto M5S 1H8, Canada.

Holland, Arthur, Commr
Rel: Salvation Army, The.
International Organizations: Territorial Commander of The Salvation Army.
Mailing address: Kowdiar, Trivandrum 695 003, Kerala State, India.

Hollander, Fritz
Rel: Jewish.
International Organizations: President of the Scandinavian Zionist Federation.
Chairman of the European Council of Jewish Community Services.
Mailing address: Nybrogatan 19, S-114 39 Stockholm, Sweden. T. (46 8) 62 60 98

Hollemweguer, Juan E, Bishop
Rel: Methodist; Ordained.
International Organizations: President of the Methodist Church of Peru.
Mailing address: Apartado 1386, Lima 100, Peru.

Hollings, Michael Richard, Rev
Rel: Roman Catholic; Ordained on 08 APR 1950; b. 30 DEC 1921 in Camberley, Surrey, UK; celibate; Citizen of United Kingdom; Lang. English and French.
International Organizations: Executive (since 1984) of the Council of Christians and Jews.
Other organizations: Religious Adviser to Thames Television (since 1968); Member (since 1970), Westminster Diocesan Schools Commission; Parish Priest and Area Dean (since 1978), St Mary of the Angels, Bayswater, London.
Main field of work: Ministry, preaching and counselling. **Career:** Major, Coldstream Guards from 1941 to 1945. Assistant Priest, St Patrick's, Soho Square from 1950 to 1954. Chaplain, Westminster Cathedral from 1954 to 1958. Chaplain, Sovereign Military Order of Malta in 1957. Chaplain, London University from 1958 to 1959. Religious Adviser to ATV from 1958 to 1959. Religious Adviser to Rediffusion from 1958 to 1968. Chaplain to Roman Catholics at Oxford University from 1959 to 1970. Lay Member, Press Council from 1969 to 1975. Chaplain, National Council of Lay Apostolate from 1970 to 1974. Parish Priest, St Anselm's, Southall from 1970 to 1978. Member, Southall Council of Churches, Oxford and Cambridge Catholic Education Board from 1971 to 1978. Executive, Council of Christians and Jews from 1971 to 1979. Chaplain, Catholic Institute of International Relations from 1971 to 1980. Executive, Ealing Community Relations Council from 1973 to 1976. Member, Standing Committee, National Conference of Priests from 1974 to 1976. Member, Rampton Committee from 1979 to 1981. Member, Swann Committee from 1981 to 1984. Member of Board, Christian Aid from 1984 to 1987. **Educ.:** M.A. from University of Oxford, St Catherine's College, UK (1939-1941); Commdt. from Sandhurst Military Academy, UK (1941); Trained for Priesthood from Beda College, Rome, Italy (1946-1950). **Interests:** reading, people. **Hons.:** Military Cross (1943); Mentioned in despatches (World War II). **Pubs.:** Living Priesthood, 1977; Go in Peace, 1984 and Dying to Live, 1990.
Mailing address: St Mary of the Angels, Moorhouse Road, Bayswater, London W2 5DJ, UK. T. (44 71) 229 0487
Residence: See Mailing Address. T. See Mailing Address

Holloway, Richard Frederick, Rt Rev
Rel: Church of Scotland; Ordained in 1960; b. 26 NOV 1933 in Glasgow, Scotland, UK; m. to Jean; c. 1 boy and 2 girls; Citizen of United Kingdom.
International Organizations: Bishop (since 1986) of Edinburgh.
Other organizations: Involved in a number, agencies working for people with AIDS; Reviewer and writer for the *Church Times*; Regular contributor to the religious feature in the *Independent and Correspondent*; Frequent broadcaster on radio and television; Presenter on several BBC series; Regular retreat conductor.
Main field of work: Ministry. **Career:** Served in the Army from 1952 to 1953. Secretary to the Bishop, Accra, Ghana from 1956 to 1958. Ordained Deacon, Glasgow in 1959. Lecturer in Christian Ethics, Edinburgh Theological College from 1959 to 1963. Curate, St Ninian's, Glasgow from 1959 to 1963. Priest-in-charge, St Margaret's and St Mungo's, Gorbals from 1963 to 1968. Rector, Old St Paul's, Edinburgh from 1968 to 1980. Rector, The Church of the Advent, Boston MA from 1980 to 1984. Vicar, St Mary Magdalen's, Oxford from 1984 to 1986. Founder/First Chairman, Christian Action Housing Association,

Glasgow. Founder/First Chairman, Castle Rock Housing Association, Edinburgh. Missions in England, USA, South Africa, Australia and Canada. **Educ.:** Kelham Theological College; Edinburgh Theological College, UK; B.D. from University of London, UK (external) (1963); S.T.M. from Union Theological Seminary, New York NY, USA (1968). **Interests:** reading (biography, history, poetry, fiction), cinema, running and long-distance walking, listening to music. **Hons.:** Winifred Mary Stanford Award for *The Killing* (1984). **Mem.:** Human Fertilisation and Embriology Authority (1990) **Pubs.:** *Crossfire: Faith and Doubt in an Age of Certainty*, Collins, 1988; Editor, *The Divine Risk*, Darton Longman and Todd, 1990 and *Another Country Another King*, Collins, 1991.
Mailing address: Bishop's Office, Diocesan Centre, Walpole Hall, Chester Street, Edinburgh EH3 7EN, Scotland, UK. T. (44 31) 226 3359. Fax 225 3181
Residence: 3 Eglington Crescent, Edinburgh EH12 5DH, Scotland, UK. T. (44 31) 226 5099

Hölter, Günter

Rel: Roman Catholic.
International Organizations: Treasurer of the Euro-Caritas.
Mailing address: Rue du Commerce 70-72, B-1040 Bruxelles, Belgium. T. (32 2) 511 42 55. Tx 25 173.

Holtrop, Pieter Nanne, Rev Prof Dr

Rel: Reformed Church; Ordained on 16 JAN 1977; b. 31 JAN 1943; m. to Agatha Arnold Fina Van Berge; c. 2 girls; Citizen of Netherlands; Lang. Dutch, English, German, French and Indonesian.
International Organizations: Member of the Executive Committee (since 1989) of the World Alliance of Reformed Churches.
Other organizations: Member, External Relations Commission (since 1984), the Reformed Churches in The Netherlands; Advisor to the Missionary Board (since 1985), the Reformed Churches in The Netherlands; Professor (since 1987), Free University, Amsterdam.
Main field of work: Education. **Career:** Assistant Professor, Free University, Amsterdam from 1971 to 1976. Teacher, Theological Seminary, Eastern Indonesia from 1977 to 1982. Minister, Reformed Church, Amsterdam from 1983 to 1987. **Educ.:** Drs. Theol. from Free University, Amsterdam, Netherlands (1961-1971); Dr. Theol. from Free University, Amsterdam, Netherlands (1971-1975). **Interests:** Church history, ecumenics, mission.
Mailing address: Kampen Theological University, Koornmarkt 1, NL-8260 GA Kampen, Netherlands. T. (31 5202) 92 666
Residence: Domineeswal 6, NL-8356 DS Blokzijl, Netherlands T. (31 5272) 1620

Holtz, Bruno, S.M.B., Fr

Rel: Roman Catholic; Ordained; celibate.
International Organizations: Secretary-General of the International Catholic Union of the Press.
Member of the Bethlehem Mission Society. Advisor of the Pontifical Council for Social Communications.
Mailing address: UCIP, 37-39 rue de Vermont, CP 197, CH-1211 Genève 20, Switzerland. T. (41 22) 734 00 17 – 734 74 16. Fax 731 1051. Tx 412946 CH.

Holzer, Emmanuel, Rabbi

Rel: Jewish.
International Organizations: Chairman of the Universal Torah Registry.
Mailing address: 14 Research Way, Setauket NY 11733, USA. T. (1 516) 689 9688.

Hong, H, Rev

Rel: Christian; Citizen of Republic of Korea.
International Organizations: Moderator of the Korean Christian Church in Japan.
Mailing address: Young Ki, 3-10-7, Inabamotomachi, Amagasaki 660, Japan.

Hooker, Morna Dorothy, Prof

Rel: Methodist; b. 19 MAY 1931 in Beddington, Surrey, UK; m. to Rev Dr W David Stacey; Citizen of United Kingdom.
International Organizations: Lady Margaret's Professor of Divinity (since 1976) of the University of Cambridge.
Other organizations: Fellow (since 1977), Robinson College, Cambridge; Joint Editor (since 1985), *Journal of Theological Studies.*
Main field of work: New Testament studies. **Career:** Research Fellow at University, Durham from 1959 to 1961. Lecturer in New Testament, King's College, London from 1961 to 1970. Lecturer in Theology, University of Oxford from 1970 to 1976. Lecturer in Theology, Keble College, Oxford from 1972 to 1976. Fellow, Linacre College, Oxford from 1970 to 1976. **Educ.:** B.A. from University of Bristol, UK (1950-1953); M.A. from University of Bristol, UK (1953-1955); Ph.D. from Victoria University of Manchester, UK (1957-1958). **Hons.:** Fellow, King's College, London (1979); Hon. Fellow, Linacre College (1980). **Mem.:** Society for the Study of the New Testament (President 1988-89) **Pubs.:** Jesus and the Servant, 1979; From Adam to Christ, 1990 and A Commentary on Mark, 1991.
Mailing address: The Divinity School, St John's Street, Cambridge CB2 1TW, UK. T. (44 223) 332598

Hoppe, Wolfgang, F.F.S.C., Br

Rel: Roman Catholic; Ordained on 01 OCT 1961; b. 19 OCT 1940 in Westpreussen, Germany; celibate; Citizen of Germany; Lang. German, English and French.
International Organizations: General Secretary (since 1974) of the Franciscan Brothers of the Holy Cross.
Other organizations: Manager (since 1979), working group for pastoral vocation in Germany; Member, Board of Ministers (since 1984), Catholic Council; Member, Board of Ministers (since 1984), Pastoral Council; President, Council of the religious people (since 1991), Diocese of Trier.
Main field of work: Education of young religious. **Educ.:** M.T.A. from University of Cologne, Germany (1962-1966).
Mailing address: Hönninger Strasse 16, 53547 Hausen/Wied, Germany. T. (49 2638) 8970.
Residence: St Josefshaus, 53547 Hausen/Wied, Germany. T. (49 2638) 897154

Horrocks, Harvey

Rel: Buddhist: Mahayana.
International Organizations: Director of the Foundation for the Preservation of the Mahayana Tradition.
Mailing address: P.O. Box 1778, Soquel 95073 CA, USA. T. (1 408) 476 8435, Fax (1 408) 476 4823

Horsch, V

Rel: Mennonite.
International Organizations: Contact of the International Mennonite Organization.
Mailing address: Langendorfstrasse 29, 56564 Neuwied, Germany. T. (49 2631) 3 14 40.

Horton, Vicki M

Rel: Christian.
International Organizations: Head, Secretariat of the Ambassadors for Christ International.
Mailing address: c/o VMH Associates, 3490 Piedmont Road, Atlanta GA 30305, USA.

Horvath, Tibor, S.J., Rev

Rel: Roman Catholic; Ordained in 1956; b. 28 JUL 1927 in Banhida, Hungary; celibate.
International Organizations: Professor of Systematic Theology (since 1962) of the Regis College, University of Toronto.
Member (since 1946) of the Society of Jesus. Consultor (since 1990) of the Pontifical Council for Dialogue with Non-Believers.
Other organizations: Professor (since 1968), Toronto School of Theology; Director (since 1970), Institute for Encyclopedia of Human Ides on Ultimate Reality and Meaning; Founder and General Editor (since 1978), *Ultimate Reality and Meaning: Interdisciplinary Studies in the Philosophy of Understanding*; Founder and first Secretary-General (since 1986), International Society for Encyclopedia of Church History in Hungary.

Main field of work: Education. **Career:** Pastoral work, Germany from 1951 to 1952. Pastoral work, Germany from 1959 to 1960. Pastoral work, Spain from 1958 to 1959. Visiting Professor of Theology, St Paul's University, Ottawa from 1967 to 1969. President, International Society for the Study of URAM from 1985 to 1987. **Educ.:** Phil. Stds. from Aloysianum, Szeged, Hungary (1948); University of Innsbruck, Austria (1949); Aloysianum, Chieri, Italy (1949-1951); M.A., L.Phil. from College of Philosophy, St Albert, Louvain, Belgium (1952-1954); S.T.L. Theol. Stds. from University of Granada, Spain (1954-1958); S.T.D. from Pontifical Gregorian University, Vatican City State (1960-1962); P.D. Stds. Inf.Sc. and Comp. Sc. from University of Chicago, IL, USA (1970-1971). **Interests:** hiking, mountain climbing, concerts. **Hons.:** Recipient, Deutsche Forschungsgemeinschaft (1966). **Mem.:** American Philosophers; Systematic Theologians; Anthropologists; International Community Service; American Canadian Indians **Pubs.:** Author of several books and articles on theology, philosophy and faith.
Mailing address: Regis College, 15 St Mary St, Toronto ON M4Y 2R5, Canada.
Residence: 2 Dale Ave, Toronto ON M4W 1K4, Canada.

Horvath, Zigmund, Bishop
Rel: Reformed Church.
International Organizations: Bishop of Reformed Christian Church in Slovakia.
Mailing address: Predsednictvo Synody, Jokaiho 34, Box 78, 94501 Komarno, Slovakia.

Hosein, M K
Rel: Muslim.
International Organizations: Executive Secretary of the Islamic Missionaries Guild of the Caribbean and South America.
Mailing address: 67 Prince Street, PO Box 800, Port-of-Spain, Trinidad-Tobago. T. (1 809) 627 5797. Fax 627 5797. C. ILMGUILD, Port of Spain.

Houssiau, Albert, Mgr
Rel: Roman Catholic; Ordained on 06 FEB 1949; b. 02 NOV 1924 in Halle, Belgium; celibate.
International Organizations: Protector of the International Union of Prayer of Banneux Notre-Dame.
Other organizations: Bishop (since 1986), Liège.
Mailing address: Rue de l'Evêché 25, B-4000 Liège, Belgium. T. (32 41) 23 48 20.

Houston, Haviland, Dr
Rel: Methodist.
International Organizations: Co-Secretary-General of the United Methodist Church, Board of Church and Society.
Mailing address: 100 Maryland Ave NE, Washington DC 20002, USA. T. (202) 488 5600.

Houston, Tom, Rev
Rel: Baptist; Ordained on 11 APR 1951; b. 26 MAR 1928 in Dumbarton, Scotland, UK; m. to Elizabeth Hazle Findlay; c. 1 boy and 1 girl; Citizen of United Kingdom; Lang. English, Kiswahili and French.
International Organizations: International Director (since 1989) of the Lausanne Committee for World Evangelization.
Other organizations: Pastor Emeritus (since 1991), Nairobi Baptist Church.
Main field of work: Strategic consultancy. **Career:** Pastor, Johnstone Baptist Church, Scotland from 1951 to 1954. Chaplain, Quarriers Homes, Scotland from 1955 to 1959. Pastor, Nairobi Baptist Church from 1959 to 1971. Executive Director, British and Foreign Bible Society from 1971 to 1983. President, World Vision International from 1984 to 1988. **Educ.:** M.A. (Hons) from University of Glasgow, UK (1946-1950); B.D. from University of London, UK (1954). **Pubs.:** "Characters Around the Cross" and "King David".
Mailing address: PO Box 300, Oxford OX2 9XB, UK. T. (44 865) 749070. Fax 714324.
Residence: 134A Cumnor Hill, Oxford OX2 9PJ, UK. T. (44 865) 862616

Howard, Charles, F.M.S., Br
Rel: Roman Catholic; Ordained; celibate.
International Organizations: Superior General of the Little Brothers of Mary.
Mailing address: Piazzale M Champagnat 2, CP 10250, I-00144 Roma, Italy. T. (39 6) 592 44 56/7/8.

Howard, David M, Dr
Rel: Evangelical; Ordained in 1952; b. 28 JAN 1928 in Philadelphia PA, USA; m. to Phyllis Gibson Howard; c. 3 boys and 1 girl; Citizen of USA; Lang. English and Spanish.
International Organizations: Past International Director (since 1982) of the World Evangelical Fellowship.
Other organizations: Advisory Life Trustee (since 1975), Wheaton College, IL.
Main field of work: Administration, ministry. **Career:** Assistant General Director, Latin American Mission from 1953 to 1968. Missions Director, Inter-Varsity Christian Fellowship from 1968 to 1977. Congress Director, Lausanne Committee for World Evangelization from 1977 to 1980. **Educ.:** B.A. from Wheaton College, IL, USA (1945-1949); M.A. Theol. from Wheaton Graduate School, IL, USA (1950-1952). **Hons.:** Alumnus of the Year, Wheaton College, IL (1977); Hon. LL.D., Geneva College, Beaver Falls PA (1974); Hon. L.H.D., Taylor University, Upland IN (1978). **Pubs.:** By the Power of the Holy Spirit, 1973; Student Power in World Missions, 1979 and The Dream That Would Not Die, 1986.
Mailing address: 141 Middle Road 05-05, GSM Bldg, Singapore 0718, Singapore. T. (65) 339 7900. Fax (65) 338 3756.
Residence: 27 Balmoral Park 08-27, Singapore 1025, Singapore T. (65) 732 7058

Howe, Margaret
Rel: Buddhist.
International Organizations: Coordinator of the Buddhist Peace Fellowship.
Mailing address: PO Box 4650, Berkeley CA 94704, USA. T. (1 415) 525 8596.

Howland, Frances
Rel: Buddhist: Mahayana.
International Organizations: South Asia Regional Office of the Foundation for the Preservation of the Mahayana Tradition.
Other organizations: Director, Himalyan Yogic Institute.
Mailing address: GPO Box 817, Kathmandu, Nepal. T. (997 1) 413 094, Fax (997 1) 410 992

Howson, Justin, B.G.S., Br
Rel: Roman Catholic; Ordained; celibate.
International Organizations: Brother-General of the Piccoli Fratelli del Buon Pastore.
Mailing address: PO Box 1003, Hamilton, Ontario L8N 3R1, Canada. T. (1 416) 528 9109

Hruby, Olga S
Rel: Christian.
International Organizations: Executive Director of the Research Center for Religion and Human Rights in Closed Societies.
Mailing address: 475 Riverside Drive, Suite 448, New York NY 10015, USA. T. (1 212) 870 2481.

Huberts, H
Rel: Christian.
International Organizations: Executive Director of the ZOA Refugee Care Netherlands.
Mailing address: Sleutelbloemstaat 8, PO Box 10343, NL-7322 AG Apeldoorn, Netherlands. T. (31 55) 66 38 33. Fax 66 87 99. Tx 20010 Attn ZOA.

Huddleston, (Ernest Urban) Trevor, Most Rev

Rel: Anglican; Ordained in 1937; b. 15 JUN 1913 in Bedford, UK; Citizen of United Kingdom.

International Organizations: President (since 1983) of the National Peace Council.

President (since 1981) of the Anti-Apartheid Movement. President (since 1984) of the International Voluntary Service.

Other organizations: Trustee (since 1972), Runnymede Trust; Provost (since 1983), Selly Oak Colleges.

Career: Deacon in 1936. Curate, St Mark's, Swindon from 1936 to 1939. Professed Community of the Resurrection, Mirfield in 1941. Prior and Priest-in-Charge, Sophiatown, Orlando, and Pimville, Johannesburg in 1943. Provincial, Community of the Resurrection from 1949 to 1956. Novice-Master, Community of the Resurrection from 1956 to 1958. Prior, London Community of the Resurrection from 1958 to 1960. Bishop, Masasi, Tanzania from 1960 to 1968. Suffragan Bishop, Stepney (resigned) from 1968 to 1978. Bishop, Mauritius from 1978 to 1983. also Archbishop, the Anglican Province of the Indian Ocean from 1978 to 1983. Vice-President, Anti-Apartheid Movement from 1969 to 1981. Chairman, International Defence and Aid Fund for Southern Africa from 1983 to 1991. **Hons.:** Hon. D.D., Aberdeen University; Hon. D.Litt., Lancaster University, Warwick University (1988), City of London Polytechnic (1989); Hon. D.H.L., Denison University (1989); Order of the Grand Companion of Freedom, Zambia (1984). **Pubs.:** God's World, 1966; I Believe: Reflections on the Apostle's Creed, 1986 and Trevor Huddleston: Essays on His Life and Work, 1988.

Mailing address: House of the Resurrection, Mirfield WF14 0BN, UK.

Huet, Denis

Rel: Buddhist: Mahayana.

International Organizations: Member of the Foundation for the Preservation of the Mahayana Tradition.

Other organizations: Director, Institut Vajra Yogini.

Mailing address: Château d'en Clausade, Marzens, F-81500 Lavaur, France. T. (33 63) 581 722. Fax (33 63) 580 348

Hughes, Alex, Col

Rel: Salvation Army, The.

International Organizations: Territorial Commander of The Salvation Army.

Mailing address: Casilla 2240, 1000 Buenos Aires, Argentina.

Hughes, Catherine, SDNdeN, Sr

Rel: Roman Catholic; Ordained; celibate.

International Organizations: General Moderator of the Sisters of Notre Dame de Namur.

Mailing address: Casa Generalizia, Via Monte Altissimo 23, I-00141 Roma, Italy. T. (39 6) 892 24 73.

Huguenin, Willy, Commr

Rel: Salvation Army, The; Ordained on 19 APR 1954; b. 22 SEP 1931 in Le Locle, Neuchâtel, Switzerland; m. to Miriam Lüthi; c. 2 boys; Citizen of Switzerland; Lang. French, English, German, Kikongo, Lingala and Munukutuba.

International Organizations: Territorial Commander for Switzerland, Austria and Hungary (since 1987) of The Salvation Army.

Main field of work: Ministry. **Career:** Teaching for the Salvation Army in Zaire from 1955 to 1969. Director, School of Officers, Kinshasa from 1969 to 1971. Secretary-General, The Salvation Army, Zaire from 1971 to 1976. Director, School of Officers, Bern from 1976 to 1979. Territorial Commander for Congo/Brazzaville from 1979 to 1987. **Educ.:** Teacher from Ecole Normale, Neuchâtel, Switzerland (1950-1951).

Mailing address: Die Heilsarmee, Postfach, Laupenstrasse 5, CH-3001 Berne, Switzerland. T. (41 31) 25 05 91

Residence: Mühledorfstrasse 25, CH-3018 Bern, Switzerland T. (41 31) 55 67 22

Huhn, Martina

Rel: Lutheran.

International Organizations: Vice-President of the Lutheran World Federation.

Member, Program Committee for Theology and Studies of the Lutheran World Federation. Member, Standing Committee for Constitution of the Lutheran World Federation.

Mailing address: Dorfstrasse 6B, 04651 Hopfgarten, Germany.

Hulbert, Fiona

Rel: Christian.

International Organizations: Contact of the Ecumenical Forum of European Christian Women.

Mailing address: Centre oecuménique, Rue Joseph II 174, B-1040 Bruxelles, Belgium. T. (32 2) 230 17 32. Fax 231 14 13.

Hull, John Martin, Prof

Rel: Christian; Unordained; b. 22 APR 1935 in Corryong, Victoria, Australia; m. to Marilyn Elaine Hull Gasson; c. 3 boys and 2 girls; Citizen of Australia; Lang. English.

International Organizations: General Secretary and Founder (since 1978) of the International Seminar on Religious Education and Values.

Other organizations: Editor (since 1971), *British Journal of Religious Education*; Professor of Religious Education (since 1989), University of Birmingham; Dean, Faculty of Education and Continuing Studies (since 1990), University of Birmingham.

Main field of work: Religious education. **Career:** Assistant Teacher, Caulfield Grammar School, Melbourne from 1957 to 1959. Head, Religious Education Dept., Selhurst Grammar School for Boys, London from 1962 to 1966. Lecturer in Divinity, Westhill College, Birmingham from 1966 to 1968. Lecturer in Religious Education, University of Birmingham in 1968. Senior Lecturer in Religious Education, University of Birmingham in 1978. Reader in Religious Education, University of Birmingham in 1986. **Educ.:** B.A. from University of Melbourne, Victoria, Australia (1952-1955); Dip. Ed. from University of Melbourne, Victoria, Australia (1956); B.Ed. from University of Melbourne, Victoria, Australia (1957-1958); M.A. from University of Cambridge, UK (1959-1962); Ph.D. from University of Birmingham, UK (1966-1970). **Hons.:** Fellow, Royal Society of Arts (1991). **Mem.:** United Reformed Church (elder); Church of England (layperson); Amnesty International; Anti-Apartheid Movement; Greenpeace; Friends of the Earth; Campaign for Nuclear Disarmament; Christian Education Movement **Pubs.:** *Studies in Religion and Education*, London, Falmer Press, 1984; *Touching the Rock: An Experience of Blindness*, London, Arrow Books, 1991 (also published in various countries and translations) and *What Prevents Christian Adults from Learning?*, Philadelphia, Trinity Press International, 1991.

Mailing address: School of Education, University of Birmingham, Edgbaston, Birmingham B15 2TT, UK. T. (44 21) 414 4836. Fax 414 4865. Tx 333762 UOBHAM G

Residence: 90 Bournbrook Road, Selly Oak, Birmingham B29 7BS, UK. T. (44 21) 472 4141

Hull, Lilla, Sr

Rel: Roman Catholic; Ordained; celibate.

International Organizations: Member of the Maryknoll Sisters of Saint Dominic.

Other organizations: Member, Association of Contemplative Sisters.

Mailing address: Maryknoll Cloister, Maryknoll NY 10545, USA.

Humbert, John O, Rev

Rel: Disciples of Christ; Ordained.

International Organizations: General Minister and President of the Christian Church (Disciples of Christ) in the United States.

Mailing address: 222 South Downey Ave, PO Box 1986, Indianapolis IN 46206, USA.

Humbert, Marie

Rel: Christian.

International Organizations: Contact of the Présence et vie.

Mailing address: Résidence Lamartine, 12 rue Dublineau, F-53200 Château-Gontier, France. T. (33) 43 07 86 78.

Hume, George Basil, O.S.B., H E Card

Rel: Roman Catholic; Ordained on 23 JUL 1950; b. 02 MAR 1923 in Newcastle-upon-Tyne, UK; celibate; Citizen of United Kingdom.

International Organizations: Member (since 1976) of the College of Cardinals.

Member (since 1942) of the Benedictine Confederation. Member of the Sacred Congregation for the Sacraments and Divine Worship. Member of the Congregation for Religious and Secular Institutes. Member of the Pontifical Council for Promoting Christian Unity. Member of the Pontifical Council for Pastoral Assistance to Health Care Workers.

Other organizations: Archbishop (since 1976), Westminster; President (since 1979), Bishop's Conference of England and Wales.

Career: Senior Master in Modern Languages, Ampleforth College, York from 1952 to 1963. Housemaster, St Bede's from 1955 to 1963. Professor of Dogmatic Theology, Ampleforth Abbey from 1955 to 1963. Magister Scholarum, English Benedictine Congregation from 1957 to 1963. Abbot, Ampleforth from 1963 to 1976. Created Cardinal by Pope Paul VI in 1976. Chairman, Benedictine Ecumenical Commission from 1972 to 1976. Vice-President, Bishops' Conference of England and Wales from 1976 to 1979. Member, Council for Secretariat, International Synod of Bishops from 1978 to 1987. President, Council of European Bishops Conferences from 1979 to 1987. **Educ.:** M.A. Hist. from University of Oxford, UK (1947); S.T.L. from University of Fribourg, Switzerland (1951). **Interests:** angling. **Hons.:** Hon. D.D., University of Newcastle-upon-Tyne (1979), University of Cambridge (1979), Manhattan College (1980), Catholic University of America (1980), University of London (1980), University of Oxford (1981), University of York (1982), University of Kent (1983), University of Durham (1987), Benedictine International Athenaeum St Anselm (1988), University of Hull (1989), University of Keele (1990); Named Hon. Bencher of Inner Temple (1976), Hon. Freeman of Newcastle-upon-Tyne (1980) and London (1981). **Pubs.:** In Praise of Benedict, 1981; To Be a Pilgrim, 1984 and Towards a Civilisation of Love, 1988.
Mailing address: Archbishop's House, Westminster, London SW1P 1QJ, UK. T. (44 71) 828 6548. Fax 828 7397

Hummel, Nikolaus, Bishop

Rel: Old Catholic.
International Organizations: Bishop of Old Catholic Church in Austria.
Mailing address: Schottenring 17, A-1010 Wien, Austria.

Hun, Khoo Leong

Rel: Buddhist.
International Organizations: Vice-President of the World Fellowship of Buddhists.
Mailing address: WFB Penang Regional Centre, 51-C Loboh Pantai, 10300 Penang, Malaysia.

Hunke, Heinz

Rel: Christian.
International Organizations: Secretary-General of the International Documentation and Communication Centre.
Mailing address: Via S Maria dell'Anima 30, I-00186 Roma, Italy. T. (39 6) 686 8332. Fax 6832 76671. Tx 0402-6105943 doc lu. E-mail: IDOC@Geomail.GEO2.org.

Hunking, Willis

Rel: Protestant.
International Organizations: President of the Missionary Women International.
Mailing address: 3901 S Wayne Ave, Fort Wayne IN 46807, USA. T. (1 219) 744 1291.

Hunt, Mary Elizabeth, Dr

Rel: Roman Catholic; b. 01 JUN 1951 in Syracuse NY, USA; single; Citizen of USA; Lang. English and Spanish.
International Organizations: Co-founder and Co-director (since 1983) of the Women's Alliance for Theology, Ethics and Ritual (WATER).
Other organizations: Board of Directors, Women's International Network for Development and Democracy in El Salvador (WINDS), Upper Room AIDS Ministry (honorary board), Institute for the Study of Christianity and Sexuality (London, patron of the Institute), Catholics for a Free Choice, National Association for Science, Technology and Society; Editorial Boards, *Journal of Feminist Studies in Religion*, *Continuum*; Women's Advisory Committee, *Concilium*.
Career: Library Assistant, Graduate Theological Union Library, Berkeley CA from 1978 to 1980. Frontier Internship in Mission, Buenos Aires, Argentina: Visiting Professor of Theology, ISEDET (ecumenical seminary); co-director, Women's Project, Centro de Estudios Cristianos; Collaborator, Servicio Paz y Justicia from 1980 to 1982. Visiting Assistant Professor of Religion, Colgate University, Hamilton NY from 1986 to 1987. **Educ.:** B.A. (magna cum laude) from Marquette University, Milwaukee WI, USA (1969-1972); M.T.S. from Harvard University, Divinity School, Cambridge MA, USA (1972-1974); M.Div. from Jesuit School of Theology at Berkeley, CA, USA (1977-1979); Ph.D. from Graduate Theological Union, Berkeley CA, USA (1974-1980). **Hons.:** Ms. Magazine 'Eighty Women to Watch in the Eighties'; Isaac Hecker Award, Paulist Center, Boston; Women's Ordination Conference Prophetic Figure Award; Crossroad Women's Studies Prize (1990). **Mem.:** American Academy of Religion; Phi Sigma Tau; Alpha Sigma Nu **Pubs.:** *Fierce Tenderness: A Feminist Theology of Friendship*, New York, Crossroad, 1990; Editor, *From Woman-Pain to Woman-Vision: Writings in Feminist Theology*, by Anne McGrew Bennett, Minneapolis, Fortress Press, 1989 and "Ethics on Ice: Soul-Chilling Dilemmas in New Reproductive Technologies", in *Conscience*, Vol. X, No 5, September-October 1989, pp. 1-6, 23-24.
Mailing address: 8035 13th St, Silver Spring MD 20910, USA. T. (1 301) 589 2509 – 589 3150
Residence: 828 Richmond Ave, Silver Spring MD 20910, USA. T. (1 301) 589 1292

Hurley, Denis Eugene, O.M.I., Rev

Rel: Roman Catholic; Ordained on 09 JUL 1939; b. 09 NOV 1915 in Cape Town, South Africa; celibate; Citizen of South Africa.
International Organizations: Chairman (since 1975) of the International Commission on English in the Liturgy.
Member of the Missionary Oblates of Mary Immaculate.
Career: Curate, Emmanuel Cathedral, Durban from 1940 to 1943. Superior, St Joseph's Scholasticate Prestbury (now Cedera) from 1944 to 1946. Vicar-Apostolic, Natal with rank of Bishop in 1947. Promoted Archbishop, Durban in 1951. President, South African Catholic Bishops' Conference from 1952 to 1960. President, South African Catholic Bishops' Conference from 1981 to 1987. Attended Second Vatican Council from 1962 to 1965. President, South African Institute of Race Relations from 1965 to 1966. Founded ecumenical agency for social concern known as Diakonia in Durban in 1976. Charged with offence against South African law for making 'false statement' concerning atrocities by para-military police unit in Namibia in 1984. Acquitted in 1985. **Interests:** reading, cricket, rugby. **Hons.:** Hon. Dr.Iur., Notre Dame (1978), De Paul University (1986); Hon. D.Litt., Natal (1978); Hon. D.Hum.Litt., University of America (1982), Georgetown University (1987); Hon. D.S.T., University of Santa Clara (1986); Hon. Dr., University of Louvain (1988); Hon. D.Sc., University of Cape Town (1988); civic honours, City of Durban (1972); Knight, Legion of Honour (1975).
Mailing address: Archbishop's House, 154 Gordon Road, PO Box 47489, Greyville 40023, South Africa. T. (27 31) 3031417

Hurley, Michael Anthony, S.J., Rev

Rel: Roman Catholic; Ordained on 15 AUG 1954; b. 10 MAY 1923 in Ardmore, Co. Waterford, Ireland; celibate; Citizen of Ireland; Lang. English, French and Italian.
International Organizations: Ecumenical theologian and Ecclesiastic of the Roman Catholic Church.
Member (since 1940) of the Society of Jesus.
Other organizations: Member (since 1983), Columbanus Community of Reconciliation; Lecturer (since 1984), Irish School of Ecumenics, Dublin.
Main field of work: Ecumenism. **Career:** Lecturer in Systematic Theology, Milltown Park Jesuit Faculties, Dublin from 1958 to 1970. Director, Irish School of Ecumenics, Dublin from 1970 to 1980. Leader, Columbanus Community of Reconciliation from 1983 to 1991. Member, Methodist/Roman Catholic International Commission from 1968 to 1975. **Educ.:** B.A. from National University of Ireland, University College, Dublin, Ireland (1942-1945); Lic. Theol. from Facultés SJ St Albert, Eegenhoven-Louvain, Belgium (1951-1955); D.D. from Pontifical Gregorian University, Vatican City State (1956-1958). **Interests:** ecumenical spirituality. **Mem.:** Irish Theological Association, etc **Pubs.:** *John Wesley's Letter to a Roman Catholic*, London, 1968; Editor, *Irish Anglicanism 1869-1969*, Dublin, 1970 and Editor, *Beyond Tolerance: The Challenge of Mixed Marriage*, London, 1975.
Mailing address: 683 Antrim Road, Belfast BT15 4EG, N Ireland, UK. T. (44) 778009
Residence: See Mailing Address. T. See Mailing Address

Hürlimann, Marianne

Rel: Spiritualist.
International Organizations: President, Geneva Office of the Lucis Trust.
Mailing address: Case Postale 31, 1 rue Varembé (3e), CH-1211 Genève 20, Switzerland. T. (41 22) 734 12 52. Fax 740 09 11.

Husar, Lubomyr, M.S.U., Most Rev.
Rel: Roman Catholic; Ordained; celibate.
International Organizations: Superior General of the Monaci della Regola Studita.
Mailing address: Monastero Studion, Via dei Laghi 25 (km 9), I-00047 Roma, Italy. T. (39 6) 945 89 85

Huston, Ralph, Ms
Rel: Christian.
International Organizations: Executive Secretary of the World-Wide Prayer and Missionary Union.
Mailing address: 6821 North Ottawa Ave, Chicago IL 60631, USA. T. (312) 763 2553.

Huston, Robert W, Rev Dr
Rel: Methodist; Ordained.
International Organizations: General Secretary of the United Methodist Church.
Mailing address: 475 Riverside Drive, Room 1300, New York NY 10115, USA.

Hutagalung, S M, Rev Dr
Rel: Protestant; Ordained.
International Organizations: General Secretary of the Christian Protestant Church in Indonesia.
Mailing address: Kantor Pusat/Head Office, Jalan Kapt. MH Sitorus 13, Pematangsiantar SU, Indonesia.

Huzzey, Jim, Pastor
Rel: Seventh-Day Adventist.
International Organizations: Church Ministries Director, Trans-European Division of the General Conference of Seventh-Day Adventists.
Mailing address: 119 St Peter's Street, St Albans AL1 3EY, UK.

Hwang, John, Rev Dr
Rel: Christian.
International Organizations: President of the Asian Christian Association.
Mailing address: PO Box 210, Taipei, Taiwan (Rep. of China). T. (886 2) 391 75 55.

Hynes, Fabian, O.H., Br
Rel: Roman Catholic; celibate.
International Organizations: Procurator-General of the Hospitaller Order of St John of God.
Other organizations: Managing Director of Pharmaceutics, Directorate of Health Services, Vatican City Government.
Mailing address: Fate Bene Fratelli, Via della Nocetta 263, I-00164 Roma, Italy. T. (39 6) 663 8937

Iakovos, I, Archbishop
Also know as Demetrios Coucouzis; Rel: Greek Orthodox; Ordained in 1940; b. 29 JUL 1911 in Island of Imbros, Turkey; Citizen of USA.
International Organizations: Archbishop (since 1959) of North and South America.
Other organizations: Founder and Chairman, the Standing Conference of Canonical Orthodox Bishops in the Americas; President, Holy Cross Orthodox Theological School, MA.
Career: Ordained deacon in 1934. Invited to the USA in 1939. Served on the faculty, Holy Cross Greek Orthodox School of Theology, Pomfret CT in 1939. Dean, Cathedral of the Annunciation, Boston MA from 1942 to 1954. Elected titular bishop, Malta in 1954. Served as personal representative, the ecumenical patriarch to the World Council of Churches, Geneva from 1955 to 1959. Co-President, World Council of Churches from 1959 ,to 1968. Former Dean, Holy Cross Orthodox Theological School, MA. **Educ.:** Grad. from Patriarchal Theological School of Halki (1934); S.T.M. from Harvard University, Cambridge MA, USA (1945). **Hons.:** awarded numerous honorary doctoral degrees and other awards in the US and abroad, including: Presidential Medal of Freedom (1980), Inaugural Award, New York University (1981), Great Cross of the Holy Sepulchre, Patriarchate of Jerusalem (1982), Gold Medal, Academy of Athens (1985). **Pubs.:** Author of books in Greek, French, and English.
Mailing address: Greek Orthodox Archdiocese, 8-10 East 79th Street, New York NY 10021, USA.

Ibiam, Francis Akanu, Sir
Rel: Presbyterian; b. 29 NOV 1906; Citizen of Nigeria.
International Organizations: Vice-President (since 1972) of the United Bible Societies.
Career: Built Abiriba Hospital from 1936 to 1945. Medical Superintendent, CSM Hospital, Itu from 1945 to 1948. Medical Superintendent, CSM Hospital, Uburu from 1952 to 1957. Member, Board of Governors, Hope Waddell Training Institute from 1945 to 1957. Principal, Hope Waddell Training Institute from 1957 to 1960. Member, Administrative Committee, International Missionary Council from 1957 to 1961. Member, Legislative Council, Nigeria from 1947 to 1952. Member, Executive Council, Nigeria from 1949 to 1952. Privy Council, East Region of Nigeria from 1954 to 1959. Governor, East Nigeria from 1960 to 1966. Adviser to Military Governor, Eastern Provinces, Nigeria in 1966. President, Christian Council of Nigeria from 1955 to 1958. Chairman, Provisional Committee, All-Africa Church Council from 1958 to 1962. Chairman, Council, University Colleges, Ibadan from 1958 to 1961. One of six Presidents, World Council of Churches in 1961. Chairman, Council, United Bible Socities from 1966 to 1972. Founder and President, Bible Society of Nigeria from 1963 to 1974. Patron, Bible Society of Nigeria in 1974. Chairman, Imo State Council of Traditional Rulers. Elder, Presbyterian Church. **Hons.:** Appointed Officer of the Order of the British Empire (1949), Knight Commander of the Order of the British Empire (1951), Knight Commander of the Order of St Michael and St George (1962) and renounced these honours in 1967 in protest against British policy regarding Biafra; Hon. D.Sc., University of Ife (1966); Hon. LL.D., University of Ibadan; Order of Russian Orthodox Church; Grand Cross, Order of the Niger and other decorations.
Mailing address: Ganymede, Unwana, PO Box 240, Afikpo, Imo State, Nigeria.

Icardi, Franco, S.M.A., Fr
Rel: Roman Catholic; Ordained; celibate.
International Organizations: Procurator-General of the Society of African Missions.
Mailing address: Via della Nocetta 111, I-00164 Roma, Italy. T. (39 6) 625 77 39, (39 6) 625 77 35

Idchinkhorlo, Jamyangiin, Prof
Rel: Buddhist.
International Organizations: Vice-President of the World Fellowship of Buddhists.
Mailing address: The Centre of Mongolian Buddhists, Gangdan Tekchingling Monastery, Ulan Bator, Mongolia.

Ieremia, Lale, Rev
Rel: Congregational.
International Organizations: General Secretary of the Congregational Christian Church in Samoa.
Mailing address: PO Box 468, Apia, Samoa.

Iglesias, Francisco, O.F.M. Cap., Fr
Rel: Roman Catholic; Ordained; celibate.
International Organizations: Past Procurator-General of the Order of Friars Minor Capuchin.
Advisor to the Section for Secular Institutes of the Congregation for Religious and Secular Institutes.
Mailing address: Via Piemonte 70, I-00187 Roma, Italy. T. (39 6) 474 06 43

Ignatios, I, IV, H H
Rel: Greek Orthodox.
International Organizations: Patriach of the Antioch and All the East.
Mailing address: PO Box 5376, Beirut, Lebanon.

Ihaji, Peter Angode, Rev
Rel: African Church of the Holy Spirit.
International Organizations: Executive Secretary of the African Church of the Holy Spirit.
Mailing address: PO Box 183, Kakagema, Kenya.

Ihsanoglu, Ekmeleddin, Prof Dr
Rel: Muslim.
International Organizations: Secretary of the International Commission for the Preservation of Islamic Cultural Heritage.
Director-General of the Research Centre for Islamic History, Art and Culture.
Mailing address: Yildiz Sarayi, Seyir Köskü, Barbaros Bulvari, 80700 Besiktas, Istanbul, Turkey. T. (90 1) 160 59 88. Fax 158 43 65. C. islam yildizi. Tx 26486 isam tr.

Ikeda, Daisaku
Rel: Buddhist: Nichiren Shoshu; b. 02 JAN 1928 in Tokyo, Japan; m. to Kaneko Shiraki; c. 2 boys; Citizen of Japan.
International Organizations: President (since 1975) of the Soka Gakkai International.
Other organizations: Honorary President (since 1979), Soka Gakkai; Honorary President (since 1987), Seikyo Press.
Career: President, Soka Gakkai from 1960 to 1979. Founder, Tokyo and Osaka Soka Junior and Senior High Schools, Soka University, Min-on Concert Association, Institute of Oriental Philosophy, Tokyo and Shizuoka Fuji Art Museum (Tokyo), Sapporo Soka Kindergarten (Tokyo), and Osaka Soka Primary Schools. **Interests:** music, art, table tennis, calligraphy. **Hons.:** Dr. h.c., Moscow State University (1975), Sofia (1981), Soka (1983); Honorary Professor, National University of San Marcos (1981), Beijing University (1984), Fudan University (1984); UN Peace Award (1983); Order of the Sun of Peru with Grand Cross (1984); Kenya Oral Literature Award (1986); Chinese Peace and Friendship Trophy (1986). **Pubs.:** Before It Is Too Late, 1984; Buddhism and Cosmos, 1986 and Unlocking the Mysteries of Birth and Death, 1988.
Mailing address: 32 Shinano-machi, Shinjuku-ku, Tokyo 160, Japan. T. (81 3) 3353 7111. Fax 3353 5431. C. SOKAGAKKAI TOKYO. Tx J 33145 SKG.

Ikeda, Ryoshin
Rel: Buddhist; Ordained; b. 20 MAR 1918; Citizen of Japan.
International Organizations: Monk of Tendaishu Chorakuji.
Educ.: Taisho University, Japan.
Mailing address: 938 Ojimamachi Serada, Nitta-gun, Gunma 370-04, Japan. T. (81 276) 52 1035

Ilangyi, Bya'ene Akulu, Bishop
Rel: Free Methodist.
International Organizations: Member of the Free Methodist World Fellowship.
Mailing address: AIM, PO Box 53435, Nairobi, Kenya.

Ilia, I, II, H H and Beatitude
Rel: Orthodox.
International Organizations: Catholicos and Patriarch of All Georgia.
Mailing address: 4 Sioni Street, Tbilisi 380005, Georgia.

Ilktac, Ahmet Cemal
Rel: Muslim; Citizen of Cyprus.
International Organizations: Mufti (acting) of Cyprus.
Mailing address: POB 142, Nicosia, Cyprus.

Imathiu, Lawi, Rt Rev
Rel: Methodist.
International Organizations: Presiding Bishop of Methodist Church in Kenya.
Mailing address: PO Box 47633, Nairobi, Kenya.

Imokhai, Charles A, Rev Fr
Rel: Roman Catholic; Ordained; celibate.
International Organizations: Secretary-General of the Association of Episcopal Conferences of Anglophone West Africa.
Mailing address: PO Box 502, Monrovia, Liberia. T. (231) 262447. C. Catholic Mission.

Inaoka, Kakujun
Rel: Buddhist; Ordained; b. 09 NOV 1905; Citizen of Japan.
International Organizations: Chief of the Kinkai Komyoji Saifukuji.
Main field of work: Monk. **Career:** Chief Manager, Jodo-shu. **Educ.:** Taisho University, Japan.
Mailing address: 3-24-13 Nishi Kawaguchi, Kawaguchi, Saitama 332, Japan. T. (81 482) 51 3675

Ingles, Mary
Rel: New Age.
International Organizations: Member of the Findhorn Foundation.
Mailing address: Findhorn, Forres IV36 OTZ, UK. T. (44 309) 69 03 11. Fax 69 09 33.

Inglessis, Emile, Dr
Rel: Roman Catholic.
International Organizations: Secretary-General of the International Council of Catholic Men.
Mailing address: Palazzo San Calisto 16, I-00120 Città del Vaticano, Vatican. T. (39 6) 698 7382.

Innocenti, Antonio, H E Card
Rel: Roman Catholic; Ordained on 17 JUL 1938; b. 23 AUG 1915 in Poppi, Fiesole, Italy; celibate; Citizen of Italy.
International Organizations: Prefect (since 1986) of the Congregation for the Clergy.
Member of the Congregation for Bishops. Member of the Sacred Congregation for the Sacraments and Divine Worship. Member of the Congregation for the Causes of Saints. Member of the Congregation for the Evangelization of Peoples. Member of the Congregation for Catholic Education. Member of the Pontifical Council for the Interpretation of Legislative Texts.
Other organizations: President (since 1988), Pontifical Commission for the Conservation of the Church's Artistic and Historic Heritage; Member (since 1985), College of Cardinals; Member, Congregation for the Doctrine of the Faith; Member, Pontifical Commission for the Vatican City State.
Career: Consecrated Bishop, Titular See, Aeclanum in 1968. then Archbishop. Former Apostolic Nuncio in Spain. Created Cardinal by Pope John Paul II in 1985.
Mailing address: Piazza della Città Leonina 9, I-00193 Roma, Italy. T. (39 6) 689 6789

Inverardi, Giuseppe, I.M.C., Fr
Rel: Roman Catholic; Ordained; celibate.
International Organizations: Superior General of the Institute of the Consolata for Foreign Missionaries.
Mailing address: Viale delle Mura Aurelie 11-13, I-00165 Roma, Italy. T. (39 6) 326 67 26.

Inverardi, Giuseppe
Rel: Roman Catholic.
International Organizations: Superior General of the Institute of the Consolata for Foreign Missionaries.
Mailing address: Viale delle Mura Aurelie 11-13, I-00165 Roma, Italy. T. (39 6) 63 84 241.

Iotama, Seve, Rev
Rel: Christian.
International Organizations: President of the Church of Tuvalu.
Mailing address: Funafuti, Tuvalu.

Irenna, Y
Rel: Lutheran.
International Organizations: Representative of the Institute for Church Leadership Development in Africa.
Mailing address: Lutheran Theological Education Programme in Africa, PO Box 846, Harare, Zimbabwe.

Irvine, Donald Fredrick, Rev
Rel: Anglican; Ordained in 1963; b. 31 JUL 1934 in Vancouver, British Columbia, Canada; m. to Joyce Reid; c. 2 boys and 1 girl; Citizen of Canada.
International Organizations: Lecturer (since 1968) of the Huron College, London.
Career: Ordained Deacon in 1962. Associate Priest, St Aidan's Church, Windsor, Ontario from 1962 to 1965. Associate Priest, Church of the Holy Saviour, Waterloo, Ontario from 1965 to 1968. Member of Board and Corp., London Pastoral Institute from 1982 to 1990. Professor of Philosophy, of Religion and Ethics and Member of Corp., Huron College from 1976 to 1990. Former Academic Dean of Theology, Huron College. **Educ.:** B.A. from University of Western Ontario, Huron

College, London, Canada (1960); B.Th. from University of Western Ontario, London, Canada (1962); M.A. from University of Waterloo, Ontario, Canada (1966); Ph.D. from University of Waterloo, Ontario, Canada (1978). **Interests:** calligraphy. **Mem.:** Hastings Centre (Associate Member); Alban Institute; Canadian Philosophical Association; Association Case Teaching (Director and Fellow); Society of Christian Philosophers
Mailing address: London, Ontario N6G 1H3, Canada.
Residence: 856 Berkshire Dr, London, Ontario N6J 3S7, Canada.

Irvine, Graeme
Rel: Protestant.
International Organizations: President of the World Vision International.
Mailing address: 919 West Huntington Drive, Monrovia CA 91016, USA. T. (1 818) 303 8811. C. WORVIS. Tx 275 335 WORV UR.

Irving, Donald R, Canon
Rel: Church of England.
International Organizations: Secretary-General of the Intercontinental Church Society.
Mailing address: 175 Tower Bridge Road, London SE1 2AQ, UK. T. (44 71) 407 4588. C. Chaplaincy, London SE1 2AQ.

Iryani, Qadi Abdul Rahman, Sheikh
Rel: Muslim; b. 18 JUL 1917; Citizen of Yemen.
International Organizations: Leader of the Zaidi Community of Shi'a Muslims.
Career: Took part in abortive revolution against Imam in 1948. Imprisoned from 1948 to 1954. Took part in uprising in 1955. Member, Revolutionary Council, Yemen Arab Republic from 1962 to 1974. Minister, Justice from 1962 to 1963. Vice-President, Executive Council from 1963 to 1964. Member, Political Bureau from 1964 to 1974. Chairman, Peace Committee set up after Khamer Peace Talks in 1965. Head, State from 1967 to 1969. Chairman, Presidential Council (deposed by military coup) from 1969 to 1974. In exile from 1974 to 1981.
Mailing address: c/o Military Command Council, Sana'a, Yemen.

Isaac, Sh, Rev
Rel: Evangelical.
International Organizations: Moderator of the Synod of the Evangelical Church of Iran.
Mailing address: PO Box 1505, Teheran, Iran Islamic Rep.

Isaji, Kanetsugu
Rel: Shinto; Ordained; b. 04 JUN 1941; Citizen of Japan.
International Organizations: Chief Priest of the Kasuga Shrine.
Career: Seki City Assemblyman. **Educ.:** Kokugakuin University, Tokyo, Japan.
Mailing address: 684-1 Imojiya, Seki, Gifu 501-32, Japan. T. (81 575) 22 3485

Islamoglu, Faruk
Rel: Muslim.
International Organizations: Secretary of the Fondation religieuse islamique turque en Belgique.
Mailing address: Bd Général Jacques 38, B-1050 Bruxelles, Belgium. T. (32 2) 647 38 98. Fax 646 51 27.

Isokai, Yoichi
Rel: Shinto; Ordained; b. 09 FEB 1932; Citizen of Japan.
International Organizations: Chief Priest of the Shibahiko Shrine.
Career: Shiokago Shrine. **Educ.:** Kokugakuin University, Tokyo, Japan.
Mailing address: 1-2 Ichimoriyama, Shiogama, Miyagi 985, Japan. T. (81 22) 362 0621

Israel, Gérard
Rel: Jewish; b. 24 NOV 1928 in Oran, Algeria; m. to Danielle Rondolat; c. 1 boy; Citizen of France.
International Organizations: Adviser to the President of the Consultative Council of Jewish Organizations.
Main field of work: Writing. **Career:** Served with French Army during Algerian campaign from 1955 to 1958. Assistant Secretary-General, Alliance Israélite Universelle (AIU) from 1958 to 1960. Writer for AIU from 1960 to 1980. Director, *Les Nouveaux Cahiers*, AIU. Member, European Parliament from 1980 to 1984. **Educ.:** Lic. ès Lettres, D.E.S. Phil. from University of Paris IV (Paris-Sorbonne), France (1947-1954). **Interests:** politics, Jewish history, European construction. **Hons.:** Knight of the Legion of Honour (1978).
Mailing address: 45 rue La Bruyère, F-75009 Paris, France. T. (33 1) 42 80 35 00. C. ALLISRUN.

Israel, Jillapegu, Commr
Rel: Salvation Army, The.
International Organizations: Territorial Commander of The Salvation Army.
Mailing address: Flat No 103 – First Floor D-1 (K-84), Ashirwad Complex – Green Park, New Delhi 110 016, India.

Issayi, Youhannan Semaan, Most Rev Archbishop
Rel: Chaldean Catholic; Ordained on 03 MAR 1940; b. 27 JUN 1914; celibate; Citizen of Iran.
International Organizations: Archbishop of Teheran.
Other organizations: President, Bishops' Conference of Iran; Member, Patriachal Synod, Assembly, and Episcopal Conference of Oriental Rites.
Mailing address: Archevêché, Forsat Ave 91, Teheran, Iran Islamic Rep. T. (98 21) 823549

Itçaïna, Jean-Baptiste, M.E.P., Fr
Rel: Roman Catholic; Ordained; celibate.
International Organizations: Procurator-General of the Paris Foreign Mission Society.
Mailing address: Via Adelaide Ristori 26, I-00197 Roma, Italy. T. (39 6) 808 46 45.

Itozu, Kiminari
Rel: Shinto; Ordained; b. 04 NOV 1937; Citizen of Japan.
International Organizations: Chief Priest of the Usa Shrine.
Career: Councilor, Association of Shinto Shrine. **Educ.:** Kokugakuin University, Tokyo, Japan.
Mailing address: 446 Minami Usa, Usa, Oita 872-01, Japan. T. (81 978) 37 0001

Iustin, I, Patriach
Rel: Orthodox: Romanian.
International Organizations: Patriarch of Romanian Orthodox Church.
Mailing address: The Holy Synod Office, Str Antim no 29, 70666 Bucuresti V, Romania.

Ivory, Thomas, Rev
Rel: Roman Catholic.
International Organizations: Representative of the Order of Our Lady of Mercy.
Mailing address: Ave des Croix du Feu 22, B-1410 Waterloo, Belgium.

Iwas, Ignatius Zakka I, H H
Also know as Sanharib Bashir Iwas; Rel: Syrian Orthodox; Ordained in 1957; b. 21 APR 1933 in Mosul, Iraq; celibate; Citizen of Iraq; Lang. Syriac, Arabic and English.
International Organizations: Patriarch (since 1980) of Antioch and All the East (Supreme Head of Universal Syrian Orthodox Church).
Main field of work: Ministry. **Career:** Consecrated Metropolitan for Archdiocese, Mosul in 1963. Transferred to Archdiocese, Baghdad in 1969. Fellow, Faculty of Syriac Studies, Lutheran School of Theology, Chicago IL in 1981. Former Member, Central Committee, World Council of Churches. **Interests:** journalism, theology. **Hons.:** Hon. D.D., General Theological Seminary, NY (1962). **Mem.:** Iraq Academy of Science; Jordan Arabic Academy (Honorary Member) **Pubs.:** Several books on Church history and theology; Journey books and memoirs and articles in the Church magazine.
Mailing address: Syrian Orthodox Patriarchate, Bab Toma, PB 22260, Damascus, Syria. T. (963 11) 432401. Tx 411876
Residence: See Mailing Address. T. See Mailing Address

Iwashiro, Shuo

Rel: Buddhist; Ordained; b. 21 NOV 1911; Citizen of Japan.
International Organizations: Head Manager of the Shingonshu Godaigo-Sect.
Career: Chief, Hokaiji. **Educ.:** Shuchiin University, Japan.
Mailing address: 19 Hino Nishi Daidocho, Fishimi-ku, Kyoto 601-13, Japan. T. (81 75) 571 0024

Iyer, Rama Subramania, Sri

Rel: Hindu.
International Organizations: Chief Priest of the Sri Manikkapillayar Temple, Hatton.
Mailing address: Sri Manikkapillayar Temple, Hatton, Sri Lanka.

Jackson, David G, Rev

Rel: United Brethren in Christ; Ordained in 1968; b. 05 NOV 1936 in Derby NY, USA; single; Citizen of USA; Lang. English.
International Organizations: Secretary-General (since 1984) of the World's Christian Endeavor Union.
Educ.: B.S. from State University of New York at Buffalo, USA (1960); M.D. from Huntington College, IN, USA (1964).
Mailing address: 1221 East Broad Street, Columbus OH 43216, USA. T. (1 614) 258 9545.
Residence: See Mailing Address. T. See Mailing Address

Jackson, Jesse Louis, Rev

Rel: Baptist; Ordained in 1968; b. 08 OCT 1941 in Greenville SC, USA; m. to Jacqueline Lavinia Brown; c. 3 boys and 2 girls; Citizen of USA.
International Organizations: Founder of the Operation PUSH (People United to Save Humanity).
Main field of work: Political and religious leader. **Career:** Answered Martin Luther King, Jr.'s call to stage a march for black rights in Selma AL in 1965. As a member, King's Atlanta-based Southern Christian Leadership Conference (SCLC), joined in the fight against Chicago mayor Richard Daley's alleged discrimination against blacks in 1966. Became head, Operation Breadbasket in Chicago in 1966. Founded also some satellite organizations to deal with problems of African-American unemployment, poverty and educational opportunity, including the National Association for the Advancement of Colored People (NAACP). Following his ordination, continued working with King's SCLC for more than three years. In the late 1970s, turned his attention to politics. Became the first African-American candidate for the presidency of the US (unsuccessful) in 1984. Again candidate for the presidency of the US (unsuccessful) in 1988. **Educ.:** B.A. from North Carolina Agricultural and Technical State University, Greensboro, USA (1964); Attended (left prior to graduation) from Chicago Theological Seminary, IL, USA (1964). **Hons.:** Presidential Award, National Medical Association (1969); Humanitarian Father of the Year, National Father's Day Committee (1971). **Pubs.:** *Straight From the Heart*, Ed. by Roger D Hatch and Frank E Watkins, Fortress, 1987 and *A Time to Speak: The Autobiography of the Reverend Jesse Jackson*, Simon and Schuster, 1988.
Mailing address: 930 E 50th Street, Chicago IL 60615, USA.

Jacobi, Reinhold, Dr

Rel: Roman Catholic.
International Organizations: Regional Secretary of the International Catholic Organization for Cinema and Audiovisual.
Mailing address: Kaiserstrasse 163, 53113 Bonn, Germany.

Jacobs, Irving, Rabbi Dr

Rel: Jewish; b. 02 AUG 1938 in London, UK; m. to Ann Jacobs; c. 1 boy and 3 girls; Citizen of United Kingdom; Lang. English and Hebrew.
International Organizations: Principal (since 1990) of the Jews' College, London.
Career: Research Fellow, Jews' College, London from 1966 to 1969. Lecturer, Jews' College, London from 1969 to 1975. Senior Lecturer, Jews' College, London from 1975 to 1984. Dean, Jews' College, London from 1984 to 1990. **Educ.:** B.A. from University of London, UK (1960); Ph.D. from University of London, UK (1971); Rabb. Dip. from University of London, UK (1971). **Interests:** early Jewish Bible exegesis. **Pubs.:** "Literary Motifs in the Testament of Job", in *Journal of Jewish Studies*, Vol. 21, pp. 1 ff.; "Elements of Near Eastern Mythology in Rabbinic Aggadah", in *JJS*, Vol. 28, pp. 11 ff. and "Holiness and Kingship in the Third Benediction of the Amidah and the Yozer", in *JJS*, Vol. 141, pp. 62-74.
Mailing address: Jew's College, University of London, 44a Albert Road, London NW4 2SJ, UK. T. (44 81) 203 6427. Fax 203 6420
Residence: 28 Elmstead Ave, Wembley HA9 8NX, UK. T. (44 81) 908 3081

Jacqueline, Bernard René Henri, H E Mgr

Rel: Roman Catholic; Ordained on 12 MAR 1944; b. 13 MAR 1918 in Saint-Lô, Manche, France; celibate; Citizen of Vatican and France; Lang. French, English, Italian and Latin.
International Organizations: Apostolic Pro-Nuncio to Morocco (since 1986) of the Roman Catholic Church.
Other organizations: Titular Archbishop (since 1982), Abbir maggiore.
Main field of work: Diplomacy, research in history. **Career:** Professor of Philosophy at Saint Paul de Cherbourg from 1947 to 1951. Chaplain, Lycée Français, Rome from 1951 to 1961. Officer, Sacred Congregation for the Evangelization of Peoples from 1961 to 1973. Under-Secretary, Secretariat for Non-Believers from 1973 to 1982. Apostolic Pro-Nuncio to Burundi from 1982 to 1985. **Educ.:** Dr. ès Lettres from University of Paris IV (Paris-Sorbonne), France (1944-1946); D.Cn.L. from Institut Catholique de Paris, France (1944-1946). **Interests:** Islam, history of canon law. **Hons.:** Knight of the Legion of Honour; Officer of the National Order of Merit; Chaplain then Honorary Prelate of HH the Pope Paul VI (1972); Lauréat de l'Académie française. **Mem.:** Académie des sciences morales et politiques (corresponding member, 1989) **Pubs.:** La Bataille de Saint-Lô; Episcopat et Papauté chez Saint-Bernard de Clairvaux and Ecrits spirituels de Charles de Foucauld (edition).
Mailing address: Nonciature Apostolique, rue Béni M'tir, BP 1303, Rabat, Morocco. T. (212 7) 77 22 77. Fax 75 62 13
Residence: See Mailing Address. T. See Mailing Address

Ja'fer, Muhammed

Rel: Muslim.
International Organizations: Secretary-General of the World Federation of Islamic Missions.
Mailing address: Islamic Centre Road, B-Block, North Nazimabad, Karachi 74700, Pakistan. T. (92 21) 6644156. C. 'ALEEMIYAH' Karachi.

Jagir Singh, J, Dr

Rel: Sikh; Ordained on 23 MAR 1963; b. 04 OCT 1949 in Baprour, Rajpura, Patiala, Punjab, India; married; c. 1 boy and 1 girl; Citizen of India; Lang. Punjabi, Hindi and English.
International Organizations: Subject Expert (since 1976) of the the Punjab Schools Education Board, Chandigarh Mohali.
Other organizations: Shiromani Ragi (religious singer) (since 1966); Honorary Editor (since 1989), *Amrit Kirtan* (monthly magazine).
Main field of work: Ministry and school education. **Career:** Religious Singer, Shiromani Gurdwara Management from 1966 to 1972. Lecturer in Religion, Sri Guru Gobingh Singh College, Chandigarh from 1974 to 1976. **Educ.:** M.A. Punjabi from Punjab University, Chandigarh, India (1970-1972); M.A. Rel. Studies from Gurmat College, Patiala, India (1972-1974); Ph.D. from Punjabi University, Patiala, India (1980). **Interests:** religious music. **Pubs.:** *Gurmat Mimansa* (anthology or religious articles); *Mere Dadaji* (for children) and article on Sikh music in journal published by Punjab University, Chandigarh and Punjabi University, Patiala.
Mailing address: HN 422, Sector 15-A, Chandigarh 160015, India. T. (91) 24660
Residence: See Mailing Address. T. See Mailing Address

Jago, John, S.M., Fr

Rel: Roman Catholic; Ordained; celibate.
International Organizations: Superior General of the Society of Mary. Advisor of the Congregation for Catholic Education.
Mailing address: Villa S Maria, Via Alessandro Poerio 63, I-00152 Roma, Italy. T. (39 6) 58 99 041.

Jahan, Emmanuel

Rel: Christian.
International Organizations: Chairman of Programme Committee of the International Christian Union of Business Executives.
Mailing address: c/o Centre Nord-Sud de l'Institut de l'Entreprise, rue Clément Marot 6, F-75008 Paris, France.

Jahshan, Moris

Rel: Lutheran.
International Organizations: Director of the Middle East Lutheran Ministry.
Mailing address: PO Box 60-307, Beirut, Lebanon. T. (961 1) 402585. C. LELUMO, Beirut. Tx 40947 MELM LE.

Jain, Sudhanshu Kumar, Dr

Rel: Jain; b. 30 JUN 1926 in Amroha, Uttar Pradesh, India; m. to Satya Mital; c. 3 boys; Citizen of India; Lang. English and Hindi.
International Organizations: Chairman (since 1982) of the IUAES Commission on Ethnobotany.
Other organizations: Vice-President (since 1984), Indian Association for the Study of Tradition in Asian Medicine; Emeritus Scientist (since 1986), National Botanical Research Institute, Lucknow, India; Vice-President (since 1988), Society of Ethnobotanists.
Main field of work: Botanical research. **Career:** Assistant Professor of Botany, Meerut College, Uttar Pradesh, India from 1947 to 1979. Systematic Botanist, Meerut College, Uttar Pradesh, India from 1956 to 1960. Economic Botanist, Meerut College, Uttar Pradesh, India from 1960 to 1971. Deputy Director, Joint Director, Meerut College, Uttar Pradesh, India from 1971 to 1978. Director, Botanical Survey of India from 1978 to 1984. President, Society of Ethnobotanists from 1982 to 1988. **Educ.:** B.Sc. from University of Allahabad, Uttar Pradesh, India (1941-1943); M.Sc. from University of Allahabad, Uttar Pradesh, India (1943-1946); Ph.D. from University of Pune, Maharashtra, India (1958-1963). **Interests:** Ethnobotany, conservation of biological diversity. **Hons.:** several Gold medals and fellowships. **Mem.:** Fellow: Indian National Science Academy, Indian Academy of Sciences and others **Pubs.:** *Medicinal Plants*, 1968 (followed by many other editions and translations); *Glimpses of Indian Ethnobotany*, 1981 (revised edition in print); *An Assessment of Threatened Plants of India*, 1983; *Indian Plant Red Data Book I*, 1984; *A Manual of Ethnobotany*, 1987 and 15 more books and 250 papers.
Mailing address: National Botanical Research Institute, Rana Patrap Marg, Lucknow UP 226 001, India. T. (91 522) 236 431
Residence: A-26, Mall Avenue Colony, Lucknow UP 226001, India T. (91 522) 244 556

Jaki, Stanley L, O.S.B.

Rel: Roman Catholic; Ordained; b. 17 AUG 1924 in Györ, Hungary; Citizen of USA.
International Organizations: Member of the Benedictine Confederation.
Other organizations: Distinguished University Professor of Physics and Philosophy of Science (since 1975), Seton Hall University, South Orange NJ.
Career: Member of staff, Seton Hall University, South Orange NJ from 1965 to 1972. Professor of Physics, Seton Hall University, South Orange NJ from 1972 to 1975. Gifford Lecturer, University of Edinburgh from 1974 to 1976. Fremantle Lecturer, University of Oxford in 1977. Priest, Benedictine Order. **Hons.:** Lecomte du Nouy Prize (1970); Templeton Prize (1987). **Mem.:** Corresponding member, Académie des Sciences de Bordeaux (1985); Honorary member, Pontifical Academy of Sciences (1990) **Pubs.:** Miracles and Physics, 1989; God and the Cosmologists, 1989 and The Only Chaos and Other Essays, 1990.
Mailing address: PO Box 167, Princeton NJ 08542, USA. T. (1 609) 896 3979

Jakobovits, Immanuel, Lord

Rel: Jewish; Ordained in 1941; b. 08 FEB 1921 in Koenigsberg, Germany; m. to Amélie Munk; c. 2 boys and 4 girls; Citizen of United Kingdom; Lang. English and Hebrew.
International Organizations: Emeritus Chief Rabbi of Great Britain. President of the Conference of European Rabbis. President of the Memorial Foundation for Jewish Culture.
Main field of work: Medical ethics. **Career:** Rabbi, Brondesbury Synagogue, London from 1941 to 1944. Rabbi, South East London Synagogue from 1944 to 1947. Rabbi, Great Synagogue, London from 1947 to 1949. Chief Rabbi, Ireland from 1949 to 1958. Rabbi, Fifth Avenue Synagogue, New York from 1958 to 1967. Chief Rabbi, Great Britain from 1967 to 1991. Fellow, University College, London in 1984. **Educ.:** B.A. (Hons), Rabb. Dip. from Jews' College, London, UK; Ph.D. from University of London, UK. **Hons.:** Baron (Life Peer), created 1987, of Regents Park in Greater London; Hon. Doc., Yeshiva University, New York NY (1977); Hon. D.D., Lambeth (1987); Hon. D.Litt., City University, London (1986); Hon. Fellow, Queen Mary College (1987). **Pubs.:** Jewish Medical Ethics, 1959; The Timely and the Timeless, 1977 and If Only My People .. Zionism is My Life, 1984 (autobiography).
Mailing address: 44A Albert Road, London NW4 2SJ, UK. T. (44 81) 203 8667

Jamnicky, John Andrew, Rev

Rel: Roman Catholic; Ordained on 10 MAY 1972; b. 27 OCT 1945 in Chicago IL, USA; celibate; Citizen of USA; Lang. English and German.
International Organizations: President (since 1984) of the International Association of Civil Aviation Chaplains.
Other organizations: Chaplain (since 1982), Chicago O'Hare International Airport; President (since 1986), National Catholic Conference of Airport Chaplains; Founder/Chaplain (since 1989), Second Monday Club of O'Hare; Certified Volunteer Deputy Registrar (since 1990); Provider, an annual VITA program at O'Hare (Volunteer Income Tax Assistance Program).
Main field of work: Aviation. **Career:** Has been involved in clinical pastoral education programs at San Quentin Prison, CA and Cook County Jail, Chicago IL. Chaplain, Rush Presbyterian-St Luke's Medical Center, Chicago IL (Dept. of Psychiatry) and North Chicago Veterans Administration Hospital. Certified Field Education Supervisor for St Mary of the Lake Pontifical Seminary, Catholic Theological Union in Chicago, and Garret Evangelical United Methodist Seminary, Evanston IL. Pastor, St Martin Church and School in Chicago (for 10 years). Initiator, the Mayor's Summer Youth Program (providing jobs for poverty level adolescents at O'Hare Airport). Chaplain to the USAF (US Air Force) with the rank of Captain from 1984 to 1989. **Educ.:** M.Div. from St Mary of the Lake Pontifical Seminary, Mundelein IL, USA (1965-1971). **Pubs.:** Published articles in *American Airlines* (magazine), *American Way*, and in *On the Move* (publication of the Pontifical Commission).
Mailing address: Chicago O'Hare International Airport, PO Box 66353, Chicago IL 60666, USA. T. (1 312) 686 2636.
Residence: See Mailing Address. T. See Mailing Address

Janda, Clement, Rev

Rel: Christian.
International Organizations: Director of the Mindolo Ecumenical Foundation.
Mailing address: Box 21493, Kitwe, Zambia. T. (260 2) 214572 – 211488 – 215198 – 211269. C. 'Mincen'. Tx ZA 52050.

Janki, Brahma Kumari

Rel: Raja Yoga; Ordained in 1938; b. in Sind, Pakistan; single; Citizen of India; Lang. Hindi and Sindhi.
International Organizations: Additional Administrative Head (since 1983) of the Brahma Kumaris World Spiritual University.
Other organizations: Student (since 1937), Brahma Kumaris World Spiritual University (BKWSU); Teacher (since 1938), BKWSU; Teacher/Trainer (since 1969), BKWSU; Director, Overseas Services (since 1974), BKWSU.
Main field of work: Coordinating centres in 55 countries. **Career:** Established major branches, Brahma Kumaris World Spiritual University (BKWSU) in Bombay, Delhi, the Punjab, Maharastra and Gujrat from 1954 to 1969. Head of Maharastra Zone, BKWSU from 1969 to 1974. Moved to London and established the first, the University's Teaching Centres abroad from 1974 to 1978. Traveled to about 80 countries. Addressed several major conferences and events and met leaders from the world of politics, religion, science and many others. **Interests:** human development, world peace.
Mailing address: Global Co-operation House, 65 Pound Lane, London NW10 2HH, UK. T. (44 81) 459 1400
Residence: 98 Tennyson Road, London NW6 7SB, UK. T. See Mailing Address

Janne, Françoise

Rel: Unknown.
International Organizations: Representative of the Institut international supérieur de sciences et recherches religieuses.
Mailing address: Rue des Augustins 19, B-4000 Liège, Belgium.

Jaswinder Singh, J, Gyani

Rel: Sikh; Ordained in 1974; b. 10 APR 1958 in Banoor, Dist Patiala, Punjab, India; m. to Daljit Kaur; c. 2 boys and 3 girls; Citizen of India; Lang. Punjabi.
International Organizations: Head Granthi (since 1989) of the Gurudwara Fatehgarh Sahib.
Main field of work: Religious studies. **Interests:** social welfare, preaching. **Hons.:** Gurbani Vichar All India Roadio. **Mem.:** Station-Jullundur
Pubs.: Article on history of Fatehgarh Sahib published in *Punjabi Tribune*, 28th December 1991.
Mailing address: Gurudwara Fatehgarh Sahib, Fatehgarh Sahib, 140406, India.
Residence: Baba Jorawar Singh Fateh Singh Nagar, Sirhind City 140406, India.

Javierre Ortas, Antonio María J, S.D.B., H E Card

Rel: Roman Catholic; Ordained on 24 APR 1949; b. 21 FEB 1921 in Siétamo, Huesca; celibate.
International Organizations: Archivist and Librarian (since 1988) of the Holy Roman Church.
Member of the Society of St Francis of Sales. Member of the Congregation for Catholic Education. Member of the Pontifical Council for Promoting Christian Unity.
Other organizations: Member (since 1988), College of Cardinals; Member, Congregation for the Doctrine of the Faith.
Career: Elected in 1976, Consecrated bishop in 1976, then Archbishop, Meta. Created Cardinal by Pope John Paul II in 1988.
Mailing address: Via Rusticucci 13, I-00193 Roma, Italy. T. (39 6) 698 39 14

Jayadeva Yogendra, Hansa J, Smt

Rel: Hindu; b. 10 OCT 1947 in Bombay, India; m. to Dr Jayadeva Yogendra; c. 1 boy; Citizen of India; Lang. English, Hindi, Gujarati and Marathi.
International Organizations: Dean of Yoga Education (since 1973) of The Yoga Institute, Bombay.
Career: Commissioned by the Government of India to help prepare Yoga syllabus for schools. **Educ.:** B.Sc.; LL.B. from Bombay University, India. **Pubs.:** Growing with Yoga – Yoga Daily Planner Heart Care; Asthma, Bronchitis and Yoga and Pregnancy, Parenthood and Yoga.
Mailing address: The Yoga Institute, Prabhat Colony, Road No 1, Santa Cruz East, Bombay 400 055, India. T. (91 22) 6122185

Jayanti, Brahma Kumari

Rel: Raja Yoga; Ordained in 1968; b. 16 MAY 1949 in Poona, India; single; Citizen of United Kingdom; Lang. Hindi and English.
International Organizations: London and European Director (since 1974) of the Brahma Kumaris World Spiritual University.
NGO representative to the United Nations in Geneva (since 1980) of the Brahma Kumaris World Spiritual University. International Coordinator (since 1988) of the Global Cooperation for a Better World. International Committee (since 1989) of the World Congress of Faiths. Executive Committee (since 1991) of the World Congress of Faiths.
Other organizations: Student and teacher (since 1968), Brahma Kumaris World Spiritual University (BKWSU); Advisor to the Council for a Parliament of the World's Religions, Chicago IL (since 1990).
Main field of work: Spiritual development. **Career:** Member, Executive Committee, World Spiritual Council from 1972 to 1980. Traveled extensively in approximately 70 countries, including Europe, Australasia, the middle and far East, Africa, North, Central and South America and the Caribbean. **Interests:** creating a better world.
Mailing address: Global Co-operation House, 65 Pound Lane, London NW10 2HH, UK. T. (44 81) 459 1400
Residence: 98 Tennyson Road, London NW6 7SB, UK. T. (44 71) 328 2478. FAX 372 1337. TX 9312102333 GC G. E MAIL 87 CQQ091.

Jayasuriya, Gamini

Rel: Buddhist; Citizen of Sri Lanka.
International Organizations: Member of the Maha Bodi Society.
Main field of work: Monk.
Mailing address: Maha Bodi Society, Maligakande, Colombo, Sri Lanka.

Jayawardene, N U

Rel: Buddhist.
International Organizations: Chairman of the Seventh Committee (Socio-Economi Development) of the World Fellowship of Buddhists.
Mailing address: Sri Lanka Regional Centre of WFB, 380 Bauddhaloka Mawatha, Colombo 7, Sri Lanka.

Jemison, Theodore Judson, Rev Dr

Rel: Baptist.
International Organizations: President of the National Baptist Convention, USA, Inc.
Mailing address: Mt Zion First Baptist Church, 915 Spain Street, Baton Rouge LA 70802, USA.

Jenkins, David Edward, Most Rev

Rel: Church of England; Ordained in 1954; b. 26 JAN 1925 in London, UK; m. to Stella M Peet; c. 2 boys and 2 girls; Citizen of United Kingdom.
International Organizations: Bishop (since 1984) of Durham.
Other organizations: Director (since 1979), William Temple Foundation, Manchester.
Career: Captain, Royal Artillery from 1943 to 1947. Succentor, Birmingham Cathedral from 1953 to 1954. Fellow and Chaplain/Praelector in Theology, Queen's College, Oxford from 1954 to 1969. Director, Humanum Studies, World Council of Churches, Geneva from 1969 to 1973. Professor of Theology, Leeds University from 1979 to 1984. Trustee, Press International, Philadelphia in 1989. Director, SCM Press. **Interests:** music (opera and church music), walking, birdwatching, nature and conservation, travel, books. **Hons.:** Hon. D.D., Durham, Aberdeen, Trinity College, Toronto. **Pubs.:** Contributor to Man, Fallen and Free, 1969; God, Jesus and Life in the Spirit, 1988 and Still Living With Questions, 1990.
Mailing address: Auckland Castle, Bishop Auckland, Co. Durham DL14 7NR, UK.

Jenkins, Keith Pellow

Rel: Methodist; b. 28 DEC 1941 in Kingsbridge, UK; m. to Murphy Fran; c. 2 boys; Citizen of United Kingdom; Lang. English and French.
International Organizations: Secretary-General (since 1990) of the European Ecumenical Commission for Church and Society.
Career: Lawyer, various British local authorities from 1971 to 1981. Executive Secretary, Community and Race Relations Unit, British Council of Churches from 1981 to 1986. Assistant Secretary-General, British Council of Churches from 1986 to 1990. **Educ.:** M.A. from University of Oxford, Jesus College, UK (1960-1963). **Interests:** theology, politics, economics. **Pubs.:** The Closed Door – A Christian Critique of British Immigration Policy, 1985; An Open House? – The European Community – Looking to 1992, 1990 and essay on the situation of immigrants' children in Great Britain in *Wanderarbeiter in der EG*, 1985.
Mailing address: Ecumenical Centre, Rue Joseph II 174, B-1040 Bruxelles, Belgium. T. (32 2) 230 17 32. Fax 231 14 13.
Residence: Drève de Nivelles 174, Bte 17, B-1150 Bruxelles, Belgium. T. (32 2) 660 09 58

Jenkinson, Willie, Fr

Rel: Roman Catholic; Ordained; celibate.
International Organizations: Executive Secretary of the Documentation and Research Centre.
Mailing address: Via dei Verbiti 1, PO Box-CP 5080, I-00100 Roma, Italy. T. (39 6) 574 13 50.

Jentzsch, Heber C, Rev

Rel: Church of Scientology.
International Organizations: President of the Church of Scientology International.
Mailing address: 6331 Hollywood Boulevard, Suite 1200, Los Angeles CA 90028-6329, USA. T. (1 213) 960 3500. Fax 960 3509.

Jenzer, Carlo, Dr

Rel: Old Catholic.
International Organizations: President of the Old Catholic Church of Switzerland.
Mailing address: St Nilausstrasse 61, CH-4500 Solothurn, Switzerland.

Jep, Jalan Let
Rel: Muslim.
International Organizations: Contact of the International Islamic Organization.
Mailing address: S Parman No 66, Slipi Raya, Jakarta, Indonesia. T. (62 21) 592480 – (62 21) 543686.

Jha, Jogendra, Dr
Rel: Hindu.
International Organizations: Secretary-General of the World Hindu Federation.
Mailing address: Pashupati Kshetra, Kathmandu, Nepal. T. (977 4) 71722

Jha, Pitambar, Dr
Rel: Hindu: Sanatan; b. 25 DEC 1942 in Madhubani, Bihar, India; m. to Sarojini; c. 1 boy and 2 girls; Citizen of India; Lang. Hindi, Sanskrit and English.
International Organizations: Assistant Director of Research, Philosophy-Literary Research Dept. (since 1984) of the Kaivalyadhama Yoga Research Institute, Lonavla.
Main field of work: Research in Yoga. **Career:** Research Assistant, Kaivalyadhama Yoga Research Institute from 1970 to 1976. Research staff member, Kaivalyadhama Yoga Research Institute from 1976 to 1978. Research Officer, Kaivalyadhama Yoga Research Institute from 1978 to 1984. **Educ.:** M.A. from Bihar University, Muzaffarpur, India (1963-1965); Ph.D. from KSD Sanskrit University, Darbhanga, India (1966-1970). **Interests:** Sanskrit, philosophy, yoga, religion. **Hons.:** Yoga Martanda. **Mem.:** Indian Council for Research in Yoga, Bhopal **Pubs.:** *Yoga Kosa*; *Hathapradipika* and *Yoga Parichaya.*
Mailing address: Kaivalyadhama, Lonavla 410 403, India.

Jiménez, Jorge, C.I.M., Fr
Rel: Roman Catholic; Ordained; celibate.
International Organizations: Secretary of the Latin American Confederation of Religious Orders.
Member of the Congregation of Jesus and Mary.
Mailing address: Calle 64 Nr 140-45, 5 Piso, Apartado Aéro 90710, Bogotá, Colombia. T. (57 91) 211 66 56 – 211 66 86. C. SECLAR.

Jiménez, Julio, Fr
Rel: Roman Catholic; Ordained; celibate.
International Organizations: Director of the Movement for a Better World.
Mailing address: Via San Saba 9, I-00153 Roma, Italy. T. (39 6) 574 39 20.

Jin, Yong Ting
Rel: Christian.
International Organizations: Regional Representative for Asia-Pacific of the World Student Christian Federation.
Mailing address: Kiu Kin Mansion, 12/F – 568 Nathan Road, Kowloon, Hong Kong. T. (852 3) 852550 – 856467. Fax 823980.

Jinno, Ryuko
Rel: Buddhist; Ordained; b. 01 JUL 1925; Citizen of Japan.
International Organizations: Monk of Shingonshu Gokurakuji.
Mailing address: 4-36 Ofuki, Saijo, Ehime 799-12, Japan. T. (81 897) 55 0011

John, Emmy Lou
Rel: Methodist.
International Organizations: Treasurer of the World Federation of Methodist Women.
Mailing address: 411 Grand Avenue, Aurora IL 60506, USA.

John Paul, J, II, H H Pope
Also know as Karol Wojtyla; Rel: Roman Catholic; Ordained on 01 NOV 1946; b. 18 MAY 1920 in Wadowice, Poland; celibate; Citizen of Poland.
International Organizations: Pope (since 1978) of the Roman Catholic Church.
Other organizations: Bishop, Rome; Vicar, Jesus Christ; Successor, the Prince of the Apostles; Supreme Pontiff, the Universal Church; Patriarch, the West; Primate, Italy; Archbishop and Metropolitan, the Roman Province; Sovereign, the Vatican City State.
Career: Worked as a laborer and performed with an avant-garde theater group in German-occupied Cracow during the early years of World War II. Professor of Theology and Ethics, Universities of Lublin and Cracow from 1954 to 1958. Titular Bishop, Ombi in 1958. Auxiliary Bishop, Cracow from 1958 to 1964. Vicar-Capitular in 1962. Archbishop and Metropolitan, Cracow from 1964 to 1978. Elected Cardinal in 1967. Elected Pope in 1978. Made four major tours: to Latin America, Poland, Ireland, and the United States in 1979. Survived an assassination attempt in St Peter's Square in 1981. Through the 1980s, continued many tours around the world. **Educ.:** Studied Lit., Polish Lang. and Acting from Jagiellonian University, Cracow, Poland (1938-1939); Studied for the priesthood from an underground seminary (1942-1946); D.D. from Pontificio Ateneo Angelicum, Vatican City State (1948). **Mem.:** Formerly Member, Congregations Pro Institutione Catholica, Pro Sacramentis et Cultu Divino, and Pro Clero **Pubs.:** *The Future of the Church*, 1979; *Easter Vigil and other poems*, 1979 and *Collected Poems*, 1982 (English translation by Jerzy Peterkiewicz).
Mailing address: Apostolic Palace, 00120 Città del Vaticano, Vatican.

Johnson, Adrian
Rel: Christian.
International Organizations: President of The Gospel's Message.
Mailing address: PO Box 180, N-1851 Mysen, Norway. T. (47 9) 89 72 92.

Johnson, D Bernard, OCSO, Fr
Rel: Roman Catholic.
International Organizations: Procurator-General of the Cistercian Order of the Strict Observance.
Mailing address: Viale Africa 33, I-00144 Roma, Italy. T. (39 06) 592 41 55.

Johnson, Hugh G, Rev Dr
Rel: Protestant.
International Organizations: President of the Protestant Church of Algeria.
Mailing address: 31 rue Reda Houhou, Algiers, Algeria. T. (213 2) 66 22 16

Johnson, Paul V, Dr
Rel: Spiritualist.
International Organizations: President of the Spiritual Advisory Council.
Mailing address: 14345 SE 103rd Terrace, Summerfield FL 32691, USA.

Johnston, Allen Howard
Rel: Anglican; Ordained; b. 02 SEP 1912 in Auckland, New Zealand; m. to Joyce Rhoda Grantley; c. 4 girls; Citizen of New Zealand.
International Organizations: Chaplain and Sub-Prelate (since 1974) of Order of St John of Jerusalem.
Career: Assistant Curate, St Mark's, Remuera in 1935. Vicar, Northern Wairoa in 1937. Vicar, Otahuhu in 1944. Vicar, Whangarei in 1949. Archdeacon, Waimate in 1949. Bishop, Dunedin from 1953 to 1969. Bishop, Waikato from 1969 to 1980. Primate and Archbishop, New Zealand from 1972 to 1980. Chairman, New Zealand Church Pension Board from 1973 to 1988. Member, Royal Commission of Inquiry into Arthur Allan Thomas Case in 1979. **Interests:** gardening. **Hons.:** Companion of the Order of St Michael and St George; Hon. LL.D., Otago. **Mem.:** Fellow, St John's College, Auckland
Mailing address: 3 Wymer Terrace, Hamilton, New Zealand. T. (64 71) 553 238

Johnston, Geoffry

Rel: Christian.

International Organizations: Chairman of the Ecumenical Coalition for Economic Justice.

Mailing address: 11 Madison Avenue, Toronto ON M5R 2S2, Canada. T. (1 416) 921 46 15.

Johnston, John, F.S.C., Fr

Rel: Roman Catholic; Ordained; celibate.

International Organizations: Superior General of the Brothers of the Christian Schools.

Member of the Congregation for Religious and Secular Institutes.

Mailing address: Casa Generalizia, Via Aurelia 476, I-00165 Roma, Italy. T. (39 6) 662 15 91 – 662 15 94 – 62 01 01. C. LASALLE-ROMA. Tx 622429 TELSAL I.

Johnston, William Bryce, Very Rev Dr

Rel: Church of Scotland; Ordained on 06 JUL 1945; b. 16 SEP 1921 in Edinburgh, Scotland, UK; m. to Ruth Margaret Cowley; c. 1 boy and 2 girls; Citizen of United Kingdom; Lang. English and German.

International Organizations: Ecclesiastic of the Church of Scotland.

Other organizations: Chairman, Judicial Commission (since 1989), Church of Scotland.

Main field of work: Ministry. **Career:** Chaplain to H M Forces from 1945 to 1948. Minister, St Andrew's Boness from 1949 to 1955. Minister, St George's Greenock from 1955 to 1964. Minister, Colinton Church, Edinburgh from 1964 to 1991. Moderator, General Assembly, Church of Scotland from 1980 to 1981. Visiting Lecturer in Social Ethics, Heriot-Watt University from 1966 to 1989. Cunningham Lecturer, New College, Edinburgh from 1968 to 1971. **Educ.:** M.A. from University of Edinburgh, UK (1938-1942); B.D. from University of Edinburgh, UK (1942-1945). **Interests:** organ-playing, bowls. **Hons.:** Hon. D.D., Aberdeen University (1980); Hon. D.Litt., Heriot-Watt University (1989). **Pubs.:** Ethics and Defence (Davis, Ed.); Calvin's Commentaries – Hebrews and 1/2 Peter (translation) and Karl Barth – Dogmatics, Vol. 1/1 (translation).

Mailing address: 15 Elliot Road, Edinburgh EH13 0JR, UK. T. (44 31) 441 2315

Residence: See Mailing Address. T. See Mailing Address

Johnston, William

Rel: Anglican.

International Organizations: Director of the Episcopal Churchpeople for a Free Southern Africa.

Mailing address: 339 Lafayette Street, New York NY 10012, USA. T. (1 212) 477 0066. Fax (1 212) 979 1013.

Jones, Derwyn Morris, Rev

Rel: Christian.

International Organizations: General Secretary of the Union of Welsh Independents.

Mailing address: Ty John Penry, 11 St Helen's Road, Swansea SA1 4AL, Wales, UK.

Jones, G, Rev Canon

Rel: Christian.

International Organizations: Secretary-General of the Missions to Seamen.

Mailing address: St Michael Paternoster Royal, College Hill, London EC4R 2RL, UK. T. (44 71) 248 52 02 – 248 74 22. Fax 248 47 61.

Jones, Richard Ditzel, Rev

Rel: Methodist; Ordained; b. 17 SEP 1906 in Elizabeth NJ, USA; m. to Evelyn; c. 1 boy and 1 girl.

International Organizations: President Emeritus (since 1976) of the Canadian Council of Christian and Jews.

Career: Pastor, Methodist Church of Gladstone NJ from 1934 to 1937. Pastor, Methodist Church of Kearny NJ from 1937 to 1947. Executive Director, Canadian Council of Christians and Jews from 1947 to 1967. President, Canadian Council of Christians and Jews from 1967 to 1976. **Educ.:** B.A. from Wesleyan University, Middletown CT, USA (1928); M.A. from Boston University, School of Theology, MA, USA (1932); B.S.T. from Boston University, School of Theology, MA, USA (1933). **Interests:** fishing, travelling. **Hons.:** Beth Sholom Brotherhood Award; Award of Merit, Corp. of City of Toronto (1972); Canadian Citizen (1957); Officer, Order of Canada (1972); Award of John Diefenbaker Memorial Foundation (1984); Hon LL.D., University of Windsor, Ontario (1968) and York University, Downsview, Ontario (1975). **Mem.:** Rotary (Honorary); Kiwanis **Pubs.:** *A Life Long Pursuit of Brotherhood*, 1988 (autobiography) and contributed various articles on human relations to numerous journals.

Mailing address: 2466 Shepard Ave, Mississauga ON L5A 2H6, Canada.

Residence: See Mailing Address.

Jones, Sean

Rel: Buddhist: Mahayana.

International Organizations: Member of the Foundation for the Preservation of the Mahayana Tradition.

Other organizations: Director, Jamyang Meditation Centre.

Mailing address: 10 Finsbury Park Road, London N4 2JZ, UK. T. (44 71) 359 1394. Fax (44 71) 405 3814

Jones, William

Rel: Unknown.

International Organizations: Administrative Director of the Friends of the World Council of Churches.

Mailing address: 475 Riverside Drive, Room 1062, New York NY 10115, USA. T. (212) 870 2529.

Jonsson, Bernt Erik

Rel: Christian; Unordained; b. 31 JUL 1935 in Skellefteå, Sweden; m. to Norma Anita Ohvall; c. 2 boys and 1 girl; Citizen of Sweden; Lang. Swedish and English.

International Organizations: Director (since 1990) of the Life and Peace Institute.

Main field of work: Peace research. **Career:** Journalist at Aftonbladet from 1968 to 1972. Producer for the Swedish Television from 1972 to 1974. Political Adviser to the Swedish Ministry of Education from 1974 to 1976. Manager, Radio Uppland (local radio) from 1976 to 1982. Political Adviser to the Swedish Ministry for Foreign Affairs from 1983 to 1989. **Educ.:** B.A. from Uppsala University, Sweden (1960). **Interests:** peace, disarmament, development issues. **Mem.:** Fellowship of Reconciliation, Mission Covenant Church, Social Democratic Party **Pubs.:** *Peace Policy for the 90s.*

Mailing address: Sysslomansgatan 4, PO Box 297, S-751 05 Uppsala, Sweden. T. (46 18) 16 95 00. Fax 69 30 59. Tx 76094 LUTHER S

Residence: Rotegatan 17, S-753 37 Uppsala, Sweden T. (46 18) 26 10 29.

Jordan, Fred, Rev

Rel: Spiritualist.

International Organizations: President of the International General Assembly of Spiritualists.

Mailing address: 1809 E Bayview Blvd, Norfolk VA 23505, USA.

Jorgenson, Melvin E

Rel: Assembly of God.

International Organizations: Director of the General Council of the Assemblies of God.

Mailing address: Avenue d'Italie 45 – Bte 31, B-1050 Bruxelles, Belgium.

Jornod, J-P, Rev

Rel: Protestant.

International Organizations: President of the Federation of Swiss Protestant Churches.

Mailing address: SEK/FEPS, Sulgenauweg 26, Postfach, CH-3000 Bern 23, Switzerland.

Joseph, P C, Rev

Rel: Orthodox.

International Organizations: Secretary of the Mar Thoma Syrian Church of Malabar.

Mailing address: Poolatheen, Tiruvalla, Kerala 689 101, India.

Joshi, Sudarshan U

Rel: Hindu; b. 03 OCT 1942 in Bombay, Maharashtra, India; m. to Shalini; c. 1 boy and 1 girl; Citizen of India; Lang. English, Hindi, Marathi, Kannada and Gujarati.

International Organizations: Principal (since 1966) of the Shri Ramtirth Yogashram (Yoga Institute).

Other organizations: Member, Special Executive (since 1991), Magistrate Directory Committee.

Main field of work: Medical research. **Educ.:** B.Com. from University of Mysore, India (1962-1966). **Interests:** philanthropic activities, social services. **Hons.:** International European Award (Shield of Merit). **Mem.:** Lions Club (1991); Bombay Manufacturers Association (1991); and business clubs **Pubs.:** Umesh Yoga Darshan, Part I and II.

Mailing address: Shri Ramtirth Yogashram, 27 Umesh Dham, L. T. Colony St No 2, Dadar, Bombay 400 014, India. T. (91 22) 414 44 81

Residence: 107/4 Atharva, 4th Lane, Hindu Colony, Dadar, Bombay 400 014, India. T. (91 22) 414 26 87

Joulia, Henri, B.M.V., Fr

Rel: Roman Catholic; Ordained; celibate.

International Organizations: Superior General of the Missionnaires de l'Immaculée-Conception de Lourdes.

Mailing address: Notre-Dame de Garaison, F-65670 Monléon-Magnoac, France. T. (33) 62 99 41 55

Jubany Arnáu, Narciso J, H E Card

Rel: Roman Catholic; Ordained on 30 JUL 1939; b. 12 AUG 1913 in Santa Coloma de Farnés, Gerona, Spain; celibate; Citizen of Spain.

International Organizations: Member (since 1973) of the College of Cardinals.

Member of the Sacred Congregation for the Sacraments and Divine Worship. Member of the Congregation for Religious and Secular Institutes.

Other organizations: Archbishop Emeritus (since 1990), Barcelona.

Career: Auxiliary Bishop, Barcelona in 1955. Bishop, Gerona from 1964 to 1971. Archbishop, Barcelona from 1971 to 1990. Created Cardinal by Pope Paul VI in 1973. President, Episcopal Commission on Liturgy.

Mailing address: Sales i Ferrer 60, E-08026 Barcelona, Spain. T. (34 3) 256 53 43

Jullien, Jacques André Marie, Mgr

Rel: Roman Catholic; Ordained on 03 APR 1954; b. 07 MAY 1929 in Brest, Finistère, France; celibate; Citizen of France; Lang. French and English.

International Organizations: Archbishop (since 1984) of Rennes.

Career: Professor of Moral Theology, Grand Séminaire de Quimper from 1957 to 1969. Curate, St Louis, Brest from 1969 to 1978. Bishop, Beauvais from 1978 to 1984. **Educ.:** Lic. Theol. from Université Catholique de l'Ouest, Angers, France (1949-1954); Lic. Soc.Sc. from Institut Catholique de Paris, France (1955-1957). **Hons.:** Knight of the Legion of Honour. **Pubs.:** Le Chrétien et la politique, DDB, 1965; Les prêtres dans le combat politique, Ed. Ouvrières, 1972 and L'homme debout, DDB, 1980.

Mailing address: Archevêché, 45 rue de Brest, F-35042 Rennes Cedex, France. T. (33) 99 54 06 06

Residence: See Mailing Address. T. See Mailing Address

Jung, Hans-Gernot

Rel: Presbyterian; Ordained; b. 10 FEB 1930 in Marburg/Lahn, Germany; m. to Nina Schwerdtfeger; c. 2 boys and 1 girl; Citizen of Germany.

International Organizations: Bishop (since 1978) of Kurhessen-Waldeck.

Career: Minister, Kreuzkirche, Kassel from 1958 to 1962. Student Chaplain, Marburg from 1962 to 1965. Director, Evangelische Akademie Hofgeismar from 1965 to 1974. Adviser, diocesan office, Evangelical Church in Kurhessen-Waldeck, Kassel from 1974 to 1978. **Pubs.:** Befreiende Herrschaft, Die politische Verkündigung der Herrschaft Christi, 1965 and Gemeinden im Bildungsprozess, 1977.

Mailing address: Burgfeldstrasse 14, 34131 Kassel-Wilh, Germany.

Jung, Park Sang, Rev

Rel: Christian.

International Organizations: Contact of the Christian Conference of Asia.

Mailing address: 6-10 Nakagawa Nishi, 2-chome, Ikuno-ku, Osaka 544, Japan. T. (81 6) 712 2719. Fax (81 6) 718 0988.

Junkkaala, Timo Tapio, Dr

Rel: Evangelical Lutheran; Ordained on 07 JUN 1973; b. 22 OCT 1948 in Helsinki, Finland; m. to Marjatta Kaarina Esko; c. 2 boys and 2 girls; Citizen of Finland; Lang. Finnish, Swedish and English.

International Organizations: Secretary-General (since 1986) of the Finnish Evangelical Lutheran Student Mission.

Other organizations: Secretary-General (since 1991), Bible-Quiz Project (Finnish Bible Celebration Year).

Main field of work: Ministry. **Career:** High School Secretary, Finnish Evangelical Lutheran Student Mission from 1973 to 1974. Student Secretary, Finnish Evangelical Lutheran Student Mission from 1974 to 1985. **Educ.:** Dr. Theol. from University of Helsinki, Finland (1986). **Pubs.:** Tunnetko Tien, 1978 (confirmation book); Hannulan Herätys, 1986 (thesis) and Urho Muroma, Herâtyssaarnaaja, 1990 (biography).

Mailing address: Pohjoisranta 4 A 8, SF-00170 Helsinki, Finland. T. (358 90) 628500

Residence: Gunillankuja 2 C 28, SF-08700 Helsinki, Finland. T. (358 90) 6984930

Juritsch, Martin, S.A.C., Fr

Rel: Roman Catholic; Ordained; celibate.

International Organizations: Rector-General of the Society of the Catholic Apostolate.

Mailing address: Piazza S Vincenzo Pallotti 204, I-00186 Roma, Italy. T. (39 6) 654 81 85 – 687 62 13

Juvent, R P Octave

Rel: Christian.

International Organizations: Representative of the Aid for the Church in Africa.

Mailing address: Rue de Bavais 211, B-7301 Hornu, Belgium.

Jyotirmayananda, J, Swami

Rel: Yoga; b. 03 FEB 1931 in Dumari Buzurg, District Saran, Bihar, India.

International Organizations: Founder/Director of the Yoga Research Foundation.

Career: Entered Sanyasa in 1953. Professor of Religion, Yoga Vedanta Forest Academy (for over 9 years). Editor, *Yoga Vedanta Journal*. Frequently lectured at the All-India Vedanta Conferences in Delhi, Amritsar, Ludhiana, and other parts of India. Came to the West and founded Sanatan Dharma Mandir in Puerto Rico in 1962. Conducted regular classes, gave weekly radio lectures in English and in Spanish, and appeared on TV, Puerto Rico from 1962 to 1969. Moved to Miami, FL, and established his ashram in 1969. **Pubs.:** books and tapes.

Mailing address: 6111 SW 74th Ave, Miami FL 33143, USA. T. (1 305) 666 2006.

Kabwe-ka-Leza, Kitobo, Bishop

Rel: Episcopal Baptist.

International Organizations: Bishop of Church of Christ in Zaire (Episcopal Baptist Community of Africa).

Mailing address: Cabinet Episcopal ECZ-CEBA, BP 3866, Lumbumbashi, Zaire.

Kada, Lajos, H E Mgr

Rel: Roman Catholic; Ordained on 10 OCT 1948; b. 16 NOV 1924 in Budapest, Hungary; celibate; Citizen of Hungary.

International Organizations: Secretary of the Sacred Congregation for the Sacraments and Divine Worship.

Advisor of the Pontifical Council for the Family.

Other organizations: Titular Bishop (since 1975), Tibica.

Educ.: Dr. Theol., J.D.

Mailing address: Palazza delle Congregazioni, Piazza Pio XII 10, I-00193 Roma, Italy. T. (39 6) 698 4201 – 698 4202 – 698 4316 – 698 4318 – 698 4326. Fax 698 3499

Residence: Via del Casaletto 481, I-00151 Roma, Italy T. (39 6) 537 4385 – 537 3319

Kadowaki, K

Rel: Unknown.

International Organizations: Representative of the Institute of Oriental Religions, Tokyo.

Mailing address: c/o Sophia University, Chiyoda-ku, Kioicho 7-1, Tokyo 102, Japan. T. (81 3) 3238 3131.

Kaganoff, Nathan M
Rel: Jewish.
International Organizations: Executive Officer of the World Council on Jewish Archives.
Mailing address: Amer. Jewish Hist. Soc., Brandeis Univ, 2 Thornton Rd, Waltham MA 02154, USA. T. (1 617) 891 8110.

Kahf, Monzer
Rel: Muslim.
International Organizations: Head, Office of the Islamic Research and Training Institute, Jeddah.
Mailing address: c/o Islamic Development Bank, PO Box 9201, Jeddah 21432, Saudi Arabia. T. (966 2) 636 14 00 – 636 00 11 – 636 20 86. Fax 636 68 71. C. BANKISLAMI. Tx 401407 – 101137.

Kahinga, Kayuwa Tshibumbu Wa, Patriarch
Rel: Community of Light.
International Organizations: Patriarch of Church of Christ in Zaire (Community of Light).
Mailing address: BP 10498, Kinshasa I, Zaire.

Kahn, Jean
Rel: Jewish.
International Organizations: President of the European Jewish Congress.
Mailing address: 78 ave des Champs-Elysées, F-75008 Paris, France. T. (33 1) 43 59 94 63. Fax 42 25 45 28. Tx 650320.

Kailasanatha Kurukkal, K K, Prof
Rel: Hindu.
International Organizations: Chief Priest of the Sri Muthu Vinayagar Devasthanam.
Other organizations: Director, Priest Training Centre.
Mailing address: Sri Muthu Vinayagar Devasthanam, Sea Street, Colombo 11, Sri Lanka.

Kajaria, Nirmala Ratanshi, Dr
Rel: Hindu; Ordained on 19 OCT 1962; b. 02 OCT 1935 in Bombay, Maharashtra, India; single; Citizen of India; Lang. Hindi, Gujarati and English.
International Organizations: Director for the Brahma Kumaris Raja Yoga Centres in Australia, New Zealand, the Asia-Pacific Region, South Africa and Sri Lanka (since 1978) of the Brahma Kumaris World Spiritual University.
Main field of work: Lectures on Raja Yoga. **Career:** Joined the Brahma Kumaris World Spiritual University in 1962. Established a Raja Yoga Centre in Bombay in 1966. Left India for London in 1971. Travelled throughout the UK, East Africa and Mauritius for the next seven years. **Educ.:** M.B.B.S. from Bombay University, India (1955-1961). **Interests:** music, travel. **Pubs.:** Editorial Board of *Positive Health*.
Mailing address: 78 Alt Street, Ashfield, Sydney NSW 2131, Australia. T. (61 2) 749 9880. Fax 716 7795. Tx 100200 YGCT AA REF RAJ YOGA. E MAIL 07 YGC001.
Residence: See Mailing Address. T. See Mailing Address

Kajitani, Sonin
Rel: Buddhist: Zen; Ordained; b. 08 JUL 1914; Citizen of Japan.
International Organizations: Chief of the Sokokuji Sect of Rinzai Zen Buddhism, Kinkakuji.
Main field of work: Monk. **Career:** Manager, Sokokuji Sect of Rinzai Zen Buddhism.
Mailing address: 1 Kinkakuji, Kita-ku, Kyoto 603, Japan. T. (81 75) 461 0013

Kakaria, Sunita
Rel: Buddhist: Mahayana.
International Organizations: Member of the Foundation for the Preservation of the Mahayana Tradition.
Other organizations: Director, Tushita Mahayana Meditation Centre.
Mailing address: C-259 Defence Colony, New Delhi 110024, India. T. (91 11) 462 9562

Kakkanattu, Benicasia, C.M.C., Sr
Rel: Roman Catholic; Ordained; b. 18 NOV 1941 in Chennamkary, Alapuzha District, Kerala, India; celibate; Citizen of India; Lang. Malayalam, Hindi and English.
International Organizations: Superior General (since 1991) of the Congregation of Mother of Carmel.
Other organizations: Member (since 1987), Kerala Higher Education Forum; Pastoral Council Member (since 1992), Archdiocese of Changanacherry.
Main field of work: University education. **Educ.:** Master's Deg. from Morgan State College, Baltimore MD, USA (1965-1966).
Mailing address: Mount Carmel Road, Aluva 683106, India. T. (91 4854) 4270
Residence: See Mailing Address. T. See Mailing Address

Kakkanattu, Samuel M, C.M.S.F., Br
Rel: Roman Catholic; Ordained; celibate.
International Organizations: Procurator-General of the Missionary Brothers of St Francis of Assisi.
Mailing address: Via di S Paolo alla Regola 6, I-00186 Roma, Italy. T. (39 6) 654 24 08

Kaldy, Zoltan, Rt Rev
Rel: Lutheran.
International Organizations: Presiding Bishop of Lutheran Church in Hungary.
Mailing address: Puskin utca 12, H-1088 Budapest, Hungary.

Kalilombe, Patrick-Augustine, Rt Rev Dr
Rel: Roman Catholic; Ordained on 03 FEB 1958; b. 28 AUG 1933 in Mua, Dedza, Malawi; celibate; Citizen of Malawi; Lang. Chichewa, English, French, Italian and German.
International Organizations: Member of the Society of Missionaries of Africa.
Other organizations: Lecturer in Mission, and responsible for running the Certificate in Theology Course (since 1985), Centre for Black and White Christian Partnership; Executive Director (since 1987), Centre for Black and White Christian Partnership; External examiner of dissertation in inter-cultural theology (since 1987), University of Birmingham.
Main field of work: University education. **Career:** Was sent for novitiate to Algiers, North Africa from 1954 to 1955. Was sent to finish theological training at the White Fathers' scholasticate in Thibar, Tunisia from 1955 to 1957. Was sent to White Fathers' scholasticate at Carthage, Tunisia from 1957 to 1958. Further studies in Rome from 1958 to 1962. Returned home and served in parish work in home Diocese, Dedza from 1962 to 1964. Rector and Lecturer, Kachebere Major Seminary, Malawi from 1964 to 1972. Bishop, Lilongwe Diocese, Malawi from 1972 to 1978. Vice-President, Association of Member Episcopal Conferences in Eastern Africa (AMECEA) from 1973 to 1976. Research Fellow in Jerusalem from 1979 to 1980. Tutor/Lecturer in Third World Theologies and Primal World Views, Selly Oak Colleges, Birmingham from 1982 to 1987. Acting Director, Centre for Black and White Christian Partnership from 1985 to 1987. **Educ.:** S.T.L. from Pontifical Gregorian University, Vatican City State (1958-1960); S.S.L. from Pontifical Biblical Institute, Vatican City State (1960-1962); Ph.D. from Graduate Theological Union, Berkeley CA, USA (1978-1983). **Mem.:** Ecumenical Association of Third World Theologians; International Association for Mission Studies **Pubs.:** *From Outstations to Small Christian Communities*, Eldoret, 1984; "Black Theology", in *The Modern Theologians*, Oxford, Blackwell, 1989 and "A Malawian Example: The Bible and Non-Literate Communities", in *Voices from the Margin*, London, SPCK, 1991.
Mailing address: Selly Oak Colleges, Birmingham B29 6LQ, UK. T. (44 21) 472 7952
Residence: Flat 10, Elizabeth Court, 107 Metchley Lane, Harborne, Birmingham B17 OJH, UK. T. (44 21) 426 1738

Kalke, David, Rev
Rel: Christian.
International Organizations: Head, Secretariat of the International Association Against Torture.
Mailing address: PO Box 1752, Cathedral Station, New York NY 10025, USA. T. (1 212) 927 9050.

Kangwana, James
Rel: Christian.
International Organizations: Regional Secretary for Africa of the World Association for Christian Communication.
Mailing address: PO Box 14205, Nairobi, Kenya.

Kannan, Aiyasamy, Dr

Rel: Hindu; Unordained; b. 19 FEB 1924 in Madras, Tamilnadu, India; m. to Sarojini Kannan; c. 2 boys and 2 girls; Citizen of India; Lang. English, Tamil and other Indian languages.

International Organizations: Secretary (since 1962) of the Theosophy Science Centre.

Main field of work: Research on science and society, spirituality. **Career:** Research Chemist, National Dairy Research Institute, Bangalore, India from 1944 to 1951. Food Analyst, Municipal Corporation, Delhi, India from 1951 to 1964. Research Associate, Agricultural Biochemistry Dept., University of Minnesota, Saint Paul, USA from 1955 to 1957. Post-Graduate Teacher, Avinasilingam Home Science College, Coimhatore, India from 1966 to 1968. Editor, Bulletin of Theosophy Science Study Group, India from 1962 to 1988. Editor, *Holistic Science and Human Values* from 1989 to 1990. **Educ.:** B.Sc. from University of Madras, India (1940-1942); M.Sc. Chem. from Andhra University, Waltair, India (1942-1944); Ph.D. Dairy Chem. from Bombay University, India (1946-1950); Further Stds. from University of Minnesota, St Paul, USA (1955-1957). **Interests:** science and spirituality (technology and human values), holism. **Mem.:** Life Member, Theosophical Society **Pubs.:** *Impact of Theosophy and Science*, TPH, Madras, 1971; *The One Life and Functional Forms*, TPH, Madras, 1973 and Editor, *Holistic Human Concern for World Welfare*, Theosophy Science Study Group, Madras, 1987.

Mailing address: c/o Theosophical Society, Adyar, Madras 600 020, India. T. Madras (91 44) 417198. C. Olcott Madras.

Kansakar, Maniharsha Jyoti

Rel: Buddhist.

International Organizations: Vice-President and Member of the Executive Council of the World Fellowship of Buddhists.

Mailing address: The Dharmodaya Sabha, PO Box 4053, Buddha Vihara, Bhrikuti Mandap, Kathmandu, Nepal.

Kao, C M, Rev

Rel: Presbyterian.

International Organizations: General Secretary of the Presbyterian Church in Taiwan.

Mailing address: 3 Roosevelt Road, Lane 269, Sec. 3, Taipei, Taiwan (Rep. of China).

Kao, Ruth

Rel: Reformed Church.

International Organizations: Member of the Executive Committee of the World Alliance of Reformed Churches.

Mailing address: 24 Alley 14, Lane 208, Zuei-An Street, Taipei 10638, Taiwan (Rep. of China). T. (886 2) 362 5282. Fax (886 2) 362 8096. Tx 20588 PRESTAI

Residence: T. (886 2) 707 7647

Kaplan, Jacob, Rabbi

Rel: Jewish; Ordained; b. 07 NOV 1895 in Paris, France; m. to Fanny Dichter; c. 3 boys and 2 girls; Citizen of France.

International Organizations: Chief Rabbi (since 1981) of Central Consistory of France.

Career: Rabbi, Mulhouse in 1922. Rabbi in Paris in 1929. Auxiliary Rabbi to Chief Rabbi, France in 1939. Chief Rabbi, Paris in 1950. Chief Rabbi, France from 1955 to 1980. **Hons.:** Grand Officer, Legion of Honour; Military Cross (twice); Grande Plaque de la Ville de Paris; Grand Cross, National Order of Merit (1984); Dr. h.c., Theological Seminary of New York; Citation by Mayor of New York for distinguished service, Grand Cross (1988). **Mem.:** Académie des Sciences Morales et Politiques, Paris (since 1967) **Pubs.:** Justice pour la Foi juive, 1977; Un Enseignement de l'Estime, 1982 and N'oublie pas, 1984.

Mailing address: 1 rue Andrieux, F-75008 Paris, France. T. (33 1) 45 22 33 97

Karasick, Joseph, Rabbi

Rel: Jewish.

International Organizations: Member of the World Jewish Congress.

Mailing address: 35 West 36th Street, New York NY 10018, USA.

Karekin (Sarkissian), K, II, H H

Rel: Armenian Apostolic Orthodox; Ordained on 28 SEP 1952; b. 27 AUG 1932 in Kessab, Lattaquieh, Syria; celibate; Citizen of Lebanon; Lang. Armenian, English and French.

International Organizations: Catholicos (since 1977) of the Cilicia. Honorary President (since 1990) of the Middle East Council of Churches.

Other organizations: Head (since 1977), Armenian Church under Catholicosate of Cilicia.

Main field of work: Theology, church history. **Career:** Ordained Deacon in 1949. Elevated to the rank, Vardapet in 1955. Appointed Dean, Cilician Seminary in 1956. Founded the Armenian Church University Student Association (ACUSA) in 1963. Elevated to the rank of Senior Archimandrite in 1963. Consecrated Bishop in 1964. Elected Prelate, Diocese of New Julfa, Isfahan (Iran) in 1971. Granted the rank of Archbishop in 1973. Appointed Pontifical Legate, Eastern Prelacy of the Armenian Apostolic Church of America in 1973. Elected Primate, Eastern Prelacy in 1974. Attended the World Council of Churches Assemblies in New Delhi, India (1961), Uppsala, Sweden (1968), Nairobi, Kenya (1975), and Vancouver, Canada (1983). Attended the Second Vatican Council as an observer from 1963 to 1965. Attended the Lambeth Conference as an observer in 1968. Elected Member of the Central and Executive Committees, World Council of Churches in 1968. Vice Moderator, World Council of Churches from 1975 to 1983. One of the initiators, the 'Ecumenical' movement within the Armenian Apostolic Church. One of the organizers, the Addis Ababa Conference of Heads of Oriental Orthodox Churches. Lectured on theology, literature, history and culture at the Palandjian College in Beirut, at the Beirut College for Women (now Beirut University College), and at the American University of Beirut. **Educ.:** B.Litt. from University of Oxford, UK (1957-1959). **Interests:** inter-church relations, ecumenical concerns. **Pubs.:** The Council of Chalcedon and the Armenian Church, SPCK, London, 1965, reprinted in New York, 1979 (in English); The Witness of Oriental Orthodox Churches, Antelias, 1968 (in English) and In Search of Spiritual Life, Antelias, 1991 (in English).

Mailing address: Armenian Catholicosate of Cilicia, Antelias, Lebanon. T. (961 1) 410 002

Residence: See Mailing Address. T. See Mailing Address

Kariamadam, Paul, Rev Dr

Rel: Roman Catholic; Ordained on 18 APR 1966; b. 30 SEP 1939 in Vazhakulam, Kerala, India; celibate; Citizen of India; Lang. Malayalam, English, Italian, French and German.

International Organizations: Superior General (since 1988) of the Vicentian Congregation.

Main field of work: Biblical research. **Career:** Professor of Sacred Scripture, Pontifical Institute Alwaye from 1979 to 1989. **Educ.:** Doc. from Pontifical Biblical Institute, Vatican City State (1969-1978). **Pubs.:** The Zachaeus Story; Vachanamalyam (in Malayalam) and The End of the Travel Narrative (Lk 18:31, 19:46).

Mailing address: Vincentian Generalate, Angamaly South, Angamaly 683 573, Kerala State, India. T. (91 4856) 2336

Residence: See Mailing Address. T. See Mailing Address

Karin, M, Sr

Rel: Roman Catholic.

International Organizations: Representative of the Ordo Sanctissimi Salvatoris S Birgittae.

Mailing address: Abbess, Sta Birgittas Kloster, Pax Mariae, Myntbacken 2, S-592 00 Vadstena, Sweden. T. (46 143) 103 82 – 109 43.

Karivalis, Damianos George

See Diodoros.

Karjian, Hovhannes, Rev

Rel: Armenian Evangelical.

International Organizations: Moderator of the Union of the Armenian Evangelical Churches in the Near East.

Mailing address: PO Box 11-2508, Beirut, Lebanon.

Karner, Peter, Rev Mag.

Rel: Evangelical.

International Organizations: Moderator (Helvetic Confession) of the Evangelical Church of the Augsburg and Helvetic Confession.

Mailing address: Evangelischer Oberkirchenrat H.B., Dorotheergasse 16, A-1010 Wien, Austria.

Karo, David H, Rev
Rel: Baptist.
International Organizations: President of the Nigerian Baptist Convention.
Mailing address: Baptist Building, PMB 5113, Ibadan, Nigeria.

Karta, K, Lama
Rel: Buddhist; celibate; Lang. French, English and Tibetan.
International Organizations: Spiritual Director of Tibetan Institutes in Belgium (Huy, Brussels, Schoten), France (Lille) and The Netherlands (Haarlem).
Main field of work: Education. **Educ.:** Monastery of the Venerable Kalu Rinpoche, Sonada, West Bengal, India.
Mailing address: Institut Yeunten Ling, Château de Fond l'Evêque, 4 Promenade Saint Jean l'Agneau, B-5201 Huy, Belgium. T. (32 85) 21 48 20
Residence: See Mailing Address. T. See Mailing Address

Karthikesa Kurukkal, K K, Brahmasri
Rel: Hindu.
International Organizations: Chief Priest of the Munneswaram Devasthanam.
Mailing address: Munneswaram Devasthanam, Munneswaram, Sri Lanka.

Karunan, Victor Prasad, Dr
Rel: Roman Catholic; b. 07 AUG 1954 in Bangalore, Karnataka State, India; m. to Wanida Karunan; Citizen of India; Lang. English, French, Tamil and Hindi.
International Organizations: Secretary-General (since 1988) of the Pax Romana, International Catholic Movement for Intellectual and Cultural Affairs.
Main field of work: Research and teaching. **Career:** National President, All-India Catholic University Federation (AICUF) from 1975 to 1976. Resident Correspondent, News Bureau of India (NBI) from 1975 to 1976. Asian Secretary, International Movement of Catholic Students (IMCS Asia) from 1979 to 1982. Editorial Board Member, Plough Publications, Hong Kong from 1979 to 1987. Asian Training Staff for Asian Workshops for Community Organisers, Indian Social Institute, Bangalore, India from 1983 to 1984. Research Staff, Rural Development Programme among Sugar Workers in Puntamba Village, Ahmednagar District, Maharastra State, India from 1983 to 1984. Research Consultation towards a PhD programme on 'Peasant Movements and Agrarian Conflicts in Asia' at Third World Studies Centre, Catholic University of Nijmegen, The Netherlands from 1984 to 1985. Visiting Lecturer in Asian Studies, Dept. of Sociology, Asian Social Institute, Manila, The Philippines from 1986 to 1988. Programme Secretary, Asian Network for Information Exchange, Documentation and Research on Transnational Corporations in Asia, for Documentation for Action Groups in Asia from 1986 to 1988. **Educ.:** B.A. Econ., Pol.Sc. and Soc. from St Joseph's College, Bangalore, India (1971-1976); M.A. Soc. from Jawaharlal Nehru University, New Delhi, India (1977-1979); Ph.D. Soc. from Catholic University of Nijmegen, Netherlands (Beginning in 1988). **Pubs.:** Books and over 40 articles, reports and conference papers, including: *If the Land Could Speak It Would Speak for Us – A History of Peasant Movements in Thailand and the Philippines*, Vol. 1, Plough Publications, Hong Kong, April 1984, 150 pp.; *A History of Peasant Protest and Agrarian Conflicts in India and Sri Lanka*, Vol. 2, Plough Publications, Hong Kong, 1988, 250 pp.; "Idealism is my Motivation: An Interview", in *Bijeen* (Third World Publication), Netherlands, Vol. 17, No 10, 1984; "Sri Lanka's State of Seige", in *Inside Asia*, London, No 5, Sept.-Oct. 1985 and *Case-Study: The Bhopal Tragedy in India*, Paper presented to Special NGO Committee on TNCs: Symposium on 'TNCs and the Protection of the Environment', United Nations, Geneva, Nov. 16, 1989.
Mailing address: 37-39 rue de Vermont, PO Box 85, CH-1211 Genève 20, Switzerland. T. (41 22) 733 67 40. Fax (41 22) 733 67 49
Residence: 38 rue du 31 Décembre, CH-1207 Genève, Switzerland T. (41 22) 736 78 43

Kasannoin, Chikatada
Rel: Shinto; Ordained; b. 03 AUG 1918; Citizen of Japan.
International Organizations: Chief Priest of the Kasuga Shrine.
Career: Member, Nara Prefecture Board of Education. **Educ.:** Kokugakuin University, Tokyo, Japan.
Mailing address: 160 Kasuganocho, Nara, Nara 630, Japan. T. (81 742) 22 7788

Kasarherou, Tell, Rev
Rel: Evangelical.
International Organizations: General Secretary of the Evangelical Church in New Caledonia and the Loyalty Isles.
Mailing address: BP 277, Noumea, New Caledonia.

Kaseje, Dan C O, Dr
Rel: Anglican; b. 29 APR 1946 in Siaya, Kenya; m. to Margaret Achien'g; c. 2 boys and 2 girls; Citizen of Kenya; Lang. English.
International Organizations: Director (since 1989) of the Christian Medical Commission.
Main field of work: Communicable disease epidemiology and primary health care. **Career:** Medical Doctor, Ministry of Health from 1975 to 1976. Medical Officer of Health, Ministry of Health from 1976 to 1977. Lecturer in Medicine, University of Nairobi, Kenya from 1976 to 1984. Project Manager, Aga Khan Health Services from 1984 to 1989. **Educ.:** M.B.Ch.B. from University of Nairobi, Kenya (1970-1975); M.P.H. from Harvard University, Graduate School of Public Health, Cambridge MA, USA (1977-1977); Ph.D. from Liverpool School of Tropical Medicine, UK (1985-1989). **Interests:** Community health. **Hons.:** Fellow, Royal College of Tropical Medicine. **Mem.:** International Epidemiology Association **Pubs.:** *Le paludisme: prévention, traitement et influence sur la mortalité. Le cas du Kenya*, 1989; *Community Based Health Care, Saradidi Experience in Social Science and Medicine*, 1988 and 17 articles on Community Based Health Care and Malaria in *Annals of Tropical Medicine and Parasitology*, Special Issue, 1987.
Mailing address: PO Box 2100, 150 route de Ferney, CH-1211 Genève 2, Switzerland. T. (41 22) 791 62 15. C. OIKOUMENE. Tx 415 730 OIKCH. Fax (41 22) 791 03 61.
Residence: 33 Chemin des Palettes, Grand Lancy, CH-1212 Genève, Switzerland T. (41 22) 794 68 54

Kasper, Walter, H E Mgr Dr
Rel: Roman Catholic; Ordained on 06 APR 1957; b. 05 MAR 1933 in Heidenheim/Brenz, Baden-Württemberg, Germany; celibate; Citizen of Germany; Lang. German, English, French and Italian.
International Organizations: Bishop (since 1989) of Rottenburg-Stuttgart.
Member of the Pontifical Council for Promoting Christian Unity. Member of the International Theological Commission.
Other organizations: President, World Church Commission (since 1991), German Conference of Bishops; Vice-Chairman, Faith Commission, German Conference of Bishops.
Career: Assistant Lecturer, Faculty of Theology, University of Tübingen from 1961 to 1964. Professor of Dogmatism, University of Münster from 1964 to 1970. Professor of Dogmatism, University of Tübingen from 1970 to 1989. Special Secretary, Synod of Bishops in 1985. Member, Faith and Order Commission, World Council of Churches. **Educ.:** Dr. Theol. from Eberhard-Karls-Universität Tübingen, Germany (1960-1961); Hab. from Eberhard-Karls-Universität Tübingen, Germany (1964). **Interests:** climbing. **Hons.:** Bundesverdienstkreuz (1st Class); 2 honorary doctorates. **Mem.:** Heidelberg Akademie der Wissenschaften; European Academy of Arts and Sciences **Pubs.:** Jesus der Christus, Mainz, 1974; Der Gott Jesu Christi, Mainz, 1982 and Theologie und Kirche, Mainz, 1987.
Mailing address: Bishop-von-Keppler-Str 7, 72108 Rottenburg a.N., Germany. T. (49 7472) 169224
Residence: See Mailing Address. T. See Mailing Address

Kasravi, Karim
Rel: Muslim: Sufi.
International Organizations: of the School of Islamic Sufism.
Mailing address: PO Box 5827, Washington DC 20016, USA. T. (1 301) 654 91 91.

Kaszuba, Tadeusz, C.R., Fr
Rel: Roman Catholic; Ordained; celibate.
International Organizations: Procurator-General of the Congregation of the Resurrection of our Lord Jesus Christ.
Mailing address: Via S Sebastianello 11, 00187 Roma, Italy. T. (39 6) 679 59 08

Katoneene, Jonah Mwesigwa, Rev
Rel: Anglican; Ordained on 01 FEB 1981; b. 03 OCT 1951 in Busheenyi, Western Uganda, Uganda; m. to Elivaida; c. 3 boys and 1 girl; Citizen of Uganda; Lang. English, Runyankore and French.
International Organizations: Executive Secretary (since 1988) of the Association of Christian Lay Centres in Africa.
Main field of work: Ministry. **Career:** Associate Provincial Youth Secretary, Church of Uganda from 1980 to 1983. Acting Chaplain, Makerere University from 1982 to 1983. Director, Lweza Training and Conference Centre, Church of Uganda from 1984 to 1988. Chairperson for the Africa Region, World Student Christian Federation (WSCF) from 1983 to 1986. Member of the Executive Committee, WSCF from 1986 to 1990. **Educ.:** B.A., Conc. Dip.Ed. from Makerere University, Kampala, Uganda (1976-1979); Dip. C.Ed. from Imanuel College of Theology, Ibadan, Nigeria (1979-1980). **Interests:** current affairs, reading. **Mem.:** World Student Christian Federation
Mailing address: AACC Bldg, Wayaki Wayi, Nairobi 2542, Kenya. T. (254 2) 742293 – (254 2) 742296. C. CHURCHCON. Tx 22175 AACC. Fax (254 2) 742352.
Residence: Kabarnet Road, Nairobi 2542, Kenya. T. (254 2) 569083

Katongo, Clément
Rel: Roman Catholic; Unordained; b. 15 SEP 1949 in Mpika, Northern Province, Zambia; m. to Doreen Kumoyo; c. 3 boys and 3 girls; Citizen of Zambia; Lang. English.
International Organizations: Organizer of the International Young Christian Workers.
Other organizations: Groups organizer (since 1985), Workers Pastoral Centre, Diocese of Ndola.
Main field of work: Education for workers' awareness. **Pubs.:** Sub-Editor, *Workers Challenge Magazine* and Sub-Editor, *Plight Magazine*.
Mailing address: Kwacha Road, PO Box 270035, c/o Zambia YCW, Kitwe, Zambia. T. (260 2) 210904
Residence: 1914 Kwacha T Ship, PO Buchi, Kitwe, Zambia. T. (260 2) 213430

Katsunuma, Tomohito
Rel: Buddhist.
International Organizations: Executive Director of the Society for Advancement of Buddhist Understanding.
Mailing address: 16925 E Gale Ave, City of Industry CA 91745, USA.

Kaudre, Bansilal Gangaram, Shri
Rel: Hindu; b. 02 FEB 1939 in Karoshi, Tal-Khed District, Maharashtra, India; m. to Anusaya Bansilal Kaudre; c. 3 boys and 3 girls; Citizen of India; Lang. Marathi, Hindi and Gujrathi.
International Organizations: Sarpanch of the Kharoshi Gaon.
Pubs.: Publisher of Bhimashankar Shakti Times.
Mailing address: Post Office – Rajgurunagar, Tal-Khed, Nehru Chowk District, Pune 410505, India.
Residence: Post Office – Bhimashankar, Taluka-Khed District, Pune, India.

Kaufmann, Frank
Rel: Unknown.
International Organizations: Executive Director of the Council for the World's Religions.
Mailing address: 4 West 43rd St, New York NY 10036, USA. T. (1 212) 869 6023. Fax 869 6424. Tx 4991393 NWERA.

Kaut, Bernd, Mgr
Rel: Roman Catholic.
International Organizations: President of the Pontifical Missionary Society.
Mailing address: Goethestrasse 43, Postfach 1110, D-52064 Aachen, Germany. T. (49 241) 75 07 00. Fax 75 07 00 335. C. MISSIO. Tx 0832719 D.

Kavanagh, Aengus J, F.S.P., Br
Rel: Roman Catholic; Ordained; celibate.
International Organizations: Superior General of the Brothers of Saint Patrick.
Mailing address: Patrician Brothers' Generalate, Tullow Hill, Co. Carlow, Ireland. T. (353 503) 51190. Fax 51190.

Kawahara, Hiromi, Dr
Rel: Christian.
International Organizations: Executive Director of the Asian Health Institute.
Mailing address: 987 Minamiyama, Nisshin, Aichi-Gun – Aichi 470-01, Japan. T. (81 5617) 3 1950. Fax 3 1990. C. ASIAHEALTH NAGOYA.

Kazemzadeh, Firuz
Rel: Bahá'í.
International Organizations: Secretary for External Affairs of the National Spiritual Assembly of the Bahá'ís of the US.
Other organizations: Professor of History, Yale University, New Haven CT; Editor, *World Order* magazine.
Educ.: Grad. from Stanford University, CA, USA; Ph.D. from Harvard University, Cambridge MA, USA. **Pubs.:** Author of several books and articles.
Mailing address: Bahá'í National Center, 536 Sheridan Rd, Wilmette IL 60091, USA.

Kearns, D Thomas
Rel: Roman Catholic.
International Organizations: Secretary-General of the Institute of Charity.
Mailing address: Via di Porta Latina 17, I-00179 Roma, Italy. T. (39 6) 70491777.

Keating, Jeremiah Columba, C.F.C., Br
Rel: Roman Catholic; Ordained; b. in Ireland; celibate; Citizen of Ireland.
International Organizations: Superior General of the Congregation of Christian Brothers.
Mailing address: Via della Maglianella 375, I-00166 Roma, Italy. T. (39 6) 696 02 53. C FRATRES ROMA.

Kedge, Peter
Rel: Buddhist: Mahayana.
International Organizations: Asia Pacific Regional Office of the Foundation for the Preservation of the Mahayana Tradition.
Other organizations: Director, SFE Systems Ltd.
Mailing address: c/o SFE Systems Ltd, PO Box 98650, Tsim Sha Tsui, Hong Kong. T. (852) 721 1974. Fax (852) 721 2942

Kee, Alexander Alistair, Prof
Rel: Church of Scotland; b. 17 APR 1937 in Alexandria, Dumbartonshire, Scotland, UK; m. to Anne Mary Paterson; c. 1 boy and 1 girl; Citizen of United Kingdom; Lang. English and French.
International Organizations: Professor (since 1988) of the New College, University of Edinburgh.
Main field of work: University education. **Career:** Lecturer, University College of Rhodesia from 1964 to 1967. Lecturer, University of Hull from 1967 to 1976. Reader, University of Glasgow from 1976 to 1988. **Educ.:** M.A. from University of Glasgow, UK (1955-1958); B.D. from University of Glasgow, UK (1958-1961); S.T.M. from Union Theological Seminary, New York NY, USA (1961-1962); Ph.D. from Union Theological Seminary, New York NY, USA (1962-1964). **Interests:** ideology and religion. **Pubs.:** The Way of Transcendence, 1971; Constantine versus Christ, 1982 and Marx and the Failure of Liberation Theology, 1990.
Mailing address: University of Edinburgh, New College, Mound Place, Edinburgh EH1 2LX, Scotland, UK. T. (44 31) 225 8400
Residence: 7 Belmont Crescent, Glasgow G12 8EU, Scotland, UK. T. (44 41) 334 9992

Keikoin, Shun
Rel: Shinto; Ordained; b. 12 MAR 1903; Citizen of Japan.
International Organizations: Chief Priest of the Ise Shrine.
Mailing address: 4 Ujiuratacho, Ise, Mie 516, Japan. T. (81 596) 22 2341

Kell, Gerhard
Rel: Buddhist: Jôdo-Shinshú.
International Organizations: Head, German Branch of the Jôdo Shin Buddhist Community – Europe.
Mailing address: Courbièrestrasse 8, 10787 Berlin, Germany. T. (49 6030) 211 91 13.

Keller, Columban, C.M.S.F., Br

Rel: Roman Catholic; Ordained; celibate.
International Organizations: Superior General of the Missionary Brothers of St Francis of Assisi.
Mailing address: Schlossstrasse 30, 96049 Bamberg, Germany. T. (39 951) 56 214

Kelly, Daniel M

Rel: Christian.
International Organizations: Executive Director of the ACROSS.
Mailing address: PO Box 21033, Nairobi, Kenya. T. (254 2) 220693 – 220655. Fax 339388. C. ACROSS, Nairobi. Tx WORCON 22165 Nairobi.

Kelly, Jerome T, F.P.M., Br

Rel: Roman Catholic; Ordained; celibate.
International Organizations: Superior General of the Presentation Brothers.
Mailing address: Mount St Joseph, Blarney Street, Cork, Ireland. T. (353 21) 43160.

Kelly, Patrick Altham

Rel: Roman Catholic; Ordained on 16 FEB 1962; b. 23 NOV 1938 in Morecambe, UK; celibate; Citizen of United Kingdom.
International Organizations: Bishop (since 1984) of Salford.
Career: Assistant Priest, Lancaster Cathedral from 1964 to 1966. Professor of Dogmatic Theology, Oscott College, Birmingham from 1966 to 1979. Rector, Oscott College, Birmingham from 1979 to 1984.
Mailing address: Wardley Hall, Worsley, Manchester M28 5ND, UK. T. (44 61) 794 2825

Kemp, Eric Waldram, Rt Rev

Rel: Church of England; Ordained in 1940; b. 27 APR 1915 in Grimsby, UK; m. to Leslie Patricia Kirk; c. 1 boy and 4 girls; Citizen of United Kingdom.
International Organizations: Bishop (since 1974) of Chichester.
Other organizations: Honorary Provincial Canon (since 1960), Cape Town.
Main field of work: Ministry. **Career:** Ordained Deacon in 1939. Curate, St Luke's, Southampton from 1939 to 1941. Librarian, Pusey House, Oxford from 1941 to 1946. Chaplain, Christ Church, Oxford from 1943 to 1946. Acting Chaplain, St John's College, Oxford from 1943 to 1945. Fellow, Chaplain, Tutor and Lecturer in Theology and Medieval History, Exeter College, Oxford from 1946 to 1969. Dean, Worcester from 1969 to 1974. Examining Chaplain to Bishop, Mon from 1942 to 1945. Examining Chaplain to Bishop, Southwark from 1946 to 1950. Examining Chaplain to Bishop, St Albans from 1946 to 1969. Examining Chaplain to Bishop, Exeter from 1949 to 1969. Examining Chaplain to Bishop, Lincoln from 1950 to 1969. Proctor in Convocation for University, Oxford from 1949 to 1969. Bishop, Oxford's Commissary for Religious Communities from 1952 to 1969. Chaplain to the Queen from 1967 to 1969. Canon and Prebendary, Caistor in Lincoln Cathedral in 1952. Bampton Lecturer from 1959 to 1960. **Educ.:** M.A. from University of Oxford, Exeter College, UK; D.D. from University of Oxford, St Stephen's House, UK. **Interests:** music, travel. **Hons.:** Hon. D.Litt., Sussex (1986); Hon. D.D., Bern (1987). **Mem.:** National Liberal (Club); Fellow, Royal Historical Society (1951) **Pubs.:** Man: Fallen and Free, 1969; Square Words in a Round World, 1980 and contributions to *English Historical Review*, *Journal of Ecclesiastical History*.
Mailing address: The Palace, Chichester, W Sussex PO19 1PY, UK. T. (44 243) 782161

Kemp, Roger Francis, Rev Dr

Rel: Baptist; Ordained in 1974; b. 26 JAN 1945 in Adelaide, South Australia, Australia; m. to Barbara Joan Jury; c. 1 boy and 1 girl; Citizen of Australia; Lang. English.
International Organizations: Secretary-General (since 1989) of the International Council of Accrediting Agencies for Evangelical Theological Education.
Other organizations: Vice-Chairman, Missionary Department (since 1986), Australian Baptist Missionary Society, NSW; Lecturer in Church History and Missiology (since 1988), Baptist Union of New South Wales.
Main field of work: Missiology. **Career:** Pastor, Baptist Union of South Australia from 1969 to 1976. Missionary, Australian Baptist Missionary Society from 1977 to 1985. **Educ.:** B.A. from University of Adelaide, Australia (1962-1973); Dip. Theol. from Melbourne College of Divinity, Australia (1969-1973); B.D. from Melbourne College of Divinity, Australia (1973-1975); Th.D. from University of South Africa, Pretoria, South Africa (1980-1987). **Interests:** theological education. **Mem.:** South Pacific Association of Missiological Studies **Pubs.:** "Jesus the Revolutionary Missionary", 1990.
Mailing address: 4A Paterson Road, Springwood NSW 2777, Australia. T. (61 47) 511 722. Fax (61 47) 515 792
Residence: See Mailing Address. T. See Mailing Address

Kempton, Wendell W

Rel: Baptist.
International Organizations: President of the Association of Baptists for World Evangelism.
Mailing address: PO Box 5000, Cherry Hill NJ 08034, USA. T. (1 609) 424 46 06 – (1 609) 424 24 24.

Kendel, Ole, Pastor

Rel: Seventh-Day Adventist.
International Organizations: Youth Director, Trans-European Division of the General Conference of Seventh-Day Adventists.
Mailing address: 119 St Peter's Street, St Albans AL1 3EY, UK.

Keningsberg, Rosita Ehrlich

Rel: Jewish.
International Organizations: President of the International Network of Children of Jewish Holocaust Survivors.
Mailing address: 19918 NE 19th Ct, North Miami Beach FL 33179, USA.

Kern, Ingrid

Rel: Evangelical.
International Organizations: Chairman of the Commission on Women's Concerns.
Mailing address: Drosselstieg 24, D-10318 Berlin, Germany. T. (49 30) 5090 171. Fax 5090 171.

Kern, Paul R

Rel: Church of Jesus Christ of Latter-Day Saints.
International Organizations: Secretary of the Esperantist Mormons.
Mailing address: Rue Haute Colline 28, B-4052 Beaufays, Belgium.

Kerr, David A

Rel: Unknown.
International Organizations: Director of the Duncan Black MacDonald Center for the Study of Islam and Christian / Muslim Relations.
Mailing address: c/o Hartford Seminary, 77 Sherman Street, Hartford CT 06105, USA. T. (1 203) 232 4451.

Keun, Kim Sang, Rev

Rel: Presbyterian.
International Organizations: General Secretary of the Presbyterian Church in the Republic of Korea.
Mailing address: 507, 136-46 Yunchi Dong, Chongno Ku, Seoul 110, Korea.

Kevehazi, Michael

Rel: Jewish.
International Organizations: Chairman of the Maccabi World Union.
Mailing address: Kfar Hamaccabiah, 52105 Ramat-Gan, Israel. T. (972 3) 715722. Fax 772059. Tx 99919 MACAB IL.

Key, John Christopher, Rev

Rel: Church of England; Ordained on 23 SEP 1962; b. 04 AUG 1936 in Newton Abbot, Devon, UK; m. to Margaret Loveday; c. 2 boys and 1 girl; Citizen of United Kingdom and Australia; Lang. English, French, Melanesian and Pidgin.

International Organizations: Director of Partnership Development (since 1991) of the World Vision International.

Main field of work: Ministry. **Career:** Rector, Church of Papua New Guinea from 1967 to 1971. Rural Dean, Church of England from 1971 to 1981. Director, World Vision International from 1981 to 1991. **Educ.:** M.A. from University of Cambridge, UK (1957-1962). **Interests:** music.

Mailing address: 919 W Huntington Drive, Monrovia CA 91016, USA. T. (1 818) 303 8811. C. Worvis. Tx 275 335 WORV UR.

Residence: 900 W Sierra Ave, Suite 127, Azusa CA 91702, USA. T. (1 818) 969 8113

Keyes, Lawrence E

Rel: Christian.

International Organizations: President of the OC International.

Mailing address: PO Box 36900, Colorado Springs CO 80936-6900, USA. T. (1 719) 592 9292. Fax 592 0693.

Khadro, Sangye, Ven

Also know as Kathleen Ann McDonald; Rel: Buddhist; Ordained on 02 MAY 1974; b. 11 MAR 1952 in Sacramento CA, USA; celibate; Citizen of USA; Lang. English, French and Tibetan (very little).

International Organizations: Member of the Foundation for the Preservation of the Mahayana Tradition.

Other organizations: Spiritual Advisor (since 1989), Amitabha Buddhist Centre (affiliated with Foundation for the Preservation of the Mahayana Tradition); Spiritual Advisor (since 1991), National University of Singapore Buddhist Society.

Main field of work: Meditation. **Career:** Director, Dorje Pamo Nunnery (France) from 1982 to 1983. Spiritual Advisor and Teacher, Buddha House (Australia) from 1984 to 1986. **Pubs.:** *How to Meditate*, Wisdom Pub. (under name Kathleen McDonald) and Editor, *Compassion and Wisdom*, ABC Pub.

Mailing address: Amitabha Buddhist Centre, 494-D Geylang Rd, Singapore 1438, Singapore. T. (65) 7458547

Residence: 40-B Ridout Rd, Singapore 1024, Singapore. T. (65) 4720451

Khalifa, Isa Muhammad Abdulla Aal, Shaikh

Rel: Muslim.

International Organizations: Secretary of the International Islamic Charitable Organization.

Mailing address: PO Box 3434, Safat 13035, Kuwait. T. (965) 2418025 – 2448786. C. AL-ISLAMIA. Tx 30494 CHARITY – KT.

Khalsa, Shakti Parwha Kaur

Rel: Hindu.

International Organizations: Executive Director of the 3HO Foundation.

Mailing address: 1620 Preuss Road, Los Angeles CA 90035, USA. T. (1 213) 552 3416.

Khamenei, Hojatoleslam Ali

Rel: Muslim; b. in Mashad, Khorassan, Iran Islamic Rep; married; c. 4 boys and 1 girl; Citizen of Iran.

International Organizations: Friday Prayer Leader (since 1989) of the Mosque in Teheran.

Main field of work: Wali Faqih (Religious Leader). **Career:** Returned to Meshed in 1964. Imprisoned six times from 1964 to 1978. Once exiled in 1978. Former personal representative, Ayatollah Khomeini, Supreme Defence Council. Secretary-General (and President, Central Committee), Islamic Republican Party from 1980 to 1987. President, Iran in 1981. Member, Revolutionary Council until its dissolution in 1979. Survived assassination attempt in 1981. Commander, Revolutionary Guards in 1980. Representative, Revolutionary, Ministry of Defence. **Interests:** traditional Persian athletics.

Mailing address: Office of the Religious Leader, Teheran, Iran Islamic Rep.

Khamis, Mar Aprim, Bishop

Rel: Apostolic-Catholic Assyrian Church.

International Organizations: Bishop of the United States and Canada of Apostolic Catholic Assyrian Church of the East.

Mailing address: 8908 Birch Avenue, Morton Grove IL 60053, USA.

Khan, Abrahim Habibulla

Rel: Muslim; b. 13 APR 1943 in Guyana; m. to Pamela O'Neill; c. 3 children.

International Organizations: Lecturer in Religious Studies and Divinity (since 1986) of the Trinity College, University of Toronto.

Other organizations: Research Associate in Arts at Trinity College (since 1986); Convenor (since 1987), Kierkegaard Circle, University of Toronto; Member, Ethics Committee (since 1990), Clarke Institute of Psychiatry; Executive Member (since 1990), North American Association for the Study of Religion.

Career: Lecturer in Religious Studies, McGill University in 1973. Research Associate in Philosophy, McGill University from 1974 to 1975. Assistant Professor, McGill University from 1979 to 1981. Lecturer in Religious Studies, Concordia University and Champlain College from 1974 to 1975. Assistant Professor of Religious Studies, University of Manitoba from 1975 to 1976. Assistant Professor of Religious Studies, University of Toronto from 1976 to 1979. Lecturer in Philosophy, Trent University in 1982. Assistant Professor, Trent University from 1983 to 1984. Associate Professor of Religious Studies, McMaster University (summer) from 1984 to 1985. Research Associate at Trinity College, University of Toronto from 1984 to 1985. Honorary Fellow and Research Reader, Centre for Religious Studies from 1985 to 1986. Board Member, Dialogue Centre Montreal from 1969 to 1975. Editor, Newsletter, Canadian Theological Society from 1984 to 1987. President, Canadian Theological Society from 1985 to 1986. Chairman, Kierkegaard Seminar, American Academy of Religion from 1985 to 1989. **Educ.:** B.S. from Howard University, Washington DC, USA (1965); B.D. from Yale University, New Haven CT, USA (1968); M.A. from McGill University, Montreal, Canada (1971); Ph.D. from McGill University, Montreal, Canada (1973). **Interests:** volleyball. **Mem.:** Centre for Bioethics, University of Toronto (Associate Member); Canadian Society for the Study of Religion; Canadian Theological Society; Association for the Advancement of Scandinavian Studies, Canada; American Academy of Religion; Canadian Philosophical Society; Søren Kierkegaard Society

Mailing address: Trinity College, University of Toronto, 6 Hoskin Ave, Toronto, Ontario M5S 1H8, Canada.

Residence: 1045 Escott Court, Pickering ON L1X 1P2, Canada.

Khan, Amanullah, Prof Dr

Rel: Muslim; b. 15 AUG 1932; married; Lang. Urdu, English, Arabic, Persian and Punjabi.

International Organizations: Professor (since 1982) of the Institute of Islamic Studies, University of the Punjab, Lahore.

Career: Asia Foundation Research Fellow from 1961 to 1963. Lecturer, University of the Punjab, Lahore from 1963 to 1970. Assistant Professor, University of the Punjab from 1970 to 1978. Associate Professor, University of the Punjab from 1978 to 1982. Government of West Pakistan Research Fellow from 1970 to 1972. Chairman, Institute of Islamic Studies, University of the Punjab from 1975 to 1976. Chairman, Institute of Islamic Studies, University of the Punjab from 1981 to 1988. Warden, University of the Punjab from 1974 to 1975. Member, Editorial Boards, *Science and Technology in the Islamic World* (Research Journal of the National Science Council of Pakistan), *The Majallah-e-Tahqiq* (University of the Punjab), *The Light of Islam* (journal), etc. Teaching at Ulama Academy (for 12 years). About 1000 television and radio broadcasts. **Educ.:** M.A. Islamic Stds.; M.A. Hist.; Ph.D. Islamic Stds./Arabic. **Mem.:** member of 20 learned and academic bodies **Pubs.:** about 70 books and research articles (in English and Urdu).

Mailing address: Institute of Islamic Studies, University of the Punjab, Quaid-e-Azam Campus, Lahore, Pakistan. T. (92 42) 867993

Khan, Elder Sardar Feroze

Rel: United Presbyterian.

International Organizations: Moderator of the United Presbyterian Church of Pakistan.

Mailing address: PO Box 395, Rawalpindi, Pakistan.

Khan, Inamullah, Dr

Rel: Muslim; b. 17 SEP 1914 in Rangoon, Burma; m. to Khadoja Khalik; c. 1 boy and 5 girls; Citizen of Pakistan.
International Organizations: Secretary-General of the World Muslim Congress.
Honorary President of the World Conference on Religion and Peace. President of the Muslim Minority Committee. President of the Commission on Islamic Specialized Organizations.
Career: Editor-in-Chief, Burma Muslim Daily (until 1942). Founder, Modern World Muslim Congress. **Interests:** reading, writing.
Mailing address: D-26, Block 8, Gulshan-e-Iqbal, Karachi 75300, Pakistan. T. (92 21) 468 737

Khan, Peter Jamel, Dr

Rel: Bahá'í; b. 12 NOV 1936 in Bowral NSW, Australia; m. to Janet Griffith; Citizen of Australia; Lang. English.
International Organizations: Member, Universal House of Justice (since 1987) of the Bahá'í International Community.
Other organizations: Editorial Board (since 1971), IEEE *Microwave Transactions*.
Main field of work: Administration. **Career:** Research Engineer, Cooley Electronics Laboratory, MI, USA from 1963 to 1967. Associate Professor, University of Michigan, USA from 1967 to 1975. Visiting Professor, University of New South Wales, Australia in 1975. Associate Professor, University of Queensland, Australia from 1976 to 1983. Counsellor, International Teaching Centre, Israel from 1983 to 1987. **Educ.:** B.Sc. from University of Sydney, New South Wales, Australia (1953-1956); B.Eng. from University of Sydney, New South Wales, Australia (1957-1959); Ph.D. from University of Sydney, New South Wales, Australia (1959-1963). **Interests:** Science and religion. **Mem.:** Fellow, Institute of Engineers (Australia), Institute of Radio and Electronics Engineers **Pubs.:** Over 100 articles in research journals on microwave theory, solid-state circuit design, and non-linear mathematics.
Mailing address: PO Box 155, 31001 Haifa, Israel. T. (972 4) 358 358. C. BAHAIFAITH, HAIFA. Tx 46626 BAYT IL. Fax (972 4) 372 425.
Residence: 10 Abraham Lincoln, 31001 Haifa, Israel T. (972 4) 244 063

Khong, Peter, Rev

Rel: Lutheran.
International Organizations: Chairperson of the Lutheran Communications in Asia.
Mailing address: c/o Glad Sounds SDN. BHD., PO Box 1019, Jalan Semangat, 46970 Petaling Jaya, Selangor, Malaysia. T. (60 3) 78 76 45 - 78 70 70

Khoraiche, Antoine Pierre, H E Card

Rel: Roman Catholic; Ordained on 12 APR 1930; b. 20 SEP 1907 in Ain Ebel, Lebanon; celibate; Citizen of Lebanon.
International Organizations: Member (since 1983) of the College of Cardinals.
Other organizations: Maronite Patriarch Emeritus (since 1986), Antioch.
Career: Church of Tarsus of Maronites from 1950 to 1957. Saida of Maronites from 1957 to 1975. Maronite Patriarch, Antioch from 1975 to 1986. Created Cardinal by Pope John Paul II in 1983.
Mailing address: c/o Patriarcat Maronite, Bkerké, Lebanon. T. (961) 93 00 11

Khoromi, Parvin

Rel: Muslim: Sufi.
International Organizations: Member of the School of Islamic Sufism.
Mailing address: PO Box 5827, Washington DC 20016, USA. T. (1 301) 654 91 91.

Khouri, Elia Khader, Rev

Rel: Anglican; Ordained; b. in West Bank; married; Citizen of Palestine.
International Organizations: Assistant Bishop in Amman of Anglican Church.
Other organizations: Member, Executive Committee (since 1987), Palestine National Council; Leader, Arab Anglican Communion in Amman.
Career: Anglican priest in several cities, expelled from the West Bank in 1969. Supporter, Fatah for many years. Representative, Palestinian Christians, Palestine Liberation Organization. Candidate for Bishopric, Jerusalem supported by a majority vote of the Arab Diocesan Council in 1974.
Mailing address: PO Box 598, Amman, Jordan.

Khumalo, Samson A, Rev

Rel: Presbyterian.
International Organizations: Chief Clerk of the General Assembly of the Presbyterian Church of Africa.
Mailing address: PO Box 54840, Umlazi 4031, South Africa.

Kiedron, Vladislav, Bishop

Rel: Evangelical.
International Organizations: Bishop of Silesian Evangelical Church of the Augsburg Confession.
Mailing address: Na Nivach 7, 73701 Cesky Tesin, Czechoslovakia.

Kienel, Paul A, Dr

Rel: Christian.
International Organizations: Executive Director of the Association of Christian Schools International.
Mailing address: PO Box 4097, Whittier CA 90607, USA. T. (213) 694 4791.

Kiggins, Thomas, Rev

Rel: Christian.
International Organizations: Executive Secretary of the Irish Missionary Union.
Mailing address: Orwell Park, Rathgar, Dublin 6, Ireland. T. (353 1) 97 17 70 - 96 54 33. Fax 96 50 29.

Kihara, Kakuei

Rel: Buddhist; Ordained; b. 27 JUN 1927; Citizen of Japan.
International Organizations: Manager of the Komyonenbutsu Shingoshoshu Honpukuji.
Educ.: Koyasan University, Japan.
Mailing address: 2120 Kiyamacho Miyaura Nakayama, Miyaki-gun, Saga 841-02, Japan. T. (81 942) 92 2451

Kilani, Mohamed Mounif

Rel: Muslim.
International Organizations: President of the Fondation européenne pour la rencontre des civilisations et le dialogue interréligieux.
Mailing address: Avenue Albert 12, B-1060 Bruxelles, Belgium. T. (32 2) 347 05 90.

Kilem-Mbila, Daniel

Rel: Christian.
International Organizations: Executive Secretary of the Association pour la coopération des Eglises, l'environnement et le développement de l'Afrique centrale.
Mailing address: BP 13798, Brazzaville, Congo.

Kim, Eunice

Rel: Christian.
International Organizations: Executive Secretary of the Asian Church Women's Conference.
Mailing address: 91-175 Shin Soo Dong, Mapo-ku, Seoul 121, Korea Rep.

Kim, Stephen Sou-Hwan, H E Card

Rel: Roman Catholic; Ordained on 15 SEP 1951; b. 08 MAY 1922 in Taegu, Kyong Buk, Korea Rep; celibate; Citizen of Republic of Korea; Lang. Korean, Japanese, German, French, Italian and English.
International Organizations: Member (since 1969) of the College of Cardinals.
Member of the Congregation for the Evangelization of Peoples. Member of the Pontifical Council for Dialogue with Non-Believers.
Other organizations: Metropolitan Archbishop (since 1968), Seoul.
Career: Pastor, Andong Parish, Taegu from 1951 to 1953. Secretary to Archbishop, Taegu from 1953 to 1955. Pastor, Kimchon Parish, Taegu from 1955 to 1956. Director, Sung-Eui Schools, Kimchon from 1955 to 1956. Sociology studies, University of Münster, Germany from 1956 to 1964. President and Editor-in-Chief, *Catholic Shibo* (weekly) from 1964 to 1966. Bishop, newly created Diocese of Masan from 1966 to 1968. Created Cardinal by Pope Paul VI in 1969. President, Follow-up Committee for Federation of Asian Bishops' Conference from 1970 to 1973. President, Bishops' Conference of Korea from 1970 to 1975.

President, Bishops' Conference of Korea from 1981 to 1987. Delegate, Bishops' Conference of Korea to the Synod (in 1967, 1971, 1974, 1980, 1983, 1985). **Educ.:** Theol. Stds. from Major Seminary of Seoul, Korea (1947-1951). **Interests:** music, literature. **Hons.:** Dr.h.c., Sogang University, Seoul (1974), Notre Dame University, IN (1977), Sophia University, Tokyo (1988), Korea University, Seoul (1990), Seton Hall University, NJ (1990).
Mailing address: Archbishop's House, 2-Ga 1, Myong-dong, Chung-ku, Korea Rep. T. (82 2) 771 7600
Residence: See Mailing Address. T. See Mailing Address

Kimba, Mutombo Dunga, Rev

Rel: Methodist.
International Organizations: President for West Africa of the World Federation of Methodist Women.
Mailing address: BP 447, Kamina, Zaire.

Kinaro, Elizabeth

Rel: Quaker.
International Organizations: Regional Representative for Africa of the Friends World Committee for Consultation.
Mailing address: Friends Int. Centre, N'gong Road, PO Box 41946, Nairobi, Kenya. T. (254 2) 567601.

King, Eileen

Rel: Christian.
International Organizations: Executive Director of the International Committee for World Day of Prayer.
Mailing address: Room 812, 475 Riverside Drive, New York NY 10115, USA. T. (1 212) 870 3049. Fax 870 2338. C. CHWOMENUN NY, USA.

King, Geoffrey, Rev Dr

Rel: Roman Catholic; Ordained on 08 DEC 1973; b. 08 JUN 1943 in Burwood NSW, Australia; celibate; Citizen of Australia; Lang. English, German and French.
International Organizations: Director (since 1989) of the East Asian Pastoral Institute.
Moderator (since 1990) of the Catholic Biblical Federation. Secretary (since 1991) of the Conference of Asian-Pacific Pastoral Institutes.
Main field of work: Higher education. **Career:** Professor, United Faculty of Theology from 1979 to 1989. Rector, Jesuit Theological College from 1981 to 1987. **Educ.:** B.A. (Hons) from University of Melbourne, Victoria, Australia (1965-1968); B.D. from Melbourne College of Divinity, Australia (1971-1973); J.C.D. from Catholic University of America, Washington DC, USA (1975-1979). **Interests:** history of canon law. **Mem.:** Catholic Biblical Federation, Canon Law Society of America **Pubs.:** *The Acceptance of Law by the Community*, Ann Arbor, 1979; "The Catholic Church in China: A Canonical Evaluation", in *Jurist* 49, pp. 69-94, 1989 and "Limits and Hopes: Catholics and Religious Freedom in the People's Republic of China", in *Pacifica* 2, pp. 175-209, 1989.
Mailing address: PO Box 221, U.P. Campus, Quezon City 1101, Philippines. T. (63 2) 98 70 18
Residence: Ateneo de Manila, Katipunan Road, Quezon City 1101, Philippines T. (63 2) 98 31 82

King, George, H E Dr Sir

Rel: Aetherius Society; Ordained in 1960; b. 23 JAN 1919 in Wellington, Shropshire, UK; m. to The Rt Rev Dr Monique Noppe; Citizen of United Kingdom; Lang. English.
International Organizations: President (since 1955) of the Aetherius Society.
Other organizations: Metropolitan Archbishop (since 1980), The Aetherius Churches worldwide; President (since 1981), The Aetherius Society College of Spiritual Sciences; International Police Chaplain (since 1984), American Federation of Police; National Chaplain, American Park Rangers Association; Deputy Director General/Member of the Advisory Board, International Biographical Association, Cambridge; Governor/Life Member, Royal National Lifeboat Institution, UK; Deputy Governor, American Biographical Institute Research Association; Presidential Commission, Republican Presidential Task Force, USA; Permanent Member, National Republican Senatorial Committee, USA; State Advisor, United States Congressional Advisory Board; Founder/President, College of Spiritual Sciences, UK and USA; Member of the Advisory Board, Florida Crime Prevention Commission, USA; Ambassador of Good Will among all People, State of West Virginia; Lieutenant-Colonel Aide-de-Camp, Governor's Staff, State of Virginia; Honorary Kentucky Colonel on staff, Governor of Kentucky; Honorary Lieutenant-Colonel Aide-de-Camp, Alabama State Militia.
Main field of work: Religious leader, author, teacher, metaphysical scientist. **Career:** Served as Special Duties Section Leader in Britain's national Fire Service (during World War II). Founded The Aetherius Society in 1955. Consecrated as a Bishop and Archbishop in 1980. Founded The Aetherius Society College of Spiritual Sciences in 1981. Became active in the work, the Government of Poland (in Exile) and served as an Ambassador-at-Large, as Minister of State and Special Advisor to the President, and as Deputy Prime-Minister from 1980 to 1990. **Educ.:** D.D. from International Theological Seminary, CA, USA (1962); Dr. Phil. Theol. from International Theological Seminary, CA, USA (1984); Prof. Metaphysics from Parthasarathy Cultural Academy, Madras, India (1985). **Interests:** photography, film-making, electronics, hi-fi, boating. **Hons.:** Count de Florina, Lord of the Manor of Allington, Sussex; Prize of Peace and Justice (Knight of Humanity) (1981); Minister of the Year, International Evangelism Crusades (1981); Freeman of the City of London and of England (1986); Empire Medal (UK); World War II Defence Medal (UK); Medal of Europe, Member of the Federation of Combatant Allies in Europe; Medal of the Secret Army of Belgium; Certificate of Appreciation, The John Edgar Hoover Memorial, American Police Hall of Fame; Man of the Year, American Biographical Institute (1990, 1991); Peace Prize, International Evangelism Crusades; Gold Medal for Outstanding Service to Humanity, Imperial House fo Byzantium; Distinguished Leadership Award, American Security Council; Humanitarian Award, California Motion Picture Council; Grant of Arms, Her Majesty's College of Arms, London; Prince Grand Master (and founder), The Mystical Order of Saint Peter (1981); Prince Grand Master, International Chivalric Order of The Knights of Justice; Honourable Lieutenant Grand Master, The Grand Sovereign Dynastic Hospitaller Order of Saint John, Knights of Malta; Order of the White Eagle; Grand Cordon, Order of Polonia Restituta; Order of Virtuti Militari. **Mem.:** Guild of Freemen of the City of London (Life Member), Association of Freemen of England (Member), Royal Society of Tropical Medicine and Hygiene, UK (Fellow), Royal Society of Health, UK (Affiliate), The Commonwealth Trust (Fellow and Foundation Member), American Police Hall of Fame and Museum (Member of the ex-officio Board of Trustees), United States National Rifle Association (Life Member) **Pubs.:** *The Nine Freedoms*; *The Twelve Blessings* and *You Are Responsible*.
Mailing address: 6202 Afton Place, Hollywood CA 90028-8298, USA. T. (1 213) 465 9652.
Residence: 6216 Afton Place, Hollywood CA 90028-8298, USA. T. (1 213) 462 2672

King, Wayne

Rel: Christian.
International Organizations: President of the Church of God – Men International.
Mailing address: PO Box 175, Clayton OH 45315, USA. T. (1 513) 233 9528.

Kinghorn, Johann, Prof

Rel: Dutch Reformed; Ordained in 1975; b. 02 DEC 1946 in Pretoria, South Africa; m. to Esmé Brink; c. 2 boys; Citizen of South Africa; Lang. Afrikaans, English and German.
International Organizations: Professor of Religious Studies and Theology (since 1991) of the University of Stellenbosch.
Other organizations: Member, Faculty of Arts and Humanities and Faculty of Theology, University of Stellenbosch; Assistant Editor, *Scriptura*, Journal for Bible and theology in Southern Africa; Secretary, Theological Society of Southern Africa.
Main field of work: University education. **Career:** Temporary Lecturer in Greek, University of Pretoria from 1969 to 1970. Lecturer (part-time) in Religious Studies, University of Stellenbosch from 1973 to 1974. Temporary Lecturer in Religious Studies, University of Stellenbosch in 1976. Permanent Lecturer in Religious Studies, University of Stellenbosch from 1977 to 1980. Permanent Lecturer in Religious Studies and Theology, University of Stellenbosch from 1981 to 1983. Senior Lecturer in Religious Studies and Theology, University of Stellenbosch from 1984 to 1990. Co-founder, Centre for Contextual Hermeneutics. Programme Director, Centre for Contextual Hermeneutics from 1987 to 1990. Head, Centre for Contextual Hermeneutics in 1991. **Educ.:** B.A. from University of Pretoria, South Africa (1966-1968); B.A. (cum laude) from University of Pretoria, South Africa (1970);

B.Th. (cum laude) from University of Stellenbosch, South Africa (1971-1973); L.Th. (cum laude) from University of Stellenbosch, South Africa (1975); Eberhard-Karls-Universität Tübingen, Germany (1974-1975); D.Th. from University of Stellenbosch, South Africa (1976-1982). **Interests:** Christianity, Christian theology and the correlation with social values. **Mem.:** American Academy of Religion; Institute for Contextual Theology, Johannesburg; South African Society for Biblical Studies; South African Academy of Science and Art **Pubs.:** Editor, with B C Lategan, C F van der Merwe, *Into Africa – Afrikaners reflect on 'coming home'*, Stellenbosch, Centre for contextual hermeneutics, 70 pp. (with S J Terreblanche, C L McCarthy, H J Kotzé, W Verwoerd, C M Pauw, B A Müller), 1989; *A Home for all. Democracy in South Africa from the perspective of social theology*, Stellenbosch, Center for contextual hermeneutics, 170 pp., 1990 (in Afrikaans) and "Reflections on the Task of Theology in Africa", in *Scriptura*, 39, 9 pp.
Mailing address: Dept. of Religious Studies, Faculty of Arts and Humanities, University of Stellenbosch, Stellenbosch 7600, South Africa. T. (27 2231) 77 2117. Fax 77 2031

Kingston, Frederick Temple, Rev

Rel: Anglican; Ordained; b. 30 DEC 1925 in Toronto ON, Canada; m. to Pauline B Smith; c. 2 boys and 2 girls; Citizen of Canada.
International Organizations: Principal (since 1965) of the Canterbury College, University of Windsor.
Other organizations: Professor of Philosophy (since 1959), University of Windsor.
Main field of work: Ministry and university education. **Career:** Professor of Systematic Theology and Philosophy of Religion, The Anglican Theological College of BC from 1953 to 1959. Visiting Professor, Laval University in 1983. National President, Humanities Association of Canada from 1985 to 1987. Priest, Diocese of Huron. Member, High Table, Van Mildert College, Durham, Christ Church, Oxford. **Educ.:** B.A. from University of Toronto, Trinity College, Canada (1947); M.A. from University of Toronto, Trinity College, Canada (1950); L.Th. from University of Toronto, Trinity College, Canada (1950); B.D. from University of Toronto, Trinity College, Canada (1952); D.Phil. from University of Oxford, Christ Church, UK (1954). **Hons.:** Honorary Fellow, Huron College, University of Western Ontario. **Mem.:** Canadian Philosophical Association; Humanities Association of Canada; Mind Association; Aristotelian Society; British Society for the History of Philosophy **Pubs.:** Author, French Existentialism: A Christian Critique, 1961; Editor and Contributor, Friendship and Dialogue Between Ontario and Quebec, 1983 and Editor, Canterbury College Lenten Lecture Booklets, 1972-1986.
Mailing address: 172 Patricia Rd, Windsor ON N9B 3B9, Canada.
Residence: 833 Kildare Rd, Windsor ON N8Y 3H3, Canada.

Kingston, L C, Rev

Rel: Congregational.
International Organizations: General Secretary of the Fellowship of Congregational Churches.
Mailing address: PO Box 47, Bexley South NSW 2207, Australia.

Kinnamon, Michael K, Rev Dr

Rel: Disciples of Christ; Ordained on 15 DEC 1976; b. 15 APR 1949 in Ottumwa IA, USA; m. to Katherine; c. 2 girls; Citizen of USA; Lang. English, French and German.
International Organizations: Dean and Associate Professor of Theology and Ecumenical Studies (since 1988) of the Lexington Theological Seminary, KY.
Other organizations: Member, Disciples General Board and Administrative Committee; Member, United Methodist Church's General Commission on Christian Unity and Interreligious Concerns.
Main field of work: Theological education. **Career:** Executive Secretary, Commission on Faith and Order, World Council of Churches from 1980 to 1983. Assistant Professor, Christian Theological Seminary from 1983 to 1988. Acting Dean, Christian Theological Seminary from 1986 to 1988. Visiting Professor, United Theological College, Bangalore (India) in 1987. Lectured widely on ecumenism in Asia, Australia, Europe and North America. **Educ.:** B.A. from Brown University, Providence RI, USA (1967-1971); M.A. from University of Chicago, Divinity School, IL, USA (1973-1976); Ph.D. from University of Chicago, Divinity School, IL, USA (1980); Tel-Aviv University, Israel (1969-1970). **Interests:** sports, traveling. **Pubs.:** With Katherine Kinnamon, Every Day We Will Bless You: A Book of Daily Prayer; Editor, Disciples of Christ in the 21st Century and articles in: *One in Christ*, *Journal of Ecumenical Studies*, *Ecumenical Review*, *Mid-Stream*, *Theology Today*, *The Disciple*, and *Encounter*.
Mailing address: Lexington Theological Seminary, 631 South Limestone St, Lexington KY 40508, USA. T. (1 606) 252 0361
Residence: 395 Eagle Creek Drive, Lexington KY 40515, USA. T. (1 606) 263 1958

Kiong, Tan Chi

Rel: Christian.
International Organizations: Secretary-General of the Asia Alliance of YMCAs.
Mailing address: 23 Waterloo Road, Kowloon, Hong Kong. T. (852 3) 7808347. C. FLAMINGO. Tx 54522 ASIAY HX. Fax (852 3) 854692.

Kirima, Nicodemus, Rt Rev

Rel: Roman Catholic; Ordained on 22 DEC 1962; b. 03 MAY 1936 in Karatina, Kenya; celibate.
International Organizations: President of the Association of Member Episcopal Conferences in Eastern Africa.
Other organizations: Bishop, Nyeri; President, Kenya Episcopal Conference.
Mailing address: Gitanga Road, PO Box 21191, Nairobi, Kenya. T. (254 2) 566506 – 566507. C. 'AMECEA' NAIROBI.

Kirkendoll, Chester A, Bishop

Rel: Methodist Episcopal.
International Organizations: Presiding Bishop of Christian Methodist Episcopal Church.
Mailing address: 11470 Northway Drive, St Louis MO 63136, USA.

Kirsch, L, Fr

Rel: Roman Catholic; Ordained; celibate.
International Organizations: Contact of the Jeunesse rurale chrétienne.
Mailing address: Rue du Château, L-4936 Bettange-Mess, Luxembourg. T. (352) 37 088.

Kirton, Allan F, Rev

Rel: Protestant.
International Organizations: Secretary-General of the Caribbean Conference of Churches.
Mailing address: PO Box 616, Bridgetown, Barbados. T. (809 42) 726 81 – (809 42) 726 82 – (809 42) 726 83. Tx 2335 CADEC WB.

Kissling, Frances

Rel: Roman Catholic; b. 15 JUN 1943 in New York NY, USA; Citizen of USA; Lang. English and Spanish.
International Organizations: President (since 1982) of the Catholics for a Free Choice.
Other organizations: Founder and Treasurer, The Global Fund for Women; Standing Committee on Women's Right, American Public Health Association; Board Member, International Women's Health Coalition; Member, Development Committee, Women's Alliance for Theology, Ethics and Ritual; Member, District of Columbia Mayor's Panel on Teen Pregnancy Prevention; Lecturer and consultant on reproductive health; Active in Women-Church Convergence (coalition of Catholic organizations).
Main field of work: Education. **Career:** Held a variety of jobs during the late 1960s, including positions with Macmillan Publishing and American Association of Psychiatric Services for Children. Director, ABC Clinics, New York City from 1970 to 1974. Consultant to international agency establishing abortion clinics in Italy, Austria and Mexico from 1974 to 1975. Founder and Executive Director, National Abortion Federation from 1976 to 1980. Director, The Youth Project, Washington DC from 1980 to 1981. **Educ.:** Attended from St John's University, Jamaica NY, USA (1961-1964); B.A. from New School for Social Research, New York NY, USA (1966). **Hons.:** Named one of the 'Eighty Women to Watch in the '80s' by Ms. magazine. **Mem.:** Women-Church Convergence **Pubs.:** Co-author, with Ellen Frankfort, *Rosie: The Investigation of A Wrongful Death*, The Dial Press, 1978 and Articles and op-ed pieces in *The Washington Post*, *The Boston Globe*, *The Christian Century*, *Christianity and Crisis*, and *The Journal of Feminist Studies in Religion*.
Mailing address: 1436 U St NW, Suite 301, Washington DC 20009-3916, USA. T. (1 202) 986 6093 – 332 7995
Residence: 2701 Connecticut Ave NW, Suite 508, Washington DC 20008, USA.

Kitagawa, Kiyoka

Rel: Shinto; Ordained; b. 29 MAR 1910; Citizen of Japan.
International Organizations: Chief Priest of the Nikko Nikozan Shrine.
Educ.: Kokugakuin University, Tokyo, Japan.
Mailing address: 2302 Yamauchi, Nikko, Tochigi 321-14, Japan. T. (81 288) 54 0413

Kitbunchu, Michael Michai, H E Card

Rel: Roman Catholic; Ordained on 20 DEC 1959; b. 25 JAN 1929 in Samphran, Nakhon Pathom, Thailand; celibate; Citizen of Thailand.
International Organizations: Member (since 1983) of the College of Cardinals.
Member of the Congregation for the Evangelization of Peoples.
Other organizations: Archbishop (since 1973), Bangkok; President, Bishops' Conference of Thailand.
Career: Created Cardinal by Pope John Paul II in 1983.
Mailing address: Assumption Cathedral, 51 Oriental Ave, Bangkok 10500, Thailand. T. (66 2) 2338712 – 2337441

Kivengere, Festo, Rt Rev

Rel: Church of Uganda; Ordained in 1967; m. to Mera Kivengere; c. 4 girls; Citizen of Uganda.
International Organizations: Ecclesiastic of the Church of Uganda.
Career: Teacher and evangelist, Tanzania from 1946 to 1959. School Supervisor, Kigezi District, Uganda from 1960 to 1962. Full-time evangelist from 1962 to 1966. Ordained Deacon in 1966. Consecrated Bishop in 1972. Bishop, Kigezi Diocese, Church of Uganda from 1972 to 1988. Team Leader, African Evangelistic Enterprise in 1971. **Hons.:** International Freedom Prize, Oslo (1977); Distinguished Public Service Award, Messiah College, PA (1977); Lindsborg Medal of Honor, KS (1980); St Augustine's Cross (1981); Hon. D.D., University of the South, Sewanee. **Pubs.:** The Spirit is Moving, 1979; Hope for Uganda and the World, 1980 and Revolutionary Love, 1981.
Mailing address: c/o PO Box 65, Kabale, Uganda.

Kivuli, John Mweresa, II, Rt Rev

Rel: Christian.
International Organizations: High Priest of the African Israel Nineveh Church.
Mailing address: Nineveh Headquarters, PO Box 701, Kisumu, Kenya.

Kizito, Robert N

Rel: Christian.
International Organizations: Acting Chief Executive of the All-Africa Press Service.
Mailing address: PO Box 14205, Nairobi, Kenya. T. (254 2) 743644 – 743773. C. CHURCHCON. Tx 22175.

Klaas, Joan, Sr

Rel: Roman Catholic; Ordained; celibate.
International Organizations: Superior General of the Sisters of the Most Precious Blood of O'Fallon.
Mailing address: 204 North Main Street, O'Fallon MO 63366, USA. T. (1 314) 272 6010.

Klein, Albert, Bishop

Rel: Evangelical.
International Organizations: Bishop of Evangelical Church of the Augsburg Confession in Romania.
Mailing address: Gen Magheru 4, 2400 Sibiu, Romania.

Kleinschmit, Nathalie

Rel: Christian.
International Organizations: Representative of the International Religious Fellowship.
Mailing address: c/o Dr Binet, 29 quai St Michel, F-75005 Paris, France. T. (33 1) 43 29 08 80.

Klemp, Harold, the Mahanta, the Living ECK Master, Sri

Rel: Eckankar; b. in Wisconsin, USA; m. to Joan; Citizen of USA.
International Organizations: Spiritual Leader (since 1981) of the Eckankar, Religion of the Light and Sound of God.
Main field of work: Ministry, writing, lecturing. **Career:** Enlisted in the Air Force. Trained as a language specialist at the University of Indiana and a radio intercept operator at Goodfellow AFB, TX. First encountered Eckankar in Japan. **Educ.:** Preministerial college in Milwaukee and Ft Wayne IN, USA. **Pubs.:** Author of many books and articles about the spiritual life: *Soul Travelers of the Far Country*; *The Book of ECK Parables*, Vols. 1, 2 and 3 and *The Living Word*.
Mailing address: PO Box 27300, Minneapolis MN 55427, USA. T. (1 612) 544 3001

Kleps, Art, Chief Boo Hoo

Rel: Original Kleptonian Neo-American Church; Ordained on 17 APR 1965; b. 17 APR 1928 in New York NY, USA; m. to Joan Whitwood; c. 1 boy and 4 girls; Citizen of USA; Lang. English.
International Organizations: Chief Boo Hoo (founder) (since 1965) of the Original Kleptonian Neo-American Church.
Main field of work: Fighting persecution. **Career:** Psychologist (schools, prisons, clinics, hospitals) from 1955 to 1965. **Educ.:** B.A., M.S. from Syracuse University, NY, USA (1949-1954). **Interests:** religion and psychedelic drugs. **Pubs.:** Millbrook; Boo Hoo Bible and Divine Toad Sweat.
Mailing address: OKNeoAC, Box 3473, Austin TX 78764, USA. T. (1 512) 443 8464
Residence: 1200 Barton Hills Drive, Suite 334, Austin TX 78704, USA. T. (1 512) 443 2991

Kleron, Bryn

Rel: Christian.
International Organizations: Contact of the One World Movement of the Ecumenical Community.
Mailing address: Elm Garth, Wetheral, Carlisle, UK.

Klietmann, K-G, Dr

Rel: Unknown.
International Organizations: Honorary Secretary of the International Orders Research Society.
Mailing address: Leibnizstrasse 48, 10629 Berlin, Germany.

Klipffel, Thérèse, Pasteur

Rel: Reformed Church.
International Organizations: President of the Reformed Church of Alsace and Lorraine.
Mailing address: 2 rue du Bouclier, F-67000 Strasbourg, France.

Kmeid, Elias, O.M.M., Fr

Rel: Roman Catholic; Ordained; celibate.
International Organizations: Procurator-General of the Antonines.
Other organizations: Superior, College of Religion, Maronita Mariamita B. M. V.
Mailing address: Piazza S Pietro in Vincoli 8, 00184 Roma, Italy. T. (39 6) 488 24 15, (39 6) 481 87 12

Knall, Dieter, Bishop Mag.

Rel: Evangelical.
International Organizations: Moderator (Augsburg Confession) of the Evangelical Church of the Augsburg and Helvetic Confession.
Mailing address: Evangelischer Oberkirchenrat AB, Severin Schreibergasse 3, A-1180 Wien, Austria.

Knorr, Nathan H

Rel: Jehovah's Witnesses.
International Organizations: Member of the Jehovah's Witnesses.
Mailing address: Asn des Témoins de Jéhovah en Suisse, BP 225, CH-3602 Thoune, Switzerland.

Knupfer, Wolfgang, O.R.C., Fr

Rel: Roman Catholic; Ordained; celibate.
International Organizations: Procurator-General of the Canons Regular of the Order of the Holy Cross.
Mailing address: Via Casale Vecchio di Aguzzano 85 (accesso da Via Cartesio), I-00137 Roma, Italy. T. (39 6) 889 67 70.

Kobia, Samuel Rodger, Rev

Rel: Methodist; Ordained in 1976; b. 20 MAR 1947 in Miathene, Meru, Kenya; m. to Ruth Njira; c. 2 boys and 2 girls; Citizen of Kenya; Lang. English, French and Kiswahili.
International Organizations: Secretary-General (since 1987) of the National Council of Churches of Kenya.
Other organizations: Patron (since 1990), Tigania Education Development Association.
Main field of work: Ecumenical ministry. **Educ.:** Dip. Theol. from St Paul's Theological College, Kenya (1969-1971); Dip. Urban Ministry from McCormick Theological Seminary, USA (1972); M.C.P. from Massachusetts Institute of Technology, Cambridge, USA (1976-1978). **Pubs.:** *Evangelization in Kenya Today – Separation without Hope*, 1978; "Elitism, Power and Wealth", 1981 and editor, *Together in Hope*, 1991.
Mailing address: National Council of Churches of Kenya, Moi Avenue, PO Box 45009, Nairobi, Kenya. T. (254 2) 33 82 11 – 33 82 12
Residence: Ole Dume Road, Kilimani, Nairobi, Kenya

Koch, Klaus, Prof Dr

Rel: Lutheran; Ordained on 31 OCT 1954; b. 04 OCT 1926 in Sulzbach, Thüringen, Germany; m. to Eva Maria; Citizen of Germany; Lang. German, English and French.
International Organizations: Professor of Old Testament and Ancient Near Eastern Religions (since 1962) of the University of Hamburg.
Main field of work: University education. **Career:** Pastor, Lutheran Church of Thüringen from 1954 to 1956. Assistant Professor of Old Testament, University of Erlangen-Nürnberg from 1956 to 1957. Assistant Professor of Old Testament, University of Hamburg from 1957 to 1960. Professor of Old Testament, Kirchliche Hochschule Wuppertal from 1960 to 1962. **Educ.:** Dr. Theol. from Ruprecht-Karls-Universität Heidelberg, Germany (1946-1953); Dr.habil. from Friedrich-Alexander-Universität Erlangen-Nürnberg, Germany (1956). **Pubs.:** The Growth of the Biblical Tradition, London/New York, 1969 (translated from German: Was ist formgeschichtez, 1964); The Rediscovery of Apocalyptic, London, 1972 (translated from German: Ratlos vor der Apokalyptik, 1972) and The Prophets, Vol. I and II, London/Philadelphia, 1982 (translated from German original, 1978, 1980).
Mailing address: Diekbarg 13A, 22397 Hamburg, Germany. T. (49 40) 6080505
Residence: See Mailing Address. T. See Mailing Address

Kochi, Eigaku

Rel: Shinto; Ordained; b. 26 MAR 1913; m. to Tetsuko; Citizen of Japan.
International Organizations: Chief Priest of the Saikoji.
Other organizations: Professor Emeritus, Komazawa University.
Educ.: Komazawa University, Tokyo, Japan. **Hons.:** Hon. D.Litt.
Mailing address: 567 Tsukakoshi, Sakado, Saitama 350-02, Japan. T. (81 492) 83 8656

Kocsis, Elemér, Dr

Rel: Reformed Church in Hungary; Ordained; Citizen of Hungary.
International Organizations: Bishop of Reformed Church in Hungary.
Mailing address: Kálvin tér 17, H-4026 Debrecen, Hungary.

Koetsveld, Hermanus, Rev

Rel: Protestant; b. 19 MAR 1923 in Rotterdam, Netherlands; m. to L C Koetsveld-Nederlof.
International Organizations: Secretary-General of the European Federation of Christian Building and Woodworkers Unions.
Main field of work: Pastor. **Career:** Naval construction drawer from 1939 to 1946. Naval construction drawer for Gebr Pot Bolnes shipyard from 1946 to 1948. Naval Constructor for the Royal Shipping Line Ms Amsterdam from 1948 to 1964. Pastor, Reformed Church in Wapenveld from 1964 to 1968. Pastor in De Lier from 1968 to 1973. Pastor in Den Helder from 1973 to 1979. Pastor in Sliedrecht from 1979 to 1983. Member of the Board, Gereformeerd Beraad. Member of the Board and Chairman, Reformed Bible School, Zeist. Member, Free Synod of Zuid-Holland Reformed Churches. **Educ.:** Naval construction from HTS (1945); Theology (1963).
Mailing address: 22 Kromme Nieuwe Gracht, Postbus 414, NL-3500 AK Utrecht, Netherlands. T. (31 30) 332531.

Kohno, Timothy Hiromichi, Rev

Rel: Anglican.
International Organizations: General Secretary of the Anglican Church in Japan.
Mailing address: 4-21 Higashi 1 chome, Shibuya-Ku, Tokyo 150, Japan.

Kok-Kwong, Yung, Rev

Rel: Church of Christ.
International Organizations: General Secretary of The Hong Kong Council Church of Christ in China.
Mailing address: Morrison Memorial Centre, 191 Prince Edward Road, Kowloon, Hong Kong.

Kolodchin, Vasyl

Rel: Ukrainian Orthodox.
International Organizations: Contact of the Ukrainian Patriarchal World Federation.
Mailing address: 4437 McKinley, Warren MI 48091, USA. T. (1 313) 755 1575.

Kolvenbach, Peter-Hans, S.J., Fr

Rel: Roman Catholic; Ordained in 1961; b. 30 NOV 1928 in Druten, Gelderland, Netherlands; celibate; Citizen of Netherlands.
International Organizations: Superior General (since 1983) of the Society of Jesus.
Member of the Congregation for the Evangelization of Peoples. Advisor of the Sacred Congregation for the Oriental Churches.
Other organizations: Vice-Grand Chancellor, Pontifical Gregorian University.
Career: Entered Society of Jesus in 1948. Professor of General Linguistics and Armenian, Institute of Oriental Languages and Faculty of Humanities, St Joseph's University, Beirut from 1968 to 1981. Professor of General Linguistics, then Director, Institute of Philosophical Studies and Professor of Hermeneutics, St Joseph's University from 1968 to 1974. Delegate to 32nd General Congregation, Society of Jesus in 1974. Provincial, Jesuit Vice-Province of the Near East from 1974 to 1981. Rector, Pontifical Oriental Institute, Rome from 1981 to 1983. **Pubs.:** In Camino verso la Pasqua; A Most Pleasant Mission and numerous articles and reviews primarily in the field of linguistics and spiritual theology.
Mailing address: Borgo S Spirito 5, I-00193 Roma, Italy. T. (39 6) 686 98 41.

Komolmas, Prathip Martin, Dr

Rel: Roman Catholic; b. 22 DEC 1933 in Bangkok, Thailand; single; Citizen of Thailand; Lang. Thai, English and French.
International Organizations: Executive Secretary (since 1969) of the Catholic Education Council, Thailand.
Other organizations: President (since 1979), Assumption University.
Main field of work: Education. **Educ.:** B.A. Econ. from Loyola College, India; M.A. from Stanford University, CA, USA; Ph.D. from Southeast Asia International Development Institute, Philippines. **Interests:** educational research. **Hons.:** many awards. **Mem.:** Phi Delta Kappa
Mailing address: Assumption University, Bangkok 10240, Thailand. T. (66 2) 319 0900
Residence: 455 Sukhumvit 55, Bangkok 10110, Thailand. T. (66 2) 391 3276

Kongtrul Rinpoche, K, H E

Rel: Buddhist: Kagyupa.
International Organizations: One of the main leaders of the Kagyupa tradition.
Mailing address: Monastery of Rumtek, PO Ranipul, Gangtok, Sikkim, India.

König, Franz, H E Card

Rel: Roman Catholic; Ordained on 29 OCT 1933; b. 03 AUG 1905 in Rabenstein, Pielach, Austria; celibate; Citizen of Austria.
International Organizations: Member (since 1958) of the College of Cardinals.
Other organizations: Archbishop Emeritus (since 1985), Vienna.

Career: Dozent, Vienna University in 1946. Professor, Faculty of Theology, Salzburg in 1948. Titular Bishop, Livias in 1952. Archbishop, Vienna from 1956 to 1985. Created Cardinal by Pope John XXIII in 1958. Former President, Secretariat for Non-Believers, Rome. **Pubs.:** Der Weg der Kirche, 1986; Lexikon der Religionen, 1987 and Juden und Christen haben eine Zukunft, 1988 (König/Ehrlich).
Mailing address: Wollzeile 2, A-1010 Wien, Austria. T. (43 222) 53 25 61

Konko, Kagamitaro

Rel: Buddhist; b. 26 APR 1909; Citizen of Japan.
International Organizations: Leader of the Konkokyo.
Mailing address: 343-1 Konkocho Otani, Asakuchi-gun, Okayama 719-01, Japan.

Koontz, Gayle Gerber

Rel: Mennonite.
International Organizations: Dean of the Associated Mennonite Biblical Seminaries.
Mailing address: 3003 Benham Ave, Elkhart IN 46517-1999, USA.

Koprowski, Andrzej, S.J., Rev Dr

Rel: Roman Catholic; Ordained on 24 JUN 1969; b. 11 MAR 1940 in Lódz, Poland; celibate; Citizen of Poland; Lang. Polish, Italian, French and Russian.
International Organizations: Co-Head, Europe Regional Office (since 1990) of the Unda, International Catholic Association for Radio and Television.
Member of the Society of Jesus. Director for Poland of the Catholic European Study and Information Centre.
Other organizations: Director, Dept. of Catholic Broadcasts (since 1989), Polish Television.
Main field of work: Mass media. **Career:** Chaplain, university students, Lublin from 1970 to 1978. Rector, Jesuit College 'Bobolanum', Warsaw from 1979 to 1983. Regional Assistant for Eastern Europe, Society of Jesus from 1984 to 1989. Visiting Professor, Institute of Social Sciences, Pontifical Gregorian University from 1984 to 1989. **Educ.:** M. Polish Philology from University of Warsaw, Poland (1956-1961); Lic. Phil. from Jesuit Faculty of Philosophy, Crakow, Poland (1963-1966); Lic. Theol. from Jesuit Faculty of Theology, Crakow, Poland (1966-1970); Dr. Hum.Sc. from Polish Academy of Sciences, Warsaw, Poland (1977). **Pubs.:** "L'Europe Occidentale – vue de Pologne", in *L'Objectif Europe*, Brussels, 2/1978; "Situacao e prospectiva da Igreja na Europa de Leste", in *Broteria*, Lisbon, 2-3/1989 and 5/1989 and "La Chiesa nell'Europa dell'Est", in *La Civilta Cattolica*, Rome, 3371/1, December 1990.
Mailing address: RTV, J P Woronicza 17, P-00 950 Warszawa, Poland. T. (48 22) 47 86 86 – 47 61 50
Residence: Rakowiecka 61, P-02 532 Warszawa, Poland. T. (48 22) 48 91 31

Kort, Leonardus H, C.I., Fr

Rel: Roman Catholic; Ordained; celibate.
International Organizations: Procurator-General of the Josephite Fathers.
Mailing address: Jozefieten College, Brusselsesteenweg 459, B-9090 Melle, Belgium. T. (32 91) 52 16 35 – 52 11 09.

Korycki, Jan, S.A.C., Rev

Rel: Roman Catholic; Ordained; celibate.
International Organizations: Procurator-General of the Society of the Catholic Apostolate.
Mailing address: Piazza S Vincenzo Pallotti 204, I-00186 Roma, Italy. T. (39 6) 654 81 85 – 687 62 13

Kos, Donald, O.F.M. Conv., Fr

Rel: Roman Catholic; Ordained; celibate.
International Organizations: Procurator-General of the Order of Friars Minor Conventual.
Mailing address: Piazza SS Apostoli 51, I-00187 Roma, Italy. T. (39 6) 679 41 78, (39 6) 678 97 16

Kosela, Krzysztof

Rel: Unknown; b. 25 JUL 1951; Citizen of Poland.
International Organizations: Member of the Council of the International Society for the Sociology of Religions.
Educ.: Ph.D.
Mailing address: University of Warsaw, Institute of Sociology, Warszawa 64, Poland. T. (48 22) 278 599
Residence: Glogera 4m46, Warszawa 02 051, Poland T. (48 2) 659 4033

Koshy, T E, Dr

Rel: Evangelical.
International Organizations: Executive Officer of the International Friendship Evangelism.
Mailing address: 862 Ostrom Avenue, Syracuse NY 13210, USA. T. (1 315) 475 9650.

Kosolkitiwong, P E Suchart, Dr

Rel: Buddhist.
International Organizations: President of the International Federation of Religions.
Senator (Thailand) of the International Parliament for Safety and Peace.
Mailing address: PO Box 9, Pak Tho, Tatchaburri 70140, Thailand. T. Tx 87394 POTS TH.

Kotooka, Mitsushige

Rel: Shinto; Ordained; b. 21 AUG 1914; Citizen of Japan.
International Organizations: Chief Priest of the Konpira Shrine.
Career: Head, Kagawa Prefecture Shrine Head Office. **Educ.:** Kokugakuin University, Tokyo, Japan.
Mailing address: 833 Kotohiracho, Nakatado-gun, Kagawa 766, Japan. T. (81 877) 75 3048

Kotsonis, Ieronymos, Archbishop

Rel: Greek Orthodox; Ordained; b. in Ysternia, Tinos, Greece; Citizen of Greece.
International Organizations: Ecclesiastic of the Greek Orthodox Church.
Career: Former Professor of Canon Law, University of Salonica. Chaplain to Greek Royal Family from 1949 to 1967. Archbishop, Athens and Primate of Greece from 1967 to 1973. President, the International of Love. **Interests:** fishing, swimming. **Pubs.:** 110 works including (in Greek): Agapism, the Only Way for Peace, 1980; Love is Not a Utopia (To avoid the Nuclear War), 1981 and Am I a Christian?, 1983.
Mailing address: Ysternia, Tinos 84201, Greece. T. (30 283) 31397

Kottayarikil, Cyriac, M.C.B.S., Fr

Rel: Roman Catholic; Ordained; celibate.
International Organizations: Superior General of the Missionary Congregation of the Blessed Sacrament.
Mailing address: M.C.B.S Generalate, Chungamvely, Alwaye 683105, India.

Koukoua, Etienne

Rel: Roman Catholic.
International Organizations: Africa Regional Secretary of the International Catholic Committee of Nurses and Medico-social Assistants.
Mailing address: 18 BP 01, Abidjan 18, Ivory Coast.

Koulouris, Antonios

Rel: Evangelical.
International Organizations: General Secretary of the Greek Evangelical Church.
Mailing address: 25 Phytagora Street, 15562 Athens 401, Greece.

Kourides, Peter

Rel: Greek Orthodox.
International Organizations: Secretary of the Greek Orthodox Archdiocesan Council of North and South America.
Mailing address: Greek Orthodox Archhouse, 8-10 East 79th St, New York NY 10021, USA.

Kowalczyk, Józef, Most Rev Archbishop Dr
Rel: Roman Catholic; Ordained on 14 JAN 1962; b. 28 AUG 1938 in Jadowniki Mokre, Tarnów, Poland; celibate; Citizen of Vatican and Poland; Lang. Polish, Italian, French, English and German.
International Organizations: Apostolic Nuncio to Poland (since 1989) of the Roman Catholic Church.
Other organizations: Titular Archbishop, Eraclea; President, Committee on Karol Wojtyla's Writings; Member, John Paul II Foundation.
Career: Consecrated Bishop in 1989. **Educ.:** J.C.D. from Pontifical Gregorian University, Vatican City State; S.T.L. from Hosianum Seminary, Olsztyn, Poland.
Mailing address: Al. I Armii Wojska Polskiego 12, 00 582 Warszawa, Poland. T. (48 22) 21 23 37. Fax 28 45 56. Tx 816493 NUWA PL
Residence: See Mailing Address. T. See Mailing Address

Kowalski, Stanislaw M Tymoteusz
Rel: Old Catholic.
International Organizations: Bishop of Old Catholic Mariavite Church in Poland.
Mailing address: ul J Wieczorka 27, Plock, Poland.

Kowo, Naomi
Rel: Christian.
International Organizations: Secretary-General of the African Christian Peace Conference.
Mailing address: Zimbabwe Council of Churches, 128 Mbaya Nehanda Street, PO Box 3566, Harare, Zimbabwe. T. (263 4) 19 12 08. Fax 52 979.

Koyama, Tsuneo
Rel: Spiritualist.
International Organizations: Vice-President of the European and African Organization for Spiritual Purification by the Light of God.
Mailing address: 9 rue Charcot, F-75013 Paris, France.

Kredel, Elmar Maria, Most Rev Archbishop
Rel: Roman Catholic; Ordained on 30 JUL 1950; b. 24 FEB 1922 in Nuremberg, Germany; celibate; Citizen of Germany.
International Organizations: Archbishop (since 1977) of Bamberg.
Career: Chaplain, Pegnitz from 1952 to 1954. Studies in Rome from 1954 to 1956. Assistant, Exeget Seminar, University of Munich from 1958 to 1962. Parish administrator and parson, Freienfels and Hollfeld from 1962 to 1967. Cathedral staff, Bamberg in 1967. Deputy Vicar-General and Canonicus theologus from 1967 to 1977. Honorary papal prelate in 1975. President, Episcopal Commission for Social and Charitable Questions, Commission VI of the German Bishop's Conference from 1977 to 1986. Roman Catholic Military Bishop, Deutsche Bundeswehr from 1978 to 1990. **Interests:** classical music. **Hons.:** Bayerische Verdienstorden; Grosses Bundesverdienstkreuz; Honorary Citizen, Bamberg (1987). **Pubs.:** various theological articles.
Mailing address: Erzbischöfliches Ordinariat, Domplatz 3, Postfach 4034, 96049 Bamberg, Germany. T. (49 951) 502 205

Kressman, O
Rel: Christian.
International Organizations: Director of the CIMADE.
Mailing address: Dept. Développement, 8 bd Bonne-Nouvelle, F-75010 Paris, France. T. (33 1) 45 23 23 77. Fax 48 24 16 48. Tx AUI 643 191 F att CIMADE.

Krey, Andrew E V, Chaplain
Rel: Lutheran.
International Organizations: President of the Lutheran Association for Maritime Ministry.
Mailing address: 140 Country club Road, Waterbury CT 06708-3313, USA.

Krisetya, Mesach
Rel: Mennonite.
International Organizations: Member, Executive Committee of the Mennonite World Conference.
Mailing address: Jl. Merbabu No 3, Salatiga, Jateng 50715, Indonesia.

Kristensen, Gunnar, Rev
Rel: Baptist.
International Organizations: General Secretary of the Baptist Union of Denmark.
Mailing address: Laerdalsgade 5, DK-2300 København S, Denmark.

Kroeger, James Henry, M.M., Rev Dr
Rel: Roman Catholic; Ordained on 17 MAY 1975; b. 04 DEC 1945 in Appleton WI, USA; celibate; Citizen of USA; Lang. English, Italian, Tagalog, Cebuano and French.
International Organizations: Asia-Pacific Assistant-General (since 1990) of the Catholic Foreign Missionary Society of America.
Other organizations: Director (since 1985), Maryknoll Institute of Language; Professor (since 1988), Loyola School of Theology, Manila.
Main field of work: Mission studies. **Career:** Board Member, Philippine Mission Program from 1980 to 1990. **Educ.:** M.A. Theol. from Loyola School of Theology, Manila, Philippines (1972-1974); M.Div. from Maryknoll Seminary, NY, USA (1974-1975); Doc. Missiology (summa cum laude) from Pontifical Gregorian University, Vatican City State (1982-1985). **Interests:** Islamics and interreligious dialogue. **Mem.:** American Society of Missiology (1985); International Association of Mission Studies (1985) **Pubs.:** *The Philippine Church and Evangelization*; *Knowing Christ Jesus* and *Mission Today: Contemporary Themes in Missiology.*
Mailing address: Maryknoll Missioners, Seminary Building, Maryknoll NY 10545, USA. T. (1 914) 941 7590
Residence: One Pinesbridge Road, Ossining NY 10545, USA. T. See Mailing Address

Krol, John Joseph, H E Card
Rel: Roman Catholic; Ordained on 20 FEB 1937; b. 26 OCT 1910 in Cleveland OH, USA; celibate; Citizen of USA.
International Organizations: Member (since 1967) of the College of Cardinals.
Other organizations: Archbishop Emeritus (since 1988), Philadelphia.
Career: Professor of Canon Law, St Mary's Seminary in 1942. Vice-Chancellor, Diocese of Cleveland from 1943 to 1951. Chancellor, Diocese of Cleveland from 1951 to 1953. Titular Bishop, Cadi, Auxiliary Bishop to Bishop of Cleveland, also Vicar-General from 1953 to 1961. Archbishop, Philadelphia from 1961 to 1987. Created Cardinal by Pope Paul VI in 1967. Member, Pontifical Commission for Mass Media Communications from 1964 to 1969. Vice-President, National Conference of Catholic Bishops and US Catholic Conference from 1966 to 1972. President, National Conference of Catholic Bishops and US Catholic Conference from 1973 to 1975. Former member, Pontifical Commission for Revision of Code of Canon Law. **Hons.:** Dr. h.c. of 16 American universities; numerous decorations.
Mailing address: 5700 City Avenue, Philadelphia PA 19131, USA. T. (1 215) 587 4501

Kron, Heinrich
Rel: Evangelical.
International Organizations: President of the Evangelical Church of the Palatinate.
Mailing address: Domplatz 5, 67346 Speyer/Rhein, Germany.

Krug, Bernard, Rev
Rel: Christian.
International Organizations: Secretary-General of the International Christian Maritime Association.
Mailing address: 81 Orwell Road, Felixstowe IP11 7PY, UK. T. (44 3942) 79911.

Krüger, Jacobus Stefanus, Prof
Rel: Quaker; b. 27 MAY 1940 in Middelburg, Transvaal, South Africa; m. to Christina Steinberg; c. 1 boy and 2 girls; Citizen of South Africa; Lang. Afrikaans, English, Dutch and German.
International Organizations: Head, Department of Religious Studies (since 1982) of the University of South Africa, Pretoria.
Other organizations: Chairman (since 1990), Association for the Study of Religion (South Africa).
Main field of work: University education. **Educ.:** B.A.; B.D. from University of Pretoria, South Africa (1958-1963); Drs. Th. from Free University, Amsterdam, Netherlands (1964-1967); D.Th. from University of South Africa, Pretoria, South Africa (1968-1970). **Pubs.:** Aandag, Kalmte en Insig, 1989; Metatheism, 1990 and Buddhism from the Buddha to Asoka, 1991.

Mailing address: UNISA, PO Box 392, Pretoria 0001, South Africa. T. (27 12) 429 4523
Residence: 292 Kandelaar Street, Pretoria 0184, South Africa T. (27 12) 83 1974

Kruse, Martin, Dr
Rel: Evangelical; Ordained; b. 21 APR 1929 in Lauenberg, Einbeck, Germany; m. to Marianne Kittel; c. 2 boys and 2 girls; Citizen of Germany.
International Organizations: Bishop (since 1977) of Berlin-Brandenburg.
Other organizations: Member (since 1979), Council of Evangelical Church in Germany; Chairman (since 1985), Council of Evangelical Church in Germany; Member (since 1983), Central Committee of Ecumenical Council of Churches.
Interests: walking, playing piano, reading. **Pubs.:** Gruppierungen in der Kirche, 1982; Hochwürden 10%, 1984 and Aufmerksamkeiten, 1984.
Mailing address: Bachstrasse 1-2, 10555 Berlin, Germany. T. (49 30) 39 09 10

Kruyswijck, A, Dr
Rel: Dutch Reformed.
International Organizations: Contact of the Reformed Churches in the Netherlands.
Mailing address: Postbus 202, Leusden, Netherlands.

Krzyzanowski, Kazimierz, M.I.C., Fr
Rel: Roman Catholic; Ordained; celibate.
International Organizations: Procurator-General of the Marian Fathers of the Immaculate Conception of the Blessed Virgin Mary.
Mailing address: Via Corsica 1, I-00198 Roma, Italy. T. (39 6) 855 4358 – 884 1885.
Residence: See Mailing Address. T. See Mailing Address

Kubo, Tsugunari, Dr
Rel: Buddhist.
International Organizations: President of the International Institute for Buddhist Studies.
Mailing address: The Reiyukai, 1-7-8 Azabudai, Minato-ku, Tokyo 106, Japan.

Kuftaro, Ahmad
Rel: Muslim; Citizen of Syria.
International Organizations: Grand Mufti of Syria.
Mailing address: No mailing address.

Kuharic, Franjo, H E Card
Rel: Roman Catholic; Ordained on 15 JUL 1945; b. 15 APR 1919 in Pribic, Zagreb, Yugoslavia; celibate; Citizen of Yugoslavia.
International Organizations: Member (since 1983) of the College of Cardinals.
Member of the Sacred Congregation for the Sacraments and Divine Worship. Member of the Congregation for the Clergy. Member of the Pontifical Council for Dialogue with Non-Believers.
Other organizations: Metropolitan Archbishop (since 1970), Zagreb; President, Bishops' Conference of Yugoslavia; Member, Council for Cardinals and Bishops.
Career: Titular Bishop, Meta and Auxiliary Bishop of Zagreb from 1964 to 1970. Created Cardinal by Pope John Paul II in 1983. President, Yugoslavia Episcopal Conference in 1983. **Educ.:** University of Zagreb, Yugoslavia.
Mailing address: Kaptol 31, PP 553, 41000 Zagreb, Yugoslavia. T. (38 41) 275 911

Kuhn, Andrew, O.F.M.Cap., Br
Rel: Roman Catholic; Ordained; celibate.
International Organizations: Member of the Order of Friars Minor Capuchin.
Other organizations: Vice-President, National Assembly of Religious Brothers.
Mailing address: 1337 West Ohio Street, Chicago IL 60622-6490, USA. T. (1 312) 829 8525

Kuijper, P J M
Rel: Roman Catholic.
International Organizations: Director of the Dutch Bishops' Lenten Campaign.
Mailing address: PO Box 587, NL-3800 AN Amersfoort, Netherlands. T. (31 33) 61 29 57. Fax 63 11 57.

Kulak, Józef, M.S.F., Fr
Rel: Roman Catholic; Ordained; celibate.
International Organizations: Procurator-General of the Congregation of the Missionaries of the Holy Family.
Mailing address: Via di Villa Troili 56, I-00163 Roma, Italy. T. (39 6) 625 85 46.

Kulananda, K, Dharmachari
Rel: Buddhist.
International Organizations: Member of the Friends of the Western Buddhist Order.
Mailing address: Padmaloka, Lesingham House, Surlingham, Norwich NR14 7AL, UK. T. (44 50) 88 310. Tx 295141 TxLINK G/105088639 (ATT MBX).

Kumaraseri, G K Alfred, H E
Rel: Buddhist; Unordained; b. 01 NOV 1942 in Kuala Lumpur, Malaysia; m. to Asoka; c. 2 boys and 1 girl; Citizen of Malaysia; Lang. English, Malay, Sinhalese, Japanese and Tamil.
International Organizations: Chairman of the Special Committee on UNESCO (since 1990) of the World Fellowship of Buddhists.
Main field of work: Diplomat. **Career:** Assistant Secretary, Ministry of Foreign Affairs in 1966. Assistant High Commissioner in Madras. Counsellor, High Commission of Malaysia, New Delhi. Counsellor, High Commission of Malaysia, Tokyo. Minister/Counsellor, Embassy of Malaysia, Washington. Chief of Inspectorate, Ministry of Foreign Affairs. **Educ.:** B.A. (Hons) from University of Malaya, Kuala Lumpur, Malaysia (1966); P.G. Int. Affairs at Bandaranaike Centre for International Studies, Colombo, Sri Lanka (1989); Ph.D. from University of Colombo, Sri Lanka (1989); Cert. Int. Law and Diplomacy from Indian Academy of International Law and Diplomacy, New Delhi, India (1974).
Mailing address: 5 Jalan Serai Wangi, Taman Cheras, 56100 Kuala Lumpur, Malaysia. T. (60 3) 9312062
Residence: See Mailing Address. T. See Mailing Address

Kumoyo, Doreen
Rel: Roman Catholic.
International Organizations: Pan-African Secretariat of the International Young Christian Workers.
Mailing address: Panafrican YCW, PO Box 46528, Nairobi, Kenya.

Kunchithapatha Kurukkal, S K, Brahmasri
Rel: Hindu.
International Organizations: Chief Priest of the Ponnambalavaneswarar.
Mailing address: Ponnambalavaneswarar, Ramanathan Mawatha, Colombo 13, Sri Lanka.

Küng, Hans C, Prof Dr
Rel: Roman Catholic; Ordained in 1954; b. 19 MAR 1928 in Sursee, Lucerne, Switzerland; celibate; Citizen of Switzerland.
International Organizations: Director (since 1963) of the Institute for Ecumenical Research, Tübingen.
Other organizations: Full Professor of Ecumenical Theology (since 1980), University of Tübingen; Consulting Editor (since 1988), Currents of Encounter: Studies on the Contact between Christianity and Other Religions, Beliefs, and Cultures, University of Amsterdam; Consultant to the Indian School of Ecumenical Theology (since 1989), Ecumenical Christian Centre, Whitefield, Bangalore, India; Member, Editorial Board (since 1989), *The Journal of Religious Pluralism*, Faculty of Religious Studies, McGill University, Montreal, Canada, and Department of Religious Studies, University of San Diego, USA; Advisor for the Grawemeyer Award in Religion (since 1989); Presidential Advisor to the Association for the Rights of Catholics in the Church (since 1990); Member, Editorial Board (since 1990), *Tendencias* (magazine), Madrid; President (since 1991), Herbert-Haag-Stiftung für Freiheit in der Kirche; Member, Executive Editorial Committee, *Concilium*, Nijmegen; Associate Editor, *Journal of Ecumenical Studies*, Philadelphia; Member, International Advisory Board, *International Journal for Philosophy of Religion*; Member, Editorial Committee, *Christian Culture Review*.

Career: Pastoral work at the Hofkirche in Lucerne from 1957 to 1959. Scientific Assistant for Dogmatic Theology, Faculty of Catholic Theology, University of Münster (Westphalia) from 1959 to 1960. Full Professor of Fundamental Theology, Faculty of Catholic Theology, University of Tübingen from 1960 to 1963. Official theological consultant (Peritus) to the Second Vatican Council, appointed by Pope John XXIII from 1962 to 1965. Full Professor of Dogmatic and Ecumenical Theology, Faculty of Catholic Theology, University of Tübingen from 1963 to 1980. Censured by the Vatican and forbidden to teach theology under Roman Catholic church auspices in 1979. Honorary President, Edinburgh University Theological Society from 1982 to 1983. Associate Editor, *Tübinger Theologische Quartalschrift* from 1960 to 1980. **Educ.:** Lic.phil. from Pontifical Gregorian University, Vatican City State (1948-1951); Lic.theol. from Pontifical Gregorian University, Vatican City State (1951-1955); Dr.theol. from Sorbonne and Institut Catholique de Paris, France (1955-1957). **Interests:** water-sport, skiing, classical music. **Hons.:** Hon. LL.D., University of St Louis MO, USA (1963); Hon. D.D. (Protestant Theology), Pacific School of Religion, Berkeley CA, USA (1966); Hon. HH.D., Loyola University, Chicago IL, USA (1970); Hon. D.D.(Protestant Theology), University of Glasgow, UK (1971); Hon. LL.D., University of Toronto, Canada (1984); Hon. D.D., University of Cambridge, UK (1985); Hon. L.H.D., University of Michigan, Ann Arbor, USA (1985); Oskar Pfister Award, American Psychiatric Association (1986); Guest Professor: Union Theological Seminary, New York City (1968), Faculty of Protestant Theology at the University of Basel (1969), University of Chicago Divinity School (1981), University of Michigan, Ann Arbor (1983), University of Toronto (1985), Rice University, Houston TX (1987); and numerous guest lectures at various universities in Europe, America, Asia and Australia. **Mem.:** PEN American Center and PEN Center of the Federal Republic of Germany (since 1978); International Buddhist-Christian Theological Encounter Group, Boston University, School of Theology; Society for Buddhist-Christian Studies, Berkeley CA **Pubs.:** Author of 30 books including: *Infallible? An Inquiry*, 1970 (English translation: Collins, London 1971, Fontana Library, London 1972, Doubleday, New York 1971, Image Book, New York 1972); *On Being a Christian*, 1974 (English translation: Collins, London 1977, Fount Paperbacks 1978, Doubleday, New York 1976) and *Does God Exist?*, 1978 (English translation: Collins, London 1980, Doubleday, New York 1980).
Mailing address: Waldhäuserstrasse 23, 72076 Tübingen, Germany. T. (49) 62646

Kuntz, Dolores, Sr
Rel: Roman Catholic; Ordained; b. 04 DEC 1925 in London, Ontario, Canada; celibate; Citizen of Canada.
International Organizations: Principal and Dean (since 1977) of the Brescia College.
Other organizations: University Teacher at Brescia College (since 1959); Member, Ursuline Community of Chatham Union.
Main field of work: Education. **Career:** High School Teacher, Grimsby, Ontario from 1947 to 1948. Teacher at The Pines, Chatham, Ontario from 1948 to 1958. Teacher at St Mary's High School, Edmonton from 1958 to 1959. Dean of Residence, Brescia College from 1960 to 1963. Psychologist at Brockville Psychiatric Hospital (summers) from 1968 to 1974. **Educ.:** B.A. from University of Western Ontario, London, Canada (1946); M.A. from University of Detroit, MI, USA (1962); Ph.D. from Queen's University at Kingston, Ontario, Canada (1968).
Mailing address: 1285 Western Rd, London, Ontario N6G 1H2, Canada.

Kuria, Manasses, Most Rev
Rel: Anglican; Ordained; b. 22 JUL 1929 in Limuru, Kenya; m. to Mary Nyambura Kuria; c. 2 boys and 4 girls; Citizen of Kenya.
International Organizations: Archbishop of Kenya and Bishop (since 1980) of Nairobi.
Career: Archdeacon, Eldoret in 1965. Assistant Bishop, Nakuru Diocese in 1970. Bishop, Nakuru from 1976 to 1980. **Pubs.:** Uwakili Katika Kristo (Stewardship of Christ), 1969.
Mailing address: 26 State House Ave, PO Box 40502, Nairobi, Kenya. T. (254 2) 333324/5

Kuria, Plawson, Rev Dr
Rel: Presbyterian.
International Organizations: Secretary-General of the Presbyterian Church of East Africa.
Mailing address: PO Box 48268, Nairobi, Kenya.

Kurtz, Robert, C.R., Fr
Rel: Roman Catholic; Ordained; celibate.
International Organizations: Superior General of the Congregation of the Resurrection of our Lord Jesus Christ.
Mailing address: Via S Sebastianello 11, I-00187 Roma, Italy. T. (39 6) 679 59 08

Kuttickattu, Sebastian J, Fr
Rel: Roman Catholic.
International Organizations: Secretary-General of the Order of the Imitation of Christ.
Mailing address: Bethany Ashram, 280 Diamond Harbour, Thakurpukur PO, Calcutta 700063, India.

Kuttickattu, Sebastian John, Rev
Rel: Roman Catholic; Ordained on 21 SEP 1972; b. 06 AUG 1942 in Venmoney (Alleppy), Kerala, India; celibate; Citizen of India; Lang. Malayalam, English, Syriac and Hindi.
International Organizations: Secretary-General (since 1980) of the Order of the Imitation of Christ.
Main field of work: Ministry.
Mailing address: Bethany Ashram, Post Bag Nr 6, Vadavathoor 686 010, Kottayam, Kerala, India. T. (91) 8494
Residence: See Mailing Address. T. See Mailing Address

Kuzmic, Peter, Rev Prof Dr
Rel: Evangelical; Ordained in 1970; b. 26 JUN 1946 in Nuskova, Slovenia, Yugoslavia; m. to Vlasta Volf; c. 3 girls; Citizen of Yugoslavia; Lang. Slovenian, Serbo-Croat, German, English and Russian.
International Organizations: Chairman, Executive Committee (since 1986) of the Theological Commission of the World Evangelical Fellowship.
Other organizations: Executive (since 1982), European Fellowship of Evangelical Theologians; President (since 1989), Evangelical Church of Yugoslavia; Editor-in-Chief (since 1990), *Izvori* (religious monthly in Serbo-Croat); President and Professor of New Testament Studies (since 1990), Evangelical Theological Seminary.
Main field of work: Ministry. **Career:** Dean, Biblical Theological Institute, Osijek from 1972 to 1976. Director, Biblical Theological Institute, Osijek from 1976 to 1990. **Educ.:** B.A. from Southern California College, Costa Mesa CA, USA (1970-1971); M.A. from Wheaton College, IL, USA (1971-1972); M.Th., D.Th. from University of Zagreb, Yugoslavia (1972-1979). **Interests:** literature, politics. **Hons.:** completed all his studies summa cum laude; numerous awards and distinctions. **Mem.:** Lausanne Committee for World Evangelization (LCWE), and numerous professional and academic associations **Pubs.:** Author of various articles and books, including: *The Gospel of John: A Study Guide* (textbook); *Biblical Hermeneutics* and *Kingdom of God.*
Mailing address: D Tucovica 32, pp 370, 54103 Osijek, Yugoslavia. T. (38 54) 556 466
Residence: See Mailing Address. T. See Mailing Address

Kverndal, Roald, Dr
Rel: Lutheran.
International Organizations: Representative of the International Coordinating Committee for Maritime Follow-up Ministry.
Mailing address: Lutheran Martime Ministry Consultancy, 2513 162nd Av NE, Bellevue, Seattle WA 98008, USA.

Kvinge, Rolf, Pastor
Rel: Seventh-Day Adventist.
International Organizations: President, West Nordic Union Conference of the General Conference of Seventh-Day Adventists.
Mailing address: Holmenkollveien 31, N-0376 Oslo 3, Norway.

Kwaw, Edmund Samuel Kwakukye, Dr
Rel: Roman Catholic; b. 15 NOV 1934 in Enchi, Ghana; m. to Elizabeth Amonoo-Neizer; c. 5 boys and 1 girl.
International Organizations: Regional Secretary for Africa (since 1984) of the International Council of Catholic Men.
Member, Board of Directors (since 1988) of the International Council of Catholic Men.

Other organizations: Medical Practitioner and Founder (since 1967), Sixth Street Clinic, Sekondi, Ghana; Chairman, Board of Governors (since 1971), Archbishop Porter Girls' Secondary School, Takoradi; Organist (since 1972), St Paul's Pro-Cathedral Choir, Sekondi; Member (since 1981), Diocesan Ordination Committee, Sekondi/Takoradi; Member (since 1984), Regional Consultative Council, Western Region; Consultant to the National Medical Ethics Commission (since 1984), the Catholic Church; Member of Council and Vice-President (since 1987), International Alliance of Catholic Knights (IACK); Member (since 1987), Diocesan Justice and Peace Commission, Sekondi/Takoradi.
Career: Secretary for the Western Region, General Medical Association from 1964 to 1966. Member, Medical Commission, Sports Council of Ghana from 1967 to 1968. Member, Board of Trustees, Centre for Civic Education from 1971 to 1972. Member, University Council, University of Cape Coast, Ghana from 1972 to 1974. Member, Board of Directors, Western Regional Development Corporation from 1975 to 1979. Chairman, Board of Directors, African Timber and Plywood Ltd from 1979 to 1980. Chairman, Bishop's Enthronment Committee, Sekondi/Takoradi in 1980. Secretary for the Western Region, General Medical Practitioners from 1984 to 1986. Vice-President, Catholic Doctors' Guild of Ghana from 1984 to 1988. Member, Diocesan Transport Committee, Sekondi/Takoradi in 1989. **Educ.:** University College, Ibadan, Nigeria (1957-1962); M.B., B.S. from University of London, UK (1962). **Interests:** Music, tennis, swimming. **Hons.:** Conferment of Knighthood of St Gregory the Great by His Holiness, Pope John Paul II (1987), Invested and Knighted (1989). **Pubs.:** Author of 2 unpublished papers: *The Catholic Doctor in Private Practice* and *The Catholic Church and Family Planning.*
Mailing address: Sixth Street Clinic, PO Box 476, Sekondi, Ghana. T. (233) 31 6686 – (233) 31 6588

Kyi, Saw Shwe Daung, Capt

Rel: Salvation Army, The.
International Organizations: Officer-in-Charge of The Salvation Army.
Mailing address: 176-178 Anawrahta Street, East Yangon 11161, Myanmar (Burma).

Kynaston, Dianne K

Rel: Theosophist.
International Organizations: Chairman of the Indo-Pacific Theosophical Federation.
Mailing address: 484 Kent Street, Sydney NSW 2000, Australia. T. (61 2) 264 7056. C. THEOSOPH, SYDNEY.

Labardakis, Augoustinos

Rel: Greek Orthodox; Ordained in 1964; b. 07 FEB 1938 in Voukoulies-Chania, Crete, Greece; Citizen of Greece.
International Organizations: Greek Orthodox Metropolitan (since 1980) of Federal Republic of Germany.
Other organizations: Exarch (since 1980), Central Europe.
Career: Worked as priest, West Berlin from 1964 to 1972. Ordained a Bishop in 1972. **Hons.:** Grosses Verdienstkreuz.
Mailing address: Greek Orthodox Metropolis of Germany, Dietrich-Bonhoeffer-Strasse 2, PO Box 300409, D-5300 Bonn, Germany. T. (49 228) 462041

La Bianca, Salvatore

Rel: Roman Catholic.
International Organizations: Supreme Secretary of the Knights of St John.
Mailing address: 6517 Charles Avenue, Parma OH 44129, USA. T. (1 216) 845 0570.

Laboa, José Sebastián, H E Mgr

Rel: Roman Catholic; Ordained on 16 APR 1949; b. 20 JAN 1923 in Pasajes de San Juan, Spain; celibate.
International Organizations: Apostolic Nuncio to Paraguay (since 1990) of the Roman Catholic Church.
Other organizations: Titular Archbishop, Zarai.
Mailing address: Calle Ciudad del Vaticano (entre 25 de Mayo y Caballero), Casilla Postal 83, Asunción, Paraguay. T. (595 21) 200 750 – 25 522. Fax 212 590

Laborda, Jose Luis

Rel: Unknown.
International Organizations: Representative of the International Association of Medical, Psychological and Religious Studies.
Mailing address: OTAC, Sepulveda 45-47, E-08015 Barcelona, Spain.

Ladrière, Jean A M G, Prof Dr

Rel: Catholic; b. 07 SEP 1921 in Nivelles, Belgium.
International Organizations: President of the World Union of Catholic Philosophical Societies.
Consultor of the Pontifical Council for Dialogue with Non-Believers.
Other organizations: Professor, Catholic University of Louvain (UCL), Belgium; President, Higher Institute of Philosophy, UCL.
Main field of work: Education. **Educ.:** Lic. Math.Sc., Ph.D. from Catholic University of Louvain (UCL), Belgium.
Mailing address: Institut Supérieur de Philosophie, chemin d'Aristote 1, B-1348 Louvain-la-Neuve, Belgium. T. (32 10) 43 46 14
Residence: Rue de Charnoy 6/208, B-1348 Louvain-la-Neuve, Belgium. T. (32 10) 41 56 16

Ladusans, Stanislaus, Prof Dr

Rel: Roman Catholic.
International Organizations: President of the Latin American Association of Catholic Philosophers.
Mailing address: Rua Marquês de Sao Vicente 293, 22451 Rio de Janeiro FJ, Brazil. T. (55 21) 274 45 96.

La Fontaine, Yves, C.M.M., Fr

Rel: Roman Catholic; Ordained on 17 MAY 1969; b. 08 NOV 1943 in Sherbrooke PQ, Canada; celibate; Citizen of Canada; Lang. French, English and Italian.
International Organizations: Superior General (since 1990) of the Congregation of Mariannhill Missionaries.
Vice-President (since 1991) of the Documentation and Research Centre. Member, Committee of 18 (since 1991) of the Union of Superiors General.
Main field of work: Administration. **Career:** Master of Novices, Congregation of Mariannhill Missionaries from 1974 to 1985. Provincial Superior, Congregation of Mariannhill Missionaries from 1977 to 1986. Founder/President, Inter-Noviciat de Québec from 1975 to 1977. **Educ.:** Bacc. ès Arts from University of Sherbrooke, Quebec, Canada (1965); Maîtrise en Théol. Pastorale from Université Laval, Québec, Canada (1970). **Interests:** religious life and education in the Catholic Church. **Mem.:** Union of Superiors General (USG); Documentation and Research Centre (SEDOS) **Pubs.:** Comme une jeune épouse (for religious communities); A la recherche de son identité (for religious communities) and Le relais institutionnel (for private schools in Quebec).
Mailing address: Via S Giovanni Eudes 91, I-00163 Roma, Italy. T. (39 6) 625 85 47.
Residence: See Mailing Address. T. See Mailing Address

Lagarense, J P, Rev

Rel: Christian.
International Organizations: President of the Christian Church in Central Sulawesi.
Mailing address: Tentena, Poso, Central Sulawesi, Indonesia.

Laghi, Pio, H E Mgr

Rel: Roman Catholic; Ordained on 20 APR 1946; b. 21 MAY 1922 in Castiglione, Italy; celibate.
International Organizations: Pro-Prefect of the Congregation for Catholic Education.
Other organizations: Titular Archbishop, Mauriana.
Mailing address: Palazzo delle Congregazioni, Piazza Pio XII 3, I-00193 Roma, Italy. T. (39 6) 698 41 67. Fax 698 41 72

Lah, Okhah Kim

Rel: Methodist.
International Organizations: President Emeritus of the World Federation of Methodist Women.
Mailing address: 1-42 Shin Moon Ro 2ka, Seoul 110, Korea Rep.

Lahey, Raymond John, Most Rev

Rel: Roman Catholic; Ordained on 13 JUN 1963; b. 29 MAY 1940 in St John's, Newfoundland, Canada; celibate; Citizen of Canada.
International Organizations: Bishop (since 1986) of St George's, Newfoundland.
Member (since 1989) of the Pontifical Council for Promoting Christian Unity.
Other organizations: Member, Permanent Council, Canadian Conference of Catholic Bishops; President, Episcopal Commission for Liturgy. **Career:** Principal, St John's College, University of Newfoundland from 1968 to 1981. Assistant Professor of Religious Studies, St John's College in 1968. Associate Professor, St John's College in 1972. Acting Head of Dept., St John's College from 1974 to 1975. Professor, St John's College from 1980 to 1981. Head of Religious Studies, St John's College from 1977 to 1981. Vicar-General, Archdiocese of St John's, Newfoundland from 1981 to 1986. Pastor, St Peter's Parish, Mount Pearl, Newfoundland from 1982 to 1986. **Educ.:** B.Th. from University of Ottawa, Ontario, Canada (1961); M.Th. from University of Ottawa, Ontario, Canada (1963); Ph.D. from University of Ottawa, Ontario, Canada (1966); P.D. Stds. from Pontifical Gregorian University, Vatican City State and University of Cambridge, UK (1966). **Hons.:** Named Prelate of Honour to His Holiness (1985). **Pubs.:** Author, James Louis O'Donel in Newfoundland 1784-1807: The Establishment of the Roman Catholic Church, 1984; Co-author, Early European Exploration and Settlements in Atlantic Canada, 1984 and contributor to *Dictionary of Canadian Biography*.
Mailing address: Bishop's Residence, 16 Hammond Dr, Corner Brook, Newfoundland A2H 2W2, Canada. T. (1 709) 639 8662
Residence: See Mailing Address. T. (1 709) 639 7073

Lai, Chiu-Nan

Rel: Buddhist: Mahayana.
International Organizations: Member of the Foundation for the Preservation of the Mahayana Tradition.
Other organizations: Director, Land of Medicine Buddha.
Mailing address: 5800 Prescott Rd, Soquel CA 95073, USA. T. (1 408) 462 8383. Fax (1 408) 462 8380

Laine, Dan

Rel: Buddhist: Mahayana.
International Organizations: Member of the Foundation for the Preservation of the Mahayana Tradition.
Other organizations: Director, Shantideva Center.
Mailing address: HCR 4660, Keaau, Hawaii 96749, USA. T. (1 808) 966 6877, (1 808) 966 4076

Laine, Molly

Rel: Buddhist: Mahayana.
International Organizations: Member of the Foundation for the Preservation of the Mahayana Tradition.
Other organizations: Director, Shantideva Center.
Mailing address: HCR 4660, Keaau, Hawaii 96749, USA. T. (1 808) 966 6877, (1 808) 966 4076

Lak, Yeow Choo, Dr

Rel: Christian.
International Organizations: Executive Director of the Association for Theological Education in South East Asia.
Mailing address: 324 Onan Road, Singapore 1542, Singapore. T. (65) 344 7316. Fax 344 7316. C. THEOLOGY SINGAPORE.

Lalitananda, L, Swami

Rel: Yoga.
International Organizations: Vice-President of the Yoga Research Foundation.
Other organizations: Editor, *International Yoga Guide* (monthly), Yoga Research Foundation.
Mailing address: 6111 SW 74th Ave, Miami FL 33143, USA. T. (1 305) 666 2006.

Lall, Piyara, Rev

Rel: United Presbyterian.
International Organizations: Secretary of the United Presbyterian Church of Pakistan.
Mailing address: PO Box 395, Rawalpindi, Pakistan.

Lamb, E M

Rel: Roman Catholic.
International Organizations: Europe Regional Secretary of the International Catholic Committee of Nurses and Medico-social Assistants.
Mailing address: 18 Shipton Close, East Boldon, Tyne and Wear NE35 9JL, UK.

Lambert, James

Rel: Disciples of Christ.
International Organizations: Contact of the Christian Churches, International Convention.
Mailing address: Box 1986, Indianapolis IN 46206, USA. T. (317) 353 1491.

Lambert, Jean-Marie, Rev

Rel: French Reformed; b. 05 DEC 1925 in Kaiserslautern, Germany; m. to Martine de Montmollin; c. 2 boys and 1 girl; Citizen of France; Lang. French, English and Spanish.
International Organizations: Executive Director (since 1986) of the International Coordinating Committee for NGOs on the Question of Palestine.
Representative to United Nations (since 1986) of the Parliamentary Association for Euro-Arab Cooperation.
Main field of work: Information, development and advocacy. **Career:** Pastor, French Reformed Church from 1952 to 1963. Program Director, World Council of Churches from 1963 to 1978. Director, Palestine Refugees Program, Middle East Council of Churches from 1979 to 1982. Official Representative, Catholic Committee Against Hunger and for Development, CIMADE, Parliamentary Association for Euro-Arab Cooperation from 1983 to 1986. **Educ.:** Expt. Compt. from Collège Turgot, Paris, France (1946-1948); Lic. Theol. from Faculty of Protestant Theology, Paris, France and Geneva, Switzerland (1948-1952). **Interests:** Middle East, Africa, Third World problems.
Mailing address: PO Box 2100, 150 Route de Ferney, CH-1211 Genève 2, Switzerland. T. (41 22) 791 64 81
Residence: 9 rue Vautier, Carouge, CH-1227 Genève, Switzerland T. (41 22) 43 73 41

Lamont, Donal Raymond, O.Carm., H E Mgr

Rel: Roman Catholic; Ordained on 11 JUL 1937; b. 27 JUL 1911 in Ballycastle, Co. Antrim, Ireland; celibate; Citizen of Ireland; Lang. English, Italian, Latin and Shona.
International Organizations: Emeritus Bishop of Mutare.
Member of the Order of Brothers of the Blessed Virgin Mary of Mount Carmel.
Career: Professed in Carmelite Order in 1930. First Superior, Carmelite Mission to Rhodesia in 1946. Prefect Apostolic, Umtali (now Mutare, Zimbabwe) in 1953. Bishop, Umtali from 1957 to 1981. Attended all Sessions, Second Vatican Council from 1962 to 1965. Member, Vatican Secretariat for the Promotion of Christian Unity from 1962 to 1975. President, Rhodesia Catholic Bishop's Conference from 1970 to 1972. Representative, Rhodesia Catholic Bishop's Conference at Roman Synods (in 1969, 1971, 1974). Sentenced to prison and deported from Rhodesia back to Ireland in 1977. Returned to Rhodesia after Independence in 1980. **Educ.:** B.A. (Hons) from National University of Ireland, University College, Dublin, Ireland (1930-1933); S.T.L. from Collegio S Alberto (Carmelite International College), Rome, Italy (1933-1938); Higher Dip. Ed. from National University of Ireland, University College, Dublin, Ireland (1938-1939); M.A. (Hons) from National University of Ireland, University College, Dublin, Ireland (1941). **Interests:** reading, poetry. **Hons.:** Honorary LL.D., Universities of Notre Dame (IN) (1977), Seton Hall (NJ), Mount St Mary's (MD), and Marymount (NY); People of God Award, Washington Theological Union. **Pubs.:** Purchased People (pastoral letter) and Speech from the Dock, London, Kevin Mayhew, Publishers, 1977.
Mailing address: Terenure College, Dublin 6W, Ireland. T. (353 1) 904621
Residence: See Mailing Address. T. See Mailing Address

Landázuri Ricketts, Juan L, O.F.M., H E Card

Rel: Roman Catholic; Ordained on 16 MAY 1939; b. 19 DEC 1913 in Arequipa, Peru; celibate; Citizen of Peru.
International Organizations: Member (since 1962) of the College of Cardinals.
Member of the Order of Friars Minor. Member of the Congregation for the Clergy. Member of the Congregation for Religious and Secular Institutes.

Other organizations: Archbishop Emeritus (since 1989), Lima. **Career:** Titular Archbishop, Roina in 1952. Created Cardinal by Pope John XXIII in 1962. Archbishop of Lima. **Hons.:** Knight Commander, Order of Malta; and many honours.
Mailing address: Luis Espejo 1064, Urbz. Sta., Catalina, La Victoria, Lima, Peru. T. (51 14) 72 01 40

Landousies, Jean
Rel: Roman Catholic.
International Organizations: Secretary-General of the Regional Episcopal Conference of North Africa.
Mailing address: 13 rue Khelifa Boukhalfa, 16000 Algiers, Algeria. T. (213 2) 63 42 44.

Lane, David John, Rev
Rel: Anglican; Ordained on 23 DEC 1962; b. 12 JUN 1935 in Huddersfield, Yorkshire, UK; single; Citizen of United Kingdom.
International Organizations: Principal (since 1991) of the College of the Resurrection.
Main field of work: Ordinand education. **Career:** Associate Professor, University of Toronto from 1971 to 1982. **Educ.:** B.A. from University of Oxford, Magdalen College, UK (1958); M.A. from University of Oxford, Magdalen College, UK (1962); B.D. from University of Oxford, Magdalen College, UK (1989). **Pubs.:** Editor, Leiden Peshitta, Leviticus; Editor, Leiden Peshitta, Ecclesiastes and various articles in *Vetus Testamentum, Orientalia Christiana Analecta.*
Mailing address: College of the Resurrection, Mirfield WF14 OBW, UK.

Lanfranchi Mantovani, Guido L, F.N., Fr
Rel: Roman Catholic; Ordained; celibate.
International Organizations: Procurator-General of the Congrégation de la Sainte Famille de Nazareth.
Mailing address: Via Casal del Marmo 110, I-00135 Roma, Italy. T. (39 6) 339 27 87

Lanigan, Aileen, Sr
Rel: Roman Catholic; Ordained; celibate.
International Organizations: Contact of the Missionary Sisters of the Society of Mary.
Mailing address: Via Cassia 1243, I-00189 Roma, Italy. T. (39 6) 376 78 67. C. TRINITA.

Lapierre, François, P.M.E., Fr
Rel: Roman Catholic; Ordained; celibate.
International Organizations: Superior General (since 1991) of the Society of Foreign Missions of the Province of Quebec.
Mailing address: 180 Place Juge-Desnoyers, Pont-Viau, Laval H7G IA4, Canada. T. (1 514) 667 4190.

Lapp, James M
Rel: Mennonite.
International Organizations: Executive Secretary of the Mennonite Church General Board.
Mailing address: 421 S Second Street, Ste 600, Elkhart IN 46516-3243, USA.

Lapp, John A, Dr
Rel: Mennonite; b. 15 MAR 1933 in Lansdale PA, USA; m. to Alice Weber; c. 1 boy and 2 girls; Citizen of USA; Lang. English and German.
International Organizations: Past Executive Secretary (since 1985) of the Mennonite Central Committee.
Observer to the 7th General Assembly (since 1991) of the World Council of Churches.
Main field of work: Church administration. **Career:** Professor of History, Eastern Mennonite College, Harrisonburg VA from 1960 to 1969. Executive Secretary, Peace Section, Mennonite Central Committee from 1969 to 1972. Professor of History, Dean and Provost, Goshen College, IN from 1972 to 1985. **Educ.:** B.A. from Eastern Mennonite College, Harrisonburg VA, USA (1950-1954); M.A. from Case Western Reserve University, Cleveland OH, USA (1955-1958); Ph.D. from University of Pennsylvania, Philadelphia, USA (1958-1968). **Interests:** church and society. **Mem.:** Conference on Faith and History **Pubs.:** *The Mennonite Church in India 1897-1962*, Herald Press, 1972; *The View from East Jerusalem*, Herald Press, 1980 and "Human Rights: An Agenda for the Church", to appear in Wm. Ball, Ed., *In Search of a National Morality*, 1993.

Mailing address: 21 South 12th Street, PO Box 500, Akron PA 17501-0500, USA. T. (1 717) 859 1151. C. MENCENCOM. Tx 90-2210
Residence: 13 Knollwood, Akron PA 17501, USA. T. (1 717) 859 4412

Larnaud, Jean André
Rel: Roman Catholic; Unordained; b. 04 MAY 1926 in Bourg-la-Reine, France; single; Lang. French, English, Italian and Spanish.
International Organizations: Secretary-General (since 1951) of the International Catholic Centre for UNESCO.
Head, International Catholic Center for UNESCO (since 1967) of the Conference of International Catholic Organizations. Consulting Member of the Pontifical Council for the Laity.
Career: Member, Pontifical Council for Culture from 1981 to 1987. **Educ.:** Dip. Droit from Institut Catholique de Paris, France. **Hons.:** Knight of the Arts and Letters (France); Commander of the Order of Saint Gregory the Great (Holy See); UNESCO Silver Medal. **Pubs.:** Les chrétiens et la vie internationale.
Mailing address: 9 rue Cler, F-75007 Paris, France. T. (33 1) 47 05 17 59. Fax (33 1) 45 56 90 92. Tx 270644 F.
Residence: 66 ave Breteuil, F-75007 Paris, France.

Larosa, B, Rev
Rel: Protestant.
International Organizations: General Secretary of the Nias Protestant Christian Church.
Mailing address: Gunungsitoli, Nias, North Sumatra, Indonesia.

Laroussi, Troudi, Dr
Rel: Muslim.
International Organizations: UNESCO Representative of the Muslim World League.
Mailing address: 92 rue d'Alésia, F-75014 Paris, France.

Larson, Rebecca
Rel: Unknown.
International Organizations: Representative of the World Council of Churches.
Mailing address: PO Box 2100, Route de Ferney 150, CH-1211 Genève 2, Switzerland. T. (41 22) 791 61 11. C. OIKOUMENE GENEVA. Tx 415 730 OIK CH. Fax (41 22) 791 03 61.

Larsson, John Alfred, Commr
Rel: Salvation Army, The; Ordained on 24 MAY 1957; b. 02 APR 1938 in Malmo, Sweden; m. to Freda Turner; c. 2 boys; Citizen of Sweden; Lang. English, Spanish and Swedish (some).
International Organizations: Territorial Commander for UK and Republic of Ireland (since 1990) of The Salvation Army.
Main field of work: Spiritual and administrative leadership. **Career:** Salvation Army (SA) corps, youth and training work from 1957 to 1980. SA Chief Secretary for Chile, Peru and Bolivia from 1980 to 1984. Principal, William Booth Training College, USA from 1984 to 1988. SA Assistant to the Chief of Staff, International Headquarters from 1988 to 1990. **Educ.:** B.D. from University of London, UK (1965). **Pubs.:** *Spiritual Breakthrough*, 1983; *The Man Perfectly Filled with the Spirit*, 1985 and *How Your Corps Can Grow*, 1988.
Mailing address: PO Box 249, 101 Queen Victoria Street, London EC4P 4EP, UK. T. (44 71) 236 5222
Residence: 23 South View, Bromley, Kent, UK

Lasso de la Vega Miranda, Juan Manuel L, C.SS.R., Fr
Rel: Roman Catholic; Ordained; celibate.
International Organizations: Superior General of the Congregation of the Most Holy Redeemer.
Mailing address: Via Merulana 31, CP 2458, I-00185 Roma, Italy. T. (39 6) 731 58 41.

Lataste-Dorolle, C
Rel: Quaker.
International Organizations: UNESCO Representative of the Friends World Committee for Consultation.
Mailing address: 162 Bd Montparnasse, F-75014 Paris, France. T. (33 1) 43 54 75 22.

Latour, Dan, Rev
Rel: Non-denominational Spiritist; Ordained on 02 JUL 1984; b. 23 SEP 1952 in Lafayette LA, USA; m. to Melissa; Citizen of USA; Lang. English.
International Organizations: Active Minister and Co-Director (since 1984) of the National Alliance for Spiritual Growth (USA).
Main field of work: Counseling, hypnotherapy. **Educ.:** St John's University, Springfield LA, USA. **Pubs.:** Editor, *Within and Beyond* (quarterly).
Mailing address: PO Box 2683, Lafayette LA 70502, USA. T. (1 318) 235 6535
Residence: See Mailing Address. T. (1 318) 235 6539

Latour, Melissa, Rev
Rel: Non-denominational Spiritist; Ordained on 21 NOV 1986; b. 16 JUL 1962 in Lafayette LA, USA; m. to Dan Latour; Citizen of USA; Lang. English.
International Organizations: Active Minister and Co-Director (since 1984) of the National Alliance for Spiritual Growth (USA).
Main field of work: Counseling, hypnotherapy. **Educ.:** St John's University, Springfield LA, USA. **Pubs.:** Editor, *Within and Beyond* (quarterly).
Mailing address: PO Box 2683, Lafayette LA 70502, USA. T. (1 318) 235 6535
Residence: See Mailing Address. T. (1 318) 235 6539

Lätsch, Walter
Rel: Roman Catholic.
International Organizations: Head, Europe – Permanent Regional Secretariat of the Caritas Internationalis.
Mailing address: Caritas Europa, c/o Schweizerische Caritas, Löwenstrasse 3, CH-6002 Luzern, Switzerland. T. (41 41) 50 11 50. Tx 862825 SCV CH.

Lau, James, Lt-Col
Rel: Salvation Army, The.
International Organizations: Commanding Officer of The Salvation Army.
Mailing address: Orchard Point, PO Box 198, Singapore 9123, Singapore.

Lauber, Samuel
Rel: Christian.
International Organizations: President of the Europäischer Verband Freikirchlicher Diakoniewerke.
Mailing address: 1 bd J Preiss, F-67082 Strasbourg, France.

Lauby, Paul T, Dr
Rel: Protestant.
International Organizations: Executive Director of the United Board for Christian Higher Education in Asia.
Mailing address: 475 Riverside Dr, New York NY 10027, USA. T. (1 212) 870 2608.

Laurent, Gisèle
Rel: Christian.
International Organizations: Secretary of the International Ecumenical Fellowship.
Mailing address: Drève des Bruyères 27, B-1630 Linkebeek, Belgium. T. (32 2) 380 78 11.

Laurie, Margaret Elizabeth, Rev
Rel: United Reformed Church; Ordained on 16 JUN 1966; b. 07 JAN 1933 in Wanstead, Essex, UK; single; Citizen of United Kingdom; Lang. English.
International Organizations: Director, England Region (since 1989) of the International Association of Women Ministers.
Other organizations: Minister (since 1966), United Reformed Church and prior to its formation in the Congregational Church.
Main field of work: Ministry. **Educ.:** Cert. of Proficiency in Religious Knowledge from University of London, UK (1965). **Interests:** hospital chaplaincy, modern religious art.
Mailing address: 11 Walscombe Close, Stroke-sub-Hamdon TA14 6QJ, UK. T. (44 935) 822 176
Residence: See Mailing Address. T. See Mailing Address

Lauritzen, Grete
Rel: Roman Catholic.
International Organizations: Chairman of the Action Children Aid, Mother Theresa's Co-workers.
Mailing address: Baunegåardsvej 41, 2, DK-2820 Gentofte, Denmark. T. (45 1) 65 66 50.

Lautrey, Pierre
Rel: Catholic.
International Organizations: Secretary of the International Federation of Married Catholic Priests.
Mailing address: 2 square du Docteur Courcoux, F-93260 Les Lilas, France. T. (33 1) 43 61 69 43.

Lauzon, Georges, M.Afr., Fr
Rel: Roman Catholic; Ordained; celibate.
International Organizations: Procurator-General of the Society of Missionaries of Africa.
Mailing address: Via Aurelia 269, CP 9078, I-00165 Roma, Italy. T. (39 6) 63 23 18 – 63 23 14.

Lavarenne, Christian
Rel: Unknown.
International Organizations: Secretary-Treasurer of the Asocio de Studado Internacia pri Spiritaj kaj Teologiaj Instruoj.
Mailing address: 7 rue Parmentier, F-18000 Bourges, France.

La Vine, Mona
Rel: Spiritualist.
International Organizations: Contact of the Tara Center.
Mailing address: PO Box 6001, North Hollywood CA 91603, USA. T. (1 818) 785 6300.

Law, Bernard Francis, H E Card
Rel: Roman Catholic; Ordained on 21 MAY 1961; b. 04 NOV 1931 in Torreón, Mexico; celibate; Citizen of USA.
International Organizations: Member (since 1985) of the College of Cardinals.
Member of the Congregation for the Evangelization of Peoples. Member of the Congregation for Religious and Secular Institutes. Member of the Pontifical Council for the Pastoral Care of Migrants and Itinerant People.
Other organizations: Archbishop (since 1984), Boston.
Career: Bishop, Springfield-Cape Girardeau in 1973. Created Cardinal by Pope John Paul II in 1985.
Mailing address: 2101 Commonwealth Ave, Brighton MA 02135, USA. T. (1 617) 782 2544

Lawrence, Wilhelmina
Rel: Methodist.
International Organizations: Vice-President for North America of the World Federation of Methodist Women.
Mailing address: 1845 48nd Street SE, Washington DC 20020, USA.

Lawton, Leon R, Rev
Rel: Baptist: Seventh-Day; Ordained in 1951; b. in Battle Creek MI, USA; m. to Dorothy G; c. 3 boys and 1 girl; Citizen of USA; Lang. English.
International Organizations: Executive Vice-President (since 1970) of the Seventh Day Baptist Missionary Society.
Main field of work: Missions administration, witnessing. **Career:** Pastor, Los Angeles Seventh-Day Baptist Church (SDB) from 1950 to 1956. Missionary, SDB Missionary Society in Jamaica from 1956 to 1964. Director of Extension, SDB Missionary Society in the USA from 1964 to 1969. Pastor, Denver CO SDB Church from 1969 to 1970. **Educ.:** B.A. from Salem College, WV, USA (1948); B.D., M.Div. from Seminary of the West Covina, CA, USA (1951). **Hons.:** Netherlands Gold Medal of Honor, WWII. **Pubs.:** Missions – for Seventh Day Baptists.
Mailing address: 119 Main Street, Westerly RI 02891, USA. T. (1 401) 596 4326. Fax (1 401) 596 9568.
Residence: 73 Granite Street, Westerly RI 02891, USA. T. (1 401) 596 6149.

Lazerus, A, Rev
Rel: Lutheran.
International Organizations: UNESCO Representative of the Christian Peace Conference.
Mailing address: Eglise Luthérienne, 12-14 rue Kilford, F-92400 Courbevoie, France.

Leanza, Giuseppe, H E Mgr

Rel: Roman Catholic; Ordained on 17 JUL 1966; b. 02 JAN 1943 in Cesarò, Italy; celibate.
International Organizations: Member of the Synod of Bishops.
Other organizations: Apostolic Nuncio (since 1990), Haiti; Titular Archbishop, Lilibeo.
Mailing address: Morne Calvaire, Pétionville, BP 326, Port-au-Prince, Haiti. T. (509 1) 73 411 – 76 308

Lebel, Robert, Most Rev

Rel: Roman Catholic; Ordained on 18 JUN 1950; b. 08 NOV 1924 in Trois-Pistoles, Quebec, Canada; celibate; Citizen of Canada.
International Organizations: Bishop (since 1976) of Valleyfield.
Other organizations: President (since 1986), Conference of Catholic Bishops of Canada.
Career: Teacher of Theology and Patristic, Séminaire de Rimouski from 1951 to 1965. Rector, Séminaire de Rimouski from 1963 to 1965. Rector, College of Rimouski from 1965 to 1968. Professor of Religious Sciences, University of Rimouski from 1970 to 1974. Auxiliary Bishop, St Jean-de-Québec. **Educ.:** B.A. from University of Ottawa, Ontario, Canada (1946); D.Th. from Pontifical University San Tommaso d'Aquino, Vatican City State (1951). **Mem.:** Catholic Conference of Canadian Bishops (President); Assemblée des Evêques du Québec; Synod of Catholic Bishops, Rome (1980); Société Canadienne de Théologie **Pubs.:** various articles on religion and pastorals.
Mailing address: 11 rue de l'Eglise, Valleyfield PQ 16T 1J5, Canada. T. (1 514) 373 8122

Le Bret, Marie-Michèle

Rel: Catholic.
International Organizations: Representative at UNESCO of the World Organization of Former Pupils of Catholic Education.
Mailing address: 19 rue Nélaton, F-75015 Paris, France.

Lebrun Moratinos, José Alí L, H E Card

Rel: Roman Catholic; Ordained on 19 DEC 1943; b. 19 MAR 1919 in Puerto Cabella, Valencia, Venezuela; celibate; Citizen of Venezuela.
International Organizations: Member (since 1983) of the College of Cardinals.
Member of the Congregation for Bishops. Member of the Congregation for Catholic Education.
Other organizations: Archbishop (since 1980), Caracas, Santiago de Venezuela; Member, Council of Cardinals for the Study of Organizational and Economic Problems of the Holy See; Member, Council of Cardinals and Bishops.
Career: Titular Bishop, Arado in 1956. Bishop, Maracay from 1958 to 1962. Bishop, Valencia from 1962 to 1980. Given title of Archbishop in 1972. Created Cardinal by Pope John Paul II in 1983.
Mailing address: Quinta Silvestre, Calle Otama, Valle Arriba, Caracas 1060, Venezuela. T. (58 2) 92 23 19

Lecoq, André, Mgr

Rel: Roman Catholic; Ordained on 29 JUN 1947; b. 02 FEB 1924 in St Quentin-les-Chardonnets, France; celibate; Citizen of France; Lang. French, Latin and English.
International Organizations: Rector of the Sanctuary (since 1969) of the Oeuvre Notre-Dame de Montligeon.
Main field of work: Management. **Hons.:** Honorary Prelate of His Holiness (1971). **Pubs.:** *Chemins d'Eternité* (6 times a year).
Mailing address: Chapelle Montligeon, F-61400 Mortagne-au-Perche, France. T. (33) 33 83 83 55
Residence: See Mailing Address. T. See Mailing Address

Leddy, Mary Jo, NDS, Sr

Rel: Roman Catholic; Ordained; b. 01 FEB 1946 in Toronto, Ontario, Canada; celibate; Citizen of Canada.
International Organizations: Member of the Congregation of Our Lady of Sion.
Career: High School Teacher from 1968 to 1970. Campus Minister at Newman Centre, Toronto from 1976 to 1977. Founder and Editor, *Catholic New Times* in 1976. Editor, *Catholic New Times* from 1976 to 1984. Editorial Team, CNT from 1985 to 1988. Part-time University Lecturer, University College (Toronto), Maryknoll Seminary (Higham MA), Regis College (Toronto). Regular Columnist, *Toronto Star* Religion Page from 1979 to 1986. Board of Directors, Newman Centre, Toronto from 1980 to 1981. Provincial Council, Sisters of Sion from 1973 to 1979. Provincial Coordinator, Sisters of Sion, North America Province from 1983 to 1988. Moderator, Niagara Institute in 1982. Commentator, Papal Visit CBC TV (1984 and 1987). **Educ.:** B.A. from University of Saskatchewan, Saskatoon, Canada (1968); B.Ed. from University of Saskatchewan, Saskatoon, Canada (1970); M.A. from University of Toronto, Ontario, Canada (1972); Ph.D. from University of Toronto, Ontario, Canada (1980). **Hons.:** Received Governor General Bronze Medal (1963); Canadian University Presidents' Award: One of Outstanding Young Women (1978); Canadian Church Press Award: Best Editorial (1978), Best Editorial: Best Feature Article (1979, 1988); Catholic Press Association Award for Best National Catholic Newspaper in North America (1982); College of Theology Society Award for best book (*Faith That Transforms*) (1988); Ida Nudel Humanitarian Award (1984); Canadian Jewish Congress Humanitarian Award (1987); Honorary Director, Shalom Institute, Vancouver. **Mem.:** Holocaust Remembrance Committee, Toronto; Catholic Press Association; Canadian Church Press; Canadian Religious Conference, Ontario (Executive Council); Interfaith Committee on Soviet Jewry **Pubs.:** Author of numerous articles in *Catholic New Times*, *Toronto Star*, *Catalyst*, *Grail*, *Compass*, *National Catholic Reporter*, *Prairie Messenger*, etc., 1976-88; Introduction to *Quiet Heroes* by Andre Stein and *The Faith That Transforms: Essays in Honor of Gregory Baum*, 1987.
Mailing address: 15 Oakmount Rd, Toronto, Ontario M6P 2M3, Canada.

Ledger, Christine

Rel: Christian.
International Organizations: Co-Secretary-General of the World Student Christian Federation.
Mailing address: Ecumenical Centre, 5 route des Morillons, Grand-Saconnex, CH-1218 Genève, Switzerland.

Lee, Allan Wren, Rev

Rel: Protestant; Ordained on 04 SEP 1949; b. 03 JUN 1924 in Yakima WA, USA; m. to Mildred Elaine Ferguson; c. 1 girl; Citizen of USA; Lang. English and Spanish.
International Organizations: Secretary-General (since 1971) of the World Convention of Churches of Christ.
Other organizations: Board Member (since 1974), European Evangelistic Society; Member, Board of Trustees (since 1985), Northwest Christian College, Eugene OR; TV Panel Member (since 1986), Religious Town Hall, Dallas TX; Pastor (since 1990), Melissa Christian Church, TX.
Main field of work: World ministry, ecumenism, convention director.
Career: Pastor, Eagle Mills Church of Christ (Disciples), Troy NY from 1950 to 1951. Pastor, Trinity Christian Church, Dallas TX from 1955 to 1958. Pastor, First Christian Church, Bremerton WA from 1958 to 1960. Pastor, First Christian Church, Seattle WA from 1960 to 1971. President, Seattle Civitan Club from 1962 to 1963. President, Washington-No. Idaho Convention of Christian Churches in 1966. Member, General Board, Christian Church (Disciples of Christ) from 1970 to 1973. Men's Executive Committee, Japan International Christian University from 1971 to 1973. **Educ.:** B.A. from Phillips University, Enid OK, USA (1946-1949); M.A. from Vanderbilt University, George Peabody College for Teachers, Nashville TN, USA (1951-1953); B.D. from Texas Christian University, Brite Seminary, Fort Worth, USA (1953-1955); D.D. from Texas Christian University, Fort Worth, USA (1968). **Hons.:** Young Man of the Year and Distinguished Service Award by Bremerton WA Junior Chamber of Commerce (1959); Distinguished Service Citation, Children's Home Society, Seattle WA (1967); Jamaica Tourist Board, Distinguished Citation (1984); Mayor's Citation, Montego Bay, Jamaica (1983). **Mem.:** Religious Conference Management Association (Vice-President), Disciples of Christ Historical Society (Life Member) **Pubs.:** *Under the Shadow of the Nine Dragons*, 1969; *One Great Fellowship*, 1974 and *Fifty Years of Faith and Fellowship*, 1980, and over 60 published articles.
Mailing address: First City Bank Center, 100 N Central Expwy., Suite 804, Richardson TX 75080, USA. T. (1 214) 480 0118.
Residence: 2112 Stone Creek Drive, Plano TX 75075, USA. T. (1 214) 596 3785

Lee, Bona So Woo, Dr

Rel: Roman Catholic.
International Organizations: Asia Regional Secretary of the International Catholic Committee of Nurses and Medico-social Assistants.
Mailing address: CNAK – Dept. of Nursing, College of Medicine, Seoul University, 28 Yunkeun-dong, Chongo-Gu, Seoul, Korea Rep.

Lee, Glenda
Rel: Buddhist: Mahayana.
International Organizations: Member of the Foundation for the Preservation of the Mahayana Tradition.
Other organizations: Director, Hospice of Mother Tara.
Mailing address: 13 Picton Crescent, Bunbury WA 6230, Australia. T. (61 97) 219 497

Lee, J B, Rev
Rel: Uniting Church in Australia.
International Organizations: Secretary of the Uniting Church World Mission.
Mailing address: PO Box E 266, St James NSW 2000, Australia. T. (61 2) 287 0900. Fax (2 61) 261 4359.

Lee, Simon
Rel: Christian.
International Organizations: Chairman of the International Council for Christian Leadership.
Mailing address: G PO Box 2592, Kowloon, Hong Kong. T. (852 3) 943073.

Lee, Soo-Min, Dr
Rel: Protestant; b. 08 FEB 1927 in Seoul, Korea Rep; m. to Mihwa Lee; c. 1 boy and 1 girl; Citizen of Republic of Korea; Lang. Korean and English.
International Organizations: Secretary-General (since 1985) of the World Alliance of Young Men's Christian Associations.
Educ.: B.D. from Han-Kuk Theological College, Republic of Korea; B.A. from Kon-Kuk University, Seoul, Republic of Korea; YMCA Secretary from YMCA Training College, Republic of Korea; S.T.M. from Union Theological Seminary, New York NY, USA.
Mailing address: Quai Wilson 37, CH-1201 Genève, Switzerland. T. (41 22) 732 31 00. C. FLEMGO GENèVE. Tx 412 332 YMCA CH. Fax (41 22) 738 40 15
Residence: Genève, Switzerland

Lee, William David
Rel: Christian.
International Organizations: Representative of the International Media Ministries.
Mailing address: Chaussée de Waterloo 45, B-1640 Rhode-Saint-Genèse, Belgium.

Lefère, Willem, C.R.S.P., Fr
Rel: Roman Catholic; Ordained; celibate.
International Organizations: Contact of the Clerics Regular of St Paul.
Mailing address: Av Brugmann 121, B-1060 Bruxelles, Belgium. T. (32 2) 344 35 25.

Leffell, Abraham Bernard, Rabbi
Rel: Jewish; Ordained on 13 JUN 1954; b. 12 JUN 1926 in Montreal, Quebec, Canada; m. to Freda Deckelbaum; c. 4 boys; Citizen of Canada; Lang. English, Hebrew and Yiddhish.
International Organizations: Rabbi Emeritus (since 1991) of Shaare Zedek Congregation, Montreal.
Main field of work: Ministry. **Career:** Rabbi, Shaare Zadek Congregation, Montreal from 1954 to 1991. President, Montreal Board of Jewish Affairs from 1965 to 1968. Chairman of Religious Affairs Committee, Eastern Region, Canadian Jewish Congress from 1965 to 1968. Executive Council, Rabbinical Assembly from 1975 to 1978. Honorary Chaplain, Royal Canadian Legion from 1978 to 1991. Chairman, Jewish Chaplaincy Board of Montreal from 1979 to 1990. President, Montreal Board of Jewish Affairs from 1985 to 1987. **Educ.:** B.A. from McGill University, Montreal, Canada (1943-1947); Rabb. Dip., M.H.L. from Jewish Theological Seminary of America, New York NY, USA (1950-1954). **Hons.:** Hon. D.D., Jewish Theological Seminary, NY (1980). **Mem.:** Montefiore Club, Montreal
Mailing address: Shaare Zedek Congregation, 5305 Rosedale Ave, Montreal, Quebec H4V 2H7, Canada. T. (1 514) 484 1122
Residence: 5511 Robinson Ave, Montreal, Quebec H4V 2P4, Canada. T. (1 514) 488 3783

Lefringhausen, M
Rel: Christian.
International Organizations: Representative of the Gemeinsame Konferenz Kirche und Entwicklung.
Mailing address: Adenauerallee 80, D-53113 Bonn, Germany. T. (49 228) 21 83 01.

Légaré, Henri Francis, O.M.I., Most Rev
Rel: Roman Catholic; Ordained on 29 JUN 1943; b. 20 FEB 1918 in Willow Bunch, Saskatchewan, Canada; celibate.
International Organizations: Archbishop (since 1972) of Grouard-McLennan.
Member of the Missionary Oblates of Mary Immaculate.
Career: Bishop, Labrador-Schefferville from 1967 to 1972. National Chaplain and Executive Director, Catholic Hospital Association of Canada from 1952 to 1957. Professor of Sociology, Laval University, Quebec in 1946. Treasurer and Professor, Grand Seminary, St Norbert, Manitoba from 1947 to 1948. Professor of Sociology, Grand Seminary, St Boniface, Manitoba from 1950 to 1952. Professor of Sociology, University of Ottawa in 1951. Professor of Medical Ethics, Faculty of Medicine, University of Ottawa from 1955 to 1958. Dean, Faculty of Social, Economic and Political Science, University of Ottawa from 1954 to 1958. Second Vice-Rector, University of Ottawa from 1955 to 1958. Rector, University of Ottawa from 1958 to 1964. Director of Studies, Oblate, Province of Manitoba and Provincial Superior in 1966. Consecrated Bishop in 1967. Assistant Editor, *La Liberté et le Patriote* (French weekly) from 1950 to 1952. Editor, *L'Ami du Foyer* (magazine) from 1950 to 1952. Director, School of Social Action, St Boniface, Manitoba from 1951 to 1952. President, National Conference of Canadian Universities and Colleges from 1960 to 1962. Chairman, Canadian Universities Foundation from 1960 to 1962. President, Western Catholic Conference of Bishops from 1974 to 1980. President, Canadian Catholic Conference of Bishops from 1981 to 1983. **Mem.:** 'Le conseil de la vie française en Amérique' (Honorary Member); International Academy of Political Science; Canadian Welfare Council (French Commission); Ontario Cancer Treatment and Research Foundation
Mailing address: Box 388, McLennan, Alberta T0H 2L0, Canada. T. (1 403) 324 3002

Léger, Paul Emile, P.S.S., H E Card
Rel: Roman Catholic; Ordained on 25 MAY 1929; b. 26 APR 1904 in Valleyfield, Canada; celibate; Citizen of Canada.
International Organizations: Member (since 1953) of the College of Cardinals.
Member of the Society of the Priests of St Sulpice.
Other organizations: Archbishop Emeritus (since 1968), Montreal.
Career: Professor, Issy Séminaire (near Paris) from 1931 to 1932. Assistant Master, Novices, Saint-Sulpice, Paris from 1932 to 1933. Founder-Superior, Fukuoka Seminary, Japan from 1933 to 1939. Professor, Montreal Seminary of Philosophy from 1939 to 1940. Vicar-General and Cathedral Pastor, Valleyfield from 1940 to 1947. Rector, Canadian College, Rome from 1947 to 1950. Elected to Episcopal See in 1950. Enthroned as Archbishop, Montreal in 1950. Resigned as Archbishop, Montreal to become missionary in leper colonies in 1967. Created Cardinal with title of St Mary of the Angels by Pope Pius XII in 1953. Papal legate to closing the Marian Year at Lourdes in 1954. Papal legate to crowning, St Joseph at Montreal in 1955. Papal legate to St Anne de Beaupré Tercentenary PQ in 1958. Parish priest, St Madeleine Sophie Barat in 1974. Resigned in 1975. **Hons.:** numerous honorary degrees; Knight Grand Cross, Legion of Honour; Equestrian Order of Holy Sepulchre (Jerusalem); Sovereign Order of Malta; Order of Canada Medal; Grand Cross de Benemerencia (Portugal); Humanitarian Award of Variety Clubs International; Pearson Medal for Peace, Canada Association UN; Commemorative plate for exceptional contribution to human relations, Canadian Council for Christians and Jews; Velan Foundation Award.
Mailing address: 130 rue de l'Epée, CP 1500 – Succ A, Montréal PQ H3C 2Z9, Canada. T. (1 514) 495 2421

Legge, Russell, Dr
Rel: Disciples of Christ.
International Organizations: Moderator of the Christian Church (Disciples of Christ) in Canada.
Mailing address: 39 Arkell Road, RR No 2, Guelph ON N1H 6H8, Canada.

Lehman, Robert L, Rabbi
Rel: Jewish.
International Organizations: President of the American Federation of Jews from Central Europe.
Mailing address: 570 Seventh Ave, New York NY 10018, USA. T. (1 212) 921 3871. C. AMFEDERA, NEW YORK.

Leich, Werner, Bishop
Rel: Evangelical Lutheran.
International Organizations: Bishop of Evangelical Lutheran Church in Thuringia.
Mailing address: Dr Moritz-Mitzenheim-Strasse 2a, 99817 Eisenach, Germany.

Leipold, Winfried, E.C.M.C., Fr
Rel: Roman Catholic; Ordained; celibate.
International Organizations: Major Superior of the Camaldolese Hermits of the Congregation of Monte Corona.
Mailing address: Eremo di Monte Rua, I-35038 Torreglia, Italy. T. (39 49) 521 10 41

Lejeune, Geneviève
Rel: Roman Catholic.
International Organizations: Representative at Council of Europe of the International Catholic Child Bureau.
Mailing address: 65 Rue de Lausanne, CH-1202 Genève, Switzerland. T. (41 22) 731 32 48. Fax 731 77 93.

Lekisch, K Peter, Dr
Rel: Jewish.
International Organizations: Past President of the American Federation of Jews from Central Europe.
Mailing address: 570 Seventh Ave, New York NY 10018, USA. T. (1 212) 921 3871. C. Amfedera, New York.

Lema, Anza A, Dr
Rel: Lutheran.
International Organizations: Associate Secretary-General of the Lutheran World Federation.
Mailing address: Route de Ferney 150, PO Box 2100, CH-1211 Genève 2, Switzerland. T. (41 22) 791 61 11. Fax 798 86 16. C. LUTHWORLD GENEVA. Tx 415730 OIK CH.

Lemaître, Henri, H E Mgr
Rel: Roman Catholic; Ordained on 28 JUL 1946; b. 17 OCT 1921 in Mortsel, Antwerp, Belgium; celibate.
International Organizations: Apostolic Pro-Nuncio to Denmark (since 1987) of the Roman Catholic Church.
Other organizations: Apostolic Pro-Nuncio to Finland, Iceland, Norway and Sweden (since 1985); Titular Archbishop, Tongeren.
Mailing address: Immortellevej 11, DK-2950 Vedbaek, Denmark. T. (45) 42 89 35 36 – 42 89 36 37. Fax 45 66 17 71

Lemhényi, Zoltán, Dr
Rel: Yoga.
International Organizations: Secretary-General of the United Nations of Yoga.
Mailing address: Box 21089, S-100 31 Stockholm 21, Sweden. T. (46 0) 31 30 77.

Lemire, Gérard, S.M.M., Fr
Rel: Roman Catholic; Ordained; celibate.
International Organizations: Superior General of the Missionaries of the Company of Mary.
Other organizations: Vice-President, Pontifical Academy of the Immaculate.
Mailing address: Viale dei Monfortani 65, I-00135 Roma, Italy. T. (39 6) 305 23 32.

Lenssen, Frans, C.M.M., Fr
Rel: Roman Catholic; Ordained; celibate.
International Organizations: Procurator-General of the Congregation of Mariannhill Missionaries.
Mailing address: Via S Giovanni Eudes 91, I-00163 Roma, Italy. T. (39 6) 625 85 47, (39 6) 625 85 60

Leonard, Graham Douglas, Rt Rev and Rt Hon
Rel: Anglican; Ordained in 1948; b. 08 MAY 1921 in Greenwich, UK; m. to Vivien Priscilla Swann; c. 2 boys; Citizen of United Kingdom.
International Organizations: Bishop (since 1981) of London.
Other organizations: Chairman (since 1981), Churches Main Committee; Dean (since 1981), H M's Chapel Royal; Prelate (since 1981), Order of British Empire; Episcopal Canon (since 1981), St George's Cathedral, Jerusalem; Honorary Master (since 1981), Bench of Middle Temple; Chairman (since 1981), Jerusalem and Middle East Church Association; Prelate (since 1986), Imperial Society of Knights Bachelor; Member (since 1989), Polytechnics and Colleges Funding Council.
Career: Captain, Oxon and Bucks. Light Infantry from 1941 to 1945. Captain, Army Operational Research Group from 1944 to 1945. Ordained Deacon in 1947. Curacies from 1948 to 1952. Vicar, Ardleigh from 1952 to 1955. Director, Religious Education, Diocese of St Albans, and Canon, St Albans Cathedral from 1955 to 1958. Secretary-General, National Society and Secretary-General of Church of England Schools Council from 1958 to 1962. Archdeacon, Hampstead, and Rector of St Andrew Underschaft and St Mary Axe from 1962 to 1964. Bishop Suffragan, Willesden from 1964 to 1973. Bishop, Truro from 1973 to 1981. Member, Anglican/Orthodox Joint Doctrinal Commission from 1974 to 1981. Delegate to 5th Assembly, World Council of Churches, Nairobi in 1975. Chairman, Church of England Board for Social Responsibility from 1976 to 1983. Chairman, Board of Education from 1983 to 1989. Chairman, BBC and IBA Central Religious Advisory Committee from 1984 to 1989. **Educ.:** University of Oxford, Balliol College, UK; University of Cambridge, Westcott House, UK. **Interests:** reading. **Hons.:** Hon. D.D., Episcopal Theological Seminary, KY (1974), Westminster College, Fulton (1987); Hon. D.C.L., Nashotah House (1983); Hon. S.T.D., Siena College (1984); Hon. LL.D., Simon Greenleaf School of Law (1987); Hon. D.Litt., Council for National Academic Awards (1989); John Findlay Green Foundation Lecture, Fulton (1987); Hon. Bencher, Middle Temple (1981); Hon. Fellow, Balliol College, Oxford; Episcopal Canon of Jerusalem (1982); Privy Councillor (1981). **Pubs.:** Firmly I Believe and Truly, 1985; Life in Christ, 1986 and contributions to several theological works.
Mailing address: London House, 8 Barton Street, Westminster, London SW1P 3NE, UK. T. (44 71) 222 8661

Leonardi, Giuseppe, Fr
Rel: Roman Catholic; Ordained; celibate.
International Organizations: Superior General of the Congrégation des Prêtres des Ecoles de Charité.
Mailing address: Via Casilina 600, I-00177 Roma, Italy. T. (39 6) 27 16 336.

Leppington, Diane, Rev
Rel: Christian.
International Organizations: Secretary of the European Contact Group on Urban Industrial Mission.
Mailing address: 52 Newton Court, Leeds LS8 2PH, UK. T. (44 532) 48 50 11.

Le Renard, Pierre-Yves
Rel: Protestant.
International Organizations: Treasurer of the Medico-social Association of French-Speaking Protestants.
Mailing address: 68 rue Marcel Bourdarias, F-94140 Alfortville, France. T. (33 1) 48 99 27 83 – (33 1) 43 76 06 66. Fax (33 1) 48 98 03 34.

Lerer, Shmuel
Rel: Jewish; b. in Jerusalem, Israel.
International Organizations: Officiant at High Holy Day prayer services of the Montreal.
Career: Served as chief cantor, the Ramah Synagogue, Tel Aviv and of the Great Synagogue in Ramat Gan. Served as chief cantor, the Marble Arch Synagogue, London from 1963 to 1968. Chief cantor in the central Bet Midrash in Johannesburg from 1968 to 1978, then returned to Israel. Produced several records of cantorial music.
Mailing address: No mailing address.

Lerman, Anthony
Rel: Jewish.
International Organizations: Co-Director of the Institute of Jewish Affairs.
Mailing address: 11 Hertford Street, London W1Y 7DX, UK. T. (44 71) 491 3517. C. IJA STUDIES LONDON W1. Tx 21633. Fax (44 71) 493 58 38.

Lesaffre, Stéphane, Fr

Rel: Roman Catholic; Ordained; celibate.
International Organizations: Member of the International Commission of General Prison Chaplains.
Mailing address: 106 rue du Bac, F-75341 Paris, France.

Lesueur, Michel

Rel: Catholic.
International Organizations: President of the World Organization of Former Pupils of Catholic Education.
Mailing address: 21 C rue du Général Faidherbe, F-94130 Nogent-sur-Marne, France.

Leuluaialii, Siatua, Rev

Rel: Methodist.
International Organizations: General Secretary of the Methodist Church in Samoa.
Mailing address: PO Box 199, Apia, Samoa.

Leva, Angelo, F.M.I., Fr

Rel: Roman Catholic; Ordained; celibate.
International Organizations: Procurator-General of the Sons of Mary Immaculate – Pavonians.
Mailing address: Piazza dei Geografi 15, I-00176 Roma, Italy. T. (39 6) 29 57 66.

Lévesque, Georges-Henri, OP, Rev Fr

Rel: Roman Catholic; Ordained on 15 APR 1928; b. 16 FEB 1903 in Roberval, Quebec, Canada; celibate; Citizen of Canada; Lang. French and English.
International Organizations: Ecclesiastic (retired) of the Roman Catholic Church.
Member (since 1923) of the Order of Preachers.
Other organizations: Adviser/Honorary President (since 1971), National University of Rwanda.
Career: In Europe in 1930. Professor during 1933. Professor, Faculty of Social Sciences, Collège des Dominicains d'Ottawa from 1935 to 1938. Professor of Social Philosophy, University of Montreal from 1936 to 1938. Founder and first Dean, Laval University from 1938 to 1962. Founder-President, Faculty of Social Sciences, Laval University from 1939 to 1955. Founder and first Director, Quebec Co-operation Council from 1939 to 1944. Social Advisor to the Agronomists' Guild, *Ensemble!* Magazine from 1940 to 1944. Founder, Editor from 1941 to 1949. Member, Labour Council, *Les Cahiers de la Faculté de Sciences Sociales* in 1941. Member, Economic Council, Quebec Province from 1943 to 1951. Member, Quebec Province from 1943 to 1945. Predicator Generalis, Court of Inquiry on the Canadian Youth from 1943 to 1946. Member, his Order in 1949. President, Royal Commission on the development of arts, sciences and letters in Canada (Massey-Lévesque Commission) during 1951. Contributed to the foundation, Canadian Political Science Association from 1953 to 1952. Member of Board, Faculty of Social Sciences, University of Salamanca in 1955. Founder and first Director, Union of International Associations, Brussels from 1955 to 1960. Vice-President, La Maison Montmorency, Quebec (cultural, social and religious centre) from 1957 to 1963. Vice-President, Canadian Arts Council from 1960 to 1962. Vice-President, Association canadienne française pour l'avancement des sciences (ACFAS) from 1962 to 1961. Founder and first Rector, Canadian Royal Society during 1963. Co-Founder, National University of Rwanda until 1972. Association for Adult Education. **Educ.:** Lecteur en Théol. from Collège des Dominicains, Ottawa, Canada (1924-1930); Dip. Sup. en Sc.Soc. from Fédération Universitaire et Polytechnique de Lille, France (1931-1933). **Hons.:** Dr. h.c.: Universities of Vancouver (1948), Manitoba (1950), Toronto (1952), Antigonish (1953), Western Ontario (1953), Saint Joseph, Moncton (1954), Saskatchewan (1961), Ottawa (1961), Laval (1963), Sherbrooke (1967), McGill (1974), Montreal (1976), National University of Rwanda (1977), University of Quebec at Chicoutimi (1985); Named General Preacher, Dominican Order (1943); Ordre du Mérite coopératif (1948); Knight, Legion of Honour (1950); Médaille Archambault, Association canadienne française pour l'avancement des sciences (ACFAS) (1959); International Cooperation Medal (1964); Honorary President, Institute canadien d'outremer (1965); Molson Prize, Canadian Arts Council (1966); Medal of the Centenary of the Canadian Confederation (1967); Commander, Ordre national des Mille Collines, Rwanda (1977); Medal, Canadian Cooperation Council (1978); Medal, Education Association of Quebec (1979); Medal (1967), Officer (1972), Companion (1979), Order of Canada; Knight, Order international de La Pléiade, International Assembly of French-Speaking Parliamentarians (1980); Officer, Compagnie des Cents-Associés (1981); Royal Bank of Canada Award (1982); Pearson Medal for Peace (1983); Medal, Canadian Association for Adult Education (1985); Officer, National Order of Quebec (1985); Institute for International Affairs Prize (1986); Fondation Edouard Montpetit Prize (1986); Named "Grand Québécois" (1989); Named "Robervalois émérite" (1989); Creation of the Seminary G.-H. Lévesque at Laval University (1990). **Mem.:** Société des Ecrivains (1945-); Royal Society of Canada (1949-); International Union for Social Studies (1949-55); World Brotherhood (1951-); Polish Institute of Arts and Science (1963); Institute of Man and Science, University of New York (1966) **Pubs.:** La première décennie de la Faculté des Sciences sociales à Laval, presented at Colloque de la Société royale: Continuité et rupture dans les sciences sociales de 1935 à 1985, 1981; "Les valeurs éducatives et l'évolution du Québec", in *Vivre en français* (revue de l'ACELF), 1982 and *Souvenances*, Vol. I, 1983, Vol. II 1988, Vol. III 1989, Ed. La Presse.
Mailing address: 2715 ch de la Côte Sainte-Catherine, Montréal, Quebec H3T 1B6, Canada. T. (1 514) 738 3157
Residence: See Mailing Address. T. See Mailing Address

Levi, Bruno

Rel: Jewish.
International Organizations: Co-Chairman of the B'nai B'rith International.
Mailing address: Rua Rio de Janeiro 67, Apt 71, Sao Paulo, Brazil.

Levi, William Haim, Rabbi

Rel: Jewish: Messianic; Ordained in 1987; b. 19 SEP 1926 in Medellin, Colombia; m. to Rachelle Sottu Levi; c. 2 boys and 3 girls; Citizen of USA; Lang. English, Spanish and Portuguese.
International Organizations: South East USA and Latin America Regional Director (since 1987) of the International Alliance of Messianic Congregations and Synagogues.
Other organizations: Executive Committee (since 1988), International Messianic Jewish Alliance, London; Vice-President and International Outreach Committee Chairman (since 1991), Messianic Jewish Alliance of America (MJAA).
Main field of work: Teaching. **Career:** Started Beth Israel Messianic Synagogue, Tampa Bay FL in 1978. Established Etz Chayim Yeshiva (Messianic Jewish Institute) in 1980. **Educ.:** M.Th. from Etz Chayim Yeshiva, Clearwater, Tampa FL, USA (1981-1982); D.Th. from Etz Chayim Yeshiva, Clearwater, Tampa FL, USA (1982-1983). **Pubs.:** *Golden Verses for Your Life* and *Judaizing or Gentalizing: What is the Issue?* (forthcoming).
Mailing address: Beth Israel, 13137 North Dale Mabry Highway, Tampa FL 33618, USA. T. (1 813) 264 1957
Residence: 5006 Country Hills Drive, Tampa FL 33624, USA T. (1 813) 960 7756

Levine, Lee I, Rabbi Prof

Rel: Jewish; Ordained on 14 JUN 1965; b. 01 FEB 1939 in Bangor ME, USA; m. to Mira E; c. 2 boys and 2 girls; Citizen of Israel and USA; Lang. English and Hebrew.
International Organizations: Fellow (since 1989) of the World Union of Jewish Studies.
Other organizations: Dean/Director (since 1987), Seminary of Judaic Studies; Lecturer/Professor (since 1971), Hebrew University of Jerusalem.
Main field of work: University education. **Career:** Instructor, Jewish Theological Seminary of America from 1969 to 1971. **Educ.:** Rabbinic Ordination from Jewish Theological Seminary of America, New York NY, USA (1961-1965); Ph.D. from Columbia University, New York NY, USA (1967-1970). **Mem.:** Fellow, World Union of Jewish Studies (1989); Fellow, Israel Exploration Society (1986) **Pubs.:** The Rabbinic Class of Roman Palestine in Late Antiquity; Caesarea under Roman Rule and Roman Caesarea: An Archeological-Topographical Study.
Mailing address: Seminary of Judaic Studies, PO Box 8600, Jerusalem 91083, Israel. T. (972 2) 617155
Residence: 27 Mevo Hamaavag French Hill, Jerusalem 97877, Israel. T. (972 2) 813164

Lévy, Jacques

Rel: Jewish.
International Organizations: Director of the Alliance israélite universelle.
Mailing address: 45 rue La Bruyère, F-75425 Paris CEDEX 09, France. T. (33 1) 42 80 35 00. Fax 48 74 51 33.

Lewin, Isaac, Dr Rabbi

Rel: Jewish; b. 14 JAN 1906 in Wieliczka, Poland; widowed Peppy Sternheim; c. 2 boys; Citizen of USA; Lang. English, Hebrew, Polish and Yiddish.

International Organizations: Member of Presidium (since 1980) of the Agudath Israel World Organization.
Representative to United Nations (since 1948) of the Agudath Israel World Organization.
Main field of work: Professor and author. **Career:** Member, City Council of Lodz, Poland during 1936. Member, City Council of Lodz, Poland during 1938. Professor of History, Bernard Revel Graduate School, Yeshiva University, New York, NY, USA from 1944 to 1976. Principal, Central High Schools for Girls in Brooklyn and Manhattan, Yeshiva University, New York, NY, USA from 1948 to 1972. Chairman, American Section of World Executive, Agudath Israel World Organization from 1950 to 1980. **Educ.:** LL.M. from Lvov Ivan Franko State University, USSR (1928); Absolutorium of Phil. from Vienna University, Austria (1931); Dr. Law from University of Wilno (1937). **Hons.:** United Nations Medal for Peace (1980). **Mem.:** Polish Institute of Arts and Sciences in America, Memorial Foundation for Jewish Culture (Executive Committee and Board of Trustees), American Jewish Joint Distribution Committee (Board of Directors) **Pubs.:** Author of many history books in Polish, Hebrew, Yiddish and English including: *War on War*, (English), New York, 1969; *Ten Years of Hope*, (English), New York, 1971; *Towards International Guarantees for Religious Freedom*, (English), 1981; *From Morning to Evening*, (Hebrew), Jerusalem, 1982 and *The Jewish Community in Poland*, New York, 1985.
Mailing address: 471 West End Ave, New York NY 10024, USA.
Residence: 258 Riverside Drive, New York NY 10025, USA T. (1 212) 222 6839

Lewis, Ronald Hugh, Rev

Rel: United Reformed Church and Messianic Jew; Ordained on 28 JUL 1959; b. 13 JUN 1930 in London, UK; m. to Doreen Rose MacDonald; c. 1 boy and 1 girl; Citizen of United Kingdom; Lang. English, French (little) and German (little).

International Organizations: Executive Secretary (since 1979) of the International Messianic Jewish /Hebrew Christian/ Alliance.
Main field of work: Ministry. **Career:** Presbyterian Church Adviser on Christian-Jewish relations. **Educ.:** B.A. from University of Wales, University College of Cardiff, UK (1948-1952); Exit Cert. from Westminster College, Cambridge, UK (1956-1959).
Mailing address: 'Shalom', Brockenhurst Road, Ramsgate CT11 8ED, UK. T. (44 843) 592669.
Residence: 70 Bank Hey Lane South, Blackburn BB1 5RQ, UK. T. (44 254) 68284

Lezama, José, C.S.V., Fr

Rel: Roman Catholic; Ordained; celibate.

International Organizations: Procurator-General of the Clerics of St Viator.
Mailing address: Via Padre Angelo Paoli 41, I-00144 Roma, Italy. T. (39 6) 529 1603.

Libanio Christo, Carlos Alberto L, Br

Also know as Frei Betto; Rel: Roman Catholic; Ordained; b. 25 AUG 1944 in Belo Horizonte, Minas Gerais, Brazil; celibate; Citizen of Brazil; Lang. Portuguese, Spanish and French.

International Organizations: Brother of the Order of Preachers.
Other organizations: Assistant (since 1979), Workers' Pastoral of Sao Bernardo do Campo; Popular education, Instituto Cajamar (since 1988); Adviser to Central de Movimentos Populares (since 1989); Adviser to comunidades eclesiais de base (since 1974).
Main field of work: Writer. **Career:** National leader, Catholic Young Students from 1962 to 1964. Political prisoner in 1964. Newspaper and magazine editor from 1966 to 1969. Political prisoner from 1969 to 1973. Organizer, basic Church communities from 1973 to 1979. **Educ.:** Studied Phil. from Dominican's School, São Paulo, Brazil (1965-1968); Studied Theol. from Dominican's School, São Paulo, Brazil (1969-1973). **Interests:** literature, religion. **Hons.:** Intellectual of the Year, Brazil (1985). **Mem.:** Ecumenical Association of Third World Theologians (EATWOT) **Pubs.:** L'Eglise des Prisons; Against Principalities and Powers and Fidel and Religion.
Mailing address: Rua Atibaia 420, 01235 Sao Paulo SP, Brazil. T. (55 11) 65 14 73
Residence: See Mailing Address. T. See Mailing Address

Libbey, Scott S

Rel: Christian.

International Organizations: Executive Vice-President of the United Church Board for World Ministries.
Mailing address: 475 Riverside Drive, 16th Floor, New York NY 10115, USA. T. (1 212) 870 26 37. C. FERNSTALK, NY. Tx 420 423 UBWM UI.

Libouban, Jean-Baptiste

Rel: Roman Catholic; b. 24 FEB 1935 in Paris, France; m. to Jeannine; c. 4 girls; Citizen of France; Lang. French and Spanish.

International Organizations: In charge (since 1989) of the Communities and Movement of the Ark (Lanza del Vasto).
Career: In charge of the school of the head-community of Borie Noble (for many years). Participated in the campaigns for non-violence lead by Lanza del Vasto. Collaborated with Pierre Parodi to the Rassemblement oecuménique européen, Basle in 1989. Organized numerous seminars and training sessions on non-violence. **Educ.:** Ecole Supérieure de Commerce, Strasbourg, France (1954-1957); Tchr. Dip. (1962). **Pubs.:** Non-Violence, Actualité, No 132-137; Accueil, Rencontre, No 123 and articles in *Les Nouvelles de l'Arche*.
Mailing address: Communauté de l'Arche, La Flayssière, F-34650 Joncels, France. T. (33) 67 23 72 00
Residence: See Mailing Address. T. See Mailing Address

Lidell, Elisabeth, Rev

Rel: Evangelical Lutheran.

International Organizations: Secretary on Inter-Church Relations of the Evangelical Lutheran Church of Denmark.
Mailing address: Nørregade 11, DK-1165 København K, Denmark.

Liebich, Hartwig, Rev

Rel: Christian.

International Organizations: Head, Secretariat of the Ecumenical Documentation and Information Centre for Eastern and Southern Africa.
Mailing address: PO Box H 94, Hatfield, Harare, Zimbabwe. T. (263 0) 50311 – (263 0) 50312. Tx 26206 EDIC ZW.

Liebling, Mordechai Eliahv, Rabbi

Rel: Jewish: Reconstructionist; Ordained on 14 JUN 1985; b. 29 OCT 1948 in Brooklyn NY, USA; m. to Devorah Bartnoff; c. 2 boys and 2 girls; Citizen of USA; Lang. English, Hebrew, French and Yiddish.

International Organizations: Executive Director (since 1986) of the Federation of Reconstructionist Congregations and Havurot.
Other organizations: Editorial Board (since 1986), *Reconstructionist Magazine*; Board (since 1989), Shomrei Adamahv Guardians of the Earth; Board (since 1989), US Inter-religious Committee for Peace in the Middle East; Editorial Council (since 1992), Alban Institute.
Main field of work: Administration. **Career:** Rabbi, Manhattan Reconstructionist Havurot from 1982 to 1984. Executive Director, P'nai Or Religious Fellowship from 1984 to 1985. **Educ.:** B.A. from Cornell University, Ithaca NY, USA (1965-1969); M.A. from Brandeis University, Waltham MA, USA (1970-1972); M.H.L.; Rabbinic Ordination from Reconstructionist Rabbinical College, Wyncok PA, USA (1980-1985). **Interests:** environment, working for peace. **Mem.:** Reconstructionist Rabbinical Association **Pubs.:** Children of Abraham – Passover Haggadah.
Mailing address: Church Road and Greenwood Ave, Wyncote PA 19095, USA.
Residence: 130 W Johnson Street, Philadelphia PA 19144, USA T. (1 215) 844 0687

Lightfoot, Neil Roland, Dr

Rel: Church of Christ; Ordained on 21 OCT 1947; b. 22 SEP 1929 in Waco TX, USA; m. to Ollie Robinson; c. 3 girls; Citizen of USA; Lang. French, German, Greek and Hebrew.

International Organizations: Frank Pack Distinguished Professor of New Testament (since 1987) of the Abilene Christian University, TX.
Main field of work: University education. **Career:** Assistant Professor, Abilene Christian University, TX from 1958 to 1961. Associate Professor, ACU from 1961 to 1967. Professor, ACU from 1967 to 1987. **Educ.:** B.A. from Baylor University, Waco TX, USA (1950-1952); M.A. from Baylor University, Waco TX, USA (1952-1955); Ph.D. from Duke University, Durham NC, USA (1955-1958). **Hons.:** Trustees' Award – University Teacher of the Year (1978). **Mem.:** Society of Biblical Literature **Pubs.:** *Jesus Christ Today: A Commentary*

on the Book of Hebrews; *How We Got the Bible*, 2nd ed. and *Lessons From the Parables*.
Mailing address: ACU Station, Box 8446, Abilene TX 79699, USA. T. (1 915) 674 3738
Residence: 1093 Cedar Crest, Abilene TX 79601, USA T. (1 915) 677 3274

Lill, Helma
Rel: Christian.
International Organizations: Secretary-General and Representative of the German-Speaking Region of the International Secretariat of Christian Artists.
Mailing address: Innstrasse 9, 81679 München, Germany. T. (49 89) 98 07 25.

Lim, Ah-Ang, Col
Rel: Salvation Army, The.
International Organizations: Territorial Commander of The Salvation Army.
Mailing address: PO Box 3830, Manila 1099, Philippines.

Linden, Ian
Rel: Roman Catholic.
International Organizations: Secretary-General of the Catholic Institute for International Relations.
Mailing address: 22 Coleman Fields, London N1 7AF, UK. T. (44 71) 354 0883. Fax 359 0017. Tx 21118 CIIR G.

Lindgren, Juhani, Rev Dr
Rel: Evangelical Lutheran; Ordained in 1978; b. 25 DEC 1937 in Helsinki, Finland; m. to Sinikka Hirvonen; c. 2 boys and 2 girls; Citizen of Finland.
International Organizations: Secretary, Missionary Training and Christian Education of the Finnish Lutheran Overseas Mission.
Other organizations: Chairman (since 1990), Finnish Missionary Council; Chairman (since 1991), Nordic Missionary Council.
Educ.: Doc. from University of Helsinki, Finland (1990).
Mailing address: Onnelantie 17A, SF-13100 Hämeenlinna, Finland. T. (358 917) 533 484

Lindner, Louis, Fr
Rel: Roman Catholic; Ordained; celibate.
International Organizations: Permanent Secretary of the Episcopal Conference of the Pacific.
Mailing address: CEPAC, PO Box 289, 14 Williamson Road, Suva, Fiji. T. (679) 300340. Fax (679) 303143.

Lindsay Schimpf, Freda L
Rel: Christian; m. to Gordon Lindsay.
International Organizations: President (since 1973) of the Christ for the Nations.
Educ.: Graduate from Aimee Semple McPherson's L.I.F.E. Bible College.
Mailing address: PO Box 769000, Dallas TX 75376, USA. T. (214) 376 1711.

Lindsey, Stephanie, Sr
Rel: Roman Catholic; Ordained; celibate.
International Organizations: Executive Secretary of the Christian Peace Conference in Latin America and the Caribbean.
Mailing address: Apto 20 – 227, San Angel, 01000 México D F, Mexico.

Lingelser, G
Rel: Catholic.
International Organizations: UNESCO Representative of the World Union of Catholic Teachers.
Mailing address: 57 rue Saint-Paul, F-59420 Mouvaux, France.

Lingscheid, Rainer
Rel: Christian.
International Organizations: Representative of the Association of the Churches Development Services.
Mailing address: Kniebisstr 29, D-70188 Stuttgart, Germany. T. (49 711) 282021. C. AGKED STUTTGART. Email GeoNet GEOD:AGKED.

Lini, Walter Hadye, Fr
Rel: Anglican; Ordained in 1970; b. in Pentecost, Vanuatu; m. to Mary B Ketu; c. 4 boys and 2 girls; Citizen of Vanuatu.
International Organizations: Prime Minister (since 1980) of the Vanuatu.
Career: Ordained Deacon in 1968. Deputy Chief Minister and Minister, Social Services, New Hebrides (now Vanuatu) in 1979. Chief Minister and Minister, Justice from 1979 to 1980. President, Vanuaaku Pati (VP), formerly the National Party. **Hons.:** Commander of the Order of the British Empire. **Pubs.:** Beyond Pandemonium, 1980.
Mailing address: Office of the Prime Minister, PO Box 110, Port Vila, Vanuatu.

Linn, Robert
Rel: Roman Catholic.
International Organizations: America Regional Secretary of the International Catholic Conference of Scouting.
Mailing address: Cerro Largo 1000, CEP 11 100, PO Box 1027, Montevideo, Uruguay.

Linnemann, H M
Rel: Evangelical.
International Organizations: President of the Evangelical Church of Westphalia.
Mailing address: Altstädter Kirchplatz 5, 33602 Bielefeld, Germany.

Linyard, Fred, Rev
Rel: Moravian.
International Organizations: Chairman and General Secretary of the Moravian Church in Great Britain and Ireland.
Mailing address: 5 Muswell Hill, London N10 3TJ, UK.

Lion, Felix, Dr
Rel: Christian.
International Organizations: Chairperson of the World Inter-faith Education Association, Canada.
Mailing address: PO Box 7384, Station 'D', Victoria BC V9B 5B7, Canada. T. (1 604) 382 3012 – 478 8508. Fax 478 6421.

Littman, David G
Rel: Jewish: Progressive.
International Organizations: Geneva Representative of the World Union for Progressive Judaism.
Mailing address: Villa Itea, CH-1196 Gland, Switzerland. T. (41 22) 64 12 05.

Llurba Ferrer, Juan L
Rel: Buddhist: Mahayana.
International Organizations: Member of the Foundation for the Preservation of the Mahayana Tradition.
Other organizations: Director, Nagarjuna C.,E.T. Barcelona.
Mailing address: Rosselon 298 Pral 2a, E-08037 Barcelona, Spain. T. FAX AND TEL (34 93) 257 0788

Loane, Edith M, Dr
Rel: Methodist.
International Organizations: President of the World Federation of Methodist Women.
Mailing address: Inglenook, Royal Terrace Lane, Dun Laoghaire (Co. Dublin), Ireland.

Loane, Marcus Lawrence, Most Rev

Rel: Anglican; Ordained; b. 14 OCT 1911 in Waratah, Tasmania, Australia; m. to Patricia Knox; c. 2 boys and 2 girls; Citizen of Australia.
International Organizations: Ecclesiastic of the Anglican Church.
Career: Resident Tutor and Chaplain, Moore Theological College, Sydney from 1935 to 1958. Vice-Principal, Moore Theological College, Sydney from 1939 to 1953. Principal, Moore Theological College, Sydney from 1954 to 1958. Canon, St Andrew's Cathedral from 1949 to 1958. Bishop Co-adjutor, Diocese of Sydney from 1958 to 1966. Archbishop, Sydney and Metropolitan of New South Wales from 1966 to 1982. Primate, Australia from 1978 to 1982. **Interests:** walking. **Hons.:** Hon. D.D., Wycliffe College, Toronto (1958); Knight of the Order of the British Empire. **Pubs.:** This is My Son, 1977; The God who Acts, 1978 and Makers of Puritan History, 1981.
Mailing address: 18 Harrington Ave, Warrawee NSW 2074, Australia.

Lobato Casado, Abelardo L, Prof Dr Rev

Rel: Roman Catholic; Ordained on 16 APR 1949; b. 20 JAN 1925 in San Pedro de la Viña, Zamora, Spain; celibate; Citizen of Spain; Lang. Spanish, Italian, French, German, English and others.
International Organizations: Director (since 1978) of the Société internationale Thomas d'Aquin.
Other organizations: Professor, Faculty of Philosophy (since 1963), Pontificia Università San Tommaso d'Aquino; Delegate (since 1982), Holy See to the Steering Committee for Human Rights, Council of Europe; Dean, Faculty of Philosophy (since 1986), Pontificia Università San Tommaso d'Aquino; Director (since 1986), San Tommaso Institute.
Main field of work: University teaching. **Educ.:** Lic. Theol. from Univeristy of Salamanca, Spain (1948-1950); Dr. Phil. from University of Rome, Italy (1950-1952); Lic. Phil. and Lett. from University of Madrid, Spain (1958-1960). **Interests:** Philosophy. **Mem.:** Società San Tommaso d'Aquino **Pubs.:** *Ser y belleza*, Herder, 1963; *La pregunta por la mujer*, Sigueme, 1965 and *La dignidad del nombre*, Madrid, 1982.
Mailing address: Pontificia Università San Tommaso, Largo Angelicum 1, I-O0184 Roma, Vatican. T. (39 6) 6702 281. Fax (39 6) 679 0407
Residence: See Mailing Address. T. See Mailing Address

Locke, Kevin

Rel: Bahá'í.
International Organizations: Member of the National Spiritual Assembly of the Bahá'ís of the US.
Mailing address: National Spiritual Assembly of the Bahá'ís of the US, Bahá'í National Center, 536 Sheridan Rd, Wilmette IL 60091, USA.

Loewen, Charles Nelson

Rel: Mennonite; b. 19 JAN 1954 in Steinbach MB, Canada; m. to Cheryl; c. 3 boys; Citizen of Canada; Lang. English.
International Organizations: Chairman (since 1991) of the Mennonite Economic Development Associates.
Other organizations: President (since 1990), Loewen Windows.
Main field of work: Business. **Educ.:** B.A. from University of Winnipeg, Manitoba, Canada.
Mailing address: Loewen Windows, Box 2260, Steinbach MB R0A 2A0, Canada. T. (1 204) 326 6446
Residence: Lot 8, Homestead Crescent, Steinbach MB R0A 2A0, Canada T. (1 204) 326 2755

Loftus, Margaret F

Rel: Catholic.
International Organizations: Director of the International Mission Network.
Mailing address: US Catholic Mission Assn., 3029 Fourth St NE, Washington DC 20017, USA. T. (1 202) 832 3112.

Logvinenko, V E

Rel: Evangelical Baptist.
International Organizations: President of the Union of Evangelical Christian Baptists.
Mailing address: GPO Box 520, Moscow, Russia.

Lohse, Eduard, Prof Bishop Emer

Rel: Lutheran; Ordained on 06 JAN 1952; b. 19 FEB 1924 in Hamburg, Germany; m. to Roswitha Flitner; c. 2 boys and 1 girl; Citizen of Germany; Lang. German, English and French.
International Organizations: President (since 1988) of the United Bible Societies.
Other organizations: Abbot (since 1977), Loccum.
Main field of work: New Testament theology. **Career:** Professor of New Testament, University of Kiel from 1956 to 1964. Professor of New Testament, University of Göttingen from 1964 to 1971. Bishop, Hannover from 1971 to 1988. President, Council of the Evangelical Church in Germany from 1979 to 1985. **Educ.:** Dr. Theol. from Georg-August-Universität Göttingen, Germany (1949). **Interests:** music. **Hons.:** Dr. Theol. h.c., University of Mainz (1961), University of Glasgow (1984). **Mem.:** Academy of Sciences, Göttingen **Pubs.:** A Commentary on the Epistles to the Colossians and Philemon, Philadelphia, 1971; The New Testament Environment, London, 1976 and Theological Ethics of the New Testament, Minneapolis, 1991.
Mailing address: Kirchenkreisamt, Düstere Strasse 19, 37073 Göttingen, Germany. T. (49 551) 496123
Residence: Ernst-Curtius-Weg 7, 37075 Göttingen, Germany. T. (49 551) 42424

Loipfinger, Anton, C.PP.S., Rev

Rel: Roman Catholic; Ordained; celibate.
International Organizations: Moderator-General of the Society of the Precious Blood.
Mailing address: Viale di Porta Ardeatina 66, I-00154 Roma, Italy. T. (39 6) 574 16 56.

Lok, Kim Bong, Bishop

Rel: Methodist.
International Organizations: Bishop of Korean Methodist Church.
Mailing address: Methodist Building, KPO Box 285, Seoul, Korea Rep.

Lokamitra, L, Dharmachari

Also know as Jeremy Goody; Rel: Buddhist; Ordained in 1974; b. 10 SEP 1947; m. to Ranjana Goody; c. 1 boy and 1 girl; Lang. English and Marathi.
International Organizations: President of all the centres in India of the Friends of the Western Buddhist Order.
Career: Worked for the Friends of the Western Buddhist Order in London (for 4 years). Moved to India to help establish activities of the Movement there in 1978.
Mailing address: c/o TBMSG, Raja Harishchandra Road, Dapodi, Pune 411012, India. T. (91) 52919. Tx OI45333 MCCI

Lokiec, Cyna

Rel: Unknown.
International Organizations: Representative of the Ordre de Melchisedec.
Mailing address: Centre Interplanétaire d'extraterrestres, d'Ovnis, d'Elohims et d'Anges, 20 rue Jules Vallès, F-750011 Paris, France.

Long, Z Bradford, Rev Dr

Rel: Presbyterian.
International Organizations: Executive Director of the Presbyterian and Reformed Renewal Ministries International.
Mailing address: PRRMI, 115 Richardson Blvd, PO Box 429, Black Mountain NC 28711-0429, USA. T. (1 704) 669 7373. Fax 669 4880.

Lønning, Inge Johan, Prof Dr

Rel: Lutheran; Ordained on 26 JUN 1963; b. 20 FEB 1938 in Bergen, Norway; m. to Kari B Andersen; c. 2 boys and 2 girls; Citizen of Norway; Lang. Norwegian (and other Scandinavian languages), English and German.
International Organizations: Rectro (since 1985) of the University of Oslo.
Other organizations: Professor of Theology (since 1971), University of Oslo; Chairman of Board (since 1985), Norsk Korrespondanseskole; Chairman of Board (since 1986), Universitetsf Orlaget; President (since 1989), National Council of Universities.

Main field of work: Theologian and university education. **Career:** Navy Chaplain, Church of Norway from 1964 to 1965. Assistant Professor, University of Oslo from 1965 to 1970. Member, Oslo City Council from 1972 to 1976. Dean, Faculty of Theology, University of Oslo from 1977 to 1981. Chairman of Board, Norwegian Research Council for Science and Humanities from 1980 to 1984. **Educ.:** Cand. Theol. from University of Oslo, Norway (1958-1962); Pastoral Sem. from University of Oslo, Norway (1962-1963); Dr. Theol. from University of Oslo, Norway (1971). **Interests:** fishing, skiing. **Hons.:** Dr. h.c., Lutheran College, Decorah (USA); Commander, Royal Norwegian Order of St Olav; Commander with Star, Order of Merit of Federal Republic of Germany. **Mem.:** Norwegian Academy of Science and Letters; Royal Norwegian Society of Science and Letters; International Catholic-Lutheran Dialogue Commission (1986) **Pubs.:** Kanon im Kanon. Zum Grundlagen Problem des Neutestamentlichen Kanons, Munich, 1972; Editor, Martin Luther: Selected Writings, 6 Vols., 1978-1983 (in Norwegian) and Editor, *Kirke og Kultur* (journal).
Mailing address: Rector's Office, University of Oslo, PO Box 1072, Blindern, N-0316 Oslo 3, Norway. T. (47 2) 856303
Residence: Skollerudstubben 22, N-1188 Oslo, Norway. T. (47 2) 289512

Lønning, Per, Rt Rev Dr

Rel: Evangelical Lutheran; Ordained on 04 JUL 1951; b. 24 FEB 1928 in Bergen, Norway; m. to Ingunn Bartz-Johannessen; c. 3 boys and 1 girl; Citizen of Norway; Lang. English, German, French and all Scandinavian languages.
International Organizations: Bishop (since 1987) of Bergen.
Other organizations: Member, Council for Worship (since 1988), Church of Norway; Chairperson, Theological Committee for Ecumenical Relations (since 1990), Church of Norway.
Main field of work: Ministry. **Career:** Assistant Pastor, Lilleborg Church, Oslo from 1951 to 1953. Lecturer, Oslo Teachers' Training College from 1954 to 1957. Member, Norwegian Parliament from 1957 to 1965. Dean, Bergen Cathedral from 1964 to 1969. Bishop, Borg, Fredrikstad from 1969 to 1975. Member, Central Commission, World Council of Churches from 1974 to 1983. Professor of History of Christian Thought, University of Oslo from 1977 to 1981. Research Professor at Institute for Ecumenical Research, Strasbourg from 1981 to 1987. **Educ.:** Cand. Theol. from Independent Faculty of Theology, Oslo, Norway (1949); Dr. Theol. from University of Oslo, Norway (1955); Dr. Phil. from University of Oslo, Norway (1959). **Interests:** skiing, swimming. **Hons.:** Dr.Litt. h.c., St Olaf's College, Northfield MN (1986); Pax Christi Award, St John's University, Collegeville MN (1975). **Mem.:** Royal Norwegian Society of Humanities and Science; Norwegian Academy of Humanities and Science **Pubs.:** Cet effrayant pari – Une pensée pascalienne et ses critiques, Vrin, Paris, 1980; Der begreiflich Unergreifbare, Vandenhoeck and Ruprecht, Göttingen, 1986 and Creation – An Ecumenical Challenge?, Mercer University Press, 1989.
Mailing address: Bjørgvin Bispedømme, Kalvedasveien 45A, N-5018 Bergen, Norway. T. (47 5) 31 24 04
Residence: Landaaslien 78, N-5030 Bergen, Norway. T. (47 5) 28 02 87

Looper, G Kirk

Rel: Baptist: Seventh-Day.
International Organizations: Executive Vice-President of the Seventh-Day Baptist Missionary Society.
Mailing address: 119 Main Street, Westerly RI 02891, USA. T. (1 401) 596 4326. Fax 596 4326.

Loorthusamy, Angy

Rel: Christian.
International Organizations: Regional Secretary for Asia of the World Association for Christian Communication.
Mailing address: ASI Comm Centre, 1518 Leon Guinto St Malate, Manila, Philippines.

Lopes Coelho, Armindo L, Rt Rev

Rel: Roman Catholic; Ordained on 01 AUG 1954; b. 16 FEB 1931 in Regilde, Porto, Portugal; celibate.
International Organizations: Bishop (since 1982) of Viana do Castelo.
Mailing address: Câmara Eclesiástica, Rua de Góis Pinto, P-4900 Viana do Castelo, Portugal. T. (351 58) 322 615

Lopez, Alejandro

Rel: Christian; b. 24 DEC 1965 in Puebla, Mexico; single; Citizen of Mexico; Lang. Spanish, English and Danish.
International Organizations: Latin American Coordinator (since 1989) of the Federation of National Committees in the International Christian Youth Exchange.
Other organizations: President (since 1988), International Service for Youth Exchange, A.C.
Main field of work: Executive Secretary.
Mailing address: Apartado Postal 1586, 72 000 Puebla, Mexico. T. (52 22) 37 68 94
Residence: Av Hidalgo 512B, Col Libertad, 72 130 Puebla, Mexico T. (52 22) 48 43 39

Lopez, Edwin I, Rev

Rel: Baptist.
International Organizations: Secretary-Treasurer of the Asian Baptist Federation.
Mailing address: PO Box 10057, Quezon City, Manila, Philippines. T. (63 2) 615811.

López-Gay, Jesús, Rev Prof Dr

Rel: Roman Catholic; Ordained on 18 MAR 1960; b. 24 MAR 1929 in Almería, Spain; celibate; Citizen of Spain; Lang. Spanish, Japanese, Italian, English and French.
International Organizations: Professor of Japanese Buddhism and Oriental Art (since 1965) of the Pontifical Gregorian University.
Other organizations: Honorary Professor, University of Buenos Aires, Argentina.
Main field of work: Research in history. **Career:** Dean, Missiology Faculty, Pontifical Gregorian University from 1969 to 1972. Dean, Missiology Faculty, Pontifical Gregorian University from 1972 to 1975. Dean, Missiology Faculty, Pontifical Gregorian University from 1987 to 1992. Consultor to Pontifical Council for Inter-religious Dialogue from 1973 to 1983. Consultor to Congregation for the Evangelization of Peoples from 1973 to 1992. **Educ.:** M.Phil. from Chamartín, Madrid, Spain (1950-1953); M.Th. from Sophia University, Tokyo, Japan (1956-1961); Dr. Missiology from Pontifical Gregorian University, Vatican City State (1962-1965). **Pubs.:** El Catecumenado en la misión japonesa del s. XVI, Rome, 1966 (translated into Japanese); La liturgia en la misión del Japón del s. XVI, Rome, 1970 and La Mística del Budismo, Madrid, 1975.
Mailing address: Pontifical Gregorian University, Piazza della Pilotta 4, I-00187 Roma, Italy. T. (39 6) 67015268
Residence: See Mailing Address. T. See Mailing Address

López Montes, Clímaco L, O.A.R., Rev

Rel: Roman Catholic; Ordained on 21 DEC 1963; b. 25 JAN 1938 in Filadelfia, Caldas, Colombia; celibate; Citizen of Colombia; Lang. Spanish, Italian, English, French and Latin.
International Organizations: Procurator-General (since 1986) of the Order of the Augustinian Recollects.
Main field of work: Ministry. **Educ.:** Dip. from Pontifical Faculty of Theology 'Marianum', Vatican City State (1987-1989); Lic. from Pontifical University San Tommaso d'Aquino, Vatican City State (1989-1991). **Interests:** reading, sport, psychology.
Mailing address: Viale dell'Astronomia 27, I-00144 Roma, Italy. T. (39 6) 592 65 34.

Lopez-Penalver, Virginia

Rel: Protestant.
International Organizations: Secretary of the Foyer international protestant 'David Livingstone'.
Mailing address: Rue Blaes 60, B-1000 Bruxelles, Belgium.

López Trujillo, Alfonso L, H E Card

Rel: Roman Catholic; Ordained on 13 NOV 1960; b. 08 NOV 1935 in Villahermosa, Ibague, Colombia; celibate; Citizen of Colombia.
International Organizations: President (since 1990) of the Pontifical Council for the Family.
Member of the Pontifical Commission for Latin America. Member of the Congregation for the Evangelization of Peoples. Member of the Pontifical Council for the Pastoral Care of Migrants and Itinerant People.
Other organizations: Member (since 1983), College of Cardinals; Archbishop Emeritus (since 1991), Medellín; Member, Sacred Congregation for the Doctrine of the Faith.
Career: Archbishop, Medellín from 1979 to 1990. Created Cardinal by Pope John Paul II in 1983.
Mailing address: Piazza S Calisto 16, I-00153 Roma, Italy.

Lorencin, Jovan, Pastor

Rel: Seventh-Day Adventist.
International Organizations: President, Yugoslavian Union Conference of the General Conference of Seventh-Day Adventists.
Mailing address: Bozidara Adzije 4, 11000 Belgrade, Yugoslavia.

Loreti, Antonio Toto, Rev

Rel: Roman Catholic; Ordained on 21 DEC 1950; b. 18 DEC 1925 in Castel di Tora, Rieti, Italy; celibate; Citizen of Italy; Lang. Italian, French and English.
International Organizations: Superior General (since 1989) of the Little Mission for the Deaf-Mute.
Pubs.: Article in *Effeta* (monthly).
Mailing address: Via Nosadella 49, I-40123 Bologna, Italy. T. (39 51) 330 552.
Residence: See Mailing Address. T. See Mailing Address

Lorscheider, Aloisio, O.F.M., H E Card

Rel: Roman Catholic; Ordained on 22 AUG 1948; b. 08 OCT 1924 in Picada Geraldo, Porto Alegre, Brazil; celibate; Citizen of Brazil.
International Organizations: Member (since 1976) of the College of Cardinals.
Member of the Order of Friars Minor. Member of the Congregation for Religious and Secular Institutes.
Other organizations: Archbishop (since 1973), Fortaleza.
Career: Bishop from 1962 to 1973. Created Cardinal by Pope Paul VI in 1976. **Hons.:** Entitled S Pietro in Montorio.
Mailing address: Arcebispado, Rua Senador Almino 310, CP D-6, 60001 Fortaleza CE, Brazil. T. (55 85) 226 2545

Lo Schiavo, John L, S.J., Rev

Rel: Roman Catholic; Ordained on 10 JUN 1955; b. 25 FEB 1925 in San Francisco CA, USA; celibate; Citizen of USA; Lang. English and Italian.
International Organizations: Chancellor (since 1991) of the University of San Francisco, CA.
Member of the Society of Jesus.
Other organizations: Director (since 1990), St Mary's Hospital; Director (since 1991), Convent Schools.
Main field of work: University education. **Career:** Vice-President, University of San Francisco, CA from 1966 to 1968. President, Bellarmine College Preparatory from 1968 to 1975. President, University of San Francisco, CA from 1977 to 1991. **Educ.:** B.A. from Gonzaga University, Spokane WA, USA (1946-1948); M.A. from Gonzaga University, Spokane WA, USA (1948-1949); S.T.L. from Alma College, Los Gatos CA, USA (1952-1956). **Interests:** philosophy. **Hons.:** Dr. Hum. Letters; Torch of Liberty. **Mem.:** Bohemian Club; Olympic Club
Mailing address: University of San Francisco, 2130 Fulton Street, San Francisco CA 94117, USA. T. (1 415) 666 6215
Residence: 650 Parker Ave, San Francisco CA 94118, USA. T. (1 415) 666 0123

Lossuy, Nicolas

Rel: Christian.
International Organizations: Director of the Higher Institute of Ecumenical Studies.
Mailing address: 21 rue d'Assas, F-75270 Paris CEDEX 06, France. T. (33 1) 44 39 52 56.

Lotz, Denton, Rev Dr

Rel: Baptist; Ordained; b. 18 JAN 1939 in Flushing NY, USA; m. to Janice Robinson; c. 3 children.
International Organizations: Secretary-General/Treasurer (since 1988) of the Baptist World Alliance.
Career: First Lieutenant, US Marine Corps, Okinawa from 1961 to 1963. Assistant Minister, Memorial Church, Harvard University from 1963 to 1966. Missionary to Central and Eastern Europe with the Board of International Ministries, American Baptist Churches from 1969 to 1980. Served as Associate Professor of Missions and Homiletics at the Baptist Theological Seminary, Rüschlikon, Switzerland, and founded the Summer Institute of Theological Education (SITE) from 1970 to 1980. Wilkerson Lecturer, Northern Baptist Theological Seminary in 1975. Visiting Scholar, Southern Baptist Theological Seminary in 1975. Staley Lecturer at numerous colleges and universities. Vice-President, CARE from 1980 to 1981. Director, Baptist World Aid from 1980 to 1981. Director, Evangelism and Education Division, Baptist World Alliance (BWA) from 1980 to 1988. Director, Youth Department, BWA from 1980 to 1990. Deputy to Secretary-General, BWA from 1980 to 1988. Acting General Secretary/Treasurer, BWA (from March to July 1988). **Educ.:** B.A. from University of North Carolina, Chapel Hill, USA (1961); S.T.B. from Harvard University, Divinity School, Cambridge MA, USA (1966); Dr. Theol. from University of Hamburg, Germany (1970). **Hons.:** Member, Order of the Golden Fleece, University of North Carolina, Chapel Hill (1957-1961); Freedom's Foundation Award (1963); President, student government, Harvard Divinity School (1963-1966). **Pubs.:** Editor, *Spring has Returned .. Listening to the Church in China* and author of numerous articles on church-state in Eastern Europe, as well as on Baptist mission and evangelism.
Mailing address: 6733 Curran Street, McLean VA 22101, USA. T. (1 703) 790 8980. Fax 893 5160.

Lourdusamy, Simon D, H E Card

Rel: Roman Catholic; Ordained on 21 DEC 1951; b. 05 FEB 1924 in Kalleri, Pondicherry, India; celibate; Citizen of India.
International Organizations: Prefect (since 1985) of the Sacred Congregation for the Oriental Churches.
Member of the Congregation for the Causes of Saints. Member of the Congregation for the Evangelization of Peoples. Member of the Congregation for Catholic Education. Member of the Pontifical Council for Promoting Christian Unity. Member of the Pontifical Council for the Family. Member of the Pontifical Council for Inter-Religious Dialogue. Member of the Pontifical Committee for International Eucharistic Congresses.
Other organizations: Member (since 1985), College of Cardinals; Member, Congregation for the Doctrine of the Faith.
Career: Consecrated Bishop (Titular Church of Sozusa, Libya) in 1962. Titular Archbishop, Philippi in 1964. Archbishop, Bangalore from 1968 to 1971. Secretary, Congregation for the Evangelization of Peoples from 1973 to 1985. Created Cardinal by Pope John Paul II in 1985.
Mailing address: Via dei Corridori 64, I-00193 Roma, Italy. T. (39 6) 698 4281

Lowenthal, Jeffrey

Rel: Jewish: Messianic.
International Organizations: Member of the Messianic Jewish Alliance of America.
Mailing address: PO Box 417, Wynnewood PA 19096, USA. T. (1 215) 896 5812

Lubachivsky, Myroslav Ivan, H E Card

Rel: Roman Catholic; Ordained on 21 SEP 1938; b. 24 JUN 1914 in Dolyna, Ukraine; celibate; Citizen of Ukraine.
International Organizations: Member (since 1985) of the College of Cardinals.
Member of the Sacred Congregation for the Oriental Churches. Member of the Pontifical Council for Dialogue with Non-Believers.
Other organizations: Archbishop Major (since 1984), Lviv of the Ukrainians; President, Synod of the Ukrainian Catholic Hierarchy.
Career: Consecrated Archbishop Metropolitan, Philadelphia of the Ukrainians in 1979. Coadjutor to Archbishop Major, Lviv of the Ukrainians in 1980. Created Cardinal by Pope John Paul II in 1985.
Mailing address: Piazza Madonna dei Monti 3, I-00184 Roma, Italy. T. (39 6) 474 40 68. Fax 474 38 23

Lubich Silvia, Chiara L

Rel: Roman Catholic; b. 22 JAN 1920; Citizen of Italy.
International Organizations: Founder/President (since 1943) of the Focolare Movement.
Consultor of the Pontifical Council for the Laity.
Career: Participated in extraordinary Synod of Bishops in 1985. Participated in Synod on the Vocation and Mission of the Laity in 1987. **Hons.:** Cross, Order of St Augustine of Canterbury (1981); Byzantine Cross, Phanar, Istanbul (1984); Plaque of St Catherine of Siena (1987); Templeton Foundation Prize for Progress in Religion (1977); Festival of Peace of Augsburg Prize (1988). **Pubs.:** To the Gen 3, I 1974, II 1976; Why Have You Forsaken Me, the Key to Unity, 1984 and Encounters with the Orient, 1986.
Mailing address: Centro dell'Opera, Via di Frascati 306, I-00040 Rocca di Papa (Roma), Italy. T. (39 6) 949 386. Fax (39 6) 949 320.

Lucas, Angel, CC.SS., Fr
Rel: Roman Catholic; Ordained; celibate.
International Organizations: Procurator-General of the Congregation of the Sacred Hearts of Jesus and Mary.
Mailing address: Via Rivarone 85, I-00166 Roma, Italy. T. (39 6) 680 55 38, (39 6) 680 55 83

Lucchetti Bingemer, María Clara L, Prof Dr
Rel: Roman Catholic; b. 19 MAY 1949 in Rio de Janeiro, Brazil; m. to Einardo F G Bingemer; c. 1 boy and 2 girls; Citizen of Brazil; Lang. Spanish, Portuguese, French, English, Italian and German.
International Organizations: Assistant Professor of Theology (since 1982) of the Pontifical Catholic University of Rio de Janeiro.
Other organizations: Assistant Professor (since 1981), University Santa Ursula, Rio de Janeiro; Researcher at Centro Joao XXIII (since 1990); World Vice-President (since 1991), Christian Life Communities.
Main field of work: University education, research. **Educ.:** Soc. Communication from Pontifícia Universidade Católica do Rio de Janeiro, Brazil (1968-1975); Theol. from Pontifícia Universidade Católica do Rio de Janeiro, Brazil (1976-1982); M.Th. from Pontifícia Universidade Católica do Rio de Janeiro, Brazil (1982-1985); Ph.D. Theol. from Pontifical Gregorian University, Vatican City State (1986-1989). **Interests:** philosophy of religion, inter-religious dialogue. **Mem.:** Ecumenical Association of Third World Theologians (EATWOT); PT; SOTER; CLC **Pubs.:** With I Gebara, *Maria mae de Deus e mae dos pobres*, Petrópolis, Vozes, 1987 (translated into English, Spanish, German and Italian); *Em tudo amar e servir. Mística trinitária e práxis crista em Santo Inácio de Loyola*, Loyola, Sao Paulo, 1990 and *O segredo feminino do Mistério*, Petrópolis, Vozes, 1991.
Mailing address: Centro Joao XXIII, Rua Bambino 115, Rio de Janeiro 22251, Brazil. T. (55 21) 286 8522
Residence: Rua Almirante Salgado 51, Rio de Janeiro 22240, Brazil. T. (55 21) 225 0874

Lucci, Antonio, S.D.V., Rev
Rel: Roman Catholic; Ordained; celibate.
International Organizations: Procurator-General of the Society of Divine Vocations.
Mailing address: Via di Grottarossa 193, I-00189 Roma, Italy. T. (39 6) 366 06 62

Lucey, Gregory F, S.J., Rev
Rel: Roman Catholic; Ordained; celibate.
International Organizations: Provincial of the Society of Jesus.
Mailing address: Wisconsin Province of the Society of Jesus, 1434 West State Street, Milwaukee WI 53233, USA.

Lucke, Hartmut, Rev
Rel: Reformed Church.
International Organizations: Secretary for the European Area of the World Alliance of Reformed Churches.
Mailing address: Route de Malagnou 54A, CH-1208 Genève, Switzerland. T. (41 31) 46 25 11. Fax (41 31) 45 54 18
Residence: T. (41 22) 735 07 72

Luckhoff, Johann
Rel: Christian.
International Organizations: Director of the International Christian Embassy Jerusalem.
Mailing address: PO Box 1192, Brenner Street 10, Jerusalem 91010, Israel. T. (972 2) 669 823 – (972 2) 699 389. Tx 25378 ICEJI IL. Fax (972 2) 669 970.

Lugo, Olga Rosa
Rel: Roman Catholic.
International Organizations: CENCAMEX Secretary of the International Catholic Committee of Nurses and Medico-social Assistants.
Mailing address: Villa Grillascac, Muñoz Rivera Avenue 2A, Ponce PR 00731, Puerto Rico.

Luigi Santieri, Roberto L, Rev
Rel: Roman Catholic; Ordained; celibate.
International Organizations: President of the Congregation of Regular Clerics.
Career: Rector, University of Milan from 1976 to the present.

Lukass, G Laszlo
Rel: Roman Catholic.
International Organizations: President of the International Catholic Union of the Press – Europe Region.
Mailing address: Milkszath, ter 1, 1088 Budapest, Hungary.

Lull, David John, Dr
Rel: United Methodist; Ordained in 1968; b. 14 OCT 1944 in Mount Kisco NY, USA; m. to Karen Lee Cloyed; Citizen of USA; Lang. English, German and French.
International Organizations: Executive Director (since 1987) of the Society of Biblical Literature.
Career: Assistant Professor, Yale University, Divinity School, New Haven CT from 1981 to 1987. **Educ.:** B.A. from Iowa Wesleyan College, Mount Pleasant IA, USA (1962-1966); B.D. from Southern Methodist University, Dallas TX, USA (1966-1969); Ph.D. from Claremont Graduate School, CA, USA (1971-1978). **Mem.:** American Academy of Religion; Society of Biblical Literature **Pubs.:** Co-author, Biblical Preaching on the Death of Jesus; The Spirit in Galatia: Pneuma as Divine Power and Co-editor, New Testament Interpretation from a Process Perspective.
Mailing address: 1549 Clairmont Rd, Suite 204, Decatur GA 30033, USA. T. (1 404) 636 4744. Fax 248 0815
Residence: 2221 Shasta Way NE, Atlanta GA 30345, USA. T. (1 404) 636 2071

Lumbantobing, A, Rt Rev Dr
Rel: Protestant.
International Organizations: Bishop of Christian Protestant Church in Indonesia.
Mailing address: Kantor Pusad/Head Office, Jalan Kapt MH Sitorus 13, Pematangsiantar SU, Indonesia.

Lumenta, D J, Rev
Rel: Protestant.
International Organizations: Chairman of the Protestant Church in Indonesia.
Mailing address: Jln Medan Merdeka Timur 10, Jakarta-Pusat, Indonesia.

Lumpe, Larry E
Rel: Lutheran.
International Organizations: Executive Director of the International Lutheran Laymen's League.
Mailing address: 2185 Hampton Ave, St Louis MO 63139, USA. T. (1 314) 647 4900.

Lung'aho Ganira, Thomas L
Rel: Quaker; Ordained on 23 JUL 1939; b. 24 JUN 1919 in Bumbo Kakamega, Kenya; m. to Leah Kayere; c. 6 boys and 4 girls; Citizen of Kenya; Lang. English.
International Organizations: General Secretary (since 1991) of the Central Yearly Meeting of Friends.
Other organizations: Member (since 1984), District Education Board.
Main field of work: Education. **Career:** Teaching since 1942. **Hons.:** Head of State Commendation.
Mailing address: PO Box 48, Shinyalu, Kenya. T. (254) 20625 – 41138

Lusakweno-Vangu, K N, Rev
Rel: Baptist.
International Organizations: General Secretary of the Church of Christ in Zaire (Baptist Community of Western Zaire).
Mailing address: BP 4728, Kinshasa 2, Zaire.

Lüsis, Arnolds, Most Rev
Rel: Evangelical Lutheran.
International Organizations: Archbishop of Evangelical Lutheran Church of Latvia in Exile.
Mailing address: 5 Valleymede Road, Toronto, Ontario M6S 1G8, Canada.

Lustiger, Jean-Marie, H E Card

Rel: Roman Catholic; Ordained on 17 APR 1954; b. 17 SEP 1926 in Paris, France; celibate; Citizen of France.

International Organizations: Member (since 1983) of the College of Cardinals.

Member of the Synod of Bishops. Member of the Sacred Congregation for the Oriental Churches. Member of the Sacred Congregation for the Sacraments and Divine Worship. Member of the Congregation for Bishops. Member of the Congregation for the Clergy. Member of the Congregation for Religious and Secular Institutes.

Other organizations: Archbishop (since 1981), Paris; Metropolitan, France; Member, Council of Cardinals and Bishops.

Career: Chaplain to the students, Sorbonne, Paris from 1954 to 1969. Director, Centre Richelieu, Paris from 1959 to 1969. Vicar, Sainte-Jeanne-de-Chantal Parish, Paris from 1969 to 1979. Consecrated Bishop in 1979. Bishop, Orléans from 1979 to 1981. Created Cardinal by Pope John Paul II in 1983. **Pubs.:** Le Choix de Dieu, 1987; La Messe, 1988 and Nous avons rendez-vous avec l'Europe, 1991.

Mailing address: Maison diocésaine, 8 rue de la Ville-l'Evêque, F-75384 Paris CEDEX 08, France. T. (33 1) 49 24 11 11

Lwaminda, Peter

Rel: Roman Catholic.

International Organizations: Secretary-General of the Association of Member Episcopal Conferences in Eastern Africa.

Mailing address: Gitanga Road, PO Box 21191, Nairobi, Kenya. T. (254 2) 566506 – (254 2) 566507. C. 'AMECEA' Nairobi.

Lyall, David, Rev Dr

Rel: Church of Scotland; Ordained on 10 MAR 1965; b. 15 OCT 1936 in Dunfermline, Fife, Scotland, UK; m. to Margaret; c. 1 boy and 1 girl; Citizen of United Kingdom.

International Organizations: Lecturer in Christian Ethics and Practical Theology (since 1990) of the University of Edinburgh.

Member, Coordinating Committee (since 1977) of the International Council for Pastoral Care and Counselling.

Main field of work: University education. **Career:** Minister, Park Parish Church, Ardrossan from 1965 to 1971. Chaplain, Edinburgh Northern Hospitals from 1971 to 1987. Part-time Lecturer, University of Edinburgh from 1971 to 1987. Lecturer, University of St Andrews from 1987 to 1990. Secretary/Treasurer, International Council for Pastoral Care and Counselling from 1987 to 1991. **Educ.:** B.Sc. from University of Edinburgh, UK (1954-1958); B.D. from University of Edinburgh, UK (1960-1963); S.T.M. from Yale University, New Haven CT, USA (1963-1964); Ph.D. from University of Edinburgh, UK (1974-1979). **Pubs.:** With J Foskett, Helping the Helpers: Supervision and Pastoral Care, SPCK, 1988; *British Pastoral Care Movement* (Dictionary of Pastoral Care and Counseling), Abingdon, 1990 and Editor, *Contact – The Interdisciplinary Journal of Pastoral Studies*, 1984-1992.

Mailing address: New College, The Mound, Edinburgh EH1 2LU, UK. T. (44 31) 225 8400.

Residence: 1 North Meggetland, Edinburgh EH14 1XG, UK. T. (44 31) 443 7640

Lynch, Sharon L, Rev

Rel: Spiritualist.

International Organizations: Secretary of the National Spiritual Association of Churches.

Mailing address: PO Box 217, Lily Dale NY 14752, USA. T. (1 716) 595 2000. Fax (1 716) 595 2020

Maartens, Pieter Jacobus, Prof Dr

Rel: Christian; Ordained in 1981; b. 27 JAN 1944 in Johannesburg, Transvaal, South Africa; m. to Jeanne; c. 2 boys; Citizen of South Africa; Lang. English.

International Organizations: Professor and Head (since 1981) of the Department of Biblical Literature.

Other organizations: Professor and Dean, Faculty of Theology (since 1992), University of Durban-Westville.

Main field of work: University education. **Career:** Lecturer, University of Port Elizabeth from 1976 to 1981. Senior Lecturer, University of Durban-Westville from 1982 to 1985. Associate Professor, University of Durban-Westville from 1986 to 1991. **Educ.:** B.A. from University of the Witwatersrand, Johannesburg, South Africa (1965-1968); B.D. from University of Pretoria, South Africa (1969-1972); D.D. from University of Pretoria, South Africa (1980-1985). **Interests:** New Testament. **Hons.:** Merit Award, University of Pretoria (1972). **Mem.:** NISSA; FICLA; MSBL **Pubs.:** "The 'Son of man' as a Composite Metaphor in Mark 14:62", in J H Petzer and P J Hartin, Eds., *A South African Perspective on the New Testament*, E J Brill, Leiden, 1986, pp. 76-III; "'Sign' and 'Significance' in the Theory and Practice of Ongoing Literary Critical Interpretation with reference to Mark 4:24 and 25", in J H Petzer and P J Hartin, Eds., *Text and Interpretation*, E J Brill, Leiden, 1988 and "Critical Dialogue in Theory and Practice of Literary Interpretation: A Study of Semiotic Relations in Matthew 5:11 and 12", in *Linguistica Biblica*, 65, 1991, pp. 5-24.

Mailing address: PO Box 44, Sarnia, Pinetown 3615, South Africa. T. (27 31) 8202290

Residence: 11 Douglas Road, Cowies Hill 3610, Pinetown 3610, South Africa. T. (27 31) 723629

Mabud, Shaikh Abdul, Dr

Rel: Muslim; b. 28 APR 1950 in Bogra, Bangladesh; m. to Shajeda Khanam Shaikh; c. 1 boy and 3 girls; Citizen of Bangladesh; Lang. Bengali, English, Urdu, Arabic and Persian.

International Organizations: Deputy Director-General (since 1983) of the Islamic Academy, Cambridge.

Other organizations: Assistant Editor (since 1983), *Muslim Educational Quarterly*.

Main field of work: Islamic education and philosophy of science. **Career:** Lecturer, University of Rajshahi from 1973 to 1978. Assistant Professor, University of Rajshahi from 1978 to 1981. Research Instructor, University of Utah from 1981 to 1983. **Educ.:** B.Sc. from University of Rajshahi, Bangladesh (1967-1971); M.Sc. from University of Rajshahi, Bangladesh (1971-1972); Ph.D. from University of Cambridge, UK (1974-1978). **Interests:** Arabic calligraphy. **Hons.:** 2 gold medals from University of Rajshahi. **Mem.:** Member, The Institute of Physics, London (1984); Fellow, Cambridge Philosophical Society (1984) **Pubs.:** *Theory of Evolution: An Assessment from the Islamic Point of View*, The Islamic Academy, Cambridge and The Islamic Academy of Science, Kuala Lumpur, 1991; "Curriculum Designing for Natural Sciences from an Islamic Point of View", in *Muslim Education Quarterly*, Cambridge, Vol. 5, No 2, pp. 19-33, 1988 and "A Muslim Response to the Education Reform Act 1988", in *British Journal of Religious Education*, London, Vol. 14, No 2, pp. 88-98, 1992.

Mailing address: 23 Metcalfe Road, Cambridge CB4 2DB, UK. T. (44 223) 350976.

Residence: 67 Darwin Drive, Cambridge CB4 3HQ, UK. T. (44 223) 321061

Mabuluki, Kwangwa, Rev

Rel: Christian.

International Organizations: Regional Representative for Africa of the World Student Christian Federation.

Mailing address: 7th Floor of Church House, Moi Avenue, Nairobi, Kenya. T. (254 2) 27334. Fax 21516.

MacBride, Malachy, S.D.S., Fr

Rel: Roman Catholic; Ordained; celibate.

International Organizations: Superior General of the Society of the Divine Saviour.

Mailing address: Via della Conciliazione 51, I-00193 Roma, Italy. T. (39 6) 654 76 41.

MacBride, Malachy, Fr

Rel: Roman Catholic.

International Organizations: Superior General of the Society of the Divine Saviour.

Mailing address: Via della Conciliazione 51, I-00193 Roma, Italy. T. (39 6) 654 76 41.

Mac Cullen, T R P Richard

Rel: Roman Catholic.

International Organizations: Superior General of the Congregation of the Mission.

Mailing address: Via di Bravetta 159, I-00164 Roma, Italy. T. (39 6) 62 23 241 – (39 6) 62 25 941 – (39 6) 62 29 597.

MacDonald, Donald C, Rev Dr
Rel: Presbyterian.
International Organizations: Clerk of the Assembly of the Presbyterian Church in Canada.
Mailing address: 50 Wynford Drive, Don Mills ON M3C 1J7, Canada.

MacDonald, Donald F M, Rev
Rel: Church of Scotland.
International Organizations: Principal Clerk of the Assembly of the Church of Scotland.
Mailing address: 121 George Street, Edinburgh EH2 4YN, Scotland, UK.

MacDonald, Fergus, Rev
Rel: Free Church of Scotland (Presbyterian); Ordained in 1960; b. 02 MAR 1936 in Evanton, Ross-Shire, Scotland, UK; m. to Dolina M Mackay; c. 1 boy and 4 girls; Citizen of United Kingdom; Lang. English and Spanish.
International Organizations: Secretary-General (since 1981) of the National Bible Society of Scotland.
Secretary (since 1987) of the Lausanne Committee for World Evangelization. Chairman, European Production Fund Board (since 1987) of the United Bible Societies.
Other organizations: Chairman (since 1987), ANGLOSEC (Major English-Speaking Bible Societies); Chairman.
Main field of work: Ministry. **Career:** Assistant Minister, Hope Street Free Church, Glasgow from 1960 to 1962. Minister, St Andrew's Evangelical Presbyterian Church, Lima from 1962 to 1967. Minister, Cumbernauld Free Church of Scotland from 1968 to 1981. Moderator, General Assembly, Free Church of Scotland from 1987 to 1988. Editor, *The Instructor* from 1973 to 1981. **Educ.:** M.A. from University of Edinburgh, UK (1954-1957); B.D. from Free Church College and University of Edinburgh, UK (1957-1960). **Interests:** gardening, walking. **Pubs.:** With Peter Brierley, *Prospects for Scotland*, 1985; *World Evangelism*, 1990 and "Bible Societies", in *New 20th-century Encyclopedia of Religious Knowledge*.
Mailing address: 7 Hampton Terrace, Edinburgh EH12 5XU, UK. T. (44 31) 337 9701
Residence: 113 St Alban's Road, Edinburgh EH9 2PQ, Scotland, UK. T. (44 31) 667 1546

Mac Gillis, Miriam, O.P., Sr
Rel: Roman Catholic; Ordained; celibate.
International Organizations: Member of the Dominican Sisters.
Mailing address: Genesis Farm, Box 622, Blairstown NJ 07825, USA.

Macharski, Franciszek, H E Card
Rel: Roman Catholic; Ordained on 02 APR 1950; b. 20 MAY 1927 in Cracow, Poland; celibate; Citizen of Poland.
International Organizations: Member (since 1979) of the College of Cardinals.
Member (since 1979) of the Congregation for the Clergy. Member (since 1981) of the Congregation for Catholic Education. Member (since 1983) of the Congregation for Bishops. Member (since 1989) of the Congregation for Religious and Secular Institutes.
Other organizations: Archbishop (since 1979), Cracow; Vice-President (since 1979), Polish Bishops' Conference; Vice-Chairman (since 1981), Scientific Council of Polish Episcopate; Vice-Chairman (since 1981), Episcopate Committee for Ministry of Working People; Chairman (since 1981), Episcopate Committee for Laity; Chairman (since 1983), Episcopate Committee for Catholic Science; Member, Council of Cardinals and Bishops.
Career: Engaged in pastoral work in Kozy near Bielsko-Biala from 1950 to 1956. Professor of Pastoral Theology, Pontifical Theological Academy, Cracow from 1962 to 1978. Rector, Pontifical Theological Academy, Cracow from 1970 to 1978. Created Cardinal by Pope John Paul II in 1979. Entitled S Giovanni a Porta Latina in 1979. **Educ.:** Jagiellonian University, Cracow, Poland (1945-1950); Dr. Theol. from University of Fribourg, Switzerland (1956-1960).
Mailing address: Ul Franciszkanska 3, 31 004 Cracow, Poland. T. (48 12) 21 15 33 – 22 98 27

Machiels, Cécile Marie Paule, Sr
Rel: Roman Catholic; Ordained; b. 31 MAY 1932 in Brussels, Belgium; celibate; Citizen of Belgium; Lang. French.
International Organizations: Superior General (since 1985) of the Sisters of St Ursula of the Blessed Virgin.
Main field of work: Ministry. **Career:** Teacher, Institut St Ursule from 1952 to 1954. Teacher, Institut St Ursule from 1957 to 1985. **Educ.:** Agr. de l'ens. sec. inf. from Ecole Normale Secondaire, Brussels, Belgium (1950-1952).
Mailing address: Maison Sainte Ursule, BP 338, F-39104 Dôle, France. T. (33) 84 72 21 03
Residence: Avenue des Armures 36, B-1190 Bruxelles, Belgium. T. (32 2) 345 73 42

Maciel Degollado, Marcial M, L.C., Fr
Rel: Roman Catholic; Ordained; celibate.
International Organizations: Superior General of the Legionaries of Christ.
Mailing address: Via Aurelia 677, I-00165 Roma, Italy. T. (39 6) 680 65 54.

Macini, Marcello
Rel: Buddhist: Mahayana.
International Organizations: Member of the Foundation for the Preservation of the Mahayana Tradition.
Other organizations: Director, Centro Terra di Unificazione Ewam.
Mailing address: Via Cortine 98, I-50021 Barberino Val d'Elsa, Italy. T. (39 55) 807 5732

Macioti, Maria Immacolata
Rel: Unknown; b. 08 DEC 1942; Citizen of Italy.
International Organizations: Member of the Council of the International Society for the Sociology of Religions.
Other organizations: Associate Professor of Sociology, Università degli Studi di Roma La Sapienza.
Pubs.: *La disgregazione di una comunità urbana*, SIARES, Roma, 1988.
Mailing address: Università, Dip di sociologia, Via V E Orlando 75, I-00185 Roma, Italy. T. (39 6) 481 8367 – 481 7868
Residence: Corso Vittorio Emanuele 24, I-00186 Roma, Italy T. (39 6) 678 6760

Mackey, James Patrick, Prof
Rel: Roman Catholic; Ordained in 1958; b. 09 FEB 1934 in Ireland; m. to Noelle Quinlan; c. 1 boy and 1 girl; Citizen of Ireland.
International Organizations: Thomas Chalmers Professor of Theology (since 1979) of the University of Edinburgh.
Career: Assistant Lecturer, then Lecturer in Philosophy, Queen's University of Belfast from 1963 to 1966. Lecturer in Philosophy and Theology, St John's College, Waterford from 1966 to 1969. Associate Visiting Professor, Catholic University of America, Washington DC in 1968. Associate Professor of Philosophical and Systematic Theology, University of San Francisco from 1969 to 1973. Professor of Philosophical and Systematic Theology, University of San Francisco from 1973 to 1979. Visiting Professor, University of California, Berkeley in 1974. Dean, Faculty of Divinity, University of Edinburgh from 1984 to 1988. Visiting Professor, Dartmouth College, NH in 1989. Scripted and presented TV series 'The Hall of Mirrors' (Channel Four) in 1984. Scripted and presented TV series 'Perpectives' (BBC Belfast) from 1986 to 1987. Scripted and presented TV series 'The Gods of War' (Channel Four) in 1987. **Educ.:** B.A. from National University of Ireland, Dublin, Ireland; L.Ph., B.D., S.T.L., D.D. from Pontifical University Maynooth; Ph.D. from Queen's University of Belfast, UK. **Interests:** sailing. **Mem.:** Catholic Theological Society of Great Britain; Clubs: Edinburgh University, Saint Brendan's Cruising **Pubs.:** Editor, Religious Imagination, 1986; Modern Theology: A Sense of Direction, 1987 and Editor, An Introduction to Celtic Christianity, 1989.
Mailing address: New College, Mound Place, Edinburgh EH1 2LX, UK. T. (44 31) 225 8400
Residence: 10 Randolph Crescent, Edinburgh EH3 7TT, UK. T. (44 31) 225 9408

MacLean, James, Rt Rev

Rel: Anglican; Ordained; b. 18 JUN 1922 in Magdalen Islands, Quebec, Canada; m. to Jean Black; Citizen of Canada.

International Organizations: Assistant Bishop (since 1990) of Montreal.

Career: Assistant, St Peter's Church from 1953 to 1955. Rector, St Ignatius Church from 1955 to 1961. Rector, Church of the Good Sheperd from 1961 to 1970. Rector, St George's Church from 1970 to 1984. Suffragan Bishop, Montreal from 1984 to 1990. **Educ.:** B.Th. from McGill University, Montreal, Canada (1953); D.D. from Diocesan College (1985). **Hons.:** Order of St Lazarus; Knight Grand Cross, St John of Jerusalem.

Mailing address: 1444 Union Ave, Montréal PQ H3A 2B8, Canada.

Residence: 440 Church Street, Beaconsfield PQ H9W 3S2, Canada.

MacNeil, Joseph Neil, Most Rev

Rel: Roman Catholic; Ordained on 23 MAY 1948; b. 15 APR 1924 in Sydney, Nova Scotia, Canada; celibate; Citizen of Canada.

International Organizations: Archbishop (since 1973) of Edmonton.

Other organizations: Chairman (since 1973), Alberta Bishops' Conference, Edmonton; Board Chairman (since 1973), Newman Theological College, Edmonton; Board Chairman (since 1973), St Joseph's College, University of Alberta, Edmonton; Member, Board of Management (since 1973), Edmonton General Hospital; Member, Commission on Ecumenism (since 1983), Canadian Conference of Catholic Bishops; Member, Board of Directors (since 1985), Centre for Human Development, Toronto.

Career: Bishop, St John, New Brunswick from 1969 to 1973. Chancellor, University of St Thomas from 1969 to 1973. Member, Executive, Canadian Council on Rural Development from 1965 to 1975. Director, Program and Planning Agency, Nova Scotia Government in 1969. Executive Committee, Atlantic Provinces Economic Council from 1968 to 1973. Vice-Chairman, Nova Scotia Voluntary Economic Planning Board from 1965 to 1979. Vice-President for Adult Education, and Director, Extension Dept., St Francis Xavier University from 1961 to 1969. President, Canadian Conference of Catholic Bishops from 1979 to 1981. Member, Board of Directors, Futures Secretariat in 1981. **Educ.:** B.A. from Saint Francis Xavier University, Antigonish, Nova Scotia, Canada (1944); J.C.D. from Pontifical University San Tommaso d'Aquino, Vatican City State (1958). **Hons.:** Hon. LL.D., St Francis Xavier University (1978), St Thomas University (1980), University of Alberta (1982); Named Domestic Prelate with title of Mgr. by Pope Paul VI for work in the field of social action (1964). **Mem.:** Institute for Research on Public Policy (Founding Member 1968-80); National Commission for Canadian Participation in Habitat (1976)

Mailing address: 8421-101st Ave, Edmonton, Alberta T6A 0L1, Canada. T. (1 403) 469 1010

MacQuarrie, John, Rev

Rel: Anglican; Ordained; b. 27 JUN 1919 in Renfrew, Scotland, UK; m. to Jenny Fallow Welsh; c. 2 boys and 1 girl; Citizen of United Kingdom.

International Organizations: Governor (since 1970) of the St Stephen's House, Oxford.

Main field of work: Professor of Theology. **Career:** British Army Chaplain from 1945 to 1948. Incumbent, St Ninian's Church, Brechin from 1948 to 1953. Lecturer in Systematic Theology, University of Glasgow from 1953 to 1962. Professor of Systematic Theology, Union Theological Seminary, New York from 1962 to 1970. Honorary Curate, St Mary's, Manhattan, NY from 1965 to 1970. Lady Margaret Professor of Divinity, University of Oxford and Canon of Christ Church, Oxford from 1970 to 1986. Gifford Lecturer, University of St Andrews from 1983 to 1984. Consultant to Lambeht Conferences (in 1968, 1978). Governor, Pusey House, Oxford in 1975. **Interests:** hill walking, the language and literature of Scottish Gaelic. **Hons.:** Territorial Decoration. **Mem.:** Fellow of the British Academy **Pubs.:** In Search of Humanity, 1982; In Search of Deity, 1984 and Theology, Church and Ministry, 1986.

Mailing address: 206 Headley Way, Oxford OX3 7TA, UK. T. (44 865) 61889

Madani, Ghazi Obaid, Dr

Rel: Muslim; b. in Medina, Saudi Arabia; Citizen of Saudi Arabia.

International Organizations: Director of the Centre for Research in Islamic Economics.

Main field of work: Academic. **Career:** Supervisor, Organization, later Director, Coordination and Control Dept., Petromin from 1964 to 1965. Lecturer, Faculty of Economics and Administration, King Abdulaziz University, Jeddah, Saudi Arabia in 1973. Assistant Professor in 1975. Vice-Dean, Faculty of Economics and Administration from 1974 to 1978. Associate Professor in 1979. Dean, Faculty of Economics and Administration from 1978 to 1981. Associate Professor, Dept. of Business Administration, Faculty of Economics and Administration, and Vice-Rector for Administration Affairs. Member of Board, Al-Madina for Press and Publishing, Dar Al Ilm Company, National Methonal Company, SABIC; Member: Associate of American Schools of Business Administration. **Educ.:** B.A. Bus. Admin.; M.B.A., Ph.D. Bus. and Fin. **Interests:** Chess, table tennis. **Pubs.:** Several articles and research papers.

Mailing address: King Abdulaziz University, PO Box 1540, Jeddah 21474, Saudi Arabia. T. (966 2) 687 90 33. C. JAMEATABDULAZIZ. Tx 601141 KAUNI SJ.

Residence: Jeddah, Saudi Arabia T. (966 2) 687 14 77

Madigan, Bresa, Sr

Rel: Roman Catholic.

International Organizations: Secretary-General of the Soeurs de l'Enfant-Jésus.

Mailing address: Via Girolamo Nisio 21/D, I-00135 Roma, Italy. T. (39 6) 30 54 751. Fax 30 54 808.

Magalit, Isabelo, Rev Dr

Rel: Evangelical; Ordained on 24 MAR 1991; b. 19 NOV 1940 in Capiz, Philippines; m. to Leticia Abakan; c. 2 boys and 2 girls; Citizen of Philippines; Lang. English and Filipino.

International Organizations: President (since 1989) of the Asian Theological Seminary.

Chairman, Philippine Home Council (since 1984) of the Overseas Missionary Fellowship. Vice-Chairman for Asia (since 1989) of the Lausanne Committee for World Evangelization. Executive Committee (since 1991) of the Lausanne Committee for World Evangelization.

Main field of work: Theological education, preaching. **Career:** Staff worker, Inter-Varsity Christian Fellowship of the Philippines from 1964 to 1966. Secretary-General, Inter-Varsity Christian Fellowship of the Philippines from 1966 to 1973. Associate Secretary-General, International Fellowship of Evangelical Students from 1972 to 1982. Pastor, Diliman Bible Church from 1982 to 1989. **Educ.:** A.A. (pre medicine) from University of the Philippines, Iloilo College, Iloilo, Philippines (1956-1959); M.D. from University of the Philippines, PGH, Manila, Philippines (1959-1964). **Pubs.:** Why Wait Till Marriage and Other Questions; How to Share Jesus and First Steps in Discipleship.

Mailing address: ATS, PO Box 461, Manila, Philippines.

Residence: 101 Naranghita, Project 2, Quezon City, Philippines. T. (63 2) 921 6059

Magermanss, Jacques

Rel: Roman Catholic.

International Organizations: Secretary-General of the Société des Auxiliaires des Missions.

Mailing address: Chaussée de Waterloo 244, B-1060 Bruxelles, Belgium. T. (32 2) 537 23 23.

Magnichever, Lucien

Rel: Jewish.

International Organizations: Representative at UNESCO of the Agudath Israel World Organization.

Mailing address: 53 Bd Sérurier, F-75019 Paris, France. T. (33 1) 42 05 76 91.

Magnoni, Antonio, H E Mgr

Rel: Roman Catholic; Ordained on 11 APR 1943; b. 13 JUN 1919 in Nonantola, Modena, Italy; celibate; Citizen of Vatican and Italy; Lang. Italian, French, English, Spanish and Latin.

International Organizations: Apostolic Pro-Nuncio to Egypt (since 1989) of the Roman Catholic Church.

Other organizations: Titular Archbishop, Boseta.

Main field of work: Ministry. **Educ.:** Dr. Law from University of Modena, Italy (1947-1951); D.Cn.L. from Lateran Pontifical University, Vatican City State (1952-1954).

Mailing address: Apostolic Nunciature, 5 Sharia Mohamed Mazhar, Zamalek, Cairo, Egypt. T. (20 2) 340 22 50. Fax 340 61 52

Residence: Via della Conciliazione 10, I-00193 Roma, Italy. T. (39 6) 654 1764

Magor, Murray C, Rev

Rel: Anglican; b. 17 FEB 1930 in Montreal, Quebec, Canada; m. to Susan Smith; c. 8 children; Citizen of Canada.
International Organizations: Rector (since 1989) of the Trinity Church, St Bruno.
Career: Lawyer, Heward, Holden, Hutchinson from 1954 to 1956. Dean's Vicar, Christ Church Cathedral, Montreal from 1959 to 1961. Rector, Trinity Church, St Agathe from 1961 to 1964. Rector, Christ Church, Beaconsfield from 1964 to 1970. Rector, St Peter's Church, TMR from 1970 to 1974. Archdeacon, St Andrew's, Montreal from 1972 to 1974. Teacher at Centennial Regina High School from 1975 to 1977. Vice-Principal, Centennial Academy from 1977 to 1981. Principal, St George's School, Montreal from 1981 to 1986. Headmaster, Neuchâtel Junior College, Switzerland from 1986 to 1989. **Educ.:** B.A. from McGill University, Montreal, Canada (1951); B.C.L. from McGill University, Montreal, Canada (1954); B.D. from McGill University, Montreal, Canada (1959); S.T.M. from McGill University, Montreal, Canada (1970); M.A. from McGill University, Montreal, Canada (1981).
Interests: skiing, hiking, cycling.
Mailing address: 2075 De Bedford Street, St Bruno, Quebec J3V 4A9, Canada.
Residence: See Mailing Address.

Maharaj Ji, M, Shri

Also know as Prem Pal Singh Rawat; Rel: Hindu; Citizen of India.
International Organizations: 'Perfect Master' of the Divine Light Mission.
Career: Was proclaimed 'perfect Master' upon his father's death (founder of the Divine Light Missions) in 1966. Made his first international tour in 1971.
Mailing address: 21 rue de Pondichéry, F-75015 Paris, France.

Mahase, K M J, Rev

Rel: Evangelical.
International Organizations: Executive Secretary of the Lesotho Evangelical Church.
Mailing address: PO Box MS 260, Maseru 100, Lesotho.

Mahathera, M Vajiragnana, Ven Dr

Rel: Buddhist.
International Organizations: Monk of London Buddhist Vihara.
Mailing address: London Buddhist Vihara, 5 Heathfield Gardens, Chiswick, London W4 4JU, UK.

Mahathera, Nyanasamvara, Ven

Rel: Buddhist: Theravada.
International Organizations: Supreme Patriarch of Thailand.
Mailing address: Wat Bovornives Vihara, Banglumpoo, Bangkok 10200, Thailand.

Maher, Helen, Sr

Rel: Roman Catholic; Ordained; celibate.
International Organizations: Superior General of the Sisters of St John of God.
Mailing address: Spawell Road, Wexford, Ireland. T. (353 53) 42 396.

Maheswara Kurukkal, S M, Brahmasri

Rel: Hindu.
International Organizations: Chief Priest of the Sri Nagabooshani Amman Kovil.
Mailing address: Sri Nagabooshani Amman Kovil, Nainativu, Sri Lanka.

Maillefer, Eric

Rel: Evangelical.
International Organizations: European Director of the World Evangelical Fellowship.
Mailing address: 2 Molenstraat, B-1560 Hoeilaart, Belgium. T. (32 2) 657 18 01.

Maina, John Boaz Y

Rel: Brethren.
International Organizations: General Secretary of the Brethren Church of Nigeria.
Mailing address: c/o Kulp Bible School, PO Box 1, Mubi, Gongola State, Nigeria.

Majewski, Tadeusz R, Most Rev

Rel: Catholic.
International Organizations: Prime Bishop of Polish Catholic Church in Poland.
Mailing address: ul Wilcza 31, Warszawa, Poland.

Majima, Jyokei

Rel: Buddhist.
International Organizations: Member of the International Network of Engaged Buddhists.
Mailing address: Niiya, 88 Tomita-Chiyo, Nakagawa-Ku, Nagoya-Shi, Aichi-Ken, Japan.

Mäkelä, Raimo Olavi, Rev

Rel: Evangelical Lutheran; Ordained on 06 NOV 1963; b. 09 AUG 1940 in Helsinki, Finland; m. to Seija Anneli Rantanen; c. 2 boys and 2 girls; Citizen of Finland; Lang. Finnish, Swedish, Norwegian, Danish, English and German.
International Organizations: General Secretary (since 1980) of the Finnish Bible Institute Foundation.
Other organizations: Editor-in-Chief (since 1980), *Herdä Valvomaan* magazin; Chairman (since 1986), Finnish Lutheran Overseas Mission; Member (since 1987), Nordic Council for Inner Mission.
Main field of work: Management and preaching. **Career:** Youth Pastor, a Lutheran state church congregation in Helsinki from 1964 to 1968. General Secretary, Finnish Evangelical Lutheran Student Mission (affiliated to IFES) from 1968 to 1980. **Educ.:** M.Th. from University of Helsinki, Finland (1958-1963); M.Sc.S. from University of Helsinki, Finland (1975-1979). **Pubs.:** 10 books/booklets and 30 translations of books.
Mailing address: POB 15, SF-02700 Kauniainen, Finland. T. (358 0) 5050177
Residence: Valkonauhantie 6 B 10, SF-02700 Kauniainen, Finland. T. (358 0) 5052095 – 5050097

Makhulu, Walter Paul Khotso, Most Rev

Rel: Anglican; Ordained; b. 02 JUL 1935 in Johannesburg, South Africa; m. to Rosemary; c. 1 boy and 1 girl; Citizen of United Kingdom.
International Organizations: Archbishop (since 1980) of Central Africa.
Other organizations: Bishop (since 1979), Botswana.
Career: Area Secretary for Eastern Africa and African Refugees, Commission on Inter-Church Aid Refugee and World Service, World Council of Churches (WCC) from 1975 to 1979. President, All Africa Conference of Churches from 1981 to 1986. President, WCC in 1983.
Interests: music, international affairs. **Hons.:** Honorary Curate, Holy Trinity, Geneva; Hon. D.D., Kent (1988); Officer, Ordre des Palmes Académiques (1981).
Mailing address: PO Box 769, Gaborone, Botswana. T. (267) 353779

Makuzwa, Caspen C, Rev

Rel: Methodist.
International Organizations: President of the Methodist Church in Zimbabwe.
Mailing address: PO Box 8298, Causeway, Harare, Zimbabwe.

Malan, Hugo, Rev

Rel: Waldensian.
International Organizations: Moderator of the Waldensian Regional Board.
Mailing address: Avenida Armand Ugon, 70 202 Valdense (Colonia), Uruguay.

Malenge, Zablon Isaac

Rel: Quaker; Unordained; b. 05 MAY 1947 in Kakamega, Western Province, Kenya; m. to Margaret Afamdi; c. 6 girls; Citizen of Kenya; Lang. English, Kiswahili and Luhya.

International Organizations: Executive Secretary for Africa (since 1979) of the Friends World Committee for Consultation.

Other organizations: Honorary Secretary-General (since 1985), Young Quakers Association.

Main field of work: Church ministry, administrative consultation. **Educ.:** Cert. from Daystar University, Nairobi, Kenya (1983); Cert. from Amani Counselling Society, Nairobi, Kenya (1989). **Interests:** writing, preaching, farming. **Hons.:** Commonwealth Centenary Year Essay Award. **Mem.:** Young Men's Club (YMCA), Institute of Bankers, Amani Counselling Society **Pubs.:** Editor, *FWCC Friends in Africa Newsletter*; Editor, *YQA Newsletter* and Editor, *Friends Swahili Newsletter*.

Mailing address: Friends Int. Centre, Ngong Road, PO Box 41946, Nairobi, Kenya. T. (254 2) 567601 – 568706

Residence: See Mailing Address. T. See Mailing Address

Malinsky, Abraham Joseph, Rabbi Prof

Rel: Jewish; Ordained on 10 AUG 1953; b. 03 APR 1930 in Jerusalem; m. to Rachel Silberberg; c. 2 boys and 1 girl; Citizen of Belgium; Lang. Hebrew, Yiddish, Aramean, English, French, Dutch and German.

International Organizations: Rabbi responsible (since 1972) of Main Synagogue Bouwmeester, Antwerp.

Other organizations: Professor (since 1981), Theological Faculty, Antwerp; Lecturer (since 1988), Rabbinical Seminary, Amsterdam; Government and Communities Inspector, Jewish Education, Belgium.

Career: Founded, Association for the Propagation of Jewish Values in 1986. **Educ.:** Talmudic Academy Yeshivah, Jerusalem, Israel (1945-1953); Hebrew University of Jerusalem, Israel (1948-1953). **Interests:** Jewish history, Biblical exegesis, comparative philology. **Hons.:** Medal of Distinction from Belgian Government (1984). **Mem.:** Bnei Brith; VINRO (organisation of Flemish inspectors) **Pubs.:** Introduction to the Jewish Religion (in English); Dr Mengele and 'the Twins of Auschwitz' and History of the Jews in Belgium.

Mailing address: Diksmuidelaan 6 bus 8, B-2600 Berchem, Belgium. T. (32 3) 236 03 36

Residence: See Mailing Address. T. See Mailing Address

Malley, John, O.C., Fr

Rel: Roman Catholic; Ordained; celibate.

International Organizations: Prior-General of the Order of Brothers of the Blessed Virgin Mary of Mount Carmel.

Mailing address: Via Giovanni Lanza 138, I-00184 Roma, Italy. T. (39 6) 481 41 67, (39 6) 488 28 18, (39 6) 481 39 28

Malloy, C J, Jr, Rev

Rel: Baptist.

International Organizations: General Secretary of the Progressive National Baptist Convention, Inc.

Mailing address: 601 50th St NE, Washington DC 20019, USA.

Maloney, Terry

Rel: Unknown.

International Organizations: Executive Director of the Honorable Order of the Blue Goose International.

Mailing address: 12940 Walnut Rd, Elm Grove WI 53122, USA. T. (1 414) 782 7658.

Ma Lourdes G Mendiola, M

See Ma Ramona Mendiola.

Maluit, Thomas, Rev

Rel: Presbyterian.

International Organizations: General Secretary of the Presbyterian Church in the Sudan.

Mailing address: PO Box 40, Malakal, Sudan.

Mampra, Thomas, C.M.I., Fr

Rel: Roman Catholic; Ordained; celibate.

International Organizations: Prior-General of the Carmelites of Mary Immaculate.

Mailing address: St Joseph's, Monastery Road, Ernakulam, Cochin 682011 Kerala, India. T. (91 484) 36 01 37

Manalai, Zainulla

Rel: Muslim.

International Organizations: President, Islamic Research Center of the Ministry of Islamic Affairs, Afghanistan.

Mailing address: Ministry of Islamic Affairs and Pious Foundation Organisation, Kabul, Afghanistan.

Mandemaker, G W, Pastor

Rel: Seventh-Day Adventist.

International Organizations: President, Netherlands Union Conference of the General Conference of Seventh-Day Adventists.

Mailing address: Biltseweg 14, NL-3735 Bosch en Duin, Netherlands.

Mangaleswara Kurukkal, K M, Brahmasri

Rel: Hindu.

International Organizations: Chief Priest of the Elathu Sithamparam.

Mailing address: Elathu Sithamparam, Karainagar, Sri Lanka.

Manino, Line

Rel: Roman Catholic.

International Organizations: Head of the International Catholic Organizations Center.

Head, New York International Catholic Organizations Information Centre of the Conference of International Catholic Organizations.

Mailing address: 323 East 47th Street, New York NY 10017, USA. T. (1 212) 355 5557.

Manjit Singh, M, Singh Sahib Jathedar

Rel: Sikh; b. 12 APR 1952 in Distt. Ropar, Punjab, India; m. to Sardani Kuljit Kaur; c. 4 boys; Citizen of India; Lang. Punjabi, Hindi and English.

International Organizations: Jathedar (Head Minister) (since 1990) of the Takhat Sri Kesgarh Sahib, Anandpur.

Other organizations: Member, Advisory Board (since 1990), Inter-religious Federation for World Peace; President (since 1991), Prof Sahib Singh Memorial Trust; Chairman (since 1992), 300th Khalsa Sajna Utsuv Committee.

Main field of work: University education. **Career:** Head Priest, Sri Guru Singh Sabha, Bombay from 1973 to 1977. Principal, Guru Nanak Gurmat College, Nainital from 1978 to 1980. **Educ.:** Gyani from Punjabi University, Patiala, India (1975); B.Ed. from Panjab University, Chandigarh, India (1985); M.A. from Punjabi University, Patiala, India (1989). **Interests:** music, missionary service for youth. **Pubs.:** Places of Worship of Sikhism in Multifaith India; Contemporary Consciousness and Religious Values: A Sikh Perspective and Patshahi Bhagat Jana Kauditani.

Mailing address: Takhat Sri Kessarh Sahib, Disst. Ropar, Punjab, India.

Residence: House No 5155, Phase II, Urban Estate, Patiala 147 002, Punjab, India T. (91 175) 822095

Manjuvajra, M, Dharmachari

Rel: Buddhist; Ordained on 25 NOV 1973; b. 13 AUG 1947 in Chatham, Kent, UK; single; Citizen of United Kingdom; Lang. English.

International Organizations: Chairman (since 1981) of the Friends of the Western Buddhist Order.

Director of Retreat Center (since 1985) of the Friends of the Western Buddhist Order.

Main field of work: Teaching and management. **Career:** Director/Teacher, Cornwall Buddhist Centre from 1973 to 1976. Meditation Teacher, London Buddhist Centre from 1976 to 1981. **Educ.:** B.Sc. from University of London, Imperial College of Science, Technology and Medicine, UK (1965-1968); M.Phil. from University of London, University College, UK (1978-1980). **Pubs.:** Article in *Buddhist America*, Ed. Morreale, John Muir Publications, 1988; Chapter in *Education for Peace*, Ed. Gordon and Grob, Orbis, 1987 and *Individualism and Conformity*, Windhorse Publications, 1976.

Mailing address: 'Aryaloka', Heartwood Circle, Newmarket NH 03857, USA. T. (1 603) 659 5456

Residence: See Mailing Address. T. See Mailing Address

Manoukian, Ardak, H E Mgr

Rel: Armenian Apostolic Church; Ordained on 24 JUN 1951; b. 22 JAN 1931 in Beirut, Lebanon; celibate; Citizen of Iran; Lang. Armenian, French, English, Arabic and Persian.

International Organizations: Prelate (Archbishop) (since 1960) of Armenian Diocese of Teheran.

Career: Archimandrite in 1953. Consecrated Bishop in 1956. Armenian Catholicossat, Cilicie, Antelias from 1956 to 1960. Editor, *Hassk* Review. **Educ.:** Facultés Catholiques de Lyon, France (1954-1956). **Interests:** literature. **Hons.:** Order of Hamayoun II of Iran (1968). **Pubs.:** Author of 20 books in Armenian, including: *Religion and Culture*; *Passage to Armenia* and *With Open Arms* (poetry).

Mailing address: Armenian Prelacy, Avenue Ostad Nejatollahi 311, 15988 Teheran, Iran Islamic Rep. T. (98 21) 891634/5

Residence: See Mailing Address.

Manzur, Gholam Sarwar, Mawlani

Rel: Muslim: Hanafi; Ordained on 02 MAR 1979; b. 17 MAY 1939 in Jalalabad, Nengarhar Province, Afghanistan; married; c. 5 boys and 4 girls; Citizen of Afghanistan; Lang. Arabic, Pashto and Dari.

International Organizations: Deputy Minister of the Ministry of Islamic Affairs, Afghanistan.

Main field of work: Islamic research. **Career:** Director General, High Council of Ulema and Clergy, Ministry of Islamic Affairs. **Educ.:** Dip. from Institute of Najmul Madaress, Nengarhar, Afghanistan (1932-1938). **Pubs.:** Adalat of Islam; Nezafat and Men Aiat-i Rahman.

Mailing address: Ministry of Islamic Affairs and Pious Foundation Organisation, Kabul, Afghanistan.

Residence: New City, Kabul, Afghanistan.

Marcelino, Antonio Baltasar, Mgr

Rel: Roman Catholic.

International Organizations: Representative of the European Episcopal Committee on the Media.

Mailing address: Curia Diocesana, Rua Cândido Dos Reis 107, Apartado 541, 3808 Aveiro, Portugal.

Marc'hadour, Germain Pierre, Fr

Rel: Roman Catholic; Ordained; b. 16 APR 1921 in Langonnet, France; celibate.

International Organizations: Founder and International Secretary (since 1962) of the Amici Thomae Mori.

Other organizations: Professor, English Philology (since 1952), Université Catholique de l'Ouest, Angers, France.

Career: Ordained Priest in 1944. Professor at the Collège Saint-Ivy, Pontivy from 1945 to 1952. Numerous stays as a researcher or lecturer in UK, USA and the military chaplaincy of Germany FR from 1948 to 1952. Director, 'Moreana', Angers in 1963. **Educ.:** D.E.S. Litt.angl. from Catholic Universities of Angers and Lyon, France; Dr. ès Lett. from Faculties of Letters, Rennes and Paris, France; Dip. Int. Phonetics from University of London, UK. **Hons.:** Knight, Palmes Académiques. **Mem.:** Life Member, Association Guillaume-Budé and Renaissance Society of America **Pubs.:** *The Bible in the Works of St Thomas More*, 5 Vols., 1969-1972; *Thomas More*, 1971; *Erasme de Rotterdam et Thomas More: Correspondance*, 1985; Contribution to the *Encyclopeadia Britannica* and Numerous articles in about twenty reviews in several coutries.

Mailing address: Moreanum, BP 808, F-49005 Angers CEDEX, France. T. (33) 41 88 83 98.

Residence: 29 rue Volney, F-49000 Angers, France

Marchetto, Agostino, H E Mgr Dr

Rel: Roman Catholic; Ordained on 28 JUN 1964; b. 28 AUG 1940 in Vicenza, Italy; celibate; Citizen of Vatican and Italy; Lang. Italian, French, German, English, Spanish and Portuguese.

International Organizations: Apostolic Pro-Nuncio to Tanzania (since 1990) of the Roman Catholic Church.

Other organizations: Titular Archbishop (since 1985), Astigi.

Main field of work: Diplomacy. **Career:** Secretary, Apostolic Nunciature in Zambia from 1968 to 1971. Secretary, Apostolic Nunciature in Cuba from 1971 to 1975. Auditor, Apostolic Nunciature in Algeria from 1975 to 1978. Auditor, Apostolic Nunciature in Portugal from 1978 to 1981. Counsellor, Apostolic Delegation to Mozambique from 1981 to 1985. Apostolic Pro-Nuncio to Madagascar and Mauritius from 1985 to 1990. **Educ.:** D.Cn.L. from Lateran Pontifical University, Vatican City State (1964-1967); L.Th. from Lateran Pontifical University, Vatican City State (1968); Dip. Past.Sc. from Lateran Pontifical University, Vatican City State (1967-1968); Dip. Diplomacy from Pontifical Ecclesiastical Academy, Vatican City State (1967-1968). **Interests:** history of the Middle Ages. **Pubs.:** Episcopat et Primat Pontifical dans les Décrétales Pseudo-Isidoriennes; Mgr J B Montini fut-il un historien ? and Le succès d'une falsification: le Pseudo-Isidor et le nouveau Code de Droit Canon.

Mailing address: Apostolic Nunciature, PO Box 480, Dar es Salaam, Tanzania UR. T. (255 51) 68 403. Fax 41 943

Residence: Plot 462 (off Mawenzi Rd), Musasani Peninsula, Dar es Salaam, Tanzania UR. T. See Mailing Address

Marciel Degollado, L.C., Fr

Rel: Roman Catholic; Ordained on 26 NOV 1944; b. 10 MAR 1920 in Cotija, Mexico; celibate.

International Organizations: Founder and Superior General of the Legionaries of Christ.

Mailing address: Via Aurelia 677, I-00165 Roma, Italy. T. (39 6) 680 65 55

Marcinkus, Paul Casimir, Most Rev

Rel: Roman Catholic; Ordained on 03 MAY 1947; b. 15 JAN 1922 in Cicero IL, USA; celibate; Citizen of USA.

International Organizations: Advisor of the Pontifical Commission for the Vatican City State.

Other organizations: Titular Archbishop (since 1981), Orta.

Career: Assistant Pastor, St Cristina's Parish, Chicago from 1947 to 1952. Joined Vatican State Secretariat in 1952. Served as papal diplomatist in Canada and Bolivia. General Manager, Istituto per le Opere di Religioni (Vatican Bank) from 1969 to 1971. Chairman, Istituto per le Opere di Religioni from 1971 to 1989. Pro-President, Pontifical Commission for the Vatican City State in 1981.

Mailing address: Via della Nocetta 63, I-00164 Roma, Italy.

Maréchal, Claude, A.A., Fr

Rel: Roman Catholic; Ordained; celibate.

International Organizations: Superior General of the Augustinians of the Assumption.

Mailing address: Via San Pio V 55, I-00165 Roma, Italy. T. (39 6) 62 23 998.

Margéot, Jean, H E Card

Rel: Roman Catholic; Ordained on 17 DEC 1938; b. 03 FEB 1916 in Quatre-Bornes, Mauritius; celibate; Citizen of Mauritius; Lang. French, English and Italian.

International Organizations: Member (since 1988) of the College of Cardinals.

Member (since 1988) of the Pontifical Council for the Pastoral Care of Migrants and Itinerant People. Member (since 1988) of the Pontifical Council for the Family. Member of the Congregation for the Evangelization of Peoples.

Other organizations: Bishop (since 1969), Port-Louis.

Career: Vicar-General in 1956. Bishop in 1969. Created Cardinal by Pope John Paul II in 1988. Member, Bishops' Conference of Indian Ocean Islands in 1985. **Educ.:** Lic. Phil. and Theol. from Pontifical Gregorian University, Vatican City State. **Hons.:** Knight of the National Order of the Legion of Honour (1988); Named Honorary Citizen of Quatre-Bornes (1986); Honorary Citizen of Beau-Bassin/Rose Hill (1986); Honorary Citizen, District of Grand Port/Savanne (1989). **Pubs.:** contributed numerous letters and articles to journals.

Mailing address: Evêché, 13 rue Mgr Gonin, Port-Louis, Mauritius. T. (230) 208 3068

Residence: See Mailing Address. T. See Mailing Address

Maria, Francis

Rel: Antiochian Orthodox; b. 01 JAN 1913; single.

International Organizations: Chairman, Ecumenical Committee (since 1982) of the Antiochian Orthodox Christian Archdiocese of North America.

Educ.: B.A. from Boston University, College of Liberal Arts, MA, USA; M.A. from Boston University, Graduate School, MA, USA.

Mailing address: PO Box 46, Warner NH 03278, USA.

Marie, Michael, Sr

Rel: Roman Catholic; Ordained; celibate.

International Organizations: Member of the Franciscan Sisters of Alegany.

Other organizations: Member, Association of Contemplative Sisters.

Main field of work: Contemplative.

Mailing address: c/o Franciscan Srs of Aleganny, St Elizabeth's Motherhouse, Alegany NY 14706, USA.

Mariella, M, A.C., Sr

Also know as Marie Colaco; Rel: Roman Catholic; Ordained; b. 14 NOV 1924 in Madras, India; celibate; Citizen of India; Lang. English, Hindi, Tamil and Konkani.

International Organizations: Superior General (since 1984) of the Sisters of the Apostolic Carmel.

Member (since 1944) of the Sisters of the Apostolic Carmel.

Career: Principal from 1963 to 1964. Pro-Provincial Superior, St Ann's B.Ed. College, Mangalore from 1972 to 1974. Provincial Superior, Sisters of the Apostolic Carmel from 1974 to 1978. Assistant-General, Sisters of the Apostolic Carmel from 1978 to 1984. Secretary, Women's Section, Sisters of the Apostolic Carmel from 1973 to 1976. President, Women's Section, Conference of Religious of India (CRI) from 1976 to 1978. President, Women's Section, CRI from 1988 to 1990. Delegate for India to International Union of Superiors General, CRI in 1986. Councillor for India from 1987 to 1989. Executive Member, International Union of Superiors General, Rome from 1984 to 1988. Vice-President, Sudeep Training Institute, Bangalore from 1988 to 1990. Delegate, Sudeep Training Institute, Bangalore in 1975. Delegate, CRI to the Bishop's Institute for Social Action (BISA) III, Kuala Lumpur in 1976. Invited by the Holy Father to be present as an 'Auditrix' at the VIII Synod of Bishops, CRI to the Asian Meeting of Religious (AMOR) IV, Manila in 1990. **Educ.:** B.Th., M.Sc.

Mailing address: Apostolic Carmel Generalate, PO Box 4107, Jayanagar 'T' Block PO, Bangalore 560041, India. T. (91 812) 641393.

Marin, Carlos

Rel: Christian.

International Organizations: Assistant Director of the Corporation for Cultural and Social Development.

Mailing address: Apdo Aereo 20439, Bogotá, Colombia. T. (57 1) 287 70 07 – 287 85 40. Fax 287 87 01.

Mariner, Cheryl

Rel: Jewish; b. 07 MAR 1945 in Melbourne, Victoria, Australia; m. to Rodney Mariner; c. 1 boy and 1 girl; Citizen of Australia; Lang. English.

International Organizations: Executive Director (since 1986) of the CBF World Jewish Relief.

Executive Secretary of the International Council on Jewish Social and Welfare Services.

Other organizations: Hon. Life President (since 1989), Israel Haemophilia Association.

Mailing address: Drayton House, Gordon Street, London WC1H OAN, UK. T. (44 71) 387 3925. Fax 383 4810.

Mariner, John Elwyn

Rel: Presbyterian; Unordained; b. 16 MAR 1941 in Schnectady NY, USA; m. to Sally Anne Aucock; c. 3 girls; Citizen of USA; Lang. English and Urdu.

International Organizations: Executive Secretary (since 1976) of the World Witness, Associate Reformed Presbyterian Church.

Board Member of the Evangelical Foreign Missions Association. Member, Board of Reference of the World Relief Corporation.

Other organizations: Board Member, Nancy Fulwood Hospital Society.

Career: Staff Engineer, Westvaco, New York City NY from 1962 to 1965. Project Engineer, Westvaco, Covington VA from 1966 to 1968. Administrator, Nancy Fulwood Hospital (World Witness), Pakistan from 1968 to 1976. Ordained Ruling Elder, Devenger Road Associate Reformed Presbyterian Church. **Educ.:** M.E. from Stevens Institute of Technology, Hoboken NJ, USA (1958-1962); M.B.A. from Stevens Institute of Technology, Hoboken NJ, USA (1962-1965); Grad. Crses. in Man. from University of Virginia, Charlottesville, USA (1966); P.G. Crses. in Eng. at North Carolina State University at Raleigh, USA (1967); Informal Residency in Hospital Administration from Charles Choat Memorial Hospital, Woburn MA, USA (1972).

Mailing address: One Cleveland Street, Greenville SC 29601, USA. T. (1 803) 233 5226

Residence: T. (1 803) 268 4041

Marini, Francesco, S.X., Fr

Rel: Roman Catholic; Ordained; celibate.

International Organizations: Superior General of the Society of St Francis Xavier for the Foreign Missions.

Mailing address: Viale Vaticano 40, I-00165 Roma, Italy. T. (39 6) 638 46 09 – 638 16 93. Fax 326 65 71. Tx 5042018 GISA.

Marín Salcedo, Jesús M, S.C., Br

Rel: Roman Catholic; Ordained; celibate.

International Organizations: Superior General of the Institute of the Brothers of the Sacred Heart.

Member of the Congregation for Religious and Secular Institutes.

Other organizations: Member, Commission for the Catechism and the Catechists.

Mailing address: Piazza del Sacro Cuore 3, I-00152 Roma, Italy. T. (39 6) 53 31 29.

Marín Salcedo, Jesús M, Fr

Rel: Roman Catholic.

International Organizations: Superior General of the Institute of the Brothers of the Sacred Heart.

Mailing address: Piazza del Sacro Cuore 3, I-00152 Roma, Italy. T. (39 6) 53 31 29.

Maris, Jacob Cornelia, Dr

Rel: Reformed Church; b. 06 FEB 1910 in Harlingen, Netherlands; m. to Adriana Paulina Baars; c. 2 boys and 4 girls; Citizen of Netherlands; Lang. Dutch, French, German and English.

International Organizations: Secretary-General (since 1954) of the International Council of Christian Churches.

Main field of work: Religion. **Educ.:** Dip. from Theological University, Apeldorn, Netherlands (1931-1937). **Pubs.:** Ed. "The Reformation Review" and Ed. "Getrouw".

Mailing address: PO Box 10, NL-3886 ZG Garderen, Netherlands. T. (31 5776) 1100. C. Intcouncil Garderen.

Residence: Oud Milligensweweg 26, NL-3886 ME Garderen, Netherlands T. See Mailing Address

Markham, Iris

Rel: Spiritualist.

International Organizations: Program Coordinator of The Pathwork.

Mailing address: PO Box 66, Phoenicia NY 12464, USA. T. (1 914) 688 22 11.

Marks, G T, Rev

Rel: Baptist.

International Organizations: General Superintendent of the Baptist Union of New Zealand.

Mailing address: PO Box 27-390, Wellington, New Zealand.

Marks, Thomas I, Dr

Rel: Protestant.

International Organizations: USA Director of the WEC International.

Mailing address: PO Box 1707, Fort Washington PA 19034, USA. T. (1 215) 646 2322.

Ma'rkus, Miha'ly, Dr

Rel: Reformed Church; Ordained on 15 SEP 1969; b. 26 AUG 1943 in Koma'rom, Hungary; m. to Dr Ma'rkus Miha'lyné Eördegh Ildikó; c. 4 boys and 2 girls; Citizen of Hungary; Lang. Hungarian, German and French.

International Organizations: Bishop (since 1991) of Reformed Church of the Transdanubian Region.

Other organizations: Bishop (since 1991), Tata/Veszprém; Secretary-General (since 1991), Hungarian Reformed World Federation; Vice-President (since 1991), Synod of the Hungarian Reformed Church.

Main field of work: Church leadership. **Career:** Pastor, Reformed Church of Szentgyörgyvölgy from 1966 to 1970. Pastor, Reformed Church of Pa'pa from 1970 to 1975. Pastor, Reformed Church of Tata in 1975. **Educ.:** Studied Theol. from Reformed Theological Academy, Budapest, Hungary (1961-1966); Dr. Theol. from Reformed Theological Academy, Budapest, Hungary (1980); Theol. from Westfälische Wilhelmsuniversität Münster, Germany (1982). **Interests:** Church history. **Mem.:** Hungarian Society of Literature-History **Pubs.:** *Jézus*, Budapest, 1978-1991; Bible stories for children, Budapest, 1987 and Komàromi csipkès györgy, Budapest, 1989 (thesis).

Mailing address: Dózsa Gy.u. 26, H-8200 Veszprém, Hungary. T. (36 80) 24 100

Residence: Kocsi u. 15, H-2890 Tata, Hungary. T. (36 34) 80 616

Marlier, Michel, Rev
Rel: Evangelical Lutheran.
International Organizations: President of the Evangelical Lutheran Church of France.
Mailing address: 29 rue de Damas, Vandoncourt, F-25230 Seloncourt, France.

Marr, Helen S
Rel: Jewish; b. 13 MAY 1934 in Lindsay ON, Canada; m. to Gerald Marr; c. 2 boys and 1 girl; Citizen of Canada; Lang. English.
International Organizations: President of the International Council of Jewish Women.
Other organizations: Vice-President (since 1987), Baycrest Centre for Geriatric Care.
Career: President, National Council of Jewish Women of Canada from 1979 to 1983. **Educ.:** B.A. from University of Toronto, Ontario, Canada (1951-1954). **Interests:** Music, art. **Hons.:** National Community Leadership Award, National Council of Jewish Women of Canada (1986).
Mailing address: 1110 Finch Avenue W, Suite 518, Downsview ON M3J 2T2, Canada. T. (1 416) 665 8251
Residence: 2 Penwood Crescent, Don Mills ON M3B 2B9, Canada

Marr, Robyn
Rel: Buddhist: Mahayana.
International Organizations: Member of the Foundation for the Preservation of the Mahayana Tradition.
Other organizations: Director, Buddha House.
Mailing address: PO Box 93, Eastwood SA 5063, Australia. T. (61 8) 799 153

Marr, Ronald James
Rel: Baptist; Ordained in 1964; b. 14 MAR 1933 in Caledonia ON, Canada; m. to Ruth Viola Peters; c. 2 girls; Citizen of Canada.
International Organizations: Chairman (since 1986) of the Worldwide Revival Prayer Fellowship.
Main field of work: Publishing. **Career:** Editor and Publisher, *Christian Enquirer* from 1970 to 1986. Founder and President, International Christian Communications Inc. from 1971 to 1987. Founder and President, International Christian Alert Network from 1983 to 1986. Publisher, *Prophecy Newsletter* from 1984 to 1986. Publisher, *Freedom Alert* from 1984 to 1986. Director, Greater Winnipeg Youth for Christ from 1955 to 1960. Director, Montreal Youth for Christ from 1960 to 1967. Interim Pastor, Chomedy Baptist Church, Laval, Quebec from 1965 to 1966. Youth Editor, *Evangelical Christian Magazine* from 1965 to 1968. Director of Development, Richmond College from 1968 to 1969. Vice-President, Christian Writers of Canada from 1969 to 1970. President, Christian Writers of Canada from 1971 to 1972. Vice-President, Overseas Students' Mission from 1969 to 1971. Pastor, Calvin Baptist Church, Toronto from 1969 to 1973. **Educ.:** Grad. Dip. from Prairie Bible Institute, Three Hills, Alberta, Canada (1955). **Pubs.:** The Education Time Bomb, 1979.
Mailing address: Box 508, Niagara Falls, Ontario L2E 6V2, Canada.

Marsaud, Jean-Paul, F.C., Fr
Rel: Roman Catholic; Ordained; celibate.
International Organizations: Procurator-General of the Fils de la Charité.
Mailing address: 10 rue Louis-Blanc, 75010 Paris, France. T. (33) 442 01 95 27

Marshall, J Murray
Rel: Protestant.
International Organizations: Chairman of the Latin America Mission.
Mailing address: PO Box 52-7900 Miami FL 33152, USA. T. (1 305) 884 8400.

Martelli, Claudio, Pastor
Rel: Methodist.
International Organizations: President of the Methodist National Board.
Mailing address: Via Firenze 38, I-00184 Roma, Italy.

Martello, Léo Luis, Dr
Rel: Unknown.
International Organizations: Representative of the Witches International Craft Association.
Mailing address: 1 B, 153 W 80th Street, New York NY 10024, USA.

Martin, Douglas
Rel: Bahá'í.
International Organizations: Director-General of the Bahá'í International Community.
Mailing address: Bahá'í World Centre, PO Box 155, 31001 Haifa, Israel. T. (972 4) 372 433 – 510 344. Fax 372 425. C. BAHAIFAITH, HAIFA. Tx 46626 BAYT IL.

Martin, George William, Mgr
Rel: Roman Catholic; Ordained; celibate.
International Organizations: Vicar-General of the Saint John, New Brunswick.
Mailing address: One Bayard Dr, Saint John NB E2L 5G3, Canada.

Martin, Jacques, H E Card
Rel: Roman Catholic; Ordained on 14 OCT 1934; b. 26 AUG 1908 in Amiens, France; celibate; Citizen of France.
International Organizations: Member (since 1988) of the College of Cardinals.
Career: Elected Bishop, Nabulus, Palestine and consecrated in 1964. Archbishop in 1986. Created Cardinal by Pope John Paul II in 1988. Prefect, Casa Pontificia from 1969 to 1986. **Hons.:** Commander, Legion of Honour. **Pubs.:** Dom Innocent le Masson, le Louis XIV des Chartreux, 1976; Heraldry in the Vatican, 1987 and Le Vatican inconnu, 1988.
Mailing address: 00120 Città del Vaticano, Vatican. T. (39 6) 698 37 18

Martín, Mercedes, STJ, Sr
Rel: Roman Catholic; Ordained; celibate.
International Organizations: Secretary-General of the Society of St Teresa of Jesus.
Mailing address: Via Valcannuta 134, I-00166 Roma, Italy. T. (39 6) 622 15 42.

Martindale, L Craig, Rev
Rel: Christian.
International Organizations: President of the Way International.
Mailing address: PO Box 328, 5555 Wierwille Rd, New Knoxville OH 45871, USA. T. (1 419) 753 2523. Tx 241275.

Martinelli, Antinio, Rev
Rel: Roman Catholic; Ordained; celibate.
International Organizations: Councillor-General of the Centro Internazionale Salesiano per le Communicazioni Sociali.
Mailing address: Via della Pisana 1111, CP 9092, I-00163 Roma, Italy. T. 693 13 41.

Martinez, C Angel
Rel: Roman Catholic.
International Organizations: Member of the Order of the Augustinian Recollects.
Mailing address: Viale dell'Astronomia 27, I-00144 Roma, Italy. T. (39 6) 592 65 34.

Martinez, Emilio Lage, O.S.S.R., Fr
Rel: Roman Catholic; Ordained; celibate.
International Organizations: Contact of the Order of the Most Holy Redeemer.
Mailing address: c/o Redemptoristine Nuns, CP 2458, Via Merulana 31, I-00100 Roma, Italy. T. (39 6) 7315 841. Fax 4466 012.

Martínez Somalo, Eduardo M, H E Card

Rel: Roman Catholic; Ordained on 19 MAR 1950; b. 31 MAR 1927 in Baños de Río Tobía, Spain; celibate; Citizen of Spain.
International Organizations: Prefect (since 1988) of the Sacred Congregation for the Sacraments and Divine Worship.
Member of the Pontifical Commission for Latin America. Member of the Congregation for the Causes of Saints. Member of the Congregation for Bishops. Member of the Congregation for the Evangelization of Peoples. Member of the Congregation for the Clergy. Member of the Congregation for Religious and Secular Institutes. Member of the International Association for Rural Development.
Other organizations: Member (since 1988), College of Cardinals; Member, Council of Cardinals and Bishops; Member, Institute for Religious Works.
Career: Elected Bishop, Tagora and consecrated in 1975, then Archbishop. Created Cardinal by Pope John Paul II in 1988.
Mailing address: 00120 Città del Vaticano, Vatican. T. (39 6) 698 31 55

Martinho, Ernesto Pedro

Rel: Roman Catholic.
International Organizations: Lusophone Africa Coordinator of the African Caritas.
Mailing address: Caritas Moçambicana, Caixa Postal 286, Maputo, Mozambique. T. (258) 74 40 09. Tx 6101 CARIT MO.

Martini, Carlo Maria, S.I., H E Card

Rel: Roman Catholic; Ordained on 13 JUL 1952; b. 15 FEB 1927 in Turin, Italy; celibate; Citizen of Italy.
International Organizations: President of the Council of European Bishops' Conferences.
Member of the Society of Jesus. Member of the Sacred Congregation for the Sacraments and Divine Worship. Member of the Congregation for Bishops. Member of the Congregation for Religious and Secular Institutes. Member of the Congregation for Catholic Education.
Other organizations: Archbishop (since 1979), Milan; Member (since 1983), College of Cardinals; Member, Council of Cardinals and Bishops; Member, Congregation for the Doctrine of the Faith.
Career: Created Cardinal by Pope John Paul II in 1983. President, Council of European Bishops' Conferences (CCEE) in 1987. **Educ.:** Theol. Deg. from Pontifical Gregorian University, Vatican City State (1958).
Mailing address: Palazzo Arcivescovile, Piazza Fontana 2, I-20122 Milan, Italy. T. (39 2) 855 62 66. Fax 855 62 20
Residence: See Mailing Address. T. See Mailing Address

Martins, Hermengarda Alves, Sr

Rel: Roman Catholic; Ordained; celibate.
International Organizations: Secretary-General of the Latin American Confederation of Religious Orders.
Mailing address: Calle 56 No 10-45 Piso 5, Apartado Aéreo 90710, Bogotá 2, Colombia. T. (57 1) 211 66 56 – (57 1) 211 68 58. C. SECLAR.

Martins, José Saraiva, Mgr

Rel: Roman Catholic.
International Organizations: Secretary of the Congregation for Catholic Education.
Mailing address: Palazzo delle Congregazioni, Piazza Pio XII 3, I-00193 Roma, Italy.

Martos, Javier

Rel: Catholic.
International Organizations: UNESCO Representative of the Pax Romana, International Movement of Catholic Students.
Mailing address: 171 rue de Rennes, F-75006 Paris, France. T. (33 1) 45 44 70 75.

Marty, François, H E Card

Rel: Roman Catholic; Ordained on 28 JUN 1930; b. 18 MAY 1904 in Pachins, France; celibate; Citizen of France.
International Organizations: Member (since 1969) of the College of Cardinals.
Other organizations: Archbishop Emeritus (since 1981), Paris.
Career: Priest, Villefranche and Rodez from 1932 to 1940. Curate, Bournazel, Rieupeyroux and Millau from 1940 to 1951. Vicar-General, Rodez from 1951 to 1952. Bishop, Saint-Flour in 1952. Archbishop, Reims from 1960 to 1968. Archbishop, Paris from 1968 to 1981. Created Cardinal by Pope Paul VI in 1969. President, French Bishops' Conference from 1969 to 1975. **Hons.:** Knight, Legion of Honour. **Pubs.:** Prophètes de la joie, 1978; L'évêque dans la ville, 1979 and Cardinal Marty, Chronique vécue de l'église de France, 1980.
Mailing address: Couvent des Dominicaines, Monteils, F-12200 Villefranche-de-Rouergue, France. T. (33) 65 29 62 70

Marty, Martin, Dr

Rel: Lutheran; Ordained on 13 JUL 1952; b. 05 FEB 1928 in West Point NB, USA; m. to (1st) Elsa Schumacher (died), (2nd) Harriet Julia Lindemann; c. 5 boys and 2 girls; Citizen of USA; Lang. English, German (reading), French (reading), Classical Greek and Latin.
International Organizations: Editor, now Senior Editor (since 1956) of the *Christian Century.*
Other organizations: Fairfax M Cone Distinguished Service Professor (since 1963), University of Chicago, IL; Senior Scholar in Residence (since 1986), The Park Ridge Center for the Study of Health, Faith and Ethics.
Main field of work: Historian, teacher. **Career:** Lutheran Minister from 1952 to 1963. President, American Society of Church History in 1971. President, American Catholic History Association in 1981. President, American Academy of Religion in 1988. **Educ.:** S.T.M. from Lutheran School of Theology, Chicago IL, USA (1954); Ph.D. from University of Chicago, IL, USA (1956). **Interests:** baroque music, calligraphy. **Hons.:** National Book Award for *Righteous Empire* (1972); 44 honorary degrees. **Pubs.:** 40 books including: *Modern American Religion – The Irony of It All*, Vol. I, *Modern American Religion – The Noise of Conflict*, Vol. II; *Righteous Empire: The Protestant Experience in America* and numerous articles on religious history, theology and cultural criticism.
Mailing address: University of Chicago Divinity School, 1025 E 58th St, Chicago IL 60637, USA.
Residence: 239 Scottswood Rd, Riverside IL 60546, USA.

Marusyn, Miroslav Stefan, H E Mgr

Rel: Roman Catholic; Ordained on 02 MAY 1948; b. 22 JAN 1924 in Kniaze, Ukraine; celibate.
International Organizations: Secretary of the Sacred Congregation for the Oriental Churches.
Advisor of the Pontifical Council for the Laity. Advisor of the Pontifical Council for Promoting Christian Unity. Member of the Pontifical Council for the Pastoral Care of Migrants and Itinerant People. Member of the Pontifical Council for Pastoral Assistance to Health Care Workers.
Other organizations: Titular Bishop, Cadi; Member, Pontifical Commission for Russia; Member, Commissioni Interdicasteriali Permanenti.
Mailing address: Palazzo del Bramante, Via della Conciliazione 34, I-00193 Roma, Italy. T. (39 6) 698 4282
Residence: Via dei Corridori 64, I-00193 Roma, Italy T. (39 6) 698 5253

Maruyama, Tadataka

Rel: Christian; b. 27 NOV 1939; m. to Meredith; Citizen of Japan.
International Organizations: President of the Tokyo Christian University.
Career: President, Tokyo Junior College. **Educ.:** Tokyo Gakugei University, Japan; Princeton University, NJ, USA.
Mailing address: 6-17-12 Minami-Nagareyama, Nagareyama, Chiba 270-01, Japan. T. (81 471) 50 0947

Maruyama, Teruo

Rel: Buddhist: Nichiren; Ordained on 08 APR 1950; b. 27 APR 1932 in Minobu, Yamanashi-Ken, Japan; m. to Sachiko Yamada; c. 1 girl; Citizen of Japan; Lang. Japanese.
International Organizations: President (since 1972) of the Society for the Study of Nichiren.

Main field of work: Journalism. **Educ.:** Rissho University, Tokyo, Japan (left before graduation) (1951-1954). **Mem.:** International Network of Engaged Buddhists (INEB) **Pubs.:** Social Activity of Buddhism; Compiler, The Religion for Men of Today, Vol. 10 and The Religious Situation of Present Japan.
Mailing address: INEB, Komyo-Ji, Nai Motoyashiki, Minowa-Chyo, Anjiyo-Shi, Aichi 446, Japan. T. (81 566) 76 2486
Residence: Marutaka Bldg 3F, 4-23-16 Seta, Setagaya-ku, Tokyo 158, Japan. T. (81 3) 3708 0379. SAME FOR FAX

Mary of the Rosary, M, Sr

Rel: Roman Catholic; Ordained; celibate.
International Organizations: Member of the Association of Contemplative Sisters.
Mailing address: Blessed Sacrament Monastery, 23 Park Ave, Yonkers NY 10703, USA.

Mary-Rousselière, Guy, O.M.I., Rev

Rel: Roman Catholic; Ordained; b. 06 JUL 1913 in Le Mans, Sarthe, France; celibate.
International Organizations: Missionary among Inuit (since 1944) of the Roman Catholic Church.
Other organizations: Editor (since 1953), *Eskimo* magazine; Archeological work in the Arctic (since 1958).
Career: Came to Canada in 1938. Missionary among Chippewyan Indians (Déné), N Manitoba (until 1944). **Educ.:** B. ès L. from University of Paris, France (1931); M.A. from University of Montreal, Quebec, Canada (1965). **Hons.:** National Science Award (1988); 1st Prize Missionary Film, Brussels (1964) for 'Light in the Darkness' (1957). **Pubs.:** *Les jeux de ficelle des Arviligjuarmiut*, 1969 and *Qitlarssuaq: l'histoire d'une migration polaire*, 1980.
Mailing address: Catholic Mission, Pond Inlet NWT X0A 0S0, Canada.

Masal, E, Rev

Rel: Evangelical.
International Organizations: President of the Kalimatan Evangelical Church.
Mailing address: Jalan Jenderal Sudirman 8, PO Box 86, Banjarmasin, Kalimantan Selatan, Indonesia.

Mascall, Eric Lionel, Rev

Rel: Church of England; Ordained in 1932; b. 12 DEC 1905 in Sydenham, UK; Citizen of United Kingdom.
International Organizations: Canon Emeritus (since 1984) of the Truro Cathedral.
Other organizations: Professor Emeritus, King's College, London.
Career: Parochial work (until 1937). Sub-Warden, Lincoln Theological College from 1937 to 1945. Student and Tutor, Christ Church, Oxford from 1945 to 1962. Professor of Historical Theology, King's College, London from 1962 to 1973. Honorary Canon, Truro Cathedral from 1974 to 1984. Bampton Lecturer, Oxford in 1956. Bampton Lecturer, Columbia University, NY in 1958. Gifford Lecturer, Edinburgh University from 1970 to 1971. **Interests:** reading. **Hons.:** Hon. D.D., St Andrews (1967). **Mem.:** Oratory of the Good Sheperd; Fellow, King's College, London (since 1968); Fellow, British Academy **Pubs.:** Jesus, Who He is and How We Know Him, 1985; Compliments of the Season, 1985 and The Triune God, 1986.
Mailing address: Saint Mary's House, Kings Mead, East Blatchington, Seaford BN25 2ET, UK.

Masferrer Kan, Elio M

Rel: Unknown; b. 07 DEC 1946; Citizen of Mexico.
International Organizations: Member of the Council of the International Society for the Sociology of Religions.
Other organizations: Professor Investigator, Escuela Nacional de Antropologia e Historia; Coordinator, Projecto Multidisciplinaria Naturaleza Sociedad y Cultura en la Sierra Norte de Puebla.
Educ.: Maestria Antropologia e historia.
Mailing address: Escuela Nacional de Antropologia e Historia, Periferico Sur y Zapote, 14030 Tlalpan, Mexico. T. (52 5) 655 24 79 – 655 25 04

Masiero, Giuseppe, F.D.P., Rev

Rel: Roman Catholic; Ordained; celibate.
International Organizations: Superior General of the Little Work of Divine Providence.
Mailing address: Via Etruria 6, I-00183 Roma, Italy. T. (39 6) 757 55 37, (39 6) 77 85 88

Masilo, Benjamin

Rel: Reformed Church.
International Organizations: Vice-President of the World Alliance of Reformed Churches.
Works with All-Africa Conference of Churches.
Mailing address: Lesotho Evangelical Church, PO Box 260, Maseru 100, Lesotho. T. (266) 323 942

Mason, Peter Ralph, Rev

Rel: Anglican; Ordained in 1967; b. 30 APR 1943 in Brome, Quebec, Canada; m. to Carmen Ruddock; c. 1 boy and 2 girls; Citizen of Canada.
International Organizations: Principal (since 1985) of the Wycliffe College.
Career: Curate, St Matthews Church, Montreal from 1967 to 1969. Incumbent, St Luke's Church, Hemmingford, Quebec from 1968 to 1971. Rector, St Clement's Church, Verdun, Quebec from 1971 to 1975. Rector, St Peter's Church, Town of Mount Royal, Quebec from 1975 to 1980. Rector, St Paul's Church, Halifax from 1980 to 1985. Lecturer in Homiletics and Liturgics, Atlantic School of Theology from 1983 to 1985. Provincial Chaplain, St John's Ambulance, Nova Scotia. **Educ.:** B.A. from McGill University, Montreal, Canada (1964); B.D. from McGill University, Montreal, Canada (1967); M.A. from McGill University, Montreal, Canada (1971); D.Min. from Princeton Theological Seminary, NJ, USA (1983). **Interests:** golf, skiing. **Hons.:** Hon. D.D., Montreal Diocese Theological College (1988).
Mailing address: 5 Hoskin Ave, Toronto, Ontario M5S 1H7, Canada.
Residence: 20 Queen's Park Cres., W. Toronto, Ontario M5S 2W2, Canada.

Mason, Richard

Rel: Christian.
International Organizations: President of the Mission Doctors Association.
Mailing address: 1531 W Ninth St, Los Angeles CA 90015, USA. T. (1 213) 251 3222.

Masri, Basheer Ahmad, Al-Hafiz

Rel: Muslim; Unordained; b. 12 MAY 1914 in Dist. Gurdaspur, Punjab, Pakistan; m. to Salima Begum; c. 2 boys and 1 girl; Citizen of United Kingdom; Lang. English, Urdu, Hindustani, Punjabi, Arabic and Kiswahili.
International Organizations: Founder/President (since 1990) of the International Muslim Association for Animals and Nature.
Advisory Director (since 1991) of the World Society for the Protection of Animals.
Main field of work: Ministry. **Career:** Emigrated to East Africa in 1941. Held various responsible positions in voluntary social organizations in East Africa from 1941 to 1961. Secretary, Tanzania Society for the Prevention of Cruelty to Animals (SPCA) from 1944 to 1950. Treasurer, then Secretary, Uganda SPCA from 1951 to 1961. Emigrated to UK in 1961. Joint Editor, *The Islamic Review*, London from 1963 to 1968. Imam, the Shah Jehan Mosque, Woking from 1964 to 1968. Toured European, Middle Eastern, North African and Asian countries from 1968 to 1971. **Educ.:** Degree in Hifz-Al-Qur'an from Madrasah Islamia, Dist. Gurdaspur, Punjab, Pakistan (1919-1927); B.A. Arabic (Hons) from University of the Punjab, Lahore, Pakistan (1928-1938); Dip. Jour. from Polytechnic of Central London, UK (1962-1963); Refreshers Crse. from Al-Azhar University, Cairo, Egypt (1969-1970). **Interests:** welfare of animals and environment from the religious point of view. **Pubs.:** *Islamic Concern for Animals* (text in English and Arabic), 1987; *Animals in Islam* (English) and Creatures of God (video – available in English and Arabic).
Mailing address: 7 Hurst Road, East Molesey, Surrey KT8 9AQ, UK. T. (44 81) 979 2368.
Residence: See Mailing Address. T. See Mailing Address

Mataio, Tere
Rel: Christian.
International Organizations: General Secretary of the Cook Islands Christian Church.
Mailing address: PO Box 93, Takamoa, Rarotonga, Cook Is.

Matanic, Atanasio, O.F.M., Rev
Rel: Roman Catholic; Ordained; celibate.
International Organizations: Member of the Order of Friars Minor Capuchin.
Advisor of the Congregation for the Causes of Saints.
Mailing address: Via Merulana 124, I-00185 Roma, Italy.

Mate, Martin, Rt Rev
Rel: Anglican; Ordained in 1953; b. 12 NOV 1929 in Port Rexton, Newfoundland, Canada; m. to Florence Mabel Hooper; c. 2 boys and 3 girls; Citizen of Canada.
International Organizations: Bishop (since 1980) of East Newfoundland and Labrador.
Career: Ordained Deacon in 1952. Curate, Cathedral of St John the Baptist, St John's from 1952 to 1953. Rector, Pushthrough Parish, Newfoundland from 1953 to 1958. Incumbent, St Anthony Mission from 1958 to 1964. Rural Dean, St Barbe from 1958 to 1964. Rector, Cookshire, Quebec from 1964 to 1967. Rector, Catalina, Newfoundland from 1967 to 1972. Rural Dean, Bonavista Bay from 1970 to 1972. Rector, Pouch Cove/Torbay Parish from 1972 to 1976. Treasurer, Diocesan Synod of East Newfoundland and Labrador from 1976 to 1980. **Educ.:** L.Th. from Queen's College, St John's, Canada (1953); B.A. from Bishop's University, Lennoxville, Quebec, Canada (1966); M.A. from Bishop's University, Lennoxville, Quebec, Canada (1967). **Interests:** carpentry, hunting, fishing, camping. **Pubs.:** Pentateuchal Criticism, 1967.
Mailing address: 19 King's Bridge Rd, St John's, Newfoundland A1C 3K4, Canada.
Residence: 42 Dundas St, St John's, Newfoundland A1B 1X1, Canada.

Mathebula, George
Rel: Roman Catholic.
International Organizations: Pan-African Secretariat of the International Young Christian Workers.
Mailing address: Panafrican YCW, PO Box 46528, Nairobi, Kenya.

Mathies, Ronald John Richard, Dr
Rel: Mennonite; b. 24 JUN 1940 in New Hamburg ON, Canada; m. to Gudrun Linda Wohlgemut; c. 1 boy and 2 girls; Citizen of Canada; Lang. English and German.
International Organizations: Chairman (since 1989) of the Mennonite Central Committee.
Other organizations: Director, Peace and Conflict Studies (since 1986), Conrad Grebel College.
Main field of work: University education. **Career:** Country Representative, Mennonite Central Committee, Malawi from 1970 to 1973. Country Representative, Mennonite Central Committee, Swaziland from 1978 to 1981. Teacher in Waterloo County secondary schools (1962-1964), (1967-1970), (1973-1978), (1981-1986). **Educ.:** B.A. from University of Waterloo, Ontario, Canada (1959-1962); M.Sc. from University of Guelph, Ontario, Canada (1974-1978); Ed.D. from University of Toronto, Ontario, Canada (1981-1987). **Interests:** development education, non-governmental development organization. **Mem.:** Comparative and International Education Society **Pubs.:** "Global Education Through a Church Related College", in *Educating for the Kingdom? Church Related Colleges in English-Speaking Canada*, Waterloo, St Jeromes and Conrad Grebel College, 1990 and "The Horn of Africa Project: Modelling Alternate Conflict Resolution", in *Aid As Peacemaker*, Ottawa, Parliament Art Centre, 1992.
Mailing address: Westmount Road, Waterloo ON N2L 3G6, Canada. T. (1 519) 885 0220
Residence: 19 Mourning Dove Crescent, Elmira ON N3B 1E2, Canada. T. (1 519) 669 5222

Mathur, J S
Rel: Hindu.
International Organizations: Director of the India Institute of Gandhian Thought and Peace Studies.
Mailing address: Gandhi Bhawan, Univ. of Allahabad, Allahabad 211002, India. T. (91 532) 54900 – (91 532) 51412 (home).

Matsbara, Taido
Rel: Buddhist; Ordained; b. 23 NOV 1907; Citizen of Japan.
International Organizations: Governor of the Namuno-kai.
Career: Chief, Ryugenji. **Educ.:** Waseda University, Tokyo, Japan.
Mailing address: 5-9-23 Mita, Minato-ku, Tokyo 108, Japan. T. (81 3) 3451 1853

Matsudaira, Nagayoshi
Rel: Shinto; Ordained; b. 08 MAR 1914; Citizen of Japan.
International Organizations: Chief Priest of the Yasukuni Shrine.
Mailing address: 5-7-28-101 Nakameguro, Meguro-ku, Tokyo 153, Japan. T. (81 3) 3712 5630

Matsumoto, Daien
Rel: Buddhist; Ordained; b. 05 AUG 1922; Citizen of Japan.
International Organizations: Honorary Manager of the Kiyomizudera.
Career: Chairman, Japanese-Chinese Friendship Buddhist Federation. **Educ.:** Ryukoku University, Kyoto, Japan.
Mailing address: 1-294 Kiyomizu, Higashiyama-ku, Kyoto 605, Japan. T. (81 75) 551 1234

Matsumoto, Jitsudo
Rel: Buddhist; Ordained; b. 22 MAY 1904; Citizen of Japan.
International Organizations: Superior of the Saidaiji.
Main field of work: Monk. **Career:** Manager, Shingonritsu-Sect. Chief, Ikomosan Hozanji.
Mailing address: 1-1-5 Saidaijshibamachi, Nara 631, Japan. T. (81 742) 45 4700

Matsunaga, Shinichi
Rel: Unknown; b. 19 DEC 1927; m. to Kuniko; Citizen of Japan.
International Organizations: President of the Seiwa University.
Career: Lecturer, Assistant Professor, then Professor, Seiwa University. President, Japan Society of New Testament Studies. Trustee, Nishinomiya UNESCO. **Educ.:** Dr. Theol. from Kwansei Gakuin University Graduate School, Hyogo, Japan.
Mailing address: 10-50 Nondomishimachi, Nishinomiya, Hyogo 662, Japan. T. (81 798) 52 1427

Matsunaga, Yukei
Rel: Buddhist; b. 21 JUL 1929; married; Citizen of Japan.
International Organizations: Professor of the Koyasan University.
Career: President, Institute of Esoteric Buddhism and Buddhist Culture. **Educ.:** Tohoku University, Sendai, Japan.
Mailing address: 31-3 Takanocho Koyasan, Ito-gun, Wakayama 648, Japan. T. (81 736) 56 3138

Matsuura, Kanden
Rel: Buddhist; Ordained; b. 24 AUG 1931; Citizen of Japan.
International Organizations: Chief of the Komyoji.
Main field of work: Monk. **Career:** Manager, Kenjo-Sect.
Mailing address: 749-3 Nakanoshocho, Innoshima, Hiroshima 722-22, Japan. T. (81 8452) 4 0078

Matthews, Gordon Elton
Rel: Quaker; Unordained; b. 21 MAY 1955 in Portsmouth, Hampshire, UK; single; Citizen of United Kingdom; Lang. English, German and French.
International Organizations: Executive Secretary (since 1990) of the Church and Peace.
Main field of work: Church and peace. **Career:** Assistant to House Mother in Lutheran children's home, Waiern, Kärnten, Austria from 1973 to 1974. Experimental Assistant, ICI Pollution Control Systems Ltd. from 1976 to 1977. Lecturer at Stevenage College from 1979 to 1980. Team Member, Quaker Peace Action Caravan in 1981. Working Bursar at Woodbrooke College, Selly Oak, Birmingham in 1981. Development and Youth Secretary, British Fellowship of Reconciliation from 1981 to 1986. Senior Volunteer with the Quaker Council for European Affairs at Quaker House in Brussels in 1987. Resident Friend at Charney Manor (Quaker Conference Centre) from 1987 to 1990. **Educ.:** Intermediate Dip. German from Institute of Linguists, London, UK (1976); B. Technology (Hons) from University of Bradford, School of Environmental Science, UK (1974-1978); P.G. Cert. in Ed. at University of Birmingham, UK (1978-1979). **Interests:** travel,

writing, walking, cycling, singing, circle dancing, photography. **Mem.:** Konfliktlösungsnetzwerk; 'Projekt Mediation'; Justice, Peace and the Integrity of Creation Working Group of London Yearly Meeting of the Religious Society of Friends; Quaker Peace and Service Peace Committee; Council on Christian Approaches to Defence and Disarmament; Fellowship of Reconciliation; Christian Campaign for Nuclear Disarmament; Oxford Against the Poll Tax; Youth Hostels Association **Pubs.:** "Mixing Prayer and Politics", in *The Friend*, June 9, 1989; "The Fruits of the Spirit: A reflection on Galatians 5:22", in *Quaker Monthly*, January 1989, Vol. 68-1 and "This beautiful planet", p. 122, in *Prayers for Peacemakers: A handbook of worship resources*, Valerie Flessati, Ed., compiled by John Ansell, Jenny Hunt, Gerry McFlynn and Gordon Matthews, 1989, Kevin Mayhew.
Mailing address: Kalsmuntstr 21, 35641 Schoeffengrund, Germany. T. (49 6445) 5588.
Residence: Wiesenstrasse 21, 35641 Schoeffengrund, Germany T. (49 6445) 5998

Matthias, Lesley
Rel: Christian; Unordained; b. 29 JUN 1949 in Birmingham, UK; single; Citizen of United Kingdom; Lang. English.
International Organizations: Director (since 1990) of the World Congress of Faiths.
Other organizations: Lecturer in religious education (since 1978), Cambridgeshire County Council.
Main field of work: Initiating, facilitating, coordinating inter-faith events. **Educ.:** B.A. (Hons) from Portsmouth Polytechnic, UK (1968-1972); P.G. Cert. Ed. at Portsmouth College of Education, UK (1973-1976); M.Theol. from University of Nottingham, UK (1988-1990).
Mailing address: 28 Powis Gardens, London W11 1JG, UK. T. (44 71) 727 2607.
Residence: Cobden Street, Peterborough PE1 2HF, UK

Matthieu, Robert, Fr
Rel: Roman Catholic; Ordained; celibate.
International Organizations: Director of the Euro-Children.
Mailing address: Mutsaertstraat 32, B-2000 Antwerpen, Belgium. T. (32 3) 231 78 35. Fax (32 3) 232 31 67.

Maurel, Gaston
Rel: Roman Catholic.
International Organizations: Head of the International Catholic Centre in Geneva.
Mailing address: 3 rue de Varembé, CP 43, CH-1211 Genève 20, Switzerland. T. (41 22) 734 14 65 – 734 08 75.

Maury, M, H E Mgr
Rel: Roman Catholic.
International Organizations: Representative of the Coetus Internationalis Ministrantium.
Mailing address: Archevêque de Reims, Cathédrale, F-51000 Reims, France.

Maxwell, Earle, Commr
Rel: Salvation Army, The.
International Organizations: Territorial Commander of The Salvation Army.
Mailing address: PO Box 6015, Wellington 1, New Zealand.

Maxwell, May Sutherland
See Rúhíyyih Rabbani.

May, Bernard André
Rel: Protestant; b. 23 JUL 1932 in Marcus Hook PA, USA; m. to Nancy Joyce Bostwick; c. 3 boys; Citizen of USA; Lang. English and Spanish.
International Organizations: Director (since 1980) of the Wycliffe Bible Translators.
Other organizations: Board of Directors (since 1972), Wycliffe Bible Translators International; Board of Directors (since 1980), Jungle Aviation and Radio Service Inc.; Board of Directors, Summer Institute of Linguistics.
Main field of work: Administration. **Career:** Pilot for Wycliffe, Peru and Ecuador. Director, JAARCS. **Educ.:** B.A. from King's College, New York, USA (1950-1954); Northrop University, Inglewood CA, USA (1954-1956). **Hons.:** Alumnus of the Year, King's College. **Pubs.:** Climbing on Course; Under His Wing and Learning to Trust.
Mailing address: PO Box 2727, Huntington Beach CA 92647, USA. T. (1 714) 969 4600
Residence: 9782 Scanlan Court, Fountain Valley CA 92708, USA. T. (1 714) 962 4114

Mayeda, Egaku, Prof Dr
Rel: Buddhist: Pure Land; Ordained in 1932; b. 29 NOV 1926 in Nagoya, Aichi-ken, Japan; m. to Ryu; Citizen of Japan; Lang. Japanese.
International Organizations: President (since 1986) of the Society for the Study of Pali and Buddhist Culture.
Other organizations: High Priest (since 1963), Sokunenji Temple, Pure Land Buddhism; Professor (since 1972), Aichi Gakuin University; Head (since 1991), Graduate School of Letters, Aichi Gakuin University.
Main field of work: Buddhist studies. **Career:** Research Assistant, University of Tokyo from 1956 to 1961. Professor, Women's College of Tokaigakuen from 1964 to 1969. Professor, Meijo University from 1969 to 1972. **Educ.:** Bungakushi from University of Tokyo, Japan (1947-1951). **Hons.:** The Royal Prize of the Japan Academy. **Mem.:** Science Council of Japan (1991) **Pubs.:** A History of Formation of Original Buddhist Texts, Sankibo, Tokyo, 1964; Sakyamuni Buddha, Sankibo, Tokyo, 1972 and Contemporary Buddhism in Sri Lanka, Sankibo, Tokyo, 1986.
Mailing address: N° 904 Mayeda-nishi-machi, 1-chome, Nakagawa-ku, Nagoya 454, Japan. T. (81 52) 301 8593
Residence: See Mailing Address. T. See Mailing Address

Mayer, Paul Augustin, O.S.B., H E Card
Rel: Roman Catholic; Ordained on 25 AUG 1935; b. 23 MAY 1911 in Altötting, Germany; celibate; Citizen of Germany.
International Organizations: President (since 1988) of the Pontifical Commission 'Ecclesia Dei'.
Member of the Benedictine Confederation. Member of the Congregation for Bishops. Member of the Congregation for Religious and Secular Institutes. Member of the Pontifical Council for the Interpretation of Legislative Texts.
Other organizations: Member (since 1985), College of Cardinals; Member, Council of Cardinals and Bishops; Member, Congregation for the Doctrine of the Faith.
Career: Consecrated Bishop (Titular See of Satrianum) in 1972, then Archbishop. Created Cardinal by Pope John Paul II in 1985.
Mailing address: Via Rusticucci 13, I-00193 Roma, Italy. T. (39 6) 698 41 31

Maynard, G Oliver, Rt Rev
Rel: Moravian.
International Organizations: Chairman of the Moravian Church, Eastern West Indies Province.
Mailing address: Cashew Hill, PO Box 504, Antigua-Barbuda.

Mayoraz, Emile, M.S.F.S., Fr
Rel: Roman Catholic; Ordained; celibate.
International Organizations: Superior General of the Missionnaires de Saint François de Sales d'Annecy.
Mailing address: Chemin de Proupeine, F-74000 Annecy, France. T. (33) 50 45 53 17

Mazaburu Atibasay, Moise M, Rev
Rel: Evangelical; Ordained on 10 AUG 1986; b. 13 MAR 1952 in Akataka/Wamba, Zaire; m. to Obee Beyayumoni Neema; c. 5 boys and 1 girl; Citizen of Zaire; Lang. Swahili, French, English and Sango.
International Organizations: President and Legal Representative (since 1987) of the Communauté évangélique du Christ au coeur de l'Afrique.
Other organizations: Visiting Professor (since 1989), Institut Supérieur Théologique, Bunia.
Main field of work: Ministry and higher education. **Career:** Professor, Institut d'Ibambi from 1980 to 1982. Professor, Institut Supérieur Théologique, Bunia from 1985 to 1987. Administrative Secretary, Institut Supérieur Théologique, Bunia from 1987 to 1988. **Educ.:** Grad. Theol. from Institut Supérieur Théologique, Bunia, Zaire (1978-1980);

M.Theol. from Faculty of Theology, Bangui, Central African Republic (1982-1985). **Interests:** theology and exegesis of the Ancient Testament. **Mem.:** Association of Evangelicals of Africa and Madagascar **Pubs.:** *Le Chrétien et l'Observation du Sabbat*; *Préparatifs au mariage Chrétien* and *L'Orthodoxie Evangélique*, Editions Evangéliques de Nyakunde, Bunia (forthcoming).
Mailing address: BP 500, Isiro, Zaire.
Residence: Mission CECCA, 16 Gamba, Isiro, Zaire.

Mazzuccato, L
Rel: Christian.
International Organizations: Director of the Collegio Universitario Aspiranti Medici Missionari.
Mailing address: Via San Francesco 126, I-35100 Padova, Italy. T. (39 49) 311 06.

Mbang, Sunday C, His Pre-eminence
Rel: Methodist.
International Organizations: Patriarch of Methodist Church of Nigeria.
Mailing address: Wesley House, 21/22 Marina, PO Box 2011, Lagos, Nigeria.

Mbeba, Joram Magayi, Rt Rev
Rel: Mennonite; Ordained on 09 FEB 1986; b. 15 DEC 1930 in Majita Musoma, Tanzania; m. to Ludia Nyangeta; c. 3 boys and 2 girls; Citizen of Tanzania; Lang. Kiswahili, Kijita and English.
International Organizations: Member, Executive Committee (since 1990) of the Mennonite World Conference.
Other organizations: Secretary (since 1987), Council of Eastern Africa Mennonite Churches (CEAMC); Chairman (since 1988), Leadership Training Commission, CEAMC; Chairman (since 1990), Africa Mennonite Brethren in Christ Fellowship; Bishop (since 1990), Mennonite Church of Tanzania, Eastern Diocese.
Main field of work: University education, biblical theology. **Career:** Headmaster, Ministry of Education from 1970 to 1972. Regional Officer, Ministry of Education from 1972 to 1974. District Dev. Director, Prime Minister's Office from 1974 to 1983. District Executive Director, Ministry of Local Government from 1984 to 1985. Church Secretary, Mennonite Church, Southern Diocese from 1986 to 1989. **Educ.:** Dip. Libr. Arts and Theol. from Hesston College, Kansas, USA (1963-1965); B.Sc. from Bethel College, Kansas, USA (1965-1967); M.Sc. from Kansas State University of Agriculture and Applied Science, Manhattan, USA (1967-1968). **Interests:** Tanzania Red Cross. **Mem.:** Tanzania Bible Society; Evangelical Fellowship of Tanzania.
Mailing address: PO Box 2435, Dar es Salaam, Tanzania UR. T. (255 51) 71358

Mbenda, Emmanuel, Rev
Rel: Baptist.
International Organizations: General Secretary of the Union of Baptist Churches of Cameroon.
Mailing address: BP 6007, New Bell, Douala, Cameroon. T. (237) 42 41 06

Mbila, Kilem
Rel: Christian.
International Organizations: Representative of the Africa Network on Churches Participation in Development.
Mailing address: BP 13798, Brazzaville, Congo.

Mboungou, Jean, Rev
Rel: Evangelical.
International Organizations: President of the Evangelical Church of the Congo.
Mailing address: BP 3205, Bacongo-Brazzaville, Congo. T. (242) 81 43 64

Mbugua, Judy W
Rel: Evangelical.
International Organizations: Continental Coordinator of the Pan-African Christian Women Alliance.
Mailing address: c/o AEAM, PO Box 49332, Nairobi, Kenya. T. (254 2) 72 27 69 – 71 49 77. Fax 71 02 54. C. KOINONIA. Tx 23041 TRNKE.

McAllister, Stuart
Rel: Evangelical.
International Organizations: Secretary-General of the European Evangelical Alliance.
Mailing address: Postfach 23, A-1037 Wien, Austria. T. (43 1) 713 34 12. Fax 713 83 82.

McBrien, Richard Peter, Prof
Rel: Roman Catholic; Ordained on 02 FEB 1962; b. 19 AUG 1936 in Hartford CT, USA; celibate; Citizen of USA; Lang. English.
International Organizations: Crowley-O'Brien-Walter Professor of Theology (since 1980) of the University of Notre Dame, IN.
Main field of work: Theology, university education. **Career:** Associate Pastor, Our Lady of Victory Church, West Haven CT from 1962 to 1963. Professor/Dean of Studies, Pope John XXIII National Seminary, Weston MA from 1965 to 1970. Professor of Theology, Boston College, Chestnut Hill MA from 1970 to 1980. Director, Institute of Religious Education and Pastoral Ministry, Boston College from 1975 to 1980. Chairman, Dept. of Theology, University of Notre Dame, IN from 1980 to 1991. **Educ.:** M.A. from St John's Seminary, Boston MA, USA (1956-1962); S.T.D. from Pontifical Gregorian University, Vatican City State (1963-1965). **Interests:** ecclesiology, religion, politics. **Hons.:** Hon. D.Litt., Quincy College, IL (1989); John Courtney Award for 'Outstanding and distinguished achievement in Theology', Catholic Society of America (1976); Christopher Award (1981) for *Catholicism* (2 Vols., Winston Press, 1980); Catholic Press Association of the United States and Canada Award for weekly syndicated column (1975, 1977, 1978, 1984, 1986); Certificate of Appreciation from The Catholic Press Association and *The Catholic Transcript* in recognition of 25 years of outstanding contribution to the Catholic press, Tempe AZ (1991). **Mem.:** Catholic Theological Society of America (Board of Directors 1970-76, Vice-President 1972-73, President 1973-74); American Academy of Religion; College Theology Society; American Association of University Professors; Authors League of America **Pubs.:** *Catholicism*, San Francisco, Harper-Collins, 1980; *Caesar's Coin: Religion and Politics in America*, New York, Macmillan, 1987 and *Ministry: A Theological, Pastoral Handbook*, San Francisco, Harper-Collins, 1987.
Mailing address: Dept. of Theology, University of Notre Dame, Notre Dame IN 46556, USA. T. (1 219) 239 5151

McCann, Owen, H E Card
Rel: Roman Catholic; Ordained on 21 DEC 1935; b. 29 JUN 1907 in Cape Town, South Africa; celibate; Citizen of South Africa.
International Organizations: Member (since 1965) of the College of Cardinals.
Other organizations: Archbishop Emeritus (since 1984), Cape Town.
Career: Titular Bishop, Stettorio in 1950. Archbishop, Cape Town from 1951 to 1984. Named Assistant at the Papal Throne in 1960. Created Cardinal by Pope Paul VI in 1965. **Hons.:** Hon. D.Litt., Cape Town; Hon. D.Hum.Litt., Maine (1984); Freeman, City of Cape Town (1984).
Mailing address: Cathedral Place, 12 Bouquet Street, Cape Town 8001, South Africa. T. (27 21) 45 2419

McCants, Jack E
Rel: Bahá'í.
International Organizations: Member of the National Spiritual Assembly of the Bahá'ís of the US.
Other organizations: Executive Director, Beckman Center, Greenwood SC.
Career: Has directed programs in the public health and mental health fields in the US and abroad for 27 years.
Mailing address: Bahá'í National Center, 536 Sheridan Rd, Wilmette IL 60091, USA.

McCarthy, M Paula, VHM, Sr
Rel: Roman Catholic; Ordained; celibate.
International Organizations: Member of the Visitation Nuns.
Other organizations: Member, Association of Contemplative Sisters.
Mailing address: Visitation Convent, 2455 Visitation Dr, Mendota Hts MN 55120, USA.

McCleary, Paul F, Dr
Rel: Christian.
International Organizations: Executive Director of the Christian Children's Fund.
Mailing address: 203 East Cary St, PO Box 26484, Richmond VA 23219-6484, USA. T. (1 804) 644 4654. C. CHILDFUND. Tx 6844270 CCF USA. Fax (1 804) 225 8385.

McClellan, Joel
Rel: Quaker.
International Organizations: Quaker Representative of the Quaker United Nations Office.
Mailing address: 13 av du Mervelet, CH-1209 Genève, Switzerland. T. (41 22) 733 33 93. C. AFSERCO-GENEVE.

McClinton, Wendell C
Rel: Christian.
International Organizations: Executive Director of the Gideons International.
Mailing address: 2900 Lebanon Road, Nashville TN 37214, USA. T. (1 615) 883 8533. Fax 889 6877. C. TWX No 810-371 1268 GIDEON NAS.

McCollough, Frederick
Rel: Mason.
International Organizations: Representative of the Universal Masonic Brotherhood.
Mailing address: 15 Vose Ave, South Orange NJ 07079, USA. T. (1 201) 763 1780.

McCotter, Bryan, Rev
Rel: Spiritualist.
International Organizations: President of the Ministerial Association (National Spiritual Association of Churches).
Mailing address: 313 E Center St, Leroy IL 61752, USA.

McCullen, Richard, C.M., Fr
Rel: Roman Catholic; Ordained; celibate.
International Organizations: Superior General of the Congregation of the Mission.
Member of the Pontifical Council 'Cor Unum'.
Mailing address: Via Bravetta 159, I-00164 Roma, Italy. T. (39 6) 666 37 30

McCullough, Stewart
Rel: Christian.
International Organizations: Executive Secretary of the World Conference – a Family of Christian Youth Organizations.
Mailing address: c/o The World Conference, Church House, Fisherwick Place, Belfast BT1 6DW, Ireland. T. (44 232) 23 20 24. Fax 24 83 77. C. Assembly, Belfast.

McDonald, Katheen Ann
See Sangye Khadro.

McDonald, Kevin, Mgr
Rel: Roman Catholic; Ordained; celibate.
International Organizations: Catholic Co-Secretary of the Anglican-Roman Catholic International Commission.
Member of the Pontifical Council for Promoting Christian Unity.
Mailing address: c/o Pontifical cl for Promoting Christian Unity, I-00120 Città del Vaticano, Vatican.

McErlean, Joan
Rel: Christian.
International Organizations: Contact of the Volunteer Missionary Movement.
Mailing address: Shenley Lane, London Colney AL2 1AR, UK. T. (44 727) 24 853.

McGavran, Donald A
Rel: Disciples of Christ; b. in India; m. to Mary.
International Organizations: Founder of the Church Growth Movement.
Career: Served as missionary in India with the United Christian Missionary Society (Disciples of Christ) and held posts in education, administration and village evangelism from 1923 to 1954. Became a mission administrator in 1933. After leaving India, spent five years lecturing and doing research on church growth in various countries. Established the Institute of Church Growth, Eugene OR in 1961. Became the founding dean, School of World Mission and Institute of Church Growth at Fuller Theological Seminary, Pasadena CA in 1965. Continued to teach until 1980. **Educ.:** Butler College; Yale University, Divinity School, New Haven CT, USA; Ph.D. from Columbia University, New York NY, USA. **Pubs.:** *The Bridges of God*, 1955; *How Churches Grow*, 1959 and *Understanding Church Growth*, 1970.
Mailing address: No mailing address.

McGill, Maurice, M.H.M., Very Rev
Rel: Roman Catholic; Ordained on 30 JUN 1968; b. 27 AUG 1943 in Ardara, Co. Donegal, Ireland; celibate; Citizen of Ireland; Lang. English, Irish, Italian, French and German.
International Organizations: Superior General (since 1988) of the St Joseph's Missionary Society of Mill Hill.
Main field of work: Ministry. **Career:** Teacher, Bamenda Diocese, Cameroon from 1973 to 1982. Rector, St Joseph's College, Mill Hill from 1982 to 1988. **Educ.:** B.A. from National University of Ireland, University College, Dublin, Ireland (1961-1964); S.T.L. from Pontifical Gregorian University, Vatican City State (1968-1970); L.S.S. from Pontifical Biblical Institute, Vatican City State (1970-1973).
Mailing address: St Joseph's College, Lawrence Street, Mill Hill, London NW7 4JX, UK. T. (44 81) 959 3222. Fax 959 8493. C. Milhillian London NW7. Tx 914991 MHMISS.
Residence: See Mailing Address. T. See Mailing Address

McGillicuddy, Frances
Rel: Catholic.
International Organizations: Representative at UNO-ECOSOC of the St Joan's International Alliance.
Mailing address: 435 W 119th Street, New York NY 10027, USA.

McGrath, Marcos Gregorio, C.S.C., Most Rev
Rel: Roman Catholic; Ordained on 09 JUN 1949; b. 10 FEB 1924 in Panamá, Panama; celibate; Citizen of Panama; Lang. Spanish, English, French, German, Italian and Latin.
International Organizations: Archbishop (since 1969) of Panamá.
Member (since 1984) of the Pontifical Council for Promoting Christian Unity. Member of the Congregation of Holy Cross.
Other organizations: President, Panama Catholic Bishops' Conference (since 1969 on and off until present day); Member, Vatican Council for Religious Relations with Jews.
Main field of work: Religious pastoral and social work. **Career:** Dean, Faculty of Theology, Catholic University of Chile, Santiago from 1958 to 1961. Auxiliary Bishop, Panamá from 1961 to 1964. Bishop, Panamá from 1964 to 1969. Member, Doctrinal Commission, Vatican Council II from 1962 to 1965. Member, World Council of Religions for Peace from 1965 to 1970. Secretary-General, Council of Latin American Bishops from 1966 to 1968. Vice-President, Council of Latin American Bishops from 1966 to 1972. **Educ.:** B.A. from University of Notre Dame, IN, USA (1940-1945); S.T.L. from Holy Cross College, Washington DC, USA and Institut Catholique de Paris, France (1945-1950); S.T.D. from Pontifical University San Tommaso d'Aquino, Vatican City State (1950-1953). **Hons.:** 6 honorary doctorates: Notre Dame, Georgetown, Louvain, etc. **Mem.:** Inter-American Dialogue **Pubs.:** Doctoral thesis on Vatican Council I, 1968; Collected speeches and articles and theological and social articles in US and Latin American reviews.
Mailing address: Apartado 6386, Panamá 5, Panama. T. (507) 62 4858
Residence: Calle Belen 48, Panamá, Panama. T. (507) 26 4792

McIntire, Carl, Dr
Rel: Bible Presbyterian; b. in Ypsilanti MI, USA.
International Organizations: President of the International Council of Christian Churches.
Main field of work: Militant fundamentalist and anti-Communist. **Career:** Ordained minister, the Presbyterian Church (until he was defrocked by his presbytery in 1935). Joined the newly founded church that eventually became known as the Orthodox Presbyterian Church in 1936. Led a group that separated to form the Bible Presbyterian Church. Pastored the Bible Presbyterian Church in Collingswood NJ. Helped found the denomination's Faith Theological Seminary in 1937. Founded the American Council of Churches in 1941. Formed the International Council of Christian Churches in 1948. Established the 'Twentieth-Century Reformation Hour' (daily half-hour radio broadcast) in 1957. **Educ.:** Grad. from Park College, Parkville MO, USA (1927); Grad. from Westminster Theological Seminary (1931).
Mailing address: 756 Haddon Avenue, Collingswood NJ 08108, USA. T. (1 609) 858 0700. C. INTCOUNCIL COLLINGSWOOD.

McIntyre, John, Very Rev Prof

Rel: Presbyterian; Ordained on 03 JUL 1941; b. 20 MAY 1916 in Glasgow, Scotland, UK; m. to Jessie Brown Buick; c. 2 boys and 1 girl; Citizen of United Kingdom; Lang. English and French.

International Organizations: Ecclesiastic of the Presbyterian Church.

Main field of work: University education. **Career:** Minister, locum tenens, Glenorchy and Inishail from 1941 to 1943. Minister in charge, Fenwick Parish Church, Kilmarnock from 1943 to 1945. Hunter Baillie Professor of Theology, St Andrew's College, University of Sydney from 1946 to 1956. Principal, St Andrew's College, University of Sydney from 1950 to 1956. Professor of Divinity, University of Edinburgh from 1956 to 1986. Principal Warden, Pollock Halls of University Residence from 1960 to 1971. Warfield Lecturer, Princeton Theological Seminary in 1966. Dean, Faculty of Divinity and Principal, New College, University of Edinburgh from 1968 to 1974. Acting Principal and Vice-Chancellor, University of Edinburgh from 1973 to 1974. Acting Principal and Vice-Chancellor, University of Edinburgh in 1979. Extraordinary Chaplain to H M The Queen in Scotland in 1974. Chaplain in Ordinary to H M The Queen in Scotland from 1974 to 1986. Dean, Order of the Thistle from 1974 to 1989. Moderator, General Assembly, Church of Scotland in 1982. Vice-President, Royal Society of Edinburgh from 1983 to 1986. Visiting professorships in various universities and colleges in USA, Canada and Australia. **Educ.:** M.A. (Hons) from University of Edinburgh, UK (1938); B.D. (Dist.) from University of Edinburgh, UK (1941); D.Litt. from University of Edinburgh, UK (1953). **Hons.:** DD hc, University of Glasgow (1961); DHL hc, College of Wooster, OH (1983); Dr. h.c., University of Edinburgh (1987); Commander of the Royal Victorian Order (1985). **Mem.:** Fellow, Royal Society of Edinburgh (since 1977) **Pubs.:** *St Anselm and His Critics*, 1954; *The Shape of Christology*, 1967 and *Faith, Theology and Imagination*, 1987.

Mailing address: 22/4 Minto Street, Edinburgh EH9 1RQ, UK. T. (44 31) 667 1203

Residence: See Mailing Address. T. See Mailing Address

McKenzie, Azariah

Rel: Baptist.

International Organizations: Regional Secretary of the Caribbean Baptist Fellowship.

Mailing address: PO B 98, Ligueanea, Kingston 6, Jamaica.

McKenzie, John Lawrence, Rev

Rel: Roman Catholic; Ordained; b. 09 OCT 1910 in Brazil; Citizen of USA.

International Organizations: Ecclesiastic of the Roman Catholic Church.

Career: Teacher, St John's College, Toledo from 1935 to 1936. Teacher, West Baden College from 1942 to 1960. Teacher, Loyola University, IL from 1960 to 1965. Teacher, University of Chicago, IL from 1965 to 1966. Teacher, University of Notre Dame, IN from 1966 to 1970. Teacher, DePaul University, Chicago IL from 1970 to 1978. **Hons.:** John Courtney Murray Theological Award; Catholic Press Award. **Pubs.:** Publication of the Bible, 1966; Second Isaiah, 1968 and A Theology of the Old Testament, 1974.

Mailing address: 447 West Tenth Street, Claremont CA 91711, USA. T. (1 714) 626 0431

McKimm, Inta Benita

Rel: Buddhist: Tibetan Mahayana; Ordained on 18 SEP 1975; b. 18 APR 1926 in Riga, Latvia; celibate; c. 1 girl; Citizen of Australia; Lang. English, Latvian, French and German.

International Organizations: Member of the Foundation for the Preservation of the Mahayana Tradition.

Other organizations: Director (and Psychology Counsellor) (since 1983), Chenrezig City Centre (FPMT); Meditation Teacher and Lecturer (since 1983), FPMT and C G Jung Society; President (since 1987), C G Jung Society of Queensland.

Main field of work: Meditation, retreats, psychology workshops, lectures. **Career:** Research Officer, Audio-Visual Aids Dept., University of Melbourne from 1961 to 1964. Psychologist, Mental Health Dept., University of Melbourne from 1965 to 1970. Acupuncturist, Chinese Medical Institute from 1979 to 1981. **Educ.:** B.A. from University of Melbourne, Victoria, Australia (1959-1963); Dip. Chinese Medicine from Chinese Medical Institute, Melbourne, Australia (1977-1980). **Interests:** Jungian psychology, Eastern medicine. **Mem.:** Foundation for the Preservation of the Mahayana Tradition; C G Jung Society of Queensland; Victorian Psychology Council.

Mailing address: 51 Enoggera Road, Newmarket, Brisbane QLD 4051, Australia. T. (61 7) 356 9523

Residence: See Mailing Address. T. See Mailing Address

McLane, Jack, Rev

Rel: Brethren in Christ.

International Organizations: Executive Director of the Brethren in Christ World Missions.

Mailing address: 431 Grantham Road, PO Box 390, Grantham PA 17027, USA. T. (1 717) 697 2634.

McLaughlin, Helen, CSI, Sr

Rel: Roman Catholic; Ordained; celibate.

International Organizations: Superior General of the Society of the Sacred Heart of Jesus.

Member of the Council to Report between the Congregation and the International Union of Superiors General and Superiors General of the Congregation for Religious and Secular Institutes. President of the International Union of Superiors General.

Mailing address: Via Adolfo Gandiglio 27, I-00151 Roma, Italy. T. (39 6) 53 86 71.

McLaughlin, Joseph, S.S.E., Fr

Rel: Roman Catholic; Ordained; celibate.

International Organizations: Superior General of the Society of St Edmund.

Mailing address: Fairholt-So. Prospect St, Burlington VT 05401-3599, USA. T. (1 802) 864 7040

McLaurin, Edna

Rel: Unknown.

International Organizations: Representative of the Supreme Assembly of the International Order of Rainbow for Girls.

Mailing address: Box 788, McAlester OK 74501, USA. T. (1 918) 423 1328.

McLean, George Francis, Prof

Rel: Catholic; b. 29 JUN 1929 in Lowell MA, USA.

International Organizations: Secretary-General (since 1973) of the World Union of Catholic Philosophical Societies.

Secretary-General (since 1974) of the International Society for Metaphysics.

Other organizations: Secretary (since 1980), Research Council in Values and Philosophy.

Main field of work: Clergyman and Educator. **Career:** Joined the Order of the Oblates of Mary Immaculate in 1949. Ordained Priest of the Roman Catholic Church in 1955. National Secretary, American Catholic Philosophical Association from 1965 to 1980. Research Scholar, University of Paris, France in 1970. Secretary, Inter-University Committee for Research and Policy Studies from 1975 to 1977. Secretary, Catholic Learned Societies and Scholars from 1975 to 1977. Research Scholar, University of Madras, India from 1977 to 1978. Director, International Federation of Philosophical Societies (FISP). **Educ.:** Ph.L. from Pontifical Gregorian University, Vatican City State (1952); S.T.L. from Pontifical Gregorian University, Vatican City State (1956); Ph.D. from Catholic University of America, Washington DC, USA (1958). **Mem.:** American Catholic Philosophy Association **Pubs.:** *The Nature of Metaphysical Knowledge*, 1987; *Moral Values and the Socio-Historical Context*, 1987; *Culture Human Rights and Peace in Central Africa*, 1987; *On Reading the Philosophers*, 1987 and Area Editor, *New Catholic Encyclopaedia.*

Mailing address: School of Philosophy, Catholic University of America, Washington DC 20064, USA. T. (1 202) 635 5636.

Residence: 391 Michigan Avenue NE, Washington DC 20017, USA

McLean, Walter Franklin, Rev

Rel: Presbyterian; b. 26 APR 1936 in Leamington ON, Canada; m. to Barbara Scott; c. 4 boys; Citizen of Canada.

International Organizations: Member (since 1979) of the Parliament. Co-founder of the CUSO.

Main field of work: Religious official and politician. **Career:** Alderman for City from 1976 to 1979. Spokesperson for Secretary, State in 1981. Secretary, State from 1984 to 1985. Minister, State responsible for the Status of Women from 1984 to 1986. Former Executive Director, Manitoba Centennial Corporation. Former Minister, Knox Presbyterian

Church, Waterloo. Former Member, Standing Committee on Communications and Culture, on External Affairs, on National Defence. Former Member, Special Sub-Committee on Latin America and Caribbean. Former spokesperson for status of women. **Hons.:** Privy Councillor. **Mem.:** Progressive Conservative Party
Mailing address: House of Commons, Ottawa ON K1A 0A2, Canada. T. (1 613) 731 8725

McLelland, Joseph Cumming, Rev

Rel: Presbyterian; Ordained in 1949; b. 10 SEP 1925 in Port-Glasgow, Scotland, UK; m. to Audrey Mary Brunton; c. 3 boys and 1 girl; Citizen of Canada.
International Organizations: McConnel Professor of Philosophy of Religion (since 1964) of the McGill University.
Career: Dean of Religious Studies, McGill University from 1975 to 1985. Moderator, 111th General Assembly, Presbyterian Church in Canada in 1985. Professor of History and Philosophy of Religion and Christian Ethics, The Presbyterian College, Montreal from 1957 to 1964. Associate Professor of Philosophy of Religion, McGill University from 1959 to 1964. Member, Reformed-Lutheran Consultation (North America) from 1963 to 1967. Co-Chairman, Reformed-Orthodox Consultation (North America) from 1968 to 1971. Member, Orthodox-Reformed Consultation (International) from 1979 to 1985. Sometime member, Faith and Order Commission (Canadian Council of Churches), and dialogue groups. **Educ.:** B.A. from McMaster University, Hamilton, Ontario, Canada (1946); B.D. from University of Toronto, Knox College, Canada (1951); D.D. from University of Toronto, Knox College, Canada (1976); M.A. from University of Toronto, Ontario, Canada (1949); Ph.D. from University of Edinburgh, New College, UK (1953); D.D. from Montreal Diocesan Theological College, Canada (1973). **Interests:** Scottish country dancing. **Mem.:** Canadian Theological Society (Past President); Canadian Society for the Study of Religion **Pubs.:** Celebration and Suffering, 1984; Prometheus Rebound, 1988 and over 60 articles on philosophy of religion and history of theology.
Mailing address: 3520 University St, Montréal PQ H3A 2A7, Canada.
Residence: 121 Alston Rd, Pointe Claire PQ H9R 3E2, Canada.

McManus, Eugène P, S.S.J., Fr

Rel: Roman Catholic; Ordained; celibate.
International Organizations: Superior General of the St Joseph's Society of the Sacred Heart.
Mailing address: 1130 N Calvert Street, Baltimore MD 21202, USA. T. (1 301) 727 33 86.

McMillan, Keith A, Rt Rev

Rel: Christian.
International Organizations: Provincial Secretary of the Church in the Province of the West Indies.
Mailing address: PO Box 535, Belize City, Belize, West Indies.

McMillan, Kenneth George, Very Rev

Rel: Presbyterian; Ordained; b. 28 MAR 1916 in Mount Forest ON, Canada; m. to Isobel Islay McCannel; c. 2 girls; Citizen of Canada.
International Organizations: Minister-at-Large (since 1983) of the World Vision Canada.
Career: Canadian Patron, Global Cooperation for a Better World in 1988. Secretary-General, Canadian Bible Society from 1957 to 1983. Chairman, General Committee, United Bible Societies of the World from 1976 to 1980. Minister, Burgoyne and Dunblane from 1942 to 1943. Minister, Drummond Hill Church, Niagara Falls, Ontario from 1944 to 1950. Minister, St Andrew's Church, Guelph, Ontario from 1950 to 1957. Moderator, Presbyterian Church in Canada from 1979 to 1980. Past Moderator, Presbyterian Synod of Toronto and Kingston. **Educ.:** B.A. from University of Toronto, University College, Canada (1939); Grad. from University of Toronto, Knox College, Canada (1942); M.Div. from University of Toronto, Knox College, Canada (1946); D.D. from University of Toronto, Knox College, Canada (1961); D.D. from University of Toronto, Wycliffe College, Canada (1974); D.D. from Acadia University, Wolfville, Nova Scotia, Canada (1978). **Interests:** camping, woodwork. **Hons.:** Member, Order of Canada (1984). **Pubs.:** What But Thy Grace and Against the Tide.
Mailing address: Apt 604, 80 Inverlochy Blvd, Thornhill ON L3T 4P3, Canada.
Residence: See Mailing Address.

McMorrow, Kevin, S.A., Fr

Rel: Roman Catholic; Ordained; celibate.
International Organizations: Procurator-General of the Franciscan Friars of the Atonement.
Procurator-General of the Brothers of the Poor of St Francis.
Mailing address: Piazza S Onofrio 2, I-00165 Roma, Italy. T. (39 6) 686 3498.

McMurry, John Edward, S.S., Rev

Rel: Roman Catholic; Ordained on 16 DEC 1956; b. 19 MAR 1931 in Nashville TN, USA; celibate; Citizen of USA; Lang. English, Latin (some), German (some) and French (some).
International Organizations: Spiritual Director (since 1989) of the St Mary's Spiritual Center.
Member of the Servants of the Blessed Sacrament.
Educ.: A.B. from St Ambrose College, Davenport IA, USA (1949-1953); S.T.B. from Pontifical Gregorian University, Vatican City State (1953-1955); S.T.L. from Pontifical Gregorian University, Vatican City State (1955-1957); Ph.D. from Graduate Theological Union, Berkeley CA, USA (1970-1977); Cert. from Shalem Institute for Spiritual Formation, Washington DC, USA (1981-1982); Cert. from Center for Religious Development, Cambridge MA, USA (1988). **Pubs.:** "Spirituality for a Secular Culture: a Christological Interpretation of Progoff's *Intensive Journal* Method", in *Review for Religious*, Vol. 47, No 3, May/June 1988; "Spirituality and Creativity: Ira Progoff's *Intensive Journal* Method", in *Bulletin de Saint-Sulpice*, Vol. 16, 1990 and "What About Spiritual Direction?", in *The Priest*, Vol. 46, No 11, November 1990.
Mailing address: 4408 Liberty Heights Ave, Baltimore MD 21207-7593, USA. T. (1 410) 542 0445
Residence: See Mailing Address. T. See Mailing Address

McNamara, Gertrude

Rel: Methodist.
International Organizations: President for Great Britain and Ireland of the World Federation of Methodist Women.
Mailing address: 23 Ralaly Drive, Dundrum, Dublin 16, Ireland.

McNamara, William, Rev

Rel: Roman Catholic; Ordained on 29 SEP 1951; b. 14 FEB 1926 in Providence RI, USA; celibate; Citizen of USA; Lang. English.
International Organizations: Founder Abbot (since 1960) of the Spiritual Life Institute.
Main field of work: Writing. **Educ.:** B.A. Theol. and Phil. from Carmelite Seminary, WI, USA; M.A. Psy. from Boston College, Chestnut Hill MA, USA. **Pubs.:** *The Human Adventure*; *Christian Mysticism* and *Art of Being Human.*
Mailing address: PO Box 219, Crestone CO 81131, USA. T. (1 719) 256 4778
Residence: See Mailing Address. T. See Mailing Address

McNeill, Tim

Rel: Buddhist: Mahayana.
International Organizations: Member of the Foundation for the Preservation of the Mahayana Tradition.
Other organizations: Director, Wisdom Publications Inc., USA.
Mailing address: 361 Newbury Street, Boston MA 02115, USA. T. (1 617) 536 3358, USA ONLY 800-272 4050. Fax (617) 536 1897

McSweeney, Anthony, S.S.S., Fr

Rel: Roman Catholic; Ordained; celibate.
International Organizations: Superior General of the Congregation of the Blessed Sacrament.
Member of the Congregation for Religious and Secular Institutes. President of the Union of Superiors General.
Mailing address: Via G B de Rossi 46, I-00161 Roma, Italy. T. (39 6) 844 24 30.

Meadows, Peter Charles

Rel: Christian; b. 04 OCT 1941 in Lydney, Gloucester, UK; m. to Rosemary Glayne; c. 4 boys and 1 girl; Citizen of United Kingdom; Lang. English.
International Organizations: Chief Executive (since 1984) of the Bridgehead Communications.
Communications Secretary (since 1984) of the Evangelical Alliance.
Other organizations: Board Member (since 1977), Spring Harvest; Board Member (since 1991), CNI Institute Inc. (USA); Acts as a communications consultant with a range of Christian societies and mission agencies, including Campus Crusade for Christ, Operation Mobilisation, Every Home for Christ (UK), Focus on the Family and Spurgeon's College; Conducts seminars on a range of subjects, including 'Communications', 'Culture' and 'Stress'.
Main field of work: Communication. **Career:** Creative Director, Musical Gospel Outreach from 1965 to 1974. Publishing Director, Elmhouse Christian Communication from 1974 to 1982. Executive Director, Luis Palau Mission to London from 1982 to 1984. Co-Founder, Spring Harvest. Co-Founder, Tear Fund (evangelical relief agency). Founder, Brainstormers (training event for Christian youth workers). Launched and served as Chairman, Campaign to Free the Siberian Seven (now Jubilee Campaign – acts for persecuted Christians throughout the world). **Interests:** communication, culture, broadcasting. **Pubs.:** *Pressure Points*, Kingsway (UK) and *Finding Peace under Pressure*, Here's Life Publishers (USA), Spring 1990.
Mailing address: Bridgehead Communications, 186 Kennington Park Road, London SE11 4BT, UK. T. (44 71) 582 0408
Residence: 10 Marston Ave, Chessington KT9 2HF, UK. T. (44 81) 397 1666

Meares, John L, Rev Dr

Rel: Evangelical.
International Organizations: President of the International Evangelical Church.
Mailing address: 610 Rhode Island Ave NE, Washington DC 20002, USA.

Mechsner, Josef, F.F.S.C., Br

Rel: Roman Catholic; Ordained on 12 JUN 1990; b. 08 JUN 1939 in Mittelwalde, Nordrhein-Westfalen, Germany; celibate; Citizen of Germany; Lang. German.
International Organizations: Superior General (since 1990) of the Franciscan Brothers of the Holy Cross.
Main field of work: Ministry.
Mailing address: St Josefshaus, 53547 Hausen-Linz, Germany. T. (49 2638) 1000.
Residence: See Mailing Address. T. See Mailing Address

Meck, Lynette Youndt

Rel: Mennonite.
International Organizations: Executive Secretary of the Mennonite Central Committee.
Mailing address: 21 South 12th Street, Akron PA 17501, USA.

Medini, Lise

Rel: Buddhist: Mahayana.
International Organizations: Member of the Foundation for the Preservation of the Mahayana Tradition.
Other organizations: Director, Kalachakra Center.
Mailing address: 6 Rue Baron, 75017 Paris, France. T. (33 1) 4229 7125. Fax (33 1) 4222 7744

Meegan, Gerald, S.D.B., Br

Rel: Roman Catholic; Ordained; celibate.
International Organizations: Member of the Society of St Francis of Sales.
Other organizations: Secretary, National Assembly of Relgious Brothers.
Mailing address: 1337 West Ohio Street, Chicago IL 60622-6490, USA. T. (1 312) 829 8525

Mehervan Singh, M, Dr

Rel: Sikh; Unordained; b. 08 JUL 1919 in Bhasaur Village, Patiala State, Punjab, India; m. to Bahadur Kaur; c. 3 boys and 2 girls; Citizen of Singapore; Lang. English, Punjabi and Malay.
International Organizations: Secretary-General (since 1976) of the Asian Conference on Religion and Peace.
Participated in five international conferences, and many Governing Board meetings (since 1970) of the World Conference on Religion and Peace.
Other organizations: Accountant (qualified) with Mehervan Singh and Co. (since 1958); President (since 1990), Inter-Religious Organisation, Singapore.
Main field of work: Accountancy. **Career:** Service with British Air Ministry, Indian Independence League (Indian National Army), Singapore Inland Revenue from 1939 to 1952. Public Accountant (unqualified) from 1952 to 1958. Honorary Secretary, Inter-Religious Organisation, Singapore from 1964 to 1983. **Interests:** religion and peace. **Hons.:** Public Service Star, Singapore National Day Honours (1966); Hon. Ph.D., International University Foundation in Malta (1988). **Mem.:** Fellow, Association of Certified Accountants **Pubs.:** Sikhism – Its Impact, 1973; India – For a Visit, 1987 and From Village to City, 1988.
Mailing address: 5001 Beach Road 07-24, Golden Mile Complex, Singapore 0719, Singapore. T. (65) 2922 510.
Residence: 71 Greenfield Drive, Singapore 1545, Singapore. T. (65) 4493 325.

Mehta, Priti Kiritkumar, Dr

Rel: Hindu; b. 02 JUN 1963 in Koth, India; Citizen of India.
International Organizations: Research Assistant (since 1991) of the LD Institute of Indology, Ahmedabad.
Career: Research Assistant in Prakrit Jain Vidya Vikas Fund at Gujarat University, Ahmedabad from 1988 to 1991. **Educ.:** F.Y.B.A. from Gujarat University, Ahmedabad, India (1981); S.Y.B.A. from Gujarat University, Ahmedabad, India (1982); T.Y.B.A. from Gujarat University, Ahmedabad, India (1983); M.A. from Gujarat University, Ahmedabad, India (1984); Ph.D. from Gujarat University, Ahmedabad, India (1986); M.A. from Gujarat University, Ahmedabad, India (1990). **Hons.:** academic distinctions. **Pubs.:** Thesis on Rsikas in the Vedas and conference papers.
Mailing address: 5 Gnanmandir Society, River Side, Vasna, Ahmedabad 380 007, India.

Meier, Gerhard

Rel: Roman Catholic.
International Organizations: Secretary-General of the Caritas Internationalis.
Mailing address: Palazzo San Calisto, I-00120 Città del Vaticano, Vatican. T. (39 6) 698 7197. C. INTERCARITAS ROMA. Tx (504) 2014 (CIVA). Fax (39 6) 698 72 37.

Meijering, Eginhard Peter, Dr

Rel: Remonstrant Reformed; Ordained on 27 OCT 1968; b. 02 JAN 1940 in Groningen, Netherlands; separated; c. 1 boy and 1 girl; Citizen of Netherlands; Lang. Dutch, German and English.
International Organizations: Reader in Patristic Literature (since 1976) of the State University of Leiden.
Other organizations: Minister (since 1987), Remonstrant Parish of Oude Wetering.
Main field of work: History of Christian thought. **Educ.:** D.Div. from State University at Leiden, The Netherlands (1968). **Interests:** religion, politics, sports. **Mem.:** Dutch Royal Society **Pubs.:** Die Hellenisierung des Christentums im Urteil; Athanasius: Das incarnatione Verb (Einleitung, Übersetzung, Kommentar) and Reformierte Scholastik und patristische Theologie.
Mailing address: Ridderspoorlaan 103, NL-2343 TX Oegstgeest, Netherlands. T. (31 71) 172111

Meisner, Joachim, H E Card

Rel: Roman Catholic; Ordained on 22 DEC 1962; b. 25 DEC 1933 in Breslau, Germany; celibate; Citizen of Germany.
International Organizations: Member (since 1983) of the College of Cardinals.
Member of the Sacred Congregation for the Sacraments and Divine Worship. Member of the Pontifical Council for Inter-Religious Dialogue. Member of the Pontifical Council for Dialogue with Non-Believers.

Other organizations: Archbishop (since 1989), Cologne and Primate of Germany; Member, Council of Cardinals for the Study of Organizational and Economic Problems of the Holy See.
Career: Chaplain, St Ägidien, Heiligenstadt and St Crucis, Erfurt from 1963 to 1966. Rector, Caritas, Diocese of Erfurt from 1966 to 1975. Suffragan Bishop, Erfurt from 1975 to 1980. Bishop, Berlin from 1980 to 1989. Created Cardinal by Pope John Paul II in 1983. **Educ.:** Dr. Theol. (1974). **Pubs.:** Das Auditorium Coelicum am Dom zu Erfurt, 1960; Nachreformatorische katholische Frömmigkeitsformen in Erfurt, 1971 and various articles in magazines.
Mailing address: Kardinal-Frings-Strasse 10, 50668 Köln, Germany. T. (49 221) 164 22 04

Meissner, Herbert, Rev Dr

Rel: Lutheran; Ordained on 20 APR 1969; b. 06 OCT 1940 in Berlin, Germany; m. to Ute; c. 3 girls; Citizen of Germany; Lang. German, English, Afrikaans, Sesotho and French.
International Organizations: Director of the Protestant Association for World Mission.
Other organizations: General Secretary (since 1988), Evangelisches Missionwerk in Deutschland (Association of Protestant Churches and Missions in Germany); Member (since 1988), various Governing Boards in mission and church development Service.
Main field of work: Ministry, administration. **Career:** Minister, Lutheran Church in Bavaria from 1966 to 1970. Minister, Lutheran Church in Southern Africa from 1970 to 1979. Secretary for Africa, Berliner Missionswerk, Berlin from 1979 to 1985. Oberkonsistorialrat (OKR), Evangelical Church Berlin Brandenburg (West) from 1985 to 1988. **Educ.:** 1st Theol. Exam. from Universities of Erlangen and Hamburg, Germany (1961-1966); 2nd Theol. Exam. from Universities of Erlangen and Hamburg, Germany (1969).
Mailing address: Mittelweg 143, 20148 Hamburg, Germany. T. (49 40) 41174 101. C. MISSIONSWERK HAMBURG. Tx 02/14504 EWEMI.
Residence: Remstedtstrasse 40, 22143 Hamburg, Germany.

Melada, Petar Pavao, O.F.M., Rev Dr

Rel: Roman Catholic; Ordained on 23 JUN 1940; b. 15 MAY 1916 in Kastel Stafilic, Croatia; celibate; Citizen of Croatia; Lang. Croatian, Latin, Italian, French, German, English and Spanish.
International Organizations: President (since 1976) of the Pontifical Marian International Academy.
Member of the Order of Friars Minor. Member of the Pontifical Committee for International Eucharistic Congresses.
Main field of work: Ministry. **Career:** Secretary, Pontifical International Marian Academy from 1946 to 1976. **Educ.:** Pontificio Ateneo Antonianum, Vatican City State (1938-1944); Pontifical Biblical Institute, Vatican City State (1944-1945). **Mem.:** Pontifical Roman Theological Academy; various national mariological societies
Mailing address: Via Merulana 124, I-00185 Roma, Italy. T. (39 6) 70373267.

Melchers, B, Sr

Rel: Roman Catholic; Ordained; celibate.
International Organizations: Superior General of the Society of Jesus, Mary, Joseph.
Mailing address: Jagersboschlaan 17, NL-5262 LS Vught, Netherlands. T. (31 73) 56 17 21.

Melchior, Bent, Chief Rabbi

Rel: Jewish; Ordained in 1963; b. 24 JUN 1929 in Beuthen, Germany; m. to Lilian Weisdorff; c. 4 boys; Citizen of Denmark.
International Organizations: Chief Rabbi (since 1970) of Denmark.
Other organizations: Chairman (since 1970), Denmark's Committee for Soviet Jewry; Member of Presidium (since 1971), World Conference for Soviet Jewry.
Career: Came to Denmark in 1933. Refugee in Sweden from 1943 to 1945. Chairman, B'nai Akiva in Denmark from 1945 to 1948. Cofounder, Danish Zionist Federation in 1946. Volunteer to Haganah and Israel Defence Forces from 1948 to 1959. Editor, Israel Magazine from 1949 to 1954. Teacher at Jewish Day School and Talmud Torah in Copenhagen from 1949 to 1958. Chairman, Jewish Youth Organization from 1954 to 1955. Rabbi, the Jewish Community in Copenhagen from 1963 to 1970. President, Keren Hayesod in Denmark from 1968 to 1970. Vice-President, World Council of Synagogues from 1970 to 1979. Lecturer in Jewish Literature, University of Copenhagen from 1971 to 1984. President, Denmark-Lodge B'nai B'rith from 1979 to 1981. President, Denmark-Lodge B'nai B'rith from 1983 to 1984. President, Council for Scandinavian Lodges of B'nai B'rith from 1984 to 1988. Vice-President, Central Committee District 19 from 1985 to 1987. **Educ.:** Rabb. Dip. from Jews' College, London, UK (1963). **Hons.:** Recipient of Israeli Fighters' Order and Danish Order of Knighthood; Certificates of honour and awards from various Jewish organizations and institutions. **Pubs.:** Author of a Danish version of the history of the first part of the Bible and a text book on Judaism for colleges and seminars; Translated the Pentateuch and the Pesach Hagaddah into Danish and Author of numerous articles and scientific papers.
Mailing address: Frederiksborggade 27, DK-1360 København K, Denmark. T. (45) 33 13 82 82. Fax 33 32 66 99

Mencia, Valentin

Rel: Buddhist: Mahayana.
International Organizations: Member of the Foundation for the Preservation of the Mahayana Tradition.
Other organizations: Director, Thubten Ling.
Mailing address: Aptdo 162, 03400 Villena, Alicante, Spain. T. (34 96) 580 8962

Mendelson, Jakob Ben-Zion

Rel: Jewish; b. in New York, USA.
International Organizations: Cantor of the Cantor Temple Israel, White Plains NY.
Career: Served as cantor in Riverdale, New York, at the Beth Torah Synagogue in Miami FL, in the congregation Shaarai Tefila in Flushing, NY. Appeared in concerts and in prayer services throughout the United States, and also at the Jerusalem congress of Cantors' Assembly in honor of the 20th anniversary of the unification of Jerusalem. Was teacher of the traditional style of prayer at the Hebrew Union College and at the Cantorial School of the Jewish Theological Seminary in New York. Produced a record of selections of the prayer service.
Mailing address: No mailing address.

Mendelson, Solomon

Rel: Jewish; b. in New York, USA.
International Organizations: President of the Cantors' Assembly of the United States.
Other organizations: Cantor (since 1954), Beth Shalom Synagogue, Long Beach NY; Member of the administration, Cantorial School, Jewish Theological Seminary; Teacher, the traditional style of the prayer services; Initiates and organizes concerts of cantorial music.
Career: Served in the US Army with the rank of captain. Assisted in the writing of new compositions in the areas of Jewish and cantorial music.
Mailing address: No mailing address.

Mendiola, Ma Ramona, ICM, Sr

Also know as Ma Lourdes G Mendiola; Rel: Roman Catholic; Ordained on 12 APR 1954; b. 10 NOV 1927 in Caloocan City, Metro Manila, Philippines; celibate; Citizen of Philippines; Lang. English, Tagalog, Spanish, French, Dutch and Italian.
International Organizations: Superior General (since 1980) of the Missionary Sisters of the Immaculate Heart of Mary.
Main field of work: Education. **Career:** Consultor, Pontifical Commission on Justice and Peace from 1982 to 1989. President, Agrimissio from 1986 to 1991. **Educ.:** A.B., B.S.E. from St Theresa's College, Manila, Philippines (1945-1949); M.A. Ed. from Saint Louis University, Baguio City, Philippines (1960-1961). **Pubs.:** Confucian Humanism: A Stepping-Stone to Christian Humanism.
Mailing address: Via di Villa Troili 30, I-00163 Roma, Italy. T. (39 6) 6641 71 56. C. ICMAR – 00163 Roma.
Residence: See Mailing Address. T. See Mailing Address

Menezes, George

Rel: Roman Catholic.
International Organizations: Regional Secretary for Asia of the International Council of Catholic Men.
Mailing address: 64 Mount Unique, Pedder Road, Bombay 400 026, India.

Menonkary, John, Fr

Rel: Roman Catholic; Ordained; celibate.
International Organizations: Secretary of the Carmelites of Mary Immaculate.
Mailing address: Prior General's House, PO Box 1056, Ernakulam, Cochin 682011, India. T. 360137.

Meram, Thomas, Most Rev Archbishop
Rel: Chaldean Catholic; Ordained on 11 JUN 1967; b. 06 AUG 1943 in Tall Kayf, Iraq; celibate; Citizen of Iran.
International Organizations: Archbishop (since 1984) of Urmia (Rezayeh).
Other organizations: Bishop (since 1983), Salmas (Shahpour).
Mailing address: Khalifagari Kaldani Katholiq, POB 338, Orumiyeh 57135, Iran Islamic Rep. T. (98 441) 22739

Mercado, Norberto
Rel: Christian.
International Organizations: International Coordinator of the Christians for the Upliftment of the Poor.
Mailing address: PO Box 141, Quezon City 3001, METRO Manila, Philippines.

Mercier, T
Rel: Christian.
International Organizations: Secretary-General of the Centre international des UCJG-YMCA.
Mailing address: F-77008 Melun-La-Rochette, France. T. (33) 64 37 12 32.

Merians, Melvin
Rel: Jewish; b. 26 OCT 1928 in New York NY, USA; m. to Elaine Blum; c. 1 boy and 2 girls; Citizen of USA; Lang. English.
International Organizations: Chairman of the Board (since 1991) of the Union of American Hebrew Congregations.
Other organizations: Associate Treasurer, Jewish Museum; Member of Board, Hebrew Union College-Jewish Institute of Religion, Cincinnati OH.
Main field of work: Investor-Consultant. **Career:** Founder, YM/YWCA of Mid-Westchester County. Board Member, Associated Y's of Greater New York. Vice-President, Larchmont Temple, NY. Chairman of the Board, Friends of the Neuberger Museum. Member of the Board and National Budget Chairman, American Jewish Committee. **Educ.:** B.Sc. from University of Pennsylvania, Philadelphia, USA.
Mailing address: 838 Fifth Ave, New York NY 10021, USA. T. (1 212) 249 0100
Residence: 10 Bonnie Briar Lane, Larchmont 10538, USA. T. (1 914) 834 0235

Merminod, Gérard, Rev
Rel: Protestant.
International Organizations: Executive Secretary of the European Ecumenical Commission for Church and Society.
Mailing address: 18 rue Charles Appell, F-67000 Strasbourg, France. T. (33) 88 36 90 26. Fax 88 36 90 08.

Merrill, Joseph H, Rev
Rel: Spiritualist; b. 16 JAN 1903 in Norway ME, USA.
International Organizations: President (since 1973) of the National Spiritualist Association of Churches.
Career: Secretary, First Spiritualist Church, Lynn MA. Secretary, Massachusetts State Association of Spiritualists from 1951 to 1961. Secretary, National Spiritualist Association of Churches (NSAC) from 1961 to 1971. Vice-President, NSAC from 1971 to 1973. Represented the NSAC at the International Spiritualist Federation (ISF) in 1975. Vice-President, ISF (for 6 years). Successively Secretary, Vice-President and President, and Board Member at different times, Lily Dale Assembly. Travelled for 5 seasons with the Jane Case Concert Company from 1921 to 1926. Employed in photography following his musical engagements. **Interests:** music, photography. **Pubs.:** Author of the Mini Course on Mediumship and Healing.
Mailing address: PO Box 217, 13 Cottage Row, Lily Dale, New York NY 14752, USA. T. (1 716) 595 2000. Fax (1 716) 595 2020
Residence: 13 Cleveland Ave, Lily Dale, New York NY 14752, USA.

Mertes, Josef M
Rel: Roman Catholic.
International Organizations: Secretary-General of the International Christian Union of Business Executives.
Symposium Secretary of the Churches-Transnational Corporations.
Mailing address: Place des Barricades 2, B-1000 Bruxelles, Belgium. T. (32 2) 218 31 14. C. Uniapac-Bruxelles. Fax (32 2) 219 70 37.

Mescher, Sebastian, F.F.S.C., Br
Rel: Roman Catholic; Ordained; celibate.
International Organizations: Superior General of the Franciscan Brothers of the Holy Cross.
Mailing address: St Josefshaus, 53547 Hausen/Wied, Germany. T. (49 2638) 4055

Messakh, Th A, Rev
Rel: Evangelical.
International Organizations: President of the Protestant Evangelical Church in Timor.
Mailing address: Jenderal A Yani 35, Kupang-NTT, Indonesia.

Messas, Chalom, Rabbi
Rel: Jewish; Citizen of Morocco.
International Organizations: Grand Rabbi of Casablanca.
Other organizations: President, Rabbinical Court of Casablanca.
Mailing address: Palais de Justice, Place des Nations Unies, Casablanca, Morocco.

Mészáros, István, Dr
Rel: Reformed Church in Hungary; Ordained; Citizen of Hungary.
International Organizations: Bishop of Reformed Church in Hungary.
Mailing address: Kossuth u. 17, Miskolc H-3501 PF 155, Hungary.

Metcalf, Hattie
Rel: Christian.
International Organizations: General Director of the International Christian Leprosy Mission.
Mailing address: 6917 SW Oak Drive, Portland OR 97223, USA. T. (1 503) 244 5935.

Metral, Constance, Sr
Rel: Roman Catholic; Ordained; celibate.
International Organizations: Member of the Sisters of the Cross of Chavanod.
Mailing address: Le Grand Pré, F-74650 Chavanod, France. T. (33) 50 69 10 58.

Metz, Johann Baptist
Rel: Roman Catholic; b. 05 AUG 1928 in Auerbach, Germany; Citizen of Germany.
International Organizations: Professor of Fundamental Theology (since 1963) of the University of Münster.
Career: Member, Founding Commission, University of Bielefeld in 1966. Consultant to Papal Secretariat Pro Non Credendibus from 1968 to 1973. Adviser to German Diocesan Synod from 1971 to 1975. **Hons.:** Awards from University of Innsbrück and Boston College, MA. **Pubs.:** several books on theological themes.
Mailing address: Katholisch-Theologische Fakultät, Seminar für Fundamentaltheologie, Johannisstrasse 8-10, 48143 Münster, Germany. T. (49 251) 83 2631
Residence: Kapitelstrasse 14, 48145 Münster, Germany. T. (49 251) 36662

Metzler, Edgar
Rel: Christian.
International Organizations: Executive Director of the United Mission to Nepal.
Mailing address: Box 126, Kathmandu, Nepal. T. (977) 212179 – 215573. C. UMNEPAL. Tx 2315 UMNEPA NP.

Meyendorff, John
Rel: Orthodox; Ordained in 1959; b. in France.
International Organizations: Dean (since 1983) of the St Vladimir's Orthodox Theological Seminary.
Main field of work: Theologian. **Career:** Emigrated to the US in 1959. Professor of church history, patristics and canon law at the St Vladimir's Orthodox Theological Seminary. Professor of Byzantine history at Fordham University, Bronx NY. Lecturer at Dumbarton Oaks Center for Byzantine Studies, Washington DC. Leading clergyman in the Russian Orthodox Greek-Catholic Church of America (Metropolia). Served as Chairman, Faith and Order Commission, World Council of Churches, and Department of External Affairs, Orthodox Church in America from 1967 to 1976. Editor, *The Orthodox Church* (monthly newspaper of the Orthodox Church in America) from 1965 to 1984. **Educ.:** St Serge Orthodox Theological Institute, Paris, France; Sorbonne, Paris, France. **Pubs.:** *The Orthodox Church*, 1962; *Orthodoxy and Catholicity*, 1966 and *Christ in Eastern Christian Thought*, 1975.
Mailing address: No mailing address.

Meyer, Albert J
Rel: Mennonite.
International Organizations: Executive Secretary of the Mennonite Board of Education.
Mailing address: PO Box 1142, Elkhart IN 46515, USA.

Meyer, Arthur
Rel: Quaker.
International Organizations: Contact of the Friends World College, Latin American Center.
Mailing address: Plover Lane, Huntington NY 11743, USA. T. (1 516) 549 1102.

Meyers, Robert J
Rel: Christian.
International Organizations: President of the Carnegie Council on Ethics and International Affairs.
Mailing address: Merrill House, 170 East 64th St, New York NY 10021, USA. T. (1 212) 838 4120. Fax 752 2432.

Meyieres, Jean-François
Rel: Roman Catholic.
International Organizations: Secretary of the International Secretariat for Catholic Engineers, Agronomists and Industry Officials.
Mailing address: 18 rue de Varenne, F-75007 Paris, France. T. (33 1) 42 22 18 56.

Mhelan, Muhammad, Sheikh
Rel: Muslim; Citizen of Jordan.
International Organizations: Chief Justice and President of the Supreme Muslim Secular Council.
Mailing address: No mailing address.

Miaglia, Emmanuel, Lt-Col
Rel: Salvation Army, The.
International Organizations: Officer Commanding of The Salvation Army.
Mailing address: Via Dei Marrucini 40, I-00185 Roma, Italy.

Michael, Joy
Rel: Reformed Church.
International Organizations: Member of the Executive Committee of the World Alliance of Reformed Churches.
Mailing address: 121 Madan Lal Block, Asiad Village, New Delhi 110049, India.
Residence: T. (91 212) 641 3143

Michalko, Jan, Prof Dr
Rel: Evangelical.
International Organizations: Presiding Bishop of Slovak Evangelical Church of the Augsburg Confession in the CSSR.
Mailing address: Palisady 46, Bratislava 81106, Slovakia.

Michaud, Marie-Kisito, Sr
Rel: Roman Catholic; Ordained; celibate.
International Organizations: Superior General of the Sisters of the Holy Family.
Mailing address: 30 rue Bories, F-12200 Villefranche-de-Rouergue, France. T. (33) 65 45 14 85.

Michelman, Henry D, Rabbi
Rel: Jewish; Ordained.
International Organizations: Executive Vice-President of the Synagogue Council of America.
Mailing address: 327 Lexington, Room 4210, New York NY 10016, USA. T. (1 212) 686 8670.

Michels, M Gregoris, SCC, Sr
Rel: Roman Catholic; Ordained; celibate.
International Organizations: Superior General of the Sisters of Christian Charity.
Mailing address: Via di Boccea 761, I-00166 Roma, Italy. T. (39 6) 696 08 29.

Michman, Dan, Prof
Rel: Jewish; Unordained; b. 28 JUN 1947 in Amsterdam, Netherlands; m. to Bruria; c. 3 boys and 3 girls; Citizen of Israel; Lang. Hebrew, Dutch, English, German, Yiddish and French.
International Organizations: Chairman of the Arnold and Leona Finkler Institute of Holocaust Research.
Other organizations: Member, Academic Committee (since 1989), Yad Vashem Remembrance Authority; Associate Professor (since 1989), Bar-Ilan University, Ramat-Gan.
Main field of work: Holocaust research. **Career:** Lecturer in Jewish History at Bar-Ilan University, Ramat-Gan from 1979 to 1983. Senior Lecturer at Bar-Ilan University, Ramat-Gan from 1983 to 1989. **Educ.:** B.A. from Hebrew University of Jerusalem, Israel (1967-1972); Ph.D. from Hebrew University of Jerusalem, Israel (1972-1978). **Interests:** Jewish religious life during the holocaust. **Pubs.:** *Het Liberale Jodendom in Nederland 1929-1943*, Amsterdam, 1988; Pinkas Hakehillot: Holland *Encyclopedia of Jewish Communities in The Netherlands*, Jerusalem, 1985 and "Research on the Life of Religious Jewry during the Holocaust", in *Historiography of the Holocaust*, Jerusalem, 1988.
Mailing address: Bar-Ilan University, Ramat-Gan 52900, Israel. T. (972 3) 34 03 33.
Residence: Oholei Keidar 59, POB 284, Elkana, DN Efraim 44814, Israel. T. (972 3) 936 22 56

Mietto, Paolo, C.S.I., Fr
Rel: Roman Catholic; Ordained; celibate.
International Organizations: Superior General of the Pious Congregation of St Joseph.
Mailing address: Via del Belvedere Mondello 77, I-00166 Roma, Italy. T. (39 6) 624 28 51 – 624 34 00. Fax 62 44 64.

Mifsud, P Antonio
Rel: Roman Catholic.
International Organizations: Director of the Latin American Institute of Social Theory and Social Studies.
Mailing address: Almirante Barraso 6, Casilla 14446 Correo 21, Santiago, Chile. T. (56 2) 698 0046. Fax 696 4798. Tx 294123 ILADES CL.

Migliaccio, Lucio, O.M.D., Fr
Rel: Roman Catholic; Ordained; celibate.
International Organizations: Procurator-General of the Clerks Regular of the Mother of God.
Mailing address: Piazza S Maria in Campitelli 9, I-00186 Roma, Italy. T. (39 6) 654 39 78, (39 6) 686 81 35

Miguel, María Pilar, CM, Sr
Rel: Roman Catholic; Ordained; celibate.
International Organizations: Superior General of the Carmelite Missionaries.
Mailing address: Via del Casaletto 115, I-00151 Roma, Italy. T. (39 6) 53 54 72.

Mihing, E, Rev
Rel: Evangelical.
International Organizations: General Secretary of the Kalimantan Evangelical Church.
Mailing address: Jalan Jenderal Sudirman 8, PO Box 86, Banjarmasin, Kalimantan Selatan, Indonesia.

Miki, Takahito
Rel: Church of Perfect Liberty; b. 02 DEC 1957; Citizen of Japan.
International Organizations: Chairman of the Church of Perfect Liberty.
Educ.: University of Tokyo, Japan.
Mailing address: 2172-1 Shindo, Tondabayashi, Osaka 584, Japan. T. (81 721) 24 1111

Miles, Frank
Rel: Quaker.
International Organizations: General Secretary of the Canadian Yearly Meeting of the Religious Society of Friends.
Mailing address: 60 Lowther Ave, Toronto M5R 1C7, Ontario, Canada.

Milingo, Emmanuel, Most Rev

Rel: Roman Catholic; Ordained on 31 AUG 1951; b. 13 JUN 1930 in Mnukwa; celibate; Citizen of Zambia.

International Organizations: Special Delegate (since 1983) of the Pontifical Council for the Pastoral Care of Migrants and Itinerant People.

Other organizations: Archbishop Emeritus (since 1983), Lusaka.

Career: Curate, Minga Mission from 1958 to 1961. Studied Pastoral Sociology (Diploma), Rome from 1961 to 1962. Studied at University College, Dublin (Dip. Ed.) from 1962 to 1963. Parish Priest, St Ann's Mission, Chipata from 1963 to 1966. Secretary for Mass Media, Zambia Episcopal Conference from 1966 to 1969. Archbishop, Lusaka from 1969 to 1983. **Interests:** writing, preaching. **Pubs.:** My Prayers are not heard; Precautions in the Ministry of Deliverance and Against Satan.

Mailing address: Piazza San Calisto 16, I-00153 Roma, Italy. T. (39 6) 6987275 – 698 7193. Fax 698 7111. Tx 0504 2028.

Miller, Donald E, Dr

Rel: Brethren; Ordained in 1958; b. 02 DEC 1929 in Dayton OH, USA; m. to Phyllis Louise Gibbel; c. 2 boys and 1 girl.

International Organizations: Secretary-General (since 1986) of the Church of the Brethren.

Career: Directed material aid for Brethren Service Commission in Europe from 1952 to 1954. Teacher of Social Studies at Madison Township High School, Trotwood OH from 1954 to 1955. Social worker in Dayton OH from 1954 to 1956. Teacher at Gregory School, Chicago IL from 1957 to 1958. Interim pastor, Salem Church of the Brethren, Dayton OH in 1959. Graduate Assistant at Harvard University, Cambridge MA, Bethany Theological Seminary, Oak Brook IL from 1960 to 1961. Associate Professor of Christian Education and Ethics, Bethany Theological Seminary, Oak Brook IL from 1961 to 1970. Professor of Christian Education and Ethics, Bethany Theological Seminary, Oak Brook IL from 1970 to 1982. Director of Graduate Studies, Bethany Theological Seminary, Oak Brook IL from 1973 to 1986. Brightbill Professor of Ministry Studies, Bethany Theological Seminary, Oak Brook IL from 1982 to 1986. Director of self-study for Chicago Cluster of Theological Schools from 1980 to 1981. Helped design and carry out Education for a Shared Ministry program from 1976 to 1986. Faith and Order Commission, National Council of Churches from 1976 to 1981. Executive Coordinating Committee, National Council of Churches in 1990. **Educ.:** Attended from Manchester College, North Manchester IN, USA (1947-1949); M.A. from University of Chicago, IL, USA (1952); Graduate study from United Theological Seminary, Dayton OH, USA (1955-1956); M.Div. from Bethany Theological Seminary, Oak Brook IL, USA (1958); Ph.D. from Harvard University, Cambridge MA, USA (1962); Postdoctoral study from Yale University, New Haven CT, USA (1968-1969); Postdoctoral study from University of Cambridge, UK (1975-1976). **Hons.:** Fellow of American Association of Theological Schools (1968-69), and Summer Workshop of the Case Study Institute (1972). **Mem.:** Association of Professors and Researchers in Religious Education (President, 1968); Association for Professional Education for the Ministry (President, 1976); American Theological Society; American Society of Christian Ethics; Professor and Research, National Council of Churches; Religious Education Association **Pubs.:** Author of numerous articles, contributing author to 4 books, and author of more than 10 books including: *Story and Context: An Introduction to Christian Education*, Abingdon, 1986; *The Gospel and Mother Goose*, Brethren Press, 1987 and Editor of *Yearbook of Association for Professional Education for the Ministry*, 1972.

Mailing address: 1451 Dundee Avenue, Elgin IL 60120, USA. T. (1 312) 742 5100.

Miller, John Harold, Rev

Rel: Roman Catholic; Ordained on 22 DEC 1951; b. 29 DEC 1925 in New Orleans LA, USA; celibate; Citizen of USA.

International Organizations: Editor of *Social Justice Review* (since 1986) of the Catholic Central Union of America.

Other organizations: Priest (since 1943), Congregation of the Holy Cross.

Main field of work: Ministry, university education. **Career:** Professor of Theology, Holy Cross College, Washington DC from 1955 to 1960. Lecturer in Liturgy, Catholic University of America, Washington DC from 1957 to 1960. Assistant Professor of Theology, Notre Dame University, IN from 1960 to 1963. Staff Editor, *New Catholic Encyclopedia*, Washington DC from 1963 to 1965. Associate Professor of Theology, Notre Dame University, IN from 1965 to 1967. Professor of Theology, Loyola University, New Orleans LA from 1967 to 1975. **Educ.:** Ph.B. from Pontifical University San Tommaso d'Aquino, Vatican City State (1947-1948); S.T.B., S.T.L. from Pontifical University San Tommaso d'Aquino, Vatican City State (1948-1952); S.T.D. from University of Trier, Germany (1953-1955). **Interests:** liturgy. **Pubs.:** Signs of Transformation in Christ, Prentice-Hall, 1963; Called by Love, Daughters of St Paul, 1990 and Love Responds, Central Bureau, CCUA, 1990.

Mailing address: 3835 Westminster Place, St Louis MO 63108, USA. T. (1 314) 371 1653

Residence: See Mailing Address. T. See Mailing Address

Miller, John Michael, Rev

Rel: Roman Catholic; Ordained on 29 JUN 1975; b. 09 JUL 1946 in Ottawa ON, Canada; celibate; Citizen of Canada; Lang. English, Spanish, Italian, French, German (spoken), Latin (written) and Greek (written).

International Organizations: Professor of Theology (since 1979) of the University of St Thomas, Houston TX.

Other organizations: Vice-President for Academic Affairs (since 1990), University of St Thomas.

Main field of work: University education. **Career:** Educational Advisor to Task Force on the Future of Basilian Colleges and Universities from 1983 to 1985. Educational Advisor to Standing Committee on Colleges and Universities from 1984 to 1986. Educational Advisor to Casa San Basilio from 1987 to 1988. **Educ.:** B.A. (Hons) from University of Toronto, Canada (1964-1969); M.A. from University of Wisconsin – Madison, USA (1969-1970); M.Div. from University of Toronto, St Michael's College, Canada (1971-1974); S.T.L. from Pontifical Gregorian University, Vatican City State (1974-1976); S.T.D. from Pontifical Gregorian University, Vatican City State (1976-1979). **Mem.:** Catholic Theological Society of America **Pubs.:** Divine Right of the Papacy in Recent Ecumenical Theology; What Are They Saying about Papal Primacy? and Paul VI: Vision and Theology of the Ministry of Peter for the Church (in preparation).

Mailing address: University of St Thomas, 3800 Montrose Blvd, Houston TX 77006, USA. T. (1 713) 522 7911

Residence: 4019 Yoakum Blvd, Houston TX 77006, USA. T. (1 713) 522 8515

Miller, Larry

Rel: Mennonite.

International Organizations: Executive Secretary of the Mennonite World Conference.

Mailing address: 7 ave de la Forêt-Noire, F-67000 Strasbourg, France. T. (33) 88 61 49 27. Tx 889000. Fax (33) 88 61 57 17.

Miller, Marlin E

Rel: Mennonite.

International Organizations: President of the Associated Mennonite Biblical Seminaries.

Mailing address: 3003 Benham Ave, Elkhart IN 46517-1999, USA.

Miller, Ray E

Rel: Protestant.

International Organizations: President of the Fellowship of Companies for Christ International.

Mailing address: 2920 Brandywine Rd, Suite 150, Atlanta GA 30341, USA. T. (1 404) 457 9700.

Milligan, William John, Canon

Rel: Anglican; b. 11 JUN 1928 in Lytham, Lancashire, UK; m. to Constance Evelyn Jessamine; c. 3 girls; Citizen of United Kingdom; Lang. English and French.

International Organizations: Associate Staff Member (since 1990) of the European Ecumenical Commission for Church and Society.

Other organizations: Chaplain (since 1986), Anglican Diocese in Europe; Anglican Representative at the European Institutions (since 1990), Church of England.

Main field of work: Information, study, contacts concerning the European Institutions for the Church of England. **Career:** Parish Priest, Church of England from 1955 to 1979. Cathedral Canon, Church of England from 1979 to 1986. Executive Secretary, European Ecumenical Commission for Church and Society from 1986 to 1990. **Educ.:** B.A. from University of Cambridge, Emmanuel College, UK (1946-1950); General Ordination from Cuddesdon College, Oxford, UK (1952-1954). **Pubs.:** *The New Nomads*, World Council of Churches, Risk Books, 1984.

Mailing address: 4 rue Stoeber, F-67000 Strasbourg, France. T. (33) 88 35 03 40

Residence: See Mailing Address. T. See Mailing Address

Mills, Howard McIlroy, Rev

Rel: United Church of Canada; Ordained in 1959; b. 20 OCT 1935 in Toronto, Ontario, Canada; m. to Virginia Lee Epes; c. 1 boy and 2 girls; Citizen of Canada.

International Organizations: General Secretary of General Council (since 1987) of the United Church of Canada.

Other organizations: President (since 1990), Churches' Council on Theological Education.

Career: Pastor, Lac Jeannine United Church, Ganonville, Quebec from 1960 to 1963. University Chaplain, Mount Allison University from 1965 to 1968. Doctoral studies and part-time teaching from 1963 to 1965. Doctoral studies and part-time teaching from 1968 to 1969. Associate Professor of Christian Ethics, St Stephen's College, Edmonton from 1969 to 1972. Pastor, St James-Bond United Church, Toronto from 1972 to 1974. General Secretary, Division of Ministry Personnel and Education, United Church of Canada from 1974 to 1982. President, United Theological Seminary of the Twin Cities, MN from 1983 to 1987. Visiting Fellow, Institute of Development Studies, University of Sussex in 1981. Official Adviser on Education to the 6th Assembly, World Council of Churches (WCC) in 1983. Moderator, Program on Theological Education, WCC from 1980 to 1983. Member, PTE Commission from 1977 to 1991. International Planning Team, 1st, 2nd and 3rd International Conference on 1981 UN Declaration on Religious Tolerance, MN (1986), Warsaw (1989), and New Delhi (1991). Registered Observer, 7th Assembly, WCC, Canberra, Australia in 1991. **Educ.:** B.A. from University of Toronto, Victoria University, Canada (1956); D.D. from University of Toronto, Victoria University, Canada (1988); M.D. from University of Toronto, Emmanuel College, Canada (1959); S.T.M. from Union Theological Seminary, New York NY, USA (1960); Ph.D. from Union Theological Seminary, New York NY, USA (1970); D.D. from United Theological Seminary of the Twin Cities, MN, USA (1987). **Interests:** golf, bridge, skiing, handiwork. **Mem.:** Canadian Theological Society; Canadian Association of Pastoral Education; Society of Christian Ethics; Canadian Diabetes Association; Club: Newmarket Golf and Country **Pubs.:** Co-author, College Education in the North West Territories, 1976.

Mailing address: 85 St Clair Ave E, Toronto, Ontario M4T 1M8, Canada.

Residence: 163 Lascelles Blvd., Toronto, Ontario M5P 2E7, Canada.

Milo, Lene, Rev

Rel: Methodist.

International Organizations: President of the Methodist Church in Samoa.

Mailing address: PO Box 199, Apia, Samoa.

Minami, Yoshinobu

Rel: Unknown.

International Organizations: Secretary-General of the Niwano Peace Foundation.

Mailing address: Shamvilla Catherina 5F, 1-16-9 Shinjuku, Shinjuku-ku, Tokyo 160, Japan. T. (81 3) 3226 4371. Fax 3226 1835.

Minari, Jackie

Rel: Buddhist: Mahayana.

International Organizations: Member of the Foundation for the Preservation of the Mahayana Tradition.

Other organizations: Director, Do Ngak Sung Juk Center.

Mailing address: Kyonan Manshion 212, 4-15-3 Kyonan-Cho, Musashino-Shi, Tokyo 180, Japan. T. (81 4) 2233 9405

Ming, Edith

Rel: Methodist.

International Organizations: President of the World Federation of Methodist Women.

Mailing address: 7100 Grey Oaks Drive, New Orleans LA 70126, USA. T. (1 504) 246 6562.

Mirejovsky, Lubomir, Rev Dr

Rel: Christian; b. in Teplice, Czechoslovakia; Citizen of Czech Republic.

International Organizations: Secretary-General (since 1979) of the Christian Peace Conference.

Main field of work: Clergyman. **Career:** Scholarship for study in the USA from 1945 to 1947. Parson in Karlove Vary from 1947 to 1949. Vicar and Superintendent, Church of Bohemian Brothers in Tabor from 1952 to 1965. Chairman, Middle European Division, Christian Peace Conference from 1965 to 1979. **Educ.:** Charles University, Prague, Czechoslovakia (1945-1947).

Mailing address: Jungmannova 9, PO Box 192, 111 21 Praha 1, Czech Rep. T. (42 2) 236 02 89 – (42 2) 236 02 90. C. EKUMRADA Praha. Tx 12 33 63 cpc c.

Mirghani, Osman al

Rel: Muslim; b. in Sudan.

International Organizations: Religious Leader of the Union Democratic Party.

Career: Religious Head, Khatmiya Sect in Sudan from 1949 to 1968.

Mailing address: Khartoum, Sudan.

Miri, Everino, S.F., Fr

Rel: Roman Catholic; Ordained; celibate.

International Organizations: Procurator-General of the Sons of the Holy Family.

Mailing address: Via Giovanni Giolitti 154, I-00185 Roma, Italy. T. (39 6) 73 13 362.

Misijuk, Wzodzimerz

Rel: Christian.

International Organizations: Representative for Eastern Europe Project of the World Student Christian Federation.

Mailing address: Ul Antoniuk Fabryczny 13, PL-15 762 Bialystok, Poland. T. (48 85) 516320 – 51236. Fax 512514.

Mitchell, Glenford E

Rel: Bahá'í.

International Organizations: Member, Universal House of Justice of the Bahá'í International Community.

Mailing address: PO Box 155, 31001 Haifa, Israel. T. (972 4) 372 433 – 510 344. C. BAHAIFAITH HAIFA. Tx 46626 BAYT IL. Fax (972 4) 372 425.

Mitchell, Susan Jones

Rel: Spiritual Healer; b. 24 FEB 1951 in Oakland CA, USA; m. to Paul David Mitchell; c. 1 boy and 1 girl; Citizen of USA; Lang. English.

International Organizations: Executive Director (since 1988) of the Reiki Alliance.

Other organizations: Teacher (since 1985), Reiki, the Usui system of natural healing.

Career: Teacher, College of Marin from 1977 to 1980. Teacher, Shoshone School District from 1980 to 1984. **Educ.:** B.A. from University of San Francisco, CA, USA (1968-1972); Schiller University, Heidelberg, Germany (1969-1970); CA State Teaching Credential from University of California, Berkeley, USA (1976-1977). **Interests:** Gardening, skiing. **Hons.:** Friedrich Schiller Award for Academic Excellence. **Mem.:** Reiki Outreach International, World Wildlife Fund.

Mailing address: E 33135 Canyon Road, PO Box 41, Cataldo ID 83810-1041, USA. T. (1 208) 682 3535. Fax 682 4848.

Mitton, Harold Logan, Rev

Rel: Baptist; Ordained; b. 20 DEC 1919 in Moncton, New Brunswick, Canada; m. to Marguerite McKay; c. 2 boys and 2 girls; Citizen of Canada.

International Organizations: Professor Emeritus and Director of Supervised Field Education (since 1986) of the Acadia Divinity College, Wolfville NS.

Career: Principal and Dean, Acadia Divinity College from 1975 to 1986. President, Atlantic Un. Baptist Convention in 1962. 1st Vice-President, Baptist Federation of Canada in 1968. Appointed to Canadian Baptist Overseas Mission Board in 1971. **Educ.:** B.A. from Acadia University, Wolfville, Nova Scotia, Canada (1944); B.D. from Acadia University, Wolfville, Nova Scotia, Canada (1946); D.D. from Acadia University, Wolfville, Nova Scotia, Canada (1966). **Interests:** swimming. **Mem.:** Academy of Homiletics of US and Canada **Pubs.:** *Facing Today's World* and contributions to various denominational journals.

Mailing address: 18 Grandview Dr, Wolfville NS B0P 1X0, Canada.

Miyabara, Sunao

Rel: Buddhist.
International Organizations: Vice-President and Member of the Executive Council of the World Fellowship of Buddhists.
Mailing address: Royal Thai Consulate General, 1 North King Street, Suite 200, Honolulu HI 96817, USA.

Miyake, Toshio, Rev Dr

Rel: Konko-Kyo; b. 26 MAR 1903 in Wakayama City, Japan; m. to Tsuneko Miyake; c. 3 boys; Citizen of Japan; Lang. Japanese.
International Organizations: President of Asian Center and Honorary Chairman of Council of the World Federalist Movement.
President of the World Conference on Religion and Peace. Council Member of the International Association for Religious Freedom. Vice-President of the Association of World Citizens.
Other organizations: Chief Senior Minister (since 1927), Konko Church of Izuo; Chairman, International Committee, Temple of Understanding (TOU); President, Japanese-Korean Religious Association; Vice-President, United World Federalists of Japan.
Main field of work: Religion. **Educ.:** Ph.D. (Hons), D.D. (Hons). **Hons.:** Albert Schweitzer Award; Honorary Lecturer of Cornell University, Ithaca, New York, USA. **Pubs.:** *Nandemo no Negai* (in Japanese) and *Heiwa o Ikiru* (in Japanese).
Mailing address: 3-8-21 Sangenya Nishi, Taishyo-ku, Osaka 551, Japan. T. (81 6) 551 0035
Residence: See Mailing Address. T. See Mailing Address

Mizutani, Kosho

Rel: Buddhist; b. 01 APR 1928; Citizen of Japan.
International Organizations: President (Buddhism) of the Bukkyo University.
Educ.: Ryukoku University Graduate School, Kyoto, Japan.
Mailing address: 356 Hatsukashi Kamogawacho, Fushimi-ku, Kyoto 612, Japan. T. (81 75) 921 2459

Mkhatshwa, Smangaliso, Rev

Rel: Roman Catholic; Ordained on 26 JUN 1965; b. 26 JUN 1939 in Barberton, Transvaal, South Africa; celibate; Citizen of South Africa; Lang. English, Zulu, Sotho and Afrikaans.
International Organizations: Secretary-General (since 1988) of the Institute for Contextual Theology.
Main field of work: Theology research. **Career:** Secretary-General, SACB from 1980 to 1988. Patron, United Democratic Front in 1983. **Educ.:** B.Theol, M.Theol. from Catholic University of Louvain, Belgium (1971-1973). **Interests:** tennis, music, theatre. **Hons.:** Ph.D. h.c., Georgetown, USA (1988); D.Th. h.c., Tübingen, Germany (1989). **Pubs.:** articles on philosophy, theology and politics.
Mailing address: Institute for Contextual Theology, 185 Smit St, PO Box 32047, Braamfontein 2017, South Africa. T. (27) 339 2513/21/38
Residence: 1916 F, Soshanguve 00152, South Africa. T. (27 1214) 2120

Mkuu, Sebastian Kolowa, Rt Rev

Rel: Evangelical Lutheran.
International Organizations: Presiding Bishop of Evangelical Lutheran Church in Tanzania.
Mailing address: PO Box 10, Lushoto, Tanzania.

Moberg, David Oscar

Rel: Christian; b. 13 FEB 1922 in Montevideo MN, USA; m. to Helen H Heitzman; c. 3 boys and 1 girl; Citizen of USA.
International Organizations: Coordinator (since 1971) of the Association for the Development of Religious Information Systems (ADRIS).
Other organizations: Member, Advisory Board (since 1983), Ecumenical Ministry with Mature Adults; Associate Editor (since 1983), *Religious Research Review*; Resource Scholar (since 1985), Christianity Today Institute; Co-editor (since 1986), *Research in the Social Scientific Study of Religion*; Professor Emeritus (since 1991), Marquette University, Milwaukee WI.
Career: Associate Instructor, University of Washington, Seattle from 1948 to 1949. Faculty, Bethel College, St Paul MN from 1949 to 1968. Professor of Sociology, Bethel College from 1959 to 1968. Chairman, Dept. of Social Sciences, Bethel College from 1952 to 1968. Professor of Sociology, Marquette University, Milwaukee WI from 1968 to 1991. Chairman, Dept. of Sociology and Anthropology, Marquette University from 1968 to 1977. Consultant to National Liberty Foundation from 1970 to 1971. Consultant to National Interfaith Coalition on Aging from 1973 to 1975. Adjunct Professor, San Francisco Theological Seminary, CA from 1964 to 1973. Adjunct Professor, McCormick Theological Seminary from 1975 to 1978. Adjunct Professor, McCormick Theological Seminary from 1981 to 1982. Visiting Professor, University of Southern California in 1979. Visiting Professor, Princeton Theological Seminary, NJ in 1979. Visiting Professor, Southern Baptist Theological Seminary in 1982. Editor, *International Directory of Religious Information Systems* in 1971. Editor, *Spiritual Well-Being: Sociological Perspectives* in 1979. Editor, *Religious Research Review* from 1968 to 1972. Contributing Editor, *Religious Research Review* from 1973 to 1977. Editor, *Journal of American Scientific Affiliation* from 1962 to 1964. Editor, *Adris Newsletter* from 1971 to 1976. **Educ.:** A.A. from Bethel Jr. College, St Paul MN, USA (1942); A.B. from Seattle Pacific College, WA, USA (1947); M.A. from University of Washington, Seattle, USA (1949); Ph.D. from University of Minnesota, Minneapolis, USA (1952). **Mem.:** American Sociological Association; International Sociological Association (Sociology of Religion Research Commission 1972-); Wisconsin Sociological Association (President 1969-71); Midwest Sociological Association (Wisconsin Board of Directors 1971-73); Association of Humanist Sociology; Association for the Development of Religious Information Systems; Religious Research Association (Board of Directors 1959-61, 1968-72, President 1981-82, H Paul Douglass lecturer 1986); Association for the Sociology of Religion (Executive Council 1971-73, President 1976-77); Society for the Scientific Study of Religion (Executive Council 1971-74); Evangelicals for Social Action (Planning Commission 1973-75); Christian Sociological Society (Steering Commission 1973-81); Family Research Council (Associate 1985-88, Research Network 1989-); Psychologists Interested in Religious Issues (Professional Affiliate 1984-); Midwest Council for Social Research on Aging (Fellow 1961-64, 1987-); American Society on Aging; Forum on Religion and Aging; Fellow, American Scientific Affiliation (Publications Commission 1984-, Social Ethics Commission 1985-88); Fellow, Gerontological Society of America **Pubs.:** The Church as A Social Institution, 1962, 2nd Edn., 1984; The Great Reversal: Evangelism and Social Concern, 1972 and Wholistic Christianity, 1985.
Mailing address: Dept. of Social and Cultural Science, Marquette University, Milwaukee WI 53233, USA. T. (1 414) 288 6838
Residence: 2619 E Newberry Blvd, Milwaukee WI 53211, USA. T. (1 414) 964 3465

Mochida, Tetsuya

Rel: Christian.
International Organizations: President of the Asia Evangelistic Fellowship.
Mailing address: Shimochoja Machi Agaru, Muro hachi dori, Kamigyo KV, Kyoto 602, Japan. T. (81 45) 432 1025.

Molinaro, Vincenzo, O.M.D., Fr

Rel: Roman Catholic; Ordained; celibate.
International Organizations: Rector-General of the Clerks Regular of the Mother of God.
Mailing address: Piazza S Maria in Campitelli 9, I-00186 Roma, Italy. T. (39 6) 654 39 78, (39 6) 686 81 35

Molineaux, David

Rel: Roman Catholic.
International Organizations: Secretary-General of the International Documentation and Communication Centre.
Mailing address: IDOC International, Via S Maria dell'Anima 30, Box GE02, I-00186 Roma, Italy. T. (39 6) 68 68 332. Tx 933524 GEONET G.

Monachino, Vincenzo, Fr

Rel: Roman Catholic.
International Organizations: President of the Ecclesiastic Archivistics Association.
Mailing address: Piazza San Calisto 16, I-00153 Roma, Italy. T. (39 6) 67015228.

Mondal, Barnabas Dwijen, Rt Rev

Rel: Christian.
International Organizations: Bishop of Dhaka.
Mailing address: St Thomas Church, 54 Johnson Road, Dhaka I, Bangladesh.

Mondin, Battista, Prof

Rel: Roman Catholic; Ordained on 29 MAR 1952; b. 29 JUL 1926 in Monte di Malo, Italy; celibate; Citizen of Italy; Lang. Italian, French, English, Spanish and German.

International Organizations: Dean, Faculty of Philosophy (since 1978) of the Urbanian Pontifical University, Vatican.
President of the Société internationale Thomas d'Aquin.
Other organizations: Full Professor (since 1972), Urbanian Pontifical University; President, Associazione Docenti Italiani di Filosofia (ADIF).
Main field of work: University education. **Career:** Assistant Professor in philosophical anthropology, Urbanian Pontifical University from 1967 to 1972. Director, Istituto Superiore per lo Studio dell'Ateismo in 1977. Dean, Faculty of Philosophy, Urbanian Pontifical University in 1978. Vice-Rector, Urbanian Pontifical University in 1988. **Educ.:** M.A. from Boston College, Chestnut Hill MA, USA (1954); Ph.D. from Harvard University, Cambridge MA, USA (1959); Libera Docenza from Catholic University of the Sacred Hearth, Milan, Italy (1966). **Mem.:** Pontificia Accademia di San Tommaso d'Aquino **Pubs.:** Author of over 60 books, including: *The Principle of Analogy in Protestant and Catholic Theology*, Nijhoff, The Hague, 2nd edn, 1967; *Nuova Teologia Cattolica*, Logos, Rome, 1978 and *Il pensiero filosofico di San Tommaso*, Massimo, Milan, 1985.
Mailing address: Pontificia Università Urbaniana, Via Urbano VIII 16, I-00165 Roma, Italy.
Residence: Via Aurelia 287, I-00165 Roma, Italy T. (39 6) 638 08 07

Monet, Jacques, S.J., Fr

Rel: Roman Catholic; Ordained; celibate.

International Organizations: Professor of the Regis College, Toronto.
Member of the Society of Jesus.
Mailing address: Regis College, 15 St Mary, Toronto, Ontario M4Y 2R5, Canada.

Monfette, Gérard, C.F.S., Fr

Rel: Roman Catholic; Ordained; celibate.

International Organizations: Superior General of the Congregation of the Sacerdotal Fraternity.
Mailing address: 500 Ave Claremont, Montréal H3Y 2N5, Canada. T. (1 514) 483 2030.

Monica, M, Sr

Rel: Roman Catholic; Ordained; celibate.

International Organizations: Superior General of the Sisters of St Joseph of Annecy.
Mailing address: 10 place au Bois, F-74000 Annecy, France. T. (33) 50 52 74 99.

Montacute, Paul

Rel: Baptist; Unordained; b. 10 JUL 1946 in Chippenham, UK; m. to Judith Lynne; c. 2 boys.

International Organizations: Director, Division of Baptist World Aid and BWA Youth Department (since 1990) of the Baptist World Alliance.
Director of the Baptist World Aid.
Other organizations: Director, CARE, USA.
Career: Chairman, Baptist World Alliance (BWA) Youth Committee from 1985 to 1988. Member, BWA General Council, Executive Committee, the Commission on the Ministry of the Laity, the Christian Education work group, and the Promotion and Development Commission from 1985 to 1990. Member, European Baptist Federation (EBF) Congress Resolutions Committee in 1989. Member, EBF Youth Committee from 1983 to 1989. Ecumenical Youth Ministry, Yate New Town in 1979. National Youth Officer, The Baptist Union of Great Britain from 1982 to 1990. Office holder in the Boy's Brigade (working with churches, training leaders and young people), Wales from 1972 to 1976. Office holder in the Boy's Brigade, Canada from 1976 to 1977. Office holder in the Boy's Brigade, Wales from 1977 to 1979. Inland Revenue – Tax Collector from 1962 to 1970. **Educ.:** Cert. in Youth and Community Work from Leicester College of Education, UK (1970-1972). **Interests:** international and development issues. **Mem.:** Church of England Children's Society Adoption and Fostering panel; Church Action with the Unemployed charity (founding member); Vienna Baptist Church, Vienna VA **Pubs.:** Contributions to: *Sixty Second Thoughts – epilogues for young people*, 1985; *When You Pray with Young People*, 1988 and articles and features for the *Baptist Times*.
Mailing address: 6733 Curran Street, McLean VA 22101, USA. T. (1 703) 790 8980. Fax 893 5160.

Montalvo, Gabriel, H E Mgr

Rel: Roman Catholic; Ordained on 18 JAN 1953; b. 27 JAN 1930 in Bogota, Colombia; celibate.

International Organizations: Apostolic Pro-Nuncio to Yugoslavia (since 1986) of the Roman Catholic Church.
Other organizations: Titular Archbishop, Celene.
Mailing address: Svetog Save 24, 11000 Beograd, Yugoslavia. T. (38 11) 432 822

Montazeri, Hussein Ali, Ayatollah

Rel: Muslim; b. in Najafabad, Isfahan, Iran Islamic Rep; Citizen of Iran.

International Organizations: Religious leader of Islam.
Career: Teacher of Science and Philosophy at Theological School, Qom. Arrested after riots over Shah's land reform in 1963. Visited Ayatollah Khomeini in Iraq in 1964. Arrested several times and exiled to rural parts of Iran from 1964 to 1974. Imprisoned from 1974 to 1978. Leading Ayatollah, Teheran from 1979 to 1980. Returned to Qom in 1980. Named Grand Ayatollah in 1984. Resigned as successor to Ayatollah Khomeini in 1989. Reported under arrest.
Mailing address: Madresseh Faizieh, Qom, Iran Islamic Rep.

Montcalm, Mary Beth, Sr

Rel: Roman Catholic; Ordained; b. 11 APR 1946 in Cochrane ON, Canada; celibate; Citizen of Canada.

International Organizations: General Superior (since 1990) of the Sisters of St Joseph of Toronto.
Other organizations: Member (since 1965), Sisters of St Joseph of Toronto.
Career: Assistant Professor of Political Studies, University of Manitoba in 1983. Associate Professor, University of Manitoba in 1989. Director, Fontbonne Health System (St Michael's Hospital and St Joseph's Health Centre). **Educ.:** B.A. from York University, Downsview, Ontario, Canada (1972); M.A. from Carleton University, Ottawa, Canada (1976); Ph.D. from Carleton University, Ottawa, Canada (1983). **Interests:** jogging, gardening. **Hons.:** Recipient Dr and Mrs H H Saunderson Award for Excellence in Teaching, University of Manitoba (1989). **Pubs.:** Co-author, Beyond The Quiet Revolution, 1990 and author and co-author of several articles on Quebec and Canadian politics.
Mailing address: 3377 Bayview Ave, Willowdale ON M2M 3S4, Canada.
Residence: 18 Sawley Dr, Willodale ON M2K 2J1, Canada.

Montefiore, Hugh William, Rt Rev Dr

Rel: Church of England; Ordained in 1950; b. 12 MAY 1920 in London, UK; m. to Elisabeth Mary Macdonald Paton; c. 3 girls; Citizen of United Kingdom; Lang. English and French.

International Organizations: Assistant Bishop (since 1987) of Diocese of Southwark.
Main field of work: University education and episcopal ministry. **Career:** Served during the war from 1940 to 1945. Ordained Deacon in 1949. Curate, St George's, Jesmond, Newcastle from 1949 to 1951. Chaplain and Tutor, Westcott House, Cambridge from 1951 to 1953. Vice-Principal, Westcott House, Cambridge from 1953 to 1954. Examining Chaplain to the Bishop, Newcastle from 1953 to 1970. Examining Chaplain to the Bishop, Worcester from 1957 to 1960. Examining Chaplain to the Bishop, Coventry from 1957 to 1970. Examining Chaplain to the Bishop, Blackburn from 1966 to 1970. Fellow and Dean, Gonville and Caius College from 1954 to 1963. Lecturer in New Testament, University of Cambridge from 1959 to 1963. Vicar, Great Saint Mary's, Cambridge from 1963 to 1970. Canon Theologian, Coventry from 1959 to 1970. Honorary Canon, Ely from 1969 to 1970. Bishop Suffragan, Kingston-upon-Thames from 1970 to 1978. Bishop, Birmingham from 1978 to 1987. Member, Archbishops' Commission on Christian Doctrine from 1967 to 1976. Chairman, General Synod Board for Social Responsibility from 1983 to 1987. Chairman, Independent Commission on Transport in 1973. **Educ.:** M.A. from University of Oxford, St John's College, UK; M.A., B.D. from University of Cambridge, Westcott House, UK. **Interests:** walking, water colour painting. **Hons.:** Hon. D.D., University of Aberdeen (1976) and University of Edinburgh (1985); Hon. Fellow, St John's College, Oxford (1981). **Pubs.:** The Probability of God, 1985; So Near And Yet So Far, 1986 and Communicating the Gospel in a Scientific Age, 1988.
Mailing address: White Lodge, 23 Bellevue Road, Wandsworth Common, London SW17 7EB, UK. T. (44 81) 672 6697
Residence: See Mailing Address. T. See Mailing Address

Monteiro de Castro, Manuel M, H E Mgr

Rel: Roman Catholic; Ordained on 09 JUL 1961; b. 29 MAR 1938 in Santa Eufemia de Prazins, Guimares, Portugal; celibate; Citizen of Portugal; Lang. Portuguese, English, French, Italian, Spanish and German.

International Organizations: Apostolic Nuncio to El Salvador and Honduras (since 1990) of the Roman Catholic Church.

Other organizations: Titular Archbishop, Benevento.

Main field of work: Diplomacy. **Career:** Entered the Diplomatic Service, Holy See in 1967. Diplomatic missions include: Panama from 1967 to 1969. Guatemala from 1969 to 1971. Viet Nam and Cambodia from 1972 to 1975. Australia and Papua New Guinea from 1975 to 1978. Mexico from 1978 to 1981. Belgium from 1981 to 1985. Apostolic Pro-Nuncio to Trinidad-Tobago, Barbados, Grenada, St Lucia, Dominica, Jamaica, Bahamas, Belize. Apostolic Delegate in the Antilles. **Educ.:** D.C.L. from Pontifical Gregorian University, Vatican City State (1961-1967); Dip. Diplomatic Studies from Pontifical Academy for Diplomats, Vatican City State (1965-1967); Judge from Supreme Court of the Roman Rota (1969). **Interests:** Religion, law, social matters. **Hons.:** Great Cross, Ordem do Infante D. Henrique (Portugal); Commandeur, Ordre du Mérite (Luxembourg); Commandeur, Ordre du Mérite (Belgium).

Mailing address: 87 Avenida Norte y 7 Calle Poniente, Colonia Escalón, Apartado Postal 01-95, San Salvador, El Salvador. T. (503) 23 2454. Fax 23 7607

Residence: See Mailing Address. T. See Mailing Address

Montesi, Giovanni, S.X., Fr

Rel: Roman Catholic; Ordained; celibate.

International Organizations: Procurator-General of the Society of St Francis Xavier for the Foreign Missions.

Mailing address: Viale Vaticano 40, I-00165 Roma, Italy. T. (39 6) 638 46 09 – 638 16 93. Fax 326 65 71. Tx 5042018 GISA.

Montgomery, John Warwick, Dr

Rel: Lutheran; Ordained in 1958; b. 18 OCT 1931 in Warsaw NY, USA; m. to (1st) Joyce Ann Bailer, (2nd) Lanalee de Kant; c. 1 boy and 2 girls; Citizen of USA; Lang. English and French.

International Organizations: Principal Lecturer in Law (since 1991) of the Luton College of Higher Education, UK.

Other organizations: Member (since 1978), Virginia Bar; Member (since 1979), California Bar; Member (since 1985), District of Columbia Bar; Member (since 1990), Washington Bar; Member (since 1981), Bar of the Supreme Court of the United States; Member (since 1984), Bar of the Supreme Court of the United Kingdom.

Main field of work: University education. **Career:** Librarian, University of California Library, Berkeley from 1954 to 1955. Instructor in Biblical Hebrew, Hellenistic Greek and Medieval Latin, Wittenberg University and Hamma Divinity School, Springfield OH from 1956 to 1959. Head Librarian, Swift Library of Divinity and Philosophy and member, Federated Theological Faculty, University of Chicago, IL from 1959 to 1960. Associate Professor, Chairman, Dept. of History, Waterloo Lutheran (now Wilfrid Laurier) University, Waterloo, Ontario from 1960 to 1964. Professor of Church History and Director of Studies, Trinity Evangelical Divinity School, Deerfield IL from 1964 to 1974. Professor of Law and Theology, International School of Law, Washington DC from 1974 to 1975. Theological Consultant to Christian Legal Society from 1975 to 1976. Director of Studies, International Institute of Human Rights, Strasbourg from 1979 to 1981. Served on the Human Rights Committee, California Bar Association from 1980 to 1983. Dean, Director of Library, Professor of Jurisprudence, Simon Greenleaf School of Law, Anaheim CA from 1980 to 1987. Honorary Professor of Theology and Law, Faith Evangelical Lutheran Seminary, Tacoma WA from 1988 to 1991. Visiting Professor, Concordia Theological Seminary, Springfield IL from 1964 to 1967. Visiting Professor, DePaul University, Chicago IL from 1967 to 1970. General Editor, *Evangelical Perspectives* series from 1970 to 1973. **Educ.:** A.B. Phil. (Dist.) from Cornell University, Ithaca NY, USA (1952); B.L.S. from University of California, Berkeley, USA (1954); M.A. from University of California, Berkeley, USA (1958); B.D. from Wittenberg University, Springfield OH, USA (1958); M.S.T. from Wittenberg University, Springfield OH, USA (1960); Ph.D. from University of Chicago, IL, USA (1962); Doc. d'univ., mention Théol. Prot. from University of Strasbourg II, France (1964); LL.B. from LaSalle Extension University, USA (1977); Dip. (cum laude) from International Institute of Human Rights, Strasbourg, France (1978); M.Phil. Law from University of Essex, Colchester, UK (1983). **Interests:** Christian apologetics, law and theology, human rights, piano, travel, antique cars, book collecting. **Hons.:** Angel Award, National Religious Broadcasters (1989). **Mem.:** American Scientific Affiliation (Fellow); Victoria Institute (Fellow); Royal Society of Arts (UK); Folio Club, Wig and Pen Club, London; Presbyterian Historical Society, Belfast; Middle Temple and Lincoln's Inn (UK); International Bar Association; International Real Estate Federation; World Society of Law Professors; International Union of Lawyers; American Bar Association; American Association of Law Libraries; American Library Association; American Theological Library Association; American Historical Association; Medical Library Association; Evangelical Theological Society; Society for Reformation Research; Creation Research Society; Tolkien Society of America; Phi Beta Kappa; Académie de gastronomie Brillat-Savarin, Paris **Pubs.:** Author of more than 40 major theological works in English, French, Spanish and German, including: *Evidence for faith: How to decide the God question*, 1991; Author of over 100 scholarly journal articles and editor and translator of many books.

Mailing address: 2 rue de Rome, F-67000 Strasbourg, France; also: Flat 9, 4 Crane Court, London EC4, UK.

Monthienvichienchai, Raphael Chainarong

Rel: Roman Catholic.

International Organizations: President of the Unda, International Catholic Association for Radio and Television.

Member of the Pontifical Council for Social Communications.

Mailing address: St John's College-Ladprao, Bangkok 10900, Thailand.

Mooi, R J, Rev Dr

Rel: Dutch Reformed.

International Organizations: General Secretary of the Netherlands Reformed Church.

Mailing address: Overgoo 11, NL-2260 AK Leidschendam, Netherlands.

Moolaveetil, Louis, O.I.C., Fr

Rel: Roman Catholic; Ordained; celibate.

International Organizations: Superior General of the Order of the Imitation of Christ.

Mailing address: Bethany Ashram, Vadavathoor PB 6, Kottayam 686010 Kerala, India. T. (91 481) 84 94

Moon, Sun Myung, Rev

Rel: Holy Spirit Association for the Unification of World Christianity; Citizen of Korea.

International Organizations: Founder (since 1954) of the Holy Spirit Association for the Unification of World Christianity.

Mailing address: 4 West 43rd Street, New York NY 10036, USA. T. (1 212) 997 0050.

Moons, Hubert, Rev

Rel: Roman Catholic; Ordained; celibate.

International Organizations: Member of the Order Secular of Servants of Mary.

Mailing address: Servi di Maria, Piazza S Marcello, al Corso 5, I-00187 Roma, Italy. T. (39 6) 678 0888 – 78 9995.

Moore, James Harry Manson

Rel: Deist; Unordained; b. 16 DEC 1929 in Saltash, Cornwall, UK; single; Citizen of United Kingdom; Lang. English and German.

International Organizations: Member – Group Leader and Apologist (since 1956) of the Gurdjieff Society.

Main field of work: Writing. **Interests:** life and teaching of George Ivanovitch Gurdjieff. **Mem.:** Fellow: Royal Society of Arts (1963) and Royal Asiatic Society (1974) **Pubs.:** *Gurdjieff and Mansfield*, RKP Ltd., 1980; *Gurdjieff: The Anatomy of A Myth*, Element Books, 1991 and "Gurdjieff: The Man and the Literature", in *Resurgence*, 1983.

Mailing address: c/o Vanessa Holt Associates Ltd, 59 Crescent Road, Leigh-on-Sea, Essex, UK.

Morales, Elfo Eloy

Rel: Roman Catholic.

International Organizations: Regional Secretary for Latin America of the International Council of Catholic Men.

Mailing address: Avda de Mayo 621, Piso 1, 1084 Buenos Aires, Argentina.

Morand, Christiane Blandine, Sr

Rel: Roman Catholic; Ordained; b. 17 MAR 1932 in Charvonnex-Annecy, France; celibate; Citizen of France; Lang. French and English.
International Organizations: Superior General of the Sisters of the Cross of Chavanod.
Main field of work: Education and management. **Career:** Professor, Ecole Secondaire et Technique from 1956 to 1969. Director, Ecole Secondaire et Technique from 1969 to 1977. **Educ.:** Lic. ès Lettres, Lic. Phil. from University of Lyon III, France (1950-1955).
Mailing address: Maison Générale, Le Grand Pré, F-74650 Chavanod, France. T. (33) 50 69 10 58.
Residence: See Mailing Address. T. See Mailing Address

Morand, Paul

Rel: Catholic.
International Organizations: Administrator of the Conference of International Catholic Organizations.
Mailing address: Chemin de Chantebise 13, CH-1226 Thônex, Switzerland. T. (41 22) 48 52 90.

Morandini, Giovanni Battista, H E Mgr

Rel: Roman Catholic; Ordained on 22 JUL 1962; b. 30 JUN 1937 in Bienno, Brescia, Italy; celibate.
International Organizations: Apostolic Nuncio to Guatemala (since 1990) of the Roman Catholic Church.
Other organizations: Titular Archbishop, Numida.
Mailing address: 10 Calle 4/47, Zona 9, Guatemala, Guatemala. T. (502 2) 32 42 74. Fax 34 19 18

Morden, John Grant, Ven

Rel: Anglican; b. 17 AUG 1925 in London, UK; m. to Elizabeth Grace Tannahill; c. 1 boy and 2 girls; Citizen of Canada.
International Organizations: Archdeacon Emeritus (since 1990) of the Diocese of Huron.
Other organizations: Member, Sub-Committee on World Religions, General Synod (since 1980), Anglican Church of Canada.
Career: Archdeacon, Diocese of Huron from 1968 to 1990. Professor of Theology, Huron College from 1962 to 1990. Principal, Huron College from 1962 to 1984. Principal Emeritus, Huron College from 1984 to 1990. Member, Corporation and Executive Board, Huron College from 1961 to 1990. Member, General Synod, Anglican Church of Canada from 1961 to 1988. With Cassidy's Ltd. as Clerk from 1940 to 1943. Assistant Curate, Annunciation and All Soul's Parish, Toronto from 1951 to 1953. Assistant Curate, St Bartholomew's, White Plains, NY from 1953 to 1956. Rector, St Matthews, Toronto from 1956 to 1957. Registrar and Assistant Professor of Theology, Huron College in 1957. Professor of Theology and Vice Principal, Huron College in 1961. Served with RCAF (Royal Canadian Air Force) from 1943 to 1946. **Educ.:** B.A. from University of Toronto, Canada (1949); L.Th. from University of Toronto, Wycliffe College, Canada (1952); B.D. from University of Toronto, Wycliffe College, Canada (1953); D.D. from University of Toronto, Wycliffe College, Canada (1963); S.T.M. from Union Theological Seminary, New York NY, USA (1954); D.Th. from General Theological Seminary, New York NY, USA (1961); D.D. from University of Western Ontario, Huron College, London, Canada (1985); D.D. from University of Toronto, Trinity College, Canada (1986); D.D. from St John's College, Winnipeg, Canada (1989). **Mem.:** Corporation of Huron College (Lifetime Honorary Member)
Mailing address: 286 Steele St, London, Ontario N6A 2L1, Canada.

Moreau, O

Rel: Roman Catholic.
International Organizations: UNESCO Representative of the International Federation of Catholic Universities.
Mailing address: 7 bis Sentier des Basses Pointes, F-92100 Meudon, France.

Moreira Neves, Lucas M, O.P., H E Card

Rel: Roman Catholic; Ordained on 09 JUL 1950; b. 16 SEP 1925 in Sao Joao del Rei, Brazil; celibate; Citizen of Brazil.
International Organizations: Member (since 1988) of the College of Cardinals.
Member of the Order of Preachers.
Other organizations: Archbishop (since 1987), São Salvador da Bahia.
Career: Elected to the Titular Church, Feradi Maggiore in 1967. Consecrated bishop in 1967. Consecrated Archbishop in 1979. Transferred to the Titular Church, Vescovio in 1987. Created Cardinal by Pope John Paul II in 1988.
Mailing address: CP 1907, Av 7 de Setembro 309, Campo Grande, 40120 Salvador BA, Brazil. T. (55 71) 321 5572 – 321 5433

Moreni, Gian Vincenzo, H E Mgr

Rel: Roman Catholic; Ordained on 04 APR 1959; b. 29 JAN 1932 in Montichiari, Brescia, Italy; celibate; Citizen of Vatican and Italy; Lang. Italian, French, Spanish and English.
International Organizations: Apostolic Nuncio to the Philippines (since 1990) of the Roman Catholic Church.
Other organizations: Titular Archbishop, Torre di Mauritania.
Main field of work: Diplomacy. **Career:** Apostolic Pro-Nuncio to Tanzania from 1982 to 1990. **Educ.:** S.T.D. from Pontifical Gregorian University, Vatican City State (1957-1960); S.T.L. from Institute of Moral Theology (Alphonsianum), Vatican City State (1960-1962); J.C.D. from Lateran Pontifical University, Vatican City State (1962-1963).
Mailing address: Apostolic Nunciature, PO Box 3604, 1099 Manila, Philippines. T. (63 2) 521 03 06. Fax 521 12 35. Tx 65079 NUNCIO PN
Residence: 2140 Taft Ave, 1004 Manila, Philippines. T. See Mailing Address

Moreno, Pierino, C.R.S., Fr

Rel: Roman Catholic; Ordained; celibate.
International Organizations: Superior General of the Clerics Regular of Somasca.
Mailing address: Piazza Tempio di Diana 14, I-00153 Roma, Italy. T. (39 6) 574 25 92.

Moretti, Giovanni, H E Mgr

Rel: Roman Catholic; Ordained on 28 JUN 1947; b. 20 NOV 1923 in Meina, Italy; celibate.
International Organizations: Apostolic Nuncio to Belgium, Luxembourg and the European Communities (since 1989) of the Roman Catholic Church.
Other organizations: Titular Archbishop, Vartana.
Mailing address: Av des Franciscains 5-9, B-1150 Bruxelles, Belgium. T. (32 2) 762 20 05. Fax 762 20 32

Morgan, John Hanly, Rev

Rel: Unitarian; Ordained; b. 28 NOV 1918 in New Albany IN, USA; m. to Jeannette Mutzfeld; c. 3 boys and 1 girl.
International Organizations: Honorary President (since 1986) of the Canadian Peace Congress.
Career: Secretary, Unitarian Ministers Association from 1956 to 1960. Vice-President, Unitarian Fellowship for Social Justice from 1958 to 1959. Chairman, Toronto Committee for Survival from 1960 to 1964. Co-Chairman, Canadians for Middle East Understanding from 1968 to 1970. President, Canadian Peace Congress from 1972 to 1986. Vice-President, World Peace Council from 1983 to 1987. **Educ.:** B.A. from Ball State University, Muncie IN, USA (1943); B.D. from Harvard University, Divinity School, Cambridge MA, USA (1946); M.A. from University of Michigan, Ann Arbor, USA (1955); Ph.D. from University of South Florida, Sarasota, USA (1979). **Hons.:** International Lenin Peace Prize (1983); Joliot-Curie Gold Medal for Peace (1984). **Pubs.:** Poetry: *Kangaroo City*, 1980; Poetry and prose: *That We Shall Live*, 1985 and *Journeys from Breakfast*, 1987.
Mailing address: 895 Ellesmere Ave, Peterborough ON K9H 7A8, Canada.
Residence: See Mailing Address.

Morimoto, Kyojun

Rel: Buddhist; Ordained; b. 17 NOV 1902; Citizen of Japan.
International Organizations: Superior of the Toshodaiji.
Main field of work: Monk. **Career:** Manager, Ritsu-shu. **Educ.:** Otani University, Japan.
Mailing address: 1-1-14 Gojo, Gojo, Nara 637, Japan. T. (81 7472) 2 2630

Morris, George E

Rel: Methodist.
International Organizations: Director of the World Methodist Evangelism Institute.
Mailing address: Candler School of Theology, Emory University, Atlanta GA 30322, USA. T. (1 404) 727 6344.

Morris, Michael

Rel: Evangelical.
International Organizations: Director of the Evangelical Alliance.
Mailing address: Whitefield House, 186 Kennington Park Rd, London SE11 4BT, UK. T. (44 71) 582 02 28. Fax 582 62 21.

Morris, Ted, Col

Rel: Salvation Army, The; b. 10 SEP 1933; m. to Lt Louise Holmes.
International Organizations: Chief Secretary in Japan of The Salvation Army.
Career: Entered Training College from New Orleans, LA. Commissioned Salvation Army Officer in 1956. Odessa TX Assistant in 1956. 2nd Lieutenant in 1957. Borger TX Commanding Officer in 1957. 1st Lieutenant in 1959. Captain in 1959. Odessa TX Commanding Officer in 1960. Transferred to Overseas Department in 1961. Macon GA Commanding Officer in 1961. Transferred to Tokyo, Japan at the Territorial Commander's Office in 1961. After language study, appointed Territorial Music Director. Attended International College for Officers in 1966. Assistant Training Principal, Japan in 1967. Transferred to USA: Atlanta Temple Commanding Officer in 1971. Major in 1972. Divisional Secretary, Maryland and Northern West Virginia in 1972. Transferred to Japan: Special Efforts and Public Relations Secretary in 1974. Training Principal, Japan in 1975. Transferred to USA: Divisional Secretary KT Division in 1978. General Secretary, Texas Division in 1981. Assistant Divisional Commander, Texas Division in 1985. Transferred to Japan: Chief Secretary in 1986. Lieutenant-Colonel in 1986. Social Secretary, Japan in 1989. Relieved as Social Secretary in 1990. Colonel in 1990. **Educ.:** Grad. from Salisbury State College, MD, USA (1953); B.A. Phil. and Rel. from Asbury College, Wilmore KY, USA (1955). **Hons.:** Outstanding Young Man of the Year, Junior Chamber of Commerce; LSS (35 years) (1991).
Mailing address: 17 2-chome Kanda Jimbocho, Chiyodaku, Tokyo 101, Japan. T. (88 3) 3237 0881-5. Tx J26431 YULAW PAT

Morris, William James, Very Rev

Rel: Church of Scotland; Ordained in 1951; b. 22 AUG 1925 in Cardiff, Wales, UK; m. to Jean Daveena Ogilvy Howie; c. 1 boy; Citizen of United Kingdom; Lang. English and French.
International Organizations: Chaplain in Ordinary to HM The Queen in Scotland (since 1969) of Church of Scotland.
Other organizations: Minister (since 1967), Glasgow Cathedral; Justice of the Peace (since 1971), County of the City of Glasgow; Chaplain in The Priory of Scotland (since 1990), the Most Venerable Order of the Hospital of St John of Jerusalem; Dean (since 1991), Chapel Royal in Scotland; President (since 1991), Scottish Church Society; President (since 1992), Saint Andrew Society, Glasgow; Chairman, Iona Cathedral Trust; Force Chaplain, Strathclyde Police.
Main field of work: Ministry. **Career:** Minister, Presbyterian Church of Wales from 1951 to 1953. Minister (Church of Scotland), St David's, Buckhaven, Fife from 1953 to 1957. Minister (Church of Scotland), Peterhead Old Parish Church from 1957 to 1967. Chaplain, Peterhead Prison from 1963 to 1967. Justice of the Peace, County of Aberdeen from 1963 to 1967. Sub-Chaplain in the Priory of Scotland, the Most Venerable Order of the Hospital of St John of Jerusalem from 1978 to 1991. Member, Independent Broadcasting Authority representing Scottish interests from 1979 to 1984. Governor, Jordanhill College of Education in 1983. **Educ.:** B.A. from University of Wales, UK (1946); B.D. from University of Wales, UK (1949); Dr. Phil. from University of Edinburgh, UK (1954). **Interests:** swimming. **Hons.:** Hon. LL.D., Strathclyde (1974); Hon. D.D., Glasgow (1979); Hon. Fellow, Royal College of Physicians and Surgeons, Glasgow (1983); Hon. Chaplain, Royal Scottish Automobile Club and West of Scotland Engineers Association; Hon. President, Glasgow Society of Social Services. **Mem.:** Rotary Clubs of Denniston and Glasgow (Hon. Member) **Pubs.:** A Walk Through Glasgow Cathedral, 1986.
Mailing address: 94 St Andrews Drive, Glasgow G41 4RX, UK. T. (44 41) 427 2757
Residence: See Mailing Address. T. See Mailing Address

Morrisey, Francis G, O.M.I., Rev

Rel: Roman Catholic; Ordained in 1961; b. 13 FEB 1936 in Charlottetown, Prince Edward Island, Canada; celibate; Citizen of Canada.
International Organizations: Professor of Canon Law (since 1970) of the St Paul University.
Member of the Missionary Oblates of Mary Immaculate.
Other organizations: Editor (since 1967), *Studia Canonica*; Titular Professor (since 1977), St Paul University; Consultor to Pontifical Council for the Interpretation of Legislative Texts (since 1985).
Career: Secretary, Ecclesiastical Faculties, University of Ottawa from 1963 to 1965. Registrar, University of Ottawa from 1965 to 1971. Vice-Rector, St Paul University Seminary from 1971 to 1983. Lecturer, Faculty of Canon Law, St Paul University from 1970 to 1972. Associate Professor, St Paul University from 1972 to 1977. Dean of Canon Law, St Paul University from 1972 to 1984. Secretary-Treasurer, Canadian Canon Law Society from 1966 to 1973. President, Canadian Canon Law Society from 1973 to 1975. Executive Secretary, Canadian Canon Law Society from 1975 to 1977. Consultor to Canadian Canon Law Society from 1977 to 1983. International Lecturer on Canon Law and Canadian Religious History. **Educ.:** B.A., B.Ph. from University of Ottawa, Ontario, Canada (1955); Ph.L. from University of Ottawa, Ontario, Canada (1957); S.T.B. from University of Ottawa, Ontario, Canada (1960); S.T.L. from University of Ottawa, Ontario, Canada (1962); M.A. from University of Ottawa, Ontario, Canada (1963); J.C.B. from University of Ottawa, Ontario, Canada (1965); M.D.C. from University of Ottawa, Ontario, Canada (1966); Ph.D. Canon Law from University of Ottawa, Ontario, Canada (1971); J.C.D. from University of Ottawa, Saint Paul University, Ontario, Canada (1971). **Interests:** philately. **Mem.:** Canadian Canon Law Society (Honorary Life Member); Canon Law Society of Great Britain and Ireland, Australia and New Zealand, America (Honorary Life Member) **Pubs.:** Numerous articles in national and international journals.
Mailing address: 223 Main St, Ottawa ON K1S 1C4, Canada.
Residence: 175 Main St, Ottawa ON K1S 1C3, Canada.

Morrison, Alexander B

Rel: Church of Jesus Christ of Latter-Day Saints; b. 22 DEC 1930 in Edmonton, Alberta, Canada; m. to Shirley Brooks; c. 4 boys and 4 girls; Citizen of Canada.
International Organizations: General Authority (since 1987) of the Church of Jesus Christ of Latter-Day Saints.
Career: Chemist, Mead Johnson and Co., Evansville, IN from 1956 to 1959. Various technical and managerial positions, Dept of National Health and Welfare, Ottawa from 1959 to 1972. Assistant Deputy Minister in charge of Health Protection Br. from 1972 to 1984. Professor and Chairman, Dept. of Food Science, University of Guelph from 1984 to 1987. Frequent Consultant to World Health Organization on human health in developing countries, tropical diseases, nutrition, essential drugs. Former Director, Connaught Laboratories. Visiting Professor of Pharmacology, University of Toronto. Member, Technical and Executive Committees, Thrasher Research Fund, Salt Lake City, UT. **Educ.:** B.Sc. from University of Alberta, Edmonton, Canada (1951); M.Sc. from University of Alberta, Edmonton, Canada (1952); Ph.D. from Cornell University, Ithaca NY, USA (1956); M.S. from University of Michigan, Ann Arbor, USA (1966). **Interests:** family. **Hons.:** Recipient, Queen's Jubilee Medal; David M Kennedy International Service Award, Brigham Young University. **Mem.:** Royal Society of Medicine (Fellow); Chemical Institute of Canada (Fellow) **Pubs.:** Author of a number of technical reports and papers on nutrition, biochemistry, toxicology, public health.
Mailing address: 47 East South Temple, Salt Lake City UT 84150, USA.
Residence: 3852 South Monarch Dr, Bountiful UT 84010, USA.

Morrow, John Watson, Rev

Rel: Presbyterian; Ordained; b. 28 JUN 1931 in Belfast, N. Ireland, UK; m. to Shirley H E Duncan; c. 3 boys and 1 girl; Citizen of United Kingdom.
International Organizations: Leader (since 1980) of the Corrymeela Community (a community dedicated to reconciliation).
Career: Minister, Kilmatree Church, Dunmurry from 1960 to 1967. Chaplain to Overseas Students, Glasgow from 1967 to 1971. Chaplain to Presbyterian Students, Dublin from 1971 to 1975. Chaplain, Queen's University, Belfast from 1975 to 1980. **Interests:** reading, music, walking, ecology. **Pubs.:** The Captivity of the Irish Churches and Laity under Pressure.
Mailing address: Corrymeela House, 8 Upper Crescent, Belfast BT7 1NT, UK.

Mose David, M
See David Brandt Berg.

Moshe, Nancy
Rel: Unknown.
International Organizations: Representative of the International Coordinating Committee on Religion and the Earth.
Mailing address: Wainwright House, 260 Stuyvesant Av, Rye, New York NY 10580, USA. T. (1 914) 967 6080. Fax 967 6114.

Moss, Cath
Rel: Protestant.
International Organizations: Secretary-General of the Ecumenical Youth Council in Europe.
Mailing address: PO Box 464, Belfast BT4 2DE, Ireland. T. (44 232) 651134. Fax 651134.

Motoyama, Hiroshi, Dr
Rel: New Religion.
International Organizations: Founder-Director of the International Association for Religion and Parapsychology.
Mailing address: Inokashira 4-11-7, Mitaka-shi, Tokyo 181, Japan. T. (81 422) 48 3535. Fax 48 3548.

Motte, Maurice, Lt-Col
Rel: Salvation Army, The.
International Organizations: Officer Commanding of The Salvation Army.
Mailing address: Nouveau Marché aux Grains 34, B-1000 Bruxelles, Belgium.

Mottu, Henry, Prof
Rel: Protestant; Ordained in 1963; b. 23 AUG 1939 in Geneva, Switzerland; m. to Liliane Weber; c. 2 boys; Citizen of Switzerland; Lang. French, German and English.
International Organizations: Ordinary Professor of Practical Theology, Faculty of Theology (since 1988) of the University of Geneva.
Main field of work: University education. **Career:** Visiting Professor, Union Theological Seminary, New York NY, USA from 1970 to 1972. Pastor, National Protestant Church of Geneva (ENPG) from 1972 to 1988. **Educ.:** Lic. Theol. from University of Basel, Switzerland (1959-1963); Dr. Theol. from University of Geneva, Switzerland (1971). **Interests:** ecclesiology, ecumenism, sacraments, biblical sciences (apocalyptic). **Pubs.:** La manifestation de l'Esprit selon Joachim de Fiore, Delachaux, Paris-Neuchâtel, 1977; Les Confessions de Jérémie. Une protestation contre la souffrance, Labor et Fides, Genève, 1985 and Editor, La Réforme, un ferment dans l'Eglise universelle, Labor et Fides, Genève, 1987.
Mailing address: Faculté de Théologie, Université de Genève, 3 place de l'Université, CH-1211 Genève 4, Switzerland. T. (42 22) 705 74 75
Residence: 29 chemin des Pinsons, Thônex, CH-1226 Genève, Switzerland. T. (42 22) 48 19 10

Moubad, M
Rel: Zoroastrian; Citizen of Iran.
International Organizations: Leader of the Zoroastrians.
Mailing address: No mailing address.

Moule, Charles Francis Digby, Rev
Rel: Church of England; Ordained in 1934; b. 03 DEC 1908 in Hangchow, China; single; Citizen of United Kingdom.
International Organizations: Ecclesiastic of the Church of England.
Main field of work: University education and research. **Career:** Ordained Deacon in 1933. Tutor, Ridley Hall from 1933 to 1934. Curate, Church of England from 1934 to 1936. Vice-Principal, Ridley Hall (Church of England Seminary) from 1936 to 1944. Lecturer, University of Cambridge from 1944 to 1951. Professor, University of Cambridge from 1951 to 1976. **Educ.:** University of Cambridge, Emmanuel College, UK; Ridley Hall, Cambridge, UK. **Interests:** christology. **Hons.:** Commander of the Order of the British Empire; D.D. h.c., St Andrews University and Cambridge University; Fellow of the British Academy (awarded Burkitt Medal, 1970). **Mem.:** Studiorum Novi Testamenti Societas; Society of Biblical Literature **Pubs.:** *An Idiom Book of New Testament Greek*, 1953; *The Birth of the New Testament*, 1962, 1981 and *The Origin of Christology*, 1977.
Mailing address: 1 King's Houses, Pevensey BN24 5RJ, UK. T. (44 323) 762436

Mounayer, Eustache Joseph, H E Mgr
Rel: Syrian Catholic; Ordained in 1949; b. in Qatana, Damascus, Syria; celibate; Citizen of Syria.
International Organizations: Archbishop (since 1978) of Damascus.
Career: Sacred Archbishop in 1971.
Mailing address: PO Box 2129, Damascus, Syria.

Mourad, George, Rev
Rel: Reformed Church.
International Organizations: Member of the Executive Committee of the World Alliance of Reformed Churches.
Mailing address: Middle East Council of Churches, PO Box 4259, Limassol, Cyprus. T. (961 1) 411100 – 411179. Tx (VIA MECC – BEIRUT) OIK 22662 LE
Residence: PO Box 70890, Antelias, Lebanon T. (961 1) 402 565

Moya, Rafael, Fr
Rel: Roman Catholic.
International Organizations: Procurator-General of the Order of Preachers.
Mailing address: Piazza Pietro d'Illiria 1, I-00153 Roma, Italy. T. (39 6) 574 34 59 – 574 35 75.

Moyo, Ambrose, Dr
Rel: Evangelical Lutheran; Ordained in 1969; b. 22 JUL 1943 in Mberengwa, Zimbabwe; m. to Iris Themba; c. 3 children; Citizen of Zimbabwe.
International Organizations: Executive Secretary (since 1989) of the Conference of African Theological Institutions.
Other organizations: Lutheran Chaplain to the University of Zimbabwe (since 1981); Senior Lecturer (since 1988), University of Zimbabwe; Bishop's Deputy (since 1992), Evangelical Lutheran Church in Zimbabwe.
Career: Parish Pastor, Evangelical Lutheran Church in Zimbabwe from 1969 to 1971. Teaching Fellow at Harvard University, Cambridge MA from 1975 to 1976. Assistant Professor (part-time), Boston State University, MA from 1977 to 1978. Teaching Fellow at Harvard University, Cambridge MA from 1977 to 1978. Lecturer, University of Botswana from 1979 to 1981. Lecturer, University of Zimbabwe from 1981 to 1987. Chairman, Dept. of Religious Studies, Classics and Philosophy, University of Zimbabwe from 1985 to 1990. Deputy Dean, Faculty of Arts, University of Zimbabwe from 1986 to 1987. Member, Church Council, Evangelical Lutheran Church in Zimbabwe from 1982 to 1986. Chairman, Conference of African Theological Institutions from 1986 to 1989. Chairman, Lutheran World Federation Advisory Committee on Theological Education in Africa, and Director of LWF sponsored Institutes on Church Leadership Development in Africa from 1987 to 1991. Chairman, Association of Theological Institutions in Central and Southern Africa from 1985 to 1990. **Educ.:** Dip. Theol. from Umpumolo Theological College, Zimbabwe (1968); B.A. Theol. from University of South Africa, Pretoria, South Africa (1971); M.Th. from Harvard University, Cambridge MA, USA (1973); M.A. from Harvard University, Cambridge MA, USA (1975); Ph.D. from Harvard University, Cambridge MA, USA (1978). **Interests:** New Testament and Christian origins. **Mem.:** Society of Biblical Literature; International Association of Mission Studies; Ecumenical Association of Third World Theologians (EATWOT); Ecumenical Association of African Theologians **Pubs.:** Co-editor (with A Pero), *Theology and the Black Experience*, Augsburg, 1988; Co-editor (with CF Hallencreutz), *Church and State in Zimbabwe*, Mambo Press, 1988; *Essays on African Theology: Towards a Meaningful Christian Encounter with African Traditional Religions in Zimbabwe*, 1991 and several articles and conference papers.
Mailing address: PO Box MP 36, Mount Pleasant, Harare, Zimbabwe. T. (263 4) 30 32 11 EXT 1248
Residence: 20 Dublin Road, Emerald Hill, Harare, Zimbabwe. T. (263 4) 39 157

Moyse, François
Rel: Jewish; b. in Luxembourg, Luxembourg; single; Citizen of Luxembourg; Lang. French, English, German and Luxemburgese.
International Organizations: President (since 1991) of the European Union of Jewish Students.
Member of the Praesidium (since 1989) of the European Coordination Bureau of International Youth Organizations.
Other organizations: Barrister, Luxembourg Bar.

Main field of work: Youth politics. **Career:** Member of the Praesidium, European Union of Jewish Students from 1989 to 1991. **Educ.:** Maîtrise en Droit from University of Strasbourg III, France; Maîtrise en Sc.pol. from University of Strasbourg III, France.
Mailing address: Chaussée de Vleurgat 89, B-1050 Bruxelles, Belgium. T. (32 2) 647 72 79. Tx 20 625 (EUJS B). Fax (32 2) 648 24 31.
Residence: Rue du Conseil 30 Bte 17, B-1050 Bruxelles, Belgium. T. (32 2) 513 61 27

Mshana, Eliewaha E, Bishop

Rel: Lutheran.
International Organizations: Chairperson of the All-Africa Lutheran Churches Information and Coordination Centre.
Mailing address: ELCT-Bldg, PO Box 314, Arusha, Tanzania. T. (255 57) 2318. C. AFRILUTHCO. Tx 42054 LUTH T2.

Muaca, Eduardo Andre

Rel: Roman Catholic.
International Organizations: Representative of the Conferencia Episcopal de Angola e São Tomé.
Mailing address: CP 1230, Luanda, Angola. T. (244 2) 34640. TX 4064 CARITAS AN ATTN CEAST.

Mubarak, M A, Moulavi

Rel: Muslim.
International Organizations: Principal of the Gaffooriya Arabic College, Maharagama.
Mailing address: Gaffooriya Arabic College, Maharagama, Sri Lanka.

Mukhtar, Ali Mohamad Abbas

Rel: Muslim; b. in Mecca, Saudi Arabia; Citizen of Saudi Arabia.
International Organizations: Assistant Secretary-General of the Higher International Council of Mosques of the Muslim World League.
Main field of work: Official. **Career:** Secretary-General, Directorate of Petroleum and Minerals. Director, Immigration, Jeddah, Saudi Arabia. Director General, Office of the Minister of Hajj. Deputy Minister of Hajj and Endowments. Founder, Haram Benevolent Fund, Mecca, Saudi Arabia. Attended several conferences in Egypt and Beirut pertinent to tourism. Other conferences in various Islamic countries in South East Asia, Europe and Africa, for the propagation of Islam. Adviser to Mecca Cultural Club. **Educ.:** Dip. General Administration from American University of Beirut, Lebanon. **Interests:** Swimming, reading, travel. **Mem.:** Mecca Welfare Society **Pubs.:** *The Mosque Strategy in Islam* and *Pilgrim's Guide.*
Mailing address: P O Box 537 – 538, Ummul Joud, Makkah, Saudi Arabia. T. (966 2) 5446700. Fax 5436202. Tx 540009 RABITA SJ.
Residence: Makkah, Saudi Arabia T. (966 2) 574 35 73

Mulder, Edwin George, Rev Dr

Rel: Reformed Church; Ordained in 1954; b. 25 MAR 1929 in Raymond MN, USA; m. to Luella Rozeboom; c. 2 boys and 2 girls; Citizen of USA.
International Organizations: Member of the Executive Committee (since 1990) of the World Alliance of Reformed Churches.
Central Committee (since 1991) of the World Council of Churches.
Other organizations: Secretary-General (since 1983), Reformed Church in America; Executive Committee (since 1991), National Council of Churches.
Main field of work: Clergy. **Career:** Pastor, Reformed Church in America (RCA) from 1954 to 1983. Vice-President, Particular Synod (NJ), RCA from 1975 to 1976. President, Particular Synod (NJ), RCA from 1976 to 1977. Vice-President and President, General Synod, RCA from 1978 to 1980. **Educ.:** B.A. from Central College, Pella IA, USA (1951); B.D. from Western Theological Seminary, Holland MI, USA (1954); D.D. from Central College, Pella IA, USA (1979).
Mailing address: c/o Reformed Church in America, 18th Floor, 475 Riverside Drive, New York NY 10115, USA. T. (1 212) 870 2841. Fax (1 212) 870 2499
Residence: T. (1 212) 532 3674

Muldowney, Anita, Sr

Rel: Roman Catholic.
International Organizations: Secretary of the Sisters of St Louis Juilly-Monaghan.
Mailing address: St Louis Generalate, 5 Grosvenor Road, Rathgar, Dublin 6, Ireland. T. (353 1) 977974. Fax 977774.

Müller, Christian, Rev

Rel: Moravian.
International Organizations: Contact of the European Continental Province of the Moravian Church – District Herrnhut.
Mailing address: Vogtshof, Zittauer Strasse 20, 02747 Herrnhut, Germany.

Müller, Gerhard, Dr

Rel: Lutheran; Ordained on 15 APR 1956; b. 10 MAY 1929 in Marburg/Lahn, Hessen, Germany; m. to Ursula Herboth; c. 2 boys; Citizen of Germany; Lang. German and English.
International Organizations: Evangelical-Lutheran Bishop (since 1982) of Brunswick.
Other organizations: Professor Emeritus (since 1983), Georg-August-Universität Göttingen, Germany.
Career: Priest in Hanau/Main from 1956 to 1957. Deutsche Forschungsgemeinschaft scholarship, Italy from 1957 to 1959. Assistant, Ecumenical Seminar, University of Marburg from 1959 to 1961. Docent, Faculty of Theology, University of Marburg from 1961 to 1966. Guest Lecturer, German Historical Institute, Rome from 1966 to 1967. Professor of Historical Theology (Modern Church History), University of Erlangen from 1967 to 1982. **Educ.:** 1st Theol. Exam. from Kirche Kurhessen-Waldeck, Marburg, Germany (1954); Dr. Theol. from Philipps-Universität Marburg, Germany (1955); 2nd Theol. Exam. from Kirche Kurhessen-Waldeck, Marburg, Germany (1956); Habil. from Philipps-Universität Marburg, Germany (1960). **Hons.:** DD hc, St Andrews University (1980). **Mem.:** Mainz and Netherlands Academies **Pubs.:** Die Rechtfertigungslehre, 1977; Zwischen Reformation und Gegenwart, 2 Vols., 1983 and 1988 and Causa Reformationis, 1989.
Mailing address: Landeskirchenamt, Neuer Weg 88, 38302 Wolfenbüttel, Germany. T. (49 5331) 802 100
Residence: Salzdahlumer Strasse 43, 38302 Wolfenbüttel, Germany. T. (49 5331) 802 333

Muller, M, Mr

Rel: Unknown.
International Organizations: Representative of the Maharishi European Research University.
Mailing address: Station 24, NL-6063 MP Vlodrop, Netherlands.

Mullins, Patricia, Sr

Rel: Roman Catholic; Ordained; b. 25 JAN 1930 in Boston MA, USA; celibate; Citizen of Canada.
International Organizations: General Assistant/Treasurer (since 1988) of the Sisters of Charity of St Vincent-de-Paul.
Member (since 1951) of the Sisters of Charity of St Vincent-de-Paul.
Career: High School Teacher, Nova Scotia from 1951 to 1965. Assistant Professor of Chemistry, Mount St Vincent University in 1968. Associate Professor, Mount St Vincent University in 1972. Professor of Chemistry, Mount St Vincent University from 1978 to 1988. Dean of Humanities and Sciences, Mount St Vincent University from 1979 to 1988. Professor of Chemistry, Mount St Vincent University from 1978 to 1988. Visiting Professor, University of Western Ontario (summer) in 1974. Visiting Post-doctoral Research Associate, University of Notre Dame Radiation Laboratory from 1975 to 1976. **Educ.:** B.A. from Mount St Vincent College, Halifax, Canada (1951); M.Sc. from University of Notre Dame, IN, USA (1963); Ph.D. from University of Notre Dame, IN, USA (1968). **Interests:** reading, music. **Pubs.:** Author or co-author of various articles in chemical publications.
Mailing address: Mount St Vincent Motherhouse, 150 Bedford Highway, Halifax NS B3M 3J5, Canada. T. (1 902) 443 4620.

Mumemthaler, Mariette

Rel: Roman Catholic; b. 06 FEB 1947 in St Moritz, Switzerland; single; Citizen of Switzerland; Lang. French, German and English.
International Organizations: Executive Secretary of the International Catholic Conference of Guiding.
Main field of work: Professor.
Mailing address: Case Postale 178, CH-9004 St Gallen, Switzerland. T. (41 39) 26 76 29.
Residence: Recrêtes 18, CH-2300 Chaux de Fonds, Switzerland T. (41 39) 26 76 29

Munce, Frederick L, Rev

Rel: Methodist.
International Organizations: Representative of the Methodist Church in Ireland World Development and Relief Fund.
Mailing address: 12 Lester Park, Magherafelt Co Londonderry, BT45 6HP, UK.

Mungovan, James, Br
Rel: Roman Catholic; Ordained; celibate.
International Organizations: Superior General of the Franciscan Brothers of the Third Order Regular, Mountbellew.
Mailing address: Franciscan Monastery, Mountbellew, Co. Galway, Ireland. T. (353 905)79 295

Muñoz, Juan Antonio
Rel: Roman Catholic.
International Organizations: Procurator-General of the Missions Etrangères de Sainte Marie de Guadaloupe.
Mailing address: Insurgentes Sur 4135, 14000 México DF, Mexico. T. (52 5) 733000.

Muñoz Vega, Paolo M, S.I., H E Card Dr
Rel: Roman Catholic; Ordained on 25 JUL 1933; b. 23 MAY 1903 in Mira, Carchi, Ecuador; celibate; Citizen of Ecuador; Lang. Castilian, Italian, French, English, German, Latin and Greek.
International Organizations: Member (since 1969) of the College of Cardinals.
Member of the Society of Jesus. Member of the Congregation for Catholic Education. Member of the Congregation for Institutes of Consecrated Life and for Societies of Apostolic Life.
Other organizations: Archbishop Emeritus (since 1986), Quito.
Career: Professor of Philosophy, Pontifical Gregorian University from 1938 to 1945. Professor of Theology, Pontifical Gregorian University from 1945 to 1950. Rector, Pontificio Colegio Pio-Latino-Americano from 1955 to 1957. Rector, Pontifical Gregorian University from 1957 to 1964. Titular Bishop, Ceramo from 1964 to 1967. Archbishop, Quito from 1967 to 1986. Created Cardinal by Pope Paul VI in 1969. **Educ.:** Dr. Phil. from Faculty of Theology, Oña, Spain (1931); Lic. and Doctor Magister Aggregatus from Pontifical Gregorian University, Vatican City State (1938). **Hons.:** decorations from the governments of Ecuador, Spain and Venezuela. **Pubs.:** Causalidad filosofica y determinismo cientifico, 1946; Los Problemas de la experiencia mística a la luz del pensamiento agustiniano, en Augustinus Magister, 1954 and Nuestra fe en la Eucaristia, 1988.
Mailing address: Apartado 17-02-5222, 'Urbanización La Armenia', Quito, Ecuador. T. (593 2) 322982/3/4
Residence: See Mailing Address. T. See Mailing Address

Munthe, A, Rev
Rel: Protestant.
International Organizations: Ephorus of the Simalungun Protestant Christian Church.
Mailing address: Jln Jend Sudirman 14, Kotak Pos 2, Pematang Siantar, Sumatra Utara, Indonesia.

Munyanshongore, Martin, F.F.S.I., Br
Rel: Roman Catholic; Ordained; celibate.
International Organizations: Superior General of the Frères Fils de Saint Joseph du Rwanda.
Mailing address: BP 37 Gitarama, Kabgayi, Rwanda. T. RWANDA 62 110

Murphy, Colm, S.S.C.M.E., Fr
Rel: Roman Catholic; Ordained; celibate.
International Organizations: Secretary-General of the Unda, International Catholic Association for Radio and Television.
Advisor of the Pontifical Council for Social Communications. Member of the Missionary Society of St Columban.
Mailing address: Rue de l'Orme 12, B-1040 Bruxelles, Belgium. T. (32 2) 734 97 08 – 734 97 09. Fax 734 70 18. C. UNDABRUSSELS. Tx Une England – Code 051 – 933524 – GEONE 1 G, Quote in first line Box GEO 2 UNDA. Electronic mail GEONE 1 BOX GEO 2 UNDA.

Murphy, David
Rel: Unknown.
International Organizations: Representative of the International Gay Association – International Coordination and Information Centre on Religion.
Mailing address: c/o RFSL Box 350, S-10124 Stockholm, Sweden. T. (46 8) 84 80 50.

Murphy, John Aloysius, Most Rev
Rel: Roman Catholic; Ordained on 21 MAR 1931; b. 21 DEC 1905 in Birkenhead, UK; celibate; Citizen of United Kingdom.
International Organizations: Archbishop Emeritus (since 1983) of Cardiff.
Career: Co-adjutor Bishop, Shrewsbury in 1948. Bishop, Shrewsbury from 1949 to 1961. Archbishop, Cardiff from 1961 to 1983. **Hons.:** Hon. D.D.
Mailing address: 'Ty-Mair', St Joseph's Nursing Home, Harding Ave, Malpas, Newport Gwent NP9 6ZE, UK. T. (44 633) 85 04 59

Murphy, John Patrick
Rel: Buddhist: Mahayana.
International Organizations: Member of the Foundation for the Preservation of the Mahayana Tradition.
Other organizations: Director, Osel Shen Phen Ling.
Mailing address: PO Box 7604, Missoula MT 59807, USA. T. (1 406) 721 3155

Murphy-O'Connor, Cormac, Rt Rev
Rel: Roman Catholic; Ordained in 1956; b. 24 AUG 1932 in Reading, UK; celibate; Citizen of United Kingdom.
International Organizations: Catholic Co-Chairman (since 1983) of the Anglican-Roman Catholic International Commission.
Other organizations: Bishop (since 1977), Arundel and Brighton; Chairman (since 1985), TVS Religious Advisers Panel.
Career: Parish Priest, Portswood, Southampton from 1970 to 1971. Rector, Venerable English College, Rome from 1971 to 1977. First Chairman, Bishops' Committee for Europe from 1979 to 1982. **Educ.:** Ph.L., S.T.L. from Pontifical Gregorian University, Vatican City State. **Interests:** music, walking, reading. **Pubs.:** The Family of the Church, 1984.
Mailing address: St Joseph's Hall, Greyfriars Lane, Storrington, Pulborough RH20 4HE, UK. T. (44 903) 742172

Murray, Edward M K
Rel: Christian.
International Organizations: President of the World Association of Daily Vacation Bible Schools.
Mailing address: 141 Linkside Circle, Ponte Vedra Beach FL 32082, USA. T. (1 904) 285 6647.

Murray, George W
Rel: Christian.
International Organizations: Executive Director of the Bible Christian Union.
Mailing address: PO Box 410, Hatfield PA 19440, USA. T. (1 215) 361 0500.

Murray, Jon Garth
Rel: Atheist; b. 16 NOV 1956 in Baltimore MD, USA; single; Citizen of USA; Lang. English.
International Organizations: President (since 1984) of the American Atheists.
Other organizations: Director (since 1975), American Atheist General Headquarters; Anchorperson (since 1982), American Atheist Television Forum; Board Member (since 1980), Charles E Stevens American Atheist Library; President (since 1984), Society of Separationists Inc.
Career: Secretary, Society of Separationists Inc from 1975 to 1984. Chapter Coordinator, American Atheists from 1978 to 1982. **Educ.:** B.A. from University of Texas at Austin, USA (1972-1975). **Interests:** Separation of state and church, civil rights of atheists. **Pubs.:** *Essays on American Atheism*, 2 Vols; *Essays of an Activist Atheist* and editor of *An Atheist's Bertrand Russell.*
Mailing address: PO Box 140195, Austin TX 78714-0195, USA. T. (1 512) 458 1244
Residence: 7215 Cameron Road, Austin TX 78752, USA

Murray, Nicholas, S.S.C.M.E., Fr
Rel: Roman Catholic; Ordained; celibate.
International Organizations: Superior General of the Missionary Society of St Columban.
Mailing address: St Columban's, Grange Road, Donaghmede, Dublin 13, Ireland. T. (353 1) 47 66 47. Fax 47 48 53.

Murray, Robert, S.J., Rev Dr
Rel: Roman Catholic; Ordained; celibate.
International Organizations: Member of the Society of Jesus.
Mailing address: Heythrop College, 11-13 Cavendish Square, London W1M OAN, UK.

Musconi, Marino, Fr
Rel: Roman Catholic; Ordained; celibate.
International Organizations: Procurator-General of the Little Mission for the Deaf-Mute.
Mailing address: Via Vincenzo Monti 1, I-00152 Roma, Italy. T. (39 6) 581 08 86

Muytjens, Johannes Joseph, F.I.C., Br Dr
Rel: Roman Catholic; Ordained on 18 AUG 1959; b. 07 MAR 1939 in Heer-Maastricht, Limburg, Netherlands; celibate; Citizen of Netherlands; Lang. Dutch, English, German and French.
International Organizations: Superior General (since 1982) of the Brothers of the Immaculate Conception of the Blessed Virgin Mary.
Main field of work: Spiritual and organizational leadership. **Career:** Teacher, Primary School from 1960 to 1962. Councilor, Provincial Council Holland from 1970 to 1982. Chairman, Office for spiritual and psychological counselling (CAPER) from 1975 to 1991. Chairman, Dutch Section, European Forum of Catholic Lay Organizations from 1976 to 1984. **Educ.:** Doc. Chem. and Physics from Catholic University of Nijmegen, Netherlands (1964-1970). **Pubs.:** Apostolic Spirituality, Maastricht, 1985 and The Future of our Tradition, 1991.
Mailing address: Brusselsestraat 38, NL-6211 PG Maastricht, Netherlands. T. (31 43) 21 21 14. FAX 21 80 46
Residence: See Mailing Address. T. See Mailing Address

Muzzi, Latino, F.S.M.I., Fr
Rel: Roman Catholic; Ordained; celibate.
International Organizations: Procurator-General of the Fils de Sainte Marie Immaculée.
Mailing address: Via del Mascherone 55, I-00186 Roma, Italy. T. (39 6) 687 59 15

Mveng, Engelbert, S.J., Prof
Rel: Roman Catholic; Ordained on 07 SEP 1963; b. 09 MAY 1930 in Enam-Ngal, Southern Province, Cameroon; celibate; Citizen of Cameroon; Lang. Ewondo, Kikongo, French, English, Italian, Spanish, German, Latin and Ancient Greek.
International Organizations: Secretary-General (since 1983) of the Association oecuménique des théologiens africains.
Member of the Society of Jesus.
Other organizations: Secretary-General (since 1980), Mouvement des intellectuels chrétiens africains (MICA); Professor (since 1965), University of Yaoundé.
Main field of work: University education. **Career:** Professor, Collège Libermann, Douala from 1957 to 1960. Consultant, UNESCO from 1963 to 1966. Head, Dept. of Cultural Affairs, Ministry of Education and Culture from 1966 to 1974. Pan-African Co-ordinator, Ecumenical Association of Third World Theologians from 1981 to 1986. Rapporteur General, 1st Festival of Negro Art, Dakar in 1966. Rapporteur General, 2nd Festival of Negro Art, Lagos in 1977. Founded Collège Le Sillon, Yaoundé in 1974. Founded Musée d'Art Nègre, A. Diop in 1966. Founded Congrégation des Béatitudes in 1980. **Educ.:** Lic. ès Lettres from University of Dakar, Senegal and University of Paris, France (1957-1960); Doc. 3ème Cycle from Universities of Paris and Lyon, France (1961-1964); Doc. d'Etat from University of Paris IV (Paris-Sorbonne), France (1965-1970); Lic. Phil. from University of Lyon III, France (1954-1957); Lic. Théol. from University of Lyon III, France (1961-1964). **Interests:** art, history, archaeology, spirituality. **Hons.:** Lauréat de l'Académie Française. **Mem.:** Academy of Overseas Sciences (Paris) **Pubs.:** Over 30 works on the history of Cameroon, African art, theology, and spirituality, including: *L'Histoire du Cameroun*, 2 Vols., 1984; *Le Cameroun. Cent ans d'évangélisation*, 1990 and *L'Afrique dans l'Eglise*, 1986.
Mailing address: PO Box 1539, Yaoundé, Cameroon. T. (237) 22 09 56
Residence: See Mailing Address. T. See Mailing Address

Mvoe, Pascal
Rel: Roman Catholic.
International Organizations: Head, French-Speaking Africa Regional Office of the Unda, International Catholic Association for Radio and Television.
Mailing address: Serv. des Informations Catholiques, Archidiocèse de Yaoundé, BP 185, Yaoundé, Cameroon.

Mwakasege, Shadrack, Rev
Rel: Moravian.
International Organizations: General Secretary of the Joint Board of the Moravian Church in Tanzania.
Mailing address: PO Box 377, Mbeya, Tanzania.

Mwangi, Samuel Mugo, Rev
Rel: Christian.
International Organizations: General Secretary of the African Christian Church and Schools.
Mailing address: PO Box 1365, Thika, Kenya.

Mwingira, Mary
Rel: Roman Catholic.
International Organizations: Head, Africa Regional Secretariat of the Pax Romana, International Catholic Movement for Intellectual and Cultural Affairs.
Mailing address: Christian Professionals in Tanzania, PO Box 4836, Dar es Salaam, Tanzania.

Myers, Paul
Rel: Mennonite.
International Organizations: Director of SELFHELP Crafts of the Mennonite Central Committee.
Mailing address: 21 South 12th Street, Akron PA 17501, USA.

Myung, Chu Ke, Rev
Rel: Presbyterian.
International Organizations: General Secretary of the Presbyterian Church of Korea (Tong-Hap).
Mailing address: CPO Box 1125, 135 Yunji-dong, Chongro-ku, Seoul 110, Korea.

Nababan, Soritua A E, Rt Rev Dr
Rel: Lutheran.
International Organizations: Chairperson, Program Committee for World Service of the Lutheran World Federation.
Other organizations: Bishop, Protestant Christian Batak Church.
Mailing address: Pearaja-Tarutung 22413, Tapanuli Utara, Sumatra, Indonesia. T. (62 633) 21707. Fax 21693. C. KANTOR PUSAT HKBP, TARUTUNG
Residence: T. (62 633) 21022

Nabil, R G
Rel: Roman Catholic.
International Organizations: Regional Secretary for Near and Middle East of the Catholic International Education Office.
Mailing address: Collège de la Ste Famille, BP 73, Fagallah, 11000 Cairo, Egypt.

Nabulivou, Inoke, Rev
Rel: Methodist.
International Organizations: General Secretary of the Methodist Church in Fiji.
Mailing address: GPO Box 357, Suva, Fiji.

Naddair, Kaledon
Rel: Unknown.
International Organizations: Secretary-General of the European Pan-Keltic League.
Mailing address: Kaer Eidyn, 4A Minto Street, Edinburgh EH9 1RG, UK. T. (44 31) 667 57 88.

Nadolski, Boguslaw, T.Chr., Fr
Rel: Roman Catholic; Ordained; celibate.
International Organizations: Superior General of the Società di Cristo per gli Emigrati della Polonia.
Mailing address: ul Panny Marii 4, Poznan 61-120, Poland. T. (48 61) 543 83 – 790 171

Nagabodh, Dharmachari
See Terence Pilchick.

Naganuma, Motoyuki
Rel: Buddhist; b. 10 NOV 1923; Citizen of Japan.
International Organizations: Chairman of the Risshokoseikai.
Career: Chairman, Niwano Peace Foundation.
Mailing address: 1-43-13 Wada, Suginami-ku, Tokyo 166, Japan. T. (81 3) 3383 0601

Nagar, Maharishi
Rel: Hindu.
International Organizations: Contact of the Maharishi World Centre of Ayur-Ved.
Mailing address: A 214 New Friends Colony, New Delhi 110065, India.

Nagy, Gyula, Bishop
Rel: Reformed Church.
International Organizations: Bishop of Reformed Church of Romania.
Mailing address: Eparphia Reformata Cluj, Str 23 August nr 51, 3400 Cluj-Napoca, Romania.

Naidu, Vijaykumar Joseph, S.J., Rev
Rel: Roman Catholic; Ordained; b. 24 AUG 1941 in Bangalore, Karnataka, India; celibate; Citizen of India; Lang. English.
International Organizations: Secretary-Treasurer (since 1984) of the OCIC/Asia.
Member, Executive Committee (since 1987) of the International Catholic Organization for Cinema and Audiovisual. Member of the Society of Jesus.
Other organizations: Director (since 1984), Jesuit Media Centre.
Main field of work: Production, training, research. **Career:** Secretary, International Jesuit Centre for Social Communications (JESCOM), South Asia from 1974 to 1981. Secretary, UNDA/OCIC, India from 1982 to 1988. **Educ.:** B.Sc. from Saint Joseph's College, Bangalore, India (1959-1962); B.Phil. from Sacred Heart College, Bangalore, India (1966-1968); B.Th. from De Nobili College, Pune, India (1969-1973); P.G. Dip. in Theory and Practise of AVA for Dev. Countries at University of London, UK (1975-1976). **Interests:** photography. **Mem.:** Asian Mass Communication Research and Information Centre (AMIC), Singapore; World Association for Christian Communication (WACC), London **Pubs.:** *Audiovisual Mobile Units and the Church in India*, 1973; Use of Modern Media for Evangelization (monograph), 1990 and Communication Technology in India and the Right to Communicate (monograph), 1991.
Mailing address: 96 Lavelle Road, Bangalore 560 001, India. T. (91 812) 214 805
Residence: See Mailing Address. T. (91 812) 212 268

Najjar, Rauf Salim, Mgr
Rel: Roman Catholic; Ordained; b. in Haifa, Israel; celibate; Citizen of Jordan.
International Organizations: President (since 1962) of the Ecclesiastic Tribunal.
Other organizations: Editor and Director (since 1968), *Voice of the Holy Land* (monthly review); Director General (since 1968), Al-Wassifiya Vocational Centre.
Career: Assistant, Roman Catholic Bishop in Amman. Chancellor, Apostolic Delegation in Jordan in 1977. **Interests:** music, sports, law. **Hons.:** Commander, Order of St John of Jerusalem, Order of Independence (1973); Commandeur, Ordre International de l'Etoile de la Paix.
Mailing address: PO Box 1317, Amman, Jordan.

Nakajima, John M, Rev
Rel: United Church of Christ.
International Organizations: General Secretary of the United Church of Christ in Japan.
Mailing address: 31 Japan Christian Centre, 2-3-18 Nishi-Waseda, Shinjuku-ku, Tokyo 160, Japan.

Nakamura, Koryu
Rel: Buddhist; Ordained; Citizen of Japan.
International Organizations: Monk of Jodshu Zojoji.
Other organizations: Honorary Professor, Taisho University.
Educ.: Taisho University, Japan.
Mailing address: 12-19 Shimizu, Shimizu, Shizuoka 424, Japan. T. (81 543) 52 4029

Nakano, Toshihiro
Rel: New Religion.
International Organizations: Secretary-General of the Organization for Industrial, Spiritual and Cultural Advancement International.
Mailing address: 6-12 Izumi 3-chome, Suginami-ku, Tokyo 168, Japan. T. (81 3) 3322 5161/4. Fax 3499 4968. C. WHQOISCA. Tx J29397 OISCA.

Nakata, Shokan
Rel: Buddhist; Ordained; b. 20 JUN 1917; Citizen of Japan.
International Organizations: Monk of Shin-Yakushiji.
Mailing address: 1352 Takahata Fukuicho, Nara 630, Japan. T. (81 742) 22 3736

Nakhjavani, Ali-Yullah
Rel: Bahá'í; b. 19 SEP 1919 in Haifa, Israel; m. to Violette Banani; c. 1 boy and 1 girl; Citizen of Iran; Lang. English, French, Persian and Arabic.
International Organizations: Member, Universal House of Justice (since 1963) of the Bahá'í International Community.
Main field of work: Administration. **Career:** Member, National Spiritual Assembly of the Bahá'ís of Iran from 1950 to 1951. Chairman, National Spiritual Assembly of the Bahá'ís of Central and East Africa from 1956 to 1961. President, International Bahá'í Council from 1961 to 1963. **Educ.:** B.A. from American University of Beirut, Lebanon (1935-1939). **Pubs.:** A number of talks printed in Bahá'í publications.
Mailing address: Baha'i World Center, Golomb Ave, 31001 Haifa, Israel. T. (972 4) 35 83 58
Residence: 6A Bethlehem Street, Haifa 31001, Israel. T. (972 4) 33 55 05

Nambiaparambil, Albert, Fr
Rel: Roman Catholic.
International Organizations: Secretary of the World Fellowship of Inter-Religious Councils.
Mailing address: CBCI Centre, Ashok Place, New Delhi 110001, India. T. (91 11) 344470 – 344453.

Nanayakkara, Irène
Rel: Buddhist.
International Organizations: Contact of the International Buddhist Women's Centre.
Mailing address: 4 Sarasavi Udyanaya, Nugegoda, Sri Lanka.

Napier, Albert
Rel: Roman Catholic; Unordained; b. 10 SEP 1956 in Victoria, Seychelles; m. to Stella Maryse; c. 1 girl; Citizen of Seychelles; Lang. French and English.
International Organizations: Pastoral Worker, Seychelles (since 1991) of the International Young Christian Workers.
Other organizations: Collaborator (since 1991), Mouvement d'Action Chrétienne.
Career: Youth Coordinator for Africa, International Young Christian Workers from 1983 to 1989.
Mailing address: c/o Seychelles YCW, Centre Action Chrétienne, rue Francis Rachel, Victoria-Mahé, Seychelles.
Residence: Mont Bux Ton, Victoria-Mahé, Seychelles. T. (248) 21519

Napier, Michael, d.O., Fr
Rel: Roman Catholic; Ordained; celibate.
International Organizations: Delegate for the Oratory of the Confederation of the Oratory of St Philip Neri.
Mailing address: Via Parione 33, I-00186 Roma, Italy. T. (39 6) 687 52 89.

Napier, Wilfrid Fox, Rt Rev

Rel: Roman Catholic; Ordained on 25 JUL 1970; b. 08 MAR 1941 in Matatiele, Natal, South Africa; celibate; Citizen of South Africa; Lang. English, Xhosa, French and Afrikaans.
International Organizations: President (since 1987) of the Southern African Catholic Bishops' Conference.
Other organizations: Bishop (since 1981), Kokstad.
Main field of work: Ministry. **Career:** Assistant Pastor, St Anthony's Parish, Lusikisiki from 1971 to 1973. Parish Priest, St Francis Parish, Tabankulu from 1973 to 1980. Vice-President, South African Catholic Bishops' Conference from 1984 to 1987. Vice-President, Inter-Regional Meeting of Bishops of Southern Africa from 1984 to 1988. **Educ.:** B.A. from National University of Ireland, University College, Galway, Ireland (1961-1964); B.Phil. from Catholic University of Louvain, Belgium (1964-1966); B.Th. from Catholic University of Louvain, Belgium (1966-1970). **Interests:** gardening, fishing.
Mailing address: 97 St John's Street, Kokstad 4700, South Africa. T. (27 372) 2239
Residence: See Mailing Address. T. See Mailing Address

Narayama, Daiten

Rel: Buddhist; Unordained; b. 18 MAY 1914; Citizen of Japan.
International Organizations: Monk of Sodoshu Hokoji.
Career: President, Tokyo Grand Hotel. **Educ.:** Komazawa University, Tokyo, Japan.
Mailing address: 20 Nagawamachi Oaza Hokoji Aza Hokoji, Sannohe-gun, Aomori 039-05, Japan. T. (81 178) 76 2506

Narayanan, Sanguni Nair

Rel: Hindu; b. 03 JUL 1928 in Kuala Lipis, Pahang, Malaysia; m. to Padma Narayanan; c. 1 boy; Citizen of Malaysia; Lang. Malay, English, Tamil and Malayalam.
International Organizations: President (since 1989) of the Indo-Pacific Theosophical Federation.
Other organizations: Lawyer (since 1988), Narayanan Mohanadas Associates.
Main field of work: Law. **Career:** Teacher, Ministry of Education, UK from 1949 to 1956. Officer, Ministry of Labour, UK from 1956 to 1969. Manager, Personnel Dept., ICI, UK from 1969 to 1976. Director, Main Board, ICI, Malaysia from 1976 to 1983. Executive Director, Malaysian Employers Federation from 1984 to 1986. **Educ.:** LL.B. from University of London, UK (1962-1964); Bar. at Law from Lincoln's Inn, London, UK (1963-1964).
Mailing address: 6 Jalan Tinggi, Petaling Jaya 46000, Malaysia. T. (60 3) 7924 950
Residence: See Mailing Address. T. See Mailing Address

Naré, Laurent, Fr

Rel: Roman Catholic.
International Organizations: Director of the Catholic Biblical Centre for Africa and Madagascar.
Mailing address: PO Box 24215, Karen, Nairobi, Kenya. T. (254 2) 882448.

Narzynski, Janusz, Rt Rev

Rel: Evangelical.
International Organizations: Bishop of Evangelical Church of the Augsburg Confession in the People's Republic of Poland.
Mailing address: ul Miodowa 21, 00 246 Warszawa, Poland.

Nash, Nancy

Rel: Buddhist; single; Citizen of USA.
International Organizations: Originator and International Coordinator of the Buddhist Perception of Nature.
Advisor to 'Health of the Planet' Survey of the Gallup International Research Institute, Seoul. Hong Kong Coordinator of the Global Network on Responsibilities to Future Generations.
Other organizations: Freelance reporter (since 1963); Consultant for hotel companies including Hilton International, other international hotel groups, airlines, fashion houses, motion picture companies, travel agencies and conservation organizations (since 1974); Far East Representative, Society for Wildlife Art for the Nations (UK); Asia Advisor, Assisi Nature Council; Honorary Advisor to Earth Ethics Research Group (International).
Main field of work: International consultant: marketing, public affairs and public relations. **Career:** Area Director of Public Relations (Asia/Pacific), Hilton International Co. from 1966 to 1974. Originator and Designer, the successful proposal to Chinese authorities for first international NGO contact (WWF/China) from 1979 to 1981. Negotiator for the first foreign cooperation in the Chinese Giant Panda conservation programme (joint China/WWF Project) from 1979 to 1981. Editing and creative service for the WWF/IUCN/UNEP-designed World Conservation Strategy, Coordinator for the three Asian cities (Peking, Hong Kong, Bangkok) for the actual launch, On-Site Director in Peking from 1979 to 1980. Designed, managed and found funds for the inauguration, WWF Hong Kong from 1981 to 1984. Co-designer and co-organizer for the first ever exhibitions for Hong Kong and Peking on Wildlife Conservation from 1981 to 1983. Originator and a chief implementator, China Wildlife Conservation Association (CWCA) from 1982 to 1984. International Advisor and Guest Speaker at environment events. **Hons.:** Laureate, one of the five 1987 Rolex Awards for Enterprise (International) for the Buddhist Perception of Nature Project. **Mem.:** Friends of the Earth (Hong Kong) (Life Member); Friends of SOS Childrens Villages (USA); International Society for Environmental Ethics (USA); IUCN Working Group on Ethics, the Arts and Humanities (International Union for the Conservation of Nature and Natural Resources, Education Commission); Society of Woman Geographers (International); Siam Society (International); The Wildfowl and Wetlands Trust (UK) (Life Member); Worldwide (Women for Nature) (International); WWF-World Wide Fund for Nature (Hong Kong) (a founder) **Pubs.:** Author of more than 150 newspaper and magazine articles (some 25 cover stories) for journals including *McCall's, Asiaweek, Asia Magazine, International Herald Tribune, Harper's and Queen, Newsweek, The World Book Year Book, Cosmopolitan, People* (World Resources Institute), and news agencies; 3 books as sole author (1 as illustrator) and contribution to 11 books.
Mailing address: 5H Bowen Road, 1st Floor, Hong Kong, Hong Kong. T. (852) 523 3464. Fax 869 1619. Tx 72149 SIDAN HX.

Nasrallah, Joseph Moussa, Fr

Rel: Roman Catholic; Ordained in 1934; b. 11 OCT 1911 in Nabk, Saudi Arabia; celibate; Citizen of Syria.
International Organizations: Vicar Patriarchal in France (since 1961) of Roman Catholic Church.
Career: Superior, Yabroud College from 1934 to 1941. Teacher at the Patriarchal College, Damascus from 1941 to 1944. Teacher of History and Geography at St Paul Seminary, Harissa from 1944 to 1950. Rector, St Julien le Pauvre of Paris, Vicar-General in 1950. President, Comparative Ecclesiastical History Commission. Syrian Delegate to the French Prehistoric Association. **Interests:** manuscripts. **Hons.:** Knight of the Legion of Honour; Commander, Academic Palms; Grand Cross, Order of St Lazarus; Prix Paul Verrières de l'Académie Française (1954). **Mem.:** French Institute of Damascus **Pubs.:** Author of several books, including: *Marie dans la Sainte et Divine Liturgie Byzantine*, 1955; *Voyageurs et Pélerins au Qalamoun*, 1943 and *Catalogue des Manuscrits au Liban*, 5 Vols., 1959.
Mailing address: 17 rue Petit Pont, Paris V, France.

Nasseef, Abdullah Bin Omar, Dr

Rel: Muslim; b. 05 JUL 1939 in Jeddah, Saudi Arabia; married; c. 2 boys and 2 girls; Citizen of Saudi Arabia; Lang. Arabic and English.
International Organizations: Past Secretary-General of the Muslim World League.
Member, Educational Methods Group and World Programme Committee of the World Scout Foundation. Vice-President of the World Assembly of Muslim Youth. President, Board of Trustees of the East-West University. Secretary-General of the World Supreme Council for Masajid.
Other organizations: International Commissioner (since 1979), Saudi Arabian National Boy-Scout Association; Member, Board of Governors, Faisal Islamic Bank, Khartoum, Sudan; Member, Board of Governors, Islamic Bank of Egypt, Cairo, Egypt; Member, Planning and Development Committee, Islamic University, Islamabad, Pakistan; Chairman, Board of Governors, Caribbean Educational Institute, Jamaica, West Indies; Member, Board of Trustees, Dar Al-Islam, New Mexico, USA; Member, Advisory Body, Islamic Arts Foundation, London, UK; Member, International Advisory Board, Institute of Policy Studies, Islamabad, Pakistan.
Career: Assistant Professor, Geology Dept., King Saud University, Riyadh, Saudi Arabia from 1971 to 1973. Assistant Professor and Head of Geology Dept., King Abdul Aziz University, Jeddah, Saudi Arabia from 1973 to 1974. Secretary-General, King Abdul Aziz University

from 1974 to 1976. Associate Professor, Faculty of Earth Sciences, King Abdul Aziz University from 1977 to 1983. Vice-President, King Abdul Aziz University from 1976 to 1980. President, King Abdul Aziz University from 1980 to 1983. **Educ.:** B.Sc. Geol. and Chem. from King Saud University, Riyadh, Saudi Arabia (1964); Ph.D. Geol. from University of Leeds, UK (1971). **Hons.:** Wood Badge (1975), Arab Scout Medal (1981). **Mem.:** Fellow, Geological Society, London, UK; Fellow, Geological Society of America, Boulder CO, USA; Member, Royal Academy of Morocco, Rabat; Member, International Commission on Muslim Minorities, UK; **Pubs.:** Publications include: With I G Gass, *Arabian Shield Granite Traverse*, Institute of Applied Geology, Jeddah, Bull. 3, v. 4, 1980; With I G Gass, *Geo-chemistry and Geo-chronology of Granites*; *Science, Shariah and Education*, World Centre for Islamic Education, Makkah, Saudi Arabia, 1981 (in English and Arabic); *Islam and Communism*, Research and Documentation, London, 1981 and With I R Al-Faruqi, *Social and Natural Sciences*, Hodder and Stoughton, London, 1981.
Mailing address: P O Box 537 -538, Makkah, Saudi Arabia. T. (966 2) 537 0011 (DIRECT) – (966 2) 536 4477 – (966 2) 536 7373. Tx 440390 MWL SJ. Fax (996 2) 544 46700.
Residence: Jeddah, Saudi Arabia T. (966 2) 660 7070 – (966 2) 660 7979

Nasseef, Abdullah Omar, Dr
Rel: Muslim.
International Organizations: Secretary-General of the World Supreme Council for Masajid.
Mailing address: PO Box 537-538, Makkah, Saudi Arabia. T. (966 2) 5445335. Fax 54446700. Tx 540009 Rabita SJ.

Natale, Tommaso
Rel: Catholic.
International Organizations: Representative at FAO of the World Organization of Former Pupils of Catholic Education.
Mailing address: Via della Pisana IIII, I-00163 Roma, Italy.

Naters, James, SSJE, Fr
Rel: Church of England; Ordained.
International Organizations: Superior General of the Society of St John the Evangelist.
Mailing address: St Edwards House, 22 College Street, London SW1P 3QA, UK. T. (44 71) 222 9234.

Natho, Eberhard
Rel: Evangelical.
International Organizations: President of the Evangelical Church of Anhalt.
Mailing address: Friedrichstrasse 22, 06844 Dessau, Germany.

Natowski, Damian, OCD, Sr
Rel: Roman Catholic; Ordained; celibate.
International Organizations: Member of the Discalced Carmelite Nuns.
Other organizations: Member, Association of Contemplative Sisters.
Main field of work: Archaeology.
Mailing address: Carmelite Monastery, "Mt. Carmel Project", 5714 Hollady Blvd, Salt Lake City UT 84121, USA.

Naudé, Beyers, Rev Dr
Rel: Christian; Ordained; m. to Ilse Weder; c. 3 boys and 1 girl; Citizen of South Africa.
International Organizations: Co-founder of the Multi-Racial Christian Institute of South Africa.
Career: Given seven-year Government banning order in 1977. Secretary-General, South African Council of Churches from 1985 to 1987. **Hons.:** Hon. D.Theol., Amsterdam (1972); Hon. LL.D., Witwatersrand (1974), Notre Dame, USA (1985), Limburg (1989); Hon. D.Litt., Cape Town (1983); many international humanitarian awards, including Bruno Kreisky Award, Austria (1979), Robert F Kennedy Human Rights Award (1985), Herbert Haag Prize, Germany.
Mailing address: 7th Floor, Auckland House, 185 Smit St, Braamfontein 2017, South Africa.

Naudé, Jacobus Adriaan, Prof
Rel: Protestant; Ordained in 1967; b. 26 NOV 1941 in Heilbron, Orange Free State, South Africa; m. to Maria Magdalena Loubser; c. 1 boy and 1 girl; Citizen of South Africa; Lang. Afrikaans, English, German and Dutch.
International Organizations: Senior Lecturer (since 1969) of the Rand Afrikaans University.
Other organizations: Head (since 1980), Centre for Islamic Studies; Editor (since 1981), *Journal for Islamic Studies*; Chairman (since 1982), Department of Semitic Languages; National Chairman (since 1991), Southern African Society for Semitics.
Main field of work: University education. **Career:** Lecturer, University of South Africa from 1966 to 1969. **Educ.:** B.D. from University of Pretoria, South Africa (1962-1965); M.A. from University of Pretoria, South Africa (1967-1967); D.Litt. from University of Pretoria, South Africa (1968-1971). **Interests:** semitic languages, Islam. **Mem.:** Middle East Studies Association (USA) **Pubs.:** Revelation and Reality in Islam, 1979; Translation and Interpretation, 1983 and Significance of Style in Quran, 1985.
Mailing address: Rand Afrikaans University, PO Box 524, 2000 Johannesburgh, South Africa. T. (27 11) 489 2747
Residence: 4 Ysterhout Ave, 2194 Randpark Extention 1, South Africa. T. (27 11) 793 3062

Navamani, Peter
Rel: Protestant.
International Organizations: Chairperson of the International Committee for World Day of Prayer.
Mailing address: 107 4th Cross 19th Main, 2 Block Koramangala, Bangalore 560-034, India.

Navone, John Joseph, S.J., Rev Prof
Rel: Roman Catholic; Ordained on 16 JUN 1962; b. 19 OCT 1930 in Seattle WA, USA; celibate; Citizen of USA and Italy; Lang. English, Italian, Latin, Greek, Spanish, French, German and Portuguese.
International Organizations: Professor of Biblical Theology (since 1967) of the Pontifical Gregorian University.
Member of the Society of Jesus.
Main field of work: University education, writing. **Career:** Entered the Society of Jesus in 1949. Teacher at Seattle Preparatory School, WA from 1956 to 1959. **Educ.:** M.Phil. from Gonzaga University, Spokane WA, USA (1956); S.T.M. from Saint Mary's University, Halifax, Canada (1959-1963); Th.D. from Pontifical Gregorian University, Vatican City State (1965). **Interests:** drama, films, literature, music, art, European history, sociology, international relations, geo-politics, communications. **Pubs.:** Author of 15 books on narrative theology, including: With Thomas Cooper, *Tellers of the Word*, formerly Le Jacq, now published by Jesuit Educational Center for Human Development, Cambridge MA, 1981, pp. 341; *Seeking God in Story*, The Liturgical Press, Collegeville MN, 1990, pp. 335 and 200 articles in such journals as: *Chicago Studies, Concilium, Dizionario della Spiritualità, Gregorianum, La Civiltà Cattolica, Irish Theological Quarterly, Philosophy and Phenomenological Research, Revue de l'University d'Ottawa, Studies in Religion, Scottish Journal of Theology, Thought*, and *Theology*.
Mailing address: Pontifical Gregorian University, Piazza della Pilotta 4, I-00187 Roma, Italy. T. (39 6) 67011
Residence: See Mailing Address. T. (39 6) 6701 5312

Nazir-Ali, Michael James, Rt Rev
Rel: Anglican; Ordained; b. 19 AUG 1949 in Karachi, Pakistan; m. to Valerie Cree; c. 2 boys; Citizen of United Kingdom and Pakistan.
International Organizations: Member of the Anglican Consultative Council.
Career: Tutorial Supervisor in Theology, University of Cambridge from 1974 to 1976. Tutor then Senior Tutor, Karachi Theological College from 1976 to 1981. Associate Priest, Holy Trinity Cathedral, Karachi from 1976 to 1979. Priest-in-Charge, St Andrew's Akhtar Colony, Karachi from 1979 to 1981. Provost, Lahore Cathedral from 1981 to 1984. Bishop, Raiwind from 1984 to 1986. Former Visiting Lecturer, Centre for the Study of Islam and Muslim-Christian Relations, Birmingham. Assistant to Archbishop, Canterbury. Coordinator of Studies for Lambeth Conference in 1988. President, Horticulture Society, Cambridge in 1975. Secretary-General, Church Missionary Society. Director-in-Residence, Oxford Centre for Mission Studies. Director, Christian Aid. Trustee, Traidcraft. **Interests:** cricket, hockey, table tennis, reading. **Hons.:** Radio Pakistan Prize for English Language and

Literature (1964); Burney Award, Cambridge (1973, 1975); Oxford Society Award for Graduates (1973). **Pubs.:** Islam, a Christian Perspective, 1982; Frontiers in Christian-Muslim Encounter, 1985 and numerous articles on Islam and Christianity.
Mailing address: 157 Waterloo Road, London SE1 8UT, UK. T. (44 71) 620 1110 – 620 1111. Fax 620 1070. C. COMPASROSE LONDON SW1. Tx 8950907 ANGCOM G. ELECTRONIC MAIL TELECOM GOLD 81 CEL002, BT/TYMNET QNI 3005.

Nazzàro, Giuseppe, O.F.M., Fr

Rel: Roman Catholic; Ordained; celibate.
International Organizations: Custos of the Custody of the Holy Land. Member of the Order of Friars Minor.
Mailing address: Custodia di Terra Santa, Terra Sancta Convent, PO Box 186, 91001 Jerusalem, Israel. T. (972 2) 28 23 54 – 27 31 11. C. TERRASANTA. Tx 26389 JCT.

Ndandali, Justin, Most Rev

Rel: Christian.
International Organizations: Contact of the Church of the Province of Burundi, Rwanda and Zaire.
Mailing address: PO Box 225, Butare, Rwanda.

N'Dayen, Joachim, Most Rev Archbishop

Rel: Roman Catholic; Ordained on 22 JUL 1961; b. 22 DEC 1934 in Loko, Central African Republic; celibate.
International Organizations: President of the Association of the Episcopal Conferences of the Central African Region.
Other organizations: Archbishop (since 1970), Bangui; President, Episcopal Conference of Central Africa.
Mailing address: BP 1518, Bangui, Central African Rep.

Ndeukoya, Augustine, Mgr

Rel: Roman Catholic; Ordained; celibate.
International Organizations: Rector of the Catholic University of Eastern Africa.
Mailing address: PO Box 62157, Nairobi, Kenya. T. (254 2) 89 16 01 – 89 16 02.

Ndungane, Njongonkulu Winston Hugh, Rt Rev

Rel: Anglican; Ordained on 03 JUL 1974; b. 02 APR 1941 in Kokstad, Cape Province, South Africa; m. to (1st) Nosipo Ngeelwane (died), (2nd) Nomahlubi Vokwana; c. 1 boy and 1 girl; Citizen of South Africa; Lang. Xhosa and English.
International Organizations: Bishop of Diocese of Kimberley and Kuruman.
Career: Associate, Federal Theological Seminary Fedsem, Alice from 1971 to 1973. Rector, St Nicholas Church, Matroosfontein, Cape Town from 1980 to 1981. Principal Liaison Officer, Johannesburg from 1982 to 1984. Principal, St Bede's Theological College, Umtata from 1985 to 1986. **Educ.:** B.D. from University of London, King's College, UK (1975-1978); M.Theol. from University of London, King's College, UK (1979). **Interests:** music, walking. **Hons.:** Honorary Provincial Canon.
Mailing address: Diocese of Kimberley and Kuruman, 129 Dutoitspan Road, Kimberley 8301, South Africa. T. (27 531) 32433
Residence: Bishopsgarth, 2 Bishops Avenue, Kimberley 8301, South Africa. T. (27 531) 812871

Neff, Robert W, Dr

Rel: Church of the Brethren.
International Organizations: General Secretary of the Church of the Brethren.
Mailing address: 1451 Dundee Ave, Elgin IL 60120, USA.

Nehmé, Abraham, O.Bas. Soarita, H E Mgr

Rel: Greek-Melkite Catholic; Ordained on 03 JAN 1954; b. 10 OCT 1927 in Kaïtouly, South Lebanon, Lebanon; celibate; Citizen of Lebanon; Lang. French, Arabic, Italian, Latin and Ancient Greek.
International Organizations: Metropolitan (since 1986) of Eparchies of Homs, Hama and Yabroud.
Member of the Sacred Congregation for the Oriental Churches.
Other organizations: Member, Ordre Basilien Chouerite.
Career: Director, college and seminary, Khomchara, Lebanon from 1956 to 1962. Teaching at college and seminary, Khomchara from 1962 to 1969. Director of Novices, St John Convent from 1969 to 1972. Director, Oriental College, Zahle, Lebanon from 1972 to 1978. Vicar-General, Eparchy of Beyrouth from 1972 to 1981. **Educ.:** Lic. Phil., Lic. Theol. from Pontifical Gregorian University, Vatican City State (1948-1956); Lic. Lettres Mod. from Ecole de Lettres, Beirut, Lebanon (1962-1965). **Interests:** translation. **Hons.:** Chevalier du Cèdre, Lebanon. **Pubs.:** "Le Problème du mal chez St Jean Damascene selon le dialogue contre le Manichéen"; Articles on religious art in *Al-Yanbou* and translation into Arabic *Religieux au coeur du monde* by P J Tillard (not published).
Mailing address: Greek-Catholic Bishopric, Bustan El-Diwan, Homs 1525, Syria. T. (963 31) 21587
Residence: See Mailing Address. T. See Mailing Address

Neiditch, Michael, Dr

Rel: Jewish.
International Organizations: Director of the B'nai B'rith International Commission on Continuing Jewish Education.
Mailing address: 1640 Rhode Island Ave NW, Washington DC 20036, USA. T. (1 202) 857 6580.

Neki, Jaswant Singh, Prof

Rel: Sikh; b. 27 AUG 1925 in Murid, Punjab, Pakistan; m. to Kanwerjit; c. 1 boy and 1 girl; Citizen of India; Lang. English, Hindi, Punjabi, Urdu, Persian and Braji.
International Organizations: Member, Governing Council of the Bhai Vir Singh Sahit Sadan.
Other organizations: Vice-President (since 1992), Punjabi Sahit Academy, Ludhiana, Panjab; Member, Governing Council (since 1992), Punjabi Academy, New Delhi.
Main field of work: Psychiatry. **Career:** Assistant Professor of Psychiatry, Medical College, Amritsar from 1958 to 1967. Associate Professor of Psychiatry, Medical College, Rohtak, Amritsar from 1965 to 1967. Specialist, Federal Ministry of Health, Lagos from 1967 to 1968. Professor and Chairman, Dept. of Psychiatry, All-India Institute of Medical Sciences, New Delhi from 1968 to 1978. Director, Postgraduate Institute of Medical Education and Research, Chandigarh from 1978 to 1981. Consultant to World Health Organization from 1981 to 1986. Consultant to United Nations Development Program in 1990. Expert to United Nations International Drug Control Programme in 1991. **Educ.:** M.B., B.S. from Punjab University, Solan, India (1943-1948); M.A. Psych. from Aligarh Muslim University, India (1953-1955); D.P.M. from University of Mysore, India (1956-1958). **Interests:** writing poetry. **Hons.:** Bhai Vir Singh Award (1978); Sahitya Akademi Award (1979); International Punjabi Conference Award (1979); Asan Memorial Award (1981); Sahit Samikhya Board Award (1982); Honoured by Punjabi Academy, Delhi (1984, 1986); Kartar Singh Dhaliwal Award (1987); Shiromani Sahitkar Award (1988); Waris Shah Award (1991); Prof Puran Singh Memorial Award (1991); Ph.D. h.c., Guru Nanak Dev University, Amritsar (1990); Robe of Honour conferred at the Sikh Educational Conference (1981, 1991); Distinguished Aluminus Award, Medical College, Amritsar. **Mem.:** Royal College of Psychiatrists, London (Founding Fellow, 1971); World Psychiatric Association (Honorary Member); Australian and New Zealand College of Psychiatrists (Honorary Fellow); American Psychiatric Association (Corresponding Fellow); American Academy of Medical Sciences (Fellow) **Pubs.:** Ardas (The Sikh Congregational Prayer), published by Singh Brothers, Amritsar, 1989; Haumai Diragh Rog Hai (The Sense of Egoism), published by Punjabi Sahit Academy, Ludhiana, 1987 and Kal Te Ausar (Chronos and Kairos), published by Guru Nanak Dev University, Amritsar.
Mailing address: 52 Hemkunt Colony, New Delhi 110048, India. T. (91 11) 643 2690
Residence: See Mailing Address. T. See Mailing Address

Nelly, Marie, Sr

Rel: Roman Catholic; Ordained; celibate.
International Organizations: Member of the International Federation for Family Life Promotion.
Mailing address: 7626 Route de la Pyrotechnie, Mermoz, Dakar, Senegal.

Nelson, Dorothy W

Rel: Bahá'í.
International Organizations: Vice-Chairman of the National Spiritual Assembly of the Bahá'ís of the US.
Mailing address: Bahá'í National Center, 536 Sheridan Rd, Wilmette IL 60091, USA.

Nelson, H P
Rel: Roman Catholic.
International Organizations: Secretary-General of the Pan-African Laity Coordination Board.
Mailing address: c/o SECAM, 4 Senchi Street, PO Box 9156, Airport Residential Area, Accra, Ghana. T. (233) 777706 – 774724. C. SECAM ACCRA. Tx 2587 SECAM GH.

Nelson, James F
Rel: Bahá'í.
International Organizations: Chairman of the National Spiritual Assembly of the Bahá'ís of the US.
Career: Los Angeles Municipal Court Judge. Presiding Judge over the court system.
Mailing address: Bahá'í National Center, 536 Sheridan Rd, Wilmette IL 60091, USA.

Nelson, John, Col
Rel: Salvation Army, The.
International Organizations: Territorial Commander of The Salvation Army.
Mailing address: PO Box 242, Lahore 54000, Pakistan.

Nemes, José Luiz, C.S.S., Fr
Rel: Roman Catholic; Ordained; celibate.
International Organizations: Superior General of the Congregation of the Sacred Stigmata.
Mailing address: Via Mazzarino 16, I-00184 Roma, Italy. T. (39 6) 488 19 94, (39 6) 482 79 82. Fax (39 6) 482 87 24

Neri, Ferdinando
Rel: Unitarian.
International Organizations: Director of the Unitarian Mission Cooperation Centre.
Mailing address: Via Bacilieri 1A, I-37139 Verona, Italy. T. (39 45) 890 03 29 – 890 02 99. Fax 890 31 99.

Neto, Joao Severino, Rev
Rel: Evangelical Presbyterian.
International Organizations: President of the Evangelical Presbyterian Church of Portugal.
Mailing address: Avenida do Brasil, 92-2° Dto, Lisbon 5, Portugal.

Neufeld, Elmer
Rel: Mennonite.
International Organizations: Chairman of the Council of Mennonite Colleges.
Mailing address: Bluffton College, Bluffton OH 45817, USA. T. (1 419) 358 8015.

Neufeld, John H
Rel: Mennonite.
International Organizations: President of the Canadian Mennonite Bible College.
Mailing address: 600 Shaftesbury Blvd, Winnipeg MB R3P OM4, Canada.

Neuhaus, Richard John, Rev
Rel: Roman Catholic; Ordained on 07 SEP 1991; b. 14 MAY 1936 in Pembroke, Ontario, Canada; celibate; Citizen of USA; Lang. English.
International Organizations: President of the Institute on Religion and Public Life.
Other organizations: Editor-in-Chief, *First Things: A Monthly Journal of Religion and Public Life*.
Career: Senior (Lutheran) Pastor, a low-income Black parish in Brooklyn, NY (for 17 years). Has played a leadership role in organizations dealing with civil rights, international justice, and ecumenism. **Educ.:** Grad. from Concordia Theological Seminary, MO, USA. **Hons.:** Recipient of numerous honours from universities and other institutions, including the John Paul II Award for Religious Freedom. **Pubs.:** *The Naked Public Square: Religion and Democracy in America*; With Rabbi Leon Klenicki, *Believing Today: Jew and Christian in Conversation* and *America Against Itself*, Notre Dame Press, 1992.
Mailing address: 156 Fifth Ave, Suite 400, New York NY 10010, USA. T. (1 212) 627 2288
Residence: 338 E 19th St, New York NY 10003, USA. T. (1 212) 260 8340

Neuman, Thomas
Rel: Jewish.
International Organizations: Executive Vice-President of the B'nai B'rith International.
Executive Secretary of the Coordinating Board of Jewish Organizations.
Mailing address: 1640 Rhode Island Ave NW, Washington DC 20036, USA. T. (1 202) 857 6600. Tx 710-822-0068.

Neves, Joyce
Rel: Buddhist: Mahayana.
International Organizations: Member of the Foundation for the Preservation of the Mahayana Tradition.
Other organizations: Director, Dorje Jig Je Centro.
Mailing address: Rua Ribeiro de Almeida 50, Laranjeiras, 22240 Rio de Janeiro, Brazil. T. FAX AND TEL (55 21) 205 0583

Neville, William F
Rel: Roman Catholic.
International Organizations: President of the Pax Romana, International Catholic Movement for Intellectual and Cultural Affairs.
Mailing address: 37-39 rue de Vermont, PO Box 85, CH-1211 Genève 20-CIC, Switzerland. T. (41 22) 733 67 40.

Newbigin, Lesslie, Rt Rev
Rel: Church of Scotland; Ordained in 1936; b. 08 DEC 1909 in Newcastle upon Tyne, UK; m. to Helen S Henderson; c. 1 boy and 3 girls; Citizen of United Kingdom.
International Organizations: Missionary of the Church of Scotland.
Career: Secretary, Student Christian Movement, Glasgow from 1931 to 1933. Missionary in Madras, South India from 1936 to 1939. Missionary in charge, Kancheepuram from 1939 to 1946. Bishop, Church of South Indian, Madura from 1947 to 1959. Secretary-General, International Missionary Council from 1959 to 1961. Associate Secretary-General, World Council of Churches, and Director, Division of World Mission and Evangelism from 1961 to 1965. Bishop, Madras from 1965 to 1974. Lecturer, Selly Oak Colleges from 1974 to 1979. Moderator, United Reformed Church of England and Wales from 1978 to 1979. Minister, Winson Green United Reformed Church from 1980 to 1988. **Interests:** music, climbing mountains. **Hons.:** Commander, Order of the British Empire. **Pubs.:** Unfinished Agenda, 1985; Foolishness to the Greeks, 1986 and The Gospel in a Pluralist Society, 1989.
Mailing address: Selly Oak College, Bristol Road, Birmingham 29, UK.

Newborn, Janice R
Rel: Christian.
International Organizations: Executive Secretary of the International Christian Women's Fellowship.
Mailing address: PO Box 1986, Indianapolis IN 46206, USA. T. (1 317) 353 1491.

Newby, William, Fr
Rel: Roman Catholic.
International Organizations: Secretary-General of the Order of St Vincent.
Mailing address: PO Box 1548, Joplin MO 64802, USA. T. (1 417) 623 6893. Fax 623 7870.

Ng, Yar-Chag Pal-Bum-Tsang
See Dhargyey.

Ngabu, Faustin, Rt Rev
Rel: Roman Catholic; Ordained on 21 DEC 1963; b. in Lokwa, Zaire; celibate.
International Organizations: Past President of the Association des conférences épiscopales de l'Afrique Centrale.
Member of the Congregation for the Evangelization of Peoples.
Other organizations: Bishop (since 1974), Goma, Zaire.
Mailing address: BP 20511, Kinshasa, Zaire.

Ngalamu, Elinana J, Most Rev
Rel: Episcopal.
International Organizations: Archbishop of Province of the Episcopal Church of the Sudan.
Mailing address: c/o Across, PO Box 44838, Nairobi, Kenya.

Ngcobo, Samuel, Rev
Rel: Presbyterian Reformed.
International Organizations: General Secretary of the Reformed Presbyterian Church in Southern Africa.
Mailing address: 48 Eagle Street, Umtata, Transkei, South Africa.

Ngeiyamu, Joel
Rel: Evangelical Lutheran.
International Organizations: General Secretary of the Evangelical Lutheran Church in Tanzania.
Mailing address: PO Box 3033, Arusha, Tanzania.

Ngobe, J Sidney, Rev
Rel: Evangelical Presbyterian.
International Organizations: General Secretary of the Evangelical Presbyterian Church in South Africa.
Mailing address: PO Box 31961, Braamfontein 2017, South Africa.

Ngoyagoye, Evariste, Rt Rev
Rel: Roman Catholic; Ordained on 06 JAN 1966; b. 03 JAN 1942 in Jenda, Burundi; celibate.
International Organizations: President of the Association des conférences épiscopales de l'Afrique Centrale.
Other organizations: Bishop (since 1980), Bubanza.
Mailing address: Evêché de Bubanza, DS 213, Bujumbura, Burundi. T. BURUNDI (042) 2148

Nias, Ibrahim
Rel: Muslim.
International Organizations: Head, West Africa Regional Office of the World Muslim Congress.
Mailing address: Shaikhul Islam, PO Box 1 Madina, Kaolack, Senegal.

Nichols, Doug
Rel: Christian.
International Organizations: Director of the Action International Ministries.
Mailing address: PO Box 490, Bothell WA 98041, USA. T. (1 206) 485 1967.

Nichols, Susan
Rel: Christian.
International Organizations: Representative at UN, New York of the International Association for Religious Freedom.
Mailing address: UU – UN Office, 777 UN Plaza – Room 7 D, New York NY 10017, USA.

Nicol, Iain Garden, Rev
Rel: Church of Scotland; Ordained in 1964; b. 07 DEC 1934 in Cove, Scotland, UK; widowed Ute Hofmann; c. 1 boy and 1 girl.
International Organizations: Professor of Systematic Theology (since 1975) of the Knox College, Toronto.
Career: Director, Toronto School of Theology from 1980 to 1987. Served with British Merchant Navy in 1951. RAF (Royal Air Force) National Service from 1952 to 1954. Held various positions prior to university studies from 1954 to 1956. Assistant Minister, Wellington Church, Glasgow from 1963 to 1965. Assistant becoming full-time Lecturer in Systematic Theology, University of Glasgow from 1965 to 1975. **Educ.:** M.A. from University of Glasgow, UK (1959); B.D. from University of Glasgow, UK (1962); Ph.D. from University of Glasgow, UK (1971); P.G. Stds. at Philipps-Universität Marburg, Germany (1971). **Interests:** tennis, fishing. **Hons.:** Various academic awards, including Duke of Hamilton Award for Post-Graduate Study, Marburg, and Prince's Award for Study, Oxford. **Mem.:** American Academy of Religion; Association of Theological Schools of USA and Canada; Faith and Order Committee, Canadian Council of Churches; various committees of the Presbyterian Church in Canada **Pubs.:** Author of various articles in theological books and journals.
Mailing address: Knox College, 59 St George St, Toronto ON M5S 2E6, Canada.
Residence: 55 Stibbard Ave, Toronto, Ontario M4P 2B9, Canada.

Nicolaos, N, VI, His Beatitude
Rel: Greek Orthodox.
International Organizations: Pope and Patriarch of Alexandria and All Africa.
Mailing address: PO Box 2006, Alexandria, Egypt.

Nicole, Jacques Henri Georges, Prof Dr
Rel: Reformed Church; Ordained on 26 DEC 1965; b. 26 NOV 1937 in Lausanne, Vaud, Switzerland; m. to Eugenia Thérèse Verschoor van Nisse; c. 5 boys; Citizen of Switzerland; Lang. French, English, Italian, Tahitian and German.
International Organizations: Director (since 1990) of the Ecumenical Institute Bossey.
Director, Programme for Theological Education (since 1990) of the World Council of Churches.
Main field of work: Teaching postgraduate ecumenics. **Career:** Pastor, Reformed Church, Canton de Vaud from 1963 to 1968. Youth Chaplain, Reformed Church, Canton de Vaud from 1968 to 1975. Lecturer, Evangelical Church, Tahiti from 1975 to 1977. Professor and Deputy Principal, Pacific Theological College, Suva, Fiji from 1978 to 1990. **Educ.:** Lic. Theol. from Faculté de Théologie de l'Eglise Evangélique Libre, Lausanne, Switzerland (1957-1962); Doc. Theol. from University of Lausanne, Switzerland (1985-1988). **Interests:** biblical studies, missiology, ecumenics. **Pubs.:** "Au Pied de la Lettre", Editions Haere Po, Tahiti, 1988; *The First Missionary Text in a Polynesian Language*, in *The Journal of Pacific History*, Vol. 22, April 1987 and "Gunkel, Father of Biblical Narratology", in *After Narrative*, Ed. Subramani, University of South Pacific, 1990.
Mailing address: Château de Bossey, CH-1298 Céligny, Switzerland. T. (41 22) 776 25 31. C. INSTITUTCELIGNY GENEVA. Fax (41 22) 776 01 69.
Residence: See Mailing Address. T. (41 22) 776 01 38

Niederman, David, Rabbi
Rel: Jewish.
International Organizations: Executive Director of the Rav Tov International Jewish Rescue Organization.
Mailing address: 500 Bedford Av, Brooklyn NY 11211, USA. T. (1 718) 935 97 71.

Niehaus, Ferd
Rel: Roman Catholic.
International Organizations: Regional Secretary for North America of the International Council of Catholic Men.
Mailing address: 5664 Midforest Lane, Cincinnati OH 45238, USA.

Nielsen, Jørgen S
Rel: Unknown.
International Organizations: Contact of the Centre for the Study of Islam and Christian-Muslim Relations.
Mailing address: c/o Selly Oak Colleges, 996 Bristol Road, Birmingham B29 6LQ, UK. T. (44 21) 472 42 31. Fax 472 8852. Tx 334 349 SELLYO G.

Nielsen, S, Rt Rev
Rel: Moravian.
International Organizations: Contact of the Moravian Church in South Africa.
Mailing address: P Bag 524, PO Cedarville, East Griqualand, South Africa.

Niessen, Jan, Dr
Rel: Protestant; b. 08 OCT 1951 in Netherlands; married; Citizen of Netherlands; Lang. Dutch, English, German and French.
International Organizations: Executive Secretary (since 1989) of the Churches' Committee for Migrants in Europe.
Main field of work: Information and networking. **Educ.:** Doc. from Free University, Amsterdam, The Netherlands (1970-1978); Ph.D. from Free University, Amsterdam, The Netherlands (1987).
Mailing address: Rue Joseph II 174, B-1040 Bruxelles, Belgium. T. (32 2) 230 20 11. Fax (32 2) 231 14 13.

Nieto, Evaristo Martin, Rev
Rel: Roman Catholic; Ordained; celibate.
International Organizations: Member of the International Commission of General Prison Chaplains.
Mailing address: Nogal 4, Mostoles, E-Madrid, Spain.

Nieuwenhuijsen, J W A
Rel: Remonstrant Reformed.
International Organizations: General Secretary of the Remonstrant Brotherhood.
Mailing address: Nieuwe Gracht 23, Utrecht, Netherlands.

Nijmé, Elias, B.A., H E Mgr
Rel: Greek Catholic; Ordained on 20 JUL 1944; b. 16 AUG 1920 in Aleppo, Syria; celibate; Citizen of Syria.
International Organizations: Archbishop (since 1978) of Tripoli.
Other organizations: Member, Ordre Basilien Alépin.
Career: Secretary to Patriarch Maximos IV Saïgh from 1962 to 1965. Professor, St Esprit University. **Hons.:** Lebanese Medal (1958).
Mailing address: Archevêché Grec-Catholique, BP 72, Tripoli, Lebanon. T. (961 6) 43 16 02 – 61 01 20

Nikkel, Ronald Wilbert
Rel: Anglican; Unordained; b. 08 JUN 1946 in Lethbridge, Alberta, Canada; m. to Celeste; Citizen of Canada; Lang. English, German and Spanish.
International Organizations: President (since 1988) of the Prison Fellowship International.
Chairman (since 1988) of the Alliance of NGOs on Crime Prevention and Criminal Justice.
Other organizations: Advisory Board (since 1988), Love in Action.
Main field of work: International ministry development. **Career:** Director, YFC/Youth Guidance, Toronto from 1973 to 1976. National Director, YFC/Youth Guidance, Canada from 1976 to 1978. National Director, YFC/Youth Guidance, USA from 1978 to 1982. Executive Director, Prison Fellowship International from 1982 to 1988. **Educ.:** B.A. from University of Winnipeg, Manitoba, Canada (1967-1970); M.P.S. from Loyola University, Chicago IL, USA (1980-1982). **Interests:** international affairs. **Pubs.:** Volunteer Motivation, Mobilization and Management.
Mailing address: PO Box 17434, Washington DC 20041, USA. T. (1 703) 481 0000. Fax 481 0003. Tx USA 903048 PFI HDQRS.
Residence: 1856 Old Reston Ave, Reston VA 22091, USA.

Niles, H Clifton, Rev
Rel: Methodist.
International Organizations: Secretary of the Conference of the Methodist Church in the Caribbean and the Americas.
Mailing address: Belmont, PO Box 9, St John's, Antigua-Barbuda. T. (1 809) 46 21234.

Nilon, Marie Thérèse, Sr
Rel: Roman Catholic; Ordained; celibate.
International Organizations: Superior General of the Little Company of Mary.
Mailing address: 28 Trinity Crescent, Tooting Bec, London SW17 7AE, UK.

Nilson, Donald Richard
Rel: Quaker.
International Organizations: Director of the Friends World College.
Mailing address: North American Center, Plover Lane, Lloyd Harbor, Huntington NY 11743-1098, USA. T. (1 516) 549 5000. Fax 549 5346.

Ninan, George, Rev
Rel: Christian.
International Organizations: Contact of the Christian Conference of Asia.
Mailing address: 6-10 Nakagawa Nishi, 2-chome, Ikuno-ku, Osaka 544, Japan. T. (81 6) 712 2719. Fax (81 6) 718 0988.

Nineham, Dennis Eric, Rev Prof Dr
Rel: Church of England; Ordained on 30 SEP 1945; b. 27 SEP 1921 in Southampton, UK; m. to Ruth Corfield Miller; c. 2 boys and 2 girls; Citizen of United Kingdom; Lang. English, French and German.
International Organizations: Ecclesiastic of the Church of England.
Main field of work: University education. **Career:** Fellow and Tutor, Queen's College, Oxford from 1945 to 1954. Professor of Biblical and Historical Theology, University of London from 1954 to 1958. Professor of Divinity, University of London from 1958 to 1964. Regius Professor of Divinity, University of Cambridge from 1964 to 1969. Warden, Keble College, Oxford from 1969 to 1979. Professor, University of Bristol from 1980 to 1986. **Educ.:** M.A. Lit. Hum. and Theol. (1st) from University of Oxford, Queen's College, UK (1939-1945); B.D., B.D. Cantab. from University of Oxford, Queen's College, UK (1964); D.D. Oxon from University of Oxford, Queen's College, UK (1977); Lincoln Theological College (1944-1945). **Interests:** walking, reading. **Hons.:** Honorary D.D., University of Birmingham (1972); Hon. B.D.S., Yale University (1965); Honorary Fellow, Keble College (1980) and Queen's College (1991), Oxford; Honorary Canon Emeritus, Bristol Cathedral (1983); Fellow, King's College, London (1963). **Pubs.:** The Gospel of St Mark, 1963; The Use and Abuse of the Bible, 1966 and Explorations in Theology, 1977.
Mailing address: 4 Wootten Drive, Iffley Turn, Oxford OX4 4DS, UK. T. (44 865) 715941
Residence: See Mailing Address. T. See Mailing Address

Nishitakatsuji, Nobuyoshi
Rel: Shinto; Ordained; b. 06 OCT 1953; Citizen of Japan.
International Organizations: Chief Priest of the Dazaifu Tenmangu.
Educ.: Keio University, Tokyo, Japan.
Mailing address: 4-7-7 Dazaifu, Dazaifu, Fukuoka 818-01, Japan. T. (81 92) 922 3555

Nissen, Karsten, Very Rev
Rel: Lutheran.
International Organizations: Chairperson, Program Committee for Mission and Development of the Lutheran World Federation.
Mailing address: Sct Leonisstraede 1, DK-8800 Viborg, Denmark. T. (45) 86 62 04 37. Fax 86 60 12 40

Niwa, Renbo
Rel: Buddhist; Ordained; b. 23 FEB 1905; Citizen of Japan.
International Organizations: Monk of Soto Zen Sect Eiheiji.
Mailing address: Eiheiji, Eiheijicho Shii, Yoshida-gun, Fukui 910-12, Japan. T. (81 776) 63 3102

Niwano, Nikkyo
Rel: Buddhist; b. 15 NOV 1906 in Tokamachi-City, Niigata Prefecture, Japan; m. to Sai Abé; c. 3 boys and 3 girls; Citizen of Japan; Lang. Japanese.
International Organizations: Founder-President (since 1938) of the Rissho Kosei-Kai.
Chairman (since 1977) of the Asian Conference on Religion and Peace. President (since 1978) of the Niwano Peace Foundation. Honorary President (since 1979) of the World Conference on Religion and Peace. Honorary President (since 1987) of the International Association for Religious Freedom.
Other organizations: Chairman (since 1965), Shinshuren (Federation of the New Religious Organizations of Japan); Trustee (since 1967), Japan Religions League; President (since 1972), Japanese Committee for World Conference on Religion and Peace.
Main field of work: Religious cooperation, peace activity. **Interests:** calligraphy, ink painting. **Hons.:** Imperial Household Agency Medal of Honour with Dark Navy Blue Ribbon (1958); Hon. LL.D., Meadville Lombard Theological School, Chicago IL (1975); Templeton Foundation Prize for Progress in Religion (1979); United Nations Plaque (1983). **Pubs.:** Buddhism for Today, 5 Vols., 1959-60 (in Japanese); A Modern Interpretation of the Threefold Lotus Sutra, 10 Vols., 1964-68 (in Japanese) and A Buddhist Approach to Peace, 1977 (in English).
Mailing address: Rissho Kosei-kai, 2-11-1 Wada, Suginami-ku, Tokyo 166, Japan. T. (81 3) 3383 1111

Njike, Charles E, Rev
Rel: Evangelical.
International Organizations: Secretary of the Evangelical Church of Cameroon.
Mailing address: BP 89, Douala, Cameroon. T. (237) 423611

Njuguna, John Macharia, Rt Rev
Rel: Christian.
International Organizations: Moderator of the African Christian Church and Schools.
Mailing address: PO Box 1365, Thika, Kenya.

Noakes, George, Most Rev
Rel: Anglican; Ordained in 1952; b. 13 SEP 1924 in Bwlchllan, Dyfed, Wales, UK; m. to Jane Margaretta Davies; Citizen of United Kingdom.
International Organizations: Archbishop (since 1987) of Wales.
Other organizations: Bishop (since 1982), St Davids.
Career: Ordained Deacon in 1950. Curate, Lampeter from 1950 to 1956. Vicar, Eglwyswrw with Meline from 1956 to 1959. Vicar, Tregaron from 1959 to 1967. Vicar, Eglwys Dewi Sant Cardiff from 1967 to 1976. Rector, Aberystwyth from 1976 to 1980. Canon, St Davids Cathedral from 1977 to 1979. Archdeacon, Cardigan from 1979 to 1982. Vicar, Llanychearn with Llanddeiniol from 1980 to 1982. Bishop, St Davids in 1982. **Educ.:** B.A. from University of Wales, University College of Wales, Aberystwyth, UK; Wycliffe Hall, Oxford, UK. **Interests:** angling.
Mailing address: Llys Esgob, Abergwili, Carmarthen, Dufed SA31 2JG, UK. T. (44 267) 236597

Noble, Ted
Rel: Christian.
International Organizations: President of the Greater Europe Mission.
Mailing address: PO Box 668, Wheaton IL 60189-0668, USA. T. (1 708) 462 8050.

Noguchi, Zenyu, Rev
Rel: Buddhist.
International Organizations: Chairman of the First Committee (Finance) of the World Fellowship of Buddhists.
Mailing address: c/o Japan Buddhist Federation – Meisho-Kaikan 2 F, 4-7-4 Shiba-koen, Minato-ku, Tokyo 105, Japan.

Noirot-Nerin, M
Rel: Roman Catholic.
International Organizations: Unesco Representative of the Caritas Internationalis.
Mailing address: Secours catholique, 106 rue du Bac, F-75341 Paris CEDEX 07, France.

Noirot-Nerin, M, Sr
Rel: Roman Catholic; Ordained; celibate.
International Organizations: UN Representative of the Caritas Internationalis.
Mailing address: 106 rue du Bac, F-75341 Paris CEDEX 07, France.

Nolan, Nancy, Sr
Rel: Roman Catholic; Ordained; celibate.
International Organizations: Superior General of the Sisters of Providence of St Mary-of-the-Woods.
Mailing address: St Mary-of-the-Woods IN 47876-1089, USA.

Noller, Charles Geoffrey, Rev Dr
Rel: Uniting Church in Australia; Ordained on 20 AUG 1968; b. 13 JUN 1934 in Sydney, New South Wales, Australia; m. to Dr Patricia Tate; c. 1 boy and 3 girls; Citizen of Australia; Lang. English.
International Organizations: Secretary-General (since 1987) of the Life Line International.
Main field of work: Ministry, teaching. **Career:** Trainee (Chemical Engineering) with Bristol Myers from 1951 to 1955. Chemical Engineer with S C Johnson from 1955 to 1960. Parish appointments (Uniting Church in Australia) from 1968 to 1970. Director, Life Line, Brisbane from 1971 to 1979. Executive Officer, Life Line Australia from 1978 to 1983. **Educ.:** B.Sc. from University of New South Wales, Sidney, Australia (1950-1957); B.A. from University of Sydney, New South Wales, Australia (1960-1963); Ph.D. from Australian National University, Canberra, Australia (1964-1967). **Interests:** impact of alcohol abuse on families, counseling practice. **Hons.:** USA Bicentennial Travelling Fellow (1976).
Mailing address: 34 Middle Street, West End, Brisbane 4101, Australia. T. (61 7) 844 1767
Residence: See Mailing Address. T. See Mailing Address

Nomoto, Shinya, Rev Prof Dr
Rel: United Church of Christ; Ordained on 05 DEC 1967; b. 12 OCT 1935 in Okayama, Japan; m. to Masuyo Ikenaga; c. 1 boy and 1 girl; Citizen of Japan; Lang. Japanese and German.
International Organizations: Professor, Dean (since 1979) of the School of Theology, Doshisha University.
Other organizations: Pastor (since 1967), Kamo Church, The United Church of Christ in Japan.
Main field of work: University education. **Educ.:** M.Th. from Doshisha University, Kyoto, Japan (1958-1960); Dr. Theol. from University of Hamburg, Germany (1962-1967). **Pubs.:** Co-author, Introduction to the Old Testament (in Japanese).
Mailing address: School of Theology, Doshisha University, Kavasuma Imadegawa, Kamikyo-ku, Kyoto 602, Japan. T. (81 75) 251 3330
Residence: 1-1 Yanaidacho, Matsugasaki, Sakyo-ku, Kyoto 606, Japan. T. (81 75) 701 6754

Nordholt, Gerhard, Dr
Rel: Evangelical Reformed.
International Organizations: Landessuperintendent of the Evangelical Reformed Church in North-West Germany.
Mailing address: Saarstrasse 6, 26789 Leer/Ostfriesland, Germany.

Norman, Kenneth Roy
Rel: Unknown; b. 21 JUL 1925; m. to Pamela Raymont; c. 1 boy and 1 girl.
International Organizations: President of the Pali Text Society.
Other organizations: Reader in Indian Studies (since 1978), University of Cambridge, UK.
Career: Fellow and Tutor, Downing College, University of Cambridge, UK from 1952 to 1964. Lecturer in Indian Studies (Prakrit), University of Cambridge, UK from 1955 to 1968. **Educ.:** M.A. from University of Cambridge, Downing College, UK (1954). **Interests:** Reading, walking. **Mem.:** Foreign Member, Royal Danish Academy of Sciences and Letters (1983); Fellow, British Academy (1985) **Pubs.:** Translator of *Jain Cosmology*, 1981; *Pali Literature*, 1983; *The Group of Discourses (Sutta-nipata)*, 1984; Editor, *Pali Tipitakath Concordance*, Vol. II 4-9, 1963-73 and Editor, *Critical Pali Dictionary*, Vol. II, 11-16, 1981-89.
Mailing address: 73 Lime Walk, Headington, Oxford OX3 7AD, UK. T. (44 865) 74 21 25

Norris, William P, Col
Rel: Salvation Army, The.
International Organizations: Territorial Commander of The Salvation Army.
Mailing address: PO Box 34352, Lusaka 10101, Zambia.

Norvila, Julius, Rev
Rel: Reformed Church.
International Organizations: Member of the Executive Committee of the World Alliance of Reformed Churches.
Mailing address: Vilnius Box 1243, Vilnius 1, Lithuania. T. Tx 269 857 22 (C/O VYTAUTAS MAGNUS UNIV.)
Residence: T. (7 95) 959 (IN KAUNAS)

Notelé, Frans Armand, Rev
Rel: Roman Catholic; Ordained on 14 SEP 1968; b. 12 AUG 1938 in Hoboken, Antwerp, Belgium; celibate; Citizen of Belgium; Lang. Dutch, French, German and Italian (some).
International Organizations: Chaplain (since 1988) of International Union of Prayer of Banneux Notre-Dame.
Mailing address: Rue de l'Esplanade 57, B-4141 Louveigné, Belgium. T. (32 41) 60 82 26.
Residence: Rue des Fawes 62, B-4141 Louveigné, Belgium. T. (32 41) 60 82 34

Nott, Peter John, Rt Rev
Rel: Anglican; Ordained in 1962; b. 30 DEC 1933 in Belfast, N Ireland, UK; m. to Elizabeth May Maingot; c. 1 boy and 3 girls; Citizen of United Kingdom.
International Organizations: Bishop (since 1985) of Norwich.
Other organizations: Vice-Chairman (since 1988), Archbishops' Commission on Rural Areas.

Career: Served in Regular Army, commissioned Royal Artillery from 1951 to 1955. Ordained Deacon in 1961. Curate, Harpenden from 1961 to 1964. Chaplain and Fellow, Fitzwilliam College, Cambridge from 1964 to 1969. Chaplain, New Hall, Cambridge from 1966 to 1969. Rector, Beaconsfield from 1969 to 1977. Bishop, Taunton from 1977 to 1985. **Educ.:** Royal Military Academy, Sandhurst, UK; M.A. from University of Cambridge, Fitzwilliam College, UK; Westcott House, Cambridge, UK. **Interests:** gardening, music, sketching, sport. **Mem.:** Royal Commonwealth Society, Norfolk (Norwich)
Mailing address: Bishop's House, Norwich NR3 1SB, UK. T. (44 603) 629001

Nottingham, William Jesse, Dr

Rel: Disciples of Christ; Ordained on 21 OCT 1945; b. 22 NOV 1927 in Sharon PA, USA; m. to Patricia Clutts; c. 2 boys and 2 girls; Citizen of USA; Lang. English, Spanish, French and German.
International Organizations: President, Division of Overseas Ministries (since 1984) of the Christian Churches, International Convention. President of the United Christian Missionary Society.
Other organizations: General Board (since 1984), National Council of Christian Churches, USA (NCCCUSA).
Main field of work: World mission administration. **Career:** Associate Minister, National City Christian Church, Washington DC from 1954 to 1958. Fraternal Worker, French Protestant Church from 1958 to 1965. Youth Dept. Staff, World Council of Churches from 1965 to 1968. Executive Secretary for Latin America and the Caribbean, Division of Overseas Ministries from 1968 to 1976. Executive Secretary for East Asia and the Pacific, Division of Overseas Ministries from 1976 to 1984. Chairperson, Europe/USSR Committee, NCCCUSA from 1988 to 1991. **Educ.:** B.A. from Bethany College, WV, USA (1945-1949); B.D. from Union Theological Seminary, New York NY, USA (1950-1953); Ph.D. from Columbia University, New York NY, USA (1962). **Interests:** literature. **Hons.:** Hon. D.D. (two). **Mem.:** Democratic Party **Pubs.:** The Practice and Preaching of Liberation; God's Underground (translated from French) and Christian Faith and Secular Action, an Introduction to the Life and Thought of Jacques Maritain.
Mailing address: Box 1986, Indianapolis IN 46206, USA. T. (1 317) 353 1491.
Residence: 50 North Kenmore Road, Indianapolis IN 46219, USA. T. (1 317) 359 7322

Novak, Michael

Rel: Roman Catholic; b. 09 SEP 1933 in Johnstown PA, USA; m. to Karen Laub; c. 1 boy and 2 girls; Citizen of USA; Lang. English, Italian, French (some) and Spanish (some).
International Organizations: George Frederick Jewett Chair in Religion and Public Policy/Director of Social and Political Studies (since 1988) of the American Enterprise Institute.
Other organizations: Member (since 1984), Board for International Broadcasting; Columnist (since 1989), *Forbes Magazine*; Serves on the editorial boards, several publications in the US and abroad.
Main field of work: Research and education. **Career:** Teaching Fellow at Harvard University, Cambridge MA from 1961 to 1963. Assistant Professor of Humanities at Stanford University, CA from 1965 to 1968. Professor at the newly formed experimental College at SUNY Old Westbury from 1968 to 1973. Ledden-Watson Distinguished Professor of Religion at Syracuse University, NY from 1976 to 1979. Visiting Professor at the University of Notre Dame, IN from 1987 to 1988. Launched the new humanities program at the Rockfeller Foundation from 1973 to 1974. Founder, Ethnic Millions Political Action Committee (EMPAC) in 1974. Advisor to EMPAC during President Ford and President Carter administrations. Joined the American Enterprise Institute as a Resident Scholar in 1978. Columnist, *National Review* from 1979 to 1986. Head, the US Delegation to the United Nations Human Rights Commission in Geneva (in 1981, 1982). Appointed to the Presidential Task Force on Project Economic Justice in 1985. Head (with the rank of ambassador), the US Delegation to the Experts' Meeting on Human Contacts at the Conference on Security and Cooperation in Europe in 1986. Judge, National Book Awards and DuPont Awards in Broadcast Journalism (for ten years). Co-founder, *This World*, *Crisis* and *First Things*. **Educ.:** B.A. (Summa Cum Laude) from Stonehill College, North Easton MA, USA (1956); B.T. (Cum Laude) from Pontifical Gregorian University, Vatican City State (1958); Theol. Stds. from Catholic University of America, Washington DC, USA (1958); M.A. Hist. and Phil. of Rel. from Harvard University, Cambridge MA, USA (1966). **Hons.:** numerous awards including: Freedom Award, Coalition for a Democratic Majority (1979); George Washington Honor Medal, Freedoms Foundation (1984); Award of Excellence, Religion and Media, 8th Annual Angel Awards (1985); first US member, Argentine National Academy of Sciences, Moral and Politics (1985); Ellis Island Medal of Honor (1986); and ten honorary degrees. **Mem.:** Knights of Malta; Council of Scholars **Pubs.:** Author of over 20 books in the area of philosophy, theology, politics, economics and culture, including: *The Spirit of Democratic Capitalism*, 1982; *Will It Liberate? Questions About Liberation Theology*, 1986 and author/editor of numerous monographs and over 200 articles and reviews published in popular journals including *The New Republic*, *The Atlantic*, and in scholarly journals as *Theology Today*, *The Yale Law Review*, *The Journal of Ecumenical Studies*, *The Review of Politics*.
Mailing address: 1150 17th Street NW, 10th Floor, Washington DC 20036, USA. T. (1 202) 862 5839.
Residence: 3211 Northampton St NW, Washington DC 20015, USA. T. (1 202) 966 4990

Novak, Miroslav, Rt Rev Dr

Rel: Hussite Church.
International Organizations: Patriarch of Czechoslovak Hussite Church.
Mailing address: V V Kujbyseva 5, 16626 Praha 6-Dejvice, Czech Rep.

Nowak, Edward, H E Mgr

Rel: Roman Catholic; Ordained on 13 JAN 1963; b. 21 FEB 1940 in Zmigród, Poland; celibate.
International Organizations: Secretary and President of the Study of the Congregation for the Causes of Saints.
Other organizations: Titular Archbishop, Luni.
Mailing address: Palazzo delle Congregazioni, Piazza Pio XII 10, I-00193 Roma, Italy. T. (39 6) 698 4244
Residence: Via dei Delfini 16, I-00186 Roma, Italy T. (39 6) 678 0546

Noyens, H, Fr Abbot

Rel: Roman Catholic; Ordained; celibate.
International Organizations: Contact of the Archconfraternity of the Mass of Reparation.
Mailing address: Norbertijnen-Abdij Tongerlo, Abdijstraat 40, B-2260 Westerlo, Belgium.

Nter, Antoine, Rev

Rel: Protestant.
International Organizations: Director-General of the African Protestant Church.
Mailing address: BP 26, Lolodorf, Cameroon.

Ntetome, Emile, Rev

Rel: Evangelical.
International Organizations: Secretary of the Evangelical Church of Gabon.
Mailing address: BP 10080, Libreville, Gabon.

Nthamburi, Zablon, Rev Br

Rel: Christian.
International Organizations: Treasurer of the Conference of African Theological Institutions.
Mailing address: PO Box 50784, Nairobi, Kenya. T. (254 2) 74 22 81 – 74 22 93 – 74 24 10 Ext 234. C. CHURCHCON. Tx 22175 AACC.

Ntontolo, K Lukombo, Rev

Rel: Evangelical.
International Organizations: President of the Church of Christ in Zaire (Evangelical Community).
Mailing address: BP 36, Luozi, Zaire.

Nuchelmans, Jan, Prof

Rel: Roman Catholic.
International Organizations: Head, Europe Regional Secretariat of the Pax Romana, International Catholic Movement for Intellectual and Cultural Affairs.
Mailing address: 40 Viesseltsebaan, NL-6564 CS Nijmegen, Netherlands.

Nukaga, Hirooki

Rel: Shinto; Ordained; b. 05 APR 1908; Citizen of Japan.
International Organizations: Chief Priest of the Nikko Toshogu.
Educ.: Jingu Kogakukan University, Japan.
Mailing address: 2329 Yamauchi, Nikko, Tochigi 321-14, Japan. T. (81 288) 54 0033

Nurrie, Rick
Rel: Theosophist.
International Organizations: Head, Secretariat of the Theosophical Network.
Mailing address: PO Box 155, Muskogee OK 74402, USA.

Nuss, Joseph
Rel: Jewish.
International Organizations: Co-President of the Consultative Council of Jewish Organizations.
Other organizations: President, Canadian Friends of the Alliance Israélite Universelle.
Mailing address: Canadian Friends of the AIU, BOP 578, Victoria Station, Montréal H3Z 2Y6, Canada.

Nussbaumer, Louise
Rel: Mennonite.
International Organizations: Member, Executive Committee of the Mennonite World Conference.
Mailing address: 2 rue des Aubépines, F-67120 Molsheim, France.

Nutter, Harold Lee, Most Rev
Rel: Church of Canada; Ordained in 1947; b. 29 DEC 1923 in Welsford NB, Canada; m. to Edith M Carew; c. 1 boy and 1 girl; Citizen of Canada.
International Organizations: Ecclesiastic of the Church of Canada.
Career: Rector, Parish of Simonds and Upham from 1947 to 1951. Rector, Parish of Woodstock from 1951 to 1957. Rector, Parish of St Mark, Saint John, NB from 1957 to 1960. Dean, Fredericton from 1960 to 1971. Bishop, Fredericton from 1971 to 1980. Archbishop, Fredericton from 1980 to 1989. Metropolitan, Ecclesiastical Province of Canada from 1980 to 1989. Examining Chaplain to Bishop, Fredericton from 1960 to 1971. Co-Chair, New Brunswick Task Force on Social Development from 1970 to 1971. Vice-Chair, New Brunswick Police Commission in 1988. **Educ.:** M.A., M.S.Litt. from Mount Allison University, Sackville, New Brunswick, Dalhousie University, Halifax, and University of King's College, Halifax, Canada. **Interests:** gardening, skiing. **Hons.:** Hon. D.D., King's College, Halifax, Montreal Diocesan College, Wycliffe College, Toronto, Trinity College, Toronto; Hon. LL.D., Mount Allison University. **Pubs.:** Co-author, *New Brunswick Report on Social Development*, 1971.
Mailing address: No mailing address.

Nyanja, Peter Nathaniel, Rt Rev
Rel: Christian; Ordained; b. 10 JUN 1940 in Malawi; m. to Irene Matrida Kayamba; c. 7 boys and 1 girl; Citizen of Malawi.
International Organizations: Diocesan Bishop (since 1978) of Diocese of Lake Malawi.
Career: Primary school teacher from 1963 to 1967. Parish priest from 1972 to 1977. **Interests:** gardening.
Mailing address: Diocese of Lake Malawi, PO Box 30349, Capital City, Lilongwe 3, Malawi. T. (265) 731966
Residence: T. (265) 722670

Nzeyimana, Noah, Bishop
Rel: Free Methodist.
International Organizations: Member of the Free Methodist World Fellowship.
Mailing address: BP 667, Bujumbura, Burundi.

Obaid, Soliman Mohammad
Rel: Muslim.
International Organizations: Director of the Islamic Call Service.
Career: Clerk, Public Notary Department, Mecca in 1956. Radio Announcer from 1963 to 1971. Director, Nida Al Islam (The Call of Islam) and the Holy Koran Radio, Jeddah in 1972. Member, Saudi Arabian Delegation to the Conference of the Islamic States Broadcasting Services in 1975.
Mailing address: Jeddah Broadcasting Service, PO Box 570, Jeddah, Saudi Arabia.

Obaikol, Charles, Rev Canon
Rel: Roman Catholic.
International Organizations: Provincial Secretary of the Church of Uganda.
Mailing address: PO Box 14123, Kampala, Uganda.

Obando Bravo, Miguel O, S.D.B., H E Card
Rel: Roman Catholic; Ordained on 10 AUG 1958; b. 02 FEB 1926 in La Libertad (Chontales), Nicaragua; celibate; Citizen of Nicaragua.
International Organizations: Member (since 1985) of the College of Cardinals.
Member of the Society of St Francis of Sales. Member of the Pontifical Commission for Latin America. Member of the Congregation for the Clergy.
Other organizations: Archbishop, Managua.
Career: Consecrated Bishop (Titular Church of Puzia di Bizacena) in 1968. Archbishop in 1970. Created Cardinal by Pope John Paul II in 1985. Chairman, National Reconciliation Commission for Nicaragua in 1987.
Mailing address: Arzobispado, Apartado 3058, Managua, Nicaragua. T. (505 2) 71 754

Obinah-Aguh, Philip
Rel: Roman Catholic.
International Organizations: Member of the Symposium of Episcopal Conferences of Africa and Madagascar.
Mailing address: 4 Senchi Street, PO Box 9156, Airport, Accra, Ghana. T. (233) 777706 – 774724. C. SECAM ACCRA. Tx 2587 SECAM GH.

O'Brien, John, C.P.M., Fr
Rel: Roman Catholic; Ordained; celibate.
International Organizations: Superior General of the Congregation of the Priests of Mercy.
Mailing address: 14480 Bowling Green Rd, South Union Owensboro KY 42283, USA. T. (1 502) 542 4146

O'Brien, Keith Michael Patrick, Most Rev
Rel: Roman Catholic; Ordained on 03 APR 1965; b. 17 MAR 1938 in Ballycastle, Co. Antrim, Northern Ireland, UK; celibate; Citizen of United Kingdom; Lang. English.
International Organizations: Archbishop (since 1985) of St Andrews and Edinburgh.
Main field of work: Ministry. **Career:** Teacher/Chaplain, St Columba's Secondary School, Dumferline from 1966 to 1971. Assistant Priest, St Patrick's Church, Kilsyth from 1972 to 1975. Assistant Priest, St Mary's Catholic Church, Bathgate from 1975 to 1978. Spiritual Director, St Andrew's College, Drygrange from 1978 to 1980. Rector, Blairs College, Aberdeen from 1980 to 1985. **Educ.:** B.Sc. from University of Edinburgh, UK (1955-1959); St Andrew's College, Drygrange, UK (1959-1965); Dip. Ed. from University of Edinburgh, UK (1965-1966).
Mailing address: Archdiocesan Offices, 106 Whitehouse Loan, Edinburgh EH9 1BD, UK. T. (44 31) 452 8244
Residence: St Bennet's, 42 Greenhill Gardens, Edinburgh EH10 4BJ, UK. T. (44 31) 447 3337

Ochs, M Roberta, O.S.B., Sr
Rel: Roman Catholic; Ordained; b. 17 DEC 1933 in Lucinda PA, USA; celibate; Citizen of USA; Lang. English.
International Organizations: Directress, International Secretariat (since 1985) of the World Apostolate of Fatima.
Other organizations: Benedictine Sister (since 1952), Benedictine Sisters of Elk County, St Marys PA.
Main field of work: Religious retreat work, ministry. **Career:** Elementary Education Teacher, Diocese of Erie PA from 1956 to 1967. Secondary Education Teacher, Diocese of Erie PA from 1968 to 1972. Religious Education Teacher, Diocese of Erie PA from 1973 to 1985. **Educ.:** M.Counselor Ed. from Duquesne University, Pittsburgh PA, USA (1974-1978). **Interests:** church music.
Mailing address: Avenue Beato Nuno, Apartado 38, P-2496 Fatima, Portugal. T. (351 49) 532 931
Residence: See Mailing Address. T. See Mailing Address

O'Collins, Gerald Glynn, S.J., Rev Prof
Rel: Roman Catholic; Ordained on 02 JAN 1961; b. 02 JUN 1931 in Melbourne, Victoria, Australia; celibate; Citizen of Australia; Lang. English, Italian, German, Latin and Greek.
International Organizations: Professor of Fundamental Theology (since 1974) of the Pontifical Gregorian University.

Main field of work: University education, research. **Career:** Research Fellow at Pembroke College, Cambridge from 1967 to 1969. Honoured Visitor at Pembroke College, Cambridge from 1973 to 1974. Dean of Theology, Pontifical Gregorian University from 1985 to 1991. **Educ.:** M.A. (1st) from University of Melbourne, Victoria, Australia (1955-1958); S.T.L. (1st) from Heythrop College, Oxford, UK (1965-1967); Ph.D. from University of Cambridge, UK (1965-1968). **Hons.:** Malipiero Prize (1987); Hon. Doc., University of San Francisco (1991). **Mem.:** Society of Jesus (1963) **Pubs.:** Fundamental Theology, New York, 1981; Interpreting Jesus, London, 1983 and Jesus Risen, New York, 1987.
Mailing address: Pontificia Università Gregoriana, Piazza della Pilotta 4, I-00187 Roma, Italy. T. (39 6) 6701 5360

O'Connor, Jay Francis, Rev

Rel: Roman Catholic; Ordained on 14 DEC 1974; b. 15 DEC 1948 in Baltimore MD, USA; celibate; Citizen of USA; Lang. English, French and Latin.
International Organizations: Executive Director (since 1991) of the National Conference of Diocesan Vocation Directors.
Main field of work: Vocation ministry. **Educ.:** B.A. from Saint Mary's Seminary and University, Baltimore MD, USA (1966-1970); M.S.T. from Saint Mary's Seminary and University, Baltimore MD, USA (1970-1974); M.Sc. from Johns Hopkins University, Baltimore MD, USA (1982-1985).
Mailing address: 1603 South Michigan Ave, Suite 400, Chicago IL 60616, USA. T. (1 312) 663 5456
Residence: 2311 North Southport Ave, Chicago IL 60614, USA. T. (1 312) 281 7317

O'Connor, John Joseph, H E Card

Rel: Roman Catholic; Ordained on 15 DEC 1945; b. 15 JAN 1920 in Philadelphia PA, USA; celibate; Citizen of USA.
International Organizations: Member (since 1985) of the College of Cardinals.
Member of the Congregation for Bishops. Member of the Congregation for the Evangelization of Peoples. Member of the Pontifical Council for the Pastoral Care of Migrants and Itinerant People. Member of the Pontifical Council for Pastoral Assistance to Health Care Workers. Member of the Pontifical Council for Social Communications.
Other organizations: Archbishop (since 1984), New York; Member, Council of Cardinals for the Study of Organizational and Economic Problems of the Holy See; Member, Council of Cardinals and Bishops; Member, Institute for Religious Works.
Career: Served in Chaplain Corps, beginning as lieutenant major, US Navy in 1952. Retired as Rear Admiral in 1979. Elevated to Monsignor in 1966. Consecrated Bishop in 1979. Auxiliary Bishop, Vicar-General (Military Vicariate) from 1979 to 1983. Bishop, Scranton in 1983. Created Cardinal by Pope John Paul II in 1985. **Educ.:** M.A. Adv. Ethics and Clin. Psy. from St Charles Borromeo Seminary, USA (1949); Catholic University of America, Washington DC, USA (1954); Ph.D. Pol.Sc. from Georgetown University, Washington DC, USA (1970); D.R.E. from Villanova University, PA, USA (1976). **Hons.:** Legion of Merit Award. **Pubs.:** A Chaplain Looks at Vietnam, 1969; In Defense of Life, 1981 and With Edward I Koch, His Eminence and Hizzoner, 1989.
Mailing address: Archdiocese of New York, 452 Madison Ave, New York NY 10022, USA. T. (1 212) 371 1000

Oddi, Silvio, H E Card

Rel: Roman Catholic; Ordained on 21 MAY 1933; b. 14 NOV 1910 in Morfasso, Italy; celibate; Citizen of Italy.
International Organizations: Member (since 1969) of the College of Cardinals.
Career: Titular Archbishop, Mesembria in 1953. Apostolic Delegate to Palestine, Jerusalem and Cyprus from 1953 to 1956. Apostolic Internuncio in Egypt and UAR from 1956 to 1962. Apostolic Nuncio in Belgium and Luxembourg from 1962 to 1969. Created Cardinal by Pope Paul VI in 1969. Pontifical Delegate for the Basilica of Assisi. Prefect, Sacred Congregation for the Clergy.
Mailing address: Via Pompeo Magno 21, I-00192 Roma, Italy. T. (39 6) 321 3329

O'Donnell, Brian Phillip, O.H., Br

Rel: Roman Catholic; Ordained; b. 07 OCT 1933 in Forbes NSW, Australia; celibate; Citizen of Australia; Lang. English and Italian.
International Organizations: Superior General (since 1988) of the Hospitaller Order of St John of God.
Other organizations: President (since 1989), Fate Bene Fratelli International Foundation.
Main field of work: Administration. **Career:** Provincial Superior, St John of God Brothers, Australasia from 1974 to 1982. **Educ.:** Teachers Dip. from Bouverie St College, Melbourne, Australia (1956-1957). **Pubs.:** The Charism of St John of God in the Europe of Tomorrow; St John of God – Servant and Prophet and St John of God Lives On.
Mailing address: Fate Bene Fratelli, Via della Nocetta 263, I-00164 Roma, Italy. T. (39 6) 663 8937.
Residence: See Mailing Address. T. See Mailing Address

O'Dwyer, Malachy, O.P., Fr

Rel: Roman Catholic; Ordained; celibate.
International Organizations: Procurator-General of the Order of Preachers.
Mailing address: Piazza Pietro d'Illiria 1, I-00153 Roma, Italy. T. (39 6) 579 45 55. Fax 575 06 75.

Ogarepke, Mba Ogarepke, Rt Rev

Rel: Presbyterian.
International Organizations: Moderator of the Presbyterian Church of Nigeria.
Mailing address: 2 Effiwatt St, Box 207, Calabar, Cross River State, Nigeria.

Ogilvie, Renate

Rel: Buddhist: Mahayana; b. 01 MAY 1949 in Weimar, Thüringen, Germany; m. to David John Ogilvie; c. 1 boy; Citizen of Germany; Lang. German and English.
International Organizations: Member of the Foundation for the Preservation of the Mahayana Tradition.
Other organizations: Psychotherapist, London and Sydney (since 1988); 1991, Shakyamuni Buddha Hospice; Director.
Main field of work: Training of counsellors. **Career:** Literary Agent, London from 1977 to 1985. Editor, London from 1985 to 1987. **Educ.:** M.A. from Johann Wolfgang Goethe-Universität Frankfurt, Germany (1969-1975); M.A. from Antioch University, London, UK (1987-1989). **Interests:** Mahayana Buddhism. **Mem.:** Society for Psychoanalytic Psychotherapy (USA); Society for Existential Counselling (UK); Australia and New Zealand Association of Psychotherapy
Mailing address: PO Box A 453, Sydney South NSW 2000, Australia. T. (61 2) 967 2020. Fax (61 2) 906 5086

Ogle, Robert Joseph, Fr

Rel: Roman Catholic; Ordained in 1953; b. 24 DEC 1928 in Rosetown, Saskatchewan, Canada; celibate; Citizen of Canada.
International Organizations: Founder (since 1985) of the Broadcasting for International Understanding.
Career: Various pastoral activities, Diocese of Saskatoon in 1953. Missionary in N.E. Brazil (for 6 years). Elected to House of Commons for Saskatoon-East (NDP) from 1979 to 1984. Former President, Western Conference of Priests. **Educ.:** B.A. from University of Western Ontario, London, Canada (1949); J.C.D. from University of Ottawa, Ontario, Canada (1957); D.D. from University of Saskatchewan, Saskatoon, Canada (1989); D.D. from University of Ottawa, Saint Paul University, Ontario, Canada (1989). **Interests:** friends. **Hons.:** Member, Order of Canada (1989). **Pubs.:** When the Snake Bites the Sun, 1977; North-South Calling, 1986 and A Man of Letters, 1990.
Mailing address: 249 Main St, Ottawa, Ontario K1S 1C5, Canada.
Residence: 220, 2902 Louise Street, Saskatoon, Saskatchewan S7J 3M2, Canada.

Ogushi, Kazuo

Rel: Shinto; Ordained; Citizen of Japan.
International Organizations: Chief Priest of the Tado Shrine.
Career: Chief Priest, Nogi Shrine. **Educ.:** Kogakukan University, Japan.
Mailing address: 1612 Tadocho Tado, Kuwana-gun, Mie 511-01, Japan. T. (81 594) 48 2065

O'Hair, Madalyn Evalyn, Dr

Rel: Atheist; b. 13 APR 1919 in Pittsburgh PA, USA; widowed Richard Franklin O'Hair; c. 2 boys and 1 girl; Citizen of USA; Lang. English and French.

International Organizations: President of the United World Atheists.

Other organizations: Founder (since 1959), American Atheist Organizations; Founder (since 1963), American Atheists, Inc.; Founder (since 1972), Charles E Stevens American Atheist Library; Editor Emeritus (since 1985), American Atheist Press.

Main field of work: Law (state/church separation) and research (atheism). **Career:** President, American Atheists, Inc. from 1963 to 1985. Editor-in-Chief, *American Atheist* Magazine from 1959 to 1981. **Educ.:** B.A. from Ashland University, OH, USA (1948); LL.B., J.D. from South Texas School of Law, Houston TX, USA (1952); MPSW from Howard University, Washington DC, USA (1956). **Interests:** atheism, civil rights of atheists. **Hons.:** Honoree, National Women's History Week (1982). **Pubs.:** *An Atheist Epic*, 1990; *Why I Am an Atheist*, 1991 and Editor, *The Atheist Viewpoint*, Arno Press (New York Times).

Mailing address: PO Box 140195, Austin TX 78714-0195, USA. T. (1 512) 458 1244. Fax 467 9525.

Residence: 7215 Cameron Rd, Austin TX 78752, USA. T. See Mailing Address

O'Hare, Joseph Aloysius, S.J., Rev

Rel: Roman Catholic; Ordained; b. 12 FEB 1931 in New York, USA; celibate; Citizen of USA.

International Organizations: President (since 1984) of the Fordham University, NY.

Member of the Society of Jesus.

Career: Instructor in Humanities at Ateneo de Manila University from 1955 to 1958. Associate Professor in Philosophy, Ateneo de Manila University from 1967 to 1972. Associate Editor, *America Magazine*, New York from 1972 to 1975. Editor-in-Chief, *America Magazine*, New York from 1975 to 1984. **Interests:** contemporary fiction, Irish folk music. **Hons.:** Hon. D.H.L., Fairfield University, Rockhurst College; Hon. D.L., College of New Rochelle.

Mailing address: Fordham University, Bronx NY 10458, USA. T. (1 212) 579 2003

Ohashi, Shunyu

Rel: Buddhist; Ordained; b. 11 FEB 1913; Citizen of Japan.

International Organizations: President of the Kasei Gakuen.

Career: Vice-Chairman, Japan Junior College Association. **Educ.:** Taisho University, Japan.

Mailing address: 34 Ayakojidori Omiya Nishi Hairu, Shijo Omiyacho, Shimogyo-ku, Kyoto 600, Japan. T. (81 75) 841 2088

Ohtani, Monshu Koshin

Rel: Buddhist: Jôdo-Shinshû; Ordained in 1960; b. 12 AUG 1945 in Kyoto, Japan; m. to Noriko Tanaka; c. 2 boys and 2 girls; Citizen of Japan.

International Organizations: President (since 1988) of the All-Japanese Buddhist Federation.

Career: Monshu (Ecclesiastic Patriarch) Apparent from 1970 to 1977. President, All-Japanese Buddhist Federation from 1978 to 1980. **Interests:** study of literature, skiing.

Mailing address: Horikawa-dori, Hanayacho-sagaru, Shimogyo-ku, Kyoto 600, Japan. T. (81 75) 371 5181

Okamoto, Kenji

Rel: Shinto; Ordained; b. 04 MAR 1924; Citizen of Japan.

International Organizations: Chief Priest of the Atsuta Shrille.

Career: Executive Director, Association of Shinto Shrine. Head, Aichi Prefecture Sint Shrine Agency. **Educ.:** Jingokogakukan University, Japan.

Mailing address: 4-83 Takemicho, Mizuho-ku, Nagoya, Aichi 467, Japan. T. (81 52) 832 9813

Okano, Shokan

Rel: Buddhist; b. 22 DEC 1920; Citizen of Japan.

International Organizations: Governor of the Kodo kyodan.

Career: Chairman, International Buddhism Exchange Center. **Educ.:** Waseda University, Tokyo, Japan.

Mailing address: 38 Torigoe, Kanagawa-ku, Yokohama, Kanagawa 221, Japan. T. (81 45) 432 1201

Oke, Philip, Dr

Rel: Christian.

International Organizations: UN Representative of the Christian Peace Conference.

Mailing address: c/o NACPC, 777 United Nations Plaza, Room 1, New York NY 10017-3521, USA. T. (1 516) 223 1880. Fax 223 1880.

O'Keeffe, Augustine, Br

Rel: Christian.

International Organizations: Executive Director of the Christian Relief and Development Association.

Mailing address: PO Box 5674, Addis Ababa, Ethiopia. T. (251 1) 167100 – 163046. Fax 652280. C. CRDA. Tx 21307 CRDA ET.

Okogie, Anthony Olubunmi, H E Mgr

Rel: Roman Catholic; Ordained on 11 DEC 1966; b. 16 JUN 1936 in Lagos, Nigeria; celibate; Citizen of Nigeria.

International Organizations: President (since 1988) of the Catholic Bishops' Conference of Nigeria.

Other organizations: Archbishop (since 1973), Lagos; National President (since 1988), Christian Association of Nigeria.

Career: Acting Parish Priest, St Patrick's Church, Idumagbo, Lagos. Assistant Priest, Holy Cross Cathedral, Lagos. Religious Instructor, King's College, Lagos. Director of Vocations, Archdiocese of Lagos. Manager, Holy Cross Group of Schools, Lagos. Master of Ceremonies, Holy Cross Cathedral, Lagos. Broadcaster, religious programmes, NBC/TV. Auxiliary Bishop, Oyo Diocese from 1971 to 1972. Auxiliary Bishop to the Apostolic Administration, Archdiocese, Lagos from 1972 to 1973. Vice-President, Catholic Bishops' Conference of Nigeria from 1985 to 1988. **Interests:** reading. **Mem.:** Prerogative of Mercy, Religious Advisory Council

Mailing address: Archdiocese of Lagos, 19 Catholic Mission Street, PO Box 8, Lagos, Nigeria. T. (234 1) 633841 – 635729

Okuda, Munehiro

Rel: New Religion.

International Organizations: Director of the Universal Love and Brotherhood Association.

Mailing address: Kameoka-City, Kyoto 621, Japan. T. (81 75) 1232145.

Okure, Teresa, Sr Prof Dr

Rel: Roman Catholic; Ordained; b. 25 OCT 1941 in Anua, Uyo, Akwa Ibom State, Nigeria; celibate; Citizen of Nigeria; Lang. Ibo, English, French, Italian, German and others.

International Organizations: Executive Secretary (since 1986) of the Ecumenical Association of Third World Theologians.

Head of Department and Academic Dean (since 1989) of the Catholic Institute of West Africa.

Other organizations: Member, Editorial Board, Research and Publications Board (since 1989), Catholic Theological Association of Nigeria (CATHAN).

Main field of work: Professor of Sacred Scripture. **Career:** Head of Library, Catholic Institute of West Africa from 1988 to 1989. **Educ.:** B.A. (Hons) from National University of Ireland, University College, Dublin, Ireland (1967-1970); M.A., Ph.D. from Fordham University, Bronx NY, USA (1977-1985); Elève Titulaire from Biblical School, Jerusalem, Israel (1981-1983). **Interests:** People, music, reading, languages. **Mem.:** Phi Beta Kappa, Society of Biblical Literature (SBL), Ecumenical Association of Third World Theologians (EATWOT), Catholic Theological Association of Nigeria (CATHAN), Pontifical Council for Justice and Peace (PACE), AIUW, CBA, EAAT, NABIS **Pubs.:** *The Johannine Approach to Mission*, WUNT, 2/31, Tubingen; "Inculturation: Biblical/Theological Bases", *Spearhead*, 112-14; "Leadership in the New Testament", *NJT*, 1, 5, 1990; "Women in the Bible", *With Passion and Compassion* and "Gustavo Gutierrez", *Future of Liberation Theology.*

Mailing address: PO Box 499, Port Harcourt, Nigeria. T. (234 84) 33 09 61. Fax (234 84) 33 29 52

Residence: See Mailing Address. T. See Mailing Address

Okuyama, Minoru, Rev

Rel: Evangelical.

International Organizations: Secretary-General of the Asia Missions Association.

Mailing address: PO Box 1, Takaku, Nasu 325-03, Japan. T. (81 287) 78 1585. Fax 78 2532.

Olang', Festo Habakkuk, Most Rev
Rel: Anglican; Ordained; b. 14 NOV 1914 in Maseno, Kenya; m. to Eseri; c. 4 boys and 8 girls; Citizen of Kenya.
International Organizations: Ecclesiastic of the Anglican Church.
Career: Teacher, Maseno Secondary School from 1936 to 1939. On staff, Butere Girls' School from 1940 to 1945. Ordained Deacon in 1945. Consecrated Bishop in 1955. Bishop, Maseno from 1961 to 1970. Archbishop, Kenya from 1970 to 1979.
Mailing address: PO Box 1, Maseno, Kenya.

Olaniyi, David Olawale, Rev Dr
Rel: Baptist; Ordained on 04 OCT 1970; b. 10 SEP 1937 in Ara Via Ede, Osun State, Nigeria; m. to Ruth Olufunmilayo; c. 2 boys and 2 girls; Citizen of Nigeria; Lang. English, Hausa and Yoruba.
International Organizations: Director of Sunday School (since 1986) of the Nigerian Baptist Convention.
Other organizations: Member, Current Affairs Panel of Discussion (since 1986), Federal Radio Corporation of Nigeria; Member, Executive Committee (since 1988), Christian Association of Nigeria; Member, Executive Committee (since 1990), Nigerian Baptist Convention.
Main field of work: New Testament studies. **Career:** Headmaster, Primary Schools in Ede, and Teacher at Secondary (Modern) School, Ede from 1962 to 1964. Pastor, Songbe Baptist Church, Ede from 1964 to 1966. Pastor, Mayin and Ago-Ofa Baptist Churches, Ogbomoso from 1966 to 1968. Pastor, Kano First Baptist Church from 1969 to 1972. Manager, Baptist Schools, Kano from 1969 to 1975. Education Officer, Ministry of Education, Kano from 1975 to 1977. Principal, Joint Baptist Pastors' School, Oyo from 1977 to 1982. Lecturer (part-time), Federal College of Education, Oyo from 1978 to 1982. Lecturer (part-time), Oyo State College of Education, Ilesa from 1981 to 1982. Lecturer (full-time), Federal College of Education, Oyo from 1982 to 1983. Pastor, Idi-Ape Baptist Church, Ejigbo, Osun from 1983 to 1986. Member, Executive Committee, Nigerian Baptist Convention from 1984 to 1986. Conference Secretary, Oyo East Baptist Conference, Osogbo from 1984 to 1986. Secretary, Ministerial Board, Nigerian Baptist Convention from 1986 to 1990. Interim Pastor, University of Ibadan Baptist Church in 1990. Acting Secretary General, Nigerian Baptist Convention from 1990 to 1991. **Educ.:** B.Th. from Nigerian Baptist Theological Seminary, Ogbomoso, Nigeria (1964-1968); B.A. Ed. from Ahmadu Bello University, Zaria, Nigeria (1972-1975); M.A. Rel. Stds. from University of Ife, Ile-Ife, Nigeria (1981-1982); Ph.D. from Obafemi Awolowo University, Ado-Ekiti, Nigeria (1984-1990). **Interests:** writing, reading. **Mem.:** Christian/Muslim Peace Movement of Nigeria, Ibadan (1990) **Pubs.:** *Seven Profitable Psalms*, Baptist Press Limited, 1987, pp. 52; *Titus: The Faithful Servant*, Baptist Press Limited, 1988, pp. 39 and contribution to several books, including: *Seven Famous Women of the Bible: Sarah, Ruth, Deborah, Esther, Eve, Hannah – The Virtuous Women*, Dr W R Ola Ojo, Ed., Baptist Press Limited, 1980.
Mailing address: Baptist Building, PMB 5113, Ibadan, Nigeria. T. (234 22) 412308 – 412146
Residence: Molete-Challenge, Ibadan 5113, Nigeria. T. (234 22) 316578

Olayiwola, David Omoleke Folagbade, Rev Dr
Rel: Evangelical; Ordained on 24 APR 1981; b. 10 MAY 1950 in Ikole-Ekiti, Ondo State, Nigeria; m. to Rachel Lola Bosede; c. 1 boy and 1 girl; Citizen of Nigeria; Lang. Yoruba and English.
International Organizations: Senior Lecturer, Department of Religious Studies (since 1981) of the Obafemi Awolowo University, Ile-Ife.
Main field of work: University education. **Educ.:** University of Ife, Ile-Ife, Nigeria (1977-1986). **Interests:** music. **Mem.:** International Religious Foundation **Pubs.:** *Aladura Spirituality in an Urban Setting: The Impact of African Independent Churches on Ijesa-Yormba Community* (forthcoming); "The Social Impact of New Religious Movement", in *Africana Marburgensia* and "Church and the Healing Ministry in Nigeria", in *Asia Journal of Theology*, Singapore.
Mailing address: Dept. of Religious Studies, Obafemi Awolowa University, Ile-Ife, Nigeria.
Residence: 9-11 Olayemi Street, Off Fadahunsi Ave, Ilesa, Nigeria.

Olckers, Roy, Commr
Rel: Salvation Army, The.
International Organizations: Territorial Commander of The Salvation Army.
Mailing address: PO Box 1018, Johannesburg 2000, South Africa.

Oles, Marian, H E Mgr
Rel: Roman Catholic; Ordained on 09 JUL 1961; b. 08 DEC 1935 in Miastkowo; celibate.
International Organizations: Apostolic Pro-Nuncio to Iraq and Kuwait (since 1987) of the Roman Catholic Church.
Other organizations: Titular Archbishop, Raziara.
Mailing address: PO Box 2090, Saadoun Street 904/2/46, Baghdad, Iraq. T. (964 1) 719 51 83

Oliver, Victor
Rel: Evangelical.
International Organizations: Chairman of the Living Bibles International.
Mailing address: 2413 Windon Court, Doraville GA 30340, USA. T. (1 404) 391 9781 – (1 404) 457 5024.

Olivera, Bernardo, O.C.R., Fr Abbot
Rel: Roman Catholic; Ordained; celibate.
International Organizations: Abbot-General of the Cistercian Order of the Strict Observance.
Mailing address: Viale Africa 33, I-00144 Roma, Italy. T. (39 06) 592 41 55

Ollarves Sanchez, Felix Octavio O
Rel: Roman Catholic; Unordained; b. 20 NOV 1956 in Barquisimeto, Venezuela; m. to Moncia Leon; c. 1 boy; Citizen of Venezuela; Lang. Spanish, French and English.
International Organizations: International President (since 1987) of the International Young Christian Workers.
Main field of work: Education. **Pubs.:** *Info Internacional.*
Mailing address: Rue Plantin 11, B-1070 Bruxelles, Belgium. T. (32 2) 521 69 83. C. JOCINT BRUXELLES.
Residence: 34 rue Peter Benoit, B-1040 Bruxelles, Belgium T. (32 2) 640 90 05

Ols, Daniel Jean Henri, Prof Dr
Rel: Roman Catholic; Unordained on 19 SEP 1971; b. 25 JUN 1942 in Paris, France; celibate; Citizen of France; Lang. French, Italian and German.
International Organizations: Secretary (since 1985) of the Société internationale Thomas d'Aquin.
Other organizations: Professor of Theology (since 1974), Pontifical University S. Tommaso d'Aquino, Vatican City State and others; Member, Academic Council (since 1987), Pontifical Roman Academy of Theology.
Main field of work: Theology education. **Educ.:** Lettres classiques from University of Paris IV (Paris-Sorbonne), France (1959-1963); Dr.theol. from Pontifical University San Tommaso d'Aquino, Vatican City State (1971-1973). **Interests:** Theology. **Mem.:** Ordinary Member, Pontifical Roman Academy of Theology **Pubs.:** *Cristología*, Buenos Aires, CEIOP, 1981; "Diversités et communion", in *Angelicum*, 60, pp.122-150, 1983 and "Scorciatoie ecumeniche", in *L'Osservatore Romano*, pp. 1-5, Feb. 1985.
Mailing address: Piazza della Minerva 42, I-00186 Roma, Italy. T. (39 6) 699 06 72
Residence: See Mailing Address. T. See Mailing Address

Olschanksy, Pedro J
Rel: Jewish.
International Organizations: Contact of the Latin American Jewish Congress.
Mailing address: Larrea 744, 1030 Buenos Aires, Argentina. T. (54 1) 961 4534 – 962 5028. Fax 961 4534. Tx 18490 WJC AR.

Olson, James
Rel: Unitarian.
International Organizations: Representative at the UN of the Unitarian Universalist Association.
Mailing address: 777 United Nations Plaza, Room 7D, New York NY 10017, USA. T. (212) 986 5265.

Olson, Jarle M
Rel: Lutheran.
International Organizations: Director of the Lutheran Brethren World Missions.
Mailing address: 1007 Westside Drive, PO Box 655, Fergus Falls MN 56537, USA. T. (1 218) 739 3336.

Oltra Vidal, José O, TC, Fr
Rel: Roman Catholic; Ordained; celibate.
International Organizations: Superior General of the Third Order Capuchin of Our Lady of Sorrows.
Member of the Third Order Capuchin of Our Lady of Sorrows.
Mailing address: Via Bernardo Blumenstihl 28-36, I-00135 Roma, Italy. T. (39 6) 305 59 31.

Olufosoye, Timothy Omotayo, Most Rev
Rel: Anglican; Ordained on 15 DEC 1946; b. 31 MAR 1918 in Ondo, Ondo State, Nigeria; m. to Joan; c. 2 boys and 3 girls; Citizen of Nigeria; Lang. English.
International Organizations: Archbishop (since 1979) of Nigeria.
Career: Bishop, Gambia and Rio Pongas from 1965 to 1970. Bishop, Ibadan in 1970. **Educ.:** University of British Columbia, Vancouver, Canada. **Interests:** poultry, reading, writing, travelling. **Hons.:** Officer, Order of the Niger; Knight Commander of African Redemption; Justice of the Peace. **Pubs.:** Glossary of Ecclesiastical Terms; My Memoirs and Ibadan Ecclesia Anglicana.
Mailing address: 27 Jimekun St, PO Box 1666, Ondo, Nigeria.
Residence: 27 Jimekun St, PO Box 675, Ondo, Nigeria. T. (234) 610634

Olutimayin, Nathaniel L
Rel: Protestant.
International Organizations: President of the Evangelical Churches of West Africa.
Mailing address: PO Box 63, Jos, Nigeria. T. (234 73) 54482.

O'Meara, Noel P, Rev
Rel: Roman Catholic; Ordained; b. 22 DEC 1937 in Gort Galway, Ireland; celibate; Citizen of Ireland; Lang. English, French, Portuguese, Spanish and Italian.
International Organizations: Secretary-General (since 1990) of the Congregation of the Holy Spirit.
Main field of work: Ministry. **Career:** Dean, Templeogue College, Dublin from 1966 to 1972. Assistant Provincial, Congregation of the Holy Spirit (CSSp) from 1973 to 1977. Pastor in Santos, Brazil from 1978 to 1983. Director, Brazilian Spiritan Missions, New York from 1983 to 1987. Pastor in Sao Paulo, Brazil from 1987 to 1990. **Educ.:** B.A. Ed. from National University of Ireland, University College, Dublin, Ireland (1958-1962); B.D. from Kimmage College, Dublin, Ireland (1962-1966); M.Ed. from University of Dublin Trinity College, Ireland (1971-1973); Ph.D. from Fordham University, Bronx NY, USA (1984-1987).
Mailing address: Clivo di Cinna 195, I-00136 Roma, Italy. T. (39 6) 34 82 47. Fax 345 06 76.
Residence: See Mailing Address. T. See Mailing Address

O'Neill, John Cochrane, Rev Prof
Rel: Church of Scotland; Ordained on 22 NOV 1960; b. 08 DEC 1930 in Melbourne, Victoria, Australia; m. to Judith Beatrice Lyall; c. 3 girls; Citizen of United Kingdom; Lang. English and German.
International Organizations: Professor of New Testament Language, Literature and Theology (since 1985) of The University of Edinburgh.
Main field of work: University education. **Career:** Dunn Professor of New Testament Language, Literature and Theology, Westminster College, Cambridge from 1964 to 1985. **Educ.:** B.A. (Hons.) from University of Melbourne, Victoria, Australia (1948-1951); B.D. from Melbourne College of Divinity, Victoria, Australia (1952-1954); Ph.D. from University of Cambridge, UK (1956-1959). **Interests:** swimming, gardening, walking. **Mem.:** Society for New Testament Studies (SNTS) **Pubs.:** Paul's Letter to the Romans, Harmondsworth, Penguin, 1975; Messiah: Six Lectures on the Ministry of Jesus, Cambridge, Cochrane, 1980, 2nd Edn. 1984 and The Bible's Authority: A Portrait Gallery of Thinkers from Lessing to Bultruann, Edinburgh, T and T Clark, 1991.
Mailing address: Dept. of New Testament, The University of Edinburgh, New College, The Mound, Edinburgh EH1 2LX, Scotland, UK. T. (44 31) 225 8400
Residence: 9 Lonsdale Terrace, Edinburgh EH3 9HN, UK. T. (44 31) 229 6070

Ono, Kaen
Rel: Buddhist; Ordained; b. 10 SEP 1917; Citizen of Japan.
International Organizations: Chief of the Horyuji.
Educ.: Ryukoku University, Kyoto, Japan.
Mailing address: 1-1 Ikaruga Horyuji, Ikoma-gun, Nara 636-01, Japan. T. (81 7457) 5 2555

Oommen, Tharaileth Koshy, Prof Dr
Rel: Christian; Unordained; b. 23 OCT 1937 in Venmoney, Kerala, India; m. to Josephine John; c. 2 boys; Citizen of India; Lang. English, Malayalam and Hindi.
International Organizations: Member of the Council (since 1989) of the International Society for the Sociology of Religions.
Vice-Moderator, Church and Society (since 1987) of the World Council of Churches. President (since 1990) of the International Sociological Association.
Other organizations: Professor (since 1976), Jawaharlal Nehru University, New Delhi; Social Science Consultant to the National Commission on Theological Education, Serampore, Calcutta.
Main field of work: University education. **Career:** Lecturer, University of Delhi from 1964 to 1970. Reader, University of Delhi from 1970 to 1971. Associate Professor, Jawaharlal Nehru University, New Delhi from 1971 to 1976. **Educ.:** B.A. from Trivandrum, Kerala, India (1955-1957); M.A. from University of Pune, Maharashtra, India (1958-1960); Ph.D. from University of Pune, Maharashtra, India (1961-1964). **Interests:** sociology of religion, political sociology. **Hons.:** VKRV Rao Prize (1981), G S Gurye Award (1985). **Mem.:** International Sociological Association, Indian Sociological Society **Pubs.:** *Charisma, Stability and Change: An Analysis of Bhoodan Gramdan Movement in India*, Thompson Press, New Delhi, 1972; *From Mobilization to Institutionalization: The Dynamics of Agrarian Movement in 20th Century Kerala*, Popular Prakashan, Bombay, 1985 and *Protest and Change: Studies in Social Movements*, Sage Publications, New Delhi, 1990.
Mailing address: Ctr. for the Study of Social Systems, Sch. of Social Sc., 5 Jawaharlal Nehru University New Campus, New Mehrauli Road, New Delhi 110 067, India. T. (91 11) 686 4242
Residence: See Mailing Address. T. See Mailing Address

Opocensky, Milan, Rev Dr
Rel: Reformed Church; Ordained on 06 MAR 1955; b. 05 JUL 1931 in Hradec Kralové, Czech Rep.; m. to Jana Juránková; c. 2 boys and 1 girl; Citizen of Czech Rep.; Lang. Czech, English, German, Russian and French.
International Organizations: Secretary-General (since 1989) of the World Alliance of Reformed Churches.
Other organizations: Moderator (since 1984), Board of Ecumenical Institute in Bossey, Geneva.
Main field of work: University education. **Career:** Lecturer, Comenius Faculty of Protestant Theology, Prague, Czechoslovakia from 1954 to 1967. European Secretary, World Student Christian Federation (WSCF) from 1967 to 1973. Professor of Theology, Comenius Faculty of Protestant Theology, Prague, Czechoslovakia from 1973 to 1990. Chairman, European Region, WSCF from 1974 to 1979. Member, Board of Bossey from 1976 to 1983. **Educ.:** M.Div. from Comenius Evangelical Faculty of Theology, Prague, Czechoslovakia (1950-1954); Dr. Theol. from Comenius Evangelical Faculty of Theology, Prague, Czechoslovakia (1965). **Hons.:** Hon. Dr. Theol., University Faculty of Protestant Theology, Brussels (1984); Hon. D.D., College of Wooster, OH (1986). **Pubs.:** *Christians and Revolutions*, Geneva, 1977; *Widerstand und Revolution*, Munich, 1982 and Editor, *Sprung über die Mauer*, Wuppertal, 1990.
Mailing address: 150 route de Ferney, PO Box 2700, CH-1211 Genève 2, Switzerland. T. (41 22) 791 62 37 – (41 22) 791 62 38. C. WARC Geneva. Tx 415 730 OIK CH (WARC). Fax (41 22) 791 65 05 WARC Reformierter Weltbund.
Residence: 4 place de l'Etrier, Chêne-Bougeries, CH-1224 Genève, Switzerland. T. (41 22) 48 46 24

Orbegozo Jáuregui, José Agustin O, C.P., Rev
Rel: Roman Catholic; Ordained on 02 APR 1960; b. 01 FEB 1937 in Goyaz, Guipúzcoa, Spain; celibate; Citizen of Spain; Lang. Basque, Spanish, Italian, English, French and Portuguese.
International Organizations: Superior General (since 1988) of the Congregation of the Passion.
Main field of work: Ministry, administration. **Career:** Professor of Holy Scripture, Teologado de Urretxu, Guipúzcoa from 1964 to 1968. Professor of Holy Scripture, Seminario de San Sebastián from 1969 to 1971. Superior in Urretxu from 1969 to 1973. Provincial Councillor, Bilbao from 1973 to 1977. Director, Seminary Goiz-Argi from 1975 to 1977. Professor of Holy Scripture, Seminario de San Sebastián from 1978 to 1979. Professor of Holy Scripture, Facultad de Teologia de Bilbao from 1978 to 1979. Provincial Superior, Bilbao from 1981 to 1988. Member, team for the translation of the Bible into Basque from 1970 to 1975. **Educ.:** Lic. Theol. from Pontifical University San Tommaso d'Aquino, Vatican City State (1960-1962); Lic. Bible from Pontifical

Biblical Institute, Vatican City State (1962-1964). **Interests:** Religious life, Holy Scriptures, Basque culture. **Pubs.:** Various articles in *Zeruko Argia, Goiz-Argi, Orleti, Redención.*
Mailing address: Piazza SS Giovanni e Paolo 13, I-00184 Roma, Italy. T. (39 6) 700 57 58.
Residence: See Mailing Address. T. See Mailing Address

Orcasitas, Gómez Miguel Angel, O.S.A., Fr

Rel: Roman Catholic; Ordained; celibate.
International Organizations: Prior-General of the Order of St Augustine.
Mailing address: Via del S Uffizio 25, I-00193 Roma, Italy. T. (39 6) 686 45 14

Ord, John, Commr

Rel: Salvation Army, The.
International Organizations: Territorial Commander of The Salvation Army.
Mailing address: Box 6866, St Olavs Plass, N-0130 Oslo 1, Norway.

O'Reilly, Maureen

Rel: Roman Catholic.
International Organizations: International Coordinator of the Association internationale Foi et lumière.
Mailing address: 3 rue du Laos, F-75015 Paris, France. T. (33 1) 47 34 33 08.

Ornauer, Helmut

Rel: Roman Catholic.
International Organizations: Director of the Coordination Centre for the International Development Promotion of the Austrian Bishops' Conference.
Mailing address: Türkenstrasse 3, A-1090 Wien, Austria. T. (43 1) 34 03 21 – (43 1) 34 03 22.

Orsmond, Reginald Joseph, Rt Rev

Rel: Roman Catholic; Ordained on 08 DEC 1954; b. 18 JUN 1931 in Johannesburg, South Africa; celibate.
International Organizations: Bishop (since 1984) of Synod of Bishops.
Mailing address: PO Box 17054, Hillbrow 2038, South Africa. T. (27 11) 648 6451

Ortega-Montoya, Ofelia, Rev

Rel: Presbyterian; Ordained on 14 FEB 1967; b. 01 JUN 1936 in Cardenas, Mantanzas, Cuba; m. to Daniel Montoya; c. 1 girl; Citizen of Cuba; Lang. Spanish, English and French.
International Organizations: Director, Programme on Theological Education (since 1988) of the World Council of Churches.
Other organizations: Pastor (since 1967), Presbyterian-Reformed Church in Cuba; Organizer (since 1986), Women's seminars at the Ecumenical Institute, Bossey.
Main field of work: Pastoral and feminist theology. **Career:** Leader, Student Christian Movement in Cuba from 1960 to 1965. Director, Study Center in Cuba from 1968 to 1969. Professor, Theological Faculty in Matanzas, Cuba from 1960 to 1984. Member, Commission 'Renewal in Education', World Council of Churches (WCC). Delegate to the IV General Assembly, WCC, Uppsala, Sweden in 1968. Adviser to the V General Assembly, WCC, Nairobi, Kenya in 1975. Member, Commission of International Affairs, WCC from 1975 to 1983. Adviser to several Central Committee meetings, WCC. Secretary-General, Programme Agency, Presbyterian Church in Cuba from 1962 to 1985. Editor, a new Curriculum for the Churches in Cuba from 1975 to 1984. Lecturer at the Ecumenical Institute, Ecumenical Institute of the WCC, Bossey, Switzerland from 1985 to 1988. Professor, Ecumenical Institute, Bossey from 1989 to 1991. Coordinator of Section III 'Unity and Reconciliation', General Assembly, WCC in 1991. Bible Study leader in different international meetings in the USA, Europe, Latin America and the Caribbean. Organizer, the first Association of Women in Theology and Pastoral Work, Buenos Aires, Argentina in 1989. Lecturer in international meetings in the USA, Europe, Latin America and the Caribbean. Associate Convener of Unit I: Unity and Renewal, WCC. **Educ.:** B.A. from Colegio 'La Progresiva', Cuba (1951-1956); B.Ch.Ed. from Faculty of Theology, Matanzas, Cuba (1953-1959); B.Th. from Faculty of Theology, Matanzas, Cuba (1964-1966); P.G. Theol. at The Ecumenical Institute, Bossey, Switzerland (1967-1968); M.Div. from Faculty of Theology, Matanzas, Cuba (1979-1980); M.Ed. from University of Havana, Cuba (1989). **Pubs.:** "Women Discovering Mutuality: Luke 1:34-45, 56", in *Search for New Community*, Edited by Thomas F Best, WCC Publications, Bossey, 1987; "Jesucristo: Su Resurrección y Ascensión", in *Dialogo Intermundos*, CELEPA (Centró Evangélico Latinoamericano de Estudios Pastorales), San José, Costa Rica, 1989 and "Lateinamerikanische Theologie der Befreiung", in *Andere Christus*, Verlag der Ev.-Luth. Mission, Erlangen, 1991.
Mailing address: PO Box 2100, Route de Ferney 150, CH-1211 Genève 2, Switzerland. T. (41 22) 791 63 33. Fax 791 03 61. C. OIKOUMENE GENEVA. Tx 415 730 OIK CH.
Residence: The Ecumenical Institute, CH-1298 Céligny, Switzerland. T. (41 22) 776 15 12.

Ortiz, Angel Miguel, Rev

Rel: Christian; b. 19 JAN 1959 in Matanzas, Cuba; single; Citizen of Cuba; Lang. Spanish and English.
International Organizations: Executive Secretary for Latin America and the Caribbean (since 1988) of the World Student Christian Federation.
Main field of work: Academic community. **Career:** Chairman, Student Christian Movement in Cuba from 1985 to 1988. Pastor, The Christian Reformed Church in Cuba from 1984 to 1988. **Educ.:** Eng. Animal Production from University of Havana, Cuba (1978-1984); Theology from Theological Seminary, Matanzas, Cuba (1982-1985). **Interests:** Relations between religion and politics.
Mailing address: 9 de Octubre 158 y Patria, Oficina 402, Quito, Ecuador. T. (593 2) 541 204 – (593 2) 544 804. Fax (593 2) 568 664
Residence: 6 de Diciembre y Eloy Alfaro (Torres del Norte), Quito, Ecuador T. (593 2) 440 104

Oruba, Véronique

Rel: Roman Catholic.
International Organizations: President and Secretary-General of the European Coordination, International Young Catholic Students – International Movement of Catholic Students.
Mailing address: Rue du Marteau 19, B-1040 Bruxelles, Belgium. T. (32 2) 218 54 37. Fax 219 83 96.

Osborne, James, Commr

Rel: Salvation Army, The; Ordained on 02 JUN 1947; b. 03 JUL 1927 in Toledo OH, USA; m. to Ruth Campbell; c. 1 girl; Citizen of USA; Lang. English.
International Organizations: National Commander (since 1986) of The Salvation Army.
Board Member (since 1989) of the Christian Children's Fund.
Other organizations: Board Member (since 1989), Christian Management Association; Board Member (since 1990), National Association of Evangelicals.
Main field of work: Ministry, administration. **Career:** Commander for The Salvation Army's Southern Territory. Divisional Secretary for Kentucky-Tennessee. General Secretary for North and South Carolina. Public Relations Secretary at the Territorial Headquarters. Divisional Commander for Maryland and Northern West Virginia, National Capital/Virginias, and Florida. Served on various national commissions, The Salvation Army. Chairman, The Salvation Army National Planning and Development Commission (for 2 terms). Chief Secretary, The Salvation Army's Western Territory. National Chief Secretary (second in command to the National Commander), The Salvation Army at the National Headquarters in Verona NJ from 1984 to 1986. **Educ.:** Officer from School for Officers' Training, Atlanta GA, USA (1947); Studied from International College for Officers, London, UK (1964). **Hons.:** Hon. Dr. of Public Service, Gordon College, Wenham MA (1991).
Mailing address: 615 Slaters Lane, PO Box 269, Alexandria VA 22348, USA. T. (1 703) 684 5505
Residence: 8328 Fort Hunt Road, Alexandria VA 22308-1815, USA. T. (1 703) 781 0254

Osburn, Carroll Duane, Prof

Rel: Christian; b. 02 SEP 1941 in Arkansas City KS, USA; m. to Linda Carol; c. 2 girls; Citizen of USA; Lang. English, German, French, Greek, Syriac, Latin, Coptic, Ethiopic, Georgian and Hebrew.
International Organizations: Carmichael Distinguished Professor of New Testament (since 1987) of the Abilene Christian University, TX.

Main field of work: Graduate education. **Career:** Professor of New Testament, Harding Graduate School of Religion from 1973 to 1983. Professor of Religion, Pepperdine University from 1983 to 1987. Chairman, New Testament Textual Christian Section, Society of Biblical Literature from 1984 to 1990. Editor, International Project on the Text of Acts in 1985. Vice-Chairman, North American Committee, International Greek New Testament Project in 1989. **Educ.:** D.Div. from Vanderbilt University, Nashville TN, USA (1968-1970); Ph.D. from University of St Andrews, UK (1970-1973). **Interests:** aerobatic aviation. **Hons.:** Educator of the Year (1983); 20th Century Christian. **Mem.:** Society of Biblical Literature; International Association of Patristic Studies **Pubs.:** "The Interpretation of Romans 8:28", in *Westminster Theological Journal*, 44, 1982, pp. 99-109; "The Historical Present in Mark as a Text Critical Criterion", in *Biblica*, 64, 1983, pp. 486-500 and "1 En 80:2-8 (67:5-7) and Jude 12-13", in *Catholic Biblical Quarterly*, 47, 1985, pp. 296-303.
Mailing address: ACU Station, Box 8425, Abilene TX 79699, USA. T. (1 915) 674 3799
Residence: 1210 Nottingham Road, Abilene TX 79602, USA.

Osei, Joseph, Rev Fr
Rel: Roman Catholic; Ordained; celibate.
International Organizations: Chairman of the Symposium of Episcopal Conferences of Africa and Madagascar.
Mailing address: 4 Senchi Street, PO Box 9156, Airport, Accra, Ghana. T. (233) 777706 – 774724. C. SECAM ACCRA. Tx 2587 SECAM GH.

Osel Rinpoche, Tenzin O, Lama
Rel: Buddhist: Mahayana; b. 12 FEB 1985.
International Organizations: Reincarnation of Lama Thubten Yeshe, founder of the Foundation for the Preservation of the Mahayana Tradition.
Mailing address: Jetsa Osel Labrang, House No 103, Sera Monastery, PO Bylakuppe 571 104, Mysore Karnataka, India.

Oshita, Isshin
Rel: Buddhist; Ordained; b. 02 JUL 1948; Citizen of Japan.
International Organizations: Monk of Rinzaishu Zuisenji.
Educ.: Komazawa University, Tokyo, Japan.
Mailing address: 710 Nikaido, Kamakura, Kanagawa 248, Japan. T. (81 467) 22 1191

Oskam, Cees
Rel: Christian.
International Organizations: Director-General of the Interchurch Organization for Development Cooperation.
Mailing address: Zusterplein 22A, PO Box 151, 3700 AD Zeist, Netherlands. T. (31 3404) 24844. Tx 47644. C. Develop. Fax (31 3404) 256 14.

Osorno Gil, Jesús Emilio O, M.X.Y., Fr
Rel: Roman Catholic; Ordained; celibate.
International Organizations: Superior General of the Foreign Missions of Yarumal.
Mailing address: Carrera 81 No 52 B-120, Medellín, Colombia. T. (57 4) 234 09 21.

Osswald, Suzanne
Rel: Theosophist.
International Organizations: Head, European Office of the World Theosophical Youth Federation.
Mailing address: Enstone Guest House, Lennox Avenue, Sidmouth 4444, UK.

Osswald, Walter, Prof
Rel: Roman Catholic; b. 20 SEP 1928 in Porto, Portugal; m. to Domingas Branco Osswald; c. 2 boys and 4 girls; Citizen of Portugal; Lang. Portuguese.
International Organizations: Secretary-General (since 1986) of the International Federation of Catholic Medical Associations.
Other organizations: Full Professor (since 1972), University of Porto, Portugal; Chairman, Dept. of Pharmacology (since 1988), University of Porto Medical School, Portugal.
Main field of work: Medical research and teaching. **Career:** Professor, University of Porto, Portugal from 1968 to 1972. **Educ.:** M.D. from University of Porto, Portugal (1945-1951); Ph.D. (1958). **Interests:** Literature, music, history. **Hons.:** several Portuguese scientific prizes. **Mem.:** Member of Scientific Societies in Portugal and abroad **Pubs.:** Co-editor of the only Portuguese textbook on Pharmacology and Therapeutics; Author of several chapters in textbooks and Author of almost 200 scientific articles and a number of publications dealing with ethical and moral aspects of the practice of medicine.
Mailing address: Rua do Passeio Alegre 822-2T, P-4100 Porto, Portugal. T. (351 2) 68 40 01.
Residence: See Mailing Address. T. See Mailing Address

Otani, Kocho
Rel: Buddhist; Ordained; b. 01 OCT 1903; Citizen of Japan.
International Organizations: Chief of the Higashi-Monganji.
Educ.: Otani University, Japan.
Mailing address: 754 Karasumadori Hichijoagaru Tokiwacho, Shimogyo-ku, Kyoto 600, Japan. T. (81 75) 371 2922

Otani, Koshin
Rel: Buddhist; Ordained; b. 12 AUG 1945; Citizen of Japan.
International Organizations: Chief of the Nishi-Honganji.
Main field of work: Monk. **Career:** Manager, Honpa Honganji. **Educ.:** University of Tokyo, Japan.
Mailing address: Horikawadori Hanayashiki Shita, Shimogyo-ku, Kyoto 600, Japan. T. (81 75) 371 5181

Ottaviano, Antonio, S.d.C., Fr
Rel: Roman Catholic; Ordained; celibate.
International Organizations: Procurator-General of the Servants of Charity.
Mailing address: Vicolo Clementi 41, I-00148 Roma, Italy. T. (39 6) 655 24 54

Otunga, Maurice Michael, H E Card
Rel: Roman Catholic; Ordained on 03 OCT 1950; b. in Chebukwa, Kenya; celibate; Citizen of Kenya.
International Organizations: Member (since 1973) of the College of Cardinals.
Member of the Congregation for the Evangelization of Peoples. Member of the Congregation for Religious and Secular Institutes. Member of the Congregation for Catholic Education.
Other organizations: Archbishop (since 1971), Nairobi; Member, Council of Cardinals for the Study of Organizational and Economic Problems of the Holy See.
Career: Consecrated Titular Bishop, Tacape in 1957. Bishop, Kisii in 1961. Titular Archbishop, Bomarzo in 1969. Created Cardinal by Pope Paul VI in 1973.
Mailing address: Archbishop's House, PO Box 14231, Nairobi, Kenya. T. (254 2) 49 620

Ouazan, E
Rel: Jewish.
International Organizations: UNESCO Representative of the International Council of Jewish Women.
Mailing address: 131 Bd Gen Koenig, F-92200 Neuilly, France. T. (33 1) 46 24 00 15.

Outen, George H
Rel: Methodist.
International Organizations: Co-Secretary-General of the United Methodist Church, Board of Church and Society.
Mailing address: 160 Maryland Avenue NE, Washington DC 20002, USA. T. (202) 488 5600.

Overman, Helen Mary, SND, Sr

Also know as Sister Mary Joell; Rel: Roman Catholic; Ordained; b. 30 SEP 1938 in Covington KY, USA; celibate; Citizen of USA; Lang. English and German.
International Organizations: Superior General (since 1986) of the Sisters of Notre Dame.
Career: Administrator, St Claire Medical Centre, Morehead KY from 1969 to 1983. Provincial Superior, Sisters of Notre Dame, Covington KY from 1983 to 1986. **Educ.:** Ed.B. Secondary from Thomas More College, Crestview Hills, KY, USA (1958-1963); M.H.A. from Saint Louis University, MO, USA (1967-1969). **Interests:** collaboration in the Church, ethics, health care. **Hons.:** Thomas More Medallion Award.
Mailing address: Via della Camilluccia 687, I-00135 Roma, Italy. T. (39 6) 327 91 44.
Residence: See Mailing Address. T. See Mailing Address

Owen, D H, Rev

Rel: Presbyterian.
International Organizations: Clerk of the General Assembly of the Presbyterian Church of Wales.
Mailing address: 53 Richmond Road, Cardiff CF2 3UP, Wales, UK.

Owens, Joseph, Rev

Rel: Roman Catholic; Ordained; b. 17 APR 1908 in St John, New Brunswick, Canada; celibate; Citizen of Canada.
International Organizations: Professor, School of Graduate Studies (since 1961) of the University of Toronto.
Other organizations: Staff Member (since 1954), Pontifical Institute of Medieval Studies, Toronto.
Career: Lecturer in Philosophy, St Alphonsus Seminary, Woodstock ON from 1936 to 1940. Missionary to Dawson Creek BC from 1940 to 1944. Lecturer in Philosophy, St Alphonsus Seminary, Woodstock ON from 1948 to 1951. Lecturer, Accademia Alfonsiana, Rome from 1952 to 1953. Assistant to Associate Professor, University of Toronto from 1954 to 1961. **Educ.:** M.S.D. from Pontifical Institute of Medieval Studies, Toronto, Canada (1951). **Interests:** swimming. **Hons.:** Hon. D.Litt., Mount Allison (1975); Hon. D.Hum.Litt., Catholic University (1984); Hon. LL.D., St Francis Xavier University, Antigonish (1988). **Pubs.:** *St Thomas Aquinas on the Existence of God*, 1980; *Human Destiny*, 1985 and *Towards a Christian Philosophy*, 1990.
Mailing address: St Patrick's Rectory, 141 McCaul St, Toronto, Ontario M5T 1W3, Canada.

Oyer, Juanita

Rel: Unknown.
International Organizations: President of the Licentiate Ministers and Certified Mediums Society.
Mailing address: 6216 SE 10th St, Oklahoma City OK 73110, USA.

Pacaut, Marcel, Prof

Rel: Roman Catholic; b. 30 OCT 1920 in Lyon, France; m. to Anne-Marie Lanier; Citizen of France; Lang. French, English and Italian.
International Organizations: President (since 1955) of the International Commission of Comparative Church History.
Director of the European Centre on Religious Orders and Congregations.
Other organizations: Professor (since 1956), University of Lyon II, France.
Main field of work: Higher education. **Career:** Assistant (history), University of Paris IV from 1948 to 1952. Seconded, Centre National de la Recherche Scientifique from 1952 to 1953. Professor, Lycées, Paris from 1953 to 1956. **Educ.:** Agrég. Hist. from University of Lyon II, France (1940-1946); D. ès L. from University of Paris IV (Paris-Sorbonne), France (1948-1954). **Hons.:** Croix des Combattants Volontaires de la Résistance. **Pubs.:** *Alexandre III*, 1956; *Louis VII et son royaume*, 1964; *La Papauté de l'origine à nos jours*, 1976; *L'ordre de Cluny*, 1986 and *La théocratie*, 1989.
Mailing address: Université Lumière, 88 rue Pasteur, F-69006 Lyon, France. T. (33) 78 69 24 45
Residence: 3 rue Duquesne, F-69006 Lyon, France. T. (33) 78 93 17 73

Pacciolla, Aureliano

Rel: Roman Catholic.
International Organizations: Member of the Carmelite Third Order.
Mailing address: Via Sforza Pallavicini 10, I-00193 Roma, Italy.

Pace, Enzo, Prof

Rel: Roman Catholic; b. 03 FEB 1944 in Chieti, Italy; married; c. 1 boy; Citizen of Italy; Lang. Italian, French and English.
International Organizations: Secretary-General (since 1990) of the International Society for the Sociology of Religions.
Main field of work: University teaching. **Educ.:** Law from University of Padova, Italy (1963-1967); Phil. from University of Padova, Italy (1968-1970). **Pubs.:** *Asceti e Mistici in una Società Secolarizzata*, Venezia, 1983; *Sapere Technico e Sapere Concreto*, Bologna, 1984; *La Società Parallela*, Milano, 1985; *Dizionario di Sociologia*, Assisi, 1985 and *Tensioni e Tendenze delli America di Reagan*, Padova, 1989.
Mailing address: University of Padova, Dept. of Sociology, Via Andreini 12, I-35100 Padova, Italy. T. (39 49) 657 508.
Residence: Via Tre Garofami 49, I-35100 Padova, Italy T. (39 49) 685 958

Padilla, Osvaldo, H E Mgr

Rel: Roman Catholic; Ordained on 20 FEB 1966; b. 05 AUG 1942 in Sogod, Philippines; celibate.
International Organizations: Apostolic Nuncio to Panama (since 1990) of the Roman Catholic Church.
Other organizations: Titular Archbishop, Pia.
Mailing address: Punta Paitilla, Apartado 4251 – Zona 5, Panama. T. (507) 693 138 – 692 102. Fax 642 116

Padiyara, Anthony, H E Card

Rel: Roman Catholic; Ordained on 19 DEC 1945; b. 11 FEB 1921 in Changanacherry, India; celibate; Citizen of India.
International Organizations: Member (since 1988) of the College of Cardinals.
Member of the Sacred Congregation for the Oriental Churches. Member of the Pontifical Council for Promoting Christian Unity.
Other organizations: Archbishop, Ernakulam; President, Episcopal Conference Siro-Malabarese.
Career: Elected to Ootacamund in 1955. Consecrated bishop in 1955. Prefect to Changanacherry dei Siro-Malabaresi in 1970. Transferred to Ernakulam in 1985. Created Cardinal by Pope John Paul II in 1988.
Mailing address: The Archdiocesan Curia, Post Bag N° 2580 Ernakulam, Cochin 682031, Kerala, India. T. (91 484) 352 629 – 353 756

Page, Rose, OCD, Sr

Rel: Roman Catholic; Ordained; celibate.
International Organizations: Member of the Discalced Carmelite Nuns.
Other organizations: Member, Association of Contemplative Sisters; Editor, *Living Prayer.*
Mailing address: Carmelite Monastery, Beckley Hill, Barre VT 05641, USA.

Pagura, Federico, Bishop

Rel: Evangelical Methodist.
International Organizations: Bishop of Evangelical Methodist Church of Argentina.
Mailing address: Rivadavia 4044, 1205 Buenos Aires, Argentina.

Paige, John R, C.S.C., Br

Rel: Roman Catholic; Ordained; celibate.
International Organizations: Member of the Congregation of Holy Cross.
Other organizations: President, National Assembly of Religious Brothers.
Mailing address: 1337 West Ohio Street, Chicago IL 60622-6490, USA. T. (1 312) 829 8525

Paik, Mary, Rev

Rel: Reformed Church.
International Organizations: Member of the Executive Committee of the World Alliance of Reformed Churches.
Mailing address: 27075 W Nine Mile Road, Southfield MI 48034, USA. T. (1 313) 356 4488
Residence: T. (1 313) 350 1580

Paisley, Ian Richard Kyle, Rev

Rel: Presbyterian; Ordained in 1946; b. 06 APR 1926; m. to Eileen Emily Cassells; c. 2 boys and 3 girls; Citizen of United Kingdom.
International Organizations: President (since 1979) of the Whitefield College of the Bible Laurencetown, Co. Down.
Other organizations: Minister (since 1946), Martyrs Memorial Free Presbyterian Church, Belfast; Moderator (since 1951), Free Presbyterian Church of Ulster; Leader (co-founder) (since 1971), Democratic Unionist Party; Member (since 1974), Parliament for North Antrim; Member (since 1979), European Parliament.
Career: Founded *The Protestant Telegraph* in 1966. Member, Parliament (Protestant Unionist) for Bannside, Co. Antrim from 1970 to 1974. Leader, the Opposition and Chairman of Public Accounts Committee in 1972. Member, Northern Ireland Association from 1973 to 1974. Elected to Second Northern Ireland Association in 1982. Member, Constitutional Convention from 1975 to 1976. Chairman, Agricultural Committee and Committee of Privileges in 1983. Co-Chairman, World Congress of Fundamentalists in 1978. Member, Political Committee, European Parliament. **Educ.:** South Wales Bible College, UK; Reformed Presbyterian Theological College, Belfast, N Ireland, UK. **Mem.:** International Cultural Society of Korea (1977); Fellow, Royal Geographical Society **Pubs.:** Ulster the Facts, 1981; No Pope Here, 1982 and Paisley's Pocket Preacher, 1987.
Mailing address: The Parsonage, 17 Cyprus Avenue, Belfast BT5 5NT, UK.

Pal, Raj Sawindar, Mahatma

Rel: Hindu; b. 01 AUG 1922 in Lyallpur, Punjab, India (now in Pakistan); celibate; Citizen of India; Lang. Hindi, Sanskrit and English.
International Organizations: Head (since 1980) of the Metaphysical Research Ashram.
Main field of work: Education and research. **Career:** Joined the Indian Army in 1947. Renounced to the world in 1948. Studied the leading religions of the world and travelled extensively visiting ashrams and meeting holy men. Gave up home, wealth and marriage in 1950. Developed the philosophy of 'Samata Yoga'. **Educ.:** B.A. from University of the Punjab, Lahore, Pakistan (1942-1946); M.Sc. Physics (uncompleted) from University of the Punjab, Lahore, Pakistan (1946-1947). **Hons.:** Highest Order in Religion. **Pubs.:** *Samata Yoga* (in English and in Hindi), Vision Books, Delhi.
Mailing address: Devi Place, Barsana, District Mathura, Uttar Pradesh, India.
Residence: See Mailing Address.

Palau, Luis

Rel: Protestant; Ordained; b. 27 NOV 1934 in Buenos Aires, Argentina; m. to Patricia Marilyn Scofield; c. 4 boys; Citizen of USA.
International Organizations: Founder (since 1978) of the Luis Palau Evangelistic Association.
Career: Member, staff, Bank of London, Buenos Aires and Cordoba from 1952 to 1959. Moved to USA in 1960. Worked as interpreter for Billy Graham in 1962. Began Spanish radio broadcasts as missionary in Colombia in 1967. Began evangelistic ministry as part, Overseas Crusades in 1968. Made crusades broadcasts to all Latin America in 1975. Named President, Overseas Crusades in 1976. First major crusade in USA, San Diego in 1981. Crusades on four continents in 1982. Launched 12-year crusade in India in 1988. **Interests:** family. **Hons.:** Dr. h.c., Talbot Theological Seminary (1977), Wheaton College (1985). **Pubs.:** Time To Stop Pretending, 1985; So You Want To Grow, 1986 and 17 books and booklets in Spanish.
Mailing address: PO Box 1173, Portland OR 97207, USA. T. (1 503) 643 0777

Palazzini, Pietro, H E Card

Rel: Roman Catholic; Ordained on 06 DEC 1934; b. 19 MAY 1912 in Piobbico, Pesaro, Italy; celibate.
International Organizations: Member (since 1973) of the College of Cardinals.
Member of the Sacred Congregation for the Oriental Churches. Member of the Sacred Congregation for the Sacraments and Divine Worship. Member of the Congregation for the Clergy. Member of the Pontifical Council for the Interpretation of Legislative Texts.
Career: Advocate, Sacred Roman Rota. Former Assistant Director, Pontificio Seminario Maggiore. Professor of Moral Theology, Lateran Pontifical University from 1945 to 1956. Under-Secretary, Sacred Congregation for Religious and Secular Institutes from 1956 to 1958. Secretary, Sacred Congregation for the Clergy in 1962. Prefect, Congregation for the Causes of Saints. Created Cardinal by Pope Paul VI in 1973. **Pubs.:** *Vita Sacramentale*, 3 Vols., 1972-86; *Dizionario dei concili*, 6 Vols., 1963-67 and *Dictionarium morale et canonicum*, 4 Vols., 1962-68.
Mailing address: Via Proba Petronia 83, I-00136 Roma, Italy. T. (39 6) 345 25 55

Palgye, Thubten, Ven.

Also know as Marcel Bertels; Rel: Buddhist: Mahayana; Ordained.
International Organizations: Contact of the International Mahayana Institute.
Mailing address: PO Box 1981, Kathmandu, Nepal. T. Fax (977) 1 410 992

Pallhuber, Heinrich, M.H.M., Fr

Rel: Roman Catholic; Ordained; celibate.
International Organizations: Procurator-General of the St Joseph's Missionary Society of Mill Hill.
Other organizations: Rector, College for Religious of St Joseph's Missionary Society of Mill Hill.
Mailing address: Via Innocenzo X 16, I-00152 Roma, Italy. T. (39 6) 581 03 30

Palmer, Martin

Rel: Unknown.
International Organizations: Director of the International Consultancy on Religion, Education and Culture.
Mailing address: Manchester Polytechnic, Didsbury Site, Wilmslow Road, Manchester M20 8RR, UK.

Palmerini, Umberto, C.P., Fr

Rel: Roman Catholic; Ordained; celibate.
International Organizations: Procurator-General of the Congregation of the Passion.
Mailing address: Piazza SS Giovanni e Paolo 13, I-00184 Roma, Italy. T. (39 6) 700 57 58.

Palmiero, Antonio, M.SS.CC., Fr

Rel: Roman Catholic; Ordained; celibate.
International Organizations: Superior General and Procurator-General of the Missionaries of the Sacred Hearts of Jesus and Mary.
Mailing address: Via dei Falegnami 23, 00186 Roma, Italy. T. (39 6) 686 16 79

Palombi, Luigi, C.R., Fr

Rel: Roman Catholic; Ordained; celibate.
International Organizations: Procurator-General of the Congregation of Regular Clerics.
Mailing address: Piazza Vidoni 6, I-00186 Roma, Italy. T. (39 6) 68 61 339.

Panciroli, Romeo, M.C.C.I., H E Mgr

Rel: Roman Catholic; Ordained on 11 JUN 1949; b. 21 NOV 1923 in Reggio nell'Emilia, Italy; celibate.
International Organizations: Apostolic Pro-Nuncio to Gambia and Liberia (since 1984) of the Roman Catholic Church.
Member of the Comboni Missionaries of the Heart of Jesus.
Other organizations: Apostolic Delegate to Sierra Leone (since 1984); Apostolic Pro-Nuncio to Guinea (since 1987); Titular Archbishop, Noba.
Career: Teacher of Natural Ethics and Sociology from 1951 to 1959. Vatican Secretary, State in 1960. Attaché to Apostolic Nunciature, Nigeria from 1961 to 1964. Secretary to Pontifical Commission for Social Communications from 1965 to 1976. Consultant to Pontifical Committee for Revision of Canon Law from 1975 to 1983. Director, Vatican Press Office from 1976 to 1984. Information Officer and member, Papal Suite with Pope John Paul II in His journeys to 54 countries from 1978 to 1984. **Interests:** reading, fine arts, classical music, travelling. **Pubs.:** Words on the Faith and Papal Messages in Asia and Australia.
Mailing address: Apostolic Nunciature, PO Box 4211, Monrovia, Liberia. T. (231) 26 29 48

Pandit, M P, Sri
Rel: Hindu; b. 14 JUN 1918 in Sirsi, Karnataka, India; Citizen of India; Lang. English, French, Hindi, Kannada, Konkani, Marathi and Sanskrit.
International Organizations: Chairman of the World Union.
Educ.: Bombay University, India. **Pubs.:** *Project Universal Man*; *Traditions in Occultism*; *Traditions in Mysticism*; *Dynamics of Yoga* and *Kundalini Yoga*.
Mailing address: Sri Aurobindo Ashram, Pondicherry 605002, India. T. (91 413) 24834. C. AUROBINDO PONDICHERRY-2
Residence: 4 St Gilles, Pondicherry 605001, India

Panjabi, Mohini, Sr
Rel: Hindu; b. 10 APR 1941 in Karachi, Pakistan; single; Citizen of India; Lang. Hindi and English.
International Organizations: President and Director (since 1968) of the Brahma Kumaris World Spiritual Organization.
Other organizations: Permanent Representative (since 1980), Brahma Kumaris at UN; Coordinator (since 1986), Million Minutes of Peace Project; International Coordinator (since 1988), Global Cooperation for a Better World International Project.
Main field of work: Ministry. **Career:** Director, Raja Yoga Center, Delhi from 1960 to 1974. President, Brahma Kumaris World Spiritual Organization, America from 1980 to 1992. Member, ECOSOC and UNICEF from 1980 to 1992. **Educ.:** B.A. Hist. and Pol.Sc. (1957-1962); Dip. Jour. (1957-1962). **Pubs.:** Articles in *Purity* (monthly), *World Renewal* (monthly).
Mailing address: 41-30 Fosley St, Elmurst NY 11373, USA. T. (1 718) 565 5733

Pankratz, James
Rel: Mennonite.
International Organizations: President of the Mennonite Brethren Bible College.
Mailing address: 1-169 Riverton Ave, Winnipeg MB R2L 2E5, Canada.

Paññavaro, Douglas
Rel: Buddhist.
International Organizations: President of the International Buddhist Forum.
Mailing address: Int Buddhist Forum Project, PO Box 11 1568, Bangkok 11, Thailand.

Pantane, Nino, Br
Rel: Roman Catholic; Ordained; b. 11 DEC 1931 in Rome, Italy; celibate; Citizen of Italy; Lang. Italian, French and Spanish.
International Organizations: Procurator-General (since 1984) of the Little Brothers of Jesus.
Mailing address: CP 13195, I-00184 Roma, Italy. T. (39 6) 435 88796

Panteghini, Antonio Giovanni, S.C.J., Fr
Rel: Roman Catholic; Ordained; celibate.
International Organizations: Superior General of the Priests of the Sacred Heart of Jesus.
Mailing address: Via Casale S Pio V 20, I-00165 Roma, Italy. T. (39 6) 66 38 809 – 66 38 825.

Papadimitriou-Demaître, Georges
Rel: Catholic; Unordained; b. 24 MAR 1925 in Antwerp, Belgium; Citizen of Greece; Lang. French, Spanish, English, German and Dutch.
International Organizations: President and interreligious lecturer (since 1974) of the Gandhi Centre in the Service of Man and Life.
Main field of work: Popularization of non-violence. **Career:** International Civil Servant, United Nations Relief and Rehabilitation Administration (UNRRA) from 1945 to 1948. Regional Secretary, World Alliance of Young Men's Christian Associations (YMCA) from 1948 to 1949. Film producer for the Radio Télévision Belge (RTBF) from 1960 to 1966. Journalist for various Belgian, French and Spanish newspapers from 1950 to 1980. **Educ.:** Ecole Française de Droit, Cairo, Egypt (1950-1951); Business School for Administrative and Economic Sciences, Brussels, Belgium (1951-1953). **Interests:** peace, justice, integrity of creation, harmony. **Mem.:** L'Arche (Lanza del Vasto), World Conference on Religion and Peace (WCRP), Pax Christi – International Catholic Peace Movement **Pubs.:** *Mutations* and hundreds of articles, lectures, radio and TV broadcasts.
Mailing address: Rue de l'Aqueduc 36, B-1050 Bruxelles, Belgium. T. (32 2) 539 18 62.
Residence: See Mailing Address. T. See Mailing Address

Papadopoulos, Dimitrios
See Dimitrios.

Papp, Laszlo, Bishop
Rel: Reformed Church.
International Organizations: Bishop of Reformed Church of Romania.
Mailing address: Eparphia Reformata Cluj, Str 23 August nr 51, 3400 Cluj-Napoca, Romania.

Pappalardo, Salvatore, H E Card
Rel: Roman Catholic; Ordained on 12 APR 1941; b. 23 SEP 1918 in Villafranca, Italy; celibate; Citizen of Italy.
International Organizations: Member (since 1973) of the College of Cardinals.
Member of the Congregation for Bishops. Member of the Sacred Congregation for the Oriental Churches. Member of the Congregation for the Clergy.
Other organizations: Archbishop (since 1970), Palermo.
Career: Counsellor, Vatican Secretariat of State from 1947 to 1965. Apostolic Pronuncio, Indonesia from 1965 to 1969. President, Pontifical Ecclesiastical Academy, Rome from 1969 to 1970. Created Cardinal by Pope Paul VI in 1973.
Mailing address: Arcivescovado, Via Matteo Bonello 2, I-90134 Palermo, Italy. T. (39 91) 58 34 42

Papworth, John, Revd
Rel: Anglican; b. 12 DEC 1921 in London, UK; m. to Marcelle Coralie Florentine Fouquet; c. 2 boys and 1 girl; Citizen of United Kingdom; Lang. English and German.
International Organizations: Assistant Priest (non-stipendiary) (since 1984) of the St Mark's Anglican Church, St John's Wood.
Other organizations: Editor/Founder (since 1965), *Resurgence*; Editor/Director (since 1981), *Fourth World Review* (Fourth World General Council).
Main field of work: Journalism. **Career:** Cook with the RAF (Royal Air Force) from 1940 to 1946. Business administration and teaching experience. Parliamentary candidate in 1955. Personal Assistant to the President, Zambia from 1971 to 1980. **Educ.:** B.Sc. Econ. from University of London, UK (1949-1954). **Interests:** politics, community, economics. **Pubs.:** *Economics of Humanism*, 1978; *New Politics*, 1980 and *The Monster*, 1992.
Mailing address: 24 Abercorn Place, London NW8 9XP, UK. T. (44 71) 286 4366
Residence: See Mailing Address. T. See Mailing Address

Paquette, Jacques, Rev
Rel: Roman Catholic; Ordained; celibate.
International Organizations: Co-Head, North America Regional Office of the Unda, International Catholic Association for Radio and Television.
Mailing address: Off. des Communications Sociales, 4005 rue de Bellechasse, Montréal H1X 1J6, Canada.

Parabirsing, B J
Rel: Moravian.
International Organizations: General Secretary of the Moravian Church in Surinam.
Mailing address: Maagdenstraat 50, PO Box 219/1811, Paramaribo, Surinam.

Parini, Pierre
Rel: Bahá'í.
International Organizations: Secretary of the Bahá'í Esperanto League.
Mailing address: 8 rue de la Fontaine, L-4447 Soleuvre, Luxembourg. T. (352) 59 44 20.

Paris, Angelo, C.S.F., Fr
Rel: Roman Catholic; Ordained; celibate.
International Organizations: Superior General of the Congrégation de la Sainte Famille de Bergame.
Mailing address: Via Incoronata 1, I-24057 Martinengo, Italy. T. (39 363) 98 80 98

Park, Dong Ki
Rel: Buddhist.
International Organizations: Vice-President of the World Fellowship of Buddhists.
Mailing address: The WFB Korea Regional Centre, CPO Box 7672, Seoul, Korea Rep.

Park, Sun Ai Lee
Rel: Christian.
International Organizations: Coordinator of the Asian Women's Resource Centre for Culture and Theology.
Mailing address: 134-5 Nokbun-dong, Eunpyong-ku, Seoul 122 020, Korea Rep. T. (82 2) 358 8283. Fax 358 8284.

Parke, Anthony
Rel: Roman Catholic.
International Organizations: Secretary of the Caribbean Association of Catholic Teachers.
Mailing address: Morne Jaloux, St George's, Grenada.

Parker, Christopher, Maj
Rel: Salvation Army, The.
International Organizations: Officer of The Salvation Army.
Mailing address: Apartado 5-165, 1704 Lisboa, Portugal.

Parkerson, B H, Pastor
Rel: Seventh-Day Adventist.
International Organizations: Secretary, South African Union of the General Conference of Seventh-Day Adventists.
Mailing address: PO Box 468, Orange Free State, Bloemfontein 9300, South Africa.

Parkin, Geoffrey
Rel: Salvation Army, The.
International Organizations: Administrative Officer of the Salvation Army William Booth Memorial Training College.
Member of the Salvation Army, The.
Mailing address: Denmark Hill, London SE5 8BQ, UK. T. (44 71) 733 1191. C. Traincoll, Camber, London, UK. Fax (44 71) 733 1191.

Parks, Lauren A
Rel: Religious Science.
International Organizations: Executive Secretary of the Religious Science International.
Mailing address: W 1516 Riverside, Spokane WA 991210-2152, USA. T. (1 509) 624 7000.

Parrilla, Luis, Rev
Rel: Church of the Disciples of Christ.
International Organizations: President of the Church of the Disciples of Christ.
Mailing address: Juan A Garcia 2044, 1416 Buenos Aires, Argentina.

Parrot, J Edward
Rel: Christian.
International Organizations: Secretary of the Foundation for a Christian Civilization.
Mailing address: McLain St and Guard Hill Rd, Bedford NY 10506, USA. T. (1 914) 241 1733.

Parteli Keller, Carlos P, Most Rev
Rel: Roman Catholic; Ordained on 15 APR 1933; b. 08 MAR 1910 in Rivera, Uruguay; celibate; Citizen of Uruguay.
International Organizations: Archbishop Emeritus (since 1985) of Montevideo.
Career: Bishop, Tacuarembo in 1960. Coadjutor Archbishop, Apostolic Administrator Sede Plena, Montevideo in 1966. Archbishop, Montevideo from 1976 to 1985.
Mailing address: Arzobispado, Calle Treinta y Tres 1368, Montevideo, Uruguay. T. (598 2) 95 8127 – 95 8926 – 95 8879

Parthenios, P, III, H B
Rel: Greek Orthodox; Ordained; b. in Port Said, Egypt.
International Organizations: Pope and Patriarch of Alexandria and all Africa.
Mailing address: Bureau de Presse, Patriarcat Greek Orthodox, PO Box 1307-2006, Alexandria, Egypt.

Pasco, Tito E, Rt Rev
Rel: Christian.
International Organizations: General Secretary of the Philippine Independent Church.
Mailing address: 1500 Taft Ave, Ermita, PO Box 2484, Metro Manila, Philippines.

Pascual, Antonio
Rel: Buddhist: Mahayana.
International Organizations: Member of the Foundation for the Preservation of the Mahayana Tradition.
Other organizations: Director, Nagarjuna C.E.T. Madrid.
Mailing address: Costanilla de los Angeles, N.2/escalera izda./3-dcha, 28013 Madrid, Spain. T. (34 91) 541 3755. Fax (34 91) 429 6708

Paskai, László, O.F.M., H E Card Dr
Rel: Roman Catholic; Ordained on 03 MAR 1951; b. 08 MAY 1927 in Szeged, Hungary; celibate; Citizen of Hungary.
International Organizations: Member (since 1988) of the College of Cardinals.
Member (since 1949) of the Order of Friars Minor. Member of the Sacred Congregation for the Oriental Churches. Member of the Congregation for Religious and Secular Institutes. Member of the Pontifical Council for Justice and Peace. Member of the Pontifical Council for the Interpretation of Legislative Texts.
Other organizations: Archbishop (since 1987), Esztergom and Primate of Hungary.
Career: Assumed diocesan service in 1950. Episcopal liturgist, Szeged from 1952 to 1955. Professor of Philosophy, Theological Academy, Szeged from 1955 to 1965, simultaneously Apostolic Prefect from 1955 to 1962. Spiritual from 1962 to 1965. Spiritual Prefect, Central Seminary of Budapest from 1965 to 1969. Commissioned lecturer from 1965 to 1967. Leading Professor of Philosophy, Theological Academy, Budapest from 1967 to 1978. Rector, Seminary from 1973 to 1978. Appointed Titular Bishop, Bavagaliana and Apostolic Governor of Veszprém in 1978. Diocesan Bishop, Veszprém in 1979. Coadjutor with right of succession to Archbishop, Kalocsa in 1982.
Mailing address: Mindszenty hercegprímás tér 2, H-2501 Esztergom, Hungary. T. (36 33) 11 288 – 13 690. Fax 11 085

Pasleau, Pierre-Percy, Mgr
Rel: Roman Catholic; Ordained; celibate.
International Organizations: President of the Comité international de liaison et d'information des ordres, académies, instituts et confréries. Contact of the Ordre Souverain et Militaire du Temple de Jerusalem.
Mailing address: Rue de la Reine 19, B-4500 Huy, Belgium. T. (32 85) 21 26 71.

Pasquali, Pietro, S.d.C., Fr
Rel: Roman Catholic; Ordained; celibate.
International Organizations: Superior General of the Servants of Charity.
Mailing address: Vicolo Clementi 41, I-00148 Roma, Italy. T. (39 6) 655 24 54

Passa, Sailali, Rev
Rel: Evangelical.
International Organizations: President of the Evangelical Church in New Caledonia and the Loyalty Isles.
Mailing address: BP 277, Noumea, New Caledonia.

Pastò, Gianluigi, Rev
Rel: Roman Catholic; Ordained on 16 MAY 1971; b. 25 MAR 1943 in Bagnoli di Sopra, Padova, Italy; celibate; Citizen of Italy; Lang. Italian and French.
International Organizations: Vicar/Procurator-General (since 1984) of the Congregazione di Gesù Sacerdote.
Main field of work: Ministry. **Educ.:** Studied Theol. from Seminary, Verona, Italy (1967-1971).
Mailing address: Via dei Giardini 36, I-38100 Trento, Italy. T. (39 461) 98 38 63.
Residence: See Mailing Address. T. See Mailing Address

Pastor Alvarez, Manuel P, M.D., Fr
Rel: Roman Catholic; Ordained; celibate.
International Organizations: Procurator-General of the Order of Discalced Mercedarians.
Mailing address: Santa María de la Merced, Urbanización Club de Golf, Calle Cabo Mayor 1, 28290 Las Matas (Madrid), Spain. T. (34 91) 630 30 81.

Pastore, Pierfranco, Mgr
Rel: Roman Catholic.
International Organizations: Secretary of the Pontifical Council for Social Communications.
Mailing address: Palazzo S Carlo, I-00120 Città del Vaticano, Vatican. T. (39 6) 698 83197 – 698 83597. Fax 698 85373. Tx 2019 PCCS VA.

Pasztor, Janos Dezso, Rev Prof Dr
Rel: Reformed Church; Ordained on 28 NOV 1952; b. 28 MAY 1925 in Budapest, Hungary; m. to Judith Kádár; c. 2 boys and 1 girl; Citizen of Hungary; Lang. Hungarian, English, German and French.
International Organizations: Professor of Ecumenical Studies (since 1989) of the Reformed Theological Academy, Budapest.
Other organizations: Pastor (since 1988), Budafok Reformed Church.
Main field of work: Ministry, theology. **Career:** Pastor, Szentendre Reformed Church in Hungary from 1952 to 1970. Tutor in Theology, St Paul's United Theological College, Kenya from 1970 to 1976. Professor of Theology, Reformed Theological Academy, Debrecen from 1976 to 1987. Professor of Theology, Columbia Theological Seminary, Decatur GA from 1987 to 1988. **Educ.:** Reformed Theological Academy, Budapest, Hungary (1943-1947); University of Edinburgh, UK (1947-1948); Victoria University of Manchester, UK (1948-1949); M.Div. from Reformed Theological Academy, Budapest, Hungary (1949-1950); Th.M. from Princeton Theological Seminary, NJ, USA (1964-1965); Th.D. from Reformed Theological Academy, Budapest, Hungary (1966-1969). **Interests:** biblical and theological research. **Pubs.:** Liturgika, Debrecen, 1985 (in Hungarian); Homiletika, Debrecen, 1986 (in Hungarian) and "God's Glory and Humanity's Future", in *Pacific Theological Review*, 1979 Special Issue.
Mailing address: Ráday utca 28, H-1092 Budapest IX, Hungary. T. (36 1) 117 2403
Residence: Kárpitos utca 2, H-1221 Budapest XXII, Hungary. T. (36 1) 226 0178

Pat, Leva Kila, Rev
Rel: United Church; Ordained in 1979; b. 21 OCT 1951 in Hula, Central Province, Papua New Guinea; m. to Lepo Ume; c. 1 child; Citizen of Papua New Guinea; Lang. Hula and English.
International Organizations: Secretary-General (since 1986) of the Melanesian Council of Churches.
Other organizations: Director (since 1990), National Housing Cooperation.
Main field of work: Administration. **Career:** Parish priest, Daru United Church from 1978 to 1979. Parish priest, Hula United Church in 1980. Lecturer, Rarong Theological College, Rabaul from 1983 to 1985. **Educ.:** B.D. from Pacific School of Religion, Berkeley CA, USA (1972-1977); M.A. from Rorongo Theological College, Rabaul, Papua New Guinea (1980-1982). **Interests:** Peace and environment activism.
Mailing address: PO Box 1015, Boroko, Papua New Guinea. T. (675) 25 99 61. C. MELCON BOROKO. TX NE22213 C/O WANTOK.
Residence: Korobosea Drive, Sec 92, Lot 3, Port Moresby, Papua New Guinea

Patanduk, P, Rev
Rel: Christian.
International Organizations: Secretary of the Toraja Church.
Mailing address: Jln Ratulangi, 58 Rantepao, South Sulawesi, Indonesia.

Patterson, Cecil John, Rt Rev
Rel: Anglican; Ordained in 1932; b. 09 JAN 1908 in London, UK; Citizen of United Kingdom.
International Organizations: Ecclesiastic of the Anglican Church.
Career: Ordained Deacon in 1931. London curacy from 1931 to 1934. Missionary, Nigeria from 1934 to 1969. Assistant Bishop from 1942 to 1945. Bishop on the Niger from 1945 to 1969. Archbishop, West Africa from 1961 to 1969. Representative for Archbishops, Canterbury and York for Community Relations from 1970 to 1972. Honorary Assistant Bishop, London from 1970 to 1976. **Hons.:** Hon. D.D., University of Nigeria (1963), Lambeth (1963); Commander, Federal Republic of Nigeria (1965); Companion of the Order of St Michael and St George; Commander of the Order of the British Empire.
Mailing address: 6 High Park Road, Kew, Richmond TW9 4BH, UK. T. (44 81) 876 1697

Paul, Bill E
Rel: Unknown.
International Organizations: Representative of the International Society of Bible Collectors.
Mailing address: PO Box 30526, Seattle WA 98103, USA.

Paul, P, Archbishop
Rel: Orthodox.
International Organizations: Archbishop of Karelia and All Finland.
Mailing address: Karjalankatu 1, SF-70300 Kuopio, Finland.

Paul, Pierre, O.M.V., Fr
Rel: Roman Catholic; Ordained; celibate.
International Organizations: Procurator-General of the Oblats de la Vierge Marie.
Mailing address: Viale Trenta Aprile 17, I-00153 Roma, Italy. T. (39 6) 589 10 71, (39 6) 589 05 36

Paul, William E
Rel: Church of Christ; Ordained on 02 JUL 1967; b. 12 APR 1927 in Alliance OH, USA; m. to Bethel Roslyn Carpenter; c. 5 boys and 1 girl; Citizen of USA; Lang. English.
International Organizations: Evangelist (since 1978) of the Crown Hill Church of Christ, Seattle WA.
Other organizations: Professor (since 1978), Seattle Christian College, WA; Editor (since 1992), *Bible Collectors' World.*
Main field of work: Ministry. **Career:** Editor, *News and Truths, Impact* from 1955 to 1981. Professor, Puget Sound Christian College in 1991. **Educ.:** B.Sacred Lit. from Midwestern School of Evangelism, Ottumwa IA, USA (1951-1955); B.Th. from Midwestern School of Evangelism, Ottumwa IA, USA (1977); M.Sacred Lit. from Midwestern School of Evangelism, Ottumwa IA, USA (1978). **Mem.:** International Society of Bible Collectors; Toastmasters Club (Past President) **Pubs.:** *Telling On God*; *A Christian View of Armed Warfare* and articles in: *Christian Standard, Restoration Herald, Voice of Evangelism, The Sword and Staff, One Body*, etc.
Mailing address: Crown Hill Church, PO Box 30526, Seattle WA 98103, USA. T. (1 206) 789 4954
Residence: 8720 Phinney Ave N, Suite 13, Seattle WA 98103, USA. T. (1 206) 782 3992

Paulose, Kallapara V, Rev Prof
Rel: Syrian Orthodox; b. 20 APR 1940 in Aluva, Kerala, India; m. to Prof Elizabeth Kuruvilla; c. 1 boy and 1 girl; Citizen of India; Lang. Malayalam, English, Hindi and Syriac.
International Organizations: Secretary-General (since 1989) of the Asian-Pacific Christian Peace Conference.
Member, International Commissions on Development, Theology, Anti-Racism (since 1976) of the Christian Peace Conference. Member, International Continuation Committee (since 1985) of the Christian Peace Conference.
Other organizations: Lecturer (since 1963), Baselius College, Kottayam, India; Head of the Dept. of Economics (since 1967), Baselius College, Kottayam, India; President (since 1982), Centre for Development Action; Chairman, Board of Studies (Economics) (since 1990), Mahatma Ghandi University; Chairman, District Literacy Committee (BJVJ – Sponsored by th Ministry of Human Resources Development, Government of India).
Main field of work: University education. **Career:** Secretary-General, Christian Peace Conference, India from 1983 to 1990. **Educ.:** B.A. Econ. from University of Kerala, India (1958-1961); M.A. Econ. from University of Kerala, India (1961-1963). **Interests:** Literacy work, social action, development work. **Hons.:** Won'Raja Annamalai Chettiar

Proficiency, Prize in Economics from Madras Christian College. **Pubs.:** Economics Textbook for first and second year pre-degree (joint work); *Christian and Churches in Socialist Countries* (joint work) and Several articles related to peace, development, justice, gandhism, ecology, etc. **Mailing address:** Kallapara House, Chelliozhukom Street, Kottayam 686001, India. T. (91) 3066 (KOTTAYAM). Tx 888 227 OLYM IN
Residence: See Mailing Address. T. See Mailing Address

Paulsen, Jan, Dr

Rel: Seventh-Day Adventist.
International Organizations: President, Trans-European Division of the General Conference of Seventh-Day Adventists.
Mailing address: 119 St Peter's Street, St Albans AL1 3EY, UK.

Paulson, Hank

Rel: Christian.
International Organizations: General Director of the Eastern European Bible Mission.
Mailing address: PO Box 110, Colorado Springs CO 80901, USA. T. (1 719) 577 4450.

Paupini, Giuseppe, H E Card

Rel: Roman Catholic; Ordained on 19 MAR 1930; b. 25 FEB 1907 in Mondavio, Italy; celibate; Citizen of Italy.
International Organizations: Member (since 1969) of the College of Cardinals.
Career: Titular Archbishop, Sebastopolis in Abasgia in 1956. Apostolic Nuncio in Colombia in 1959. Created Cardinal by Pope Paul VI in 1969. Grand Penitentiary from 1973 to 1984.
Mailing address: Via Rusticucci 13, I-00193 Roma, Italy.

Pauw, Christoff Martin, Dr

Rel: Christian; Ordained on 04 JUL 1965; b. 06 FEB 1940 in Chipata, Eastern Province, Zambia; m. to Ilse Heese; c. 3 boys and 1 girl; Citizen of South Africa; Lang. Afrikaans, Chewa, Dutch, English, German (some) and Zulu (some).
International Organizations: Senior Lecturer in Missiology (since 1983) of the University of Stellenbosch.
Main field of work: Ministry, university education. **Career:** Minister of Religions, Church of Central Africa Presbyteria, Malawi from 1965 to 1973. Lecturer, Theological College, Church of Central Africa, Malawi from 1968 to 1973. Lecturer, Theological College, Reformed Church in Zambia from 1975 to 1983. **Educ.:** B.A. from University of Pretoria, South Africa (1958-1960); B.Th. from University of Stellenbosch, South Africa (1961-1963); Lic. Theol. from University of Stellenbosch, South Africa (1964); M.Th. from University of Stellenbosch, South Africa (1968); D.Th. from University of Stellenbosch, South Africa (1980). **Interests:** History of Christianity in Africa, African independent churches. **Mem.:** Southern African Missiological Society; International Association for Mission Studies; South African Institute for Race Relations; Africa Institute of South Africa **Pubs.:** Mission and Church in Malawi: The History of the Nkhoma Synod of the Church of Central Africa Presbyteria 1889-1962; The Holiness of the Church in Africa, Missionalia, 1974 and The Challenge of the Power Encounter for Reformed Theology, Ecumenical Council, 1988.
Mailing address: Theological Faculty, 171 Dorp Street, Stellenbosch 7600, South Africa. T. (27 2231) 773264
Residence: 29 Kommandeurs Ave, Stellenbosch 7600, South Africa T. (27 2231) 71944

Pavan, Pietro, H E Card

Rel: Roman Catholic; Ordained on 08 JUL 1928; b. 30 AUG 1903 in Treviso, Italy; celibate; Citizen of Italy.
International Organizations: Member (since 1985) of the College of Cardinals.
Other organizations: Deacon, San Francesco da Paola ai Monti.
Career: Created Cardinal by Pope John Paul II in 1985.
Mailing address: Via della Magliana 1240, Ponte Galeria, I-00050 Roma, Italy. T. (39 6) 647 00 69

Payne, S Stewart

Rel: Anglican; Ordained; b. 06 JUN 1932 in Fogo, Newfoundland, Canada; m. to Selma Carlson Penney; c. 2 boys and 2 girls; Citizen of Canada.
International Organizations: Archbishop (since 1990) of Diocese of Western Newfoundland.
Other organizations: Metropolitan (since 1990), Ecclesiastical Province of Canada.
Career: School Teacher (in 1949, 1950, 1951). Parish Priest, Happy Valley, Labrador from 1957 to 1965. Parish Priest, Bay Roberts, Newfoundland from 1965 to 1970. Parish Priest, St Anthony, Newfoundland from 1970 to 1978. Bishop, Diocese of Western Newfoundland from 1978 to 1990. **Educ.:** B.A. from Memorial University of Newfoundland, St John's, Canada (1958); L.Th. from Queen's University at Kingston, Queen's Theological College, Ontario, Canada (1958); B.D. (by correspondence) from General Synod (1968). **Interests:** cross-country skiing, gardening, walking, reading. **Hons.:** Hon. D.D., King's College (1981). **Mem.:** numerous committees at national, provincial and diocesan level
Mailing address: Newfoundland and Labrador Light and Power Bldg, 83 West St, Corner Brook, Newfoundland A2H 2Y6, Canada.
Residence: 13 Cobb Lane, Corner Brook, Newfoundland A2H 2V3, Canada.

Peale, Norman Vincent

Rel: Methodist Episcopal; Ordained in 1922; b. in Bowersville OH, USA; m. to Loretta Ruth Stafford; Citizen of USA.
International Organizations: Minister of the Methodist Episcopal Church.
Other organizations: Promoter, positive thinking.
Career: Pastor, a small church in Berkeley RI in 1922. Pastor, a small congregation in Brooklyn NY from 1924 to 1927. Pastor, University Methodist Church in Syracuse NY. Pastor, Marble Collegiate Church in New York City NY – a move that required him to join the Reformed Church in America (from 1932 until the end of his career). Established a psychiatric clinic, the American Foundation of Religion and Psychiatry, the Foundation for Christian Living, and *Guideposts* (monthly magazine). **Educ.:** Grad. from Ohio Wesleyan University, Delaware OH, USA (1920); B.S.T., M.A. from Boston University, MA, USA (1921-1924). **Pubs.:** *A Guide to Confident Living*, 1948; *The Art of Real Happiness*, 1950 and *The Power of Positive Thinking*, 1952.
Mailing address: No mailing address.

Pearce, (John) Brian

Rel: Christian; b. 25 SEP 1935; m. to Michelle Etcheverry; c. 4 boys.
International Organizations: Director (since 1987) of the Inter Faith Network for UK.
Career: National Service with RAF (Royal Air Force) from 1954 to 1956. Pilot Officer from 1955 to 1956. Ministry of Power from 1959 to 1960. Colonial Officer from 1960 to 1967. Dept. of Economic Affairs from 1967 to 1969. Civil Service Dept. from 1969 to 1981. HM Treasury from 1981 to 1984. **Educ.:** B.A. from University of Oxford, Brasenose College, UK. **Interests:** reading (mainly theology), travel.
Mailing address: Tavistock Place, London WC1, UK. T. (44 71) 388 0008
Residence: 124 Court Lane, London SE21 7EA, UK.

Pecqueux, Pierre-Yves, Rev

Rel: Christian.
International Organizations: Executive Officer of the Liaison Committee of Christian International Cooperation Organizations.
Mailing address: 9-11 rue Guyton de Morveau, F-75013 Paris, France. T. (33 1) 45 65 96 65. Tx 206293 F COOPCAT. Fax (33 1) 45 81 30 81.

Pécriaux, François, Fr

Rel: Roman Catholic; Ordained; celibate.
International Organizations: Superior General of the Institute of Prado.
Mailing address: 13 rue du Père Chevrier, F-69007 Lyon, France. T. (33) 78 72 41 67

Peel, Shitoku Adriaan, Rev Prof

Rel: Buddhist: Jôdo-Shinshû; Ordained on 15 OCT 1978; b. 28 MAY 1927 in Antwerp, Belgium; Citizen of Belgium; Lang. Dutch, French, English and German.
International Organizations: Dean and Managing Board Secretary (since 1980) of the Faculty for the Comparative Study of Religions, Antwerp.
Other organizations: Kyoshi (Sutra-Master) (since 1977), Jikoji Temple, Antwerp; Chairman (since 1982), Federation of Buddhist Associations of Belgium.
Main field of work: Education. **Educ.:** D.Div. from Southern Baptist University, Fort Walton FL, USA (1979). **Pubs.:** Syllabus Boeddhisme I, FVG, 2nd Edn.; Aldus heb ik gehoord (Translation of Buddhist Texts) and Het Reine-Landboeddhisme.
Mailing address: Centrum voor Shin Boeddhisme VZW, Pretoriastraat 68, B-2600 Antwerpen-Berchem, Belgium.
Residence: Leemputstraat 10, B-2600 Berchem, Belgium. T. (32 3) 218 73 63

Peers, Michael Geoffrey, Most Rev

Rel: Anglican; Ordained in 1960; b. 31 JUL 1934 in Vancouver, British Columbia, Canada; m. to Dorothy Elizabeth Bradley; c. 2 boys and 1 girl; Citizen of Canada.
International Organizations: Primate (since 1986) of the Anglican Church of Canada.
Career: Ordained Deacon in 1959. Served as Assistant Curate, 2 Ottawa parishes from 1959 to 1965. Diocesan Chaplain, Carleton University and University of Ottawa from 1961 to 1962. Instructor, Ottawa Teachers' College from 1962 to 1966. Instructor, St Paul's High School, Winnipeg from 1967 to 1969. Rector, St Bede's Church, Winnipeg from 1966 to 1972. Incumbent, St Martin's, Winnipeg, St Paul's, Middlechurch and Priest Superior of River N. Anglican Parishes from 1972 to 1974. Archdeacon, Winnipeg from 1969 to 1974. Rector, St Paul's Cathedral, Regina and Dean of Diocese of Qu'Appelle from 1974 to 1977. Consecrated Bishop, Diocese of Qu'Appelle from 1977 to 1982. Archbishop, Ecclesiastical Province of Rupert's Land from 1982 to 1986. Member, Vancouver Assembly, World Council of Churches in 1983. Co-leader, ecumenical Easter visit to Churches in Soviet Union in 1985. Chairman, Ecumenical Section, Lambeth Conference in 1988. Former Chairman of Council, College of Emmanuel and St Chad, Saskatoon. **Educ.:** Interpreter's Cert. from Ruprecht-Karls-Universität Heidelberg, Germany (1955); B.A. (Hons) from University of British Columbia, Vancouver, Canada (1956); L.Th. from University of Toronto, Trinity College, Canada (1959). **Hons.:** Hon. D.D., Trinity College, Toronto (1977), St John's College, Winnipeg (1981), Wycliffe College, Toronto (1987), University of Kent at Canterbury (1988), Montreal Diocesan College (1989), College of Emmanuel and St Chad, Saskatoon (1990).
Mailing address: 600 Jarvis St, Toronto, Ontario M4Y 2J6, Canada. T. (1 416) 924 9192
Residence: 195 Westminster Ave, Toronto, Ontario M6R 1N9, Canada.

Pellabeuf, René

Rel: Roman Catholic.
International Organizations: President of the Fraternitas Sancti Benedicti pro Europa Christiana.
Mailing address: 4-L'Oliveraie, 126 cours Gambetta, F-13100 Aix-en-Provence, France. T. (33) 42 26 47 87.

Pellegrino, Carlo, C.R.S., Fr

Rel: Roman Catholic; Ordained; celibate.
International Organizations: Procurator-General of the Clerics Regular of Somasca.
Mailing address: Piazza Tempio di Diana 14, I-00153 Roma, Italy. T. (39 6) 574 25 92.

Pellman Good, Phyllis P

Rel: Mennonite.
International Organizations: Recording Secretary of the Mennonite World Conference.
Mailing address: 223 E Walnut Street, Lancaster PA 17602, USA.

Pelzer, Anne Marie Augustine Clementine

Rel: Catholic; b. 24 JUN 1926 in Liège, Belgium; single; Citizen of Belgium; Lang. French and English.
International Organizations: President (since 1989) of the St Joan's International Alliance.
Main field of work: Christian feminism. **Career:** Researcher in Physiology, University of London, School of Hygiene and Tropical Medicine, UK in 1961. Assistant Professor, University of Toronto, Ontario, Canada. Professor, University of Montreal, Quebec, Canada until 1968. **Educ.:** Ph.D. Human Physiology from University of London, UK (1961-1965); B.A. Theol. from University of Cambridge, UK (1971-1973); B.Th. from Catholic University of Louvain (UCL), Belgium (1974-1976); M.Th. from University of Cambridge, UK (1977). **Pubs.:** Many articles in the *Catholic Citizen* and *L'Alliance* (periodicals from St Joan's International Alliance).
Mailing address: Quai Churchill 19/061, B-4020 Liège, Belgium. T. (32 41) 42 04 71
Residence: See Mailing Address. T. See Mailing Address

Peñaranda, Miguel A

Rel: Christian.
International Organizations: Executive Secretary of the Federation of National Committees in the International Christian Youth Exchange.
Mailing address: Goethestrasse 85-87, 10623 Berlin, Germany. T. (49 30) 313 70 46. C. INYOUTHEX BERLIN. Tx 183 279 icye d.

Pende, Thubren

Rel: Buddhist: Mahayana.
International Organizations: Member of the Foundation for the Preservation of the Mahayana Tradition.
Other organizations: Manager, Nalanda Monastery.
Mailing address: Rouzegas, Labastide Saint-Georges, F-81500 Lavaur, France. T. (33 63) 580 225

Pender, Dinsdale, Commr

Rel: Salvation Army, The.
International Organizations: Territorial Commander for Australian Southern Territory of The Salvation Army.
Mailing address: GPO Box 1287K, Melbourne 3001, Victoria, Australia.

Penney, Alphonsus Liguori, Most Rev

Rel: Roman Catholic; Ordained on 29 JUN 1949; b. 17 SEP 1924 in St John's, Newfoundland, Canada; celibate; Citizen of Canada.
International Organizations: Archbishop (since 1979) of St John's, Newfoundland.
Other organizations: Member (since 1979), Catholic Education Committee.
Career: Served with RCAF (Royal Canadian Air Force) from 1952 to 1957. Assistant Priest becoming Parish Priest from 1957 to 1972. Named Vicar Forane in 1960. Vicar-General in 1971. Prelate of Honour in 1971. Bishop, Grand Falls, Newfoundland from 1973 to 1979. **Educ.:** L.Ph., L.Th. from Saint Paul's Seminary and University of Ottawa, Ontario, Canada; LL.D. from Memorial University of Newfoundland, St John's, Canada (1980).
Mailing address: PO Box 37, St John's Newfoundland A1C 5H5, Canada.

Pennick, Nigel

Rel: Unknown.
International Organizations: Secretary of the International Institute of Practical Geomancy.
Mailing address: 142 Pheasant Rise, Bar Hill, Cambridge CB3 8SD, UK. T. (44 954) 780 932.

Penoukou, Benu Efoevi, Fr

Rel: Roman Catholic; Ordained; celibate.
International Organizations: Francophone Africa Coordinator of the African Caritas.
Mailing address: Sec des Caitas d'Afrique Francophone, PB 8395, Lomé, Togo.

Penrose, Ramanus, OSB, Sr
Rel: Roman Catholic; Ordained; celibate.
International Organizations: Member of the Benedictine Nuns.
Other organizations: Member, Association of Contemplative Sisters.
Mailing address: Benedictine Monastery, 8300 Morganford Rd, St Louis MO 63123, USA.

Penttinen, Pirjo-Liisa, Rev
Rel: Christian.
International Organizations: Secretary of the European Fellowship.
Mailing address: PTK-Helsinki, Vuorikatu 22A, PL 345, SF – 00101 Helsinki, Finland. T. (358 0) 17 7334. Fax 65 7654.

Pepper, Stewart
Rel: Christian.
International Organizations: Director of the Far East Broadcasting Association.
Mailing address: Box 234, Mahé, Seychelles. T. (248) 24449. Fax 248 25171. C. FEBCOM. Tx 2287 FEBCOM SZ. Fax via Bureau 22555 FEBCOM.

Pepping, Johannes Wilhelmus, F.N.D.L., Br
Rel: Roman Catholic; Ordained on 08 DEC 1948; b. 12 DEC 1930 in Heiloo, Noord Holland, Netherlands; celibate; Citizen of Netherlands; Lang. Dutch and English.
International Organizations: Superior General (since 1985) of the Brothers of Our Lady of Lourdes.
Educ.: Master Nursing from Nijmegen, Netherlands (1964-1977).
Mailing address: Sint Jozefstraat 1, B-9041 Oostakker, Belgium. T. (32 91) 51 01 85
Residence: See Mailing Address. T. See Mailing Address

Pereira de Queiroz, Maria Isaura P, Prof Dr
Rel: Unknown; b. 26 AUG 1918; Citizen of Brazil.
International Organizations: Member of the Council of the International Society for the Sociology of Religions.
Educ.: Dr. Sociology.
Mailing address: Centro e Est. Rurais e Urb., Departamento de Sociologia, Fac. Fil. Letras e C. Humanas, Universidade de Sao Paulo, Brazil.
Residence: Rua Buri 131, 01246 Sao Paulo SP, Brazil T. (55 11) 258 5946

Pereira Neves, Jubal P, Rev
Rel: Episcopal.
International Organizations: General Secretary of the Episcopal Church of Brazil.
Mailing address: Caixa Postal 965, 90 000 Porto Alegre RS, Brazil.

Perera, Vincent, Col
Rel: Salvation Army, The.
International Organizations: Territorial Commander of The Salvation Army.
Mailing address: PO Box 193, Colombo, Sri Lanka.

Perez, Pablo E, Dr
Rel: Protestant.
International Organizations: Executive Secretary of the Commission on Church Renewal.
Mailing address: Apartado Postal 20-839, 01000 México DF, Mexico.

Perez-Albert, Mercedes
Rel: Buddhist: Mahayana.
International Organizations: Member of the Foundation for the Preservation of the Mahayana Tradition.
Other organizations: Director, Nagarjuna C.,E.T. Barcelona.
Mailing address: Rosselon 298 Pral 2a, 08037 Barcelona, Spain. T. FAX AND TEL (34 93) 257 0788

Pergola, Philippe, Prof Dr
Rel: Roman Catholic; b. 01 JUN 1953 in Corte, Corse, France; m. to Vismara Cinzia; Citizen of France; Lang. Italian, French, German, English and Spanish.
International Organizations: Professor of Topography of the Orbis Christianus Antiquus (since 1984) of the Pontifical Institute of Christian Archeology.
Other organizations: Chargé de recherche (since 1985), Centre National de la Recherche Scientifique (CNRS), Paris.
Main field of work: University education, research. **Career:** Librarian, Centre d'Etudes St Louis de France, Rome from 1974 to 1976. Librarian, Pontifical Institute of Christian Archeology, Rome from 1976 to 1981. Member, Ecole française de Rome from 1981 to 1984. **Educ.:** Laurea from University of Rome, Italy (1971-1976); Doc. from Pontificio Istituto di Archeologia Cristiana, Rome, Italy (1971-1975); Doc. 3ème cycle from University of Aix-Marseille III, Aix-en-Provence, France (1975-1982). **Pubs.:** "Architecture religieuse et topographie de la Corse médiévale: deux cas concrets: S Mariona de Talcini (Corte) et S Ghjuvanni d'Altiani, in *Etudes Corses*, XV, 1980, pp.93-124; "Le catacombe romane: miti e realtà (a proposito del cimitero di Domitilla)", in *Società Romana e impero tardoantico – Roma: politica, economia, paesaggio urbano*, Vol II, Roma-Bari, 1986, pp. 333-350, 484, 490 and "Mensores frumentarii christiani", in *Revista di Archeologia Cristiana*, LXVI, pp. 167-1984.
Mailing address: Via Napoleone III 1, I-00185 Roma, Italy. T. (39 6) 4465 574. Fax 4469 197.
Residence: Viale delle Province 101, I-00162 Roma, Italy T. (39 6) 428 694

Perino, Renato, S.S.P., Rev
Rel: Roman Catholic; Ordained; b. in Front Canavese, Torino, Italy; celibate; Citizen of Italy.
International Organizations: Ex-Superior General (since 1992) of the Society of St Paul.
Career: Superior General, Society of St Paul from 1980 to 1992. **Educ.:** Dr. Theol. from Pontifical St Anselmo Athenaeum, Vatican City State.
Mailing address: Via della Fanella 39, I-00148 Roma, Italy. T. (39 6) 522 2927.

Perna, Franco
Rel: Quaker.
International Organizations: Regional Representative for Europe and the Near East of the Friends World Committee for Consultation.
Mailing address: 1 rue B Haal, L-1711 Luxembourg, Luxembourg. T. (352) 44 87 20. Fax 44 02 01.

Perpiñan, Mary Soledad, Sr
Rel: Roman Catholic; Ordained; b. 18 MAY 1937 in Manila, Philippines; celibate; Citizen of Philippines; Lang. English, Filipino and Spanish.
International Organizations: Coordinator, Ruah Community (since 1990) of the Sisters of Our Lady of Charity of the Good Shepherd. Coordinator (since 1980) of the Third World Movement Against the Exploitation of Women. Contact person of the Campaign Against Military Prostitution International.
Other organizations: Corporate Secretary (since 1988), IBON Databank and Research Center; Council Member (since 1990), Asia-Pacific Peace Research Association.
Main field of work: Social services. **Career:** Founder/Editor, *IBON Facts and Figures* from 1978 to 1982. Founder/Editor, *BALAI Asian Journal* from 1980 to 1986. Columnist, *Daily Globe* from 1989 to 1990. Founded Belen sa Ermita, Malate in 1988. Founded Belen sa Subic in 1988. Founded Nazareth Growth Home in 1990. Founded Belen sa Batangas in 1992. **Educ.:** A.B. (Summa cum Laude) from St Theresa's College, Manila, Philippines (1953-1956); B.S.E. (Summa cum Laude) from St Theresa's College, Manila, Philippines (1953-1957); M.S. from Fordham University, Bronx NY, USA (1958-1959). **Interests:** women. **Pubs.:** *Yielding* and *Thrust.*
Mailing address: Good Shepherd Sisters, 1043 Aurora Blvd, Quezon City 1108, Philippines. T. (63 2) 922 1314
Residence: See Mailing Address. T. See Mailing Address

Perry, Troy D, Rev
Rel: Christian.
International Organizations: Moderator of the Universal Fellowship of Metropolitan Community Churches.
Mailing address: 5300 Santa Monica Boulevard, Suite 304, Los Angeles CA 90029, USA. T. (213) 464 5100.

Persson, Walter, Rev Dr
Rel: Evangelical.
International Organizations: President of the International Federation of Free Evangelical Churches.
Mailing address: Tegnergatan 8, Box 6302, S-113 81 Stockholm, Sweden.

Pessoa Cámara, Helder P, Most Rev
Rel: Roman Catholic; Ordained on 15 AUG 1931; b. 07 FEB 1909 in Fortaleza, Brazil; celibate; Citizen of Brazil.
International Organizations: Theologian of the Roman Catholic Church.
Other organizations: Archbishop Emeritus (since 1985), Olinda and Recife.
Career: Consecrated Bishop in 1952. Titular Archbishop, Salde from 1955 to 1964. Archbishop, Olinda and Recife from 1964 to 1984. Became a champion of the poor and called for peaceful but radical social transformation and an end to 'internal colonialism'. Helped found the Brazilian Conference of Bishops and the Latin American Conference of Bishops and played a major role in the radicalization of the Latin American clergy. **Hons.:** René Sande Award (1962); Via International Peace Prize, Italy (1970); Martin Luther King Jr Award; John XXIII Memorial Award for Pax Christi, Spain (1970); People's Peace Prize (1974); Niwano Peace Prize (1983); Christopher Award (1987); Dr. h.c., Universities of St Louis, Louvain, Münster, Fribourg and Sorbonne. **Pubs.:** *Revolution Through Peace*, 1971; *The Conversions of a Bishop*, 1979 and *Hoping Against All Hope*, 1984.
Mailing address: Rua Henrique Dias, 208-Derbi 50070, Recife, Brazil. T. (55 81) 222 0400

Péter, János
Rel: Reformed Church; Ordained; Citizen of Hungary.
International Organizations: Politician and Ecclesiastic of the Reformed Church.
Other organizations: Member (since 1953), National Assembly.
Career: Bishop Transtibiscan Synod, the Reformed Church from 1949 to 1956. President, Institute of Cultural Relations from 1957 to 1958. Deputy Speaker from 1973 to 1988. First Deputy Minister for Foreign Affairs from 1958 to 1961. Minister from 1961 to 1973. Member, Central Committee, Hungarian Socialist Workers' Party from 1966 to 1988. **Hons.:** Banner, Order 1st degree of the Hungarian People's Republic (1970); Red Banner of Labour Order of Merit; State Prize (1985). **Pubs.:** History of Hungarian Soviet Diplomatic Relations 1939-1941, 1980.
Mailing address: Országgyülés, Kossuth Lajos tér, H-1357 Budapest, Hungary.

Peter, Navamani
Rel: Methodist.
International Organizations: Vice-President of the World Federation of Methodist Women.
Mailing address: Bishop's House, 27 1st Main Road, Cambridge Lay Out, Unoor, Bangalore 360008, India.

Peters, Douglas C
Rel: Christian.
International Organizations: Secretary of the World Radio Missionary Fellowship.
Mailing address: 20201 NW 37th Ave, Opa-Locka FL 33055, USA. T. (1 305) 624 42 52. Fax 621 2333.

Peters, Greg
Rel: Christian.
International Organizations: Representative of the Bible Memory Association International.
Mailing address: Box 12000, Ringgold LA 71068, USA. T. (1 318) 894 9154.

Peters, Guillaume
Rel: Buddhist: Mahayana.
International Organizations: Member of the Foundation for the Preservation of the Mahayana Tradition.
Other organizations: Director, Land of Medicine Buddha.
Mailing address: 5800 Prescott Rd, Soquel CA 95073, USA. T. (1 408) 462 8383. Fax (1 408) 462 8380

Peters, Hans Gerhard
Rel: Mennonite.
International Organizations: Member, Executive Committee of the Mennonite World Conference.
Mailing address: Rua Gabriel C Domingues 1000, Curitiba, Parana 81670, Brazil.

Peterson, Birdie, Rev
Rel: Spiritualist.
International Organizations: President of the Universal Church of the Masters.
Mailing address: 501 Washington Street, Santa Clara CA 95050, USA.

Petraitis, Donald Stanley, Rt Rev
Rel: Roman Catholic; Ordained on 09 MAY 1964; b. 13 AUG 1937 in Chicago IL, USA; celibate; Citizen of USA; Lang. English, Italian, Lithuanian and Portuguese.
International Organizations: Superior General (since 1984) of the Marian Fathers of the Immaculate Conception of the Blessed Virgin Mary.
Ecumenical Commission (since 1989) of the International Union of Superiors General. Postulator-General (since 1991) of the Marian Fathers of the Immaculate Conception of the Blessed Virgin Mary.
Other organizations: Trustee (since 1978), Pontifical College San Casimiro.
Main field of work: Ministry and administration. **Career:** Dean of Students/Assistant Headmaster, Marianapolis Preparatory School, Thompson CT from 1964 to 1969. Headmaster, Marianapolis Preparatory School from 1969 to 1975. Priests' Senate, Diocese of Norwich, CT. Liturgical Commission, Diocese of Norwich, CT from 1965 to 1968. Vicar-General, Marian Fathers from 1975 to 1984. **Educ.:** B.A. from Illinois Benedictine College, Lisle, USA (1953-1954); St Procopius Abbey, Lisle IL, USA (1958-1960); M.A. from Assumption College, Worcester MA, USA (summers) (1960-1964). **Interests:** reading, history. **Hons.:** Commander, Polonia Restituta; Knight of Columbus (4th degree). **Mem.:** Union of Superiors General; Knights of Columbus
Mailing address: Via Corsica 1, I-00198 Roma, Italy. T. (39 6) 855 4358 – 884 1885.
Residence: See Mailing Address.

Petranker, Jack
Rel: Buddhist.
International Organizations: Vice-President of the World Fellowship of Buddhists.
Mailing address: Tibetan Nyingma Meditation Centre, 2425 Hillside Avenue, Berkeley CA 94704, USA.

Petrowski, Chari
Rel: Hindu.
International Organizations: Head, Secretariat of the Krishnamurti Foundation of America.
Mailing address: PO Box 1560, Ojai CA 93023, USA. T. (1 805) 646 27 26.

Pettiti, Louis Edmond
Rel: Roman Catholic; b. 14 JAN 1916 in Asnières, France; m. to Zina Cyrin; c. 2 boys; Citizen of France.
International Organizations: President of the International Movement of Catholic Lawyers.
Judge (since 1980) of the European Court of Human Rights.
Main field of work: Barrister. **Career:** Joined the Bar, Paris in 1935. Vice-President, Société de législation comparée in 1977. President, Paris Bar from 1978 to 1979. President, Société des Prisons from 1985 to 1988. Member, National Committee of Human Rights from 1985 to 1989. French Branch, UNESCO from 1985 to 1989. Bicentenary Mission from 1987 to 1989. **Educ.:** Lic. Law from Faculty of Law, Paris, France. **Interests:** Fencing. **Hons.:** Officer of the Legion of Honour; Commander, National Order of Merit (France) and Order of the Italian Republic; Officer of the Order of the Republic of Senegal; Officer of the Phoenix (Greece); Commander of the Order of St Gregory the Great; UNESCO Prize for Human Rights (1979). **Mem.:** International Academy of Law, Mexico; Association of European Jurists (Vice-President) **Pubs.:** *Le Procès Charansky* (joint authorship), Grasset, 1979; *Mission en Iran*, 1982; *Le Rôle actuel du Conseil de l'Europe*, 1983; *La Cour européenne des droits de l'homme*, 1983 and *Liberté d'expression*, 1985.

Mailing address: 4 square La Bruyère, F-75009 Paris, France. T. (33 1) 42 80 49 54.
Residence: See Mailing Address. T. See Mailing Address

Peyrovan, Abdosalam, Dr

Rel: Muslim: Sufi.
International Organizations: of the School of Islamic Sufism.
Mailing address: PO Box 5827, Washington DC 20016, USA. T. (1 301) 654 91 91.

Peyton, Patrick, Rev

Rel: Roman Catholic; Ordained; celibate.
International Organizations: Director of the Family Rosary.
Mailing address: Executive Park Drive, Albany NY 12203, USA. T. (1 518) 489 8900.

Pezzullo, Lawrence A

Rel: Roman Catholic.
International Organizations: Executive Director of the Catholic Relief Services – United States Catholic Conference.
Mailing address: 209 West Lafayette Street, Baltimore MD 21201, USA.

Pfortmiller, Gene, Rev

Rel: Spiritualist; Ordained on 08 JAN 1992; b. 01 MAY 1936 in Bloomingdale IL, USA; m. to Sandra Pfortmiller; c. 3 boys; Citizen of USA; Lang. English.
International Organizations: President, Healers League (since 1990) of the National Spiritualist Association of Churches.
Other organizations: Missionary, medium and healer, National Spiritualist Association of Churches.
Main field of work: Ministry. **Educ.:** Morris Pratt Institute, Milwaukee WI, USA (1985-1990). **Interests:** antique cars, sports, camping, nature. **Pubs.:** *The NSAC Spiritual Manual*; *The NSAC Lyceum Manual* and *The National Spiritualist – Summit* (magazine).
Mailing address: 2020 W Turney Ave, Phoenix AZ 85015, USA. T. (1 602) 265 6413
Residence: See Mailing Address. T. See Mailing Address

Pfortmiller, Sandra Peterson, Rev

Rel: Spiritualist; Ordained on 08 JAN 1992; b. in Minneapolis MN, USA; m. to Gene Pfortmiller; c. 2 boys and 1 girl; Citizen of USA; Lang. English.
International Organizations: Secretary-Treasurer (since 1988) of the Licentiate Ministers and Certified Mediums Society.
Other organizations: Editor (since 1990), *The National Spiritualist – Summit*; Missionary, medium, healer, National Spiritualist Association of Churches.
Main field of work: Ministry, writing. **Educ.:** Stephens College, Columbia MO, USA (1957-1959); Phoenix College, AZ, USA (Beginning in 1980); Morris Pratt Institute, Milwaukee WI, USA (1985-1990). **Interests:** painting, nature, camping, antique cars. **Pubs.:** *The NSAC Spiritual Manual*; *The NSAC Lyceum Manual* and *The National Spiritualist – Summit* (magazine).
Mailing address: 2020 W Turney Ave, Phoenix AZ 85015, USA. T. (1 602) 265 6413
Residence: See Mailing Address. T. See Mailing Address

Philadelphia, P, H E Metropolitan Meliton

Rel: Orthodox.
International Organizations: Chief Secretary of the Holy Synod of the Ecumenical Patriarchate of Constantinople.
Mailing address: Chief Secretary of the Holy Synod, Halic-Fener, Istanbul, Turkey.

Phillips, Brian Winfred Gilbert

Rel: Quaker; b. 28 MAR 1924 in London, UK; m. to Maureen Elizabeth Phillips; c. 2 boys and 2 girls; Citizen of United Kingdom; Lang. English, Esperanto, French and Malagasy.
International Organizations: President of the Quaker Esperanto Society.
Career: Worker, various farms and garden from 1942 to 1950. Teacher, State junior school from 1950 to 1952. Missionary, Friends Service Council (Quakers) from 1953 to 1962. Teacher of handicapped children, state schools from 1962 to 1987. Diagnostic Radiographer, state hospitals in 1971. **Educ.:** Tchr. Cert. from Kirkby Teacher Training College, Lancashire, UK (1950-1951); Dip. for teaching handicapped children from Manchester Teacher Training College, UK (1963-1964); Dip. Soc. of Radiographers from School of Radiography, Ipswich, UK (1970-1971). **Interests:** Quakerism, ecumenism, esperanto, environmental issues. **Mem.:** Green Party, UK
Mailing address: 60 Tithe Barn Rd, Strafford ST16 3PL, UK. T. (44 785) 44 320
Residence: See Mailing Address. T. See Mailing Address

Phillips, Keith, Dr

Rel: Baptist; Ordained on 20 AUG 1971; b. 21 OCT 1946 in Portland OR, USA; m. to Mary Katherine; c. 3 boys; Citizen of USA; Lang. English, Latin, German and Hebrew.
International Organizations: Founder/President (since 1965) of the World Impact Services.
Other organizations: Chairman (since 1986), Mayor's Prayer Book Committee for Los Angeles; Lecturer (Staley lectures) (since 1988); Editorial Advisor to Zondervan Publications (since 1990); Member, Executive Board (since 1992), Christian Community Development Association.
Main field of work: Ministry. **Educ.:** B.A. from University of California, Los Angeles, USA (1965-1968); M.Div. from Fuller Theological Seminary, Pasadena CA, USA (1968-1971); Dr. of Ministry from Fuller Theological Seminary, Pasadena CA, USA (1968-1971). **Hons.:** Hon. D.Litt., Hon. D.H.L., John Brown University, Siloam Springs AR (1991). **Pubs.:** They Dare to Love the Ghetto; The Making of a Disciple and No Quick Fix.
Mailing address: 2001 S Vermont Ave, Los Angeles CA 90007, USA. T. (1 213) 735 1137. Fax 735 2576.
Residence: See Mailing Address. T. See Mailing Address

Philpot, Ford

Rel: Evangelical.
International Organizations: President of the Ford Philpot Evangelical Association.
Mailing address: PO Box 3000, Lexington KY 40533-9989, USA. T. (1 606) 223 0260. Fax 223 9441.

Piano, Lino, S.S.C., Rev

Rel: Roman Catholic; Ordained; celibate.
International Organizations: Procurator-General of the Société Sacerdotale de Saint Joseph-Benoît Cottolengo.
Member of the Congregation for Catholic Education.
Mailing address: Via di Villa Alberici 14, I-00165 Roma, Italy. T. (39 6) 63 09 03.

Piatti, Franco

Rel: Buddhist: Mahayana.
International Organizations: Member of the Foundation for the Preservation of the Mahayana Tradition.
Other organizations: Director, Shine.
Mailing address: Via Poggiberna 9, I-56040 Pomaia, Italy. T. (39 50) 685 774. Fax (39 50) 685 768

Piazza, Giuseppe, P.I.M.E., Fr

Rel: Roman Catholic; Ordained; celibate.
International Organizations: Procurator-General of the Pontifical Institute for Foreign Missions.
Mailing address: Via F D Guerrazzi 11, I-00152 Roma, Italy. T. (39 6) 58 97 941.

Picachy, Lawrence Trevor, S.I., H E Card

Rel: Roman Catholic; Ordained on 21 NOV 1947; b. 07 AUG 1916 in Darjeeling, India; celibate; Citizen of India.
International Organizations: Member (since 1976) of the College of Cardinals.
Member (since 1934) of the Society of Jesus.
Other organizations: Archbishop Emeritus (since 1986), Calcutta.

Career: Headmaster, St Xavier's School, Calcutta from 1954 to 1960. Principal, St Xavier's College, Calcutta from 1954 to 1958. Rector, St Xavier's College, Calcutta from 1954 to 1960. Parish Priest, Basanti village, West Bengal from 1960 to 1962. First Bishop, Jamshedpur from 1962 to 1969. Attended four sessions, II Vatican Council from 1962 to 1965. Chairman, Social Communications (Mass Media) Commission, Catholic Bishops' Conference of India in 1966. World tour studying Mass Media in 1968. Archbishop, Calcutta from 1969 to 1987. Vice-President, Catholic Bishops' Conference of India from 1972 to 1976. President, Catholic Bishops' Conference of India from 1976 to 1981. Created Cardinal by Pope Paul VI in 1976.
Mailing address: St Xavier's College, 30 Park Street, Calcutta 700016, India. T. (91 33) 44 5995

Picazo Gálvez, Melecio P, M.Sp.S., Fr

Rel: Roman Catholic; Ordained; celibate.
International Organizations: Superior General of the Missionaries of the Holy Ghost.
Mailing address: Ave Universidad 1702, 04010 México DF, Mexico. T. (52 5) 658 74 33

Pick, Philip L

Rel: Jewish.
International Organizations: President of the International Jewish Vegetarian and Ecological Society.
Mailing address: 855 Finchley Road, Golders Green, London NW11 8LX, UK. T. (44 81) 455 0692.

Pickard, David

Rel: Evangelical.
International Organizations: Director General, Singapore of the Overseas Missionary Fellowship.
Mailing address: 2 Cluny Road, Singapore 1025, Singapore. T. (65) 473 5755. Fax 472 2398.

Piederis, Jan, Rev

Rel: Christian.
International Organizations: General Secretary of the Batak Christian Community Church.
Mailing address: JL Hos Cokroaminoto 96, PO Box 96/MT, Jakarta 10310, Indonesia.

Piercy, Jessie

Rel: Protestant.
International Organizations: President of the International Prayer Fellowship.
Mailing address: PO Box 1236, Lake Junaluska NC 28745, USA. T. (1 704) 456 4454.

Pierli, Francesco, M.C.C.I., Fr

Rel: Roman Catholic; Ordained; celibate.
International Organizations: Superior General of the Comboni Missionaries of the Heart of Jesus.
Member, Council for Reporting to the Congregation and the International Union of Superior Generals and the Union of Superior Generals of the Congregation for the Evangelization of Peoples. Advisor of the Pontifical Council for Inter-Religious Dialogue.
Mailing address: Via Luigi Lilio 80, CP 10733, I-00143 Roma, Italy. T. (39 6) 519 1224.

Pierre, Marirose

Rel: Roman Catholic.
International Organizations: Supreme Secretary of the Supreme Ladies Auxiliary Knights of St John.
Mailing address: 832 W Franklin, Evansville IN 47720, USA. T. (1 812) 424 4493.

Pierre, Maurice

Rel: Unknown.
International Organizations: Secretary-General of the International Federation of Magical Societies.
Mailing address: FISM, 35 rue Vineuse, F-75116 Paris, France. T. (33 1) 47 27 17 68.

Pierson, Paul E

Rel: Evangelical.
International Organizations: Chairman of the Latin America Mission.
Mailing address: PO Box 52-7900 Miami FL 33152, USA. T. (1 305) 884 8400.

Pieterse, Jude, P.F.M., Br

Rel: Roman Catholic; Ordained; b. 01 SEP 1939 in Cape Town, South Africa; celibate; Citizen of South Africa; Lang. English and Afrikaans.
International Organizations: Secretary-General (since 1988) of the Southern African Catholic Bishops' Conference.
Member of the Little Brothers of Mary.
Other organizations: Trustee (since 1986), Community-based Education Programme; Trustee (since 1988), Independent Examinations Board; Vice-Provincial (since 1989), Marist Brothers, South African Province.
Main field of work: Education. **Career:** Principal, Marist Brothers High School, Port Elizabeth from 1970 to 1974. Provincial, Marist Brothers, South Africa from 1974 to 1983. Director, Catholic Institute of Education from 1985 to 1988. **Educ.:** B.Sc. from University of Natal, Pietermaritzburg, South Africa (1960-1962); M.R.E. from Seattle University, WA, USA (1984-1987). **Interests:** education, human rights issues. **Pubs.:** Author of several articles and papers on the desegregation of Catholic schools in South Africa (1976-91).
Mailing address: Khanya House, 140 Visagie Street, Pretoria 0002, South Africa. T. (27 12) 323 6458. C. CHURCHWELL PRETORIA. Tx 320776 SA. Fax (27 12) 211795.
Residence: Sacred Heart College, Eckstein Street, Observatory, Johannesburg 2143, South Africa. T. (27 11) 648 1501

Pilarczyk, Daniel Edward, Rev

Rel: Roman Catholic; Ordained on 20 DEC 1959; b. 12 AUG 1934 in Dayton OH, USA; celibate; Citizen of USA.
International Organizations: Archbishop (since 1982) of Cincinnati.
Other organizations: Trustee (since 1974), St Gregory Seminary; President (since 1986), National Conference of Catholic Bishops.
Career: Assistant Chancellor, Archdiocese of Cincinnati from 1961 to 1963. Faculty, Athenaeum of Ohio (St Gregory Seminary) from 1963 to 1974. Vice-President/Rector, St Gregory Seminary from 1968 to 1974. Synodal Judge, Archdiocesan Tribunal from 1971 to 1982. Director, Archdiocesan Educational Services from 1974 to 1982. Auxiliary Bishop, Cincinnati from 1974 to 1982. **Hons.:** Hon. LL.D., Xavier University (1975), Calumet College (1982), University of Dayton (1990), Margrette University (1990). **Pubs.:** Twelve Tough Issues, 1988 and numerous articles in newspapers and journals.
Mailing address: 100 East Eighth Street, Cincinnati OH 45202, USA. T. (1 513) 421 3131. Fax 381 2242
Residence: 29 East Eighth Street, Cincinnati OH 45202, USA.

Pilchick, Terence

Also know as Dharmachari Nagabodh; Rel: Buddhist; Ordained on 10 JAN 1974; b. 08 JUN 1948 in London, UK; single; Citizen of United Kingdom; Lang. English and French.
International Organizations: Dharma Worker (full time) (since 1974) of the Friends of the Western Buddhist Order.
President (Auckland, Wellington, Boston) (since 1990) of the Friends of the Western Buddhist Order.
Other organizations: Chairman (since 1975), Windhorse Publications; Editor, *Golden Drum Magazine* (quarterly); Managing Trustee, Karuna Trust.
Main field of work: Publishing, buddhist teaching. **Career:** Film editing for BBC from 1970 to 1974. **Educ.:** B.A. Eng. and American Litt. (Hons) from University of Kent at Canterbury, UK (1967-1970). **Pubs.:** Jai Bhim: Dispatches from a Peaceful Revolution.
Mailing address: 354 Crookesmoor Road, Sheffield S10 1BH, UK. T. (44 742) 684775
Residence: See Mailing Address. T. See Mailing Address

Pillers, Lauritz P

Rel: Christian.
International Organizations: President of the International Christian Leprosy Mission.
Mailing address: PO Box 23353, Portland OR 97223-0043, USA. T. (1 503) 244 5935.

Pimen, (Sergey Mikhailovich Isvekov), Patriarch
Rel: Orthodox: Russian; Ordained in 1932; b. 23 JUL 1910 in Bogorodsk, Moscow, Russia; Citizen of Russia.
International Organizations: Patriarch (since 1971) of Moscow and All Russia.
Career: Became monk in 1927. Bishop, Balta in 1957. Archbishop, Balta in 1960. Metropolitan, Leningrad from 1960 to 1963. Metropolitan, Krutitsky and Kolomna from 1963 to 1971. **Hons.:** Grand Cordon, Order of the Cedar of Lebanon (1972); Order of St Sergius of Radonezh, USSR (1979); Order of Friendship of the Peoples (1980); Order of the Red Banner of Labour (1985).
Mailing address: Danilov Monastery, Danilovsky Val 22, Moskva, Russia.

Pimenta, Simon Ignatius, H E Card
Rel: Roman Catholic; Ordained on 21 DEC 1949; b. 01 MAR 1920 in Bombay, India; celibate; Citizen of India.
International Organizations: Member (since 1988) of the College of Cardinals.
Member of the Synod of Bishops. Member of the Congregation for the Evangelization of Peoples. Member of the Pontifical Council for Promoting Christian Unity. Member of the Pontifical Council for Inter-Religious Dialogue.
Other organizations: Archbihsop (since 1978), Bombay.
Career: Elected to the Titular Church, Bocconia in 1971. Consecrated bishop in 1971. Coadjutor bishop in 1977. Created Cardinal by Pope John Paul II in 1988.
Mailing address: Archbishop's House, 21 Nathalal Parekh Marg, Bombay 400039, India. T. (91 22) 204 96 96

Pinto, Roger
Rel: Jewish; Unordained; b. 01 JUN 1933 in Oran, Algeria; m. to Mireille Gras; c. 2 boys; Citizen of France; Lang. French, Hebrew and Arabic.
International Organizations: President of the International Committee for the Freedom of Syrian Jewry.
Other organizations: Member of the Council (since 1970), Consistory of Paris and France; President, Conseil National des Organismes de Formation Professionnelle, Paris; Director-General, Asprofoc – Formation Professionnelle; Vice-President, Conseil Représentatif des Institutions Juives de France (CRIF); President, Commission 'Communautés Juives en Péril', CRIF; President, Fédération Sépharade Européenne; President, Siona.
Main field of work: Defence of human rights. **Educ.:** Dip. d'Etudes Arabes. **Interests:** Human Rights supporter. **Hons.:** Knight of the National Order of Merit. **Pubs.:** Many press articles in various papers and magazines.
Mailing address: 52 rue Richer, F-75009 Paris, France. T. (33 1) 42 46 04 34
Residence: 153 bd Haussmann, F-75008 Paris, France T. (33 1) 43 02 84 94

Piovanelli, Silvano, H E Card
Rel: Roman Catholic; Ordained on 13 JUL 1947; b. 21 FEB 1924 in Ronta di Mugello, Italy; celibate; Citizen of Italy.
International Organizations: Member (since 1985) of the College of Cardinals.
Member of the Congregation for Catholic Education. Member of the Pontifical Council for Dialogue with Non-Believers.
Other organizations: Archbishop (since 1983), Florence.
Career: Consecrated Bishop (Titular Church of Tubune, Mauritania) in 1982. Created Cardinal by Pope John Paul II in 1985.
Mailing address: Arcivescovado, Piazza S Giovanni 3, I-50129 Firenze, Italy. T. (39 55) 29 88 13

Pipaón Monreal, José Javier P, O.A.R., Fr
Rel: Roman Catholic; Ordained; celibate.
International Organizations: Prior-General of the Order of the Augustinian Recollects.
Mailing address: Viale dell'Astronomia 27, I-00144 Roma, Italy. T. (39 6) 592 65 34

Pires da Silveira, Abival P, Rev Dr
Rel: Reformed Church.
International Organizations: Vice-President of the World Alliance of Reformed Churches.
Mailing address: Rua Maranhao 236, Apto 32, 01240 Sao Paulo SP, Brazil. T. (55 11) 255 6111. Fax (55 11) 259 0009
Residence: T. (55 11) 256 8159

Piroelle, Pierre
Rel: Christian.
International Organizations: President of the Missions et développement.
Mailing address: 3 rue Turgot, F-21000 Dijon, France. T. (33) 80 30 92 86.

Pironio, Eduardo Francisco, H E Card
Rel: Roman Catholic; Ordained on 05 DEC 1943; b. 03 DEC 1920 in Nueve de Julio, Argentina; celibate; Citizen of Argentina.
International Organizations: President (since 1984) of the Pontifical Council for the Laity.
Member of the Sacred Congregation for the Oriental Churches. Member of the Congregation for the Causes of Saints. Member of the Congregation for Bishops. Member of the Congregation for Catholic Education. Member of the Pontifical Council for the Interpretation of Legislative Texts.
Other organizations: Titular Archbishop (since 1975), Tiges; Member (since 1976), College of Cardinals; Member, Council of Cardinals and Bishops.
Career: Titular Bishop, Ceciri from 1964 to 1972. Bishop, Mar del Plata from 1972 to 1975. Created Cardinal by Pope Paul VI in 1976. Prefect, Sacred Congregation for Religious Orders and Secular Institutes from 1976 to 1984.
Mailing address: Piazza del S Uffizio 11, I-00193 Roma, Italy. T. (39 6) 698 4264

Pirson, Joseph
Rel: Roman Catholic.
International Organizations: Secretary-General of the International Federation of Rural Adult Catholic Movements.
Mailing address: Rue Africaine 92, B-1050 Bruxelles, Belgium. T. (32 2) 538 78 42.

Pittaluga, Agnese, MC, Sr
Rel: Roman Catholic; Ordained; celibate.
International Organizations: Secretary of the Consolata Missionary Sisters.
Mailing address: Corso G Allamano 137, I-10095 Grugliasco, Italy. T. (39 11) 703 703.

Pitts, Glen R, Rev
Rel: Protestant.
International Organizations: International Vice-President of the Every Home for Christ International.
Mailing address: 20232 Sunburst Street, Chatsworth CA 91311-6286, USA.

Pityana, Nyameko Barney, Rev Dr
Rel: Anglican; b. 07 AUG 1945 in Uitenhage, South Africa; m. to Nosidima Dimza Songca; c. 1 girl; Citizen of United Kingdom; Lang. English, Xhosa, Zulu and Afrikaans.
International Organizations: Director (since 1988) of the Programme to Combat Racism.
Main field of work: Church. **Career:** Secretary-General, South African Students Organization from 1970 to 1973. Assistant Curate, Wonghton Ecumenical Parish from 1982 to 1985. Vicar, Immanuel Hightex Heath from 1985 to 1988. **Educ.:** B.A. (uncompl.) from University of Fort Hare, Alice, Ciskei, South Africa (1966-1968); B.A. from University of South Africa, Pretoria, South Africa (1973-1975); B.Proc. from University of South Africa, Pretoria, South Africa (1976-1978); B.D. from University of London, King's College, UK (1978-1981); Cert. Theol. from Ripon College Cuddesdon, Oxford, UK (1981-1982).
Mailing address: PO Box 2100, Route de Ferney 150, CH-1211 Genève 2, Switzerland. T. (41 22) 791 61 11. C. OIKOUMENE Geneva. Tx 415 730 OIK CH. Fax (41 22) 791 03 61.

Platt, Albert T, Dr
Rel: Christian.
International Organizations: President of the CAM International.
Mailing address: 8625 La Prada Dr, Dallas TX 75228, USA. T. (214) 327 8206.

Platten, Stephen George, Rev Canon

Rel: Anglican; Ordained on 26 SEP 1975; b. 17 MAY 1947 in Southgate, Middlesex, UK; m. to Rosslie Thompson; c. 2 boys; Citizen of United Kingdom; Lang. English and German.
International Organizations: Anglican Co-Secretary (since 1990) of the Anglican-Roman Catholic International Commission.
Other organizations: Minister Provincial for Europe (since 1990), Society of Saint Francis (Third Order); Archbishop (since 1990), Canterbury's Secretary for Ecumenical Affairs; Honorary Canon (since 1990), Canterbury Cathedral.
Main field of work: Ecumenical affairs. **Career:** Assistant Curate, St Andrew's Headington Oxford from 1975 to 1978. Chaplain and Tutor, Lincoln Theological College from 1978 to 1982. Residentiary Canon, Portsmouth Cathedral from 1983 to 1989. Director of Ordinands and Director of Ministerial Training, Diocese of Portsmouth from 1983 to 1989. **Educ.:** B.Ed. (Hons) from University of London, UK (1968-1972); Dip. Theol. from University of Oxford, Trinity College, UK (1973-1974); Cert. Theol. from Cuddesdon Theology College, Oxford, UK (1973-1975). **Interests:** English literature, music, walking, Christian ethics, Christian doctrine. **Mem.:** Society for the Study of Christian Ethics **Pubs.:** Series Editor, *Ethics and our Choices*, 1990; Contributor to *Spirituality and Psychology*, 1990 and Contributor to *Say one for me*, A W Carr, Ed., 1991.
Mailing address: Dept. of Ecumenical Affairs, Lambeth Palace, London SE1 7JU, UK. T. (44 71) 928 8282
Residence: See Mailing Address. T. See Mailing Address

Plourde, Joseph Aurèle, Most Rev

Rel: Roman Catholic; Ordained on 07 MAY 1944; b. 12 JAN 1915 in St François, Madawaska Co., New Brunswick, Canada; celibate; Citizen of Canada.
International Organizations: Archbishop Emeritus (since 1989) of Ottawa.
Career: Served with RCA Chaplain Corps during World War II. Former Professor of Philosophy, St Louis College of Edmunston and Diocesan Director of social works. Pastor, St Leonard, Madawaska from 1955 to 1959. Consecrated Bishop in 1964. Apostolic Administrator, Alexandria in 1964. President, Canadian Conference of Catholic Bishops from 1969 to 1971. Archbishop, Ottawa from 1967 to 1989. **Educ.:** Lic. Social Stds. from University of Ottawa, Ontario, Canada. **Hons.:** Knight of Malta; Knight of the Holy Sepulchre; Hon. Ph.D. in Ed., University of Moncton. **Pubs.:** Author of numerous publications and articles in ecclesiology, spirituality and Catholic education.
Mailing address: 1247 Kilborn Place, Ottawa ON K1H 6K9, Canada. T. (1 613) 738 1239

Pobee, John Samuel, Prof Dr

Rel: Anglican; Ordained in 1988; b. 09 JUL 1937 in Cape Coast, Ghana; single; Citizen of Ghana; Lang. English.
International Organizations: Associate Director, Programme on Ecumenical Theological Education (since 1984) of the World Council of Churches.
President (since 1988) of the International Association for Mission Studies.
Main field of work: University education. **Career:** Lecturer-Professor, University of Ghana from 1966 to 1984. Claus Petri Professor, University of Uppsala in 1983. **Educ.:** University of Ghana, Legon (1957-1961); University of Cambridge, UK (1961-1966). **Pubs.:** Towards an African Theology, Abingdon, 1979; Persecution and Martyrdom in the Theology of Paul, Academic Press, 1985 and Kwame Nkrumah and the Church in Ghana, Asempa, 1987.
Mailing address: Château de Bossey, CH-1298 Céligny, Switzerland. T. (41 22) 776 25 31. Fax 776 01 69. C. INSTITUTCELIGNY GENEVA.
Residence: 17 rue de Zurich, CH-1201 Genève, Switzerland. T. (41 22) 732 37 93

Poggi, Luigi, H E Mgr

Rel: Roman Catholic; Ordained on 28 JUL 1940; b. 25 NOV 1917 in Piacenza, Italy; celibate.
International Organizations: Apostolic Nuncio to Italy (since 1986) of the Roman Catholic Church.
Other organizations: Titular Archbishop, Forontoniana.
Mailing address: Via Po 27-29, I-00198 Roma, Italy. T. (39 6) 854 62 87 – 855 20 92. Fax 854 97 25

Pohl, Rudolf, Mgr

Rel: Roman Catholic; Ordained; celibate.
International Organizations: Executive Officer of the Consociatio Internationalis Musicae Sacrae.
Mailing address: Kaüberg 18, B-4720 Kelmis, Belgium.

Poisson, André, Fr

Rel: Roman Catholic; Ordained; celibate.
International Organizations: Minister-General of the Carthusian Order.
Mailing address: Monastère Grande Chartreuse, F-38380 St Laurent-du-Pont, France. T. (33) 76 88 60 30.

Poletti, Ugo, H E Card

Rel: Roman Catholic; Ordained on 29 JUN 1938; b. 19 APR 1914 in Omegna, Novara, Italy; celibate; Citizen of Italy.
International Organizations: Member (since 1973) of the College of Cardinals.
Member of the Congregation for Bishops. Member of the Sacred Congregation for the Oriental Churches. Member of the Sacred Congregation for the Sacraments and Divine Worship. Member of the Congregation for the Clergy. Member of the Congregation for Religious and Secular Institutes. Member of the Pontifical Council for the Laity.
Other organizations: High Priest (since 1991), Patriarcale Basilica Liberiana de Santa Maria Maggiore.
Career: Consecrated Titular Bishop, Medeli in 1958. Promoted to Spoleto in 1967. Consecrated Archbishop, Cittanova in 1969. Created Cardinal by Pope Paul VI in 1973. Vicar-General, Rome from 1973 to 1991.
Mailing address: Piazza S Giovanni in Laterano 6, I-00184 Roma, Italy. T. (39 6) 698 64 36

Poli, Paul

Rel: Catholic.
International Organizations: Representative of the Comité laïque pour l'éducation au développement – Educateurs sans frontières.
Mailing address: 28 boulevard Bonne Nouvelle, F-75010 Paris, France. T. (33 1) 45 23 10 81. Fax 48 01 08 69.

Polkinghorne, John Charlton, Rev Dr

Rel: Church of England; Ordained in 1982; b. 16 OCT 1930 in Weston-super-Mare, Somerset, UK; m. to Ruth Isobel Martin; c. 2 boys and 1 girl; Citizen of United Kingdom; Lang. English.
International Organizations: Ecclesiastic of the Church of England.
Other organizations: Honorary Professor of Theoretical Physics (since 1984), University of Kent; President (since 1989), Queens' College, Cambridge; Member (since 1989), Church of England Doctrine Commission; Member (since 1990), General Synod of the Church of England.
Career: Fellow, Trinity College, Cambridge from 1954 to 1986. Lecturer, University of Edinburgh from 1956 to 1958. Lecturer, University of Cambridge from 1958 to 1965. Reader, University of Cambridge from 1965 to 1968. Professor of Mathematical Physics, University of Cambridge from 1968 to 1979. Professor, Westcott House, Cambridge from 1979 to 1981. Ordained Deacon in 1981. Curacies in Cambridge and Bristol from 1981 to 1984. Vicar, Blean, Kent from 1984 to 1986. Fellow, Dean and Chaplain, Trinity Hall, Cambridge from 1986 to 1989. Honorary Fellow, Trinity Hall, Cambridge in 1989. Member, Science Research Council from 1975 to 1979. Chairman, Nuclear Physics Board from 1978 to 1979. Chairman, Committee to Review the Guidance on the Research Use of Fetuses and Fetal Material from 1988 to 1989. **Educ.:** M.A. from University of Cambridge, UK (1955); Ph.D. from University of Cambridge, UK (1956); Sc.D. from University of Cambridge, UK (1974). **Interests:** gardening. **Mem.:** Fellow, Royal Society (1974) **Pubs.:** *One World*, 1986; *Science and Providence*, SPCK, 1989, New Science Library, 1989 and *Rochester Roundabout*, Longman, 1989.
Mailing address: The President's Lodge, Queens' College, Cambridge CB3 9ET, UK. T. (44 223) 335532
Residence: See Mailing Address. T. See Mailing Address

Polok, W, Pastor

Rel: Seventh-Day Adventist.
International Organizations: President, Polish Union Conference of the General Conference of Seventh-Day Adventists.
Mailing address: Foksal 8, Warszawa 00 366, Poland.

Poloma, Margaret, Prof

Rel: Evangelical; b. 27 AUG 1943 in Los Angeles CA, USA; Citizen of USA; Lang. English.

International Organizations: Professor of Sociology (since 1970) of the University of Akron.

Other organizations: Council (since 1991), Society for the Scientific Study of Religion.

Main field of work: Teaching and research. **Educ.:** Ph.D. from Case Western Reserve University, Cleveland OH, USA (1965-1970). **Interests:** effects of spirituality on mental health. **Pubs.:** The Assemblies of God at the Crossroads, University of Tennessee Press; Varieties of Prayer, Trinity Press and Exploring Neglected Dimensions of Religion in Quality of Life Research, Edwin Mellen Press.

Mailing address: University of Akron, Olin Hall, Sociology Dept., Akron OH 44225-1905, USA. T. (1 216) 972 6837

Residence: 2872 Silver Lake Blvd, Silver Lake OH 44229, USA.

Pomberger, Elisabeth, Dr

Rel: Roman Catholic.

International Organizations: Representative at UN, Vienna of the Pax Romana, International Catholic Movement for Intellectual and Cultural Affairs.

Mailing address: Ramperstorffergasse 12/I/14, A-1050 Wien, Austria.

Pomeroy, David Wardell, Rev

Rel: United Church of Christ; Ordained on 22 NOV 1964; b. 29 SEP 1939 in South Bend IN, USA; m. to D Ann F Krick; c. 2 boys and 2 girls; Citizen of USA; Lang. English.

International Organizations: Director of Electronic Media (since 1972) of the National Council of Churches.

Other organizations: Associate Pastor (since 1986), United Church of Spring Valley; Chairman (since 1989), Interfaith Broadcasting Commission.

Main field of work: Electronic communication. **Career:** Pastor, The People's Church, Long Beach NY from 1964 to 1968. Minister to Youth, First Congregational Church, Chappaqua NY from 1968 to 1972. Interim Pastor, New Hyde Park Community Church, New Hyde Park NY from 1973 to 1974. Interim Pastor, Briarcliff Congregational Church, Briarcliff NY from 1979 to 1980. Interim Pastor, Community Church of the Pelhams, Pelham NY from 1982 to 1983. Chairman, Dept. of Radio and Television, The Council of Churches of the City of New York from 1970 to 1972. President, Tri-State Media Ministry from 1976 to 1978. Chairman, North American Broadcast Section, World Association for Christian Communication (1980 and 1991). Member, Executive Committee, North American Regional Association, World Association for Christian Communication from 1990 to 1991. **Educ.:** B.A. from DePauw University, Greencastle IN, USA (1956-1960); M.Div. from Union Theological Seminary, New York NY, USA (1960-1964). **Interests:** film reviewing. **Hons.:** Emmy ("Duty Bound"); Blue Ribbon, American Film Festival ("Into the Mouths of the Babes"). **Mem.:** National Association of Educational Broadcasters; Association for Education Communications and Technology; The National Academy of Television Arts and Sciences; American Civil Liberties Union **Pubs.:** *The Vision of Ingmar Bergman* (unpublished); "Computers and Communion(s)", in *Church Bytes* (magazine) and film reviews in various magazines, including: *The Christian Century, Film Information, Theology Today.*

Mailing address: National Council of Churches, Room 856, 475 Riverside Drive, New York NY 10115, USA. T. (1 212) 870 2574

Residence: 684 Union Road, Spring Valley NY 10977, USA. T. (1 914) 362 1977

Ponnau, Geneviève, Sr

Rel: Roman Catholic; Ordained; celibate.

International Organizations: Superior General of the Sisters of St Joseph of Tarbes.

Mailing address: Maison St Joseph, Cantaous Cidex 501, F-65160 Saint Laurent de Neste, France. T. (33) 62 98 92 70 – 62 98 94 02. Fax 62 98 91 78.

Popescu, G, Rev Dr

Rel: Orthodox: Romanian.

International Organizations: Priest of the St Mary's Romanian Orthodox Church.

Mailing address: St Mary's Romanian Orthodox Parish, 64 Linthorn Ave, Enfield South NSW 2133, Australia.

Popkes, Wiard Udo, Dr

Rel: Baptist; Ordained on 01 SEP 1966; b. 30 JUN 1936 in Ihren, Ostfriesland, Germany; m. to Irma Blask; c. 1 boy and 1 girl; Citizen of Germany; Lang. German, English, Dutch and French.

International Organizations: Executive Officer (since 1976) of the European Baptist Theological Teachers' Conference.

Other organizations: Lecturer (since 1969), Hamburg Baptist Theological Seminary; Outside Lecturer (since 1987), Hamburg University; Chairman, Board of Trustees (since 1989), Rüschlikon Baptist Seminary.

Main field of work: University and Seminary education. **Career:** Pastor, Baptist Church, Münster from 1966 to 1969. **Educ.:** B.D. from Baptist Seminary, Rüschlikon, Germany (1958-1962); Dr. Theol. from University of Zürich, Switzerland (1960-1965). **Interests:** New Testament, international relations. **Mem.:** Wiss. Gesellschaft für Theologie, SNTS **Pubs.:** Adressaten, Situation und Form des Zabusbriefes; Christus Traditus and Gemeinde – Raum des Vertrauens.

Mailing address: Theologisches Seminar Hamburg, Rennbahnstrasse 115, 22111 Hamburg, Germany. T. (49 40) 65 58 50. Fax (49 40) 65 58 51 34.

Residence: Oberförsterkoppel 10, 21521 Aumühle, Germany. T. (49 4104) 2294

Porte, Bernard

Rel: Catholic; b. 07 MAR 1938 in Nantes, France; m. to Marie-Paule Ferras; c. 2 boys and 3 girls; Citizen of France; Lang. French, English and Italian.

International Organizations: President (since 1989) of the International Federation of Catholic Dailies.

Other organizations: Chairman of Board (since 1985), Bayard Presse; President, Centre Nationale de Presse Catholique; President, La Sauvegarde de l'enfance des Yvelines.

Main field of work: Press. **Career:** Director, Fondation pour la recherche sociale from 1965 to 1967. Secretary General, L'Union Nationale des Centres d'Etudes et D'Actions Sociales from 1967 to 1976. Directeur Central, Bayard Presse from 1976 to 1979. Managing Director, Bayard Presse from 1979 to 1982. Directeur General, Bayard Presse from 1982 to 1985. **Educ.:** D.E.S. Droit Public from University of Paris II (University of Law, Economics and Social Sciences), France; D.E.S. Sc. Pol. from University of Paris II (University of Law, Economics and Social Sciences), France. **Pubs.:** L'initiative bénévole, une société réinventée, 1977.

Mailing address: UCIP, CP 197, CH-1211 Genève 20, Switzerland. T. (41 22) 734 74 16

Residence: Route de Rueil 15, F-78150 Le Chesnay, France T. (33 1) 39 54 56 87

Portigliatti, Bruno

Rel: Buddhist.

International Organizations: President of the Buddhist Union of Europe.

Mailing address: 58 Eccleston Square, London SW1Y 1PH, UK. T. (44 71) 828 9242.

Portillo, Alvaro del P, Most Rev Mgr

Rel: Roman Catholic; Ordained on 25 JUN 1944; b. 11 MAR 1914 in Madrid, Spain; celibate; Citizen of Spain; Lang. Spanish, Italian and English.

International Organizations: Prelate (since 1982) of the Prelature of the Holy Cross and Opus Dei.

President General of the Priestly Society of the Holy Cross. Consultor (since 1966) of the Congregation for the Clergy. Consultor (since 1982) of the Congregation for the Causes of Saints. Consultor (since 1984) of the Pontifical Council for Social Communications.

Other organizations: Titular Bishop, Vita; Honorary Academician (since 1982), Pontificia Academia Theologica Romana; Grand Chancellor, the Universities of Navarra (Spain), Piura (Peru), and La Sabana (Colombia), and of the Roman Athenaeum of the Holy Cross (Italy).

Main field of work: Ministry. **Career:** Secretary-General, Opus Dei from 1940 to 1947. Procurator-General, Opus Dei from 1947 to 1956. Secretary-General, Opus Dei from 1956 to 1975. President General, Opus Dei from 1975 to 1982. Counsellor, Italy, Opus Dei from 1947 to 1950. Consultor, Sacred Congregation for Religious from 1954 to 1966. Consultor, Sacred Congregation of Council from 1959 to 1966. Consultor, Postconciliar Commission on Bishops in 1966. President, Antepreparatory Commission for the Laity of Vatican Council II from 1960 to 1961. Member, Preparatory Commission, Vatican Council II on

States of Perfection from 1961 to 1962. Examiner, Sacred Congregation of the Holy Office in 1960. Judge, Sacred Congregation of the Holy Office in 1964. Consultor, Pontifical Commission for the Revision of the Code of Canon Law from 1963 to 1983. Consultor, Sacred Congregation for the Doctrine of the Faith from 1966 to 1983. Member of the Secretariat, Synod of Bishops in 1983. Member, Synod of Bishops (1987 and 1990). Vatican Council II (from 1962 to 1965): Council Expert, Secretary of Commission on the Discipline of the Clergy and of the Christian People, Consultor of Commission on Bishops, Commission on Religious, Commission on the Discipline of the Faith, and Mixed Commission on Associations of the Faithful. **Educ.:** Dr. Civil Eng. from University of Madrid, Spain (1941); Dr. Phil. and Letters from University of Madrid, Spain (1944); Dr. Canon Law from Pontifical University San Tommaso d'Aquino, Vatican City State (1948). **Hons.:** Knight of Honour and Devotion of the Supreme Order of Malta (1958); Grand Cross of St Raymond of Pennafort (1967). **Pubs.:** *Discovery and Explorations of the Coasts of California*, 1947 (latest Edn. 1982); *Faithful and Laity in the Church*, 1969 (translated into 6 languages) and *On Priesthood*, 1970 (translated into 5 languages).
Mailing address: Viale Bruno Buozzi 73, I-00197 Roma, Italy. T. (39 6) 8078741.

Poser, Klaus, Dr

Rel: Christian.
International Organizations: Director of the Commission on Inter-Church Aid, Refugee and World Services.
Mailing address: WCC, PO Box 2100, CH-1211 Genève 2, Switzerland. T. (41 22) 791 61 11. C. OIKUMENE Geneva. Tx 415 730 OIK CH. Fax (41 22) 791 03 61.

Post, Avery D, Rev Dr

Rel: United Church of Christ.
International Organizations: President of the United Church of Christ.
Mailing address: 105 Madison Ave, New York NY 10016, USA.

Poswick, Reginald-Ferdinand, OSB, Br

Rel: Roman Catholic; Ordained on 13 FEB 1957; b. 04 JAN 1937 in Brussels, Belgium; celibate; Citizen of Belgium; Lang. French, English, Dutch, German, Italian, Spanish, Latin, Hebrew and Greek.
International Organizations: Secretary-General (since 1982) of the International Association Bible and Computers.
Member of the Benedictine Confederation. Board of Directors (since 1986) of the Catholic Biblical Federation.
Other organizations: Founder and Manager (since 1980), Centre: Informatique et Bible.
Main field of work: Computer applications for Bible and religious studies. **Career:** Member, Executive Committee, World Catholic Federation for the Biblical Apostolate from 1984 to 1990. **Educ.:** Lic. Theol. from Pontifical St Anselmo Athenaeum, Vatican City State (1964); Cert. Anal.-Prog. from IBM Education Center, Belgium (1971). **Interests:** computer culture and Christian message. **Hons.:** International Association for Literary and Linguistic Computing (ALLC); International Organization for Septuagint and Cognate Studies (IOSCS); Society of Biblical Literature (SBL); JRI. **Mem.:** Society of Biblical Literature **Pubs.:** *Table Pastorale de la Bible*, Lethielleux, Paris, 1974; *Dictionnaire Encyclopédique de la Bible*, Brepols, Turnhout, 1987 and Actes des Colloques de l'AIBI: Louvain-la-Neuve, 1985, Jerusalem, 1988, Tübingen, 1991, Slatkine, Genève.
Mailing address: Abbaye de Maredsous, B-5537 Denée, Belgium. T. (32 82) 69 96 47. Fax 22 32 69
Residence: See Mailing Address. T. (32 82) 69 91 55

Potgieter, Pieter Cornelius, Prof

Rel: Dutch Reformed; Ordained on 20 JAN 1966; b. 10 FEB 1940 in Bloemfontein, Orange Free State, South Africa; m. to Diana De Vries; c. 2 boys and 1 girl; Citizen of South Africa; Lang. Afrikaans, English, Dutch and German.
International Organizations: Moderator (since 1990) of the General Synod, Dutch Reformed Church of South Africa.
Other organizations: Professor (since 1980), University of the Orange Free State, Bloemfontein; Dean, Faculty of Theology (since 1989), University of the Orange Free State, Bloemfontein.
Main field of work: University education. **Career:** Minister, Dutch Reformed Church from 1966 to 1978. Associate Professor, University of Pretoria from 1978 to 1980. **Educ.:** B.A. (Hons) from University of Stellenbosch, South Africa (1958-1961); M.Th. from University of Stellenbosch, South Africa (1962-1965); D.Th. from University of Stellenbosch, South Africa (1973). **Interests:** ecumenism, research. **Mem.:** Theological Society of South Africa **Pubs.:** Skrif, Dogma en Verkondiging; Personal Faith in the Theology of Karl Barth and The Providence of God in Calvin's Correspondence.
Mailing address: PO Box 28782, Danhof, Bloemfontein 9310, South Africa. T. (27 51) 401 2667
Residence: 63 Albrecht Street, Bloemfontein 9301, South Africa. T. (27 51) 311 343

Potter, David

Rel: Unknown.
International Organizations: Honorary Secretary of the World Congress of Faiths.
Mailing address: 28 Powis Gardens, London W11 1JG, UK. T. (44 71) 727 2607.

Potter, Philip Alford, Rev Dr

Rel: Methodist; Ordained; b. 19 AUG 1921 in Dominica, West Indies; m. to (1st) Doreen Cousins (died), (2nd) Baerbel van Wartenburg; Citizen of West Indies.
International Organizations: Chaplain (since 1985) of University of the West Indies.
Other organizations: Professor (since 1985), United Theological College of the West Indies.
Career: Missionary and Overseas Secretary, Student Christian Movement of Great Britain from 1948 to 1950. Minister, Methodist Church, Haiti from 1950 to 1954. Secretary, Youth Dept., World Council of Churches (WCC) from 1954 to 1957. Executive Secretary, Youth Dept., WCC from 1957 to 1960. Director, Commission on World Mission and Evangelism, WCC from 1967 to 1972. Secretary-General, WCC, Geneva from 1972 to 1984. Secretary for West Africa and West Indies, Methodist Missionary Society, London from 1961 to 1966. **Interests:** hiking, geology. **Hons.:** Hon. Th.D., University of Hamburg (1971), University of Geneva (1976), Theological Institute of Romanian Orthodox Church (1977), Humboldt University, Berlin (1982), University of Uppsala (1984), University of Birmingham (1985); Hon. LL.D., University of the West Indies (1974); Niwano Peace Prize, Japan (1986). **Pubs.:** Living the Christian Year, 1975; Life in all its Fullness, 1981 and numerous articles in *Ecumenical Review*, *International Review of Mission*, *Student World*.
Mailing address: 3A York Castle Ave, Kingston 6, Jamaica.

Poupard, Paul, H E Card

Rel: Roman Catholic; Ordained on 18 DEC 1954; b. 30 AUG 1930 in Bouzillé, Maine et Loire, France; celibate; Citizen of France; Lang. French, Italian and English.
International Organizations: President (since 1985) of the Pontifical Council for Dialogue with Non-Believers.
President (since 1988) of the Pontifical Council for Culture. Member (since 1959) of the Pontifical Council for Inter-Religious Dialogue. Member (since 1980) of the Congregation for the Evangelization of Peoples. Member (since 1985) of the Sacred Congregation for the Sacraments and Divine Worship. Member of the Congregation for Catholic Education.
Other organizations: Member (since 1985), College of Cardinals.
Main field of work: Culture, unbelief. **Career:** Attaché, Centre National de la Recherche Scientifique (CNRS) from 1958 to 1959. Attaché, Secretariat of State of the Holy See from 1959 to 1971. Rector, Institut Catholique de Paris from 1971 to 1981. Auxiliary Bishop, Paris from 1979 to 1980. Titular Bishop, Usula in 1979. Archbishop in 1980. Created Cardinal by Pope John Paul II in 1985. **Educ.:** Dip. Sc.Rel. from Ecole Pratique des Hautes Etudes, Paris, France (1958-1959); Dr. Theol. from Université Catholique de l'Ouest, Angers, France (1959); Dr. Hist. from University of Paris IV (Paris-Sorbonne), France (1960). **Hons.:** Prix du Roi René d'Anjou; Grand Prix Cardinal-Grente de l'Académie Française; Médaille de Vermeil, Paris; Officier de la Légion d'Honneur; Prix Journalistique international, Rome; Commander of the National Orders of Benin and Madagascar; Grand Cross of the Equestrian Order of the Holy Sepulchre of Jerusalem; Dr. h.c., Universities of Aix-en-Provence, Fu Jen, Louvain, Quito; Robert Schuman Prize for Europe; Medal from Pio Manzù International Research Centre on the Environment; Grand Cross, Order of Merit of the Federal Republic of Germany. **Mem.:** Académie des Belles-Lettres, Sciences et Arts d'Angers **Pubs.:** *Connaissance du Vatican*, Beauchesne (translated into 7 languages); *Dictionnaire des Religions*, PUF, 1st Ed. 1984, 1856 p., 2nd Ed. 1985 and 1990 (translated into Italian and Spanish) and *Le Pape*, *Le Vatican*,

La Foi catholique, Le Concile Vatican II, Les Religions, Col. Que sais-je, PUF.
Mailing address: Piazza San Calisto 16, I-00153 Roma, Italy. T. (39 6) 698 7393. Fax 698 7368. Tx C/O 2013 CULTVAT VA.

Powathil, Joseph, Most Rev

Rel: Roman Catholic; Ordained on 03 OCT 1962; b. 14 AUG 1930 in Kurumpanadam, India; celibate; Citizen of India.
International Organizations: Archbishop (since 1985) of Changanacherry.
Career: Lecturer in Economics, St Berchmans' College, Changanacherry from 1963 to 1972. Auxiliary Bishop, Changanacherry, Titular Bishop Caesarea Philipi in 1972. Consecrated Bishop in 1972. 1st Bishop, Kanjirappally Diocese, Kerala from 1977 to 1986. Chairman, CBCI Commission for Education and Clergy, KCBC Commission for Education, KCBC Commission for Development, Justice and Peace, SMBC Commission for Ecumenism. **Mem.:** Pontifical Commission for Dialogue with the Orthodox Syrian Church.
Mailing address: Archbishop's House, Changanacherry 686 101, Kerala, India. T. (91 4824) 22540

Powell, G, Rev

Rel: Church of Christ.
International Organizations: President of the Churches of Christ in Australia.
Mailing address: 7 Elm Road, Glen Iris, Victoria 3146, Australia.

Powell, James Matthew, Prof

Rel: Roman Catholic; Unordained; b. 09 JUN 1930 in Cincinnati OH, USA; m. to Judith Catherine Davidorf; c. 4 boys and 2 girls; Citizen of USA; Lang. English, Italian, Latin, French, German and Greek.
International Organizations: Secretary-General (since 1989) of the Society for the Study of the Crusades and the Latin East.
Other organizations: Professor of Medieval History (since 1972), Syracuse University, NY.
Main field of work: University education. **Career:** Instructor, Kent State University, OH from 1959 to 1961. Assistant Professor, University of Illinois, Urbana from 1961 to 1965. Assistant Professor, Syracuse University, NY from 1965 to 1967. Associate Professor, Syracuse University, NY from 1967 to 1972. **Educ.:** B.A. from Xavier University, Cincinnati OH, USA (1949-1953); M.A. from Xavier University, Cincinnati OH, USA (1953-1955); Ph.D. from Indiana University, Bloomington, USA (1957-1960). **Interests:** crusades, medieval society and culture. **Hons.:** Awarded John Gilmary Shea Prize for *Anatomy of a Crusade, 1213-1221*, American Catholic Historical Association (1987). **Mem.:** American Historical Association, Medieval Academy of America, etc. **Pubs.:** *Leopold von Ranke and the Shaping of the Historical Discipline*, edited with George G Iggers, Syracuse University Press, Syracuse, 1989; *Muslims under Latin Rule, 1100-1300*, edited with essays by David Abulafia, Robert I Burns, Benjamin Z Kedar, Joseph O'Callaghan, and James M Powell (also wrote an introduction, conclusion, and chapter on "The Papacy and the Muslim Frontier"), Princeton University Press, Princeton, 1990 and *Albertanus of Brescia: The Pursuit of Happiness in the Early Thirteenth Century*, University of Pennsylvania Press, Philadelphia, 1992 (forthcoming).
Mailing address: Syracuse University, Dept. of History, 320 Maxwell Hall, Syracuse NY 13244, USA. T. (1 315) 433 2210
Residence: 114 Doll Parkway, Syracuse NY 13214, USA. T. (1 315) 446 8402

Powell, James R

Rel: Christian.
International Organizations: President of the International Bible Society.
Mailing address: PO Box 62970, Colorado Springs CO 80962-2970, USA. T. (1 719) 528 19 00. Fax 488 3840. Tx 312284.

Powell, John Sidney

Rel: Christian; b. 28 JUL 1957 in Suva, Lau, Fiji; m. to Joann Caroline Duthie; c. 1 boy and 4 girls; Citizen of Fiji; Lang. English and Fijian.
International Organizations: Secretary for Finance (since 1992) of the Pacific Conference of Churches.
Treasurer (since 1988) of the South Pacific Association of Theological Schools.
Other organizations: Lay preacher (since 1981), Methodist Church in Fiji.
Career: Cashier/clerk, Methodist Church in Fiji from 1976 to 1978. Assistant Accountant, Methodist Church in Fiji from 1979 to 1982. Accountant, Methodist Church in Fiji from 1983 to 1987. Secretary for Youth, Pacific Conference of Churches from 1988 to 1991. **Educ.:** Fiji Institute of Technology, Suva, Fiji (1979-1982). **Interests:** Church finances, youth work.
Mailing address: 4 Thurston Street, PO Box 208, Suva, Fiji. T. (679) 302332. C. PACIFICONS, SUVA.
Residence: 12 Gardiner Road, Suva, Fiji T. (679) 311231

Powell, Philip L, Rev

Rel: Assembly of God.
International Organizations: National General Secretary of the Assemblies of God in Australia.
Mailing address: PO Box 145, Campbelltown SA 5074, Australia.

Poyntz, Samuel Greenfield, Rt Rev Dr

Rel: Church of Ireland; Unordained on 01 JUL 1950; b. 04 MAR 1926 in Deloraine MB, Canada; m. to Noreen Henrietta Armstrong; c. 1 boy and 2 girls; Citizen of Ireland.
International Organizations: Bishop (since 1987) of Diocese of Connor.
Main field of work: Ministry. **Career:** Curate, St George's, Dublin from 1950 to 1952. Curate, Bray from 1952 to 1955. Curate, St Michael and St Paul, Dublin from 1955 to 1959. Incumbent, St Stephen, Dublin from 1959 to 1967. Incumbent, St Ann, Dublin from 1967 to 1970. Incumbent, St Ann with St Stephen, Dublin from 1970 to 1978. Archdeacon, Dublin from 1974 to 1978. Examining Chaplain, Archbishop, Dublin from 1974 to 1978. Bishop, Cork, Cloyne and Ross from 1978 to 1987. Bishop, Connor from 1987 to 1990. Chairman, Irish Council of Churches from 1986 to 1988. Vice-President, British Council of Churches from 1987 to 1990. **Educ.:** B.A. from University of Dublin Trinity College, Ireland (1948); M.A. from University of Dublin Trinity College, Ireland (1951); B.D. from University of Dublin Trinity College, Ireland (1953); Ph.D. from University of Dublin Trinity College, Ireland (1960). **Pubs.:** The Exaltation of the Blessed Virgin Mary, 1953; Journey Towards Union, 1975 and Our Church – Praying with our Church Family, 1983.
Mailing address: Church House, 12 Talbot Street, Belfast BT, UK.
Residence: Bishop's House, 22 Deramore Road, Belfast BT9 5JU, UK. T. (44) 668442

Prakash, Narain, Dr

Rel: Yoga.
International Organizations: Head of Yoga Department, Sports College of the Haryana Agricultural University.
Mailing address: Yoga Dept., Sports College, Haryana Agricultural University, Hissar 125 004, India.

Prakashmani, Brahma Kumari

Rel: Raja Yoga; Ordained in 1937; b. 01 JUN 1922 in Hyderabad, Sind, Pakistan; single; Citizen of India; Lang. Hindi and Sindhi.
International Organizations: Administrative Head (since 1969) of the Brahma Kumaris World Spiritual University.
President (since 1988) of the Global Cooperation for a Better World.
Main field of work: Spiritual teacher, meditation, spiritual development, spiritual values, women's development. **Career:** Leader, a delegation to the 2nd World Religious Congress in Japan in 1954. Organised India's Unity Youth Marches in 1985. Organised the All-India Health Awareness Campaign in 1989. President, Annual Universal Peace Conferences, Mount Abu from 1983 to 1991. Has chaired and spoken at international conferences in over 50 countries. **Hons.:** Key to the City of Los Angeles (1984); City medal of Sao Paulo (1984); City medal of Frankfurt (1984); UN International Year of Peace medal (1986); International Peace Messenger Award (1987).
Mailing address: Global Co-operation House, 65 Pound Lane, London NW10 2HH, UK. T. (44 81) 459 1400. Fax (44 81) 451 6480
Residence: PO Box No 2, Mount Abu 307501 (Rajasthan), India. T. (91) 248 – 249 – 251. C. BAPDADA. TX 03165476 OM IN.

Pratzner, Ferdinand, S.S.S., Rev Fr

Rel: Roman Catholic; Ordained; celibate.
International Organizations: Secretary of the Pontifical Committee for International Eucharistic Congresses.
Procurator-General of the Congregation of the Blessed Sacrament.
Mailing address: Piazza San Calisto 16, I-00120 Città del Vaticano, Vatican. T. (39 6) 698 7366. Fax 698 7214.

Preheim, Vern Quincy
Rel: Mennonite; Unordained; b. 27 JUN 1935 in Hurley SD, USA; m. to Marion Keeney; c. 2 boys and 3 girls; Citizen of USA; Lang. English and French.
International Organizations: General Secretary (since 1980) of the Mennonite World Conference.
Main field of work: Church administration. **Career:** Africa-Mid East Director, Mennonite Central Committee (MCC) from 1965 to 1972. Zaire Director, MCC from 1972 to 1974. Africa Secretary, MCC from 1974 to 1975. Asia Secretary, MCC from 1975 to 1980. Recording Secretary, Board of Christian Service, General Conference of the Mennonite Church (GCMC) from 1965 to 1968. Recording Secretary, Commission on Overseas Mission, GCMC from 1968 to 1972. Chairman, Commission on Overseas Mission, GCMC from 1974 to 1980. Member, General Board, GCMC from 1974 to 1980. Board Member, Habitat for Humanity International from 1976 to 1983. Secretary, Council of Moderators and Secretaries from 1981 to 1987. **Educ.:** B.A. from Bethel College, Newton KS, USA (1955-1957); B.D. from Mennonite Biblical Seminary, Elkhart IN, USA (1957-1960).
Mailing address: 722 Main Street, Box 347, Newton KS 67114, USA. T. (1 316) 283 5100
Residence: 1112 Lorna Lane, Newton KS 67114, USA T. (1 316) 283 2412

Premakantha Kurukkal, S P, Sri
Rel: Hindu.
International Organizations: Chief Priest of the Norwood Sri Kathivelayutha Swamy Kovil.
Mailing address: Norwood Sri Kathivelayutha Swamy, Kovil, Norwood, Sri Lanka.

Preus, David W, Dr
Rel: Lutheran.
International Organizations: Presiding Bishop of American Lutheran Church.
Mailing address: 422 South Fifth St, Minneapolis MN 55415, USA.

Price, Peter Bryan, Rev Canon
Rel: Anglican; Ordained on 30 JUN 1974; b. 17 MAY 1944 in Exmouth, Devon, UK; m. to Edith; c. 4 boys; Citizen of United Kingdom; Lang. English.
International Organizations: Secretary-General (since 1992) of the World Mission Association.
Other organizations: Honorary Canon (since 1991), Southwark Cathedral.
Main field of work: Mission. **Career:** Head of Religious Studies, Cordeaux School, Louth from 1970 to 1972. Community Chaplain, Crookhorn Church, Portsmouth from 1974 to 1978. Chaplain, Scargill House, Yorkshire from 1978 to 1980. Vicar, St Mary Magdalene, Addiscombt, Cromdon from 1980 to 1988. Chancellor, Southwark Cathedral, Diocese of Southwark from 1988 to 1991. **Educ.:** Cert. Ed. from University of Bristol, UK (1963-1966); Dip. Pastoral Stds. from Trent Polytechnic, Nottingham and Oakhill College, London, UK (1972-1974); Gen. Ordination Exams (1972-1974). **Interests:** mission studies, base ecclesial communities. **Pubs.:** Church as the Kingdom.
Mailing address: Partnership House, 157 Waterloo Road, London SE1 6XA, UK. T. (44 71) 928 86 81. C. GOSPELIZE LONDON SE1.
Residence: 14 Oakwood Road, Wimbledon, London 8W20 0PW, UK. T. (44 81) 946 3814

Prigione, Girolamo, H E Mgr
Rel: Roman Catholic; Ordained on 18 MAY 1944; b. 12 OCT 1921 in Castellazzo Bórmida, Italy; celibate.
International Organizations: Apostolic Delegate to Mexico (since 1978) of the Roman Catholic Church.
Other organizations: Titular Archbishop, Lauriaco.
Mailing address: Calle Juan Pablo II 118, Col Guadalupe Inn, Deleg Alvaro Obregón, 01020 México DF, Mexico. T. (52 5) 660 0542 – 524 2467. Fax 660 0571. Tx 1772396 DASS ME

Primatesta, Raúl Francisco, H E Card
Rel: Roman Catholic; Ordained on 25 OCT 1942; b. 14 APR 1919 in Capilla del Señor, Argentina; celibate; Citizen of Argentina.
International Organizations: Member (since 1973) of the College of Cardinals. Member of the Congregation for the Clergy. Member of the Congregation for Religious and Secular Institutes.
Other organizations: Archbishop (since 1965), Córdoba.
Career: Consecrated Titular Bishop, Tanais in 1957. Bishop, San Rafael from 1961 to 1965. Created Cardinal by Pope Paul VI in 1973.
Mailing address: Arzobispado, Avenida H, Irigoyen 98, 5000 Córdoba, Argentina. T. (54 51) 38 942

Primeau, Liette, Sr
Rel: Roman Catholic; Ordained; celibate.
International Organizations: Superior General of the Sisters of the Sacred Heart of Jesus.
Mailing address: Villa des Otages 8, 85 Rue Haxo, F-75020 Paris, France. T. (33 1) 43 64 03 59.

Prince, Bernard A, Rev
Rel: Roman Catholic.
International Organizations: Secretary-General of the Pontifical Society for the Propagation of Faith.
Mailing address: Via di Propaganda 1-C, I-00187 Roma, Italy.

Prine, Margaret L
Rel: Baptist.
International Organizations: President of the American Baptist Churches in the USA.
Mailing address: PO Box 851, Valley Forge PA 19482-0851, USA.

Pringarbe, Robert
Rel: Roman Catholic.
International Organizations: Secretary-General of the International Catholic Federation of Physical Education and Sports.
Mailing address: 22 rue Oberkampf, F-75011 Paris, France. T. (33 1) 47 66 03 23.

Printz, Othon, Dr
Rel: Christian.
International Organizations: Secretary-General of the International Federation for Inner Mission and Christian Social Work.
Mailing address: 22 rue d'Oberhoffen, F-67240 Bischwiller, France. T. (33) 88 63 25 31.

Prinz, Richard
Rel: Buddhist: Mahayana.
International Organizations: Member of the Foundation for the Preservation of the Mahayana Tradition.
Other organizations: Director, Vajrapani Institute.
Mailing address: PO Box "I", Boulder Creek CA 95006, USA. T. (1 408) 338 6654. Fax (1 408) 338 3666

Probst, W
Rel: Protestant.
International Organizations: General Secretary of the Federation of Swiss Protestant Churches.
Mailing address: SEK/FEPS, Sulgenauweg 26, Postfach, CH-3000 Bern 23, Switzerland.

Prophet, Elizabeth Clare, Rev
Rel: Non-denominational; Ordained on 05 JUL 1964; b. 08 APR 1939 in Red Bank NJ, USA; m. to Edward L Francis; c. 1 boy and 3 girls; Citizen of USA; Lang. English and French.
International Organizations: President (since 1962) of the Church Universal and Triumphant.
Founder/President (since 1970) of the International Montessori Association.
Other organizations: President (since 1971), Summit University.
Main field of work: Ministry, writing, lecturing. **Educ.:** Antioch College, Yellow Springs OH, USA (1958-1960); B.A. Pol.Sc. from Boston University, MA, USA (1960-1962); George Washington University, Washington DC, USA (1964). **Interests:** hiking, swimming. **Pubs.:** The Lost Years of Jesus; The Lost Teachings of Jesus and Saint Germain On Alchemy.
Mailing address: Church Universal and Triumphant, Box A, Livingston MT 59047, USA. T. (1 800) 245 5445
Residence: Highway 89, Box A, Corwin Springs MT 59021, USA. T. (1 406) 848 7441

Provoost, Constant

Rel: Christian.
International Organizations: Secretary-General of the Missionary Co-operation Service for Development.
Mailing address: Bd Leopold II 195, B-1080 Bruxelles, Belgium. T. (32 2) 427 47 20. Fax 425 90 31. Tx 65 172.

Prowse, Margaret

Rel: Roman Catholic; Unordained; b. 27 APR 1954 in Olney, Buckinghamshire, UK; Citizen of United Kingdom; Lang. English.
International Organizations: General Co-ordinator (since 1991) of the Volunteer Missionary Movement.
Main field of work: Overseas missions.
Mailing address: Shenley Lane, London Colney AL2 1AR, UK. T. (44 727) 824 853.

Prudence, Marie-Antoinette

Rel: Roman Catholic.
International Organizations: Executive Officer of the International Movement of Apostolate of Children.
Mailing address: 8 rue Duguay-Trouin, F-75006 Paris, France. T. (33 1) 42 22 84 81.

Puchu, Zhao

Rel: Buddhist.
International Organizations: Vice-President of the World Fellowship of Buddhists.
Mailing address: The Buddhist Association of China, No 25 Fuchengmennei Dajie, Beijing 100034, China.

Pudaite, Rochunga

Rel: Christian.
International Organizations: President of the Bibles for the World.
Mailing address: PO Box 805, Wheaton IL 60189-0805, USA. T. (1 708) 668 7733. Fax 668 6348.

Puente, Pablo, H E Mgr

Rel: Roman Catholic; Ordained on 02 APR 1956; b. 16 JUN 1931 in Colindres, Santander, Spain; celibate.
International Organizations: Apostolic Nuncio to Lebanon (since 1989) of the Roman Catholic Church.
Other organizations: Titular Archbishop, Macri.
Mailing address: Rue Georges Picot, BP 1882, Beirut, Lebanon. T. (961 1) 33 83 88

Puig, Enrique Capo, Rev

Rel: Evangelical.
International Organizations: Secretary of the Spanish Evangelical Church.
Mailing address: C Tallers 26, E-08001 Barcelona, Spain.

Pulgram-Arthen, Lucia Deirdre, Rev

Rel: Pagan; Ordained on 12 JUN 1980; b. 05 NOV 1956 in Atlanta GA, USA; m. to Andras Corban-Arthen; c. 1 boy and 1 girl; Citizen of USA; Lang. English.
International Organizations: Director (since 1983) of the Earth Spirit Community.
Other organizations: Beldame (since 1982), Glainn Sidhr Order; Program Coordinator (since 1989), Center for Community Resources; Public Information Officer (since 1991), Covenant of the Goddess.
Main field of work: Outreach ministry. **Educ.:** B.A. Drama and Hist. from Tufts University, Medford MA, USA (1974-1978); M.A. Counseling Psy. from Lesley College, Cambridge MA, USA (1984-1986). **Interests:** music, children's education, environmental issues. **Mem.:** Covenant of the Goddess; La Leche League International
Mailing address: Earth Spirit, PO Box 365, Medford MA 02155, USA. T. (1 617) 395 1023
Residence: 195 Spring St, Medford MA 02155, USA. T. See Mailing Address

Pulh, Jacques

Rel: Roman Catholic.
International Organizations: Secretary-General of the World Movement of Christian Workers.
Member of the Pontifical Council for the Laity.
Mailing address: Rue des Palais 90, B-1210 Bruxelles, Belgium. T. (32 2) 216 56 96. Fax 215 19 89.

Pung, Robert Edward, S.V.D., Rev

Rel: Roman Catholic; Ordained on 15 AUG 1944; b. 08 JAN 1916 in Westphalia MI, USA; celibate; Citizen of USA; Lang. English, German, Italian and Spanish.
International Organizations: Procurator-General (since 1985) of the Society of the Divine Word.
Advisor (since 1990) of the Congregation for the Clergy.
Other organizations: Postulator-General (since 1991), SVD Generalate, Rome; Editor (since 1991), SVD Word in the World.
Main field of work: Ministry. **Career:** Assistant Prefect, Professor, Vocation and Lay Retreat Director in Bay St Louis MO from 1945 to 1950. Editor, *St Augustine's Messenger* from 1946 to 1958. Editor, *Shepherd Magazine* from 1950 to 1958. Rector, Divine Word Major and Minor Seminary, Bay St Louis MO from 1952 to 1955. Provincial, Southern Province, Bay St Louis MO from 1955 to 1958. General Consultor at SVD Generalate, Rome from 1958 to 1968. Rector, SVD Renewal Center, Rome from 1968 to 1971. Administrative Secretary, Pontifical Council for Justice and Peace from 1971 to 1985. **Educ.:** B.Phil. from St Mary's Major Seminary, Techny IL, USA (1937-1939); M.Th. from St Mary's Major Seminary, Techny IL, USA (1939-1945); Dip. Accounting (Correspondence crse.) from University of Illinois, Urbana, USA (1967-1970).
Mailing address: Via dei Verbiti 1, CP 5080, I-00153 Roma, Italy. T. (39 6) 575 4021
Residence: See Mailing Address. T. See Mailing Address

Pupko, Isaac, Rabbi

Rel: Jewish.
International Organizations: Chairman of the International Rabbinic Committee for the Safety of Israel.
Mailing address: One Penn Plaza, Suite 100, New York NY 10119, USA. T. (1 212) 279 6666.

Purcell, Borden Conrad, Rev Canon

Rel: Anglican; Ordained; b. 08 NOV 1928 in Athens, Ontario, Canada; m. to Carter Dorwart; c. 2 boys and 1 girl; Citizen of Canada.
International Organizations: Canon (since 1972) of the Christ Church Cathedral, Ottawa.
Other organizations: Convention Refugee Determination Division (since 1988), Immigration and Refugee Board of Canada; Member (since 1958), Council of Associate Parishes.
Career: Curate, Grace Church, Hamilton from 1954 to 1955. Rector, Christ Church, St Catharines from 1955 to 1959. Rector, St Hilda's, Oakville from 1959 to 1963. Rector, St John's, Thorold from 1963 to 1969. Rector, St John's, Ottawa from 1969 to 1981. Chairman, Anglican Book Society and Canterbury Book Stores from 1969 to 1978. Consultative Committee on Human Rights, Canadian Council of Churches in 1982. Chairman, Ontario Human Rights Commission from 1982 to 1988. National Council, Canadian Human Rights Foundation in 1985. President, Canadian Association of Statutory Human Rights Agencies in 1986. Delegate to International Consultation on Human Rights, Ireland in 1978. Canadian Representative for other worldwide seminars and conventions. **Educ.:** B.A. from Bishop's University, Lennoxville, Quebec, Canada (1951); S.T.B. from University of Toronto, Trinity College, Canada (1954). **Interests:** gardening, water sports, skiing, travel. **Mem.:** Toronto Interfaith Council on Soviet Jewry (Founding Member, 1985) **Pubs.:** Author of several papers and articles.
Mailing address: 70 University Ave, Toronto, Ontario M5J 2M5, Canada.
Residence: 'The New Town of York', 135 George St S, Ste 903, Toronto, Ontario M5A 4E8, Canada.

Purnell, David Lyle

Rel: Quaker; b. 07 APR 1941 in Sydney NSW, Australia; divorced; c. 1 boy and 2 girls; Citizen of Australia; Lang. English.
International Organizations: Presiding Clerk (since 1992) of the Religious Society of Friends in Australia.

Other organizations: National Administrator (since 1989), United Nations Association of Australia.
Career: Secretary, Life Line Canberra from 1971 to 1980. Member of Executive, Australian Council of Churches from 1980 to 1986. President, Conflict Resolution Service from 1988 to 1991. **Educ.:** B.A. (Hons) from Australian National University, Canberra, Australia (1958-1961). **Interests:** non-violence, peace education, music. **Mem.:** Conflict Resolution Network **Pubs.:** Creative Conflict, 1988; Effective Peacemaking, 1988 and Resolving International Conflict, 1990.
Mailing address: 10 O'Sullivan St, Higgins ACT 2615, Australia. T. (61 6) 254 1111
Residence: See Mailing Address. T. See Mailing Address

Pye, Edward Michael, Prof

Rel: Anglican; b. 22 APR 1939 in Newport, Shropshire, UK; m. to Christine Roether; c. 2 boys and 2 girls; Citizen of United Kingdom; Lang. English, French, German and Japanese.
International Organizations: Secretary-General (since 1985) of the International Association for the History of Religions.
Other organizations: Professor (since 1991), Lancaster University.
Main field of work: University education. **Career:** Lecturer, Lancaster University from 1968 to 1973. Lecturer, Leeds University from 1973 to 1982. Professor, Marburg University (Germany) from 1982 to 1991. **Educ.:** B.A. Theol. (Hons) from University of Cambridge, Clare College, UK (1958-1961); Ph.D. from University of Leeds, UK (1977). **Pubs.:** *The Cardinal Meaning*, 1973; *Skilful Means, A Concept in Mahayana Buddhism*, 1978 and *Emerging from Meditation*.
Mailing address: Dept. of Religious Studies, Lancaster University, Lancaster LA1 4YG, UK. T. (44 524) 65201. Tx 65111 LANCUL G.

Qabbani, Muhammad Rashid, Sheikh

Rel: Muslim: Sunni; Citizen of Lebanon.
International Organizations: Grand Mufti (acting) of Lebanon.
Mailing address: Dar el-Fatwa, Ilewi Rushed St, Beirut, Lebanon.

Qaisari, Abdul Aziz, Mawlawi

Rel: Muslim; b. 13 FEB 1945 in Qaisar, Qaisar District, Faryab Province, Afghanistan; m. to Gul Zarin; c. 4 boys and 3 girls; Citizen of Afghanistan; Lang. Pashtoo, Dari, Iranian, Persian, Arabic, Uzbik and Azarbaijani.
International Organizations: Chairman, Propaganda Board (since 1991) of the Ministry of Islamic Affairs and Awqaf.
Other organizations: Member (since 1980), Directorship of Islamic Affairs; Vice-President (since 1986), High Council of Ulema and Clergy Men.
Main field of work: Islamic scientific research and propaganda. **Educ.:** Mawlawi from Madrasa of Barnabad Ghorian, Herat. **Interests:** study and compilation of religious books. **Hons.:** two medals and one order of the P.F. **Mem.:** Islamic People's Party of Afghanistan (Member of Executive Committee)
Mailing address: Wazarat Shaqwn Islami Wa Awqaf, Shri-Now, Char-Rahi Fiz Mohammad, Kabul, Afghanistan.
Residence: Kabul, Afghanistan.

Quaisin, Michel

Rel: Protestant.
International Organizations: Representative of the Gideons International.
Mailing address: Rue d'Aunale 38, B-1070 Bruxelles, Belgium.

Quaranta, Ciro, R.C.I., Fr

Rel: Roman Catholic; Ordained; celibate.
International Organizations: Procurator-General of the Rogationistes du Coeur de Jésus.
Advisor of the Congregation for Catholic Education.
Mailing address: Via Tuscolana 167, I-00182 Roma, Italy. T. (39 6) 702 0751.

Quartier, Jacques

Rel: Roman Catholic.
International Organizations: Secretary-General of the International Sporting Federation of Catholic Education.
Mailing address: 277 rue Saint Jacques, F-75005 Paris, France. T. (33 1) 43 25 49 67.

Quek, Kiok Chiang, Rev Dr

Rel: Presbyterian; Ordained on 12 AUG 1956; b. 10 JUL 1916 in Swatow, Guandung, China; m. to Tan Mui Keow; c. 4 boys; Citizen of Singapore; Lang. English, Mandarin, Teochew and 2 other Chinese dialects.
International Organizations: Executive Secretary (since 1979) of the International Council of Christian Churches.
Other organizations: President, Singapore Council of Christian Churches (since the 1960s); Vice-President, Singapore Christian Evangelistic League (since the 1960s); Pastor (since 1960), Faith Bible-Presbyterian Church; President (since 1973), Bible-Presbyterian Welfare Services, Singapore; Chairman (since 1983), Ling Kwang Home for Senior Citizens.
Main field of work: Christian ministries. **Interests:** Christian ministries, welfare services. **Hons.:** DD, Faith Theological Seminary, USA (1983); PBM Public Service Medal for Social and Community Service; Singapore Government Award (1991). **Pubs.:** Editor and Publisher, *Southeast Asia Christian* (Chinese monthly), since the 1950s and Editor and Publisher, *Far Eastern Beacon* (English monthly), since 1968.
Mailing address: ICCC Far Eastern Office, 5 Tavistock Avenue, Singapore 1955, Singapore. T. (65) 2884786
Residence: 21 Olive Road, Singapore 1129, Singapore. T. (65) 2875466 – 2803338

Quilici, Oriano, H E Mgr

Rel: Roman Catholic; Ordained on 08 DEC 1954; b. 29 NOV 1929 in San Pietro a Vico, Italy; celibate.
International Organizations: Apostolic Nuncio to Venezuela (since 1990) of the Roman Catholic Church.
Other organizations: Titular Archbishop, Tabla.
Mailing address: Apartado 54 597, Caracas 1052-A, Venezuela. T. (58 2) 781 8939 – 781 3101. Fax 793 2403. Tx 24040 NUNAP VC

Quint, Eduard, Rev Br

Rel: Roman Catholic; Ordained; b. 06 APR 1941 in Nieuwer-Amstel, Netherlands; celibate; Citizen of Netherlands; Lang. Dutch, English, German and French.
International Organizations: Superior General (since 1987) of the Brothers of the Immaculate Conception of the Blessed Virgin Mary, Mother of God.
Main field of work: Ministry. **Educ.:** Tchr. Cert. from Training College, Breda, Netherlands; Lang. Deg. from Free University, Amsterdam, Netherlands.
Mailing address: Generalate CFH, Boomstraat 7, NL-4635 CX Huijbergen, Netherlands. T. (31 1644) 2750.
Residence: See Mailing Address. T. See Mailing Address

Quintero, Manuel

Rel: Protestant.
International Organizations: Co-Secretary-General of the World Student Christian Federation.
Mailing address: Ecumenical Centre, 5 route des Morillons, Grand-Saconnex, CH-1218 Genève, Switzerland.

Qutub, Musa Y, Imam Prof Dr

Rel: Muslim; Unordained; b. 06 FEB 1940 in Jerusalem, Israel; m. to Ghladeer; Citizen of USA; Lang. English and Arabic.
International Organizations: President (since 1983) of the Islamic Information Center of America.
President (since 1989) of the International Association for the Advancement of Earth and Environmental Sciences.
Other organizations: Professor (since 1979), Northeastern Illinois University, Chicago IL.
Main field of work: University education. **Career:** Instructor, Iowa State University of Science and Technology, Ames IA from 1966 to 1969. Assistant Professor then Associate Professor, Northeastern Illinois University, Chicago IL from 1969 to 1977. Expert, Ministry of Planning, Saudi Arabia from 1977 to 1979. President, National Association of Geology Teachers from 1975 to 1979. **Educ.:** B.A. from Simpson College, Indianola IA, USA (1960-1964); M.S. from Colorado State University, Fort Collins, USA (1964-1966); Ph.D. from Iowa State University of Science and Technology, Ames IA, USA (1966-1969). **Interests:** tennis, track. **Pubs.:** 55 articles on Islam; 70 articles on science and 1 book.
Mailing address: 720 Madelyn Dr, Des Plaines IL 60016, USA. T. (1 708) 297 1383.
Residence: See Mailing Address. T. See Mailing Address

Rabbani, Rúhíyyih

Also know as May Sutherland Maxwell; Rel: Bahá'í; b. 08 AUG 1910 in New York NY, USA; widowed Shoghi Effendi Rabbani (great grandson of Bahá'ú'lláh, World Head of the Bahá'í Faith and Guardian of the Cause); Citizen of USA; Lang. English, French, German and Persian.
International Organizations: High ranking officer of the Bahá'í International Community.
Other organizations: Member (since 1973), International Teaching Centre, Haifa; Honorary President (since 1989), Sacred Literature Trust, Manchester; Member, International Support Committee (since 1991), International Centre for Peace and its associated Peace Museum in Verdun.
Main field of work: Ministry. **Career:** Named Hand of the Cause, Bahá'í Faith in 1952. Opened to public worship all 7 continental Bahá'í Houses of Worship. Official representative, Universal House of Justice at formation of 12 of present 155 National Spiritual Assemblies. Presented (on behalf of Universal House of Justice) special Bahá'í peace plan 'The Promise of World Peace' to the Secretary-General of the United Nations, Javier Pérez de Cuellar in 1985. Representative at Canadian Bahá'í Conference for Indians and Eskimos in 1986. Presented to H R H the Prince Philip, International President of WWF, a compilation of Bahá'í writings on ecology and environmental protection in 1987. Served as delegate to the 31st Annual Meeting, International Society for General Systems Research, Budapest in 1987. Principal speaker at European Bahá'í Women's Conference, Madrid in 1990. Travelled and lectured in 167 countries and at many international conferences.
Educ.: McGill University, Montreal, Canada (part-time student). **Pubs.:** *A Manual for Pioneers*, Bahá'í Publishing Trust, New Delhi, 1974 (in English); *The Desire of the World*, George Ronald, Publisher, 1982 (in English and in Portuguese) and *The Guardian of the Bahá'í Faith*, Bahá'í Publishing Trust, London, 1988 (in English).
Mailing address: Bahá'í World Center, PO Box 155, 31001 Haifa, Israel. T. (972 4) 358358. Fax 372 425. C. BAHAIFAITH, HAIFA. Tx 46626 BAYT IL.
Residence: See Mailing Address. T. (972 4) 358632

Rabinovicz, Pinhas

Rel: Jewish; b. in Lodz, Poland.
International Organizations: Chief Cantor of the the largest Orthodox congregation in the Western US.
Career: Immigrated to Israel (at the age of three). Began to serve as a cantor in the Gevurat Yisrael Synagogue in Tel Aviv in 1968. Appeared in concerts and on radio and television programs. Moved to Montreal in 1973. Appeared in broadcasts, the Canadian Broadcasting Authority. Served as a cantor in the Beth Ora Synagogue in Montreal. Was President, the Cantors' Association. Moved to Los Angeles in 1983.
Mailing address: No mailing address.

Rader, Paul A, Commr

Rel: Salvation Army, The.
International Organizations: Territorial Commander of The Salvation Army.
Mailing address: 30840 Hawthorne Boulevard, Rancho Palos Verdes CA 90274, USA.

Radhakrishna, K S

Rel: Hindu.
International Organizations: Secretary of the Gandhi Peace Foundation.
Mailing address: 221/223 Deen Dayal, Upadhyaya Marg, New Delhi 110 002, India. T. (91 11) 331 74 91 - (91 11) 331 74 93 - (91 11) 331 17 34. C. SATYAGRAHA NEW DELHI.

Rahula, Walpola, Ven Dr

Rel: Buddhist: Theravada.
International Organizations: Monk of Theravada Buddhism.
Mailing address: 126 Buddhist Institute Ave, Parliament Rd, Kotte, Sri Lanka.

Raï, Boutros, B.A., H E Mgr

Rel: Greek-Melkite Catholic; Ordained on 26 JUN 1948; b. 08 NOV 1922 in Aleppo, Syria; celibate; Citizen of Syria.
International Organizations: Archbishop (since 1988) of Nuestra Señora del Paraíso en México.
Other organizations: Member, Ordre Basilien Alépin; Apostolic Exarch to the Melkite Greek Catholics in Venezuela; Apostolic Visitor to the Melkite Greek Catholics in Argentina.
Career: Professor, Saint Anne Seminary. Secretary, late Maximos IV (Patriarch). Superior, Basilian Congregation from 1965 to 1968.
Mailing address: Venustiano Carranza 107, México 06060 DF, Mexico. T. (52 5) 542 0225

Raimbault, Philippe

Rel: Catholic.
International Organizations: President of the Union mondiale de la Retraite.
Mailing address: 3 rue du Bois Jacques, BCA, F-95600 Eaubonne, France.

Räisänen, Heikki Martti, Prof

Rel: Lutheran; Ordained on 03 MAR 1965; b. 10 DEC 1941 in Helsinki, Finland; m. to Leena Marjatta Wright; c. 3 boys and 1 girl; Citizen of Finland; Lang. Finnish, Swedish, English and German.
International Organizations: Full Professor in New Testament Exegesis (since 1975) of the University of Helsinki.
Other organizations: Research Professor at the Academy of Finland (since 1984); Member, Editorial Board (since 1984), *Teologinen Aikakauskirja*; Chief Editor (since 1989), *Vartija.*
Main field of work: University education. **Career:** Minister in the Lutheran congregation, Kanta-Espoo from 1965 to 1966. Assistant Professor in Exegesis from 1968 to 1970. Fulbright Visiting Scholar, Harvard from 1970 to 1971. Research Fellow at the Finnish Academy from 1971 to 1972. Docent in New Testament Exegesis, University of Helsinki from 1969 to 1974. Acting Associate Professor in Biblical Languages, University of Helsinki in 1971. Acting Associate Professor in Biblical Languages, University of Helsinki from 1972 to 1974. Vice Dean, Theological Faculty, University of Helsinki in 1977. Vice Dean, Theological Faculty, University of Helsinki from 1979 to 1980. Humboldt Visiting Scholar, Tübingen from 1980 to 1982. Committee Member, Novi Testamenti Societas from 1985 to 1988. Member, Managing Board, Finnish Exegetical Society from 1969 to 1985. Chairman, Finnish Exegetical Society from 1980 to 1985. Leader, the research project 'Redaction Criticism of the Synoptic Gospels', funded by the Finnish Academy from 1983 to 1986. Leader, the research project 'The Early Church and Judaism', funded by the Finnish Academy from 1986 to 1989. Leader, the research project 'Expectation of the End and Early Christian Identity', funded by the Finnish Academy from 1990 to 1992.
Educ.: B.D. from University of Helsinki, Finland (1965); Lic. Theol. from University of Helsinki, Finland (1967); M.A. Greek, Oriental and Roman Lit. from University of Helsinki, Finland (1968); Dr. Theol. from University of Helsinki, Finland (1969). **Hons.:** Hon. D.D., University of Edinburgh (1990). **Mem.:** Finnish Academy of Sciences (1978)
Pubs.: *Paul and the Law*, 2nd revised and enlarged edition, Tübingen, 1987.; *The Torah and Christ* (Essays in German and English on the Problem of the Law in Early Christianity), Publications of the Finnish Exegetical Society 45, Helsinki, 1986. and *Beyond New Testament Theology* (A Story and a Programme), SCM Press, London, and Trinity Press International, Philadelphia.
Mailing address: Theological Faculty, University of Helsinki, Neitsytpolku 1b, SF-00140 Helsinki, Finland.
Residence: Vantaanjänne 1 B 11, SF-01730 Vantaa, Finland.

Raiser, Elisabeth

Rel: Christian.
International Organizations: Co-President of the Ecumenical Forum of European Christian Women.
Mailing address: Dr Kolbestrasse 13, 58452 Witten, Germany. T. (49 2302) 32799. Fax 3527.

Raiser, Konrad, Dr

Rel: Unknown.
International Organizations: Secretary-General of the World Council of Churches.
Mailing address: PO Box 2100, Route de Ferney 150, CH-1211 Genève 2, Switzerland. T. (41 22) 791 61 11. Fax 791 03 61. C. OIKOUMENE Geneva. Tx 415 730 OIK CH.

Rajanani Kurukkal, K R, Brahmasri
Rel: Hindu.
International Organizations: Chief Priest of the Sri Muthukumarasamy Devasthanam.
Mailing address: Sri Muthukumarasamy Devasthanam, Trincomalee, Sri Lanka.

Rajaratnam, K, Dr
Rel: United Evangelical Lutheran.
International Organizations: Executive Secretary of the United Evangelical Lutheran Church in India.
Mailing address: Luther House, Gurukul Campus, Thambuswamy Rd, Kilpauk, Madras 600010, India.

Rall, M
Rel: Roman Catholic.
International Organizations: European Council Representative of the International Association of Charities.
Mailing address: 3 rue Stimmer, F-67000 Strasbourg, France.

Ralston, Helen
Rel: Christian; b. 07 MAR 1948; Citizen of Canada.
International Organizations: Member of the Council of the International Society for the Sociology of Religions.
Educ.: Ph.D. from Carleton University, Ottawa, Canada (1973). **Pubs.:** *Christian Ashrams: A New Religious Movement in Contemporary India*, Lewiston, NY/Queenston, Ontario, The Edwin Mellen Press, 1987; "Strands of Research on Religious Movements in India", in *Sciences religieuses/Studies in Religion*, 17, 3, 1988, pp. 257-277 and "The Construction of Authority in the Christian Ashram Movement", in *Archives de Sciences Sociales des Religions*, 67/1 (janv-mars), 1989, pp. 53-75.
Mailing address: St Mary's University, Dept. of Sociology, Halifax NS B3H 3C3, Canada. T. (1 902) 420 5878
Residence: 5820 Spring Garden Road, Halifax NS B3H 1X8, Canada T. (1 902) 422 3693

Ramadhani, John Acland, Most Rev
Rel: Anglican; Ordained; b. 01 AUG 1932 in Zanzibar, Tanzania; Citizen of Tanzania.
International Organizations: Archbishop (since 1984) of Province of Tanzania.
Other organizations: Bishop (since 1980), Zanzibar and Tanga.
Career: Principal, St Andrew's Teacher Training College, Korogwe from 1967 to 1969. Warden, St Mark's Theological College, Dar es Salaam from 1977 to 1979. **Interests:** reading.
Mailing address: Diocese of Zanzibar and Tanga, PO Box 35, Korogwe, Tanzania.

Raman, Mannampurath Kavu, Master
Rel: Hindu; b. 17 AUG 1927 in Kerala, India; celibate; Citizen of India; Lang. Malayalam and English.
International Organizations: Chief Instructor of the Kavil Bhavan Physical Culture Institute, Nileshwar.
Main field of work: Yoga teaching and research. **Educ.:** S.S.L.C.; T.T.C. **Mem.:** Conferred the title of Yogacharya (1988) **Pubs.:** *Yoga Margam*; *Jnana Margam* and *Yoga and Nature Cure-text.*
Mailing address: Yoga Therapy Centre, Nileshwar 671314, Kasaragod District, Kerala State, India.

Ramanatha Kurukkal, K P R, Brahmasri
Rel: Hindu.
International Organizations: Chief Priest of the Sri Kailasanatha Swamy Kovil.
Mailing address: Sri Kailasanatha Swamy Kovil, Captain's Garden, Maradana, Sri Lanka.

Ramanatha Kurukkal, K R, Brahmasri
Rel: Hindu.
International Organizations: Chief Priest of the Kandasamy Kovil.
Mailing address: Kandasamy Kovil, Inuvil, Sri Lanka.

Ramanbasoa, Joseph J, Rev
Rel: Church of Jesus Christ.
International Organizations: President of the Church of Jesus Christ in Madagascar.
Mailing address: 'Ifanomezantsoa', Analakely, BP 623, Antananarivo, Madagascar.

Ramasamy Adigal, Santhalinga R, Rev Kailaimunivar
Rel: Hindu; Ordained on 03 MAR 1950; b. 01 SEP 1926 in Mudalipalayam (Coimbatore), Tamilnadu, India; celibate; Citizen of India; Lang. Tamil and English.
International Organizations: President of the Indian Saints Association.
Career: Academic Council, Senate Member, Chairman Board of Studies, Madras University. Head, transcription of ancient palm manuscripts. Organizor, Tamilnadu Universal Hindu Organization (VHP). Member, Tamil Poets Congregation. Member, Tamilnadu Deiveeka Peravai. **Pubs.:** Essays on Tamil Morals; Avirotha Yundiar and Kolai Maruthal.
Mailing address: Santhalinga Swamigal Muth, Perur, Coimbatore 641 010, Tamilnadu, India.

Ramasso, Venanzio M, O.S.M., Fr
Rel: Roman Catholic; Ordained; celibate.
International Organizations: Procurator-General of the Order of the Servants of Mary.
Mailing address: Piazza S Marcello al Corso 5, I-00187 Roma, Italy. T. (39 6) 678 08 88 – 678 99 95.

Ramcharan, Michael
Rel: Christian.
International Organizations: Coordinating Secretary of the Collaboration for Ecumenical Planning and Action in the Caribbean.
Mailing address: c/o St Paul's AC Rectory, 12 Harris Promenade, San Fernando, Trinidad-Tobago. T. (809) 652 4489. Fax 652 4489.

Ramino, Paul A, Rev
Rel: Church of Jesus Christ.
International Organizations: General Secretary of the Church of Jesus Christ in Madagascar.
Mailing address: 'Ifanomezantsoa', Analakely, BP 623, Antananarivo, Madagascar.

Ramírez Piñón, Crescencio R, M.J., S.S.I., Fr
Rel: Roman Catholic; Ordained; celibate.
International Organizations: Procurator-General of the Missionaries of St Joseph of Mexico.
Mailing address: Via di S Alessio 23, I-00153 Roma, Italy. T. (39 6) 574 13 75.

Rams, Konrad, F.M.M.A., Br
Rel: Roman Catholic; Ordained; celibate.
International Organizations: Procurator-General of the Frères de la Miséricorde de Sainte Marie-Auxiliatrice.
Mailing address: Via delle Sette Chiese 280, I-00147 Roma, Italy. T. (39 6) 513 39 56, (39 6) 513 51 93. Fax (39 6) 513 54 61

Ramsey, Arthur Michael
Rel: Church of England; Ordained in 1928; Citizen of United Kingdom.
International Organizations: Clergyman of the Church of England.
Career: Served at various parishes and colleges as a teacher and vicar in the years before World War II. Named Professor of Divinity at Cambridge in 1940. Named Bishop, Durham in 1942. Transferred to the archbishopric, York in 1956. Archbishop, Canterbury from 1961 to 1974. Met with Pope Paul VI (first encounter between the leaders of the Roman Catholic and Anglican churches since their separation in 1534) in 1966. **Educ.:** Grad. from University of Cambridge, Magdalene College, UK (1927). **Pubs.:** *From Gore to Temple*, 1960; *Canterbury Pilgrim*, 1974 and *Be Still and Know*, 1982.
Mailing address: No mailing address.

Ranaghan, Kevin M
Rel: Unknown.
International Organizations: Co-founder of the People of Praise.
Mailing address: 237 North Michigan, South Bend IN 46601, USA.

Ranaivojaona-Razafimantsoa, R, Rev
Rel: Lutheran.
International Organizations: President of the Malagasy Lutheran Church.
Mailing address: BP 1741, Antananarivo, Madagascar.

Ranatunga, George, Col
Rel: Buddhist; b. 22 DEC 1922 in Mawanella, Kegalle, Sri Lanka; m. to Beryl Rodrigo; c. 1 boy and 1 girl; Lang. Sinhalese and English.
International Organizations: Founder/President (since 1990) of the Asian Buddhist Congress.
Other organizations: Member (since 1990), Human Rights Commission.
Career: Colonel, Sri Lanka Army from 1950 to 1979. **Educ.:** Eng. Trained Teachers' Cert. (1st Class) from Teachers Training College, Colombo, Sri Lanka (1946-1948). **Interests:** social and humanitarian work. **Mem.:** All-Ceylon Buddhist Congress; Young Men's Buddhist Association **Pubs.:** Words of Command for the Thres Services in Sinhala and Athletic Training in Sinhala.
Mailing address: 380 Bauddaloka Mawatha, Colombo 7, Sri Lanka. T. (94 1) 691695
Residence: 104 Allen Ave, Dehiwala, Sri Lanka. T. (94 71) 3102

Randerath, Manuela, CPS, Sr Mother
Rel: Roman Catholic; Ordained; celibate.
International Organizations: Superior General of the Missionary Sisters of the Precious Blood.
Mailing address: Via S Giovanni Eudes 93, I-00163 Roma, Italy. T. (39 6) 625 85 59.

Ranft, Eckart
Rel: Evangelical.
International Organizations: President of the Bremen Evangelical Church.
Mailing address: Franziuseck 2-4, 28199 Bremen, Germany.

Ransom, Constance A
Rel: Spiritualist.
International Organizations: Executive Officer of the International Spiritualist Federation.
Mailing address: 90 Bamburgh Close, Reading RE2 7UD, UK. T. (44 734) 86 33 98.

Rao, P Jayachandra
Rel: Baptist.
International Organizations: General Secretary of the Samavesam of Telugu Baptist Churches.
Mailing address: CAM Higschool Compound, Nellore 524 003 Andhra Pradesh, India.

Raper, Mark, SJ, Rev
Rel: Roman Catholic; Ordained; celibate.
International Organizations: Director of the Jesuit Refugee Service.
Mailing address: 5, CP 6139, I-00195 Roma, Italy. T. (39 6) 686 9841.

Raphael, Yitchak, Dr
Rel: Jewish.
International Organizations: Secretary-General of the World Centre for Torah and Judaic Studies.
Mailing address: Maimon Quarter, PO Box 642, Jerusalem, Israel. T. (972 2) 526231.

Rapisarda, Alfio, H E Mgr
Rel: Roman Catholic; Ordained on 14 JUL 1957; b. 03 SEP 1933 in Zafferana Etnea, Italy; celibate.
International Organizations: Apostolic Pro-Nuncio to Zaire (since 1985) of the Roman Catholic Church.
Other organizations: Titular Archbishop, Canne.
Mailing address: Avenue Goma 81, BP 3091, Kinshasa-Gombe, Zaire. T. (243 12) 31 419. Tx 21527 NAPOST ZR

Rapoo, D P, Rt Rev
Rel: Evangelical Lutheran.
International Organizations: Presiding Bishop of Evangelical Lutheran Church in Southern Africa.
Mailing address: Box 536, Rustenburg 0300, South Africa.

Rapp, James
Rel: Jewish.
International Organizations: Chairman of the International League for the Repatriation of Russian Jews.
Mailing address: 2 Fountain Lane, Ste 2J, Scarsdale NY 10583, USA. T. (1 800) 448 1866.

Rasoaharison Fégeat, Liliane R
Rel: Reformed Church; Unordained; b. 17 AUG 1942 in Diego-Suarez, Madagascar; m. to Jean-Louis Fégeat; c. 2 boys and 2 girls; Citizen of France; Lang. French, English and Malagasy.
International Organizations: Member of the Executive Committee (since 1989) of the World Alliance of Reformed Churches.
Coordinator (since 1987) of the France-Libertés Fondation Danielle Mitterrand.
Other organizations: President (since 1988), Reformed Church of the Reunion.
Career: Association de Formation des Cadres (AFCA) in 1966. CHS (psychiatry for adults) in 1974. Regional Delegation to the State Department for Women's Rights in 1990. **Educ.:** Lic. Psy. from University of Paris (Paris-Sorbonne), France; Dip. Psy. and Pathology from University of Paris (Paris-Sorbonne), France. **Interests:** humanitarian policy.
Mailing address: 52 rue Labourdonnais, 97400 Saint Denis, Réunion, France. T. (262) 219 125

Rathbone, Thomas
Rel: Christian.
International Organizations: Contact of the International Religious Fellowship.
Mailing address: 38 Gilbert Road, Cambridge CB4 EPE, UK. T. (44 223) 35 45 42.

Rathke, Heinrich Karl Martin Hans, Dr
Rel: Evangelical Lutheran; Ordained on 19 APR 1955; b. 12 DEC 1928 in Mölln, Kreis Malchin, Germany; m. to Marianne Rusam; c. 6 boys and 1 girl; Citizen of Germany; Lang. German, Latin, Greek, Hebrew and English.
International Organizations: Pastor in Ruhestand (since 1991) of the Evangelical Lutheran Church.
Other organizations: Vice-President (since 1984), Gustav-Adolf-Werkes; Bishop Emeritus (since 1984).
Main field of work: Ministry, people's mission. **Career:** Parish priest, Althof bei Bad Doberan/Mecklenburg from 1954 to 1955. Parish priest, Warnkenhagen, Kreis Teterow/Mecklenburg from 1955 to 1962. Parish priest, Rostock Südstadt/Mecklenburg from 1962 to 1970. Priest in charge, community service and people's mission, Mecklenburg (Güstrow) from 1970 to 1971. Bishop, the Evangelical-Lutheran Church of Mecklenburg from 1971 to 1984. Presiding Bishop, the United Evangelical Lutheran Church of the GDR from 1977 to 1981. Pastor in Crivitz/Mecklenburg from 1984 to 1991. **Educ.:** Christian-Albrechts-Universität Kiel, Friedrich-Alexander-Universität Erlangen-Nürnberg, and Eberhard-Karls-Universität Tübingen, Germany (1949-1953); Dr. Theol. from Wilhelm-Pieck-Universität Rostock, Germany (1956). **Interests:** theology, Russian Orthodox Church, Evangelical churches in the USSR. **Pubs.:** Ignatius von Antiochien und die Paulusbriefe, 1967; Gemeinde heute und morgen, 1979 and Einstehen für Gemeinschaft in Christus, 1980.
Mailing address: Münzstrasse 8, 19055 Schwerin, Germany.
Residence: Schleifmühlenweg 11, 19061 Schwerin/Mecklenburg, Germany.

Rato, Khyongla, Ven Geshe
Rel: Buddhist: Tibetan Mahayana; Ordained in 1936; b. 04 JUN 1923 in Tibet; single; Citizen of USA; Lang. Tibetan and English.
International Organizations: Spiritual Head (since 1975) of the Tibet Center, New York City.
Other organizations: Spiritual Head (since 1986), Rato Drabang Foundation.

Educ.: Highest Deg. of Lharampa Geshe (Beginning in 1947). **Mem.:** Rato Monastery, Mundgod (India) **Pubs.:** *My Life and Lives* (autobiography).
Mailing address: Tibet Center, 359 Broadway, New York NY 10013, USA. T. 966 8504
Residence: 220 E 36th, New York NY 10016, USA

Ratzinger, Joseph Alois, H E Card
Rel: Roman Catholic; Ordained on 29 JUN 1951; b. 16 APR 1927 in Marktl am Inn, Passau, Germany; celibate; Citizen of Germany.
International Organizations: President (since 1981) of the International Theological Commission.
President (since 1981) of the Pontifical Biblical Commission. Patron of the Pontifical Marian International Academy. Member of the Synod of Bishops. Member of the Sacred Congregation for the Oriental Churches. Member of the Sacred Congregation for the Sacraments and Divine Worship. Member of the Congregation for Bishops. Member of the Pontifical Commission for Latin America. Member of the Congregation for the Evangelization of Peoples. Member of the Congregation for the Clergy. Member of the Congregation for Catholic Education. Member of the Pontifical Council for Promoting Christian Unity.
Other organizations: Member (since 1977), College of Cardinals; Prefect (since 1981), Congregation for the Doctrine of Faith; Member, Council for Cardinals and Bishops.
Career: Chaplain in 1951. Professor of Theology, Freising in 1958. Professor of Theology, Bonn in 1959. Professor of Theology, Münster in 1963. Professor of Theology, Tübingen in 1966. Professor of Theology, Regensburg in 1969. Archbishop in 1977. Created Cardinal by Pope Paul VI in 1977. **Pubs.:** books and articles on theological matters.
Mailing address: 00120 Città del Vaticano, Vatican. T. (39 6) 698 3296.

Ravichandra Kurukkal, S R, Brahmasri
Rel: Hindu.
International Organizations: Chief Priest of the Sri Badrakali Ambal Devasthanam.
Mailing address: Sri Badrakali Ambal Devasthanam, Trincomalee, Sri Lanka.

Rawddara, Amara
Rel: Buddhist.
International Organizations: Secretary-General of the World Fellowship of Buddhist Youth.
Mailing address: 33 Sukhumvit Road, Bangkok 10110, Thailand. T. (66 2) 2511188. Fax 2557363.

Ray, Chaplain
Rel: Christian.
International Organizations: Director of the International Prison Ministry.
Mailing address: PO Box 63, Dallas TX 75221, USA. T. (1 214) 494 2302.

Raya, Joseph M, Most Rev
Rel: Greek Catholic; Ordained on 20 JUL 1941; b. 20 JUL 1917 in Zahleh, Lebanon; celibate; Citizen of Lebanon.
International Organizations: Titular Archbishop of Scythopolis.
Career: Head, Greek Catholic Church in Birmingham from 1950 to 1967. Archbishop, St John of Acre, Haifa and Nazareth in 1968. **Pubs.:** The Gospel (translation of Les Evangiles), 1966.
Mailing address: Madonna House, 685 Pape Ave, Toronto ON M4K 3S4, Canada. T. (1 416) 465 8909

Rayner, John Desmond, Rabbi Dr
Rel: Jewish; Ordained on 21 JUN 1953; b. 30 MAY 1924 in Berlin, Germany; m. to Jan Priscilla Heilbronn; c. 2 boys and 1 girl; Citizen of United Kingdom; Lang. English, German, French and Hebrew.
International Organizations: Emeritus Rabbi of Liberal Jewish Synagogue, London.
Other organizations: Lecturer (since 1966), Leo Baeck College, London; Vice-President (since 1969), Leo Baeck College, London; Chairman (since 1989), Council of Reform and Liberal Rabbis; Co-President (since 1990), London Society of Jews and Christians.
Main field of work: Teaching. **Career:** Rabbi, South London Liberal Synagogue from 1953 to 1957. Associate Rabbi, Liberal Jewish Synagogue, London from 1957 to 1961. Senior Rabbi, Liberal Jewish Synagogue, London from 1961 to 1989. Chairman, European Board, World Union of Liberal and Progressive Judaism from 1973 to 1976. Chairman, Council of Reform and Liberal Rabbis of Great Britain from 1969 to 1971. Chairman, Council of Reform and Liberal Rabbis of Great Britain from 1982 to 1984. **Educ.:** B.A., M.A. from University of Cambridge, Emmanuel College, UK (1947-1953); Rabbinical Ord. from Hebrew Union College-Jewish Institute of Religion, Cincinnati OH, USA (1963-1965). **Hons.:** Hon. D.D., Hebrew Union College-Jewish Institute of Religion, Cincinnati OH (1980). **Pubs.:** Co-editor, *Service of the Heart* and *Gate of Repentance* (prayer books); Co-author, *The Jewish People: Their History and their Religion* and Co-author, *Judaism for Today.*
Mailing address: Liberal Jewish Synagogue, 28 St John's Wood Road, London NW8 7HA, UK. T. (44 71) 286 5181
Residence: 37 Walmington Fold, London N12 7LD, UK. T. (44 81) 446 6196

Rayner, Keith, Most Rev Dr
Rel: Anglican; Ordained on 21 DEC 1953; b. 22 NOV 1929 in Brisbane, Queensland, Australia; m. to Audrey Evelyn Fletcher; c. 1 boy and 2 girls; Citizen of Australia; Lang. English.
International Organizations: Primate (since 1991) of the Anglican Church of Australia.
Other organizations: Archbishop (since 1991), Melbourne.
Career: Bishop, Wangaratta from 1969 to 1975. Archbishop, Adelaide from 1975 to 1990. **Educ.:** Ph.D. from University of Queensland, Brisbane, Australia (1963). **Hons.:** Officer, Order of Australia; Th.D. h.c., Australian College of Theology, Sydney (1990).
Mailing address: St Paul's Cathedral Buildings, 209 Flinders Lane, Melbourne VIC 3000, Australia. T. (61 3) 653 4220
Residence: 120 Clarendon St, East Melbourne VIC 3002, Australia. T. See Mailing Address

Razafimahatratra, Victor, S.I., H E Card
Rel: Roman Catholic; Ordained on 28 JUL 1956; b. 08 SEP 1921 in Ambanitsilena-Ranomasina, Fianarantsoa, Madagascar; celibate; Citizen of Madagascar.
International Organizations: Member (since 1976) of the College of Cardinals.
Member of the Society of Jesus. Member of the Congregation for the Evangelization of Peoples.
Other organizations: Archbishop (since 1976), Antananarivo.
Career: Bishop, Farafangana from 1971 to 1976. Created Cardinal by Pope Paul VI in 1976. Entitled S Croce in Gerusalemme.
Mailing address: Archevêché, Andohalo, 101 Antananarivo, Madagascar. T. (261 2) 20 726

Read, Allan Alexander, Rt Rev
Rel: Anglican; Ordained; b. 19 SEP 1923 in Toronto, Ontario, Canada; m. to Beverly Roberts; c. 2 boys and 2 girls; Citizen of Canada.
International Organizations: Bishop (since 1981) of Ontario, Kingston.
Career: Curate, Parish of Mono from 1947 to 1954. Curate, Parish of Barrie from 1954 to 1971. Archdeacon, Simcoe from 1960 to 1971. Suffragan Bishop, Diocese of Toronto from 1972 to 1981. Member, Barrie School Board from 1954 to 1963. **Educ.:** B.A. from University of Toronto, Ontario, Canada; D.D. from University of Toronto, Trinity College and Wycliffe College, Canada; S.T.D. from Laurentian University of Sudbury, Thorneloe University, Ontario, Canada. **Hons.:** Citizen of the Year, City of Barrie (1968).
Mailing address: Anglican Diocesan Centre, 90 Johnson St, Kingston, Ontario K4L 1X7, Canada.
Residence: 39 Riverside Dr, RR 1, Kingston, Ontario K7L 4V1, Canada.

Read, John Edward, Commr
Rel: Salvation Army, The; b. 01 OCT 1927 in Amherst NS, Canada; m. to Louisa Doris Harrison; c. 2 boys and 3 girls; Citizen of Canada; Lang. English.
International Organizations: Principal (since 1987) of the Salvation Army International College for Officers.
Chairman of the Advisory Council to the General (since 1988) of the Salvation Army.

Main field of work: Education and ministry. **Career:** Officer, Salvation Army in 1946. Training Principal, Salvation Army from 1967 to 1977. Field Secretary, Salvation Army from 1977 to 1982. Territorial Commander, Caribbean, Salvation Army from 1982 to 1986. **Educ.:** Comm. Off. from College for Officers Training, Toronto, Canada (1945-1947); Memorial University of Newfoundland, St John's, Canada (1969). **Interests:** Writing. **Pubs.:** Studies in Sanctification; Burning, Always Burning; I Believe in the Dawn; Heart to God, Hand to Man and Training – and Retraining – Timothy.
Mailing address: The Cedars, 34 Sydenham Hill, London SE26 6LS, UK. T. (44 81) 693 3290.
Residence: 95 Longton Ave, London SE26, UK T. (44 81)699 21 00

Reardon, John Patrick, Rev

Rel: United Reformed Church; Ordained on 12 DEC 1962; b. 15 JUN 1933 in Gravesend, Kent, UK; m. to Molly Pamela Young; c. 4 boys and 1 girl; Citizen of United Kingdom; Lang. English.
International Organizations: General Secretary (since 1990) of the Council of Churches for Britain and Ireland.
Board Member (since 1983) of the Christian Aid. Trustee (since 1985) of the Action for World Development Fund.
Main field of work: Ministry and ecumenical leadership. **Career:** Teacher, schools in London and Kent from 1958 to 1961. Minister, Congregational Church, Horsham from 1961 to 1968. Minister, Congregational Church, St Albans from 1968 to 1972. Secretary of Dept., United Reformed Church from 1972 to 1990. Deputy General Secretary, United Reformed Church from 1983 to 1990. **Educ.:** B.A. from University of London, UK (1954-1957); P.G. Cert. Ed. at University of London, UK (1957-1958). **Interests:** ecumenism, overseas development, church and society. **Pubs.:** Leaves From The Tree of Peace (edited) and Threads of Creation (edited).
Mailing address: 35-41 Lower Marsh, London SE1 7RL, UK. T. (44 71) 620 4444
Residence: 27 Vanda Crescent, St Albans AL1 5EX, UK. T. (44 727) 52921

Reardon, Martin, Rev Canon

Rel: Church of England; Ordained on 01 JUN 1958; b. 03 OCT 1932 in Purley, Surrey, UK; m. to Dr Ruth Slade; c. 1 boy and 1 girl; Citizen of United Kingdom; Lang. English.
International Organizations: General Secretary (since 1990) of the Churches Together in England.
Other organizations: Co-Chairman (since 1968), Association of Interchurch Families.
Career: Secretary, Sheffield Council of Churches from 1961 to 1970. Sub-Warden, Lincoln Theological College from 1971 to 1978. Secretary, Board for Mission and Unity, Church of England from 1978 to 1989. **Educ.:** M.A. from University of Cambridge, Selwyn College, UK (1951-1955); Dip. Oecuménique from University of Geneva, Graduate School of Ecumenical Studies, Bossey, Switzerland (1955-1956); G.O.E. from Cuddesdon College, Oxford, UK (1956-1958). **Pubs.:** Christian Unity in Sheffield, 1967; What on Earth is the Church For, 1985 and Christian Initiation: A Policy for the Church of England, 1990.
Mailing address: Churches Together in England, Interchurch House, 35-41 Lower Marsh, London SE1 7RL, UK. T. (44 81) 620 4444

Rech, Roland, Rev

Rel: Buddhist: Zen; Ordained in 1974; b. 20 JUN 1944 in Paris, France; single; Citizen of France; Lang. French.
International Organizations: President (since 1982) of the International Zen Association.
Career: Commercial Executive, CFR from 1968 to 1971. Commercial Executive, Rhône Poulenc from 1972 to 1981. **Educ.:** Institut d'Etudes Politiques de Paris, France (1963-1966); UER de langues orientales from University of Paris VII, France (1981-1983); UER de Sc.Hum. Cliniques from University of Paris VII, France (1981-1985). **Interests:** Soto Zen.
Mailing address: 17 rue Keller, F-75011 Paris, France. T. (33 1) 43 79 89 73
Residence: Raymond Losserand 108, F-75014 Paris, France

Redmond, Francis J, F.S.P., Br

Rel: Roman Catholic; Ordained; celibate.
International Organizations: Procurator-General of the Brothers of Saint Patrick.
Mailing address: Patrician Brothers' Generalate, Tullow Hill, Co. Carlow, Ireland. T. (353 503) 51190. Fax 51190.

Redrado Marchite, José Luis R, O.H., Rev Fr

Rel: Roman Catholic; Ordained; celibate.
International Organizations: Secretary of the Pontifical Council for Pastoral Assistance to Health Care Workers.
Member of the Hospitaller Order of St John of God.
Mailing address: Via della Conciliazione 3, I-00193 Roma, Italy. T. (39 6) 689 68 45 – (39 6) 689 67 98. Fax (39 6) 689 68 41. Tx 2031 SANITPC VA
Residence: Villa S Famiglia, Largo O Respighi 6, I-00135 Roma, Italy T. (39 6) 329 42 28

Reece, Donald James, Most Rev

Rel: Roman Catholic; Ordained on 03 JAN 1971; b. 13 APR 1934 in Kingston, Jamaica; celibate; Citizen of Jamaica; Lang. English and Spanish.
International Organizations: Secretary (since 1984) of the Antilles Episcopal Conference.
Member (since 1985) of the Pontifical Council for Promoting Christian Unity.
Other organizations: Bishop of St-John's – Basseterre (since 1981), Roman Catholic Church.
Main field of work: Adminstration and pastoral. **Career:** Member, Joint Working Group from 1985 to 1990. **Educ.:** B.A.; S.T.B.; M.A. from Catholic University of America, Washington DC, USA (1964-1971).
Mailing address: PO Box 836, St John's, Antigua-Barbuda. T. (1 809) 461 1135. Fax (1 809) 462 2383
Residence: Catholic Chancery Office, Church Street, St Johns, Antigua-Barbuda T. See Mailing Address

Rees, Paul Stromberg

Rel: Evangelical Covenant Church; Ordained in 1921; b. in Rhode Island, USA.
International Organizations: Minister of the Evangelical Covenant Church.
Main field of work: Ministry, evangelism, writing. **Career:** Served in the Pilgrim Tabernacle, Pasadena CA from 1920 to 1923. Served in the Holiness Tabernacle, Detroit MI from 1928 to 1932. Pastor, First Covenant, Minneapolis MN from 1938 to 1958. Moderator, Evangelical Covenant in 1948. Vice-President, World Evangelical Fellowship from 1950 to 1955. President, National Association of Evangelicals from 1952 to 1954. Minister to pastors in the Billy Graham Crusades from 1954 to 1959. Advisor to the World Council of Churches assembly in New Delhi in 1961. Vice-President, World Vision from 1958 to 1975. Editor, *World Vision Magazine* from 1964 to 1972. Associate Editor, *Herald* from 1955 to 1975. Trustee, Asbury College from 1933 to 1965. Trustee, Asbury Theological Seminary from 1967 to 1980. **Educ.:** B.A. from University of Southern California, Los Angeles, USA (1923). **Hons.:** awarded several honorary doctorates. **Pubs.:** author of numerous books on devotional themes, the practical Christian life and Bible exposition.
Mailing address: No mailing address.

Reeves, Paul Alfred, Most Rev Sir

Rel: Anglican; Ordained in 1960; b. 06 DEC 1932; m. to Beverley Watkins; c. 3 girls; Citizen of New Zealand.
International Organizations: Ecclesiastic of the Anglican Church.
Career: Ordained Deacon in 1958. Curate, Tokoroa, New Zealand from 1958 to 1959. Curate, St Mary the Virgin, Oxford from 1959 to 1961. Curate, Kirkley St Peter, Lowestoft from 1961 to 1963. Vicar, St Paul, Okato, New Zealand from 1964 to 1966. Lecturer in Church History, St John's College, Auckland from 1966 to 1969. Director of Christian Education, Diocese of Auckland from 1969 to 1971. Bishop, Waiapu from 1971 to 1979. Bishop, Auckland in 1979. Primate and Archbishop, New Zealand from 1980 to 1985. Governor-General, New Zealand from 1985 to 1990. Chairman, Environmental Council from 1974 to 1976. **Interests:** swimming, sailing, jogging. **Hons.:** Hon. D.C.L., Oxon (1985); Knight of the Order of St John of Jerusalem (1986); Honorary Fellow, St Peter's College, Oxford (1980); Knight Grand Cross of the Order of St Michael and St George; Knight Grand Cross of the Royal Victorian Order.
Mailing address: c/o Government House, Private Bag, Wellington, New Zealand.

Regazzoni, Giuseppe, F.M.I., Fr

Rel: Roman Catholic; Ordained; celibate.
International Organizations: Procurator-General of the Sons of Mary Immaculate – Pavonians.
Mailing address: Piazza dei Geografi 15, I-00176 Roma, Italy. T. (39 6) 29 57 66.

Regehr, John, Rev

Rel: Mennonite Brethren; Ordained in 1957; b. 29 DEC 1925 in Mexico; m. to Mary Unger; c. 2 boys and 2 girls; Citizen of Canada.
International Organizations: Co-Director (since 1990) of the Recovery of Hope counseling service.
Career: Instructor, Mennonite Brethren Bible College, Winnipeg, Manitoba from 1968 to 1991. Acting President, Mennonite Brethren Bible College, Winnipeg from 1972 to 1974. Began teaching in Jr. and Sr. High School, West Kildonan, Winnipeg in 1950. Principal, H.C. Avery Elementary School, Winnipeg in 1956. Pastor, Silver Lake Mennonite Brethren Church, Marion, SD in 1957. Instructor, Mennonite Brethren College Institute from 1959 to 1966. Assistant Pastor, Elmwood Mennonite Brethren Church from 1962 to 1966. Pastor, Georgetown E.U.B. Church, IN from 1966 to 1968. Member, Board, Interfaith Pastoral Institute from 1972 to 1974. **Educ.:** B.A., B.Ed. from University of Manitoba, Winnipeg, Canada (1950); B.D. from Mennonite Brethren Bible College (1950); M.A. from Winona Lake School of Theology (summers) (1950); North American Baptist Seminary (part-time) (1957-1959); Th.D. from Southern Baptist Theological Seminary (part-time) (1970). **Pubs.:** Articles and sermons published in numerous theological magazines and journals.
Mailing address: 1-169 Riverton Ave, Winnipeg MB R2L 2E5, Canada.
Residence: 275 Bredin Drive, Winnipeg MB R2K 1N7, Canada.

Reich, Seymour D

Rel: Jewish.
International Organizations: President of the B'nai B'rith International.
Mailing address: 1640 Rhode Island Ave NW, Washington DC 20036, USA. T. (1 202) 857 6545. Tx 710-822-0068.

Reid, William Stanford

Rel: Presbyterian; Ordained; b. 13 SEP 1913 in Montreal, Canada; m. to Priscilla Lee; Citizen of Canada.
International Organizations: Professor Emeritus (since 1979) of the University of Guelph.
Career: Minister, Fairmount Taylor Presbyterian Church, Montreal from 1941 to 1945. Founder and Minister, Presbyterian Church of Town of Mount Royal, Montreal from 1945 to 1951. Lecturer, subsequently Professor of History, McGill University, Montreal from 1941 to 1965. Professor and first Chairman, Dept. of History, University of Guelph from 1965 to 1970. Professor of History, University of Guelph from 1970 to 1978. Visiting Lecturer, Presbyterian Theological Hall, Melbourne from 1982 to 1984. **Hons.:** Hon. L.H.D., Wheaton College (1975); Hon. D.D., Presbyterian College, Montreal (1978). **Pubs.:** Trumpeter of God, 1974; The Scottish Tradition in Canada, 1976 and John Calvin: His Influence on the Western World, 1983.
Mailing address: Dept. of History, University of Guelph, Guelph N1G 2W1, Canada. T. (1 519) 824 4120 EXT 3887
Residence: No 906, 19 Woodlawn Road W, Guelph ON N1H 7B1, Canada. T. (1 519) 822 7032

Reimer, A James

Rel: Mennonite.
International Organizations: Professor of Theology of the Conrad Grebel College.
Mailing address: Westmount Rd North, Waterloo ON N2L 3G6, Canada.

Reimer, Reginald E

Rel: Christian.
International Organizations: Chairman of the Inter-Church Relief and Development Alliance.
Mailing address: c/o World Relief, PO Box 874, Station B, Willowdale ON M2K 2R1, Canada.

Reinich, Rodolfo R, Rev

Rel: Evangelical.
International Organizations: President of the Evangelical Church of the River Plate.
Mailing address: Esmeralda 162, 1035 Buenos Aires, Argentina.

Remmerswaal, James, O.S.C., Fr

Rel: Roman Catholic; Ordained; celibate.
International Organizations: Procurator-General of the Order of the Holy Cross.
Procurator-General of the Frères de Saint Louis de Gonzague.
Mailing address: Via del Velabro 19, I-00186 Roma, Italy. T. (39 6) 679 33 35.

Rempel, Erwin Henry, Rev

Rel: Mennonite; Ordained on 20 JUL 1969; b. 08 AUG 1944 in Los Angeles CA, USA; m. to Angela Marie Albrecht; c. 1 boy and 2 girls; Citizen of USA; Lang. English and Portuguese.
International Organizations: Executive Secretary, Commission on Overseas Mission (since 1982) of the General Conference of the Mennonite Church.
Main field of work: Administration. **Career:** Pastor, Indian Valley Mennonite Church from 1968 to 1975. Missionary in Brazil from 1975 to 1982. **Educ.:** B.A. from Grace College, Omana NB, USA (1962-1966); M.Div. from Westminster Theological Seminary, Philadelphia PA, USA (1968-1970); Th.M. from Eastern Baptist Theological Seminary, Philadelphia PA, USA (1971-1973). **Mem.:** American Society of Missiology
Mailing address: 722 Main Street, Newton KS 67114, USA. T. (1 316) 283 5100
Residence: 1701 West 4th, Newton KS 67114, USA. T. (1 316) 283 7762

Rempel, Ron

Rel: Mennonite.
International Organizations: Editor of the *Mennonite Reporter*.
Mailing address: 3-312 Marsland Drive, Waterloo ON N2J 3Z1, Canada.

Renaud, Etienne, M.Afr., Fr

Rel: Roman Catholic; Ordained; celibate.
International Organizations: Superior General of the Society of Missionaries of Africa.
Vice-Grand Chancellor of the Pontifical Institute for Arabic and Islamic Studies.
Mailing address: Via Aurelia 269, CP 9078, I-00165 Roma, Italy. T. (39 6) 63 23 18 – 63 23 14.

Renggli-Enderle, Claire

Rel: Roman Catholic; b. in St Gallen, Switzerland; m. to Karl Renggli; Citizen of Switzerland; Lang. German, French, Italian, English and Spanish.
International Organizations: Secretary-General (since 1986) of the International Catholic Conference of Guiding.
Main field of work: Librarian.
Mailing address: Metzgergasse 5, CH-9004 St Gallen, Switzerland. T. (41 71) 23 41 30
Residence: See Mailing Address. T. See Mailing Address

Resler, Loel L, Dr

Rel: Christian; Ordained on 04 FEB 1974; b. 02 FEB 1941 in Sterling CO, USA; m. to Nita Marie; c. 4 boys and 1 girl; Citizen of USA; Lang. English.
International Organizations: Chancellor (since 1984) of the Westbrook University.
Main field of work: University education. **Educ.:** D.Th. from Faith Bible College and Seminary, Denver CO, USA (1970-1974); D.Psy. from Pacific International University, Riverside CA, USA (1980-1985); D.Ph. from Faith Bible College and Seminary, Denver CO, USA (1985-1987).
Mailing address: World Congregational Fellowship Church, 404 N Mesa Verde, Aztec NM 87410, USA. T. (1 505) 334 1115
Residence: See Mailing Address. T. See Mailing Address

Resner, André, Prof

Rel: Church of Christ; Ordained on 01 AUG 1983; b. 21 JUL 1957 in Killeen TX, USA; m. to Mary Larner; c. 2 boys and 1 girl; Citizen of USA; Lang. English, Hebrew, Greek, German and French.
International Organizations: Assistant Professor of Hebrew Homiletics (since 1991) of the Abilene Christian University, TX.
Other organizations: Corporate Board (since 1992), Restoration Quarterly.
Main field of work: Ministry of grieving. **Educ.:** B.A. from Pepperdine University, Malibu CA, USA (1978-1981); M.S. from Pepperdine University, Malibu CA, USA (1981-1985); M.Div. from Princeton Theological Seminary, Princeton NJ, USA (1985-1988); Ph.D. from Princeton Theological Seminary, Princeton NJ, USA (1990-1992). **Pubs.:** "Lament: Faith's Response to Loss", in *Restoration Quarterly*, 32, pp. 129-142, 1990; "The Death and Rebirth of Praise", in *Praising God: Themes from the Psalms*, ACU Press, pp 30-55, 1991 and "A Year of Lament", in *Princeton Seminary Bulletin*, 9, pp. 159-162, 1988.
Mailing address: ACU Station, Box 8403, Abilene TX 79699, USA. T. (1 915) 674 3740
Residence: 1726 Cedar Crest, Abilene TX 79601, USA T. (1 915) 677 5723

Restrepo, Jesús Angel, C.I.M., Fr

Rel: Roman Catholic; Ordained; celibate.
International Organizations: Procurator-General of the Congregation of Jesus and Mary.
Mailing address: Via dei Querceti 15, I-00184 Roma, Italy. T. (39 6) 731 57 89 – 731 53 84.

Restrepo, Jesús, Fr

Rel: Roman Catholic.
International Organizations: Procurator-General of the Congregation of Jesus and Mary.
Mailing address: Via dei Querceti 15, I-00184 Roma, Italy. T. (39 6) 731 57 89.

Rètailleau, Michel, F.C., Fr

Rel: Roman Catholic; Ordained; celibate.
International Organizations: Superior General of the Fils de la Charité.
Mailing address: 10 rue Louis-Blanc, 75010 Paris, France. T. (33) 442 01 95 27

Revivi, Menachem

Rel: Jewish.
International Organizations: Executive Director of the World Confederation of Jewish Community Centers.
Mailing address: 12 Hess Street, Jerusalem 94185, Israel. T. (972 2) 231371. Fax 247767. Tx 26528 JWB-IL

Revollo Bravo, Mario R, H E Card

Rel: Roman Catholic; Ordained on 31 OCT 1943; b. 15 JUN 1919 in Genoa, Italy; celibate; Citizen of Italy.
International Organizations: Member (since 1988) of the College of Cardinals.
Member of the Sacred Congregation for the Sacraments and Divine Worship. Member of the Pontifical Commission for Latin America. Member of the Congregation for Religious and Secular Institutes.
Other organizations: Archbishop, Bogotá.
Career: Elected to the Titular Church, Tinisa di Numidia in 1973. Consecrated bishop in 1973. Prefect at Nueva Pamplona in 1978. Transferred to Bogotá in 1984. Created Cardinal by Pope John Paul II in 1988.
Mailing address: Arzobispado, Carrera 7a N 10-20, Bogotá DE, Colombia. T. (57 1) 334 5500

Ribadeneira, Maria Cristina

Rel: Christian.
International Organizations: Regional Representative for Latin America and the Caribbean of the World Student Christian Federation.
Mailing address: Casilla 17-21-967, Quito, Ecuador. T. (593 2) 541 204 Ext 7. Fax 568 664.

Ribeiro, António, H E Card

Rel: Roman Catholic; Ordained on 05 JUL 1953; b. 21 MAY 1928 in Gandarela de Basto, Braga, Portugal; celibate; Citizen of Portugal.
International Organizations: Member (since 1973) of the College of Cardinals.
Member of the Congregation for the Clergy.
Other organizations: Patriarch (since 1971), Lisbon; President, Episcopal Conference of Portugal.
Career: Professor, University of Lisbon from 1965 to 1967. Auxiliary Bishop, Braga in 1967. Auxiliary Bishop, Lisbon in 1969. Created Cardinal by Pope Paul VI in 1973. **Interests:** sports, philately, travel. **Pubs.:** The Aevum According to Thomas Aquinas, 1958 and The Socialization, 1964.
Mailing address: Campo dos Mártires da Pátria 45, P-1198 Lisboa CODEX, Portugal. T. (351 1) 56 39 01

Ribush, Alison

Rel: Buddhist: Tibetan Mahayana.
International Organizations: Director, Tara Institute of the Foundation for the Preservation of the Mahayana Tradition.
Mailing address: 3 Mavis Ave, Brighton East, Melbourne VIC 3187, Australia. T. (61 3) 596 8900. FAX (61 3) 596 4856

Ribush, Nick

Rel: Buddhist: Mahayana.
International Organizations: Member of the Foundation for the Preservation of the Mahayana Tradition.
Other organizations: Director, Kurukulla Center.
Mailing address: PO Box 67262, Chestnut Hill MA 02167, USA. T. (1 617) 421 9668. Fax (1 617) 536 1897

Ricciardi, Salvatore, Rev

Rel: Reformed Church.
International Organizations: Member of the Executive Committee of the World Alliance of Reformed Churches.
Mailing address: Via della Signora 6, I-20122 Milano, Italy. T. Fax (39 2) 791 518 (C/O LIBERIA CLAUDIANA CHURCH BOOKSHOP)
Residence: T. (39 2) 7600 2654

Rice, Robert F, Rev

Rel: Christian.
International Organizations: Director of the Literacy and Evangelism International.
Mailing address: 1800 South Jackson, Tulsa OK 74107, USA. T. (1 918) 585 3826. Fax 585 3224.

Richard, D John, Rev Dr

Rel: Protestant.
International Organizations: Chairman of the Commission on Church Renewal.
Mailing address: E894 Chittaranjan Park, New Delhi 110 019, India.

Richman, M L

Rel: Jewish.
International Organizations: Director of the Jewish Colonization Association.
Mailing address: ORT Resource Centre, Westfiedl College, Kidderpore Avenue, London NW3 7ST, UK. T. (44 71) 431 5002.

Richter, Manfred

Rel: Protestant.
International Organizations: President of the Protestant Association for Adult Education in Europe.
Mailing address: c/o Ev Bildungswerk, Goethestrasse 26-30, 10625 Berlin, Germany. T. (49 30) 31 91 223.

Ridala, Esmo

Rel: Evangelical Lutheran.
International Organizations: General Secretary of the Estonian Evangelical Lutheran Church.
Mailing address: Box 450 74, S-104 30 Stockholm 45, Sweden.

Riehm, Magda

Rel: Christian.
International Organizations: Contact of the International Fellowship for Research in Hymnology.
Mailing address: Beethovenstr 2, 69121 Heidelberg, Germany. T. (49 6221) 45775.

Rieth, Theobald

Rel: Christian.
International Organizations: Secretary-General of the Association of Christian Centres in Europe.
Mailing address: Lothringerstrasse 69, 52070 Aachen, Germany. T. (49 241) 50 90 47.

Rieutord, Maurice Georges Gabriel, Fr

Rel: Roman Catholic; Ordained; b. 12 AUG 1932 in Paris, France; celibate; Citizen of France; Lang. French, Italian and English (reading).
International Organizations: Secretary-General of the Robert Schuman Institute for Europe.
Vice-President of the Liaison Committee of NGOs enjoying Consultative Status with the Council of Europe. Delegate to Strasbourg and Vice-President of the Catholic European Study and Information Centre. Member (since 1966) of the Society of Jesus. General Delegate for the Associated Bodies to the Managing Board of the Centre for the Europe of the Citizens and the Human Rights. Member, Orientation Council of the Observatoire européen de géopolitique, Paris.
Other organizations: Director, External Relations (since 1962), Centre Culturel “Les Fontaines”; Secretary-General, Centre Européen de Recherche Interdisciplinaire de Chantilly (CERIC); Organizer, sessions, conferences, symposiums, etc; Committee Member, Association Française pour les Nations Unies, Paris; Member, Scientific Council, UPCS/INFAC Group, Nogent-sur-Marne.
Main field of work: Culture. **Career:** Vicar, Chaplain, apostolic movements, colleges, and in charge of the relations with the authorities for the parish of Saint-Louis d'Hyères (Var) from 1962 to 1965. Vicar, parish of Saint-Georges du Cap Brun, Toulon (Var) from 1965 to 1966. Spiritual Director, Collège Saint-Michel, Saint-Etienne from 1967 to 1969. In charge, the catechesis at the School Centre of Saint-Marc, Lyon and at two other schools of the diocese of Lyon from 1971 to 1974. Assistant to the Director-General, School Centre of Saint-Marc, Lyon from 1972 to 1974. Deputy Director, the Catholic Education, Diocese of Lyon from 1975 to 1980. Professor of Christian Education, Centre de Formation Pédagogique de l'Enseignement Spécialisé (CFPES), Catholic Institute of Paris from 1982 to 1987. Secretary-General, Centre Culturel 'Les Fontaines' from 1980 to 1988. Vice-President and Founder, Foundation for ACP-EEC Cultural Cooperation in 1985. Vice-President and Founding Member, National Liaison Committee 'Etre Citoyen', Paris in 1985. **Educ.:** Phil. from Grand Séminaire de Francheville and Grand Séminaire International d'Issy-les-Moulineaux, France; Theol. from Grand Séminaire International d'Issy-les-Moulineaux, Grand Séminaire Interdiocésain d'Aix-en-Provence, France and Pontifical Gregorian University, Vatican City State. **Interests:** Painting, drawing, decoration, swimming, tennis, parlor games. **Hons.:** Knight of the Order of the Academic Palms (France), Silver Medal of the European Merit Foundation. **Mem.:** Centre of Studies and Initiatives for the Europe of the Citizens and the Human Rights; Founding Member, Robert Schuman Institute for Europe (1982); European Cultural Center; Centre Européen de Recherche Interdisciplinaire de Chantilly – CERIC (since 1980); Founding Member, Association Internationale de Recherche et d'Animation Pédagogiques (1972); Member, Association de Paris de l'Institut des Hautes Etudes de Défense Nationale (IHEDN); Les Amis de Robert Schuman, Scy-Chazelles (1957); Les Amis de la Maison de l'Europe de Paris; Centre Unité Chrétienne, Lyon; Conférence Mondiale des Religions pour la Paix, Paris; Fondation Cziffra, Senlis; Fondation de France, Paris; Unicef, Paris **Pubs.:** *Culture et Mécénat*, Coll. Europe-Cultures, Robert Schuman Institute for Europe and Several articles and conferences on Europe and culture.
Mailing address: Les Fontaines, BP 219, F-60631 Chantilly CEDEX, France. T. (33) 44 57 19 26 – (33) 44 57 24 60.
Residence: See Mailing Address. T. See Mailing Address

Rigali, Justin Francis, H E Mgr

Rel: Roman Catholic; Ordained on 25 APR 1961; b. 19 APR 1935 in Los Angeles CA, USA; celibate.
International Organizations: Secretary of the Congregation for Bishops.
Member of the Pontifical Council for the Pastoral Care of Migrants and Itinerant People. Advisor of the Pontifical Council for the Laity.
Other organizations: Titular Archbishop (since 1985), Bolsena, Volsinium; Secretary, College of Cardinals; Advisor, Congregazione per la Dottrina della Fede; Member, Pontifical Commission for Latin America.
Mailing address: Palazzo delle Congregazioni, Piazza Pio XII 10, I-00193 Roma, Italy. T. (39 6) 698 4217. Fax 698 5303
Residence: Piazza della Minerva 74, I-00186 Roma, Italy T. (39 6) 686 5850 – 687 4941

Righi-Lambertini, Egano, H E Card

Rel: Roman Catholic; Ordained on 25 MAY 1929; b. 22 FEB 1906 in Casalecchio di Reno, Bologna, Italy; celibate; Citizen of Italy.
International Organizations: Member (since 1979) of the College of Cardinals.
Career: Titular Archbishop, Doclea from 1960 to 1979. Created Cardinal by Pope John Paul II in 1979.
Mailing address: Piazza della Città Leonina 9, I-00193 Roma, Italy. T. (39 6) 687 94 63

Rights, Graham H, Rev

Rel: Moravian.
International Organizations: President of the Moravian Church in America (Southern Province).
Mailing address: Drawer O, Salem Station, Winston-Salem NC 27108, USA.

Rigsel, Geshe Lhundrup

Rel: Buddhist: Mahayana.
International Organizations: Member of the Foundation for the Preservation of the Mahayana Tradition.
Other organizations: Director, Kopan Gompa.
Mailing address: GPO Box 817, Kathmandu, Nepal. T. (977 1) 226 717

Riley-Smith, Jonathan S C, Prof

Rel: Unknown.
International Organizations: President of the Society for the Study of the Crusades and the Latin East.
Mailing address: 5 Claremont Road, Windsor SL4 3AX, UK.

Rime, Paul, M.S.F.S., Fr

Rel: Roman Catholic; Ordained; celibate.
International Organizations: Procurator-General of the Missionnaires de Saint François de Sales d'Annecy.
Mailing address: Via Federico Torre 21, I-00153 Roma, Italy. T. (39 6) 581 75 89

Ríos Zalapa, Alejandro R, M.G., Fr

Rel: Roman Catholic; Ordained; celibate.
International Organizations: Superior General of the Missions Etrangères de Sainte Marie de Guadaloupe.
Mailing address: Córdoba 17, 06700 México DF, Mexico. T. (52 5) 533 65 60

Rivera, Alberto R, Dr

Rel: Christian.
International Organizations: Director of the AIC International Christian Ministries.
Mailing address: PO Box 31340, Los Angeles CA 90031, USA.

Rivera, Cornelius A

Rel: Christian.
International Organizations: Executive Director of the Spanish World Gospel Mission.
Mailing address: PO Box 542, Winona Lake IN 46590, USA. T. (1 219) 267 8821.

Riviera, Luis N, Prof

Rel: Christian.
International Organizations: President of the Christian Peace Conference in Latin America and the Caribbean.
Mailing address: POB 23098, University of Puerto Rico, Río Piedras 00931, Puerto Rico.

Ro, Bong Rin, Dr

Rel: Presbyterian; b. 23 JAN 1935 in Sunchon, Korea Rep; m. to Alma L. Ro; c. 2 boys and 1 girl; Citizen of USA; Lang. Korean, English, Chinese (Mandarin) and Japanese.

International Organizations: Executive Secretary (since 1990) of the Theological Commission of the World Evangelical Fellowship. Executive Secretary (since 1974) of the Asia Theological Association. Member, Executive Committee (since 1982) of the Evangelical Fellowship of Asia. Member, Executive Committee (since 1984) of the International Council of Accrediting Agencies for Evangelical Theological Education.

Other organizations: Professor of Church History (Part-time) (since 1984), Central Taiwan Theological Seminary; Dean (since 1984), Asia Graduate School of Theology; Editor (since 1976), *Asia Theological News* (Quarterly) and *Asian Perspective* Booklets; Member, Editorial Board (since 1983), *Transformation.*

Main field of work: Church education. **Career:** Professor of Church History, Discipleship Training Centre in Singapore and Singapore Bible College from 1970 to 1974. Visiting Professor of Asian Studies, Wheaton Graduate School from 1974 to 1975. Professor of Asian Church History (Part-time), China Evangelical Seminary, Taipei from 1978 to 1979. Professor of Chinese Religions, Missions and Doctrines (Part-time), Wheaton College and Trinity Evangelical Divinity School from 1979 to 1980. Professor of Comparative Religions (Part-time), Tung Hai University of Taiwan from 1981 to 1984. Chairman, Executive Committee, International Council of Accrediting Agencies for Evangelical Theological Education from 1980 to 1984. **Educ.:** Seoul National University, Republic of Korea (1955-1956); B.A. Bib. Ed. from Columbia College, SC, USA (1956-1960); B.A. Hist. from Wheaton College, IL, USA (1961-1962); B.D. from Covenant Seminary, St Louis MO, USA (1960-1961); M.Th. from Concordia Lutheran Seminary, St Louis MO, USA (1964-1969). **Pubs.:** A number of books including: ed. *Christian Alternatives to Ancestor Practices*, Taichung, Taiwan: ATA, 1984, 370p; With Ruth Eshenaur, eds., *The Bible and Theology in Asian Contexts*, Taichung, Taiwan: ATA, 1984, 404p; Several articles including: "The Inspiration of Scripture among the 17th Century Reformed Theologians," *The Living and Active Word of God*, eds. M Inch and R Youngblood, Winona Lake, Indiana: Eisenbrauns, 1983, pp. 207-24; "Historical Perspective on Evangelism and Social Responsibility", *In Word and Deed*, eds. K Kantzer and B Nicholls, Exeter, UK: Paternoster Press, 1984 and a video tape program: "Korean Church Growth History", produced by ATA in Taipei, 1976.

Mailing address: PO Box 94 Choong Jongno, Seoul 120-751, Korea Rep. T. (82 2) 363 3247. Fax (82 2) 393 3789.

Residence: 17-4 Jongdong, Choong-ku, Seoul, Korea Rep T. (82 2) 393 9895

Ro, Chung-Hyun, Prof

Rel: Reformed Church.

International Organizations: Vice-President of the World Alliance of Reformed Churches.

Mailing address: 120-2 Daeshin-dong, Sudaemoon-ku, Seoul, Korea Rep. T. (82 2) 392 2460. Fax (82 2) 392 0618

Residence: T. (82 2) 392 0658

Roberts, Granville Oral

Rel: United Methodist; Ordained in 1968 (Pentecostal Holiness Church), 1978 (United Methodist Church); b. in Pontotoc County OK, USA; Citizen of USA.

International Organizations: Founder of the Oral Roberts University, Tulsa OK.

Main field of work: Evangelist, faith healer and television minister. **Career:** Had a conversion experience and experienced a divine healing from tuberculosis in 1935. Ordained in the Pentacostal Holiness Church in 1936. Launched an independent ministry based on divine healing in 1947. Founded the Oral Roberts Evangelical Association in 1947. Conducted huge tent revivals and pioneered sophisticated methods of mass mailing. Broadened his influence by televising his healing crusades in 1954. Stopped holding crusades in 1968. Canceled his television program in 1967. Returned in a series of prime-time specials that featured well-known personalities from the entertainment field in 1969. Left the Pentecostal Holiness Church and was ordained a minister in the United Methodist Church in 1968. **Pubs.:** *If You Need Healing, Do These Things*, 1952; *God Is a Good God*, 1960 and *Don't Give Up*, 1980.

Mailing address: No mailing address.

Roberts, Richard Frederick Anthony, Rev

Rel: Anglican; Ordained in 1988; b. 12 MAY 1932; m. to Melvern Hollis (Bain) Roberts; c. 1 boy and 2 girls; Citizen of Bahamas.

International Organizations: Assistant Priest (since 1988) of the St Barnabas Anglican Church, Nassau, Bahamas.

Career: Personnel Officer, Bahamas Airways from 1963 to 1967. Executive Director and Partner, Venn Livingstone Roberts (public relations) from 1967 to 1968. Personnel Director, New Providence Development Co. Ltd. in 1968. Member, Parliament for Centreville constituency from 1968 to 1977. Parliamentary Secretary to Ministry of Finance from 1969 to 1972. Parliamentary Secretary to Ministry of Agriculture from 1971 to 1972. Minister, Agriculture and Fisheries from 1972 to 1973. Minister, Home Affairs in 1973. Minister, Agriculture, Fisheries and Local Government from 1974 to 1977. High Commissioner in UK from 1979 to 1984. Ordained Deacon in 1987. Previous activities include: President, Airline Workers Union. Secretary-General, Amalgamated Building Constructional Engineering Trade Union. Assistant Secretary-General, Bahamas Federation of Labour. President, Bahamas TUC. Secretary-General, Bahamas TUC. Assistant Secretary-General, Progressive Liberal Party (PLP). First Vice-Chairman, PLP. Chairman, Maritime Board. Vice-Chairman, Bahamas Agricultural Corporation. Secretary, Methodist Preachers Committee. **Interests:** fishing, reading, sports, religion.

Mailing address: Carmichael Road, PO Box 565, Nassau, Bahamas.

Robertson, Marion Gordon 'Pat'

Rel: Baptist: Southern; b. in Lexington VA, USA.

International Organizations: Founder of the Christian Broadcasting Network (CBN).

Main field of work: Religious television broadcaster and politician. **Career:** Became a Christian and turned toward a full-time commitment to ministry in 1956. While a student, became a pioneer in the modern charismatic movement. Purchased a small UHF television station in Norfold VA, and became a Southern Baptist associate minister in a nearby church. Established the '700 Club' in 1963. Moved his headquarters to Virginia Beach and established a full graduate university. Was a candidate for the US presidency in 1988. Withdrew before the summer Republican Convention and returned to religious broadcasting. **Educ.:** Grad. (cum laude) from Washington and Lee University, Lexington VA, USA (1950); Law Degree from Yale University, Law School, New Haven CT, USA (1952-1955); New York Theological Seminary, USA (1952-1955). **Pubs.:** *Shout It from the Housetop*, 1972.

Mailing address: No mailing address.

Robinson, Donald William Bradley, Most Rev

Rel: Anglican; Ordained; b. 09 NOV 1922 in Australia; m. to Marie E Taubman; c. 3 boys and 1 girl; Citizen of Australia.

International Organizations: Archbishop (since 1982) of Sydney.

Career: Curate, St Matthew's, Manly from 1950 to 1951. Curate, St Philip's, Sydney from 1952 to 1953. Lecturer, Moore Theological College from 1952 to 1953. Senior Lecturer, Moore Theological College from 1954 to 1959. Vice-Principal, Moore Theological College from 1959 to 1972. Acting Principal, Moore Theological College in 1968. Visiting Lecturer, Union Biblical Seminary, Yeotmal, India in 1969. Consecrated Assistant Bishop, Sydney in 1973. Bishop in Parramatta from 1973 to 1982. **Hons.:** Dr h.c., Australian College of Theology.

Mailing address: PO Box Q190, Queen Victoria Building, Sydney NSW 2000, Australia.

Robinson, Gnana, Rev Prof Dr

Rel: Protestant; Ordained in 1960; b. 17 MAR 1935 in Gnanapuram-Tamil Nadu, Tamil Nadu, India; m. to Florence Robinson; c. 1 boy and 1 girl; Citizen of India; Lang. Tamil, English, German, Hebrew (working knowledge), Greek (working knowledge) and Syriac (working knowledge).

International Organizations: President (since 1985) of the Centre for Peace and Justice, India.

Other organizations: President (since 1978), Peoples' Association for Social Action; Guest Professor for Ecumenical Studies (since 1987), Preachers Seminary of the Westfalen Church in Germany.

Main field of work: Theology. **Career:** Professor for Old Testament, Tamilnadu Theological Seminary/College from 1962 to 1978. Principal, Tamilnadu Theological Seminary, Madurai, South India from 1978 to 1987. Honorary President, Serampore University from 1986 to 1988. President, Partners in Justice Concerns, India from 1983 to 1987. **Educ.:** B.Sc. from Travencore University (1952-1956); B.D. from Serampore University (1957-1960); M.Th. from Serampore University

(1961-1963); S.T.M. from Union Theological Seminary, New York NY, USA (1964-1965); Dr. Theol. from University of Hamburg, Germany (1972-1975). **Interests:** justice issues. **Hons.:** The World Christian Tamil Academy. **Mem.:** Honoured with the title of 'Narchaithy Selvar' (1983) **Pubs.:** *The Origin and Development of the Old Testament Sabbath*, Peter Lang, 1988; *Struggle is Life*, C.L.S., Madras, 1988 and *Siding With the Poor*, C.L.S., Madras, 1989.
Mailing address: Im Schäferkamp 6a, 59494 Soest, Germany. T. (49 2921) 12760
Residence: See Mailing Address. T. See Mailing Address

Robinson, Joseph, Rev Canon

Rel: Church of England; Ordained in 1953; b. 23 FEB 1927; m. to Anne Antrobus; c. 2 boys and 2 girls.
International Organizations: Master (since 1980) of The Temple.
Career: Ordained Deacon in 1952. Curate, Tottenham from 1952 to 1955. Minor Canon, St Paul's Cathedral from 1955 to 1968. Lecturer in Old Testament Studies, King's College, London from 1959 to 1968. Canon, Canterbury from 1968 to 1980. **Educ.:** B.D., M.Th. from University of London, King's College, UK. **Interests:** reading, gardening. **Hons.:** Hon. LL.D., Simon Greenleafe Law School, Anaheim CA. **Mem.:** Clubs: Athenaeum; Fellow, King's College, London **Pubs.:** Cambridge Bible Commentary on 1 Kings, 1972 and Cambridge Bible Commentary on 2 Kings, 1976.
Mailing address: Temple Church, The Temple, London EC4Y 7BB, UK. T. (44 71) 353 8559

Robinson, Margorie, OCD, Sr

Rel: Roman Catholic; Ordained; celibate.
International Organizations: Member of the Discalced Carmelite Nuns.
Other organizations: Member, Association of Contemplative Sisters; President, Metropolitan Association of Contemplative Communities.
Mailing address: Carmelite Monastery, 139 Depuyster Ave, Beacon NY, USA.

Robinson, Philippus Johannes, Prof

Rel: Christian; Ordained in 1959; b. 28 MAY 1932 in Hobhouse, Orange Free State, South Africa; m. to Ammerensia Susanna Odendaal; c. 2 boys and 3 girls; Citizen of South Africa; Lang. Afrikaans, English, Xhosa and Dutch.
International Organizations: Professor (since 1976) of the University of the Western Cape.
Main field of work: University education. **Career:** Minister, Dutch Reformed Church in South Africa from 1960 to 1968. Minister, Dutch Reformed Church in Africa (Xhosa) from 1968 to 1972. Lecturer, Theological School, Decoligny, Umtata from 1972 to 1976. **Educ.:** B.A. and B.A. (Hons) from University of the Orange Free State, Bloemfontein, South Africa (1951-1955); B.D. from University of Stellenbosch, South Africa (1956-1959); D.Th. from University of Stellenbosch, South Africa (1982). **Mem.:** South African Missiological Society (1973); American Academy of Religion (1991); American Society of Biblical Literature (1991) **Pubs.:** With H L Pretorius, Reflecting on Mission in the African Context, 1986; "Mission as ethos/Ethos as mission", in *Missionalia*, 18:1, 1990, pp.156-167 and "Integrity of creation and Christian Mission", in *Missionalia*, 19:2, 1991, pp. 144-153.
Mailing address: University of the Western Cape, Modderdam Rd (Private Bag X17), Bellville 7535, South Africa. T. (27 21) 959 2206
Residence: 10 Skool Street, Kuils River 7580, South Africa. T. (27 21) 903 4264

Robinson, Samuel E

Rel: Christian.
International Organizations: President of the Christian Missions in Many Lands.
Mailing address: PO Box 13, Spring Lake NJ 07762, USA. T. (1 212) 267 0306.

Robitaille, Jean

Rel: Christian.
International Organizations: Coordinator of the Christian Committee for Human Rights in Latin America.
Mailing address: 25 Jarry Ouest, Montréal PQ H2P 1S6, Canada. T. (1 514) 387 2541. Fax (1 514) 387 2541.

Robles Díaz, Luis R, H E Mgr

Rel: Roman Catholic; Ordained on 14 APR 1963; b. 06 MAR 1938 in El Grullo; celibate.
International Organizations: Apostolic Pro-Nuncio to Uganda (since 1990) of the Roman Catholic Church.
Other organizations: Titular Archbishop, Stefaniaco.
Mailing address: PO Box 7177, Kampala, Uganda. T. (256 41) 221 167 – 221 774

Roch, Raymond, Rev Fr

Rel: Roman Catholic; Ordained; celibate.
International Organizations: Chaplain of Association Our Lady of the Nations: Group of Catholic International Civil Servants.
Mailing address: 5 chemin des Tulipiers, CH-1208 Geneva, Switzerland.

Rochette, Jean-François, Dr

Rel: Reformed Church; Unordained; b. 15 MAR 1939 in Paris, France; m. to Véronique Turrettini; c. 1 boy and 2 girls; Citizen of Switzerland; Lang. French, English and German.
International Organizations: General Treasurer (since 1971) of the World Alliance of Reformed Churches.
Other organizations: General Staff Officer (Colonel) (since 1978), Swiss Army; Colonel (since 1985), Geneva Regiment; Director-General (since 1983), United Overseas Bank.
Main field of work: Banking. **Career:** Assistant Director, Lombard Odler et Cie from 1969 to 1981. Director, Cramer et Cie from 1981 to 1983. **Educ.:** Lic. ès sc.pol. from University of Geneva, Switzerland (1963); Lic. ès sc.écon. from University of Geneva, Switzerland (1964); Dip. from University of Finance, New York NY, USA (1968-1969). **Mem.:** Parti Libéral (Geneva), Golf Club Bessinge, Club du Lundi, Bureau des Familles, Fondation des Enfants Surdi-Mutite
Mailing address: United Overseas Bank, 11 quai des Bergues, CH-1200 Genève, Switzerland. T. (41 22) 731 9841. Fax (41 22) 732 6733. Tx 412100
Residence: 'La Lezardière', 290B route de Lausanne, CH-1293 Bellevue, Switzerland T. (41 22) 774 2643

Roddy, Joan, D.M.I., Sr

Rel: Roman Catholic; Ordained; b. 28 APR 1946 in Ballinamore, Leitrim, Ireland; celibate; Citizen of Ireland; Lang. English, French and Italian.
International Organizations: Superior General (since 1989) of the Daughters of Mary and Joseph.
Executive Member (since 1992) of the International Union of Superiors General.
Educ.: Dip. Theol. from Maynooth College, Ireland (1968-1969); B.Soc.Sc. from National University of Ireland, University College, Cork, Ireland (1969-1972). **Interests:** social justice.
Mailing address: Via Rivarone 100, Int 9, I-00166 Roma, Italy. T. (39 6) 6641 5629.
Residence: See Mailing Address. T. See Mailing Address

Rodrigo Mendizábal, Iván Fernando R

Rel: Roman Catholic; b. 26 AUG 1961 in La Paz, Bolivia; m. to Catalina Vaca Espín; Citizen of Bolivia; Lang. Spanish and English.
International Organizations: Executive Secretary (since 1990) of the International Catholic Organization for Cinema and Audiovisual – Latin America.
Vice-President for Bolivia (since 1987) of the International Catholic Organization for Cinema and Audiovisual.
Other organizations: Production Chief (since 1985), Canal 13 Televisión Universitaria; Production and Programming Chief (since 1989), Canal 13 Televisión Universitaria.
Main field of work: Communications. **Educ.:** Lic. Comunicación Social from Universidad Católica Boliviana, La Paz, Bolivia (1985-1989). **Interests:** Identification with the institutional projects. **Hons.:** Video prizes. **Mem.:** Press Syndicate, Bolivia; International Catholic Organization for Cinema and Audiovisual, Bolivia **Pubs.:** Articles in reviews: *Contacto* and *Educación Popular*; "Video Nacional", in *Hoy* (1984-1989) – video critic and "Video Boliviano", in *Presencia* (1990) – video critic.
Mailing address: OCIC-AL, Selva Alegre 169 y Av. 10 de Agosto, Apartado 21-178, Quito, Ecuador. T. (593 2) 548 046. Fax (593 2) 501 658.
Residence: See Mailing Address. T. See Mailing Address

Rodriguez, Ester, Ms
Rel: Spiritualist.
International Organizations: President, New York Office of the Lucis Trust.
Mailing address: 113 University Place, 11th Floor, PO Box 722 – Cooper Station, New York NY 10276, USA. T. (1 212) 982 87 70.

Rodriguez, Ignacio Ortega, S.S.J., Fr
Rel: Roman Catholic; Ordained; celibate.
International Organizations: Superior General of the Missionaries of St Joseph of Mexico.
Mailing address: Cerrada de Carracci 13, Apartado 19-243, 03730 México D F, Mexico. T. (52 5) 563 33 44

Rodriguez, Norberto
Rel: Protestant.
International Organizations: Secretary-General of the Latin American Confederation of YMCAs.
Mailing address: Culpina 272, 1406 Buenos Aires, Argentina. T. (54 1) 613 3747. Fax 613 3747. Tx 17125.

Rodríguez Maradiaga, Oscar Andrés R, S.D.B., H E Mgr
Rel: Roman Catholic; Ordained on 28 JUN 1970; b. 29 DEC 1942 in Tegucigalpa, Honduras; celibate.
International Organizations: Past Secretary-General of the Latin American Episcopal Council.
Member of the Pontifical Commission for Latin America. Member of the Pontifical Council for Justice and Peace. Member of the Pontifical Council 'Cor Unum'. Member of the Society of St Francis of Sales.
Other organizations: Titular Bishop, Pudenziana; Vicar-General, Tegucigalpa.
Mailing address: CELAM, Carrera 5 118-31, Usaquén, Apartado Aéreo 5278, Bogotá, Colombia. T. (57) 612 1495 – (57) 612 1929 – (57) 612 1889. Fax (57) 612 1929.

Roehrs, Zahir
Rel: Muslim: Sufi.
International Organizations: Representative of the Sufi Order International.
Mailing address: 23 rue de la Tuilerie, F-92150 Suresnes, France.

Roeroe, Wilhelmus A, Rev Dr
Rel: Reformed Church.
International Organizations: Member of the Executive Committee of the World Alliance of Reformed Churches.
Mailing address: Tromolpos 18, Jalan Lelepoan 10, Kakaskasen III, Tomohon 95362 Sul-Ut, Indonesia.

Roest, Crollius Arij, S.J., Rev Fr
Rel: Roman Catholic; Ordained; celibate.
International Organizations: Director of the Centre for Coordination of Research of the International Federation of Catholic Universities. Advisor of the Pontifical Council for Inter-Religious Dialogue. Member of the Society of Jesus.
Mailing address: Palazzo Frascara, Piazza della Pilotta 4, I-00187 Roma, Italy. T. (39 6) 678 62 53. C. PUGI ROMA. Fax (39 6) 678 62 53.

Roger-Machart, M
Rel: Roman Catholic.
International Organizations: UNESCO Representative of the Unda, International Catholic Association for Radio and Television.
Mailing address: 12 rue Malher, F-75004 Paris, France. T. (33 1) 49 83 27 20.

Rogers, Anthony, Br
Rel: Roman Catholic; Ordained; celibate.
International Organizations: Zone Coordinator, Asia and Oceania of the Caritas Internationalis.
Mailing address: Office for Human Devel (OHD), Dona Matilda Bldg 301, PO Box EA-12, Ermita, Manila, Philippines.

Rogers, Edward W, Fr
Rel: Roman Catholic; Ordained; celibate.
International Organizations: Director of the Inter-Regional Meeting of Bishops of Southern Africa.
Mailing address: 4 Bayswater Road, Highlands, PO Box 230 BE Belvedere, Harare, Zimbabwe. T. (263 4) 791053. Fax 724971. Tx 22390 GENSEC ZW Attn IMBISA.

Rogestam, Christina
Rel: Lutheran; b. 31 DEC 1943 in Linköping, Sweden; m. to Lennart Widing; Citizen of Sweden; Lang. Swedish and English.
International Organizations: Treasurer (since 1990) of the Lutheran World Federation.
Chairperson, Program Committee for Administrative Services of the Lutheran World Federation. Director-General of the Nordic Immigration Committee.
Other organizations: Director-General (since 1988), Swedish Immigration Board.
Career: Member, Swedish Parliament from 1973 to 1982. Director, University of Gothenburg from 1982 to 1987. **Educ.:** M.S.S.S. from University of Gothenburg, Sweden (1963-1970).
Mailing address: c/o Swedish Immigration Board, Postfack 6113, S-600 06 Norrköping, Sweden. T. (46 11) 15 63 00. Tx 50940. C. STIVERK.
Residence: Tösse prästgård, Pl 5169, S-66200 Åmål, Sweden. T. (46 532) 20145

Rogues, Jean
Rel: Christian.
International Organizations: Director of the Higher Institute of Ecumenical Studies.
Mailing address: 21 rue d'Assas, F-75270 Paris CEDEX 06, France. T. (33 1) 42 22 41 80.

Roll, Bertin, Rev
Rel: Roman Catholic; Ordained; celibate.
International Organizations: USA Contact of the Archconfraternity of Christian Mothers.
Mailing address: 220 37th Street, Pittsburgh PA 15201, USA. T. (412) 683 2400.

Rome, Brian
Rel: Christian.
International Organizations: Director of the Far East Broadcasting Association.
Mailing address: Box 234, Mahé, Seychelles. T. (248) 24449. Fax 248 25171. C. FEBCOM. Tx 2287 FEBCOM SZ. Fax via Bureau 22555 FEBCOM.

Romeo, Paolo, H E Mgr
Rel: Roman Catholic; Ordained on 04 OCT 1970; b. 20 FEB 1938 in Acireale, Sicily, Italy; celibate.
International Organizations: Apostolic Nuncio to Colombia (since 1990) of the Roman Catholic Church.
Other organizations: Titular Archbishop, Vulturia.
Mailing address: Carrera 15, N 36-33, Apartado Aéreo 3740, Bogóta, Colombia. T. (57 1) 2454 260. Fax 2851 817. Tx 42 455 NAPBO CO

Rondo, Kelly H, Rev
Rel: Evangelical.
International Organizations: General Secretary of the Christian Evangelical Church in Minahasa.
Mailing address: Kantor Sinode GMIM, Tomohon, Minahasa, Sulawesi Utara, Indonesia.

Root, Michael, Prof
Rel: Lutheran.
International Organizations: Director of the Institute for Ecumenical Research, Strasbourg.
Mailing address: 8 rue Gustave-Klotz, F-67000 Strasbourg, France. T. (33) 88 36 10 57.

Roozen, Nico, Drs
Rel: Christian.
International Organizations: Contact of the Interchurch Action for Latin America.
Mailing address: Laan van Meerdervoort 148, NL-2587 BE Den Haag, Netherlands. T. (31 70) 65 52 07 – (31 70) 45 89 46.

Rosa, Alfred Christopher Henricus, Rev Dr
Rel: Methodist; Ordained on 01 MAY 1948; b. 01 DEC 1922 in Jaela, Western Province, Sri Lanka; widowed Joyce Vimalawathy; c. 3 girls; Citizen of Sri Lanka; Lang. Sinhala and English.
International Organizations: Vice-President (since 1989) of the Asian-Pacific Christian Peace Conference.
Deputy Secretary-General (since 1958) of the Christian Peace Conference.
Educ.: B.D. from United Theological College, Bangalore, India (1944-1948); M.A. from Columbia University, New York NY, USA (1951-1952). **Interests:** global peace enterprise. **Hons.:** Dr. h.c., Comenius Faculty of Theology, Prague (1989). **Pubs.:** Bad Times for the Third World; Energy Problems in Developing Countries with Urgent Summons for Alternate Sources of Energy and The Gulf War Syndrome.
Mailing address: Shalom 33, Lourdes Mawatha, Jaela, Sri Lanka. T. (94) 536 839.
Residence: See Mailing Address. T. See Mailing Address

Rose, A Moshe, Rabbi
Rel: Jewish; Ordained.
International Organizations: Secretary-General of the Conference of European Rabbis.
Mailing address: PO Box 5324, Jerusalem, Israel. T. (972 2) 81 28 59.

Rose, Maurice, Rabbi
Rel: Jewish; b. 07 SEP 1925 in Birmingham, UK; m. to Cynthia Rose; c. 4 boys; Citizen of United Kingdom; Lang. Hebrew and English.
International Organizations: Director (since 1962) of the Conference of European Rabbis.
Other organizations: Director (since 1975), Israel Council of Young Israel.
Educ.: B.A. (Hons); M.A. from University of London, UK (1952-1956); Rabbinical Dip. from Jews College, London, UK (1953-1960).
Mailing address: PO Box 5324, Jerusalem 91052, Israel. T. (972 2) 81 28 59

Rosen, David Shlomo, Rabbi
Rel: Jewish; Ordained in 1972; b. 18 AUG 1951 in Newbury, Berkshire, UK; m. to Sharon Rothstein; c. 3 girls; Citizen of Israel and United Kingdom; Lang. English and Hebrew.
International Organizations: Director, Inter-faith Relations at Israel Office (since 1986) of the Anti-Defamation League of B'nai B'rith.
Executive (since 1988) of the International Council of Christians and Jews.
Other organizations: Executive (since 1989), Rabbis for Human Rights; Executive (since 1991), Inter-religious Co-ordinating Council in Israel; Professor of Jewish Studies (since 1990), Jerusalem Center for Near East Studies, Mt Scopus.
Main field of work: Inter-faith relations, university education. **Career:** Director, South Africa Federation of Jewish Student Organizations from 1973 to 1975. Judge, Cape Ecclesiastical Court from 1975 to 1979. Senior Rabbi, Sea Point, Cape Town from 1975 to 1979. Lecturer on Jewish-Christian Relations, Irish School of Ecumenics from 1982 to 1985. Chief Rabbi, Jewish Communities of Ireland from 1979 to 1985. Editor, *Irish-Jewish Yearbook* from 1979 to 1985. **Educ.:** Doc. Rec. from University of Dublin Trinity College, Ireland (1982-1985). **Mem.:** International Jewish Vegetarian Society (President of Israel Chapter) **Pubs.:** Inter-faith Relations in Israel Today; Salvation and Redemption in Judaism and Guide to the Sabbath Home Service.
Mailing address: Anti-Defamation League, Israel Office, 30 King David Street, Jerusalem 94101, Israel. T. (972 2) 251 171
Residence: 20 Dan Street, Baka'a, Jerusalem, Israel T. (972 2) 724 596

Rosen, Jeremy, Rabbi Prof
Rel: Jewish; Ordained on 01 JUN 1968; b. 11 SEP 1942 in Manchester, UK; m. to Suzanne Kaszirer; c. 2 boys and 2 girls; Citizen of United Kingdom; Lang. English, Hebrew and French.
International Organizations: Professor of Judaism (since 1991) of the Faculty for the Comparative Study of Religions, Antwerp.
Other organizations: Trustee (since 1980), Yakar Foundation; Rabbi (since 1986), Western Marble Arch Synagogue, London.
Main field of work: Consultant. **Career:** Rabbi, Giffnock, Glasgow from 1969 to 1971. Principal, Carmel College, UK from 1971 to 1985. Advisor on Interfaith to Chief Rabbi, England from 1987 to 1991. **Educ.:** B.A. (Hons) from University of Cambridge, UK (1962-1965); M.A. from University of Cambridge, UK (1965-1967); Rabb. Dip. from Mir Yeshiva, Jerusalem, Israel (1965-1968).
Mailing address: Kaszirer, 78 Pelikaanstraat, B-2018 Antwerpen, Belgium. T. (32 3) 234 36 30
Residence: Mechelsesteenweg 213, B-2018 Antwerpen, Belgium.

Rosenhager, Ursel, Rev
Rel: Reformed Church.
International Organizations: Programme Secretary of the World Alliance of Reformed Churches.
Mailing address: 150 route de Ferney, PO Box 2100, CH-1211 Genève 2, Switzerland. T. (41 22) 791 62 40. C. WARC GENEVA. Tx 415 730 OIK CH (WARC). Fax (41 22) 791 65 05 WARC REFORMIERTER WELTBUND.
Residence: T. (41 22) 788 0480

Rosenstock, Werner, Dr
Rel: Jewish; Unordained; b. 10 APR 1908 in Berlin, Germany; m. to Susanne Philipp; c. 1 boy; Citizen of United Kingdom; Lang. English and German.
International Organizations: Honorary Secretary (since 1968) of the Council of Jews from Germany.
Career: Organiser, Central Verein, Berlin from 1934 to 1938. Director, Association of Jewish Refugees from 1949 to 1982. **Educ.:** Dr.jur. from Universities of Berlin and Freiburg, Germany (1926-1931). **Pubs.:** *Exodus 1933-1939* (Jewish emigration from Germany) and *The Jewish Youth Movement.*
Mailing address: 9 Adamson Road, London NW3 3HX, UK. T. (44 71) 483 2536.
Residence: 77D Campagne Gardens, London NW6 3RT, UK.

Rosner, Regine
Rel: Roman Catholic.
International Organizations: Secretary-General of the International Movement of Catholic Agricultural and Rural Youth.
Mailing address: Tiensevest 68, B-3000 Leuven, Belgium. T. (32 16) 22 83 12.

Ross, Ian
Rel: Christian.
International Organizations: Director of the International Christian Relief.
Mailing address: PO Box 180, 16 St John's Hill, Sevenoaks TN13 1AS, UK.

Rossello, Lorenzo
Rel: Buddhist: Mahayana.
International Organizations: Member of the Foundation for the Preservation of the Mahayana Tradition.
Other organizations: Manager, Takden Shedrup Targye Ling.
Mailing address: Via Poggiberna 9, 56040 Pomaia (Pisa), Italy. T. (39 50) 685 976

Rossi, Agnelo, H E Card
Rel: Roman Catholic; Ordained on 27 MAR 1937; b. 04 MAY 1913 in Joaquim Egidio, Campinas, Brazil; celibate; Citizen of Brazil.
International Organizations: Member (since 1965) of the College of Cardinals.
Member of the Sacred Congregation for the Oriental Churches. Member of the Congregation for the Causes of Saints. Member of the Congregation for Bishops. Member of the Congregation for the Clergy. Member of the Congregation for Religious and Secular Institutes. Member of the Congregation for Catholic Education.
Other organizations: Titular Bishop, Ostia and Sabina-Poggio Mirteto; Member, Council of Cardinals and Bishops.
Career: Bishop, Barra do Pirai in 1956. Archbishop, Sao Paulo in 1964. Created Cardinal by Pope Paul VI in 1965. Prefect, Sacred Congregation for the Evangelization of the People from 1970 to 1984. Titular Bishop, Ostia and Sabina-Poggio Mirteto. President, Administration of the Patrimony of the Holy See.
Mailing address: 00120 Città del Vaticano, Vatican.

Rossi, Gianfranco
Rel: Unknown.
International Organizations: Secretary General of the International Association for the Defence of Religious Liberty.
Mailing address: Schosshaldenstrasse 17, CH-3006 Bern, Switzerland. T. (41 31) 44 62 62. Fax 44 62 66. Tx 845-912950.

Rossi, Opilio, H E Card

Rel: Roman Catholic; Ordained on 11 MAR 1933; b. 14 MAY 1910 in New York, USA; celibate.
International Organizations: Member (since 1976) of the College of Cardinals.
Other organizations: President, Commission of Cardinals for the Pontifical Sanctuaries of Pompei, Loreto and Bari.
Career: Created Cardinal by Pope Paul VI in 1976.
Mailing address: Via della Scrofa 70, I-00186 Roma, Italy. T. (39 6) 683 37 88

Rossner, John Leslie, Rt Rev Abbot, Prof Dr

Rel: Anglican; Ordained on 21 DEC 1958; b. 08 SEP 1930 in Norwood MA, USA; m. to Prof Marilyn Zwaig; Citizen of USA; Lang. English, French, Spanish (some), German (some), Russian (some), Ancient languages: Latin, Greek and Hebrew.
International Organizations: President (since 1976) of the International Institute of Integral Human Sciences.
Co-President (since 1980) of the World Fellowship of Religions. Rector (since 1988) of the International Council of World Religions and Cultures. Abbot-General (since 1989) of the Order of the Transfiguration.
Other organizations: Priest (non-parochial) (since 1958), Episcopal Church/USA, Diocese of Massachusetts; From Lecturer to Full Professor of Comparative Religion and Culture (since 1962), Concordia University (Sir George William Campus), Montreal; Adjunct Professor of Comparative Religion and Culture, World University, Ojai, CA, USA.
Main field of work: University education. **Career:** US Air Force Officer – Active duty from 1952 to 1954. Lecturer in New Testament, Nashotah House, WI (Episcopal Seminary) from 1955 to 1958. Tutor/Lecturer in Ancient History, Montreal Diocesan Theological College from 1958 to 1960. Served as priest in parishes in Montreal, Newport RI, New York City from 1958 to 1962. **Educ.:** B.A. from Trinity College, Hartford CT, USA (1948-1952); M.A. from Brown University, Providence RI, USA (1954-1955); B.D., M.Div. from Nashotah House, WI, USA (1955-1958); S.T.M. from McGill University, Montreal, Canada (1958-1960); Ph.D. from Heed University, Hollywood FL, USA (1975-1979). **Interests:** religions of the Ancient World and primitive Christianity, religion and psychical research, new sciences of consciousness and healing, furthering the cause of the introduction of such subjects into educational systems for interreligious understanding and world peace. **Hons.:** Honorary Canon, Cathedral St Cyr sur Mer (1978); Knight Conv. Chaplain, Sovereign Military Order of St John of Jerusalem, Malta (1980); D.D. h.c., The New Seminary, New York City NY (1988); D.Litt. h.c., World University of America, Ojai CA (1989); D.Sc. h.c., Open International University, Colombo, Sri Lanka (1992). **Mem.:** Academy of Religion and Psychical Research (1974); Catholic Theological Society of America (1990); Canadian Society for the Study of Religion; American Academy of Religion; Society for the Scientific Study of Religion **Pubs.:** Numerous books, articles, and reviews: *From Ancient Magic to Future Technology*, The University Press of America, 1979; *From Ancient Religion to Future Science*, The University Press of America, 1979 and *The Primordial Tradition and the Cosmic Christ*, 1989.
Mailing address: PO Box 1387, Station H, Montréal, Québec, Canada. T. (1 514) 937 8359.
Residence: PO Box 460, Station H, Montréal PQ H3G 2N3, Canada.

Rossoondyal, N

Rel: Hindu.
International Organizations: Secretary-General of the International Council of Hindu Youth.
Mailing address: 12 Bourbon Street, Rm 4, PO Box 382, Port Louis, Mauritius. T. (230) 89964. Tx 4647 ATTN:ICHY.

Roth, Alice M

Rel: Mennonite.
International Organizations: Vice-President of the Mennonite Board of Missions.
Mailing address: Box 370, Elkhart IN 46515-0370, USA.

Roth, Donald

Rel: Evangelical Mennonite.
International Organizations: President of the Evangelical Mennonite Church.
Mailing address: 1420 Kerrway Court, Fort Wayne IN 46805, USA.

Rouhe, Eino Olavi, Pastor

Rel: Seventh-Day Adventist; Ordained on 07 JUN 1969; b. 30 APR 1934 in Salo, Finland; m. to Anna-Liisa Helle; c. 3 boys and 1 girl; Citizen of Finland; Lang. Finnish, Swedish and English.
International Organizations: President, Finland Union Conference (since 1986) of the General Conference of Seventh-Day Adventists.
Main field of work: Ministry. **Career:** Book Editor, Seventh-Day Adventists (SDA) Publishing House Kirjatoimi from 1961 to 1967. Editor-in-Chief, SDA Publishing House Kirjatoimi from 1967 to 1981. President, West-Finland Conference of the SDA Church from 1981 to 1983. Editor, SDA Publishing House Kirjatoimi from 1983 to 1984. Secretary-General, Union Conference of the SDA Church in Finland from 1984 to 1986. **Pubs.:** *Jumalan Israel*, 1961; *Yksi Lauma Ja Yksi Paimen*, 1968; *Raportti Jeesuksen Uskosta*, 1976 and other books and hundreds of articles.
Mailing address: Pl 2, SF-20721 Turku 72, Finland. T. (358 21) 365100
Residence: Voudinkatu 6 E 31, SF-20780 Kaarina, Finland T. (358 21) 434398

Roussel, François-Xavier

Rel: Roman Catholic.
International Organizations: Contact of the Secrétariat International Spiritain de Communications Sociales par l'Audio-visuel.
Mailing address: 30 rue Lhomond, F-75005 Paris, France. T. (33 1) 43 31 88 62.

Roux, Alain

Rel: Christian.
International Organizations: Representative of the Comité laïque pour l'éducation au développement – Educateurs sans frontières.
Mailing address: 28 boulevard Bonne-Nouvelle, F-75010 Paris, France. T. (33 1) 45 23 10 81. Fax 48 01 08 69.

Roux, Albert G, Fr

Rel: Roman Catholic.
International Organizations: National Director of the Marian Movement of Priests.
Mailing address: PO Box 8, St Francis ME 04774, USA. T. (207) 398 3375.

Rovida, Edoardo, H E Mgr

Rel: Roman Catholic; Ordained on 29 JUN 1950; b. 26 AUG 1927 in Alexandria, Egypt; celibate.
International Organizations: Apostolic Nuncio to Switzerland (since 1985) of the Roman Catholic Church.
Other organizations: Apostolic Nuncio to Liechtenstein (since 1987); Titular Archbishop, Taormina.
Mailing address: Thunstrasse 60, Postfach 199, CH-3000 Bern 16, Switzerland. T. (41 31) 44 60 40. Fax 44 50 64

Rowinski, Francis C, Most Rev

Rel: Catholic.
International Organizations: Prime Bishop of Polish National Catholic Church.
Mailing address: 115 Lake Scranton Road, Scranton PA 18505, USA.

Roy, David K

Rel: New Age.
International Organizations: Secretary of the New Age World Religious and Scientific Research Foundation.
Mailing address: 62091 Valley View Circle, Joshua Tree CA 92252, USA. T. (1 619) 366 2833.

Roy, Jacques

Rel: Roman Catholic.
International Organizations: Secretary of the Secular Institute of Pius X.
Mailing address: 1645, 80e Rue Est, C P 1815, Charlesbourg G1K 7K7, Canada. T. (1 418) 626 5882.

Roy Choudhury, Punyabrata R

Rel: Brahmo Samaj; b. 26 DEC 1928 in Calcutta, West Bengal, India; m. to Sunanda; c. 1 boy and 1 girl; Citizen of India; Lang. English, Bengali, Hindi and Russian.

International Organizations: President (since 1990) of the International Association for Religious Freedom.

Other organizations: Examiner and paper setter (since 1978), University of Calcutta; Consultant to the jute industry (since 1978); Member of the Governing Board (since 1983), City College, Calcutta; President (since 1989), Hari Navi Brahmo Samaj, Calcutta.

Main field of work: Consultancy. **Career:** Production Manager, Jardine Henderson Ltd. from 1954 to 1966. Head of Quality Control, Indian Jute Industries Research Association from 1967 to 1970. Joint Director for Textiles, Ministry of Commerce, India from 1970 to 1977. **Educ.:** B.Sc. (Hons) from University of Calcutta, India (1945-1949); P.G. Dip. in Textile Technology at Dundee College of Technology, UK (1952-1954). **Interests:** social service. **Hons.:** Fellow of the Institute of Standards Engineers. **Mem.:** Calcutta Club

Mailing address: 78 B Sarat Bose Road, Calcutta 700025, India. T. (91 33) 75 3088

Residence: C-3, Government Housing Estate, 98 Karaya Road, Calcutta 700019, India. T. (91 33) 47 6469

Royer, Therese, Ms

Rel: Catholic.

International Organizations: Representative at UNESCO of the St Joan's International Alliance.

Mailing address: 261 Avenue du Général Leclerc, F-94700 Maisons-Alfort, France.

Ruangskul, Prasert

Rel: Buddhist; b. 17 AUG 1914 in Singburi Province, Thailand; m. to Varaporn Ruangskul; c. 2 boys and 2 girls; Citizen of Thailand; Lang. Thai and English.

International Organizations: Honorary Secretary-General and Member of the Executive Council (since 1976) of the World Fellowship of Buddhists.

Career: Teacher, Government Schools from 1939 to 1944. Section Chief, Public Relations Dept., Bank of Thailand from 1944 to 1950. Chief, Governor's Office and Secretary, Court of Directors, Bank of Thailand from 1950 to 1974. **Educ.:** Math. and Hist. from University of Oxford, UK (1933-1937). **Pubs.:** *Modern Europe*, 1962 (Thai translation from english); *The Word of the Buddha*, 1974 (Thai translation from english); *The Revolutionary Ideas in France*, 1975 (Thai translation from english) and *Bismarck*, 1979 (Thai version).

Mailing address: 33 Sukhumvit Road, Bangkok 10110, Thailand. T. (66 2) 251 1188 (66 2) 251 1190. C. WORLBUDIST BANGKOK. Fax (66 2) 255 7363.

Residence: 209 Sukhumvit Road 31, Bangkok 10110, Thailand T. (66 2) 258 9132

Rüegg, François, Dr

Rel: Greek Catholic; b. 23 MAR 1948 in Bern, Switzerland; m. to Melinda Coelho; c. 1 boy and 2 girls; Citizen of Switzerland; Lang. French, English, German, Italian, Spanish (passive) and Romanian.

International Organizations: Secretary-General (since 1985) of the International Catholic Child Bureau.

Other organizations: Secretary (since 1989), Centre International de Formation Chrétienne; Treasurer (since 1989), Centre Catholique d'Etudes; Board Member (since 1989), Association Frères et Soeurs d'Espérance.

Main field of work: Child welfare. **Career:** Lecturer, Section of Psychology and Educational Sciences, University of Geneva, Switzerland from 1974 to 1975. Lecturer, Université Laval, Québec, Canada from 1979 to 1980. Permanent Representative to the United Nations Centre for Social Development and Humanitarian Affairs, International Catholic Child Bureau from 1980 to 1984. Correspondent, Conference of International Catholic Organizations, Vienna from 1980 to 1984. Assistant Secretary-General, International Catholic Child Bureau from 1984 to 1985. **Educ.:** Lic. ès Lett. from University of Geneva, Switzerland (1968-1970); Maîtrise spéc. Ethn. from Ecole des Hautes Etudes en Sciences Sociales, Paris, France (1970-1972); P.G. at University of Paris VII, France (1972-1980); Doc. d'Etat ès L., Sc.Hum. from University of Montpellier, France (1986). **Pubs.:** "Les maisons bourgeoisiales dans l'espace traditionnel anniviard (Suisse)", in *Etudes Rurales*, 83, July-Sept. 1981, pp. 69-95 and "Drogue: Pour une approche globale de la prévention", in *Bulletin de la Division des Stupéfiants*, United Nations, Vol. XXXVII, Nos 2 and 3.

Mailing address: 65 Rue de Lausanne, CH-1202 Genève, Switzerland. T. (41 22) 731 32 48. Fax (41 22) 731 77 93.

Residence: Chemin Chantebise 13, CH-1226 Thonex, Switzerland

Ruegg-Saillen, Rudolf

Rel: Roman Catholic.

International Organizations: Administrator of the Conference of International Catholic Organizations.

Mailing address: Chemin de Chantebise 13, CH-1226 Thônex, Switzerland. T. (41 22) 48 52 90.

Ruff, Francis, Fr

Rel: Roman Catholic; Ordained; celibate.

International Organizations: Superior General of the Glenmary Home Missioners.

Mailing address: PO Box 465618, Cincinnati OH 45246, USA. T. (1 513) 874 8900.

Rugambwa, Laurean, H E Card

Rel: Roman Catholic; Ordained on 12 DEC 1943; b. 12 JUL 1912 in Bukongo, Tanzania; celibate; Citizen of Tanzania.

International Organizations: Member (since 1960) of the College of Cardinals.

Member of the Congregation for the Causes of Saints.

Other organizations: Archbishop (since 1969), Dar-es-Salaam.

Career: Bishop, Rutabo from 1952 to 1960. Created Cardinal by Pope John XXIII in 1960. Bishop, Bukoba from 1960 to 1969. **Hons.:** Hon. LL.D., Notre Dame (1961), St Joseph's College, Philadelphia (1961), Rosary Hill College, Buffalo (1965); Hon. D.H.L., College of New Rochelle (1961); Hon. Dr. Civil and Canon Law, Georgetown University (1961); Giving of the Scroll, Catholic University of America (1961). **Mem.:** Knights of Columbus; Knights of St Peter Claver

Mailing address: Archbishop's House, PO Box 167, Dar es Salaam, Tanzania. T. (255 51) 22 031

Rügg, François

Rel: Roman Catholic.

International Organizations: Secretary-General of the International Catholic Child Bureau.

Mailing address: 65 Rue de Lausanne, CH-1202 Genève, Switzerland. T. (41 22) 731 32 48. Fax 731 77 93.

Ruhe, David S, Dr

Rel: Bahá'í.

International Organizations: Member, Universal House of Justice of the Bahá'í International Community.

Mailing address: PO Box 155, 31001 Haifa, Israel. T. (972 4) 372 433 – 510 344. C. BAHAIFAITH HAIFA. Tx 46626 BAYT IL. Fax (972 4) 372 425.

Ruhumuriza, Aaron, Bishop

Rel: Free Methodist; Ordained in 1971; b. in Rwamatamu, Kibuye, Rwanda; m. to Edith Kankindi; c. 3 boys and 1 girl; Citizen of Rwanda; Lang. Kinyarwanda, French, English, Kirundi and Swahili.

International Organizations: Bishop (since 1985) of Free Methodist Church in Rwanda.

Other organizations: Vice Legal Representative (since 1982), Protestant Council of Churches in Rwanda; Vice-Chairman (since 1984), Africa Evangelical Enterprise; Chairman of Sub-Committee, Francophone Africa Countries (since 1985), Scripture Union; Coordinating Committee Chairman (since 1989), Central Africa Protestant University Project.

Main field of work: Ministry. **Educ.:** B.D. from Union Biblical Seminary, Yeotmal, India (1964-1970); Doc. Stds. from Trinity Evangelical School, IL, USA (Beginning in 1986). **Interests:** pulpit preaching and pastoral counselling. **Hons.:** Hon. Doctorate, Greenville College, USA. **Mem.:** Bible Society of Rwanda, Scripture Union, Africa Evangelical Entreprise

Mailing address: BP 1668, Kigali, Rwanda. T. (250 7) 3047

Residence: Gikondo, Kigali, Rwanda. T. See Mailing Address

Ruiz, Alejandro, Bishop

Rel: Methodist.

International Organizations: Bishop of Methodist Church of Mexico.

Mailing address: Miravalle 209, Col Portales, México 03570 DF, Mexico.

Russell, Gary

Rel: Protestant.

International Organizations: Contact of the Transport for Christ International.

Mailing address: PO Box 303, Denver PA 17517, USA. T. (1 215) 267 2444.

Russell, Philip Welsford Richmond

Rel: Anglican; Ordained in 1951; b. 21 OCT 1919 in Durban, South Africa; m. to Violet Eirene Hogarth; c. 1 boy and 3 girls; Citizen of South Africa.

International Organizations: Ecclesiastic of the Anglican Church.

Career: Ordained Deacon in 1950. Curate, St Peter's Parish, Pietermaritzburg from 1950 to 1954. Incumbent, Parish of Greytown from 1954 to 1957. Incumbent, Parish of Ladysmith from 1957 to 1961. Incumbent, Parish of Kloof from 1961 to 1966. Archdeacon, Pinetown from 1962 to 1966. Bishop Suffragan, Diocese of Cape Town from 1966 to 1970. Bishop, Port Elizabeth from 1970 to 1974. Bishop, Natal from 1974 to 1981. Archbishop, Cape Town from 1981 to 1986. **Interests:** climbing, jogging. **Pubs.:** With Bishop Lawrence Zulu, Tools for the Job.

Mailing address: 400 Currie Road, Durban, Natal, South Africa.

Rutt, Cecil Richard, Rt Rev

Rel: Anglican; Ordained in 1952; b. 27 AUG 1925 in Langford, Beds, UK; m. to Joan Mary Ford; Citizen of United Kingdom.

International Organizations: Bishop (since 1979) of Leicester.

Other organizations: Chairman (since 1980), Advisory Council for Relations of Bishops and Religious Communities.

Career: With RNVR (Royal Naval Volunteer Reserve) from 1943 to 1946. Curate, St George's, Cambridge from 1951 to 1954. Church of England Mission to Korea from 1954 to 1974. Rector, St Michael's Seminary, Oryu Dong, Seoul from 1965 to 1966. Consecrated Bishop in 1966. Bishop, Taejon, Republic of Korea from 1968 to 1974. Bishop Suffragan, St Germans, Cornwall, and Honorary Canon of St Mary's Cathedral, Truro from 1974 to 1979. Member, Anglican/Orthodox Joint Doctrinal Discussions from 1983 to 1989. **Educ.:** M.A. from University of Cambridge, Pembroke College, UK. **Hons.:** Bard of the Gorseth of Cornwall (1976); Chaplain, Order of St John of Jerusalem (1978); Hon. D.Litt., Confucian University, Seoul (1974); Tasan Cultural Award for writings on Korea (1964); Order of Civil Merit, Peony Class, Korea (1974); Commander of the Order of the British Empire (1973). **Mem.:** Clubs: United Oxford and Cambridge **Pubs.:** The Bamboo Grove, 1971; Virtuous Women, 1974 and A History of Hand Knitting, 1987.

Mailing address: Bishop's Lodge, 10 Springfield Road, Leicester LE2 3BD, UK. T. (44 533) 708985

Ruygrok, E

Rel: Christian.

International Organizations: President of the Society for Interests of Active Missionaries in Asia, Africa and America.

Mailing address: 105 Gerard Brandtstraat, NL-2332 AK Leiden, Netherlands. T. (31 71) 760951 – 768989. Fax 317031 – 323454. Tx 39222 Siama NL.

Ryder of Warsaw, R, Lady

Rel: Roman Catholic; b. 03 JUL 1923 in Leeds, UK; widowed Group Captain Lord Cheshire of Woodhall; c. 1 boy and 1 girl; Citizen of United Kingdom; Lang. English, Polish and German.

International Organizations: Founder of the Sue Ryder Foundation.

Other organizations: Trustee, Ryder-Cheshire Mission for the Relief of Suffering; Trustee, Leonard Cheshire Foundation.

Hons.: Order of the British Empire (1957); Companion of the Order of St Michael and St George (1976); Life Peer – title: Baroness Ryder of Warsaw in Poland and of Cavendish in the County of Suffolk (1978); Order of Polonia Restituta (1965); Medal of Yugoslav Flag with Gold Wreath and Diploma (1971); Golden Order of Merit, Poland (1976); Order of Smile, Poland (1980); several honorary degrees from various British universities. **Pubs.:** *Child of my Love* (autobiography) and *Remembrance* (annual foundation magazine).

Mailing address: Sue Ryder Home, Cavendish, Sudbury C010 8AY, UK. T. (44 787) 280252

Residence: See Mailing Address. T. See Mailing Address

Sabattani, Aurelio, H E Card

Rel: Roman Catholic; Ordained on 26 JUL 1935; b. 18 OCT 1912 in Casal Fiumanese, Imola, Italy; celibate; Citizen of Italy.

International Organizations: Member (since 1983) of the College of Cardinals.

Member of the Sacred Congregation for the Oriental Churches. Member of the Pontifical Council for the Interpretation of Legislative Texts.

Other organizations: President (since 1983), Fabbrica di San Pietro; Vicar-General (since 1991), His Holiness for the Vatican City.

Career: Former Prelate, Loreto (until 1971). Former Prefect, Holy See's Supreme Tribunal. Created Cardinal by Pope John Paul II in 1983.

Mailing address: Palazzo del Tribunale, Piazza S Marta, 00120 Città del Vaticano, Vatican. T. (39 6) 698 46 15

Sabbah, Michel, His Beatitude

Rel: Roman Catholic; Ordained on 29 JUN 1955; b. 19 MAR 1933 in Nazareth, Israel; celibate; Citizen of Jordan; Lang. Arabic, Hebrew, Latin, French, English and Italian.

International Organizations: Latin Patriarch (since 1988) of Jerusalem.

President of the Conference of Latin Bishops in Arab Regions. Member of the Sacred Congregation for the Oriental Churches. Member of the Pontifical Council for Justice and Peace. Member of the Pontifical Council for Inter-Religious Dialogue. Member of the Synod of Bishops. Member of the Equestrian Order of the Holy Sepulchre of Jerusalem.

Main field of work: Ministry. **Career:** Director, Schools, Latin Patriarchate from 1955 to 1965. Curate, Misdar Parish, Amman in 1970. Rector, University of Bethlehem from 1980 to 1987. **Educ.:** Dr. Arabic Phil. from University of Paris IV (Paris-Sorbonne), France.

Mailing address: Latin Patriarchate, Jerusalem 14152, Israel. T. (972 2) 282323

Residence: See Mailing Address. T. (972 2) 283992

Sabir, Khalid

Rel: Muslim.

International Organizations: Secretary of the World Muslim Minorities Commission.

Mailing address: PO Box 420, Safat, Kuwait.

Sacks, Jonathan Henry, Rabbi Dr

Rel: Jewish; b. 08 MAR 1948; m. to Elaine Taylor; c. 1 boy and 2 girls; Citizen of United Kingdom.

International Organizations: Chief Rabbi (since 1990) of United Hebrew Congregations of the British Commonwealth of Nations.

Career: Lecturer in Moral Philosophy, Middlesex Polytechnic from 1971 to 1973. Lecturer in Jewish Philosophy, Jews' College, London from 1973 to 1976. Lecturer in Talmud and Jewish Philosophy, Jews' College from 1976 to 1982. Chief Rabbi Lord Jakobovits Professor (first incumbent) in Modern Jewish Thought, Jews' College in 1982. Director, Rabbinic Faculty, Jews' College from 1983 to 1990. Principal, Jews' College from 1984 to 1990. Sherman Lecturer, University of Manchester in 1989. Reith Lecturer, University of Manchester in 1990. Visiting Professor of Philosophy, University of Essex from 1989 to 1990. Rabbi, Golders Green Synagogue, London from 1978 to 1982. Rabbi, Marble Arch Synagogue, London from 1983 to 1990. Editor, *L'Eylah. A Journal of Judaism Today* from 1985 to 1990. **Educ.:** M.A. from University of Cambridge, Gonville and Caius College, UK; Ph.D. from University of Oxford, New College, UK. **Interests:** walking. **Mem.:** Careers Research and Advisory Council (CRAC) **Pubs.:** Tradition and Transition: Essays Presented to Chief Rabbi Sir Immanuel Jakobovits to Celebrate Twenty Years in Office, 1986; Traditional Alternatives, 1989 and Tradition in an Untraditional Age, 1990.

Mailing address: Adler House, Tavistock Square, London WC1H 9HN, UK. T. (44 71) 387 1066. Fax 383 4920

Sadiki, Hujatul Islam Mohamed Ewaz

Rel: Muslim.

International Organizations: Deputy Minister of the Ministry of Islamic Affairs, Afghanistan.

Mailing address: Ministry of Islamic Affairs and Pious Foundation Organisation, Kabul, Afghanistan.

Sahiouny, Salim, Rev Dr

Rel: Evangelical.

International Organizations: Executive Secretary of the National Evangelical Synod of Syria and Lebanon.

Mailing address: PO Box 110-235, Beirut, Lebanon.

Saier, Oskar, Most Rev Archbishop
Rel: Roman Catholic; Ordained on 02 JUN 1957; b. 12 AUG 1932 in Wagensteig, Germany; celibate; Citizen of Germany.
International Organizations: Archbishop (since 1978) of Freiburg im Breisgau.
Member (since 1984) of the Congregation for the Clergy.
Career: Curate in 1957. Assistant, Kanonist Institut, University of Munich in 1963. Bishop, Freiburg in 1972. Chairman, Pastoral Commission, German Conference of Bishops in 1979. Second Chairman, German Conference of Bishops in 1987. **Pubs.:** Communio in der Lehre des Zweiten Vatikanischen Konzils, 1973.
Mailing address: Herrenstrasse 35, 79098 Freiburg im Breisgau, Germany. T. (49 761) 34 545. Tx 7721854 EBOF D

Sainte-Beuve, Michel, Br
Rel: Roman Catholic; Ordained; celibate.
International Organizations: Prior-General of the Little Brothers of Jesus.
Mailing address: 97 Sudbourne Road, London SW2 5AF, UK. T. (44 71) 733 75 83

Saint-Pierre, Valère, C.F.S., Fr
Rel: Roman Catholic; Ordained; celibate.
International Organizations: Procurator-General of the Congregation of the Sacerdotal Fraternity.
Mailing address: 500 Ave Claremont, Montréal H3Y 2N5, Canada. T. (1 514) 483 2030.

Saint-Roch, Patrick
Rel: Roman Catholic.
International Organizations: Member of the Pontifical Institute of Christian Archeology.
Mailing address: Via Napoleone III 1, I-00185 Roma, Italy. T. (39 6) 4465 574. Fax 4469 197.

Sáinz de Baranda, Felipe S, O.C.D., Fr
Rel: Roman Catholic; Ordained; celibate.
International Organizations: Preposito Generale of the Order of Discalced Brothers of the Blessed Virgin Mary of Mount Carmel.
Other organizations: Grand Chancellor, Pontifical Faculty of Theology and Pontifical Institute of Spirituality 'Teresianum'.
Mailing address: Corso d'Italia 38, I-00198 Roma, Italy. T. (39 6) 841 65 78 – 884 09 58.

Sainz Muñoz, Faustino S, H E Mgr
Rel: Roman Catholic; Ordained on 19 DEC 1964; b. 05 JUN 1937 in Almadén, Spain; celibate.
International Organizations: Apostolic Pro-Nuncio to Cuba (since 1988) of the Roman Catholic Church.
Other organizations: Titular Archbishop, Novaliciana.
Mailing address: Calle 12 No 514, Miramar Marianao 13 (Habana), Cuba. T. (53 7) 29 5700 – 2 5296. Tx 512267 NUNAP CU

Saionji, Masami
Rel: New Religion.
International Organizations: Chairperson of the World Peace Prayer Society.
Mailing address: 800 Third Avenue, 31st Floor, New York NY 10022, USA. T. (1 212) 755 4755. Fax (1 212) 935 1389.

Sairin, Weinata, Rev
Rel: Christian.
International Organizations: General Secretary of the Pasundan Christian Church.
Mailing address: Jalan Pasirkaliki 121-123, PO Box 178, Bandung, West Java, Indonesia.

Sajid, Abduljalil, Imam Prof Dr
Rel: Muslim; Ordained on 26 MAR 1965; b. 01 NOV 1949 in Brighton, Sussex, UK; m. to Jamila; c. 2 boys and 3 girls; Citizen of United Kingdom; Lang. Arabic, English, Bengali and Urdu.
International Organizations: Imam (since 1976) of the Brighton Islamic Mission.
Other organizations: Director (since 1976), Sussex Muslim Society; Justice of the Peace, Brighton Bench.
Main field of work: Muslim family laws. **Educ.:** B.A. (Hons) from University of the Punjab, Lahore, Pakistan (1964); M.A. Arabic from University of the Punjab, Lahore, Pakistan (1966); M.A. Islamic Studies from University of the Punjab, Lahore, Pakistan (1967); M.A. Pol.Sc. from University of Dhaka, Bangladesh (1969); Ph.D. from Islamic University, Islamabad, Pakistan (1972). **Pubs.:** Muslim Family Laws, 1978; The Status of Women, 1980 and Muslims in Non-Muslim Society, 1985.
Mailing address: Brigthon Islamic Mission, 8 Caburn Road, Hove, Sussex BN3 6EF, UK. T. (44 273) 722 438 – 505 247
Residence: See Mailing Address. T. See Mailing Address

Sakala, Richard
Rel: Christian.
International Organizations: Director of the Africa Church Information Service.
Mailing address: PO Box 14205, Nairobi, Kenya. T. (254 2) 743644 – (254 2) 743773. C. CHURCHCON. Tx 22175.

Sakurai, Katsunoshin
Rel: Shinto; b. 28 OCT 1909; m. to Fumi; Citizen of Japan.
International Organizations: Chairman of the Kogakukan University.
Career: President, Shrine Head Office. Professor, Ashiya University.
Mailing address: 30-23 Sakuragicho, Ise, Mie 516, Japan. T. (81 596) 22 0201

Sakurai, Shuyu
Rel: Buddhist; b. 11 SEP 1916; Citizen of Japan.
International Organizations: Professor (Buddhism) of the Komazawa University.
Career: President, Komazawa University. **Educ.:** Tohoku University, Sendai, Japan.
Mailing address: 2-5-3 Kamata, Setagaya-ku, Tokyo 157, Japan. T. (81 3) 3700 1824

Salamate, E J, Rev
Rel: Evangelical.
International Organizations: Chairman of the Synod of the Evangelical Church of Sangir-Talaud.
Mailing address: Kantor Sinode GMIST, Tahuna, Sangir Talaud, Sulawesi Utara, Indonesia.

Salame, Joseph, H E Mgr
Rel: Maronite Catholic; Ordained in 1940; b. 24 DEC 1914 in Antelias, Lebanon; Citizen of Lebanon.
International Organizations: Bishop (since 1967) of Aleppo.
Career: General Vicar for the Diocese, Cyprus from 1940 to 1966. Dean, Theological League of Lebanon. Judge, Patriarchal Higher Court (Clerical Personal Status Tribunal) (for 7 years).
Mailing address: Maronite Bishopric of Aleppo, Syria.

Salaverri Aranegui, José María S, S.M., Fr
Rel: Roman Catholic; Ordained; celibate.
International Organizations: Superior General of the Society of Mary.
Other organizations: Member, Commission for the Catechism and the Catechists.
Mailing address: Via Latina 22, I-00179 Roma, Italy. T. (39 6) 757 58 92.

Salazar, Rubén, Rev
Rel: Roman Catholic; Ordained; celibate.
International Organizations: Zone Coordinator, 'Bolivariana' of the Caritas Internationalis.
Mailing address: Sec Nacional de Pastoral Social, Apartado 12309, Bogotá DE, Colombia.

Salazar López, José S, H E Card
Rel: Roman Catholic; Ordained on 26 MAY 1934; b. 12 JAN 1910 in Ameca, Guadalajara, Mexico; celibate; Citizen of Mexico.
International Organizations: Member (since 1973) of the College of Cardinals.
Other organizations: Archbishop Emeritus (since 1987), Guadalajara.
Career: Consecrated Titular Bishop, Prusiade in 1961. Transferred to Zamora in 1967. Archbishop, Guadalajara from 1970 to 1987. Created Cardinal by Pope Paul VI in 1973.
Mailing address: Garibaldi 770, 44290 Guadalajara, Jal, Mexico. T. (52 36) 26 39 60

Saliba, Philip
Rel: Antiochian Orthodox; b. in Lebanon.
International Organizations: Archbishop of New York.
Career: Came to the US in 1956. Formed a united Antiochian Diocese in 1975. Established the Antiochian Village in 1978.
Mailing address: No mailing address.

Salisse, John
Rel: Unknown.
International Organizations: Honorary Secretary of the Magic Circle.
Mailing address: 12 Hampstead Way, London NW11 7LS, UK. T. (44 81) 455 1569.

Salomon, S, Ms
Rel: Jewish.
International Organizations: Head, Montreal Office of the Consultative Council of Jewish Organizations.
Mailing address: 5711 Edgemore Avenue, Cote Saint Luc, Montréal H4W 1V8, Canada. T. 487 1243.

Salon, Mario, CC.RR.MM., Fr
Rel: Roman Catholic; Ordained; celibate.
International Organizations: Superior General of the Clerics Regular Minor.
Mailing address: Via Alpi Apuane 1, I-00141 Roma, Italy. T. (39 6) 890 055, (39 6) 896 721

Salter, Arthur Thomas John, Rev
Rel: Church of England; Ordained on 23 MAY 1961; b. 22 NOV 1934 in Tong-Norton, Shropshire, UK; single; Citizen of United Kingdom; Lang. English and French.
International Organizations: Chairman (since 1990) of the Anglican and Eastern Churches Association.
Other organizations: Vicar (since 1970), St Sila's with All Saints, Pentonville, London; Priest-in-charge (since 1979), St Dunstan-in-the-West with St Thomas of Canterbury within the Liberty of the Rolls, London.
Career: Served with the Intelligence Corps from 1954 to 1955. Served with the Royal Army Medical Corps from 1955 to 1956. Assistant Priest, St Peter's, Mount Park, Ealing from 1961 to 1965. Assistant Priest, St Stephen with St Thomas the Apostle, Shepherd's Bush from 1965 to 1966. Assistant Priest, St Alban the Martyr, Holborn with St Peter, Saffron Hill from 1966 to 1970. Priest-in-charge, St Clement, Barnsbury and St Michael the Archangel, Islington from 1970 to 1979. Chaplain, 36 Signal Regiment from 1975 to 1980. **Educ.:** AKC from University of London, King's College, UK (1957-1960); St Boniface's Theological College, Warminster, UK (1960-1961). **Interests:** Eastern churches, genealogy, reading. **Hons.:** Honorary Archimandrite's Cross of Byelo-Russian Autocephalic Orthodox Church-in-Exile (1979); Archpriest's Cross of Ethiopian Catholic Uniate Church (Eparchy of Asmara, Eritrea) (1980); Archpriest's Cross, Exarchate of Pope Shenouda III (Coptic Orthodox Patriarchate of Alexandria) (1981); Honorary Knight, Order of St Michael of the Wing (Royal House of Braganza, Portugal) (1984); Companion of Honour, Order of Orthodox Hospitallers (Cyprus) (1985); Territorial Decoration (1988). **Mem.:** Clubs: Army and Navy, City Livery, Polish Hearth **Pubs.:** Anglicans and the Holy See, 1967 and Should I Receive Auction?, 1968.
Mailing address: 87 Richmond Ave, London N1 0LX, UK. T. (44 71) 607 2865 – 405 1929
Residence: See Mailing Address. T. See Mailing Address

Salti, Yacoub
Rel: Christian.
International Organizations: Executive Officer of the Near East Council of Churches Committee for Refugee Work.
Mailing address: PO Box 1295, Amman, Jordan. T. (962 6) 236 58. Tx 23658 NECCRW JO.

Samaha, Basilios, H E Mgr
Rel: Greek Orthodox; Ordained; b. in Syria; Citizen of Syria.
International Organizations: Archbishop of Hauran and Mount Druze.
Mailing address: c/o Melkite-Greek Catholic Patriarchate, PO Box 22249, Damascus, Syria.

Sambi, Pietro, H E Mgr
Rel: Roman Catholic; Ordained on 14 MAR 1964; b. 27 JUN 1938 in Sogliano al Rubicone, Forli, Italy; celibate; Citizen of Vatican; Lang. Italian, French, English and Spanish.
International Organizations: Apostolic Pro-Nuncio to Indonesia of the Roman Catholic Church.
Other organizations: Titular Archbishop, Belcastro.
Career: Apostolic Nunciature in Cameroon from 1969 to 1971. Apostolic Delegation in Jerusalem from 1971 to 1975. Apostolic Nunciature in Cuba from 1975 to 1978. Apostolic Nunciature in Algeria from 1978 to 1979. Apostolic Nunciature in Nicaragua from 1979 to 1981. Apostolic Nunciature in Belgium from 1981 to 1983. Apostolic Nunciature in India from 1983 to 1985. Apostolic Pro-Nuncio to Burundi from 1985 to 1991. **Educ.:** Dr. Hist. from Lateran Pontifical University, Vatican City State (1965); Dr. Cn.L. from Lateran Pontifical University, Vatican City State (1969).
Mailing address: Apostolic Nunciature, Jl. Merdeka Timur No 18, PO Box 4227, Jakarta-Pusat, Indonesia. T. (62 21) 381 0736
Residence: See Mailing Address. T. (62 21) 384 1142

Samdhong Rinpoche, S, Prof
Rel: Buddhist; Ordained in 1946; b. 05 NOV 1939 in Nadgud Village, Kham Province, Tibet; single; Citizen of Tibet; Lang. Tibetan, Hindi, English and Sanskrit.
International Organizations: Director (since 1988) of the Central Institute of Higher Tibetan Studies, Sarnath.
Main field of work: Teaching. **Career:** Principal, Central Institute of Higher Tibetan Studies, Sarnath from 1971 to 1977. Principal/Director, Central Institute of Higher Tibetan Studies, Sarnath from 1977 to 1988. **Educ.:** Rabjampa Deg. in Buddhist Phil. from Drepung Monastery, Tibet (1951-1959); Geshe Lharampa Deg. from Drepung Monastery, India (1959-1968); Geshe Ngagrampa Deg. from Gyuto Tantric College, Dalhousie, India (1968-1970).
Mailing address: Central Institute of Higher Tibetan Studies, Sarnath, Varanasi 221 007, Uttar Pradesh, India. T. (91) 42881
Residence: T. (91) 45036

Samson, Pierre, P.M.E., Fr
Rel: Roman Catholic; Ordained; celibate.
International Organizations: Superior General of the Society of Foreign Missions of the Province of Quebec.
Mailing address: 180 Place Juge-Desnoyers, Pont-Viau, Laval PQ H7G 1A4, Canada. T. (1 514) 667 4190

Samuel, J V, Rev
Rel: Christian.
International Organizations: Secretary-General of the Christian Conference of Asia.
Mailing address: 2 Jordan Road, Kowloon, Hong Kong. T. (852 3) 722 1137 722 1214. Fax 721 6007. Tx 37618 URM HX. C. CHRISCONAS, Hong Kong.

Samuel, Prasana Kumari
Rel: Lutheran.
International Organizations: Chairperson, Program Committee for Theology and Studies of the Lutheran World Federation.
Mailing address: Gurukul Campus, 94 Purasawalkam High Road, Kilpauk, 600010 Madras, India. T. (91 44) 66 18 70 – 47 16 76 – 66 10 07. Tx 416613 NIEO IN. C. CHRISNIEO MADRAS-6

Samy, Ama, SJ, Fr
Rel: Roman Catholic; Ordained; celibate.
International Organizations: Member of the Society of Jesus.
Mailing address: Dhyana Ashram, 13 Madha Church Road, Madras 600 028, India.

Sánchez, Arturo, Bishop
Rel: Reformed Episcopal.
International Organizations: Bishop of Spanish Reformed Episcopal Church.
Mailing address: Beneficencia 18, E-28004 Madrid, Spain.

Sánchez, José T, Most Rev Archbishop Emer
Rel: Roman Catholic; Ordained on 12 MAY 1946; b. 17 MAR 1920 in Pandan, Philippines; celibate.
International Organizations: Secretary of the Congregation for the Evangelization of Peoples.
Member of the Pontifical Committee for International Eucharistic Congresses. Member of the Pontifical Council for Pastoral Assistance to Health Care Workers. Member of the Pontifical Council for the Pastoral Care of Migrants and Itinerant People. Advisor of the Pontifical Council for Promoting Christian Unity. Advisor of the Pontifical Council for the Laity.
Other organizations: Archbishop Emeritus, Nueva Segovia; Vice Chancellor, Pontifical University Urbaniana; President, Commissione di Theologica, Spititualità e Animazione Missionaria; President, Commissione Pastorale per lo Studio; President, Commissione per la Catechesi e per i Catechisti.
Mailing address: Palazzo di Propaganda Fide, Piazza di Spagna 48, I-00187 Roma, Italy. T. (39 6) 679 6941. Fax 679 3906
Residence: See Mailing Address. T. (39 6) 679 4796 – 679 6941

Sanders, Hadijah
Rel: Subud.
International Organizations: Director of the Subud International Cultural Association.
Mailing address: 13701 Bel-Red Road, Suite B, Bellevue WA 98005, USA. T. (1 206) 643 1904.

Sándor, Cserháti, Dr
Rel: Lutheran; Ordained on 29 JUN 1953; b. 21 JUN 1930 in Mohács, Hungary; m. to Judit Gellén; c. 1 boy and 1 girl; Citizen of Hungary; Lang. Hungarian and German.
International Organizations: Professor of New Testament (since 1983) of the Lutheran Theological Faculty, Budapest.
Other organizations: President (since 1987), Committee for Bible Translation in Hungary.
Main field of work: University education. **Career:** Pastor, Lutheran Church in Hungary from 1953 to 1983. **Educ.:** Lutheran Pastor from Lutheran Theological Faculty, Budapest, Hungary (1948-1953); Dr. Theol. from Lutheran Theological Faculty, Budapest, Hungary (1973). **Interests:** Pauline theology. **Mem.:** Society for the Study of the New Testament **Pubs.:** Commentaries on Paul's Letter: to the Philippians, 1975; to the Colossians and to Philemon, 1979 and to the Galatians, 1982.
Mailing address: Rózsavölgyi-Köz 3, H-1141 Budapest, Hungary. T. (36 1) 163 64 51
Residence: Damjanich-U 28/b, H-1071 Budapest, Hungary. T. (36 1) 142 68 15

Sándor, Kostya
Rel: Unknown.
International Organizations: Secretary of the METEM-International Society of Toronto for Hungarian Church History.
Mailing address: Regis College, 15 St Mary Street, Toronto ON M4Y 2R5, Canada. T. (1 416) 922 2476. Fax 922 2898.

Sangharakshita, S, Ven
Rel: Buddhist.
International Organizations: Head and Founder of the Friends of the Western Buddhist Order.
Mailing address: Padmaloka, Lesingham House, Surlingham, Norwich NR14 7AL, UK. T. (44 50) 88 310. Tx 295141 TxLINK G/105088639 (ATT MBX).

Sanharib Bashir Iwas, S
See Ignatius Zakka I Iwas.

Sanjo, Saneharu
Rel: Shinto; Ordained; Citizen of Japan.
International Organizations: Chief Priest of the Heian Shrine.
Career: Chief Priest, Kasuga Shrine. **Educ.:** University of Tokyo, Japan.
Mailing address: 251-19 Iwakura Kinocho, Sakyo-ku, Kyoto 601-11, Japan. T. (81 75) 741 2788

Sansbury, Cyril Kenneth, Rt Rev
Rel: Anglican; Ordained in 1929; b. 21 JAN 1905; m. to Ada Ethelreda Mary Wamsley; c. 1 boy and 2 girls; Citizen of United Kingdom.
International Organizations: Ecclesiastic (retired) of the Anglican Church.
Career: Ordained Deacon in 1928. Curate, Diocese of Southwark from 1928 to 1932. Missionary in Japan from 1932 to 1941. Professor, Central Theological College, Tokyo, and Chaplain of St Andrew's Church, Tokyo from 1934 to 1941. Chaplain, Royal Canadian Air Force from 1941 to 1945. Warden, Lincoln Theological College from 1945 to 1952. Canon and Prebendary, Lincoln Cathedral from 1948 to 1953. Warden, St Augustine's College, Canterbury from 1952 to 1961. Honorary Canon, Canterbury from 1953 to 1961. Bishop, Singapore and Malaya from 1961 to 1966. Secretary-General, British Council of Churches from 1966 to 1973. Assistant Bishop, Diocese of London from 1966 to 1973. Priest-in-charge, St Mary in the Marsh, Norwich from 1973 to 1984.
Mailing address: 67C The Close, Norwich NR1 4DD, UK. T. (44 603) 618808

Sanson, Henri, S.J., Fr
Rel: Roman Catholic; Ordained; celibate.
International Organizations: Member of the Society of Jesus.
Mailing address: 1 impasse Ahmed Kara, Hydra, Algiers, Algeria.

Santer, Mark, Rt Rev
Rel: Anglican; Ordained on 22 SEP 1963; b. 29 DEC 1936 in Bristol, UK; m. to Henriette Cornelia Weststrate; c. 1 boy and 2 girls; Citizen of United Kingdom; Lang. English, Dutch and German.
International Organizations: Anglican Co-Chairman (since 1983) of the Anglican-Roman Catholic International Commission.
Other organizations: Bishop (since 1987), Birmingham.
Main field of work: Episcopal work. **Career:** Assistant Curate, All Saints Cuddesdon, Oxford from 1963 to 1967. Tutor, Cuddesdon College, Oxford from 1963 to 1967. Dean and Fellow, Clare College, Cambridge from 1967 to 1972. Principal, Westcott House, Cambridge from 1973 to 1981. Bishop, Kensington from 1981 to 1987. **Educ.:** M.A. from University of Cambridge, Queens' College, UK (1957-1961); Wescott House, Cambridge, UK (1961-1963). **Hons.:** Honorary Fellow, Clare College and Queen's College, Cambridge. **Pubs.:** With M F Wiles, Documents in Early Christian Thought, 1975 and Their Lord and Ours, 1982.
Mailing address: Bishop's Croft, Old Church Road, Harborne, Birmingham B17 OBG, UK. T. (44 21) 427 1163
Residence: See Mailing Address. T. See Mailing Address

Santram, Pritam B, Rev
Rel: Church of North India.
International Organizations: General Secretary of the Church of North India.
Mailing address: Wesley Lodge, 16 Pandit Pant Marg, New Delhi 110 001, India.

Sarasvati, Divyanand, Dr
Rel: Hindu; b. 01 JUL 1939 in Sirsa Ganj, Uttar Pradesh, India; celibate; Citizen of India; Lang. Hindi and Sanskrit.
International Organizations: Director (since 1976) of the Patanjal Yog Dham.
Other organizations: Director of Research, Education and Training (since 1983), Yoga Institute of Research in Education and Advanced Studies.

Main field of work: University education. **Career:** Visiting Professor at Gurukula Kangri University, Hardwar from 1976 to 1982. Visiting Professor at International Preahus College, Tankara from 1983 to 1984. Visiting Professor at Gurukula Arya Nagar, Hissar from 1985 to 1986. Visiting Professor at Gurukula Jhajjar, Rohtak from 1991 to 1992. Visiting Professor at Ramjas Foundation Lajpat Nagar, Delhi in 1992. **Educ.:** M.A. Vedas from Gurukula Kangri Vishwavidyalaya, Hardwar, India (1972-1974); M.A. Phil. from Gurukula Kangri Vishwavidyalaya, Hardwar, India (1974-1976); Ph.D. from Gurukula Kangri Vishwavidyalaya, Hardwar, India (1976-1980). **Interests:** Yoga. **Hons.:** Awarded Merit Certificate and Medal, World Conference of Yoga by President of India, New Delhi (1985); Awarded Shield, All India Yoga Conference at Swai Man Singh College, Jaipur (1991). **Mem.:** Life Member: Yog Dham; Atam Smmddhi Ashram; Gyan Uday Yoga Centre, Sirsaganj, Firozabad UP **Pubs.:** Vedon main Yog Vidya and Yog Darpan.
Mailing address: Yoga Dham, Arya Nagar, Jwalapur 249 407, Uttar Pradesh, India.

Saraswati, Kumari Shiva, Shiva Saraswati

Rel: Hindu; b. 12 MAR 1967 in Kota, Rajasthan, India; single; Citizen of India; Lang. Hindi, Sanskrit and English.
International Organizations: Secretary-General (since 1988) of the Matra Mandal, Moji Baba Kee Gufa, Kota.
Other organizations: Organizer (since 1982), Shri Shiv Joti Bal Vidhyani Ketan Talwandi Kota; Central Board Member (since 1990), Vishwa Hindu Parishad.
Educ.: B.A. from University of Rajasthan, Jaipur, India (1986); M.A. Sanskrit from University of Rajasthan, Jaipur, India (1988). **Pubs.:** Articles in *Hindu Chetna*(magazine), *Dainik Jagrana* (newspaper), *AAJ Newspaper*, etc.
Mailing address: Sabarmati Colony, House No Chh-4, Kota 324006, India.

Sardella, Giovanna Lia, Sr

Rel: Roman Catholic; Ordained; celibate.
International Organizations: Superior General of the Soeurs Maîtresses de Sainte Dorothée, 1838.
Mailing address: Via Raffaele Conforti 25, I-00166 Roma, Italy. T. (39 6) 662 40 41.

Sardjonan, S, Rev

Rel: Christian.
International Organizations: Moderator of the East Java Christian Church.
Mailing address: Jln S Supriadi 18, Malang, East Java, Indonesia.

Sarei, Alexis Holyweek, Sir

Rel: Roman Catholic; Ordained; b. 25 MAR 1934 in Buka, Papua New Guinea; m. to Claire Dionne; c. 3 boys and 3 girls; Citizen of Papua New Guinea.
International Organizations: Diplomatist and Ecclesiastic of the Roman Catholic Church.
Career: Priest from 1966 to 1971. Private Secretary to Chief Minister from 1972 to 1973. District Commander from 1973 to 1975. Warden, University of Papua New Guinea in 1972. Premier, North Solomons Province Government from 1976 to 1980. High Commander in UK (also Ambassador to other European and Middle Eastern countries) from 1980 to 1983. Permanent Representative to UN and Ambassador to USA from 1983 to 1986. **Interests:** music, all sports, writing letters, swimming, hunting. **Hons.:** Papua New Guinea Independence Medal. **Pubs.:** The Practice of Marriage Among the Solos, 1974.
Mailing address: c/o Ministry of Foreign Affairs, Port Moresby, Papua New Guinea.

Sarkis, Paul Antoine, S.J., Fr

Rel: Roman Catholic; Ordained in 1964; b. 18 SEP 1933 in Cairo, Egypt; celibate.
International Organizations: Rector (since 1973) of the Collège de la Sainte Famille, Cairo.
Member of the Society of Jesus.
Career: Invigilator and teacher at Collège de la Sainte Famille from 1957 to 1961. Vice-Principal, Collège de la Sainte Famille from 1965 to 1970. Vice-Principal, Collège de la Sainte Famille from 1972 to 1973.
Mailing address: Collège de la Sainte Famille, Cairo, Egypt.

Sarma, B Balasunthara, Sri

Rel: Hindu.
International Organizations: Member of the Sri Badrakali Amman Kovil.
Mailing address: Sri Badrakali Amman Kovil, Mayura Place, Colombo 6, Sri Lanka.

Sarma, B Shanmugartna, Brahmasri

Rel: Hindu.
International Organizations: Chief Priest of the Sri Bala Selva Vinayagar Kovil.
Mailing address: Sri Bala Selva Vinayagar Kovil, Captain's Garden, Maradana, Sri Lanka.

Sartorius, Willi, Rev

Rel: Evangelical Reformed; Ordained on 26 JUN 1952; b. 12 AUG 1928 in Zürich, Switzerland; m. to Ursula Blankart; c. 2 boys and 3 girls; Citizen of Switzerland; Lang. German, Spanish, English, French and Italian.
International Organizations: President (since 1987) of the European Evangelical Alliance.
Other organizations: Member of the Board (since 1958), the Swiss Indian Mission; Member of the Board, the Swiss Evangelical Alliance.
Career: Youth Pastor in Germany from 1952 to 1953. Pastor, Swiss Reformed Church in Präz from 1953 to 1957. Pastor, Swiss Reformed Church in Arosa from 1957 to 1962. Director/President, Swiss Indian Mission from 1958 to 1967. Pastor, Swiss Reformed Church in Aaurau from 1967 to 1978. Pastor, Swiss Reformed Church in Basel from 1978 to 1990. **Educ.:** M.Th. from University of Basel, Switzerland (1948-1952). **Interests:** politics, social and missiological concerns. **Mem.:** Member of a Swiss regional parliament; Member of the Board of various missions **Pubs.:** Mission als Zeichen der Zeit and Der Absolutheitsanspruch der biblischen Offenbarung gegenüber den Weltreligionen.
Mailing address: Höfjiweg 27, CH-7250 Klosters, Switzerland. T. (41 81) 69 52 27.
Residence: See Mailing Address. T. See Mailing Address

Sarwar Manzur, Alhaj Gholam S

Rel: Muslim; Citizen of Afghanistan.
International Organizations: General Director of the High Council of Ulema and Clergy of Afghanistan.
Mailing address: No mailing address.

Sasingen, O, Rev

Rel: Evangelical.
International Organizations: General Secretary of the Evangelical Christian Church in Halmahera.
Mailing address: Synod Office, Tobelo, Halmahera, Maluku Utara, Indonesia.

Satini, Louisie, Dr

Rel: Roman Catholic; Ordained on 15 AUG 1952; b. 30 JAN 1928 in Klaten, Central Java, Indonesia; celibate; Citizen of Indonesia; Lang. Indonesian, English and Dutch.
International Organizations: Superior General (since 1988) of the Sisters of Charity of St Charles Borromeo.
Educ.: Dr. Hist. (1975).
Mailing address: Postbus 206, NL-6200 AE Maastricht, Netherlands. T. (31 43) 21 92 41.
Residence: St Servaas Klooster 14, NL-6211 TE Maastricht, Netherlands. T. See Mailing Address

Satowaki, Joseph Asajiro, H E Card

Rel: Roman Catholic; Ordained on 17 DEC 1932; b. 01 FEB 1904 in Shittsu, Nagasaki, Japan; celibate; Citizen of Japan.
International Organizations: Member (since 1979) of the College of Cardinals.
Other organizations: Archbishop Emeritus (since 1990), Nagasaki.
Career: Bishop, Kagoshima from 1955 to 1968. Archbishop, Nagasaki from 1968 to 1990. Created Cardinal by Pope John Paul II in 1979.
Educ.: Nagasaki Theological School, Japan; Urbanian Pontifical University, Vatican City State; Catholic University of America, Washington DC, USA.
Mailing address: Archbishop's House, 5-3 Minami Yamatema-cho, Nagasaki 850, Japan. T. (81 958) 23 2934

Satyapati Parivrajak, S, Swami

Rel: Hindu: Arya-Samaji; Ordained in 1947; single; Citizen of India; Lang. Sanskrit and Hindi.
International Organizations: President of the Vishaw Kalyan Dharmarth Nyas.
Other organizations: Vice-Chancellor, Darshan Yoga Maha Vidhyalaya.
Mailing address: Arya Van Vikas Shetra, Rojar, PO Sangpur 383 307, Sabar Kantha, Gujrat, India.

Saukkonen, (Veikko Antti) Juhani

Rel: Christian; Ordained; b. 22 JUN 1937 in Viipuri, Finland; m. to Taimi Tuulikki Tervonen; Citizen of Finland.
International Organizations: Senior Clergyman (since 1977) of the Parikkala Parish.
Other organizations: Member (since 1972), Parliament.
Career: Curate, Parikkala Parish from 1966 to 1977. Minister, Defence from 1982 to 1983. **Mem.:** Centre Party
Mailing address: c/o Ministry of Defence, Et Makasiinikatu 8, SF-00130 Helsinki 13, Finland. T. (358 0) 625 801

Saukkonen, Paula

Rel: Christian.
International Organizations: Regional Representative for Europe of the World Student Christian Federation.
Mailing address: Prins Hendriklaan 37, NL-1075 BA Amsterdam, Netherlands. T. (31 20) 717 7654. Fax 675 57336.

Saurma, Adalbert

Rel: Unknown; b. 13 MAY 1939; Citizen of Switzerland.
International Organizations: Member of the Council of the International Society for the Sociology of Religions.
Other organizations: Assistant Lecturer, ETS Basel, University of Zurich.
Educ.: Dr. phil. hist. **Pubs.:** "Das Politische, Heilige, Identische und der 1 August", in *Basler Magazin*, 31/87 and "Die Endinger Stiftung 'La Colonia Tovar'", in Endingen a. K., *Die Geschichte der Stadt*, Endingen, 1988, pp. 538-543.
Mailing address: Schützenmattstrasse 63, CH-4051 Basel, Switzerland. T. (41 61) 22 98 09

Savarimuthu, John Gurubatham, Rt Rev Tan Sri

Rel: Anglican; Ordained on 08 JUN 1952; b. 29 NOV 1925 in Kulasekharapatnam, India; widowed Puan Sri Catherine Edith Yesadian; c. 2 boys and 1 girl; Citizen of Malaysia.
International Organizations: Bishop (since 1973) of West Malaysia.
Other organizations: Vice-President (since 1978), Assunta Hospital Foundation Board of the Roman Catholic Church; Member (since 1980), National Unity Board under the Prime Minister's Department; Episcopal Chairman (Primate) (since 1987), Council of the Church of East Asia (CCEA).
Career: Ordained Deacon in 1951. Assistant Curate, St Mark's Church, Seremban from 1952 to 1955. Vicar, Negri Sembilan, St Mark's Church, Seremban from 1956 to 1962. Curate, St Mary's Church, Portsmouth, UK in 1963. Curate, St George's Cathedral, Kingston, Ontario, Canada in 1963. Vicar, South Johor, St Christopher's Church, Johor Bahru from 1964 to 1970. Examining Chaplain to the Bishop, Singapore and Malaya from 1964 to 1970. Visiting Lecturer, St Peter's Hall, Trinity College, Singapore from 1964 to 1970. Promoted as Archdeacon, Southern Archdeaconry of the Diocese of West Malaysia in 1970. Vicar, St James' Church, Kuala Lumpur, Archdeacon of Southern and Central Archdeaconries in 1971. Director of Theological Training, Diocese of West Malaysia from 1971 to 1972. Appointed as full-time Archdeacon, three Archdeaconries in the Diocese of West Malaysia in 1972. Appointed as Vicar-General, Diocese of West Malaysia in 1972. Consecrated Bishop in 1973. National Secretary, Church Union Negotiating Committee of Malaysia and Singapore from 1970 to 1973. National Vice-President, Council of Churches of Malaysia and Singapore from 1973 to 1975. National President, Council of Churches of Malaysia (CCM) from 1979 to 1983. Provincial Secretary, Council of the Church of East Asia (CCEA) from 1979 to 1983. Chairman, CCEA Theological Commission from 1979 to 1987. Chairman, CCEA from 1987 to 1991. **Educ.:** University of Madras, Hindu College, India (1943-1947); B.D. from Serampore College, India (1948-1952); Dip. Theol. from St Augustine's College, Canterbury, UK (1963). **Interests:** reading, swimming, walking. **Hons.:** Pangalima Setia Mahota (1980). **Mem.:** Anglican-Lutheran International Commission (1986); Anglican International Network on Refugee and Migrant Ministry (1988)
Mailing address: Bishop's House, 14 Pesiaran Stonor, 50450 Kuala Lumpur, Malaysia. T. (60 3) 2427303 – 2417116. Fax 2416460
Residence: See Mailing Address. T. (60 3) 2421850

Sawatsky, Rodney

Rel: Mennonite.
International Organizations: President of the Conrad Grebel College.
Mailing address: Westmount Road N, Waterloo ON N21 3G6, Canada.

Sawatsky, Walter

Rel: Mennonite.
International Organizations: Professor of Church History of the Associated Mennonite Biblical Seminaries.
Mailing address: 3003 Benham Ave, Elkhart IN 46517, USA.

Sawen, P, Rev

Rel: Evangelical.
International Organizations: President of the Evangelical Christian Church in West Irian.
Mailing address: Jln Argapura, PO Box 14, Jayapura, Irian Jaya, Indonesia.

Saxena, Kabir

Rel: Buddhist: Mahayana.
International Organizations: Member of the Foundation for the Preservation of the Mahayana Tradition.
Other organizations: Director, Root Institute.
Mailing address: Bodhgaya (Gaya Dist), Bihar 824231, India. T. (91 631) 81214

Saxer, Victor, Mgr

Rel: Roman Catholic; Ordained; celibate.
International Organizations: Rector (since 1982) of the Pontifical Institute of Christian Archeology.
Other organizations: Professor Emeritus (since 1989), Pontifical Institute of Christian Archeology; President, Pontifical Committee of Historical Science; Representative, Holy See to International Committee of Historical Sciences (ICHS); Censor, Pontifical Roman Academy of Archeology.
Main field of work: History of Christian worship. **Career:** Professor, Pontifical Institute of Christian Archeology from 1966 to 1989. Rector, Pontifical Institute of Christian Archeology from 1970 to 1973. **Hons.:** Knight of the National Order of Merit (France), Honorary Prelate of His Holiness. **Mem.:** Pontifical Roman Academy of Archeology **Pubs.:** Le culte de Marie Madeleine en Occident, Paris, 1959; Vie quotidienne et liturgique à Carthage au IIIè siècle, Rome, 1969 and Les rites de l'initiation chrétienne (IIè – VIè siècle), Spoleto, 1988.
Mailing address: Via Napoleone III 1, I-00185 Roma, Italy. T. (39 6) 4465 574. Fax 4469 197.

Say, Richard David, Rt Rev Dr

Rel: Church of England; Ordained on 20 DEC 1940; b. 04 OCT 1914 in London, UK; m. to Irene Frances Rayner; c. 1 boy and 2 girls; Citizen of United Kingdom; Lang. English.
International Organizations: Assistant Bishop (since 1988) of Canterbury.
Other organizations: Sub-Prelate (since 1961), Order of St John; Honorary Chaplain (since 1968), Pilgrims of Great Britain; Pro-Chancellor (since 1983), University of Kent.
Main field of work: Ministry. **Career:** General Secretary, Church of England Youth Council from 1944 to 1947. General Secretary, British Council of Churches from 1947 to 1955. Rector, Hatfield, Diocese of St Albans from 1955 to 1961. Honorary Canon, St Albans Abbey from 1957 to 1961. Bishop, Rochester from 1961 to 1988. Member, House of Lords from 1969 to 1988. Deputy Pro-Chancellor, University of Kent from 1977 to 1983. Chairman, Age Concern England from 1986 to 1989. **Educ.:** M.A. from University of Cambridge, Christ's College, UK (1934-1938); Ridley Hall, Cambridge, UK (1938-1939). **Interests:** Ecumenical movement, university education, history, travel. **Hons.:** Lord High Almoner to H M The Queen (1970-88); Hon. D.C.L., Kent (1987); Knight Commander of the Royal Victorian Order (1988); Freeman, City of London (1953); Hon. Freeman, Borough of Tonbridge and Malling (1987), Borough of Rochester Upon Medway (1988). **Mem.:** Institute of Royal Engineers (Honorary Member, 1986); Friends of Kent Churches (President, 1988); Clubs: United Oxford and Cambridge
Mailing address: 23 Chequers Park, Wye, Ashford TN25 5BB, UK. T. (44 233) 812720
Residence: See Mailing Address. T. See Mailing Address

Sayegh, Selim, H E Mgr

Rel: Roman Catholic; Ordained on 28 JUL 1959; b. 15 JAN 1935 in Ermenim; celibate; Citizen of Jordan.

International Organizations: Bishop and Vicar-General (since 1982) of Latin Patriarchate in Jordan.

Other organizations: Titular Bishop, Acque di Proconsolare.

Career: President, Latin Patriarchal Court in Jerusalem from 1967 to 1979. Rector, Latin Patriarchal Seminary from 1976 to 1981. **Interests:** table tennis, volley ball, chess, history, ecclesiastical law. **Hons.:** Commander, Equestrian Order of the Holy Sepulchre of Jerusalem. **Pubs.:** Le Statu Quo des Lieux Saints, 1971 and The Christian Family's Guidebook.

Mailing address: Latin Vicariate, Jebel Lweibdeh, PO Box 1317, Amman, Jordan. T. (962 6) 637 440

Sbarbaro, Eugenio, H E Mgr

Rel: Roman Catholic; Ordained on 11 JUN 1960; b. 26 JUN 1934 in Belpiano; celibate.

International Organizations: Apostolic Pro-Nuncio to Zambia and Malawi (since 1985) of the Roman Catholic Church.

Other organizations: Titular Archbishop, Tiddi.

Mailing address: Hussein Saddam Boulevard, PO Box 31445, Lusaka, Zambia. T. (260 1) 250 786 – 251 033 – 253 466. Tx 40403 ANL ZA

Scannell, Anthony, Rev

Rel: Roman Catholic; Ordained; celibate.

International Organizations: Contact of the Franciscan Communications Center.

Mailing address: 1229 S Santee Street, Los Angeles CA 90015, USA.

Schadl, Claudia

Rel: Buddhist: Tibetan.

International Organizations: Representative of the Tibetan Monastic Institute, Rikon.

Mailing address: Wildbergstrasse, CH-8486 Rikon ZH, Switzerland. T. (41 52) 35 17 29 – 35 26 98.

Schaeffer, Paul, Fr

Rel: Roman Catholic; Ordained; celibate.

International Organizations: Secretary-General of the Commission of the Episcopacies of the European Community.

Mailing address: Ave Père Damien 13, B-1150 Bruxelles, Belgium. T. (32 2) 771 36 78. Fax (32 2) 771 36 78.

Schaer, Edith

Rel: Yoga.

International Organizations: Contact of the Corona Mundi.

Mailing address: Rue de Lyon 61, CH-1203 Genève, Switzerland. T. (41 22) 744 66 53.

Schaller, Stefan, Rev

Rel: Evangelical Lutheran.

International Organizations: President of the Evangelical Lutheran Church in Chile.

Mailing address: Avenida Ricardo Lyon 1483, Casilla 15167, Santiago, Chile.

Schalück, Herman, Fr

Rel: Roman Catholic; Ordained; celibate.

International Organizations: Grand Chancellor of the Pontificio Ateneo Antonianum.

Mailing address: Via Merulana 124, I-00185 Roma, Italy. T. (39 6) 70 37 35 02.

Scharf, Kurt Franz Wilhelm

Rel: Protestant; Ordained; b. 21 OCT 1902 in Landsberg/Warthe, Germany; m. to (1st) Ingeborg Sommerwerck (died), (2nd) Renate; c. 1 boy and 4 girls; Citizen of Germany.

International Organizations: Ecclesiastic (retired) of the Protestant Church in Germany.

Career: Pastor, Friesack in 1928. Pastor, Sachsenhausen in 1933. Präses, Brotherhood of Brandenburg in 1933. Chairman, Conference of Brotherhoods of Confessional Churches in Germany in 1938. Several times arrested and forbidden to preach and publish by Nazi regime. Präses, Brandenburg Confessional Synod in 1935. Provost, Berlin-Brandenburg Province in 1945. Chairman, United Evangelical Church in 1955. Chairman, Protestant Church in Germany from 1961 to 1967. Deputy Chairman, Protestant Church in Germany from 1967 to 1979. Bishop, Berlin-Brandenburg (Protestant Church) from 1966 to 1976. Vice-President, United World Bible Society from 1966 to 1969. Member, Central Committee, World Council of Churches from 1968 to 1975. Editor, *Rundbriefe der Bekennenden Kirche* (illegal 1935-45) from 1933 to 1945. **Interests:** chess, poetry. **Hons.:** Hon. Th.D., Humboldt University, Berlin; Hon. D.D., Eden Seminary, Webster Groves MA, Christliche Akademie, Warsaw; Buber-Rozenzweig Medal; Jüdisch-Christlicher Koordinierungsrat; Kopernikusmedaille; Ernst-Reuter-Medaille der Stadt Berlin West; Gustav-Heinemann Preis; Dr Leopold-Lucas Preis der Universität Tübingen; Goldenes Athos-Kreuz des Ökumenischen Patriarchates, Istanbul; Augustin Cross of Archbishop of Canterbury. **Pubs.:** Brücken und Breschen, 1977; Streit mit der Macht, 1983 and Widerstehen und Versohnen, 1987.

Mailing address: Am Hirschsprung 35, 14195 Berlin, Germany. T. (49 30) 831 65 00

Scheele, Paul-Werner, H E Mgr

Rel: Roman Catholic; Ordained on 29 MAR 1952; b. 06 APR 1928 in Olpe, Germany; celibate.

International Organizations: Co-Chairperson of the Lutheran-Roman Catholic Joint Commission.

Member of the Pontifical Council for Promoting Christian Unity.

Other organizations: Bishop (since 1979), Würzburg.

Mailing address: c/o SPCU, I-00120 Città del Vaticano, Vatican.

Scheidegger, Maddalena

Rel: Roman Catholic.

International Organizations: Secretary of the International Catholic Society for Girls.

Mailing address: 37-39 rue de Vermont, CP 22, CH-1211 Genève 20, Switzerland. T. (41 22) 734 96 85. C. ACISJF.

Schelhoth, George, H E Mgr

Rel: Syrian Catholic; Ordained in 1945; b. in Aleppo, Syria; Citizen of Syria.

International Organizations: Archbishop (since 1971) of Damascus.

Career: Secretary to Archbishop, Aleppo. Director, St Joseph School, Aleppo. Priest in Gerablos then in Aleppo (St Ephrem Church). Founder, YCW in Syria and many other movements. Founder, Secular Institute of Jesus the Worker. **Pubs.:** Author of many books in Arabic.

Mailing address: Foyer Jésus Ouvrier, Quartier al-Chahba, Aleppo, Syria.

Schembri, Salvatore, M.S.P., Rev

Rel: Roman Catholic; Ordained on 22 MAR 1947; b. 23 MAY 1922 in Siculiana, Agrigento, Italy; celibate; Citizen of Italy; Lang. Spanish and French.

International Organizations: Procurator-General of the Missionnaires Serviteurs des Pauvres.

Main field of work: Ministry. **Interests:** Research in history. **Pubs.:** P Salvatore Gambino – Una vita per il Vangelo e per i Poveri, La Cartà, Parlerme; I Missionari Servi dei Poveri – 100 anni tra cronaca e storia, La Carità, Palerme and Missionari Italiani in Messico, Déhoniani, Rome.

Mailing address: Via della Pisana 95, I-00163 Roma, Italy. T. (39 6) 622 37 62

Residence: See Mailing Address. T. See Mailing Address

Scherer, Alfredo Vicente, H E Card

Rel: Roman Catholic; Ordained on 03 APR 1926; b. 05 FEB 1903 in Bom Princípio, Rio Grande do Sul, Brazil; celibate; Citizen of Brazil.

International Organizations: Member (since 1969) of the College of Cardinals.

Other organizations: Archbishop Emeritus (since 1981), Porto Alegre.

Career: Private Secretary to Archbishop, Porto Alegre from 1927 to 1933. Organizer, Parishes of Tapes and Barra do Ribeiro from 1933 to 1935. Parish Priest, Sao Geraldo, Porto Alegre from 1935 to 1946. Auxiliary Bishop, Porto Alegre in 1946. Archbishop, Porto Alegre from 1946 to 1981. Created Cardinal by Pope Paul VI in 1969.

Mailing address: CP 546, Rua da Gruta 145, 91700 Porto Alegre RS, Brazil. T. (55 512) 36 6100

Schindler, Alexander Moshe, Rabbi

Rel: Jewish; Ordained in 1953; b. 04 OCT 1925 in Munich, Germany; m. to Rhea Rosenblum; c. 2 boys and 3 girls; Citizen of USA; Lang. English, German, Hebrew and Yiddish.

International Organizations: President (since 1973) of the Union of American Hebrew Congregations.

Vice-President of the World Jewish Congress. Vice-President of the Memorial Foundation for Jewish Culture. Member, Governing Board of the Jewish Agency for Israel.

Career: Came to the US in 1937. Assistant Rabbi, Temple Emmanuel, Worcester MA from 1953 to 1956. Associate Rabbi, Temple Emmanuel, Worcester MA from 1956 to 1959. Director, New England Council, Union of American Hebrew Organizations from 1959 to 1963. National Director of Education, Union of American Hebrew Organizations from 1963 to 1967. Vice-President, Union of American Hebrew Organizations from 1967 to 1972. Chairman, Conference of Presidents of Major American Jewish Organizations from 1976 to 1978. **Educ.:** B.S. from City University of New York, City College, USA; M.H.L. from Hebrew Union College-Jewish Institute of Religion, Cincinnati OH, USA. **Interests:** tennis, opera, music, reading, theater. **Hons.:** Bublik Prize, Hebrew University (1978); Harris Medal, City College, New York (1979); Hon. D.D., University of South Carolina, Lafayette and Hamilton College. **Pubs.:** From Discrimination to Extermination, 1950.

Mailing address: 838 Fifth Ave, New York NY 10021, USA. T. (1 212) 249 0100

Residence: 6 River Lane, Westport CT 06880, USA.

Schindler-McGraw, Kevin, Rev

Rel: Roman Catholic; Ordained on 27 DEC 1984; b. 11 OCT 1954 in Cleveland OH, USA; celibate; Citizen of USA; Lang. English, German, Italian and Spanish.

International Organizations: International General Assistant (since 1989) of the Secular Franciscan Order.

Other organizations: Coordinator (since 1989), Centro Francescano Internazionale per il Dialogo; President (since 1990), Conference of General Assistants (OFS).

Main field of work: Administration and spiritual counsel. **Career:** Instructor, Bishop Montgomery High School, Torrance CA from 1983 to 1984. Associate Pastor, St Thomas Paris, Riverside CA from 1984 to 1985. **Educ.:** B.A. from Purdue University, West Lafayette IN, USA (1975-1976); M.A. from St Paul Seminary, MN, USA (1977-1982); M.Div. from Graduate Theological Union of Berkeley, Franciscan School of Theology, CA, USA (1981-1983); Doc. Candidate from Theologische Fakultät, Luzern, Switzerland (Beginning in 1985). **Interests:** Biblica and Judaic studies, ecumenism, liturgical and sacramental theology. **Mem.:** International Council of Christians and Jews (1986); Society of Biblical Literature; Catholic Biblical Association

Mailing address: Via Fortini 24, I-06081 Assisi PG, Italy.

Residence: Sacro Convento, Piazza Inferiore di S Francesco, I-06082 Assisi PG, Italy. T. (39 75) 81 22 38

Schlegel, John P, S.J., Rev

Rel: Roman Catholic; Ordained on 22 JUN 1973; b. 31 JUL 1943 in Dubuque IA, USA; celibate; Citizen of USA.

International Organizations: President (since 1991) of the University of San Francisco, CA.

Member of the Society of Jesus.

Career: Lecturer, Creighton University, Omaha NB from 1969 to 1970. Research Associate, Commonwealth Institute, Oxford from 1975 to 1976. Assistant Professor, Creighton University, Omaha NB from 1976 to 1979. Associate Professor, Creighton University, Omaha NB in 1979. Assistant Academic Vice-President, Creighton University, Omaha NB from 1978 to 1982. Fellow in Academic Administration, American Council on Education from 1982 to 1983. Academic Dean and Dean of the College of Arts and Sciences, Rockhurst College, Kansas City MO from 1982 to 1984. Dean, College of Arts and Sciences, Marquette University, Milwaukee WI from 1984 to 1988. Academic Vice-President, John Carroll University, Cleveland OH from 1988 to 1991. Executive and Academic Vice-President, John Carroll University, Cleveland OH from 1989 to 1991. **Educ.:** B.A. from Saint Louis University, MO, USA (1969); M.A. from Saint Louis University, MO, USA (1970); B.D. from University of London, UK (1973); D.Phil. from University of Oxford, UK (1977). **Mem.:** American Association of Higher Education; Council of Liberal Learning; Academy of Political Science; American Conference of Academic Deans; Association of Canadian Studies in the US; Western Social Science Association; Phi Delta Kappa; Delta Phi Alpha; Pi Gamma Mu; Alpha Sigma Nu; Phi Alpha Theta; Pi Sigma Alpha; Phi Kappa Psi **Pubs.:** Reviews, professional papers, articles and chapters including: "San Francisco's First University", in *Current Issues in Catholic Higher Education*, Vol 12:2, Winter 1992; "20 Years of Policy Evolution: Canada, The USA and South Africa", in *The Round Table*, London, Jan 1987 and "Patterns of Diplomacy: Canada and Australia in the Third World", in *Australian Journal of History and Politics*, 30-1, Winter 1984.

Mailing address: Univ. of San Francisco, 2130 Fulton St, San Francisco CA 94117-1080, USA. T. (1 415) 666 6762. FAX 666 2303

Schmaus, Michael

Rel: Roman Catholic; Ordained in 1922; b. 17 JUL 1897 in Oberbaar, Bavaria, Germany; celibate; Citizen of Germany.

International Organizations: Professor Emeritus (since 1965) of the University of Münster.

Career: Lecturer, Philosophische-theologische Hochschule und Seminar, Freising from 1924 to 1929. Lecturer, University of Munich from 1928 to 1929. Professor of Dogmatic Theology, German University in Prague from 1929 to 1933. Founder, Grabmann Institut zur Erforschung der mittelalterlichen Philosophie und Theologie in 1954. Professor, University of Münster from 1933 to 1945. Professor, University of Munich from 1946 to 1965. Visiting Professor, Chicago from 1969 to 1970. Rector, University of Munich from 1951 to 1952. Peritus, Second Vatican Council from 1960 to 1965. Supernumerary Pronotary Apostle in 1984. **Hons.:** Günther Kunze Prize (1983); Commander of Greek Order of Phoenix; Commander of Spanish Order of Civil Merit; Bayerischer Verdienstorden; Bayerische Maximiliansorden für Wissenschaft und Kunst (1984); Grosses Bundesverdienstkreuz; Päpstlicher Ehrenprälat; Erzbischöflicher Geistlicher Rat. **Mem.:** Bavarian Academy of Sciences **Pubs.:** Die Denkform Augustins in seinem Werk 'De Trinitate', 1962; Wahrheit als Heilsbegegnung, 1964 and articles, translations, several series.

Mailing address: Junkerstrasse 5, 82131 Gauting bei München, Germany. T. (49 89) 850 2800

Schmidel, David, Rabbi

Rel: Jewish.

International Organizations: President of the World Rabbinic Committee for the Preservation of Ancient Tombs in Tiberias.

Mailing address: 183 Wilson Street, Brooklyn NY 11211, USA. T. (718) 782 5910.

Schmidt, M G

Rel: Roman Catholic.

International Organizations: UNESCO Representative of the International Christian Union of Business Executives.

Mailing address: 56 rue Cambronne, F-75015 Paris, France. T. (33 1) 47 34 50 85.

Schmidt, Paula, OSSR, Sr

Rel: Roman Catholic; Ordained; celibate.

International Organizations: Member of the Order of the Most Holy Redeemer.

Other organizations: Member, Association of Contemplative Sisters.

Mailing address: Monastery of Our Lady of Perpetual Help, Esopus NY 12329, USA.

Schmidt-Brabant, M

Rel: Spiritual Science.

International Organizations: President of the General Anthroposophical Society.

Mailing address: The Goetheanum, CH-4143 Dornach, Switzerland. T. (41) 72 42 42.

Schmidt-Clausen, Kurt Hermann

Rel: Evangelical Lutheran; Ordained; b. 01 OCT 1920 in Hanover, Germany; m. to Erika Rokahr; c. 1 boy and 1 girl; Citizen of Germany.
International Organizations: International Official of the Evangelical Lutheran Church.
Career: Pastor, City Parish, Hanover from 1951 to 1952. Member, Administrative Staff, Evangelical Lutheran Church of Hanover from 1952 to 1955. Pastor, Wunstorf Suburban Parish (near Hanover) from 1955 to 1960. Assistant Executive Secretary, later Acting Executive Secretary, Lutheran World Federation (LWF) from 1960 to 1961. Executive Secretary, LWF from 1961 to 1964. Secretary-General, LWF from 1964 to 1965. Member, Executive Committee, LWF from 1970 to 1977. Chairman, Ecumenical Commission, Evangelical Lutheran Church of Hanover. Editor, *Lutherische Monatshefte* (monthly). Regional Bishop, Osnäbruck Diocese from 1970 to 1982. **Interests:** Roman archaeology, history, liturgical music. **Pubs.:** Vorweggenommene Einheit, 1964; Reformation als ökumenisches Ereignis, 1970 and Geschichte des Lutherischen Weltkonvents, 1975.
Mailing address: Lemförderstrasse 8A, 30169 Hannover, Germany. T. (49 511) 887347

Schmiegelow, P K L, Deaconess

Rel: Anglican.
International Organizations: Secretary-General of the Intercontinental Church Society.
Mailing address: 175 Tower Bridge Road, London SE1 2AQ, UK. T. (44 71) 407 4588. Fax 378 0541. C. Chaplaincy, London SE1 2AQ.

Schneemelcher, Wilhelm

Rel: Evangelical; b. 21 AUG 1914 in Berlin, Germany; m. to Eva Ackermann; c. 3 boys and 1 girl; Citizen of Germany.
International Organizations: Professor Emeritus of the University of Bonn.
Main field of work: Theologian. **Career:** Pastor, Sülbeck, Kreis Einbeck from 1947 to 1949. Teaching appointment, Göttingen from 1946 to 1949. Professor in 1953. Professor of New Testament and History of the Ancient Church, Bonn in 1954. Dean, Evangelical Theology Faculty from 1958 to 1959. Dean, Evangelical Theology Faculty from 1963 to 1964. Rector, University of Bonn from 1967 to 1968. President, Rhineland-Westphalia Academy of Sciences, Düsseldorf from 1982 to 1985. **Hons.:** Grosses Bundesverdienstkreuz mit Stern; Dr. h.c. **Pubs.:** Martin Luther, sein Leben und sein Werk, 1983; Co-editor, Evangelisches Staatslexikon, 1987 and Editor, Neutestamentliche Apokryphen, 2 Vols., 1987-89.
Mailing address: Böckingstrasse 1, 53604 Bad Honnef, Germany. T. (49 2224) 5450

Schneerson, Menachem M, Rabbi

Rel: Jewish; Ordained; b. 14 APR 1902 in Nikolaev, Russia; m. to Chaya Moussia; Citizen of U.S.A.
International Organizations: World Leader (Rebbe) (since 1950) of the Chabad-Lubavitcher Movement of Judaism.
Other organizations: Head, Merkos L'Inyonei Chinuch; Head, Machine Israel; Head, Kehot Publishing Society.
Career: Established Lubavitch institutions in over 100 cities in the US, Canada and 11 other countries. Established Tzivos Hashem for boys and girls. Established Tifereth Zekeinim Colel Levy Yitzchak for elderly. Established Lubavitcher Mitzah-Mobiles. Established educational centers for alienated youth to bring them back to Judaism. **Educ.:** University of Berlin; University of Paris IV (Paris-Sorbonne), France. **Pubs.:** notation to Chassidic and Kabbalistic treatises.
Mailing address: Chabad Research Ctr, 770 Eastern Pkwy., Brooklyn NY, USA.

Schneider, Michel Herbert, O.F.M., Rev Dr

Rel: Roman Catholic; Ordained on 07 AUG 1965; b. 08 JUN 1938 in Dockweiler/Eifel, Rheinland-Pfalz, Germany; celibate; Citizen of Germany; Lang. German, English, French and Italian.
International Organizations: Procurator-General (since 1991) of the Brothers of Mercy of Montabaur.
Main field of work: Education, leading. **Educ.:** University of Cologne, Germany (1967-1972). **Interests:** education, spirituality, social matters. **Pubs.:** Solidarische Gesellschaft als Zukunft, No 23, Walter Raymond-Stiftung, Cologne; Priester – Ein Lebensentwurf, Pustet, Regensburg and Fragen an die Orden – Eine Antwort, Johannes-Verlag, Leutesdorf.
Mailing address: Via S Maria Mediatrice 25, I-00165 Roma, Italy. T. (39 6) 68 49 19.
Residence: See Mailing Address. T. See Mailing Address

Schneider, Rudolf

Rel: Spiritualist.
International Organizations: Secretary-General of the Institute for Planetary Synthesis.
Mailing address: PO Box 128, CH-1211 Genève 20, Switzerland. T. (41 22) 733 88 76. Fax 733 66 49. Tx 427993 TxC CH, atth IPS.

Schneider, Silvio, Rev

Rel: Lutheran.
International Organizations: General Coordinator of the Lutherans United in Communication.
Mailing address: CP 2876, 90 000 Porto Alegre RS, Brazil. T. (55 512) 213 433. Fax 257 244. Tx 512-332 LUTE BR

Schneier, Arthur, Rabbi

Rel: Jewish.
International Organizations: Member of the World Jewish Congress.
Mailing address: Park East Synagogue, 163 East 67th Street, New York NY 10021, USA.

Schochet, J Immanuel, Rabbi

Rel: Jewish; b. 27 AUG 1935 in Switzerland; m. to Jettie Elzas; c. 4 children; Citizen of Canada.
International Organizations: Professor of Philosophy and Religion (since 1971) of the Humber College.
Other organizations: Rabbi (since 1959), Kielcer Congregation.
Career: Visiting Professor of Jewish Philosophy and Jewish Mysticism, Maimonides College from 1980 to 1986. Frequent lecturer and commentator worldwide on social, ethical and religious issues. **Educ.:** University of Windsor, Ontario, Canada; University of Toronto, Ontario, Canada; McMaster University, Hamilton, Ontario, Canada; Ph.D. from University of Waterloo, Ontario, Canada (1974); Rabbinical College of Lubavich, Brooklyn NY, USA (1974). **Pubs.:** Likkutei Sichot, 5 Vols., 1980-1990; Who is a Jew?, 1987 and The Mystical Dimension, 1990.
Mailing address: 55 Charleswood Dr, Downsview ON M3H 1X5, Canada.
Residence: See Mailing Address.

Schoenberg, Harris, Dr

Rel: Jewish.
International Organizations: Director of the Coordinating Board of Jewish Organizations.
Mailing address: 823 UN Plaza, Suite 400, New York NY 10017, USA. T. (1 212) 557 9008.

Scholtz, Marge

Rel: Evangelical.
International Organizations: Secretary of the International Evangelical Church.
Mailing address: PO Box 127, Addis Ababa, Ethiopia.

Schoneveld, Jacobus, Dr

Rel: Dutch Reformed; Ordained in 1963; b. 27 JAN 1934 in Alphen aan den Rijn, Zuid-Holland, Netherlands; m. to Ellen Tuyt; c. 4 boys and 4 girls; Citizen of Netherlands; Lang. Dutch, English, German, Hebrew and French.
International Organizations: Secretary-General (since 1980) of the International Council of Christians and Jews.
Member, Consultation on the Church and the Jewish People (since 1975) of the World Council of Churches.
Other organizations: Adviser to the Council for the Relationship between the Church and Israel (since 1989), Netherlands Reformed Church.
Main field of work: Interreligious relations. **Career:** Vicar, Netherlands Reformed Church in Zeist from 1961 to 1963. Minister, Netherlands Reformed Church in Stavoren from 1963 to 1967. Theological Adviser to the Netherlands Reformed Church in Jerusalem from 1967 to 1980. Executive Secretary, Ecumenical Institute for Theological Research, Israel from 1972 to 1980. **Educ.:** Cand. Theol. from State University at Leiden, The Netherlands (1952-1955); Ecclesiastic Exam. from State University at Leiden, Netherlands (1952-1959); Doctorandus Theol. from State University at Leiden, The Netherlands (1966); Dr. Theol. from State University at Leiden, Netherlands in conjunction with the Hebrew University of Jerusalem, Israel (1976). **Pubs.:** *The Bible in*

Israeli Education, Amsterdam, 1976 (in English), 1987 (German translation); "Torah in the Flesh" (on the prologue to the gospel of John) and several articles on theological issues in the field of Christian-Jewish relations.
Mailing address: Martin Buber House, Werlestrasse 2, 64646 Heppenheim an der Bergstrasse, Germany. T. (49 6252) 5041. Fax 68331.
Residence: Neckarsteinacher Strasse 4, 64646 Heppenheim, Germany. T. (49 6252) 3834

Schonfeld, Fabian, Rabbi

Rel: Jewish; Ordained on 05 JUN 1952; b. 14 DEC 1923 in Zawiercie, Poland; m. to Ruth; c. 2 boys and 4 girls; Citizen of USA; Lang. French, English, German and Yiddish.
International Organizations: Member of the World Jewish Congress. Chairman for Inter-Religious Affairs (since 1987) of the Synagogue Council of America.
Other organizations: Rabbi, Young Israel of Kew Gardens Hills, NY.
Main field of work: Ministry. **Career:** President, Rabbinic Alumni, Yeshiva University, New York NY. President, Rabbinical Council of America. **Educ.:** B.A. from University of London, UK; Rabb. Ord. from Yeshiva University, New York NY, USA. **Interests:** inter-religious affairs, zionist affairs. **Pubs.:** Articles in *Hadarom Tradition*, *Jewish Observer* on Hassidic thought.
Mailing address: Young Israel of Kew Gardens Hills, 150-05 70th Road, Flushing NY 11367, USA. T. (1 718) 261 9723
Residence: 70-41 153 Street, Flushing NY 11367, USA. T. (1 718) 544 1040

Schönig, Heinzpeter, Pater

Rel: Christian.
International Organizations: President of the International Group of Pastors for Circus and Showmen of All Confessions.
Mailing address: Kesterstrasse 14, D-86153 Augsburg, Germany. T. (49 821) 3127 58.

Schotte, Jan, C.I.C.M., Most Rev

Rel: Roman Catholic; Ordained on 03 AUG 1952; b. 29 APR 1928 in Beveren-Leie, Belgium; celibate; Citizen of Belgium.
International Organizations: Secretary-General (since 1985) of the Synod of Bishops.
Contact (since 1989) of the Special Assembly for Africa of the Synod of Bishops. Councillor of the Pontifical Commission for Latin America. Member of the Congregation of the Immaculate Heart of Mary.
Other organizations: Titular Archbishop (since 1985), Silli; Consultor to the Congregation for Bishops (since 1986); President (since 1989), Holy See Labour Office.
Career: Entered the Congregation, the Immaculate Heart of Mary, Brussels in 1946. Promoted to Titular Bishop, Silli in 1983. Ordained a Bishop by Pope John Paul II in 1984. Vice-Rector, CICM Theological Seminary in Louvain from 1956 to 1962. Rector, Immaculate Heart Mission Seminary in Washington DC from 1963 to 1966. Assistant Professor of Canon Law at the Higher Institute of Religious Sciences, Catholic University of Louvain from 1957 to 1962. Secretary-General, Congregation of the Immaculate Heart of Mary in Rome from 1967 to 1972. Served as Attaché in the International Organizations Division, Secretariat of State of Pope Paul VI from 1972 to 1978. Served as Attaché in the International Organizations Division, John Paul I in 1978. Served as Attaché in the International Organizations Division, John Paul II from 1978 to 1980. Secretary, Pontifical Commission on Justice and Peace from 1980 to 1983. Member, Pontifical Commission on Tourism and Migration from 1980 to 1985. Consultor to the Pontifical Commission on Latin America from 1980 to 1985. Consultor to the Pontifical Council for the Laity from 1980 to 1985. Consultor to the Secretariat for the Non-Christians from 1981 to 1985. Member, Joint Consultative Group on Social Thought and Action of the Holy See and the World Council of Churches from 1981 to 1985. Chairman, Vatican Labour Relations Commission from 1982 to 1989. Staff Member, Special Synod of Bishops of the Netherlands in 1980. Press Officer (French language), Synod of Bishops on the Family in 1980. Press Officer, Synod of Bishops on Penance and Reconciliation in 1983. Chairman, Drafting Committee, Holy See's 'Charter of the Rights of the Family' from 1981 to 1983. Vice-President, Pontifical Commission on Justice and Peace from 1983 to 1985. Member, the Papal Suite during various Apostolic Voyages of H H John Paul II. Served on the Holy See Delegations to various international meetings. **Educ.:** Studied Canon Law from Catholic University of Louvain (KUL), Belgium (1953-1956); Studied Canon Law from Catholic University of America, Washington DC, USA (1962-1963). **Hons.:** Knight Commander with Star, Equestrian Order of the Holy Sepulchre of Jerusalem; Knight Commander, Order of Leopold II (Belgium); Officer, Legion of Honour (France). **Mem.:** Joint Working Group of the Holy See and the World Council of Churches (since 1981); Pontifical Commission for the Universal Catechism (since 1986)
Mailing address: CICM, Piazza Pio XII 3, I-00193 Roma, Italy. T. (39 6) 698 4821 – 698 4324. Fax 698 3392.
Residence: Via di Porta Angelica, I-00193 Roma, Italy. T. (39 6) 686 89 70

Schrock, J Roger

Rel: Brethren.
International Organizations: Executive Secretary of the World Ministries Commission of the Church of the Brethren.
Mailing address: 1451 Dundee Avenue, Elgin IL 60120, USA. T. (1 312) 742 51 00.

Schröder, Hermann

Rel: Buddhist.
International Organizations: President of the League of Buddhist Esperantists.
Mailing address: Freudenstädter Strasse 43, 72270 Baiersbronn, Germany.

Schrotenboer, Paul G, Dr

Rel: Christian.
International Organizations: Contact of the Reformed Ecumenical Synod.
Mailing address: 1677 Gentian Drive SE, Grand Rapids MI 49508, USA.

Schuller, Robert Harold

Rel: Reformed Church in America; Ordained in 1950; b. in Alton IA, USA; m. to Arvella De Haan.
International Organizations: Minister of the Reformed Church in America.
Main field of work: Televangelist. **Career:** Minister at Ivanhoe Reformed Church in Riverside, IL from 1950 to 1955. Accepted a call to begin a mission church in Orange County, Southern California, and held the first service, the Garden Grove Community Church in the Orange Drive-in Theatre in 1955. Constructed a new building in 1961. Build the Tower of Hope in 1968. Established th Robert H Schuller Institute for Successful Church Leadership in 1969. Began broadcasting over television, his Sunday worship service known as 'Hour of Power' in 1970. Construction of the Crystal Cathedral from 1977 to 1980. **Educ.:** B.A. from Hope College, Holland MI, USA (1947); B.D. from Western Theological Seminary, Holland MI, USA (1950). **Pubs.:** *Move Ahead with Possibility Thinking*, 1967; *Your Church Has Real Possibilities!*, 1974 and *Self-Esteem: The New Reformation*, 1982.
Mailing address: No mailing address.

Schultz, Joseph, Rabbi Prof

Rel: Jewish; Ordained on 12 JUN 1955; b. 02 DEC 1928 in Kansas City MO, USA; m. to Bella E Intrater; c. 1 boy and 2 girls; Citizen of USA; Lang. English, Yiddish and Hebrew.
International Organizations: Oppenstein Brothers Distinguished Professor of Judaic Studies (since 1973) of the University of Missouri-Kansas City.
Main field of work: University education. **Career:** Assistant Rabbi, Congregation Kehillath Israel from 1955 to 1957. Rabbi, Beth Jacob Synagogue from 1957 to 1960. Rabbi, Temple Beth Shalom, Cambridge MA from 1960 to 1973. Lecturer, Boston University, MA from 1963 to 1968. Assistant Professor of Religion, Boston University, MA from 1968 to 1973. **Educ.:** B.A. from Yeshiva University, New York NY, USA (1949-1951); M.H.L., Rabb. Ord. from Jewish Theological Seminary of America, New York NY, USA (1951-1955); Ph.D. from Brandeis University, Waltham MA, USA (1956-1962). **Interests:** comparative study of religion. **Hons.:** Hyman B Enelow Award, Jewish Theological Seminary of America (1976). **Pubs.:** Judaism and the Gentile Faiths; Mid-America's Promise and Editor with introduction, Ze'ena U-Re'enah Book of Genesis (translation), Horowitz.
Mailing address: University of Missouri-Kansas City, 5100 Rockhill Rd, Kansas City MO 64114, USA. T. (1 816) 235 2538
Residence: 1104 W 100th St, Kansas City MO 64114, USA.

Schultz, Walter Arthur

Rel: Lutheran; b. 16 JAN 1928 in Winnipeg MB, Canada; m. to Dorothea Brethauer; c. 2 boys and 2 girls; Citizen of Canada.

International Organizations: Director of Finance and Administration (since 1990) of the Lutheran World Federation.

Career: Audit Group Head Taxation Division Revenue Canada from 1955 to 1963. Executive Secretary, Division of Information Services, Canadian Lutheran Council from 1963 to 1967. Executive Secretary, Lutheran Council in Canada from 1967 to 1976. Treasurer, Canadian Secretariat, Lutheran Church in America from 1963 to 1985. Member, Commission on World Service, Lutheran World Federation (LWF), Geneva from 1977 to 1986. Assistant Comptroller, LWF from 1986 to 1990. Director, Lutheran Life Insurance Society of Canada from 1972 to 1986. Member, Finance Committee, Canadian Council of Churches from 1964 to 1986. Executive Director, Lutheran Council in Canada from 1976 to 1986. **Educ.:** Accounting Cert. from Angus School of Commerce (1947); C.A. from University of Manitoba, Winnipeg, Canada (1955). **Interests:** photography, jogging. **Hons.:** Recipient of Distinguished Service Award, Lutheran Brotherhood Insurance, Minneapolis (1967).

Mailing address: Route de Ferney 150, PO Box 2100, CH-1211 Genève 2, Switzerland. T. (41 22) 791 61 11. Fax 798 86 16. C. LUTHWORLD GENEVA. Tx 415730 OIK CH.

Residence: 52 ch Ami-Argand, CH-1290 Versoix, Switzerland.

Schulz, William Frederick, Rev

Rel: Unitarian Universalist; Ordained on 09 NOV 1975; b. 14 NOV 1949 in Pittsburgh PA, USA; separated; Citizen of USA; Lang. English.

International Organizations: President (since 1985) of the Unitarian Universalist Association.

Board Member of the International Association for Religious Freedom.

Other organizations: Board Member, People for the American Way; Board Member, Planned Parenthood of America.

Main field of work: Administration, ministry. **Career:** Executive Vice-President, Unitarian Universalist Association (UUA) from 1975 to 1978. Acting Director, Department of Ministerial and Congregational Service, UUA from 1978 to 1979. Director of Social Responsibility, UUA in 1979. Minister, First Parish in Bedford MA from 1979 to 1985. **Educ.:** A.B. from Oberlin College, OH, USA (1969-1971); M.A. Theol. Stds. from Meadville/Lombard Theological School, Chicago IL, USA (1971-1973); M.A. Phil. from University of Chicago, IL, USA (1974); D.Min. from Meadville/Lombard Theological School, Chicago IL, USA (1975). **Hons.:** Phi Beta Kappa. **Mem.:** ACLU **Pubs.:** Contributing Editor to *Transforming Words: Six Essays on Preaching*, Skinner House, 1984; Contributor to *Dictionary of the Ecumenical Movement*, WCC Publications, 1991 and "Goal is Human Dignity", in *USA Today*, October 6, 1986.

Mailing address: 25 Beacon Street, Boston MA 02108, USA. T. (1 617) 742 2100.

Residence: 9 Louisburg Square, Boston MA 02108, USA. T. (1 617) 720 0527

Schupp, M

Rel: Roman Catholic; Ordained; celibate.

International Organizations: Representative of the Congregation of Mariannhill Missionaries.

Mailing address: Via S Giovanni Eudes 91, I-00163 Roma, Italy. T. (39 6) 625 85 47.

Schurink, Reinder J, Commr

Rel: Salvation Army, The.

International Organizations: Territorial Commander of The Salvation Army.

Mailing address: Damstraat 15, 1012 JL Amsterdam, Netherlands.

Schüssler Fiorenza, Elisabeth S, Dr

Rel: Catholic.

International Organizations: Krister Stendhal Professor of Divinity (since 1988) of the Harvard Divinity School, Cambridge MA.

Other organizations: Co-Founder/Editor, *The Journal of Feminist Studies in Religion*; Editor, *Religious Studies Review* and *Concilium* (international journal).

Main field of work: Lecturing, teaching, research. **Career:** Taught (for almost 15 years) in the Dept. of Theology, University of Notre Dame, IN. Talbot Professor of New Testament Studies, Episcopal Divinity School, Cambridge MA. Harry Fosdick Visiting Professor at Union Theological Seminary, NY. Visiting Professor at the Catholic Theological Faculty, Universities of Tübingen and Luzern. Served as President, Society of Biblical Literature (SBL). Served on the boards, the major biblical societies. Served on the editorial boards, the *Journal of Biblical Literature, The Catholic Biblical Quarterly, Semeia, Horizons, Cross Currents, The Journal of Religion and Law*, and *Kirche und Israel*. First Co-Chair, AAR/SBL Women's Caucus. Served on numerous task forces on women in the church and ministry. Former member of the board, The Women's Alliance for Theology, Ethics, and Ritual, and the Women's Ordination Conference. Chaired the SBL Women in the Biblical World Section. Served on the Advisory Committee, the Women's Studies in Religion Program at Harvard Divinity School and of the Research Project on Women in Christianity at the University of Tübingen. **Educ.:** M.Div., Lic. Practical Theol., Doc. New Testament Studies. **Interests:** Biblical interpretation, feminist theology, hermeneutics. **Hons.:** Named 'Catholic of the Year' by *US Catholic* magazine (1987); Service Award, Women's Ordination Conference. **Pubs.:** Numerous publications (both in English and German): *In Memory of Her*; *Bread Not Stone* and *Revelation: Vision of a Just World*, Augsburg Fortress Press, 1991.

Mailing address: Harvard Divinity School, Cambridge MA 02138, USA.

Schwarz, Fred C, Dr

Rel: Baptist; Unordained; b. 15 JAN 1913 in Brisbane, Australia; Citizen of Australia.

International Organizations: President of the Christian Anti-Communism Crusade.

Career: High School Teacher. Appointed to the evening staff, Queensland Teachers College. Established a general medical practice in Sydney. Served as a part-time psychiatrist at the New South Wales Community Hospital and the Cathedral Marriage Guidance Clinic. Engaged in a debate with Max Julius, a leading Australian Communist in 1940. Made a profound study of basic communist texts. Has repeatedly challenged communist leaders to open debate on any aspect of their philosophy and program and has participated in many debates with leading communist spokesmen. Toured the US and Europe lecturing on communism in 1950. Continued to make tours which included, not only the US and Europe, but Asian countries such as India, the Philippines, Formosa, and Japan. Formed the Christian Anti-Communism Crusade in the USA in 1953. Closed his medical practice in Sydney in 1955. **Educ.:** Grad. Sc. from University of Queensland, Brisbane, Australia (1933); Grad. Arts from University of Queensland, Brisbane, Australia (1933); Deg. Med. and Surg. from University of Queensland, Brisbane, Australia (1944). **Pubs.:** Author of several books and numerous pamphlets including: *You Can Trust the Communists (to be Communists)*; *Why I Am Against Communism* and *Why Communism Kills – The Legacy of Karl Marx*.

Mailing address: PO Box 890, 227 E Sixth St, Long Beach CA 90801, USA. T. (213) 437 0941.

Schwarz, Karen Anne

Rel: Roman Catholic; Unordained; b. 15 MAR 1957 in San Francisco CA, USA; single; Citizen of USA; Lang. English and Spanish.

International Organizations: National Advisor (since 1991) of the Women's Ordination Conference.

Other organizations: Coordinator (since 1987), San Francisco Bay Area Women's Ordination Conference; Steering Committee (since 1988), Mercy Justice Coalition, Burlingame; Psychotherapist (since 1991), Foster City Counseling Services.

Main field of work: Clinical psychology. **Career:** Parish Minister, St Timothy Church, San Mateo CA from 1976 to 1977. Campus Minister, Archdiocese of Los Angeles CA from 1979 to 1980. Clinical Intern, Family Counseling Services, Elkhart IN from 1980 to 1981. Psychotherapist, Center for the Whole Person, San Carlos CA from 1981 to 1988. Clinical Intern, San Francisco General Hospital Substance Abuse from 1987 to 1988. Psychology Intern, Garfield Geropsychiatric Hospital, Oakland CA from 1989 to 1990. Psychotherapist, Robert Kaye MD and Associates, San Mateo CA from 1988 to 1991. Psychology Intern, Adolescent Unit, Valley Medical Center, San Jose CA from 1990 to 1991. **Educ.:** B.A. Psy.; B.A. Rel.St. from Mount St Mary's College, Los Angeles CA, USA (1977-1980); M.A. Couns. Psy. from University of Notre Dame, IN, USA (1980-1981); Ph.D. Candidate in Clin. Psy. from California School of Professional Psychology, Berkeley CA, USA (from 1986 on) (1980-1981). **Interests:** Feminist psychoanalytic theory, feminist theology, spirituality, reform of the Roman Catholic church. **Mem.:** American Psychological Association; Association for Women in Psychology; American Association for Counseling and Development **Pubs.:** "Called to a New Form of Mercy?", in *Mercy Association for Scripture and Theology Journal*, Fall 1990; "Alternative Membership

in Religious Congregations", in *Review for Religious*, July-August 1991 and *Paradigm in Religious Life: Alternative Membership as a Model for the Future*, Conference Proceedings, Women in Evangelization Symposium, Barry University FL, March 1992.
Mailing address: 1289 E Hillsdale Blvd, Foster City CA 94404-1219, USA. T. (1 415) 345 4006
Residence: 1813 Ashton Ave, Burlingame CA 94010, USA

Schwenzer, Gerhard, SS.CC., Fr

Rel: Roman Catholic; Ordained on 18 JUL 1964; b. 01 OCT 1938 in Lorch, Germany; celibate.
International Organizations: Secretary-General of the Scandinavian Episcopal Conference.
Member of the Congregation of the Sacred Hearts of Jesus and Mary.
Other organizations: Bishop (since 1983), Oslo.
Mailing address: Oslo Katolske Bispedømme, Akersveien 5, PB 8270, Hammersborg, N-0177 Oslo 1, Norway. T. (47 2) 20 72 26. Fax (47 2) 20 48 57.

Scobie, Richard S

Rel: Unitarian.
International Organizations: Executive Director of the Unitarian Universalist Service Committee.
Mailing address: 78 Beacon St, Boston MA 02108, USA. T. (1 617) 742 2120. C. UNSERCOM. Fax (1 617) 742 6891.

Scott, Edward Walter, Most Rev

Rel: Anglican; Ordained in 1942; b. 30 APR 1919 in Edmonton, Alberta, Canada; m. to Isabel Florence Brannan; c. 1 boy and 3 girls; Citizen of Canada.
International Organizations: Archbishop and Primate (since 1971) of Anglican Church of Canada.
Career: Rector, St Peter's, Seal Cove, Prince Rupert, British Columbia from 1942 to 1945. Secretary-General, Student Christian Movement, University of Manitoba, also part-time lecturer at St John's College from 1945 to 1949. Rector, Church of St John the Baptist, Fort Garry from 1949 to 1955. Rector, St Jude's, Winnipeg from 1955 to 1960. Director, Social Service, and Priest in charge of Indian Work for Diocese of Rupert's Land from 1960 to 1964. Helped establish first Indian-Metis Friendship Centre. Associate Secretary, Council for Social Service, National Office of Anglican Church from 1964 to 1966. Bishop, Kootenay from 1966 to 1971. Moderator, Central Committee, World Council of Churches from 1975 to 1983. Member, Commonwealth Group on South Africa from 1985 to 1986. **Educ.:** B.A. from University of British Columbia, Vancouver, Canada; L.Th. from Theological College, Vancouver, Canada (1942). **Interests:** carpentry. **Hons.:** Human Relations Award, Canadian Council of Christians and Jews (1987); Companion of the Order of Canada (1978); 14 honorary degrees.
Mailing address: 600 Jarvis St, Toronto ON M4Y 2J6, Canada.

Scremin, Lorenzo, C.S., Fr

Rel: Roman Catholic; Ordained; celibate.
International Organizations: Procurator-General of the Pious Society of the Missionaries of St Charles.
Mailing address: Via Calandrelli 11, I-00153 Roma, Italy. T. (39 6) 589 88 27 – (39 6) 589 88 28.

Seaman, Keith Douglas, Rev Sir

Rel: Methodist; Ordained; b. 11 JUN 1920 in McLaren Vale, Australia; m. to Joan Isabel Birbeck; c. 1 boy and 1 girl; Citizen of Australia.
International Organizations: State Governor and Ecclesiastic of the Methodist Church.
Career: South Australia (S.A.) Public Service from 1937 to 1954. Entered Methodist ministry in 1954. Minister, Renmark from 1954 to 1958. Central Methodist Mission from 1958 to 1977. Director, 5KA, 5AU, 5RM Broadcasting Cos. from 1960 to 1977. Secretary, Christian TV Association S.A. from 1959 to 1973. Member, Executive, World Association of Christian Broadcasting from 1963 to 1970. RAAF (Royal Australian Air Force) Overseas HQ, London from 1941 to 1945. Superintendent, Adelaide Central Methodist Mission from 1971 to 1977. Member, Australian Government Social Welfare Commission from 1973 to 1976. Governor, South Australia from 1977 to 1982. **Interests:** gardening, reading. **Hons.:** Knight Commander of the Royal Victorian Order; Knight of the Order of St John of Jerusalem; Officer of the Order of the British Empire.
Mailing address: 31 Heggerton Street, Victor Harbor, South Australia 5211, Australia. T. (61 85) 523535

Seater, Hilda Frank

Rel: Protestant.
International Organizations: International Director of the International Association of Women Ministers.
Mailing address: 338 S 8th Ave, West Bend WI 53095, USA.

Sebastiani, Sergio, H E Mgr

Rel: Roman Catholic; Ordained on 15 JUL 1956; b. 11 APR 1931 in Montemonaco, Italy; celibate.
International Organizations: Apostolic Pro-Nuncio to Turkey (since 1985) of the Roman Catholic Church.
Other organizations: Titular Archbishop, Cesarea di Mauritania.
Mailing address: PO Box 33, Cankaya-Ankara, Turkey. T. (90 4) 139 00 41. Fax 140 29 00

Secher, Paul

Rel: Jewish.
International Organizations: Director of the Commonwealth Jewish Council.
Mailing address: Somerset House, 37 Fortress Road, London NW5 1AD, UK. T. (44 71) 267 7991. Fax 267 6394.

Secondo, Louis J, Fr

Rel: Roman Catholic; Ordained; celibate.
International Organizations: Procurator-General of the Third Order Regular of St Francis.
Procurator-General of the Society of St Edmund.
Mailing address: Basilica Ss Cosma i Damiano, Via dei Fori Imperiali 1, I-00186 Roma, Italy. T. (39 6) 679 28 00.

Seebeck, Douglas

Rel: Christian Reformed Church.
International Organizations: Director of the Christian Reformed World Relief Committee.
Mailing address: PO Box 14205, Nairobi, Kenya. T. (254 2) 751039.

Seelaus, Vilma, OCD, Sr

Rel: Roman Catholic; Ordained; celibate.
International Organizations: Member of the Discalced Carmelite Nuns.
Other organizations: Member, Association of Contemplative Sisters.
Mailing address: Carmelite Forum, 1318 Dulany Valley Rd, Baltimore MD 21204-1399, USA.

Sehested, Ken

Rel: Baptist; Ordained on 17 OCT 1981; m. to Nancy Hastings; c. 2 girls; Citizen of USA.
International Organizations: Executive Director (since 1984) of the Baptist Peace Fellowship of North America.
Career: Co-Editor and Director of Education, *SEEDS* (a magazine and educational ministry on world hunger issues) from 1978 to 1984. Associate Pastor, Manhattan Baptist Church, New York NY. Circulation Manager, *Christianity and Crisis* magazine, New York NY. **Educ.:** Baylor University, Waco TX, USA (1969-1971); B.A. (cum laude) from New York University, NY, USA (1973); M.Div. from Union Theological Seminary, New York NY, USA (1978). **Pubs.:** Magazine articles in *SEEDS, Sojourners, Christianity and Crisis, Baptist Peacemaker, The Ecumenist, SBC Today, The Canadian Baptist, PeaceWork, The Christian Century, The Christian Ministry, Home Missions Magazine, Union Seminary Quarterly Review, Faith and Mission* (Southeastern Baptist Seminary); Book chapters in *Alternate Celebrations Catalog* (Pilgrim Press), *'Seek Peace and Pursue It': Proceedings from the 1988 International Baptist Peace Conference* (Institute for Baptist and Anabaptist Studies, Rüschlikon, Switzerland, Baptist Seminary and Baptist Peace Fellowship of North America), and *Beyond Bandaides* (Meyer-Stone Books) and author of *Trust and Obey: Notes Toward a Spirituality of Justice*, Baptist Peacemaker International Spirituality Series, 1988.
Mailing address: 499 Patterson Street, Memphis TN 38111, USA. T. (1 901) 324 7675.

Seiple, Robert A
Rel: Christian; b. 06 DEC 1942 in Easton NJ, USA; m. to Margaret Ann; c. 2 boys and 1 girl; Citizen of USA.
International Organizations: Past President (since 1987) of the World Vision International.
Career: Served in US Marine Corps (attaining the rank of captain) from 1966 to 1969. Field sales representative for Boise Cascade Corp. from 1969 to 1971. Became Assistant Director, Athletics for Brown University, Providence RI in 1972. Became Director, Athletics for Brown University in 1975. Appointed Vice-President for Development and Director, fund-raising campaign for Brown University in 1979. President, Eastern College and Eastern Baptist Theological Seminary, St Davids PA from 1983 to 1987. **Educ.:** A.B. Am.Lit. from Brown University, Providence RI, USA (1965). **Hons.:** Hon. L.H.D., Alderson-Broaddus College, Philippi WV (1986); Hon. LL.D., Sioux Falls College, SD (1986); Vietnamese Campaign Medal with five battle stars; Navy Commendation Award; 28 air medals; Distinguished Flying Cross.
Mailing address: 919 West Huntington Drive, Monrovia CA 91016, USA. T. (1 818) 357 7979 – 357 1111. C. WORVIS. Tx 675341 WORVIS MROV – 275335 WORV UR.
Residence: 5219 Bubbling Well Lane, La Canada CA 91011, USA.

Seki, Bokuo
Rel: Buddhist; Ordained; b. 15 APR 1903; Citizen of Japan.
International Organizations: Monk of Rinzaishu Tenryujiha.
Educ.: Keio University, Tokyo, Japan.
Mailing address: 68 Sagatenryu Susukinobabacho, Ukyo-ku, Kyoto 616, Japan. T. (81 75) 881 1235

Sekine, Masao, Prof Dr
Rel: Christian; Unordained; b. 14 AUG 1912 in Tokyo, Japan; m. to Yasuko Shirai; c. 2 boys; Citizen of Japan; Lang. Japanese.
International Organizations: Theologian of the Tokyo Kyoiku University.
Main field of work: Old Testament study. **Career:** Became a Christian in 1929. Assistant Professor of Semitic Languages and Old Testament, Tokyo Kyoiku University in 1954. Professor, Tokyo Kyoiku University from 1955 to 1976. Director, Japanese Biblical Institute from 1958 to 1988. President, Japanese Society of Old Testament Study from 1970 to 1979. **Educ.:** Dr. Theol. from Martin-Luther-Universität Halle-Wittenberg, Halle, Germany (1944). **Interests:** Japanese short poetry. **Hons.:** Dr. Theol. h.c., University of Erlangen-Nürnberg (1970); Dr. Litt. h.c., University of Tokyo (1976); Purple Prize for Christian Theology, Government of Japan (1977). **Mem.:** Japan Academy (Christian Theology) (1982) **Pubs.:** Selected Works of Masao Sekine, Vol. 1-20, 1979-1989; History of Old Testament Literature, 2 Vols., 1978-1980 and The Thinkers of Ancient Israel, 1977.
Mailing address: 5-11-14 Nishiogi-kita, Suginamiku, Tokyo 167, Japan. T. (81 3) 3390 6538
Residence: See Mailing Address. T. See Mailing Address

Sekiya, Ayako
Rel: Christian; Citizen of Japan.
International Organizations: Chairman of the Tokyo YWCA.
Mem.: Committee of Seven to Appeal for World Peace
Mailing address: 4-22-5 Jindaiji Kitamachi, Chofu, Tokyo 182, Japan. T. (81 424) 85 6743

Selby, William R
Rel: Mason.
International Organizations: Secretary-General of the General Grand Chapter of Royal Arch Masons.
Mailing address: 1084 New Circle Rd NE, Lexington KY 40505, USA. T. (1 606) 252 4618.

Selby Wright, Ronald William Vernon S, Very Rev Dr
Rel: Church of Scotland; Ordained in 1936; b. 12 JUN 1908 in Scotland, UK; single; Citizen of United Kingdom.
International Organizations: Ecclesiastic of the Church of Scotland.
Main field of work: Ministry (youth work). **Career:** Minister, Canongate, Edinburgh from 1936 to 1977. Moderator, Church of Scotland from 1972 to 1973. Chaplain to H M Forces from 1942 to 1947. Chaplain to H M The Queen from 1961 to 1977. **Educ.:** M.A. from University of Edinburgh, UK (1928-1936). **Hons.:** Commander of the Royal Victorian Order; Hon. D.D., University of Edinburgh (1956). **Mem.:** Fellow, Royal Society of Edinburgh; Athenaeum Club; London New Club **Pubs.:** *Asking them Questions*, 5 Vols., Oxford, 1936-1973; *Take Up God's Armour*, Oxford, 1968 and *Another Home* (autobiography), Blackwood, 1980.
Mailing address: The Queen's House, 36 Moray Place, Edinburgh EH3 6BX, UK. T. (44 31) 226 5566
Residence: See Mailing Address. T. See Mailing Address

Selg, Ottfried
Rel: Roman Catholic; b. 04 AUG 1940 in Mannheim, Germany; divorced; c. 3 boys; Citizen of Germany; Lang. German, English and French.
International Organizations: Secretary-General (since 1983) of the European Colloquy of Parishes.
Educ.: Sociology and Theology from Albert-Ludwigs-Universität Freiburg, Eberhard-Karls-Universität Tübingen, and Bayerische-Julius-Maximilians-Universität Würzburg, Germany (1960-1967). **Pubs.:** *Seelsorge in der Stadt Heidelberg*, 1968; *Pfarrverband – ein Koopertives Gemeindemodell*, 1974; *Glaubenszeugnis im Gespräch*, 1979; *Menschen in der heutigen Welt*, 1989 and *Wege zu gelebten Werten*, 1990.
Mailing address: Römerstr 9, 86356 Neusäss, Germany. T. (49 821) 482001.
Residence: See Mailing Address. T. (49 821) 482002

Sélis, Claude, Prof Rev
Rel: Roman Catholic; Ordained on 11 NOV 1981; b. 26 APR 1951 in Anderlecht, Brussels, Belgium; celibate; Citizen of Belgium; Lang. French, Dutch, English, German and Greek.
International Organizations: Chairman of the International Association Bible and Computers.
Other organizations: General Chaplain (since 1988), Federation of Catholic Scouts of Belgium.
Main field of work: Research in orientalism. **Career:** Teacher, several secondary schools from 1975 to 1983. Director, secondary school from 1983 to 1988. **Educ.:** Lic. Philologie. bibl.; Lic. Philologie et Hist. orientales from Catholic University of Louvain (UCL), Belgium (1969-1973). **Interests:** Orientalism, bible exegesis. **Mem.:** Order of the Dominicans **Pubs.:** Contribution to *Dictionnaire Encyclopédique de la Bible*, Brepols, Turnhout, 1987 and *Les Syriens orthodoxes et catholiques*, Col. Fils d'Abraham, Brepols, Turnhout, 1989.
Mailing address: Philoxène, Avenue de la Renaissance 40, B-1040 Bruxelles, Belgium. T. (32 2) 733 72 47
Residence: See Mailing Address. T. See Mailing Address

Selle, Murri
Rel: Christian.
International Organizations: Regional Secretary for Europe of the World Association for Christian Communication.
Mailing address: Lachnerstrasse 20, 80639 München, Germany.

Seman, Isabel, Sr
Rel: Roman Catholic; Ordained; celibate.
International Organizations: Representative of the Sisters Mercedarian Missionaries of Bérriz.
Mailing address: Viale Pola 10, I-00198 Roma, Italy.

Semple, Ian Chalmers
Rel: Bahá'í; b. 02 DEC 1928 in New Barnet, Hertfordshire, UK; m. to Louise Gloor; c. 2 boys and 1 girl; Citizen of United Kingdom; Lang. English, German and French.
International Organizations: Member, Universal House of Justice (since 1963) of the Bahá'í International Community.
Main field of work: Bahá'í administration. **Career:** Recruit and 2nd Lieutenant, Royal Corps of Signals (Military Service) from 1947 to 1949. Undergraduate, Oxford University, UK from 1949 to 1952. Apprentice and Senior, Accounting firms in London and Edinburgh, UK from 1952 to 1958. Secretary, National Bahá'í Assembly, UK from 1958 to 1961. Assistant Secretary, International Bahá'í Council from 1961 to 1963. **Educ.:** M.A. from University of Oxford, Pembroke College, UK (1949-1952). **Mem.:** Fellow (retired), Institute of Chartered Accountants in England and Wales (1955)
Mailing address: PO Box 155, 31001 Haifa, Israel. T. (972 4) 35 83 58
Residence: 39 Harofe Street, 34367 Haifa, Israel T. (972 4) 24 45 10

Sen, Noel Sushil, Rev

Rel: Church of North India; Ordained on 28 OCT 1973; b. 08 AUG 1939 in Simla, India; single; Citizen of India; Lang. Hindi and English.
International Organizations: Secretary-General (since 1990) of the Church of North India Synod.
Main field of work: Ministry and administration. **Career:** Management Trainee, Philips from 1960 to 1963. Assistant Sales Manager, Gramophone Co. of India Ltd. from 1965 to 1969. Presbytery, Diocese of Calcutta from 1972 to 1990. **Educ.:** B.A. from St Xavier's College, Calcutta, India (1957-1960); Cert. Music from Yale University, New Haven CT, USA (1963-1965); B.D. from Serampore College, India (1969-1972); Dip. Theol. from Bishop's College, Calcutta, India (1969-1972). **Interests:** ecumenical dialogue, church music. **Hons.:** Diplomas for playing the piano and violin. **Mem.:** Hislop Education Society (1990), Calcutta Club, Ramakrishna Mission Institute of Culture
Mailing address: CNI Bhavan, 16 Pandit Pant Marg, New Delhi 110 001, India. T. (91 11) 371 6513
Residence: See Mailing Address. T. (91 11) 371 0346

Senge, Mitsuhiko

Rel: Unknown; b. 23 NOV 1922; Citizen of Japan.
International Organizations: Governor/Manager of the Izumooyashirokyo.
Mailing address: Taishamachi Omiyadori, Hikawa-gun, Shimane 699-07, Japan. T. (81 853) 53 2078

Senoo, Hilary, Mgr

Rel: Roman Catholic; Ordained on 27 JUL 1963; b. 20 JAN 1938 in Gbi-Bla (Hohoe), Volta Region, Ghana; celibate; Citizen of Ghana; Lang. Ewe, English, French and Italian.
International Organizations: Secretary-General (since 1988) of the Symposium of Episcopal Conferences of Africa and Madagascar.
Educ.: Phil. and Theol. from St Peter's Major Seminary, Cape Coast (1957-1963); Dr. Canon Law from Urbanian Pontifical University, Vatican City State (1964-1967); P.G. Teacher's Cert. at University of London, St Mary's College, UK (1967-1968); University of Notre Dame, IN, USA (1984-1985).
Mailing address: 4 Senchi Street, Airport Residential Area, PO Box 9156, Airport Accra, Ghana. T. (233) 777706. C. SECAM ACCRA. Tx C/O 2471 NCS GH, ATTENTION SECAM.

Sensi, Giuseppe Maria, H E Card

Rel: Roman Catholic; Ordained on 21 DEC 1929; b. 27 MAY 1907 in Cosenza, Italy; celibate; Citizen of Italy.
International Organizations: Member (since 1976) of the College of Cardinals.
Career: Diplomatic service for the Holy See from 1934 to 1976. Created Cardinal by Pope Paul VI in 1976.
Mailing address: 16 Piazza S Calisto, I-00153 Roma, Italy. T. (39 6) 698 72 65

Senturias, Erlinda, Dr

Rel: Christian.
International Organizations: Member of the Christian Medical Commission.
Mailing address: PO Box 2100, 150 route de Ferney, CH-1211 Genève 2, Switzerland. T. (41 22) 791 61 11. Fax 791 03 61. C. Oikoumene. Tx 415 730 OIKCH.

Seper, Franjo, H E Card

Rel: Roman Catholic; Ordained; celibate.
International Organizations: Representative of the Pontifical Biblical Commission.
Mailing address: Palazzo S Congregazione, per la Dottrina della Fede, Piazza del S Uffizio, I-00193 Roma, Italy. T. (39 6) 698 4886 – 698 3736.

Sepulveda Barra, Narciso S, Rev

Rel: Pentecostal; Ordained.
International Organizations: President of the Pentecostal Mission Church.
Mailing address: Av Pedro Montt 1473, Casilla 5391, Santiago 3, Chile.

Seraphim, S, His Beatitude Archbishop

Rel: Greek Orthodox; Ordained in 1938; b. in Artesianon, Karditsa, Thessaly, Greece; Citizen of Greece.
International Organizations: Archbishop (since 1974) of Athens and All Greece.
Career: Participated in National Resistance against Nazi occupation. Clerk, later Secretary, Holy Synod. Metropolitan, Arta in 1949. Metropolitan, Ioannina in 1958. Participated in First Pan-Orthodox Conference of Rhodes.
Mailing address: c/o Holy Synod of the Church of Greece, Athens, Greece.

Sergey, John M, Rev

Rel: Evangelical; Ordained.
International Organizations: President of the World Fellowship of Slavic Evangelical Christians.
Mailing address: PO Box 35220, Chicago IL 60635, USA. T. (1 708) 453 7997.

Serina, Mercurio M, Bishop

Rel: United Church of Christ; Ordained.
International Organizations: General Secretary of the United Church of Christ in the Philippines.
Mailing address: PO Box 718, Manila 2801, Philippines.

Serour, G I, Prof

Rel: Muslim.
International Organizations: Director of the International Islamic Centre for Population Studies and Research.
Mailing address: 11651 PO Box 1894, c/o Al Azhar University, Cairo, Egypt. T. (20 2) 932406 – (20 2) 932749. C. Azpopcent. Tx Jasmin 92615 Cairo.

Serrini, Lanfranco, O.F.M. Conv., Fr

Rel: Roman Catholic; Ordained; celibate.
International Organizations: Minister-General of the Order of Friars Minor Conventual.
Member of the Congregation for Religious and Secular Institutes.
Other organizations: Grand Chancellor, Pontifical Faculty of Theology 'San Bonaventura'.
Mailing address: Piazza SS Apostoli 51, I-00187 Roma, Italy. T. (39 6) 679 41 78 – 678 97 16.

Setae, Susana

Rel: United Church; b. 07 JUL 1946 in Port Moresby, Papua New Guinea; m. to Miri Setae; c. 3 boys and 1 girl; Citizen of Papua New Guinea; Lang. English.
International Organizations: Vice-President for South Pacific of the World Federation of Methodist Women.
Other organizations: Women's Work Secretary and Womens worker (since 1985), United Church Womens Fellowship.
Main field of work: Administration and coordination. **Career:** School Teacher, Church Agent Schools from 1964 to 1967. Lecturer to Christian leaders from 1970 to 1974. Community worker for community development from 1980 to 1982. **Educ.:** Cert. from Teachers College, Central (1963-1963); Cert. from Christian Leadership, Rabaul, Papua New Guinea (1968-1970). **Interests:** Singing. **Hons.:** Member of the Order of the British Empire. **Mem.:** National Council of Women, Pacific Region Womens Organization **Pubs.:** NIUS PNG Womens Newsletter; Share United Church Womens Newsletter and The Health of Women in Papua New Guinea.
Mailing address: United Church Assembly, Douglas Street, PO Box 1401, Port Moresby, Papua New Guinea. T. (675) 211 744
Residence: Boroko Drive, Mairi Place, PO Box 1401, Port Moresby, Papua New Guinea T. See Mailing Address

Sevi, David

Rel: Jewish; b. in Istanbul, Turkey.
International Organizations: Cantor of the Beth Jacob and Neveh Shalom synagogues, Istanbul.
Career: Participated and performed at the 40th congress, the Cantors' Assembly of the US in Jerusalem.
Mailing address: No mailing address.

Seybold, Klaus Dieter, Prof Dr

Rel: Evangelical; Ordained on 26 MAR 1961; b. 28 APR 1936 in Heidenheim, Baden-Württemberg, Germany; m. to Gisela Vetterlein; c. 2 boys; Citizen of Germany; Lang. German, English and French.
International Organizations: Professor, Faculty of Theology (since 1979) of the University of Basel.
Other organizations: Editor (since 1980), *Theologische Zeitschrift Basel.*
Main field of work: Bible research, university education. **Educ.:** 1 Theol. Exam. from Eberhard-Karls-Universität Tübingen, Germany (1956-1961); Dr. Theol. from Christian-Albrechts-Universität Kiel, Germany (1964-1968); Dr. Theol. Habil. from Christian-Albrechts-Universität Kiel, Germany (1968-1972). **Mem.:** Rotary Club **Pubs.:** Der aaronitische Segen. Studien zu Numeri 6, 22-27, Neukirchen, 1977; With U Müller, Krankheit und Heilung, Stuttgart, 1978 and Die Psalmen. Eine Einführung, Stuttgart, 1991.
Mailing address: Theologische Fakultät, Nadelberg 10, CH-4051 Basel, Switzerland. T. (41 61) 267 29 00/3
Residence: Bruderholzrain 62, CH-4102 Binningen BL, Switzerland. T. (41 61) 421 03 67

Seymour, Christine Janet

Also know as Dharmacharini Sanghadevi; Rel: Buddhist; Ordained on 06 MAY 1977; b. 07 JAN 1954 in London, UK; single; Citizen of United Kingdom; Lang. English.
International Organizations: Chairwoman of Taraloka Buddhist Retreat Centre for Women (since 1985) of the Friends of the Western Buddhist Order.
Preceptor for Women (in consultation with the head and other preceptors) (since 1989) of the Western Buddhist Order.
Main field of work: Buddhist teacher. **Educ.:** B.Sc. Biol.Sc. from University of Leicester, UK (1972-1975). **Pubs.:** several taped lectures available on various Buddhist themes.
Mailing address: Taraloka, Cornhill Farm, Bettisfield, Near Whitchurch, North Shropshire SY13 2LV, UK. T. (44 94) 875 646 – 875 654
Residence: See Mailing Address. T. See Mailing Address

Sfeir, Antoine, O.M.M., Fr Abbot

Rel: Roman Catholic; Ordained; celibate.
International Organizations: Superior General of the Antonines.
Mailing address: Ordine Mariamita Maronita, Couvent N D de Louaizé, Zouk Mosbeh 500, Lebanon.

Sfeir, Nasrallah-Pierre, His Beatitude

Rel: Maronite Catholic; Ordained on 07 MAY 1950; b. 15 MAY 1920 in Rayfoun, Kesrouan, Lebanon; celibate; Citizen of Lebanon; Lang. Arabic, French, Latin, Italian, English and Syriac.
International Organizations: Patriarch (since 1986) of Antioch and All the East.
Member of the Sacred Congregation for the Oriental Churches.
Other organizations: Head, Maronite Patriarchal Synod; President, Assembly of Catholic Patriarchs and Bishops of Lebanon.
Educ.: L.Ph., L.Th. from Saint Joseph's University, Beirut, Lebanon. **Hons.:** Legion of Honour. **Pubs.:** Homélies de Dimanche, 4 Vols.; Des visages qui ne sont plus and Des sources de l'Evangile.
Mailing address: Patriarcat Maronite, Bkerké, Lebanon. T. (961 9) 915441
Residence: See Mailing Address. T. See Mailing Address

Shafer, Raymond Donald

Rel: Brethren in Christ; Ordained in 1959; b. 22 MAY 1936 in Waynesboro PA, USA; m. to Ceola Marlene Engel; c. 1 boy and 1 girl; Citizen of USA; Lang. English.
International Organizations: General Secretary (since 1984) of the Brethren in Christ General Conference.
Other organizations: Board of Trustees (since 1972), Messiah College; Board (since 1984), Mennonite Central Committee.
Main field of work: Administration. **Career:** Board of Trustees, Western Evangelical Seminary from 1972 to 1984. Board of Directors, Azusa Pacific University, CA from 1972 to 1984. **Educ.:** B.A. from Messiah College, Grantham PA, USA (1958); Lancaster Theological Seminary, PA, USA (1958-1959); B.D. from Eastern Baptist Seminary, Philadelphia PA, USA (1960-1962); Dr. of Ministry from Fuller Theological Seminary, Pasadena CA, USA (1975-1979). **Pubs.:** Numerous articles in the *Evangelical Visitor.*
Mailing address: PO Box 245, Upland CA 91785, USA. T. (1 714) 946 0088
Residence: 865 Sharon Way, Upland CA 91786, USA T. (1 714) 982 4794

Shakarian, Demos

Rel: Armenian Pentecostal; b. in Los Angeles CA, USA.
International Organizations: Founder-President (since 1953) of the Full Gospel Business Men's Fellowship International.
Career: Prospered in the California dairy industry during the 1940s. Began sponsoring summer revivals in 1940. Helped organize Oral Robert's Los Angeles campaign in 1951. **Hons.:** recipient of many awards.
Mailing address: 3150 Bear Street, Costa Mesa CA 92626, USA. T. (1 714) 754 1400. Fax (1 714) 757 9916.

Shamar Rinpoche, S, H E

Rel: Buddhist: Kagyupa.
International Organizations: One of the main leaders of the Kagyupa tradition.
Mailing address: Monastery of Rumtek, PO Ranipul, Gangtok, Sikkim, India.

Shambuyi, Nkumbi Mudiayi

Rel: Mennonite.
International Organizations: Member, Executive Committee of the Mennonite World Conference.
Mailing address: 32 Compagnons Bâtisseurs, BP 440, Mbujimayi, Kasai Oriental, Zaire.

Shankar, Ganesh, Dr

Rel: Hindu; b. 11 FEB 1958 in Khurja, Uttar Pradesh, India; m. to Pinki; c. 1 boy and 1 girl; Citizen of India; Lang. Hindi, English and Sanskrit.
International Organizations: Head and Chairman, Yoga Department (since 1985) of the Sagar University, Madhya Pradesh.
Main field of work: Research and teaching. **Career:** Held various positions in the field of yoga at national and international level. **Educ.:** M.A. Phys. Educ. from Punjabi University, Patiala, India (1978-1979); Ph.D. Phys. Educ. from Punjabi University, Patiala, India (1985-1987); Dip. Gym. (1980-1981); Post Grad. Dip. Yoga (1981-1982). **Interests:** reading. **Pubs.:** author of 1 book and about 20 research papers.
Mailing address: Sagar University, Sagar 470 003, Madhya Pradesh, India.
Residence: Quarter No BDS-1, University Campus, Sagar 470 003, Madhya Pradesh, India

Shanmuganatha Kurukkal, T S S, Maharaja Sri

Rel: Hindu; Ordained.
International Organizations: Chief Priest of the Mavaddapuram Randasamy Kovil.
Mailing address: Mavaddapuram Randasamy Kovil, Maviddapuram, Tellippallai, Sri Lanka.

Shapasa, Kabangy Djeke, Rev

Rel: Mennonite.
International Organizations: General Secretary of the Church of Christ in Zaire (Mennonite Community).
Mailing address: BP 18, Tshikapa, Zaire.

Shapiro, Avraham, The Chief Rabbi

Rel: Jewish: Ashkenazi; Ordained; Citizen of Israel.
International Organizations: Head of the Ashkenazi Community.
Mailing address: No mailing address.

Sharif, Mohamed Ahmed

Rel: Muslim.
International Organizations: Secretary-General of the Afro-Islamic Call Coordination Council.
Mailing address: PO Box 2549, Tripoli, Libya.

Sharon, A

Rel: Jewish.
International Organizations: Head, Bombay, India and Far East International Secretariat of the Jewish Agency for Israel.
Mailing address: Directeur, 137 Mahatma Ghandi Road, PO Box 1979, Fort Bombay, India. T. (91 22) 25 2055.

Shashaani, Avideh

Rel: Muslim: Sufi.
International Organizations: East Coast Relations Representative of the School of Islamic Sufism.
Other organizations: Taught and lectured on Sufi concentration and meditation, and corporate wellness programs at US Senate, World Bank, MCI, Bureau of Public Debt, George Washington University, Sacramento University, INTELSAT, CONTEL-ASG, AFSCME, Blue Cross Blue Shield, Pan American Health Organization, Group Health Association, American Heart Association, Global Fund for Women, Federal Communications Commission, Washington DC Convention and Visitor's Association, Mortgage Bankers Association, US Small Business Administration, and other locations in the Washington area (since 1982); Chair, Refugee Women in Development; Literary Advisor and Vice-President, Literary Friends of The DC Public Library; Member, Poetry Committee of the Greater Washington Area; East Coast Director, Wayfinders, Inc. stress management corporation; Member, Research Board of Advisors, American Biographical Institute.
Career: Consultant for the United Nations, World Health Organization, and International Labour Organization. Co-Director, International Institute for Rehabilitation in Developing Countries. Consultant to developing country governments for rehabilitation programs. Appointed by UN Secretary-General to UN Expert Group Meeting on Socio-Economic Implications of Investments in Rehabilitation of the Disabled. Guest Speaker at Harvard University, Georgetown University, George Washington University, University of Maryland, Sacramento State University, Stanford University, etc. and various radio programs. **Educ.:** Grad. Deg. Ed. Planning and Management; Grad. Deg. Sufism and Philosophy; Studied Sufism under the direct guidance of H H E, Hazrat Pir, Molana Salaheddin Ali Nader Shah Angha. **Pubs.:** 7 translated books on Sufism published by the University Press of America and articles in the area of Sufism and rehabilitation in US and international journals.
Mailing address: PO Box 5827, Washington DC 20016, USA. T. (1 301) 654 9191.

Shaughnessy, Maureen, Sr

Rel: Roman Catholic; Ordained; celibate.
International Organizations: Superior General of the Sisters of Charity of Saint Elisabeth.
Mailing address: Convent St Elizabeth, Convent Station NJ 07961, USA. T. (1 201) 539 1600.

Shaukat, Jamila, Prof Dr

Rel: Muslim; b. 01 FEB 1941; single; Citizen of Pakistan.
International Organizations: Chairman (since 1988) of the Institute of Islamic Studies, University of the Punjab, Lahore.
Other organizations: Professor (since 1990), Institute of Islamic Studies, University of the Punjab, Lahore.
Career: Lecturer, Institute of Islamic Studies, University of the Punjab, Lahore from 1969 to 1976. Assistant Professor, Institute of Islamic Studies from 1976 to 1986. Associate Professor, Institute of Islamic Studies from 1986 to 1990. **Educ.:** B.A. (Hons) Arabic from University of the Punjab, Lahore, Pakistan; M.A. Arabic from University of the Punjab, Lahore, Pakistan (1963); M.A. Islamic Stds. from University of the Punjab, Lahore, Pakistan (1964); M.A. Hist. from University of the Punjab, Lahore, Pakistan (1975); B.A. Eng. Compulsory, Arabic, Econ., Persian from University of the Punjab, Lahore, Pakistan (1976); Ph.D. from University of Cambridge, UK (1984); Dip. in Arabic from Umm Al-Qura University, Mecca, Saudi Arabia (1984). **Hons.:** Gold Medal in M.A. Islamic Studies (1964); Roll of Honour, M.A. Arabic (1963). **Pubs.:** Author of more than 20 research articles published in English and Urdu journals: "Khawateen ke Haqooq wa Hassiyat Hazoor ki Nazar Main" (in Urdu), in *Urdu Encyclopedia of Islam*, University of the Punjab, Lahore; "Al-Shaikh al-Maraghi aur unki Tafseeri Khidmat" (in Urdu), in *Al-Ma'rif*, Lahore, Shumara Khasoosi, 4, 1986 and "Musnad Literature in the Second and Third Centuries AH", in *Journal of the Research Society of Pakistan*, Vol XXIII, No 2, April 1986.
Mailing address: Institute of Islamic Studies, Quaid-e-Azam Campus, University of the Punjab, Lahore, Pakistan. T. (92 42) 867993
Residence: 3/B, Islamic Centre, Quaid-e-Azam Campus, University of the Punjab, Lahore, Pakistan.

Shea, Leo Joseph, S.J., Fr

Rel: Roman Catholic; Ordained; b. 28 DEC 1903 in Massachusetts, USA; celibate.
International Organizations: Member of the Society of Jesus.
Mailing address: Collège de la Sainte Famille, 1 Bustan et Maksy Street, PO Box 73, Faggalla, Cairo, Egypt.

Sheerin, Philomena, Sr

Rel: Roman Catholic; Ordained; celibate.
International Organizations: Superior General of the Medical Missionaries of Mary.
Mailing address: 'Greenbank', Mell, Drogheda, Ireland. T. (353 41) 31028.

Shen, C T

Rel: Unknown.
International Organizations: Executive Officer of the Institute for Advanced Studies of World Religions.
Mailing address: RD 2, Rt 301, Carmel NY 10512, USA. T. (1 914) 225 0447.

Sheng, Pai, Ven

Rel: Buddhist.
International Organizations: Contact of the World Buddhist Sangha Council.
Mailing address: 6 Shaoshing North Street, Taipei, Taiwan (Rep. of China).

Shenk, Gerald

Rel: Mennonite.
International Organizations: Professor of Social Ethics of the Eastern Mennonite Seminary.
Mailing address: 461 Lee Ave, Harrisonburg VA 22801, USA.

Shenouda, Anba, III, Pope

Rel: Coptic Orthodox; Ordained; b. 03 AUG 1923; Citizen of Egypt.
International Organizations: Patriarch (since 1985) of the See of St Mark of Egypt, the Near East and All Africa.
Other organizations: 117th Pope (since 1985), Alexandria; Leader, Coptic Orthodox Church.
Career: Theological teacher and writer. Former Bishop and Professor of Theology, Orthodox Clerical College. 1st Chairman, Association of Theological Colleges in the Near East. 117th Pope, Alexandria and Patriarch of the See of St Mark of Egypt, the Near East and All Africa from 1971 to 1981. Removed from post by President Sadat and banished to desert monastery Wai Natroun in 1981. Released in 1985.
Mailing address: Coptic Orthodox Patriarchate, Anba Ruess Bldg., 222 Ramses Street, Abbasiya, POB 9035, Cairo, Egypt. T. Fax (20 2) 2825983. Tx 23281

Shepard, David Stuart, Rt Rev

Rel: Anglican; Ordained in 1955; b. 06 MAR 1929 in UK; m. to Grace Isaac; c. 1 girl; Citizen of United Kingdom.
International Organizations: Bishop (since 1975) of Liverpool.
Other organizations: Chairman (since 1989), Central Religious Advisory Committee; Chairman (since 1991), Church of England Board for Social Responsibility.
Career: Curate, St Mary Islington from 1955 to 1957. Warden, Mayflower Family Centre, Canning Town from 1957 to 1969. Bishop, Woolwich from 1969 to 1975. **Educ.:** M.A. Hist. from University of Cambridge, Trinity Hall, UK (1949-1953); University of Cambridge, Ridley Hall, UK (1953-1955). **Interests:** family, reading, music, painting, theatre. **Hons.:** Hon. LL.D., University of Liverpool (1983); Hon. D. Tech., Liverpool Polytechnic (1987). **Pubs.:** Built as a City, 1974; Bias to the Poor, 1983 and With Archbishop Derek Worlock, Better Together, 1988.
Mailing address: Church House, 1 Hannover Street, Liverpool L1 3DW, UK. T. (44 51) 708 9480. Fax 709 2885
Residence: Bishop's Lodge, Woolton Park, Liverpool L25 6DT, UK. T. See Mailing Address

Shepard, Lorenzo, Rev

Rel: Baptist.
International Organizations: President of the Progressive National Baptist Convention, Inc.
Mailing address: 42nd and Wallace St, Philadelphia PA 19104, USA.

Sherer, Moshe, Rabbi

Rel: Jewish; Ordained.
International Organizations: Chairman, New York Office of the Agudath Israel World Organization.
Mailing address: Abraham Hirsch, Hacherut Square, POB 326, Jerusalem, Israel. T. (972 2) 384357. Fax 383 634. C. Worldaguda, Jerusalem.

Sheridan, Terence, Rev
Rel: Roman Catholic; Ordained on 01 JUL 1960; b. 27 NOV 1935 in Coventry, UK; celibate; Citizen of United Kingdom; Lang. English and French.
International Organizations: Superior General (since 1987) of the Prêtres du Sacré-Coeur de Jésus de Bétharram.
Mailing address: Via Angelo Brunetti 27, I-00186 Roma, Italy. T. (39 6) 321 91 44.
Residence: See Mailing Address. T. See Mailing Address

Shibutani, Yuki
Rel: Buddhist; Ordained; b. 04 FEB 1917; Citizen of Japan.
International Organizations: Monk of Shinshu Bukkoji.
Mailing address: Bukkojidori Takakura Nishi Hairu, Shimogyo-ku, Kyoto 600, Japan. T. (81 75) 351 0135

Shik, Kim Kun, Rev
Rel: Christian.
International Organizations: General Secretary of the Korean Christian Church in Japan.
Mailing address: Japan Christian Centre, Room 55, 2-3-18 Nishi-Waseda, Shinjuku-ku, Tokyo 160, Japan.

Shikii, Syujo
Rel: Buddhist; Ordained; b. 12 NOV 1905; Citizen of Japan.
International Organizations: Chief of the Saikyoji.
Main field of work: Monk. **Career:** Manager, Tendai Masaworshu. **Educ.:** Kyoto University, Japan.
Mailing address: 5-13-1 Sakamoto, Otsu, Shiga 520-01, Japan. T. (81 775) 78 1163

Shimizu, Kosho
Rel: Buddhist; Ordained; b. 03 JAN 1911; m. to Toshiko; Citizen of Japan.
International Organizations: Monk Superior of Todaiji.
Educ.: Ryukoku University, Kyoto, Japan.
Mailing address: 406-1 Zoshicho, Nara, Nara 630, Japan. T. (81 742) 22 3371

Shinoda, Yasuo
Rel: Shinto; b. 03 JAN 1908; Citizen of Japan.
International Organizations: Chairman of the Kogakukan University.
Other organizations: Honorary Priest, Atsuta Shrine; Honorary Chairman, Jinja Shinpo.
Mailing address: 671-13 Fujisatocho, Ise, Mie 516, Japan. T. (81 596) 23 2074

Shiozawa, Daijo
Rel: Buddhist; Ordained; b. 24 NOV 1919; Citizen of Japan.
International Organizations: Monk of Rinzai-shu Nanzenji.
Mailing address: Nanzenji Nanzenji Fukuchicho, Sakyo-ku, Kyoto 606, Japan. T. (81 75) 771 0365

Shiragaki, Konin
Rel: Buddhist; Ordained; b. 12 OCT 1924; Citizen of Japan.
International Organizations: Chief of the Kuramadera.
Main field of work: Nun. **Career:** Manager, Kurama Kokyo.
Mailing address: 1074 Kuramahoncho, Sakyo-ku, Kyoto 601-11, Japan. T. (81 75) 741 2003

Shirai, Eiji
Rel: Shinto; Ordained; b. 10 DEC 1915; Citizen of Japan.
International Organizations: Chief Priest of the Tsuruokahachimangu.
Career: Vice-Governor, Association of Shinto Shrine. **Educ.:** Kokugakuin University, Tokyo, Japan.
Mailing address: 2-18-18 Yukinoshita, Kamakura, Kanagawa 248, Japan. T. (81 467) 22 0315

Shirayanagi, Peter Seiichi, H E Mgr
Rel: Roman Catholic; Ordained on 21 DEC 1954; b. 17 JUN 1928 in Tokyo, Japan; celibate; Citizen of Japan; Lang. Japanese.
International Organizations: Archbishop (since 1970) of Tokyo.
Other organizations: President (since 1983), Catholic Bishops' Conference of Japan.
Career: Priest from 1954 to 1966. Auxiliary Bishop, Tokyo from 1966 to 1969. Archbishop Coadjutor, Tokyo from 1969 to 1970. **Educ.:** Studied Phil. from Sophia University, Tokyo, Japan (1947-1951); Studied Theol. from Sophia University, Tokyo, Japan (1951-1953); Dr. from Urbanian Pontifical University, Vatican City State (1957-1960).
Mailing address: Archbishop's House, 3-16-15 Sekiguchi, Bunkyo-ku, Tokyo 112, Japan. T. (81 3) 3943 2301
Residence: See Mailing Address. T. See Mailing Address

Shitemi, Simeon
Rel: Quaker; b. 14 OCT 1931 in Kakamega, Western Province, Kenya; m. to Beatrice Monyani; c. 4 boys and 1 girl; Citizen of Kenya; Lang. Luhya (Bantu dialect), Swahili and English.
International Organizations: Former Clerk of the Friends World Committee for Consultation.
Deputy Director General, Flying Doctor Service (since 1988) of the African Medical and Research Foundation.
Other organizations: Vice-Chairman (since 1991), Maseno University Council; Adviser to International Co-operative Alliance (ICA).
Career: District Officer, Office of the President (Kenya) from 1961 to 1966. Under Secretary, Office of the President from 1975 to 1977. Counsellor, Kenya Mission to United Nations from 1978 to 1982. Permanent Secretary for Tourism and Wildlife from 1984 to 1985. Permanent Secretary, Ministry of Health (Kenya) from 1985 to 1986. Clerk, Friends World Committee for Consultation from 1986 to 1991. **Educ.:** B.A. from University of London, UK (1962-1965); Dip. from University of Leeds, UK (1967-1968); Dip. from Makerere University, Kampala, Uganda (1963-1964). **Interests:** church activities, disarmament issues. **Hons.:** Award from Black American Women for Cultural Enlightment. **Mem.:** United Kenya Club; Pen Club, London **Pubs.:** *Memoirs of a Kenyan Civil Servant* (forthcoming) and various articles.
Mailing address: PO Box 30985, Nairobi, Kenya.
Residence: Jacaranda Ave, Nairobi, Kenya. T. (254 2) 562048

Shogreen, Andy, Rev
Rel: Moravian; Ordained.
International Organizations: Superintendent of the Moravian Church in Nicaragua.
Mailing address: Puerto Cabezas, Zelaya Norte, Nicaragua.

Shola, Ugyen
Rel: Buddhist: Mahayana.
International Organizations: Member of the Foundation for the Preservation of the Mahayana Tradition.
Other organizations: Director, Tushita Mahayana Meditation Centre.
Mailing address: C-259 Defence Colony, New Delhi 110024, India. T. (91 11) 462 9562

Shotwell, J Ralph
Rel: Christian.
International Organizations: Executive Director of the International Council of Community Churches.
Mailing address: 7808 College Drive, Suite 2SE, Palos Heights IL 60463, USA. T. (1 708) 361 2600.

Shoults, Harold E, Commr
Rel: Salvation Army, The.
International Organizations: Territorial Commander of The Salvation Army.
Mailing address: 860 North Dearborn Street, Chicago IL 60610, USA.

Shubitz, Amy
Rel: Christian.
International Organizations: Director of the Tucson Ecumenical Council Task Force for Central America.
Mailing address: 317 West 23rd Street, Tucson AZ 85713, USA. T. (1 602) 628 7525.

Siahaan, G H M, Rev
Rel: Protestant.
International Organizations: Ephorus of the Batak Protestant Christian Church.
Mailing address: Kantor Pusat HKBP, Pearaja-Tarutung, Tapanuli Utara, Indonesia.

Sidarouss, Stephanos, I, H E Card
Rel: Coptic Catholic; Ordained in 1939; b. in Cairo, Egypt; celibate; Citizen of Egypt.
International Organizations: Past Patriarch of Alexandria.
Other organizations: Spiritual Leader, Catholic Coptic Community of Egypt.
Career: Barrister. Left his practice for the Seminary in 1933. Principal and Professor, Evreux Seminary, Dax Seminary, Beauvais Seminary. Rector, Catholic Coptic Seminary at Tanta from 1947 to 1953. Rector, Catholic Coptic Seminary at Maadi from 1947 to 1958. Bishop in 1948. Patriarch in 1958. Created Cardinal in 1965.
Mailing address: 34 Iben Sandar Street, Qubbah, Cairo, Egypt.

Siddiqi, Bashir Ahmad, Prof Dr
Rel: Muslim; b. 01 SEP 1928; married; Citizen of Pakistan; Lang. English, Arabic, Urdu, Punjabi and Persian.
International Organizations: Professor of the Institute of Islamic Studies, University of the Punjab, Lahore.
Career: Vice-President, Anjuman Arabi University Oriental College, Lahore in 1961. Member, Academic Committee, Board of Intermediate and Secondary Education, Lahore from 1970 to 1972. Member, Sub-Committee for Syllabus Reform, Education Dept., Government of Pakistan in 1971. Representative, University of the Punjab in the Action Committee on the Development of Jamia Islamia Bahawalpur in 1972. Chairman, Dept. of Islamic Studies, University of the Punjab from 1973 to 1974. Chairman, Dept. of Islamic Studies, University of the Punjab from 1979 to 1980. Member, Mosque Committee, University of the Punjab from 1975 to 1982. President, Old Boys Association, Dept. of Islamic Studies, University of the Punjab from 1981 to 1982. Professor and Chairman, Dept. of Islamic Studies, Islamia University, Bahawalpur from 1987 to 1988. Member, Editorial Board, *Light of Islam* (monthly journal), Lahore. Convenor, Board of Islamic Studies, Board of Intermediate and Secondary Education, Lahore. Member, Board of Islamic Studies, University of Multan. Member, Academic Council, Islamia University, Bahawalpur. Member, New Campus Management Committee, University of the Punjab (1 term). President, Mosque Committee, Masjid-o-Madrassah, E-Block, New Campus (3 terms). Delivered more than 1000 extension lectures and speeches at various colleges, centres of learning, public meetings and on radio and television, Pakistan. Delivered lectures in Ulema Academy, Auqaf Dept., Lahore on Fiqh-o-Qanoon and comparative study of religions. Contributed articles and speeches to various All Pakistan Conferences and other functions of national and religious importance. **Educ.:** M.A. Islamic Stds. (1st); M.A. Arabic (1st); Ph.D. Islamic Stds. **Pubs.:** Contributed to research articles to the *Encyclopedia of Islam* (in Urdu), University of the Punjab, Lahore; Contributed to research articles in various journals and revision of the translation and translation of several books, including: *Usul-al-Shashi* (Islamic Jurisprudence).
Mailing address: Institute of Islamic Studies, University of the Punjab, Quaid-e-Azam Campus, Lahore, Pakistan. T. (92 42) 867993

Sidiqui, Sulaiman, Dr
Rel: Muslim.
International Organizations: Head of the Department of Islamic Studies, Osmania University, Hyderabad.
Mailing address: Dept. of Islamic Studies, Osmania University, Hyderabad 500007, India.

Siegman, Henry, Rabbi
Rel: Jewish; Ordained in 1951; b. 12 DEC 1930 in Germany; m. to (1st) Selma Goldberger (divorced), (2nd) Miriam Cantor; c. 1 boy and 2 girls; Citizen of USA.
International Organizations: Executive Director (since 1978) of the American Jewish Congress.
Other organizations: National Vice-Chairman (since 1966), Religion in American Life; Executive Committee (since 1966), Interreligious Commission on Peace.
Career: Came to the US in 1942. Naturalized in 1948. Served to 1st Lieutenant Chaplain with AUS (Army of the United States) from 1952 to 1954. National Director, Community Activities Division, Union of Orthodox Jewish Congregations of America from 1953 to 1959. Executive Secretary, American Association for Middle East Studies from 1959 to 1964. Director for International Affairs, National Community Relations Advisory Council, New York City from 1964 to 1965. Executive Vice-President, Synagogue Council of America, New York City from 1965 to 1978. An organizer, White House Conference on Civil Rights in 1967. Chairman, Interreligious Commission of General Secretaries (National Council of Churches-US Catholic Conference-Synagogue Council of America) in 1973. Editor, Middle East Studies from 1959 to 1964. Guest Lecturer, University of Illinois, Columbia NY, Williams College MA. Steering Committee, Executive Committee, National Urban Coalition. **Educ.:** B.A. from New School for Social Research, New York NY, USA (1961); P.G. Stds. (1961-1964). **Hons.:** Decorated Bronze Star; Designated Distinguished American by President of the USA (1970). **Mem.:** AAUP Council on Foreign Relations; National Conference of Jewish Communal Service; Association of Jewish Community Relations Workers **Pubs.:** Contributed articles to professional journals.
Mailing address: 15 E 84th St, New York NY 10028, USA.
Residence: 685 West End Ave, New York NY 10025, USA.

Siegrist-Rey, Denise Marcelle Emilie Antoinette
Rel: Protestant; b. 31 JUL 1946 in Liège, Belgium; m. to Jean-Paul Siegrist; c. 2 boys and 1 girl; Citizen of Belgium and France; Lang. French, English, German and Dutch.
International Organizations: Secretary-General/Coordinator (since 1981) of the European YWCAs.
Main field of work: Sociocultural. **Educ.:** ISSD, Brussels, Belgium. **Interests:** women and Europe. **Mem.:** Belgian Liberal Party (PRL)
Mailing address: ELG Secretariat, Avenue Brugmann 94, B-1060 Bruxelles, Belgium. T. (32 2) 344 98 61.
Residence: Avenue du Cor de Chasse 19, B-1410 Waterloo, Belgium

Sieunarine, Everson T, Rt Rev
Rel: Presbyterian.
International Organizations: Moderator of the Presbyterian Church in Trinidad and Grenada.
Mailing address: PO Box 92, Paradise Hill, San Fernando, Trinidad-Tobago.

Sievers, Wilhelm, Dr
Rel: Evangelical Lutheran.
International Organizations: Bishop of Evangelical Lutheran Church in Oldenburg.
Mailing address: Postfach 1709, Philosophenweg 1, 26121 Oldenburg, Germany.

Sigurgeirsson, Petur, Rt Rev
Rel: Evangelical Lutheran.
International Organizations: Contact of the Evangelical Lutheran Church of Iceland.
Mailing address: Sudurgata 22, IS-101 Reykjavik, Iceland.

Sihombing, P M, Rev
Rel: Protestant.
International Organizations: General Secretary of the Batak Protestant Christian Church.
Mailing address: Kantor Pusat HKBP, Pearaja-Tarutung, Tapanuli Utara, Indonesia.

Sikazwe, Godfrey Mwansa, Rev
Rel: United Church.
International Organizations: Synod Moderator of the United Church of Zambia.
Mailing address: PO Box 50122, Nationalist Road at Burma Road, Lusaka, Zambia.

Silberman, Curt C
Rel: Jewish.
International Organizations: Chairman of the Board of the American Federation of Jews from Central Europe.
Mailing address: 570 Seventh Ave, New York NY 10018, USA. T. (1 212) 921 3871. C. AMFEDERA, NEW YORK.

Silva, Cèlia Junker

Rel: Methodist.

International Organizations: Vice-President for Latin America of the World Federation of Methodist Women.

Mailing address: José Anibal Colleoni 211, 09735 Audueramos SBC, Sao Paulo, Brazil.

Silva Henríquez, Raúl S, S.D.B., H E Card

Rel: Roman Catholic; Ordained on 03 JUL 1938; b. 27 SEP 1907 in Talca, Chile; celibate; Citizen of Chile.

International Organizations: Member (since 1962) of the College of Cardinals.

Member of the Society of St Francis of Sales.

Other organizations: Archbishop Emeritus (since 1983), Santiago de Chile.

Career: Diocesan Bishop, Valparaíso from 1959 to 1961. Archbishop, Santiago de Chile from 1961 to 1983. Created Cardinal by Pope John XXIII in 1962. **Hons.:** Hon. LL.D., University of Notre Dame; Latin America Jewish Conference Human Rights Prize (1972); UN Prize (1978); Bruno Kreisky Foundation (1979). **Pubs.:** La Misión Social del Cristianismo: Conflicto de Clases o Solidaridad Cristiana.

Mailing address: Calle Los Pescadores 2260, Santiago, Chile. T. (56 2) 223 30 35

Silvestrini, Achille, H E Card

Rel: Roman Catholic; Ordained on 13 JUL 1946; b. 25 OCT 1923 in Brisighella, Italy; celibate; Citizen of Italy.

International Organizations: Prefect (since 1988) of the Supreme Tribunal of the Apostolic Signatura.

Member of the Sacred Congregation for the Oriental Churches. Member of the Congregation for the Causes of Saints. Member of the Congregation for Bishops. Member of the Congregation for the Evangelization of Peoples. Member of the Pontifical Council for Promoting Christian Unity. Member of the Pontifical Council for the Interpretation of Legislative Texts.

Other organizations: Member (since 1988), College of Cardinals; Member, Council of Cardinals and Bishops; President, Supreme Court of Appeal, Vatican City State.

Career: Elected Archbishop, Novaliciana, Mauritania in 1979. Consecrated Bishop in 1979. Secretary, Council for the Public Affairs of the Church in 1979. Created Cardinal by Pope John Paul II with title of St Benedict outside the Gate of St Paul in 1988.

Mailing address: Palazzina della Zecca, 00120 Città del Vaticano, Vatican. T. (39 6) 698 48 38

Simaan, Nemeh Khalil, H E Mgr

Rel: Roman Catholic; Ordained; b. in Ramleh; Citizen of Jordan.

International Organizations: Latin Bishop of Amman.

Career: Assistant Priest in Madaba from 1932 to 1933. Vicar in the Parish, Hossan, Jordan from 1933 to 1940. Vicar-General for the Patriarch and President, Ecclesiastic Court of Justice in 1940. Auxiliary Bishop in Amman for the East Bank, Jordan in 1965. **Hons.:** Renaissance Medal; Star of Jordan; Independence Medal; Commander of the Cedars, Lebanon; Commendatore, Italy; Equestrian Order of the Holy Sepulchre in Jerusalem; Order of St John of Jerusalem. **Mem.:** Pontifical Mission for Palestine **Pubs.:** Religious and political articles in local newspapers and reviews.

Mailing address: PO Box 1317, Amman, Jordan.

Simangunsong, H, Rev

Rel: Christian.

International Organizations: General Secretary of the Indonesian Christian Church.

Mailing address: Kantor Pucuk Pimpinan (Centre), Jalan Marihat 109-111, Pematangsiantar, Sumatra, Indonesia.

Simms, George Otto, Most Rev

Rel: Church of Ireland; Ordained in 1936; b. 01 JUN 1910 in Dublin, Ireland; m. to Mercy Felicia; c. 3 boys and 2 girls; Citizen of Ireland.

International Organizations: President (since 1964) of The Leprosy Mission International.

President (since 1983) of the Society for Promoting Christian Knowledge.

Career: Ordained Deacon in 1935. Curate Assistant, St Bartholomew's, Dublin from 1935 to 1938. Chaplain, Lincoln Theological College from 1938 to 1939. Dean of Residence, Trinity College, Dublin from 1939 to 1952. Chaplain, Church of Ireland Training College from 1943 to 1952. Dean, Cork in 1952. Bishop, Cork from 1952 to 1956. Archbishop, Dublin, Bishop of Glendalough and Kildare, Primate of Ireland from 1956 to 1969. Archbishop, Armagh and Primate of Ireland from 1969 to 1980. Chairman, Irish Council of Churches from 1972 to 1974. Co-Chairman, Anglican-Roman Catholic International Commission on the Theology of Marriage from 1968 to 1975. **Hons.:** DD hc, Huron, Canada (1963); Hon DCL, Kent (1978); DLitt hc, New University of Ulster (1981); Honorary Fellow, Trinity College, Dublin (1978). **Mem.:** Member, Royal Irish Academy (1957); Honorary Life Member, Royal Dublin Society (1984) **Pubs.:** Editor, Rathmichael: A Parish History by Kathleen Turner, Dublin, 1986; Angels and Saints, 1988 and Exploring the Book of Kells, 1988.

Mailing address: 62 Cypress Grove Road, Dublin 6W, Ireland. T. (353 1) 905594

Simonis, Adrianus Johannes, H E Card

Rel: Roman Catholic; Ordained on 15 JUN 1957; b. 26 NOV 1931 in Lisse, Rotterdam, Netherlands; celibate; Citizen of Netherlands.

International Organizations: Member (since 1985) of the College of Cardinals.

Member of the Pontifical Council for Promoting Christian Unity.

Other organizations: Archbishop (since 1983), Utrecht; President, Bishops' Conference of The Netherlands.

Career: Consecrated Bishop, Rotterdam in 1971. Created Cardinal by Pope John Paul II in 1985.

Mailing address: Aartsbisdom, BP 14019, 3508 SB Utrecht, Netherlands. T. (31 30) 31 69 56

Simonnet, R P

Rel: Roman Catholic.

International Organizations: Representative of the Society of St Edmund.

Mailing address: F-89110 Aillant-sur-Tholon, France.

Sin, Ho Chee, Bishop

Rel: Methodist; Ordained.

International Organizations: Bishop of Methodist Church in Singapore.

Mailing address: 10 Mount Sophia, Singapore 0922, Singapore.

Sin, Jaime L, H E Card

Rel: Roman Catholic; Ordained on 03 APR 1954; b. 31 AUG 1928 in New Washington, Aklan, Philippines; celibate; Citizen of Philippines; Lang. English and Spanish.

International Organizations: Member (since 1976) of the College of Cardinals.

Permanent Member (since 1977) of the Synod of Bishops. Member (since 1983) of the Congregation for Bishops. Member (since 1983) of the Congregation for the Evangelization of Peoples. Member (since 1983) of the Congregation for Catholic Education. Member (since 1984) of the Pontifical Council for Social Communications. Member (since 1984) of the Sacred Congregation for the Sacraments and Divine Worship. Member (since 1984) of the Congregation for the Clergy.

Other organizations: Metropolitan Archbishop (since 1974), Manila; Chairman (since 1982), CBCP Commission on Colegio Filippino; Member (since 1982), CBCP Commission on Mass Media; Member (since 1981), Council of Cardinals for the Study of Organization and Economic Questions Relating to the Holy See; Member (since 1983), Secretariat for Non-Christians; Member (since 1984), Council for Public Affairs of the Church; Member (since 1987), Prefecture of Economic Affairs of the Holy See.

Career: Missionary Priest, Archdiocese of Capiz from 1954 to 1957. First Rector, St Pius X Seminary, Roxas City from 1957 to 1967. Domestic Prelate, Pope John XXIII in 1960. Titular Bishop, Obba and Auxiliary Bishop of Jaro in 1967. Apostolic Administrator, Sede Plena, Archdiocese of Jaro in 1970. Titular Archbishop, Massa Lubrense and Coadjutor Archbishop of Jaro, with the right of succession in 1972. Metropolitan Archbishop, Jaro in 1972. Vice-President, Catholic Bishops' Conference of the Philippines (CBCP) in 1974. Elected President, CBCP in 1977. Re-elected President, CBCP in 1979. Created Cardinal by Pope Paul VI in 1976. **Educ.:** B.Sc. Ed. from Immaculate Conception College, Roxas City, Philippines (1959). **Interests:**

music (Bach, Chopin, Wagner, Strauss), reading, writing. **Hons.:** Hon. LL.D., Adamson University (1975), Angeles University (1978), Manhattan College, Bronx NY (1984), Seton Hall University, South Orange NJ (1986); Hon. D.H.L., De la Salle University, Manila (1975), Brandeis University, Boston MA (1984), University of San Francisco, CA (1986), Georgetown University, Washington DC (1986), St Michael's College, Winooski VT (1987), University of San Diego, CA (1988); Hon. D.S.T., University of Santo Tomas, Manila (1977), Boston College, Chestnut Hills, MA (1986), Santa Clara University, San José CA (1987); Hon. D.Hum., University of San Augustin, Iloilo City (1980), Stonehill College, North Easton MA (1986), University of Portland, OR (1987), Lewis University, Romeville IL (1987), University of the East, Manila (1989); Hon. D.Phil., Ateneo de Manila University, Quezon City and Fu Jen Catholic University, Taipeh (1980); Hon. Dr.Ped., Assumption University San Fernando, Pampanga (1981); Hon. D.Litt., Republic Central Colleges, Angeles City (1982); Hon. Dr.Soc., Aquinas University, Legaspi City (1985); Hon. D.D., Yale University, New Haven CT (1986); Hon. Dr.Soc.Phil., Loras College, Dubuque IA (1989); Distinguished Son of Iloilo Award (1974); Honored Son of Capiz Award (1976); Outstanding Aklanon Award (1979); Grand Cross, Knights of the Holy Sepulchre of Jerusalem (1976); Distinguished Son of Manila (1976); Rajah Soliman Award for Distinguished Citizenry, Manila (1976); Gran Cruz de Isabel la Catolica of the King of Spain, Manila (1977); Knight Grand Cross of Rizal, Manila (1978); Outstanding Citizens' Award for Religion, Manila (1979); Bailiff Grand Cross of Honor and Devotion, Sovereign Military Order of Malta, Rome (1979); International League of Human Rights Award, New York (1986); Received numerous Plaques, Certificates and Resolutions of Appreciation from various government, civic, and religious organizations, from the Philippines and abroad. **Pubs.:** "The Christian Basis of Human Rights", 1978 (delivered before the Philosophical Society of the Philippines, July 29, 1978), Published in *L'Osservatore Romano*, 1978, *The Fookien Times*, 1981-1982; Discipline, Discipleship and Discerning Service, The Making of Men for Others, 1980 and over 200 papers, articles in periodicals.
Mailing address: Arzobispado de Manila, Arsobispo Street, Intramuros, Manila, Philippines. T. (63 2) 49 21 61
Residence: Villa San Miguel, Shaw Boulevard, Mandaluyong, Metro Manila, Philippines. T. (63 2) 79 26 71

Sinaga, L H, Rev
Rel: Christian.
International Organizations: Ephorus/Bishop of Batak Christian Community Church.
Mailing address: JL Hos Cokroaminoto 96, PO Box 96/MT, Jakarta 10310, Indonesia.

Singer, Israel
Rel: Jewish; b. 29 JUL 1942 in New York City NY, USA; m. to Evelyne Kuhl; c. 2 boys and 3 girls; Citizen of USA; Lang. English, French, German, Hebrew and Yiddish.
International Organizations: Secretary-General (since 1980) of the World Jewish Congress.
Main field of work: Politics. **Career:** Professor, City University of New York, USA from 1966 to 1969. Professor, Bar-Ilan University, Israel from 1969 to 1971. Professor, City University of New York, USA from 1971 to 1977. Assistant to Mayor Lindsay, New York in 1971. Assistant to President Ford for intergroup affairs in 1976. **Educ.:** Rabbinical Ordination from Yeshiva Torah vo Daas, New York NY, USA (1960-1964). **Interests:** Judaism, art.
Mailing address: 501 Madison Avenue, New York NY 10022, USA. T. (1 212) 755 5770. Tx 236129. Fax (1 212) 755 5883.
Residence: 112-14 Jewel Avenue, Forest Hills NY 11375, USA T. (1 718) 544 3225

Singer, Shirley
Rel: Jewish.
International Organizations: Executive Director of the Emunah World Religious Zionist Women's Organization.
Mailing address: Seven Penn Plaza, New York NY 10001, USA. T. (1 212) 564 9045.

Singh Rawat, Prem Pal S
See Maharaj Ji.

Sinnemäki, Maunu, Rev
Rel: Evangelical Lutheran.
International Organizations: Secretary-General for the Council for Foreign Affairs of the Evangelical Lutheran Church of Finland.
Mailing address: Satamakatu 11, SF-00160 Helsinki 16, Finland.

Sirait, W J, Rev
Rel: Christian.
International Organizations: President of the Indonesian Christian Church.
Mailing address: Kantor Pucuk Pimpinan (Centre), Jalan Marihat 109-111, Pematangsiantar, Sumatra, Indonesia.

Sirat, René-Samuel, Rabbi
Rel: Jewish; b. 13 NOV 1930 in Bône (now Annaba), Algeria; m. to (1st) Colette Salamon, (2nd) Nicole Holzman; c. 1 boy and 2 girls; Citizen of France.
International Organizations: Director (since 1985) of the Ecole des Hautes Etudes de Judaisme.
Other organizations: Director of Hebrew Studies, Professor (since 1965), Institut National des Langues et Civilisations Orientales (INALCO, formerly ENLOV); Director (since 1974), Centre de Documentation et Recherches des Etudes Juives modernes et contemporaines.
Career: Rabbi, Toulouse from 1952 to 1955. Chaplain, Jeunesse juive from 1955 to 1963. Professor, Ecole Rabbinique de France from 1965 to 1970. Professor, Ecole Rabbinique de France from 1977 to 1980. Inspector-General of Hebrew, Ministry of Education from 1972 to 1980. President, Hebrew Examining Board, Certificate of Professional Aptitude and Higher Studies from 1973 to 1978. President, Hebrew Examining Board, Agrégation from 1978 to 1980. President, Centre Universitaire Rachi, Troyes in 1989. Chief Rabbi, France from 1981 to 1987. Chief Rabbi, Consistoire Central in 1988. **Hons.:** Dr. h.c., Yeshiva University, USA (1985); Knight, Legion of Honour; Officer, National Order of Merit; Officer, Palmes Académiques; Jerusalem Prize; Commander of the Arts and Letters. **Pubs.:** Co-Editor, Omer Hasikha, 1973; Co-Editor, Mélanges A Neher, 1974 and Co-Editor, Mélanges Vajda, 1974-80.
Mailing address: INALCO, 106 quai de Clichy, F-92110 Clichy, France.

Sisse, Mustapha
Rel: Muslim.
International Organizations: Chairman of the Afro-Islamic Call Coordination Council.
Mailing address: Embassy of Senegal, Riyadh, Saudi Arabia.

Sisson Filho, Arnaldo S
Rel: Theosophist.
International Organizations: Head, Permanent Secretariat and President of the World Theosophical Youth Federation.
Mailing address: Caixa Postal 10360, Porto Alegre 90 000 RS, Brazil.

Sister Mary Joell, S
See Helen Mary Overman.

Sitadevi Yogendra, S, Smt
Rel: Hindu; b. 01 JUN 1912 in Mysore State, India; Citizen of India; Lang. English, Hindi and Gujarati.
International Organizations: Secretary (since 1930) of The Yoga Institute, Bombay.
Mailing address: The Yoga Institute, Prabhat Colony, Santacruz East, Bombay 400 055, India. T. (91 22) 6122185

Sitruk, Joseph
Rel: Jewish: Sephardi; b. 16 OCT 1944 in Tunis, Tunisia; married; c. 9 children; Citizen of France.
International Organizations: Chief Rabbi (since 1988) of France.
Career: Began to serve as the rabbi, Strasbourg in 1970. Rabbi in charge, the Marseilles community from 1975 to 1987.
Mailing address: Consistoire Central, Union des Communautés Juives de France, 17 rue Saint Georges, F-75009 Paris, France.

Situ Rinpoche, S, H E
Rel: Buddhist: Kagyupa.
International Organizations: One of the main leaders of the Kagyupa tradition.
Mailing address: Monastery of Rumtek, PO Ranipul, Gangtok, Sikkim, India.

Sjoberg, Donald W, Rev

Rel: Evangelical Lutheran.
International Organizations: President of the Evangelical Lutheran Church in Canada.
Mailing address: 500-365 Hargrave St, Winnipeg MB R3B 2K3, Canada.

Skaaheim, Anfin

Rel: Lutheran.
International Organizations: Secretary-General of the Nordic Home-mission Council.
Mailing address: Det norske luthersle Indremisjonsselskap, Box 6830 St Olavs plass, N-0130 Oslo 1, Norway. T. (47 2) 11 13 15. Fax 38 14 58.

Skelskey, Peter

Rel: Eckankar.
International Organizations: President of the Eckankar, Religion of the Light and Sound of God.
Mailing address: PO Box 27300, Minneapolis MN 55427-0300, USA.

Sköld, Margareta

Rel: Christian.
International Organizations: Member of the Christian Medical Commission.
Mailing address: PO Box 2100, 150 route de Ferney, CH-1211 Genève 2, Switzerland. T. (41 22) 791 61 11. Fax 791 03 61. C. Oikoumene. Tx 415 730 OIKCH.

Skop, Gizela Maria

Rel: Roman Catholic.
International Organizations: Contact of the Lumière et Vie.
Mailing address: Evangelisations-Zentrum, Licht-Leben, Kreuzweg 28, 67316 Carlsberg, Germany.

Skouras Oldknow, Dina S

Rel: Greek Orthodox.
International Organizations: President of the Greek Orthodox Ladies Philoptochos Society.
Mailing address: 345 E 74th St, New York NY 10021, USA. T. (1 212) 744 4390.

Skulason, Olafur, Rt Rev

Rel: Lutheran.
International Organizations: President of the Lutheran Foundation for Interconfessional Research.
Council Member of the Lutheran World Federation. Member, Program Committee for World Service of the Lutheran World Federation. Member, Standing Committee for Membership of the Lutheran World Federation.
Other organizations: Presiding Bishop, Evangelical Lutheran Church – The National Church of Iceland.
Mailing address: 8 rue Gustave-Klotz, F-67000 Strasbourg, France. T. (33) 88 36 10 57.

Sladkevicius, Vincentas, H E Card

Rel: Roman Catholic; Ordained on 25 MAR 1944; b. 20 AUG 1920 in Zasliai, Kaisiadorys, USSR; celibate; Citizen of Lithuania.
International Organizations: Member (since 1988) of the College of Cardinals.
Member of the Sacred Congregation for the Sacraments and Divine Worship. Member of the Congregation for Catholic Education. Member of the Pontifical Council for Justice and Peace. Member of the Pontifical Council for Dialogue with Non-Believers.
Other organizations: Archbishop, Kaunas; President, Episcopal Conference of Lithuania.
Career: Elected Bishop, Abora in 1957. Consecrated in 1957. Created Cardinal by Pope John Paul II in 1988.
Mailing address: Vilnius gatvé 4, 233000 Kaunas, Lithuania. T. (7) 22 21 97

Slaghenauffi, Daniele, Fr

Rel: Roman Catholic; Ordained; celibate.
International Organizations: Superior General of the Company of Mary for the Education of Deaf-Mutes.
Mailing address: Via Antonio Provolo 20, I-37123 Verona, Italy. T. (39 45) 8007419. Fax 575396.

Slater, C Peter R L

Rel: Anglican; Ordained in 1958; b. 24 MAR 1934 in Newcastle-upon-Tyne, UK; m. to Joanne McWilliam; c. 3 girls; Citizen of Canada.
International Organizations: Professor of Theology (since 1982) of the Toronto School of Theology Graduate Faculty.
Other organizations: Professor (since 1983), Centre for Religious Studies; Honorary Assistant (since 1983), Parish of St Clements, Eglinton; Co-Editor (since 1985), *Toronto Journal of Theology*.
Main field of work: Professor of Systematic Theology. **Career:** Assistant Professor of Religion, Haverford College, PA from 1964 to 1970. Associate Professor of Religion Sir George Williams, University of Montreal from 1970 to 1971. Associate and Full Professor of Religion, Carleton University, Ottawa from 1971 to 1982. Professor of Theology, Wycliffe College from 1982 to 1985. Dean of Divinity, Trinity College, University of Toronto from 1985 to 1990. **Educ.:** B.A. from McGill University, Montreal, Canada (1954); B.A. from University of Cambridge, Queens' College, UK (1957); M.A. from University of Cambridge, Queens' College, UK (1963); Ph.D. from Harvard University, Cambridge MA, USA (1964). **Interests:** sailing, theatre. **Hons.:** Recipient, Gold Medal Philosophy, McGill (1954). **Mem.:** Canadian Society for the Study of Religion (President, 1978-80); American Academy of Religion (Regional President, 1975); Society for the Values of Higher Education; American Theological Society; American Society for the Study of Religion; Canadian Theological Society **Pubs.:** The Dynamics of Religion, 1978; Editor, Religion and Culture in Canada/Religion et Culture au Canada, 1977 and Co-editor, Traditions in Contact and Change, 1983.
Mailing address: Toronto, Ontario M5S 1H8, Canada.

Slonim, Reuben, Rabbi

Rel: Jewish; Ordained; b. 27 FEB 1914 in Winnipeg, Manitoba, Canada; m. to Rita Short; c. 1 girl.
International Organizations: President of the Association for the Living Jewish Spirit, Toronto.
Career: Civilian Chaplain, RCAF (Royal Canadian Air Force) from 1940 to 1943. Rabbi, Community Temple, Cleveland OH from 1943 to 1944. Rabbi, Temple Beth El, Troy NY from 1944 to 1947. Rabbi, McCaul St Synagogue, Toronto from 1937 to 1940. Rabbi, McCaul St Synagogue, Toronto from 1947 to 1954. Jewish Chaplain, New York State Assembly from 1944 to 1947. Rabbi, Beth Tzedec Congregation, Toronto from 1954 to 1955. Chairman, Synagogue Council, State of Israel Bonds from 1955 to 1960. Rabbi, Congregation Habonim, Toronto from 1960 to 1983. Executive Member, Ontario Zionist Region and Canadian Jewish Congress, Ontario Region from 1947 to 1960. National Speakers' Bureau, Zionist Organization of America. President, Ministry Association of Capitol District, New York State in 1946. President, Toronto Zionist Council from 1947 to 1951. President, Assembly of Jewish Organizations, Toronto from 1947 to 1952. Co-Chairman, Interfaith Committee of Metropolitan Toronto Community Chest from 1953 to 1955. Chaplain, Variety Club, Toronto from 1952 to 1956. Associate Editor, *The Telegram*, Toronto from 1955 to 1971. **Educ.:** B.S.A.S. from Illinois Institute of Technology, Chicago IL, USA (1933); M.H.L. from Jewish Theological Seminary, New York NY, USA (1937); Albany Law School, New York, USA (1945-1947). **Mem.:** Rabbinical Assembly of America; World Society of Skippers of the Flying Dutchmen **Pubs.:** Family Quarrel, 1977; Grand to be and Orphan, 1983 and To Kill A Rabbi, 1987.
Mailing address: 625 Roselawn Ave, Toronto, Ontario M5N 1K7, Canada.
Residence: See Mailing Address.

Smellie, S H, Rev

Rel: United Church.
International Organizations: General Secretary of the United Church of Jamaica and Grand Cayman.
Mailing address: 12 Carlton Crescent, PO Box 359, Kingston 10, Jamaica.

Smet, Joachim
Rel: Roman Catholic.
International Organizations: Member of the Carmelite Third Order.
Mailing address: Via Sforza Pallavicini 10, I-00193 Roma, Italy.

Smidt, Wolf-Udo, Rev
Rel: Evangelical.
International Organizations: Secretary of the Bremen Evangelical Church.
Mailing address: Franziuseck 2-4, 28199 Bremen, Germany.

Smit, Herman, Pastor
Rel: Seventh-Day Adventist.
International Organizations: President, Greek Mission of the General Conference of Seventh-Day Adventists.
Mailing address: Keramikou 18, Athens 10437, Greece.

Smith, Dwight
Rel: Christian.
International Organizations: President of the United World Mission.
Mailing address: Box 250, Union Mills NC 28167-0250, USA. T. (1 704) 287 8996.

Smith, Nico J, Dr
Rel: Dutch Reformed; Ordained in 1952; b. 11 APR 1929 in Kroonstad, Orange Freestate, South Africa; m. to Helena Faul; c. 3 girls; Citizen of South Africa; Lang. Afrikaans, English, Dutch and German.
International Organizations: Pastor (since 1982) of the Dutch Reformed Church in Africa (Black).
Other organizations: Patron (since 1985), Koinonia Southern Africa; Advisor to Moderamen, Dutch Reformed Church in Africa (since 1987).
Main field of work: Ministry. **Career:** Pastor, Dutch Reformed Church, Benoni and Potchefstrom from 1953 to 1956. Missionary, Dutch Reformed Church amongst Venda People from 1956 to 1963. Mission Secretary, Dutch Reformed Church from 1963 to 1966. Professor of Missiology and History of Religions, University of Stellenbosch from 1966 to 1982. **Educ.:** B.A. from University of Pretoria, South Africa (1946-1948); B.D. from University of Pretoria, South Africa (1949-1951); D.D. from University of Pretoria, South Africa (1972). **Interests:** socio-political developments in South Africa. **Hons.:** Honorary doctoral degrees from Withworth and North Park; South African Toastmakers Award for Communicator of the Year; Ecumenical Consultancy Recognition and Affirmation Award. **Mem.:** International Association for Mission Studies; South African Association for Mission Studies **Pubs.:** numerous books (in Afrikaans).
Mailing address: PO Box 972, Silverton, 0127 Pretoria, South Africa. T. (27 12) 8055070
Residence: Block U-7000, Mamelodi, 0001 Pretoria, South Africa. T. See Mailing Address

Smith, R
Rel: Christian.
International Organizations: Secretary-General of the Association of Missionary Societies.
Mailing address: Overseas House, 3 Belgrave Road, Dublin 6, Ireland. T. 97 09 31 – 97 09 39.

Smith, Richard F, S.J., Rev Dr
Rel: Roman Catholic; Ordained; celibate.
International Organizations: Member of the Society of Jesus.
Other organizations: Editor, Newsletter, Association for the Development of Religious Informations Systems.
Mailing address: 3601 Lindell Blvd, St Louis MO 63108-3393, USA.

Smith, Robert Frederick, Very Rev
Rel: United Church of Canada; Ordained; b. 25 MAY 1934 in Montreal, Quebec, Canada; m. to Margaret Elen Maguire; c. 3 boys and 2 girls; Citizen of Canada.
International Organizations: Clergyman of the United Church of Canada.
Career: Minister, St Luke's, Fort St John, BC from 1958 to 1961. Minister, Trinity, Edmonton, Alberta from 1961 to 1965. Minister, Memorial Congregational Church, North Quincy MA from 1965 to 1968. Minister, Richmond Hill United Church from 1968 to 1974. Minister, Eglinton United Church, Toronto from 1974 to 1982. Adjunct Professor of Homiletics, Victoria College, University of Toronto from 1979 to 1981. Moderator, United Church of Canada from 1984 to 1986. **Educ.:** B.A. from University of Alberta, Edmonton, Canada (1956); B.D. from University of Alberta, Edmonton, Canada (1964); Th.D. from Boston University, MA, USA (1973). **Hons.:** DD hc, Victoria University (1985).
Mailing address: 1550 West 33rd Ave, Vancouver BC V6M 1A7, Canada.
Residence: 1590 West 33rd Ave, Vancouver BC V6M 1A7, Canada.

Smith, Steven Verney
Rel: Buddhist; Ordained; b. 19 APR 1947 in Honolulu HI, USA; m. to Michele McDonald; c. 1 girl; Citizen of USA; Lang. English and Pali.
International Organizations: Guiding Teacher (since 1991) of the Insight Meditation Society.
Other organizations: Founder and Guiding Teacher, Vipassana Hawaii; Guiding Teacher, Te Moata Meditation Center, Tairua, New Zealand; Guiding Teacher, Buddha Sassana Center of Australia.
Main field of work: Education. **Career:** Ordained Theravada Buddhist monk in Burma in 1982. Ordained Theravada Buddhist monk at Mahasi Meditation Center, Rangoon, under Ven Mahasi Sayadaw in 1983. **Educ.:** B.A. from Linfield College, McMinville OR, USA (1965-1967); University of Hawaii, Honolulu, USA (1967-1970). **Interests:** Integrating spiritual practice in daily life, meditation, relationships as spiritual practice, men's healing, generativity and spirituality, liberating mythologies, spiritual leadership, surfing and whitewater kayaking. **Pubs.:** "Sharing the Path: Relationships as Spiritual Practice", in *IMS Newsletter*, Spring 1992 and *Transformation and Transcendence: The Path of Love and Understanding* (forthcoming).
Mailing address: Vipassana Hawaii, 94 Niuiki Circle, Honolulu HI 96821, USA.
Residence: 380 Portlock Rd, Honolulu HI 96825, USA T. (1 808) 395 5301

Smits, Gérard, M.O., Fr
Rel: Roman Catholic; Ordained; celibate.
International Organizations: Superior General of the Aumôniers du Travail.
Mailing address: Bd Lambermont 17, B-1030 Bruxelles, Belgium. T. (32 2) 215 13 00

Snow, Charles Hugh, Rabbi
Rel: Jewish; Ordained on 29 MAR 1987; b. 01 JUN 1952 in El Paso TX, USA; m. to Ruth Fleischer; c. 1 boy; Citizen of USA; Lang. English.
International Organizations: Committee Member (since 1988) of the International Alliance of Messianic Congregations and Synagogues. Committee Member (since 1988) of the International Messianic Jewish /Hebrew Christian/ Alliance.
Other organizations: Director (since 1984), Ha Devar Publishing and Distribution; Rabbi (since 1988), London Messianic Congregation; Board of Ref. (since 1990), Manna Music; Director (since 1991), Messianic Ministries; Board of Ref. (since 1992), New Jerusalem Dancers.
Main field of work: Ministry. **Career:** Organizer, Annual Messianic European Conference 'Yeshua' from 1991 to 1992. **Educ.:** Montgomery Co. Tech., PA, USA (1971-1972); Spiro Institute, London, UK (1990-1991). **Interests:** Music, sports.
Mailing address: 61 Kings Close, Hendon, London NW4 2JU, UK. T. (44 81) 203 7500
Residence: See Mailing Address. T. See Mailing Address

Snow, R F, Rev
Rel: Protestant.
International Organizations: President of the United Protestant Church.
Mailing address: Fortkerk, Fort Amsterdam, Curaçao, Neth. Antilles.

So, Erlinda
Rel: Roman Catholic.
International Organizations: Co-Head, Asia Regional Office of the Unda, International Catholic Association for Radio and Television.
Mailing address: Overseas Program Manager, PO Box 939, Manila, Philippines.

Soares, Anjou P T
Rel: Roman Catholic.
International Organizations: President of the South Asian Catholic Press Association.
Mailing address: Catechetical Center, 104 Mohan Terrace, Parr Street, Karach 74400, Pakistan.

Soares, Fernando L, Rt Rev
Rel: Catholic-Apostolic Evangelical Church.
International Organizations: Contact of the Lusitanian Catholic-Apostolic Evangelical Church.
Mailing address: Rua de Elias Garcia 107-1 Dto, P-4400 Vila Nova de Gaia, Portugal.

Sodano, Angelo, Card
Rel: Roman Catholic.
International Organizations: President of the Fondazione Pio XII per l'Apostolato dei Laici.
Mailing address: I-00120 Città del Vaticano, Vatican.

Soejana, Koernai Atje, Rev
Rel: Christian.
International Organizations: President of the Pasundan Christian Church.
Mailing address: Jalan Jatinegara Timur 66, Jakarta 13310, Indonesia.

Solassol, Alain
Rel: Catholic.
International Organizations: Secretary General of the World Organization of Former Pupils of Catholic Education.
Mailing address: 5 rue Jules Guesde, F-92130 Issy-les-Moulineaux, France.

Soliman, Doaa Fouad
Rel: Roman Catholic.
International Organizations: Co-Secretary-General of the International Movement of Apostolate of Children.
Mailing address: 8 rue Duguay-Trouin, F-75006 Paris, France. T. (33 1) 42 22 84 81.

Solomon, Lorraine Ellen
Rel: Methodist; b. 26 DEC 1933 in Johannesburg, South Africa; m. to Kenneth Cyril; c. 2 boys and 1 girl; Citizen of South Africa; Lang. English and Afrikaans.
International Organizations: Secretary (since 1986) of the World Federation of Methodist Women.
Executive (since 1986) of the World Methodist Council.
Other organizations: Executive (since 1964), Womens Association Connexion and District.
Main field of work: Sales representative in the industrial chemical field. **Educ.:** B.A. (uncompl.) from University of South Africa, Pretoria, South Africa.
Mailing address: 4 Meersig Road, Elfindale, Diep River 7945, South Africa. T. (27 21) 52 6010
Residence: See Mailing Address. T. (27 21) 75 5713

Somers, Jeffrey Ian
Rel: Jewish; Unordained; b. 01 AUG 1936 in London, UK; m. to Nobuko Hirose; Citizen of United Kingdom.
International Organizations: Writer for *Religion Today* of the INTERACT Research Centre.
Other organizations: Director (since 1979), Fine Books Oriental; Administrator (since 1985), The English Language Society; Advisor to Element Books (since 1989); Advisor to the Museum of Music, Dushanbe, Tadjikistan.
Main field of work: Writing. **Educ.:** Studied Sociology from University of London, Birkbeck College, UK. **Interests:** Gurdjieff. **Hons.:** Fellow, Royal Asiatic Society. **Mem.:** Royal Society for Asian Affairs; The Japan Society **Pubs.:** Japanese Buddhism in Britain; Tibetan Buddhism in Britain and Theravada Buddhism in Britain.
Mailing address: 94 Addison Gardens, London W14 0DE, UK. T. (44 71) 603 9797
Residence: See Mailing Address. T. See Mailing Address

Soper, Donald Oliver, Rev Lord
Rel: Methodist; Ordained on 02 AUG 1929; b. 31 JAN 1903 in London, UK; m. to Marie Gertrude Dean; c. 4 girls; Citizen of United Kingdom; Lang. English and French.
International Organizations: Supernumary (since 1978) of the West London Mission.
Other organizations: Labour Peer (since 1965), House of Lords; President, Christian Socialist Movement; Chairman, League Against Cruel Sports.
Main field of work: Ministry. **Career:** Minister to the South London Mission from 1926 to 1929. Minister to the Central London Mission from 1929 to 1936. Superintendent, West London Mission, Kingsway Hall from 1936 to 1978. President, Methodist Church in GB from 1953 to 1954. **Educ.:** M.A. from University of Cambridge, St Catharine's College, UK (1914-1921); Wesley House, Cambridge, UK (1921-1924); Ph.D. from University of London, UK (external) (1924-1926). **Interests:** music, most games. **Hons.:** Hon. D.D., University of Cambridge (1988); World Methodist Council Peace Award (1981-82); Baron (Life Peer), created 1965, of Kingsway. **Pubs.:** All His Grace; Tower Hill, 12.30 and Calling for Action (autobiography).
Mailing address: 19 Thayer Street, London W1M 5LJ, UK. T. (44 71) 935 6179
Residence: 17 Bigwood Road, London NW11 7BG, UK.

Soplantila, A J, Rev
Rel: Protestant.
International Organizations: General Secretary of the Protestant Church in the Moluccas.
Mailing address: Jalan Imam Bonjol 2, Ambon, Indonesia.

Sorenson, Diane
Rel: Eckankar.
International Organizations: Contact of the Eckankar, Religion of the Light and Sound of God.
Mailing address: PO Box 27300, Minneapolis MN 55427-0300, USA. T. (1 612) 474 0700. Fax 474 1127.

South, Colin
Rel: Christian.
International Organizations: Secretary-General of the Christians Abroad.
Mailing address: 1 Stockwell Green, London SW9 9HP, UK. T. (44 71) 737 7811.

South, Liz
Rel: Christian.
International Organizations: Secretary of the Scottish Churches Action for World Development.
Mailing address: 41 George IV Bridge, Edinburgh EH1 1EL, UK. T. (44 31) 225 1772.

Souza, Enilson Rocha, Rev
Rel: Evangelical Pentecostal.
International Organizations: Secretary of the Evangelical Pentecostal Church 'Brazil for Christ'.
Mailing address: Caixa Postal 041, 40 000 Salvador BA, Brazil.

Sowy-Lavergne, Françoise, Sr
Rel: Roman Catholic.
International Organizations: Secretary of the Union of European Conferences of Superiors General.
Mailing address: Lakenwesstraat, 30 rue des Drapiers, B-1050 Bruxelles, Belgium. T. (32 2) 511 77 04. Fax 511 77 04.

Spanu, Paolo
Rel: Evangelical Baptist.
International Organizations: President of the Evangelical Baptist Union of Italy.
Mailing address: Piazza S Lorenzo in Lucina 35, I-00186 Roma, Italy.

Spencer, Peter, Rev
Rel: Protestant.
International Organizations: President of the Evangelical Association of the Caribbean.
Mailing address: PO Box 123, Kingston 8, Jamaica.

Spengler, Helmut, Prof
Rel: Evangelical.
International Organizations: President of the Evangelical Church in Hessen and Nassau.
Mailing address: Paulusplatz 1, 64285 Darmstadt, Germany.

Spinder, Hans, Drs
Rel: Dutch Reformed.
International Organizations: Secretary of the Inter-European Commission on Church and School.
Mailing address: Center for Education of the Netherlands Reformed Church, Postbus 1100, NL-3970 BC Driebergen, Netherlands. T. (31 3438) 15725 - 18334. Fax 23250.

Spindler, Marc Robert, Rev Prof Dr
Rel: Protestant; Ordained on 12 JUL 1959; b. 25 AUG 1930 in Mulhouse, Alsace, France; m. to Antoinette Theis; c. 2 boys and 2 girls; Citizen of France; Lang. French, English, German, Dutch and Malagasy.
International Organizations: Director (since 1974) of the Interuniversity Institute for Missiological and Ecumenical Research.
Other organizations: Professor, Faculty of Theology (since 1974), State University at Leiden; President (since 1987), Centre de Recherches et d'Echanges sur la Diffusion et l'Inculturation du Christianisme, Lyon.
Main field of work: University education and research. **Career:** Parish Minister, Reformed Church of Alsace (ERAL) from 1957 to 1960. Missionary, Paris Evangelical Missionary Society from 1960 to 1971. Missionary, Evangelical Community of Apostolic Action (CEVAA) from 1972 to 1974. **Educ.:** Centre de Hautes Etudes sur l'Afrique et l'Asie Modernes, Paris, France (1960-1961); Doc. d'Etat en Théol. from University of Strasbourg, France (1950-1967); Lic. ès Lett. from University of Montpellier III, France (1974). **Interests:** publishing. **Hons.:** Knight of the National Malagasy Order. **Mem.:** International Association for Mission Studies; Malagasy Academy **Pubs.:** La mission combat pour le salut du monde, 1967; Pour une théologie de l'espace, 1968 and Oecumenische inleiding in de missiologie, 1988.
Mailing address: Afdeling Missiologie, Rapenburg 61, NL-2311 GJ Leiden, Netherlands. T. (31 71) 144 248.
Residence: Van Diepenburchstraat 25, NL-2597 PR Den Haag, Netherlands T. (31 70) 324 1588

Spital, Hermann Josef Silvester, Rt Rev
Rel: Roman Catholic; Ordained on 06 AUG 1952; b. 31 DEC 1925 in Münster, Germany; celibate; Citizen of Germany.
International Organizations: Bishop (since 1981) of Trier.
Career: Vicar-General, Münster in 1973. Consecrated Bishop in 1980.
Mailing address: Liebfrauenstrasse 1, Postfach 3444, 54290 Trier, Germany. T. (49 651) 7105209. Fax (49 651) 710 54 98. Tx 472553

Spolti, Luiz, S.D.S., Fr
Rel: Roman Catholic; Ordained; celibate.
International Organizations: Procurator-General of the Society of the Divine Saviour.
Mailing address: Via della Conciliazione 51, I-00193 Roma, Italy. T. (39 6) 654 76 41.

Spong, John Shelby, Rt Rev
Rel: Christian; Ordained; b. 16 JUN 1931 in Charlotte NC, USA; widowed Joan Lydia Ketner; c. 3 girls; Citizen of USA.
International Organizations: Bishop (since 1976) of Diocese of Newark, NJ.
Career: Rector, St Joseph's, Durham NC from 1955 to 1957. Rector, Calvary Church, Tarboro NC from 1957 to 1965. Rector, St John's Church, Lynchburg VA from 1965 to 1969. Rector, St Paul's Church, Richmond VA from 1969 to 1976. **Hons.:** Roger Baldwin Award. **Pubs.:** Into the Whirlwind, 1983; Beyond Moralism, 1986 and Survival and Consciousness, 1987.
Mailing address: 24 Rector Street, Newark NJ 07102, USA. T. (1 201) 622 3876

Sponselee, Theodor, Br
Rel: Roman Catholic; Ordained; celibate.
International Organizations: Superior General of the Frères de Saint Louis de Gonzague.
Mailing address: PB 24, NL-4730 AA Oudenbosch, Netherlands. T. (31 1652) 12 437

Sporre, Karin, Rev
Rel: Reformed Church.
International Organizations: Member of the Executive Committee of the World Alliance of Reformed Churches.
Mailing address: Teologiska Seminariet, Kottlavägen 116, S-18141 Lidingo, Sweden. T. (46 8) 765 2605. Fax (46 8) 767 0714
Residence: T. (46 8) 767 6181

Spradlin, Byron
Rel: Christian.
International Organizations: Director of the Church Planting International.
Mailing address: 9521 Business Centre Drive, PO Box 1002, Cucamonga CA 91730, USA. T. (1 714) 945 1264. Fax 944 7583.

Sprondel, Walter M
Rel: Unknown; b. 28 MAR 1938; Citizen of Germany.
International Organizations: Member of the Council of the International Society for the Sociology of Religions.
Other organizations: Full Professor, Soziologisches Seminar, Eberhard-Karls-Universität Tübingen.
Mailing address: Soz. Seminar, Wilhelmstrasse 36 (Hegelbau), 72074 Tübingen, Germany. T. (49 7071) 296892
Residence: T. (49 7071) 23334

Sprunger, Arthur L, Rev Dr
Rel: Protestant.
International Organizations: Executive Director of the United Association of Christian Counselors International.
Mailing address: 41 Short St, Harrisburg PA 171093731, USA. T. (1 717) 652 7688.

Squicciarini, Donato, H E Mgr
Rel: Roman Catholic; Ordained on 12 APR 1952; b. 24 APR 1927 in Altamura, Italy; celibate.
International Organizations: Apostolic Nuncio to Austria (since 1989) of the Roman Catholic Church.
Other organizations: Titular Archbishop, Tiburnia; Permanent Representative, the Holy See to the International Atomic Energy Agency (IAEA); Permanent Observer, the Holy See to the United Nations Office in Vienna; Permanent Observer, the Holy See to the United Nations Industrial Development Organization (UNIDO).
Mailing address: Theresianumgasse 31, A-1040 Wien, Austria. T. (43 222) 505 13 27. Fax 505 61 40 75

Srisang, Koson, Dr
Rel: Christian.
International Organizations: Executive Secretary of the Ecumenical Coalition on Third World Tourism.
Mailing address: PO Box 24 Chorakhebua, Bangkok 10230, Thailand. T. (66 2) 510 7287. C. Countours Bangkok. Fax (66 2) 510 7287.

Stäbler, Martin, Rev

Rel: Christian.
International Organizations: Representative of the Third World Tourism European Ecumenical Network.
Mailing address: ZEB, Nikolaus-Otto-Strasse 13, D-70771 Leinfelden-Echterdingen, Germany. T. (49 711) 7989-0 – 7898 281. Fax 7989 123.

Stålsett, Gunnar Johan, Dr

Rel: Christian; b. 10 FEB 1935 in Nordkapp, Norway; m. to Unn Eriksen Staalsett; c. 1 boy and 1 girl; Citizen of Norway; Lang. Norwegian, English, German, French and Spanish.
International Organizations: Secretary-General (since 1985) of the Lutheran World Federation.
Contact person of the Committee on Christian Communication in Africa.
Main field of work: Church, peace, human rights. **Career:** Student Pastor, Christian Student Movement from 1962 to 1964. Professor of Theology, Norwegian Mission Society from 1964 to 1970. Secretary-General, Church of Norway Council on Foreign Relations from 1970 to 1977. State Secretary, Norwegian Government from 1972 to 1973. Party Chairman, Centre Party from 1977 to 1979. Dean, Church of Norway from 1979 to 1982. Secretary-General, Norwegian Bible Society from 1982 to 1985. Member, Nobel Peace Prize Committee from 1984 to 1991. **Educ.:** University of Oslo, Norway (1954-1957); B.D. from Bethany Lutheran Seminary, Mankato MN, USA (1957-1959); Cand. Theol. from Free Faculty of Theology, Church of Norway, Oslo, Norway (1959-1962). **Interests:** International affairs. **Hons.:** DLit, DD, DTheol hc. **Mem.:** Non-voting member of the Nobel Peace Prize Committee (since 1991)
Mailing address: 150 route de Ferney , PO Box 2100, CH-1211 Genève 2, Switzerland. T. (41 22) 791 63 61. C. LUTHWORLD. Tx 23423 OIK CH. Fax (41 22) 798 86 16.
Residence: 13 Champ-d'Anier, CH-1209 Genève, Switzerland T. (41 22) 798 36 79

Stamm, Ute

Rel: Unknown.
International Organizations: Head, Secretariat of the Standing Conference of Jews, Christians and Muslims in Europe.
Mailing address: Hedwig Dransfeld Haus, 56170 Bendorf-Rhein, Germany.

Stanislaw, Turek

Rel: Roman Catholic.
International Organizations: Procurator-General of the Order of St Paul the First Hermit.
Mailing address: Via Alcamo 12/a, I-00182 Roma, Italy. T. (39 6) 7014569.

Staquet, Michel

Rel: Roman Catholic.
International Organizations: Secretary-General of the European Union of Former Pupils of Catholic Education.
Mailing address: 24 rue Henner, F-68000 Colmar, France. T. (33) 89 23 22 99.

Stäubli, Max

Rel: Baptist.
International Organizations: Secretary-General of the European Baptist Mission.
Mailing address: Missionszentrale, Friedbergerstrasse 101, 61350 Bad Homburg, Germany. T. (49 6172) 80 04 40 – 80 04 46. Tx 41 09 56 ebm d. Fax (49 6172) 80 04 36.

Steffans, Marion, OCD, Sr

Rel: Roman Catholic; Ordained; celibate.
International Organizations: Member of the Discalced Carmelite Nuns.
Other organizations: President, Carmelite Communities Associated; Member, Association of Contemplative Sisters.
Mailing address: Carmelite Monastery, 25 Watson Ave, Barrington RI 02806, USA.

Steg, Adolphe, Prof

Rel: Jewish.
International Organizations: Co-President of the Consultative Council of Jewish Organizations.
President of the Alliance israélite universelle.
Mailing address: 45 rue La Bruyère, F-75425 Paris CEDEX 09, France. T. (33 1) 42 80 35 00. Fax (33 1) 48 74 51 33.

Steicke, Lance Graham, Rev Dr

Rel: Lutheran; Ordained on 18 DEC 1955; b. 19 FEB 1933 in Murray Bridge, South Australia, Australia; m. to Leah Gweneth Briese; c. 2 boys and 2 girls; Citizen of Australia; Lang. English.
International Organizations: General President (since 1987) of the Lutheran Church of Australia.
Main field of work: Ministry. **Career:** Secretary, Evangelical Lutheran Church of New Zealand from 1960 to 1963. Vice-President, Evangelical Lutheran Church of New Zealand from 1963 to 1964. President, Evangelical Lutheran Church of New Zealand from 1964 to 1979. Director, Board of Radio and TV (LCA) from 1979 to 1987. 2nd Vice-President, Lutheran Church of Australia from 1981 to 1987. **Educ.:** B.Th. from Concordia Seminary, Adelaide, Australia (1951-1955); D.D. from Lutheran Church-Missouri Synod, St Louis MO, USA (1989). **Pubs.:** *A New Relationship*, 1980; *Jesus Comes*, 1982 and devotional articles.
Mailing address: 58 O'Connell Street, North Adelaide SA 5006, Australia. T. (61 8) 267 4922. FAX 239 0173
Residence: 22a Thames Street, Klemzig SA 5087, Australia T. (61 8) 261 2036

Stein, Greig

Rel: Roman Catholic.
International Organizations: Representative at UN, New York of the Pax Romana, International Catholic Movement for Intellectual and Cultural Affairs.
Mailing address: Phi Kappa Theta National Foundation, 3901 W 86th Street, Suite 290, Indianapolis IN 46268, USA.

Stella, Beniamino, H E Mgr

Rel: Roman Catholic; Ordained on 19 MAR 1966; b. 18 AUG 1941 in Pieve di Soligo, Italy; celibate.
International Organizations: Apostolic Pro-Nuncio to Chad (since 1989) of the Roman Catholic Church.
Other organizations: Apostolic Pro-Nuncio to Congo and the Central African Republic (since 1987); Titular Archbishop, Midila.
Mailing address: Avenue Boganda, BP 1447, Bangui, Central African Rep. T. (236) 612 654

Stelpstra, William, Rev

Rel: Protestant.
International Organizations: President of the World for Christ Crusade.
Mailing address: 1005 Union Valley Road, West Milford NJ 07480, USA. T. (1 201) 728 3267.

Stephanos, Francis, Ato

Rel: Lutheran.
International Organizations: Vice-President of the Lutheran World Federation.
Member, Program Committee for World Service of the Lutheran World Federation.
Other organizations: President, Ethiopian Evangelical Church Mekane Yesus.
Mailing address: PO Box 2087, Addis Ababa, Ethiopia. T. (251 11) 11 12 00. 55 29 66. C. EECMY ADDIS ABABA. Tx 21528 EECMY ET

Stephens, Alice

Rel: Roman Catholic; Unordained; b. 24 SEP 1911 in Carlisle, Cumbria, UK; single; Citizen of United Kingdom; Lang. English and French.
International Organizations: Founder Member (with Bryn Stephens) (since 1960) of the Ecumenical Community of Christian Love.
Other organizations: Religious Affairs Correspondent, British World Federalist Association.

Main field of work: Lecturer in philosophy. **Career:** Teacher in various schools. **Educ.:** B.A. from University of London, Royal Holloway and Bedford New College, UK (1934-1937); B.A. Phil. (Hons) from University of London, Birkbeck College, UK (1951-1954); M.A. Phil. from University of London, Birkbeck College, UK (1955-1958); M.Th. from Maryknoll, New York, USA (1988-1990); Ph.D. (not yet completed) from University of Lancaster, UK (1988-1990). **Pubs.:** article in *Azan* (magazine).
Mailing address: "Bryn-Kieron", 8 Elm Garth, Carlisle CA4 8LB, UK. T. (44) 561 384
Residence: See Mailing Address. T. See Mailing Address

Stephens, B, Rev

Rel: Christian.
International Organizations: Contact of the Christian World Service.
Mailing address: 165 Lichfield Street, Christchurch, New Zealand. T. (64 3) 669 274. Tx 4249 Tattler NZ.

Stephens, Jacob S A, Rev

Rel: Methodist.
International Organizations: President of the Conference of the Methodist Church, Ghana.
Mailing address: PO Box 403, Accra, Ghana.

Stephens, Sarah Jane, Rev

Rel: Presbyterian; Ordained on 12 JUL 1986; b. 10 APR 1955 in Sacramento CA, USA; single; Citizen of USA; Lang. English.
International Organizations: Executive Secretary for Co-operation and Witness (since 1990) of the World Alliance of Reformed Churches.
Main field of work: Human rights advocacy. **Career:** Associate, Presbyterian Church Committee on Social Witness Policy (USA) from 1985 to 1990. **Educ.:** B.A. Pol. Stds. from Whitworth College, Spokane WA, USA (1973-1978); M.D. from San Francisco Theological Seminary, San Anselmo CA, USA (1980-1985).
Mailing address: 150 route de Ferney, PO Box 2100, CH-1211 Genève 2, Switzerland. T. (41 22) 791 62 36. C. WARC GENEVA. Tx 415 730 OIK CH (WARC). Fax (41 22) 791 65 05 WARC REFORMIERTER WELTBUND.
Residence: Route de Treléx, CH-1261 Genolier/Vaud, Switzerland T. See Mailing Address

Stephens, William Peter, Rev Prof

Rel: Methodist; Ordained on 07 JUL 1960; b. 16 MAY 1934 in Penzance, Cornwall, UK; single; Citizen of United Kingdom; Lang. English, French, German and Swedish.
International Organizations: Professor of Church History (since 1986) of the University of Aberdeen.
Central Committee (since 1974) of the Conference of European Churches.
Main field of work: History and theology of the Reformation. **Career:** Assistant Tutor, Hartley Victoria College from 1958 to 1961. Minister and University Chaplain, Nottingham from 1961 to 1965. Minister, Shirley Methodist Church, Croydon from 1967 to 1971. Chair of Church History, Hartley Victoria College, Manchester from 1971 to 1973. Chair of Historical and Systematic Theology, Wesley College, Bristol from 1973 to 1980. Lecturer in Church History, The Queen's College, Birmingham from 1980 to 1986. Member, Bristol City Council from 1976 to 1983. **Educ.:** B.A., M.A., B.D. from University of Cambridge, UK (1952-1957); Dr. ès Sc.Rel. from University of Strasbourg, France (1965-1967). **Pubs.:** The Holy Spirit in the Theology of Martin Bucer, Cambridge, 1970; The Theology of Huldrych Zwingli, Clarendon, Oxford, 1986 and Zwingli: An Introduction to his Thought, Clarendon, Oxford, 1992.
Mailing address: Faculty of Divinity, University of Aberdeen, Aberdeen AB9 2UB, Scotland, UK. T. (44 224) 272383
Residence: Morefield House, 15 Richmond Hill Gardens, Aberdeen AB2 4LW, UK. T. (44 224) 638107

Stern, Jean, M.S., Fr

Rel: Roman Catholic; Ordained; celibate.
International Organizations: Procurator-General of the Missionaries of Our Lady of La Salette.
Member, Academic Council of the Pontifical Marian International Academy.
Mailing address: Piazza Madonna della Salette 3, I-00152 Roma, Italy. T. (39 6) 53 04 51.

Stevens, Clifford, Fr

Rel: Roman Catholic; Ordained; celibate.
International Organizations: Priest of the Roman Catholic Church.
Mailing address: Tintern Monastery, Monastery Road, Oakdale NB 68761, USA.

Stevens, David

Rel: Christian.
International Organizations: Director of the World Medical Mission.
Mailing address: PO Box 3000, Boone NC 28607, USA. T. (1 704) 262 1980. Fax 262 0175.

Stewart, J

Rel: Roman Catholic.
International Organizations: Director of the Archbishop of Sydney's Overseas Relief and Aid Fund.
Mailing address: PO Box Q190, Queen Victoria Bldg., Sydney NSW 2000, Australia. T. 26 23 71.

St George, John A S

Rel: Roman Catholic.
International Organizations: Executive Director of the Multimedia International.
Mailing address: Borgo S Spirito 5, I-00193 Roma, Italy. T. (39 6) 68 97 72 91. Fax 68 97 72 91.

St George, John A S

Rel: Roman Catholic.
International Organizations: Executive Director of the Multimedia International.
Mailing address: Borgo S Spirito 5, I-00193 Roma, Italy. T. (39 6) 68 97 72 91. Fax 68 97 72 91.

Stickler, Alfons Maria, S.D.B., H E Card

Rel: Roman Catholic; Ordained on 27 MAR 1937; b. 23 AUG 1910 in Neunkirchen, Vienna, Austria; celibate; Citizen of Austria.
International Organizations: Member (since 1985) of the College of Cardinals.
Member of the Society of St Francis of Sales.
Other organizations: Deacon, S Giorgio of Velabro.
Career: Consecrated Archbishop (Titular See of Volsinium) in 1983. Created Cardinal by Pope John Paul II in 1985. Former Chief Vatican Librarian and Archivist.
Mailing address: Piazza del S Uffizio 11, I-00193 Roma, Italy. T. (39 6) 698 33 25

Stockwood, (Arthur) Mervyn, Rt Rev

Rel: Anglican; Ordained in 1936; b. 27 MAY 1913; Citizen of United Kingdom.
International Organizations: Ecclesiastic of the Anglican Church.
Other organizations: Member, Council (since 1980), Bath University.
Career: Curate, St Matthew Moorfields, Bristol, and Blundell's School Missioner from 1936 to 1941. Vicar, St Matthew Moorfields, Bristol from 1941 to 1945. Honorary Canon, Bristol from 1953 to 1955. Vicar, St Mary the Great, Cambridge from 1955 to 1959. Bishop, Southwark from 1959 to 1980. Church Commissioner from 1972 to 1979. Labour Member, Bristol City Council from 1946 to 1955. Labour Member, Cambridge City Council from 1956 to 1959. **Interests:** fishing, walking. **Hons.:** Freedom of City of London (1976); Hon. D.D., Lambeth University (1959), Bucharest University (1977); Hon. D.Litt., Sussex University (1963). **Pubs.:** The Cross and the Sickle, 1978; From Strength to Strength, 1980 and Chanctonbury Ring, 1982 (autobiography).
Mailing address: 15 Sydney Buildings, Bath BA2 6BZ, UK. T. (44 225) 63978

Stojcevic, Pavle, H H

Rel: Orthodox; Ordained; b. 11 SEP 1914 in Kucanci, Donji Miholjac, Yugoslavia; celibate; Citizen of Yugoslavia; Lang. Serbian, Russian and Greek.
International Organizations: Patriarch (since 1990) of Serbian Orthodox Church.

Career: Monk in 1948. Deacon, Raca monastery from 1949 to 1955. Archimandrite in 1957. Ordained Bishop, Raska-Prizren in 1957. Diocesan Bishop, Raska-Prizren (until 1990). President, Commission of the Holy Synod of the Serbian Church for the newly revised officially accepted translation of the New Testament. **Educ.:** School of Theology, Sarajevo, Yugoslavia (1930-1936); Faculty of Theology of the Serbian Orthodox Church in Belgrade, Yugoslavia (1936-1940); P.G. Stds. at National Capodistrian University of Athens, Faculty of Theology, Greece (1955-1957). **Interests:** liturgical theology and biblical studies. **Hons.:** Dr. Theol. h.c., Faculty of Theology, Belgrade (1991). **Pubs.:** numerous scientific and research works in liturgical theology and biblical studies.
Mailing address: Serbian Patriarchate, 7 July N° 5, PO Box 182, 11001 Belgrade, Yugoslavia. T. (38 11) 638 161. Fax 182 780
Residence: See Mailing Address. T. See Mailing Address

Stoll, Karlheinz, Bishop

Rel: Evangelical Lutheran.
International Organizations: Presiding Bishop of North Elbian Evangelical Lutheran Church.
Mailing address: Nordebisches Kirchenamt, Dänische Strasse 21/35, 24103 Kiel, Germany.

Stoner, John K

Rel: Mennonite.
International Organizations: Secretary of the Mennonite Central Committee – Overseas Peace Office.
Mailing address: 21 S 12th Street, Akron PA 17501, USA. T. (1 717) 859 11 51. C. MENCENCOM. Tx 90 2210.

Storero, Luciano, H E Mgr

Rel: Roman Catholic; Ordained on 29 JUN 1949; b. 26 SEP 1926 in Dubbione di Pinasca, Torino, Italy; celibate; Lang. Italian, French, English and Spanish.
International Organizations: Apostolic Pro-Nuncio to Greece (since 1990) of the Roman Catholic Church.
Other organizations: Titular Archbishop, Tigimma.
Main field of work: Diplomatic service. **Career:** Attaché, Apostolic Nunciature in Egypt from 1953 to 1955. Secretary, Apostolic Nunciature in Japan from 1955 to 1959. Auditor, Apostolic Nunciature in Ireland from 1959 to 1962. Auditor, Apostolic Delegation in the USA from 1962 to 1966. In charge of the Apostolic Delegation in Sri Lanka from 1967 to 1971. Episcopal Ordination in 1970. Apostolic Nuncio to the Dominican Republic and Apostolic Delegate to Puerto Rico from 1971 to 1973. Apostolic Pro-Nuncio to Cameroon and Gabon and Apostolic Delegate to Ecuatorial Guinea from 1973 to 1976. Apostolic Pro-Nuncio to India from 1976 to 1981. Apostolic Nuncio to Venezuela from 1981 to 1990. **Educ.:** D.Cn.L. from Lateran Pontifical University, Vatican City State (1949-1952); Dip. Eccl. Diplomacy from Pontifical Ecclesiastical Academy, Vatican City State (1951-1953).
Mailing address: Apostolic Nunciature, Odós Mavili 2, Paleo Psychicó, Athinai 154 52, Greece. T. (30 1) 647 35 98. Fax 647 28 49
Residence: See Mailing Address. T. See Mailing Address

St-Pierre, Valère, Fr

Rel: Roman Catholic.
International Organizations: Procurator-General of the Congregation of the Sacerdotal Fraternity.
Mailing address: 500 Ave Claremont, Montréal H3Y 2N5, Canada. T. (1 514) 483 2030.

Stradling, Leslie Edward, Rt Rev

Rel: Anglican; Ordained; b. 11 FEB 1908; Citizen of United Kingdom.
International Organizations: Ecclesiastic (retired) of the Anglican Church.
Career: Curate in London from 1933 to 1938. Vicar from 1938 to 1945. Bishop, Masasi from 1945 to 1952. Bishop, South West Tanganyika from 1952 to 1961. Bishop, Johannesburg from 1961 to 1974. **Pubs.:** A Bishop on Safari, 1960; A Bishop at Prayer, 1971 and Praying the Psalms, 1977.
Mailing address: Braehead House, Auburn Road, 7700 Kenilworth, South Africa. T. (27 21) 761 6251

Stramare, Tarcisio, O.S.I., Fr

Rel: Roman Catholic; Ordained; celibate.
International Organizations: Procurator-General of the Congregation of the Oblates of St Joseph.
Advisor of the Congregation for the Causes of Saints.
Mailing address: Via Boccea 364, I-00166 Roma, Italy. T. (39 6) 622 33 47

Strange, John, Prof Dr

Rel: Lutheran; Ordained on 25 JUL 1979; b. 20 JUL 1934 in Gentofte, Denmark; m. to Lisbeth Laursen; c. 3 boys and 2 girls; Citizen of Denmark; Lang. Danish, English, German and French.
International Organizations: Pro-Rector (since 1991) of the University of Copenhagen.
Other organizations: Army Chaplain (since 1984), The Citadel, Copenhagen.
Main field of work: University administration, archaeological research and excavation. **Career:** Research Fellow, University of Copenhagen from 1962 to 1967. Assistant Professor, University of Copenhagen from 1967 to 1972. Associate Professor, University of Copenhagen in 1972. Dean (Theology), University of Copenhagen from 1984 to 1991. Curate, Jægersborg Sogn, Danish Church from 1979 to 1986. **Educ.:** Cand. Theol. from University of Copenhagen, Denmark (1962); Dr. Theol. from University of Copenhagen, Denmark (1980); Hebrew University of Jerusalem, Israel (1963); British School of Archaeology in Jerusalem, Israel (1964-1967). **Interests:** biblical archaeology and geography. **Hons.:** Gold Medal, University of Copenhagen (1960). **Mem.:** Collegium Biblicum, Denmark **Pubs.:** Caphtor/Keftiu, Acta Theologica Danica XIV, Leiden, 1980 (dissertation); Bibelselskabets Bibelatlas, 1988 (Finnish, German, Italian translations 1989 and 1990) and Theology and Politics in Architecture and Iconography, Sjot, 1991, pp. 23-44.
Mailing address: University of Copenhagen, Nørregade 10, PO Box 2177, DK-1017 København K, Denmark. T. (45) 33910828
Residence: Trondhjemsgade 5, 2., DK-2100 København 0, Denmark. T. (45) 31380406

Stranges, Frank Ernest, Dr

Rel: Christian; Ordained on 06 JUN 1951; b. 10 JUN 1927 in Brooklyn NY, USA; m. to Julie Ann; c. 2 boys; Citizen of USA; Lang. English, Italian and Korean (some).
International Organizations: President of the International Evangelism Crusades/International Theological Seminary of California.
Other organizations: President, National Investigations Committee on Unidentified Flying Objects; Author and Lecturer on the subject of UFO's; Deputy Assistant, California Marshals Association.
Main field of work: Ministry, preaching, teaching. **Career:** Pioneered a number of churches in Minnesota, Massachusetts and California. Ministered in the churches of various denominations, seminaries, Bible colleges, as well as conventions and crusades around the US and other parts of the world. Founded the International Evangelism Crusades (IEC) in 1959. Established the International Theological Seminary of California in 1976. Former Member, Mayor's Community Advisory Committee, City of Los Angeles. **Educ.:** B.Th., M.Th., Th.D., Ph.D. from Tennessee State Christian University, Chattanooga, USA; Dr. Int.Law from Union University, Los Angeles CA, USA. **Interests:** space, science. **Hons.:** Recipient FBI Gold Medal Award. **Mem.:** United Science Federation (FL); Planetary Society (CA); American Federation of Police (FL); Association of Chiefs of Police (VA); World Chinese Martial Arts Association (Life Member); Alpha Psi Omega **Pubs.:** articles in: IEC International Newsletter, many newspapers and other religious magazines.
Mailing address: 14617 Victory Blvd, Suite 4, Van Nuys CA 91411, USA. T. (1 818) 989 5942 (SAME FOR Fax)

Strauss, Stephen John, Rev

Rel: Evangelical; Ordained in 1980; b. 10 JUL 1955 in Dallas TX, USA; m. to Marcia Dawn; c. 2 boys and 1 girl; Citizen of USA; Lang. English and Amharic.
International Organizations: Pastor and Bible Institute Director (since 1982) of the International Evangelical Church.
Main field of work: Teaching and school administration. **Career:** Instructor at Dallas Seminary from 1980 to 1981. Missionary, Society of International Missionaries (SIM) from 1982 to 1991. **Educ.:** B.A. from Bryan College, Dayton TN, USA (1972-1976); Th.M. from Dallas Seminary, Dallas TX, USA (1976-1980). **Interests:** contextualization and hermeneutics. **Pubs.:** Book reviews in *Africa Journal of Evangelical Theology*, 1989-91.

Mailing address: PO Box 127, Addis Ababa, Ethiopia. T. (251 1) 710063
Residence: See Mailing Address. T. See Mailing Address

Struharik, Juraj, Rt Rev
Rel: Evangelical.
International Organizations: Bishop of Slovak Evangelical Church of the Augsburg Confession in Yugoslavia.
Mailing address: Karadziceva 2, 21000 Novi Sad, Yugoslavia.

Stuart, Ian D
Rel: Episcopal.
International Organizations: Secretary of the Scottish Episcopal Church.
Mailing address: 21 Grosvenor Crescent, Edinburgh EH12 5EE, Scotland, UK.

Stubs, Alan Charles, Very Rev
Rel: Presbyterian; Ordained on 29 APR 1960; b. 31 MAR 1924 in Launceston, Tasmania, Australia; m. to Esther Sophia Bain; c. 3 boys and 4 girls; Citizen of Australia; Lang. English.
International Organizations: Past Moderator-General (since 1992) of the Presbyterian Church of Australia.
Main field of work: Ministry. **Career:** Moderator, Presbyterian Church of Tasmania from 1973 to 1975. Moderator, Presbyterian Church of Tasmania from 1977 to 1979. Moderator, Presbyterian Church of Victoria from 1983 to 1984. Moderator, Presbyterian Church of South Australia from 1986 to 1987. Moderator-General, Presbyterian Church of Australia from 1988 to 1992. Lecturer, Presbyterian Theological Hall, Melbourne from 1979 to 1984. **Educ.:** B.A. from University of Tasmania, Hobart, Australia (1943-1954); B.D., Dip. Rel.Ed. from Melbourne College of Divinity, Australia (1957-1960); M.Th. from University of Edinburgh, UK (1974-1975). **Interests:** music (organ and choral work), cricket. **Mem.:** Christian Research Association **Pubs.:** The Pastoral Care of the Aged.
Mailing address: 91 Cecil Ave, Castle Hill NSW 2154, Australia. T. (61 2) 634 2911
Residence: See Mailing Address. T. See Mailing Address

Stückelberger, Hansjürg, Rev
Rel: Christian.
International Organizations: President of the Christian Solidarity International.
Mailing address: Forchstrasse 280, PO Box 881, CH-8029 Zürich, Switzerland. T. (41 1) 422 66 46. Fax 422 67 14.

Stückelberger, Johannes Georg, Rev
Rel: Reformed Church; Ordained on 10 MAY 1956; b. 06 DEC 1930 in Schiers, Switzerland; m. to Christa Haarbeck; c. 2 boys and 3 girls; Citizen of Switzerland; Lang. German, French, English and Spanish.
International Organizations: President/Founder (since 1975) of the Christian Solidarity International.
Other organizations: Minister (part-time), Reformed Church of Balgrist, Zürich.
Career: President, Christian East Mission. President, Christian Radio Mission. **Educ.:** Studied from University of Zürich and University of Basel, Switzerland, Georg-August-Universität Göttingen, Germany, and Free Scientological Seminary, Paris, France (1950-1956). **Interests:** further respect and love among Christian leaders. **Pubs.:** Christen in Not; Hand in Hand and Solidaritäts fonds.
Mailing address: Forchstrasse 280, Postfach 52, CH-8029 Zürich, Switzerland. T. (41 1) 55 66 46
Residence: Lenggstrasse 71, CH-8008 Zürich, Switzerland. T. (41 1) 53 94 94

Studer, Gerald C, Rev
Rel: Mennonite; Ordained on 31 OCT 1947; b. 31 JAN 1927 in Smithville OH, USA; m. to Marilyn Ruth Kreider; c. 2 girls; Citizen of USA; Lang. English.
International Organizations: President (since 1988) of the International Society of Bible Collectors.
Other organizations: Conference Minister (since 1990), Atlantic Coast Mennonite Conference.
Career: Pastor, 3 successive congregations, The Mennonite Church of North America from 1947 to 1990. **Educ.:** B.A. from Goshen College, IN, USA (1945-1949); B.Th. from Goshen Biblical Seminary, Goshen IN, USA (1949-1950); B.D. from Goshen Biblical Seminary, Goshen IN, USA (1956-1957); M.Div. from Goshen Biblical Seminary, Elkhart IN, USA (1971). **Pubs.:** *Christopher Dock: Colonial Schoolmaster*; *After Death, What?* and *Toward a Theology of Servanthood.*
Mailing address: 1260 Orchard Lane, Lansdale PA 19446, USA. T. (1 215) 368 8455
Residence: See Mailing Address. T. See Mailing Address

Stuker, Andrea
Rel: Reformed Church.
International Organizations: Member of the Executive Committee of the World Alliance of Reformed Churches.
Mailing address: Albisstrasse 10, CH-8134 Adliswil, Switzerland.

Sturzaker, D L
Rel: Kabbalist.
International Organizations: Principal of the International Order of Kabbalists.
Mailing address: 25 Circle Gardens, Merton Park, London SW19 3JX, UK. T. (44 81) 542 3611.

Sturzaker, J
Rel: Kabbalist.
International Organizations: Principal of the International Order of Kabbalists.
Mailing address: 25 Circle Gardens, Merton Park, London SW19 3JX, UK. T. (44 81) 542 3611.

Stylianos, S, H E Archbishop
Rel: Greek Orthodox.
International Organizations: Primate of the Greek Orthodox Archdiocese of Australia.
Mailing address: 242 Cleveland Street, Redfern NSW 2016, Australia.

Suarez, Guillermo Santa Maria
Rel: Catholic.
International Organizations: Regional Secretary for Latin America of the Pax Romana, International Catholic Movement for Intellectual and Cultural Affairs.
Mailing address: La Alborada, Xetapa, Manzana 103, Villa 15, CP 15067, Guayaquil, Ecuador.

Subhuti, S, Dharmachari
Rel: Buddhist.
International Organizations: Member of the Friends of the Western Buddhist Order.
Mailing address: Padmaloka, Lesingham House, Surlingham, Norwich NR14 7AL, UK. T. (44 50) 88 310. Tx 295141 TxLINK G/105088639 (ATT MBX).

Suchy, Wladyslaw, C.S.M.A., Fr
Rel: Roman Catholic; Ordained; celibate.
International Organizations: Procurator-General of the Congregation of the Archangel St Michael.
Mailing address: Pontificio Santuario Maria SS ad Rupes, I-01030 Castel Sant'Elia (Viterbo), Italy. T. (39 761) 55 77 29.

Suenens, Leo Jozef, H E Card
Rel: Roman Catholic; Ordained on 04 SEP 1927; b. 16 JUL 1904 in Ixelles, Brussels, Belgium; celibate; Citizen of Belgium.
International Organizations: Member (since 1962) of the College of Cardinals.
Other organizations: Archbishop Emeritus (since 1979), Malines-Brussels.

Career: Teacher at Institut Sainte-Marie, Brussels in 1929. Lecturer in Philosophy, Malines Seminary from 1930 to 1940. Vice-Rector, Catholic University of Louvain from 1940 to 1945. Private Chamberlain to Pope in 1941. Auxiliary Bishop and Vicar-General, Archdiocese of Malines from 1945 to 1961. Archbishop, Malines-Brussels and Primate of Belgium from 1961 to 1979. Created Cardinal by Pope John XXIII in 1962. Moderator, Vatican Council from 1962 to 1965. Member, Pontifical Commission for Revision of Canon Law in 1962. President, Belgian Bishops' Conference. **Hons.:** Templeton Foundation Prize (1976). **Pubs.:** A New Pentecost? Ecumenism and Charismatic Renewal; With H Camara, Charismatic Renewal and Social Action: A Dialogue and Renewal and the Powers of Darkness, Nature and Grace, 1986.
Mailing address: Boulevard de Smet de Nayer 570, B-1020 Bruxelles, Belgium. T. (32 2) 479 19 50

Suharto, S H, Rev
Rel: Christian.
International Organizations: General Secretary of the East Java Christian Church.
Mailing address: Jln S Supriadi 18, Malang, East Java, Indonesia.

Suka, A Ginting, Rev
Rel: Protestant.
International Organizations: Moderator of the Karo Batak Protestant Church.
Mailing address: Jln Kapt Pala Banun, 90 Kabanjahe, North Sumatra, Indonesia.

Sukhu, Rawle, Rev
Rel: Presbyterian.
International Organizations: General Secretary of the Presbyterian Church in Trinidad and Grenada.
Mailing address: PO Box 92, Paradise Hill, San Fernando, Trinidad-Tobago.

Sumah, Rose, Sr
Rel: Roman Catholic.
International Organizations: Representative of the Missionary Sisters of Our Lady of the Apostles.
Mailing address: Via Guiseppe Ghislieri 15, I-00152 Roma, Italy. T. (39 6) 53 5818. Fax 53 70155.

Summers, Sarah, Sr
Rel: Roman Catholic; Ordained; celibate.
International Organizations: Superior General of the Medical Mission Sisters.
Mailing address: 41 Chatsworth Gardens, Acton, London W3 9LP, UK. T. (44 81) 992 6444.

Summey, David
Rel: Christian.
International Organizations: Representative of the Help International Ministries.
Mailing address: Rt 1, Box 171D, Harlem GA 30814, USA. T. (1 706) 556 3408.

Summit, Jeffrey
Rel: Jewish.
International Organizations: Executive Director of the Association of Hillel-Jewish Campus Professionals.
Mailing address: B'nai B'rith Hillel Foundation, Curtis Hall, 474 Boston Ave, Medford MA 02155, USA. T. (1 617) 381 3242.

Sundby, (Carl) Olof (Werner)
Rel: Lutheran; Ordained in 1943; b. 06 DEC 1917 in Karlskoga, Sweden; m. to Birgitta Nordfors; c. 1 boy and 3 girls; Citizen of Sweden.
International Organizations: Ecclesiastic (retired) of the Lutheran Church.
Other organizations: Chairman (since 1984), Swedish Committee for the Life and Peace Institute, Uppsala.
Career: Parish Curate. Military Chaplain. Prison Chaplain. Diocesan Youth Pastor, Karlstad in 1944. Senior Lecturer on Theological Ethics, University of Lund in 1959. Rector, St Peter, Lund in 1960. Bishop, Växjö in 1970. Archbishop, Uppsala from 1972 to 1983. President, Swedish Ecumenical Council. Member, Central Committee and Executive Committee, World Council of Churches (WCC). Member, Presidium, WCC from 1975 to 1983. Chairman of Board, Kyrkans Tidning from 1982 to 1988. **Interests:** tennis, skiing. **Pubs.:** Lutheran Conception of Marriage, 1959 and Pastoral Letter to the Diocese of Växjö, 1970.
Mailing address: Gilleskroken 8, S-222 47 Lund, Sweden.

Sunshine, Catherine A
Rel: Christian.
International Organizations: Director of the Ecumenical Program on Central America and the Caribbean.
Mailing address: 1470 Irving St NW, Washington DC 20010, USA. T. (1 202) 332 02 92.

Benthareswara Kurukkal, E S, Sivasiri
Rel: Hindu.
International Organizations: Chief Priest of the Durga Devasthanam.
Mailing address: Durga Devasthanam, Tellippallai, Sri Lanka.

Sunthareswara Kurukkal, S S, Brahmasri
Rel: Hindu.
International Organizations: Chief Priest of the Vannai Venkadeswara Devasthanam.
Mailing address: Vannai Venkadeswara Devasthanam, Jaffna, Sri Lanka.

Suquía Goicoechea, Angel S, H E Card
Rel: Roman Catholic; Ordained on 07 JUL 1940; b. 02 OCT 1916 in Zaldivia, San Sebastián, Spain; celibate; Citizen of Spain.
International Organizations: Member (since 1985) of the College of Cardinals.
Member of the Congregation for Bishops. Member of the Congregation for the Evangelization of Peoples. Member of the Congregation for Catholic Education.
Other organizations: Archbishop (since 1983), Madrid; President, Episcopal Conference of Spain; Member, Council of Cardinals for the Study of Organization and Economic Questions Relating to the Holy See; Member, Council of Cardinals and Bishops.
Career: Consecrated Bishop, Almeria in 1966. Bishop, Málaga in 1969. Archbishop, Santiago de Compostela in 1973. Created Cardinal by Pope John Paul II in 1985.
Mailing address: San Justo 2, 28005 Madrid, Spain. T. (34 91) 266 56 01

Sutton, Norma, Rev
Rel: Mennonite; Ordained on 27 MAY 1984; b. 22 APR 1946 in Three Rivers MI, USA; divorced; Citizen of USA; Lang. English, German, French and Hebrew.
International Organizations: Theological Librarian (since 1979) of the North Park Theological Seminary.
Other organizations: Board of Directors (since 1990), American Theological Library Association.
Main field of work: Theological librarianship. **Career:** President, Chicago Area Theological Library Association from 1991 to 1992. **Educ.:** B.S. from Goshen College, IN, USA (1966-1968); M.Div. from Associate Mennonite Biblical Seminaries, Elkhart IN, USA (1969-1976); M.A. from University of Notre Dame, IN, USA (1976-1978); M.A. from Emory University, Atlanta GA, USA (1978-1979). **Interests:** printing history. **Mem.:** American Theological Library Association (ATLA); CATLA; AAR **Pubs.:** "The Influence of Radical Pietism on Russian Mennonites", in *Covenant Quarterly*, 38, November 1980, pp. 19-26 and "Arise for it is Day: John Day Imprints and the English Reformation, 1545-1559", in *The Journal of Religious and Theological Information*, 1, Spring 1992.
Mailing address: North Park Theological Seminary, 3225 W Foster Ave, Chicago IL 60625, USA. T. (1 312) 583 2700
Residence: 5218 N Sawyer Ave, Chicago IL 60625, USA. T. (1 312) 463 3044

Sutton, Peter Alfred, O.M.I., Most Rev

Rel: Roman Catholic; Ordained on 22 OCT 1960; b. 18 OCT 1934 in Chandler, Gaspé, Canada; celibate.
International Organizations: Archbishop (since 1986) of Keewatin-Le Pas.
Accompanying Bishop (since 1983) of the Arche Internationale. Member of the Missionary Oblates of Mary Immaculate.
Other organizations: Secretary-Treasurer, Western Catholic Conference.
Career: Entered Congregation of Missionary Oblates Mary Immaculate in 1953. Final profession as Oblate in 1957. Teacher at St Patrick's College High School, Ottawa in 1961. Teacher at Catholic Central High School, London, Ontario in 1963. Bishop, Labrador-Schefferville from 1974 to 1986. **Educ.:** M.Ed. from Loyola University, Chicago IL, USA.
Mailing address: PO Box 270, The Pas MB R9A 1K4, Canada. T. (1 204) 623 2503

Suvajra, S, Dharmachari

Rel: Buddhist.
International Organizations: Member of the Friends of the Western Buddhist Order.
Mailing address: Padmaloka, Lesingham House, Surlingham, Norwich NR14 7AL, UK. T. (44 50) 88 310. Tx 295141 TxLINK G/105088639 (ATT MBX).

Suzuki, Koei

Rel: Buddhist; Ordained; b. 09 AUG 1924; Citizen of Japan.
International Organizations: Chief of the Jofukuji.
Main field of work: Monk. **Career:** Manager, Shigisan Shingonshu. **Educ.:** Koyasan University, Japan.
Mailing address: Heguricho Shigisan, Ikoma-gun, Nara 636, Japan. T. (81 745) 72 2581

Suzuki, Ryowa

Rel: Buddhist.
International Organizations: Chairman of the International Network of Engaged Buddhists.
Mailing address: Komyo-Ji, Motoyashiki Minoya-Chyo, Anjiyo-Shi, Aichi-Ken, Japan.

Suzuki, Shunon

Rel: Buddhist; Ordained; b. 31 MAR 1930; Citizen of Japan.
International Organizations: Monk of Nichirenshu Hoonji.
Other organizations: President, Japan University of Social Welfare.
Educ.: Waseda University Graduate School, Tokyo, Japan.
Mailing address: 3-6 Komagatacho, Showa-ku, Nagoya, Aichi 466, Japan. T. (81 52) 831 7135

Swaggart, Jimmy Lee

Rel: Pentecostal; Ordained in 1964; b. in Louisiana, USA; m. to Frances Swaggart; c. 1 boy.
International Organizations: Founder of the the Family Worship Center and the Jimmy Swaggart Bible College in Baton Rouge LA.
Main field of work: Evangelist and gospel singer. **Career:** Committed himself to full-time ministry in 1958. Began to preach extended revivals in his denomination's churches in 1964. Launched 'The Camp Meeting Hour' (radio show) in 1969. Began holding crusades in civic auditoriums in 1972. Began 'The Jimmy Swaggart Telecast' (weekly television show) in 1973. Launched an attack on fellow televangelist Jim Bakker accusing him of adultery in 1987. Was himself accused of voyeuristic engagements with a prostitute in 1987. His ordination was withdrawn by his denomination but he continued his ministry in 1988. **Pubs.:** With R P Lamb, *To Cross a River*, 1977.
Mailing address: No mailing address.

Swamy, Jeevananda, Sri

Rel: Hindu.
International Organizations: Chief Priest of the Sri Manikkapillayar Temple, Hatton.
Mailing address: Sri Manikkapillayar Temple, Hatton, Sri Lanka.

Swapp, Bruce, Rev

Rel: Methodist.
International Organizations: Secretary of the Conference of the Methodist Church in the Caribbean and the Americas.
Mailing address: Belmont, PO Box 9, St John's, Antigua-Barbuda.

Swartley, Willard

Rel: Mennonite.
International Organizations: Professor of New Testament of the Goshen Biblical Seminary.
Mailing address: 3003 Benham Ave, Elkhart IN 46517, USA.

Swedish, Margaret

Rel: Christian.
International Organizations: Coordinator of the Religious Task Force on Central America.
Mailing address: 1747 Connecticut Ave NW, Washington DC 20009, USA. T. (202) 387 7652.

Sweeney, Brian, F.M.S., Fr

Rel: Roman Catholic; Ordained; celibate.
International Organizations: Procurator-General of the Little Brothers of Mary.
Mailing address: Piazzale M Champagnat 2, CP 10250, I-00144 Roma, Italy. T. (39 6) 592 44 56/7/8.

Sweeney, Redempta, IHM, Sr Mother

Rel: Roman Catholic; Ordained; celibate.
International Organizations: Superior General of the Sisters Servants of the Immaculate Heart of Mary of Scranton.
Mailing address: Marywood, 2300 Adams Avenue, Scranton PA 18509-1598, USA. T. (1 717) 342 6850.

Swinburne, Richard Granville, Prof

Rel: Anglican; b. 26 DEC 1934 in Smethwick, Staffs, UK; sep. from Monica H N Holmstrom; c. 2 girls; Citizen of United Kingdom; Lang. English, Russian, French (reading), Italian (reading), Latin (reading) and Greek (reading).
International Organizations: Nolloth Professor of the Philosophy of the Christian Religion (since 1985) of the University of Oxford.
Main field of work: University education. **Career:** Lecturer in Philosophy, University of Hull from 1963 to 1972. Professor of Philosophy, University of Keele from 1972 to 1984. **Educ.:** B.A. from University of Oxford, UK (1954-1957); B.Phil. from University of Oxford, UK (1957-1959); Dip. Theol. from University of Oxford, UK (1959-1960). **Pubs.:** The Coherence of Theism, 1977; The Existence of God, 1979 and The Evolution of the Soul, 1986.
Mailing address: Oriel College, Oxford OX1 4EW, UK. T. (44 865) 276589

Swingle, L A

Rel: Jehovah's Witnesses.
International Organizations: Assistant Secretary-Treasurer of the International Bible Students Association.
Mailing address: 25 Columbia Heights, Brooklyn NY 11201, USA. T. (1 718) 625 3600.

Swords, Brian, S.F.M., Fr

Rel: Roman Catholic; Ordained; celibate.
International Organizations: Superior General of the Scarboro Foreign Mission Society.
Mailing address: 2685 Kingston Road, Scarborough ON M1M 1M4, Canada. T. (1 416) 261 7135.

Syeed, Sayyid Muhammad, Dr

Rel: Muslim; Unordained; b. 03 APR 1941 in Sopor, Kashmir, India; m. to Rafia; c. 3 boys and 3 girls; Citizen of USA; Lang. Kashmiri, Urdu, Arabic and English.
International Organizations: Director of Academic Outreach (since 1984) of the International Institute of Islamic Thought.
Secretary-General (since 1988) of the Association of Muslim Social Scientists.
Main field of work: Ministry, research. **Career:** President, Muslim Students Association of USA and Canada from 1980 to 1983. Secretary-General, International Islamic Federation of Student Organizations (IIFSO) from 1984 to 1988. One of the founders, Islamic Society of North America. Has played a prominent role in the founding, the American Muslim Council, the Kashmiri-American Council, Americans for Soviet Muslim Rights, Indian Muslim Relief Committee and the Committee for the Defense of Human Rights in India. **Educ.:**

Ph.D. from Indiana University, Bloomington, USA (1984). **Interests:** research on Islamic social sciences. **Mem.:** Middle East Studies Association **Pubs.:** Editor-in-Chief of *American Journal of Islamic Social Scientists* (AJISS).
Mailing address: 555 Grove Street, Suite 101, Box 669, Herndon VA 22070, USA. T. (1 703) 471 1133. Tx 901153 IIIT WASH. Fax (1 701) 471 3922.
Residence: 1602 Attorney Ct, Herndon VA 22070, USA

Sykes, Stephen Whitefield, Rt Rev

Rel: Church of England; Ordained in 1965; b. 01 AUG 1939 in Bristol, Somerset, UK; m. to Joy; c. 1 boy and 2 girls; Citizen of United Kingdom; Lang. English and German.
International Organizations: Member of the Anglican-Lutheran International Commission.
Other organizations: Bishop (since 1990), Ely; President, Council, St John's College, Durham; Member, Doctrine Commission, Church of England.
Main field of work: Ministry. **Mem.:** Académie internationale des sciences religieuses
Mailing address: The Bishop's House, Ely CB7 4DW, UK. T. (44 353) 662749
Residence: See Mailing Address. T. See Mailing Address

Syme, Daniel B, Rabbi

Rel: Jewish: Reform.
International Organizations: Senior Vice-President of the Union of American Hebrew Congregations.
Mailing address: 838 Fifth Ave, New York NY 10021, USA. T. (1 212) 249 0100

Syron, Rita, Sr

Rel: Roman Catholic.
International Organizations: Society Coordinator of the Medical Mission Sisters.
Mailing address: 41 Chatsworth Gardens, Acton, London W3 9LP, UK. T. (44 81) 992 6444.

Szedressy, Paul, Bishop

Rel: Evangelical Presbyterian.
International Organizations: Bishop of Evangelical Synodal Presbyterial Church of the Augsburg Confession in Romania.
Mailing address: Bulevardul Lenin 1, 3400 Cluj-Napoca, Romania.

Szigeti, Jeno, Pastor

Rel: Seventh-Day Adventist.
International Organizations: President, Hungarian Union Conference of the General Conference of Seventh-Day Adventists.
Mailing address: Szekely Bertalan u 13, H-1062 Budapest, Hungary.

Szoka, Edmund Casimir, H E Card

Rel: Roman Catholic; Ordained on 05 JUN 1954; b. 14 SEP 1927 in Grand Rapids, USA; celibate; Citizen of USA.
International Organizations: President (since 1990) of the Prefecture for Economic Affairs of the Holy See.
Member of the Congregation for the Causes of Saints. Member of the Congregation for Bishops. Member of the Congregation for the Evangelization of Peoples. Member of the Congregation for the Clergy. Member of the Congregation for Religious and Secular Institutes.
Other organizations: Member (since 1988), College of Cardinals; Archbishop Emeritus (since 1990), Detroit; Member, Council of Cardinals and Bishops.
Career: Elected to Gaylord in 1954. Consecrated Bishop, Detroit in 1971. Created Cardinal by Pope John Paul II in 1988.
Mailing address: Via Rusticucci 13, I-00193 Roma, Italy. T. (39 6) 698 42 66

Szücs, Ferenc, Prof

Rel: Reformed Church in Hungary; Ordained on 10 SEP 1967; b. 29 SEP 1942 in Örszentmiklós, Pest Co., Hungary; married; c. 3 boys and 1 girl; Citizen of Hungary; Lang. Hungarian, English and German.
International Organizations: Secretary-General (since 1990) of the "Collegium Doctorum" of the Reformed Church of Hungary.
Other organizations: Professor of Systematic Theology (since 1985), Református Teológiai Akadémia Budapest.
Main field of work: University education. **Career:** Assistant Minister, Reformed Church of Hungary from 1965 to 1969. Minister, Reformed Church of Hungary from 1970 to 1987. Spiritual Director, Theological Academy, Budapest from 1979 to 1983. Lecturer, Theological Academy, Budapest from 1983 to 1985. Dean, Theological Academy, Budapest from 1986 to 1988. **Educ.:** Dip. Ministry from Reformatus Teológiai Akadémia, Budapest, Hungary (1960-1965); P.G. Stds. at University of Edinburgh, New College, UK (1969-1970); Dr. Theol. from Reformatus Teológiai Akadémia, Budapest, Hungary (1984). **Pubs.:** Az 'Imago Dei' értelme a barthi és a Barth utáni teológiában, 1984 (Understanding of 'Imago Dei' in the Barthian and post-Barthian Theology); Dogmatikai Prolegomena, 1991 and Teológiai Etika, 1992 (Theological Ethics).
Mailing address: Ráday u. 28, Budapest H-1092, Hungary. T. (36 1) 180 816
Residence: See Mailing Address. T. See Mailing Address

Taberdo, Lettie A

Rel: Catholic.
International Organizations: Managing Director of the Asian Socio-Pastoral Communications Centre.
Mailing address: PO Box 4082, Sonolux Asia Bldg., Ateneo de Manila University, Loyola Heights, Quezon City, METRO Manila 1099, Philippines. T. (63 2) 96 59 58 – 98 25 41 to 98 25 45 Ext 314 – 363. Fax 521 81 25 – 817 97 42 Attn 67568 PIC PN. C. SONOLUX JESUITAS MANILA. Tx 67568 PIC PN.

Tabet, Paul Fouad, H E Mgr

Rel: Roman Catholic; Ordained on 22 DEC 1956; b. 28 NOV 1929 in Maarab; celibate.
International Organizations: Apostolic Pro-Nuncio to Nigeria (since 1984) of the Roman Catholic Church.
Other organizations: Titular Archbishop, Sinna.
Mailing address: 9 Anifowoshe Street, Victoria Island, PO Box 2470, Lagos, Nigeria. T. (234 1) 614 441 – 613 744. Fax 618 635. Tx 22455 NUNLAG NG

Tabibi, Aboul Makim, Dr

Rel: Muslim.
International Organizations: Permanent Representative to the UN, Geneva of the World Muslim Congress.
Mailing address: Representative of WMC, Rue de Lyon 81, CH-1203 Genève, Switzerland.

Tabick, Jacqueline Hazel, Rabbi

Rel: Jewish; Ordained on 20 MAY 1975; b. 08 OCT 1948 in Dublin, Ireland; m. to Larry; c. 2 boys and 1 girl; Citizen of United Kingdom; Lang. Hebrew and English.
International Organizations: Associate Rabbi (since 1975) of West London Synagogue.
Other organizations: Vice-Chairman (since 1990), Council of Reform and Liberal Rabbis.
Main field of work: Pastoral, education. **Educ.:** B.A. (Hons) from University of London, UK (1967-1970); Dip. Ed. from Victoria University of Manchester, UK (1970-1971); Rabb. Dip. from Leo Baeck College, London, UK (1971-1975). **Interests:** education, social issues. **Mem.:** Assembly of Rabbis **Pubs.:** Education materials.
Mailing address: West London Synagogue, 33 Seymour Place, London W1H 6AT, UK. T. (44 71) 723 4404

Tack, Martha

Rel: Buddhist: Mahayana.
International Organizations: Member of the Foundation for the Preservation of the Mahayana Tradition.
Other organizations: Director, Milarepa Center.
Mailing address: Barnet Mountain, Barnet VT 05821, USA. T. (1802) 633 4136

Tagawa, Jokaku

Rel: Buddhist; Ordained; b. 12 OCT 1935; Citizen of Japan.
International Organizations: Monk of Hososhu Kofukuji.
Educ.: Ryukoku University, Kyoto, Japan.
Mailing address: 48 Nobori Ojicho, Nara 630, Japan. T. (81 742) 22 7755

Tagliaferri, Mario, H E Mgr

Rel: Roman Catholic; Ordained on 05 AUG 1950; b. 01 JUN 1927 in Alatri, Italy; celibate.
International Organizations: Apostolic Nuncio to Spain (since 1985) of the Roman Catholic Church.
Other organizations: Titular Archbishop, Formia.
Mailing address: Avenida Pío XII 46, E-28016 Madrid, Spain. T. (34 1) 766 83 11. Fax 766 70 85. Tx 47768 NAPM E

Taher, Muhammad, Dr

Rel: Muslim.
International Organizations: Chief Librarian of the American Research Center.
Mailing address: Osmania University, Hyderabad AP 500007, India.

Taherzadeh, Adib

Rel: Bahá'í; b. 29 APR 1921 in Dublin, Ireland; m. to Lesley Elaine Gibson; c. 2 boys and 2 girls; Citizen of Ireland and Iran; Lang. Persian and English.
International Organizations: Member, Universal House of Justice (since 1988) of the Bahá'í International Community.
Main field of work: Bahá'í administration. **Career:** Chief Engineer, Messrs. Hughes and Coyle Ltd. from 1950 to 1984. Chairman, National Bahá'í Assembly, Ireland. **Educ.:** B.Sc. Eng. from University of Teheran, Iran (1939-1944). **Interests:** Writing. **Mem.:** Ex-Member, Continental Board Counsellors, Bahá'í **Pubs.:** *The Revelation of Bahá'ú'lláh*, Vols. 1, 2, 3 and 4 and *Trustees of the Merciful.*
Mailing address: PO Box 155, 31001 Haifa, Israel. T. (972 4) 35 83 58
Residence: 16 Kauder Street, Haifa, Israel T. (972 4) 33 43 53

Tahzib, Bahiyyih G

Rel: Bahá'í.
International Organizations: Secretary of the European Bahá'í Youth Council.
Mailing address: Utrechtseweg 115, NL-3702 AB Zeist, Netherlands. T. (31 3404) 18 441. Fax (31 3404) 13 534.

Takada, Koin

Rel: Buddhist; Ordained; b. 30 MAR 1924; Citizen of Japan.
International Organizations: Monk of Hososhu Yakushiji.
Educ.: Ryukoku University, Kyoto, Japan.
Mailing address: 457 Yakushiji Nishi Nokyomachi, Nara 630, Japan. T. (81 742) 33 6001

Takamatsu, Ryoshu

Rel: Buddhist; Ordained; b. 21 NOV 1914; Citizen of Japan.
International Organizations: Monk of Zenshoji.
Other organizations: President, Iida Women's Junior College.
Educ.: Kyoto University, Japan.
Mailing address: 2-31 Tenmacho, Iida, Nagano 395, Japan. T. (81 265) 24 0420

Takasawa, Shinichiro

Rel: Shinto; Ordained; b. 21 JUN 1908; Citizen of Japan.
International Organizations: Chief Priest of the Meiji Shrine.
Career: Superior, Association of Shinto Shrine. Trustee, Kokugakuin University. **Educ.:** Kokugakuin University, Tokyo, Japan.
Mailing address: 1-1-1 Yoyogi, Shibuya-ku, Tokyo 151, Japan. T. (81 3) 3370 3032

Takenaka, Masao, Dr

Rel: Christian; Unordained; b. 06 SEP 1925 in Beijing, China; married; c. 1 boy and 1 girl; Citizen of Japan; Lang. Japanese, English and German.
International Organizations: Chairman (since 1978) of the Asian Christian Art Association.
Chairman (since 1986) of the Programme for Theology and Cultures in Asia.
Other organizations: Professor of Christian Ethics and Sociology and Religion (since 1955), Doshisha University, Kyoto; Executive Board (since 1962), Nippon Christian Academy.
Main field of work: Christian ethics and sociology of religion. **Career:** Chairman, Urban Industrial Mission, Christian Conference of Asia from 1968 to 1973. Chairman, Urban Industrial Mission, World Council of Churches from 1968 to 1975. Chairman, Nishijin Community Center from 1970 to 1989. **Educ.:** M.Econ. from Kyoto University, School of Economics, Japan (1948); B.D. from Doshisha University, School of Theology, Kyoto, Japan (1950); B.D. from Yale University, Divinity School, New Haven CT, USA (1951); S.T.M. from Yale University, Divinity School, New Haven CT, USA (1952); Ph.D. from Yale University, Divinity School, New Haven CT, USA (1955). **Hons.:** Carnahan Lecture at Buenos Aires (1965), Burns Memorial Lectures, New Zealand (1966), Canadian Research Award (1990). **Mem.:** United Church of Christ in Japan **Pubs.:** God is Rice, 1986; Cross and Circle, 1990 and The Bible Through Asian Eyes, 1991.
Mailing address: School of Theology, Doshisha University, Karasuma Imadegawa, Kyoto 602, Japan. T. (81 75) 251 3338
Residence: 3 Kita Chanoki-cho, Shimogamo, Sakyo-ku, Kyoto 606, Japan. T. (81 75) 781 7598

Takeuchi, Bartholomew Kentaro, Rev

Rel: Anglican; Ordained on 21 DEC 1958; b. 05 APR 1931 in Tokyo, Japan; m. to Lydia Toshiko; c. 2 girls; Citizen of Japan; Lang. Japanese and English.
International Organizations: Moderator (since 1988) of The National Christian Council of Japan.
Main field of work: Ministry. **Educ.:** B.A. from Keio University, Tokyo, Japan (1951-1954); Central Theological College, Tokyo, Japan (1955-1957).
Mailing address: Holy Trinity Church, 2-10-11 Daizawa, Setagayaku, Tokyo 155, Japan. T. (81 3) 3414 2891
Residence: See Mailing Address. T. (81 3) 3421 3646

Talvivaara, Anu

Rel: Orthodox; Unordained; b. 20 APR 1964 in Ilomantsi, Finland; single; Citizen of Finland; Lang. Finnish, Swedish, English, Russian and French.
International Organizations: Secretary-General (since 1989) of the World Fellowship of Orthodox Youth.
Main field of work: Youth ministry. **Educ.:** M.A. from University of Joensuu, Finland (1984-1990).
Mailing address: Résidence des Pépinières, 12 Bd Desgranges, F-92230 Sceaux, France. T. (33 1) 46 60 17 74. Fax (33 1) 46 60 45 54 SYNDESMOS.
Residence: Rue du Prieuré 10B, CH-1260 Nyon, Switzerland. T. (41 22) 625 824

Tamás, Bertalan, Rev

Rel: Calvinist-Reformed; Ordained on 15 SEP 1965; b. 07 APR 1940 in Debrecen, Hungary; m. to Erzsébet; c. 1 girl; Citizen of Hungary; Lang. Hungarian and English.
International Organizations: Member of the Executive Committee (since 1989) of the World Alliance of Reformed Churches.
Member, Executive Committee (since 1990) of the International Council of Christians and Jews.
Other organizations: Head of the Ecumenical Dept (since 1979), Synodal Office of the Scottish Mission in Budapest; Pastor, Reformed Church in Hungary.
Main field of work: Ministry. **Educ.:** Divinity from Reformed Theological Seminary, Budapest, Hungary (1959-1965); King's College, Aberdeen, UK (1972-1973). **Interests:** literature, gardening. **Mem.:** Rotary Club, President of Budapest-Duna **Pubs.:** Agonia Christiana (Lev Tolstoj); Antisemitism from the Christian point of view and Christian-Jewish Dialogue.
Mailing address: Abonyi u. 21, H-1146 Budapest, Hungary. T. (36 1) 1228 479
Residence: Temesvári u. 4, H-1215 Budapest, Hungary

Tamez, Carolos, Rev

Rel: Protestant.
International Organizations: General Coordinator of the Evangelical Latin American Commission on Christian Education.
Mailing address: Apartado 5312, Managua, Nicaragua. T. (505 2) 44594.

Támez, Elsa, Prof Dr

Rel: Methodist; b. 17 SEP 1950 in Victoria, Tamaulipas, Mexico; m. to José Duque; c. 1 boy and 1 girl; Citizen of Mexico; Lang. Spanish, English, French and German.
International Organizations: Vice-Moderator, Program on Theological Education (since 1983) of the World Council of Churches.
Other organizations: Professor of Theology (since 1980), Seminario Bíblico Latinoamericano; Researcher at Departamento Ecuménico de Investigaciones (since 1978).
Main field of work: Research, theological education. **Educ.:** M.Th. from Seminario Bíblico, Costa Rica (1970-1979); M.Litt. from Universidad Nacional, Costa Rica (1978-1985); Th.D. from University of Lausanne, Switzerland (1987-1990). **Interests:** Biblical studies, liberation theology, women. **Mem.:** Ecumenical Association of Third World Theologians (EATWOT) **Pubs.:** Bible of Oppressed, Orbis Books; The Scandalous Message of James, Cross Road, 1990 and Contra toda condena. La justificación por la fe desde los excluidos, DEI, 1991 (will soon be available in English).
Mailing address: Apartado 901, San José 1000, Costa Rica. T. (506) 227555
Residence: Calle 3, Avenida 14, San José, Costa Rica. T. (506) 355922

Tan, Hup Cheng

Rel: Buddhist: Mahayana.
International Organizations: Member of the Foundation for the Preservation of the Mahayana Tradition.
Other organizations: Director, Amitabha Buddhist Centre.
Mailing address: 494-D Geylang, Singapore 1439, Singapore. T. (65) 745 8547. Fax (65) 745 0438

Tanaka, Fumikiyo

Rel: Shinto; Ordained; b. 17 APR 1905; Citizen of Japan.
International Organizations: Honorary Priest of the Iwashimizu Hachimangu.
Career: President, Kyoto Kokugakuin. Superior, Shrine Head Office.
Educ.: Kokugakuin University, Tokyo, Japan.
Mailing address: 14 Yawata aza Shibaza, Yawata, Kyota 614, Japan. T. (81 75) 981 3001

Tang, Dominic, Rt Rev

Rel: Roman Catholic; Ordained in 1941; b. 13 MAY 1908 in Hong Kong; celibate; Citizen of China.
International Organizations: Ecclesiastic of the Roman Catholic Church.
Career: Apostolic Administrator, Canton from 1951 to 1957. Imprisoned in China in 1957. Released in 1980. Appointed Archbishop, Canton by Pope John Paul II in 1981. Appointment rejected by Chinese Catholics.
Mailing address: Catholic Mission, She-Se, Guangzhou, Guangdong, China.

Tani, Kogetsu

Rel: Buddhist; Ordained; b. 22 OCT 1931; Citizen of Japan.
International Organizations: President of the Shogen Junior College.
Main field of work: Monk. **Career:** Chief, Shogenji. **Educ.:** Bukkyo University, Kyoto, Japan.
Mailing address: 872-2 Ibukacho, Minokamo, Gifu 505, Japan. T. (81 574) 29 1369

Tanner, Ian B, Rev

Rel: Uniting Church in Australia.
International Organizations: President of the Uniting Church in Australia.
Mailing address: 31 Boothby Street, Panorama 5041, Australia.

Tantawi, Muhammad Sayed Attiyah, Dr

Rel: Muslim; Citizen of Egypt.
International Organizations: Grand Mufti of Egypt.
Mailing address: No mailing address.

Taofinu'u, Pio, S.M., H E Card

Rel: Roman Catholic; Ordained on 08 DEC 1954; b. 09 DEC 1923 in Falealupo, Savaii, Samoa; celibate; Citizen of Western Samoa.
International Organizations: Member (since 1973) of the College of Cardinals.
Member of the Society of Mary. Member of the Congregation for the Causes of Saints.
Other organizations: Archbishop (since 1968), Samoa-Apia and Tokelau.
Career: Created Cardinal by Pope Paul VI in 1973. **Interests:** gardening, music, art. **Pubs.:** The Kava Ceremony is a Prophecy, 1973.
Mailing address: Archbishop's House, PO Box 532, Apia, Western Samoa. T. (684) 20 401

Taranger, Bengt

Rel: Lutheran; Ordained on 25 MAY 1960; b. 24 JUN 1932 in Gothenburg, Sweden; m. to Anita Sjögren; c. 1 boy and 2 girls; Citizen of Sweden; Lang. Swedish, English and Tamil.
International Organizations: Member, Executive Committee of the International Christian Federation for the Prevention of Alcoholism and Drug Addiction.
Other organizations: Director (since 1974), Christian Temperance Federation of Sweden; Member of the Board, Temperance Council of Scandinavia; Member of the Board, Temperance Council of Sweden.
Main field of work: Drug policy, prevention and treatment in the field of alcohol and drugs. **Career:** Director, Hospital for alcoholics from 1960 to 1962. Missionary, Church of Sweden Mission, TECC India from 1963 to 1971. Secretary, Dept. for diaconal work, Diocese of Växjö, Church of Sweden from 1971 to 1974. **Educ.:** Bachelor of Social Work from Sköndalsinstitutet, Stockholm, Sweden (1955-1960). **Interests:** expert on alcohol and drugs, treatment of alcoholics and drug addicts. **Pubs.:** Dessa mina minsta bröder. Några tankar om diakoni och dom, 1980 and numerous articles in books, magazines and papers within the field of interest.
Mailing address: DKSN, Älvsjö gårdsväg 7, S-125 30 Älvsjö, Sweden. T. (46 8) 749 12 70
Residence: Bondegårdsgatan 18, S-352 44 Växjö, Sweden. T. (46 470) 189 21

Tarter, Bill K, Rev

Rel: Christian.
International Organizations: Director-General of the International Missions.
Mailing address: PO Box 14866, Reading PA 19612-4866, USA. T. (1 215) 375 0300.

Tarzia, Antonio, S.S.P., Rev Dr

Rel: Roman Catholic; Ordained on 30 JUN 1969; b. 27 NOV 1940 in Amaroni, Catanzaro, Italy; celibate; Citizen of Italy; Lang. Italian and French.
International Organizations: Director-General (since 1984) of the Edizioni Paoline.
Other organizations: President (since 1990), Club della Famiglia.
Main field of work: Publishing. **Career:** Vice-Director, *Il Giornalino* (children weekly) from 1972 to 1976. Director, *Jesus* (monthly) from 1977 to 1984. **Educ.:** Lic. Theol. from Lateran Pontifical University, Vatican City State (1966-1970); Laurea lettere moderne from University of Milan, Italy (1972-1979). **Pubs.:** Books for children: *La storia del presepe*; *Primo incontro con Gesù* and *L'uccellino fotoreporter*.
Mailing address: Piazza Soncino 5, I-20092 Cinisello-Balsamo, Italy. T. (39 2) 66 00 621
Residence: See Mailing Address. T. (39 2) 61 75 341

Tatabe, Roask

Rel: Subud.
International Organizations: Head, Secretariat of the World Subud Association.
Mailing address: c/o Int Subud Committee, Takatsu, PO Box 40, Kawasaki-shi 213, Japan.

Tatebe, Zuiyu

Rel: Buddhist; Ordained; b. 03 JAN 1911; Citizen of Japan.
International Organizations: Monk of Chinkokuji.
Career: Chief, Ninnaji. Consultant to Shingonshu Omuro-ha. Chief, Chinkokuji.
Mailing address: Chinkokuji, Genkaimachi Yoshida, Munakata-gun, Fukuoka 811-35, Japan. T. (81 940) 62 0111

Tauvette, Liliane Irène, Sr

Rel: Roman Catholic; Ordained; b. 25 OCT 1941 in Strickland, Ontario, Canada; celibate; Citizen of Canada; Lang. French, English and Spanish.
International Organizations: Superior General (since 1991) of the Sisters of Our Lady of Charity of the Good Shepherd.
Main field of work: Administration. **Career:** Director, Administrative Depts, Centre Rose-Virginie Pelletier from 1968 to 1988. **Educ.:** Bacc. Adm. from University of Montreal, Quebec, Canada (1971).
Mailing address: Via Raffaello Sardiello 20, I-00165 Roma, Italy. T. (39 6) 6641 85 45. Fax 6641 88 64.
Residence: See Mailing Address. T. See Mailing Address

Tavard, Georges Henri, Rev

Rel: Roman Catholic; Ordained in 1947; b. 06 NOV 1922 in Nancy, France; celibate; Citizen of USA.
International Organizations: Ecclesiastic of the Roman Catholic Church.
Career: Lecturer in Theology, Capenor House, Surrey, UK from 1949 to 1951. Associate Editor, Documentation Catholique, Paris from 1951 to 1952. Lecturer in Theology, Assumption College, Worcester MA from 1953 to 1958. Chairman and Professor, Dept. of Theology, Carlow College, Pittsburgh PA from 1959 to 1966. Professor of Religious Studies, Pennsylvania State University from 1966 to 1969. Professor of Theology, Methodist Theological School, OH from 1970 to 1987. **Hons.:** Medal of St Augustine of Canterbury; Hon. D.D., Kenyon College. **Pubs.:** The Seventeenth Century Tradition, 1978; Poetry and Contemplation in St John of the Cross, 1988 and Song for Avalokita, Poems, 1979 (under pseudonym of Henri Wasser).
Mailing address: Methodist Theological School, Delaware OH 43015, USA. T. (1 614) 548 4824
Residence: 2151 Waldorf Road, Columbus OH 43229, USA. T. (1 614) 475 6811

Tawil, Joseph Elias, H E Mgr

Rel: Greek Catholic; Ordained on 20 JUL 1936; b. 25 DEC 1913 in Damascus, Syria; celibate; Citizen of Syria.
International Organizations: Archbishop Emeritus (since 1989) of Newton.
Career: Teacher at Patriarchal College, Cairo from 1936 to 1939. Headmaster, Patriarchal College, Cairo from 1939 to 1943. Principal, Patriarchal College, Cairo from 1943 to 1954. Patriarchal Vicar in Alexandria from 1954 to 1961. Elected Archbishop, Myra by Greek Catholic Synod in 1959. Patriarchal Vicar to His Beatitude Patriarch Maximox IV, Damascus from 1961 to 1969. Melkite Apostolic Exarch for the USA in 1970. Archbishop, Newton from 1976 to 1989. Contributor to *Le Lien* (periodical), Cairo from 1943 to 1954. **Pubs.:** The Three Sacraments of the Christian Initiation, 1977; St Basil the Great, 1981 and various articles in the *National Catholic Register* and other newspapers.
Mailing address: 7 Veterans of Foreign Wars Parkway, Boston MA 02131, USA. T. (1 617) 323 5242

Taylor, David W A, Rev Dr

Rel: Christian; Ordained.
International Organizations: Secretary-General of the Consultation on Church Union.
Mailing address: 151 Wall St, Princeton NJ 08540, USA. T. (1 609) 921 7866.

Taylor, Edwin L, Rev

Rel: Methodist.
International Organizations: President of the Conference of the Methodist Church in the Caribbean and the Americas.
Mailing address: PO Box 9, St John's, Antigua-Barbuda.

Taylor, Eric, Most Rev Dr

Rel: Liberal Catholic; Ordained; celibate.
International Organizations: Presiding Bishop of General Episcopal Synod of the Liberal Catholic Church.
Mailing address: Drayton House, 30 Gordon Street, London WC1, UK. T. (441) 387 3260.

Taylor, Humphrey Vincent, Rev Canon

Rel: Anglican; Ordained; b. 05 MAR 1938; m. to Anne Katharine Dart; c. 2 girls.
International Organizations: Secretary (since 1984) of the World Mission Association.
Other organizations: Honorary Canon (since 1986), Bristol Cathedral; Honorary Canon (since 1989), Province of Southern Africa.
Career: Curate, Hammersmith from 1963 to 1964. Curate, North Kensington from 1964 to 1966. Rector, Lilongue, Malawi from 1966 to 1971. Chaplain and Lecturer, Bishop Grosseteste College, Lincoln from 1972 to 1974. Secretary for chaplaincies in higher education, General Synod Board of Education from 1974 to 1980. Mission Programs Secretary, United Society for the Propagation of the Gospel (USPG) from 1980 to 1984. **Educ.:** B.A., M.A. from University of Cambridge, Pembroke College, UK; M.A. from University of London, UK.
Mailing address: Partnership House, 157 Waterloo Road, London SE1 8XA, UK. T. (44 71) 928 8681. Fax (44 71) 620 1070. C. GOSPELIZE LONDON SE1. Tx 8590 907 ANG COM G

Taylor, John Bernard, Dr

Rel: Methodist; b. 04 JUN 1937 in Birmingham, Warwickshire, UK; m. to Margaret Carol Maynard; c. 2 boys and 1 girl; Citizen of United Kingdom; Lang. English, French and German.
International Organizations: Secretary-General of the World Conference on Religion and Peace.
Educ.: B.A. (Hons), M.A. from University of Cambridge, Trinity College, UK (1957-1961); M.A., Ph.D. from McGill University, Montreal, Canada (1962-1965).
Mailing address: 14 Chemin Auguste Vilbert, Grand Saconnex, CH-1218 Genève, Switzerland. T. (41 22) 798 51 62. C. RELPEACE, GENEVA. Fax (41 22) 791 00 34.
Residence: L'Echappée, Route d'Arzier, CH-1264 St Cergue, Switzerland T. (41 22) 60 17 22

Taylor, Michael Hugh, Rev

Rel: Baptist; Ordained in 1960; b. 08 SEP 1936 in Northampton, Northamptonshire, UK; m. to Adèle May Dixon; c. 2 boys and 1 girl; Citizen of United Kingdom.
International Organizations: Director (since 1985) of the Christian Aid.
Other organizations: Chair (since 1979), Audenshaw Trustees; Vice Moderator (since 1985), WCC (World Council of Churches) Commission on Theological Education; Member of the Council, VSO (Voluntary Service Overseas) and ODI (Overseas Development Institute).
Main field of work: Ministry. **Career:** Baptist Minister, North Shields, Northumberland from 1961 to 1966. Baptist Minister, Hall Green, Birmingham from 1966 to 1969. Principal, Northern Baptist College, Manchester from 1970 to 1985. Examining Chaplain to Bishop, Manchester from 1975 to 1985. **Educ.:** B.A., B.D., M.A. from Victoria University of Manchester, UK; Union Theological Seminary, New York NY, USA. **Interests:** cooking, theatre, walking. **Hons.:** Fulbright Travel Award (1969). **Pubs.:** Sermon on the Mount, 1982; Learning to Care, 1983 and Good for the Poor, 1990.
Mailing address: 53 Woodland Rise, London N10 3UN, UK. T. (44 81) 883 7217
Residence: See Mailing Address. T. See Mailing Address

Taylor, Robert Selby, Most Rev Dr

Rel: Anglican; Ordained on 29 SEP 1931; b. 01 MAR 1909 in Darlington, Durham, UK; single; Citizen of South Africa; Lang. English.
International Organizations: Life President (since 1971) of the South African Council of Churches.
Other organizations: Archbishop Emeritus (since 1988), Cape Town.
Main field of work: Ministry. **Career:** Principal, Diocesan Theological College, Northern Rhodesia from 1939 to 1941. Bishop, Northern Rhodesia from 1941 to 1951. Bishop, Pretoria from 1951 to 1958. Bishop, Grahamstown from 1958 to 1964. Archbishop, Cape Town from 1964 to 1974. Bishop, Central Zambia from 1979 to 1984. **Educ.:**

University of Cambridge, St Catharine's College, UK (1927-1930); University of Oxford, Cuddesdon College, UK (1930-1931). **Interests:** race relations, ecumenical relations. **Hons.:** DD, Lambeth (1991); Hon DD, Rhodes University (1966); Commander, Order of the British Empire (1972). **Mem.:** Priest of the Oratory of the Good Sheperd; Honorary Fellow, St Catharine's College, Cambridge (1964)
Mailing address: Braehead House, Auburn Road, Kenilworth 7700, South Africa. T. (27 21) 762 1440
Residence: See Mailing Address. T. See Mailing Address

Taylor, Thomas F
Rel: Quaker.
International Organizations: Secretary-General of the Friends World Committee for Consultation.
Mailing address: Drayton House, 30 Gordon Street, London WC1H OAX, UK. T. (44 71) 388 0497.

Taylor, William D, Dr
Rel: Evangelical.
International Organizations: Executive Secretary of the Missions Commission of the World Evangelical Fellowship.
Mailing address: 29 Simpson Ave, Cedar Creek TX 78612, USA. T. (1 512) 321 7618. Fax (1 512) 321 0535.

Taylor-Pearce, Modupe, Dr
Rel: Evangelical.
International Organizations: Principal of the Nairobi Evangelical Graduate School of Theology.
Mailing address: c/o AEAM, PO Box 24686, Nairobi, Kenya. T. (254 2) 88 21 04 – (254 2) 88 21 05.

Teauariki, Turaki, Rev
Rel: Christian.
International Organizations: President of the Cook Islands Christian Church.
Mailing address: PO Box 93, Takamoa, Rarotonga, Cook Is.

Techera, Jorge, Rev
Rel: Catholic.
International Organizations: Secretary for Latin America of the Caritas Internationalis.
Mailing address: Apartado 1389-A, Quito, Ecuador. T. (593 2) 567309. Fax 504577. Tx 2427 CONFER ED.

Tedla, Stefanos, O.F.M. Cap., Fr
Rel: Roman Catholic; Ordained on 16 JUN 1973; b. 01 JAN 1940 in Degra Libe, Akele Guzai, Eritrea, Ethiopia; celibate; Citizen of Ethiopia; Lang. Tigrigna, Amharic, English and Italian.
International Organizations: Anglophone Africa Coordinator (since 1966) of the African Caritas.
Member of the Order of Friars Minor Capuchin. Consultant, Emergency Aid Group (since 1966) of the Pontifical Council 'Cor Unum'. Member, International Working Group on Refugees (since 1989) of the Caritas Internationalis. Member, International Working Group on AIDS (since 1991) of the Caritas Internationalis. Member (since 1984) of the Caritas Internationalis.
Main field of work: Coordination of relief and development aid, socio-pastoral activities of the Church organizations in Anglophone Africa. **Career:** Secretary-General, Ethiopian Catholic Secretariat from 1973 to 1985. **Educ.:** Dip. Social Leadership from Cody Institute, Nova Scotia, Canada (1966-1967); Lic. Theol. from Pontificio Ateneo Antonianum, Vatican City State (1970-1973). **Interests:** helping the needy affected by emergency. **Mem.:** Caritas African Regional Commission (1966)
Pubs.: Articles on the famine in Ethiopia, development and refugees in Africa.
Mailing address: Anglophone Africa Secretariat, PO Box 14954, Nairobi, Kenya. T. (254 2) 74 52 96. Tx 23297 ISEDC M (FOR CARITAS OFFICE).
Residence: See Mailing Address. T. See Mailing Address

Teeple, Howard Merle, Prof Emer Dr
Rel: Independent; b. 29 DEC 1911 in Salem OR, USA; m. to Gladys Berdean Windedahl; Citizen of USA; Lang. English, French, German, Greek (Hellenistic) and Hebrew.
International Organizations: Executive Director (since 1973) of the Religion and Ethics Institute, Inc.
Main field of work: Historical research on religion. **Career:** New Testament Research for International Greek New Testament Project (at Emory University, Atlanta GA) from 1955 to 1957. Instructor, Bexley Hall Seminary OH from 1957 to 1958. Associate Professor, West Virginia Wesleyan College from 1958 to 1961. Assistant Ref. Librarian, Northwestern University, Evanston OH from 1963 to 1969. Head Ref. Librarian, Chicago State University, IL from 1969 to 1977. **Educ.:** B.A. from Willamette University, Salem OR, USA (1931-1938); Ph.D. from University of Chicago, IL, USA (1950-1955); M.L.S. from University of Chicago, IL, USA (1962-1963). **Interests:** Bible, church origins, ethics. **Mem.:** Society of Biblical Literature; American Academy of Religion
Pubs.: Literary Origin of the Gospel of John; The Historical Approach to the Bible and How Did Christianity Really Begin?: A Historical Archaeological Approach.
Mailing address: PO Box 5510, Evanston IL 60204, USA. T. (1 708) 328 4049
Residence: 400 Main St, Apt 6B, Evanston IL 60202, USA. T. See Mailing Address

Tegyey, Gabor, Rev
Rel: Roman Catholic; Ordained; b. 22 DEC 1930 in Budapest, Hungary; celibate; Citizen of Germany; Lang. German, French, English and Hungarian.
International Organizations: Acting Director (since 1990) of the Catholic European Study and Information Centre.
Main field of work: Research, journalism, editing. **Career:** Russian Teacher, School from 1955 to 1956. Scientific Advisor, Formation de volontaires de progrès from 1964 to 1972. Lecturer in Social Ethics from 1973 to 1977. **Educ.:** Bacc.phil. from Académie Théologique, Budapest, Hungary (1950-1952); Dip. Russian Lang. from Janus Pannonius University of Pécs, Hungary (1952-1955); Lic. Theol. from Facultés Catholiques de Lyon, France (1959-1963). **Pubs.:** "Le socialisme à Madagascar"; "Le problème des réfugiés hongrois"; "Première rencontre oecuménique des évêques et dirigeants d'Eglise à Chantilly", 1978 and "L'Eglise et le Grand Marché".
Mailing address: Rue de la Loi 221, B-1040 Bruxelles, Belgium. T. (32 2) 231 06 97
Residence: Rue des Trévires, B-1040 Bruxelles, Belgium T. See Mailing Address

Teissier, Henry Antoine-Marie, Most Rev Archbishop
Rel: Roman Catholic; Ordained on 25 MAR 1955; b. 21 JUL 1929 in Lyon, France; celibate; Citizen of Algeria; Lang. French, Arabic, English, Italian and Spanish.
International Organizations: President (since 1981) of the Regional Episcopal Conference of North Africa.
Other organizations: Archbishop (since 1981), Algiers; Member, Council (since 1989), African Synod.
Main field of work: Ministry. **Career:** Director of Works, Diocese of Algiers from 1962 to 1966. Director of Study Centre, Diocese of Algiers from 1966 to 1973. Bishop, Diocese of Oran from 1973 to 1981. Vice-President, Caritas Internationalis from 1981 to 1988. **Educ.:** Lic. ès Lettres from University Mohammed V, Rabat, Morocco (1946-1948); Lic. Phil. from University of Paris IV (Paris-Sorbonne), France (1948-1950); Lic. Théol. from Institut Catholique de Paris, France (1950-1955). **Pubs.:** *Eglise en Islam*, Centurion, Paris, 1984; *La Mission de l'Eglise*, Desclée, Paris/Bruxelles, 1985 and *Histoire des Chrétiens d'Afrique du Nord* (joint authorship), Desclée, Paris, 1991.
Mailing address: 13 rue Khalifa Boukhalfa, 16000 Algiers, Algeria. T. (213 2) 63 42 44.
Residence: See Mailing Address. T. See Mailing Address

Tekeyan, Vartan, Rt Rev Dr
Rel: Armenian Catholic; Ordained on 01 JAN 1944; b. 05 MAR 1921 in Adana, Turkey; celibate; Citizen of Iran.
International Organizations: Bishop of Isfahan.
Mailing address: Armenian Catholic Bishopric, Khiaban Ghazzali 22, Teheran, Iran Islamic Rep. T. (98 21) 677204

Téllez, Thomas A

Rel: Baptist.
International Organizations: Executive Secretary of the Baptist Convention of Nicaragua.
Mailing address: Apartado 2593, Managua, Nicaragua.

Temperini, Lino, TOR, Fr

Rel: Roman Catholic; Ordained; celibate.
International Organizations: Procurator-General of the Third Order Regular of St Francis.
Mailing address: Via dei Fori Imperiali 1, I-00186 Roma, Italy. T. (39 6) 67 92 800 – 67 91 201

Temu, W T, Dr

Rel: Christian.
International Organizations: Director of the Christian Organizations Research Advisory Trust.
Mailing address: Imani House, St John's Gate, PO Box 42493, Nairobi, Kenya. T. (254 2) 33 16 98 – (254 2) 211 15. Tx 25782.

Teng, Betty

Rel: Methodist.
International Organizations: President for East Asia of the World Federation of Methodist Women.
Mailing address: PO Box 119, Sibu Sarawak 9600, Malaysia.

Tenke, Sándor, Prof

Rel: Reformed Church; Ordained on 31 OCT 1972; b. 06 MAY 1948 in Vasad, Hungary; m. to Katalin Szilágyi; c. 3 boys; Citizen of Hungary; Lang. Hungarian and English.
International Organizations: Professor of Theology (since 1988) of the Reformed Academy of Theology in Budapest.
Other organizations: Editor-in-Chief (since 1984), *Confessio* (theological quarterly); Vice-President (since 1987), Commission on Church Art; Principal (since 1990), Reformed Academy of Theology in Budapest.
Main field of work: Sociology and history. **Educ.:** Pastor from Reformed Academy of Theology, Budapest, Hungary (1966-1971); Teacher in History from Kossuth Lajos University, Debrecen, Hungary (1975-1979). **Interests:** church law, church history. **Pubs.:** "A History of the past 40 Years of the Reformed Church in Hungary", in *Confessio*, 1989.4; *'Uj Aranyhárfa' Poems for Parish Church and Christian Family*, S Tenke, Editor, Budapest, 1987 and *The Hungarian Reformed Church in the Light of the Church Constitution and of the Experiences of Church Sociology*, Biblia és egyház, Budapest, 1990.
Mailing address: 28 Ráday, H-1092 Budapest, Hungary. T. (36 1) 118 0816
Residence: See Mailing Address. T. (36 1) 118 0266

Terai, Taneshige

Rel: Shinto; Ordained; b. 27 NOV 1926; Citizen of Japan.
International Organizations: Chief Priest of the Osaka Tenmangu.
Career: Chairman, Osaka Kokugakuin University.
Mailing address: 2-5-6 Tenjinbashi, Kita-ku, Osaka, Osaka 530, Japan. T. (81 6) 356 0025

Teresa, T, Mother

Also know as Agnes Gonxha Bojaxhiu; Rel: Roman Catholic; Ordained; b. 27 AUG 1910 in Skopje, Serbia; celibate; Citizen of India.
International Organizations: Founder (since 1950) of the Missionaries of Charity.
Main field of work: Missionary. **Career:** Joined Sisters of Loretto in 1928. Worked at Loretto institutes in Ireland and India. Principal, St Mary's High School, Calcutta. Resigned as Head in 1990. Set up over fifty schools, orphanages and houses for the poor in India and other countries. Opened Nirmal Hriday (Pure Heart) Home for Dying Destitutes in Calcutta in 1952. Started a leper colony in West Bengal in 1964. **Hons.:** Padma Shri (1962); Pope John XXIII Peace Prize (1971); Templeton Foundatin Prize (1973); Nobel Peace Prize (1979); Bharat Ratna (Star of India) (1980); Hon. Dr.Med., Catholic University of Sacred Heart, Rome (1981), Catholic University of Louvain (1982); Honorary Citizen of Assisi (1982); Hon. O.M. (1983); Presidential Medal of Freedom (1985); Woman of the Year Award (1989). **Pubs.:** Gift for God, 1975.
Mailing address: 54A Acharya Jagadish Chandra Bose Road, Calcutta 700016, India. T. (91 33) 24 741150

Terme, Jacques, Rev

Rel: Reformed Church.
International Organizations: General Secretary of the Reformed Church of France.
Mailing address: 47 rue de Clichy, F-75009 Paris, France.

Terranova, Juan

Rel: Protestant.
International Organizations: Executive Officer of the Confraternity of Evangelicals in Latin America.
Mailing address: CONELA, Apartado 4949, 1000 Buenos Aires, Argentina.
Residence: Casilla de Correo 96, Sucursal 31, Buenos Aires 1431, Argentina. T. (54 1) 572 1035. Fax 572 1888.

Ter Steeg, Maria T

Rel: Catholic.
International Organizations: Regional Representative for Europe of the Unda, International Catholic Association for Radio and Television.
Mailing address: Albert Neuhuysstraat 27, NL-3583 SV Utrecht, Netherlands.

Testa, Elide, FdCC, Sr

Rel: Roman Catholic; Ordained; celibate.
International Organizations: Superior General of the Canossian Daughters of Charity.
Mailing address: Via della Stazione di Ottavia 70, I-00135 Roma, Italy. T. (39 6) 339 53 52.

Testa, Emilo, O.S.F.S., Fr

Rel: Roman Catholic; Ordained; celibate.
International Organizations: Procurator-General of the Oblates of St Francis de Sales.
Mailing address: Via Dandolo 49, I-00153 Roma, Italy. T. (39 6) 58 23 89.

Thai-arry, Phallop

Rel: Buddhist.
International Organizations: Chairman of the Sixth Committee (Youth) of the World Fellowship of Buddhists.
Mailing address: 33 Sukhumvit Road, Bangkok 10110, Thailand. T. (66 2) 2511188. Fax (66 2) 2557363.

Thangalathil, Benedict Mar Gregorios, O.I.C., Most Rev Dr

Rel: Syro-Malankara Catholic; Ordained on 24 AUG 1944; b. 01 FEB 1916 in Trivandrum, Kerala, India; celibate; Citizen of India; Lang. Malayalam, English, Latin and Syriac.
International Organizations: Archbishop (since 1955) of Trivandrum. Member of the Order of the Imitation of Christ.
Other organizations: Head (since 1955), Malankara Catholic Church.
Main field of work: Religious, educational, social, cultural, economic, agricultural and charitable activities. **Career:** Patron and Manager, three University Colleges and a large number of schools. One of the Trustees, Central Trust of Cheshire Homes of India. President, National Association of the Blind, Kerala Branch, etc. Patron, Kerala Federation of the Blind. Vice-President, Association for Social Health in India. Member, Academic Council, Kerala Agricultural University. Founder-Chairman, Cheshire Home, Trivandrum. **Educ.:** M.Econs. from St Joseph's College, Tiruchirapalli, Tamil Nadu, India (1946). **Interests:** astronomy, agriculture. **Pubs.:** Published a large number of articles in leading publications and periodicals in India and abroad on various topics relating to religion, education, Indian culture, agriculture, industry, social welfare, development, etc.
Mailing address: Archbishop's House, Trivandrum 695004, India. T. (91 471) 437331 – 437642/3
Residence: See Mailing Address. T. See Mailing Address

Thara Bhai, L T, Dr

Rel: Hindu; b. 16 SEP 1949 in Madurai, Tamilnadu, India; m. to S Raveendran Nair; c. 1 boy and 1 girl; Citizen of India; Lang. English and Malayalam.
International Organizations: Vice-President, RC 22 Sociology of Religion (Research Committee) (since 1990) of the International Sociological Association.
Other organizations: Reader in Sociology (since 1976), Madurai-Kamaraj University.
Main field of work: University education. **Educ.:** B.Sc. from University of Kerala, India (1966-1969); M.A. from Loyola College, Kerala, India (1969-1971); Ph.D. from University of Kerala, India (1971-1975). **Interests:** religion and women. **Mem.:** Indian Sociological Society (1975); Society for the Scientific Study of Religion (1986) **Pubs.:** Changing Patterns of Social Stratification in South India.
Mailing address: Dept. of Sociology, Madurai-Kamaraj University, Madurai 626021, Tamil Nadu, India.
Residence: 1/86, Staff Quarters, Palkalai Nagar, Madurai 625021, India. T. (91) 85462

Thatcher, William L

Rel: Evangelical.
International Organizations: Executive Director of the International Christian Media Commission.
Mailing address: PO Box 70632, Seattle WA 98107, USA. T. (1 206) 781 0461. Fax 781 0571.

Thelen, Bert R, S.J., Rev

Rel: Roman Catholic; Ordained; celibate.
International Organizations: Rector of the Marquette University Jesuit Community.
Member of the Society of Jesus.
Mailing address: Marquette University Jesuit Community, 1404 West Wisconsin Ave, Milwaukee WI 53233, USA.

Thenguvila, Mary, Sr

Rel: Roman Catholic; Ordained; celibate.
International Organizations: Contact of the Tutmonda Ekumena Ligo.
Mailing address: Via F Berni 9, I-00185 Roma, Italy. T. (39 6) 7588388.

Thenoz, Yves, Fr

Rel: Roman Catholic.
International Organizations: Secretary-General of the Little Brothers of Mary.
Mailing address: Piazzale M Champagnat 2, CP 10250, I-00144 Roma, Italy. T. (39 6) 592 44 56 – (39 6) 592 44 57.

Theodosios, T, H E Metropolitan

Rel: Orthodox.
International Organizations: Archbishop of Japanese Orthodox Church.
Mailing address: No 1-3, 4 Chome, Surugadai Kanda, Chiyoda-Ku, Tokyo, Japan.

Theodosius, T, His Beatitude

Rel: Orthodox.
International Organizations: Metropolitan of All America and Canada.
Other organizations: Archbishop, Washington.
Mailing address: Rte 25A, Box 675, Syosset NY 11791, USA.

Thesing, Kenneth F, M.M., Rev

Rel: Roman Catholic; Ordained on 24 MAY 1969; b. 09 MAY 1942 in Winona MN, USA; celibate; Citizen of USA; Lang. English, Swahili and Spanish.
International Organizations: Superior General (since 1991) of the Catholic Foreign Missionary Society of America.
Main field of work: Ministry. **Career:** Assistant Regional Superior, Maryknoll Fathers from 1975 to 1978. Assistant Regional Superior, Maryknoll Fathers from 1979 to 1982. Second Assistant, General Council Member, Maryknoll Fathers from 1983 to 1991. **Educ.:** B.A. Phil. from Glen Ellyn Seminary, IL, USA (1961-1964); B.D.; M.A. from Maryknoll Seminary, NY, USA (1964-1968).
Mailing address: Maryknoll Seminary, Box 303, Maryknoll NY 10545, USA. T. (1 914) 941 7590
Residence: See Mailing Address. T. See Mailing Address

Thiandoum, Hyacinthe, H E Card

Rel: Roman Catholic; Ordained on 18 APR 1949; b. 02 FEB 1921 in Poponguine, Dakar, Senegal; celibate; Citizen of Senegal.
International Organizations: Member (since 1976) of the College of Cardinals.
Member of the Congregation for the Evangelization of Peoples. Member of the Congregation for the Clergy. Member of the Congregation for Religious and Secular Institutes. Member of the Pontifical Council for Culture. Member of the Pontifical Council for Social Communications.
Other organizations: Archbishop (since 1962), Dakar.
Career: Created Cardinal by Pope Paul VI in 1976. General Reporter to the Synod of Bishops in 1987. Former President, Episcopal Conference of Senegal and Mauritania. Former President, Symposium of Episcopal Conferences of Africa and Madagascar (SECAM). **Hons.:** Hon. Chaplain with Grand Cross, Order of Malta (1972); Grand Cross, Order of Lion (1976); Commander, Legion of Honour (1980).
Mailing address: BP 5082, Villa 'Les Badamiers', Dakar-Fann, Senegal. T. (221) 23 69 18

Thielemans, Edward

Rel: Christian; b. 17 FEB 1928 in Melsbroek, Belgium; m. to Godelieve de Backer.
International Organizations: Chairman of Finances Committee of the International Christian Union of Business Executives.
Career: Director, Société Universitaire Européenne de Recherches Financières. **Educ.:** Lic. Econ.Sc. from Catholic University of Louvain (KUL), Belgium.
Mailing address: c/o Almanij S.A., Schoenmarkt 33, B-2000 Antwerpen, Belgium. T. (32 3) 234 27 95
Residence: Assesteenweg 129, B-1750 St-Kwintens-Lennik, Belgium T. (32 2) 532 47 65

Thien-Dinh, Tich, Ven

Rel: Buddhist.
International Organizations: Vice-President of the World Fellowship of Buddhists.
Mailing address: Association Culturelle Buddhique de Marseille, 3 rue de la Pagode (Vallon des Tuvas), F-13015 Marseille, France.

Tholens, Cornelius J A, O.S.B., Dom

Rel: Roman Catholic; Ordained on 21 DEC 1940; b. 30 AUG 1913 in Amsterdam, Netherlands; celibate; Citizen of Netherlands; Lang. Dutch, English, French, German, Italian and Latin.
International Organizations: Committee Member (since 1977) of the World Conference on Religion and Peace.
Other organizations: Founder and Member (since 1960), Benedictine secretariat for monks in Asia, Africa, Latin America; Founder and Member (since 1973), Benedictine secretariat for dialogue with Buddhist and Hindu monks.
Main field of work: Interreligious studies, writing, lecturing, ministry.
Career: Entered the Order of St Benedict in 1938. Superior and Abbot of Slangenburg Abbey, Doetinchem from 1945 to 1972. Stayed in Saccidananda Ashram (India) with the English monk Dom Bede Griffiths from 1972 to 1976. **Educ.:** Studied Theol. from Theologicum, Abbey of Oosterhout, Netherlands (1935-1942). **Pubs.:** Teksten om bij stil the staan uit Oost and West (Texts from East and West for meditation), Delft, 1984; Kroniek van een monnik, Delft, 1988 (Italian Edn.: Incontri di un monaco) and Von Buddha bis C G Jung, Walter-Verlag, 1990 (contributed: Beim Hinduismus in der Lehre).
Mailing address: Begijnhof 37b, NL-1012 WV Amsterdam, Netherlands. T. (31 20) 622 3829
Residence: See Mailing Address. T. See Mailing Address

Thoma, Alexander Mar, Most Rev Dr

Rel: Orthodox.
International Organizations: Metropolitan of Mar Thoma Syrian Church of Malabar.
Mailing address: Poolatheen, Tiruvalla, Kerala 689 101, India.

Thomanek, A

Rel: Roman Catholic.
International Organizations: ECOSOC Representative of the International Association of Charities.
Mailing address: Albertgasse 19/14, A-1080 Wien, Austria.

Thomas, Peter, Rev

Rel: Roman Catholic; Ordained; celibate.

International Organizations: Co-Head, Pacific Regional Office of the Unda, International Catholic Association for Radio and Television.

Mailing address: Catholic Communications, 18 Brunswick St, Fitzroy 3065, Australia.

Thomas, Ronald Stuart

Rel: Church of Wales; Ordained in 1937; b. 29 MAR 1913 in Caerdydd, Wales, UK; m. to Mildred E Eldridge; c. 1 boy; Citizen of United Kingdom.

International Organizations: Poet and Clergyman of the Church of Wales.

Career: Ordained Deacon in 1936. Curate, Chirk from 1936 to 1940. Curate, Hammer from 1940 to 1942. Rector, Manafon from 1942 to 1954. Vicar, St Michael's Eglwysfach from 1954 to 1968. Vicar, St Hywyn, Aberdaron from 1968 to 1978. Rector, Rhiw with Llanfaelrhys from 1973 to 1978. **Hons.:** Heinemann Award, Royal Society of Literature (1956) for Song at the Year's Turning; Sovereign's Gold Medal for Poetry (1964); Cholmondeley Award (1978). **Pubs.:** Selected Poems, 1946-1968; Between Here and Now, 1981 and The Echoes Return Slow, 1988.

Mailing address: Sarn y Plas, Y Rhiw, Pwllheli, Wales, UK.

Thompson, Eugene Mayne

Rel: United Baptist; b. 05 JAN 1931 in Oxford NS, Canada; m. to Rhoda Mitchell; c. 2 boys and 1 girl; Citizen of Canada.

International Organizations: Executive Minister (since 1984) of the United Baptist Convention of the Atlantic Provinces.

Other organizations: Council and Executive, Canadian Baptist Federation; Board of Directors, Atlantic Baptist Senior Citizens Homes Inc.; Board of Directors, Atlantic Baptist Foundation; Board of Governors, Atlantic Baptist College; Board of Trustees, Acadia Divinity College.

Career: Pastor, South End Baptist Church from 1954 to 1958. Associate Secretary of Christian Education, United Baptist Convention, Atlantic Provinces from 1958 to 1961. Pastor, Immanuel Baptist Church from 1961 to 1965. Pastor, Hillcrest Baptist Church from 1965 to 1968. Area Minister for Manitoba, Baptist Union of W Canada from 1968 to 1974. Pastor-at-Large for W Nova Scotia, United Baptist Convention, Atlantic Provinces from 1974 to 1984. **Educ.:** B.A. from Acadia University, Wolfville, Nova Scotia, Canada (1954); M.Div. from Acadia Divinity College, Wolfville, Canada (1976); D.Min. from Southern Baptist Theological Seminary (1979). **Interests:** music, gardening, cross-country skiing. **Mem.:** Canadian Baptist Overseas Mission Board **Pubs.:** Baptist Youth Fellowship Handbook, 1958 and New Design for a Dynamic Church, 1973.

Mailing address: 1655 Manawagonish Rd, Saint John NB E2M 3Y2, Canada.

Residence: Box 37, Site 10, Chalmers Drive, Saint John NB E2L 3W7, Canada.

Thompson, George Ralph

Rel: Seventh-Day Adventist; b. 20 MAR 1929 in Barbados, Antilles; m. to Imogene Clotilde Barker; c. 1 boy and 2 girls; Citizen of USA; Lang. English and French.

International Organizations: Secretary (since 1980) of the General Conference of Seventh-Day Adventists.

Main field of work: Church administration. **Career:** Evangelist, South Caribbean Conference of Seventh-Day Adventists from 1950 to 1953. Teacher, Caribbean Union College from 1953 to 1954. Teacher/Pastor, Caribbean Union College from 1959 to 1964. Chairman, Dept. of Theology, Caribbean Union College from 1962 to 1964. President, East Caribbean Conference of Seventh-Day Adventists from 1964 to 1970. President, Caribbean Union Conference from 1970 to 1975. Vice-President, General Conference of Seventh-Day Adventists from 1975 to 1980. **Educ.:** B.A. Rel. from Atlantic Union College, South Lancaster MA, USA (1956); M.Th.; Church Hist. from Andrews University, Berrien Springs MI, USA (1958); B.D. from Andrews University, Berrien Springs MI, USA (1962). **Hons.:** Hon. D.D., Andrews University, Berrien Springs, MI, USA (1983).

Mailing address: 12501 Old Columbia Pike, Silver Spring MD 20904, USA. T. (1 301) 680 6640. Tx 440186. Fax (1 301) 680 6635.

Residence: 10117 Phoebe Lane, Adelphi MD 20783, USA T. (1 301) 434 6914

Thompson, Henry Orrin, Rev Prof Dr

Rel: United Methodist; Ordained on 10 JUN 1956; b. 23 OCT 1931 in Northwood IA, USA; m. to Joyce Elaine Beebe; c. 2 boys; Citizen of USA; Lang. English.

International Organizations: Executive Secretary of the Global Congress of the World's Religions.
Executive Secretary of the International Association of Gandhian Studies.

Other organizations: Professor of Bible and Ministry (since 1977), Unification Theological Seminary; Adjunct Professor of Ethics (since 1980), School of Nursing, University of Pennsylvania, Philadelphia PA.

Main field of work: Biblical studies. **Career:** Professor of Old Testament, New York Theological Seminary from 1967 to 1971. Director, American Center of Oriental Research (ACOR), Amman from 1971 to 1972. Visiting Professor, ACOR from 1972 to 1973. Assistant Professor, Syracuse University from 1963 to 1967. Lecturer, Colgate-Rochester Divinity School from 1962 to 1963. **Educ.:** B.S. from Iowa State University of Science and Technology, Ames IA, USA (1949-1953); M.Div. from Drew University, Madison NJ, USA (1955-1958); Ph.D. from Drew University, Madison NJ, USA (1958-1964); M.Sc. Ed. from Syracuse University, NY, USA (1966-1971); M.A. Educational Psy. from Jersey City State College, NJ, USA (1974-1975); Dip. in School from Jersey City State College, NJ, USA (1975-1976); 24 grad. hours from Rutgers University, New Brunswick NJ, USA (1977-1980). **Interests:** biblical archaeology, world religions. **Mem.:** American Institute of Archaeology; American Schools of Oriental Research **Pubs.:** Biblical Archaeology: The World, The Mediterranean, The Bible, NY, Paragon House, 1987; World Religions in War and Peace, Jefferson, NC, McFarland, 1988 and Eternal Love: Biblical Covenants and Modern Man, Delhi, ISPCK, 1991.

Mailing address: 10 Dock Road, Barrytown NY 12507, USA. T. (1 914) 758 6881.

Residence: 2942 Valley View Dr, Doylestown PA 18901, USA. T. (1 215) 794 3724

Thompson, John, Very Rev

Rel: Presbyterian; Ordained; b. 14 JUL 1922 in Northern Ireland, UK; m. to (1st) Ann M Normiton (died), (2nd) Heather M Lewis; c. 2 boys and 1 girl; Citizen of United Kingdom.

International Organizations: Professor of Systematic Theology and Apologetics (since 1976) of the Union Theological College, Belfast.

Other organizations: External Lecturer, Queen's University, Belfast; Member, Senate, Queen's University, Belfast.

Career: Assistant Minister, St Enoch's, Belfast from 1949 to 1952. Minister, Sandymount Presbyterian Church, Dublin and Lecturer in Catechetics, Trinity College, Dublin from 1952 to 1961. Minister, Fortwilliam Park Presbyterian Church, Belfast from 1961 to 1976. Moderator, General Assembly, Presbyterian Church in Ireland in 1986. Various guest and part-time lectureships. **Interests:** sport (especially hockey, tennis, soccer), English literature, gardening. **Hons.:** Hon. D.D., Presbyterian Theological Faculty, Ireland (1986). **Pubs.:** With J L M Haire, Commentary on the Shorter Catechism, 1966; Christ in Perspective: Christological Perspectives in the Theology of Karl Barth, 1976 and essays and articles in theological journals.

Mailing address: Union Theological College, Botanic Ave, Belfast BT7 1JT, UK.

Residence: Station Road, Carnalea, Co. Down, N Ireland, UK.

Thompson, Kenneth A

Rel: Rosicrucian.

International Organizations: Secretary of the Board of the Rosicrucian Order, Ancient and Mystical Order Rosae Crucis.

Mailing address: Rosicrucian Park, 1342 Maglee Avenue, San Jose CA 95191, USA. T. (1 408) 287 9171. Fax 286 4038.

Thompson-Travis, Beverly, Rev

Rel: Christian.

International Organizations: North American Desk Liaison of the Asian Women's Institute.

Mailing address: United Church Board for World Ministries, 475 Riverside Drive, 16th Floor, New York NY 10115, USA.

Thomsen, Jørgen

Rel: Christian.

International Organizations: Secretary-General of the Conference of the Churches on Development Education.

Mailing address: Nørregade 11, DK-1165 København K, Denmark. T. (45) 33 15 59 27.

Thomson, Lyle

Rel: Assembly of God.

International Organizations: President of the International Correspondence Institute.

Mailing address: Chaussée de Waterloo 45, B-1640 Rhode-Saint-Genèse, Belgium. T. (32 2) 358 35 10.

Thomson, Robert E, Commr

Rel: Salvation Army, The; Ordained on 15 JUN 1951; b. 21 FEB 1928 in Racine WI, USA; married; c. 2 boys and 1 girl; Citizen of USA; Lang. English.

International Organizations: Eastern Territorial Commander (since 1990) of The Salvation Army.

Other organizations: Chairman of the Board (since 1990), Booth Memorial Medical Center.

Main field of work: Ministry. **Career:** Editor-in-Chief, The Salvation Army, USA from 1970 to 1972. International Secretary, The Salvation Army, London from 1988 to 1990. **Educ.:** B. Music from Saint Olaf College, Northfield MN, USA (1945-1949); Booth Memorial College, Chicago IL, USA (1950-1951); Northwestern University, Chicago IL, USA (1953-1955). **Mem.:** Evangelical Press Association, Rotary International **Pubs.:** Numerous articles for the Religious Press.

Mailing address: 440 Nyack Road, West Nyack NY 10994, USA. T. (1 914) 620 7201

Residence: 13 Washington Circle, New City NY 10956, USA. T. (1 914) 634 1225

Thorngate, Dale D, Rev

Rel: Baptist: Seventh-Day.

International Organizations: Executive Secretary of the Seventh-Day Baptist General Conference.

Mailing address: 3120 Kennedy Rd, PO Box 1678, Jonesville WI 53547, USA. T. (1 608) 752 5055.

Thorogood, Bernard George, Rev

Rel: Congregational; Ordained in 1952; b. 21 JUL 1927; widowed Jannett Lindsay Paton; c. 2 boys; Citizen of United Kingdom.

International Organizations: Secretary-General (since 1980) of The United Reformed Church in the United Kingdom.

Career: Missionary appointment under London Missionary Society in South Pacific Islands from 1953 to 1970. Secretary-General, Council for World Mission from 1971 to 1980. Moderator, Executive Committee, British Council of Churches in 1984. Member, Central Committee, World Council of Churches in 1984. **Interests:** sketching. **Pubs.:** Risen Today, 1987; The Flag and the Cross, 1988 and No Abiding City, 1989.

Mailing address: 86 Tavistock Place, London WC1H 9RT, UK. T. (44 71) 837 7661

Thottam, Eustace, C.M.I., Fr

Rel: Roman Catholic; Ordained; celibate.

International Organizations: Procurator-General of the Carmelites of Mary Immaculate.

Advisor of the Sacred Congregation for the Oriental Churches.

Mailing address: Via Martino V 26/b-c, I-00167 Roma, Italy. T. (39 6) 662 16 76

Thrower, James Arthur, Dr

Rel: None; b. 05 OCT 1936 in Guisborough, Yorkshire, UK; m. to Judith Elizabeth Gauss; c. 3 girls; Citizen of United Kingdom; Lang. English and Russian.

International Organizations: Director (since 1987) of the Center for the Study of Religions, University of Aberdeen.

Other organizations: Riddoch Lecturer in Comparative Religion (since 1987), University of Aberdeen.

Main field of work: University education. **Career:** Tutor, WEA (Eastern District) from 1962 to 1964. Lecturer, University of Ghana from 1964 to 1968. Lecturer, Bede College, Durham from 1968 to 1970. Senior Lecturer and Head, Dept. of Religious Studies, University of Aberdeen from 1970 to 1987. **Educ.:** B.A., M.A. from University of Durham, UK (1955-1959); B.Litt. from University of Oxford, UK (1959-1962); Ph.D. from University of Aberdeen, UK (1981). **Interests:** religion and rejection of religion. **Mem.:** International Association for History of Religions **Pubs.:** A Short History of Western Atheism, Remberton, London and New York, 1971; The Alternative Tradition, Routon, The Hague, 1981 and Marxist-Leninist 'Scientific Atheism' and the Study of Religion in the USSR, Routon de Gruijter, Berlin, 1983.

Mailing address: Centre for the Study of Religions, University of Aberdeen, Aberdeen AB9 2UB, Scotland, UK. T. (44 224) 272147

Residence: 233 Clifton Road, Aberdeen AB2 2ET, UK. T. (44 224) 485641

Thubten Chodron, T, Ven

Also know as Cheryl Greene; Rel: Buddhist; Ordained in 1977; b. 18 SEP 1950 in Chicago IL, USA; celibate; Citizen of USA; Lang. English, Italian and French.

International Organizations: Dharma Teacher (since 1991) of the Dharma Friendship Foundation.

Other organizations: One of the Vice-Presidents (since 1991), Inter-Buddhist Council.

Main field of work: Teaching, counselling. **Career:** Dharma Teacher, Chamtse Ling Buddhist Centre, Hong Kong in 1986. Dharma Teacher, Amitabha Buddhist Centre, Singapore from 1987 to 1988. Teaching tour in USA and Canada in 1989. Teaching tour in USA and Canada in 1992. **Educ.:** University of Southern California, Los Angeles, USA (1967-1969); B.A. Hist. from University of California, Los Angeles, USA (1969-1971); Grad. Crses. in Ed. for Teaching Cert. from University of Southern California, Los Angeles, USA (1974-1975); Rel. Ed. in Nepal, India, Italy and France with teachers including H H The Dalai Lama, Ven Serkong Rinpoche, Ven Zopa Rinpoche (1974-1975). **Interests:** psychology, history. **Pubs.:** Open Heart, Clear Mind, Snow Lion Publications; Taming the Monkey Mind, Graham Brash Publishers and I Wonder Why (forthcoming – to be published by Snow Lion Publications).

Mailing address: Dharma Friendship Foundation, 11739 Palatine Ave N, Seattle WA 98133, USA. T. (1 206) 363 1437

Residence: See Mailing Address. T. See Mailing Address

Thurman, Barbara, Rev

Rel: Spiritualist.

International Organizations: President of the National Spiritualist Teachers Club.

Mailing address: 200 Marina Vista Rd, Larkspur CA 94939, USA.

Tibaldo, María Fidelis, Sr

Rel: Roman Catholic; Ordained; celibate.

International Organizations: Executive Secretary of the International Catholic Organization for Cinema and Audiovisual.

Mailing address: Secretaría Ejecutiva, Casilla C 11194 A17, Quito, Ecuador.

Tien, Stella L

Rel: Buddhist; b. 15 MAR 1916 in Peking, China; m. to Pao-Tai Tien; c. 2 boys and 1 girl; Citizen of Taiwan (Rep. of China); Lang. Chinese and English.

International Organizations: Vice-President (since 1984) of the World Fellowship of Buddhists.

Main field of work: Buddhist philosophy. **Career:** Associate Professor, Fu Jen University from 1967 to 1970. **Educ.:** B.A. from University of Peking, China (1935-1939). **Pubs.:** Yehman's Essays; Flower and Smile and Heart Sutra.

Mailing address: International Development Corporation, PO Box 26-174, Taipei, Taiwan (Rep of China).

Residence: 5F, 145 Ta-an Road, Sec 1, Taipei 106, Taiwan (Rep. of China)

Tienou, Tite, Dr

Rel: Christian.

International Organizations: Chairman of the Accrediting Council for Theological Education in Africa.

Mailing address: PO Box 60875, Nairobi, Kenya.

Tingson, Greg, Dr

Rel: Christian.

International Organizations: President of the Asian Christian Outreach.

Mailing address: 56 S Maya Dr, Quezon City, Manila, Philippines. T. (63 2) 99 56 45 – (63 2) 99 55 04. C. EVANGELISM MANILA.

Tinsley, Ernest John, Rt Rev
Rel: Anglican; Ordained; b. 22 MAR 1919 in Maghull, Lancashire, UK; widowed Marjorie Dixon; c. 2 girls; Citizen of United Kingdom.
International Organizations: Ecclesiastic of the Anglican Church.
Career: Lecturer in Theology, University College, Hull from 1946 to 1954. Lecturer in charge of Dept. of Theology, University College, Hull from 1954 to 1961. Senior Lecturer and Head of Dept., University College, Hull from 1961 to 1962. Professor of Theology and Head of Dept., Leeds University from 1962 to 1975. Anglican Bishop, Bristol from 1976 to 1985. Special Lecturer, Bristol University from 1976 to 1984. Chairman, General Synod Board of Education from 1979 to 1982. **Interests:** gardening, travel. **Pubs.:** Dictionary of Christian Spirituality, 1983; Tragedy, Irony and Faith, 1985 and Tell it Slant, 1990.
Mailing address: 100 Acre End Street, Eynsham, Oxford OX8 1PD, UK. T. (44 865) 880 822

Tintelott, Hubert
Rel: Roman Catholic; Unordained; b. 17 OCT 1947 in Paderborn, North Rhine Westphalia, Germany; m. to Elisabeth Hajdin; c. 1 boy and 1 girl; Citizen of Germany; Lang. German and English.
International Organizations: International Secretary (since 1978) of the International Kolping Society.
Other organizations: Vice-President, Arbeitsgemeinschaft katholisches Verbände (AGKV); Treasurer, Lernen und Helfen in Übersee (LHÜ).
Main field of work: Development cooperation. **Educ.:** Dipl.-Kaufmann from Friedrich-Alexander-Universität Erlangen-Nürnberg, Germany (1969-1971); Dipl.-Kaufmann from Westfälische Wilhelmsuniversität Münster, Germany (1971-1973). **Hons.:** Distinguished Service Award (1980). **Pubs.:** "Von der sozialen Frage zur Internationalen Sozialen Frage – Herausforderung der christlich-sozialen Bewegung", Band 16 *Kolpingwerk in Staat und Gesellschaft*, 1988; *Für eine solidarische Entwicklung der Menscheit*, Verantwortung katholischer Verbände für entwicklungspolitische Bildungsarbeit, Entwicklungshilfe und Entwicklungspolitik, Dokumentation 89, Deutscher Katholikentag and Bewahrung der Schöpfung – Aktuelle Aufgaben und Handlungsperspektiven kath. Verbände, Berichte und Dokumente 77, ZdK 1991.
Mailing address: Kolpingplatz 5-11, 50667 Köln, Germany. T. (49 221) 20701-49
Residence: Dürerstrasse 12, 50389 Wesseling, Germany

Tisdale, Janet
Rel: Spiritualist.
International Organizations: Secretary-Treasurer of the Spiritualist Healers League.
Mailing address: 11457 W Hidalgo, Tolleson AZ 85353, USA.

Todd, T G
Rel: Church of Christ.
International Organizations: General Secretary of the Associated Churches of Christ in New Zealand.
Mailing address: 90A Mount Street, Nelson, New Zealand.

Todo, Kyoshun
Rel: Buddhist; Ordained; b. 19 NOV 1918; Citizen of Japan.
International Organizations: Professor of the Bukkyo University.
Career: Chief, Shinjuin. **Educ.:** Taisho University, Japan.
Mailing address: Chioninsannai, Higashiyama-ku, Kyoto 605, Japan. T. (81 75) 541 3933

Todo, Shunsho
Rel: Buddhist; Ordained; b. 01 JAN 1914; Citizen of Japan.
International Organizations: Monk of Jodoshu Zendoji.
Educ.: Taisho University, Japan.
Mailing address: 550 Zendojimachi Iida, Kurume, Fukuoka 839-11, Japan. T. (81 75) 541 3933

Toeda, Yoshiaki, Rev
Rel: United Church of Christ; Ordained on 08 DEC 1955; b. 05 FEB 1925 in Tokyo, Japan; m. to Keiko; c. 3 boys; Citizen of Japan; Lang. Japanese, English and German.
International Organizations: First Vice-President (since 1989) of the Asian-Pacific Christian Peace Conference.
Other organizations: Pastor (since 1956), United Church of Christ, Sendai; President (since 1971), Educational Development International; General Secretary (since 1972), Christian Peace Exchange Committee; President (since 1989), Japan Mission Society.
Main field of work: International education, peace. **Educ.:** Harbin Medical College, China (1943-1945); Tokyo Union Theological Seminary, Japan (1947-1951); Evangelische Prediger Seminar Essen und Bad Kreuznach, Germany (1962). **Interests:** intercultural-religious program. **Hons.:** Awards from President Reagan and California State. **Mem.:** Rotary International **Pubs.:** Encounter with Contemporary, 1978; An Encounter with the 21st Century, 1988 and Christianity in Rapid Changing World, 1991.
Mailing address: 301 Toei Building, Kotobashi 4-31-7, Sumida-ku, Tokyo 130, Japan. T. (81 3) 3631 3110. Fax 3846 2568
Residence: Wakabamachi 20-1, Taihaku-ku, Sendai-Shi 982, Japan. T. (81 22) 229 3009

Toews, John E
Rel: Mennonite.
International Organizations: Dean of the Mennonite Brethren Biblical Seminary.
Mailing address: 4824 E Butler Ave, Fresno CA 93727-5097, USA.

Toews, Reginald
Rel: Mennonite.
International Organizations: Treasurer of the Mennonite World Conference.
Mailing address: 180 Minnetonka St, Winnipeg MB R2M 3Y6, Canada.

Tokutomi, Giko
Rel: Buddhist; Ordained; b. 24 APR 1920; Citizen of Japan.
International Organizations: Chief of the Mitsugenin.
Main field of work: Monk. **Career:** Head, Takanocho. Managing Director, Koyasan Shingonshu.
Mailing address: 478 Takanocho Oaza Koyasan, Ito-gun, Wakayama 648-02, Japan. T. (81 736) 56 2202

Tolu, Giovannino, O. de M., Fr
Rel: Roman Catholic; Ordained; celibate.
International Organizations: Procurator-General of the Order of Our Lady of Mercy.
Mailing address: Via Monte Carmelo 3, I-00166 Roma, Italy. T. (39 6) 680 15 98 – 680 17 16.

Tomásek, Frantisek, H E Card
Rel: Roman Catholic; Ordained on 05 JUL 1922; b. 30 JUN 1899 in Studénka, Czechoslovakia; celibate; Citizen of Slovakia.
International Organizations: Member (since 1977) of the College of Cardinals.
Member of the Congregation for the Clergy.
Other organizations: Archbishop (since 1977), Prague, Metropolitan and Primate of Bohemia; President, Bishops' Conference of Czechoslovakia.
Career: Assistant Lecturer in Education and Catechetics, Faculty of Theology, Olomouc University in 1934. Lecturer in Education, Olomouc University in 1945. Ordinary Professor, Olomouc University in 1947. Auxiliary Bishop, Olomouc in 1949. Apostolic Administrator in Prague in 1965. Created Cardinal by Pope Paul VI in 1977. **Interests:** education, catechism. **Pubs.:** Editor, Letters on Education and author of 25 monographs on theology, education and catechism of the Catholic religion.
Mailing address: Arcibiskupsky palác, Hradcanské nám 16, 119 02 Praha I, Czech Rep. T. (42 2) 53 45 83 – 23 69 18

Tomasi, Silvio, C.S.S., Fr
Rel: Roman Catholic; Ordained; celibate.
International Organizations: Procurator-General of the Congregation of the Sacred Stigmata.
Other organizations: Member of the Administration of the Tribunal, Vicariate of Rome.
Mailing address: Via Mazzarino 16, I-00184 Roma, Italy. T. (39 6) 488 19 94, (39 6) 482 79 82. Fax (39 6) 482 87 24

Tomaskovic, Emil R, S.A., Fr
Rel: Roman Catholic; Ordained; celibate.
International Organizations: Superior General of the Franciscan Friars of the Atonement.
Mailing address: PO Box 5, Garrison, New York NY 10524, USA. T. (1 914) 424 3539.

Tomko, Jozef, H E Card

Rel: Roman Catholic; Ordained on 12 MAR 1949; b. 11 MAR 1924 in Udavské, Kosice, Czechoslovakia; celibate; Citizen of Slovakia.
International Organizations: Prefect (since 1985) of the Congregation for the Evangelization of Peoples.
Member of the Congregation for Bishops. Member of the Sacred Congregation for the Sacraments and Divine Worship. Member of the Congregation for Religious and Secular Institutes. Member of the Congregation for Catholic Education. Member of the Pontifical Council for Promoting Christian Unity. Member of the Pontifical Council for the Interpretation of Legislative Texts. Member of the Pontifical Council for Inter-Religious Dialogue. Member of the Pontifical Council for Dialogue with Non-Believers.
Other organizations: Member (since 1985), College of Cardinals; Grand Chancellor, Urbanian Pontifical University; Member, Congregation for the Doctrine of the Faith; Member, Pontifical Commission for the Vatican City State.
Career: Consecrated Bishop (Titular See of Doclea) in 1979. Created Cardinal by Pope John Paul II in 1985.
Mailing address: Via Urbano VIII 16, I-00165 Roma, Italy. T. (39 6) 687 0331

Tongdonmuan, Arun, Rev

Rel: Church of Christ.
International Organizations: Moderator of the Church of Christ in Thailand.
Mailing address: 14 Pramuan Road, Bangkok 10500, Thailand.

Tonucci, Giovanni, H E Mgr

Rel: Roman Catholic; Ordained on 16 MAR 1966; b. 04 DEC 1941 in Fano, Italy; celibate.
International Organizations: Apostolic Nuncio to Bolivia (since 1989) of the Roman Catholic Church.
Other organizations: Titular Archbishop, Torcello.
Mailing address: Avenida Arce 2990, Casilla Postal 136, La Paz, Bolivia. T. (591 2) 375 007. Fax 39 21 22

Tonui, John Arap, Rev

Rel: Christian.
International Organizations: General Secretary of the African Israel Nineveh Church.
Mailing address: Nineveh Headquarters, PO Box 701, Kisumu, Kenya.

Torrance, Thomas Forsyth, Very Rev Emer Prof

Rel: Church of Scotland; Ordained on 20 MAR 1940; b. 30 AUG 1913 in Chengdu, Sichuan, China; m. to Margaret Edith Spear; c. 2 boys and 1 girl; Citizen of United Kingdom; Lang. English and German.
International Organizations: Minister of the Church of Scotland.
Other organizations: Member, Advisory Board (since 1980), Center of Theological Inquiry, Princeton NJ; Member, Kuratorium (since 1983), Das Deutsche Institut für Bildung und Wissen, Berlin and Paderborn; Member, Kuratorium (since 1986), Europäische Akademie für Umweltfragen, Tübingen; Emeritus Professor of Christian Dogmatics.
Career: Professor of Systematic Theology, Auburn Theological Seminary, NY from 1938 to 1939. Parish Minister at Alyth, Perthshire from 1940 to 1947. Chaplaincy Service with Army, Middle East and Italy from 1943 to 1945. Minister, Beechgrove Parish Church, Aberdeen from 1947 to 1950. Professor of Church History, University of Edinburgh and New College from 1950 to 1952. Professor of Christian Dogmatics, University of Edinburgh and New College from 1952 to 1979. Convener, Church of Scotland Commission on Baptism from 1954 to 1962. Moderator, General Assembly, Church of Scotland from 1976 to 1977. **Educ.:** M.A. Class. Lang. and Phil. from University of Edinburgh, UK (1931-1934); B.D. from University of Edinburgh, UK (1934-1937); University of Basel, Switzerland (1937-1939); University of Oxford, Oriel College, UK (1939-1940); Dr. Theol. from University of Basel, Switzerland (1946); D.Litt. from University of Edinburgh, UK (1970). **Interests:** theological science, relations between theology and science, walking, golf, travel. **Hons.:** Degrees h.c.: D.D., Presbyterian College, Montreal (1950), Dr. Theol., University of Geneva and Faculté Libre, Paris (1959), D.D., St Andrews University (1960), Dr. Theol., Oslo University (1961), D.Sc., Heriot-Watt University, Edinburgh (1983), Dr. Theol., Reformed College, Debrecen (1988); Member of the Order of the British Empire (1944); Collins Religious Book Prize (1969); Cross of St Mark, Patriarchate of Alexandria (1970); Honorary Protopresbyter of the Greek Orthodox Church, Patriarchate of Alexandria (1973); Cross of Thyateira (1977); Templeton Prize for Progress in Religion (1978). **Mem.:** Scottish Church Theology Society (Founder, 1945); Church Service Society of the Church of Scotland (Member, 1946); Society for the Study of Theology (Founder, 1952); Société de l'Histoire du Protestantisme Français (Foreign Member, 1968); International Society for the Study of Medieval Philosophy (Member, 1968); International Academy of Religious Sciences (Member, 1969); Institute of Religion and Theology of Great Britain and Ireland (Founder Member, 1973); International Academy of Philosophy of Science (1976); Royal Society of Edinburgh (Fellow, 1979; Committee Member, 1985); British Academy, London (Fellow, 1982) **Pubs.:** *Theological Science*, Oxford University Press, London, 1969; *Space, Time and Incarnation*, Oxford University Press, London, 1969 and *The Trinitarian Faith. The Evangelical Theology of the Ancient Catholic Church*, T and T Clark, Edinburgh, 1988.
Mailing address: Church of Scotland, 121 George Street, Edinburgh, Lothian EH2 4YN, UK. T. (44 31) 225 5722
Residence: 37 Braid Farm Road, Edinburgh EH10 6LE, UK. T. (44 31) 447 3224

Torres, Enrique, C.S.V., Fr

Rel: Roman Catholic; Ordained; celibate.
International Organizations: Procurator-General of the Society of Mary.
Mailing address: Via Latina 22, I-00179 Roma, Italy. T. (39 6) 757 58 92.

Torres, Javier, Rev

Rel: Reformed Church.
International Organizations: Member of the Executive Committee of the World Alliance of Reformed Churches.
Mailing address: Apartado 1.148, Barquisimeto 3001, Venezuela. T. (58 51) 29 946
Residence: T. (58 51) 26 848

Torres, Nelly

Rel: Christian.
International Organizations: Secretary-General of the International Movement of Apostolate of Children.
Mailing address: 8 rue Duguay-Trouin, F-75006 Paris, France. T. (33 1) 42 22 84 81.

Tóth, Károly, Bishop Dr

Rel: Reformed Church; Ordained in 1955; b. 03 APR 1931 in Györszemere, Györ, Hungary; m. to Piroska Bán; c. 2 girls; Citizen of Hungary; Lang. Hungarian, English (preaching, lecturing), German (preaching, lecturing), French (preaching, lecturing), Russian (reading and speaking).
International Organizations: President (since 1991) of the Ecumenical Study Centre, Budapest.
Career: Leader, Foreign Dept., Reformed Church in Hungary Synodal Office from 1958 to 1977. Member, Commission of the Churches on International Affairs and Executive and Central Committee, World Council of Churches from 1968 to 1991. Bishop, Danubian District, Reformed Church in Hungary from 1977 to 1991. President of the Synod, Reformed Church in Hungary from 1987 to 1989. President, Ecumenical Council of Churches in Hungary from 1980 to 1991. Vice-President, World Alliance of Reformed Churches from 1977 to 1990. Member, Hungarian Parliament from 1988 to 1991. Lecturer at Princeton Theological Seminary, NJ. **Educ.:** Pastor's Dip. from Theological Academy in Budapest, Hungary (1955); D.E.S. Théologiques from Montpellier, France (1956); Dr. Theol. (summa cum laude) from Debrecen Theological Academy, Hungary (1966). **Interests:** ecumenism, Church-State relations. **Hons.:** Dr. h.c., Comenius Theological Faculty, Prague and Hus Theological Faculty, Prague. **Pubs.:** *Living as a Christian iin Todays World*, Prague, 1981; *A Reorientation to Peace*, New Delhi, 1985 and many articles in Hungary and abroad.
Mailing address: Ecumenical Study Centre, Bocskai ut 15 III/3, H-1114 Budapest, Hungary. T. (36 1) 1664 790. 1664 790
Residence: See Mailing Address. T. See Mailing Address

Toulouse, Mark G

Rel: Disciples of Christ; Ordained on 22 DEC 1971; m. to Jessica L Smith; c. 1 boy and 2 girls.
International Organizations: Associate Dean (since 1991) of the Brite Divinity School, Texas Christian University.
Other organizations: Associate Professor of History of Christianity (since 1989), Brite Divinity School, Texas Christian University.

Main field of work: Ministry, research in American religion. **Career:** Part-time Instructor (Religious Studies), Illinois Benedictine College from 1980 to 1982. Assistant Professor of Religious Studies, Illinois Benedictine College from 1982 to 1984. Assistant Professor of History of Christianity, Phillips University Graduate Seminary from 1984 to 1986. Assistant Professor of Church History, Brite Divinity School, Texas Christian University from 1986 to 1989. **Educ.:** B.A. from Howard Payne University, Brownwood TX, USA (1974); M.Div. from Southwestern Baptist Theological Seminary, USA (1977); Ph.D. from University of Chicago, IL, USA (1984). **Interests:** religion and American public life. **Mem.:** American Academy of Religion; American Historical Association; American Society of Church History; Disciples of Christ Historical Society; Theta Phi **Pubs.:** *The Transformation of John Foster Dulles: From Prophet of Realism to Priest of Nationalism*, Macon, Mercer University Press, 1985; *Joined in Discipleship: The Maturing of an American Religious Movement*, St Louis, Chalice Press, 1992 and journal articles, chapters in books, book reviews.
Mailing address: Brite Divinity School, TCU, 2800 S University, PO Box 32923, Fort Worth TX 76129, USA. T. (1 817) 921 7577
Residence: 4129 Alava Dr, Fort Worth TX 76133, USA. T. (1 817) 294 3965

Toumani, Ali Mohaman

Rel: Muslim.
International Organizations: Executive Director of the Islamic Solidarity Fund.
Mailing address: PO Box 178, Jeddah 21411, Saudi Arabia. T. (966 2) 6800800. C. ISLAMIAH JEDDAH. Tx 601366 ISLAMI SJ.

Tovini, Sergo

Rel: Christian.
International Organizations: President of the Tutmonda Ekumena Ligo.
Mailing address: Odos Rodou 5, Athinai 11252, Greece. T. (30 1) 8674873.

Toyama, Koshi

Rel: Buddhist; Ordained; b. 28 MAY 1933; Citizen of Japan.
International Organizations: Manager of the Ryousenji Shingonshu.
Main field of work: Monk. **Educ.:** Koyasan University, Japan.
Mailing address: 3873 Nakamachi, Nara, Nara 631, Japan. T. (81 742) 45 0081

Traer, Robert, Rev Dr

Rel: Presbyterian; Ordained on 28 SEP 1969; b. 12 JAN 1943 in Kalamazoo MI, USA; m. to Nancy; c. 2 boys and 3 girls; Citizen of USA; Lang. English and German.
International Organizations: Secretary-General (since 1990) of the International Association for Religious Freedom.
Main field of work: Ministry, education. **Career:** Faculty, St Joseph's College from 1978 to 1985. Faculty, University of San Francisco, CA from 1988 to 1990. Faculty, St Mary's College from 1987 to 1990. **Educ.:** D.Min. from University of Chicago, Divinity School, IL, USA (1965-1969); J.D. from University of California, School of Law, Davis, USA (1973-1976); Ph.D. from Graduate Theological Union, Berkeley CA, USA (1985-1988). **Mem.:** American Academy of Religion **Pubs.:** *Faith and Human Rights: Support in Religious Traditions for a Global Struggle*, Georgetown University Press, Washington DC, 1991; "Faith in the Buddhist Tradition", in *Buddhist-Christian Studies*, Vol. II, 1991, pp. 85-120 and "Freedom Now: The Challenge for Liberal Religious Communities", The Essex Hall Lecture 1991, Unitarian Publications, London, 1991.
Mailing address: Dreieichstrasse 59, 60594 Frankfurt/Main, Germany. T. (49 69) 62 87 72. Fax (49 69) 62 18 20.
Residence: Letzter Hasenpfad 125, D-6000 Frankurt/Main 70, Germany. T. (49 69) 68 25 83

Tragan, Pius, Rev Prof

Rel: Roman Catholic; Ordained on 24 SEP 1955; b. 20 AUG 1928 in Esparreguera, Barcelona, Spain; celibate; Citizen of Spain; Lang. Castillan, Catalan, German, English, French and Italian.
International Organizations: Rector (since 1990) of the Pontifical Athenaeum St Anselmo.
Main field of work: University education. **Educ.:** L.S.S. from Pontifical Biblical Institute, Vatican City State (1969); Lic. Phil. from University of Strasbourg II, France (1972); Dr. Theol. from University of Strasbourg II, France (1976). **Pubs.:** La Parabole du Pasteur: Jean 10, 1-18; El Cantic dels Cantics. Versió i comentari and Fede e Sacramenti nel Vangelo di Giovanni.
Mailing address: Piazza dei Cavalieri di Malta 5, I-00153 Roma, Italy. T. (39 6) 578 35 81

Travell, John Charles, Rev

Rel: Congregational; Ordained on 19 OCT 1966; b. 04 APR 1930 in Gloucester, Gloucestershire, UK; m. to Annette Patricia Fare; c. 1 boy and 1 girl; Citizen of United Kingdom; Lang. English.
International Organizations: Co-Chairman (since 1989) of the International Congregational Fellowship.
Main field of work: Ministry. **Career:** Minister, Christ Church and Opton Chapel Lambeth from 1966 to 1968. Minister, Penge Congregational Church from 1969 to 1989. Chairman, Penge (Community Forum) from 1973 to 1980. Member, Council for World Mission and Officer for World Mission, Congregational Federation from 1974 to 1986. Chairman, European Region, Congregational Federation in 1979. President, Congregational Federation from 1982 to 1983. **Educ.:** B.D. (Hons) from University of London, New College, UK (1962-1966); M.Th. from University of London, King's College, UK (1968-1969). **Interests:** walking, literature, music, theatre, films. **Mem.:** Fellow, Royal Society of Arts; Institute of Religion and Medicine (1964); Congregational Memorial Hall Trust (1973) **Pubs.:** "The Wisdom of Sri Ramakrishna", Vedanta Society, October 1969; "Jesus is Lord", Congregational Federation, May 1982 and "Leslie Weatherhead", (articles – current subject of research).
Mailing address: 44 Cornwall Road, Dorchester DT1 1RY, UK. T. (44 305) 26 46 81
Residence: See Mailing Address. T. See Mailing Address

Tremblay, Robert, Mgr

Rel: Roman Catholic; Ordained; b. 05 AUG 1929 in Les Eboulements PQ, Canada; celibate; Citizen of Canada; Lang. French, English, Italian, Spanish, Latin and Greek.
International Organizations: Secretary-Archivist (since 1986) of the Pontifical Council for the Family.
Member of the Pontifical Council for the Family.
Main field of work: Pastoral care of the family. **Career:** Professor, Seminary, Valleyfield, Quebec from 1954 to 1955. Member, Commission of Sacred Music, Diocese of Valleyfield from 1956 to 1984. Professor of Music, Seminary, Valleyfield from 1958 to 1969. Instructor of Religious and Greek Method, Seminary, Valleyfield from 1958 to 1959. Member of the Board, Young Musicians of Canada from 1960 to 1966. Member of the Executive Council, Young Musicians of Canada from 1962 to 1966. Professor of Music Pedagogy, University of Montreal, Canada from 1962 to 1966. Professor of Music and Rhythmic, Institut Doréa from 1962 to 1976. Member, Diocesan Commission for the Liturgical Pastorate and Commission of Sacred Music from 1964 to 1967. Responsible for Sacred Music, Diocese of Valleyfield from 1965 to 1975. Assistant to the responsible person for Liturgy, Diocese of Valleyfield from 1967 to 1969. Responsible for Liturgy, Diocese of Valleyfield from 1969 to 1975. Responsible for Sacred Art, Diocese of Valleyfield from 1969 to 1984. Pro-Chancellor, Diocese of Valleyfield from 1970 to 1972. Private Secretary, Bishop of Valleyfield from 1970 to 1975. Chancellor, Diocese of Valleyfield from 1972 to 1984. Member, Presbytery Council, Presbytery Council from 1973 to 1975. Chaplain, Knights of Columbus, Diocese of Valleyfield from 1977 to 1984. Secretary, Episcopal Committee, Diocese of Valleyfield from 1972 to 1984. Counsellor-Director, Conseil de la Société Canadienne de Droit Canonique from 1976 to 1978. President, Selection Committee, Low-cost Housing, Valleyfield from 1978 to 1983. Vicaire Episcopal, Legal Affairs from 1981 to 1984. Secretary-General, Bishop from 1981 to 1984. Honorary Prelate, His Holyness Jean-Paul II in 1987. **Educ.:** Lic. Theol. from University of Montreal, Quebec, Canada (1950-1954); Cert. Gregorian Chant, Choirmaster, Choir Conductor, Music Ped. Ward from Paris, France (1955-1958); Dip. Mus. Ped. from University of Montreal, Quebec, Canada (1962); L.D.C. from Pontifical University, Vatican City State (1984-1986). **Interests:** Church music, organ, sacred art, liturgy. **Hons.:** Honorary Prelate; Commander of the Order of the Holy Sepulchre; Knight of Colombus (4th degree). **Mem.:** Member of professional clubs
Mailing address: Piazza San Calisto 16, I-00153 Roma, Italy. T. (39 6) 698 7243.
Residence: Piazza San Giovanni in Laterano 4, I-00120 Roma, Italy T. (39 6) 698 6448

Tresoldi, Giuseppina, S.M.C., Rev Sr
Rel: Roman Catholic; Ordained; celibate.
International Organizations: Superior General of the Comboni Missionary Sisters.
Other organizations: Councillor for Reporting to the Congregations, the International Union of Superior Generals and the International Union of Superiors, Commission for the Catechism and the Catechists.
Mailing address: Via Boccea 506, I-00166 Roma, Italy. T. (39 6) 696 02 73.

Tricarico, Alberto, H E Mgr
Rel: Roman Catholic; Ordained on 06 JAN 1950; b. 10 AUG 1927 in Gallipoli, Italy; celibate.
International Organizations: Apostolic Pro-Nuncio to Singapore and Thailand (since 1987) of the Roman Catholic Church.
Other organizations: Apostolic Delegate to Malaysia, Brunei and Laos (since 1987); Apostolic Delegate to Myanmar (since 1990); Titular Archbishop, Sistroniana.
Mailing address: 55 Waterloo Street 6, Singapore 0718, Singapore. T. (65 2) 337 2466

Trickett, David
Rel: Christian.
International Organizations: Secretariat of the Washington Institute of Ecumenics.
Mailing address: 487 Michigan Ave NE, Washington DC 20017, USA.

Trigano, Shmuel, Prof
Rel: Jewish; Unordained; b. 12 SEP 1948 in Blida, Algeria; m. to Hélène Acoca; c. 1 boy and 1 girl; Citizen of France; Lang. French, Hebrew and English.
International Organizations: Director (since 1985) of the Collège des études juives.
Other organizations: Director (since 1985), *Pardès* (Revue d'Etudes Juives); Lecturer in Sociology (since 1986), University of Paris X (Paris-Nanterre).
Main field of work: Education, writing, communication. **Career:** Lecturer, University of Montpellier III, France from 1981 to 1986. **Educ.:** B.A. from Hebrew University of Jerusalem, Israel (1969-1973); Doc. 3ème Cycle from University of Paris X (Paris-Nanterre), France (1981); Hab. professorale from University of Paris X (Paris-Nanterre), France (1990). **Pubs.:** *Le récit de la disparue. Essai sur l'identité juive*, Gallimard, 1977; *La nouvelle question juive*, Gallimard, 1979 and *Philosophie de la loi. L'origine de la politique dans la Tora*, Le Cerf, 1991.
Mailing address: 45 rue de la Bruyère, F-75009 Paris, France. T. (33 1) 42 80 35 00. C. ALLISRUN.

Trinley, Tseyangla
Rel: Buddhist: Mahayana.
International Organizations: Member of the Foundation for the Preservation of the Mahayana Tradition.
Other organizations: Director, Khachoe Ghakyil Nunnery.
Mailing address: GPO Box 817, Kathmandu, Nepal. T. (977 1) 226 717

Trisolini, John Francis, Rev
Rel: Roman Catholic; Ordained on 15 APR 1967; b. 02 MAR 1937 in Hoboken NJ, USA; celibate; Citizen of USA; Lang. English, Korean, French, Italian and Spanish.
International Organizations: Chairman (since 1971) of the Archdiocesan Labor Pastoral Commission.
Main field of work: Labor ministry. **Career:** Chaplain, Young Christian Workers, Seoul from 1968 to 1982. Chaplain (National), Young Christian Workers, Korea from 1983 to 1985. Chaplain, Seoul Juvenile Correctional Institute from 1979 to 1990. Director, Salesian Seminary from 1984 to 1990. **Educ.:** B.A. from Don Bosco College, Newton NJ, USA (1956-1959); Ecole Supérieure de Théologie, Lyon, France (1964-1967). **Pubs.:** "Some Reflections on the Pastoral Mission of the Clergy to the World of Labor", in *Samok Magazine*, 48, 1976; "Pastoral Concern for Workers", in *Samok Magazine*, 85, 1983 and "The Church's Mission to the Incarcerated", in *Samok Magazine*, 113, 1987.
Mailing address: Don Bosco Youth Center, 4491 Shin Gil 6 Dong, Yong Dung Po Ku, Seoul 150-056, Korea Rep. T. (82 2) 845 7714 – 833 4010
Residence: See Mailing Address. T. See Mailing Address

Tron, Georges
Rel: Roman Catholic.
International Organizations: Contact of the Fédération européenne des associations maristes.
Mailing address: 12 avenue de Corinthe, F-13006 Marseille, France.

Trujillo, Alfonso López, Card
Rel: Roman Catholic.
International Organizations: President of the Pontifical Council for the Family.
Mailing address: Piazza San Calisto 16, I-00120 Città del Vaticano, Vatican. T. (39 6) 698 7243. Fax 698 72 72.

Trujillo López, Ildefonso T, Rev
Rel: Roman Catholic; Ordained on 17 DEC 1966; b. 14 MAY 1940 in Jaraiz de la Vera, Cáceres, Spain; celibate; Citizen of Spain; Lang. Spanish, English and Sindebele.
International Organizations: Director-General (since 1988) of the Spanish Institute for Foreign Missions.
Main field of work: Parish and social work. **Educ.:** Studies Theol. from Seminario de Misiones, Burgos, Spain (1961-1966). **Pubs.:** 3 books in Sindebele: *Ngilandela*; *Kasikhulekeni* and *Ibhayibili*.
Mailing address: Calle Ferrer del Río 17, E-28028 Madrid, Spain. T. (34 91) 355 23 42.
Residence: See Mailing Address. T. See Mailing Address

Tshihamba, M L, Dr
Rel: Presbyterian.
International Organizations: General Secretary of the Church of Christ in Zaire (Presbyterian Community).
Mailing address: PB 117, Kananga, Zaire.

Tsurumi, Shoseki
Rel: Buddhist; Ordained; b. 15 JAN 1917; Citizen of Japan.
International Organizations: Monk of Naritasan Shinshoji.
Mailing address: 1 Narita, Narita, Chiba 286, Japan. T. (81 476) 22 2111

Tuckett, C M, Prof
Rel: Christian.
International Organizations: Secretary of the Society for the Study of the New Testament.
Mailing address: Faculty of Theology, University of Manchester, Manchester M13 9PL, UK.

Tuckmann, Frederick Augustus (Fred)
Rel: Jewish; b. 09 JUN 1922; m. to Patricia Caroline Myers; c. 2 boys and 1 girl.
International Organizations: Co-President of the Consultative Council of Jewish Organizations.
Other organizations: President (since 1989), Anglo-Jewish Association.
Career: Departmental Manager, Marks and Spencer from 1950 to 1954. Co-Secretary and Personnel Manager, British and International Address Ltd. from 1955 to 1963. Assistant Co-Secretary, Temple Press from 1963 to 1965. Management Consultant to Hay Group from 1965 to 1985. Partner, Hay Associates from 1975 to 1985. Member, European Parliament (European Democratic Group) for Leicester from 1979 to 1989. Conservative spokesman on Social and Employment Committee from 1984 to 1989. Led Conservatives in Latin American Delegation from 1982 to 1987. Chairman, International Group of Small Business in 1985. **Educ.:** B.Sc. from University of London, London School of Economics and Political Science, UK. **Interests:** reading, music, walking, discussion, family. **Hons.:** Officer of the Order of the British Empire. **Mem.:** Fellow, Chartered Institute of Secretaries; Fellow, Institute of Personnel Management; Fellow, British Institute of Management; Member, Institute of Marketing; Clubs: Carlton, Athenaeum
Mailing address: Woburn House, Upper Woburn Place, London WC1H OEP, UK. T. (44 71) 387 5937 – (44 71) 387 5938. C. ANJEWGLO.

Tucunan, E P
Rel: Evangelical.
International Organizations: General Secretary of the Evangelical Church of Sangir-Talaud.
Mailing address: Kantor Sinode GMIST, Tahuna, Sangir Talaud, Sulawesi Utara, Indonesia.

Tumansi, Kititwa
Rel: Christian.
International Organizations: Vice-President for Africa of the International Christian Union of Business Executives.
Mailing address: c/o CADICEC-Uniapac-Zaïre, 81 av des 3 Z, Zone de la Gombe, BP 3417, Kinshasa, Zaire.

Tumi, Christian Wiyghan, H E Card
Rel: Roman Catholic; Ordained on 17 APR 1966; b. 15 OCT 1930 in Kikaikelaki, Cameroon; celibate; Citizen of Cameroon.
International Organizations: President of the Symposium of Episcopal Conferences of Africa and Madagascar.
Member of the Synod of Bishops. Member of the Congregation for the Evangelization of Peoples. Member of the Pontifical Council for Inter-Religious Dialogue.
Other organizations: Member (since 1988), College of Cardinals; Archbishop, Garoua; President, National Episcopal Conference of Cameroon.
Career: Elected to Yagoua in 1979. Consecrated Bishop in 1980. Coadjutor Bishop in 1982. Diocesan Bishop in 1984. Created Cardinal by Pope John Paul II in 1988.
Mailing address: Archevêché, BP 272, Garoua, Cameroon. T. (237) 271 353

Tung, Sophia
Rel: Lutheran.
International Organizations: Vice-President of the Lutheran World Federation.
Member, Program Committee for World Service of the Lutheran World Federation.
Mailing address: 86 Section 3 Hsin Sheng South Road, 10769 Taipei, Taiwan (Rep of China). T. (886 2) 362 1623. Fax 363 9401

Turek, Stanislaw, O.S.P.P.E., Fr
Rel: Roman Catholic; Ordained; celibate.
International Organizations: Procurator-General of the Order of St Paul the First Hermit.
Mailing address: Via Alcamo 12/a, I-00182 Roma, Italy. T. (39 6) 701 45 69.

Turoldo, D Ercole, C.R.L., Fr
Rel: Roman Catholic; Ordained; celibate.
International Organizations: Procurator-General of the Cononici Regolari della Congregazione del Ss.mo Salvatore Lateranense.
Mailing address: Piazza S Pietro in Vincoli 4/a, I-00184 Roma, Italy. T. (39 6) 481 89 69, (39 6) 488 28 65

Tutu, Desmond Mpilo, Most Rev
Rel: Anglican; Ordained in 1961; b. 07 OCT 1931 in Klerksdorp, Transvaal, South Africa; m. to Leah Nomalizo Shenxane; c. 1 boy and 3 girls; Citizen of South Africa.
International Organizations: Archbishop (since 1986) of Cape Town.
Other organizations: Metropolitan (since 1986), Church of the Province of Southern Africa.
Career: Teacher at Johannesburg Bantu High School in 1954. Teacher at Munsieville High School, Krugersdorp from 1955 to 1958. Ordained Deacon in 1960. Part-time curate, St Alban's from 1962 to 1965. Part-time curate, St Mary's, Blechingley, Surrey in 1966. Member of staff (Lecturer), Federal Theological Seminary, Alice, Cape, and Chaplain, University of Fort Hare from 1967 to 1969. Lecturer, Dept. of Theology, UBLS Roma, Lesotho from 1970 to 1972. Associate Director, Theological Education Fund, World Council of Churches from 1972 to 1975. Dean, Johannesburg from 1975 to 1976. Bishop, Lesotho from 1977 to 1978. Secretary-General, South African Council of Churches from 1978 to 1985. Bishop, Johannesburg from 1985 to 1986. President, All-Africa Conference of Churches in 1987. Chancellor, University of the Western Cape in 1988. **Educ.:** Teacher's Dip. from Pretoria Bantu Normal College, South Africa (1951-1953); B.A. from University of South Africa, Pretoria, South Africa (1954); Ord. Trng., Lic. Theol. from St Peter's Theological College, Rosettenville, Johannesburg, South Africa (1958-1960); M.Th. from University of London, UK (1966). **Hons.:** Prix d'Athènes, Onassis Foundation (1980); The Family of Man Gold Medal Award (1983); Martin Luther King Jr Humanitarian Award (1984); Nobel Peace Prize (1984); International Integrity Award, John-Roger Foundation, Los Angeles (1986); USA President's Award, Glassboro State College, NJ (1986); Martin Luther King Jr Peace Award, Atlanta GA (1986); Order of the Southern Cross, Brazil (1987); Order of Merit, Brazil (1987); Albert Schweitzer Humanitarian Award, Emmanuel College, Boston MA (1988); Joint recipient of the Third World Prize awarded jointly to the Most Rev D M Tutu, Rev Dr F Chikane, Rev Dr A Boesak and Dr B Naude (1989); Distinguished Peace Leadership Award, Nuclear Age Peace Foundation (1990); Freedom of the City of Florence, Italy (1985); Freedom of the City of Methyr Tydfil, UK (1986); Freedom of the City of Durham, UK (1987); Freedom of the City of Hull, UK (1989); Freedom of Borough of Lewisham, UK (1990); Freedom of the City of Kinshasa (1990); Fellow of King's College, London (1978); numerous honorary degrees (D.D., LL.D., Th.D., D.C.L., S.T.D., D.H., D.Hum.Litt.). **Mem.:** International Social Prospects Academy (1983); National Association for Advancement of Coloured People (Life Member – 1985) **Pubs.:** Collections of sermons and speeches: Crying in the Wilderness, 1982 and Hope and Suffering, 1983.
Mailing address: Bishopscourt, Claremont, Cape Town 7700, South Africa. T. (27 21) 761 2531. Fax 761 4193. Tx 524320 CPSA

Twagirayesu, Michel, Rev
Rel: Presbyterian.
International Organizations: President of the Presbyterian Church of Rwanda.
Mailing address: BP 56, Kigali, Rwanda.

Tzadua, Paulos, H E Card
Rel: Roman Catholic; Ordained on 12 MAR 1944; b. 25 AUG 1921 in Addifini; celibate.
International Organizations: Member (since 1985) of the College of Cardinals.
Member of the Sacred Congregation for the Oriental Churches. Member of the Pontifical Council for Dialogue with Non-Believers.
Other organizations: Archbishop, Addis Ababa; Member, Council of Cardinals for the Study of Organization and Economic Questions Relating to the Holy See; President, Ethiopian Episcopal Conference.
Career: Created Cardinal by Pope John Paul II in 1985.
Mailing address: PO Box 21903, Addis Ababa, Ethiopia. T. (251 1) 12 24 54

Tzelios, Menalaos
Rel: Unknown.
International Organizations: Secretary-General of the International Federation for the Protection of the Rights of Ethnic, Religious, Linguistic and Other Minorities.
Mailing address: 11-25 30th Avenue, Long Island City NY 11102, USA.

Uesugi, Chisato
Rel: Shinto; Ordained; b. 03 APR 1923; Citizen of Japan.
International Organizations: Chief Priest of the Chinzei Taisha Suwa Shrine.
Educ.: Kokugakuin University, Tokyo, Japan.
Mailing address: 3 Kaminishiyamamachi, Nagasaki, Nagasaki 850, Japan. T. (81 958) 24 0445

Uhac, Giuseppe, H E Mgr
Rel: Roman Catholic; Ordained on 16 APR 1949; b. 20 JUL 1924 in Bresec; celibate.
International Organizations: Apostolic Nuncio to Germany (since 1984) of the Roman Catholic Church.
Other organizations: Titular Archbishop, Tharros.
Mailing address: Turmstrasse 29, 53175 Bonn, Germany. T. (49 228) 37 69 01. Fax 37 91 80. Tx 8869794 APNU D

Ukpong, Justin S
Rel: Roman Catholic.
International Organizations: Contact of the Catholic Institute of West Africa.
Mailing address: PO Box 499, Port Harcourt, Nigeria.

Umeda, Shinryu
Rel: Buddhist; Ordained; b. 22 FEB 1906; Citizen of Japan.
International Organizations: Chief of the Sojiji.
Main field of work: Monk. **Career:** Manager, Soto Zen Sect. **Educ.:** Kyushu University Graduate School, Fukuoka-shi, Japan.
Mailing address: 2-1-1 Tsurumi, Tsurumi-ku, Yokohama, Kanagawa 230, Japan. T. (81 45) 581 6021

Uranga, Washington
Rel: Roman Catholic.
International Organizations: Second Vice-President of the Unda, International Catholic Association for Radio and Television.
Mailing address: Estados Unidos 2057, 1227 Buenos Aires, Argentina.

Uria, Paul, Fr
Rel: Roman Catholic; Ordained; celibate.
International Organizations: Regional Secretary of the International Catholic Organization for Cinema and Audiovisual.
Mailing address: Secretary AMECEA, PO Box 21191, Nairobi, Kenya.

Uribe Guzmán, Francisco Javier U, Fr
Rel: Roman Catholic; Ordained; celibate.
International Organizations: Superior General of the Misioneros de la Natividad de María.
Mailing address: Madero 724 Apartado 447, León Gto., Mexico. T. (52 471) 3 36 43

Urich, Francis Xavier, S.I.C., Sir
Rel: Roman Catholic; Ordained; celibate.
International Organizations: Secretary of the Conference of Major Superiors of the Antilles.
Mailing address: Alvernia, PO Box 124, Montego Bay, Jamaica.

Urrisk, B E M Rolf M
Rel: Roman Catholic.
International Organizations: Secretary-General of the International Military Apostolate.
Mailing address: Armeekommando, Hütteldorferstrasse 126, A-1140 Wien, Austria.

Ursi, Corrado, H E Card
Rel: Roman Catholic; Ordained on 25 JUL 1931; b. 26 JUL 1908 in Andria, Bari, Italy; celibate; Citizen of Italy.
International Organizations: Member (since 1967) of the College of Cardinals.
Other organizations: Archbishop Emeritus (since 1987), Naples.
Career: Bishop, Nardo in 1951. Archbishop, Acorensa in 1961. Archbishop, Naples from 1966 to 1987. Created Cardinal by Pope Paul VI in 1967.
Mailing address: Via Capodimonte 13, I-80136 Napoli, Italy. T. (39 81) 741 39 85

Ushiroku, Toshio, Rev
Rel: United Church of Christ.
International Organizations: Moderator of the United Church of Christ in Japan.
Mailing address: 31 Japan Christian Centre, 2-3-18 Nishi-Waseda, Shinjuku-ku, Tokyo 160, Japan.

Uthman, Mustafa, Dr
Rel: Muslim.
International Organizations: Secretary-General of the International Islamic Federation of Student Organizations.
Mailing address: 555 Grove Street, PO Box 448, Herndon VA 22070, USA. T. (1 703) 471 3922. Fax 471 3922. Tx 901153 IIIT WASH.

Utia, Marurai, Rev
Rel: Evangelical.
International Organizations: President of the Evangelical Church of French Polynesia.
Mailing address: BP 113, Papeete, Tahiti, Polynesia Fr.

Utterback, Bertil, Pastor
Rel: Seventh-Day Adventist.
International Organizations: President, Swedish Union Conference of the General Conference of Seventh-Day Adventists.
Mailing address: PO Box 536, S-10130 Stockholm, Sweden.

Uwadi, Rogers O, Rt Rev
Rel: Methodist.
International Organizations: Secretary of Conference of the Methodist Church Nigeria.
Mailing address: Wesley House, 21/22 Marina, PO Box 2011, Lagos, Nigeria.

Vaccaro, Gabriel O, Rev Dr
Rel: Church of God.
International Organizations: President of the Church of God.
Mailing address: Miralla 435, 1408 Buenos Aires, Argentina.

Vachon, Louis-Albert, H E Card
Rel: Roman Catholic; Ordained on 11 JUN 1938; b. 04 FEB 1912 in St Frédéric-de-Beauce, Quebec, Canada; celibate; Citizen of Canada.
International Organizations: Member (since 1985) of the College of Cardinals.
Member (since 1985) of the Congregation for the Clergy.
Other organizations: Member, Administrative Board (since 1981), Canadian Conference of Catholic Bishops; Member, Administrative Board (since 1985), Ordre national du Québec; Archbishop Emeritus (since 1990), Quebec.
Career: Professor of Philosophy, Laval University from 1941 to 1947. Professor of Theology, Laval University from 1949 to 1955. Superior, Grand Séminaire de Québec from 1955 to 1959. General Superintendent, Grand Séminaire de Québec from 1960 to 1977. Domestic Prelate in 1958. Vice-Rector, Laval University from 1959 to 1960. Rector, Laval University from 1960 to 1972. Vicar-General, Diocese of Quebec from 1960 to 1981. Protonotary Apostolic from 1963 to 1977. Auxiliary Bishop, Quebec from 1977 to 1981. Archbishop, Quebec and Primate of Canada from 1981 to 1990. President, Assemblée des évêques du Québec from 1981 to 1985. Delegate, Canadian Conference of Catholic Bishops (CCCB) to Synod of Bishops on Reconciliation and Penance in the Mission of the Church in 1983. Created Cardinal by Pope John Paul II in 1985. Governor, Canadian Bible Society in 1985. **Educ.:** B.A. from Québec Seminary, Canada (1934); D.Ph. from Université Laval, Québec, Canada (1947); D.Th. from Pontifical University San Tommaso d'Aquino, Vatican City State (1949). **Interests:** reading, fine art. **Hons.:** Officer, Ordre de la Fidélité française (1963); Centennial Medal of Confederation (1967); Companion of the Order of Canada (1969); Hon. Fellow, Royal College of Physicians and Surgeons, Canada (1972); Prix '3 juillet 1608' du Conseil de la langue française (1983); Knight Grand Cross, Equestrian Order of the Holy Sepulchre of Jerusalem (1985); Officer, National Order of Quebec (1985); Honorary President, Laval University Fund Drive (1986); Bailli Great Cross and Devotion, Knights of Malta, Canada (1987); Officer, Legion of Honour (1988); Hon. degrees from Montreal, McGill, Victoria (1964), Guelph (1966), Moncton (1967), Bishop's, Queen's, Strasbourg (1968), Notre Dame (1971), Carleton (1972), Laval (1982). **Mem.:** Entraide universitaire mondiale du Canada (President); Conférence (1965-68) des Recteurs et des Principaux des Universités du Québec; Association des Universités et Colleges du Canada (President 1965-66); International Federation of Catholic Universities (IFCU) (Administrative Council); Association of Partially or Wholly French-Language Universities (AUPELF) (Administrative Council); Canadian Education Association; Société canadienne de l'histoire de l'Eglise catholique; Catholic Theological Society of America; Société canadienne pour l'histoire et la philosophie des sciences; World Union of Catholic Teachers (WUCT); Association canadienne des éducateurs de langue française (ACELF) (President); Association of Commonwealth Universities; Société des écrivains canadiens; Ordre des francophones d'Amérique; Société Royale du Canada **Pubs.:** Responsabilité collective des universitaires, 1964; Les humanités d'aujourd'hui, 1966 and Excellence et loyauté des universitaires, 1969.
Mailing address: 1 rue des Remparts, Québec G1R 5L7, Canada.

Vaerge, Vilhelm, Rev
Rel: Lutheran; Unordained.
International Organizations: Chairman of the Lutheran European Commission on the Church and the Jewish People.
Mailing address: Grove praestegard, DK-7540 Haderup, Denmark. T. (45 97) 143073.

Vaerge, Vilhelm, Rev
Rel: Lutheran.
International Organizations: Chairman of the Lutheran European Commission on the Church and the Jewish People.
Mailing address: Grove praestegard, DK-7540 Haderup, Denmark. T. (45) 97 14 30 73.

Valabek, Francis J
See Redemptus Maria Valabek.

Valabek, Redemptus Maria, O. Carm., Rev Prof
Also know as Francis J Valabek; Rel: Roman Catholic; Ordained on 25 JUL 1960; b. 03 JUL 1934 in Shelton CT, USA; celibate; Citizen of USA; Lang. English, Italian, Spanish, French and Czech.
International Organizations: Delegate (since 1984) of the Carmelite Third Order.
Dean, English Section (since 1975) of the Pontifical Institute 'Regina Mundi'. Contact person of the Confraternity of Our Lady of Mount Carmel.
Other organizations: Editor (since 1971), *Carmel in the World*; Postulator General (since 1981), Carmelite Order; Apostolic Examiner, the Roman Clergy, Diocese of Rome; Professor of Theology in various Roman institutions.
Main field of work: University education. **Educ.:** Lic. Theol. from Collegio Internazionale San Alberto and Lateran Pontifical University, Vatican City State (1957-1962); Dr. Theol. from Lateran Pontifical University, Vatican City State (1962-1965). **Interests:** Carmelite lay movements, mariology, liturgy. **Pubs.:** Prayer Life in Carmel, 1982; Mary, Mother of Carmel, 2 Vols., Rome, 1986, 1987 and Editor, Essays on Titus Brandsma, Rome, 1985.
Mailing address: Via Sforza Pallavicini 10, I-00193 Roma, Italy. T. (39 6) 65 43 513. C. COLLALBERT.
Residence: See Mailing Address. T. See Mailing Address

Valle, Carlos A
Rel: Christian.
International Organizations: General Secretary of the World Association for Christian Communication.
Mailing address: 357 Kennington Lane, London SE11 5QY, UK. T. (44 71) 582 91 39. Fax 735 03 40. C. WACC LONDON SE11. Tx 8812669 WACC G.

Vanackere, Hans L
Rel: Roman Catholic.
International Organizations: Secretary-General of the Christian Academy for European Dialogue.
Mailing address: Abdij van 't Park, Abdijdreef 7A, B-3001 Heverlee, Belgium. T. (32 16) 20 71 91. Fax 20 72 07.

Van Asperdt, G
Rel: Christian.
International Organizations: Secretary-General of the Missionary Cooperation Service for Development.
Mailing address: 195 Bd Leopold II, B-1080 Bruxelles, Belgium. T. (32 2) 427 47 20. Tx 65 172.

Vanasse, Roman, O.Praem., Fr
Rel: Roman Catholic; Ordained; celibate.
International Organizations: Member of the Order of the Canons Regular of Prémontré.
Mailing address: Postfach 1209, 61462 Königstein, Germany.

Van Chau, André N, Dr
Rel: Roman Catholic.
International Organizations: Secretary-General of the International Catholic Migration Commission.
Mailing address: 37-39 rue de Vermont, CP 96, CH-1211 Genève 20, Switzerland. T. (41 22) 33 41 50. C. Cathmicom. Tx 28 100 ICMC CH. Fax (41 22) 734 7929.

Van Culin, Samuel, Rev Canon
Rel: Church of England; Ordained; b. 20 SEP 1930 in Honolulu, Hawaii; Citizen of USA.
International Organizations: Secretary-General (since 1983) of the Anglican Consultative Council.
Secretary of the Lambeth Conference of Bishops of the Anglican Communion.
Other organizations: Canon, Canterbury, St George's Cathedral Jerusalem, Ibadan, Nigeria; Canon, Province of Southern Africa.
Career: Curate, St Andrew's Cathedral Parish, Honolulu from 1955 to 1956. Canon Precentor, St Andrew's Cathedral Parish, Honolulu from 1956 to 1958. Assistant Rector, St John's Church, Lafayette Square, Washington DC from 1958 to 1960. Secretary-General, Laymen International, Washington DC from 1960 to 1961. Assistant Secretary, Overseas Mission Society of the US Church from 1962 to 1968. Executive Secretary for Africa and the Middle East, Episcopal Church Center, New York from 1968 to 1976. World Secretary, Unit for National and World Mission, Episcopal Church Center, New York from 1976 to 1983. Appointed Executive for World Mission, Episcopal Church USA in 1976. Secretary, Lambeth Conference in 1988. **Educ.:** A.B. from Princeton University, NJ, USA (1953); D.D. from Virginia Theological Seminary, USA (1955). **Interests:** music, travel. **Hons.:** Hon. D.D., Virginia Theological Seminary (1977). **Mem.:** Club: Athenaeum
Mailing address: Partnership House, 157 Waterloo Road, London SE1 8UT, UK. T. (44 71) 620 1110 – (44 71) 620 1111. C. Compasrose London SW1. Tx 8950907 ANGCOM G. Fax (44 71) 620 1070. Electronic Mail Telecom Gold 81 CEL002, BT/TYMNET QNI 3005.

van den Akker, Frans
Rel: Buddhist: Mahayana.
International Organizations: Member of the Foundation for the Preservation of the Mahayana Tradition.
Other organizations: Director, Stichting De Stroom.
Mailing address: Postbus 340, 5000 AH Tilburg, Netherlands. T. (31 13) 421 720

Van den Audenaerde, Jef
Rel: Roman Catholic.
International Organizations: Secretary of the International Federation of Catholic Parochial Youth Communities.
Mailing address: FIMCAP, Kipdorp 30, B-2000 Antwerpen, Belgium. T. (32 3) 231 07 95. Fax 232 51 62.

Vandeputte, Germain
Rel: Catholic.
International Organizations: Secretary of the Church for the World.
Mailing address: Route de Barvaux 30, B-6940 Wéris, Belgium.

Van der Bruggen, C
Rel: Roman Catholic.
International Organizations: UNESCO Representative of the Pax Romana, International Movement of Catholic Students.
Mailing address: Rue de la Faisanderie, 40, B-1150 Bruxelles, Belgium. T. (32 2) 771 14 68.

Vander Elst, R
Rel: Mason.
International Organizations: Representative of the Universal Freemasons League.
Mailing address: Avenue F D Roosevelt 12, Bte 14, B-1050 Bruxelles, Belgium. T. (32 2) 648 14 74.

Van der Kooij, A, Prof
Rel: Christian.
International Organizations: Secretary of the International Organization for the Study of the Old Testament.
Mailing address: Faculteit der Godgeleerdheid, Postbus 9515, NL-2300 RA Leiden, Netherlands.

van der Ness, H J
Rel: Seventh-Day Adventist.
International Organizations: Treasurer, South African Union of the General Conference of Seventh-Day Adventists.
Mailing address: PO Box 468, Orange Free State, Bloemfontein 9300, South Africa.

Vander Stricht, Jean-Marie, C.I., Fr
Rel: Roman Catholic; Ordained; celibate.
International Organizations: Superior General of the Josephite Fathers.
Mailing address: Jozefieten College, Brusselsesteenweg 459, B-9090 Melle, Belgium. T. (32 91) 52 16 35 – 52 11 09.

Vandertuin-Roy, Victoria Elva Claudia Belot, Rev
Rel: New Age; Ordained on 14 JUN 1964; b. 16 OCT 1933 in New Belford MA, USA; m. to David Kent Roy; c. 1 boy and 2 girls; Citizen of USA; Lang. English.
International Organizations: Minister of the New Age World Religious and Scientific Research Foundation.
Main field of work: Ministry and writing. **Pubs.:** *The Power and Wisdom of the Universe* and *The Polaris Newsletter*.
Mailing address: 62091 Valley View Circle, Joshua Tree CA 92252, USA. T. (1 619) 366 2833.
Residence: See Mailing Address. T. See Mailing Address

Van de Ven, Marcel V, O.Praem., Fr Abbot
Rel: Roman Catholic; Ordained; celibate.
International Organizations: Abbot-General of the Order of the Canons Regular of Prémontré.
Member, Council for Reporting to the Congregation and the International Union of Superiors and Superiors General of the Congregation for Religious and Secular Institutes.
Educ.: Ph.d.
Mailing address: Viale Giotto 27, I-00153 Roma, Italy. T. (39 6) 574 69 79 – 574 23 97.

Van De Wyngard, Viola
Rel: Spiritualist.
International Organizations: President of the International Spiritual Alliance.
Mailing address: Calle Sazié 2078, Casilla 2941, Santiago, Chile. T. (56 2) 695 1170.

van Espen, Daniel
Rel: Roman Catholic.
International Organizations: Director of General Secretariat of the International Catholic Organization for Cinema and Audiovisual.
Secretary of the EUROCIC.
Mailing address: OCIC, Rue de l'Orme 8, B-1040 Bruxelles, Belgium. T. (32 2) 734 42 94. C. OCIC BRUXELLES. Tx 21275 CIPINF B/OCIC. Fax (32 2) 734 32 07.

van Frankenhuijsen, Wim, S.M.A., Fr
Rel: Roman Catholic; Ordained; celibate.
International Organizations: Councillor-General of the Society of African Missions.
Executive Secretary of the Commission on Justice and Peace.
Mailing address: Via della Nocetta 111, I-00164 Roma, Italy. T. (39 6) 62 57 739 – 62 57 735.

van Geene, Harrie, Fr
Rel: Roman Catholic.
International Organizations: Superior General of the Brothers of Our Lady, Mother of Mercy.
Mailing address: Gasthuisring 54, 5041 DT Tilburg, Netherlands. T. (31 13) 432 777. Fax 441 405. C. GENERALAAT TILBURG.

Vanhengel, Marcel, O.P., Rev Dr
Rel: Roman Catholic; Ordained; celibate.
International Organizations: Executive Director of the Catholic Media Council.
Member of the Order of Preachers. Consultor of the Pontifical Council for Social Communications.
Mailing address: Anton Kurze Allee 2, 52074 Aachen, Germany. T. (49 241) 73081. Fax 73462. C. MEDIENPLANUNG. Tx 0832719 mira d.

Vanier, Jean
Rel: Roman Catholic.
International Organizations: Past International Coordinator of the Association internationale Foi et lumière.
Mailing address: 3 rue du Laos, F-75015 Paris, France. T. (33 1) 47 34 33 08.

van Kleef, G A, Rt Rev
Rel: Old Catholic.
International Organizations: Secretary of the Old Catholic Church of the Netherlands.
Mailing address: Bishop's House, Papenpad, NL-1506 GR Zaandam, Netherlands.

van Kooten, Karl-Heinz, Dr
Rel: Protestant.
International Organizations: Director of the International Federation for the Pastoral Care of Persons with Defective Hearing.
Mailing address: Krokusstr 1, D-48527 Nordhorn, Germany. T. (49 5921) 70312. Fax 77702.

van Litsenborgh, Dirk
Rel: Roman Catholic.
International Organizations: Secretary-General of the International Federation for Catholic Associations of the Blind.
Mailing address: FIDACA, Avenue Dailly 90, B-1030 Bruxelles, Belgium. T. (32 2) 216 06 15. Fax 215 88 21.

Van Looy, Lucas Karel Leo August, SDB, Rev
Rel: Roman Catholic; Ordained on 12 SEP 1970; b. 28 SEP 1941 in Tielen, Belgium; celibate; Citizen of Belgium; Lang. Dutch, Korean, English, Italian, German, Spanish, Portuguese and French.
International Organizations: Secretary-General of the Salesian Youth Movement.
General Councillor for Youth Ministry (since 1990) of the Society of St Francis of Sales.
Main field of work: Ministry, coordination. **Career:** Professor of Religious Education, Salesian High School, Kwanedu from 1972 to 1974. Director, Catholic Students, Archdiocese of Seoul from 1974 to 1978. Salesians of Don Bosco, Korea (Provincial Salesians) from 1979 to 1984. General Councillor for Missions, Salesians of Don Bosco, Rome from 1984 to 1990. **Educ.:** Lic. Music from Hanyang University, Seoul, Republic of Korea (1964-1968); Lic. Missiology from Catholic University of Louvain (KUL), Belgium (1967-1972). **Interests:** youth animation through music. **Pubs.:** "The Evangelizer, 12 Steps"; "Evangelization in Latin America" and "Salesian Youth Movement".
Mailing address: Via della Pisana 1111, I-00163 Roma, Italy. T. (39 6) 6592915. Fax 6592929. Tx 628527 SDB ROM. C. SALESIANI PISANA ROMA.
Residence: See Mailing Address. T. See Mailing Address

van Munster, Hendrikus Antonius, O.F.M., Rev Dr
Rel: Roman Catholic; Ordained in 1951; b. 17 NOV 1925 in Gouda, Netherlands; celibate; Citizen of Netherlands; Lang. Dutch, French, English, German, Danish (some) and Swedish (some).
International Organizations: President (since 1991) of the Pontifical Council for Justice and Peace.
Member (since 1944) of the Order of Friars Minor.
Career: Lecturer in Philosophy, Studium Philosophicum, Venray from 1955 to 1967. Regent, RK Lyceum for boys, Venray from 1963 to 1967. Rector for the training, priests, Venray from 1966 to 1967. Lecturer then Professor of Philosophy, Catholic Theological Faculty of Utrecht from 1967 to 1970. Rector, Catholic Theological Faculty of Utrecht from 1968 to 1970. President, Curatorium, Catholic Theological Faculty of Utrecht from 1970 to 1984. Vicar-General, Archdiocese of Utrecht from 1970 to 1981. Secretary-General, Bishops' Conference of the Netherlands from 1981 to 1991. Member, Board for the Netherlands, Franciscans from 1964 to 1970. Vice-Provincial for the Netherlands, Franciscans from 1970 to 1974. President, Stichting Alfrinkfonds from 1975 to 1991. **Educ.:** Studied Phil. from Venray (1945-1947); Studied Theol. from Wychen en Weert (1947-1951); Dr. Phil. from Catholic University of Louvain (KUL), Belgium (1951-1955). **Interests:** cycling, literature. **Hons.:** Officer, Orde van Oranje Nassau (1976); Knight, Orde van de Nederlandse Leeuw (1991). **Pubs.:** Drie prioriteiten voor de Kerk van Europa, 1983; Kantelt de koets wel? Verschuivende denkpatronen in kerkelijk spreken, 1989 and De katolieke school voor primair en secundair onderwijs op weg naar de volgende eeuw, 1989.

Mailing address: Piazza San Calisto 16, 00120 Città del Vaticano. T. (39 6) 698 71 91. Fax 698 72 05.
Residence: Jan van Scorelstraat 75, NL-3583 CL Utrecht, Netherlands.

van Oossanen, Karel Cornelis, Rev

Rel: Seventh-Day Adventist; Ordained on 14 MAY 1959; b. 20 DEC 1931 in Breda, Netherlands; m. to K E Steens; c. 2 boys and 2 girls; Citizen of Netherlands; Lang. Dutch, English, German and French.
International Organizations: Secretary, Trans-European Division (since 1987) of the General Conference of Seventh-Day Adventists.
Main field of work: Ministry, church and administration. **Career:** Ministerial intern, Seventh-Day Adventists (SDA) Church in The Hague from 1951 to 1952. Entered ministry in Zeeland, Southwestern Province, Netherlands, then served as a minister in The Hague, Delft, Leiden and Gouda from 1953 to 1959. Elected Youth Director, South Netherlands Conference of SDA in 1960. Elected Youth Director, Netherlands Union Conference of SDA in 1963. President, South Netherlands Conference of SDA from 1969 to 1971. President, Netherlands Union Conference of SDA from 1971 to 1987. **Educ.:** B.A. Theol. from Adventist Junior College, Oud-Zandbergen, Zeist, Netherlands; Study in Pastoral Psy. (part-time) from State University of Utrecht, Netherlands. **Interests:** Eastern Europe culture, special icons. **Pubs.:** Author of a few children's books published in Netherlands.
Mailing address: 119 St Peter's Street, St Albans AL1 3EY, UK. T. (44 727) 860331
Residence: 41 Cunningham Hill Road, St Albans AL1 5BX, UK. T. (44 727) 862153

Van Schoote, Jean-Pierre

Rel: Catholic.
International Organizations: President of the Latin American College.
Mailing address: Tervuursestraat 56, B-3000 Leuven, Belgium. T. (32 16) 22 58 45.

van Straaten, Werenfried, O.Praem., Fr

Rel: Roman Catholic; Ordained on 27 JUL 1940; b. 17 JAN 1913 in Mijdrecht, Netherlands; celibate; Citizen of Netherlands; Lang. Dutch, French, English, Italian, German and Spanish.
International Organizations: Founder (since 1947) of the Aid to the Church in Need.
Member of the Order of the Canons Regular of Prémontré.
Other organizations: Editor (since 1958), *Mirror.*
Main field of work: Ministry, fund raising. **Career:** General Moderator, Aid to the Church in Need from 1964 to 1981. Ecclesiastical Assistant, Aid to the Church in Need from 1981 to 1988. **Interests:** Writing. **Hons.:** Bundesverdienstkreuz (1982). **Pubs.:** *They call me the Bacon Priest* (available in 6 languages) and *Where God weeps* (available in 6 languages).
Mailing address: Bischof Kaller Strasse 3, 61462 Königstein-im-Taunus, Germany. T. (49 6174) 291-0

van Straten, E R, Rev

Rel: Mennonite.
International Organizations: Secretary of the Mennonite World Conference.
Other organizations: Secretary, General Mennonite Society.
Mailing address: Singel 454, Amsterdam, Netherlands.

van Veldhoven, Ad

Rel: Roman Catholic.
International Organizations: Director of the Dutch Bishops' Lenten Campaign.
Mailing address: Huize O L Vrouw Ter Eem, Daam Fockemalaan 22, NL-3818 KG Amersfoort, Netherlands. T. (31 33) 61 29 57. Tx 79 326 vaned nl.

Van Vught, Jan, Rev

Rel: Christian; Ordained.
International Organizations: Secretary-General of the International Commission of General Prison Chaplains.
Mailing address: Treublaan 2, NL-2555 HHS Den Haag, Netherlands.

Van Winden, Jacobus Cornelis Maria, Rev Prof Dr

Rel: Roman Catholic; Ordained on 07 MAR 1948; b. 10 OCT 1922 in Schipluiden, Zuid Holland, Netherlands; celibate; Citizen of Netherlands; Lang. Dutch, English, German and French.
International Organizations: Ecclesiastic of the Roman Catholic Church.
Other organizations: Editor-in-Chief (since 1977), *Vigilae Christianae* (a review of early Christian life and language); Professor Emeritus (since 1987).
Main field of work: Study of Greek Fathers. **Career:** Assistant, State University of Leiden from 1954 to 1956. Teacher of Latin and Greek, Rotterdam Secondary School from 1956 to 1966. Reader, State University of Leiden from 1966 to 1979. Professor of Patristics, State University of Leiden from 1979 to 1987. **Educ.:** Class. Philology from State University at Leiden, The Netherlands (1949-1954); Dr. (1959). **Interests:** Latin and Greek patristics and philosophy. **Mem.:** Franciscan Order (1941); Royal Academy of Sciences **Pubs.:** *Calcidius on Matter. His Doctrine and Sources*, Diss., Leiden, 1959; *An Early Christian Philosopher. Justin Martyr's Dialogue*, chap. 1-9, Leiden, 1971 and With J H Waszink, *Tertullianus, De Idolatria*, Ed., translation, commentary, Leiden, 1987.
Mailing address: Haarlemmerstraat 106b, NL-2312 GD Leiden, Netherlands. T. (31 71) 120401
Residence: See Mailing Address. T. See Mailing Address

van Zadel, Th L

Rel: Evangelical Lutheran.
International Organizations: Secretary of the Evangelical Lutheran Church in the Kingdom of the Netherlands.
Mailing address: Jan de Bakkerstraat 13-15, NL-3441 ED Woerden, Netherlands.

van Zyl, Daniel Cornelis, Dr

Rel: Dutch Reformed; Ordained on 12 FEB 1977; b. 18 JUN 1951 in Bredasdorp, Cape Province, South Africa; m. to Cornelia Linde; c. 1 boy and 3 girls; Citizen of South Africa; Lang. English, Afrikaans and Xhosa.
International Organizations: Lecturer, Department of Theology (since 1990) of the University of Western Cape, Bellville.
Other organizations: Founder (since 1991), 'Sokhanya' theological school for African independent churches in the Cape peninsula; Lecturer (since 1991), Hugenote Bible Institute; Tutorial Coordinator, Dept. of Theology (since 1992), University of Western Cape.
Main field of work: Theological education. **Career:** Minister of Religion, Dutch Reformed Church in Africa from 1977 to 1984. **Educ.:** B.A. from University of Stellenbosch, South Africa (1970-1972); M.Th. from University of Stellenbosch, South Africa (1984-1984); D.Th. from University of Stellenbosch, South Africa (1985-1989). **Interests:** Missions, community development. **Mem.:** Missiological Society of South Africa **Pubs.:** Teks en konteks in perspektief uit Exodus 21:2-6, NGTT, 1991 and Cosmology, ecology and missiology – A Perspective from Genesis 1, Missionalia, 1991.
Mailing address: University of Western Cape, PO Box 182, Bellville 7530, South Africa. T. (27) 9592 632
Residence: 57 Governor Street, Welgemoed, Bellville 7530, South Africa T. (27) 9131 937

Varah, Edward Chad, Prebendary Dr

Rel: Church of England; Ordained on 27 DEC 1936; b. 12 NOV 1911 in Barton on Humber, South Humberside, UK; m. to Doris Susan Whanslaw; c. 4 boys and 1 girl; Citizen of United Kingdom; Lang. English, Dutch, French, German, Spanish and Russian (some).
International Organizations: Founder (since 1974) of the Befrienders International.
Other organizations: Rector (since 1953), Lord Mayor's Parish Church of St Stephen Walbrook, City of London; Prebendary (Honorary Canon) (since 1975), St Paul's Cathedral, London.
Main field of work: Ministry. **Career:** Vicar, Holy Trinity, Blackburn from 1942 to 1949. Vicar, St Paul, Clapham Junction, London from 1949 to 1953. Scriptwriter/Visualiser for *Eagle* and *Girl* magazines from 1950 to 1962. Founded The Samaritans in 1953. Chairman, The Samaritans Inc. Chairman, Befrienders International. President, Befrienders International. **Educ.:** M.A. from University of Oxford, Keble College, UK (1930-1933); G.O.E. from Lincoln Theological College, UK (1934-1935). **Interests:** sex therapy, suicide prevention. **Hons.:** Roumanian Patriarchal Cross (1968); Officer, Order of the British Empire (1969);

Albert Schweitzer Gold Medal (1972); Louis Dublin Award, American Association for Suicide Prevention (1974); Prix de l'Institut de la Vie, Paris (1978); Hon. LL.D., Leicester (1979). **Mem.:** International Association for Suicide Prevention **Pubs.:** *Notny Sbornik Russkogo Pravoslavnogo Tserkovnaya Penia*, 1962; *The Samaritans: befriending the suicidal*, Constable, 1984, 2nd Ed., 1988 and *Before I Die Again* (autobiography), Constable, 1992.
Mailing address: St Stephen's Church, Walbrook, London EC4N 8BN, UK. T. (44 71) 626 8242
Residence: 39 Walbrook, London EC4N 8BP, UK. T. (44 71) 283 4444

Varah, Susan
Rel: Church of England; m. to Edward Chad Varah; c. 4 boys and 1 girl.
International Organizations: Past President of the Anglican Mother's Union.
Mailing address: 42 Hillersdon Ave, London SW13 0EF, UK.

Varela, Ana
Rel: Roman Catholic.
International Organizations: Head, Latin America Regional Secretariat of the Pax Romana, International Catholic Movement for Intellectual and Cultural Affairs.
Mailing address: Plaza Cagancha 1368/12, Montevideo, Uruguay.

Varelopoulos, Nicolaos, VI, H B
Rel: Greek Orthodox; Ordained; b. in Istanbul, Turkey; Citizen of Greece.
International Organizations: Pope and Patriarch of Alexandria and all Africa.
Mailing address: Bureau de Presse, Patriarcat Greek Orthodox, PO Box 1307-2006, Alexandria, Egypt.

Varga, Mary Alice, SDR, Sr
Rel: Roman Catholic; Ordained; celibate.
International Organizations: Secretary of the Sisters of the Divine Redeemer.
Mailing address: Via Casale Piombino 14, I-00135 Roma, Italy. T. (39 6) 338 18 12.

Varqa, Ali Muhammad, Dr
Rel: Bahá'í.
International Organizations: Member of the National Spiritual Assembly of the Bahá'ís of Canada.
Mailing address: c/o National Spiritual Assembly of the Bahá'ís of Canada, 7200 Leslie St, Tornhill ON L3T 6L8, Canada.

Varsamis, Thomas, Rev Archimandrite
Rel: Catholic (Byzantine rite); Citizen of Turkey.
International Organizations: Vicar Delegate of the Istanbul.
Mailing address: Apostolic Exarchate of Istanbul, Hamalbasi Cad. 44, 80070 Beyoglu, Istanbul, Turkey. T. (90 1) 1497104

Vasey, William L, Rev
Rel: Christian.
International Organizations: General Director of the International Mission Board.
Mailing address: 7872 Glenwood Ave, Youngstown OH 44512, USA. T. (1 216) 726 2643.

Vasken, V, I, H H
Rel: Armenian Apostolic Church.
International Organizations: Supreme Patriach and Catholicos of the All Armenians.
Mailing address: Holy See, Etchmiadzin, Armenia.

Vasquez, Carlos M Lee
Rel: Roman Catholic.
International Organizations: Zone Coordinator, Central America of the Caritas Internationalis.
Mailing address: Depto Nacional de Pastoral, Social y Caritas, Apartado 1149, Panamá 9A, Panama.

Vassallo, Laura
Rel: Buddhist: Mahayana.
International Organizations: Member of the Foundation for the Preservation of the Mahayana Tradition.
Other organizations: Director, Istituto Lama Tzong Khapa.
Mailing address: Via Poggiberna 9, I-56040 Pomaia, Italy. T. (39 50) 685 654. Fax (39 50) 685 768
Residence: T. (39 50) 685 976

Vassaux, P, Rev
Rel: Unitarian.
International Organizations: UNESCO Representative of the International Association for Religious Freedom.
Mailing address: 4 rue de l'Oratoire, F-75001 Paris, France.

Vassilaqui, Alejandro
Rel: Protestant.
International Organizations: President of the World Alliance of Young Men's Christian Associations.
Mailing address: John R Mott House, Quai Wilson 37, CH-1201 Genève, Switzerland. T. (41 22) 732 31 00. Fax 738 40 15. C. FLEMGO Genève. Tx 412332 YMCA CH.

Vaughn, John, OFM, Rev Fr
Rel: Roman Catholic; Ordained; celibate.
International Organizations: Grand Chancellor of the Pontificio Ateneo Antonianum.
Minister General of the Order of Friars Minor. Member of the Congregation for the Evangelization of Peoples.
Mailing address: Via Merulana 124, I-00185 Roma, Italy. T. (39 6) 77 97 49.

Vazken, V, I, Catholicos of Armenia
Also know as L K Baldzhian; Rel: Armenian Church; Ordained; b. in Romania.
International Organizations: Catholicos (since 1955) of the All Armenia.
Career: Priest in Armenian Church in Romania, later Head, Armenian Church in Romania and Bulgaria.
Mailing address: Residence of Catholicos of Armenia, Etchmiadzin, Armenia.

Vedanti, Brahma Kumari, Sr
Rel: Hindu; Unordained; b. 13 SEP 1947 in Umrath, Gujarat, India; single; Citizen of India; Lang. Gujarati, Hindi and English.
International Organizations: Regional Head for Africa (since 1968) of the Brahma Kumaris World Spiritual University.
Main field of work: Spiritual education. **Educ.:** M.A. Psy. from Ahmedabad University, India.
Mailing address: Parkland Road, Mawa Close, PO Box 12349, Nairobi, Kenya. T. (254 2) 741239
Residence: See Mailing Address. T. (254 2) 743572

Veem, Konrad, Rev
Rel: Evangelical Lutheran.
International Organizations: Archbishop of Estonian Evangelical Lutheran Church.
Mailing address: Box 450 74, S-104 30 Stockholm 45, Sweden.

Vegliò, Antonio Maria, H E Mgr
Rel: Roman Catholic; Ordained on 18 MAR 1962; b. 03 FEB 1938 in Macerata, Italy; celibate.
International Organizations: Apostolic Pro-Nuncio to Cape Verde, Guinea-Bissau, Mali and Senegal (since 1989) of the Roman Catholic Church.
Other organizations: Apostolic Delegate to Mauritania (since 1989); Titular Archbishop, Eclano.
Mailing address: BP 5076, Dakar, Senegal. T. (221) 24 26 74. Fax 24 19 31

Veilleux, Armand J P, O.C.R., Rev Dr

Rel: Roman Catholic; Ordained on 11 AUG 1963; b. 10 NOV 1937 in Québec, Canada; celibate; Citizen of USA; Lang. English, French, Italian, Spanish and German.

International Organizations: Procurator-General (since 1990) of the Cistercian Order of the Strict Observance.

Other organizations: Rector, College of Religion for Cistercian Order of Strict Observance.

Career: Abbot, Cistercian Abbey, Québec from 1969 to 1976. Abbot, Cistercian Abbey, Georgia from 1983 to 1990. **Educ.:** D.Theol. from Pontifical St Anselmo Athenaeum, Vatican City State (1962-1967). **Interests:** patristics, gnosticism.

Mailing address: Viale Africa 33, I-00144 Roma, Italy. T. (39 06) 592 41 55.

Residence: See Mailing Address. T. See Mailing Address

Veitch, J A, Rev Dr

Rel: Presbyterian.

International Organizations: Joint Secretary for Council for Mission and Ecumenical Cooperation of the Presbyterian Church of New Zealand.

Mailing address: PO Box 10-000, The Terrace, Wellington, New Zealand.

Vekemans, Roger, S.J., Fr

Rel: Roman Catholic; Ordained in 1949; b. 12 DEC 1921 in Brussels, Belgium; celibate; Citizen of Belgium; Lang. French, Dutch, Spanish, English, German, Italian, Latin and Portuguese.

International Organizations: President (since 1985) of the Aid to the Church in Need.

Member of the Society of Jesus.

Other organizations: Executive Director (since 1970), CEDIAL; President (since 1986), IIHJ.

Main field of work: Research. **Educ.:** Studied Phil., Namur, Belgium (1939-1944); Studied Soc., Louvain, Belgium (1944-1945); Studied Theol., Louvain, Belgium (1946-1951); Studied Soc., Nimègue, Belgium (1952-1955). **Interests:** socio-theology. **Pubs.:** Doctrina, ideologia, política; La pre-revolución latino-americana and Teologia de la liberación y cristianos por el socialismo.

Mailing address: CEDIAL, Providencia 2250, 3 Piso, Santiago, Chile. T. (56 2) 231 6735

Residence: Las Catalpas 1367, Vitacura, Santiago, Chile. T. (56 2) 208 3575

Venanzio, Ramasso, Fr

Rel: Roman Catholic.

International Organizations: Procurator-General of the Order of the Servants of Mary.

Mailing address: Piazza S Marcello al Corso 5, I-00187 Roma, Italy. T. (39 6) 678 08 88 – 678 99 95.

Vence, Ilda, Rev

Rel: Evangelical Methodist.

International Organizations: President of the Evangelical Methodist Church in Uruguay.

Mailing address: San José 1457, Montevideo, Uruguay.

Vencer, Agustin B, Jr, Rev Dr Atty

Rel: Evangelical; Ordained in 1985; b. 11 FEB 1946 in Barotac Viejo, Iloilo, Philippines; m. to Annabella Castillo; c. 3 boys and 1 girl; Citizen of Philippines; Lang. English.

International Organizations: International Director (since 1992) of the World Evangelical Fellowship.

Member, Executive Council (since 1979) of the World Evangelical Fellowship. Country Director (since 1980) of the World Relief International. Member, Executive Committee (since 1983) of the Evangelical Fellowship of Asia.

Other organizations: Secretary-General/Chief Executive Officer (since 1978), Philippine Council of Evangelical Churches; Secretary-General (since 1980), Philippine Relief and Development Services (PHILRADS); Chairman (since 1983), Discipling a Whole Nation Philippines DAWN AD2000; Pulpit Minister (since 1984), Alliance Fellowship Church; Chairman, Coordinating Committee (since 1989), Philippines Congress for World Evangelization (PCWE); Executive Vice-President, Dept. of Foreign Mission (since 1990), Christian and Missionary Alliance of the Philippines (CAMACOP); President (since 1991), Philippines Alliance College of Theology; Vice-Chairman (since 1992), National Ecumenical Consultative Committee (NECCOM).

Main field of work: Management, training, fund raising. **Career:** Chairman of Production, Philippine Bible Society from 1974 to 1975. Business Legal Officer, Christian and Missionary Alliance of the Philippines (CAMACOP), Zamboanga from 1974 to 1977. Chairman, Coordinating Committee, Billy Graham Crusade from 1977 to 1978. Founding Chairman, Interchurch Relief and Development Committee from 1979 to 1980. Associate Director for Church Relations, World Vision Philippines from 1979 to 1980. Country Director, Compassion Philippines from 1981 to 1988. Founding Chairman, Evangelical Fellowship of Asia from 1982 to 1983. Executive Chairman, Asia Missions Congress 1990 from 1989 to 1990. Chairman, Coordinating Committee, Say Yes for Christ Movement from 1990 to 1991. **Educ.:** B.S. Preparatory Med. from Central Philippine University, Iloilo, Philippines (1965-1969); B.S. Gen. Sc., A.B. Lit. from Southern Island Colleges, Iloilo, Philippines (1969-1970); B.L. from University of the East, Manila, Philippines (1970-1972). **Interests:** counselling, management, theological and development issues. **Hons.:** Hon. D.Litt., Toccoa Falls College, GA (1990); D.Div., Ebenezer Bible College and Seminary (1992); Dedicated Christian Service, Alliance Men International. **Mem.:** Inter-Church Relief and Development Agency, Evangelical Fellowship of Asia, Quezon City Sports Club, PAL Mabuhay Club, NECCOM-National Ecumenical and Consultative Committee **Pubs.:** Time and the Christian Worker, PCEC, 1990, Revised Edn.; A Biblical Framework for an Evangelical Response to the Current Socio-Political Unrest in the Philippines and The Church and Relief.

Mailing address: 62 Molave St Project 3, Quezon City 1102, Philippines. T. (63 2) 922 9821/2/3

Residence: 34 Ferdinand St, Kingsville Subd, Antipolo, Philippines.

Vendley, William, Dr

Rel: Unknown.

International Organizations: Secretary-General ad interim of the World Conference on Religion and Peace.

Mailing address: 777 United Nations Plaza, New York NY 10017, USA. T. (1 212) 687 2163. Fax 983 0566.

Vera Olvera, Pablo V, M.Sp.S., Fr

Rel: Roman Catholic; Ordained; celibate.

International Organizations: Procurator-General of the Missionaries of the Holy Ghost.

Other organizations: Procurator-General, Misioneros de la Natividad de Maria.

Mailing address: Piazza S Salvatore in Campo 57, I-00186 Roma, Italy. T. (39 6) 683 37 73.

Verdieu, Ernst Jacques Nicolas, Dr

Rel: Roman Catholic; b. 09 DEC 1929 in Port-au-Prince, Haiti; m. to Agathe Chevalier; c. 1 boy; Citizen of Haiti; Lang. Creole, French, English, Spanish (read) and Italian (read).

International Organizations: Zone Coordinator, Caribbean of the Caritas Internationalis.

Other organizations: Director General (since 1987), National Caritas of Haiti, Port-au-Prince.

Career: Professor, Collège Saint-Martial, Port-au-Prince from 1955 to 1969. National Chaplain, Young Catholic Students from 1956 to 1969. Director, Institut de Catéchèse, Port-au-Prince from 1966 to 1969. Secretary-General, Diocèse des Gonaïves from 1966 to 1968. Director, Petit Séminaire Diocésain, Port-au-Prince from 1968 to 1969. Assistant-General, Maison Généralice, Holy Spirit Fathers, Rome, Italy from 1971 to 1973. Lecturer, Faculty of Theology, University of Montreal, Canada from 1970 to 1971. Lecturer, Faculty of Theology, University of Montreal, Canada from 1974 to 1979. Director, Research and Documentation Dept., International Study and Cooperation Centre (CECI), Montreal, Canada from 1974 to 1981. Chairman of the Board, Solidarity-Haiti from 1979 to 1982. General Coordinator, Interregional Council for Haitian Refugees from 1982 to 1986. President, Quebec Chapter, Society for International Development from 1981 to 1986. Director, Development Dept., Catholic Relief Services from 1986 to 1987. **Educ.:** B.A. Theol. from University of Strasbourg II, France (1961-1962); Lic.soc. from University of Strasbourg II, France (1961-1963); D.Sc.Rel. from University of Strasbourg II, France (1970). **Pubs.:** Co-author, *Les Prêtres Noirs s'interrogent*, 1954; *Les Voies du Tiers-Monde*, 1981; *Essai sur la Condition Haïtienne*, CECI, 1981; Co-author, *Haïti, briser les chaînes*, 1984 and *La Caritas en Haïti, Quel Présent, Quel Avenir*, 1988.

Mailing address: BP 13-197 Delmas, Caritas Nationale d'Haïti, Port-au-Prince, Haiti. T. (509 1) 50 128 – (509 1) 55 347. Tx 0001 PP BOOTH HI (POUR CARITAS).
Residence: 7 rue Biassou, Delmas 31, Port-au-Prince, Haiti T. (509 1) 65 318

Verdot, Charles, Fr
Rel: Roman Catholic; Ordained; celibate.
International Organizations: Superior General of the Institut de l'Oeuvre de la Jeunesse Jean-Joseph Allemand.
Mailing address: 41 rue St Savournin, F-13392 Marseille CEDEX 5, France. T. (33) 91 42 49 30

Verebalaire, Sern
Rel: Christian.
International Organizations: Regional Secretary for Pacific of the World Association for Christian Communication.
Mailing address: LPP, PO Box 208, Suva, Fiji.

Vereeck, Jan
Rel: Catholic.
International Organizations: Treasurer of the European Organization of the International Group of Catholic Parents' Associations.
Mailing address: Rozestraat 9, B-2860 Sint-Katelijne-Waver, Belgium.

Verfaillie, Maurice, Pastor
Rel: Seventh-Day Adventist; Ordained on 10 APR 1965; b. 05 JAN 1934 in Kinshasa, Zaire; m. to Irène Nave Leal; c. 2 girls; Citizen of Belgium; Lang. French, English and Portuguese.
International Organizations: Vice-President (since 1987) of the International Association for the Defence of Religious Liberty.
Other organizations: Director (since 1983), Department of Religious Liberty; Secretary-General (since 1987), French-Belgian Union of Adventist Churches.
Main field of work: Study and promotion of the freedom of conscience. **Career:** Professor of History, Adventist College of Tananarive, Madagascar from 1958 to 1960. President, Adventist Mission, Reunion from 1965 to 1968. President, North-African Union of Adventist Churches from 1968 to 1971. Pastor, Parish of Marseille, Federation of Adventist Churches of the South of France from 1971 to 1979. Pastor, Parishes in Brussels, Federation of Belgian-Luxemburg Adventist Churches from 1979 to 1987. **Educ.:** Lic. Theol. from Faculté Adventiste, Collonges, France (1954-1948); Lic. Hist. from University of Aix-Marseille III, Aix-en-Provence, France (1975-1978); Lic. Hist. du Christ. from Free University of Brussels (ULB), Belgium (1987); Doc. en Hist. du Christ. (not yet completed) from Free University of Brussels (ULB), Belgium (1987). **Interests:** research on the history of religious freedom. **Pubs.:** "Les Galériens pour la foi", in *Conscience et liberté*, n° 13, 1977 and "Situation dans les relations entre l'école et la religion dans quelques pays de l'Europe occidentale", in *Conscience et liberté* (forthcoming), n° 43, 1992.
Mailing address: 684 avenue de la Libération, F-77350 Le Mée-sur-Seine, France. T. (33 1) 60 68 62 32
Residence: 762 rue de la Noue, F-77350 Le Mée-sur-Seine, France. T. (33 1) 64 09 66 50

Verlinde, A
Rel: Roman Catholic.
International Organizations: Secretary-General of the International Catholic Committee of Nurses and Medico-social Assistants.
Mailing address: CICIAMS, Square Vergote 43, B-1040 Bruxelles, Belgium. T. (32 2) 732 10 50. Fax (32 2) 734 84 60.

Verschuren, Willem J G, Br
Rel: Roman Catholic; Ordained; celibate.
International Organizations: Superior General of the Brothers of Our Lady, Mother of Mercy.
Mailing address: Gasthuisring 54, 5041 DT Tilburg, Netherlands. T. (31 13) 432 777. C. GENERALAAT TILBURG.

Vessels, John
Rel: Roman Catholic.
International Organizations: Contact person of the Apostleship of Prayer.
Contact person of the Eucharistic Youth Movement.
Mailing address: Borgo Santo Spirito 5, Casella Postale 6139, I-00193 Roma, Italy. T. (39 6) 656 98 41.

Vethathiri, Yogiraj, Maharishi
Rel: World Order of Holistic Unity; b. 14 AUG 1911 in Guduvancheri, Tamil Nadu, India; m. to Logambal; Citizen of India; Lang. Tamil and English.
International Organizations: Founder/President (since 1958) of the World Community Service Centre.
Other organizations: Received (during a time of intense solitary meditation) knowledge of the origin and functions of the cosmos and all life, as Nature revealed to his inner vision (since 1984), Vethathiri Maharishi Kundalini Yoga and Kaya Kalpa Research Foundation, Pollachi (Tamil Nadu); Wrote over 2000 Tamil poems to celebrate, commemorate and convey his vision (since 1986), International Educational Research Foundation for Vethathiri Law of Unified Force, Munster IN; Dedicated his life to serve his fellow-beings by conveying his understanding of God and universe through his speech and writings (since 1986), Ayurveda and Siddha Medical Education and Research Foundation, Madras; Founder/Trustee (since 1989), World Order of Holistic Unity, California.
Main field of work: Writing and lecturing. **Educ.:** Ayurvedic Dr from University of Delhi, Ayurvedic and Unani Tabbia College, India (1936-1939); Dip. Homeopathy from Hahnemann Institute, Thiruvarur, Tamil Nadu, India (1960). **Interests:** spiritual service to humanity. **Pubs.:** All publications by Vethathiri Publications, Erode on behalf of World Community Service Centre, Madras: Ulaga Samadhanam (in Tamil) (World Peace), 1957, Reprinted with English Commentary, 1990; The World Order of Holistic Unity (in English), 1990 and Philosophy of Magnetism (in Tamil and English), 1990.
Mailing address: World Community Service Centre, 20 Second Seaward St, Thiruvanmiyur, Madras 600 041, India. T. (91 44) 411692
Residence: See Mailing Address. T. See Mailing Address

Vezignol, Michel
Rel: Protestant.
International Organizations: Treasurer of the Medico-social Association of French-Speaking Protestants.
Mailing address: AMSP, Ecole des Diaconesses, Ecole d'Infirmières, 95 rue de Reuilly, F-75012 Paris, France. T. (33 1) 43 41 41 41.

Vialard, Antonio Vazquez
Rel: Roman Catholic.
International Organizations: Zone Coordinator, 'Southern Cone' of the Caritas Internationalis.
Mailing address: Caritas Argentina, Hipolito Yrigoyen 785, 5 piso, 1086 Buenos Aires, Argentina.

Vichta, Kathy
Rel: Buddhist: Mahayana.
International Organizations: Member of the Foundation for the Preservation of the Mahayana Tradition.
Other organizations: Director, Chenrizig Institute.
Mailing address: PO Box 41, Eudlo QLD 4554, Australia. T. (61 74) 459 047. Fax (61 74) 459 720

Vicidomini, Immacolata M, Sr
Rel: Roman Catholic; Ordained; celibate.
International Organizations: Superior General of the Congregation of the Sisters of St John the Baptist.
Mailing address: Circonvallazione Cornelia 65, I-00165 Roma, Italy. T. (39 6) 62 164 45.

Viczian, Janos, Rev
Rel: Baptist.
International Organizations: President of the Baptist Union of Hungary.
Mailing address: Aradi utca 48, H-1062 Budapest VI, Hungary.

Vidal, Ricardo J, H E Card

Rel: Roman Catholic; Ordained on 17 MAR 1956; b. 06 FEB 1931 in Mogpoc, Marinduque, Philippines; celibate; Citizen of Philippines.
International Organizations: Member (since 1985) of the College of Cardinals.
Member of the Synod of Bishops. Member of the Congregation for the Evangelization of Peoples. Member of the Pontifical Council for Pastoral Assistance to Health Care Workers.
Other organizations: Archbishop (since 1982), Cebu.
Career: Consecrated Bishop (Titular Church of Claterna) in 1971. Archbishop, Lipa from 1973 to 1982. Created Cardinal by Pope John Paul II in 1985. **Interests:** walking.
Mailing address: D Jakosalem Street, PO Box 52, Cebu City 6000, Philippines. T. (63 32) 621 20 – 72 124

Viganò, Egidio D, S.D.B., Rev

Rel: Roman Catholic; Ordained on 31 MAY 1947; b. 26 JUL 1920 in Sondrio, Italy; celibate; Citizen of Italy; Lang. Italian, Spanish, French and Latin.
International Organizations: Superior General (since 1977) of the Society of St Francis of Sales.
Consultor (since 1982) of the Pontifical Council for the Family. Consultor (since 1985) of the Pontifical Council for the Laity. Member (since 1989) of the Congregation for the Evangelization of Peoples.
Other organizations: Chancellor (since 1977), Salesian Pontifical University.
Main field of work: Administration. **Career:** Lecturer, Faculty of Theology, Catholic University of Chile in 1971. Expert at Council Vatican II from 1962 to 1965. Provincial, Salesians in Chile from 1968 to 1971. Member, General Council, Salesians of Don Bosco from 1971 to 1977. President, Union of Superiors General, Rome from 1983 to 1985. **Educ.:** D.D. from Catholic University of Chile, Santiago (1948). **Interests:** ecclesiology and theology of the religious life. **Hons.:** Dr. h.c. Ed.Sc., University of Parma (1988); Honorary citizenship of several cities in Italy, Latin America and Israel. **Pubs.:** Mistero e Storia, SEI, Turin, 1986; Interioridad Apostólica, Fortín Mercedes (Argentina), 1988 and Don Bosco Ritorna, Paoline, Milan, 1992.
Mailing address: Casa Generalizia Salesiana, POB 9022, I-00163 Roma, Italy. T. (39 6) 65 92 915. Fax 65 92 929
Residence: Via della Pisana 1111, CP 9092, I-00163 Roma, Italy.

Vignolo, Roque

Rel: Christian.
International Organizations: Secretary-General of the Caribbean Alliance of YMCAs.
Mailing address: Apartado 705, Santo Domingo, Dominican Rep.

Vikström, John Edvin

Rel: Evangelical Lutheran; Ordained in 1956; b. 01 OCT 1931 in Kronoby, Finland; m. to Birgitta Hellberg; c. 2 boys and 1 girl; Citizen of Finland.
International Organizations: Archbishop (since 1982) of Evangelical-Lutheran Church of Finland.
Member, Central Committee (since 1983) of the World Council of Churches.
Career: Pastor for Porvoo (Borgå) Diocese from 1957 to 1961. Assistant in Theological Faculty, Åbo Akademi, Turku from 1963 to 1965. Instructor, Åbo Akademi from 1966 to 1970. Associate Professor, Åbo Akademi in 1970. Bishop, Porvoo (Borgå) Diocese from 1970 to 1982. **Interests:** sports. **Hons.:** Dr. h.c., Theological Academy of Leningrad (1985), Theological Academy of Budapest (1987); Great Award, Swedish Culture Foundation (1981); Groszes Bundesverdienstkreuz, Federal Republic of Germany (1989); Great Cross, Order of the White Rose, Finland (1987). **Pubs.:** Ihmisen usko, 1982; Kuitenkin, 1983 and Uusi rohkeus elää, 1985.
Mailing address: Agricolankatu 2, SF-20500 Turku 50, Finland.

Villegas, George

Rel: Roman Catholic.
International Organizations: Head, Asia Regional Secretariat of the Pax Romana, International Catholic Movement for Intellectual and Cultural Affairs.
Mailing address: c/o ACES, No 12 -11th Avenue, Murphy Cubao, Manila, Philippines. T. (63 2) 79 70 53.

Villena Villegas, Rolando V, Rt Rev

Rel: Evangelical Methodist.
International Organizations: Contact of the Evangelical Methodist Church in Bolivia.
Mailing address: Casilla 356, La Paz, Bolivia.

Vimercati, Giuseppe, C.F.I.C., Fr

Rel: Roman Catholic; Ordained; celibate.
International Organizations: Procurator-General of the Sons of the Immaculate Conception.
Mailing address: Via della Luce 46, I-00153 Roma, Italy. T. (39 6) 58 23 31

Vincent, John James, Rev Dr

Rel: Methodist; Ordained on 04 JUL 1956; b. 29 DEC 1929 in Sunderland, UK; m. to Grace Johnston Stafford; c. 2 boys and 1 girl; Citizen of United Kingdom; Lang. English, German, French, Hellenistic Greek, Latin and Hebrew.
International Organizations: Minister of the Methodist Church.
Other organizations: Founder, leader (since 1967), Ashram Community; Director (since 1970), Urban Theology Unit, Sheffield; Superintendent Minister (since 1970), Sheffield Inner City Ecumenical Mission; Coordinator (since 1990), British Liberation Theology Project; Honorary Lecturer, Dept. of Biblical Studies (since 1990), University of Sheffield.
Main field of work: Urban mission and theology, politics. **Career:** Sergeant with the RAMC (Royal Army Medical Corps) from 1947 to 1949. Minister, Manchester and Salford Mission from 1956 to 1962. Superintendent Minister, Rochdale Mission from 1962 to 1969. Adjunct Professor, New York Theological Seminary from 1979 to 1988. President, Methodist Church of Great Britain from 1989 to 1990. Secretary, Regional Working Party, World Council of Churches Faith and Order Dept from 1958 to 1963. Member, British Council, Churches Commission on Defence and Disarmament from 1963 to 1965. Member, British Council, Churches Commission on Defence and Disarmament from 1969 to 1972. Member, Theological Commission, Christian Peace Conference, Prague from 1965 to 1969. Chairman, North West Campaign for Nuclear Disarmament from 1957 to 1965. Founding Member, Editor, Methodist Renewal Group from 1961 to 1970. Chairman, Alliance of Radical Methodists from 1970 to 1976. Chairman, Urban Mission Training Association of Great Britain from 1976 to 1977. Chairman, Urban Mission Training Association of Great Britain from 1984 to 1990. Council, Christian Organisations for Social, Political and Economic Change from 1981 to 1989. Executive, Association of Centres of Adult Theological Education from 1984 to 1990. **Educ.:** B.D. (1st) from University of London, Richmond College, UK (1950-1954); S.T.M. (Summa cum laude) from Drew University, Madison NJ, USA (1954-1955); University of Basel, Switzerland (1955-1956); Dr. Theol. (Insigni cum laude) from University of Basel, Switzerland (1960). **Interests:** writing, jogging. **Pubs.:** *Secular Christ: A Contemporary Interpretation of Jesus*, Lutterworth Press, 1968 (also Abingdon Press); *Radical Jesus*, Marshall Pickering, 1986 and *Britain in the 90's*, Methodist Publishing House, 1989.
Mailing address: 210 Abbeyfield Road, Sheffield S4 7AZ, UK. T. (44 742) 435342
Residence: 178 Abbeyfield Road, Sheffield S4 7AY, UK. T. (44 742) 436688

Viquar, Syed Nasiruddin

Rel: Muslim; b. 05 JUL 1948 in Hyderabad, Andhra Pradesh, India; m. to Yusuf; c. 3 boys and 1 girl; Citizen of India; Lang. English and Urdu.
International Organizations: Librarian (since 1984) of the King Faisal Centre for Research and Islamic Studies.
Other organizations: President (since 1982), Imamul Uloom Educational Society, India; Editor (since 1983), *Treasure Hunt*, Hyderabad; Honorary Director (since 1987), New Era Mission School, Hyderabad.
Main field of work: Islamic education. **Educ.:** B.A. from Osmania University, Hyderabad, India (1965-1968); M.A. Islamic Stds. from Osmania University, Hyderabad, India (1969-1971); B.L.S. from Osmania University, Hyderabad, India (1971-1972). **Pubs.:** 5000 Years Gregorian Calendar, 1982 and Glossary of Islamic Terms (in press).
Mailing address: PO Box 1460, Saidabad, Hyderabad AP 500659, India. T. (91) 527081
Residence: 16-2-842/3 Saidabad, Hyderabad AP 500659, India. T. (91) 527381

Visvanatha Kurukkal, S V, Brahmasri
Rel: Hindu.
International Organizations: Chief Priest of the Old Kathiresan Kovil.
Mailing address: Old Kathiresan Kovil, Bambalapitiya, Colombo 4, Sri Lanka.

Visvanatha Kurukkal, Sami V, Brahmasri
Rel: Hindu.
International Organizations: Chief Priest of the Sri Vishnu Kovil.
Mailing address: Sri Vishnu Kovil, Nedimalai, Dehiwela, Sri Lanka.

Vittachi, Varindra
Rel: Subud.
International Organizations: Chairperson of the World Subud Association.
Mailing address: IO Waterside Plaza, Apt 22J, New York NY 10010, USA.

Vives Aguilella, Juan Antonio V, T.C., Rev Dr
Rel: Roman Catholic; Ordained on 07 DEC 1973; b. 15 SEP 1948 in Onda, Castellón, Spain; celibate; Citizen of Spain; Lang. Castellan, Italian, French, Latin, Greek and Hebrew.
International Organizations: Procurator-General (since 1983) of the Third Order Capuchin of Our Lady of Sorrows.
Main field of work: Ministry. **Educ.:** Lic. Theol. from Universidad Pontificia de Comillas, Madrid, Spain (1971-1976); Dr.Theol. from Pontificio Ateneo Antonianum, Vatican City State (1983-1985). **Pubs.:** *Un hombre que se fió de Dios – Luis Amigón su vida y obra*, Rome 1984; *Testigos del Amor de Cristo* (doctoral thesis), Rome, 1986 and *En la Casa del Padre – Necrologio Amigoniano*, Madrid, 1988.
Mailing address: Via Bernardo Blumenstihl 28-36, I-00135 Roma, Italy. T. (39 6) 305 59 31.

Vizzari, Domenico, P.O.C.R., Fr
Rel: Roman Catholic; Ordained; celibate.
International Organizations: Procurator-General of the Pieux Ouvriers Catéchistes Ruraux.
Mailing address: Via della Lungara 45, I-00165 Roma, Italy. T. (39 6) 65 43 169.

Vlug, René J
Rel: Christian.
International Organizations: Director of The European Missionary Association.
Mailing address: PO Box 420, NL-3770 AK Barreveld, Netherlands.

Vogel, André
Rel: Protestant.
International Organizations: Executive Secretary of the Solidarité protestante.
Mailing address: Rue du Champ de Mars 5, B-1050 Bruxelles, Belgium. T. (32 2) 511 44 71.

Vonck, Christiaan Jan Germana, Rev Dr
Rel: Christian; b. 21 MAR 1940 in Antwerp, Belgium; married; c. 1 boy; Citizen of Belgium; Lang. Dutch and English.
International Organizations: Rector (since 1980) of the Faculty for the Comparative Study of Religions, Antwerp.
Mailing address: c/o Faculty for the Comparative Study of Religions, Bist 164, B-2610 Antwerpen, Belgium. T. (32 3) 830 51 58
Residence: Tolbareel 60, B-2930 Brasschaat, Belgium. T. (32 3) 651 57 44

Von Holzen, Walter, Fr
Rel: Roman Catholic.
International Organizations: Director of the Documentation and Research Centre.
Mailing address: Via dei Verbiti 1, I-00154 Roma, Italy. T. (39 6) 574 13 50. Fax 574 13 50.

von Keler, Hans
Rel: Evangelical Lutheran; Ordained; b. 12 NOV 1925 in Bielitz, Germany; m. to Brigitte Holland; c. 1 boy and 2 girls; Citizen of Germany.
International Organizations: Ecclesiastic of the Evangelical Lutheran Church.
Career: Assistant Minister, Haigstkirchen Stuttgart from 1951 to 1953. Parish Minister, Wildenstein/Hohenlohe from 1953 to 1957. Director, YWCA work, Württemberg from 1957 to 1964. Parish Minister, Neuenstein/Hohenlohe from 1964 to 1969. Executive Director, Herrenberger Diakonieschwesternschaft from 1969 to 1976. General Superintendent, Ulm District, Evangelical-Lutheran Church from 1976 to 1979. Presiding Bishop, Evangelical Lutheran Church of Württemberg from 1979 to 1988. **Interests:** studies in history. **Hons.:** Dr. Theol. h.c. **Pubs.:** more than 50 articles in the field of youth work, pastoral counselling, etc.
Mailing address: Heinrich Schütz Str 13, 71083 Herrenberg, Germany.

von Oeyen, Paul F
Rel: Christian.
International Organizations: Executive Director of the International Christian Education Association.
Mailing address: PO Box 69006, Pleasant Ridge MI 48069-0006, USA. T. (1 313) 399 6500.

Vos, Frans
Rel: Christian.
International Organizations: Secretary of the International Association of Christian Education.
Mailing address: Acacialaan 15, NL-8181 GG Heerde, Netherlands. T. (33 1) 31 82 37 53.

Waddaulab, Hassanal Bolkiah Muizzaddin, Paduka Seri Baginda Sultan
Rel: Muslim; b. 15 JUL 1946 in Istana Darussalam, Bandar Seri Begawan, Brunei Darussalam.
International Organizations: Sultan (since 1967) of the Brunei.
Other organizations: Prime Minister, Minister of Finance and Minister of Home Affairs (since 1984), Brunei.
Career: Qualified for admission as an officer cadet at Sandhurst Military Academy, UK in 1966. Commissioned a captain in 1967.
Mailing address: No mailing address.

Wadud-Muhsin, Amina, Dr
Rel: Muslim; b. 25 SEP 1952 in Bethesda MD, USA; m. to Nadir Mushin; c. 2 boys and 3 girls; Citizen of USA; Lang. English, Arabic, Malay and Spanish.
International Organizations: Lecturer, Department of Revealed Knowledge and Heritage (since 1989) of the International Islamic University, Petaling Jaya.
Main field of work: Research and university education. **Educ.:** B.Sc. from University of Pennsylvania, Philadelphia, USA (1970-1975); M.A. from University of Michigan, Ann Arbor, USA (1980-1982); Ph.D. from University of Michigan, Ann Arbor, USA (1983-1988). **Interests:** women and religion. **Mem.:** Sisters in Islam **Pubs.:** "Women as Minority", in *Journal of Muslim Minority Affairs*, January 1989; *Are Women and Men Equal Before Allah*, co-author with and published by Sisters in Islam and *Qur'an and Woman*, Fajar Bakh Publ., Malaysia, April 1992.
Mailing address: 66 Seaton Place NW, Washington DC 20001, USA. T. (1 202) 265 1383
Residence: 81 Jalan Athinahapan Satu, Taman Tun Dr Ismail, Kuala Lumpur 60000, Malaysia. T. (60 3) 718 5292

Wahlström, Jarl Holger
Rel: Salvation Army, The; b. 09 JUL 1918 in Helsinki, Finland; m. to Maire Helfrid Nyberg; c. 2 boys and 1 girl; Citizen of Finland.
International Organizations: Officer (retired) of The Salvation Army.
Career: Salvation Army Field Officer, Finland from 1939 to 1945. Youth Officer from 1945 to 1952. Youth Officer from 1954 to 1960. Private Secretary to Territorial Commander from 1952 to 1954. Division Commander from 1960 to 1963. Training College Principal from 1963 to 1968. Chief Secretary, Finland from 1968 to 1972. Chief Secretary, Canada and Bermuda from 1972 to 1976. Territorial Commander, Finland from 1976 to 1980. Territorial Commander, Sweden in 1981. International Leader, The Salvation Army from 1981 to 1986. **Interests:**

music. **Hons.:** Hon. D.Hum.Litt., Western Illinois University; Knight, Order of Lion of Finland; Finnish Liberty Cross (Fourth Class); Order of Civil Merit; Mugunghwa Medal, Republic of Korea; Commander, Order of Whiter Rose of Finland (1989). **Pubs.:** Autobiography: Matkalaulu (in Finnish) and En Vallfartssång (in Swedish), 1989.
Mailing address: Borgströminkuja 1A10, SF-00840 Helsinki 84, Finland. T. (358 0) 698 2413

Wahlström, Per Åke, Rev
Rel: Baptist.
International Organizations: Chairman of the Nordic Committee for Central Africa.
Mailing address: Baptist Union, Ekensbergsvägen 128 plan 6, S-172 30 Sundbyberg, Sweden. T. (46) 764 64 80. C. Baptist Stockholm. Tx 12442 Baptist Stockholm.

Waite, Terrence Hardy
Rel: Anglican; b. 31 MAY 1939 in Bollington, UK; m. to Helen Frances Watters; c. 1 boy and 3 girls; Citizen of United Kingdom.
International Organizations: Adviser to Archbishop (since 1980) of Canterbury on Anglican Communion Affairs.
Career: Lay Training Adviser to Bishop and Diocese, Bristol from 1964 to 1968. Adviser to Archbishop, Uganda, Rwanda and Burundi from 1968 to 1971. International consultant working with Roman Catholic Church from 1972 to 1979. Libyan hostages mission in 1985. Disappeared in Beirut in 1987. Released in 1991. Member, Church of England National Association from 1966 to 1968. Coordinator, Southern Sudan Relief Project from 1969 to 1971. Council, Worldlife Fund. Council, Uganda Society for Disabled Children. Chairman, YMCA International Relief and Development Committee. Patron, Strode Park Foundation for the Disabled, Hearne, Kent. **Interests:** music, walking, travel, Jungian studies, international affairs and politics. **Hons.:** UK 'Man of the Year' (Radar Award) (1985); Hon DCL, Kent (1986), City of London; Hon. D.L.L., Liverpool; Member, Order of the British Empire. **Mem.:** Royal Institute of International Affairs (1980); Butler Trust (Prison Officers Award Programme)
Mailing address: c/o Lambeth Palace, London SE1 7JU, UK.

Wakefield, Wesley Halpenny
Rel: Bible Holiness Movement; b. 22 AUG 1929 in Vancouver, British Columbia, Canada; m. to Mildred June Shouldice.
International Organizations: International Leader (Bishop-General) (since 1949) of Bible Holiness Movement.
Other organizations: Editor (since 1949), *Truth on Fire* (now *Hallelujah* magazine); Editor (since 1980), *Wesleyan Annotated Edition Bible*; Editor (since 1988), *Miniature Railways* quarterly.
Career: Director, *Bible Broadcast* from 1952 to 1956. Editor, *Christian Social Vanguard* from 1960 to 1961. Researcher on effects of marijuana on youth in 1969. Editor, *Canadian Church and State* from 1977 to 1990. Vice-President, Canadians United for Separation of Church and State from 1977 to 1990. Chairman, Religious Freedom Conference from 1978 to 1979. Editor, *Hallelujah* (hymnal) from 1981 to 1983. Advisory Committee, Wesley Bible from 1988 to 1990. Board Member, Canadian Council of Japan Evangelistic Band. National Committee Member, Christian Holiness Association. West Canada representative, Canadians for the Protection of Religious Liberty. Chairman, Religious Information Centre. President and Founder, Double Dial Lock Co., Manual Offset Printing Press Co. President, Imperial Security Guard Service Ltd. Board Member, Agarwal Resources Ltd. President, Cumo Resources Ltd. Secretary, NP Energy Corp. Manager, Liberty Press, Evangelistic Book Service, Aldersgate Advertising Agency. **Interests:** model railways. **Hons.:** Honoured twice for activity in religious liberty (Religious Freedom Crusade); Honoured twice with International Community Service Award (UK). **Mem.:** Anti-Slavery Society; Canadian Bible Society; Creation Science Association; Evangelicals for Social Action; National Association for the Advancement of Coloured People; International Platform Association **Pubs.:** Bringing Back the Ark, 1987; John Wesley – The Burning Heart, 1988 and Antinominism, the Curse of the Ages, 1990.
Mailing address: PO Box 223, Postal Stn A, Vancouver BC V6C 2M3, Canada.

Waldenfels, Hans, Prof Dr
Rel: Christian.
International Organizations: Representative of the Institute for Scientific Research on Missions.
Mailing address: Sem. für Fundamentaltheologie, Universität Bonn, 53113 Bonn, Germany.

Waldstein, Ernst, Dr
Rel: Roman Catholic.
International Organizations: Contact of the European Forum of National Laity Committees.
Mailing address: Waaggassee 18, A-9013 Klagenfurt, Austria. T. (43 463) 5877.

Walker, Alan, Rev Dr Sir
Rel: Uniting Church in Australia; Ordained in 1935; b. 04 JUN 1911 in Sydney NSW, Australia; m. to Winifred Garrard Channon; c. 3 boys and 1 girl; Citizen of Australia; Lang. English.
International Organizations: Director of World Evangelism (since 1978) of the World Methodist Council.
Principal (since 1989) of the Pacific College for Evangelism.
Career: Associate Director, Methodist Young People's Dept., New South Wales Conference from 1936 to 1938. Minister, Cessnock, New South Wales from 1939 to 1944. Superintendent, Waverley Methodist Mission, Sydney from 1944 to 1954. Director, Mission to the Nation for the Methodist Church of Australasia from 1953 to 1956. Visiting Professor of Evangelism, Boston School of Theology, MA, USA from 1956 to 1957. Superintendent, Central Methodist Mission, Sydney from 1958 to 1978. Delegate to First Assembly, World Council of Churches (WCC), Amsterdam in 1948. Appointed Adviser to Australian Delegation at UN by the Australian Government in 1949. Adviser to Third Assembly, WCC, New Delhi in 1962. Founded Sydney Life Line Telephone Counselling Centre in 1963. Appointed President, Life Line International in 1966. Adviser to Fourth Assembly, WCC, Uppsala in 1968. President, NSW Methodist Conference in 1971. Has undertaken many lecturing and preaching engagements in USA and elsewhere. **Educ.:** B.A. from University of Sydney, New South Wales, Australia (1936); M.A. from University of Sydney, New South Wales, Australia (1943). **Interests:** swimming, gardening. **Hons.:** Order of the British Empire (1955); Hon. D.D., Bethany Biblical Seminary, Chicago IL (1956); Received (jointly with Lady Winifred Walker) World Methodist Peace Award (1986). **Pubs.:** Author of 30 books including: *Caring for the World*, William Collins, London, 1979; *Life in the Holy Spirit*, Discipleship Resources, Nashville, 1986 and *Try God*, 1990.
Mailing address: 14 Owen Stanley Ave, Beacon Hill NSW 2100, Australia. T. (61 2) 451 3923
Residence: See Mailing Address. T. See Mailing Address

Walker, Alan, Rev
Rel: Methodist.
International Organizations: Principal of the Pacific College for Evangelism.
Mailing address: 6 Lincluden Place, North Parramatta 2151, Australia. T. (61 2) 633 0777. Fax 630 0678.

Walker, Eric
Rel: Quaker.
International Organizations: Secretary of the Quaker Esperanto Society.
Mailing address: Downham Reach Cottage, Nacton, Ipswich IP10 0JZ, UK.

Wall, James M
Rel: United Methodist; Ordained; b. 27 OCT 1928 in Monroe GA, USA; m. to Mary Eleanor Kidder; c. 3 boys.
International Organizations: Editor (since 1972) of the *Christian Century* Magazine.
Other organizations: President, Christian Century Foundation; Adjunct Professor of Religion and Culture, Claremont School of Theology, CA; Board of Trustees, Elmhurst Public Library, IL; Board of Directors, Chicago Center for Peace Studies; Advisory Board, MidAmerica Leadership Foundation, Chicago; North American representative and one of two international Vice-Presidents, Inter Film; Board of Directors, Greater Chicago Broadcast Ministries; Member, Communications Commission of the National Council of Churches; Contributing Editor for two weekly television programs in Chicago.
Career: Sports writer with two Atlanta newspapers, the *Journal* and the *Constitution* from 1946 to 1950. Lieutenant in the US Air Force from 1950 to 1952. Associate Pastor, East Lake Methodist Church, Atlanta GA from 1952 to 1953. Pastor, Moreland and Luthersville Methodist churches in the North Georgia Annual Conference from 1955 to 1957. Pastor, Bethel (Mt Greenwood) United Methodist Church, Chicago IL from 1957 to 1959. Democratic Party nominee for Congress in Illinois'

14th District in 1972. Elected as delegate to Democratic National Conventions (in 1972, 1976 and 1980). Elected as delegate to Democratic National Conferences (in 1974 and 1978). Appointed Jimmy Carter's representative to the Illinois delegations (in 1976 and 1980). Chairman, Jimmy Carter presidential campaign in Illinois (in 1976 and 1980). Campaign manager, Paul Simon's senatorial primary campaign in 1984. Chairman, Al Gore's presidential campaign in Illinois in 1988. Member from Illinois, the Democratic National Committee from 1976 to 1980. Appointed by President Carter to be a member, the President's Commission on White House Fellowships from 1977 to 1981. Elected a member, the Illinois Democratic State Central Committee from the 14th and 6th congressional districts (in 1974, 1978 and 1982). Member, Program Review Committee for Battelle from 1980 to 1987. Member, Department of Energy's Office of Nuclear Waste Isolation from 1980 to 1987. Chairman, National Council of Churches's Film and Industry Committee (several terms). Managing Editor, *Christian Advocate* from 1960 to 1962. Editor, *Christian Advocate* from 1962 to 1972. Editor for *Together*. **Educ.:** B.A. from Emory University, Alanta GA, USA (1949); B.D. from Emory University, Alanta GA, USA (1955); M.A. from University of Chicago, IL, USA (1961). **Hons.:** honorary degrees from Emory University, MacMurray College, Willamette University, Ohio Northern University and Coe College. **Mem.:** Northern Illinois Annual Conference **Pubs.:** Author and/or Editor of several books, including: *Church and Cinema*, Eerdmans, 1971; *Three European Directors*, Eerdmans, 1972 and *Winning the War, Losing Our Soul*, Christian Century Press, 1991.
Mailing address: 407 South Dearborn, Chicago IL 60605, USA. T. (1 312) 427 5380

Wallace, S L, Rev
Rel: Methodist.
International Organizations: Secretary of Conference of the Methodist Church Sierra Leone.
Mailing address: Wesley House, George Street, PO Box 64, Freetown, Sierra Leone.

Wallin, Ingvar
Rel: Lutheran; b. 11 JUL 1936 in Lund, Sweden; m. to Christiane Wallin; c. 2 boys and 2 girls; Citizen of Sweden; Lang. Swedish, English, French, German and Spanish.
International Organizations: Secretary-General (since 1970) of the International Association of Y's Men's Clubs.
Main field of work: Voluntary NGO. **Educ.:** Fil. Lic. Nuclear Physics from Lunds University, Sweden (1959-1969).
Mailing address: 37 quai Wilson, CH-1201 Genève, Switzerland. T. (41 22) 732 31 00. C. FLEMGO. Tx 412332 YMCA CH. Fax (41 22) 738 40 15.
Residence: 83 rue de Lyon, CH-1203 Genève, Switzerland T. (41 22) 44 14 42

Walls, A F
Rel: Christian.
International Organizations: Editor of the Society for African Church History.
Mailing address: Centre for the Study of Christianity in the Non-Western World, University of Edinburgh, New College, Edinburgh EH1 2LU, UK.

Walsh, Dominic, C.F.A., Br
Rel: Roman Catholic; Ordained; celibate.
International Organizations: Superior General of the Congregation of Cellites.
Mailing address: 100 James Boulevard, Signal Mountain TN 37377, USA. T. (1 615) 886 2969.

Walsh, Patrick Joseph, Rt Rev
Rel: Roman Catholic; Ordained on 25 FEB 1956; b. 09 APR 1931 in Cobh, Co. Cork, Ireland; celibate; Citizen of Ireland; Lang. English, Irish and Italian.
International Organizations: Bishop (since 1991) of Down and Connor.
Main field of work: Ministry. **Career:** Teacher, St MacNissi's College from 1958 to 1964. Chaplain, Queen's University of Belfast from 1964 to 1970. Principal, St Malachy's College, Belfast from 1970 to 1983. Auxiliary Bishop, Diocese of Down and Connor from 1983 to 1991. **Educ.:** B.A. from Queen's University of Belfast, UK (1948-1952); S.T.L. from Lateran Pontifical University, Vatican City State (1952-1956); M.A. from University of Cambridge, UK (1956-1958); M.A. from Queen's University of Belfast, UK (1956-1958). **Interests:** music, Italian civilization and culture (especially Renaissance art).
Mailing address: 73 Somerton Road, Belfast BT15 4DE, UK. T. (44) 776185
Residence: See Mailing Address. T. See Mailing Address

Walsh, Thomas G, Dr
Rel: Holy Spirit Association for the Unification of World Christianity.
International Organizations: Executive Director of the International Religious Foundation.
Executive Director of the Assembly of the World's Religions.
Mailing address: JAF Box 2347, New York NY 10116, USA. T. (1 212) 869 6023. Tx 4991393 NWERA. Fax (1 212) 869 6424.

Walter, Karl-Heinz
Rel: Baptist.
International Organizations: Secretary-Treasurer of the European Baptist Federation.
Mailing address: Albertinenhaus, Süntelstrasse 11, 22457 Hamburg, Germany. T. (49 40) 550 97 23. Fax 550 97 25.

Walter, Paul R
Rel: Episcopal.
International Organizations: Executive Director of the Episcopal World Mission.
Mailing address: PO Box 490, 102 W Main Street, Forest City NC 28043, USA. T. (1 704) 248 1377.

Walters, Clem J
Rel: Christian.
International Organizations: Mission Coordinator of the Christians in Mission.
Mailing address: 107 South Greenlawn, South Bend IN 46617, USA.

Wan, Paul Oh Sae, SSP, Fr
Rel: Roman Catholic; Ordained; celibate.
International Organizations: President of the East Asia Catholic Press Association.
Member of the Society of St Paul.
Mailing address: Society of St Paul, 103-36 Mia 4 dong Dolong-gu, Seoul 132-104, Korea Rep.

Wan-tin, Wong
Rel: Buddhist.
International Organizations: Vice-President of the World Fellowship of Buddhists.
Mailing address: c/o The Pun Tak Land I and A Co Ltd, Pun Tak Building 14th Floor, 482 Lockhart Road, Hong Kong.

Ward, Clive
Rel: Church of Christ.
International Organizations: National President of the Federal Conference of Churches of Christ in Australia.
Mailing address: 77 Capel Street, West Melbourne VIC 3003, Australia.

Ward, Daniel T, Bishop
Rel: Free Methodist; b. 08 APR 1942 in Yavatmal, Maharashtra State, India; married; c. 1 boy and 1 girl; Citizen of India; Lang. Marathi, English and Hindi.
International Organizations: Bishop (since 1989) of Free Methodist Church.
Other organizations: Ordained Elder (since 1971), Free Methodist Church.
Main field of work: Church administration and ministry. **Career:** Ordained Deacon, Free Methodist Church from 1969 to 1971. District Superintendent, Free Methodist Church from 1986 to 1987. Registrar, Yavatmal College for Leadership Training from 1983 to 1989. **Educ.:** B.A. from Amolakchand College, Yavatmal, India (1960-1964); B.D. from Union Biblical Seminary, Yavatmal, India (1964-1967); M.Theol. (Hons) from Fuller Theological Seminary, Pasadena CA, USA (1979-1980). **Interests:** preaching, writing, teaching. **Pubs.:** Christ-Jyot; Ankur and The New Day.
Mailing address: Bishop House, Vanjari Fail Road, Box 27, Yavatmal 445 001, India. T. (91) 2446
Residence: See Mailing Address. T. See Mailing Address

Ward, Donald Butler, Dr

Rel: Congregational; b. 15 JUN 1919 in Boston MA, USA; m. to Vera Barbara Bantz; c. 2 boys and 2 girls.

International Organizations: Co-Chairman of the International Congregational Fellowship.

Other organizations: Minister emeritus (since 1986), First Congregational Church, Los Angeles, California, USA.

Main field of work: Clergyman. **Career:** Television Account Executive, Blair TV Associate, Chicago from 1951 to 1954. Manager, Blair TV Associate, Chicago from 1954 to 1956. Ordained to Ministry in the Congregational Church in 1959. Minister, First Congregational Church of Ravenswood, Chicago from 1958 to 1960. Minister, Kirk of Bonnie Brae, Denver from 1960 to 1962. Minister, First Congregational Church, Evanston, IL from 1970 to 1979. Vice-President, Alaska Pacific University, Anchorage from 1979 to 1980. Senior Minister, First Congregational Church of Los Angeles, CA from 1980 to 1986. Commentator, Viewpoint (daily radio program) from 1978 to 1979. Radio performer-producer from 1942 to 1949. Television performer-producer from 1949 to 1951. President, Yankton College, SD from 1962 to 1970. **Educ.:** Crses. from Yankton College, SD, USA (1937-1939); B.Sc. from Northwestern University, Evanston IL, USA (1942); M.Div. from University of Chicago, IL, USA (1959); D.D. from Lakeland College, Sheboygan WI, USA (1964); LL.D. from Morningside College, Davenport IA, USA (1964); D.D. from Yankton College, SD, USA (1981). **Hons.:** Named Man of the Year, First Congregational Church of Evanston, IL (1956). **Mem.:** Fellow, American Congregational Center; Member: Evanston Historical Society, Aircraft Owners and Pilots Association, National Speakers Association, Academy of Magical Arts, Sigma Chi; Clubs: Los Angeles Pianists **Pubs.:** *Not The Scabbard But The Blade*, 1965; *The Underground Church Is Nonsense*, 1969 and *Master Sermon Series*, 1973.

Mailing address: Park Alhambra 56, 999 East Valley Blvd, Alhambra CA 91801, USA.

Ward, John Aloysius, O.F.M., Most Rev

Rel: Roman Catholic; Ordained on 07 JUN 1953; b. 24 JAN 1929 in Leeds, Yorkshire, UK; celibate; Citizen of United Kingdom; Lang. English, Italian and Latin.

International Organizations: Metropolitan Archbishop (since 1983) of Cardiff.

Member (since 1945) of the Order of Friars Minor.

Main field of work: Focus of unity of the local church community. **Career:** Friar from 1950 to 1953. Travelling Missioner, Diocese of Menevia, Wales from 1954 to 1960. Superior (Guardian) and Parish Priest, Franciscan Friary, Peckham, London from 1960 to 1966. Provincial Definitor (Councillor), British Province of Order from 1963 to 1969. Minister Provincial from 1969 to 1970. General Definitor (Councillor), Order in Rome from 1970 to 1980. Coadjutor Bishop, Diocese of Menevia from 1980 to 1981. Bishop, Diocese of Menevia from 1981 to 1983. **Educ.:** Theological Seminaries of Franciscan Order, Olton, Crawley, UK. **Mem.:** Order of Friars Minor Capuchin

Mailing address: Archbishop's House, 43 Cathedral Road, Cardiff CF1 9HD, Wales, UK. T. (44 222) 220411

Residence: See Mailing Address. T. See Mailing Address

Ward, John Stephen Keith, Rev Canon Prof

Rel: Church of England; Ordained on 29 JUN 1972; b. 22 AUG 1938 in Hexham, Northumberland, UK; m. to Marian; c. 1 boy and 1 girl; Citizen of United Kingdom; Lang. English.

International Organizations: Chairman of the World Congress of Faiths.

Other organizations: Regius Professor of Divinity (since 1991), University of Oxford; Executive Committee, Board for Social Responsibility, Church of England; Council, Royal Institute of Philosophy; Co-Editor, *Religious Studies*; Editorial Board, *Journal of Medical Ethics, Religion Today, Studies in Interreligious Dialogue*.

Career: Served with the Royal Air Force from 1956 to 1958. Lecturer in Logic, University of Glasgow from 1964 to 1966. Lecturer in Moral Philosophy, University of Glasgow from 1966 to 1969. Lecturer in Philosophy, University of St Andrews from 1969 to 1971. Lecturer in the Philosophy of Religion, University of London, King's College from 1971 to 1976. Fellow and Dean, Director of Studies in Theology and Philosophy, Trinity Hall from 1976 to 1983. Lecturer in Divinity, University of Cambridge from 1981 to 1983. Professor of Moral and Social Theology, University of London, King's College from 1983 to 1986. Professor of the History and Philosophy of Religion, University of London, King's College from 1986 to 1991. Cadbury Lecturer, Birmingham in 1986. Teape Lecturer, India in 1989. Visiting Professor of Religion, Claremont Graduate School, CA in 1990. **Educ.:** B.A. Phil. (1st) from University of Wales, University College of Cardiff, UK (1958-1962); B.Litt. from University of Oxford, UK (1962-1964); M.A. from University of Cambridge, UK (1976). **Hons.:** Honorary Fellow: University of Wales, Cardiff College and Trinity Hall, Cambridge. **Mem.:** Vatican Observatory Research Conference on Quantum Creation and the Origin of the Laws of Nature **Pubs.:** *Divine Action*, Collins, 1990; *A Vision to Pursue*, SCM, 1991 and "Pluralism and Truth", in *Religious Studies*, 1991.

Mailing address: Christ Church, Oxford OX1 1DP, UK.

Residence: Brewery Cottage, Lower Froyle, Alton GU34 4LX, UK

Warner, John

Rel: Evangelical.

International Organizations: African Representative of the Society of International Missionaries.

Mailing address: PO Box 50770, Nairobi, Kenya. T. (254 2) 72 58 58.

Warner, John

Rel: Buddhist: Mahayana.

International Organizations: Member of the Foundation for the Preservation of the Mahayana Tradition.

Other organizations: Director, Buddha House.

Mailing address: PO Box 93, Eastwood SA 5063, Australia. T. (61 8) 799 153

Watanabe, John M, Rt Rev

Rel: Anglican.

International Organizations: Primate of the Anglican Church in Japan.

Mailing address: 4-21 Higashi 1 chome, Shibuya-Ku, Tokyo 150, Japan.

Waters, Robert, Rev

Rel: Congregational.

International Organizations: General Secretary of the Congregational Union of Scotland.

Mailing address: PO Box 189, Glasgow G1 2BX, Scotland, UK.

Watkins, Tony

Rel: Congregational.

International Organizations: Director of the Congregational World Assembly of Youth.

Mailing address: 12 Silverdale Court, Silverdale Rd, Banister P, Southampton SO1 2NJ, UK. T. (44 703) 632053.

Watson, Richard

Rel: Protestant.

International Organizations: Contact of the World Vision International.

Mailing address: 919 West Huntington Drive, Monrovia CA 91016, USA. T. (1 818) 303 8811. C. Worvis. Tx 275 335 WORV UR.

Watson, Ruth

Rel: Quaker.

International Organizations: Regional Representative for Asia-West Pacific of the Friends World Committee for Consultation.

Mailing address: RMB 9364 Tania Road, Torbay Hill Via Albany 6330, Australia. T. (61 98) 451066.

Weatherhead, James Leslie, Rev

Rel: Church of Scotland; Ordained in 1960; b. 29 MAR 1931 in Dundee, UK; m. to Anne Elizabeth Shepherd; c. 2 boys; Citizen of United Kingdom.

International Organizations: Principal Clerk (since 1985) of the General Assembly of Church of Scotland.

Career: National Service in Royal Navy from 1955 to 1956. Parish Minister, Trinity Church, Rothesay from 1962 to 1969. Parish Minister, The Old Church, Montrose from 1969 to 1985. **Interests:** music, sailing.

Mailing address: Church of Scotland Offices, 121 George Street, Edinburgh EH2 4YN, UK. T. (44 31) 225 5722

Residence: 28 Castle Terrace, Edinburgh EH1 2EL, UK. T. (44 31) 228 6460

Weaver, Cynthia

Rel: Christian.
International Organizations: Secretary-Treasurer of The Voice of the Martyrs.
Mailing address: PO Box 443, Bartlesville OK 74005-0443, USA. T. (1 918) 337 8015. Fax 337 9287.

Weder, Dölf, Rev Dr

Rel: Evangelical Reformed; Ordained on 18 NOV 1979; b. 26 DEC 1950 in Balgach, St Gallen, Switzerland; single; Citizen of Switzerland; Lang. German, English and French.
International Organizations: Secretary-General (since 1990) of the European Alliance of YMCAs.
Other organizations: Chaplain (since 1982), Swiss army.
Main field of work: Christian youth work. **Career:** Secretary-General, YMCA of St Gallen from 1979 to 1987. Church Minister, Evangelical Reformed Church Wil SG from 1987 to 1990. Co-founder and President, YMCA Wil SG from 1988 to 1990. Vice-President, European Alliance of YMCAs from 1988 to 1990. Lecturer for Religious Education, Cantonal School, St Gallen from 1989 to 1990. **Educ.:** V.D.M. from University of Zürich, Switzerland (1971-1978); Dr.Theol. from University of Zürich, Switzerland (1979). **Interests:** Practical theology. **Pubs.:** Christliche Jugendarbeit; Das Haus des Lebens and Die Aufgasse der Kirche an der Jugend.
Mailing address: Teufener Strasse 94, CH-9000 St Gallen, Switzerland. T. (41 71) 27 72 17. Fax (41 71) 27 72 19.
Residence: Goethestr 79, CH-9008 St Gallen, Switzerland T. (41 71) 25 45 48

Wee, Paul

Rel: Lutheran.
International Organizations: Secretary-General of the Lutheran World Ministries.
Mailing address: LWF Dept. of Church Cooperation, Route de Ferney 150, CH-1211 Genève 20, Switzerland.

Weier, Helen, O.C.R., Sr

Rel: Roman Catholic; Ordained; celibate.
International Organizations: Member of the Cistercian Order of the Strict Observance.
Other organizations: Member, Association of Contemplative Sisters.
Main field of work: Contemplation.
Mailing address: Monastery of St Clare, 8650 Russell Ave S, Minneapolis MN 55431, USA.

Weiland, Paul, Rev

Rel: Lutheran.
International Organizations: Chairperson of the Communication Committee of the Lutheran Minority Churches in Europe.
Mailing address: LWB, PO Box 2100, CH-1211 Genève 21, Switzerland. T. (41 22) 791 61 11. Fax 798 86 16. C. LUTHERWORLD GENèVE. Tx 415730 OIK CH.

Weirig, Paul

Rel: Roman Catholic.
International Organizations: President of the Council of European Conferences of Priests.
Mailing address: 5 avenue Marie-Thérèse, L-2132 Luxembourg, Luxembourg. T. (352) 44743 Ext 211.

Weisman, Malcolm, Rev

Rel: Jewish; Ordained in 1956; b. in London, UK; m. to Rosalie Spiro; c. 2 boys; Citizen of United Kingdom; Lang. English, French, German, Yiddish and Hebrew.
International Organizations: Senior Jewish Chaplain (since 1972) of H M Forces.
Other organizations: Religious Adviser to small Jewish communities (since 1962); National Hillel Counsellor (since 1963); Member (since 1970), Chief Rabbi's Cabinet; Honorary Officer, a number of Jewish organizations.
Career: Assistant Commissioner, British Parliamentary Boundaries from 1976 to 1985. **Educ.:** M.A. from University of Oxford, UK (1954); Chief Rabbi Cert. (1954). **Interests:** reading, music, walking. **Hons.:** Man of the Year, B'nai B'rith (1979); Honorary Chaplain, Association of Jewish Ex-Servicemen. **Pubs.:** Editor of *Menorah* magazine.
Mailing address: Woburn House, Tavistock Square, London WC1H OEP, UK. T. (44 71) 387 3081
Residence: See Mailing Address. T. See Mailing Address

Welsh, William A

Rel: Christian.
International Organizations: Executive Director of the Christian Educators Association International.
Mailing address: James F Byrnes Intl. Center, Univ. of South Carolina, Columbia SC 29208, USA. T. (1 803) 777 2933.

Wenger, Antoine, Rev Fr

Rel: Roman Catholic; Ordained; b. 02 SEP 1919 in Rohrwiller, Bas-Rhin, France; celibate; Citizen of France.
International Organizations: Advisor of the Pontifical Council for Dialogue with Non-Believers.
Member of the Augustinians of the Assumption.
Other organizations: Advisor to Council for Church and Public Affairs (since 1983), Vatican.
Career: Director of Oriental Theology, Catholic University of Lyon from 1948 to 1956. Professor, Catholic University of Lyon in 1956. Chief Editor, *La Croix* from 1957 to 1969. President, Fédération International des Directeurs de Journaux Catholiques from 1957 to 1965. Member, Pontifical Marian Academy, Rome in 1959. Professor of Ancient Christian Literature, University of Strasbourg from 1969 to 1973. Ecclesiastical Counsellor to the French Ambassador to the Holy See from 1973 to 1983. **Interests:** old books, stamps. **Pubs.:** Upsal, le Défi du Siècle aux Eglises, 1968; Rome et Moscou, 1900-1950, 1987 and Le Cardinal Jean Villot, Secrétaire d'Etat de trois Papes, 1988.
Mailing address: Via San Pio V 55, I-00165 Roma, Italy. T. (39 6) 46 38 41

Wenker, Zita, SJC, Sr

Rel: Roman Catholic; Ordained; celibate.
International Organizations: Member of the Sisters of St Joseph of Cluny.
Other organizations: Member, Association of Contemplative Sisters.
Mailing address: Regina Mundi Priory, Waterloo and Fairfield, Devon PA 19333, USA.

Wennberg, Barbro

Rel: Baptist.
International Organizations: Contact of the Scandinavian Independent Baptist Union.
Mailing address: PO Box 144, S-522 00 Tidaholm, Sweden. T. (46 502) 110 10 – (46 502) 125 26.

Werkström, Bertil, Most Rev

Rel: Christian.
International Organizations: Archbishop of Church of Sweden.
Mailing address: Box 438, S-751 06 Uppsala, Sweden.

Wessels, Antoine, Rev Prof Dr

Rel: Protestant; Ordained on 24 SEP 1972; b. 08 OCT 1937 in Lemmer, Friesland, Netherlands; m. to Catharina G J M van Hooreveld; c. 2 boys and 1 girl; Citizen of Netherlands; Lang. Dutch, English, German and French.
International Organizations: Full Professor of History of Religions/Missions (since 1978) of the Free University of Amsterdam.
Other organizations: Chairman of Board (since 1991), Institute of Ecumenical and Missionary Studies.
Main field of work: University education. **Educ.:** Doc. Exam. Theol. from Amsterdam, Netherlands (1965); Promotion from Amsterdam, Netherlands (1972). **Interests:** relationship between Islam and Christianity (particularly in the Middle East). **Pubs.:** A Modern Arabic Biography of Mohammad, Brill, Leiden, 1972; The Renaissance of Islam, Baaren, 1980 and The Images of Jesus, Grand Rapids, London, 1990.
Mailing address: Free University of Amsterdam, De Boelelaan 1105, NL-1081 HV Amsterdam, Netherlands. T. (31 20) 548 5448
Residence: Rodenburghlaan 27, NL-1181 PX Amstelveen, Netherlands. T. (31 20) 641 1107

Wessels, Marc A, Rev Dr

Rel: Christian.
International Organizations: Executive Director of the International Network for Religion and Animals.
Mailing address: PO Box 1335, North Wales PA 19454, USA. T. (1 215) 699 6067.

West, Dorothy
Rel: Mason.
International Organizations: Coordinator of the Universal Masonic Order of the Eastern Star.
Mailing address: 95 N 18th St, East Orange NJ 07017, USA.

Westwood, William John, Rt Rev
Rel: Church of England; Ordained on 21 SEP 1952; b. 28 DEC 1925 in Saul, Gloucestershire, UK; m. to Anne Shirley; c. 1 boy and 1 girl; Citizen of United Kingdom; Lang. English.
International Organizations: Bishop (since 1984) of Peterborough.
Other organizations: Member (since 1973), Church Commissioner; President (since 1985), English Churches Housing; Member (since 1988), Broadcasting Standards Council; Member (since 1988), House of Lords.
Main field of work: Ministry. **Career:** Curate, Hull from 1952 to 1957. Rector, Lowestort from 1957 to 1965. Vicar, St Peter, Mancroft, Norwich from 1965 to 1975. Bishop, Edmonton from 1975 to 1984. **Educ.:** M.A. from University of Cambridge, UK (1948-1952). **Hons.:** Hon. LL.D., University of Leicester (1991); Hon. Fellow, Emmanuel College, Cambridge.
Mailing address: The Palace, Peterborough PE1 1YA, UK. T. (44 733) 62492
Residence: See Mailing Address. T. See Mailing Address

Wetter, Friedrich, H E Card
Rel: Roman Catholic; Ordained on 10 OCT 1953; b. 20 FEB 1928 in Landau, Speyer, Germany; celibate; Citizen of Germany.
International Organizations: Member (since 1985) of the College of Cardinals.
Member of the Congregation for the Evangelization of Peoples. Member of the Congregation for Catholic Education.
Other organizations: Archbishop (since 1982), Munich and Freising.
Career: Consecrated Bishop, Speyer in 1968. Created Cardinal by Pope John Paul II in 1985.
Mailing address: Kardinal-Faulhaber-Strasse 7, 80333 München, Germany. T. (49 89) 29 69 55

Weyer, Michel, Prof
Rel: United Methodist.
International Organizations: Representative of the United Methodist Church, Board of Church and Society.
Mailing address: Bellinostrasse 35, D-72764 Reutlingen, Germany.

Whaling, Frank, Rev Dr
Rel: Methodist; Ordained on 06 JUL 1962; b. 05 FEB 1934 in Pontefract, Yorkshire, UK; m. to Patricia Hill; c. 1 boy and 1 girl; Citizen of United Kingdom; Lang. French, Hindustani and Hindi.
International Organizations: Director (since 1985) of the Religious Studies Centre, University of Edinburgh.
Other organizations: University Teacher in Theology and Religious Studies (since 1973), University of Edinburgh; Council Member (since 1973), SHAP Working Party on Religious Education; Chairman to President (since 1975), Scottish Working Party on Religion in Education; Council Member (since 1978), Farmington Institute of Religious Education; Council Member (since 1978), Christian Education Movement in Scotland; Committee Member (since 1980), British Association of the History of Religion; Chairman (since 1985), Scottish Churches China Group; Chairman (since 1987), Scottish Inter-Faith Symposium; Member (since 1988), International Hall of Leaders.
Main field of work: University education, ministry. **Career:** Methodist Minister, Birmingham Central Hall from 1960 to 1962. Minister and College Manager, Methodist Church in North India from 1962 to 1966. Methodist Minister in Eastbourne (UK) from 1966 to 1969. **Educ.:** B.A. Hist. from University of Cambridge, Christ's College, UK (1954-1957); B.Th. from University of Cambridge, Wesley House, UK (1957-1959); M.A. from University of Cambridge, UK (1961); D.Th. from Harvard University, Cambridge MA, USA (1973); Ph.D. from University of Cambridge, UK (1990). **Interests:** Interfaith dialogue, global theology, history of religions. **Hons.:** FRAS (1981); FWLA (1985); FIBA (1986); FABI (1987); MIPSP (1990). **Mem.:** International Center for Integrative Studies, New York (1973); Royal Asiatic Society; Wesley Historical Society; World Institute of Achievement; World Literary Academy **Pubs.:** *The Rise of the Religious Significance of Rama*, Motilal Banarsidass, New Delhi, 1980; *John and Charles Wesley*, Paulist Press, New York and SPCK, London, 1981 and *Contemporary Approaches to the Study of Religion: The Humanities and The Social Sciences*, Berlin, Amsterdam, New York, Mouton, 1984-85.
Mailing address: New College, Mound Place, Edinburgh EH1 2LX, UK. T. (44 31) 225 8400
Residence: 29 Ormidale Terrace, Edinburgh EH12 6EA, UK T. (44 31) 337 5119

White, Thomas A, H E Mgr
Rel: Roman Catholic; Ordained on 25 FEB 1956; b. 12 AUG 1931 in Durrow, Ireland; celibate.
International Organizations: Apostolic Pro-Nuncio to New Zealand and Fiji (since 1989) of the Roman Catholic Church.
Other organizations: Apostolic Delegate to the Pacific Ocean (since 1989); Titular Archbishop, Sabiona.
Mailing address: PO Box 14-044, 112 Queen's Drive, Wellington 3, New Zealand. T. (64 4) 873 470. Fax 878 170. Tx 30857 FOSS

Whitehead, Richard
Rel: Quaker.
International Organizations: Presiding Clerk of the Religious Society of Friends: Friends United Meeting.
Mailing address: 101 Quaker Hill Drive, Richmond IN 47374, USA.

Wickham, John, S.J., Fr
Rel: Roman Catholic; Ordained; celibate.
International Organizations: Member of the Society of Jesus.
Mailing address: 4567 West Broadway, Montreal PQ H4B 2A7, Canada.

Widmann, Wolfgang, F.M.M.A., Br
Rel: Roman Catholic; Ordained; celibate.
International Organizations: Superior General of the Frères de la Miséricorde de Sainte Marie-Auxiliatrice.
Mailing address: Nordallee 1, 54292 Trier, Germany. T. (49 651) 208 2200

Wiebe, Donald
Rel: Unknown; b. 29 APR 1943 in Niagara Falls ON, Canada; m. to Gloria Willems; c. 1 boy; Citizen of Canada.
International Organizations: Associate Director (since 1990) of the Centre for Religious Studies, University of Toronto.
Executive Board Member (since 1985) of the International Association for the History of Religions.
Other organizations: Professor (since 1987), Trinity College, University of Toronto; Executive Board (since 1987), North American Association for the Study of Religion; Editorial Advisory Board (since 1987), Peter Lang Press.
Career: Assistant Professor, Canadian Nazarene College from 1965 to 1967. Assistant Professor, Canadian Nazarene College from 1975 to 1977. Assistant Professor, Winkler Bible Institute from 1967 to 1969. Sessional Lecturer, University of Guelph from 1969 to 1970. Sessional Lecturer, Universities of Winnipeg and Manitoba from 1977 to 1980. Assistant Professor, Trinity College from 1980 to 1982. Associate Professor, Trinity College from 1982 to 1987. Editorial Board Member, *Religion, The Journal of the Institute for the Study of Religion, Numen.* **Educ.:** B.Th. from Mennonite Brethren College of Arts (1967); B.A. from Wilfrid Laurier University, Waterloo, Ontario, Canada (1967); M.A. from University of Guelph, Ontario, Canada (1970); Ph.D. from University of Lancaster, UK (1974). **Mem.:** International Association for the History of Religions (Executive Director, XIVth Quinquennial Congress); Canadian Society for the Study of Religion (Executive Board 1980-83); American Academy of Religion; North American Association for the Study of Religion (President 1986-87) **Pubs.:** Religion and Truth, 1981; The Irony of Theology and the Nature of Religious Thought, 1991 and over 50 articles.
Mailing address: 6 Hoskin Ave, Toronto ON M5S 1H8, Canada.
Residence: 9 Douglas Ave, Toronto ON M5M 1G4, Canada.

Wiesel, Elie
Rel: Jewish; b. 30 SEP 1928 in Sighet, Romania; m. to Marion E Rose; c. 1 boy; Citizen of USA.
International Organizations: Founder of the Elie Wiesel Foundation for Humanity.
Vice-President-International (since 1985) of the International Rescue Committee. Board of Trustees (since 1985) of the American Jewish World Service. Comité d'Honneur (since 1985) of the International League Against Racism and Antisemitism. Chairman, Advisory Board (since 1985) of the World Union of Jewish Students.

Other organizations: Member, Board of Directors (since 1973), American Associates of Ben-Gurion University of the Negev; Andrew W Mellon Professor in the Humanities, Boston University, MA (since 1976); Member, Board of Governors (since 1977), Haifa University; Member, Board of Directors (since 1983), National Committee on American Foreign Policy.
Career: Deported to Birkenau, Auschwitz Monowitz and Buchenwald from 1944 to 1945. Naturalized in 1963. Distinguished Professor of Judaic Studies, City College, City University of New York from 1972 to 1976. Henry Luce Visiting Scholar in the Humanities and Social Thought, Whitney Humanities Center, Yale University, New Haven CT from 1982 to 1983. Distinguished Visiting Professor of Literature and Philosophy, Florida International University in 1982. Chairman, United States President's Commission on the Holocaust from 1979 to 1980. Chairman, United States Holocaust Memorial Council from 1980 to 1986. Member of the Jury, Neustadt International Prize for Literature in 1984. Honorary President, American Gathering of Jewish Holocaust Survivors in 1985. Founded the Elie Wiesel Foundation for Humanity in 1987. Founded The National Jewish Center for Learning and Leadership (CLAL). Journalist for Israeli, French, and American newspapers. **Educ.:** University of Paris IV (Paris-Sorbonne), France. **Hons.:** numerous awards including: First Herzl Literary Award; First David Ben-Gurion Award, National UJA; First International Remembrance Award, World Federation of Bergen-Belsen Associations (1965); First Jewish Heritage Award, B'nai B'rith (1966); Prix Médicis, France (1968); Eleanor Roosevelt Award, New York UJA (1972); American Liberties Medallion, American Jewish Committee (1972); Prix Bordin de l'Académie Française (1972); Martin Luther King Medallion, City College of New York (1973); Congressional Gold Medal of Achievement (awarded unanimously by the Senate and the House of Representatives, 1984); Commandeur de la Légion d'Honneur, France (1984), elevated to the rank of Grand Officier (1990); Nobel Peace Prize (1986); etc.; and numerous honorary degrees (D.Litt., D.Hum.Litt., D.Phil., D.H.L., LL.D., D.H., D.Sc., D.Theol.). **Mem.:** Société des Auteurs, Paris; PEN; Writers Guild of America, East; The Author's Guild; Writers and Artists for Peace in the Middle East; Royal Norwegian Society of Sciences and Letters; Amnesty International; Foreign Press Association (Honorary Life Member); Albert Einstein Society, Philadelphia (Honorary Member); American Academy of Arts and Sciences (Fellow) **Pubs.:** *Against Silence: The Voice and Vision of Elie Wiesel* (collected shorter writings), Irving Abrahamson, Ed., 3 Vols., 1985; With Albert Friedlander, *The Six Days of Destruction*, 1988 and *L'Oublié* (novel), 1989 (publication in English as *The Forgotten*, Spring 1992).
Mailing address: Boston University, The University Professors, 745 Commonwealth Ave, Boston MA 02215, USA. T. (1 617) 353 4566

Wiesenthal, Simon

Rel: Jewish.
International Organizations: Contact of the Jewish Documentation Centre.
Mailing address: Salztorgasse 6/IV/5, A-1010 Wien, Austria. T. (43 1) 63 91 31 – (31 222) 63 98 05. Fax (43 1) 535 0397.

Wietzke, Joachim, Dr

Rel: Christian.
International Organizations: General Secretary of the International Association for Mission Studies.
Mailing address: Mittelweg 143, 20148 Hamburg, Germany. T. (49 40) 41174 104. Fax 45 84 79. C. Missionswerk Hamburg. Tx 402/6105429 gma lu.

Wiggins, James Bryan, Prof

Rel: Methodist; b. 24 AUG 1935 in Mexia TX, USA; m. to Kay Moeckly; c. 1 boy and 1 girl; Citizen of USA; Lang. English and German.
International Organizations: Professor of Religion, Department of Religion (since 1963) of the Syracuse University, NY.
Other organizations: Chairman, Committee on International Directions (since 1991), American Academy of Religion.
Main field of work: University education, writing. **Career:** Executive Director, American Academy of Religion from 1982 to 1991. **Educ.:** B.A. from Texas Wesleyan College, Fort Worth TX, USA (1953-1957); B.D. from Southern Methodist University, Dallas TX, USA (1957-1959); Ph.D. from Drew University, Madison NJ, USA (1960-1963). **Interests:** pluralism and diversity. **Mem.:** American Society of Church History; Society for Arts, Religion and Contemporary Culture **Pubs.:** Christianity: A Cultural Perspective; Religion as Story and Foundations of Christianity.
Mailing address: Dept. of Religion, 501 H2, Syracuse University, Syracuse NY 13244, USA. T. (1 315) 443 3861
Residence: 308 Kimber Rd, Syracuse NY 13224, USA. T. (1 315) 446 4322

Wilckens, Ulrich

Rel: Christian; Ordained; b. 05 AUG 1928 in Hamburg, Germany; m. to Inge Westermann; c. 3 girls; Citizen of Germany.
International Organizations: Bishop (since 1981) of Holstein-Lübeck.
Career: Vicar, Hinterzarten, Black Forest from 1953 to 1955. Assistant, Ökumenisches Institut, University of Heidelberg from 1955 to 1958. Docent, Marburg from 1958 to 1960. Professor of New Testament, Kirchliche Hochschule, Berlin from 1960 to 1968. Professor of New Testament, Theological Faculty, University of Hamburg from 1968 to 1981. **Pubs.:** Das Neue Testament übersetzt und kommentiert, 1970; Rechtfertigung als Freiheit, Paulusstudien, 1974 and Der Brief and die Römer, Vol. I, 1978, Vol. II, 1980, Vol. III, 1982.
Mailing address: Bäckerstrasse 3-5, 23564 Lübeck, Germany. T. (49 451) 79 71 76

Wilkins, Howell O

Rel: Christian.
International Organizations: President of the Christians Associated for Relationships with Eastern Europe.
Mailing address: 114 North Washington Street, Easton MD 21601, USA. T. (1 301) 822 3573.

Wilkins, John Anthony Francis

Rel: Roman Catholic; b. 20 DEC 1936 in Cheltenham, Glos, UK; single; Citizen of United Kingdom.
International Organizations: Editor (since 1982) of *The Tablet.*
Educ.: B.A. from University of Cambridge, Clare College, UK (1958-1961). **Hons.:** Unda Radio Prize (1973).
Mailing address: The Tablet, 1 King Street Cloisters, London W6 OQZ, UK. T. (44 81) 748 8484. FAX 748 1550
Residence: 46 Winchester Street, London SW1 4NF, UK

Wilkinson, Georgalyn

Rel: Protestant.
International Organizations: Executive Director of the Gospel Literature International.
Mailing address: PO Box 488, 1409 N Walnut Grove Ave, Rosemead CA 91770, USA. T. (1 818) 288 2812. Fax 288 8803.

Willebrands, Johannes Gerardus Maria, H E Card

Rel: Roman Catholic; Ordained on 26 MAY 1934; b. 04 SEP 1909 in Bovenkarspel, Noord Holland, Netherlands; celibate; Citizen of Netherlands; Lang. Dutch, French, English, German and Italian.
International Organizations: President Emeritus (since 1989) of the Pontifical Council for Promoting Christian Unity.
Member of the Pontifical Council for Promoting Christian Unity.
Other organizations: Member (since 1969), College of Cardinals.
Main field of work: Ecumenism. **Career:** Chaplain, Begijnhof Church, Amsterdam from 1937 to 1940. Professor of Philosophy, Major Seminary, Warmond from 1940 to 1960. Rector, Major Seminary, Warmond from 1945 to 1960. President, St Willibrord Association from 1946 to 1960. Appointed delegate for ecumenical work by the Dutch hierarchy from 1958 to 1960. Organized and was Secretary, Catholic Conference on Ecumenical Questions in 1951. Appointed Secretary, the newly created Vatican Secretariat for Promoting Christian Unity in 1960. Consecrated Titular Bishop, Mauriana in 1964. Created Cardinal Deacon by Pope Paul VI, SS Cosmas and Damian in 1969. Appointed President, Secretariat for Promoting Christian Unity in 1969. Became President, Commission for Religious Relations with the Jews in 1974. Named Archbishop, Utrecht and Cardinal-Priest of St Sebastian at the Catacombs from 1975 to 1983. Primate, the Netherlands. Chairman, Bishops' Conference of the Netherlands. Became Catholic Co-Chairman, Commission for the Theological Dialogue of the Roman Catholic Church and the Orthodox Church in 1980. Chairman, International Liaison Committee between Jews and the Catholic Church. President, Council for Promoting Christian Unity. **Educ.:** Phil. and Theol. Stds. from Major Seminary, Warmond, The Netherlands (1934); Ph.D. from Pontifical University San Tommaso d'Aquino, Vatican City State (1934-1937). **Hons.:** Hon. D.Litt.: University of St Louis, MO (1968), St Olaf College, Northfield MN (1976), St Thomas College, St Paul

MN (1979), Assumption College, Worcester MA (1980); Hon. LL.D.: University of Notre Dame, IN (1970), Catholic University of Washington, DC (1974), Seton Hall University, South Orange NJ (1987); Hon. D.Theol.: Catholic University of Louvain, Belgium (1971), Theological Academy of Leningrad, USSR (1973), Catholic University of Lublin, Poland (1985), University of Oxford, UK (1987), Hellenic College/Holy Cross Greek Orthodox Theological School, Brookline MA (1989), University of St Michael's College, Toronto, Canada (1990); D.Hum.Litt. h.c.: Bellarmine College, Louisville KY (1987); many awards – both civil and ecclesiastical – for his services to ecumenism and to Christian-Jewish relations, as well as for his overall contribution to contemporary society and culture. **Pubs.:** *Oecuménisme et Problèmes Actuels*, Ed. du Cerf, Paris, 1969; *Mandatum Unitatis*, Bonifatius Verlag, Paderborn, 1989 and numerous articles.
Mailing address: 00120 Città del Vaticano, Vatican. T. (39 6) 698 49854
Residence: Piazza Città Leonina, int. 4, I-00193 Roma, Italy.

Willemsen, Antonia
Rel: Christian.
International Organizations: Secretary-General of the Aid to the Church in Need.
Mailing address: Bischof-Kaller-Strasse 3, 61462 Königstein, Germany. T. (49 6174) 3091. Tx 410654 KINI D.

William, Carmen Dasko, Sr
Rel: Roman Catholic; Ordained; celibate.
International Organizations: Co-Head, Pacific Regional Office of the Unda, International Catholic Association for Radio and Television.
Mailing address: Mercederan Sisters, PO Box 3487, Agana 96910, Guam.

William, Stanley
Rel: Buddhist.
International Organizations: Member of the Executive Council of the World Fellowship of Buddhists.
Mailing address: PO Box 699, 3rd Floor, Lloyds Building, Sir Baron Jayatilleke Mawatha, Colombo 1, Sri Lanka.

Williams, Delores L Kennedy
Rel: Methodist Episcopal.
International Organizations: President of the Women's Missionary Society – African Methodist Episcopal Church.
Mailing address: 1134 11th St NW, Washington DC 20001, USA. T. (1 202) 371 8886.

Williams, Jimmy
Rel: Christian.
International Organizations: Executive Director of the Probe Ministries International.
Mailing address: 1900 Firman Dr, Richardson TX 75081-6796, USA.

Williams, Kathryn, Rev
Rel: Protestant.
International Organizations: President of the International Association of Women Ministers.
Mailing address: Box 7030, Indianapolis IN 46207, USA.

Williams, Theodore, Rev Dr
Rel: Methodist; Ordained on 22 SEP 1973; b. 24 FEB 1935 in Nazareth, Tamilnadu, India; m. to Esther; Citizen of India; Lang. Tamil, English and Hindi.
International Organizations: President (since 1982) of the World Evangelical Fellowship.
Chairman (since 1986) of the Missions Commission of the World Evangelical Fellowship.
Other organizations: President (since 1990), Indian Evangelical Mission; Minister at Large (since 1991), Evangelical Fellowship of India.
Main field of work: Bible teaching, preaching, mission leading. **Career:** Teacher at the South India Biblical Seminary from 1957 to 1969. Pastor, Methodist Church from 1972 to 1977. General Secretary, Indian Evangelical Mission from 1965 to 1990. Executive Secretary, World Evangelical Fellowship Missions Commission from 1979 to 1986. Secretary for Missions, Evangelical Fellowship of India from 1977 to 1990. **Educ.:** M.A. from University of Madras, India (1950-1955); B.D. from University of Serampore, United Theological College, Bangalore, India (1961-1964). **Interests:** writing. **Hons.:** Hon. D.D., Asbury College, Wilmore KY (1983). **Pubs.:** God Alive – A Study in Psalms; The Local Church and Mission and Mission Ignored – A Study on Hosea.
Mailing address: 7 Langford Road, Bangalore 560 025, Karnataka, India. T. (91 812) 213 080
Residence: See Mailing Address. T. (91 812) 210 183

Williams, Thomas Stafford, H E Card
Rel: Roman Catholic; Ordained on 20 DEC 1959; b. 20 MAR 1930 in Wellington, New Zealand; celibate; Citizen of New Zealand.
International Organizations: Member (since 1983) of the College of Cardinals.
Member of the Congregation for the Evangelization of Peoples.
Other organizations: Archbishop (since 1979), Wellington and Metropolitan of New Zealand.
Career: Assistant, St Patrick's Parish, Palmerston North in 1963. Director of Studies, Catholic Enquiry Centre, Wellington in 1965. Parish Priest, St Anne's, Leulumoega, Western Samoa in 1971. Parish Priest, Holy Family Parish, Porirua East, Wellington in 1976. Created Cardinal by Pope John Paul II in 1983.
Mailing address: Archbishop's House, Private Bag 1937, Wellington 1, New Zealand. T. (64 4) 496 1795

Williamson, Clark Murray, Prof
Rel: Disciples of Christ; Ordained on 11 JUN 1961; b. 03 NOV 1935 in Memphis TN, USA; m. to Barbara Unger; c. 1 boy; Citizen of USA; Lang. English, French and German.
International Organizations: Indiana Professor of Christian Thought (since 1991) of the Christian Theological Seminary, Indianapolis IN.
Other organizations: Editor, *Encounter* (theological quarterly); Member, Church Relations Committee, US Holocaust Memorial Council.
Main field of work: Systematic theology. **Career:** Served as President, Christian Study Group on Judaism, Association of Disciples for Theological Discussion, Indiana Academy of Religion, and Alumni Council of Disciples Divinity House of University of Chicago. Served as Teaching Assistant to Paul Tillich and Editor/Advisor to the 3rd Volume, Tillich's *Systematic Theology*. Visiting Professor at the School of Theology in Claremont, CA, and at the Graduate School, the Ecumenical Institute (Bossey), Switzerland. Lectured widely and took part in numerous workshops, seminars and scholarly symposia. **Educ.:** B.A. from Transylvania University, Lexington KY, USA (1953-1957); M.A., B.D., Ph.D. from University of Chicago, Divinity School, IL, USA (1957-1964). **Hons.:** Noted as Distinguished Disciple Scholar by the *Disciples Theological Digest* (1990). **Mem.:** Christian Study Group on Judaism; Association of Disciples for Theological Discussion; American Academy of Religion; American Theological Society; Theological Commission of the Christian Church (Disciples of Christ) **Pubs.:** Author of several books, including the following most recent: With Ronald J Allen, *Interpreting Difficult Texts: Anti-Judaism and Christian Preaching*, London, SCM, 1989 and Philadelphia, Trinity Press International, 1989; With Ronald J Allen, *A Credible and Timely Word: Process Theology and Preaching*, St Louis, Chalice Press, 1991 and chapters in books, papers in: *Religion and Intellectual Life, Impact, Mid-Stream, Quarterly Review, Process Studies, The Journal of Religion*, the *Chicago Theological Seminary Register, Encounter, The Evangelical Quarterly, Quest, The Journal of Ecumenical Studies.*
Mailing address: Christian Theological Seminary, 100 W 42nd St, Indianapolis IN 46208, USA. T. (1 317) 924 1331
Residence: 5261 Boulevard Place, Indianapolis IN 46208, USA. T. (1 317) 251 7309

Wilson, C
Rel: Christian.
International Organizations: Representative of the Church Missionary Society, UK.
Mailing address: Overseas House, 12 Talbot Street, Belfast BT1 2LE, UK. T. 24581.

Wilson, Donald P

Rel: Unknown.

International Organizations: Grand Exalted Ruler of the Improved Benevolent Protective Order of Elks of the World.

Mailing address: PO Box 159, Winston NC 27986, USA. T. (1 919) 358 7661.

Wilson, Henry Steward, Rev Dr

Rel: Protestant; Ordained on 19 NOV 1978; b. 30 DEC 1946 in Mangalore, Karnataka, India; m. to Eva Beryl Unger; c. 2 girls; Citizen of India; Lang. Tulu, Kannada and English.

International Organizations: Executive Secretary, Department of Theology (since 1989) of the World Alliance of Reformed Churches.

Main field of work: Theology. **Career:** Lecturer, United Theological College, India from 1977 to 1980. Director, Board of Theological Education, India from 1980 to 1989. **Educ.:** B.D. from Serampore University, India (1967-1970); M.A. from Karnatak University, Karnataka, India (1970-1972); Th.M. from Princeton Seminary, NJ, USA (1972-1973); Ph.D. from Drew University, Madison NJ, USA (1974-1977). **Interests:** history of christianity. **Mem.:** International Association for Mission Studies **Pubs.:** Editor, *Church on the Move: An Affirmation of Biblical Faith*, Madras, CLS, 1988; *Theological Education and Addiction Concerns in India*, Bangalore, BTE, 1989 and Editor, *Christian Community in a Changing Society*, Geneva, WARC, 1991.

Mailing address: 150 route de Ferney, PO Box 2100, CH-1211 Genève 2, Switzerland. T. (41 22) 791 62 41. C. WARC GENEVA. Tx 415 730 OIK CH (WARC). FA (41 22) 791 65 05 WARC REFORMIERTER WELTBUND.

Residence: 72b rue du Village, CH-1214 Vernier, Switzerland T. (41 22) 41 25 19

Wilson, Lloyd Lee

Rel: Quaker.

International Organizations: General Secretary of the Religious Society of Friends: Friends General Conference.

Mailing address: 1520-B Race Street, Philadelphia PA 19102, USA.

Wilson, Lois Miriam, Very Rev Dr

Rel: United Church of Canada; Ordained on 09 JUN 1965; b. 08 APR 1927 in Winnipeg MB, Canada; m. to Rev Dr Roy F Wilson; c. 2 boys and 2 girls; Citizen of Canada; Lang. English and French.

International Organizations: McGeachy Senior Scholar (since 1989) of The United Church of Canada.

Other organizations: Board Member (since 1989), Federal Government Environmental Review of Nuclear Waste Disposal; Vice-President (since 1989), Civil Liberties Association of Canada; Chancellor (since 1991), Lakehead University.

Main field of work: Ministry. **Career:** Minister, First Church United, Thunder Bay, Ontario from 1965 to 1969. Minister, First Church United, Hamilton, Ontario from 1969 to 1978. Minister, Chalmers United Church, Kingston, Ontario from 1978 to 1980. Officer, Ontario Human Rights Commission from 1974 to 1975. Director, Ecumenical Forum of Canada from 1983 to 1989. President, The Canadian Council of Churches from 1976 to 1979. Moderator, The United Church of Canada from 1980 to 1982. President, World Council of Churches from 1983 to 1991. **Educ.:** B.A. from University of Winnipeg, Manitoba, Canada (1947); M.Div. from University of Winnipeg, Manitoba, Canada (1950). **Interests:** skiing, sailing, canoeing, reading. **Hons.:** Order of Canada; Order of Ontario; Pearson Peace Prize (1985); World Federalist Peace Prize (1985); Queen's Jubilee Medal; several honorary degrees. **Mem.:** Amnesty International (Commissioner, Canada section) **Pubs.:** Like a Mighty River, 1980; Turning the World Upside Down, 1989 (memoir) and Tell me a Story, 1991.

Mailing address: 482 Markham Street, Toronto ON M6G 2L3, Canada. T. (1 416) 531 1732

Residence: See Mailing Address. T. See Mailing Address

Wilson, Paul F

Rel: Christian Church.

International Organizations: Director of the International Human Rights Program of the Christian Church.

Mailing address: PO Box 1986, Indianapolis IN 46206, USA. T. (1 317) 353 1491 Ext 362.

Wilson, Roger Plumpton, Rt Rev Dr

Rel: Anglican; Ordained on 21 DEC 1935; b. 03 AUG 1905 in Bristol, Avon, UK; m. to Mabel Joyce Avery; c. 2 boys and 1 girl; Citizen of United Kingdom; Lang. English, Latin and Greek.

International Organizations: Ecclesiastic of the Church of England.

Career: Classical Master, Shrewsbury School from 1928 to 1930. Classical Master, Shrewsbury School from 1932 to 1934. Classical Master, St Andrew's College, Grahamstown, South Africa from 1930 to 1932. Ordained Deacon in 1935. Curate, St Paul's, Liverpool from 1935 to 1938. Curate, St John's, Westminster from 1938 to 1939. Vicar, South Shore, Blackpool from 1939 to 1945. Vicar, Radcliffe-on-Trent, Notts. from 1945 to 1949. Archdeacon, Nottingham from 1945 to 1949. Bishop, Wakefield from 1949 to 1958. Bishop, Chichester from 1958 to 1974. Clerk, the Closet to HM The Queen from 1963 to 1975. Member, Presidium, Conference of European Churches from 1967 to 1974. **Educ.:** University of Oxford, Keble College, UK (1924-1928). **Interests:** education, church unity. **Hons.:** DD, Lambeth (1989); Knight Commander of the Royal Victorian Order (1974).

Mailing address: Kingsett, Ropers Lane, Wrington, Bristol BS18 7NH, UK. T. WRINGTON 862464

Residence: See Mailing Address. T. See Mailing Address

Wilson, Ronald

Rel: Uniting Church.

International Organizations: Former President of the Uniting Church in Australia.

Mailing address: c/o Uniting Church Assembly Offices, PO Box E266, St James NSW 2000, Australia.

Wilson, Wes

Rel: Protestant.

International Organizations: International Executive Administrator of the Every Home for Christ International.

Mailing address: 20232 Sunburst Street, Chatsworth CA 91311, USA. T. (1 818) 341 7870.

Winant, Pierre, M.O., Fr

Rel: Roman Catholic; Ordained; celibate.

International Organizations: Procurator-General of the Aumôniers du Travail.

Mailing address: Bd Lambermont 17, B-1030 Bruxelles, Belgium. T. (32 2) 215 13 00

Winchell, Richard

Rel: Evangelical.

International Organizations: General Director of The Evangelical Alliance Mission.

Mailing address: PO Box 969, Wheaton IL 60189, USA. T. (1 312) 653 5300.

Winer, Robert, Dr

Rel: Jewish: Messianic.

International Organizations: International Relations Committee of the Messianic Jewish Alliance of America.

Mailing address: PO Box 417, Wynnewood PA 19096, USA. T. (1 215) 896 5812

Wing, Joseph, Rev

Rel: United Congregational.

International Organizations: Secretary of the United Congregational Church of Southern Africa.

Mailing address: PO Box 31083, Braamfontein 2017, South Africa.

Winning, Thomas Joseph, Most Rev

Rel: Roman Catholic; Ordained on 18 DEC 1948; b. 03 JUN 1925 in Wishaw, Lanarkshire, UK; celibate; Citizen of United Kingdom.

International Organizations: Archbishop (since 1974) of Glasgow.

Other organizations: President (since 1986), Bishops' Conference of Scotland.

Career: Assistant Priest, Chapelhall from 1949 to 1950. Assistant Priest, St Mary, Hamilton from 1953 to 1957. Assistant Priest, Cathedral, Motherwell from 1957 to 1958. Chaplain, Franciscans of the Immaculate Conception, Bothwell from 1958 to 1961. Diocesan Secretary, Motherwell from 1956 to 1961. Spiritual Director, Pontifical Scots College, Rome from 1961 to 1966. Paris Priest, St Luke's, Motherwell from 1966 to 1970. Officialis and Vicar Episcopal, Motherwell from 1966 to 1970. President, Scottish Catholic Marriage Tribunal from 1970 to 1971. Auxiliary Bishop and Vicar-General, Glasgow from 1971 to 1974. Parish Priest, Our Holy Redeemer's Clydebank from 1972 to 1974. **Interests:** golf, music. **Hons.:** Hon. D.D., Glasgow University (1983); Hon. Fellow, Educational Institute of Scotland (1986); Grand Prior of Scottish Lieutenancy of Equestrian Order of the Holy Sepulchre of Jerusalem (1989).
Mailing address: Curial Office, 196 Clyde Street, Glasgow G1 4JY, UK. T. (44 41) 226 5898
Residence: The Oaks, 40 Newlands Road, Glasgow G43 2JD, UK. T. (44 41) 649 1224

Winston, E Beulah, Dr
Rel: Unknown.
International Organizations: President of the Women's International Religious Fellowship.
Mailing address: c/o Norma E Boulevard, 1602 Webster St NW, Washington DC 20011, USA. T. (1 202) 882 4033.

Wirakotan, J H, Rev Dr
Rel: Christian.
International Organizations: General Secretary of the Indonesian Christian Church.
Mailing address: Jln Panglima Polim 1/51A, Kebayoran Baru, Jakarta Selatan, Indonesia.

Witteveen, H J, Dr
Rel: Muslim: Sufi.
International Organizations: General Representative of the International Sufi Movement.
Mailing address: PO Box 1066, NL-2240 BB Wassenaar, Netherlands.

Wodajo, Datan
Rel: Christian.
International Organizations: Director of the World Missionary Evangelism.
Mailing address: PO Box 1237, Addis Ababa, Ethiopia. T. (251 1) 18 75 21 – 18 75 22. Tx 21307 (CRDA ADDIS).

Wojtyla, Karol
See John Paul II.

Wolf, Michael, Rabbi
Rel: Jewish; Ordained.
International Organizations: Chairman of the International Alliance of Messianic Congregations and Synagogues.
Mailing address: PO Box 417, Wynnewood PA 19096, USA. T. (1 215) 896 5812.

Wong, T M
Rel: Buddhist: Mahayana.
International Organizations: Member of the Foundation for the Preservation of the Mahayana Tradition.
Other organizations: Director, Cham-Tse-Ling.
Mailing address: 50 Hoi An Street, 1st Floor Block D, Shau Ki Wan, Hong Kong. T. (852) 721 1974

Wongmo, Thubten
Rel: Buddhist: Mahayana.
International Organizations: Member of the Foundation for the Preservation of the Mahayana Tradition.
Other organizations: Director, Tushita Retreat Centre.
Mailing address: McLeod Ganj, Dharamsala, Dist. Kangra, Himachal Predesh 176219, India. T. (91 1892) 2266

Wood, Anne
Rel: Quaker.
International Organizations: Contact of the Wider Quaker Fellowship.
Mailing address: 1506 Race Street, Philadelphia PA 19102, USA. T. (1 215) 241 72 50.

Wood, Harold W, Jr
Rel: Pantheist; b. 21 NOV 1950 in Whittier CA, USA; m. to Janet Stuhr; Citizen of USA; Lang. English.
International Organizations: Secretary of the Universal Pantheist Society.
Other organizations: Editor (since 1973), *Pantheist Vision.*
Main field of work: Editing and writing. **Career:** Attorney, Tulare County Counsel in 1985. **Educ.:** B.Sc. from University of California, Davis, USA (1974); M.Sc. from University of Washington, Seattle, USA (1976); J.D. from University of Puget Sound, Tacoma WA, USA (1980). **Interests:** wilderness. **Mem.:** California State Bar Association (1984); Sierra Club; and many other environmental organizations **Pubs.:** "Modern Pantheism as an Approach to Environmental Ethics", in *Environmental Ethics*, Summer 1985; "A Proof of Pantheism Using the 'Argument by Design'", in *Pantheist Vision*, December 1990 and "Pantheism and Biodiversity", in *Pantheist Vision*, September 1991.
Mailing address: PO Box 265, Big Pine CA 93513, USA.
Residence: 2011 W Laurel Ave, Visalia CA 93277, USA.

Wood, Maurice Arthur Ponsonby, Rt Rev
Rel: Anglican; Ordained in 1940; b. 26 AUG 1916 in London, UK; m. to (1st) Marjorie Pennell (died), (2nd) M Margaret Sandford; c. 4 boys and 2 girls; Citizen of United Kingdom.
International Organizations: Chairman (since 1986) of the Order of Christian Unity.
Other organizations: Honorary Chaplain (since 1971), Royal Naval Reserve; Honorary Assistant Bishop (since 1985), Diocese of London; Honorary Assistant Bishop (since 1988), Diocese of Oxford; Resident Priest (since 1987), Englefield.
Career: Curate, St Paul's Portman Square, London from 1940 to 1943. Royal Naval Chaplain from 1943 to 1947. Rector, St Ebbe's, Oxford from 1947 to 1952. Vicar and Rural Dean, Islington from 1952 to 1961. Principal, Oak Hill Theological College, Southgate, London from 1961 to 1971. Bishop, Norwich from 1971 to 1985. Entered House of Lords in 1975. Chaplain, Commando Association, Worshipful Co. of Weavers. Governor, Monkton Combe School, Bath. **Interests:** painting. **Pubs.:** To Everyman's Door, 1968; Into the Way of Peace, 1982 and This is our Faith, 1985.
Mailing address: St Mark's House, Englefield, Nr Reading RG7 5EP, UK. T. (44 734) 302 227

Wood, Robert S, Rev
Rel: Reformed Church.
International Organizations: Member of the Executive Committee of the World Alliance of Reformed Churches.
Mailing address: 226 Church Street NW, Huntsville AL 35801, USA. T. (1 205) 536 7481
Residence: T. (1 205) 536 9073

Woodley, A K, Rev
Rel: Methodist.
International Organizations: General Secretary of the Methodist Church of New Zealand.
Mailing address: PO Box 931, Christchurch, New Zealand.

Woodring, Wayne S, Dr
Rel: Unknown.
International Organizations: Executive Director of the Religious Conference Management Association.
Mailing address: One Hoosier Dome, Suite 120, Indianapolis IN 46225, USA. T. (1 317) 632 18 88.

Woods, Frank, Archbishop Sir

Rel: Church of England; Ordained in 1931; b. 06 APR 1907 in Davos, Switzerland; m. to Jean Margaret Sprules; c. 2 boys and 2 girls; Citizen of United Kingdom; Lang. English, German and French.
International Organizations: Ecclesiastic of the Church of England.
Main field of work: Ministry. **Career:** Curate, St Mary, Portsea from 1931 to 1933. Chaplain, Trinity College, Cambridge and Examining Chaplain to Bishop of Bristol from 1933 to 1936. Vice-Principal, Wells Theological College from 1936 to 1945. Chaplain to the Forces from 1939 to 1945. Vicar and Rural Dean, Huddersfield from 1945 to 1952. Honorary Canon, Wakefield and Proctor in Convocation, Wakefield from 1947 to 1952. Chaplain to H M The King from 1951 to 1952. Suffragan Bishop, Middleton and Canon Residentiary of Manchester Cathedral from 1952 to 1957. Archbishop, Melbourne from 1957 to 1977. Chaplain, (Victoria) Order of St John of Jerusalem from 1962 to 1971. Primate, the Church of England in Australia from 1971 to 1977. **Educ.:** University of Cambridge, Trinity College, UK. **Interests:** music, walking. **Hons.:** Knight, Order of the British Empire; Hon. D.D., Lambeth (1957); Hon. LL.D., Monash (1979). **Mem.:** Melbourne Club
Mailing address: 18 Victoria Road, Camberwell VIC 3124, Australia. T. (61) 882 9945
Residence: See Mailing Address. T. See Mailing Address

Woolford, Orville, Pastor

Rel: Seventh-Day Adventist.
International Organizations: Education Director, Trans-European Division of the General Conference of Seventh-Day Adventists.
Mailing address: 119 St Peter's Street, St Albans AL1 3EY, UK.

Woollcombe, Kenneth John, Rt Rev

Rel: Anglican; Ordained; b. 02 JAN 1924 in Sutton, Surrey, UK; m. to (1st) Gwendoline R V Hodges (died), (2nd) Juliet Dearmer; c. 4 girls; Citizen of United Kingdom.
International Organizations: Assistant Bishop (since 1989) of Diocese of Worcester.
Career: Sub-Lieutenant, R.N.V.R. (Royal Naval Volunteer Reserve) from 1945 to 1946. Curate, Grimsby Parish Church from 1951 to 1953. Fellow, Chaplain and Tutor, St John's College, Oxford from 1953 to 1960. Professor of Dogmatic Theology, General Theological Seminary, New York from 1960 to 1963. Principal, Episcopal Theological College, Edinburgh and Canon of St Mary's Cathedral, Edinburgh from 1963 to 1971. Bishop, Oxford from 1971 to 1978. Assistant Bishop, London from 1978 to 1981. Canon Residentiary, St Paul's Cathedral from 1981 to 1989. Precentor, St Paul's Cathedral from 1982 to 1989. Chairman, Society for the Promotion of Christian Knowledge (SPCK) from 1973 to 1979. Chairman, Churches' Council for Covenanting from 1978 to 1982. Chairman, Committee for Roman Catholic Relations from 1985 to 1988. Member, Court of Ecclesiastical Causes Reserved from 1984 to 1989. Honorary Fellow, St John's College, Oxford in 1971. **Interests:** reading, music. **Hons.:** Hon. S.T.D., University of the South, Sewanee, TN (1963); Hon. D.D., Hartford, CT (1975). **Pubs.:** Essays on Typology (joint author), 1957 and contributions to other theological publications.
Mailing address: 4 Flax Piece, Upton Snodsbury, Worcester WR7 4PA, UK.

Worach, Piotr, Br

Rel: Roman Catholic; Ordained; celibate.
International Organizations: Superior General of the Fils de la Mère de Dieu Douloureuse.
Mailing address: Aleja Przyszlosci 8, PL-05 420 Warszawa, Poland.

Worcelo, Gail, C.P., Sr

Rel: Roman Catholic; Ordained; b. 10 SEP 1956 in Brooklyn NY, USA; celibate; Citizen of USA; Lang. English and French.
International Organizations: Member of the Passionist Nuns.
Other organizations: Director, Homecomings: Center for Ecology and Contemplation.
Main field of work: Education, psychology, ministry. **Educ.:** B.A. from St Joseph's College, NY, USA (1974-1978); M.A. from Fordham University, Bronx NY, USA (1979-1981); M.A. from Marywood College, Scranton PA, USA (1990-1992). **Interests:** ecology, spirituality. **Mem.:** Amnesty International; Riverdale Center for Religious Research; Association of Contemplative Sisters; Psy Chi (1991) **Pubs.:** "The Passion of the Earth"; "Toward a Vow of Non-Violence" and "Experiencing Monastic Formation in the 80's".
Mailing address: St Gabriels Monastery, 631 Griffin Pond Rd, Clarks Summit PA 18411, USA. T. (1 717) 586 2791
Residence: See Mailing Address. T. See Mailing Address

Worlock, Derek John Harford, Most Rev

Rel: Roman Catholic; Ordained on 03 JUN 1944; b. 04 FEB 1920 in London, UK; celibate; Citizen of United Kingdom; Lang. English and French.
International Organizations: Archbishop (since 1976) of Liverpool. Advisor of the Pontifical Council for the Laity.
Other organizations: Metropolitan (since 1976), Northern Province; Vice-President (since 1978), Bishop's Conference of England and Wales.
Career: Curate, Our Lady of Victories, Kensington from 1944 to 1945. Private Secretary to Archbishop, Westminster from 1945 to 1964. Rector and Rural Dean, Church of St Mary and St Michael, London from 1964 to 1965. Bishop, Portsmouth from 1965 to 1976. Privy Chamberlain to Pope Pius XII from 1949 to 1953. Domestic Prelate, the Pope from 1953 to 1965. Peritus at Vatican Council II from 1963 to 1965. Consultor to Council of Laity from 1967 to 1976. Episcopal Secretary to Roman Catholic Bishops' Conference from 1967 to 1976. Member, Synod Council from 1976 to 1977. Member, Committee for the Family from 1977 to 1981. English delegate to International Synod of Bishops (in 1974, 1977, 1980, 1983, 1987). **Hons.:** Hon. D.D., University of Cambridge; Hon. LL.D., University of Liverpool; Knight Commander, Holy Sepulchre of Jerusalem (1966). **Pubs.:** Better Together, 1987 (Collins Religious Book Prize 1988); With Christ in the Wilderness, 1990 and Bread Upon the Waters, 1991.
Mailing address: Archbishop's House, 87 Green Lane, Mosley Hill, Liverpool L18 2EP, UK. T. (44 51) 722 2379
Residence: See Mailing Address. T. See Mailing Address

Wright, David Frederick

Rel: Church of Scotland; b. 02 OCT 1937 in Farnborough, Kent, UK; m. to Anne-Marie MacDonald; c. 1 boy and 1 girl; Citizen of United Kingdom; Lang. English and French.
International Organizations: Senior Lecturer in Ecclesiastical History (since 1973) of the University of Edinburgh.
Other organizations: Chairman (since 1986), Tyndale Fellowship; Dean, Faculty of Divinity (since 1988), University of Edinburgh.
Main field of work: University education. **Career:** Lecturer in Ecclesiastical History, University of Edinburgh from 1964 to 1973. **Educ.:** M.A. from University of Cambridge, Christ's College, UK (1957-1961); University of Oxford, Lincoln College, UK (1961-1964). **Interests:** patristics, reformation. **Mem.:** International Association for Patristic Studies; North American Patristics Society; Ecclesiastical History Society **Pubs.:** Common Places of Martin Bucer, Appleford, 1972; Joint Editor, New Dictionary of Theology, Leicester, 1988 and Editor and Contributor, Chosen by God: Mary in Evangelical Perspective, London, 1989.
Mailing address: New College, Mound Place, Edinburgh EH1 2LX, UK. T. (44 31) 225 8400. Fax 220 0952
Residence: 5 Lockharton Gardens, Edinburgh EH14 1AU, UK. T. (44 31) 443 1001

Wright, Francis W, Rev

Rel: Christian.
International Organizations: Director of the Holy Childhood Association.
Mailing address: 1720 Massachusetts Ave NW, Washington DC 20036, USA.

Wright, Ken

Rel: Baptist.
International Organizations: National President of the Baptist Union of Australia.
Mailing address: 21 Cunningham St, Applecross WA 6153, USA.

Wright, Kenyon E, Canon

Rel: Anglican; Citizen of United Kingdom; Lang. English and German.
International Organizations: Coordinating Secretary (since 1990) of the Christian Peace Conference.
Mailing address: Jungmannova 9, PO Box 192, 111 21 Praha 1, Czech Rep. T. (42 2) 236 02 89 – 236 02 90. Fax 235 02 51. C. EKUMRADA PRAHA. Tx 12 33 63 CPC C.

Wright, Lawrence A

Rel: Christian.
International Organizations: Director of the Atkinson Foundation.
Mailing address: 10 West Orange Avenue, San Francisco CA 94080, USA. T. (1 415) 876 1559. Fax 876 1143.

Wu Cheng-chung, John Baptist W, H E Card

Rel: Roman Catholic; Ordained on 06 JUL 1952; b. 26 MAR 1925 in Shui-tsai, Kaying, China; celibate; Citizen of China.
International Organizations: Member (since 1988) of the College of Cardinals.
Member of the Congregation for the Evangelization of Peoples. Member of the Pontifical Council for Inter-Religious Dialogue. Member of the Pontifical Council for Dialogue with Non-Believers.
Other organizations: Bishop (since 1975), Hong Kong.
Career: Pastoral ministry, Refugee Centre, Tung-tao-tsuen, Kowloon from 1952 to 1953. Worked at Chancery, Archdioceses of New York, Boston and Chicago from 1956 to 1957. Pastoral ministry, Taiwan from 1957 to 1975. Created Cardinal by Pope John Paul II in 1988.
Mailing address: Caritas House, 2 Caine Road, Hong Kong, Hong Kong. T. (852 5) 524 16 33. Fax 521 87 37

Wucher, Nikolaus, S.D.S., Fr

Rel: Roman Catholic; Ordained; celibate.
International Organizations: Secretary-General of the Society of the Divine Saviour.
Mailing address: Via della Conciliazione 51, I-00193 Roma, Italy. T. (39 6) 654 76 41.

Wunderlich, Kenneth

Rel: Christian.
International Organizations: Representative of the Concordia Gospel Outreach.
Mailing address: PO Box 201, St Louis MO 63166-0201, USA. T. (1 314) 268 1363.

Yabusaki, Koichi

Rel: Church of Perfect Liberty; b. 05 AUG 1911; Citizen of Japan.
International Organizations: Chairman of the PL Gakuen.
Career: Teacher, Church of Perfect Liberty (PL).
Mailing address: 2172-1 Oaza Shindo, Tondabayashi, Osaka 584, Japan. T. (81 721) 24 1111

Yacoub, Ignatius, III, H E Mgr

Rel: Orthodox; Ordained in 1932; b. in Bartali; Citizen of Syria.
International Organizations: Patriarch (since 1957) of Orthodox Syrians of Antioch and all the East.
Career: Professor in Beirut in 1931. Superior, St Ignatius Seminary, India. Professor, St Ephrem Seminary, Mosul in 1946. Patriarchal Vicar in Syria in 1950. **Pubs.:** Author of several books on liturgy and history.
Mailing address: Orthodox Patriarchate, Beirut, Lebanon.

Yago, Bernard, H E Card

Rel: Roman Catholic; Ordained on 01 MAY 1947; b. in Pass, Abidjan, Ivory Coast; celibate; Citizen of Ivory Coast.
International Organizations: Member (since 1983) of the College of Cardinals.
Member of the Congregation for the Evangelization of Peoples. Member of the Pontifical Council for Promoting Christian Unity.
Other organizations: Archbishop (since 1960), Abidjan; President, Episcopal Conference of Ivory.
Career: Created Cardinal by Pope John Paul II in 1983.
Mailing address: Archevêché, 01 BP 1287, Abidjan 01, Ivory Coast. T. (225) 44 27 85

Yamada, Ryuko

Rel: Buddhist.
International Organizations: Member of the International Network of Engaged Buddhists.
Mailing address: Seigan-Ji, Nekoya Narumi, Midori-Ku, Nagoya-Shi, Japan.

Yamamori, Ted, Dr

Rel: Christian.
International Organizations: President of the Food for the Hungry International.
Mailing address: 7729 East Greenway Suite, Scottsdale AZ 85260, USA. T. (1 602) 951 5090. Fax 951 9035. Tx 156 1607 FHI UT.

Yan-Chiu, Lam, Dr

Rel: Unknown.
International Organizations: Director of the International Society for Study of Religion and Culture.
Mailing address: PO Box 79576, Mongkok, Hong Kong.

Yaroson, Garba A

Rel: Protestant.
International Organizations: President of the World Alliance of Young Men's Christian Associations.
Mailing address: John R Mott House, Quai Wilson 37, CH-1201 Genève, Switzerland. T. (41 22) 732 31 00. Fax 738 40 15. C. FLEMGO Genève. Tx 412332 YMCA CH.

Yashiro, James Takashi

Rel: Anglican; Ordained; b. 16 JUL 1931 in Kobe, Japan; m. to Yoko Kuyama; c. 2 boys and 2 girls; Citizen of Japan.
International Organizations: Bishop (since 1985) of Kitakanto.
Other organizations: Chancellor (since 1987), Rikkyo University.
Career: Assistant Professor of Christian Studies, Momoyama Gakuin University in 1968. Assistant Professor of Church History, Rikkyo University in 1971. Professor of Church History, Rikkyo University in 1972. Visiting Professor of Religion, Washington and Lee University, Lexington VA from 1980 to 1981. **Interests:** photography, painting. **Hons.:** Japan Society of Historical Studies of Christianity Prize (1977); Hon. D.D., Virginia Theological Seminary (1982). **Pubs.:** Christianity and World History, Studies in English Reformation History, 1978 and New Canterbury Tales: A Short History of the Anglican Communion, 1987.
Mailing address: 5-8-30 Nobidome, Niiza Saitama 352, Japan. T. (81 484) 77 5098

Yaw, M Zau

Rel: Baptist.
International Organizations: General Secretary of the Burma Baptist Convention.
Mailing address: 143 St John's Road, PO Box 506, Rangoon, Myanmar (Burma).

Yazicioglu, Mustafa Sait, Prof

Rel: Muslim; Citizen of Turkey.
International Organizations: Head of the Religious Affairs in Turkey.
Mailing address: No mailing address.

Yeo, Alfred C H, Rev

Rel: Protestant.
International Organizations: Secretary-General of the Evangelical Fellowship of Asia.
Mailing address: 7 Armenian Street, 03-03, Bible House, Singapore 0617, Singapore. T. (65) 337 8414. Tx RS 42252 WEFSIN. Fax (65) 336 4113.

Yeow, Choo-Lak, Dr

Rel: Presbyterian; Ordained in 1970; b. 21 MAY 1938; m. to Teo Geok Lian; c. 1 boy and 1 girl; Citizen of Singapore; Lang. English, Malay, Mandarin, Hokkien, Hebrew (reading), Greek (reading), Latin (reading), French (reading) and German (reading).
International Organizations: Executive Director (since 1981) of the Association for Theological Education in South East Asia.
Dean (since 1981) of the South East Asia Graduate School of Theology. Consultant to the Programme for Theological Education (since 1983) of the World Council of Churches. Member of the Standing Commission, Faith and Order Commission (since 1983) of the World Council of Churches.
Other organizations: Adjunct Professor (since 1979), San Francisco Theological Seminary, San Anselmo CA; Editor (since 1983), ATESEA Occasional Papers; Editor (since 1987), *The Asia Journal of Theology*; Vice-Chairperson (since 1987), Programme for Theology and Cultures in Asia; Secretary/Treasurer (since 1989), World Conference of Associations of Theological Institutions.
Career: Secretary-General, Cambridge University Malaysian Association in 1963. President, Cambridge University Malaysian Association in 1964. Editor, *The South East Asia Journal of Theology* from 1981 to 1982. Editor, *The East Asia Journal of Theology* from 1983 to 1986.
Educ.: L.Th. from Melbourne College of Divinity, Australia (1960);

B.Th. from Trinity Theological College, Singapore (1961); B.A. from University of Cambridge, UK (1964); M.A. from University of Cambridge, UK (1968); S.T.M. from Union Theological Seminary, New York NY, USA (1965); Ph.D. from Graduate Theological Union, Berkeley CA, USA (1969). **Pubs.:** Author of 7 books and 19 articles, and book reviews and editor of 10 books.
Mailing address: 324 Onan Road, Singapore 1542, Singapore. T. (65) 344 7316.

Yeung, David P L
Rel: Christian.
International Organizations: President of the South East Asia Association for the Promotion of Christian Church Music.
President of the World Association for Chinese Church Music.
Mailing address: 12/F Sincere Insurance Bldg, 4-6 Hennessy Road, Hong Kong, Hong Kong. T. (852 5) 295577.

Yo, Hsiang-Chou, Dr
Rel: Buddhist.
International Organizations: Chairman of the Second Committee (Publication, Publicity, Education, Culture and Arts) of the World Fellowship of Buddhists.
Mailing address: 6F Suite 4, Alley 2, Lane 421, Kwang-fu South Road, Taipei 10555, Taiwan (Rep of China).

Yoder, Ruth
Rel: Christian.
International Organizations: Secretary-Treasurer of The Voice of the Martyrs.
Mailing address: er, PO Box 443, Bartlesville OK 74005, USA. T. (1 918) 337 8015. Fax 337 9287.

Yoffie, Eric H, Rabbi
Rel: Jewish: Reform.
International Organizations: Vice-President of the Union of American Hebrew Congregations.
Mailing address: 838 Fifth Ave, New York NY 10021, USA. T. (1 212) 249 0100

Yogendra, Jayadeva, Dr
Rel: Hindu; b. 27 APR 1929 in Bombay, Maharashtra, India; m. to Hansa; c. 1 boy; Citizen of India; Lang. English, Hindi, Gujarati and Sanskrit.
International Organizations: Director (since 1989) of The Yoga Institute, Santa Cruz, Bombay.
Other organizations: Principal (since 1950), Teachers' Training Institute of Yoga; President (since 1989), International Board of Yoga.
Main field of work: Education. **Educ.:** B.A. (Hons) from Bombay University, India (1947-1950); M.A. from Bombay University, India (1950-1952); Ph.D. from Bombay University, India (1958). **Interests:** education, psychology, medicine. **Pubs.:** Editor: Yoga in Modern Life, Yoga Today and Encyclopedia of Yoga (2 Vols. published out of the initially proposed 10) and Editor of *Yoga and Total Health* Magazine.
Mailing address: The Yoga Institute, Prabhat Colony, Santa Cruz East, Bombay 400055, India. T. (91 22) 6122185
Residence: See Mailing Address. T. See Mailing Address

Yogeswara Kurukkal, S Y, Brahmasri
Rel: Hindu.
International Organizations: Chief Priest of the Nallur Kandaswamy Kovil.
Mailing address: Nallur Kandaswamy Kovil, Jaffna, Sri Lanka.

Yogi, Maharishi Mahesh
Rel: Hindu; Citizen of India.
International Organizations: Founder (since 1975) of the Maharishi European Research University.
Founder (since 1976) of the Association for the World Government of the Age of Enlightenment. Founder (since 1982) of the Maharishi University of Natural Law.
Mailing address: CH-6377 Seelisberg, Switzerland. T. (41 43) 33 11 44. Tx 866018 MERU CH.

Yolton, L William
Rel: Presbyterian.
International Organizations: Executive Secretary of the Presbyterian Peace Fellowship.
Mailing address: Box 271, Nyack NY 10960, USA. T. (914) 358 4601.

Yoshida, Shodo
Rel: Buddhist: Zen; Ordained; b. 22 FEB 1940; Citizen of Japan.
International Organizations: Chief of the Kenchoji.
Main field of work: Monk. **Career:** Manager, Kenchoji-Sect Rinzai Zen Buddhism.
Mailing address: 30 Yamanouchi, Kamakura, Kanagawa 247, Japan. T. (81 467) 22 7196

Yossef, Daniel
Rel: Jewish.
International Organizations: Executive Director of the World Union of Jewish Students.
Mailing address: PO Box 7914, 91077 Jerusalem, Israel. T. (972 2) 637 482 – 639 314. Fax 637 605. Tx 25615 Att WUJS.

Youakim, Saba, B.S., H E Mgr
Rel: Greek-Melkite Catholic; Ordained on 30 NOV 1939; b. 02 JUN 1914 in Wardieh, South Lebanon; celibate; Citizen of Lebanon.
International Organizations: Archbishop (since 1970) of Petra and Philadelphia.
Member of the Congregation of St Basil.
Career: Professor, St Sauveur Seminary from 1940 to 1953. Superior, St Sauveur Seminary (2 times). Assistant to the Salvatorian Sisters from 1951 to 1953. Parish Priest at Zahleh from 1953 to 1954. General Superior, Order of the Salvatorian Monks from 1956 to 1968.
Mailing address: Greek Catholic Melchite Diocese, BP 2435, Jabal-Amman, Amman, Jordan. T. (962 6) 624 757

Youn, Victorinus Kong-Hi, Most Rev Archbishop
Rel: Roman Catholic; Ordained on 20 MAR 1950; b. 08 NOV 1924 in Jinnampo City, Korea DPR; celibate; Citizen of Republic of Korea.
International Organizations: Episcopal Moderator (since 1979) of the Justice and Peace Committee.
Other organizations: Archbishop (since 1973), Kwangju; Representative (since 1974), Kwangju Catholic College Foundation.
Career: Assistant Priest, Cathedral of Seoul (Myong-Dong) in 1950. Chaplain, Pusan UN P.O.W. Camp in 1952. Vice-President, Catholic Library, Pusan in 1954. Teacher at Song-Shin (Holy Ghost) Middle and High School in 1956. Secretary, Catholic Conference of Korea in 1960. Ordained Bishop in 1963. Ordinary, Su-Won Diocese in 1963. Administrator, Seoul Archdiocese in 1967. Chairman, Episcopal Conference of Korea in 1975. Representative, Episcopal Committee of Bicentennial of Catholic Church in Korea in 1980. **Interests:** mountain climbing. **Pubs.:** Radio Message, 1963.
Mailing address: Archbishop's House, PO Box 28, Puk Kwang Ju, Kwangju City 500 600, Korea Rep. T. (82 62) 525 9004

Young, Andrew
Rel: Congregational; Ordained in 1955; b. 12 MAR 1932 in New Orleans LA, USA; m. to Jean Childs; c. 1 boy and 3 girls; Citizen of USA.
International Organizations: Director of the Southern Christian Leadership Conference.
Other organizations: Mayor (since 1982), Atlanta.
Career: Pastor in Thomasville, GA from 1955 to 1957. Associate Director for Youth Work, National Council of Churches from 1957 to 1961. Administrator, Citizenship Education Program, United Church of Christ from 1961 to 1964. On staff, Southern Christian Leadership Conference (SCLC) from 1961 to 1970. Executive Director, SCLC from 1964 to 1970. Executive Vice-President, SCLC from 1967 to 1970. Member, House of Representatives in 1972. Re-elected 1974, 1976. Permanent Representative to UN from 1977 to 1979. Chairman, Atlanta Community Relations Commission from 1970 to 1972. Chairman, Democratic voter registration drive in 1976. Organizer, voter registration and community development programmes for civil rights movement in 1960s. **Hons.:** Medal of Freedom (1980); Legion of Honour (1982); numerous awards and honorary degrees.
Mailing address: The Office of the Mayor, 68 Mitchell Street SW, Atlanta GA 30303, USA.

Young, Norman James, Rev Prof

Rel: Uniting Church in Australia; Ordained on 22 OCT 1959; b. 31 JUL 1930 in Greelong, Victoria, Australia; m. to Barbara Mae; c. 2 boys; Citizen of Australia; Lang. English and French.
International Organizations: Professor of Systematic Theology (since 1964) of the Queen's College, University of Melbourne.
Other organizations: Chairperson (since 1991), Uniting Church Ecumenical Affairs Commission.
Main field of work: Contemporary theology. **Career:** Deputy Master, King's College, University of Queensland from 1960 to 1963. Lecturer in Theology, University of Queensland from 1961 to 1963. Visiting Professor, Princeton Theological Seminary in 1968. Visiting Professor, Yale Divinity School in 1974. Visiting Professor, Duke Divinity School in 1991. **Educ.:** B.A. (Hons) from University of Melbourne, Victoria, Australia (1950-1953); B.D. (Summa cum laude) from Drew University, Madison NJ, USA (1954-1956); Ph.D. from Drew University, Madison NJ, USA (1956-1959). **Interests:** ecumenical dialogue. **Mem.:** Melbourne Cricket Club; Kew Golf Club **Pubs.:** History and Existential Theology; Creator, Creation and Faith and "Word and Church in the New Testament".
Mailing address: 4 Queen's College, University of Melbourne, Parkville VIC 3052, Australia.
Residence: See Mailing Address. T. See Mailing Address

Young, Ralph C

Rel: Methodist.
International Organizations: Head, Geneva Office of the World Methodist Council.
Mailing address: 150 route de Ferney, CH-1211 Genève 20, Switzerland.

Yousif, Fail, Rev

Rel: Presbytarian; Ordained.
International Organizations: Moderator of the Sudan Presbytery.
Mailing address: PO Box 57, Khartoum, Sudan.

Yrigoyen, Charles, Jr, Dr

Rel: United Methodist; Ordained on 10 MAY 1962; b. 09 DEC 1937 in Philadelphia PA, USA; m. to Jean Brittingham; c. 1 boy and 1 girl; Citizen of USA; Lang. English and French.
International Organizations: Secretary-General (since 1987) of the World Methodist Historical Society.
Executive (since 1982) of the United Methodist Church, Board of Church and Society. Executive Committee (since 1986) of the World Methodist Council.
Other organizations: Professor of Church History (since 1982), Drew University, Madison NJ; Instructor in Church History (since 1982), Union Theological Seminary, New York NY.
Main field of work: Executive officer. **Career:** Pastor, Methodist Church from 1958 to 1968. Professor of Religion, Albright College, Reading PA from 1968 to 1982. **Educ.:** B.Sc. Econ. from University of Pennsylvania, Philadelphia, USA (1955-1959); B.D. from Lancaster Theological Seminary, Lancaster PA, USA (1959-1962); Th.M. from Eastern Baptist Theological Seminary, Philadelphia PA, USA (1962-1964); Ph.D. from Temple University, Philadelphia PA, USA (1964-1972). **Interests:** church history, theology. **Hons.:** DD from Albright College, Reading PA (1987). **Mem.:** American Society of Church History, Wesley Historical Society **Pubs.:** Acts for Our Time; Reformed and Catholic: Selected Writings of Philip Schaff and Catholic and Reformed: Selected Writings of John Williamson Nevin.
Mailing address: PO Box 127, Madison NJ 07940, USA. T. (1 201) 822 2787
Residence: 2 Hemlock Lane, Morristown NJ 07960, USA. T. (1 201) 538 7434

Ysern, Juan Luis

Rel: Catholic.
International Organizations: Regional Representative for Latin America of the Unda, International Catholic Association for Radio and Television.
Mailing address: Casilla 405, Blanco Encalada 793, Ancud, Chile.

Ysern de Arce, Juan Luis Y, Rt Rev

Rel: Roman Catholic; Ordained on 29 JUN 1953; b. 02 MAY 1930 in Valencia, Spain; celibate.
International Organizations: Co-Head, Latin America Regional Office of the Unda, International Catholic Association for Radio and Television.
Other organizations: Bishop (since 1974), San Carlos de Ancud.
Mailing address: Casilla 405, Blanco Encalada 793, Ancud, Chile. T. (56 656) 3083. Fax 2831. Tx 343127 CONANC CK

Yuguchi, Takashi

Rel: Lutheran.
International Organizations: Chairman of the Lutheran Communications in Asia.
Mailing address: 201 Belvedere Kudan, 2-15-5 Fujimi, Chiyoda-ku, Tokyo 102, Japan. T. (81 3) 3261 2288. Fax 3261 6033.

Yugusuk, Benjamina W

Rel: Anglican; Citizen of Sudan.
International Organizations: Archbishop of Anglican Church.
Other organizations: Bishop, Juba.
Mailing address: Clergy House, POB 110, Juba, Sudan. T. (249) 2595

Yuyama, Akira, Dr

Rel: Buddhist.
International Organizations: Director of the International Institute for Buddhist Studies.
Regional Secretary for Asia of the International Association of Buddhist Studies.
Mailing address: 5-3-23 Toranomon, Minato-ku, Tokyo 105, Japan. T. (81 3) 3434 6953. C. BUDDHOLOGY TOKYO.

Zafar-ul-Haq, Raja Mohummed

Rel: Muslim.
International Organizations: Secretary-General of the Muslim World League.
Mailing address: PO Box 537 – 538, Ummul Joud, Makkah, Saudi Arabia. T. (966 2) 5446700. Fax 5436202. Tx 540009 Rabita SJ.

Zago, Marcello, O.M.I., Fr

Rel: Roman Catholic; Ordained on 13 SEP 1959; b. 08 AUG 1932 in Villorba, Treviso, Italy; celibate; Citizen of Italy; Lang. Italian, French, English, Spanish, Lao, German and Latin.
International Organizations: Superior General (since 1986) of the Missionary Oblates of Mary Immaculate.
Member of the Congregation for Religious and Secular Institutes. Member of the Pontifical Council for Inter-Religious Dialogue.
Main field of work: Administration. **Career:** Missionary in Laos from 1959 to 1974. Professor of Theology from 1970 to 1974. Assistant General, Missionary Oblates of Mary Immaculate from 1974 to 1980. Secretary, Vatican Secretariat for Inter-Religious Dialogue from 1983 to 1986. **Educ.:** S.T.D. from Pontifical Gregorian University, Vatican City State (1970). **Pubs.:** Riti e Ceremonie in Ambiente Lao, Pontifical Gregorian University, 1971; La Chiesa in Asia Oggi, EMI, Bologna, 1983 and Il Buddhismo, Rizoli, Milano, 1984.
Mailing address: Via Aurelia 290, Casella Postale 9061, I-00165 Roma, Italy. T. (39 6) 637 02 51. FAX 637 53 22.
Residence: See Mailing Address. T. See Mailing Address

Zahlen, Liliane

Rel: Unknown.
International Organizations: Administrative Secretary of the Lay Association for the Education and Professional Training of Adolescents Europe.
Mailing address: Avenue Ste-Anne 117, B-1640 Bruxelles, Belgium. T. (32 2) 358 11 65.

Zaia, Mar Meelis, Bishop

Rel: Catholic-Apostolic Assyrian Church; Ordained; celibate.
International Organizations: Bishop of Catholic-Apostolic Assyrian Church.
Mailing address: PO Box 621, Fairfield NSW 2165, Australia.

Zain, Dato Haj Ahmad Nordin Mohd
Rel: Muslim.
International Organizations: Secretary-General of the Regional Islamic Da'wah Council of Southeast Asia and the Pacific.
Mailing address: 5th Floor, Perkim Building, Jalan Ipoh, 51200 Kuala Lumpur, Malaysia. T. (60 3) 4428166 – 4412417. Tx MA 28104.

Zakar, Polikarp, O.Cist., Fr Abbot
Rel: Roman Catholic; Ordained; celibate.
International Organizations: Abbot-General of the Order of Cistercians.
Member of the Congregation for Religious and Secular Institutes. Advisor to the Religious Institutes Section of the Congregation for Religious and Secular Institutes.
Other organizations: Rector, College of Religion of Saint Bernard.
Mailing address: Piazza Tempio di Diana 14, I-00153 Roma, Italy. T. (39 6) 574 36 94, (39 6) 575 51 10

Zakka, Ignatius, I, H H
Rel: Syrian Orthodox.
International Organizations: Patriarch of Antioch and All the East.
Mailing address: Bab Tooma, Damascus, Syria.

Zanolini, Gualtiero, Dr
Rel: Roman Catholic.
International Organizations: Secretary-General of the International Catholic Conference of Scouting.
Mailing address: CICS, Piazza Pasquale Paoli 18, I-00186 Roma, Italy. T. (39 6) 68 65 270. Fax (39 6) 68 65 211.

Zantedeschi, Giovanni B., I.C., Fr
Rel: Roman Catholic; Ordained; celibate.
International Organizations: Superior General of the Institute of Charity.
Mailing address: Via di Porta Latina 17, I-00179 Roma, Italy. T. (39 6) 70491777.

Zappa, Mario Enrico, Rev
Rel: Roman Catholic; Ordained on 12 JUN 1965; b. 10 APR 1940 in Triuggio, Italy; celibate; Citizen of Italy; Lang. Italian and French.
International Organizations: Procurator-General of the Prêtres du Sacré-Coeur de Jésus de Bétharram.
Main field of work: Ministry. **Educ.:** Laurea Letteres e Fil. from University of Rome, Italy (1981).
Mailing address: Via Angelo Brunetti 27, I-00186 Roma, Italy. T. (39 6) 321 9144.

Zega, Leonardo, S.S.P., Rev
Rel: Roman Catholic; Ordained; celibate.
International Organizations: Director-General of the Gruppo Periodici Paolini.
Mailing address: Via Giotto 36, I-20145 Milano, Italy.

Zehr, Daniel
Rel: Mennonite.
International Organizations: Executive Director for Canada of the Mennonite Central Committee.
Mailing address: 134 Plaza Drive, Winnipeg MB R3T 5K9, Canada.

Zeitz, Susanne, Dr
Rel: Roman Catholic; b. 11 APR 1954 in Vienna, Austria; single; Citizen of Austria; Lang. German, French, English and Italian.
International Organizations: Secretary-General (since 1986) of the International Secretariat of Catholic Secondary School Teachers.
Other organizations: Member of the Board (since 1986), Teacher's Union, Austria; Teacher's Representative (since 1987), Secondary School.
Main field of work: Secondary Education. **Educ.:** Ph.D. Bio.; M.A. Eng. and Bio. from Vienna University, Austria (1972-1982). **Pubs.:** Thesis: *Oligocarbophite Bakterien in der Alten Donau*, 1981.
Mailing address: Florianigasse 30/43, A-1080 Wien, Austria. T. (43 1) 43 66 573
Residence: See Mailing Address.

Zhao Puchu, Z
Rel: Buddhist; b. in Taihu Co., Anhui Province, China; Citizen of China.
International Organizations: President (since 1979) of the Chinese Buddhist Association.
Other organizations: President (since 1984), Society for the Study of Sino-Japanese Relations; Vice-Chairman (since 1988), 7th CPPCC.
Career: Member, as a Buddhist representative, 1st CPPCC, Beijing in 1949. Vice-President and Secretary-General, China's Buddhist Association from 1953 to 1979. Vice-Chairman, Sino-Japanese Friendship Association from 1963 to 1966. Member, Presidium 4th NPC in 1975. Vice-Chairman, 6th CPPCC from 1983 to 1988. **Hons.:** Buddhist Evangelist Prize (1982); Niwano Peace Prize (1985). **Pubs.:** Selected Poetry Dishuiji, 1964 and Collection of Poems Pianshiji, 1978.
Mailing address: Standing Committee of Chinese, People's Political Consultative Conf., Taiping Qiao Road, Beijing, Rep of China.

Ziegler, Martin Arthur Erich, Rev
Rel: Evangelical Lutheran; Ordained on 27 APR 1958; b. 01 OCT 1931 in Berlin, Germany; m. to Gertraude Paetzold; c. 3 boys and 1 girl; Citizen of Germany; Lang. German and English.
International Organizations: Director (since 1991) of the Hoffnungstaler Anstalten Lobetal bei Bernau.
Main field of work: Ministry. **Career:** Priest in Grosskayna/Kr. Merseburg from 1957 to 1962. Priest in Kötzschen/Kr. Merseburg from 1962 to 1968. Cathedral priest and Superintendent, Merseburg from 1968 to 1974. Director of Internal Mission and Welfare, Evangelical Church in Berlin-Brandenburg from 1974 to 1983. Secretary-General, Federation of Evangelical Churches in GDR from 1983 to 1991. **Educ.:** Kand. Theol. from Humboldt University of Berlin, Germany (1950-1955). **Interests:** literature, horticulture. **Hons.:** Bruno-Kreisky-Preis (1990); Bundesverdienstkreuz (1991). **Mem.:** Diakonischer Rat Berlin-Brandenburg **Pubs.:** "Auf dem Weg zu einer missionierenden Gemeinde", 1964; "Bruderschaftliche Leitung", 1973 and "Leiten als Dienst", 1989.
Mailing address: Hoffnungstaler Anstalten, Bodelschwinghstrasse 4, 16321 Lobetal, Germany. T. BERNAU 66218
Residence: Schuberstrasse 13, Postfach 3412, 16552 Schildow, Germany. T. Mühlenbeck 344

Zinovieff, Maurice
Rel: Orthodox.
International Organizations: President of the Comité Orthodoxe des amitiés françaises dans le monde.
Mailing address: 61 rue de Vaugirard, F-75006 Paris, France. T. (33 1) 40 69 12 04.

Zirilli, Giuseppe, C.SS.R., Fr
Rel: Roman Catholic; Ordained; celibate.
International Organizations: Procurator-General of the Congregation of the Most Holy Redeemer.
Mailing address: Via Merulana 31, CP 2458, I-00185 Roma, Italy. T. (39 6) 731 58 41.

Zita, Amosse Baltazar, Rev
Rel: Presbyterian.
International Organizations: President of the Synodal Council of the Presbyterian Church of Mozambique.
Mailing address: Caixa Postal 21, Maputo, Mozambique.

Zitko, Howard John, Dr
Rel: Christian: Metaphysical; Ordained on 12 NOV 1955; b. 26 OCT 1911 in Milwaukee WI, USA; divorced; c. 1 boy and 2 girls; Citizen of USA; Lang. English.
International Organizations: Founder/President (since 1946) of the World University Roundtable.
President (since 1967) of the World University. Chancellor (since 1988) of the International Association of Educators for World Peace.
Other organizations: Chief Executive Officer, Desert Sanctuary Regional Campus, Benson AZ.
Main field of work: Educational administrator, metaphysical research.
Career: Vice-President in charge of cooperative development and co-founder, Lemurian Fellowship from 1936 to 1946. Minister, Lemurian Fellowship Chapel, Milwaukee WI from 1939 to 1941. Minister, Temple of the Jewelled Cross, Los Angeles CA from 1942 to 1946. Minister, Church of the Abundant Life, Huntington Park CA from 1955 to 1959.

Educ.: Studied from University of Wisconsin – Milwaukee, College of Letters and Science, USA (1929-1931); Studied Jour., Bus. Law, and Int. Rel. from University of California, Los Angeles, USA (1946-1948); D.D. from Golden State University at Los Angeles CA, USA (1949). **Interests:** peace education, esoteric studies. **Hons.:** Knight Commander, Royal Knights of Justice; Laureate Man of Letters, World Congress of Poets; Malayan Kris, Homeopaths of Malaya (now Malaysia); Silver Chalice, World University Roundtable; Gold Medal, World Congress of Poets; Dr of Indian Philosophy, The Indian Association of America, Shawnee OK (1946); Dr of Humane Letters, American World Patriarchates, New York (1972); Dr of Modern Humanities, L'Université Libre Asia, Karachi, Pakistan (1970); Dr of Letters, China Academy, Taipei, Taiwan, Rep. of China (1977); Dr of International Relations, Universal Orthodox College, Iperu Remo, Ogun State, Nigeria (1985). **Mem.:** Benson Chamber of Commerce, United Poets Laureate International, International Parliament for Safety and Peace, Sierra Club, American Association of Retired Persons (USA), United Nations Association of the United States of America, American Civil Liberties Union (NY and AZ), World University (Benson AZ), World University Roundtable (Benson AZ), American Ministerial Association (Santa Monica CA) **Pubs.:** *The Lemurian Theo-Christic Conception*, 1936 (word text in religious philosophy), originally published by the Lemurian Fellowship and currently a publication of the World University Roundtable; *New Age Tantra Yoga*, 1974, subtitled *The Cybernetics of Sex and Love* (a philosophy of marriage for the 21st century) and *World University Insights*, 1980, originally presented as *New Age Perspectives in Questions and Answers*.
Mailing address: Desert Sanctuary Campus, PO Box 2470, Benson AZ 85602, USA. T. (1 602) 586 2985.
Residence: See Mailing Address. T. See Mailing Address

Ziviani, Enzo
Rel: Christian.
International Organizations: Representative of the Association for Missionary Cooperation.
Mailing address: Galleria Risorgimento 3, I-37045 Legnago, Italy. T. (39 442) 28 333. Fax 60 01 11. Tx 351627 I.

Zoccola, Filippo G, F.S.F., Br
Rel: Roman Catholic; Ordained; celibate.
International Organizations: Procurator-General of the Institut des Frères de la Sainte-Famille de Belley.
Mailing address: Viale Aurelio Saffi 24, I-00152 Roma, Italy. T. (39 6) 581 38 41

Zodhiates, Spiros, Dr
Rel: Christian.
International Organizations: President of the AMG International.
Mailing address: 6815 Shallowford Road, Chattanooga TN 37421, USA. T. (1 615) 894 6060. Fax 894 6863.

Zoghbi, Elie, H E Mgr
Rel: Greek Catholic; Ordained on 20 JUL 1936; b. 09 JAN 1912 in Cairo, Egypt; celibate; Citizen of Egypt.
International Organizations: Archbishop Emeritus (since 1988) of Baalbek.
Career: Professor, St Anne Junior Seminary from 1937 to 1941. Parish Priest in Egypt from 1941 to 1949. Parish Priest in Alexandria from 1949 to 1951. Archimandrite in 1951. Consecrated Bishop Titular and Patriarchal Vicar-General for Egypt and Sudan in 1954. Archbishop, Baalbek from 1968 to 1988.
Mailing address: BP 58, Zouk-Mikael, Lebanon.

Zokoue, Isaac, Dr
Rel: Evangelical.
International Organizations: Principal of the Bangui Evangelical Graduate School of Theology.
Mailing address: BP 988, Bangui, Central African Rep. T. (236 61) 613330.

Zopa Rinpoche, Thubten Z, Lama
Rel: Buddhist: Mahayana; b. in Lawudo, Solo Khumbu, Nepal; single; Citizen of Nepal; Lang. Tibetan, English and Nepalese.
International Organizations: Spiritual Director (since 1984) of the Foundation for the Preservation of the Mahayana Tradition.
Main field of work: Ministry. **Career:** Spends almost all his time traveling around the world leading Buddhist courses and retreats. **Educ.:** Sera Je Monastery, Lhasa. **Interests:** religious art, poetry, translation. **Pubs.:** Wisdom Energy (I and II), Wisdom Publications, Boston; Kapan Meditation Course, 1982 and 1983, Wisdom Publications, Boston and Transforming Problems, Wisdom Publications, Boston.
Mailing address: PO Box 1778, Soquel CA 95073, USA. T. (1 408) 476 8435. Fax 476 4823.
Residence: See Mailing Address. T. See Mailing Address

Zopfi, Jakob, Rev
Rel: Pentecostal.
International Organizations: Chairman of the Pentecostal World Conference.
Secretary of the European Pentecostal Fellowship.
Mailing address: PO Box 98, CH-6376 Emmetten, Switzerland. T. (41 41) 64 25 55. Fax 64 32 55.

Zora, Hanna
Rel: Chaldean Catholic; Ordained on 10 JUN 1962; b. 15 MAR 1937; celibate; Citizen of Iran.
International Organizations: Archbishop (since 1974) of Ahwaz.
Mailing address: Archbishop's House, POB 61956, Naderi St, Ahwaz, Iran Islamic Rep. T. (98 61) 24890

Zoungrana, Paul, H E Card
Rel: Roman Catholic; Ordained on 02 MAY 1942; b. 03 SEP 1917 in Ouagadougou, Burkina Faso; celibate; Citizen of Burkina Faso.
International Organizations: Member (since 1965) of the College of Cardinals.
Member of the Sacred Congregation for the Sacraments and Divine Worship. Member of the Congregation for the Evangelization of Peoples. Member of the Congregation for Religious and Secular Institutes. Member of the Pontifical Council for Pastoral Assistance to Health Care Workers.
Other organizations: Archbishop (since 1960), Ouagadougou; Member, Council of Cardinals for the Study of Organization and Economic Questions Relating to the Holy See.
Career: Created Cardinal by Pope Paul VI in 1965.
Mailing address: Archevêché, 01 BP 1472, Ouagadougou 01, Burkina Faso. T. (226) 30 67 04

Zubitegui, Juan, M.SS.CC., Fr
Rel: Roman Catholic; Ordained; celibate.
International Organizations: Procurator-General of the Missionaries of the Sacred Hearts of Jesus and Mary.
Mailing address: Vicolo del Curato 12, I-00186 Roma, Italy. T. (39 6) 686 20 45.

Zukowski, Angela Ann, Sr
Rel: Roman Catholic; Ordained; celibate.
International Organizations: Co-Head, North America Regional Office and First Vice-President of the Unda, International Catholic Association for Radio and Television.
Mailing address: MHSH, Centre for Religious Telecom, University of Dayton, 102 Alumni Hall, Dayton OH 45469, USA.

Zülch, Tilman
Rel: Unknown.
International Organizations: President of the Society for Threatened Peoples.
Mailing address: Postfach 2024, Düstere Str 20, D-37073 Göttingen, Germany. T. (49 551) 49906 – 49900. Fax 58028.

Zulooza, Ismael, Fr
Rel: Roman Catholic.
International Organizations: President of the Jesuit Conference of East Asia.
Mailing address: PO Box 1264, 1100 Quezon City, METRO Manila, Philippines. T. (63 2) 96 59 64. C. PEOPLEMANILA. Fax (63 2) 921 75 34.

Zur, Giorgio, H E Mgr

Rel: Roman Catholic; Ordained on 10 OCT 1955; b. 15 FEB 1930 in Görlitz, Germany; celibate.
International Organizations: Apostolic Pro-Nuncio to India and Nepal (since 1990) of the Roman Catholic Church.
Other organizations: Titular Archbishop, Sesta.
Mailing address: 50-C, Niti Marg, Chanakyapuri, New Delhi 110021, India. T. (91 11) 606 520 – 606 921. Fax 687 42 86

Zyss, W

Rel: Jewish.
International Organizations: UNESCO Representative of the B'nai B'rith International.
Mailing address: 11 Villa Croix-Nivert, F-75015 Paris, France. T. (33 1) 45 67 48 25.

Survey of Countries

Name Index Arranged by Country

Chile

China

Colombia

Croatia

Cuba

Cyprus

Czech Republic

Denmark

Ecuador

Egypt

Estonia

Ethiopia

Fiji

Finland

France

Germany

Ghana

Greece

Haiti

Hungary
Bartha, Tibor
Benda, Kálmán, Prof
Hegedüs, Lóránt, Bishop, Dr
Kada, Lajos, H E Mgr
Kocsis, Elemér, Dr
Ma'rkus, Miha'ly, Dr
Mészáros, István, Dr
Paskai, László, O.F.M., H E Card Dr
Pasztor, Janos Dezso, Rev Prof Dr
Péter, János
Szücs, Ferenc, Prof
Sándor, Cserháti, Dr
Tamás, Bertalan, Rev
Tenke, Sándor, Prof
Tóth, Károly, Bishop Dr

Iceland
Gudmundsson, J Eric, Pastor

India
Acharuparambil, Daniel, O.C.D., Rev Prof
Amritanand, Joseph, Rt Rev
Andiappan, Sundara Yogasali Asana R, Prof
Arnold, Perumalla B, Rev Dr
Arokiasamy, Arul Maria, Rev Roshi
Askari, Hasan, Prof
Bakula, Kushok, Rev Rinpoche
Bali, Dev Dutta, Pundit
Basu, Samar, Sri
D'Souza, Eugene, M.S.F.S., Most Rev
D'Souza, Henry Sebastian, Most Rev Archbishop
Dastur, Meherjirana Kaikobad, Vada Dasturji
David, S Immanuel, Rev Prof Dr
Divarkar, Parmananda Roque, Dr
Fernandes, Angelo Innocent, Most Rev
Giri, Chidda, Prof Dr
Giri, Rakesh, Dr
Gorai, Dinesh Chandra, Rt Rev
Gorekar, Nizamuddin S, Prof Dr
Gregorios, Paulos Mar, Metropolitan Dr
Harbans Singh, H, Dr
Jagir Singh, J, Dr
Jain, Sudhanshu Kumar, Dr
Janki, Brahma Kumari
Jaswinder Singh, J, Gyani
Jayadeva Yogendra, Hansa J, Smt
Jha, Pitambar, Dr
Joshi, Sudarshan U
Kajaria, Nirmala Ratanshi, Dr
Kakkanattu, Benicasia, C.M.C., Sr
Kannan, Aiyasamy, Dr
Kariamadam, Paul, Rev Dr
Karunan, Victor Prasad, Dr
Kaudre, Bansilal Gangaram, Shri
Kuttickattu, Sebastian John, Rev
Lourdusamy, Simon D, H E Card
Maharaj Ji, M, Shri
Manjit Singh, M, Singh Sahib Jathedar
Mariella, M, A.C., Sr
Mehta, Priti Kiritkumar, Dr
Naidu, Vijaykumar Joseph, S.J., Rev
Neki, Jaswant Singh, Prof
Oommen, Tharaileth Koshy, Prof Dr
Padiyara, Anthony, H E Card
Pal, Raj Sawindar, Mahatma
Pandit, M P, Sri
Panjabi, Mohini, Sr
Paulose, Kallapara V, Rev Prof
Picachy, Lawrence Trevor, S.I., H E Card
Pimenta, Simon Ignatius, H E Card
Powathil, Joseph, Most Rev
Prakashmani, Brahma Kumari
Raman, Mannampurath Kavu, Master
Ramasamy Adigal, Santhalinga R, Rev Kailaimunivar
Robinson, Gnana, Rev Prof Dr
Roy Choudhury, Punyabrata R
Sarasvati, Divyanand, Dr
Saraswati, Kumari Shiva, Shiva Saraswati
Satyapati Parivrajak, S, Swami
Sen, Noel Sushil, Rev
Shankar, Ganesh, Dr
Sitadevi Yogendra, S, Smt
Teresa, T, Mother
Thangalathil, Benedict Mar Gregorios, O.I.C., Most Rev Dr
Thara Bhai, L T, Dr
Vedanti, Brahma Kumari, Sr
Vethathiri, Yogiraj, Maharishi
Viquar, Syed Nasiruddin
Ward, Daniel T, Bishop
Williams, Theodore, Rev Dr
Wilson, Henry Steward, Rev Dr
Yogendra, Jayadeva, Dr
Yogi, Maharishi Mahesh

Indonesia
Ali, H A Mukti
Christano, Peter Charles, Rev
Darmojuwono, Justin, H E Card
Satini, Louisie, Dr

Iran
Bedini, Ignazio, S.D.B., Most Rev Archbishop
Issayi, Youhannan Semaan, Most Rev Archbishop
Khamenei, Hojatoleslam Ali
Manoukian, Ardak, H E Mgr
Meram, Thomas, Most Rev Archbishop
Montazeri, Hussein Ali, Ayatollah
Moubad, M
Nakhjavani, Ali-Yullah
Taherzadeh, Adib
Tekeyan, Vartan, Rt Rev Dr
Zora, Hanna

Iraq
Bidawid, Raphael I, I, His Beatitude
Darmo, Mar Thomas, H B
Garmo, Georges, Most Rev
Iwas, Ignatius Zakka I, H H

Ireland
Caird, Donald Arthur Richard, Most Rev Dr
Callaghan, William Sydney, Rev
Casey, Eamonn, Most Rev
Comiskey, Brendan Oliver, SS.CC., Rt Rev
Daly, Edward Kevin, Rt Rev
Daly, Gabriel Conor, OSA, Fr
Donlon, Kevin Hastings, CSSR, Rev Fr
Faul, Denis O'Beirne, Rev
Griffin, Victor Gilbert Benjamin, Rev
Grimes, Edward, Rev
Hurley, Michael Anthony, S.J., Rev
Keating, Jeremiah Columba, C.F.C., Br
Lamont, Donal Raymond, O.Carm., H E Mgr
Mackey, James Patrick, Prof
McGill, Maurice, M.H.M., Very Rev
O'Meara, Noel P, Rev
Poyntz, Samuel Greenfield, Rt Rev Dr
Roddy, Joan, D.M.I., Sr
Simms, George Otto, Most Rev
Taherzadeh, Adib
Walsh, Patrick Joseph, Rt Rev

Israel
Assaf, David
Bender, Karpol, Rabbi
Bonfil, Robert, Rabbi Prof
Davis, Moshe, Prof
Dinitz, Simcha
Diodoros, D, I, His Beatitude
Eliahu, Mordechai, The Chief Rabbi
Even-Chen (Edelstein), Yaacov E, Rabbi
Fackenheim, Emil Ludwig, Rabbi Prof
Grünewald, Hans Isaak, Rabbi
Hammer, Reuven, Rabbi Prof
Levine, Lee I, Rabbi Prof
Michman, Dan, Prof
Rosen, David Shlomo, Rabbi
Shapiro, Avraham, The Chief Rabbi

Italy
Agazzi, Evandro, Prof Dr
Antonelli, Ferdinando Giuseppe, O.F.M., H E Card
Ballestrero, Anastasio Alberto, O.C.D., H E Card
Battista, Armando P Gregorio, O.Cist, Fr Abbot
Berloco, Giacinto, H E Mgr
Biffi, Giacomo, H E Card
Boaga, Emmanuele, O.Carm., Rev Prof
Borgomeo, Pasquale, S.J., Rev
Bouchard, Giorgio, Rev
Bressan, Luigi, H E Mgr Dr
Canestri, Giovanni, H E Card
Caprio, Giuseppe, H E Card
Carpino, Francesco, H E Card
Casaroli, Agostino, H E Card
Casoria, Giuseppe, H E Card
Cheli, Giovanni, Mgr
Ciappi, Mario Luigi, O.P., H E Card
Colombo, Giovanni, H E Card
Cortesi, Filipo, Dr
Curioni, Cesare, Mgr
Curis, Carlo, H E Mgr
D'Agostino, Federico
Davitti, Michele, Rev Fr
De Bonis, Donato, H E Mgr
De Stefano, Lucio, M.S.C., Rev
Dezza, Paolo, S.I., H E Card
Di Berardino, Angelo , Rev Prof Dr
Fabbri, Fabio, Mgr
Faccani, Clemente, H E Mgr
Farina, Raffaele, Rev Prof Dr
Felici, Angelo, H E Card
Giampiccoli, Franco, Rev
Giordano, Michele, H E Card
Guerri, Sergio, H E Card
Innocenti, Antonio, H E Card
Loreti, Antonio Toto, Rev
Lubich Silvia, Chiara L
Macioti, Maria Immacolata
Magnoni, Antonio, H E Mgr
Marchetto, Agostino, H E Mgr Dr
Martini, Carlo Maria, S.I., H E Card
Mondin, Battista, Prof
Moreni, Gian Vincenzo, H E Mgr
Navone, John Joseph, S.J., Rev Prof
Oddi, Silvio, H E Card
Pace, Enzo, Prof
Pantane, Nino, Br
Pappalardo, Salvatore, H E Card
Pastò, Gianluigi, Rev
Paupini, Giuseppe, H E Card
Pavan, Pietro, H E Card
Perino, Renato, S.S.P., Rev
Piovanelli, Silvano, H E Card
Poletti, Ugo, H E Card
Revollo Bravo, Mario R, H E Card
Righi-Lambertini, Egano, H E Card
Sabattani, Aurelio, H E Card
Schembri, Salvatore, M.S.P., Rev
Sensi, Giuseppe Maria, H E Card
Silvestrini, Achille, H E Card
Tarzia, Antonio, S.S.P., Rev Dr
Ursi, Corrado, H E Card
Viganò, Egidio D, S.D.B., Rev
Zago, Marcello, O.M.I., Fr
Zappa, Mario Enrico, Rev

Ivory Coast
Yago, Bernard, H E Card

Jamaica
Reece, Donald James, Most Rev
de Souza, Ernest Henriques, Rev

Japan
Abeno, Ryusho
Adachi, Daishin
Akagi, Yu
Akamatsu, Shinjo
Akihito, A, Emperor
Akiya, Einosuke
Akiyama, Yuga
Amabe, Mistuhiko
Aoki, Jiun, Ven Dr
Aonuma, Jakujin
Aoyama, Shundo
Arita, Hiroshige
Aso, Bunno
Eda, Koten
Fuji, Koei
Fujii, Jitsuo
Fukui, Kojun
Fukutomi, Settei
Funamoto, Sakao
Furuse, Nichiu
Futaba, Kenko
Hasuike, Nitto
Hasuo, Zenryu
Henjo, Daiyu
Hirayama, Shoji, Rev

Jordan

Kenya

Lebanon

Libya

Lithuania

Luxembourg

Madagascar

Malawi

Malaysia

Malta

Mauritius

Mexico

Mongolia

Morocco

Nepal

Netherlands

New Zealand

Nicaragua

Nigeria

Norway

Pakistan

Palestinian

Panama

Papua New Guinea

Peru

Philippines

Poland

Benedyktowicz, Witold Kazimierz, Rev
Bolonek, Janusz, H E Mgr
Dabrowski, Ray, Pastor
Deskur, Andrzej Maria, H E Card
Doroszkiewicz, Bazyli, H E M
Glemp, Józef, H E Card
Gulbinowicz, Henryk Roman, H E Card
John Paul, J, II, H H Pope
Koprowski, Andrzej, S.J., Rev Dr
Kosela, Krzysztof
Kowalczyk, Józef, Most Rev Archbishop Dr
Macharski, Franciszek, H E Card

Portugal
Afonso, Manuel Castro, S.M.P., Rev
Alves de Sousa, Pio G A, Prof
Alves, Manuel Isidro, Prof
da Silva Lima, José, Prof
da Silva Sebastião, Manuel Martins d, C.S.Sp., Rev
dos Santos, Alexander d, O.F.M., H E Card
Monteiro de Castro, Manuel M, H E Mgr
Osswald, Walter, Prof
Ribeiro, António, H E Card
Suenens, Leo Jozef, H E Card

Republic of Korea
Cho, Paul Yonggi, Rev Dr
Hong, H, Rev
Kim, Stephen Sou-Hwan, H E Card
Lee, Soo-Min, Dr
Moon, Sun Myung, Rev
Youn, Victorinus Kong-Hi, Most Rev Archbishop

Russia
Aleksandr, A., Archbishop
Gundiaev, Kirill (Vladimir), H E
Pimen, (Sergey Mikhailovich Isvekov), Patriarch

Rwanda
Ruhumuriza, Aaron, Bishop

Saudi Arabia
Al-Ghamdi, Ali Abdullah Mughram, Dr
Fahd Ibn Abdul Aziz, F, King
Hafiz, Omar Z A
Madani, Ghazi Obaid, Dr
Mukhtar, Ali Mohamad Abbas
Nasseef, Abdullah Bin Omar, Dr
al Sidais, Abdallah Abdulaziz a

Senegal
Thiandoum, Hyacinthe, H E Card

Seychelles
Chang-Him, French Kitchener, Most Rev
Napier, Albert

Singapore
Mehervan Singh, M, Dr
Quek, Kiok Chiang, Rev Dr
Yeow, Choo-Lak, Dr

Slovakia
Tomásek, Frantisek, H E Card
Tomko, Jozef, H E Card

South Africa
Boesak, Allan, Rev
Boshoff, Carel Willem Hendrik, Rev Dr
Buchanan, George Duncan, Rt Rev
Burden, Jasper Jacobus, Prof
Burnett, Bill Bendyshe, Most Rev Dr
Buthelezi, Manas, Rt Rev
Chikane, Frank, Rev Dr
Cloete, Glynn
Coertzen, Pieter, Prof
Combrink, Hans Jacob Bernardus, Prof
Du Preez, Jannie D, Prof
Hendrickse, Helenard Joe (Alan), Rev
Hurley, Denis Eugene, O.M.I., Rev
Kinghorn, Johann, Prof
Krüger, Jacobus Stefanus, Prof
Maartens, Pieter Jacobus, Prof Dr
McCann, Owen, H E Card
Mkhatshwa, Smangaliso, Rev
Napier, Wilfrid Fox, Rt Rev
Naudé, Beyers, Rev Dr
Naudé, Jacobus Adriaan, Prof
Ndungane, Njongonkulu Winston Hugh, Rt Rev
Pauw, Christoff Martin, Dr
Pieterse, Jude, P.F.M., Br
Potgieter, Pieter Cornelius, Prof
Robinson, Philippus Johannes, Prof
Russell, Philip Welsford Richmond
Smith, Nico J, Dr
Solomon, Lorraine Ellen
Taylor, Robert Selby, Most Rev Dr
Tutu, Desmond Mpilo, Most Rev
van Zyl, Daniel Cornelis, Dr

Spain
Abrego de Lacy, José Maria A, Prof
Abril y Castelló, Santos , H E Mgr
Angulo Quilis, José A, T.O.R., Most Rev
Blajot, Victor
Enrique y Tarancón, Vicente E, H E Card
Feliú Loriente, María del Pilar F, STJ, Sr
García del Cerro, Cesáreo G, Rev
González Martín, Marcelo G, H E Card
González-Montes, Adolfo, Rev Prof Dr
Jubany Arnáu, Narciso J, H E Card
Lobato Casado, Abelardo L, Prof Dr Rev
López-Gay, Jesús, Rev Prof Dr
Martínez Somalo, Eduardo M, H E Card
Orbegozo Jáuregui, José Agustin O, C.P., Rev
Portillo, Alvaro del P, Most Rev Mgr
Suquía Goicoechea, Angel S, H E Card
Tragan, Pius, Rev Prof
Trujillo López, Ildefonso T, Rev
Vives Aguilella, Juan Antonio V, T.C., Rev Dr

Sri Lanka
Ariarajah, Wesley, Rev Dr
Balasuriya, Stanislaus Tissa, Rev
Edirisinghe, Albert
Fernando, Nicholas Marcus, Most Rev Dr
Gurugé, Ananda Weihena Palliya, Dr
Jayasuriya, Gamini
Rosa, Alfred Christopher Henricus, Rev Dr
de Silva, Lily, Prof Dr

Sudan
Ador, Samuel Nyawelo, Rev
Ayik, Arop Yor, Prof
Elias, Nabeel Radi, Prof
Yugusuk, Benjamina W

Swaziland
Gamedze, A B, Dr

Sweden
Jonsson, Bernt Erik
Larsson, John Alfred, Commr
Rogestam, Christina
Sundby, (Carl) Olof (Werner)
Taranger, Bengt
Wallin, Ingvar

Switzerland
Bovon, François, Prof Dr
Böckle, Franz, Fr
Chevrier, Jôdô André, Rev Prof Dr
Ducor, Jérôme, Rev Prof
Elsener, Josef, Rev Fr
Eracle, Jean, Rev
Fischer, Jean Eugène
Fuchs, Eric V, Prof Dr
Gaspoz, Jean-Marc, O.F.M.Cap., Fr
Gebs, Simone
Handschin, Hanna
Huguenin, Willy, Commr
Küng, Hans C, Prof Dr
Mottu, Henry, Prof
Mumemthaler, Mariette
Nicole, Jacques Henri Georges, Prof Dr
Renggli-Enderle, Claire
Rochette, Jean-François, Dr
Rüegg, François, Dr
Sartorius, Willi, Rev
Saurma, Adalbert
Stückelberger, Johannes Georg, Rev
Weder, Dölf, Rev Dr

Syria
Capougi, Hilarion Bashir, H E Mgr
Dick, Ignace, Rev Archimandrite
Edelby, Neophytos Elias, H E Mgr
Farah, Augustin Joseph, H E Mgr
Hayek, Ignace Antoine, II, His Beatitude
Hazim, Ignace, Mgr
Kuftaro, Ahmad
Mounayer, Eustache Joseph, H E Mgr
Nasrallah, Joseph Moussa, Fr
Nijmé, Elias, B.A., H E Mgr
Raï, Boutros, B.A., H E Mgr
Samaha, Basilios, H E Mgr
Schelhoth, George, H E Mgr
Tawil, Joseph Elias, H E Mgr
Yacoub, Ignatius, III, H E Mgr

Taiwan (Republic of China)
Tien, Stella L

Tanzania
Mbeba, Joram Magayi, Rt Rev
Ramadhani, John Acland, Most Rev
Rugambwa, Laurean, H E Card

Thailand
Kitbunchu, Michael Michai, H E Card
Komolmas, Prathip Martin, Dr
Ruangskul, Prasert

Tibet
Dalai Lama, D, H H The
Dhargyey, D, Geshe
Samdhong Rinpoche, S, Prof

Tunisia
Belkhodja, Muhammad Habir

Turkey
Asseo, David, Rabbi
Varsamis, Thomas, Rev Archimandrite
Yazicioglu, Mustafa Sait, Prof

Uganda
Katoneene, Jonah Mwesigwa, Rev
Kivengere, Festo, Rt Rev

Ukraine
Hermaniuk, Maxim, Most Rev
Lubachivsky, Myroslav Ivan, H E Card

United Kingdom
Alves, Colin, M.Litt.
Anderson, Hector David, Rev
Appleton, George, Rt Rev
Ashraf, Syed Abu Nasr Ali, Prof Dr
Baker, John Austin, Rt Rev
Ball, Bryan William, Dr
Barbour of Bonskeid, Robert (Robin) Alexander Stewart B, Very Rev Prof
Barker, Eileen Vartan, Dr
Barlow, Bernard Francis, Rev
Barrington-Ward, Simon, Rt Rev
Bavin, Timothy John, Rt Rev
Beck, Brian Edgar, Rev
Beckford, James Arthur, Prof
Blanch, Stuart Yarworth, Rt Rev and Rt Hon Lord
Bleakley, David Wylie, Rt Hon
Bloom, William, Dr
Booth, William Bramwell, Col
Booth-Clibborn, Stanley Eric Francis, Rt Rev
Bourdeaux, Michael Alan, Rev Canon

Uruguay

USA

Vanuatu

Vatican

Venezuela

West Indies

Western Samoa

Yemen

Yugoslavia

Zaire

Zambia

Zimbabwe

Survey of Religions

Name Index
Arranged by Religions

Shiozawa, Daijo
Shiragaki, Konin
Smith, Steven Verney
Subhuti, S, Dharmachari
Suvajra, S, Dharmachari
Suzuki, Koei
Suzuki, Ryowa
Suzuki, Shunon
Tagawa, Jokaku
Takada, Koin
Takamatsu, Ryoshu
Tani, Kogetsu
Tatebe, Zuiyu
Thai-arry, Phallop
Thien-Dinh, Tich, Ven
Thubten Chodron, T, Ven
Tien, Stella L
Todo, Kyoshun
Todo, Shunsho
Tokutomi, Giko
Toyama, Koshi
Tsurumi, Shoseki
Umeda, Shinryu
Wan-tin, Wong
William, Stanley
Yamada, Ryuko
Yo, Hsiang-Chou, Dr
Yuyama, Akira, Dr
Zhao Puchu, Z

Buddhist: Jôdo-Shinshú
Eracle, Jean, Rev
Kell, Gerhard
Ohtani, Monshu Koshin
Peel, Shitoku Adriaan, Rev Prof

Buddhist: Kagyupa
Gyaltsab Rinpoche, G, H E
Kongtrul Rinpoche, K, H E
Shamar Rinpoche, S, H E
Situ Rinpoche, S, H E

Buddhist: Mahayana
Alongina, Xavi
Andrew, Bill
Aroncena, Isabel
Bajardi, Elia
Bajardi, Rodolfo
Bertels, Marcel
Biyrje, Paul
Bosa, Siliana
Braun, Ingrid
Bucherl, Gisela
Buscicchio, Mavi
Castles, Margaret
Cole, Owen
Corradi, Carol
De Wijs-Koolkin, Paula Beth
Drukier, Elisabeth
Emonson, Sue
Femenia, Marisa
Ferranti, Andriana
Foulkes, Garry
Fraser, Roy
Fraser, Sally
Furio, Angelo
Ghirardo, Danilo
Grants, Alnis
Green, Judy
Gyatso, Thubten, Ven Dr
Hawter, Ken
Horrocks, Harvey
Howland, Frances
Huet, Denis
Jones, Sean
Kakaria, Sunita
Kedge, Peter
Lai, Chiu-Nan
Laine, Dan
Laine, Molly
Lee, Glenda
Llurba Ferrer, Juan L
Macini, Marcello
Marr, Robyn
McNeill, Tim
Medini, Lise
Mencia, Valentin
Minari, Jackie
Murphy, John Patrick
Neves, Joyce
Ogilvie, Renate
Osel Rinpoche, Tenzin O, Lama
Palgye, Thubten, Ven.
Pascual, Antonio
Pende, Thubren
Perez-Albert, Mercedes
Peters, Guillaume
Piatti, Franco
Prinz, Richard
Ribush, Nick
Rigsel, Geshe Lhundrup
Rossello, Lorenzo
Saxena, Kabir
Shola, Ugyen
Tack, Martha
Tan, Hup Cheng
Trinley, Tseyangla
van den Akker, Frans
Vassallo, Laura
Vichta, Kathy
Warner, John
Wong, T M
Wongmo, Thubten
Zopa Rinpoche, Thubten Z, Lama

Buddhist: Nichiren
Maruyama, Teruo

Buddhist: Nichiren Shoshu
Ikeda, Daisaku

Buddhist: Pure Land
Fenzl, Friedrich R, Shaku
Mayeda, Egaku, Prof Dr

Buddhist: Soka Gakkai
Akiya, Einosuke

Buddhist: Theravada
Bodhi, Bhikkhu, Ven
Gunaratana Mahathera, Henepola G, Rev Dr
Mahathera, Nyanasamvara, Ven
Rahula, Walpola, Ven Dr

Buddhist: Tibetan
Carr, Gregory Allen
Dalai Lama, D, H H The
Doga, Lobsang Dorje, Ven Geshe
Schadl, Claudia

Buddhist: Tibetan Mahayana
Ahkön Norbu Lhamo, Jetsunma A, Tulku
Green, Ian Russell
McKimm, Inta Benita
Rato, Khyongla, Ven Geshe
Ribush, Alison

Buddhist: Zen
Adachi, Daishin
Fukutomi, Settei
Kajitani, Sonin
Rech, Roland, Rev
Yoshida, Shodo

Bulgarian Orthodox
Bulgaria, Maxim of, H H Patriarch

Byzantine-Ukrainian Catholic
Byblow, Frances Eudoxia, S.S.M.I., Sr

Calvinist
Benda, Kálmán, Prof

Calvinist-Reformed
Tamás, Bertalan, Rev

Catholic
Agazzi, Evandro, Prof Dr
Farah, Augustin Joseph, H E Mgr
Geuns, André Jozef, Dr
Grebe Lopez, Ronald G
Ladrière, Jean A M G, Prof Dr
Lautrey, Pierre
Le Bret, Marie-Michèle
Lesueur, Michel
Lingelser, G
Loftus, Margaret F
Majewski, Tadeusz R, Most Rev
Martos, Javier
McGillicuddy, Frances
McLean, George Francis, Prof
Morand, Paul
Natale, Tommaso
Papadimitriou-Demaître, Georges
Pelzer, Anne Marie Augustine Clementine
Poli, Paul
Porte, Bernard
Raimbault, Philippe
Rowinski, Francis C, Most Rev
Royer, Therese, Ms
Schüssler Fiorenza, Elisabeth S, Dr
Solassol, Alain
Suarez, Guillermo Santa Maria
Taberdo, Lettie A
Techera, Jorge, Rev
Ter Steeg, Maria T
Vandeputte, Germain
Van Schoote, Jean-Pierre
Vereeck, Jan
Ysern, Juan Luis

Catholic (Byzantine rite)
Varsamis, Thomas, Rev Archimandrite

Catholic-Apostolic Assyrian Church
Zaia, Mar Meelis, Bishop

Catholic-Apostolic Evangelical Church
Soares, Fernando L, Rt Rev

Chaldean Catholic
Bidawid, Raphael I, I, His Beatitude
Garmo, Georges, Most Rev
Issayi, Youhannan Semaan, Most Rev Archbishop
Meram, Thomas, Most Rev Archbishop
Zora, Hanna

Christian
Aagard, J, Dr
Aarflot, Andreas, Rt Rev
Abrahams, Stuart N
Adams, David M
Addy, Anthony J.
Adolf, Felipe
Albert, Christel
Albert, Michel Maurice Louis Delphin, Dr
Allen, Gary I
Altizer, Thomas
Alvarez, Carmelo

Poser, Klaus, Dr
Powell, James R
Powell, John Sidney
Printz, Othon, Dr
Provoost, Constant
Pudaite, Rochunga
Raiser, Elisabeth
Ralston, Helen
Ramcharan, Michael
Rathbone, Thomas
Ray, Chaplain
Reimer, Reginald E
Resler, Loel L, Dr
Ribadeneira, Maria Cristina
Rice, Robert F, Rev
Riehm, Magda
Rieth, Theobald
Rivera, Alberto R, Dr
Rivera, Cornelius A
Riviera, Luis N, Prof
Robinson, Philippus Johannes, Prof
Robinson, Samuel E
Robitaille, Jean
Rogues, Jean
Rome, Brian
Roozen, Nico, Drs
Ross, Ian
Roux, Alain
Ruygrok, E
Sairin, Weinata, Rev
Sakala, Richard
Salti, Yacoub
Samuel, J V, Rev
Sardjonan, S, Rev
Saukkonen, (Veikko Antti) Juhani
Saukkonen, Paula
Schönig, Heinzpeter, Pater
Schrotenboer, Paul G, Dr
Seiple, Robert A
Sekine, Masao, Prof Dr
Sekiya, Ayako
Selle, Murri
Senturias, Erlinda, Dr
Shik, Kim Kun, Rev
Shotwell, J Ralph
Shubitz, Amy
Simangunsong, H, Rev
Sinaga, L H, Rev
Sirait, W J, Rev
Sköld, Margareta
Smith, Dwight
Smith, R
Soejana, Koernai Atje, Rev
South, Colin
South, Liz
Spong, John Shelby, Rt Rev
Spradlin, Byron
Srisang, Koson, Dr
Stäbler, Martin, Rev
Stålsett, Gunnar Johan, Dr
Stephens, B, Rev
Stevens, David
Stranges, Frank Ernest, Dr
Stückelberger, Hansjürg, Rev
Suharto, S H, Rev
Summey, David
Sunshine, Catherine A
Swedish, Margaret
Takenaka, Masao, Dr
Tarter, Bill K, Rev
Taylor, David W A, Rev Dr
Teauariki, Turaki, Rev
Temu, W T, Dr
Thielemans, Edward
Thompson-Travis, Beverly, Rev
Thomsen, Jørgen
Tienou, Tite, Dr
Tingson, Greg, Dr
Tonui, John Arap, Rev
Torres, Nelly
Tovini, Sergo
Trickett, David
Tuckett, C M, Prof
Tumansi, Kititwa
Valle, Carlos A
Van Asperdt, G
Van der Kooij, A, Prof
Van Vught, Jan, Rev
Vasey, William L, Rev
Verebalaire, Sern
Vignolo, Roque
Vlug, René J
Vonck, Christiaan Jan Germana, Rev Dr
von Oeyen, Paul F
Vos, Frans
Waldenfels, Hans, Prof Dr
Walls, A F
Walters, Clem J
Weaver, Cynthia
Welsh, William A
Werkström, Bertil, Most Rev
Wessels, Marc A, Rev Dr
Wietzke, Joachim, Dr
Wilckens, Ulrich
Wilkins, Howell O
Willemsen, Antonia
Williams, Jimmy
Wilson, C
Wirakotan, J H, Rev Dr
Wodajo, Datan
Wright, Francis W, Rev
Wright, Lawrence A
Wunderlich, Kenneth
Yamamori, Ted, Dr
Yeung, David P L
Yoder, Ruth
Ziviani, Enzo
Zodhiates, Spiros, Dr

Christian Church
Wilson, Paul F

Christian Reformed Church
Seebeck, Douglas

Christian: Metaphysical
Zitko, Howard John, Dr

Church of Canada
Nutter, Harold Lee, Most Rev

Church of Christ
Adcock, W O
Allsop, Ian E, Rev
Boayen, Boonratna, Rev Dr
Fair, Ian A, Prof Dr
Ferguson, William Everett, Prof
Kok-Kwong, Yung, Rev
Lightfoot, Neil Roland, Dr
Paul, William E
Powell, G, Rev
Resner, André, Prof
Todd, T G
Tongdonmuan, Arun, Rev
Ward, Clive

Church of Christ on Earth by the Prophet Simon Kimbangu
Diangienda-Kuntima, Joseph, H E

Church of Christ, Scientist
Goodsell, James Nelson, Dr
Harris, Virginia

Church of England
Alves, Colin, M.Litt.
Appleton, George, Rt Rev
Baker, John Austin, Rt Rev
Barrington-Ward, Simon, Rt Rev
Blanch, Stuart Yarworth, Rt Rev and Rt Hon Lord
Booth-Clibborn, Stanley Eric Francis, Rt Rev
Bourdeaux, Michael Alan, Rev Canon
Braybrooke, Marcus Christopher Rossi, Rev
Brierley, Peter William
Brown, Alan Scott
Coggan, (Frederick) Donald, Rt Rev and Rt Hon Lord
Cupitt, Don, Rev Dr
Fleming, (William) Launcelot Scott, Rt Rev
Frend, William Hugh Clifford, Rev Prof
Gilbert, Patrick Nigel Geoffrey
Habgood, John Stapylton, Most Rev and Rt Hon Dr
Harries, Richard Douglas, Rt Rev
Hill, Christopher John, Rev Canon
Irving, Donald R, Canon
Jenkins, David Edward, Most Rev
Kemp, Eric Waldram, Rt Rev
Key, John Christopher, Rev
Mascall, Eric Lionel, Rev
Montefiore, Hugh William, Rt Rev Dr
Moule, Charles Francis Digby, Rev
Naters, James, SSJE, Fr
Nineham, Dennis Eric, Rev Prof Dr
Polkinghorne, John Charlton, Rev Dr
Ramsey, Arthur Michael
Reardon, Martin, Rev Canon
Robinson, Joseph, Rev Canon
Salter, Arthur Thomas John, Rev
Say, Richard David, Rt Rev Dr
Sykes, Stephen Whitefield, Rt Rev
Van Culin, Samuel, Rev Canon
Varah, Edward Chad, Prebendary Dr
Varah, Susan
Ward, John Stephen Keith, Rev Canon Prof
Westwood, William John, Rt Rev
Woods, Frank, Archbishop Sir

Church of God
Vaccaro, Gabriel O, Rev Dr

Church of Ireland
Armstrong, J W, Rt Rev
Bleakley, David Wylie, Rt Hon
Caird, Donald Arthur Richard, Most Rev Dr
Eames, Robert Henry Alexander, Most Rev
Poyntz, Samuel Greenfield, Rt Rev Dr
Simms, George Otto, Most Rev

Church of Jesus Christ
Ramanbasoa, Joseph J, Rev
Ramino, Paul A, Rev

Church of Jesus Christ of Latter-Day Saints
Benson, Ezra Taft
Kern, Paul R
Morrison, Alexander B

Church of Lippe
Haarbeck, Ako, Dr

Church of Melanesia
Bugotu, Ambrose H

Church of North India
Santram, Pritam B, Rev
Sen, Noel Sushil, Rev

Church of Norway
Bakkevig, Trond, Rev Dr

Ntontolo, K Lukombo, Rev
Okuyama, Minoru, Rev
Olayiwola, David Omoleke Folagbade, Rev Dr
Oliver, Victor
Passa, Sailali, Rev
Persson, Walter, Rev Dr
Philpot, Ford
Pickard, David
Pierson, Paul E
Poloma, Margaret, Prof
Puig, Enrique Capo, Rev
Ranft, Eckart
Reinich, Rodolfo R, Rev
Rondo, Kelly H, Rev
Sahiouny, Salim, Rev Dr
Salamate, E J, Rev
Sasingen, O, Rev
Sawen, P, Rev
Schneemelcher, Wilhelm
Scholtz, Marge
Sergey, John M, Rev
Seybold, Klaus Dieter, Prof Dr
Smidt, Wolf-Udo, Rev
Spengler, Helmut, Prof
Strauss, Stephen John, Rev
Struharik, Juraj, Rt Rev
Taylor, William D, Dr
Taylor-Pearce, Modupe, Dr
Thatcher, William L
Tucunan, E P
Utia, Marurai, Rev
Vencer, Agustin B, Jr, Rev Dr Atty
Warner, John
Winchell, Richard
Zokoue, Isaac, Dr

Evangelical Baptist
Logvinenko, V E
Spanu, Paolo

Evangelical Brethren
Hajek, Miloslav, Rev

Evangelical Congregational
Chipenda, José Belo, Rev
Francisco, Julio, Rev

Evangelical Covenant Church
Anderson, Philip James, Dr
Rees, Paul Stromberg

Evangelical Lutheran
Assur, M D
Burghoorn, A, Rev
Buthelezi, Manas, Rt Rev
Droste, Rolf, Rev
Esteban Denuncio, Raul E, Rev
Hanselmann, Johannes, Dr
Hark, Edgar, H E
Hempel, Johannes, Rt Rev
Heubach, Joachim, Bishop Dr
Junkkaala, Timo Tapio, Dr
Leich, Werner, Bishop
Lidell, Elisabeth, Rev
Lindgren, Juhani, Rev Dr
Lønning, Per, Rt Rev Dr
Lüsis, Arnolds, Most Rev
Mäkelä, Raimo Olavi, Rev
Marlier, Michel, Rev
Mkuu, Sebastian Kolowa, Rt Rev
Moyo, Ambrose, Dr
Ngeiyamu, Joel
Rapoo, D P, Rt Rev
Rathke, Heinrich Karl Martin Hans, Dr
Ridala, Esmo
Schaller, Stefan, Rev
Schmidt-Clausen, Kurt Hermann
Sievers, Wilhelm, Dr
Sigurgeirsson, Petur, Rt Rev
Sinnemäki, Maunu, Rev
Sjoberg, Donald W, Rev
Stoll, Karlheinz, Bishop
van Zadel, Th L
Veem, Konrad, Rev
Vikström, John Edvin
von Keler, Hans
Ziegler, Martin Arthur Erich, Rev

Evangelical Mennonite
Garcia, Raúl Oscar, Prof
Roth, Donald

Evangelical Methodist
Aquilante, Sergio, Rev
Castro, George F, Bishop
Diaz, Roberto, Rev
Pagura, Federico, Bishop
Vence, Ilda, Rev
Villena Villegas, Rolando V, Rt Rev

Evangelical Pentecostal
Domingos Caetano, José D, Rev
Souza, Enilson Rocha, Rev

Evangelical Presbyterian
de Sousa Campos, Manuel
Dzobo, N K, Rt Rev Prof
Neto, Joao Severino, Rev
Ngobe, J Sidney, Rev
Szedressy, Paul, Bishop

Evangelical Reformed
Brown, Harold Ogden Joseph, Rev Prof
Nordholt, Gerhard, Dr
Sartorius, Willi, Rev
Weder, Dölf, Rev Dr

Free Church
Calver, Clive Richard, Rev

Free Church of Scotland (Presbyterian)
MacDonald, Fergus, Rev

Free Evangelical
Fjeld, Bjorn Oyvind, Rev

Free Methodist
Bastian, Donald N, Bishop
Bates, Gerald E, Bishop
Gindi, Nathan, Supt
Hatano, Motoi, Bishop
Ilangyi, Bya'ene Akulu, Bishop
Nzeyimana, Noah, Bishop
Ruhumuriza, Aaron, Bishop
Ward, Daniel T, Bishop

Free Wesleyan
Havea, S A, Rev Dr

French Reformed
Lambert, Jean-Marie, Rev

Greek Catholic
Bustros, Cyrille Selim, M.S.S.P., H E Mgr
Gigacz, Stefan Robert
Haddad, Pierre Naaman, H E
Nijmé, Elias, B.A., H E Mgr
Raya, Joseph M, Most Rev
Rüegg, François, Dr
Tawil, Joseph Elias, H E Mgr
Zoghbi, Elie, H E Mgr

Greek Orthodox
Diodoros, D, I, His Beatitude
Eliopoulos, N
Harkianakis, Stylianos, Most Rev Dr
Hazim, Ignace, Mgr
Iakovos, I, Archbishop
Ignatios, I, IV, H H
Kotsonis, Ieronymos, Archbishop
Kourides, Peter
Labardakis, Augoustinos
Nicolaos, N, VI, His Beatitude
Parthenios, P, III, H B
Samaha, Basilios, H E Mgr
Seraphim, S, His Beatitude Archbishop
Skouras Oldknow, Dina S
Stylianos, S, H E Archbishop
Varelopoulos, Nicolaos, VI, H B

Greek-Melkite Catholic
Antaki, Paul, H E Mgr
Dick, Ignace, Rev Archimandrite
Edelby, Neophytos Elias, H E Mgr
Hakim, Maximos V, His Beatitude
Nehmé, Abraham, O.Bas. Soarita, H E Mgr
Raï, Boutros, B.A., H E Mgr
Youakim, Saba, B.S., H E Mgr

Hindu
Andiappan, Sundara Yogasali Asana R, Prof
Bang, Thakurdas
Basu, Samar, Sri
Chandrananda Kurukkal, P C, Sri
Chandrasekara Kurukkal, S C, Sivasri
Clements, Geoffrey, Dr
Daya Mataji, D, Sri Sri
Devananda, Vishnu, Swami
Dhandhania, Kishorilal
Giri, Chidda, Prof Dr
Giri, Rakesh, Dr
Gohil, Jamnadas Zaver Ragha, Prof
Grant, Michael
Iyer, Rama Subramania, Sri
Jayadeva Yogendra, Hansa J, Smt
Jha, Jogendra, Dr
Joshi, Sudarshan U
Kailasanatha Kurukkal, K K, Prof
Kajaria, Nirmala Ratanshi, Dr
Kannan, Aiyasamy, Dr
Karthikesa Kurukkal, K K, Brahmasri
Kaudre, Bansilal Gangaram, Shri
Khalsa, Shakti Parwha Kaur
Kunchithapatha Kurukkal, S K, Brahmasri
Maharaj Ji, M, Shri
Maheswara Kurukkal, S M, Brahmasri
Mangaleswara Kurukkal, K M, Brahmasri
Mathur, J S
Mehta, Priti Kiritkumar, Dr
Nagar, Maharishi
Narayanan, Sanguni Nair
Pal, Raj Sawindar, Mahatma
Pandit, M P, Sri
Panjabi, Mohini, Sr
Petrowski, Chari
Premakantha Kurukkal, S P, Sri
Radhakrishna, K S
Rajanani Kurukkal, K R, Brahmasri
Raman, Mannampurath Kavu, Master
Ramanatha Kurukkal, K P R, Brahmasri
Ramanatha Kurukkal, K R, Brahmasri
Ramasamy Adigal, Santhalinga R, Rev Kailaimunivar
Ravichandra Kurukkal, S R, Brahmasri
Rossoondyal, N
Sarasvati, Divyanand, Dr
Saraswati, Kumari Shiva, Shiva Saraswati
Sarma, B Balasunthara, Sri
Sarma, B Shanmugartna, Brahmasri
Shankar, Ganesh, Dr

Kabbalist
Sturzaker, D L
Sturzaker, J

Konko-Kyo
Miyake, Toshio, Rev Dr

Liberal Catholic
Burton, Ernest James, Rt Rev
Taylor, Eric, Most Rev Dr

Lutheran
Aarflot, Helge
Almen, Lowell G, Rev Dr
Appel, André
Ballentine, Rudolph Miller, Dr
Barth, Norman E
Brakemeier, Gottfried, Rev Dr
Brauer, Jerald Carl, Prof Dr
Capps, Carol A, Ms
Chilstrom, Herbert W, Rt Rev
Crumley, James Robert, Jr, Rt Rev
David, S Immanuel, Rev Prof Dr
Diggs, Ronald J, Bishop
Duda, Betty
Erickson, John David, Rev Dr
Gaétan, Mutembe
Gassmann, Günther, Rev
Gensichen, Hans-Werner, Prof Dr
Glebe-Møller, Jens, Prof Dr
Graz, Franz Friedrich
Harms, Hans Heinrich, Bishop Emer
Heintze, Gerhard
Hirschler, Horst, Landesbischof
Hoeffel, Michel, Pastor
Huhn, Martina
Irenna, Y
Jahshan, Moris
Kaldy, Zoltan, Rt Rev
Khong, Peter, Rev
Koch, Klaus, Prof Dr
Krey, Andrew E V, Chaplain
Kverndal, Roald, Dr
Lazerus, A, Rev
Lema, Anza A, Dr
Lohse, Eduard, Prof Bishop Emer
Lønning, Inge Johan, Prof Dr
Lumpe, Larry E
Marty, Martin, Dr
Meissner, Herbert, Rev Dr
Montgomery, John Warwick, Dr
Mshana, Eliewaha E, Bishop
Müller, Gerhard, Dr
Nababan, Soritua A E, Rt Rev Dr
Nissen, Karsten, Very Rev
Olson, Jarle M
Preus, David W, Dr
Räisänen, Heikki Martti, Prof
Ranaivojaona-Razafimantsoa, R, Rev
Rogestam, Christina
Root, Michael, Prof
Samuel, Prasana Kumari
Sándor, Cserháti, Dr
Schneider, Silvio, Rev
Schultz, Walter Arthur
Skaaheim, Anfin
Skulason, Olafur, Rt Rev
Steicke, Lance Graham, Rev Dr
Stephanos, Francis, Ato
Strange, John, Prof Dr
Sundby, (Carl) Olof (Werner)
Taranger, Bengt
Tung, Sophia
Vaerge, Vilhelm, Rev
Vaerge, Vilhelm, Rev
Wallin, Ingvar
Wee, Paul
Weiland, Paul, Rev
Yuguchi, Takashi

Malankara Jacobite Syrian Orthodox Church of India
Clemis, Abraham Mar, Metropolitan

Maronite Catholic
El-Hachem, Basile, Rev
Gemayel, Boutros, Most Rev
Salame, Joseph, H E Mgr
Sfeir, Nasrallah-Pierre, His Beatitude

Mason
Anderson, M J
Briggs, Betty J
Cohen, Mattie
Dahm, Gene
Heaslewood, Jeanne
McCollough, Frederick
Selby, William R
Vander Elst, R
West, Dorothy

Melkite
Coriaty, George M, Mgr

Mennonite
Arnold, Perumalla B, Rev Dr
Bender, Ross T
Brubacher, Ray
Brunk, George R, III, Dr
Burkholder, Mary
Christano, Peter Charles, Rev
Crossman, Gerald Peter
Delp, Lee
Driedger, Florence
Ens, Harold
Geiser, Daniel
Gerber, Samuel
Gingrich, Paul M
Hansulrich, Gerber
Heisey, Nancy
Hoekema, Alle
Horsch, V
Koontz, Gayle Gerber
Krisetya, Mesach
Lapp, James M
Lapp, John A, Dr
Loewen, Charles Nelson
Mathies, Ronald John Richard, Dr
Mbeba, Joram Magayi, Rt Rev
Meck, Lynette Youndt
Meyer, Albert J
Miller, Larry
Miller, Marlin E
Myers, Paul
Neufeld, Elmer
Neufeld, John H
Nussbaumer, Louise
Pankratz, James
Pellman Good, Phyllis P
Peters, Hans Gerhard
Preheim, Vern Quincy
Reimer, A James
Rempel, Erwin Henry, Rev
Rempel, Ron
Roth, Alice M
Sawatsky, Rodney
Sawatsky, Walter
Shambuyi, Nkumbi Mudiayi
Shapasa, Kabangy Djeke, Rev
Shenk, Gerald
Stoner, John K
Studer, Gerald C, Rev
Sutton, Norma, Rev
Swartley, Willard
Toews, John E
Toews, Reginald
van Straten, E R, Rev
Zehr, Daniel

Mennonite Brethren
Regehr, John, Rev

Methodist
Andrews, Elias, Rev
Ariarajah, Wesley, Rev Dr
Banana, Canaan Sodindo, Rev Dr
Baquete, Grace
Beck, Brian Edgar, Rev
Benedyktowicz, Witold Kazimierz, Rev
Bisai, Ana, Rev
Bloem, Renate
Borges, Armando Rodriguez, Bishop
Born, Ethel W
Brose, Susanne
Callaghan, William Sydney, Rev
Campos Leite, Nelson Luis C, Bishop
Casilla, Ercilla Palomino
Castro, Emilio Enrique, Rev Dr
Catchings, Rose, Ms
Charles, Nelson H, Rev
Clarke, Eric St C, Rev
Cooke, William Dennis Davison, Rev Dr
Cragg, D G L, Rev Dr
Downes, Stanley E, Rev
Eltscher, Susan M
Eyre, Charles G, Rev
Fang, C N, Bishop
Feng, Allan C J, Prof Dr
Fernando, Harold A, Rev
Fourie, Enid
Gorai, Dinesh Chandra, Rt Rev
Gordon, Margaret
Greet, Kenneth Gerald, Rev Dr
Gutiérrez, Isaias, Bishop
Hale, Joseph Rice, Dr
Henri, Harry Y, Rev
Hollemweguer, Juan E, Bishop
Hooker, Morna Dorothy, Prof
Houston, Haviland, Dr
Huston, Robert W, Rev Dr
Imathiu, Lawi, Rt Rev
Jenkins, Keith Pellow
John, Emmy Lou
Jones, Richard Ditzel, Rev
Kimba, Mutombo Dunga, Rev
Kobia, Samuel Rodger, Rev
Lah, Okhah Kim
Lawrence, Wilhelmina
Leuluaialii, Siatua, Rev
Loane, Edith M, Dr
Lok, Kim Bong, Bishop
Makuzwa, Caspen C, Rev
Martelli, Claudio, Pastor
Mbang, Sunday C, His Pre-eminence
McNamara, Gertrude
Milo, Lene, Rev
Ming, Edith
Morris, George E
Munce, Frederick L, Rev
Nabulivou, Inoke, Rev
Niles, H Clifton, Rev
Outen, George H
Peter, Navamani
Potter, Philip Alford, Rev Dr
Rosa, Alfred Christopher Henricus, Rev Dr
Ruiz, Alejandro, Bishop
Seaman, Keith Douglas, Rev Sir
Silva, Cèlia Junker
Sin, Ho Chee, Bishop
Solomon, Lorraine Ellen
Soper, Donald Oliver, Rev Lord
Stephens, Jacob S A, Rev
Stephens, William Peter, Rev Prof

Methodist Episcopal

Moravian

Muslim

Muslim: Hanafi

Muslim: Ismaili

Muslim: Shi'a

Muslim: Sufi

Muslim: Sunni

Nazarene

Nestorian

New Age

New Religion

Non-denominational

Non-denominational Spiritist

None

Old Catholic

Original Kleptonian Neo-American Church

Orthodox

Paul, P, Archbishop
Stojcevic, Pavle, H H
Talvivaara, Anu
Theodosios, T, H E Metropolitan
Theodosius, T, His Beatitude
Thoma, Alexander Mar, Most Rev Dr
Yacoub, Ignatius, III, H E Mgr
Zinovieff, Maurice

Orthodox, Church of Antioch
Habib, Gabriel Gergi

Orthodox: Ethiopian
Abebaw, Yegzaw, Ato
Ghebray, Kidane, Abba
Haimanot, Abune Tekle, H H

Orthodox: Romanian
Iustin, I, Patriach
Popescu, G, Rev Dr

Orthodox: Russian
Gundiaev, Kirill (Vladimir), H E
Pimen, (Sergey Mikhailovich Isvekov), Patriarch

Orthodox: Serbian
German, G, H H

Pagan
Pulgram-Arthen, Lucia Deirdre, Rev

Pantheist
Wood, Harold W, Jr

Pentecostal
Chavez Campos, Enrique C, Bishop
Cher, Arturo Palma, Rev
Foreman, Kenneth
Hampton, J E
Sepulveda Barra, Narciso S, Rev
Swaggart, Jimmy Lee
Zopfi, Jakob, Rev

Presbyterian
Abel, Lorry, Rev
Aboa, S K, Rev
Ador, Samuel Nyawelo, Rev
Aitken, I C, Rev
Ambadiang de Mendeng, Grégoire A, Rev Dr
Andrews, James E, Rev Dr
Awasom, Henry Anye, Rev
Bilheimer, Robert Sperry, Rev
Boyd, Robert Hugh Steele (Robin)
Brown, Carol, Rev
Camps, Carlos M, Rev Prof Dr
Carson, Eugene
Cochrane, Charles Clarke, Rev
Coffin, William Sloane, Jr
Cunha, Carlos, Rev
Douglas, Jane Dempsey, Prof Dr
Ellor, James W, Rev Dr
Farris, Stephen Charles, Rev Prof
Gatwa, Tharcisse
Ibiam, Francis Akanu, Sir
Jung, Hans-Gernot
Kao, C M, Rev
Keun, Kim Sang, Rev
Khumalo, Samson A, Rev
Kuria, Plawson, Rev Dr
Long, Z Bradford, Rev Dr
MacDonald, Donald C, Rev Dr
Maluit, Thomas, Rev
Mariner, John Elwyn
McIntyre, John, Very Rev Prof
McLean, Walter Franklin, Rev
McLelland, Joseph Cumming, Rev
McMillan, Kenneth George, Very Rev
Morrow, John Watson, Rev
Myung, Chu Ke, Rev
Ogarepke, Mba Ogarepke, Rt Rev
Ortega-Montoya, Ofelia, Rev
Owen, D H, Rev
Paisley, Ian Richard Kyle, Rev
Quek, Kiok Chiang, Rev Dr
Reid, William Stanford
Ro, Bong Rin, Dr
Sieunarine, Everson T, Rt Rev
Stephens, Sarah Jane, Rev
Stubs, Alan Charles, Very Rev
Sukhu, Rawle, Rev
Thompson, John, Very Rev
Traer, Robert, Rev Dr
Tshihamba, M L, Dr
Twagirayesu, Michel, Rev
Veitch, J A, Rev Dr
Yeow, Choo-Lak, Dr
Yolton, L William
Yousif, Fail, Rev
Zita, Amosse Baltazar, Rev

Presbyterian Evangelical
Ayik, Arop Yor, Prof

Presbyterian Reformed
Arce, Sergio, Rev
Ngcobo, Samuel, Rev

Protestant
Abraham, K C
Acker, Raymond A
Ackerman, William A
Aker, John B, Dr
Albertz, Heinrich
Alford, Timothy, Rev
Althouse, LaVonne, Rev
Amoore, F A, Rt Rev
Anderson, Roy W
Anhelm, Fritz Erich, Dr
Apuzen, Valmike, Rev Dr
Ayub, K Suyaga, Rev
Bacon, Daniel W
Ball, John M, Canon
Barnett, Ted, Dr
Bauer, Hartmut
Bax, Martin
Beam, Fred, Dr
Bell, Randall
Bennett, John Coleman, Dr
Berg, David Brandt
Beukenhorst, Martin J, Rev
Bovon, François, Prof Dr
Bright, William (Bill) Rohl
Brookes, Leonard
Brown, Florence
Brown, Robert McAfee
Brown, Stuart E, Dr
Burger, Stephen E, Rev
Cadier, Gérard
Cadier, Gérard
Caldwell, Happy
Casey, John W
Chapman, Paul K, Dr
Checkley, David
Choi, Stephen, Rev
Christiansen, Egon
Clark, R Norris
Colbert, James D, Rev
Conring, Warner
Crouse, J B
Cullmann, Oscar, Prof Dr
Cumberbatch, Edward R
Cuthbert, Rod
Dayton, Edward, Dr
Dicker, Helen Ruth
Ding Guangxun, K H D, Bishop
Dorey, Gordon, Dr
Dumke, Isolde, Dr
Durersingh, Christopher, Rev Dr
Eastman, Dick, Rev
Edsor, Albert W, Pastor
Eldred, Carl C
Ellerbee, Gene
Engelhardt, Klaus, H E
Engstrom, Ted, Dr
Eriksson, Throstur
Fairbairn, William D
Falkenberg, James R
Fuchs, Eric V, Prof Dr
Fugere, Robert
Gallimore, Gerry O
Gear, John A., Rev
George, Kondothra M, Prof Dr
Gibson, John Clark Love, Rev Prof
Gilbert, Bruce
Gillespie, John M, Rev
Gintings, E P, Rev
Girsang, H M, Rev
Gudmundsson, Bernhardur
Habib, Samuel, Rev Dr
Hardig, Warren G
Harper, C, Rev
Harris, Myrtle W
Hart, Shirley Alexander, Dr
Havens, Lani J
Heil, Hans-Joachim
Hensley, Kirby J
Hermiz, Thomas H, Dr
Hicks, Murray L
Hill, Richard, Rev
Hilton, David, Dr
Hunking, Willis
Hutagalung, S M, Rev Dr
Irvine, Graeme
Johnson, Hugh G, Rev Dr
Jornod, J-P, Rev
Kirton, Allan F, Rev
Koetsveld, Hermanus, Rev
Larosa, B, Rev
Lauby, Paul T, Dr
Lee, Allan Wren, Rev
Lee, Soo-Min, Dr
Le Renard, Pierre-Yves
Lopez-Penalver, Virginia
Lumbantobing, A, Rt Rev Dr
Lumenta, D J, Rev
Marks, Thomas I, Dr
Marshall, J Murray
May, Bernard André
Merminod, Gérard, Rev
Miller, Ray E
Moss, Cath
Mottu, Henry, Prof
Munthe, A, Rev
Naudé, Jacobus Adriaan, Prof
Navamani, Peter
Niessen, Jan, Dr
Nter, Antoine, Rev
Olutimayin, Nathaniel L
Palau, Luis
Perez, Pablo E, Dr
Piercy, Jessie
Pitts, Glen R, Rev
Probst, W
Quaisin, Michel
Quintero, Manuel
Richard, D John, Rev Dr
Richter, Manfred
Robinson, Gnana, Rev Prof Dr
Rodriguez, Norberto
Russell, Gary
Scharf, Kurt Franz Wilhelm

Quaker

Raja Yoga

Reformed

Reformed Church

Reformed Church in America

Reformed Church in Hungary

Reformed Episcopal

Religious Science

Remonstrant Reformed

Roman Catholic

Bayzelon, Jean-Paul, M.E.P., Fr
Beaulieu, Omer, F.D.M., Br
Becher, Paul, Dr
Bedard, Margaret, Dr
Bedini, Ignazio, S.D.B., Most Rev Archbishop
Bélisle, Gilles, H E Most Rev
Bellantonio, Alfredo, O.M., Fr
Bellini, Pietro, O.S.A., Fr
Bellotto, Giuseppe, C.S.I., Fr
Benedek, Jack
Berger, Anton, Fr
Berloco, Giacinto, H E Mgr
Bernard, S Francis, Sr
Bernardin, Joseph Louis, H E Card
Bernier, Paul J
Berra, Francesco, I.C., Rev
Berrigan, Daniel, S.J.
Berrigan, Philip, S.S.J.
Berry, Thomas, C.P., Fr
Bertello, Giuseppe, H E Mgr
Bertoli, Paolo, H E Card
Bertrand, Hector-L, Rev
Beya, Marie-Bernadette Mbuy, Sr
Bickel, Jean-François
Biffi, Giacomo, H E Card
Biggio, Piero, H E Mgr
Birmingham, Kieran, Fr
Bisimwa, Ethienne
Black, J Bernard, Rev
Blajot, Victor
Blanquet Gas, José Maria B, S.F., Fr
Bluyssen, Johannes Wilhelmus Maria, Most Rev
Boaga, Emmanuele, O.Carm., Rev Prof
Böckle, Franz, Fr
Boff, Leonardo Genezio Darci, Prof
Bohdanowicz, Piotr, Fr
Bolonek, Janusz, H E Mgr
Bomers, Hendrik Joseph Alois, C.M., Rt Rev
Bonata, Anna
Bonello, James (Anthony), M.S.S.P., Rev
Bonvin, Jean-Marc, O.R.C., Fr
Borgomeo, Pasquale, S.J., Rev
Borocz, Isaac, O.S.B., Fr
Borunda, Luis Díaz, Fr
Boscardin, Augusto, F.d.C.C., Fr
Bouchaud, Constant, P.S.S., Fr
Boucher, René, S.C., Br
Bourdeau, Gilles, O.F.M., Rev
Bourel de la Roncière, Philippe B
Bourne, Joan, Sr
Bowen, Michael George, Most Rev
Boyle, James M, C.F.X., Br
Bradley, Patrick, CC.SS., Fr
Brady, John Joseph, S.P.S., Fr
Brambilla, Franco, H E Mgr
Braun, Cornelis, M.S.C., Fr
Bressan, Luigi, H E Mgr Dr
Bressanelli, Virginio Domingo, S.C.J., Rev
Bringmann, Jürgen
Brown, Winifred, Sr
Browne, Bartholomew, F.P.M., Br
Bruce, Florence
Brunot, Jean-Marie
Brusco, Angelo, M.I., Fr
Brutosky, Veronica, Sr
Buckley, Francis Joseph, S.J., Rev
Bugin, Fiorenzo, O.C.D., Fr
Bukovsky, John, S.V.D., H E Mgr
Bulaitis, Giovanni, H E Mgr
Bunskoek, L J
Buonaiuto, G
Burghardt, Walter J
Burniak, Marian, T.Chr., Fr
Burtchaell, James Tunstead, Rev
Byrne, Damian Aloysius, O.P., Fr
Cabra, Pier Giordano, F.N., Fr
Cacchione, Ciro, Dr
Caccia, Sisto, C.S., Fr
Cacciavillan, Agostino, H E Mgr
Cáceda, Carlos Chipoco
Cagnasso, Franco, P.I.M.E., Fr
Cagni, Giuseppe, B., Fr
Cakpo, Barthélémy
Calabrese, Vito, O.S.I., Fr
Calabresi, Ubaldo, H E Mgr
Calderón Polo, Cipriano C, H E Mgr
Callaghan, Brendan Alphonsus, S.J., Rev
Callagy, Dennis
Campbell, P
Campeau, Lucien, S.J., Fr
Canalini, Francesco, H E Mgr
Candy, D J, Sr
Canestri, Giovanni, H E Card
Cano Manrique, Francisco C, M.D., Fr
Capetola, Nicola, CC.RR.MM., Fr
Capougi, Hilarion Bashir, H E Mgr
Cappello, Mario
Capraro, Giacomo, S.D.V., Rev
Caprio, Giuseppe, H E Card
Carberry, John Joseph, H E Card
Carbonnel, Jeannine Marie Léonie
Cardenal, Ernesto
Carew, William Aquin, H E Mgr
Carisio, Mario, S.P./C.Op., Fr
Carlson, Alexandre
Carmody, Joy, Sr
Carpino, Francesco, H E Card
Carrara, Giovannamaria, F.S.P., Sr
Carraro Flavio, Roberto C, O.F.M. Cap., Fr
Carrier, Hervé, S.J., Rev Prof
Carroll, Philip, Rev
Carter, Alexander, Most Rev
Carter, Gerald Emmett, H E Card
Carteri, Giuseppe, Fr
Carton, Jacques
Caruana, Edmund A, Fr
Carvelli, Giuseppe, P.O.C.R., Fr
Casaroli, Agostino, H E Card
Casey, Eamonn, Most Rev
Casoria, Giuseppe, H E Card
Casserly, Patrick, S.M., Rev
Cassidy, Edward Idris, H E Mgr
Castillo Lara, Rosalio José C, S.D.B., H E Card
Catoir, John, Rev
Caudron, Marc
Cavallari, Eugenio, O.A.D., Fr
Cavanaugh, Ellen
Cé, Marco, H E Card
Ceirano, Giovanni, H E Mgr
Celata, Pier Luigi, H E Mgr
Celeste, M, Mother
Cenkner, William, Prof
Cespites, Gennaro, C.PP.S., Fr
Chabot, Marie-Emmanuel, Rev Sister
Chakaipa, Patrick Fani, Most Rev Archbishop
Chan Soon Cheong, James C, Rt Rev
Chapellin, E Hernandez, Sr
Chapper, Adolfo, Fr
Charria, Manuelita, O.P. d. Pres., Sr
Charzat, J
Cheli, Giovanni, Mgr
Choné, Monique, FMM, Sr
Chong, Richard Lai Pong
Chooi, Theresa Ee
Chrapek, Jan, C.S.M.A., Fr
Ciappi, Mario Luigi, O.P., H E Card
Cifuni, Pietro, R.C.I., Fr
Cinà, Giuseppe, M.I., Fr
Clancy, Edward Bede, H E Card
Clavel, Pierre, Fr
Claverie, Pierre Lucien, O.P., Mgr
Clímaco, López Montes, OAR, Fr
Cloete, Glynn
Coff, Pascaline, OSB, Sr
Colasuonno, Francesco, H E Mgr
Coleman, John W, Fr
Collaço, Blasco Francisco, H E Mgr
Collet, Adrien, Fr
Colombo, Giovanni, H E Card
Comiskey, Brendan Oliver, SS.CC., Rt Rev
Connolly, Jerome
Connolly, Michael Augustine, C.F.C., Br
Conti, Luigi, H E Mgr
Conti, Mario Giuseppe, Rt Rev
Cooper, Roswitha
Coppa, Giovanni, H E Mgr
Corbon, Richard, R.S.V., Fr
Corcoran, Theresa, Sr
Cordeiro, Joseph, H E Card
Cordero Lanza di Montezemolo, Andrea C, H E Mgr
Corripio Ahumada, Ernesto C, H E Card
Corriveau, Ernest, M.S., Fr
Cortesi, Filipo, Dr
Coscia, Luis, O.F.M.Cap., Fr
Cottier, Georges Marie Martin, O.P., Rev
Coudreau, François, Rev Prof Dr
Couloubaly, Joseph
Couve de Murville, Maurice Noël Léon C, Most Rev
Coveney, Patrick, H E Mgr
Coventry, John Seton, S.J., Rev
Crespelle, Jean-Pierre
Crous Boatella, Filomena C
Crowe, Frederick Ernest, S.J., Rev
Crowley, Regina, OSB, Sr
Cuesta, Julio, F.D.P., Rev
Cumella, Elena, Dr
Cunegatti, Pietro, P.S.D.P., Fr
Curioni, Cesare, Mgr
Curis, Carlo, H E Mgr
Curran, Charles E, Rev Dr
Czaja, Tadeusz, O.M.I., Fr
Da Campo, Lino D, F.S.F., Br
da Cruz Policarpo, José, H E Mgr
Daily, Thomas Vose, Bishop
Daly, Edward Kevin, Rt Rev
Daly, Gabriel Conor, OSA, Fr
Daly, Mary
Damiba, Bernard
Dammertz, Viktor Joseph, Rt Rev
Danneels, Godfried, H E Card
Dardene, Myriam, O.S.C., Sr
Dargan, Joseph, S.J., Rev
Dargan, Joseph, S.J., Fr
Dario, Antonio, d.O., Fr
Darmojuwono, Justin, H E Card
da Silva Lima, José, Prof
da Silva Sebastiáo, Manuel Martins d, C.S.Sp., Rev
Daun, Christel, Sr
D'Avack, P A, Prof
Davitti, Michele, Rev Fr
De Angelis, Nicola, C.F.I.C., Fr
de Araújo Sales, Eugénio d, H E Card
De Bonis, Donato, H E Mgr
De Brandt, Edward
de Broucker, José
de Castro Cruz, Jorio Roberto
Decencière, Françoise, Sr
de Champigny, Yolande, Sr
Dechant, Virgil C
de Cosse, Pierre
de Coulomme-Labarthe, G
Decourtray, Albert, H E Card
Decraene, Michel, C.I.C.M., Rev Fr
de Diesbach, G
de Ferreira, Maria Victoria
de Furstenberg, Maximilian, H E Card
Degenhardt, Johannes Joachim, Most Rev Archbishop
de Jonghe, Etienne Nathalie Honoré
del Aguila, Jorge Mario Avila

Graus, Lambert, O.S.C., Fr
Gray, Gordon Joseph, H E Card
Grebe Lopez, Ronald G
Grech, Pierre
Grégoire, Paul, H E Card
Greiveldinger, P
Grimes, Edward, Rev
Grisez, Germain, Dr
Groër, Hans Hermann, O.S.B., H E Card
Grou, Claude, C.S.C., Fr
Grouès, Henri, Abbe
Guerri, Sergio, H E Card
Guillerm, Vincent, F.I.C.P., Br
Guillet, Jean-Paul, Fr
Gulbinowicz, Henryk Roman, H E Card
Guske, Hubertus
Gustavo, Mejia P
Gutiérrez, Gustavo
Gyürki-kis, Adam, C.Op., Fr
Haas, Pierre, C.S.Sp., Fr
Hakenewerth, Quentin, Fr
Halliden, William H, S.S.C.M.E., Fr
Hamelin, Jean-Guy, Most Rev
Hamer, Jean Jérôme, O.P., H E Card
Hamilton, Susan
Hamilton, Wayne
Hammerling, Mary Anselm, Sr
Handl, Mathilde, Sr
Hanssens, Jan, Rev
Harrington, Patrick, S.M.A., Fr
Harris, Augustine, Rt Rev
Hassel, David John, Prof
Hastings, Adrian Christopher, Rev Prof
Hayes, Eugen Joseph, O.Praem., Very Rev Dr
Hayes, James Martin, Most Rev
Healy, Timothy Stafford, S.J.
Heath, David Gordon, Rev
Heidkamp, Martha
Hell, Theo
Henau, Ernest, Rev
Hengsbach, Franz, H E Card
Henihan, Mary, SMR, Sr
Heon, Céline, PSA, Sr
Heptner, Elisabeth, S.A.C., Sr
Herkenrath, Norbert, Mgr
Hermans, A M
Hermans, Louis, SS.CC., Fr
Hernandez, Andrés Delgado, Fr
Herranz Casado, Julián H, H E Mgr
Hess, Robert, Rev
Hibbs, William E, III, Mgr
Hickey, James Aloysius, H E Card
Hickey, Thomas A, M.M., Br
Higgins, George Gilmary, Mgr
Himbeault, Claire, R.N.D.M., Sr
Hincq, L
Hoare, Liam, S.P., Fr
Hogan, William Francis, C.S.C., Rev
Hollings, Michael Richard, Rev
Hölter, Günter
Holtz, Bruno, S.M.B., Fr
Hoppe, Wolfgang, F.F.S.C., Br
Horvath, Tibor, S.J., Rev
Houssiau, Albert, Mgr
Howard, Charles, F.M.S., Br
Howson, Justin, B.G.S., Br
Hughes, Catherine, SDNdeN, Sr
Hull, Lilla, Sr
Hume, George Basil, O.S.B., H E Card
Hunt, Mary Elizabeth, Dr
Hurley, Denis Eugene, O.M.I., Rev
Hurley, Michael Anthony, S.J., Rev
Husar, Lubomyr, M.S.U., Most Rev.
Hynes, Fabian, O.H., Br
Icardi, Franco, S.M.A., Fr
Iglesias, Francisco, O.F.M. Cap., Fr
Imokhai, Charles A, Rev Fr
Inglessis, Emile, Dr
Innocenti, Antonio, H E Card
Inverardi, Giuseppe, I.M.C., Fr
Inverardi, Giuseppe
Itçaïna, Jean-Baptiste, M.E.P., Fr
Ivory, Thomas, Rev
Jacobi, Reinhold, Dr
Jacqueline, Bernard René Henri, H E Mgr
Jago, John, S.M., Fr
Jaki, Stanley L, O.S.B.
Jamnicky, John Andrew, Rev
Javierre Ortas, Antonio María J, S.D.B., H E Card
Jenkinson, Willie, Fr
Jiménez, Jorge, C.I.M., Fr
Jiménez, Julio, Fr
John Paul, J, II, H H Pope
Johnson, D Bernard, OCSO, Fr
Johnston, John, F.S.C., Fr
Joulia, Henri, B.M.V., Fr
Jubany Arnáu, Narciso J, H E Card
Jullien, Jacques André Marie, Mgr
Juritsch, Martin, S.A.C., Fr
Kada, Lajos, H E Mgr
Kakkanattu, Benicasia, C.M.C., Sr
Kakkanattu, Samuel M, C.M.S.F., Br
Kalilombe, Patrick-Augustine, Rt Rev Dr
Kariamadam, Paul, Rev Dr
Karin, M, Sr
Karunan, Victor Prasad, Dr
Kasper, Walter, H E Mgr Dr
Kaszuba, Tadeusz, C.R., Fr
Katongo, Clément
Kaut, Bernd, Mgr
Kavanagh, Aengus J, F.S.P., Br
Kearns, D Thomas
Keating, Jeremiah Columba, C.F.C., Br
Keller, Columban, C.M.S.F., Br
Kelly, Jerome T, F.P.M., Br
Kelly, Patrick Altham
Khoraiche, Antoine Pierre, H E Card
Kim, Stephen Sou-Hwan, H E Card
King, Geoffrey, Rev Dr
Kirima, Nicodemus, Rt Rev
Kirsch, L, Fr
Kissling, Frances
Kitbunchu, Michael Michai, H E Card
Klaas, Joan, Sr
Kmeid, Elias, O.M.M., Fr
Knupfer, Wolfgang, O.R.C., Fr
Kolvenbach, Peter-Hans, S.J., Fr
Komolmas, Prathip Martin, Dr
König, Franz, H E Card
Koprowski, Andrzej, S.J., Rev Dr
Kort, Leonardus H, C.I., Fr
Korycki, Jan, S.A.C., Rev
Kos, Donald, O.F.M. Conv., Fr
Kottayarikil, Cyriac, M.C.B.S., Fr
Koukoua, Etienne
Kowalczyk, Józef, Most Rev Archbishop Dr
Kredel, Elmar Maria, Most Rev Archbishop
Kroeger, James Henry, M.M., Rev Dr
Krol, John Joseph, H E Card
Krzyzanowski, Kazimierz, M.I.C., Fr
Kuharic, Franjo, H E Card
Kuhn, Andrew, O.F.M.Cap., Br
Kuijper, P J M
Kulak, Józef, M.S.F., Fr
Kumoyo, Doreen
Küng, Hans C, Prof Dr
Kuntz, Dolores, Sr
Kurtz, Robert, C.R., Fr
Kuttickattu, Sebastian J, Fr
Kuttickattu, Sebastian John, Rev
Kwaw, Edmund Samuel Kwakukye, Dr
La Bianca, Salvatore
Laboa, José Sebastián, H E Mgr
Ladusans, Stanislaus, Prof Dr
La Fontaine, Yves, C.M.M., Fr
Laghi, Pio, H E Mgr
Lahey, Raymond John, Most Rev
Lamb, E M
Lamont, Donal Raymond, O.Carm., H E Mgr
Landázuri Ricketts, Juan L, O.F.M., H E Card
Landousies, Jean
Lanfranchi Mantovani, Guido L, F.N., Fr
Lanigan, Aileen, Sr
Lapierre, François, P.M.E., Fr
Larnaud, Jean André
Lasso de la Vega Miranda, Juan Manuel L, C.SS.R., Fr
Lätsch, Walter
Lauritzen, Grete
Lauzon, Georges, M.Afr., Fr
Law, Bernard Francis, H E Card
Leanza, Giuseppe, H E Mgr
Lebel, Robert, Most Rev
Lebrun Moratinos, José Alí L, H E Card
Lecoq, André, Mgr
Leddy, Mary Jo, NDS, Sr
Lee, Bona So Woo, Dr
Lefère, Willem, C.R.S.P., Fr
Légaré, Henri Francis, O.M.I., Most Rev
Léger, Paul Emile, P.S.S., H E Card
Leipold, Winfried, E.C.M.C., Fr
Lejeune, Geneviève
Lemaître, Henri, H E Mgr
Lemire, Gérard, S.M.M., Fr
Lenssen, Frans, C.M.M., Fr
Leonardi, Giuseppe, Fr
Lesaffre, Stéphane, Fr
Leva, Angelo, F.M.I., Fr
Lévesque, Georges-Henri, OP, Rev Fr
Lezama, José, C.S.V., Fr
Libanio Christo, Carlos Alberto L, Br
Libouban, Jean-Baptiste
Linden, Ian
Lindner, Louis, Fr
Lindsey, Stephanie, Sr
Linn, Robert
Lobato Casado, Abelardo L, Prof Dr Rev
Loipfinger, Anton, C.PP.S., Rev
Lopes Coelho, Armindo L, Rt Rev
López-Gay, Jesús, Rev Prof Dr
López Montes, Clímaco L, O.A.R., Rev
López Trujillo, Alfonso L, H E Card
Loreti, Antonio Toto, Rev
Lorscheider, Aloisio, O.F.M., H E Card
Lo Schiavo, John L, S.J., Rev
Lourdusamy, Simon D, H E Card
Lubachivsky, Myroslav Ivan, H E Card
Lubich Silvia, Chiara L
Lucas, Angel, CC.SS., Fr
Lucchetti Bingemer, María Clara L, Prof Dr
Lucci, Antonio, S.D.V., Rev
Lucey, Gregory F, S.J., Rev
Lugo, Olga Rosa
Luigi Santieri, Roberto L, Rev
Lukass, G Laszlo
Lustiger, Jean-Marie, H E Card
Lwaminda, Peter
MacBride, Malachy, S.D.S., Fr
MacBride, Malachy, Fr
Mac Cullen, T R P Richard
Mac Gillis, Miriam, O.P., Sr
Macharski, Franciszek, H E Card
Machiels, Cécile Marie Paule, Sr
Maciel Degollado, Marcial M, L.C., Fr
Mackey, James Patrick, Prof
MacNeil, Joseph Neil, Most Rev
Madigan, Bresa, Sr
Magermanss, Jacques
Magnoni, Antonio, H E Mgr
Maher, Helen, Sr
Malley, John, O.C., Fr
Mampra, Thomas, C.M.I., Fr
Manino, Line

Picazo Gálvez, Melecio P, M.Sp.S., Fr
Pierli, Francesco, M.C.C.I., Fr
Pierre, Marirose
Pieterse, Jude, P.F.M., Br
Pilarczyk, Daniel Edward, Rev
Pimenta, Simon Ignatius, H E Card
Piovanelli, Silvano, H E Card
Pipaón Monreal, José Javier P, O.A.R., Fr
Pironio, Eduardo Francisco, H E Card
Pirson, Joseph
Pittaluga, Agnese, MC, Sr
Plourde, Joseph Aurèle, Most Rev
Poggi, Luigi, H E Mgr
Pohl, Rudolf, Mgr
Poisson, André, Fr
Poletti, Ugo, H E Card
Pomberger, Elisabeth, Dr
Ponnau, Geneviève, Sr
Portillo, Alvaro del P, Most Rev Mgr
Poswick, Reginald-Ferdinand, OSB, Br
Poupard, Paul, H E Card
Powathil, Joseph, Most Rev
Powell, James Matthew, Prof
Pratzner, Ferdinand, S.S.S., Rev Fr
Prigione, Girolamo, H E Mgr
Primatesta, Raúl Francisco, H E Card
Primeau, Liette, Sr
Prince, Bernard A, Rev
Pringarbe, Robert
Prowse, Margaret
Prudence, Marie-Antoinette
Puente, Pablo, H E Mgr
Pulh, Jacques
Pung, Robert Edward, S.V.D., Rev
Quaranta, Ciro, R.C.I., Fr
Quartier, Jacques
Quilici, Oriano, H E Mgr
Quint, Eduard, Rev Br
Rall, M
Ramasso, Venanzio M, O.S.M., Fr
Ramírez Piñón, Crescencio R, M.J., S.S.I., Fr
Rams, Konrad, F.M.M.A., Br
Randerath, Manuela, CPS, Sr Mother
Raper, Mark, SJ, Rev
Rapisarda, Alfio, H E Mgr
Ratzinger, Joseph Alois, H E Card
Razafimahatratra, Victor, S.I., H E Card
Redmond, Francis J, F.S.P., Br
Redrado Marchite, José Luis R, O.H., Rev Fr
Reece, Donald James, Most Rev
Regazzoni, Giuseppe, F.M.I., Fr
Remmerswaal, James, O.S.C., Fr
Renaud, Etienne, M.Afr., Fr
Renggli-Enderle, Claire
Restrepo, Jesús Angel, C.I.M., Fr
Restrepo, Jesús, Fr
Rètailleau, Michel, F.C., Fr
Revollo Bravo, Mario R, H E Card
Ribeiro, António, H E Card
Rieutord, Maurice Georges Gabriel, Fr
Rigali, Justin Francis, H E Mgr
Righi-Lambertini, Egano, H E Card
Rime, Paul, M.S.F.S., Fr
Ríos Zalapa, Alejandro R, M.G., Fr
Robinson, Margorie, OCD, Sr
Robles Díaz, Luis R, H E Mgr
Roch, Raymond, Rev Fr
Roddy, Joan, D.M.I., Sr
Rodrigo Mendizábal, Iván Fernando R
Rodriguez, Ignacio Ortega, S.S.J., Fr
Rodríguez Maradiaga, Oscar Andrés R, S.D.B., H E Mgr
Roest, Crollius Arij, S.J., Rev Fr
Roger-Machart, M
Rogers, Anthony, Br
Rogers, Edward W, Fr
Roll, Bertin, Rev
Romeo, Paolo, H E Mgr
Rosner, Regine
Rossi, Agnelo, H E Card
Rossi, Opilio, H E Card
Roussel, François-Xavier
Roux, Albert G, Fr
Rovida, Edoardo, H E Mgr
Roy, Jacques
Ruegg-Saillen, Rudolf
Ruff, Francis, Fr
Rugambwa, Laurean, H E Card
Rügg, François
Ryder of Warsaw, R, Lady
Sabattani, Aurelio, H E Card
Sabbah, Michel, His Beatitude
Saier, Oskar, Most Rev Archbishop
Sainte-Beuve, Michel, Br
Saint-Pierre, Valère, C.F.S., Fr
Saint-Roch, Patrick
Sáinz de Baranda, Felipe S, O.C.D., Fr
Sainz Muñoz, Faustino S, H E Mgr
Salaverri Aranegui, José María S, S.M., Fr
Salazar, Rubén, Rev
Salazar López, José S, H E Card
Salon, Mario, CC.RR.MM., Fr
Sambi, Pietro, H E Mgr
Samson, Pierre, P.M.E., Fr
Samy, Ama, SJ, Fr
Sánchez, José T, Most Rev Archbishop Emer
Sanson, Henri, S.J., Fr
Sardella, Giovanna Lia, Sr
Sarei, Alexis Holyweek, Sir
Sarkis, Paul Antoine, S.J., Fr
Satini, Louisie, Dr
Satowaki, Joseph Asajiro, H E Card
Saxer, Victor, Mgr
Sayegh, Selim, H E Mgr
Sbarbaro, Eugenio, H E Mgr
Scannell, Anthony, Rev
Schaeffer, Paul, Fr
Schalück, Herman, Fr
Scheele, Paul-Werner, H E Mgr
Scheidegger, Maddalena
Schembri, Salvatore, M.S.P., Rev
Scherer, Alfredo Vicente, H E Card
Schindler-McGraw, Kevin, Rev
Schlegel, John P, S.J., Rev
Schmaus, Michael
Schmidt, M G
Schmidt, Paula, OSSR, Sr
Schneider, Michel Herbert, O.F.M., Rev Dr
Schotte, Jan, C.I.C.M., Most Rev
Schupp, M
Schwarz, Karen Anne
Schwenzer, Gerhard, SS.CC., Fr
Scremin, Lorenzo, C.S., Fr
Sebastiani, Sergio, H E Mgr
Secondo, Louis J, Fr
Seelaus, Vilma, OCD, Sr
Selg, Ottfried
Sélis, Claude, Prof Rev
Seman, Isabel, Sr
Senoo, Hilary, Mgr
Sensi, Giuseppe Maria, H E Card
Seper, Franjo, H E Card
Serrini, Lanfranco, O.F.M. Conv., Fr
Sfeir, Antoine, O.M.M., Fr Abbot
Shaughnessy, Maureen, Sr
Shea, Leo Joseph, S.J., Fr
Sheerin, Philomena, Sr
Sheridan, Terence, Rev
Shirayanagi, Peter Seiichi, H E Mgr
Silva Henríquez, Raúl S, S.D.B., H E Card
Silvestrini, Achille, H E Card
Simaan, Nemeh Khalil, H E Mgr
Simonis, Adrianus Johannes, H E Card
Simonnet, R P
Sin, Jaime L, H E Card
Skop, Gizela Maria
Sladkevicius, Vincentas, H E Card
Slaghenauffi, Daniele, Fr
Smet, Joachim
Smith, Richard F, S.J., Rev Dr
Smits, Gérard, M.O., Fr
So, Erlinda
Soares, Anjou P T
Sodano, Angelo, Card
Soliman, Doaa Fouad
Sowy-Lavergne, Françoise, Sr
Spital, Hermann Josef Silvester, Rt Rev
Spolti, Luiz, S.D.S., Fr
Sponselee, Theodor, Br
Squicciarini, Donato, H E Mgr
Stanislaw, Turek
Staquet, Michel
Steffans, Marion, OCD, Sr
Stein, Greig
Stella, Beniamino, H E Mgr
Stephens, Alice
Stern, Jean, M.S., Fr
Stevens, Clifford, Fr
Stewart, J
St George, John A S
St George, John A S
Stickler, Alfons Maria, S.D.B., H E Card
Storero, Luciano, H E Mgr
St-Pierre, Valère, Fr
Stramare, Tarcisio, O.S.I., Fr
Suchy, Wladyslaw, C.S.M.A., Fr
Suenens, Leo Jozef, H E Card
Sumah, Rose, Sr
Summers, Sarah, Sr
Suquía Goicoechea, Angel S, H E Card
Sutton, Peter Alfred, O.M.I., Most Rev
Sweeney, Brian, F.M.S., Fr
Sweeney, Redempta, IHM, Sr Mother
Swords, Brian, S.F.M., Fr
Syron, Rita, Sr
Szoka, Edmund Casimir, H E Card
Tabet, Paul Fouad, H E Mgr
Tagliaferri, Mario, H E Mgr
Tang, Dominic, Rt Rev
Taofinu'u, Pio, S.M., H E Card
Tarzia, Antonio, S.S.P., Rev Dr
Tauvette, Liliane Irène, Sr
Tavard, Georges Henri, Rev
Tedla, Stefanos, O.F.M. Cap., Fr
Tegyey, Gabor, Rev
Teissier, Henry Antoine-Marie, Most Rev Archbishop
Temperini, Lino, TOR, Fr
Teresa, T, Mother
Testa, Elide, FdCC, Sr
Testa, Emilo, O.S.F.S., Fr
Thelen, Bert R, S.J., Rev
Thenguvila, Mary, Sr
Thenoz, Yves, Fr
Thesing, Kenneth F, M.M., Rev
Thiandoum, Hyacinthe, H E Card
Tholens, Cornelius J A, O.S.B., Dom
Thomanek, A
Thomas, Peter, Rev
Thottam, Eustace, C.M.I., Fr
Tibaldo, María Fidelis, Sr
Tintelott, Hubert
Tolu, Giovannino, O. de M., Fr
Tomásek, Frantisek, H E Card
Tomasi, Silvio, C.S.S., Fr
Tomaskovic, Emil R, S.A., Fr
Tomko, Jozef, H E Card
Tonucci, Giovanni, H E Mgr
Torres, Enrique, C.S.V., Fr
Tragan, Pius, Rev Prof
Tremblay, Robert, Mgr
Tresoldi, Giuseppina, S.M.C., Rev Sr
Tricarico, Alberto, H E Mgr
Trisolini, John Francis, Rev
Tron, Georges
Trujillo, Alfonso López, Card

Rosicrucian

Salvation Army, The

Seventh-Day Adventist

Shinto

Sikh

Spiritual Healer

Spiritual Science
Schmidt-Brabant, M

Spiritualist
Allen, Cosie Lilian Widrig, N.S.T., Rev
Armstrong, Beth
Brown, F Reed, Rev
Costa, Joaquín
Doyle, June
Gerling, Helene, Rev
Ghose, Chinmoy K, Sri
Guerrero, Bernardita
Haddick, Donald H, Rev
Hilborn, Harry, Rev
Hürlimann, Marianne
Johnson, Paul V, Dr
Jordan, Fred, Rev
Koyama, Tsuneo
La Vine, Mona
Lynch, Sharon L, Rev
Markham, Iris
McCotter, Bryan, Rev
Merrill, Joseph H, Rev
Peterson, Birdie, Rev
Pfortmiller, Gene, Rev
Pfortmiller, Sandra Peterson, Rev
Ransom, Constance A
Rodriguez, Ester, Ms
Schneider, Rudolf
Thurman, Barbara, Rev
Tisdale, Janet
Van De Wyngard, Viola

Subud
Sanders, Hadijah
Tatabe, Roask
Vittachi, Varindra

Swedenborgian
Buss, Peter Martin, Rt Rev

Syrian Catholic
Hayek, Ignace Antoine, II, His Beatitude
Mounayer, Eustache Joseph, H E Mgr
Schelhoth, George, H E Mgr

Syrian Orthodox
Aboodi, Timotheos Aphrem, H E Archbishop
Iwas, Ignatius Zakka I, H H
Paulose, Kallapara V, Rev Prof
Zakka, Ignatius, I, H H

Syro-Malankara Catholic
Thangalathil, Benedict Mar Gregorios, O.I.C., Most Rev Dr

Theosophist
Burnier, Radha, Dr
Farinch, Manuel
Filho, Arnaldo Sisson
Gray, Hugh, Dr
Kynaston, Dianne K
Nurrie, Rick
Osswald, Suzanne
Sisson Filho, Arnaldo S

Transcendental Meditation
Headley, Thomas A

Ukrainian Catholic
Hermaniuk, Maxim, Most Rev

Ukrainian Orthodox
Kolodchin, Vasyl

Unitarian
de Mille Cameron, Angus, Rev
Field, Frank, Prof
Morgan, John Hanly, Rev
Neri, Ferdinando
Olson, James
Scobie, Richard S
Vassaux, P, Rev

Unitarian Universalist
Schulz, William Frederick, Rev

United Baptist
Thompson, Eugene Mayne

United Brethren in Christ
Jackson, David G, Rev

United Church
Boseto, Leslie
Fenske, Elizabeth DeLoris Wall (Pat), Dr
Pat, Leva Kila, Rev
Setae, Susana
Sikazwe, Godfrey Mwansa, Rev
Smellie, S H, Rev

United Church of Canada
Catto, Charles, Rev
Cline, Philip A, Rev
Forsyth, Charles Harkness, Rev
Mills, Howard McIlroy, Rev
Smith, Robert Frederick, Very Rev
Wilson, Lois Miriam, Very Rev Dr

United Church of Christ
Camba, Eugenia Lily D, Ms
Hirayama, Shoji, Rev
Nakajima, John M, Rev
Nomoto, Shinya, Rev Prof Dr
Pomeroy, David Wardell, Rev
Post, Avery D, Rev Dr
Serina, Mercurio M, Bishop
Toeda, Yoshiaki, Rev
Ushiroku, Toshio, Rev

United Church of Papua New Guinea and Solomon Islands
Avi, Dick, Rev
Burua, Albert, Rev

United Church of Zambia
Chisanga, Joel, Rev
Chuba, Bwalya S, Rev

United Congregational
Wing, Joseph, Rev

United Evangelical
Domingos, A L, Rev

United Evangelical Lutheran
Rajaratnam, K, Dr

United Free Church of Scotland
Baird, Isabel D

United Methodist
Cobb, John Boswell, Jr, Prof
Handschin, Hanna
Lull, David John, Dr
Roberts, Granville Oral
Thompson, Henry Orrin, Rev Prof Dr
Wall, James M
Weyer, Michel, Prof
Yrigoyen, Charles, Jr, Dr

United Presbyterian
Khan, Elder Sardar Feroze
Lall, Piyara, Rev

United Reformed Church
Laurie, Margaret Elizabeth, Rev
Reardon, John Patrick, Rev

United Reformed Church and Messianic Jew
Lewis, Ronald Hugh, Rev

Uniting Church
Wilson, Ronald

Uniting Church in Australia
Barnes, Geoffrey Leon, Rev Dr
Brookes, Graham F, Rev
D'Arcy Wood, Harold D, Rev Dr
Edwards, Olive Eva
Gill, David Muir, Rev
Henderson, Gregor, Rev
Lee, J B, Rev
Noller, Charles Geoffrey, Rev Dr
Tanner, Ian B, Rev
Walker, Alan, Rev Dr Sir
Young, Norman James, Rev Prof

Unknown
Barker, Eileen Vartan, Dr
Brown, Kerry
Christinat, Ann-Marie, Mme
Conroy, Donald B
D'Agostino, Federico
De Concini, Barbara, Dr
Edmone, M
Emmerson, P R
Gatto, Santi, Dr
Gawa, Aka
Geertz, Armin W
Gómez, Luis, Prof
Hadjeddin, S A
Harkavy, Angela
Janne, Françoise
Jones, William
Kadowaki, K
Kaufmann, Frank
Kerr, David A
Klietmann, K-G, Dr
Kosela, Krzysztof
Laborda, Jose Luis
Larson, Rebecca
Lavarenne, Christian
Lokiec, Cyna
Macioti, Maria Immacolata
Maloney, Terry
Martello, Léo Luis, Dr
Masferrer Kan, Elio M
Matsunaga, Shinichi
McLaurin, Edna
Minami, Yoshinobu
Moshe, Nancy
Muller, M, Mr
Murphy, David
Naddair, Kaledon
Nielsen, Jørgen S
Norman, Kenneth Roy
Oyer, Juanita
Palmer, Martin
Paul, Bill E
Pennick, Nigel
Pereira de Queiroz, Maria Isaura P, Prof Dr
Pierre, Maurice
Potter, David
Raiser, Konrad, Dr
Ranaghan, Kevin M
Riley-Smith, Jonathan S C, Prof

Name Index
Arranged by Organizations

3HO Foundation
Khalsa, Shakti Parwha Kaur

ACROSS
Kelly, Daniel M

AIC International Christian Ministries
Rivera, Alberto R, Dr

AMG International
Zodhiates, Spiros, Dr

Accrediting Council for Theological Education in Africa
Foxall, George M, Rev
Tienou, Tite, Dr

Action Children Aid, Mother Theresa's Co-workers
Clivaz-Nielsen, Isabelle
Lauritzen, Grete

Action International Ministries
Nichols, Doug

Action for Soviet Jewry
Greenberg, Janet

Administrative Hierarchy of the Roman Catholic Church
Falcao, José Freire, H E Card

Adventist Development and Relief Agency International
Harcombe, E G, Pastor

Aetherius Society
King, George, H E Dr Sir

Africa Association for Liturgy, Music and Arts
Chuba, Bwalya S, Rev

Africa Christian Press
Brabbe, Richard

Africa Church Information Service
Sakala, Richard

Africa Evangelical Fellowship
Genheimer, Ron

Africa Inland Church - Hurri Hills Project
Andersen, Herbert Earl

Africa Inland Mission International
Alford, Timothy, Rev
Barnett, Ted, Dr
Beam, Fred, Dr
Checkley, David
Cuthbert, Rod
Dorey, Gordon, Dr

Africa Network on Churches Participation in Development
Mbila, Kilem

African Caritas
Fleurot, Robert, Rev
Martinho, Ernesto Pedro
Penoukou, Benu Efoevi, Fr
Tedla, Stefanos, O.F.M. Cap., Fr

African Christian Peace Conference
Kowo, Naomi

African Israel Nineveh Church
Kivuli, John Mweresa, II, Rt Rev
Tonui, John Arap, Rev

Afro-Islamic Call Coordination Council
Sharif, Mohamed Ahmed
Sisse, Mustapha

Aga Khan Foundation
Aga Khan, Karim A, IV, H H Prince

Agudath Israel World Organization
Abromowitz, J M, Rabbi
Aschkenazi, M, Dr
Hirsch, Abraham
Lewin, Isaac, Dr Rabbi
Magnichever, Lucien
Sherer, Moshe, Rabbi

Aid for the Church in Africa
Juvent, R P Octave

Aid to the Church in Need
van Straaten, Werenfried, O.Praem., Fr
Vekemans, Roger, S.J., Fr
Willemsen, Antonia

All Africa Baptist Fellowship
Akande, Samuel Titilola Oladele, Rev Dr
Durodola, V O

All Africa Conference of Churches
Chipenda, José Belo, Rev

All Africa Lutheran Churches Information and Coordination Centre
Gaétan, Mutembe
Mshana, Eliewaha E, Bishop

All Africa Press Service
Kizito, Robert N

All-Ukrainian Evangelical Baptist Fellowship
Harbuziuk, Olexa R, Rev

Alliance israélite universelle
Lévy, Jacques

Ambassadors for Christ International
Horton, Vicki M

American Association of Bible Colleges
Bell, Randall

American Enterprise Institute
Novak, Michael

American Federation of Jews from Central Europe
Lehman, Robert L, Rabbi
Lekisch, K Peter, Dr
Silberman, Curt C

American Islamic College
Aasi, Ghulam-Haider, Prof Dr
Delk, Robert C

American Jewish World Service
Griffer, Andrew

American Mission to the Chinese and Asian
Ching, Peter P S

Amici Thomae Mori
Marc'hadour, Germain Pierre, Fr

Ananda Marga Pracaraka Samgha
Avadhuta, Shambhushivananda, Dr

Anglican Consultative Council
Nazir-Ali, Michael James, Rt Rev
Van Culin, Samuel, Rev Canon

Anglican and Eastern Churches Association
Salter, Arthur Thomas John, Rev

Association of the Churches Development Services
Lingscheid, Rainer

Association of the Episcopal Conferences of the Central African Region
N'Dayen, Joachim, Most Rev Archbishop

Association of the International Christian Youth Exchange in Europe
Bocken, Dirk

Association pour la coopération des Eglises, l'environnement et le développement de l'Afrique centrale
Kilem-Mbila, Daniel

Atkinson Foundation
Harris, Myrtle W
Wright, Lawrence A

Augustinians of the Assumption
Augustijns, Louis, A.A., Fr
de Champigny, Yolande, Sr
Maréchal, Claude, A.A., Fr

Aumôniers du Travail
Smits, Gérard, M.O., Fr
Winant, Pierre, M.O., Fr

Australian and New Zealand Association of Theological Schools
Barnes, Geoffrey Leon, Rev Dr

B'nai B'rith Hillel Foundations
Fishman, Samuel, Rabbi

B'nai B'rith International
Bloch, Georges M
Erem, Shimon
Levi, Bruno
Neuman, Thomas
Reich, Seymour D
Zyss, W

B'nai B'rith International Commission on Continuing Jewish Education
Neiditch, Michael, Dr

B'nai B'rith Women
Binder, Elaine Kotell

B'nai B'rith Youth Organization
Clearfield, Sidney M

Bahá'í Esperanto League
Hofmann, Nadi Angelika
Parini, Pierre

Bahá'í International Community
Barrett, Ronald
Bos, Wytze
Chance, Hugh Emery, Dr
Dunbar, Hooper C
Ellis, Wilma M, Dr
Fathea'zam, Hushmand
Furutan, Ali-Akbar
Hancock, Tinai
Khan, Peter Jamel, Dr
Martin, Douglas
Mitchell, Glenford E
Nakhjavani, Ali-Yullah
Rabbani, Rúhíyyih
Ruhe, David S, Dr
Semple, Ian Chalmers
Taherzadeh, Adib

Bangui Evangelical Graduate School of Theology
Zokoue, Isaac, Dr

Baptist Bible Fellowship International
Baird, Robert Dean, Dr

Baptist Peace Fellowship of North America
Sehested, Ken

Baptist World Alliance
Lotz, Denton, Rev Dr
Montacute, Paul

Befrienders International
Varah, Edward Chad, Prebendary Dr

Benedettine Missionarie
Handl, Mathilde, Sr

Benedictine Confederation
Borocz, Isaac, O.S.B., Fr
Dammertz, Viktor Joseph, Rt Rev
Jaki, Stanley L, O.S.B.

Benedictine Nuns
Coff, Pascaline, OSB, Sr
Penrose, Ramanus, OSB, Sr

Benedictine Sisters of the Federation of Saint Gertrude
Fox, Ruth, Sr
Hammerling, Mary Anselm, Sr

Benedictine Sisters of the Federation of Saint Scholastica
Crowley, Regina, OSB, Sr

Berlin Conference of European Catholics
Guske, Hubertus

Beshara Trust
Collins, (John) Grenville
Collins, Grenville

Bethlehem Mission Society
Elsener, Josef, Rev Fr

Bible Christian Union
Murray, George W

Bible Churchmen's Missionary Society
Ball, John M, Canon

Bible Lands Society
Emmerson, P R

Bible Literature International
Falkenberg, James R

Bible Memory Association International
Peters, Greg

Bibles for the World
Pudaite, Rochunga

Biblical Ministries Worldwide
Acker, Raymond A

Brahma Kumaris World Spiritual University
Janki, Brahma Kumari
Jayanti, Brahma Kumari
Kajaria, Nirmala Ratanshi, Dr
Prakashmani, Brahma Kumari
Vedanti, Brahma Kumari, Sr

Brethren in Christ World Missions
Brubaker, A Graybill, Rev
McLane, Jack, Rev

Brothers of Charity of Ghent
Devestel, Waldebert, F.C., Br

Brothers of Christian Instruction of St Gabriel
Euzen, Robert, S.G., Br
Friant, Jean, S.G., Br

Brothers of Mercy of Montabaur
Frink, Ferdinand, F.M.M., Br
Schneider, Michel Herbert, O.F.M., Rev Dr

Brothers of Our Lady of Lourdes
Gogne, René Joseph, F.N.D.L., Br
Pepping, Johannes Wilhelmus, F.N.D.L., Br

Brothers of Our Lady, Mother of Mercy
van Geene, Harrie, Fr
Verschuren, Willem J G, Br

Brothers of Saint Patrick
Kavanagh, Aengus J, F.S.P., Br
Redmond, Francis J, F.S.P., Br

Brothers of St Francis Xavier
Boyle, James M, C.F.X., Br

Brothers of the Christian Schools
Del Hoyo, Félix D, F.S.C., Fr
Johnston, John, F.S.C., Fr

Brothers of the Immaculate Conception of the Blessed Virgin Mary
Bastiaens, Willy, F.I.C., Fr
Muytjens, Johannes Joseph, F.I.C., Br Dr

Brothers of the Immaculate Conception of the Blessed Virgin Mary, Mother of God
Quint, Eduard, Rev Br

Brothers of the Poor of St Francis
Gastel, Mark, C.F.P., Br

Buddhist Peace Fellowship
Howe, Margaret

Buddhist Perception of Nature
Nash, Nancy

Buddhist Union of Europe
Burton-Stibben, Arthur
Hodge, Stephen
Portigliatti, Bruno

CAM International
Platt, Albert T, Dr

CBF World Jewish Relief
Mariner, Cheryl

CIMADE
Kressman, O

Christian Anti-Communism Crusade
Colbert, James D, Rev
Schwarz, Fred C, Dr

Christian Blind Mission International
Brooker, Art

Christian Business Men's Association International
Guggenbühl, Ad, Dr

Christian Children's Fund
McCleary, Paul F, Dr

Christian Churches, International Convention
Lambert, James
Nottingham, William Jesse, Dr

Christian Committee for Human Rights in Latin America
Robitaille, Jean

Christian Conference of Asia
Byu, Esther
Jung, Park Sang, Rev
Ninan, George, Rev
Samuel, J V, Rev

Christian Educators Association International
Welsh, William A

Christian Family Movement
Hamilton, Susan
Hamilton, Wayne

Christian Learning Materials Centre, Nairobi
Buconyori, Elie, Dr

Christian Life Community
Cooper, Roswitha

Christian Medical Commission
Albert, Christel
Hilton, David, Dr
Kaseje, Dan C O, Dr
Senturias, Erlinda, Dr
Sköld, Margareta

Christian Mission for the United Nations Community
Allen, Gary I

Christian Mission to Buddhists
Aagard, J, Dr
Fogh-Hansen, Carl V, Rev

Christian Missions in Many Lands
Robinson, Samuel E

Christian Missions to the Communist World International
Bontrager, David, Rev

Christian Movement for Peace
Gomes, Rui

Christian Organizations Research Advisory Trust
Temu, W T, Dr

Christian Peace Conference
Andriamanjato, Richard, Rev Dr
Benedyktowicz, Witold Kazimierz, Rev
Lazerus, A, Rev
Mirejovsky, Lubomir, Rev Dr
Oke, Philip, Dr
Wright, Kenyon E, Canon

Christian Peace Conference in Latin America and the Caribbean
Lindsey, Stephanie, Sr
Riviera, Luis N, Prof

Christian Reformed Church World Literature Ministries
De Jager, John

Christian Reformed World Relief Committee
Seebeck, Douglas

Christian Relief and Development Association
O'Keeffe, Augustine, Br

Christian Research Institute International
Hanegraaf, Hendrik

Christian Schools International
Haan, Sheri D

Christian Solidarity International
Stückelberger, Hansjürg, Rev
Stückelberger, Johannes Georg, Rev

Christian Temperance Council for the Nordic Countries
Gjerdi, Ivar

Christian World Service
Stephens, B, Rev

Christians Abroad
South, Colin

Christians Associated for Relationships with Eastern Europe
Wilkins, Howell O

Christians for the Upliftment of the Poor
Mercado, Norberto

Christophers, The
Catoir, John, Rev

Church Growth International
Cho, Paul Yonggi, Rev Dr

Church Missionary Society, UK
Wilson, C

Church Monuments Society
Easter, Clive J

Church Planting International
Spradlin, Byron

Church World Service
Havens, Lani J

Church World Service / Lutheran World Relief
Capps, Carol A, Ms

Church and Peace
Matthews, Gordon Elton

Church for the World
Vandeputte, Germain

Church of Christ, Scientist
Goodsell, James Nelson, Dr
Harris, Virginia

Church of God - Men International
King, Wayne

Church of Scientology International
Jentzsch, Heber C, Rev

Church of the Brethren
Miller, Donald E, Dr

Churches Third World Information
Christensen, Birthe J

Churches' Committee for Migrants in Europe
Niessen, Jan, Dr

Churches' Development Service
Conring, Warner

Cistercian Order of the Strict Observance
Dardene, Myriam, O.S.C., Sr
Enoch, Elizabeth, OSC, Sr
Johnson, D Bernard, OCSO, Fr
Olivera, Bernardo, O.C.R., Fr Abbot
Veilleux, Armand J P, O.C.R., Rev Dr
Weier, Helen, O.C.R., Sr

Clergy and Laity Concerned
Collins, J

Clerics Regular Minor
Capetola, Nicola, CC.RR.MM., Fr
Salon, Mario, CC.RR.MM., Fr

Clerics Regular of Somasca
Moreno, Pierino, C.R.S., Fr
Pellegrino, Carlo, C.R.S., Fr

Clerics Regular of St Paul
Bassotti, Giuseppe M, CRSP, Fr
Cagni, Giuseppe, B., Fr
Lefère, Willem, C.R.S.P., Fr

Clerics of St Viator
Audet, Léonard, C.S.V., Fr
Lezama, José, C.S.V., Fr

Clerks Regular Poor of the Mother of God of the Pious Schools
Balcells Xuriach, José Maria B, S.P., Fr
Carisio, Mario, S.P./C.Op., Fr

Clerks Regular of the Mother of God
Migliaccio, Lucio, O.M.D., Fr
Molinaro, Vincenzo, O.M.D., Fr

Coetus Internationalis Ministrantium
Maury, M, H E Mgr

Congregation of Regular Clerics
Gómez, Eugenio Julio, C.R., Fr
Luigi Santieri, Roberto L, Rev
Palombi, Luigi, C.R., Fr

Congregation of St Basil
Black, J Bernard, Rev
Gallagher, John C, C.S.B., Fr

Congregation of the Archangel St Michael
Chrapek, Jan, C.S.M.A., Fr
Suchy, Wladyslaw, C.S.M.A., Fr

Congregation of the Blessed Sacrament
McSweeney, Anthony, S.S.S., Fr

Congregation of the Holy Spirit
da Silva Sebastiáo, Manuel Martins d, C.S.Sp., Rev
Haas, Pierre, C.S.Sp., Fr
O'Meara, Noel P, Rev
O'Meara, Noel P

Congregation of the Immaculate Heart of Mary
Decraene, Michel, C.I.C.M., Rev Fr
D'Heedene, Walter, C.I.C.M., Fr
Hanssens, Jan, Rev

Congregation of the Mission
Mac Cullen, T R P Richard
McCullen, Richard, C.M., Fr

Congregation of the Missionaries of the Holy Family
Färber, Egon, M.S.F., Fr
Kulak, Józef, M.S.F., Fr

Congregation of the Most Holy Redeemer
de Sousa, José, Fr
Lasso de la Vega Miranda, Juan Manuel L, C.SS.R., Fr
Zirilli, Giuseppe, C.SS.R., Fr

Congregation of the Oblates of St Joseph
Calabrese, Vito, O.S.I., Fr
Stramare, Tarcisio, O.S.I., Fr

Congregation of the Passion
Agustin, Obregozo Juaregui José, Fr
Berry, Thomas, C.P., Fr
Orbegozo Jáuregui, José Agustin O, C.P., Rev
Palmerini, Umberto, C.P., Fr

Congregation of the Priests of Mercy
O'Brien, John, C.P.M., Fr

Congregation of the Resurrection of our Lord Jesus Christ
Kaszuba, Tadeusz, C.R., Fr
Kurtz, Robert, C.R., Fr

Congregation of the Sacerdotal Fraternity
Monfette, Gérard, C.F.S., Fr
Saint-Pierre, Valère, C.F.S., Fr
St-Pierre, Valère, Fr

Congregation of the Sacred Hearts of Jesus and Mary
Bradley, Patrick, CC.SS., Fr
Hermans, Louis, SS.CC., Fr
Lucas, Angel, CC.SS., Fr

Congregation of the Sacred Stigmata
Nemes, José Luiz, C.S.S., Fr
Tomasi, Silvio, C.S.S., Fr

Congregation of the Sisters of St John the Baptist
Vicidomini, Immacolata M, Sr

Congregational World Assembly of Youth
Watkins, Tony

Congregazione di Gesù Sacerdote
Fornari, Franco, Fr
Pastò, Gianluigi, Rev

Congrégation de la Sainte Famille de Bergame
Fani, Italo, C.S.F., Fr
Paris, Angelo, C.S.F., Fr

Congrégation de la Sainte Famille de Nazareth
Cabra, Pier Giordano, F.N., Fr
Lanfranchi Mantovani, Guido L, F.N., Fr

Congrégation des Ouvriers Chrétiens de Saint Joseph Calasanz
Gyürki-kis, Adam, C.Op., Fr

Congrégation des Prêtres des Ecoles de Charité
Armeni, Antonio, Fr
Leonardi, Giuseppe, Fr

Congrégation du Sacré-Coeur de Jésus
Ferrari, Edoardo, S.C.J.;T.D., Fr
Fine, Charles, S.C.J.;T.D., Fr

Conservative Baptist Foreign Mission Society
Finzel, Hans W, Dr

Consociatio Internationalis Musicae Sacrae
Alvares, Vincent
Pohl, Rudolf, Mgr

Consolata Missionary Sisters
Pittaluga, Agnese, MC, Sr

Consortium on Religion and Ecology International
Conroy, Donald B

Consultation on Church Union
Taylor, David W A, Rev Dr

Consultative Council of Jewish Organizations
Cairnscross, G
Green, Warren
Israel, Gérard
Nuss, Joseph
Salomon, S, Ms
Steg, Adolphe, Prof
Tuckmann, Frederick Augustus (Fred)

Coopérateurs Paroissiaux du Christ Roi
Fernández-Cueto, José María, C.P.C.R., Fr

Coordinating Board of Jewish Organizations
Schoenberg, Harris, Dr

Coordination Centre for the International Development Promotion of the Austrian Bishops' Conference
Ornauer, Helmut

Corona Mundi
Schaer, Edith

Corporation for Cultural and Social Development
Marin, Carlos

Council for the World's Religions
Kaufmann, Frank

Council of European Bishops' Conferences
Fürer, Ivo, Rev Dr
Martini, Carlo Maria, S.I., H E Card

Council of European Conferences of Priests
Berger, Anton, Fr
Weirig, Paul

Council of Jews from Germany
Rosenstock, Werner, Dr

Council of Mennonite Colleges
Neufeld, Elmer

Custody of the Holy Land
Nazzàro, Giuseppe, O.F.M., Fr

Danchurchaid
Balslev-Olsen, Christian

Danish Missionary Council
Andersen, Birte

Daughters of Mary and Joseph
Roddy, Joan, D.M.I., Sr

Daughters of St Anne
Florio, A Angela, Sr

Daughters of Wisdom
Archambault, Odile, DW, Sr

Diakonia
Elmquist, Karl-Axel

Discalced Carmelite Nuns
Fadden, Marie Celeste, Sr
Fitzgerald, Constance, OCD, Sr
Grady, Laureen, OCD, Sr
Natowski, Damian, OCD, Sr
Page, Rose, OCD, Sr
Robinson, Margorie, OCD, Sr
Seelaus, Vilma, OCD, Sr
Steffans, Marion, OCD, Sr

Disciples Ecumenical Consultative Council
Crow, Paul A, Jr, Rev Dr

European Evangelical Alliance
Hees, Karl-Wilhelm
McAllister, Stuart
Sartorius, Willi, Rev

European Federation of Catholic Medical Associations
Gontard, Françoise, Dr

European Federation of Christian Building and Woodworkers Unions
Koetsveld, Hermanus, Rev

European Federation of Christian Student Associations
Apfel, Martin Josef Konrad

European Fellowship
Penttinen, Pirjo-Liisa, Rev

European Forum of National Laity Committees
Becher, Paul, Dr
Waldstein, Ernst, Dr

European Forum of the Apostolate of Laity
Derde, L

European Jewish Congress
Cwajgenbaum, Serge
Kahn, Jean

European Missionary Association, The
Vlug, René J

European Organization of the International Group of Catholic Parents' Associations
De Smet, Hubert
Vereeck, Jan

European Pan-Keltic League
Naddair, Kaledon

European Union of Former Pupils of Catholic Education
Staquet, Michel

European Union of Jewish Students
Moyse, François

European YWCAs
Siegrist-Rey, Denise Marcelle Emilie Antoinette
Siegrist-Rey, Denise

European and African Organization for Spiritual Purification by the Light of God
Koyama, Tsuneo

Europäischer Verband Freikirchlicher Diakoniewerke
Lauber, Samuel

Evangelical Alliance
Calver, Clive Richard, Rev
Earwicker, John
Morris, Michael
Morris, Michael

Evangelical Association of the Caribbean
Gerig, Zenas, Dr
Ho, David, Rev
Spencer, Peter, Rev

Evangelical Churches of West Africa
Olutimayin, Nathaniel L

Evangelical Fellowship of Asia
Yeo, Alfred C H, Rev

Evangelical Foreign Missions Association
Coggins, Wade T

Evangelical Latin American Commission on Christian Education
Tamez, Carolos, Rev

Evangelical Lutheran Church in America
Almen, Lowell G, Rev Dr

Evangelical Missionary Alliance
Davies, Stanley

Evangelical and Ecumenical Women's Caucus
Brown, Florence

Evangelization 2000
Forrest, Tom

Every Home for Christ International
Clément, Pierre
Eastman, Dick, Rev
Pitts, Glen R, Rev
Wilson, Wes

Family Life Mission
Heil, Hans-Joachim

Family Rosary
Peyton, Patrick, Rev

Far East Broadcasting Association
Pepper, Stewart
Rome, Brian

Federation of Asian Bishops' Conferences
D'Souza, Henry Sebastian, Most Rev Archbishop

Federation of Evangelical Lutheran Churches in Southern Africa
Graz, Franz Friedrich

Federation of Masons of the World
Anderson, M J

Federation of National Committees in the International Christian Youth Exchange
Lopez, Alejandro
Peñaranda, Miguel A

Fellowship of Companies for Christ International
Miller, Ray E

Fils de Sainte Marie Immaculée
Fain Binda, Luigi F, F.S.M.I., Fr
Muzzi, Latino, F.S.M.I., Fr

Fils de la Charité
Marsaud, Jean-Paul, F.C., Fr
Rètailleau, Michel, F.C., Fr

Fils de la Mère de Dieu Douloureuse
Bohdanowicz, Piotr, Fr
Worach, Piotr, Br

Findhorn Foundation
Bloom, William, Dr
Gibsone, Craig John Kenneth
Ingles, Mary

Focolare Movement
Bathellier, A, Ms
Del Zanna, Luisa D
Lubich Silvia, Chiara L

Fondation européenne pour la rencontre des civilisations et le dialogue interréligieux
Kilani, Mohamed Mounif

Fondation religieuse islamique turque en Belgique
Hasan, Okuyucu
Islamoglu, Faruk

Fondazione Pio XII per l'Apostolato dei Laici
Casaroli, Agostino, H E Card
El-Hachem, Mounged, Mgr
Sodano, Angelo, Card

Food for the Hungry International
Yamamori, Ted, Dr

Ford Philpot Evangelical Association
Philpot, Ford

Foreign Missions of Yarumal
Gustavo, Mejia P
Osorno Gil, Jesús Emilio O, M.X.Y., Fr

Foundation for a Christian Civilization
Parrot, J Edward

Foundation for the Preservation of the Mahayana Tradition
Alongina, Xavi
Andrew, Bill
Aroncena, Isabel
Bajardi, Elia
Bajardi, Rodolfo
Biyrje, Paul
Bosa, Siliana
Braun, Ingrid
Bucherl, Gisela
Buscicchio, Mavi
Carr, Gregory Allen
Castles, Margaret
Cole, Owen
Corradi, Carol
Dagpo Rinpoche, D
De Wijs-Koolkin, Paula Beth
Doga, Lobsang Dorje, Ven Geshe
Drukier, Elisabeth
Emonson, Sue
Femenia, Marisa
Ferranti, Andriana

General Grand Chapter of Royal Arch Masons
Selby, William R

General Grand Chapter, Order of the Eastern Star
Briggs, Betty J

Gideons International
McClinton, Wendell C
Quaisin, Michel

Girls' Brigade International
Fox, Marjorie Joyce

Glenmary Home Missioners
Ruff, Francis, Fr

Global Congress of the World's Religions
Thompson, Henry Orrin, Rev Prof Dr

Gospel Literature International
Wilkinson, Georgalyn

Gospel and Mission Groups
Collet, Adrien, Fr
Gérard, Yves, Mgr

Greater Europe Mission
Noble, Ted

Greek Orthodox Archdiocesan Council of North and South America
Kourides, Peter

Greek Orthodox Ladies Philoptochos Society
Skouras Oldknow, Dina S

Groupe de laïcs pour le Tiers-monde
Buonaiuto, G

Gurdjieff Society
Moore, James Harry Manson

Help International Ministries
Summey, David

High Twelve International
Dahm, Gene

Higher Institute of Ecumenical Studies
Lossuy, Nicolas
Rogues, Jean

Himalayan International Institute of Yoga Science and Philosophy
Ballentine, Rudolph Miller, Dr

Holy Childhood Association
Wright, Francis W, Rev

Holy Cross Foreign Mission Society
Fell, Arnold A, Rev

Holy Spirit Association for the Unification of World Christianity
Moon, Sun Myung, Rev

Honorable Order of the Blue Goose International
Maloney, Terry

Hospitaller Order of St John of God
Hynes, Fabian, O.H., Br
O'Donnell, Brian Phillip, O.H., Br

INTERACT Research Centre
Somers, Jeffrey Ian

INTERSERVE
Eckardt, Ralph W

IUAES Commission on Ethnobotany
Jain, Sudhanshu Kumar, Dr

Improved Benevolent Protective Order of Elks of the World
Wilson, Donald P

India Institute of Gandhian Thought and Peace Studies
Mathur, J S

Indo-Pacific Theosophical Federation
Kynaston, Dianne K
Narayanan, Sanguni Nair

Information Network Focus on Religious Movements
Beckford, James Arthur, Prof

Institut des Frères de la Sainte-Famille de Belley
Da Campo, Lino D, F.S.F., Br
Zoccola, Filippo G, F.S.F., Br

Institut international de yoga
Cors, Marie

Institut international supérieur de sciences et recherches religieuses
Janne, Françoise

Institute for Advanced Studies of World Religions
Shen, C T

Institute for Church Leadership Development in Africa
Irenna, Y

Institute for Ecumenical Research, Strasbourg
Root, Michael, Prof

Institute for Ecumenical Research, Tübingen
Küng, Hans C, Prof Dr

Institute for Planetary Synthesis
Schneider, Rudolf

Institute for Research on the Sephardi and Oriental Jewish Heritage
Bonfil, Robert, Rabbi Prof

Institute for Scientific Research on Missions
Waldenfels, Hans, Prof Dr

Institute of Charity
Berra, Francesco, I.C., Rev
Kearns, D Thomas
Zantedeschi, Giovanni B., I.C., Fr

Institute of Jewish Affairs
Lerman, Anthony

Institute of Missionary Catechesis
Giglioni, Paolo, Rev Fr

Institute of Muslim Minority Affairs
Abedin, Syed Z, Dr

Institute of Oriental Religions, Tokyo
Kadowaki, K

Institute of Prado
Pécriaux, François, Fr

Institute of the Brothers of the Sacred Heart
Boucher, René, S.C., Br
Gauthier, Aquin, S.C., Br
Marín Salcedo, Jesús M, S.C., Br
Marín Salcedo, Jesús M, Fr

Institute of the Consolata for Foreign Missionaries
Inverardi, Giuseppe, I.M.C., Fr
Inverardi, Giuseppe

Institute of the Heart of Jesus
Falquet, Claude

Institute on the Holocaust and Genocide
Charny, Israel W, Prof

Inter European Commission on Church and School
Brown, Alan Scott
Spinder, Hans, Drs

Inter-American Theosophical Federation
Farinch, Manuel

Inter-Church Committee for Refugees
Brown, Stuart E, Dr

Inter-Church Committee on Human Rights in Latin America
Fairbairn, William D

Inter-Church Fund for International Development
Fugere, Robert

Inter-Church Relief and Development Alliance
Reimer, Reginald E

Inter-Regional Meeting of Bishops of Southern Africa
Coleman, John W, Fr
Gonçalves, Jaime Pedro, H E Mgr
Rogers, Edward W, Fr

Inter-Territorial Catholic Bishops' Conference
Davitti, Michele, Rev Fr
Grimes, Edward, Rev

Inter-Territorial Pastoral and Social Development Centre
Anne, Lois

International Catholic Organizations Center
Manino, Line

International Catholic Programme of Evangelization
Cappello, Mario

International Catholic Rural Association
Cortesi, Filipo, Dr

International Catholic Society for Girls
Scheidegger, Maddalena

International Catholic Union of the Press
Brunot, Jean-Marie
Géraud, A
Holtz, Bruno, S.M.B., Fr

International Catholic Union of the Press - Europe Region
Hell, Theo
Lukass, G Laszlo

International Centre for Studies in Religious Education
Bacq, Philippe, Fr

International Centre for University Teaching of Jewish Civilization
Davis, Moshe, Prof

International Centre of Sindonology
Barberis, Bruno, Prof

International Christian Aid
Bass, Joe

International Christian Committee for Israel
Geraisy, Sami F, Dr

International Christian Education Association
von Oeyen, Paul F

International Christian Embassy Jerusalem
Luckhoff, Johann

International Christian Federation for the Prevention of Alcoholism and Drug Addiction
Gnanadason, Jonathan N
Taranger, Bengt

International Christian Leprosy Mission
Metcalf, Hattie
Pillers, Lauritz P

International Christian Maritime Association
Krug, Bernard, Rev

International Christian Media Commission
Adams, David M
Thatcher, William L

International Christian Relief
Ross, Ian

International Christian Service for Peace
Freise, Josef, Dr

International Christian Studies Association
Gruenwald, Oskar, Dr

International Christian Union of Business Executives
Albert, Michel Maurice Louis Delphin, Dr
Chargueraud, Marc André
Elizondo, Adán
Falck, Alberto
Jahan, Emmanuel
Mertes, Josef M
Schmidt, M G
Thielemans, Edward
Tumansi, Kititwa

International Christian Women's Fellowship
Newborn, Janice R

International Christian Youth
Clark, R Norris

International Clergy Council
Ching, Peter P S, DD, Dr

International Co-Freemasonry
Heaslewood, Jeanne

International Commission for the Preservation of Islamic Cultural Heritage
Ihsanoglu, Ekmeleddin, Prof Dr

International Commission of Comparative Church History
Bolton, Brenda
Pacaut, Marcel, Prof

International Commission of General Prison Chaplains
Curioni, Cesare, Mgr
Fabbri, Fabio, Mgr
Lesaffre, Stéphane, Fr
Nieto, Evaristo Martin, Rev
Van Vught, Jan, Rev

International Commission on English in the Liturgy
Hurley, Denis Eugene, O.M.I., Rev

International Committee for World Day of Prayer
King, Eileen
Navamani, Peter

International Committee for the Freedom of Syrian Jewry
Pinto, Roger

International Committee of the Order of Malta for Leprosy Relief
Genoud, Jocelyne

International Conference of Police Chaplains
DeRevere, David

International Congregational Fellowship
Travell, John Charles, Rev
Ward, Donald Butler, Dr

International Consultancy on Religion, Education and Culture
Palmer, Martin

International Convention of Faith Ministers
Caldwell, Happy

International Coordinating Committee for Maritime Follow-up Ministry
Kverndal, Roald, Dr

International Coordinating Committee for NGOs on the Question of Palestine
Lambert, Jean-Marie, Rev

International Coordinating Committee on Religion and the Earth
Moshe, Nancy

International Correspondence Institute
Thomson, Lyle

International Council for Christian Leadership
Lee, Simon

International Council for Pastoral Care and Counselling
Elom, Maarten E, Dr

International Council of Accrediting Agencies for Evangelical Theological Education
Kemp, Roger Francis, Rev Dr

International Council of Associations of Theological Libraries
Dumke, Isolde, Dr
Geuns, André Jozef, Dr

International Council of Catholic Men
Inglessis, Emile, Dr
Kwaw, Edmund Samuel Kwakukye, Dr
Menezes, George
Morales, Elfo Eloy
Niehaus, Ferd

International Council of Christian Churches
Maris, J C, Dr
Maris, Jacob Cornelia, Dr
McIntire, Carl, Dr
Quek, Kiok Chiang, Rev Dr

International Council of Christians and Jews
Schoneveld, Jacobus, Dr

International Council of Community Churches
Shotwell, J Ralph

International Council of Hindoo Youth
Rossoondyal, N

International Council of Jewish Women
Anav, N
Marr, Helen S
Ouazan, E

International Mahayana Institute
Bertels, Marcel
Palgye, Thubten, Ven.

International Media Ministries
Lee, William David

International Mennonite Organization
Horsch, V

International Messianic Jewish /Hebrew Christian/ Alliance
Lewis, Ronald Hugh, Rev

International Military Apostolate
Bringmann, Jürgen
Urrisk, B E M Rolf M

International Ministers' Wives and Widows Association
Hart, Shirley Alexander, Dr

International Ministries to Israel
Hicks, Murray L

International Mission Board
Vasey, William L, Rev

International Mission Network
Loftus, Margaret F

International Missionary Benefit Society
Chazelle, M E

International Missions
Tarter, Bill K, Rev

International Movement for the Apostolate in the Independent Social Milieux
Chapper, Adolfo, Fr
de Ferreira, Maria Victoria
de Rios, Maria Marta
Duval, Ayle
Francelle, Jean
Gebs, Simone

International Movement of Apostolate of Children
de Castro Cruz, Jorio Roberto
Prudence, Marie-Antoinette
Soliman, Doaa Fouad
Torres, Nelly

International Movement of Catholic Agricultural and Rural Youth
Rosner, Regine

International Movement of Catholic Lawyers
Eugenis, Gay Montalvo
Pettiti, Louis Edmond

International Muslim Association for Animals and Nature
Masri, Al-Hafiz
Masri, Basheer Ahmad, Al-Hafiz

International Muslim Union
Boukakaur, Dalil, Dr

International Network for Religion and Animals
Wessels, Marc A, Rev Dr

International Network of Children of Jewish Holocaust Survivors
Keningsberg, Rosita Ehrlich

International Network of Engaged Buddhists
Amada, Bunji
Majima, Jyokei
Suzuki, Ryowa
Yamada, Ryuko

International Old Catholic Bishops' Conference
Glazemaker, Antonius Jan, Mgr

International Order of Kabbalists
Sturzaker, D L
Sturzaker, J

International Order of the King's Daughters and Sons
Furman, L G Wayne
Furman, Marilyn

International Orders Research Society
Klietmann, K-G, Dr

International Organization for Septuagint and Cognate Studies
Greenspoon, Leonard Jay, Prof Dr

International Organization for the Study of the Old Testament
Emerton, John Adney, Rev Prof
Van der Kooij, A, Prof

International Partners in Prayer
Doyle, A C

International Pentecostal Press Association
Foreman, Kenneth

International Prayer Fellowship
Piercy, Jessie

International Prison Ministry
Ray, Chaplain

International Rabbinic Committee for the Safety of Israel
Pupko, Isaac, Rabbi

International Registry for Religious-Women Artists
Brutosky, Veronica, Sr

International Religious Fellowship
Dawson, Alan
Kleinschmit, Nathalie
Rathbone, Thomas

International Religious Foundation
Walsh, Thomas G, Dr

International Sacerdotal Brotherhood Saint Pius X
de Mallerais, Bernard Tissier, Fr

International Sacred Literature Trust
Brown, Kerry

International Secretariat for Catholic Engineers, Agronomists and Industry Officials
Meyieres, Jean-François

International Secretariat of Catholic Secondary School Teachers
Zeitz, Susanne, Dr

International Secretariat of Christian Artists
Gilby, Peter S
Lill, Helma

International Sikh Organization
Gormit Singh, G, Dr

International Sivananda Yoga Vedanta Center
Devananda, Vishnu, Swami

International Society for Krishna Consciousness
Grant, Michael

International Society for Study of Religion and Culture
Yan-Chiu, Lam, Dr

International Society for Yad Vashem
Arad, Yizhak, Dr

International Society for the Sociology of Religions
D'Agostino, Federico
Davie, Grace R C, Dr
Kosela, Krzysztof
Macioti, Maria Immacolata
Masferrer Kan, Elio M
Oommen, Tharaileth Koshy, Prof Dr
Pace, Enzo, Prof
Pereira de Queiroz, Maria Isaura P, Prof Dr
Ralston, Helen
Saurma, Adalbert
Sprondel, Walter M

International Society of Bible Collectors
Paul, Bill E
Studer, Gerald C, Rev

International Society of Jewish Librarians
Eisen, Sylvia

International Sociological Association
Thara Bhai, L T, Dr

International Spiritual Alliance
Guerrero, Bernardita
Van De Wyngard, Viola

International Spiritualist Federation
Doyle, June
Ransom, Constance A

International Sporting Federation of Catholic Education
Quartier, Jacques

International Sufi Movement
Witteveen, H J, Dr

Lambeth Conference of Bishops of the Anglican Communion
Carey, George Leonard, Most Rev and Rt Hon

Latin America Mission
Marshall, J Murray
Pierson, Paul E

Latin American Association of Catholic Philosophers
Ladusans, Stanislaus, Prof Dr

Latin American Bureau of Society of Jesus
García del Cerro, Cesáreo G, Rev

Latin American College
Van Schoote, Jean-Pierre

Latin American Confederation of Religious Orders
Charria, Manuelita, O.P. d. Pres., Sr
Coscia, Luis, O.F.M.Cap., Fr
Jiménez, Jorge, C.I.M., Fr
Martins, Hermengarda Alves, Sr

Latin American Confederation of YMCAs
Rodriguez, Norberto

Latin American Council of Catholic Women
Cumella, Elena, Dr

Latin American Council of Churches
Adolf, Felipe

Latin American Episcopal Council
Assis, Raymundo Damasceno, H E Mgr
Rodríguez Maradiaga, Oscar Andrés R, S.D.B., H E Mgr

Latin American Evangelical Center for Pastoral Studies
Gaitán, Jorge

Latin American Federation 'Faith and Joy'
Blajot, Victor

Latin American Institute of Social Theory and Social Studies
Mifsud, P Antonio

Latin American Jewish Congress
Olschanksy, Pedro J

Latin American Network of Catholic Lawyers
Cáceda, Carlos Chipoco

Latin American Pontifical College
Fernando, Londoño Bernal, Rev Fr

Latinitas Foundation
Caruana, Edmund A, Fr

Lausanne Committee for World Evangelization
Houston, Tom, Rev

Lay Association for the Education and Professional Training of Adolescents Europe
Zahlen, Liliane

Lay Cooperation Service for Development
Hermans, A M

Lay Dehonians
Girardi Oliviero, Giuseppe G, S.C.J., Fr

League of Buddhist Esperantists
Schröder, Hermann

Legion of Mary
Fay, Patrick

Legionaries of Christ
Arumi Blancafort, Rafael A, L.C., Fr
Maciel Degollado, Marcial M, L.C., Fr
Marciel Degollado, L.C., Fr

Leprosy Mission International, The
Edgar, W R
Simms, George Otto, Most Rev

Liaison Committee of Christian International Cooperation Organizations
Pecqueux, Pierre-Yves, Rev

Life Line International
Noller, Charles Geoffrey, Rev Dr

Life and Peace Institute
Ferris, Elizabeth G, Dr
Jonsson, Bernt Erik

Literacy and Evangelism International
Rice, Robert F, Rev

Little Brothers of Jesus
Pantane, Nino, Br
Sainte-Beuve, Michel, Br

Little Brothers of Mary
Howard, Charles, F.M.S., Br
Sweeney, Brian, F.M.S., Fr
Thenoz, Yves, Fr

Little Company of Mary
Nilon, Marie Thérèse, Sr

Little Mission for the Deaf-Mute
Loreti, Antonio Toto, Rev
Musconi, Marino, Fr

Little Sisters of the Assumption
Heon, Céline, PSA, Sr

Little Work of Divine Providence
Cuesta, Julio, F.D.P., Rev
Masiero, Giuseppe, F.D.P., Rev

Living Bibles International
Oliver, Victor

Lubavitch Movement
Hodakov, M A, Rabbi

Lucis Trust
Hürlimann, Marianne
Rodriguez, Ester, Ms

Lumière et Vie
Skop, Gizela Maria

Lutheran Association for Maritime Ministry
Krey, Andrew E V, Chaplain

Lutheran Brethren World Missions
Olson, Jarle M

Lutheran Communications in Asia
Khong, Peter, Rev
Yuguchi, Takashi

Lutheran European Commission on the Church and the Jewish People
Aarflot, Helge
Vaerge, Vilhelm, Rev
Vaerge, Vilhelm, Rev

Lutheran Foundation for Interconfessional Research
Skulason, Olafur, Rt Rev

Lutheran World Federation
Brakemeier, Gottfried, Rev Dr
Chilstrom, Herbert W, Rt Rev
Hirschler, Horst, Landesbischof
Hoeffel, Michel, Pastor
Huhn, Martina
Lema, Anza A, Dr
Nababan, Soritua A E, Rt Rev Dr
Nissen, Karsten, Very Rev
Rogestam, Christina
Samuel, Prasana Kumari
Schultz, Walter Arthur
Stålsett, Gunnar Johan, Dr
Stephanos, Francis, Ato
Tung, Sophia

Lutheran World Ministries
Wee, Paul

Lutheran World Relief
Barth, Norman E

Lutheran-Roman Catholic Joint Commission
Crumley, James Robert, Jr, Rt Rev
Scheele, Paul-Werner, H E Mgr

Lutherans United in Communication
Schneider, Silvio, Rev

MARC Europe
Brierley, Peter William

METEM-International Society of Toronto for Hungarian Church History
Sándor, Kostya

Maccabi World Union
Kevehazi, Michael

Madonna House Apostolate
Fox, Jean

Magdalena Aulina Institute
Crous Boatella, Filomena C

Magic Circle
Salisse, John

Missionary Society of St Columban
Halliden, William H, S.S.C.M.E., Fr
Murray, Nicholas, S.S.C.M.E., Fr

Missionary Society of St Paul
Bonello, James (Anthony), M.S.S.P., Rev

Missionary Women International
Hunking, Willis

Missionary Workers of the Immaculate
Carbonnel, Jeannine Marie Léonie

Missionnaires Serviteurs des Pauvres
Giorgio, Giuseppe, M.S.P., Fr
Schembri, Salvatore, M.S.P., Rev

Missionnaires de Saint François de Sales d'Annecy
Mayoraz, Emile, M.S.F.S., Fr
Rime, Paul, M.S.F.S., Fr

Missionnaires de l'Immaculée-Conception de Lourdes
Joulia, Henri, B.M.V., Fr

Missions Commission of the World Evangelical Fellowship
Taylor, William D, Dr

Missions Etrangères de Sainte Marie de Guadaloupe
Muñoz, Juan Antonio
Ríos Zalapa, Alejandro R, M.G., Fr

Missions et développement
Piroelle, Pierre

Missions to Seamen
Jones, G, Rev Canon

Missionszentrale der Franziskaner
Amendt, P

Modern Free and Accepted Masons of the World
Cohen, Mattie

Monaci della Regola Studita
Husar, Lubomyr, M.S.U., Most Rev.

Morris Cerullo World Evangelism
Cerullo, Morris, Rev

Moslem Mission
Chaudhry, Saeed Ahmad

Movement for a Better World
Jiménez, Julio, Fr

Multimedia International
St George, John A S
St George, John A S

Muslim World League
Al Harakan, Mohamad Ali A
Laroussi, Troudi, Dr
Mukhtar, Ali Mohamad Abbas
Nasseef, Abdullah Bin Omar, Dr
Zafar-ul-Haq, Raja Mohummed

Nairobi Evangelical Graduate School of Theology
Taylor-Pearce, Modupe, Dr

Nazarene World Mission Society
Gunter, Nina G

Near East Council of Churches Committee for Refugee Work
Salti, Yacoub

New Age World Religious and Scientific Research Foundation
Roy, David K
Vandertuin-Roy, Victoria Elva Claudia Belot, Rev

Niwano Peace Foundation
Minami, Yoshinobu

Non-Denominational Bible Prophecy Study Association
Burns, John Henri

Nordic Christian Mission to Buddhists
Bøckman, Peter Wilhelm

Nordic Committee for Central Africa
Wahlström, Per Åke, Rev

Nordic Council for Church Music
Eriksson, Throstur

Nordic Council for Church Studies
Gudmundsson, Bernhardur

Nordic Ecumenical Council
Borgegård, Gunnel

Nordic Home-mission Council
Skaaheim, Anfin

Nordic Seamen's Church Council
Christiansen, Egon

North America Maritime Ministry Association
Chapman, Paul K, Dr

OC International
Keyes, Lawrence E

OCIC Missionary Service
Guillet, Jean-Paul, Fr

OCIC/Asia
Naidu, Vijaykumar Joseph, S.J., Rev

OMS International
Crouse, J B

Oblates of St Francis de Sales
Balducelli, Roger, O.S.F.S., Fr
Testa, Emilo, O.S.F.S., Fr

Oblats de la Vierge Marie
Gendrot, Marcel, O.M.V., Fr
Paul, Pierre, O.M.V., Fr

Oekumenisches Begegnungszentrum
Begeman, G

Oeuvre Notre-Dame de Montligeon
Lecoq, André, Mgr

Oeuvres sociales et éducatives des jésuites au Tiers-Monde
Hincq, L

Omkarananda International Ashram
Edmone, M

One World Movement of the Ecumenical Community
Kleron, Bryn

Order Secular of Servants of Mary
Barlow, Bernard Francis, Rev
Moons, Hubert, Rev

Order of Brothers of Bethlehem
Alvárez García, Luís de la Cruz A, Fr

Order of Brothers of the Blessed Virgin Mary of Mount Carmel
Attard, Victor Mark, O.C., Rev Dr
Malley, John, O.C., Fr

Order of Cistercians
Battista, Armando P Gregorio, O.Cist, Fr Abbot
Zakar, Polikarp, O.Cist., Fr Abbot

Order of Discalced Augustines
Cavallari, Eugenio, O.A.D., Fr
Ferlisi, Gabriele, O.A.D., Fr

Order of Discalced Brothers of the Blessed Virgin Mary of Mount Carmel
Bugin, Fiorenzo, O.C.D., Fr
Florent, Bugin, Br
Sáinz de Baranda, Felipe S, O.C.D., Fr

Order of Discalced Mercedarians
Cano Manrique, Francisco C, M.D., Fr
Pastor Alvarez, Manuel P, M.D., Fr

Order of Friars Minor
Bourdeau, Gilles, O.F.M., Rev

Order of Friars Minor Capuchin
Carraro Flavio, Roberto C, O.F.M. Cap., Fr
Flavio, Carraro, Fr
Iglesias, Francisco, O.F.M. Cap., Fr
Kuhn, Andrew, O.F.M.Cap., Br
Matanic, Atanasio, O.F.M., Rev

Order of Friars Minor Conventual
Kos, Donald, O.F.M. Conv., Fr
Serrini, Lanfranco, O.F.M. Conv., Fr

Order of Minims
Bellantonio, Alfredo, O.M., Fr
Galuzzi, Alessandro M, O.M., Fr

Order of Our Lady of Mercy
Aguirre Herrera, Emilio Manuel A, O. de M., Fr
Ivory, Thomas, Rev
Tolu, Giovannino, O. de M., Fr

Order of Preachers
Byrne, Damian Aloysius, O.P., Fr
Libanio Christo, Carlos Alberto L, Br
Moya, Rafael, Fr
O'Dwyer, Malachy, O.P., Fr

Pontifical Council for Promoting Christian Unity
Cassidy, Edward Idris, H E Mgr
Duprey, Pierre, M.Afr., H E Mgr
Willebrands, Johannes Gerardus Maria, H E Card

Pontifical Council for Social Communications
Deskur, Andrzej Maria, H E Card
Foley, John Patrick, Most Rev
Pastore, Pierfranco, Mgr

Pontifical Council for the Family
López Trujillo, Alfonso L, H E Card
Tremblay, Robert, Mgr
Trujillo, Alfonso López, Card

Pontifical Council for the Interpretation of Legislative Texts
Fagiolo, Vincenzo, H E Mgr
Herranz Casado, Julián H, H E Mgr

Pontifical Council for the Laity
Pironio, Eduardo Francisco, H E Card

Pontifical Council for the Pastoral Care of Migrants and Itinerant People
Cheli, Giovanni, Mgr
Milingo, Emmanuel, Most Rev

Pontifical Institute for Arabic and Islamic Studies
Ferré, André, Fr

Pontifical Institute for Foreign Missions
Cagnasso, Franco, P.I.M.E., Fr
Piazza, Giuseppe, P.I.M.E., Fr

Pontifical Institute of Christian Archeology
Pergola, Philippe, Prof Dr
Saint-Roch, Patrick
Saxer, Victor, Mgr

Pontifical Institute of Sacred Music
Fantini, Augusto

Pontifical Marian International Academy
Melada, Petar Pavao, O.F.M., Rev Dr

Pontifical Missionary Society
Kaut, Bernd, Mgr

Pontifical Oriental Institute
Gallagher, Clarence, Fr

Pontifical Society for the Propagation of Faith
Prince, Bernard A, Rev

Pontificio Ateneo Antonianum
Schalück, Herman, Fr

Poor Servants of Divine Providence
Cunegatti, Pietro, P.S.D.P., Fr
Fiorini, Luigi, P.S.D.P., Br
Fiorini, Luigi

Portuguese Catholic Missions
Afonso, Manuel Castro, S.M.P., Rev

Prelature of the Holy Cross and Opus Dei
del Portillo, Alvaro , Rev
Portillo, Alvaro del P, Most Rev Mgr

Presbyterian Peace Fellowship
Yolton, L William

Presbyterian and Reformed Renewal Ministries International
Long, Z Bradford, Rev Dr

Presentation Brothers
Browne, Bartholomew, F.P.M., Br
Kelly, Jerome T, F.P.M., Br

Priests Eucharistic League
Bernier, Paul J

Priests of the Sacred Heart of Jesus
Bressanelli, Virginio Domingo, S.C.J., Rev
Panteghini, Antonio Giovanni, S.C.J., Fr

Prison Fellowship International
Nikkel, Ronald Wilbert

Probe Ministries International
Williams, Jimmy

Programme to Combat Racism
Pityana, Nyameko Barney, Rev Dr

Protestant Association for Adult Education in Europe
Richter, Manfred

Protestant Association for Cooperation in Development
Bauer, Hartmut

Protestant Association for World Mission
Meissner, Herbert, Rev Dr

Présence et vie
Humbert, Marie

Prêtres de Sainte Marie de Tinchebray
Douget, Yves, Fr
Douguet, Yves, P.S.M., Fr
Gaudeul, Bernard, F.I.C.P., Br
Guillerm, Vincent, F.I.C.P., Br

Prêtres de l'Oratoire de France
Clavel, Pierre, Fr
Dujardin, Jean, Fr

Prêtres de la Doctrine Chrétienne
Bagna, Mario, D.C., Fr
Gasparotto, Rinaldo, D.C., Fr

Prêtres du Sacré-Coeur de Jésus de Bétharram
Sheridan, Terence, Rev
Zappa, Mario Enrico, Rev

Pères et Frères Religieux de Saint Vincent-de-Paul
Corbon, Richard, R.S.V., Fr
Forgues, Réal, R.S.V., Fr

Quaker Council for European Affairs
Baker, Elisabeth, Ms

Quaker Esperanto Society
Phillips, Brian Winfred Gilbert
Walker, Eric

Quaker United Nations Office
McClellan, Joel

Radio Vaticana
Borgomeo, Pasquale, S.J., Rev

Rav Tov International Jewish Rescue Organization
Niederman, David, Rabbi

Recherches pour l'éducation affective des jeunes en Afrique
Duteil, Armel, Rev Fr

Reformed Ecumenical Synod
Schrotenboer, Paul G, Dr

Regional Episcopal Conference of French-Speaking West Africa
Agré, Bernard, H E Mgr

Regional Episcopal Conference of North Africa
Landousies, Jean
Teissier, Henry Antoine-Marie, Most Rev Archbishop

Regional Islamic Da'wah Council of Southeast Asia and the Pacific
Zain, Dato Haj Ahmad Nordin Mohd

Regional Secretariat of Catholic Education for Africa and Madagascar
Didi, Dibalu, Rev Fr

Reiki Alliance
Mitchell, Susan Jones

Religious Conference Management Association
Woodring, Wayne S, Dr

Religious Science International
Parks, Lauren A

Religious Society of Friends
Clark, Andrew C

Religious Task Force on Central America
Swedish, Margaret

Religious of Nazareth
Decencière, Françoise, Sr

Research Center for Religion and Human Rights in Closed Societies
Hruby, Olga S

Rissho Kosei-Kai
Niwano, Nikkyo

Robert Schuman Institute for Europe
Rieutord, Maurice Georges Gabriel, Fr

Rogationistes du Coeur de Jésus
Cifuni, Pietro, R.C.I., Fr
Quaranta, Ciro, R.C.I., Fr
Quaranta, Ciro, Fr

Rosicrucian Order, Ancient and Mystical Order Rosae Crucis
Thompson, Kenneth A

Sisters of St Joseph of Cluny
Wenker, Zita, SJC, Sr

Sisters of St Joseph of Tarbes
Ponnau, Geneviève, Sr

Sisters of St Louis Juilly-Monaghan
Muldowney, Anita, Sr

Sisters of St Ursula of the Blessed Virgin
Machiels, Cécile Marie Paule, Sr

Sisters of the Apostolic Carmel
Mariella, M, A.C., Sr

Sisters of the Cross of Chavanod
Metral, Constance, Sr
Morand, Christiane Blandine, Sr

Sisters of the Divine Redeemer
Varga, Mary Alice, SDR, Sr

Sisters of the Holy Family
Michaud, Marie-Kisito, Sr

Sisters of the Holy Family of Nazareth
Celeste, M, Mother

Sisters of the Most Precious Blood of O'Fallon
Klaas, Joan, Sr
Klaas, Joan, Sr

Sisters of the Sacred Heart of Jesus
Primeau, Liette, Sr

Sisters, Servants of the Immaculate Heart of Mary of Monroe
Glisky, Joan, Sr

Slavic Gospel Association
Aker, John B, Dr

Societas Liturgica - International Society for Liturgical Study and Renewal
Holeton, David, Rev Canon Prof Dr

Societies of African Missions
Evrard, Jean, Fr

Society for Advancement of Buddhist Understanding
Katsunuma, Tomohito

Society for African Church History
Walls, A F

Society for Interests of Active Missionaries in Asia, Africa and America
Ruygrok, E

Society for Promoting Christian Knowledge
Gilbert, Patrick Nigel Geoffrey

Society for Threatened Peoples
Zülch, Tilman

Society for the Study of the Crusades and the Latin East
Powell, James Matthew, Prof
Riley-Smith, Jonathan S C, Prof

Society for the Study of the New Testament
Elliott, J K, Dr
Tuckett, C M, Prof

Society of African Missions
Harrington, Patrick, S.M.A., Fr
Icardi, Franco, S.M.A., Fr

Society of African Missions
van Frankenhuijsen, Wim, S.M.A., Fr

Society of Divine Vocations
Capraro, Giacomo, S.D.V., Rev
Giacomo, Capraro D, SDV, Rev
Lucci, Antonio, S.D.V., Rev

Society of Foreign Missions of the Province of Quebec
Lapierre, François, P.M.E., Fr
Samson, Pierre, P.M.E., Fr

Society of International Missionaries
Warner, John

Society of Jesus
Arokiasamy, Arul Maria, Rev Roshi
Berrigan, Daniel, S.J.
Coventry, John Seton, S.J., Rev
Egaña, Francisco Javier, S.J., Fr
Kolvenbach, Peter-Hans, S.J., Fr
Lucey, Gregory F, S.J., Rev
Murray, Robert, S.J., Rev Dr
Samy, Ama, SJ, Fr
Sanson, Henri, S.J., Fr
Shea, Leo Joseph, S.J., Fr
Smith, Richard F, S.J., Rev Dr
Wickham, John, S.J., Fr

Society of Jesus, Mary, Joseph
Melchers, B, Sr

Society of Mary
Dilanni, Albert, S.M., Fr
Filippucci, Mauro, S.M., Fr

Society of Mary
Hakenewerth, Quentin, Fr

Society of Mary
Jago, John, S.M., Fr

Society of Mary
Salaverri Aranegui, José María S, S.M., Fr
Torres, Enrique, C.S.V., Fr

Society of Mary Reparatrix
Henihan, Mary, SMR, Sr

Society of Missionaries of Africa
Kalilombe, Patrick-Augustine, Rt Rev Dr
Lauzon, Georges, M.Afr., Fr
Lauzon, Georges, Fr
Renaud, Etienne, M.Afr., Fr

Society of St Edmund
McLaughlin, Joseph, S.S.E., Fr
Simonnet, R P

Society of St Francis Xavier for the Foreign Missions
Marini, Francesco, S.X., Fr
Montesi, Giovanni, S.X., Fr

Society of St Francis of Sales
Fiora, Luigi, S.D.B., Rev
Meegan, Gerald, S.D.B., Br
Viganò, Egidio D, S.D.B., Rev

Society of St John the Evangelist
Naters, James, SSJE, Fr

Society of St Paul
Gandolfo, Guido, SSP, Rev
Perino, Renato, S.S.P., Rev

Society of St Teresa of Jesus
Feliú Loriente, María del Pilar F, STJ, Sr
Martín, Mercedes, STJ, Sr

Society of St Vincent-de-Paul
de Tarrazi, Amin A

Society of the Catholic Apostolate
Juritsch, Martin, S.A.C., Fr
Korycki, Jan, S.A.C., Rev

Society of the Divine Saviour
MacBride, Malachy, S.D.S., Fr
MacBride, Malachy, Fr
Spolti, Luiz, S.D.S., Fr
Wucher, Nikolaus, S.D.S., Fr

Society of the Divine Word
Barlage, Heinrich, S.V.D., Fr
Pung, Robert Edward, S.V.D., Rev

Society of the Precious Blood
Cespites, Gennaro, C.PP.S., Fr
Loipfinger, Anton, C.PP.S., Rev

Society of the Priests of St Sulpice
Bouchaud, Constant, P.S.S., Fr
Deville, Raymond, P.S.S., Fr

Society of the Sacred Heart of Jesus
McLaughlin, Helen, CSI, Sr

Società di Cristo per gli Emigrati della Polonia
Burniak, Marian, T.Chr., Fr
Nadolski, Boguslaw, T.Chr., Fr

Société Sacerdotale de Saint Joseph-Benoît Cottolengo
Gemello, Francesco, S.S.C., Fr
Piano, Lino, S.S.C., Rev

Société des Auxiliaires des Missions
Magermanss, Jacques

Société internationale Thomas d'Aquin
Lobato Casado, Abelardo L, Prof Dr Rev
Ols, Daniel Jean Henri, Prof Dr

Soeurs Maîtresses de Sainte Dorothée, 1838
Sardella, Giovanna Lia, Sr

Soeurs de l'Enfant-Jésus
Madigan, Bresa, Sr

Unda, International Catholic Association for Radio and Television
Düsterfeld, Peter, Rev Prof Dr
Grebe Lopez, Ronald G
Grebe Lopez, Ronald G
Henau, Ernest, Rev
Koprowski, Andrzej, S.J., Rev Dr
Monthienvichienchai, Raphael Chainarong
Murphy, Colm, S.S.C.M.E., Fr
Mvoe, Pascal
Paquette, Jacques, Rev
Roger-Machart, M
So, Erlinda
Ter Steeg, Maria T
Thomas, Peter, Rev
Uranga, Washington
William, Carmen Dasko, Sr
Ysern, Juan Luis
Ysern de Arce, Juan Luis Y, Rt Rev
Zukowski, Angela Ann, Sr

Union mondiale de la Retraite
Raimbault, Philippe

Union of European Conferences of Superiors General
Dargan, Joseph, S.J., Rev
Dargan, Joseph, S.J., Fr
Delcourt, France, S.A., Sr
Sowy-Lavergne, Françoise, Sr

Union of Superiors General
Basterrechea, José Pablo, F.S.C., Fr

Unitarian Mission Cooperation Centre
Neri, Ferdinando

Unitarian Universalist Association
Olson, James
Schulz, William Frederick, Rev

Unitarian Universalist Service Committee
Scobie, Richard S

United Ancient Order of Druids
Davis, F J

United Association of Christian Counselors International
Sprunger, Arthur L, Rev Dr

United Bible Societies
Erickson, John David, Rev Dr
Ibiam, Francis Akanu, Sir
Lohse, Eduard, Prof Bishop Emer

United Board for Christian Higher Education in Asia
Lauby, Paul T, Dr

United Church Board for World Ministries
Libbey, Scott S

United Indian Missions International
Fredericks, Donald G

United Methodist Church, Board of Church and Society
Houston, Haviland, Dr
Outen, George H
Weyer, Michel, Prof

United Mission to Nepal
Metzler, Edgar

United Nations of Yoga
Lemhényi, Zoltán, Dr

United Society for Christian Literature
Gilmore, Alec, Rev

United World Atheists
O'Hair, Madalyn Evalyn, Dr

United World Mission
Smith, Dwight

Uniting Church World Mission
Brookes, Graham F, Rev
Lee, J B, Rev

Universal Fellowship of Metropolitan Community Churches
Perry, Troy D, Rev

Universal Freemasons League
Vander Elst, R

Universal Life Church
Hensley, Kirby J

Universal Love and Brotherhood Association
Okuda, Munehiro

Universal Masonic Brotherhood
McCollough, Frederick

Universal Masonic Order of the Eastern Star
West, Dorothy

Universal Pantheist Society
Wood, Harold W, Jr

Universal Spiritual School
Costa, Joaquín

Universal Torah Registry
Holzer, Emmanuel, Rabbi

Ursulines de l'Union Canadienne
Chabot, Marie-Emmanuel, Rev Sister

Vicentian Congregation
Arackal, George P, CV, Rev
Ayyamcolil, Joseph, C.V., Fr
Kariamadam, Paul, Rev Dr

Visitation Nuns
McCarthy, M Paula, VHM, Sr

Voice of China and Asia Missionary Society
Hammond, Robert B, Dr

Voice of the Martyrs, The
Weaver, Cynthia
Yoder, Ruth

Volunteer Missionary Movement
McErlean, Joan
Prowse, Margaret

WEC International
Marks, Thomas I, Dr

Washington Institute of Ecumenics
Trickett, David

Way International
Martindale, L Craig, Rev

Wesleyan World Headquarters
Brannon, Ronal Roy, Rev Dr

Wider Quaker Fellowship
Wood, Anne

Witches International Craft Association
Martello, Léo Luis, Dr

Women's Inter-Church Council
Butcher, Pauline

Women's International Religious Fellowship
Winston, E Beulah, Dr

Women's League for Conservative Judaism
Citak, Audrey

Women's Missionary Society - African Methodist Episcopal Church
Williams, Delores L Kennedy

Work and Economy Network in the European Churches
Addy, Anthony J.

World Alliance of Reformed Churches
Ador, Samuel Nyawelo, Rev
Andrews, James E, Rev Dr
Blei, Karel, Rev Dr
Boesak, Allan, Rev
Brown, Margrethe B J, Rev
Capulong, Emilio, Jr, Atty
Chang, Sang, Prof Dr
Chiphangwi, Saindi, Rev Dr
Dansokho, Samuel, Rev
Davidson, Hugh, Rev
Dilschneider, Gerhard
Douglas, Jane Dempsey, Prof Dr
Ehnes, Herbert, Dr
Eleblu, Lydia Aku
Farris, Stephen Charles, Rev Prof
Gilmour, Daphne Elizabeth
Hadfield, Anne
Holtrop, Pieter Nanne, Rev Prof Dr
Kao, Ruth
Lucke, Hartmut, Rev
Masilo, Benjamin
Michael, Joy
Mourad, George, Rev
Mulder, Edwin George, Rev Dr
Norvila, Julius, Rev
Opocensky, Milan, Rev Dr
Paik, Mary, Rev
Pires da Silveira, Abival P, Rev Dr
Rasoaharison Fégeat, Liliane R
Ricciardi, Salvatore, Rev
Ro, Chung-Hyun, Prof
Rochette, Jean-François, Dr
Roeroe, Wilhelmus A, Rev Dr
Rosenhager, Ursel, Rev
Sporre, Karin, Rev
Stephens, Sarah Jane, Rev
Stuker, Andrea

World Methodist Historical Society
Eltscher, Susan M
Yrigoyen, Charles, Jr, Dr

World Ministries Commission of the Church of the Brethren
Schrock, J Roger

World Mission Association
Price, Peter Bryan, Rev Canon
Taylor, Humphrey Vincent, Rev Canon

World Missionary Evangelism
Wodajo, Datan

World Missions Fellowship
Gillespie, John M, Rev

World Movement of Christian Workers
Pulh, Jacques

World Muslim Congress
Hamdani, Viqar Ahmad, Dr
Khan, Inamullah, Dr
Nias, Ibrahim
Tabibi, Aboul Makim, Dr

World Muslim Minorities Commission
Sabir, Khalid

World Organization of Former Pupils of Catholic Education
Le Bret, Marie-Michèle
Lesueur, Michel
Natale, Tommaso
Solassol, Alain

World Organization of Jews from Arab Countries
Haddad, Heskel M, Dr

World Peace Prayer Society
Saionji, Masami

World Rabbinic Committee for the Preservation of Ancient Tombs in Tiberias
Schmidel, David, Rabbi

World Radio Missionary Fellowship
Peters, Douglas C

World Relief Corporation
Gay, Arthur E, Dr

World Revival Crusade
Edsor, Albert W, Pastor

World Sephardi Federation
Dery, Shimon

World Sikh Organization
Basarke, Alice

World Student Christian Federation
Azar, Zahi
Gilbert, Bruce
Jin, Yong Ting
Ledger, Christine
Mabuluki, Kwangwa, Rev
Misijuk, Wzodzimerz
Ortiz, Angel Miguel, Rev
Quintero, Manuel
Ribadeneira, Maria Cristina
Saukkonen, Paula

World Subud Association
Tatabe, Roask
Vittachi, Varindra
Vittachi, Varindra

World Supreme Council for Masajid
Nasseef, Abdullah Omar, Dr

World Theosophical Youth Federation
Filho, Arnaldo Sisson
Osswald, Suzanne
Sisson Filho, Arnaldo S

World Union
Basu, Samar, Sri
Pandit, M P, Sri

World Union for Progressive Judaism
Friedlander, Albert H, Rabbi Prof Dr
Hirsch, Richard G, Rabbi Dr
Littman, David G

World Union of Catholic Philosophical Societies
Ladrière, Jean A M G, Prof Dr
McLean, George Francis, Prof

World Union of Catholic Teachers
Greiveldinger, P
Lingelser, G
Lingelser, G

World Union of Catholic Women's Organizations
de Diesbach, G
de Thoury, C

World Union of Jesuit Alumni
Cacchione, Ciro, Dr
Duminuco, Vincent J, Dr

World Union of Jewish Students
Yossef, Daniel

World Union of Jewish Studies
Assaf, David
Levine, Lee I, Rabbi Prof

World Union of Mapam Parties
Fisher, Avri

World Union of North African Jews
Ben-Simhon, Shaul

World University Roundtable
Zitko, Howard John, Dr

World Vision International
Dayton, Edward, Dr
Irvine, Graeme
Key, John Christopher, Rev
Seiple, Robert A
Watson, Richard

World Without War Council
Hill, Kent R

World Witness, Associate Reformed Presbyterian Church
Mariner, John Elwyn

World Yoga Society
Hills, Christopher, Dr

World Zionist Organization
Dinitz, Simcha

World Zoroastrian Organization
Captain, Shahpur F

World for Christ Crusade
Stelpstra, William, Rev

World of Islam Festival Trust
Duncan, Alistair

World's Christian Endeavor Union
Jackson, David G, Rev

World's Woman's Christian Temperance Union
Edwards, Olive Eva

World-Wide Prayer and Missionary Union
Huston, Ralph, Ms

Worldwide Marriage Encounter
Grace, P J, Fr

Yoga Research Foundation
Jyotirmayananda, J, Swami
Lalitananda, L, Swami

Youth for Christ International
Engstrom, Ted, Dr
Gallimore, Gerry O
Groen, H James

Youth with a Mission
de Visser, Jan Piet

ZOA Refugee Care Netherlands
Huberts, H